Principles of
MOLECULAR BIOLOGY

Jones & Bartlett Learning Titles in Biological Science

AIDS: Science and Society, Seventh Edition
 Hung Fan, Ross F. Conner, & Luis P. Villarreal

AIDS: The Biological Basis, Fifth Edition
 Benjamin S. Weeks & I. Edward Alcamo

Alcamo's Fundamentals of Microbiology, Body Systems Edition, Second Edition
 Jeffrey C. Pommerville

Alcamo's Laboratory Fundamentals of Microbiology, Tenth Edition
 Jeffrey C. Pommerville

Alcamo's Microbes and Society, Third Edition
 Benjamin S. Weeks

Biochemistry
 Raymond S. Ochs

Bioethics: An Introduction to the History, Methods, and Practice, Third Edition
 Nancy S. Jecker, Albert R. Jonsen, & Robert A. Pearlman

Bioimaging: Current Concepts in Light and Electron Microscopy
 Douglas E. Chandler & Robert W. Roberson

Biomedical Graduate School: A Planning Guide to the Admissions Process
 David J. McKean & Ted R. Johnson

Biomedical Informatics: A Data User's Guide
 Jules J. Berman

Botany: A Lab Manual
 Stacy Pfluger

Case Studies for Understanding the Human Body, Second Edition
 Stanton Braude, Deena Goran, & Alexander Miceli

Electron Microscopy, Second Edition
 John J. Bozzola & Lonnie D. Russell

Encounters in Microbiology, Volume 1, Second Edition
 Jeffrey C. Pommerville

Encounters in Microbiology, Volume 2
 Jeffrey C. Pommerville

Encounters in Virology
 Teri Shors

Essential Genetics: A Genomics Perspective, Sixth Edition
 Daniel L. Hartl

Essentials of Molecular Biology, Fourth Edition
 George M. Malacinski

Evolution: Principles and Processes
 Brian K. Hall

Exploring Bioinformatics: A Project-Based Approach
 Caroline St. Clair & Jonathan E. Visick

Exploring the Way Life Works: The Science of Biology
 Mahlon Hoagland, Bert Dodson, & Judy Hauck

Fundamentals of Microbiology, Tenth Edition
 Jeffrey C. Pommerville

Genetics: Analysis of Genes and Genomes, Eighth Edition
 Daniel L. Hartl & Maryellen Ruvolo

Genetics of Populations, Fourth Edition
 Philip W. Hedrick

Guide to Infectious Diseases by Body System, Second Edition
 Jeffrey C. Pommerville

Human Biology, Seventh Edition
 Daniel D. Chiras

Human Biology Laboratory Manual
 Charles Welsh

Human Body Systems: Structure, Function, and Environment, Second Edition
 Daniel D. Chiras

Human Embryonic Stem Cells, Second Edition
 Ann A. Kiessling & Scott C. Anderson

Laboratory Investigations in Molecular Biology
 Steven A. Williams, Barton E. Slatko, & John R. McCarrey

Lewin's CELLS, Second Edition
 Lynne Cassimeris, Vishwanath R. Lingappa, & George Plopper

Lewin's Essential GENES, Third Edition
 Jocelyn E. Krebs, Elliott S. Goldstein, & Stephen T. Kilpatrick

Lewin's GENES XI
 Jocelyn E. Krebs, Elliott S. Goldstein, & Stephen T. Kilpatrick

The Microbial Challenge: A Public Health Perspective, Third Edition
 Robert I. Krasner

Microbial Genetics, Second Edition
 Stanley R. Maloy, John E. Cronan, Jr., & David Freifelder

Molecular Biology: Genes to Proteins, Fourth Edition
 Burton E. Tropp

Neoplasms: Principles of Development and Diversity
 Jules J. Berman

Precancer: The Beginning and the End of Cancer
 Jules J. Berman

Principles of Cell Biology
 George Plopper

Principles of Modern Microbiology
 Mark Wheelis

Science and Society: Scientific Thought and Education for the 21st Century
 Peter Daempfle

Strickberger's Evolution, Fifth Edition
 Brian K. Hall

Symbolic Systems Biology: Theory and Methods
 M. Sriram Iyengar

20th Century Microbe Hunters
 Robert I. Krasner

Understanding Viruses, Second Edition
 Teri Shors

Principles of MOLECULAR BIOLOGY

BURTON E. TROPP
Professor Emeritus
Queens College
City University of New York
Flushing, New York

JONES & BARTLETT
LEARNING

World Headquarters
Jones & Bartlett Learning
5 Wall Street
Burlington, MA 01803
978-443-5000
info@jblearning.com
www.jblearning.com

Jones & Bartlett Learning books and products are available through most bookstores and online booksellers. To contact Jones & Bartlett Learning directly, call 800-832-0034, fax 978-443-8000, or visit our website, www.jblearning.com.

Substantial discounts on bulk quantities of Jones & Bartlett Learning publications are available to corporations, professional associations, and other qualified organizations. For details and specific discount information, contact the special sales department at Jones & Bartlett Learning via the above contact information or send an email to specialsales@jblearning.com.

Copyright © 2014 by Jones & Bartlett Learning, LLC, an Ascend Learning Company

All rights reserved. No part of the material protected by this copyright may be reproduced or utilized in any form, electronic or mechanical, including photocopying, recording, or by any information storage and retrieval system, without written permission from the copyright owner.

Principles of Molecular Biology is an independent publication and has not been authorized, sponsored, or otherwise approved by the owners of the trademarks or service marks referenced in this product.

Production Credits
Chief Executive Officer: Ty Field
President: James Homer
SVP, Editor-in-Chief: Michael Johnson
SVP, Chief Marketing Officer: Alison M. Pendergast
Publisher: Kevin Sullivan
Senior Acquisitions Editor: Erin O'Connor
Editorial Assistant: Rachel Isaacs
Editorial Assistant: Michelle Bradbury
Production Manager: Louis C. Bruno, Jr.
Senior Marketing Manager: Andrea DeFronzo
Production Services Manager: Colleen Lamy
Online Products Manager: Dawn Mahon Priest
V.P., Manufacturing and Inventory Control: Therese Connell
Composition: Circle Graphics, Inc.
Cover Design: Kristin E. Parker
Rights & Photo Research Associate: Lauren Miller
Assistant Permissions and Photo Researcher: Ashley Dos Santos
Illustrations: Elizabeth Morales
Cover and Title Page Image: © lculig/ShutterStock, Inc.
Printing and Binding: Courier Companies
Cover Printing: Courier Companies

About the Cover
A high-resolution image of the nucleosome core particle, the first level of eukaryotic chromatin organization. The double-stranded DNA (backbone shown as orange tubes) winds around a protein assembly consisting of an octameric protein complex (two copies each of histones H2A, H2B, H3, and H4—shown as colored ribbon structures). The DNA is compacted about sevenfold as a result of nucleosome core particle formation.

To order this product, use ISBN: 978-1-4496-8917-9

Library of Congress Cataloging-in-Publication Data
Tropp, Burton E.
  Principles of molecular biology/Burton Tropp. — 1st ed.
     p. ; cm.
  Includes bibliographical references and index.
  ISBN 978-1-4496-4791-9 (alk. paper)
  I. Title.
  [DNLM: 1. Biochemical Phenomena. 2. Genes—physiology. 3. Proteins—physiology. QU 450]
  572'.43—dc23
                                                                    2012020605
6048

Printed in the United States of America
16 15 14 13 12   10 9 8 7 6 5 4 3 2 1

To my wife Roslyn and to our family: Jonathan, Lauren, Matthew, Julie, Paul, Erica, Sarah, Rachel, Gabrielle, Katie, Gracie, Alice, Jane, and Charles and to the memory of my parents Sol and Renee Tropp.

Brief Contents

CHAPTER 1	Introduction to Molecular Biology	1
CHAPTER 2	Protein Structure and Function	24
CHAPTER 3	Nucleic Acid Structure	80
CHAPTER 4	Molecular Biology Technology	112
CHAPTER 5	Chromosomes	151
CHAPTER 6	Genetic Analysis in Molecular Biology	185
CHAPTER 7	Viruses in Molecular Biology	223
CHAPTER 8	DNA Replication	264
CHAPTER 9	DNA Damage and Repair	317
CHAPTER 10	Double-Strand Break Repair and Homologous Recombination	359
CHAPTER 11	Transposable Elements	380
CHAPTER 12	Bacterial Transcription and Its Regulation	406
CHAPTER 13	Eukaryotic Transcription	466
CHAPTER 14	RNA Polymerase II: Cotranscriptional and Posttranscriptional Processes	551
CHAPTER 15	Small Silencing RNAs	600
CHAPTER 16	Protein Synthesis	623
	Glossary	703

Image © Juan Gaertner/ShutterStock, Inc.

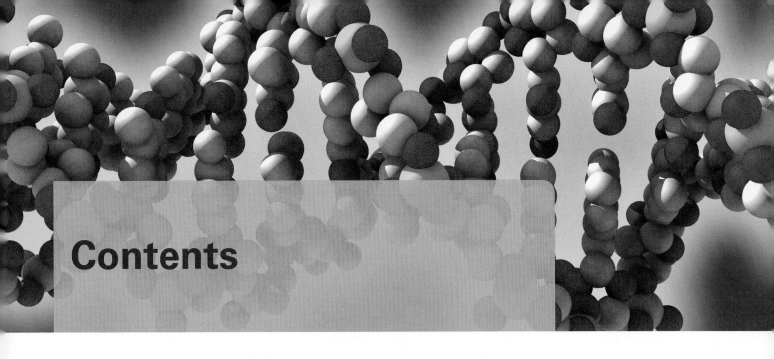

Contents

Preface xxvii

CHAPTER 1 **Introduction to Molecular Biology** 1

- **1.1** Intellectual Foundation 2
  Two studies performed in the 1860s provided the intellectual underpinning for molecular biology. 2

- **1.2** One Gene–One Polypeptide Hypothesis 4
  Each gene is responsible for the synthesis of a single polypeptide. 4

- **1.3** Introduction to Nucleic Acids 5
  DNA contains the sugar deoxyribose and RNA contains the sugar ribose. 5
  A nucleoside is formed by attaching a purine or pyrimidine base to a sugar. 6
  A nucleotide is formed by attaching a phosphate group to the sugar group in a nucleoside. 7
  DNA is a linear chain of deoxyribonucleotides. 8

- **1.4** DNA: Hereditary Material 11
  Transformation experiments led to the discovery that DNA is the hereditary material. 11
  Chemical experiments also supported the hypothesis that DNA is the hereditary material. 14

- **1.5** Watson-Crick Model 15
  Rosalind Franklin and Maurice Wilkins obtained x-ray diffraction patterns of extended DNA fibers. 15
  James Watson and Francis Crick proposed that DNA is a double-stranded helix. 15

- **1.6** Central Dogma 19
  The central dogma provides the theoretical framework for molecular biology. 19

Image © Juan Gaertner/ShutterStock, Inc.

1.7 Introduction to Recombinant DNA Technology 20
Recombinant DNA technology allows us to study complex biological systems. 20
A great deal of molecular biology information is available on the Internet. 22
Questions and Problems 22
Suggested Reading 23

CHAPTER 2 Protein Structure and Function 24

2.1 α-Amino Acids 25
α-Amino acids have an amino group and a carboxyl group attached to a central carbon atom. 25
Amino acids are represented by three-letter and one-letter abbreviations. 28

2.2 Peptide Bond 29
α-Amino acids are linked by peptide bonds. 29
Peptide chains have directionality. 30

2.3 Protein Purification 31
Protein mixtures can be fractionated by chromatography. 31
Proteins and other charged biological polymers migrate in an electric field. 35

2.4 Primary Structure of Proteins 36
The amino acid sequence or primary structure of a purified protein can be determined. 36
Polypeptide sequences can be obtained from nucleic acid sequences. 39
The BLAST program compares a new polypeptide sequence with all sequences stored in a data bank. 41

2.5 Introduction to Protein Structure 41
Proteins with just one polypeptide chain have primary, secondary, and tertiary structures, whereas those with two or more chains also have quaternary structures. 41

2.6 Weak Noncovalent Bonds 43
The polypeptide folding pattern is determined by weak noncovalent interactions. 43

2.7 Secondary Structures 46
The α-helix is a compact structure that is stabilized by hydrogen bonds. 46
The β-conformation is also stabilized by hydrogen bonds. 49
Loops and turns connect different peptide segments, allowing polypeptide chains to fold back on themselves. 49
Certain combinations of secondary structures, called supersecondary structures or folding motifs, appear in many different proteins. 50
We cannot yet predict secondary structures with absolute certainty. 50

2.8 Tertiary Structure 50
X-ray crystallography studies reveal three-dimensional structures of many different proteins. 50

Intrinsically disordered proteins lack an ordered structure under physiological conditions. 53
The primary structure of a polypeptide determines its tertiary structure. 53
Molecular chaperones help proteins to fold inside the cell. 56

2.9 Myoglobin, Hemoglobin, and the Quaternary Structure Concept 56
Differences in myoglobin and hemoglobin function are explained by differences in myoglobin and hemoglobin structure. 56
Hemoglobin subunits bind oxygen in a cooperative fashion. 60

2.10 Immunoglobulin G (IgG) and the Domain Concept 63
Large polypeptides fold into globular units called domains. 63

2.11 Enzymes 65
Enzymes are proteins that catalyze chemical reactions. 65
Molecular biologists often use colorimetric and radioactive methods to measure enzyme activity. 65
Enzymes lower the energy of activation but do not affect the equilibrium position. 66
All enzyme-catalyzed reactions proceed through an enzyme–substrate complex. 68
Molecular details for ES complexes have been worked out for many enzymes. 71
Regulatory enzymes control committed steps in biochemical pathways. 72
Regulatory enzymes exhibit sigmoidal kinetics and are stimulated or inhibited by allosteric effectors. 73
BOX 2.1. IN THE LAB: Strategies for Protein Purification 33
BOX 2.2. CLINICAL APPLICATION: Sickle Cell Anemia 58
BOX 2.3. CLINICAL APPLICATION: Clinical Significance of 2,3-Bisphosphoglycerate 61
BOX 2.4. IN THE LAB: Radioactive Tracers 67
Questions and Problems 74
Suggested Reading 77

CHAPTER 3 Nucleic Acid Structure 80

3.1 DNA Size and Fragility 81
Most eukaryotic DNA is located in a defined cell nucleus, whereas bacterial and archaeal DNA is present in the cytoplasm. 81
DNA molecules vary in size and base composition. 81
DNA molecules are fragile. 82

3.2 Recognition Patterns in the Major and Minor Grooves 82
Enzymes can recognize specific patterns at the edges of the major and minor grooves. 82

3.3 DNA Bending 83
Some base sequences cause DNA to bend. 83

3.4 DNA Denaturation and Renaturation 84
DNA can be denatured. 84
Hydrogen bonds stabilize double-stranded DNA. 86
Base stacking also stabilizes double-stranded DNA. 87
Ionic strength influences DNA structure. 87
The DNA molecule is in a dynamic state. 87
Alkali denatures DNA without breaking phosphodiester bonds. 88
Complementary single strands can anneal to form double-stranded DNA. 88

3.5 Helicases 89
Helicases are motor proteins that use the energy of nucleoside triphosphates to unwind DNA. 89

3.6 Single-Stranded DNA Binding Proteins 90
Single-stranded DNA binding proteins stabilize single-stranded DNA. 90

3.7 Topoisomers 90
DNA can exist as a circular molecule. 90
Bacterial DNA usually exists as a covalently closed, circular double-stranded DNA molecule. 91
Plasmid DNA molecules are used to study the properties of covalently closed, circular double-stranded DNA *in vitro*. 92
Covalently closed, circular double-stranded DNA molecules often have superhelical structures. 92

3.8 Topoisomerases 97
Topoisomerases catalyze the conversion of one topoisomer into another. 97

3.9 Non-B DNA Conformations 98
A-DNA is a right-handed double helix with a deep major groove and very shallow minor groove. 98
Z-DNA has a left-handed conformation. 100
Several other kinds of non-B DNA structures appear to exist in nature. 102

3.10 RNA Structure 104
RNA performs a wide variety of functions in the cell. 104
RNA secondary structure is dominated by Watson-Crick base pairs. 104
RNA tertiary structures are stabilized by interactions between two or more secondary structure elements. 105

3.11 RNA World Hypothesis 105
The earliest forms of life on earth may have used RNA as both the genetic material and the biological catalysts needed to maintain life. 105
BOX 3.1. CLINICAL APPLICATION: Plasmids and Disease 93
BOX 3.2. LOOKING DEEPER: Linking Number 95
Questions and Problems 107
Suggested Reading 108

CHAPTER 4   **Molecular Biology Technology   112**

4.1   Nucleic Acid Isolation   113
The method used to isolate DNA depends on the DNA source.   113
Great care must be taken to protect RNA from degradation during its isolation.   114

4.2   Physical Techniques Used to Study Macromolecules   115
Different physical techniques are used to study macromolecules.   115
Electron microscopy allows us to see nucleic acids and nucleoprotein complexes.   115
Centrifugation can separate macromolecules and provide information about their size and shape.   117
Gel electrophoresis separates nucleic acids based on their rate of migration in an electric field.   119
SDS-PAGE can be used to determine a polypeptide's molecular mass.   120

4.3   Enzymatic Techniques Used to Manipulate DNA   122
Nucleases are useful tools in DNA investigations.   122
Restriction endonucleases cleave within specific nucleotide sequences in DNA.   123
Restriction endonucleases can be used to construct a restriction map of a DNA molecule.   125
DNA fragments can be inserted into plasmid DNA vectors.   128
The Southern blot procedure is used to detect specific DNA fragments.   130
Northern and Western blotting are used to detect specific RNA and polypeptide molecules, respectively.   132
DNA polymerase I requires a template-primer.   132
DNA polymerase I has both $3'\rightarrow 5'$ and $5'\rightarrow 3'$ exonuclease activities.   133
DNA polymerase I can catalyze nick translation.   136
The polymerase chain reaction (PCR) is used to amplify DNA.   137
Site-directed mutagenesis can be used to introduce a specific base change within a gene.   138
The chain termination method for sequencing DNA uses dideoxynucleotides to interrupt DNA synthesis.   139
DNA sequences can be stitched together by using information obtained from a restriction map.   141
Shotgun sequencing is used to sequence long DNA molecules.   141
A new generation of DNA sequencing techniques is now being used for whole genome shotgun sequencing.   142
The human genome sequence provides considerable new information.   142
BOX 4.1. IN THE LAB: Gel Electrophoresis and Topoisomer Separation   121
BOX 4.2. IN THE LAB: Capillary Electrophoresis   122
BOX 4.3. IN THE LAB: New Generation DNA Sequencers   143
Questions and Problems   146
Suggested Reading   148

CHAPTER 5	**Chromosomes 151**
	5.1 Bacterial Chromatin 153
	Bacterial DNA is located in the nucleoid. 153
	5.2 Introduction to Eukaryotic Chromatin 155
	Eukaryotic chromatin is visible under the light microscope during certain stages of the cell cycle. 155
	5.3 Mitosis 155
	The animal cell life cycle alternates between interphase and mitosis. 155
	Mitosis allows cells to maintain the chromosome number after cell division. 156
	5.4 Meiosis 159
	Meiosis reduces the chromosome number in half. 159
	5.5 Karyotype 162
	Chromosome sites are specified according to nomenclature conventions. 162
	A karyotype shows an individual cell's metaphase chromosomes arranged in pairs and sorted by size. 163
	Fluorescent *in situ* hybridization (FISH) provides a great deal of information about chromosomes. 163
	5.6 The Nucleosome 166
	Five major histone classes interact with DNA in eukaryotic chromatin. 166
	The first level of chromatin organization is the nucleosome. 167
	X-ray crystallography reveals the atomic structure of nucleosome core particles. 168
	The precise nature of the interaction between H1 and the core particle is not known. 171
	We do not know how nucleosomes are arranged in chromatin. 171
	Condensins and topoisomerase II help to stabilize condensed chromosomes. 172
	The scaffold model was proposed to explain higher order chromatin structure. 172
	5.7 The Centromere 173
	The centromere is the site of microtubule attachment. 173
	5.8 The Telomere 175
	The telomere, which is present at either end of a chromosome, is needed for stability. 175
	BOX 5.1. CLINICAL APPLICATION: Karyotype and Diagnosis 164
	Questions and Problems 179
	Suggested Reading 180
CHAPTER 6	**Genetic Analysis in Molecular Biology 185**
	6.1 *Drosophila melanogaster* 186
	Many fundamental genetic principles were discovered by studying the common fruit fly. 186
	6.2 *Escherichia coli* 189
	E. coli is a gram-negative bacterium. 189
	Bacteria can be cultured in liquid or solid media. 191

Plating can be used to detect auxotrophs. 194
Specific notations, conventions, and terminology are used in bacterial genetics. 194
Cells with altered genes are called mutants. 195
Some mutants display the mutant phenotype under all conditions, whereas others display it only under certain conditions. 195
Mutations can be classified on the basis of the changes in the DNA. 196
Mutants have many uses in molecular biology. 197
A genetic test known as complementation can be used to determine the number of genes responsible for a phenotype. 200
E. coli cells can exchange genetic information by conjugation. 200
The F plasmid can integrate into a bacterial chromosome and carry it into a recipient cell. 204
Bacterial mating experiments can be used to produce an *E. coli* genetic map. 206
F′ plasmids contain part of the bacterial chromosome. 209
Plasmid replication control functions are usually clustered in a region called the basic replicon. 210

6.3 Budding Yeast (*Saccharomyces cerevisiae*) 211
Yeasts are unicellular eukaryotes. 211
Specific notations, conventions, and terminology are used in yeast genetics. 212
Yeast cells exist in haploid and diploid stages. 213

6.4 New Tools for Studying Human Genetics 214
Different humans may exhibit DNA sequence variations within the same DNA region. 214
Restriction fragment length polymorphisms facilitate genetic analysis in humans and other organisms. 216
Somatic cell genetics can be used to map genes in higher organisms. 219
BOX 6.1. LOOKING DEEPER: Plasmid Functions Required for Mating 204
BOX 6.2. IN THE LAB: DNA Fingerprint 218
Questions and Problems 219
Suggested Reading 220

CHAPTER 7 Viruses in Molecular Biology 223

7.1 Introduction to the Bacteriophages 225
Bacteriophages were of interest because they seemed to have the potential to serve as therapeutic agents to treat bacterial diseases. 225
Investigators belonging to the "Phage Group" were the first to use viruses as model systems to study fundamental questions about gene structure and function. 226
Bacteriophages come in different sizes and shapes. 227
Bacteriophages have lytic, lysogenic, and chronic life cycles. 228
Bacteriophages form plaques on a bacterial lawn. 228

- **7.2 Virulent Bacteriophages** 230
  - The T-even phages are tailed viruses that inject their DNA into bacteria. 230
  - Bacterial and phage RNA polymerases help to reel phage T7 DNA into the bacterial cell. 234
  - *E. coli* phage ϕX174 contains a single-stranded circular DNA molecule. 238
  - Some phages have single-stranded RNA as their genetic material. 239
- **7.3 Temperate Phages** 241
  - *E. coli* phage λ DNA can replicate through a lytic or lysogenic life cycle. 241
  - *E. coli* phage P1 can also replicate through a lytic or lysogenic life cycle. 245
- **7.4 Chronic Phages** 248
  - A chronic phage programs the host cell for continued virion particle release without killing the cell. 248
- **7.5 Animal Viruses** 249
  - Animal viruses can serve as excellent models for addressing fundamental questions in molecular biology. 249
  - Polyomaviruses contain circular double-stranded DNA. 250
  - Adenoviruses have linear blunt-ended, double-stranded DNA with an inverted repeat at each end. 253
  - Retroviruses use reverse transcriptase to make a DNA copy of their RNA genome. 255
  - BOX 7.1. LOOKING DEEPER: T-Even Phage Assembly 234
  - BOX 7.2. IN THE LAB: Generalized Transduction 246
  - BOX 7.3. IN THE LAB: Phage M13 Cloning Vector 249
  - BOX 7.4. IN THE LAB: Complementary DNA Synthesis 259
  - Questions and Problems 260
  - Suggested Reading 260

CHAPTER 8 DNA Replication 264

- **8.1 General Features of DNA Replication** 265
  - DNA replication is semiconservative. 265
  - Bacterial and eukaryotic DNA replication is bidirectional. 267
  - The DNA strand that grows in an overall 3′→5′ direction is formed by joining short fragments. 267
  - DNA ligase connects adjacent Okazaki fragments. 271
  - RNA serves as a primer for Okazaki fragment synthesis. 272
- **8.2 Bacterial DNA Replication** 273
  - The bacterial replication machinery has been isolated and examined *in vitro*. 273
  - Mutant studies provide important information about the enzymes involved in bacterial DNA replication. 275
  - The replicon model proposes that an initiator protein must bind to a DNA sequence called a replicator at the start of replication. 275
  - *E. coli* chromosomal replication begins at *oriC*. 276
  - Several enzymes act together at the replication fork. 280
  - DNA polymerase III is required for bacterial DNA replication. 280

DNA polymerase III holoenzyme has three distinct kinds of subassemblies. 281
The core polymerase has one subunit with 5'→3' polymerase activity and another with 3'→5' exonuclease activity. 283
The sliding clamp forms a ring around DNA, tethering the remainder of the polymerase holoenzyme to the DNA. 285
The clamp loader places the sliding clamp around DNA. 286
The replisome catalyzes coordinated leading and lagging strand DNA synthesis at the replication fork. 287
E. coli DNA replication terminates when the two growing forks meet in the terminus region, which is located 180 degrees around the circular chromosome from the origin. 291
The terminus utilization substance binds to *Ter* sites. 291
Topoisomerase IV and recombinase separate newly formed sister chromosomes. 291

8.3 Eukaryotic DNA Replication 293
The SV40 DNA replication system is a good model for *in vitro* eukaryotic DNA replication. 293
Eukaryotic replication machinery must replicate long linear duplexes with multiple origins of replication. 295
Eukaryotic chromosomes have many origins of replication. 296
Yeast origins of replication, called autonomously replicating sequences, determine the site of DNA chain initiation. 297
Pol δ and Pol ε are primarily responsible for copying the lagging- and leading-strand templates, respectively. 298
Studies of the *Tetrahymena* and yeast telomeres suggest that a terminal transferase-like enzyme is required for telomere formation. 300
Telomerase uses an RNA template to add nucleotide repeats to chromosome ends. 303
Telomerase plays an important role in solving the end-replication problem. 305

8.4 Replication Coupled Chromatin Synthesis 307
Chromatin disassembly and reassembly are tightly coupled to DNA replication. 307
BOX 8.1. LOOKING DEEPER: Replication Regulation at the Initiation Stage 279
BOX 8.2. IN THE LAB: Measurement of Processivity 282
BOX 8.3. LOOKING DEEPER: Helicase and Primer Coordination 288
BOX 8.4. CLINICAL APPLICATION: Telomerase in Aging and Cancer 304
Questions and Problems 310
Suggested Reading 311

CHAPTER 9 **DNA Damage and Repair 317**

9.1 Radiation Damage 319
Ultraviolet light causes cyclobutane pyrimidine dimer formation and (6-4) photoproduct formation. 319
X-rays and gamma rays cause many different types of DNA damage. 321

9.2 DNA Instability in Water 321
DNA is damaged by hydrolytic cleavage reactions. 321

9.3 Oxidative Damage 324
Reactive oxygen species damage DNA. 324

9.4 Alkylation Damage by Monoadduct Formation 325
Alkylating agents damage DNA by transferring alkyl groups to centers of negative charge. 325
Many environmental agents must be modified by cell metabolism before they can alkylate DNA. 325

9.5 Chemical Cross-Linking Agents 327
Chemical cross-linking agents block DNA strand separation. 327

9.6 Mutagen and Carcinogen Detection 331
Mutagens can be detected based on their ability to restore mutant gene activity. 331

9.7 Direct Reversal of Damage 333
Photolyase reverses damage caused by cyclobutane pyrimidine dimer formation. 333
O^6-Alkylguanine, O^4-alkylthymine, and phosphotriesters can be repaired by direct alkyl group removal by a suicide enzyme. 335

9.8 Base Excision Repair 337
The base excision repair pathway removes and replaces damaged or inappropriate bases. 337

9.9 Nucleotide Excision Repair 340
Nucleotide excision repair removes bulky adducts from DNA by excising an oligonucleotide bearing the lesion and replacing it with new DNA. 340

9.10 Mismatch Repair 345
The DNA mismatch repair system removes mismatches and short insertions or deletions that are present in DNA. 345

9.11 SOS Response and Translesion DNA Synthesis 347
Error-prone DNA polymerases catalyze translesion DNA synthesis. 347
RecA and LexA regulate the *E. coli* SOS response. 348
The SOS signal induces the synthesis of DNA polymerases II, IV, and V. 351
Human cells have at least 14 different template-dependent DNA polymerases. 352
BOX 9.1. CLINICAL APPLICATION: The Interstrand Cross-Linking Agents Cisplatin and Psoralen 328
BOX 9.2. CLINICAL APPLICATION: Xeroderma Pigmentosum 343
Questions and Problems 353
Suggested Reading 354

CHAPTER 10 **Double-Strand Break Repair and Homologous Recombination   359**

    10.1 Replication Restart After Bacterial Replication Fork Collapse   360
- Cells use homologous recombination to restart replication after a replication fork collapses.   360
- The RecBCD protein processes the broken ends of double-stranded break to form a single-strand 3'-OH tail.   361
- RecA acts as a recombinase to catalyze strand invasion.   362
- The RuvAB protein complex catalyzes branch migration in *E. coli*.   363
- RuvC helps convert the DNA with the Holliday junction into a DNA fork.   364
- PriA helps to load DnaB on the reestablished DNA fork to restart replication.   364

    10.2 Mitotic Recombination   364
- Eukaryotic cells use homologous recombination to repair double-strand breaks during the late S stage or the G_2 stage of the cell cycle.   364

    10.3 Gene Knockouts   368
- Homologous recombination can be used to create gene knockouts in mice.   368

    10.4 Nonhomologous End-Joining   371
- Nonhomologous end-joining connects broken ends that have no homology.   371

    10.5 Meiotic Recombination   372
- Homologous recombination during meiosis can result in gene conversion.   372
- Homologous recombination during meiosis requires some enzymes that are unique to it.   373

    Questions and Problems   376
    Suggested Reading   377

CHAPTER 11 **Transposable Elements   380**

    11.1 Bacterial Transposable Elements   381
- Insertion sequences (IS elements) are simple mobile genetic elements.   381
- A composite transposon consists of a pair of IS elements on either side of one or more genes.   382
- Tn*5* and Tn*10* transposons move via a cut-and-paste transposition mechanism.   383
- The Tn*3* transposon uses replicative transposition to move to a new site while maintaining a copy at the original site.   387
- Transposons are useful tools for bacterial studies.   389

    11.2 Eukaryotic Transposable Elements   391
- Some eukaryotic transposable elements move through a DNA-only mechanism and others require an RNA intermediate.   391
- The eukaryotic "*h*AT" mobile elements move by a DNA only cut-and-paste transposition mechanism.   392
- Antibody and T-cell receptor genes are formed by a mechanism that resembles the Hermes transposition mechanism.   392

A long dormant transposable element in fish called *Sleeping Beauty* has been restored to full activity by recombinant DNA technology. 396

LTR retrotransposons copy an RNA intermediate to make cDNA, which then is integrated into the genome by a cut-and-paste mechanism. 397

Non-LTR elements move using a target-primed reverse transcription mechanism. 398

Alu is the most common SINE in the human genome. 401

There is considerable debate concerning what benefits, if any, retrotransposons make to the host cells. 401

BOX 11.1. IN THE LAB: Reporter Genes 391

BOX 11.2. LOOKING DEEPER: Barbara McClintock and the Discovery of "Jumping Genes" 393

Questions and Problems 402

Suggested Reading 403

CHAPTER 12 Bacterial Transcription and Its Regulation 406

12.1 Introduction to the Bacterial RNA Polymerase Catalyzed Reaction 408

RNA polymerase requires a DNA template and four nucleoside triphosphates to synthesize RNA. 408

Bacterial RNA polymerases are large multisubunit proteins. 410

12.2 Initiation Stage 412

Bacterial RNA polymerase holoenzyme consists of a core enzyme and sigma factor. 412

A transcription unit must have an initiation signal called a promoter for accurate and efficient transcription to take place. 413

Active σ^{70} promoters have −10 and −35 boxes. 416

Genetic and biochemical studies provide additional information about bacterial promoters. 418

DNA footprinting shows that σ^{70} RNA polymerase holoenzyme binds to promoter DNA to form a closed and then an open complex. 418

Bacterial RNA polymerase crystal structures show how the enzyme is organized and provide insights into how it works. 420

Members of the σ^{70} family have four conserved domains. 424

The σ^{70} RNA polymerase holoenzyme goes through several rounds of abortive initiation before promoter escape. 425

RNA polymerase "scrunches" DNA during transcription initiation. 425

12.3 Transcription Elongation Complex 427

The transcription elongation complex is a highly processive molecular motor. 427

Pauses influence the overall transcription elongation rate. 432

RNA polymerase can detect and remove incorrectly incorporated nucleotides. 432

12.4 Transcription Termination 432

Bacterial transcription machinery releases RNA strands at intrinsic and Rho-dependent terminators. 432

12.5 Messenger RNA 434
Bacterial mRNA may be monocistronic or polycistronic. 434
Bacterial mRNA usually has a short lifetime compared with other kinds of bacterial RNA. 435
Controlling the rate of mRNA synthesis can regulate the flow of genetic information. 436
Messenger RNA synthesis can be controlled by negative and positive regulation. 436

12.6 Lactose Operon 437
The *E. coli* genes *lacZ*, *lacY*, and *lacA* code for β-galactosidase, lactose permease, and β-galactoside transacetylase, respectively. 437
The *lac* structural genes are regulated. 438
Genetic studies provide information about the regulation of *lac* mRNA. 439
The operon model explains the regulation of the lactose system. 440
Allolactose is the true inducer of the lactose operon. 442
The Lac repressor binds to the *lac* operator *in vitro*. 443
The *lac* operon has three *lac* operators. 444
The Lac repressor is a dimer of dimers, where each dimer binds to one *lac* operator sequence. 445

12.7 Catabolite Repression 448
E. coli uses glucose in preference to lactose. 448
The cAMP receptor protein combines with 3′,5′-cyclic adenylate to form a positive regulator or activator. 449
The cAMP • CRP complex binds to an activator site (AS) upstream from the *lac* promoter and activates *lac* operon transcription. 450
Glucose causes catabolite repression through cAMP modulation and inducer exclusion mechanisms. 451
cAMP • CRP activates more than 100 operons. 453

12.8 Tryptophan Operon 453
The tryptophan (*trp*) operon is regulated at the levels of transcription initiation, elongation, and termination. 453
BOX 12.1. LOOKING DEEPER: Alternate Sigma Factors 414
BOX 12.2. CLINICAL APPLICATION: Rifamycin 422
BOX 12.3. LOOKING DEEPER: RNA Polymerase Movement 430
Questions and Problems 459
Suggested Reading 460

CHAPTER 13 Eukaryotic Transcription 466

13.1 Introduction to Eukaryotic Nuclear RNA Polymerases 468
The eukaryotic cell nucleus has three different kinds of RNA polymerase. 468
RNA polymerases I, II, and III can be distinguished by their sensitivities to inhibitors. 470
Each nuclear RNA polymerase has some subunits that are unique to it and some it shares with one or both of the two other nuclear RNA polymerases. 471

13.2 RNA Polymerase II Structure 472
Yeast RNA polymerase II crystal structures help explain how the enzyme works. 472
The crystal structure has been determined for the complete 12-subunit yeast RNA polymerase II bound to a transcription bubble and product RNA. 474
Nuclear RNA polymerases have limited synthetic capacities. 474

13.3 Core Promoter for Protein-Coding Genes 476
The core promoter for protein-coding genes extends from 40 bp upstream of the transcription start site to 40 bp downstream from this site. 476

13.4 General Transcription Factors: Basal Transcription 478
RNA polymerase II requires the assistance of general transcription factors to transcribe naked DNA from specific transcription start sites. 478
TFIID or its TBP subunit must bind to a TATA core promoter before other general transcription factors can do so. 480
General transcription factors and RNA polymerase interact at the promoter to form a preinitiation complex. 482

13.5 Transcription Elongation 486
The C-terminal domain of the largest RNA polymerase subunit must be phosphorylated for chain elongation to proceed. 486
A variety of transcription elongation factors helps to suppress transient pausing during elongation. 487
Elongation factor TFIIS reactivates arrested RNA polymerase II. 487

13.6 Regulatory Promoters, Enhancers, and Silencers 487
Linker-scanning mutagenesis reveals the regulatory promoter's presence just upstream from the core promoter. 487
Enhancers stimulate transcription and silencers block transcription. 490
The upstream activating sequence regulates genes in yeast. 492

13.7 Introduction to Transcription Activator Proteins for Protein-Coding Genes 492
Transcription activator proteins help to recruit the transcription machinery. 492
A combinatorial process determines gene activity. 493
DNA affinity chromatography can be used to purify transcription activator proteins. 494
A transcription activator protein's ability to stimulate gene transcription can be determined by a transfection assay. 495

13.8 DNA-Binding Domains in Transcription Activator Proteins 496
Transcription activator proteins are commonly grouped according to the structures of their DNA-binding domains. 496

13.9 Activation Domains in Transcription Activator Proteins 511
Activation domains tend to be intrinsically disordered. 511
Gal4 has DNA-binding, dimerization, and activation domains. 511

Other proteins act along with Gal4 to regulate *GAL* genes. 512
The activation domain must associate with a DNA-binding domain to stimulate transcription. 514

13.10 Mediator 517
Squelching occurs when transcription activator proteins compete for a limiting transcription machinery component. 517
Mediator is required for activated transcription. 517
The yeast Mediator complex associates with activators at the UAS in active yeast genes. 521

13.11 Epigenetic Modifications 523
Cells remodel or modify chromatin to make the DNA in chromatin accessible to the transcription machinery. 523
Histone modification influences transcription of protein-coding genes. 525
DNA methylation plays an important role in determining whether chromatin will be silenced or actively expressed in vertebrates. 527
Epigenetics is the study of inherited changes in phenotype caused by changes in chromatin other than changes in DNA sequence. 527

13.12 RNA Polymerase I Catalyzed Transcription 528
RNA polymerase I is required to synthesize 5.8S, 18S, and 28S rRNA. 528
The rRNA transcription unit promoter consists of a core promoter and an upstream promoter element (UPE). 531
5.8S, 18S, and 28S rRNA syntheses take place in the nucleolus. 532
RNA polymerase I is a multisubunit enzyme with a structure similar to that of RNA polymerase II. 532
The upstream binding factor and selectivity factor, working together, recruit RNA polymerase I to the rDNA promoter to form a pre-initiation complex. 533
RNA polymerase I forms a transcription elongation complex, leaving UBF and SL1/TIF-1B behind. 533
RNA polymerase I transcription termination requires the assistance of a termination factor and a release protein. 534

13.13 RNA Polymerase III Catalyzed Transcription 535
RNA polymerase III transcripts are short RNA molecules with a variety of biological functions. 535
RNA polymerase III transcription units have three different types of promoters. 535
RNA polymerase III does not appear to require additional factors for transcription elongation or termination. 537
BOX 13.1. LOOKING DEEPER: The Archaeal RNA Polymerase 475
BOX 13.2. LOOKING DEEPER: TFIIB Assists in Open Promoter Formation 484
BOX 13.3. CLINICAL APPLICATION: The AP-1 Family of Transcription Activator Proteins 507
BOX 13.4. IN THE LAB: Yeast Two-Hybrid Assay 518
BOX 13.5. LOOKING DEEPER: Genomic Imprinting 529
Questions and Problems 538
Suggested Reading 541

CHAPTER 14 RNA Polymerase II: Cotranscriptional and Posttranscriptional Processes 551

14.1 Pre-mRNA 552
Eukaryotic cells synthesize large heterogeneous nuclear RNA molecules. 552
mRNA and hnRNA both have poly(A) tails at their 3'-ends. 552

14.2 Cap Formation 553
mRNA molecules have 7-methylguanosine caps at their 5'-ends. 553
5'-m^{7}G caps are attached to nascent pre-mRNA chains when the chains are 20 to 30 nucleotides long. 556
All eukaryotes use the same basic pathway to form 5'-m^{7}G caps. 556
CTD must be phosphorylated on Ser-5 to target a transcript for capping. 558

14.3 Split Genes 558
Viral studies revealed that some mRNA molecules are formed by splicing pre-mRNA. 558
Amino acid coding regions within eukaryotic genes may be interrupted by noncoding regions. 559
Exons tend to be conserved during evolution, whereas introns usually are not conserved. 564
A single pre-mRNA can be processed to produce two or more different mRNA molecules. 566
Combinations of the various splicing patterns within individual genes lead to the formation of multiple mRNAs. 568
Pre-mRNA requires specific sequences for precise splicing to occur. 569
Two splicing intermediates resemble lariats. 571
Splicing consists of two coordinated transesterification reactions. 574

14.4 Spliceosomes 575
Aberrant antibodies, which are produced by individuals with certain autoimmune diseases, bind to small nuclear ribonucleoprotein particles (snRNPs). 575
snRNPs assemble to form a spliceosome, the splicing machine that excises introns. 577
RNA and protein may both contribute to the spliceosome's catalytic site. 577
Cells use a variety of mechanisms to regulate splice site selection. 579
Splicing begins as a cotranscriptional process and continues as a posttranscriptional process. 583
mRNA splicing and export are coupled processes. 584

14.5 Cleavage/Polyadenylation and Transcription Termination 584
Poly(A) tail synthesis and transcription termination are coupled, cotranscriptional processes. 584
Transcription units often have two or more alternate polyadenylation sites. 587
Transcription termination takes place downstream from the poly(A) site. 588

RNA polymerase II transcription termination appears to involve allosteric changes and a 5′→3′ exonuclease. 589

14.6 RNA Editing 591
RNA editing permits a cell to recode genetic information. 591

14.7 The Gene Reconsidered 592
The human proteome contains a much greater variety of proteins than would be predicted from the human genome. 592
Cotranscriptional and posttranscriptional processes force us to reconsider our concept of the gene. 592
BOX 14.1. LOOKING DEEPER: Self-Splicing RNA 580
Questions and Problems 593
Suggested Reading 594

CHAPTER 15 Small Silencing RNAs 600

15.1 RNA Interference (RNAi) Triggered by Exogenous Double-Stranded RNA 601
The roundworm *Caenorhabditis elegans* is an attractive organism for molecular biology studies. 601
RNAi was discovered in *C. elegans*. 602
In vitro studies helped to elucidate the RNAi pathway. 605
Dicer cleaves long double-stranded RNA into fragments of discrete size. 607
RISC loading complex is required for siRISC formation. 609
RNAi blocks virus replication and prevents transposon activation. 609

15.2 Transitive RNAi 610
In some organisms, RNAi that starts at one site spreads throughout the entire organism. 610
SID-1, an integral membrane protein in *C. elegans*, assists in the systemic spreading of the silencing signal. 612
ERI-1, a 3′→5′ exonuclease in *C. elegans*, appears to be a negative regulator of RNAi. 612

15.3 RNAi as an Investigational Tool 612
RNAi is a powerful tool for investigating functional genomics. 612

15.4 MicroRNA Pathway 614
The miRNA pathway blocks mRNA translation or causes mRNA degradation. 614

15.5 Piwi Interacting RNAs (piRNAs) 617
piRNAs help to maintain germ line stability in animals. 617
BOX 15.1. LOOKING DEEPER: Dicer from *Giardia intestinalis* 607
Questions and Problems 618
Suggested Reading 619

CHAPTER 16 Protein Synthesis 623

16.1 Introduction to the Ribosome 625
Protein synthesis takes place on ribosomes. 625
Bacterial ribosomes are made of a large subunit with a 23S and 5S RNA and a small subunit with 16S RNA. 625

A eukaryotic ribosome also has a small and a large ribonucleoprotein subunit. 625

Eukaryotic ribosomes exist free in the cytoplasm or attached to the endoplasmic reticulum. 627

16.2 Transfer RNA 628

An amino acid must be attached to a transfer RNA before it can be incorporated into a protein. 628

All tRNA molecules have CCA_{OH} at their 3′-ends. 631

An amino acid attaches to tRNA through an ester bond between the amino acid's carboxyl group and the 2′- or 3′-hydroxyl group on adenosine. 633

Yeast tRNA^{Ala} was the first naturally occurring nucleic acid to be sequenced. 635

tRNAs have cloverleaf secondary structures. 635

tRNA molecules fold into L-shaped three-dimensional structures. 638

16.3 Aminoacyl-tRNA Synthetases 639

Some aminoacyl-tRNA synthetases have proofreading functions. 639

Ile-tRNA synthetase has a proofreading function. 639

Ile-tRNA synthetase can hydrolyze valyl-tRNA^{Ile} and valyl-AMP. 640

Each aminoacyl-tRNA synthetase can distinguish its cognate tRNAs from all other tRNAs. 641

Selenocysteine and pyrrolysine are building blocks for polypeptides. 645

16.4 mRNA and the Genetic Code 647

mRNA programs ribosomes to synthesize proteins. 647

Three adjacent bases in mRNA that specify an amino acid are called a codon. 648

The discovery that poly(U) directs the synthesis of poly(Phe) was the first step in solving the genetic code. 650

Protein synthesis begins at the amino terminus and ends at the carboxyl terminus. 652

mRNA is read in a 5′ to 3′ direction. 653

Trinucleotides promote the binding of specific aminoacyl-tRNA molecules to ribosomes. 654

Synthetic messengers with strictly defined base sequences confirmed the genetic code. 654

Three codons, UAA, UAG, and UGA, are polypeptide chain termination signals. 656

The genetic code is nonoverlapping, commaless, almost universal, highly degenerate, and unambiguous. 657

The coding specificity of an aminoacyl-tRNA is determined by the tRNA and not the amino acid. 658

Some aminoacyl-tRNA molecules bind to more than one codon because there is some play or wobble in the third base of a codon. 659

16.5 Ribosome Structure 660

Bacterial 30S (small) subunits and 50S (large) subunits each have unique structures and functions. 660

Bacterial ribosome structure has been determined at atomic resolution. 661

16.6 Four Stages of Protein Synthesis 662
Protein synthesis can be divided into four stages. 662

16.7 Initiation Stage 664
Each bacterial mRNA open reading frame has its own start codon. 664
Bacteria have an initiator methionine tRNA and an elongator methionine tRNA. 664
The 30S subunit is an obligatory intermediate in polypeptide chain initiation. 665
Initiation factors participate in the formation of 30S and 70S initiation complexes. 667
The Shine-Dalgarno sequence in mRNA interacts with the anti-Shine-Dalgarno sequence in the 16S rRNA. 668
Eukaryotic initiator tRNA is charged with a methionine that is *not* formylated. 671
Eukaryotic translation initiation proceeds through a scanning mechanism. 673
Translation initiation factor phosphorylation regulates protein synthesis in eukaryotes. 676
The translation initiation pathway in archaea appears to be a mixture of the eukaryotic and bacterial pathways. 676

16.8 Elongation Stage 677
Polypeptide chain elongation requires elongation factors. 677
The elongation factors act through a repeating cycle. 677
An EF-Tu • GTP • aminoacyl-tRNA ternary complex carries the aminoacyl-tRNA to the ribosome. 677
Specific nucleotides in 16S rRNA are essential for sensing the codon–anticodon helix. 679
EF-Ts is a GDP-GTP exchange protein. 679
The ribosome is a ribozyme. 680
The hybrid-states translocation model offers a mechanism for moving tRNA molecules through the ribosome. 684

16.9 Termination Stage 685
Bacteria have three protein release factors. 685
Mutant tRNA molecules can suppress mutations that create termination codons within a reading frame. 687

16.10 Recycling Stage 687
The ribosome release factor is required for the bacterial ribosomal complex to disassemble. 687

16.11 Nascent Polypeptide Processing and Folding 688
Ribosomes have associated enzymes that process nascent polypeptides and chaperones that help to fold the nascent polypeptides. 688

16.12 Signal Sequence 689
The signal sequence plays an important role in directing newly synthesized proteins to specific cellular destinations. 689
BOX 16.1. LOOKING DEEPER: Structural Basis for Posttransfer Editing 642
BOX 16.2. CLINICAL APPLICATION: A Neurodegenerative Disease Caused by an Editing Defect in Alanyl-tRNA Synthetase 643

BOX 16.3. LOOKING DEEPER: Bacterial Pathway for Selenocysteyl-tRNA^{Sec} Formation 647
BOX 16.4. LOOKING DEEPER: Structures of Eukaryotic and Archaeal Ribosomes 663
BOX 16.5. LOOKING DEEPER: Riboswitches 672
BOX 16.6. LOOKING DEEPER: Translocation Mechanism 686
Questions and Problems 691
Suggested Reading 693

GLOSSARY 703

INDEX 733

Preface

Principles of Molecular Biology, which describes the flow of genetic information from genes to proteins, provides an introduction to molecular biology for undergraduate students who have completed a full-year course in both college biology and general chemistry and at least a one-semester course in organic chemistry. Whenever possible the book uses a discovery approach, which introduces the reader to the experimental evidence for key concepts. This pedagogical approach provides historical and experimental background information that permits the reader to see how molecular biologists examine clues and formulate the hypotheses that lead to new advances in the field. Thus, the reader becomes part of the discovery process and begins to understand the pleasure and sense of satisfaction investigators derive from solving a molecular biology problem.

Chapter Organization

Chapter 1 provides a brief historical introduction to molecular biology including the discovery of the double helix model. This model provides a very clear example of structure and function relationships, a major theme in molecular biology and one that is emphasized throughout this book. **Chapter 2** introduces important concepts related to the primary, secondary, tertiary, and quaternary structures of proteins. It also explores structure-function relationships in proteins, using hemoglobin, myoglobin, and immunoglobulin G as illustrative examples. The final section of Chapter 2 describes enzymes and enzyme kinetics, topics that are central to understanding the processes of DNA replication, repair, and recombination, RNA synthesis and processing, and protein synthesis. **Chapter 3** builds on the information provided in Chapter 1 about DNA structure. It emphasizes conformational variations, helical stability, strand separation, helical reformation, circular DNA, and topoisomers. The last section of the chapter examines RNA structure. **Chapter 4** includes an examination of physical techniques used to isolate and characterize nucleic acids and a discussion of techniques that use enzymes to manipulate and sequence DNA in the laboratory. **Chapter 5** describes the interactions between specific proteins and DNA to form nucleoprotein complexes. It also describes events that take place during mitosis and meiosis. **Chapter 6** introduces concepts in genetic analysis that are essential for work in molecular biology. It includes a brief introduction to genetic recombination and descriptions of basic techniques used by molecular biologists to locate genes in bacteria, yeast, and higher organisms. **Chapter 7** provides basic information that molecular biologists need to know about viruses. The information presented will help to place viral systems that are discussed in subsequent chapters in perspective. Some students and instructors may decide to skip this chapter but then refer back to appropriate sections when specific viruses are mentioned in subsequent chapters. **Chapter 8** begins with a description of the general features of DNA replication. It then examines the initiation, elongation, and termination stages of bacterial DNA replication. The information gained by this examination is then applied to the study of DNA replication in eukaryotes. **Chapter 9** begins with an examination of the different types of DNA damage that take place in cells. It then explores the mechanisms cells use to reverse DNA damage, excise and replace damaged elements, or tolerate the damage. **Chapter 10** explores DNA double-strand break repair and homologous recombination. It begins by examining replication restart after a bacterial replication fork collapse and then builds on this information to explore homologous recombination and nonhomologous end-joining. **Chapter 11** examines mobile DNA sequences called transposable elements or transposons and explores the mechanism of movement. **Chapter 12** begins with an examination of bacterial RNA polymerase and its function in RNA synthesis and then studies the mechanisms that bacteria use to regulate mRNA synthesis. **Chapter 13** begins with an examination of eukaryotic RNA polymerase II (the enzyme responsible for mRNA formation) and the general transcription factors that it requires for basal transcription. It then describes two additional types of transcription factors, transcription activators and Mediators, that interact with the basal transcription machinery to form the much more efficient transcription machine that is in fact responsible for eukaryotic mRNA synthesis. It also explores how epigenetic modifications influence mRNA formation and the contribution that RNA polymerase I makes to 5.8S, 18S, and 28S ribosomal RNA synthesis and RNA polymerase III makes to 5S ribosomal RNA synthesis. **Chapter 14** investigates the various stages in cotranscriptional processing. It also describes RNA editing and messenger RNA export. **Chapter 15** examines roles that RNAi, miRNA, and piwiRNA play in regulating gene expression. **Chapter 16** begins with a description of the structure and function of transfer RNA and then examines how these molecules act along with mRNA and ribosomes to specify the amino acid sequences in proteins. The second part of the chapter examines ribosome structure and function to discover how ribosomes are able act as universal translators.

Key Features

- **Introduction**—Each chapter begins with a brief description of the material that is presented in that chapter.
- **Topic sentences**—Each section begins with a topic sentence to alert the reader to the important concept described in that section.
- **Important terms shown in bold**—Important terms within each chapter are shown in bold.
- **Looking Deeper Boxes**—These boxes present mechanisms, structure-function relationships, and other information of interest to some readers in greater detail.
- **Clinical Application Boxes**—These boxes describe important medical applications.
- **In the Lab Boxes**—These boxes describe important techniques that are used in the molecular biology laboratory.

- **Questions and Problems**—End-of-chapter problems and questions will help readers to evaluate their understanding of the material within the chapter.
- **Suggested Reading**—Each chapter ends with a suggested reading section, which is organized by topic. A reader desiring more information can refer to the references in this section.

Supplements to the Text

Jones & Bartlett Learning offers an array of ancillaries to assist instructors and students in teaching and mastering the concepts presented in this text. Additional information and review copies of the following items are available through your Jones & Bartlett Learning sales representative or by going to www.jblearning.com.

For the Student

Developed exclusively for this first edition of *Principles of Molecular Biology*, and authored by Brent Nielsen of Brigham Young University, the **Student Companion Website** offers a variety of resources to enhance understanding of molecular biology. Free access is provided with each new print copy of the text. This website contains chapter outlines, quizzes to test comprehension and retention, and an interactive glossary. The site also provides links to relevant materials such as animations, structural programs, and tutorials that are available on the Web. The URL for the website is http://biology.jbpub.com/molecular/principles.

For the Instructor

Compatible with Windows and Macintosh platforms, the Instructor's Media CD-ROM provides instructors with the following traditional ancillaries:

The **PowerPoint® Image Bank** provides the illustrations, photographs, and tables (to which Jones & Bartlett Learning holds the copyright or has permission to reproduce digitally) inserted into the PowerPoint slides. You can quickly and easily copy individual images or tables into your existing lecture outlines.

The **PowerPoint Lecture Outline** presentation package, authored by Samuel Galewsky of Millikin University, provides lecture notes and images for each chapter of *Principles of Molecular Biology*. Instructors with the Microsoft PowerPoint software can customize the outlines, art, and order of presentation.

A **Test Bank**, provided as a text file and authored by Brent Nielsen of Brigham Young University, is available to instructors through the Jones & Bartlett Learning website. Visit www.jblearning.com for more details.

Acknowledgments

I would like to acknowledge the assistance of the many people who helped to prepare *Principles of Molecular Biology*. To begin, I wish to thank the following talented scholars for reviewing one or more chapters and, in the process of doing so, correcting errors and making valuable suggestions to improve the book:

Aaron Cassill, University of Texas at San Antonio
Akif Uzman, University of Houston–Downtown
Charles L. Vigue, University of New Haven
Donna L. Pattison, University of Houston
Hannah L. Klein, New York University
Ioannis Eleftherianos, George Washington University
Joseph E. Flaherty, Coker College

Lisa M. Sardinia, Pacific University
Matthew Bahamonde, Farmingdale State College
Michael W. Persans, University of Texas–Pan American
Peter Oelkers, University of Michigan–Dearborn
R. Paul Evans, Brigham Young University
Samuel Galewsky, Millikin University
Sanghamitra Saha, Harold Washington College
Shawn E. Krosnick, Southern Arkansas University
Shere Byrd, Fort Lewis College
Subba Reddy Palli, University of Kentucky
Yevgeniya Lapik, Harold Washington College

I also wish to express my appreciation to the publisher, editors, and staff at Jones & Bartlett Learning, who contributed to the preparation of *Principles of Molecular Biology*. I am very grateful to Cathy Sether, former Publisher, for her advice and assistance throughout the project. It is a pleasure to express special thanks to Molly Steinbach, former Acquisitions Editor, for her help and encouragement during the early stages of the project and to Erin O'Conner, Senior Acquisitions Editor, for her assistance during the later stages of the project. I would also like to thank their editorial assistants, Rachel Isaacs and Michelle Bradbury, for supervising the review process. I am very grateful to Louis Bruno, Production Manager, who coordinated the copyediting, artwork, and other production activities and made the entire process run smoothly. I greatly appreciate his patience, dedication, and incredible organizational skills. Elizabeth Morales created instructive and beautiful figures, often working from crude sketches. Pamela Thomson copyedited the manuscript, making many improvements in the process. Jan Cocker served as proofreader, helping to ensure that errors did not slip through from the copyedited version to the final version that appears in this book. It is a pleasure to acknowledge Alexandra Nickerson for her work in preparing the index. I am very grateful to Ashley Dos Santos and Lauren Miller for their help in obtaining necessary permissions for the figures that appear in this book and to the authors, journal editors, and publishers for granting permissions to use copyrighted material. Every effort was made to seek such permission whenever copyrighted material was used and to make full disclosure of the source. If errors of omission have been made, they are wholly inadvertent and will be corrected at the first opportunity to do so.

Finally, I wish to thank my wife Roslyn for her encouragement, support, and understanding that made this book possible.

Burton E. Tropp

Introduction to Molecular Biology

CHAPTER OUTLINE

1.1 Intellectual Foundation
Two studies performed in the 1860s provided the intellectual underpinning for molecular biology.

1.2 One Gene–One Polypeptide Hypothesis
Each gene is responsible for the synthesis of a single polypeptide.

1.3 Introduction to Nucleic Acids
DNA contains the sugar deoxyribose and RNA contains the sugar ribose.
A nucleoside is formed by attaching a purine or pyrimidine base to a sugar.
A nucleotide is formed by attaching a phosphate group to the sugar group in a nucleoside.
DNA is a linear chain of deoxyribonucleotides.

1.4 DNA: Hereditary Material
Transformation experiments led to the discovery that DNA is the hereditary material.
Chemical experiments also supported the hypothesis that DNA is the hereditary material.

1.5 Watson-Crick Model
Rosalind Franklin and Maurice Wilkins obtained x-ray diffraction patterns of extended DNA fibers.
James Watson and Francis Crick proposed that DNA is a double-stranded helix.

1.6 Central Dogma
The central dogma provides the theoretical framework for molecular biology.

1.7 Introduction to Recombinant DNA Technology
Recombinant DNA technology allows us to study complex biological systems.
A great deal of molecular biology information is available on the Internet.

QUESTIONS AND PROBLEMS

SUGGESTED READING

Image © Juan Gaertner/ShutterStock, Inc.

The first recorded use of the term **molecular biology** appears to have been in a 1938 report prepared for the Rockefeller Foundation by Warren Weaver, the director of the Foundation's Natural Science Division. The report proposed that the Foundation should start to fund research efforts to seek molecular explanations for biological processes. Weaver used the term "molecular biology" to describe a research approach he envisioned would use the physical sciences to address fundamental biological problems. Weaver's proposal was remarkably farsighted, especially when one considers many of his contemporaries believed living cells possessed a vital force that could not be explained by chemical or physical laws that govern the inanimate world.

At the time of Weaver's report two new disciplines—genetics and biochemistry—were altering the way biologists thought about living systems. Geneticists established that the functional and physical unit of heredity is the gene. However, they did not know the chemical nature of genes, the way that hereditary information is stored in genes, how genes are replicated so they can be transmitted to the next generation, or how information stored in genes determines a specific physical trait such as eye color. Biochemists had delivered a major blow to the vital force theory by demonstrating that cell-free extracts can perform many of the same functions as intact cells. However, they knew very little about protein or nucleic acid structure and function.

Neither genetics nor biochemistry had the power to investigate the chemical basis of heredity on its own. In fact, it took an interdisciplinary effort involving specialists in many fields of the life sciences, including genetics, biochemistry, biophysics, microbiology, chemistry, x-ray crystallography, virology, developmental biology, and immunology, to solve the hereditary problem. This interdisciplinary effort resulted in the creation of a new discipline, molecular biology, which has had a remarkable impact on all areas of the life sciences from agronomy to zoology. The scope of this book is limited to an examination of the contributions that molecular biology has made to understanding the flow of information from genes to proteins.

1.1 Intellectual Foundation

Two studies performed in the 1860s provided the intellectual underpinning for molecular biology.

The earliest intellectual roots of molecular biology can be traced back to the work of two investigators in the 1860s. No connection was apparent between the experiments performed by the two investigators for more than 75 years, but when the connection finally was made, the result was the birth of molecular biology and the beginning of a scientific revolution that continues today.

The work of the first investigator, Gregor Mendel, an Austrian monk and botanist, is familiar to all biology students and so is only summarized briefly here. Mendel discovered three basic laws of inheritance by studying the way in which simple physical traits are passed on from one generation of pea plants to the next. For convenience, Mendel's laws of inheritance will be described using two modern biological terms, **gene** for a unit of heredity and **chromosome** for a structure bearing several linked genes.

1. *The law of segregation:* A specific gene may exist in alternate forms called **alleles**. An organism inherits one allele for each trait from each parent. The two alleles, which may be the same or different, segregate (or separate) during germ cell (sperm or egg) formation and combine again as a result of fertilization so that each parent transmits one allele to each offspring.
2. *The law of independent assortment:* Specific physical traits such as plant size and color are inherited independently of one another. Mendel was fortunate to have selected physical traits that were determined by genes that were on different chromosomes.
3. *The law of dominance:* An allele may be **dominant** or **recessive**. A dominant allele produces its characteristic physical trait whether it is paired with an identical allele or a recessive allele. In contrast, a recessive allele produces its characteristic physical trait only when paired with an identical allele. In pea plants, tallness is dominant and shortness recessive. When Mendel allowed pea plants with one allele for tallness and one allele for shortness to self-fertilize, he observed three times as many progeny were tall as short. Today, we know there are exceptions to the law of dominance. Sometimes neither allele is dominant. For instance, a plant that inherits a gene for a red flower and a gene for a white flower may produce an intermediate phenotype, a pink flower.

Unfortunately, scientists failed to recognize the significance of Mendel's work during his lifetime. His paper remained little-known until about 1900, when scientists rediscovered Mendel's laws of inheritance, giving birth to the science of genetics.

The second investigator, the Swiss physician Friedrich Miescher, performed experiments that led to the discovery of deoxyribonucleic acid (DNA), which we, of course, now know is the hereditary material. Miescher did not set out to discover the hereditary material but instead was interested in studying cell nuclei from white blood cells, which he collected from pus discharges on discarded bandages used to cover infected wounds. Miescher used a combination of protease (an enzyme that hydrolyzes proteins) digestion and solvent extraction to disrupt and fractionate the white blood cells. One fraction, which he called nuclein, contained an acidic material with unusually high phosphorus content. Miescher later found that salmon sperm cells, which have remarkably large cell nuclei, are also an excellent source of nuclein. In 1889 Miescher's student, Richard Altmann, separated nuclein into protein and a substance with a very high phosphorous content that he named **nucleic acid**.

1.2 One Gene–One Polypeptide Hypothesis

Each gene is responsible for the synthesis of a single polypeptide.

Mendel's experiments showed that the genetic makeup of an organism, its **genotype**, determines the organism's physical traits, its **phenotype**. However, his experiments did not show how genes are able to determine complex physical traits such as plant color or size. Archibald Garrod, an English physician, was the first to provide an explanation for the relationship between genotype and phenotype. Garrod uncovered this relationship in the early 1920s while studying alkaptonuria, a rare inherited human disorder in which the urine of affected individuals becomes very dark upon standing due to the oxidation of homogentisic acid, a breakdown product of the amino acid tyrosine. Garrod correctly proposed that alkaptonuria results from a recessive gene, which causes a deficiency in an enzyme needed to convert homogentisic acid to colorless products.

Garrod's work was generally ignored until the early 1940s, when two American scientists, George Beadle and Edward Tatum, provided additional experimental proof for a relationship between genes and enzymes. Beadle and Tatum thought that if a gene really does specify an enzyme, it should be possible to create genetic mutants that cannot carry out specific enzymatic reactions in a biochemical pathway. They decided to work with the bread mold, *Neurospora crassa*, because it could be subjected to genetic analysis. In addition, *N. crassa* has very simple nutritional requirements. It can grow in a minimal medium that contains a single carbon source such as sucrose, inorganic salts, and the vitamin biotin. Beadle and Tatum began by irradiating *N. crassa* with x-rays to generate mutants with additional nutritional growth requirements. A mutant that requires a specific nutritional supplement that is not required by the parent cell is called an **auxotroph**.

In one set of experiments Beadle and Tatum focused particular attention on arginine auxotrophs. Genetic analysis of these auxotrophs revealed that mutations in three different genes produce auxotrophs for the amino acid arginine. Each kind of auxotroph was blocked at a specific step in the arginine biosynthetic pathway and accumulated large quantities of the substance formed just before the blocked step. Beadle and Tatum thus had replicated in the bread mold the same type of situation Garrod had observed in alkaptonuria. A defective gene caused a defect in a specific enzyme that resulted in the abnormal accumulation of an intermediate in a metabolic pathway.

As a result of their work with the *N. crassa* mutants, Beadle and Tatum proposed the one gene–one enzyme hypothesis, which states that each gene is responsible for synthesizing a single enzyme. We now know that many enzymes are made of more than one type of polypeptide chain and that a single mutation may affect just one of the polypeptide chains. The original one gene–one enzyme hypothesis hence was modified to become a *one gene–one polypeptide hypothesis*. As we will see later, however, even the one gene–one polypeptide hypothesis is an oversimplification.

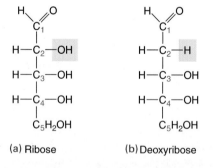

FIGURE 1.1 The two sugars present in nucleic acids. (a) Ribose and (b) deoxyribose each contain five carbon atoms and an aldehyde group and therefore belong to aldopentose family of simple sugars. The only difference between the two sugars is that ribose has a hydroxyl group at carbon-2 and deoxyribose as a hydrogen atom at carbon-2.

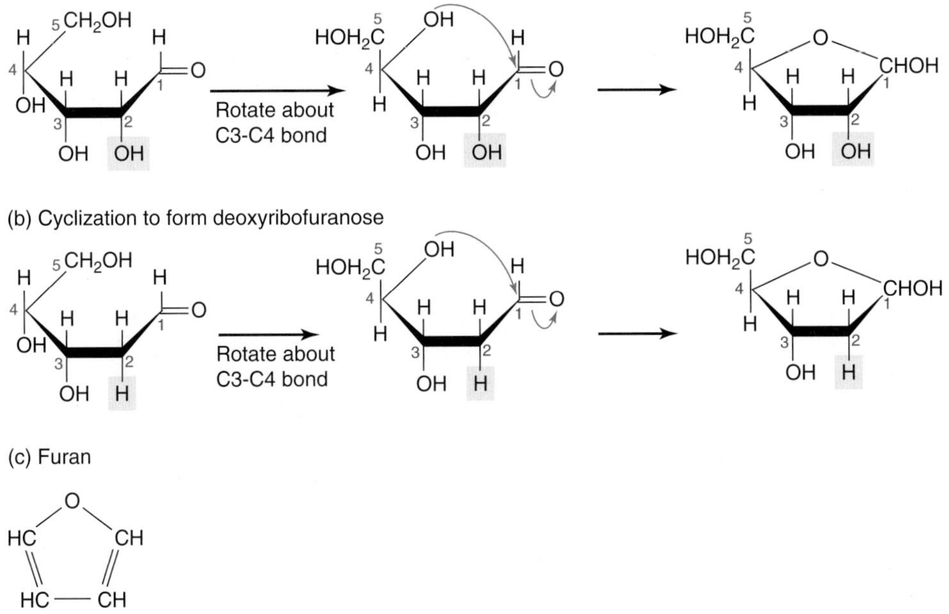

(a) Cyclization to form ribofuranose

(b) Cyclization to form deoxyribofuranose

(c) Furan

FIGURE 1.2 Conversion of straight chain ribose and deoxyribose to cyclic forms. (a) Conversion of ribose to ribofuranose and (b) conversion of deoxyribose to deoxyribofuranose. (c) Structure of furan, a five-atom ring system.

1.3 Introduction to Nucleic Acids

DNA contains the sugar deoxyribose and RNA contains the sugar ribose.

Investigators slowly came to realize that nucleic acids could be divided into two major groups: DNA and ribonucleic acid (RNA). The principal difference between DNA and RNA is that the former contains deoxyribose and the latter contains ribose (FIGURE 1.1). The two five-carbon sugars (pentoses) differ in only one substituent: deoxyribose has a hydrogen atom at carbon-2, whereas ribose has a hydroxyl group at this position. By convention, the aldehyde group in the pentose is C-1, the next carbon is C-2, and so forth.

Each pentose chain can close to form a five-carbon ring in which an oxygen bridge joins C-1 to C-4 (FIGURE 1.2ab). Because the sugar rings are derivatives of furan (Figure 1.2c) they are called **furanoses**. Ribofuranose and deoxyribofuranose are often depicted as Haworth structures (named after Walter N. Haworth, the investigator who devised the representations). As illustrated in FIGURE 1.3, a Haworth structure represents a cyclic sugar as a flat ring perpendicular to the plane of the page with the ring-oxygen in the back and C-1 to the right. The ring's thick lower edge projects toward the viewer, its upper edge projects back behind the page, and its substituents are visualized as being either above or below the plane of the ring. A line is used to

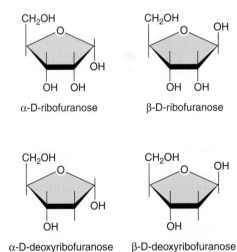

FIGURE 1.3 Haworth structures for ribofuranose and deoxyribofuranose. Haworth structure represents a cyclic sugar as a flat ring perpendicular to the plane of the page with the ring-oxygen in the back and C-1 to the right. A hydrogen atom attached to the sugar ring is represented by a line. Ring formation can lead to two different stereochemical arrangements at C-1: one in which the hydroxyl at C-1 points down and another in which it points up.

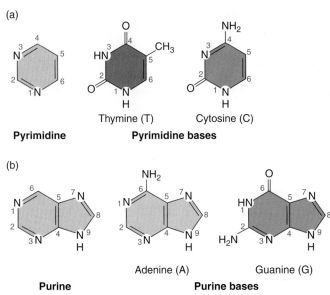

FIGURE 1.4 Pyrimidine and purine bases in DNA. (a) Pyrimidine bases and (b) purine bases.

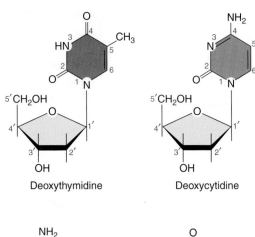

FIGURE 1.5 Uracil (U).

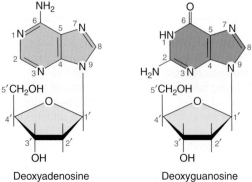

FIGURE 1.6 Deoxyribonucleosides.

represent a hydrogen atom attached to the sugar ring. Ring formation can lead to two different stereochemical arrangements at C-1. One arrangement, the α-anomer, is represented by drawing the hydroxyl group attached to C-1 below the plane of the ring and the other, the β-anomer, by drawing it above the plane of the ring. (Although ribofuranose and deoxyribofuranose actually have puckered rather than planar conformations, Haworth structures are convenient representations for the pentose rings when precise three-dimensional information is not required and, therefore, are used throughout this book.)

A nucleoside is formed by attaching a purine or pyrimidine base to a sugar.

Each deoxyribose group in DNA is attached to one of four different nitrogen heterocyclic molecules, referred to as bases because they can act as proton acceptors. Two bases, **thymine** (**T**) and **cytosine** (**C**), are derivatives of **pyrimidine** (FIGURE 1.4a) and the other two, **adenine** (**A**) and **guanine** (**G**), are derivatives of **purine** (Figure 1.4b). RNA also contains cytosine, adenine, and guanine, but the pyrimidine **uracil** (**U**) replaces thymine (FIGURE 1.5). The only difference between uracil and thymine is that the latter contains a methyl group attached to carbon-5. T, C, A, and G combine with deoxyribose to form a class of compounds called deoxyribonucleosides (FIGURE 1.6), and U, C, A, and G combine with ribose to form a related class of compounds called ribonucleosides (FIGURE 1.7). Each base is linked to the pentose ring by a bond that joins a specific nitrogen atom on the base (N-1 in pyrimidines and N-9 in purines) to C-1 on the furanose ring. This bond is termed an **N-glycosylic bond**. Because each nucleoside has two ring systems (the sugar and the base attached to it), a method is required to distinguish between atoms in each ring system. This problem is solved by adding a prime (′) after the sugar atoms. The first carbon atom in the sugar thus becomes 1′, the second 2′, and so forth.

A nucleotide is formed by attaching a phosphate group to the sugar group in a nucleoside.

Nucleosides that have a phosphate group attached to the sugar group are called nucleotides. Ribonucleoside and deoxyribonucleoside derivatives are called ribonucleotides and deoxyribonucleotides, respectively (FIGURE 1.8). The pentose carbon atom to which the phosphate group is attached is given as part of the nucleotide's name. Thus, the phosphate group is attached to C-5′ in uridine-5′-monophosphate (5′-UMP) and thymidine-5′-monophosphate (5′-dTMP) and to C-3′ in uridine-3′-monophosphate (3′-UMP) and thymidine-3′-monophosphate (3′-dTMP). As indicated in TABLE 1.1, nucleoside monophosphates have two alternative names. For example, cytidine-5′-monophosphate (5′-CMP) is also known as 5′-cytidylate. Each nucleoside monophosphate is

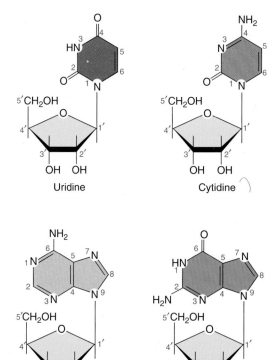

FIGURE 1.7 **Ribonucleosides.**

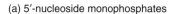
(a) 5′-nucleoside monophosphates

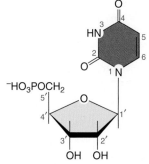
Uridine-5′-monophosphate (5′-UMP) or 5′-uridylate

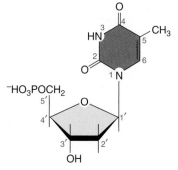
Thymidine-5′-monophosphate (5′-dTMP) or 5′-thymidylate

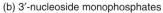
(b) 3′-nucleoside monophosphates

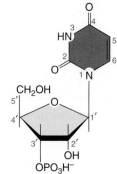
Uridine-3′-monophosphate (3′-UMP) or 3′-uridylate

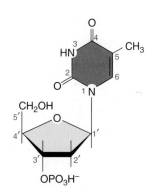
Thymidine-3′-monophosphate (3′-dTMP) or 3′-thymidylate

FIGURE 1.8 **Nucleotides.** Nucleotides are formed by adding a phosphate group to the pentose ring in a nucleoside. (a) Nucleotides formed by adding a phosphate group to the 5′-hydroxyl group in uridine or thymidine. (b) Nucleotides formed by adding a phosphate group to the 3′-hydroxyl group in uridine or thymidine.

| TABLE 1.1 Bases, Nucleosides, and Nucleotides ||||
Base	Sugar	Nucleoside	5'-Mononucleotide
Uracil (U)	ribose	uridine	Uridine-5'-monophosphate or 5'-uridylate (5'-UMP)
Cytosine (C)	ribose	cytidine	Cytidine-5'-monophosphate or 5'-cytidylate (5'-CMP)
Adenine (A)	ribose	adenosine	Adenosine-5'-monophosphate or 5'-adenylate (5'-AMP)
Guanine (G)	ribose	guanosine	Guanosine-5'-monophosphate or 5'-guanylate (5'-GMP)
Thymine (T)	deoxyribose	deoxythymidine[1]	Deoxythymidine-5'-monophosphate or 5'-deoxythymidylate (5'-dTMP)[1]
Cytosine (C)	deoxyribose	deoxycytidine	Deoxycytidine-5'-monophosphate or 5'-deoxycytidylate (5'-dCMP)
Adenine (A)	deoxyribose	deoxyadenosine	Deoxyadenosine-5'-monophosphate or 5'-deoxyadenylate (5'-dAMP)
Guanine (G)	deoxyribose	deoxyguanosine	Deoxyguanosine-5'-monophosphate or 5'-deoxyguanylate (5'-dGMP)

[1]Deoxythymidine and deoxythymidine-5'-monophosphate are also called thymidine and thymidine-5'-monophosphate, respectively. When thymine is attached to ribose, the nucleoside is called ribothymidine and the nucleotide is called ribothymidylate. This nomenclature convention follows from the fact that thymine is most frequently attached to deoxyribose.

phosphomonoester because it has a single sugar group attached to a phosphate by an ester bond.

DNA is a linear chain of deoxyribonucleotides.

The deoxyribonucleotide groups in DNA form a linear polymer in which phosphate groups join 5'- and 3'-carbons of neighboring deoxyribonucleosides (FIGURE 1.9a). The bond that joins neighboring nucleosides to a common phosphate group is called a **phosphodiester bond**. *Each linear DNA chain has a 5'- and a 3'-terminus.* This directionality will be very important in future discussions of DNA structure and function. The structure of RNA is very similar to that for DNA (FIGURE 1.10a).

Just 3 years after the discovery that nucleic acids are linear chains of nucleotides, Frederick Sanger, working in England, showed that each kind of polypeptide molecule is a linear chain of amino acids arranged in a specific order. The amino acid order is responsible for the unique properties of each type of protein, including its ability to catalyze specific reactions, protect cells from foreign organisms and substances, transport materials, or support the cellular infrastructure. The awareness that DNA molecules are made of linear chains of

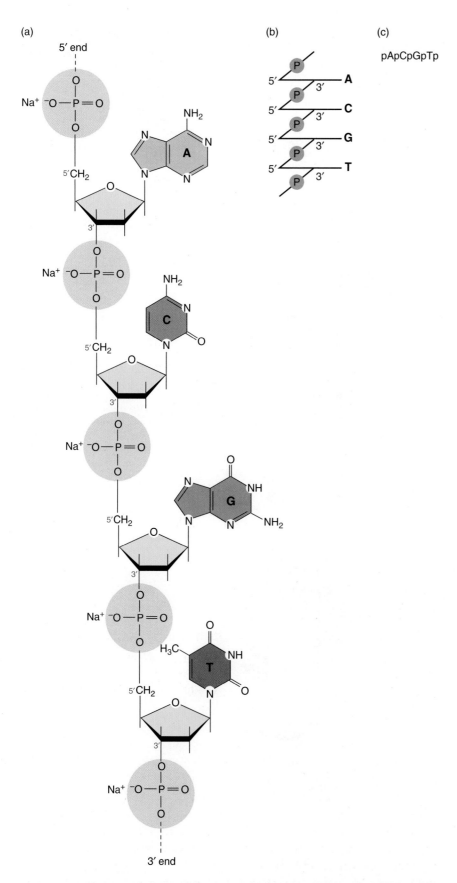

FIGURE 1.9 Segment of a polydeoxyribonucleotide. (a) Extended structure as a sodium salt, (b) stick figure structure, and (c) an abbreviated structure.

1.3 Introduction to Nucleic Acids

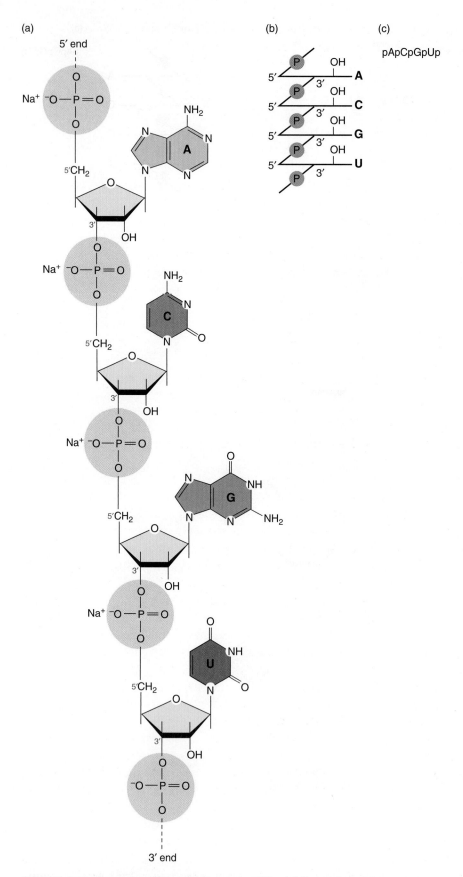

FIGURE 1.10 Segment of a polyribonucleotide. (a) Extended structure as a sodium salt, (b) stick figure structure, and (c) an abbreviated structure.

nucleotides and polypeptides are made of linear chains of amino acids led to the **sequence hypothesis,** which proposes nucleotide sequences specify amino acid sequences.

Drawing extended structures for DNA and RNA chains requires considerable time and space. It is more convenient to draw stick figure structures, which are adequate representations for many purposes (Figure 1.9b and 1.10b). The few simple conventions for drawing stick figure representations for DNA and RNA are as follows: (1) a single line (shown as horizontal in Figures 1.9b and 1.10b) represents the pentose ring; (2) the letter A, G, C, U, or T, at one end of the line, represents the purine or pyrimidine attached to C-1′ of the pentose ring; (3) the letter P, connected by short diagonal lines to adjacent lines, represents the 5′→3′ phosphodiester bond; and (4) the symbol OH represents a hydroxyl group. An even simpler method for indicating nucleotide sequence is to write the letters corresponding to the bases (Figure 1.9c and 1.10c).

1.4 DNA: Hereditary Material

Transformation experiments led to the discovery that DNA is the hereditary material.

The first step on the path leading to the discovery that genes are made of DNA was an observation made in 1928 by Fred Griffith, who was studying *Streptococcus pneumoniae,* a bacterial strain responsible for human pneumonia. This bacterium's virulence was known to depend on a surrounding polysaccharide capsule that protects the bacterium from the body's defense systems. The capsule also causes the bacterium to produce smooth-edged colonies on an agar surface. Bacteria from smooth-edged colonies, called S bacteria, normally kill mice. Griffith isolated a *S. pneumoniae* mutant that produced rough-edged colonies. Cells from rough-edged colonies, called R bacteria, proved to be both non-encapsulated and nonlethal to mice. Additional experiments revealed that although either live R or heat-killed S bacteria are nonlethal to mice, a mixture of the two is lethal (FIGURE 1.11). Furthermore, when bacteria were isolated from a mouse that had died from such a mixed infection, the bacteria were live S and R. Live R bacteria, therefore, had somehow either been replaced by or transformed to S bacteria.

Several years later investigators showed the mouse itself is not needed to mediate this transformation, because when a mixture containing live R bacteria and heat-killed S bacteria is incubated in culture medium, living S cells are produced. One possible explanation for this surprising phenomenon is that the R cells restore the viability of the dead S cells. This hypothesis was eliminated by the observation that living S cells appear even when the heat-killed S bacteria in the mixture are replaced by an S cell extract, which had been centrifuged to remove both intact cells and the capsular polysaccharide. Based on these studies it appeared that the S cell extract contains a **transforming principle** of unknown chemical composition.

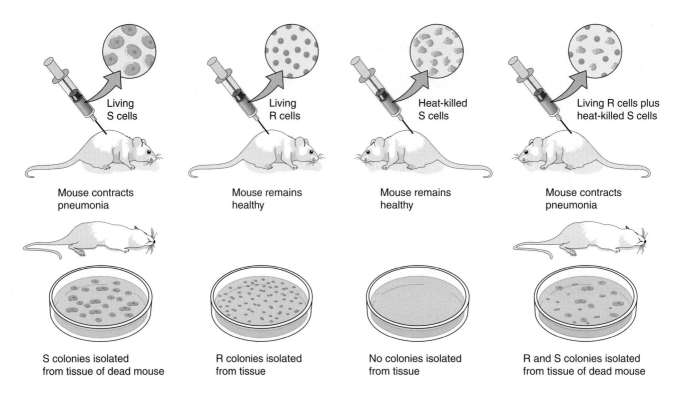

FIGURE 1.11 **Griffith's experiment demonstrating bacterial transformation.** A mouse dies from pneumonia if injected with the virulent S (smooth) strain of *Streptococcus pneumoniae*. However, the mouse remains healthy if injected with either the nonvirulent R (rough) strain or the heat-killed S strain. R cells in the presence of heat-killed S cells are transformed into the virulent S strain, killing the mouse.

The next development occurred in 1944 when Oswald Avery, Colin MacLeod, and Maclyn McCarty determined the chemical nature of the transforming principle. They did so by purifying the transforming principle from S cells and adding it to live R bacterial cultures (FIGURE 1.12). After allowing the mixture to incubate for a period of time, they placed samples on an agar surface and incubated them until colonies appeared. Some of the colonies (about 1 in 10^4) that grew were S type. To show this was a permanent genetic change, Avery and coworkers dispersed many of the newly formed S colonies and placed them on a second agar surface. The resulting colonies were again S type. If an R colony arising from the original mixture was dispersed, only R bacteria grew in subsequent generations. R colonies, hence, retained the R character, whereas the transformed S colonies bred true as S cells. Chemical analysis showed the transforming principle was a deoxyribose-containing nucleic acid, and physical measurements showed it was a highly viscous substance having the properties of DNA. Because the transforming principle contained other kinds of molecules, it was necessary to provide evidence that the transformation was actually caused by DNA and not an impurity. This evidence was provided by the following three procedures:

1. Polysaccharides isolated from S cells did not transform R cells.
2. Incubation with trypsin or chymotrypsin, enzymes that catalyze protein hydrolysis, or with ribonuclease (RNase), an enzyme that catalyzes RNA hydrolysis, did not affect

Preparation of transforming principle from S strain

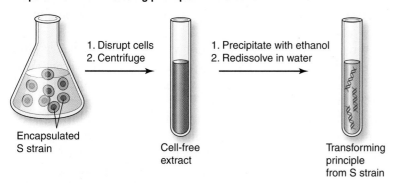

Addition of transforming principle to R strain

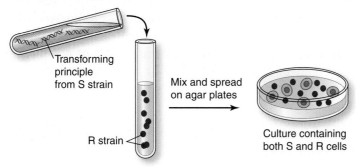

FIGURE 1.12 **Avery, MacLeod, and McCarty transformation experiment.**

transforming activity. The transforming principle thus is neither protein nor RNA.

3. Incubation with deoxyribonuclease (DNase), an enzyme that catalyzes DNA hydrolysis, inactivated the transforming principle.

In drawing conclusions from their experiments, Avery, MacLeod, and McCarty did not explicitly state that DNA is the hereditary material. One reason for their restraint may have been the crude analytical tools then available seemed to indicate that adenine, guanine, thymine, and cytosine were present in DNA in equimolar concentrations. Based on this and other evidence, many contemporary investigators mistakenly thought DNA consisted of a repeating tetranucleotide sequence and so could not possibly exhibit the variation needed by the hereditary material. Instead, the prevailing theory was that genetic information was stored in proteins because protein composition and structure were known to vary among organisms.

Investigators who supported the genes-as-protein theory posed two alternative explanations for the transformation results: (1) the transforming principle might not be DNA but rather one of the proteins invariably contaminating the DNA sample and (2) DNA somehow affected capsule formation directly by acting in the metabolic pathway for biosynthesis of the polysaccharide and permanently altering this pathway. The first point should have already been discounted by the original work because the experiments showed insensitivity to proteolytic enzymes and sensitivity to DNase. Because the DNase was not a pure enzyme, however, the possibility could not be eliminated conclusively.

The transformation experiment was repeated 5 years later by Rollin Hotchkiss using a DNA sample with a protein content that was only 0.02%, and it was found that this extensive purification did not reduce the transforming activity. This result supported the view of Avery, MacLeod, and McCarty but still did not prove it. The second alternative, however, was clearly eliminated—also by Hotchkiss—with an experiment in which he transformed a penicillin-sensitive bacterial strain to penicillin resistance. Because penicillin resistance is totally distinct from the rough-smooth character of the bacterial capsule, this experiment showed the transforming ability of DNA was not limited to capsule synthesis. Interestingly enough, most biologists still remained unconvinced that DNA was the genetic material. It was not until Erwin Chargaff showed in 1950 that a wide variety of chemical structures in DNA were possible—thus allowing biological specificity—that this idea was accepted.

Chemical experiments also supported the hypothesis that DNA is the hereditary material.

The hypothesis of a tetranucleotide structure for DNA arose for two reasons. First, in the chemical analysis of DNA, the technique used to separate the bases before identification did not resolve them very well, so the quantitative analysis was poor. Second, the DNA analyzed was usually isolated from animals, plants, and yeast in which the four bases are present in nearly equimolar concentration or from bacterial species such as *Escherichia coli* that also happen to have nearly equimolar base concentrations. Using the DNA from a wide variety of organisms, Chargaff applied new separation and analytical techniques and showed that the molar content of bases (generally called the base composition) could vary widely. The base composition of DNA from a particular organism is usually expressed as a fraction of all bases that are G • C pairs. This fraction, called the G + C content, can be expressed as follows:

$$G + C \text{ content} = ([G] + [C])/[\text{all bases}]$$

where the square brackets ([ ]) denote molar concentrations. Base compositions of DNAs from many different organisms have been determined. Generally speaking, the value of the G + C content is near 0.50 for the higher organisms, ranging from 0.49 to 0.51 for primates. For lower organisms the value of the G + C content varies widely from one genus to another. For example, for bacteria the extremes are 0.27 for the genus *Clostridium* and 0.76 for the genus *Sarcina*; *E. coli* DNA has the value 0.50. Thus, it was demonstrated that DNA could have variable composition, a primary requirement for the hereditary material.

Chargaff's studies also revealed one other remarkable fact about DNA base composition. In each of the DNA samples Chargaff studied, he found that [A] = [T] and [G] = [C]. Chargaff's findings can be summarized by two rules, known as **Chargaff's rules**, which state: (1) double stranded DNA has equimolar adenine and thymine concentrations as well as equimolar guanine and cytosine concentrations; and (2) DNA composition varies from one genus to another. Although the significance of Chargaff's rules, was not immediately apparent, they would later help to confirm the structure of the DNA molecule.

Upon publication of Chargaff's results, the tetranucleotide hypothesis quietly died and the DNA–gene idea began to catch on. Shortly afterward, workers in several laboratories found for a wide variety of organisms that somatic cells have twice the DNA content of germ cells, a characteristic to be expected of the genetic material, given the tenets of classical chromosome genetics. Although it could apply just as well to any component of chromosomes, once this result was revealed objections to the work of Avery, MacLeod, and McCarty were no longer heard, and the hereditary nature of DNA rapidly became the accepted idea.

1.5 Watson-Crick Model

Rosalind Franklin and Maurice Wilkins obtained x-ray diffraction patterns of extended DNA fibers.

At the same time chemists were attempting to learn something about the composition of DNA, crystallographers were trying to obtain a three-dimensional image of the molecule. Rosalind Franklin and Maurice Wilkins obtained some excellent x-ray diffraction patterns of extended DNA fibers in the early 1950s (FIGURE 1.13). One might predict that all x-ray diffraction patterns would look alike, but this was not the case. DNA structure and therefore x-ray diffraction patterns depend on several variables. One of the most important of these is the relative humidity of the chamber in which DNA fibers are placed.

Two types of DNA structure are of particular interest. **B-DNA** is stable at a relative humidity of about 92%, whereas **A-DNA** appears as the relative humidity falls to about 75%. Crystallographers did not know whether A-DNA or B-DNA was present in the living cell. Partly for this reason, Wilkins turned his attention toward taking x-ray diffraction pictures of DNA in sperm cells. Franklin focused her attention on x-ray diffraction patterns of A-DNA because they appeared to provide more detail. She believed that careful analyses of the detailed patterns would eventually lead to the solution of DNA's structure.

James Watson and Francis Crick proposed that DNA is a double-stranded helix.

The American biologist, James D. Watson, and the English crystallographer, Francis H. C. Crick, working together in England, took a different approach to determining DNA's structure. They tried to obtain as much information as they could from the x-ray diffraction patterns and then to build a model consistent with this information.

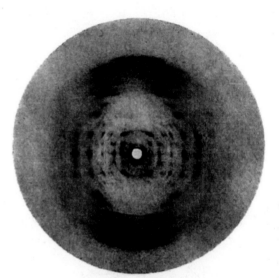
A-form DNA

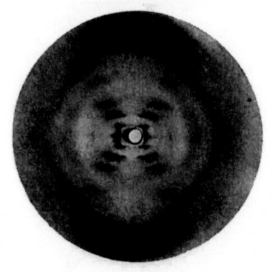
B-form DNA

FIGURE 1.13 X-ray diffraction patterns of the A and B forms of the sodium salt of DNA. (Reproduced from Franklin, R. E., and Gosling, R. G. 1953. *Acta Crystallographica* 6:673–677. Photos courtesy of International Union of Crystallography.)

The term "model" has a special meaning to scientists. A **model** is a hypothesis or tentative explanation of the way a system works, usually including the components, interactions, and sequences of events. A successful model suggests additional experiments and allows investigators to make predictions that can be tested in the laboratory. If predictions do not agree with experimental results, the model must be considered incorrect in its current form and modified. A model cannot be proved to be correct merely by showing it makes a correct prediction. If it makes many correct predictions, however, it is probably nearly, if not completely, correct.

Watson and Crick focused their attention on Franklin's x-ray diffraction patterns of B-DNA. This pattern indicated that B-DNA has a helical structure, a diameter of approximately 2.0 nm, and a repeat distance of 0.34 nm. Their model would have to account for these structural features. Watson and Crick still had to work out the number of DNA chains in a DNA molecule, the location of the bases, and the position of the phosphate and deoxyribose groups. The density of DNA seemed to be consistent with one, two, or three chains per molecule. Watson and Crick tried to build a two-chain model with hydrogen bonds (weak electrostatic attractions; see Chapter 2) holding the bases together. They were unsuccessful until their colleague, Jerry Donohue, suggested they use the *keto tautomeric* forms of T and G in their models. At first, Watson tried to link two purines together and two pyrimidines together in what he called a "like-with-like" model. A dramatic turning point occurred in 1953 when Watson realized adenine forms hydrogen bonds with thymine and guanine forms hydrogen bonds with cytosine. Watson describes this turning point in his book, *The Double Helix* (pages 194–195):

> When I got to our still empty office the following morning, I quickly cleared away the paper from my desk top so that I would have a large, flat surface on which to form pairs of bases held together by hydrogen bonds. Though I initially went back to my like-with-like prejudices, I saw all too well that they led nowhere. When Jerry [Donohue] came in I looked up, saw that it was not Francis [Crick], and began shifting the bases in and out of various other pairing possibilities. Suddenly I became aware that an adenine-thymine pair held together by two hydrogen bonds was identical in shape to a guanine-cytosine base pair held together by at least two hydrogen bonds. All the hydrogen bonds seemed to form naturally; no fudging was required to make the two types of base pairs identical in shape.

With the realization that adenine-thymine and cytosine-guanine base pairs have the same width, Watson and Crick were quickly able to construct a double helix model of DNA that fit Franklin's x-ray diffraction data (**FIGURE 1.14**).

The key features of the Watson-Crick Model for B-DNA are as follows:

1. Two polydeoxyribonucleotide strands twist about each other to form a double helix.
2. Phosphate and deoxyribose groups form a backbone on the outside of the helix.

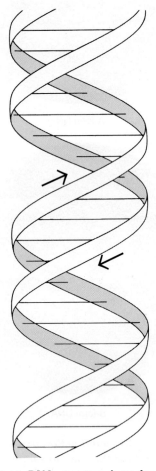

FIGURE 1.14 **DNA structure based on the 1953 paper by Watson and Crick in *Nature* showing their double-helix model for DNA for the first time.**

3. Purine and pyrimidine base pairs stack inside the helix and form planes perpendicular to the helix axis and the deoxyribose groups.
4. The helix diameter is 2.0 nm (or 20 Å).
5. Adjacent base pairs are separated by an average distance of 0.34 nm (or 3.4 Å) along the helix axis. The structure repeats itself after about 10 base pairs, or about once every 3.4 nm (or 34 Å) along the helix axis.
6. Adenine always pairs with thymine and guanine with cytosine. The original model showed two hydrogen bonds stabilizing each kind of base pair. Although this was an accurate description for A-T base pairs, later work showed that G-C base pairs are stabilized by three hydrogen bonds (FIGURE 1.15). (These base pairing relationships explained Chargaff's observation that the molar ratios of adenine to thymine and guanine to cytosine are one.)
7. The two strands are antiparallel, which means the strands run in opposite directions. That is, one strand runs 3'→5' in one direction, whereas the other strand runs 5'→3' in the same direction. Because the strands are antiparallel, a convention is needed for stating the sequence of bases of a single chain. The convention is to write a sequence with the 5' terminus at the left; for example, ATC denotes the trinucleotide 5'-ATC-3'. This is also often written as pApTpC, again using the conventions that the left side of each base is the 5'-terminus of the nucleotide and that a phosphodiester group is represented by a "p" between two capital letters.
8. A **major groove** and a **minor groove** wind about the cylindrical outer helical face. The two grooves are of about equal depth, but the major groove is much wider than the minor groove.

The Watson-Crick model indicates that when the nucleotide sequence of one strand is known, the sequence of the complementary

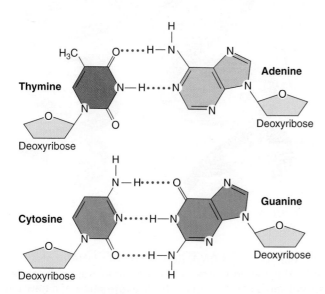

FIGURE 1.15 **Base pairs in DNA.**

strand can be predicted, providing the theoretical framework needed to understand the fidelity of gene replication. Each strand serves as a mold or **template** for the synthesis of the complementary strand (**FIGURE 1.16**). Watson and Crick ended their short paper announcing the double helix model with the following sentence that must be one of the greatest understatements in the scientific literature: "It has not escaped our notice that the specific pairing we have postulated immediately suggests a possible copying mechanism for the genetic material." The double helix showed how establishing a chemical structure can be used to understand biological function and to make predictions that guide new research.

Structure–function relationships remain a central theme of molecular biology. Knowledge of structure helps us to understand function

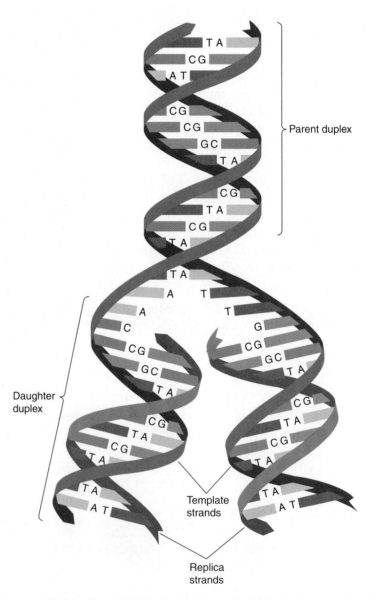

FIGURE 1.16 Replication of DNA. Replication of a DNA duplex as originally envisioned by Watson and Crick. As the parental strands separate, each parental strand serves as a template for the formation of a new daughter strand by means of A-T and G-C base pairing.

and leads us to new insights. Biological structures are sometimes quite complex and difficult to study. It is, however, worth the effort to study the structures because the reward for doing so is so great. The Watson-Crick model also serves as a powerful example of the fundamental principles that help to define molecular biology as a discipline: (1) the same physical and chemical laws apply to living systems and inanimate objects, (2) the same biological principles tend to apply to all organisms, and (3) biological structure and function are intimately related. The Watson-Crick model provided an excellent starting point for the challenging job of elucidating the chemical basis of heredity.

1.6 Central Dogma

The central dogma provides the theoretical framework for molecular biology.

Much of the research in molecular biology in the late 1950s was directed toward discovering the mechanisms by which nucleotide sequences specify amino acid sequences. In a talk given at a 1957 symposium, Crick suggested a model for information flow that provides a theoretical framework for molecular biology. The following excerpt from Crick's presentation ("On Protein Synthesis." *The Symposia of the Society for Experimental Biology* 12, (1958):138–163) states the theory known as the **central dogma** in a clear and concise fashion:

> In more detail, the transfer of information from nucleic acid to nucleic acid, or from nucleic acid to protein may be possible, but transfer from protein to protein, or from protein to nucleic acid is impossible. Information means here the precise determination of sequence, either of bases in nucleic acid or of amino acid residues in protein.

Crick thus was proposing that genetic information flows from DNA to DNA (DNA replication), from DNA to RNA (transcription), and from RNA to polypeptide (translation) (**FIGURE 1.17**).

By the mid-1960s molecular biologists had obtained considerable experimental support for the central dogma. In particular, they had discovered enzymes that catalyze replication and transcription and elucidated the pathway for translating nucleotide sequences to amino acid sequences. With some variations in detail, all organisms use the same basic mechanism to translate information from nucleotide sequences to amino acid sequences. Three major components of the translation machinery—**messenger RNA** (mRNA), **ribosomes**, and **transfer RNA** (tRNA)—play essential roles in this information transfer (**FIGURE 1.18**). Each gene can serve as a template for the synthesis of specific mRNA molecules. mRNA molecules program ribosomes (protein synthetic factories) to form specific polypeptides. tRNA molecules carry activated amino acids to programmed ribosomes, where under the direction of the mRNA the amino acids join to form a polypeptide chain.

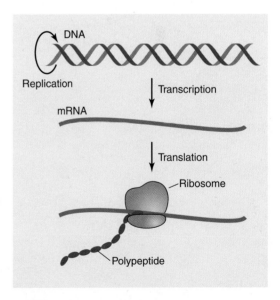

FIGURE 1.17 The "central dogma." The central dogma as originally proposed by Francis Crick postulated information flow from DNA to RNA to protein. The ribosome is an essential part of the translation machinery. Later studies demonstrated that information can also flow from RNA to RNA and from RNA to DNA (reverse transcription).

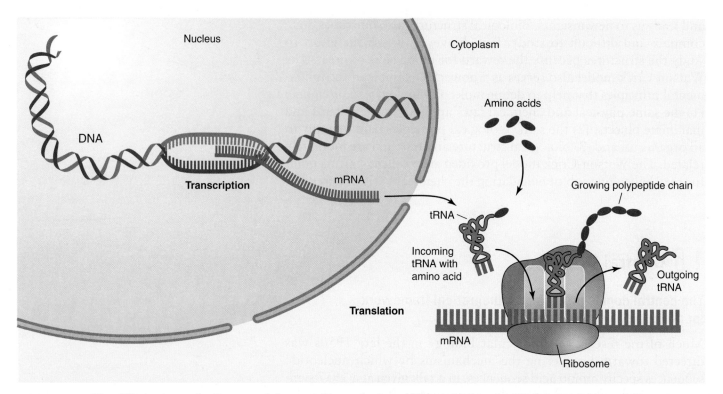

FIGURE 1.18 Simplified schematic diagram of the protein synthetic machinery in an animal, plant, or yeast cell. Transfer RNA (tRNA) carries the next amino acid to be attached to the growing polypeptide chain to the ribosome, which recognizes a match between a three nucleotide sequence in the mRNA (the codon) and a complementary sequence in the tRNA (the anticodon) and transfers the growing polypeptide chain to the incoming amino acid while it is still attached to the tRNA. (Adapted from Secko, D. 2007. *The Science Creative Quarterly 2*. http://www.scq.ubc.ca/a-monks-flourishing-garden-the-basics-of-molecular-biology-explained/, Figure 4.)

The 12 years of extraordinary scientific progress that followed the discovery of DNA's structure were capped by a series of brilliant investigations by Marshal W. Nirenberg and others that culminated in deciphering the genetic code. By 1965 investigators could predict a polypeptide chain's amino acid sequence from a DNA or mRNA molecule's nucleotide sequence. Remarkably, the genetic code was found to be nearly universal. Each particular sequence of three adjacent nucleotides or **codon** specifies the same amino acid in bacteria, plants, and animals. Could one ask for more convincing support for the uniformity of life processes?

1.7 Introduction to Recombinant DNA Technology

Recombinant DNA technology allows us to study complex biological systems.

The second major wave in molecular biology started in the late 1970s with the development of **recombinant DNA technology** (also known as **genetic engineering**). Thanks to recombinant DNA technology genes can be manipulated in the laboratory just like any other organic molecule. They can be synthesized or modified as desired and their nucleotide sequence determined. Sequence information can save molecular biologists months and perhaps even years of work. For example, when a new gene is discovered that causes a specific disease in humans, molecular biologists may be able to obtain valuable clues to the new gene's function by comparing its nucleotide sequence with sequences of all other known genes. If similarities are found with a gene of known function from some other organism, then it is likely the new gene will have a similar function. Alternatively, a segment of the nucleotide sequence of the new gene may predict an amino acid sequence that is known to have specific binding or catalytic functions.

Development of recombinant DNA technology has allowed the incredible progress in solving problems that once seemed intractable. Recombinant DNA technology has helped investigators to study cell division, cell differentiation, transformation of normal cells to cancer cells, programmed cell death (apoptosis), antibody production, hormone action, and a variety of other fundamental biological processes.

One of the most exciting applications of recombinant DNA technology is in medicine. Until relatively recently many diseases could only be studied in humans because no animal models existed. Very little progress was made in studying these diseases because of obvious ethical constraints associated with studying human diseases. Once investigators learned how to modify and transfer genes, they were able to create animal models for human diseases, thus facilitating study of the diseases. Recombinant DNA technology also has led to the production of new drugs to treat diabetes, anemia, cardiovascular disease, and cancer as well as to the development of diagnostic tools to detect a wide variety of diseases. The list of practical medical applications grows longer with each passing day.

Although recombinant DNA technology promises to change our lives for the better, it also forces us to consider important social, political, ethical, and legal issues. For example, how do we protect the interests of an individual when DNA analysis reveals the individual has alleles likely to cause a serious physical or mental disease in the future, especially if there is no cure or treatment? What impact will the knowledge have on affected individuals and their families? Should insurance companies or potential employers have access to the genetic information? If not, how do we limit access to information and how do we enforce the limitation? Rapid progress in recombinant DNA technology also raises troubling ethical issues. Germ line therapy allows new genes to be introduced into fertilized eggs and thereby alter the genetic characteristics of future generations. This technique has been used to introduce desired traits into plants and animals.

An argument can be made for using germ line therapy to correct human genetic diseases, such as Tay-Sachs disease, cystic fibrosis, or Huntington disease, so affected individuals and their families can be spared the devastating consequences of these diseases. However, we must be very careful about application of germ line therapy to humans because the technique has the power to do great harm. Who will decide which genetic characteristics are desirable and which are undesirable? Should the technique be used to change physical appearance, intelligence, or personality traits? Should anyone be permitted to make such decisions?

A great deal of molecular biology information is available on the Internet.

Recombinant DNA technology has generated so much information it would be nearly impossible to share all of it in a timely fashion with the entire molecular biology community by conventional means such as publishing in professional journals or writing books. Fortunately, there is an alternate method for sharing large quantities of rapidly accumulating information that is both quick and efficient. A worldwide network of communication networks, the Internet, allows us to gain almost instant access to the information. The Internet also provides many helpful tutorials and instructive animations. This text will include references to helpful Internet sites. Because the addresses for these Web sites tend to change over time, however, this text will refer the reader to a primary site maintained by the publisher.

Questions and Problems

1. Briefly define, describe, or illustrate each of the following terms.
   - a. Allele
   - b. Phenotype
   - c. Genotype
   - d. Dominant gene
   - e. Recessive gene
   - f. Purine
   - g. Pyrimidine
   - h. Nucleoside
   - i. N-glycosylic bond
   - j. Nucleotide
   - k. Phosphomonoester
   - l. Phosphodiester bond
   - m. Transformation
   - n. Transforming principle
   - o. Chargaff's rules
   - p. B-DNA
   - q. Model
   - r. Major groove
   - s. Minor groove
   - t. Replication
   - u. Transcription
   - v. Translation
   - w. Codon

2. Draw the four bases commonly found in DNA. Which of these bases is a purine and which is a pyrimidine?

3. Draw the base that is commonly present in RNA but not in DNA. Is this base a purine or a pyrimidine?

4. What are the chemical similarities and differences between ribose and deoxyribose?

5. Each purine and pyrimidine base in nucleic acids has distinct chemical features.
   - a. Which purine has a single amine group and no oxygen atoms?
   - b. Which pyrimidine has a single amine group and one oxygen atom?
   - c. What is the major difference between the structures of uracil and thymine?
   - d. Which pyrimidine has two oxygen atoms attached to the ring structure?

6. Although Avery, MacLeod, and McCarty did not explicitly state that DNA is the hereditary material, their experiment certainly seemed to show this was the case.
   - a. Describe their experiment.
   - b. Why were their experiments showing that the transforming principle is resistant to RNase and proteases but sensitive to DNase important?
   - c. Why were contemporary investigators slow to accept the idea that DNA is the hereditary material?

d. Why did Chargaff's studies of DNA base composition help to convince investigators that DNA is the hereditary material?
   e. How does the Watson-Crick model explain Chargaff's rules?
7. DNA and RNA are linear nucleotide chains.
   a. Draw the extended structure for a segment of a polydeoxyribonucleotide with the abbreviated structure pdTpdApdC. Indicate the 5'- and 3'-ends.
   b. Draw the stick figure structure for a segment of a polydeoxyribonucleotide with the abbreviated structure pdTpdApdC. Indicate the 5'- and 3'-ends.
   c. Draw the extended structure for a segment of a polyribonucleotide with the abbreviated structure pUpApC. Indicate the 5'- and 3'-ends.
   d. Draw the stick figure structure for a segment of a polyribonucleotide with the abbreviated structure pUpApC. Indicate the 5'- and 3'-ends.
   e. Identify two important chemical differences between a polydeoxyribonucleotide and a polyribonucleotide strand.
8. What are the major structural features of the Watson-Crick model for DNA?
9. How does the Watson-Crick model help us to understand DNA function?
10. What functions do ribosomal RNA (rRNA), messenger RNA (mRNA), and transfer RNA (tRNA) play in the cell?

Suggested Reading

General

Choudri, S. 2003. The path from nuclein to human genome: a brief history of DNA with a note on human genome sequencing and its impact on future research in biology. *Bull Sci Technol Soc* 23:360–367.

Crick, F. 1974. The double helix: a personal view. *Nature* 248:766–769.

Crick, F. 1988. *What Mad Pursuit: A Personal View of Scientific Discovery.* New York: Basic Books.

Judson, H. F. 1996. *The Eighth Day of Creation: Makers of the Revolution in Biology.* Cold Spring Harbor, NY: Cold Spring Harbor Laboratory.

Maddox, B. 2002. *Rosalind Franklin: The Dark Lady of DNA.* New York: Harper Collins.

McCarty, M. 1985. *The Transforming Principle: Discovering That Genes Are Made of DNA.* New York: W. W. Norton.

Olby, R. C. 1994. *The Path to the Double Helix: The Discovery of DNA.* New York: Dover.

Pukkila, P. J. 2001. Molecular biology: the central dogma. *Encyclopedia of Life Sciences.* pp. 1–5. London, UK: Nature.

Sayre, A. 1978. *Rosalind Franklin and DNA.* New York: W. W. Norton.

Stent, G. 1972. Prematurity and uniqueness in scientific discovery. *Sci Am* 227:84–93.

Summers, W. C. 2002. History of molecular biology. *Encyclopedia of Life Sciences.* pp. 1–8. London, UK: Nature.

Watson, J. D. 1968. *The Double Helix: A Personal Account of the Discovery of the Structure of DNA.* New York: Atheneum Books.

Wilkins, M. 2003. *The Third Man of the Double Helix: The Autobiography of Maurice Wilkins.* Oxford, UK: Oxford University Press.

Classic Papers

Avery, O. T., MacLeod, C. M., and McCarty, M. 1944. Studies on the chemical nature of the substance inducing transformation of pneumococcal types. Induction of transformation by a deoxyribonucleic acid fraction isolated from *Pneumococcus* type III. *J Exp Med* 79:137–156.

Beadle, G. W., and Tatum, E., 1941. Genetic control of biochemical reactions in Neurospora. *Proc Natl Acad Sci USA* 27:499–506.

Watson, J. D., and Crick F. H. 1953. Molecular structure of nucleic acids; a structure for deoxyribose nucleic acid. *Nature* 171:737–738.

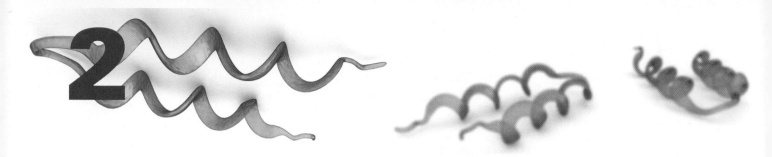

2

Protein Structure and Function

CHAPTER OUTLINE

2.1 α-Amino Acids
α-Amino acids have an amino group and a carboxyl group attached to a central carbon atom.
Amino acids are represented by three-letter and one-letter abbreviations.

2.2 Peptide Bond
α-Amino acids are linked by peptide bonds.
Peptide chains have directionality.

2.3 Protein Purification
Protein mixtures can be fractionated by chromatography.
Proteins and other charged biological polymers migrate in an electric field.

2.4 Primary Structure of Proteins
The amino acid sequence or primary structure of a purified protein can be determined.
Polypeptide sequences can be obtained from nucleic acid sequences.
The BLAST program compares a new polypeptide sequence with all sequences stored in a data bank.

2.5 Introduction to Protein Structure
Proteins with just one polypeptide chain have primary, secondary, and tertiary structures, whereas those with two or more chains also have quaternary structures.

2.6 Weak Noncovalent Bonds
The polypeptide folding pattern is determined by weak noncovalent interactions.

2.7 Secondary Structures
The α-helix is a compact structure that is stabilized by hydrogen bonds.
The β-conformation is also stabilized by hydrogen bonds.
Loops and turns connect different peptide segments, allowing polypeptide chains to fold back on themselves.
Certain combinations of secondary structures, called supersecondary structures or folding motifs, appear in many different proteins.
We cannot yet predict secondary structures with absolute certainty.

2.8 Tertiary Structure
X-ray crystallography studies reveal three-dimensional structures of many different proteins.
Intrinsically disordered proteins lack an ordered structure under physiological conditions.
The primary structure of a polypeptide determines its tertiary structure.
Molecular chaperones help proteins to fold inside the cell.

2.9 Myoglobin, Hemoglobin, and the Quaternary Structure Concept
Differences in myoglobin and hemoglobin function are explained by differences in myoglobin and hemoglobin structure.
Hemoglobin subunits bind oxygen in a cooperative fashion.

2.10 Immunoglobulin G (IgG) and the Domain Concept
Large polypeptides fold into globular units called domains.

2.11 Enzymes
Enzymes are proteins that catalyze chemical reactions.
Molecular biologists often use colorimetric and radioactive methods to measure enzyme activity.
Enzymes lower the energy of activation but do not affect the equilibrium position.
All enzyme-catalyzed reactions proceed through an enzyme–substrate complex.
Molecular details for ES complexes have been worked out for many enzymes.
Regulatory enzymes control committed steps in biochemical pathways.
Regulatory enzymes exhibit sigmoidal kinetics and are stimulated or inhibited by allosteric effectors.

BOX 2.1: IN THE LAB Strategies for Protein Purification
BOX 2.2: CLINICAL APPLICATION Sickle Cell Anemia
BOX 2.3: CLINICAL APPLICATION Clinical Significance of 2,3-Bisphosphoglycerate
BOX 2.4: IN THE LAB Radioactive Tracers
QUESTIONS AND PROBLEMS
SUGGESTED READING

Image © alexmit/123RF.

The Watson-Crick model helped to bridge a major gap between genetics and biochemistry and in so doing helped establish molecular biology as a scientific discipline. The structure–function approach, a recurring theme in molecular biology, is responsible for many advances that have been made in our understanding of the mechanisms of replication, transcription, and translation. This chapter extends the study of structure–function relationships to proteins, which catalyze specific reactions, transport materials within a cell or across a membrane, protect cells from foreign invaders, regulate specific biological processes, and support various structures.

The basic building blocks for proteins are small organic molecules called amino acids. Amino acids can combine to form long linear chains known as **polypeptides**. Each type of polypeptide chain has a unique amino acid sequence. Although a polypeptide must have the correct amino acid sequence to perform its specific biological function, the amino acid sequence alone does not guarantee that the polypeptide will be biologically active. The polypeptide must fold into a specific three-dimensional structure before it can perform its biological function(s). Once folded into its biologically active form the polypeptide is termed a protein. Proteins come in various sizes and shapes. Those with thread-like shapes, **fibrous proteins**, tend to have structural or mechanical roles. Those with spherical shapes, **globular proteins**, function as enzymes, transport proteins, or antibodies. Fibrous proteins tend to be water insoluble, whereas globular proteins tend to be water soluble.

Polypeptides are unique among biological molecules in their flexibility, which allows them to fold into characteristic three-dimensional structures with specific binding properties. Amino acid substitution that alters a protein's ability to interact with its normal molecular partners often result in a loss of protein activity. One of the most common partners of a folded polypeptide is another folded polypeptide, which may be identical to it or different. A complex that contains two, three, or more identical polypeptides is called a homodimer, homotrimer, and so forth, whereas one that contains different polypeptides is called a heterodimer, heterotrimer, and so forth.

This chapter examines the following aspects of protein structure and function: (1) the chemistry of amino acid building blocks; (2) the nature of the bond linking adjacent amino acids in a polypeptide; (3) the methods of protein purification; (4) the primary, secondary, tertiary, and quaternary levels of protein structure; (5) protein structure–function relationships; and (6) protein catalysts—the enzymes.

2.1 α-Amino Acids

α-Amino acids have an amino group and a carboxyl group attached to a central carbon atom.

The typical amino acid building block for polypeptide synthesis has a central carbon atom that is attached to an amine ($-NH_2$) group, a carboxyl ($-COOH$) group, a hydrogen atom, and a side chain ($-R$).

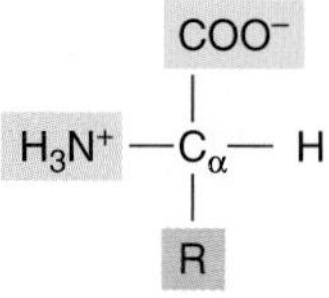

FIGURE 2.1 Structure of an α-amino acid. A typical α-amino acid in which the central carbon atom is attached to an amino (–NH$_3^+$) group, a carboxylate (–COO$^-$) group, a hydrogen atom, and a side chain (–R).

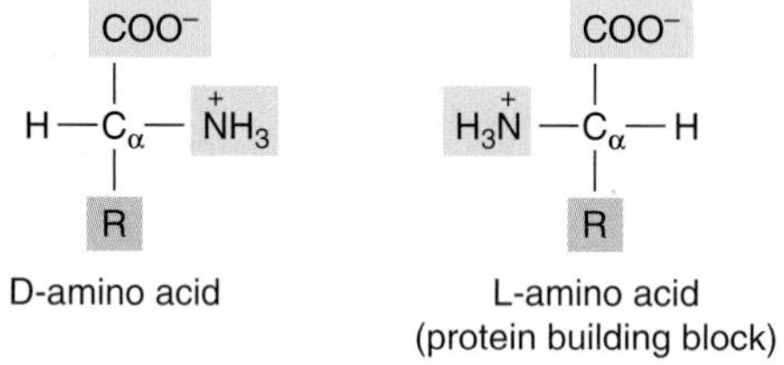

FIGURE 2.2 D- and L-amino acid configurations.

An amino acid having these properties is termed an **α-amino acid**. The α prefix is a remnant of an older chemistry nomenclature system in which atoms in a hydrocarbon chain attached to a carboxyl group were designated by Greek letters. The carbon atom closest to the carboxyl group was designated α, the next β, and so forth. At pH 7 an α-amino acid's amine group is protonated to form –NH$_3^+$ and its carboxyl group is deprotonated to form –COO$^-$ so that the amino acid has the structure shown in FIGURE 2.1.

With the exception of glycine (see below), all amino acids present in proteins have four different kinds of functional groups attached to their α-carbon atom. The four different functional groups can be arranged in different stereo configurations, forming two stereoisomers (FIGURE 2.2). The D- and L-stereoisomers are mirror images in much the same way that the right hand is a mirror image of the left hand. Cells readily distinguish between D- and L-amino acid stereoisomers and only incorporate the L-stereoisomer into proteins. Each amino acid has characteristic physical and chemical properties that derive from its unique side chain. Amino acids with similar side chains usually have similar properties. This relationship is an important consideration when comparing amino acid sequences of two different polypeptides or when considering the effect an amino acid substitution will have on protein function. Based on side chain structure, the 20 major amino acid building blocks of proteins can be divided into four different groups:

1. *Basic side chains:* **Arginine**, **lysine**, and **histidine** are called basic amino acids because their side chains are proton acceptors (FIGURE 2.3). The guanidino group in arginine's side chain is a relatively strong base. The amine group in lysine's side chain is a somewhat weaker base, and the imidazole group in histidine's side chain is the weakest of the three bases. Hence, at pH 7 arginine and lysine side chains are very likely to have positive charges, whereas histidine side chains have only about a 10% probability of having a positive charge.

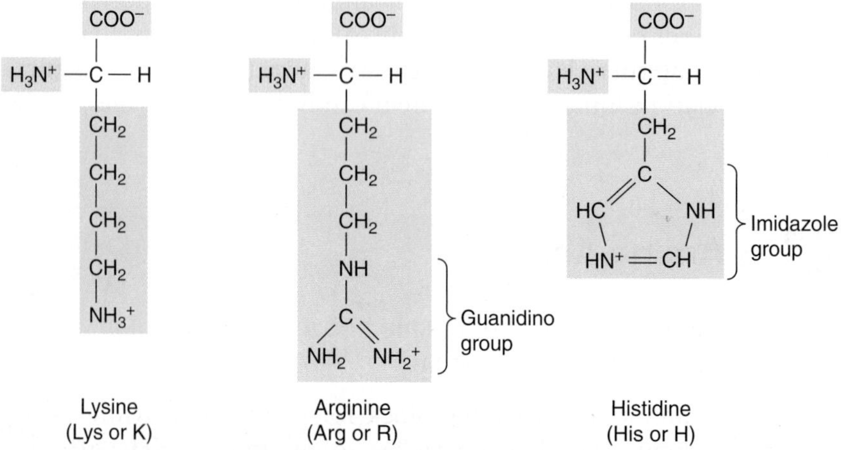

FIGURE 2.3 Amino acids with basic side chains.

CHAPTER 2 PROTEIN STRUCTURE AND FUNCTION

2. *Acidic side chains:* **Aspartic acid** and **glutamic acid** each has a carboxyl group as part of its side chain (FIGURE 2.4). Both the α-carboxyl and the side chain carboxyl groups are deprotonated and have negative charges at pH 7. The α-carboxyl group is a slightly stronger acid, however, because the α-carbon is also attached to a positively charged amino group. When the side chain is deprotonated, aspartic and glutamic acids are more appropriately called aspartate and glutamate, respectively. Because aspartic acid and aspartate refer to the same amino acid at different pH values, the names are used interchangeably. The same is true for glutamic acid and glutamate.
3. *Polar but uncharged side chains:* Six amino acids have side chains with polar groups (FIGURE 2.5). **Asparagine** and **glutamine** are amide derivatives of aspartate and glutamate, respectively. **Serine, threonine,** and **tyrosine** have side chains with hydroxyl (–OH) groups. The tyrosine side chain also has another interesting feature; it is aromatic. Cysteine is similar to serine but a sulfhydryl (–SH) group replaces the hydroxyl group. When exposed to oxygen or other oxidizing agents, sulfhydryl groups on two cysteine molecules react to form a disulfide (–S–S–) bond, resulting in the formation of **cystine** (FIGURE 2.6). Cystine, which is not a building block for polypeptide synthesis, is formed by the oxidation of cysteine side chains after the polypeptide has been formed.
4. *Nonpolar side chains:* Nine amino acids have side chains with nonpolar groups (FIGURE 2.7). **Glycine**, with a side chain consisting of a single hydrogen atom, is the smallest amino acid and the only one that lacks an asymmetric carbon atom. Because its side chain is so small, glycine can fit into tight places and tends to behave like amino acids with polar but uncharged side chains when present in a polypeptide. **Alanine, isoleucine, leucine,** and **valine** have hydrocarbon side chains. **Phenylalanine** and **tryptophan** have aromatic side chains. **Methionine** and **proline** have side chains with unique features. The methionine side chain contains a thioether (–CH$_2$–S–CH$_3$) group. Proline's side chain is part of a five-member ring that includes

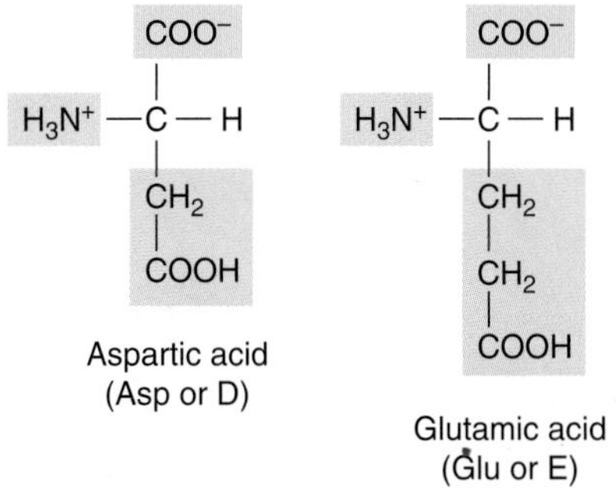

FIGURE 2.4 Amino acids with acidic side chains.

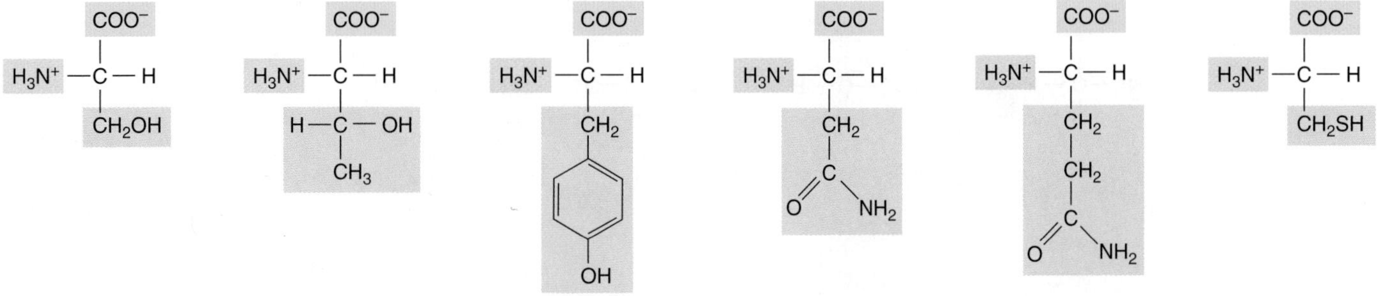

FIGURE 2.5 Amino acids with polar but uncharged side chains at pH 7.0.

2.1 α-Amino Acids

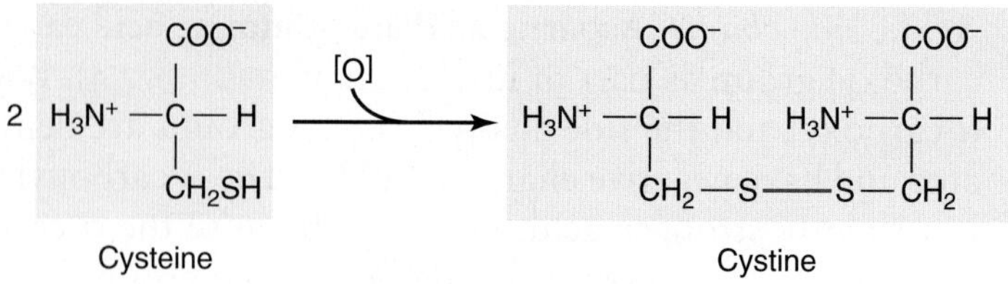

FIGURE 2.6 Oxidation of cysteine to form cystine.

the α-amino group, making the α-amino group a secondary rather than a primary amine group. The rigid ring structure can influence the way a polypeptide chain folds by introducing a kink into the structure.

Amino acids are represented by three-letter and one-letter abbreviations.

Writing the full names of the amino acids is inconvenient, especially for polypeptide chains with many amino acids. Two systems of abbreviations listed in **TABLE 2.1** offer more convenient methods for representing amino acids. In the first system each amino acid is represented by a three-letter abbreviation. For most amino acids the first three-letters of the amino acid's name are used. For example, Arg is used for arginine, Phe for phenylalanine, and Lys for lysine. But four amino acids have unusual three-letter abbreviations; aspartate and asparagine have identical first three letters and the same is true for glutamate and glutamine. So the abbreviations for aspartate and glutamate are

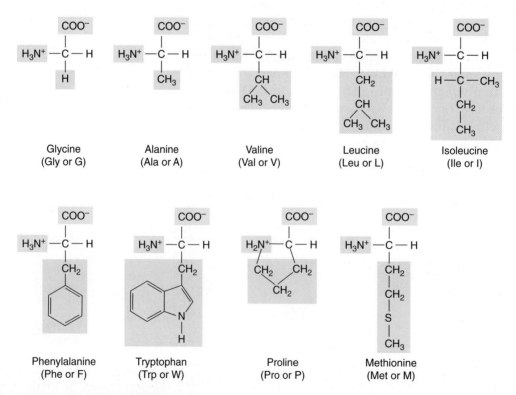

FIGURE 2.7 Amino acids with nonpolar side chains.

TABLE 2.1 Amino Acid Abbreviations		
Amino Acid	Three Letter Abbreviation	One Letter Abbreviation
Alanine	Ala	A
Arginine	Arg	R
Asparagine	Asn	N
Aspartic acid (Aspartate)	Asp	D
Cysteine	Cys	C
Glutamine	Gln	Q
Glutamic acid (Glutamate)	Glu	E
Glycine	Gly	G
Histidine	His	H
Isoleucine	Ile	I
Leucine	Leu	L
Lysine	Lys	K
Methionine	Met	M
Phenylalanine	Phe	F
Proline	Pro	P
Serine	Ser	S
Threonine	Thr	T
Tryptophan	Trp	W
Tyrosine	Tyr	Y
Valine	Val	V

the expected Asp and Glu, respectively, and those for asparagine and glutamine are Asn and Gln, respectively. The other two amino acids with unusual three-letter abbreviations are Trp for tryptophan and Ile for isoleucine. Even the three-letter abbreviation system requires too much space for many applications. So investigators devised the one-letter abbreviation system also shown in Table 2.1.

2.2 Peptide Bond

α-Amino acids are linked by peptide bonds.

The α-carboxyl group of one amino acid can react with the α-amino group of a second amino acid to form an amide bond and release water (FIGURE 2.8). Amide bonds that link amino acids are designated **peptide bonds** and the resulting molecules are called **peptides**. Peptides with two amino acids are dipeptides, those with three are tripeptides, and so forth.

Systematic physical studies by Linus Pauling and Robert R. Corey in the late 1930s provide important information about bond distances and angles in dipeptides. The results of their studies, summarized in FIGURE 2.9, show the carbon-nitrogen peptide bond is 0.133 nm long, placing it between the length of a carbon-nitrogen single bond (0.149 nm) and a carbon-nitrogen double bond (0.127 nm). The

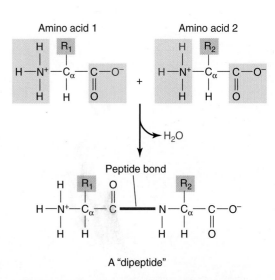

FIGURE 2.8 **Peptide bond formation in the laboratory.** Two amino acids combine with the loss of water to form a dipeptide. The peptide bond is shown in purple.

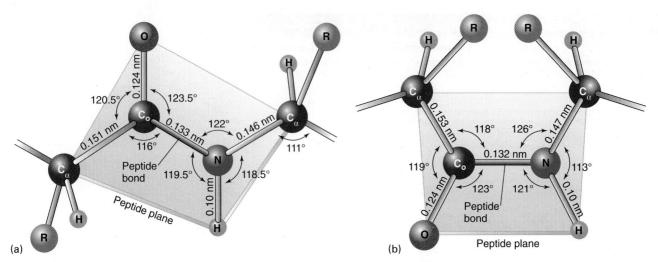

FIGURE 2.9 The peptide bond. (a) The *trans* peptide bond and (b) the *cis* peptide bond. ([a and b] Adapted from D. Voet, and J. G. Voet. 2005. *Biochemistry* (3rd ed.) John Wiley & Sons, Ltd.)

peptide bond therefore has some double-bond character, which produces an energy barrier to free rotation and imposes planarity on the peptide bond. Two configurations are possible (Figure 2.9), one in which adjacent C_α atoms are on opposite sides (the *trans* isomer) and another in which they are on the same side of the peptide bond (the *cis* isomer). The *trans* isomer is the more stable of the two because there is less physical contact between the side chains of the two amino acids forming the peptide bond. Although rare, *cis* isomers do occur in polypeptides, especially when a proline is on the carboxyl side of a peptide bond.

Peptide chains have directionality.

Like polynucleotide chains, peptide chains have directionality. A free amino group is present at one end of the peptide chain and a free carboxyl group at the other end. By convention the free amino group is drawn on the left. Once linked in a peptide chain, amino acids are called **amino acid residues** (or residues for short). Some important peptide nomenclature conventions are summarized in **FIGURE 2.10**. Note that all residues except for the C-terminal residue have a *-yl* suffix. For example, tyrosine, methionine, and glutamate residues are called tyrosyl, methionyl, and glutamyl. The C-terminal residue retains its regular name.

Rather arbitrarily, peptides are divided by size into two major groups. Those with less than 50 amino acids are **oligopeptides**, while those with 50 or more residues are polypeptides. As indicated above, the term "protein" is usually reserved for a polypeptide chain (or set of associated polypeptide chains) in the biologically active conformation. Because peptides can vary in chain length, amino acid sequence, or both, one can imagine an almost limitless variety of peptides. For example, there are 20^{50} or slightly more than 1.12×10^{65} possible sequences for polypeptides with just 50 amino acid residues.

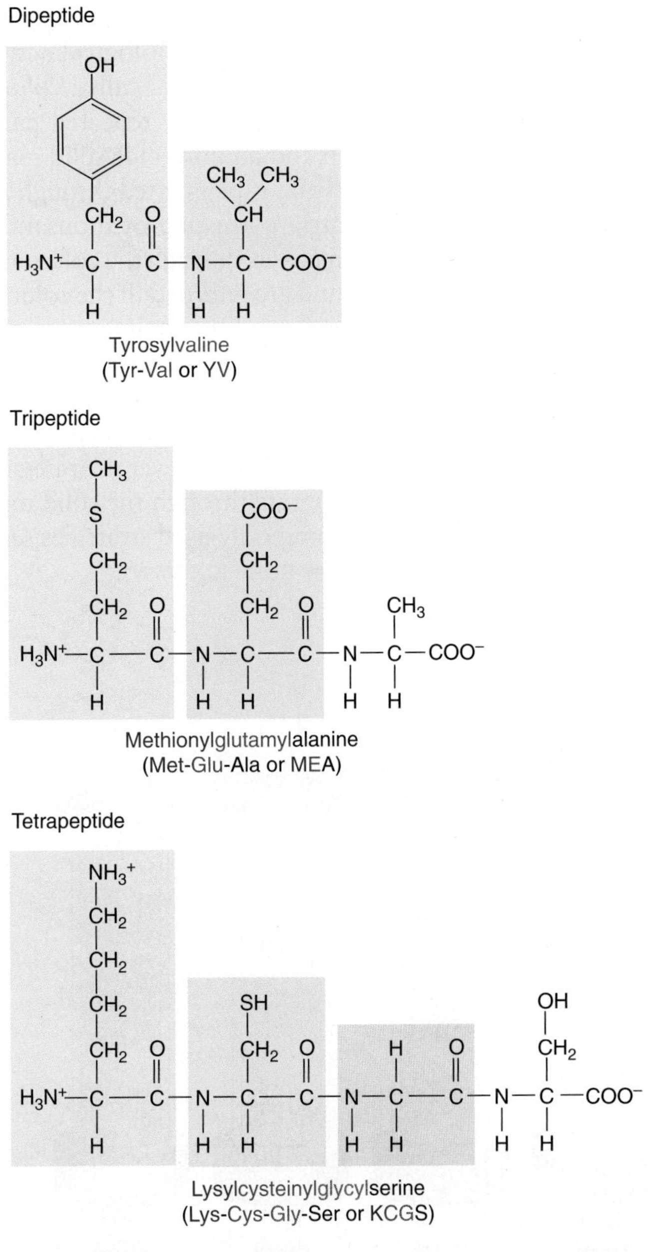

FIGURE 2.10 Conventions for drawing peptides. By convention the amino acid terminus (N-terminus) is on the left and the carboxyl terminus (C-terminus) is on the right. Peptides are named as derivatives of the carboxyl terminal amino acid.

2.3 Protein Purification

Protein mixtures can be fractionated by chromatography.

A complex process such as DNA replication or RNA synthesis requires many different proteins that must work together. Each protein makes a specific contribution to the overall process. It is difficult to examine an individual protein's structure and function(s) when the protein is in a mixture containing other proteins. Fortunately, most proteins are reasonably hardy and so retain their biological activity during purification. Nevertheless, it is usually desirable to purify proteins at 4°C and at about pH 7 to prevent the loss of biological activity.

One general method for protein purification, called **column chromatography**, separates proteins in a mixture by repeated partitioning between a mobile aqueous solution and an immobile solid matrix. The solution containing the protein mixture is percolated through a column containing the immobile solid matrix consisting of thousands of tiny beads (**FIGURE 2.11**). As the solution passes through the column, proteins interact with the immobile matrix and are retarded. If the column is long enough, it can separate proteins that have different migration rates. Proteins released from the column can be detected by an ultraviolet monitor because the aromatic amino acids (Phe, Tyr, and Trp) absorb ultraviolet light at a wavelength of about 280 nm, λ_{280}. Eluted proteins are collected in tubes by a fraction collector (Figure 2.11). Several strategies can be used to separate proteins as they migrate through the solid matrix (for a description of three of the most commonly used strategies see **BOX 2.1 IN THE LAB: STRATEGIES FOR PROTEIN PURIFICATION**).

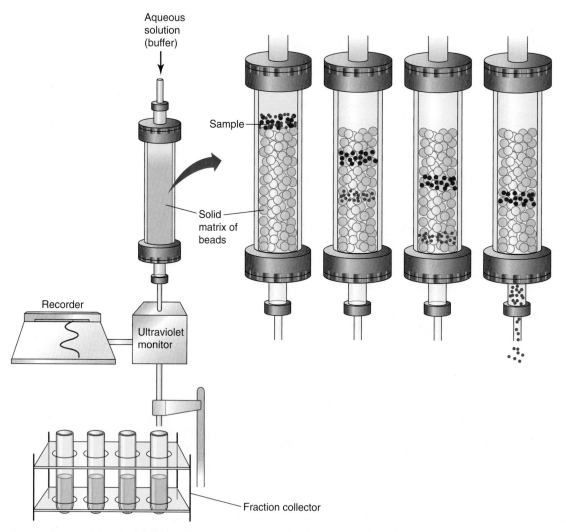

FIGURE 2.11 Schematic for column chromatography. Adapted from an illustration by Wilbur H. Campbell, Michigan Technological University (http://www.bio.mtu.edu/campbell/bl4820/lectures/lec6/482w62.htm. Accessed September 1, 2007).

BOX 2.1: IN THE LAB

Strategies for Protein Purification

Enzyme purification usually relies on some type of column chromatography. Three of the most commonly used column chromatography methods are ion exchange chromatography, gel filtration, and affinity chromatography. Investigators often use more than one of these methods to purify a protein. For instance, an investigator may obtain a partially purified protein by ion exchange chromatography and then obtain further purification by subjecting the partially purified protein to gel filtration. The basic features of the three-column chromatography purification techniques are as follows:

1. *Ion exchange chromatography*: This technique uses electrostatic interactions between the protein and the solid matrix to fractionate proteins (FIGURE B2.1a). A sample containing a mixture of proteins is allowed to percolate through a column packed with an immobile matrix, such as polysaccharide beads that are coated with positively (or negatively) charged groups. The beads' charged groups interact with the charged amino acid side chains on the protein. At pH 7, aspartate, glutamate, and carboxyl terminal residues have negative charges and interact with positively charged resins (anion-exchange chromatography). Lysine, arginine, and amino terminal residues have positive charges and interact with negatively charged resins (cation exchange resins). Proteins are released by passing aqueous solutions with progressively higher salt concentrations through the column. The salt ions displace the charged side chains from the ion exchange beads. Proteins that interact with the column most weakly migrate through the column fastest. Because proteins have both positively charged and negatively charged side chains on their surface, a specific protein may be fractionated by first using anion exchange chromatography to obtain a partially purified protein and then obtaining a further purification by subjecting the partially purified protein to cation exchange chromatography.

2. *Gel filtration*: Gel filtration (also called molecular exclusion chromatography) separates protein molecules by size. This method depends on special beads that permit small proteins to penetrate into their interior while excluding large proteins from this region (FIGURE B2.1b). A gel filtration column has two different water compartments: the internal compartment consists of the aqueous solution inside the beads and the external compartment consists of the aqueous solution outside the beads. Small protein molecules have access to both compartments, whereas large protein molecules only have access to the external compartment. The large proteins therefore appear in earlier fractions than do the small proteins (FIGURE B2.1c and 1d).

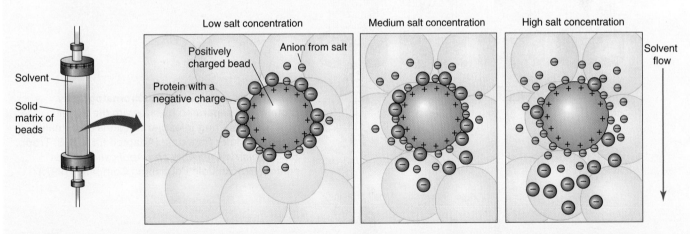

FIGURE B2.1a Anion exchange chromatography. Negatively charged groups on the proteins bind to positively charged groups on the anion exchange resin. Increasing salt concentrations produce anions that displace the proteins. Cation exchange resins work in a similar way except in this case, positively charged groups on the proteins bind to the negatively charged groups on the resin and cations displace the proteins.

(continues)

BOX 2.1: IN THE LAB

Strategies for Protein Purification *(continued)*

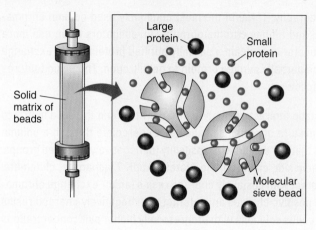

FIGURE B2.1b Gel filtration chromatography. The gel filtration column has an internal compartment consisting of the aqueous solution inside specially designed beads and an external compartment consisting of the aqueous solution outside the beads. The beads permit small proteins to penetrate into their matrix while excluding large proteins from the region; the large proteins can only get into the external compartment. (Adapted from B. E. Tropp. *Biochemistry: Concepts and Applications,* First edition. Brooks/Cole Publishing Company, 1997.)

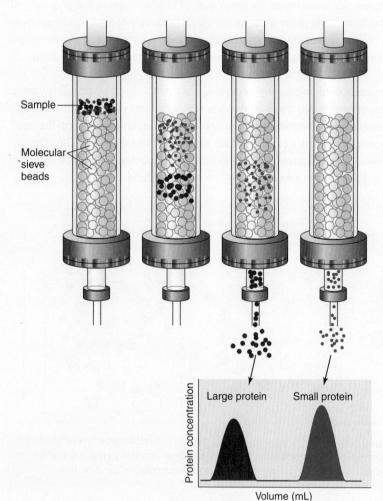

FIGURE B2.1c Gel filtration chromatography. Proteins are separated by size with the larger proteins appearing in earlier fractions and smaller proteins in later fractions. (Adapted from B. E. Tropp. *Biochemistry: Concepts and Applications,* First edition. Brooks/Cole Publishing Company, 1997.)

FIGURE B2.1d Gel filtration chromatography. Protein elution profile. (Adapted from B. E. Tropp. *Biochemistry: Concepts and Applications,* First edition. Brooks/Cole Publishing Company, 1997.)

CHAPTER 2 PROTEIN STRUCTURE AND FUNCTION

3. *Affinity chromatography*: This technique takes advantage of the fact that many proteins can bind specific small molecules, termed ligands. Affinity chromatography exploits this specificity to purify the protein. Ligands are attached to tiny beads to form affinity beads that are suspended in an aqueous buffer and poured into a column. In one of the earliest experiments the deoxythymidylic acid derivative shown in FIGURE B2.1e was attached to an insoluble polysaccharide to form affinity beads that bind nucleases (enzymes that catalyze nucleic acid hydrolysis). An aqueous solution, containing a mixture of bacterial proteins, was passed through a column packed with these affinity beads. Most of the proteins did not bind to the ligand and passed through the column. The nuclease, however, did bind to the ligand and was retained by the column. Active nuclease was recovered by washing the beads with a buffer solution at a low pH. Enzymes can also be eluted by washing the column with a solution that contains free ligand. Affinity chromatography can provide a high degree of protein purification due to the specificity of the binding step. Thanks to recombinant DNA technology, it is possible to use a variant of affinity chromatography to purify almost any protein if the DNA that codes for it has been isolated. The basic approach is to modify the DNA so the gene of interest now codes for the protein with an additional six histidine residues at its amino or carboxyl end. Specific resins have been devised that have a high affinity for proteins with a $(His)_6$ sequence. Because only the protein encoded by the modified DNA will have a $(His)_6$ sequence, the recombinant protein can be separated from all the other proteins in the mixture. Although addition of a $(His)_6$ tag usually does not alter a protein's biological properties, one must test to be certain that it does not.

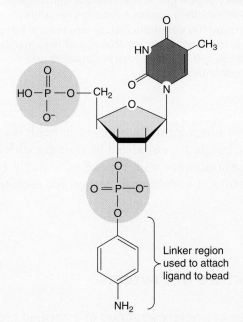

FIGURE B2.1e Affinity chromatography. Affinity chromatography exploits a protein's ability to bind to ligands. Nucleases bind to deoxyribonucleotide derivatives such as the one shown. Affinity resins are prepared by attaching the ligand to tiny water-insoluble beads. (Adapted from B. E. Tropp. *Biochemistry: Concepts and Applications*, First edition. Brooks/Cole Publishing Company, 1997.)

Proteins and other charged biological polymers migrate in an electric field.

Multiply charged macromolecules such as proteins and nucleic acids migrate through a medium in response to an electric field (FIGURE 2.12). In **gel electrophoresis** the medium is a porous matrix, such as a polyacrylamide gel saturated with buffer solution. The protein sample is applied to one end of the gel and the electric field is generated by connecting a power source to electrodes attached at either end of the gel. Proteins that migrate through the gel at the fastest rate tend to have the greatest net charge, the most compact shape, and the smallest size. The net charge can be altered by changing the pH of the medium. The pH at which a protein has no net

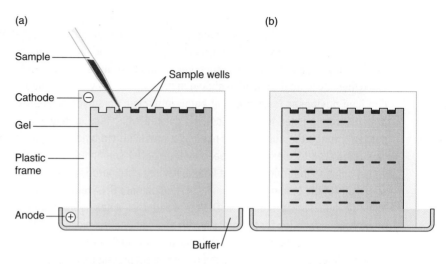

FIGURE 2.12 Gel electrophoresis. (a) Samples are applied to slots in a porous matrix such as polyacrylamide gel, and an electric field is generated. The buffer pH is adjusted to ensure the proteins have a net negative charge. (b) Proteins that migrate through the gel at the fastest rate tend to have the greatest net negative charge, the most compact shape, and the smallest size. (Adapted from B. E. Tropp. 1997. *Biochemistry: Concepts and Applications.* Brooks/Cole Publishing Company.)

charge, and therefore will not migrate in an electric field, is called its **isoelectric pH**. Protein bands are visualized by staining with dyes. A protein free of all contaminants will appear as a single band. Electrophoresis is, therefore, a very useful method for monitoring protein purity.

2.4 Primary Structure of Proteins

The amino acid sequence or primary structure of a purified protein can be determined.

Once a protein has been purified it must be characterized to learn more about its chemical and biological properties. Until recently, a degradation technique devised by Pehr Edman in 1950 was the most widely used method for amino acid sequence (also called **primary structure**) determination. *The term "degradation" indicates that covalent bonds are broken.* Edman degradation involves a series of chemical steps that remove the amino terminal residue from a polypeptide (**FIGURE 2.13**). Then the released amino acid derivative is identified and the process repeated through several rounds of amino acid removal and identification. Because cleavage efficiency is less than 100%, each successive cleavage cycle produces an increasingly heterogeneous peptide population. After about 50 cycles the peptide population is so heterogeneous it becomes nearly impossible to interpret the data.

The first step in solving the sequencing problem for long polypeptides is to cut the polypeptide into manageable fragments. This

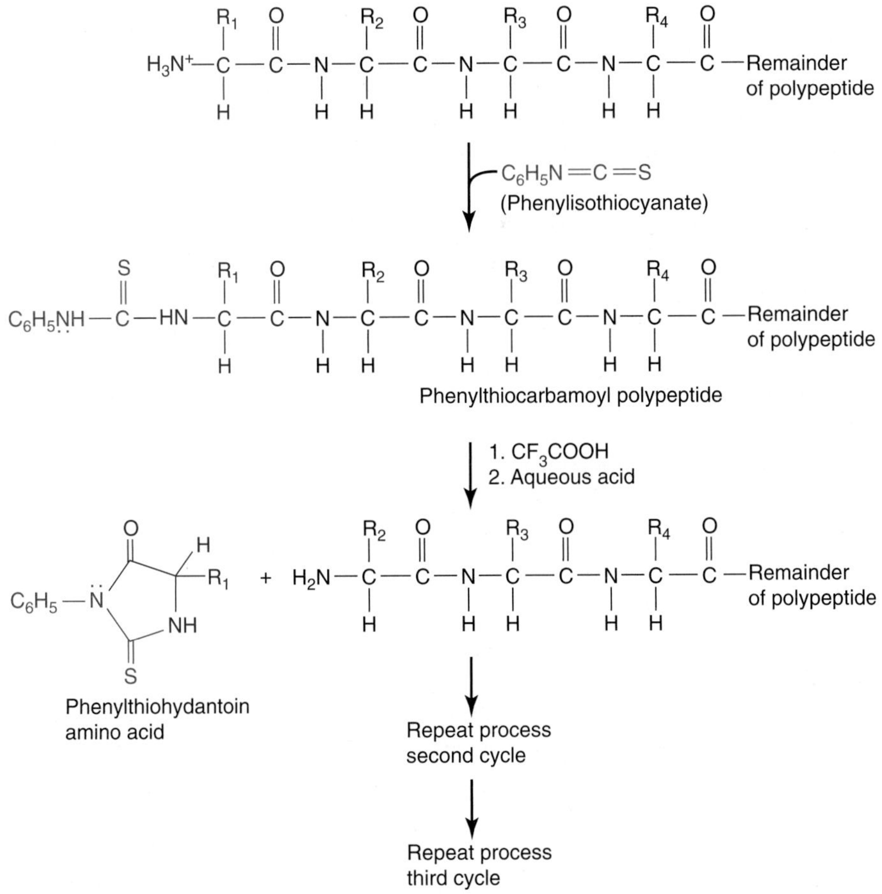

FIGURE 2.13 The Edman degradation. The Edman degradation, which selectively removes the N-terminal amino acid from a polypeptide chain, can be used to determine a polypeptide's amino acid sequence. (Adapted from B. E. Tropp. 1997. *Biochemistry: Concepts and Applications.* Brooks/Cole Publishing Company.)

task is usually accomplished by incubating the polypeptide with a specific protease (an enzyme that catalyzes protein hydrolysis) or a chemical cleaving reagent (FIGURE 2.14). The digestive enzyme trypsin, which cleaves on the carboxyl side of lysine and arginine residues, is well suited to this task. A second enzyme, chymotrypsin, cleaves peptide bonds when phenylalanine, tyrosine, or tryptophan residues are on the carboxyl side of the peptide bond. In addition to recognizing residues with aromatic side chains, however, chymotrypsin also sometimes cleaves peptide bonds in which a nonpolar residue such as leucine is on the carboxyl side of the bond. Because of this broader specificity considerably more caution must be taken when interpreting results with chymotrypsin digests than with trypsin digests. Cyanogen bromide (CNBr), a particularly useful chemical reagent for cleaving proteins, acts on peptide bonds in which methionine residues provide the carboxyl group. Although methionine is chemically altered during the cleavage process, this alteration does not interfere with subsequent sequence determination steps.

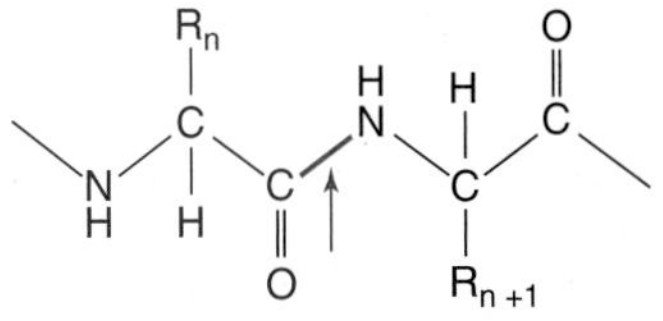

Enzyme or chemical agent	R_n
Trypsin	Lysine or arginine
Chymotrypsin	Phenylalanine, tyrosine, or tryptophan
Cyanogen bromide (CNBr)	Methionine

FIGURE 2.14 Specific cleavage. Trypsin, chymotrypsin, and cyanogen bromide (CNBr) cleave after specific amino acid residues. The arrow indicates the bond cleavage.

2.4 Primary Structure of Proteins

Before sequencing, each peptide fragment produced by trypsin, chymotrypsin, or cyanogen bromide cleavage must be isolated in the pure form by chromatography. Then the purified peptide fragments generated by a protease or CNBr are sequenced. After sequencing, the fragments must be placed in the proper order. Ordering is quite simple if the polypeptide is cut into just three fragments. Assuming the simplest possible case with no ambiguities, it is necessary to know which amino acids were present at the N- and C-termini of the intact polypeptide. The peptide fragment that begins with the residue present at the N-terminus of the intact polypeptide corresponds to the N-terminal fragment, the fragment that ends with C-terminal residue corresponds to the C-terminal fragment, and the remaining fragment is the middle one.

This simple approach to peptide fragment ordering will not work if enzymatic or chemical cleavage produces four or more peptide fragments. Then fragment order is determined by the **overlap method** (FIGURE 2.15). This method depends on having the sequences of two different fragment collections. The order of the fragments within the original polypeptide chain is determined by searching the two sets of fragments for overlapping sequences.

Despite its great value, Edman degradation also has limitations. It is time consuming, does not work when a peptide is blocked at its

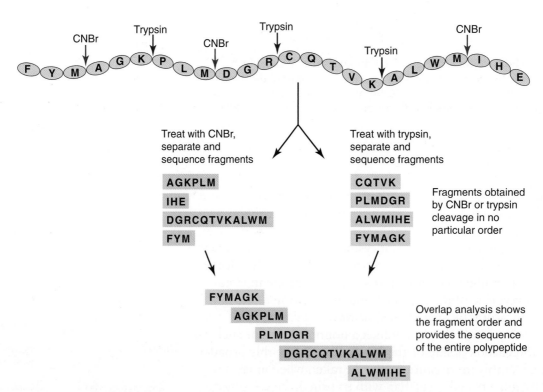

FIGURE 2.15 The overlap method for amino acid sequence determination. A polypeptide that is known to have phenylalanine (F) as its N-terminal residue is divided into two samples. One sample is treated with CNBr, which cleaves after methionine (M), and the other is treated with trypsin, which cleaves after arginine (R) or lysine (K). The CNBr cleavage sites are indicated by the red arrows and the trypsin cleavage sites by the purple arrows. Then the fragments produced from each sample are resolved by chromatography and each purified fragment is sequenced by the Edman degradation procedure. Finally, the fragments are sequenced by searching for overlaps.

amino terminus, and does not provide information about amino acid residues that have been modified. A new approach that overcomes these difficulties takes advantage of mass spectrometry, a technique in which molecules are ionized and their masses are determined by following the specific trajectories of the ionized fragments in a vacuum system. Because mass spectrometry is very sensitive, it requires very little protein. Moreover, peptide fragmentation takes place in seconds rather than hours, and sequencing is possible even if the protein is not completely pure.

FIGURE 2.16 summarizes the main steps in the mass spectrometer sequence technique. A protein population is prepared from a biological source and the individual polypeptides separated by electrophoresis. After separation the gel lane is cut into several slices. A specific protease or chemical agent is added to the gel slice of interest to digest the trapped protein, converting the protein into peptides. The peptides in the mixture that is generated by this digestion are separated by chromatography and then analyzed by the mass spectrometer. The process is repeated by cutting the same protein with another specific protease or cleavage agent and the primary structure of the protein determined by the overlap method.

Polypeptide sequences can be obtained from nucleic acid sequences.

The development of rapid DNA sequencing methods greatly accelerated the pace at which polypeptide sequences were determined. Instead of sequencing purified polypeptides, investigators determine DNA sequences of specific genes or all the genetic material in the chromosomes of a particular organism (the **genome**) and then translate this information to obtain polypeptide sequences.

Although the DNA sequencing approach is very fast and quite accurate, extrapolating to polypeptide sequences does have serious limitations. A DNA sequence does not necessarily predict the chemical nature of the biologically active protein for the following reasons. First, in eukaryotic cells (cells with defined nuclei) large precursors of messenger RNA (mRNA) are converted to mRNA by a precise splicing mechanism in which intervening RNA sequences, called **introns**, are removed with concomitant joining of flanking sequences, called **exons** (FIGURE 2.17). The resulting mRNA molecules that are missing sequences present in DNA program ribosomes form polypeptides that are shorter than those predicted from the DNA sequences. In many cases we cannot predict the sequences that will be lost during splicing and therefore cannot predict the sequence of the biologically active polypeptide. Second, many polypeptides are converted into their biologically active form by cleavage at specific sites. For instance, the polypeptide precursor to insulin, preproinsulin, is converted into the active hormone by specific peptide bond cleavages. Third, many proteins are subject to covalent modifications such as disulfide bond formation or the addition of phosphate, sugar, acetyl, methyl, lipid, or other groups. These modifications, which often influence the protein's biological activity and stability, can only be revealed by studying

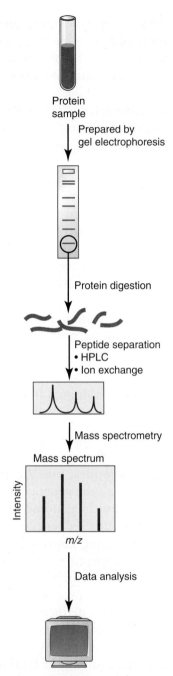

FIGURE 2.16 Mass spectrometric determination of peptide sequence. A protein population is prepared from a biological source such as a bacterial or cell culture. The protein of interest is purified; the last purification step is usually electrophoresis. The gel lane is cut to obtain the band that contains the desired protein. The protein is digested while still in the gel with specific enzymes, chemicals, or both. Then the peptide mixture generated is separated by chromatography and individual peptides are analyzed by mass spectrometry. The results are analyzed to determine the amino acid sequence within each peptide. The polypeptide sequence can be determined by digesting different samples of the same protein with different cleavage agents so that the overlap method can be used to order the proteins. (Adapted from H. Steen, and M. Mann. 2004. *Nat Rev Mol Cell Biol* 5:699–711.)

the purified protein. Finally, many polypeptides do not act alone but instead function as a part of a complex that contains other polypeptides of the same or different types. The true nature of these protein complexes can only be revealed by studying the purified complex.

The BLAST program compares a new polypeptide sequence with all sequences stored in a data bank.

The genomes of approximately 1,000 different organisms have now been sequenced. A comprehensive catalog of protein sequence information is available from the National Center for Biotechnology Information and from Uniprot. Comparison of a new sequence with sequences stored in the data bank is possible by using the BLAST (*Basic Local Alignment Search Tool*) program available at these sites.

Five different BLAST programs offer fast, sensitive, and relatively easy ways to compare specific nucleic acid or polypeptide sequences (the query sequences) with all sequences (the subject sequences) in the data bank. Each program permits a different type of search. Here we consider just one of the five BLAST programs, the *blastp program*, because it compares the amino acid query sequence with all polypeptide sequences in the data bank. The blastp program searches the data bank by first looking for every tripeptide in the data bank that is similar to tripeptides in the query polypeptide and then extends initial regions of similarity into larger alignments without gaps. Once alignments have been created, the blastp program determines and reports the probability of their arising by chance, lists the data bank sequences that are most similar to the query sequence, and shows a local alignment of the query sequence with matched data bank sequences. A newly sequenced polypeptide may be so similar to a polypeptide with a known function in another organism that we can safely conclude the two polypeptides have the same function.

A search of the data banks, however, may also reveal no similar polypeptides or polypeptides with similar sequences but with no known function. Now we are in the rather unsettling position of knowing that a polypeptide exists and knowing its sequence but without any idea about what the polypeptide actually does. Solving this "function problem" is one of the major challenges for molecular biologists in the coming decades.

2.5 Introduction to Protein Structure

Proteins with just one polypeptide chain have primary, secondary, and tertiary structures, whereas those with two or more chains also have quaternary structures.

The primary structure is just the first of four possible levels of polypeptide structures. The other three levels are determined by the way in which the polypeptide is arranged in space. The polypeptide backbone has three types of bonds, C_α—C_o, C_o—N (the peptide bond), and N—C_α (FIGURE 2.18). Although rotation around peptide bonds is severely

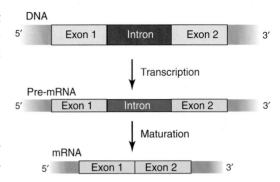

FIGURE 2.17 Split genes. Many eukaryotic genes have intervening sequences (introns) that are included in the precursor to messenger RNA (mRNA) molecules that are formed when the genes are transcribed but are removed during a maturation process in which precursor mRNA molecules are converted to mature mRNA molecules. The coding (or expressed) sequences, which are included in both the precursor mRNA and mature mRNA molecules, are called exons.

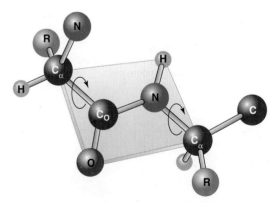

FIGURE 2.18 Rotation around C_α–C_o and N–C_α. Rotation does not take place around N–C_o but is free to take place around C_α–C_o and N–C_α. (Adapted from C. K. Mathews, et al. 2000. *Biochemistry* (3rd ed.), Prentice Hall.)

limited, rotation does occur around N—C_α and C_α—C_o, the two single bonds in the polypeptide backbone. Rotation around the single bonds permits the polypeptide to fold into biologically active proteins.

Because of this folding proteins have three levels of structure in addition to their primary structure (FIGURE 2.19a). The **secondary**

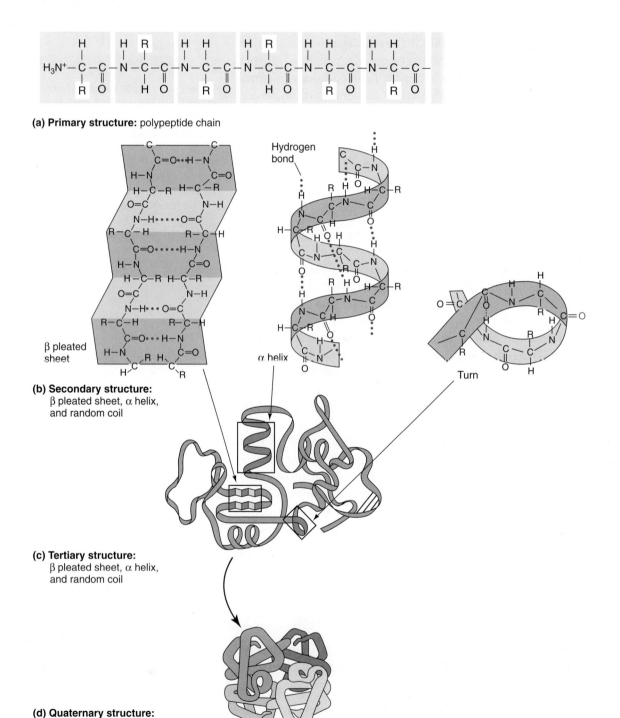

FIGURE 2.19 Protein structure. Polypeptides can be viewed at four different structural angles. The ribbon diagram (b and c) is a simple and effective way to represent secondary and tertiary structures. (Adapted from B. E. Tropp. 1997. *Biochemistry: Concepts and Applications.* Brooks/Cole Publishing Company.)

structure describes the folding pattern within a segment of a polypeptide chain containing neighboring residues. Among the many different possible secondary structures, the three most common are the α-helix, the β-conformation, and the loop or turn (Figure 2.19b). The **tertiary structure** provides a view of a protein's entire three-dimensional structure (Figure 2.19c), including spatial arrangements among different segments and among residues within the different segments. The **quaternary structure** applies only to proteins with two or more polypeptide chains and indicates the way in which the chains are arranged in space with respect to one another (Figure 2.19d).

2.6 Weak Noncovalent Bonds

The polypeptide folding pattern is determined by weak noncovalent interactions.

Secondary, tertiary, and quaternary structures are stabilized by four kinds of weak noncovalent interactions, the **hydrogen bond, ionic bond, hydrophobic interaction, and van der Waals interaction**. Because each kind of weak interaction has a range of binding energies, there is considerable energy overlap among them. Although the noncovalent interactions are weak (1–40 kJ • mol^{-1}) compared with covalent bonds (200–1,000 kJ • mol^{-1}), the combined effect of many weak interactions is sufficient to determine a protein's folding pattern. Moreover, because noncovalent interactions are weak they are easily formed and easily broken, permitting proteins to assume different conformations as they perform their functions. The four different kinds of weak interactions play such an important role in protein and nucleic acid structure that a brief description of each type is essential for studying the structure of these macromolecules (very large molecules).

Hydrogen Bond

The hydrogen bond results from an attractive interaction between an electronegative atom and a hydrogen atom attached to a second electronegative atom. Only two electronegative atoms—oxygen and nitrogen—participate in hydrogen bond formation in biological molecules. (Fluorine, which can also participate in hydrogen bond formation, is rarely present in biological molecules.) A glance at the periodic table reveals that nitrogen, oxygen, and fluorine are in period two and therefore the smallest electronegative atoms. Hence, the partially negative charge that results when these atoms pull electrons from a covalently attached carbon or hydrogen atom is concentrated in a small region of space. The partially positive charge in the hydrogen atom is concentrated in an even smaller region of space. The hydrogen bond results from an attractive force between the partially positively charged hydrogen atom attached to an oxygen or nitrogen in one group and the partially negatively charged oxygen or nitrogen atom in another group. This attractive force is considerably stronger than that resulting from other dipole–dipole interactions (see below).

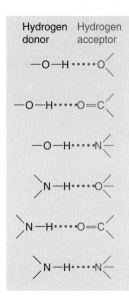

FIGURE 2.20 Some typical hydrogen bonds in proteins. The amino acid residue that supplies the hydrogen atom is designated the donor and the residue that binds the hydrogen atom is designated the acceptor.

In the strongest hydrogen bonds the two electronegative atoms and the hydrogen atom lie on a straight line. Hydrogen bond strengths range from about 8 to 40 kJ • mol^{-1}. The bond energy depends on physical environment. It is usually weaker on the protein surface than in the protein interior because it is much more likely to be subject to competing interactions with water on the protein surface than it is in the interior. Some typical examples of hydrogen bonds that occur in biological molecules are shown in FIGURE 2.20. The Watson-Crick model for DNA recognizes that hydrogen bonds between adenine-thymine base pairs and guanine-cytosine base pairs contribute to the stability of the double helix.

Ionic Bond

The ionic bond results from attraction between positively and negatively charged ionic groups. For example, the negatively charged side chain in aspartate or glutamate can form an ionic bond with the positively charged side chain in arginine or lysine. Ionic bonds can form between residue pairs near to one another in the primary structure or pairs that are far apart. The strength of the ionic bond varies depending on its surroundings. In an aqueous solution, ionic bond strength is about 2 kJ • mol^{-1}. The ionic bond strength is about 10 times greater in a protein's interior, where water molecules are very scarce. Ionic bonds are easily disrupted by pH changes, which alter the charges on interacting side chains or by high concentrations of small ions that compete with the interacting side chains.

Hydrophobic Interaction

Nonpolar molecules such as hexane, which aggregate in water, are said to be hydrophobic (Gr. *hydro* = water and *phobus* = fear). The driving force for hydrophobic interactions can be explained by considering the structural arrangement of surrounding water molecules. When a nonpolar molecule or group is placed in water, water molecules interact through hydrogen bonds to form highly ordered cages around the nonpolar molecule or group (FIGURE 2.21). According to the second law of thermodynamics, disorder is favored over order. The ordering of water molecules to form a cage around a nonpolar molecule or group is therefore an unfavorable process. That is, energy must be supplied to impose order. Placing two nonpolar molecules or groups into water requires water molecules to form two highly ordered cages. If the nonpolar molecules or groups were close together, a single water cage would suffice. For strictly geometric reasons the number of water molecules required to form a single cage around the pair of nonpolar molecules or groups is less than half the total number needed to form a cage around each molecule or group separately. In general, the number of ordered water molecules per nonpolar molecule or group is always smaller if the nonpolar molecules or groups are clustered or stacked. Hence, clustering or stacking of nonpolar molecules or groups is thermodynamically favored because a cluster or stack requires fewer water molecules to be arranged in highly ordered cages.

One can obtain an indication of the tendency of molecules or groups to aggregate in water, known as **hydrophobicity**, from their tendency to transfer from water to a nonpolar solvent. Many amino acid side chains are hydrophobic. For example, the hydrocarbon chains of alanine, leucine, isoleucine, and valine tend to form clusters and the aromatic side chains in tyrosine and tryptophan tend to form stacks in water. When interacting hydrophobic side chains are present on residues that are far apart, they bring distant parts of the polypeptide chain together. Thus, folding patterns of polypeptides are strongly influenced by hydrophobic interactions among the side chains of amino acid residues.

Van der Waals Interactions

Van der Waals interactions are weak electrostatic interactions between two polar groups, a polar group and a nonpolar group, or two nonpolar groups. Electrostatic attraction between two polar groups results from the attraction between a partially positive atom on one polar group and a partially negative atom on another polar group. Electrostatic attractions between a polar and a nonpolar group result from the polar group's ability to induce a short-lived polarity in the nonpolar group, which in turn leads to a weak attractive interaction between the oppositely charged regions on the two groups. Very weak electrostatic attractions between two nonpolar groups arise from fluctuating charge densities in the nonpolar groups. At any given time there is a small probability that a nonpolar group will have an asymmetrical electron distribution. A nonpolar group that experiences such a transitory perturbation of charge distribution can induce polarity in neighboring nonpolar groups. The combination of fluctuating and induced polarity accounts for the very weak forces of attraction that hold nonpolar molecules together.

The strength of a van der Waals interaction ranges from about 1 to 10 kJ • mol^{-1}. Because the attractive force between two atoms is proportional to $1/r^6$ (r is the distance between their nuclei), van der Waals interactions become significant only when two atoms are very near one another (0.1–0.2 nm apart). A powerful repulsive force also comes into play when the outer electron shells of the two atoms overlap. The **van der Waals radius** is defined as half the distance between two nonbonded atoms, when attractive and repulsive forces between the atoms are equal. Van der Waals radii differ from one kind of atom pair to another; some representative values are shown in TABLE 2.2.

The shape of a molecule is in essence the surface formed by the van der Waals spheres of each atom. FIGURE 2.22 shows the shapes of alanine and proline when defined in this way. Structures of this type are called **spacefill structures**. The average energy of thermal motion at room temperature is about 2.5 kJ • mol^{-1}. Therefore, the van der Waals interaction between two atoms is usually not sufficient to maintain these atoms in proximity. If the interactions of several pairs of atoms are combined, however, the cumulative attractive force can be great enough to withstand being disrupted by thermal motion. Thus, two nonpolar molecules can attract one another if several of their component atoms can mutually interact. However, because of the $1/r^6$-dependence, the intermolecular fit must be nearly perfect. Two nonpolar molecules will therefore hold together if their shapes are complementary. Likewise, two separate regions of a polymer will

(a) Water molecules in bulk phase of water

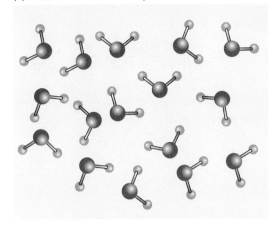

(b) Water molecules in cage around hydrocarbon

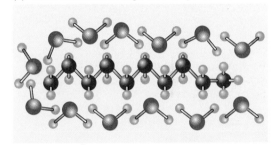

FIGURE 2.21 **The hydrophobic effect.** (a) Water molecules move randomly in the absence of nonpolar (hydrophobic) molecules. (b) Water molecules form ordered cages around hydrophobic molecules or groups. Because disorder is favored over order, the most favorable situation is one that uses the fewest possible water molecules to form cages. For strictly geometric reasons the number of water molecules required to form a cage around a group of hydrophobic molecules is less than the number required to form separate cages around each molecule. The hydrophobic molecules (or groups) therefore tend to aggregate in water. (Adapted from B. E. Tropp. 1997. *Biochemistry: Concepts and Applications*. Brooks/Cole Publishing Company.)

TABLE 2.2 van der Waals Radii	
Atom	van der Waals radii (nm)
Hydrogen	0.120
Oxygen	0.152
Nitrogen	0.155
Carbon	0.170
Sulfur	0.180
Phosphorus	0.180

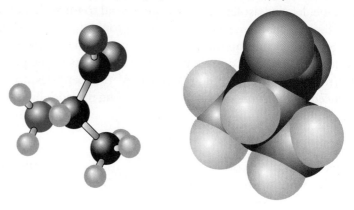

(a) Alanine shown in ball and stick (left) and van der Waals displays

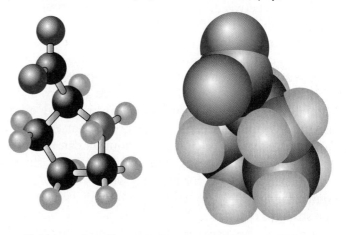

(b) Proline shown in ball and stick (left) and van der Waals displays

FIGURE 2.22 Alanine and proline as ball and stick and van der Waals displays.

hold together if their shapes match. Sometimes, the van der Waals attraction between two regions is not large enough to cause binding; however, it can significantly strengthen other weak interactions such as the hydrophobic interaction, if the fit is good.

2.7 Secondary Structures

The α-helix is a compact structure that is stabilized by hydrogen bonds.

Adjacent residues on a polypeptide chain can fold into regular secondary structures. Linus Pauling predicted the existence of the first secondary structure, the **α-helix**, in 1951 while recovering from an illness at home. To occupy his time Pauling drew a short peptide with correct bond angles and bond lengths on a piece of paper. Upon creasing the paper he noticed the peptide backbone folded into a helix. When he returned to the laboratory Pauling and his colleague, Robert Corey, constructed a more accurate three-dimensional model they named the α-helix. The model, shown in FIGURE 2.23, has the following features:

1. Each complete helical turn extends 0.54 nm along the vertical axis and requires 3.6 amino acid residues. The vertical rise per residue is 0.15 nm/residue (0.54 nm/3.6 residues).
2. Hydrogen bonds are located inside the helix, forming a regular repeating pattern.
3. The oxygen atom on the carbonyl group of residue n forms a hydrogen bond with the hydrogen atom of the N–H group of residue $n + 4$. The first and last four residues in the helix cannot form a full set of hydrogen bonds because hydrogen bond partners are not available.
4. All C=O bonds point in one direction, and all N–H bonds point in the opposite direction.

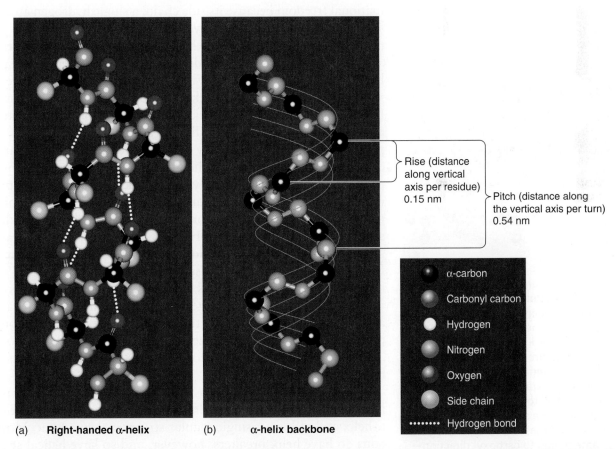

(a) Right-handed α-helix   (b) α-helix backbone

FIGURE 2.23 **The α-helix.** The N-terminal residue is at the bottom and the C-terminal residue at the top of the figure. The oxygen in each carbonyl group forms a hydrogen bond with the amide hydrogen that is four residues more toward the C-terminus, with the hydrogen bond approximately parallel to the long axis of the helix. All carbonyl groups point toward the C-terminus. In an ideal α-helix equivalent positions reappear every 0.54 nm (the pitch of the helix), each amino acid residue advances the helix by 0.15 nm along the axis of the helix (the rise), and there are 3.6 amino acid residues per turn. In a right-handed helix the backbone turns in a clockwise direction when viewed along the axis from its N-terminus. (Structures from the Protein Data Bank 1L64. From D. W. Heinz, W. A. Baase, and B. W. Matthews. 1992. *Proc Natl Acad Sci USA* 89:3751–3755. Prepared by B. E. Tropp.)

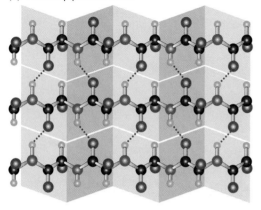

(a) Parallel β pleated sheet

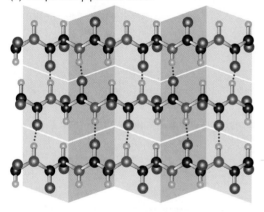

(b) Antiparallel β pleated sheet

FIGURE 2.24 **The β-conformation of polypeptide chains.** Polypeptide chains with β-conformations can line up side by side so that C=O and N–H groups on adjacent chains interact through hydrogen bonds to produce an almost flat structure called a β-sheet. The polypeptide chains' zigzag structure gives the β-sheet the appearance of being pleated with consecutive C_α atoms slightly above or below the plane of the sheet. The sheets can be organized so that (a) all peptide chains have the same amino to carboxyl direction—parallel pleated sheets or (b) successive polypeptide chains are oriented in opposite directions—antiparallel pleated sheets. The color coding is the same as in Figure 2.23. (Adapted from D. C. Nelson, and M. M. Cox. 2004. *Lehninger Principles of Biochemistry* (4th ed.). W. H. Freeman & Company.)

5. The α-helix is a right-handed helix. The right-handed nature of the helix is most easily visualized by picturing the threads of a screw. In a right-handed screw threads are shaped so that clockwise rotation produces tightening. Threads could also be in a left-handed form so that counterclockwise rotation produces tightening. Naturally occurring polypeptide chains, which are made of L-amino acids, fold into right-handed α-helices. When synthetic polypeptides are made of D-amino acids, the polypeptides fold into left-handed helices.

All amino acid residues can fit in the α-helix, but they differ in their propensity to do so. Proline and glycine have the least tendency to fit into a helix. The nitrogen atom in a proline that is part of a polypeptide lacks the substituent hydrogen atom needed to form a hydrogen bond. Moreover, the proline side chain is rigid and so does not easily fit into the α-helix. Proline is occasionally present in the first helical turn, where its side chain geometry and inability to hydrogen bond do not create a problem. Glycine's tendency not to be part of an α-helix results from an entirely different problem. Glycine has great conformational freedom because its side chain, a single hydrogen atom, is so small. Furthermore, a single hydrogen atom is insufficient to protect the hydrogen bonds inside the helix from disruption by water.

Certain combinations of adjacent residues also tend to disrupt or break the helical structure. When neighboring residues have bulky side chains, the side chains make physical contact that prevents them from fitting properly. For instance, neighboring isoleucine, tryptophan, and tyrosine residues would disrupt the helical structure. A run of positively charged or negatively charged side chains repel one another, destabilizing the helix. This phenomenon is best illustrated by comparing polyglutamate and polylysine structures at different pH values. Polyglutamate exists as an α-helix at pH 2 but not at pH 7 or pH 10. The explanation for this behavior is that the carboxylic acid side chains are uncharged at pH 2 but have negative charges at pH 7 and pH 10. Side-chain repulsion at pH 7 and pH 10 prevents helix formation. In contrast, polylysine exists as an α-helix at pH 10 but not at the two lower pH values. Once again the explanation is charge repulsion. The positively charged lysine side chains repel one another at pH 2 and pH 7. Although the α-helix is the most common helix present in proteins, it is not the only one. For instance, the so-called 3_{10} helix, which is much less common, has hydrogen bonds between residues n and $n + 3$.

A few fibrous proteins, most notably α-keratin, the major protein in hair, have no residues or combination of residues that disrupt the α-helix and so have a single regular structure throughout. Most proteins do have helix breakers, however, and so have helical segments surrounded by nonhelical segments. When present in a water-soluble globular protein, α-helical segments tend to be on the outside of the protein with one side facing the aqueous medium and the other the hydrophobic interior. The side facing the aqueous medium tends to have **hydrophilic** (acidic, basic, or polar) residues, whereas the side facing the interior has mostly hydrophobic residues. Because the α-helix has 3.6 residues per turn, such an arrangement can be achieved by switching from hydrophobic to hydrophilic side chains with a three- or four-residue periodicity.

The β-conformation is also stabilized by hydrogen bonds.

Pauling and Corey also predicted the existence of a second type of secondary structure, the **β-conformation**, in which the polypeptide chain is almost fully extended. Polypeptide chains with β-conformation can line up side by side so that C=O and N–H groups on adjacent chains interact through hydrogen bonds to produce an almost flat structure called a β-sheet (FIGURE 2.24). The polypeptide chain's zig-zag structure gives the β-sheet the appearance of being pleated with consecutive C_α atoms slightly above or below the plane of the sheet. A sheet can be organized so that all polypeptide chains have the same amino to carboxyl direction, referred to as a **parallel pleated sheet** (Figure 2.24a), or so that successive polypeptide chains are oriented in opposite directions, called an **antiparallel pleated sheet** (Figure 2.24b). In a parallel pleated sheet hydrogen bonds are evenly spaced and at angles to the long axes of the polypeptide chain. Hydrophobic side chains are present on both sides of the parallel pleated sheet. In an antiparallel pleated sheet pairs of narrowly spaced hydrogen bonds alternate with pairs of more widely spaced hydrogen bonds, with all hydrogen bonds perpendicular to the long axes of the polypeptide strands. The polypeptide chains that comprise an antiparallel pleated sheet tend to have alternating hydrophilic and hydrophobic residues, so that hydrophobic side chains tend to be present on one side of the sheet and hydrophilic residues on the other.

Loops and turns connect different peptide segments, allowing polypeptide chains to fold back on themselves.

The average globular protein has a diameter of about 2.5 nm, corresponding to about 11 residues in an α-helix and only 7 residues in the extended β-conformation. Secondary structures known as **turns** and **loops** connect α-helical and β-strand segments within a protein, allowing the polypeptide backbone to fold back on itself and reverse direction. Turns and loops usually are on the protein surface, extending into the surrounding aqueous environment. They therefore tend to be made of hydrophilic residues but also exploit the special conformational properties of glycine and proline residues to reverse direction. The most common kind of turn, the β-turn, consists of four residues and allows the polypeptide chain to reverse direction. The carbonyl oxygen of the first residue in a β-turn forms a hydrogen bond with the amino group of the fourth residue. The peptide bonds of the middle two residues do not interact through hydrogen bonds. Glycine and proline are commonly present in β-turns. Glycine's conformational flexibility allows it to fit into the tight turn. Proline's steric constraints are also well suited to the β-turn. Two of the most common types of β-turns are shown in FIGURE 2.25. Loops contain five or more residues, tend to be quite flexible, and lack a defined structure.

(a) Type I β-turn

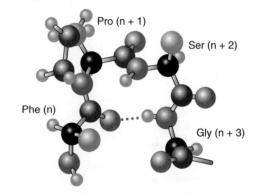

(b) Type II β-turn

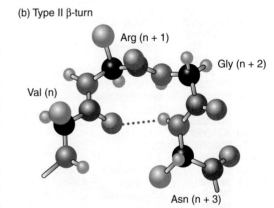

FIGURE 2.25 Two types of β-turns. The two most commonly occurring β-turns are called (a) type I and (b) type II. Type I β-turns occur with about twice the frequency of type II β-turns. In each case there is a hydrogen bond between residue n and residue $n+3$. Type II β-turns always have a glycine at position $n+2$. The color coding is the same as in Figure 2.23.

2.7 Secondary Structures

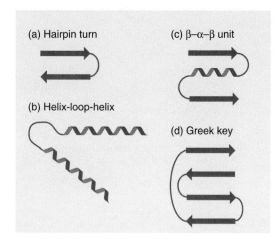

FIGURE 2.26 **A sample of supersecondary protein structures.**

Certain combinations of secondary structures, called supersecondary structures or folding motifs, appear in many different proteins.

Folding patterns for a few thousand different proteins are now available. Although at first glance these patterns look unique, more careful analysis reveals that some unifying principles do exist. One of the most important of these is that certain combinations of secondary structures, called **supersecondary structures** or **polypeptide folding motifs**, are present in many different proteins. Supersecondary structures are represented by a schematic topology diagram in which β-strands are represented by arrows pointing from the amino to the carboxyl terminus, α-helices by cylinders or helical structures, and loops and turns by ribbons. Some common supersecondary structures are shown in **FIGURE 2.26**.

We cannot yet predict secondary structures with absolute certainty.

Examination of amino acid side chain structures provides helpful insights into the way a residue may fit into a secondary structure. For example, the methyl group in alanine fits well into an α-helix, whereas the larger isopropyl group in valine experiences some steric hindrance. Additional insights come from comparing folding patterns of many different polypeptides. A statistical analysis of data from such studies allows us to predict the likelihood that a particular residue will be in an α-helix or β-strand (**FIGURE 2.27**). However, residues do not always behave as expected. For example, studies by Daniel L. Minor, Jr. and Peter S. Kim in 1996 showed that an 11-amino acid sequence (Ala-Trp-Thr-Val-Glu-Lys-Ala-Phe-Lys-Thr-Phe) folds into an α-helix when in one position in the primary sequence of a polypeptide but as a β-sheet when in another position. Peptide sequences thus can form different secondary structures when placed in different protein contexts.

2.8 Tertiary Structure

X-ray crystallography studies reveal three-dimensional structures of many different proteins.

It is much more difficult to determine the three-dimensional structure of a globular protein that contains a combination of secondary structures than it is to determine the three-dimensional structure of a regular protein that is all α-helix or all β-conformation. **X-ray crystallography** is used to elucidate the three-dimensional structure of proteins and nucleoproteins (protein–nucleic acid complexes) at atomic detail.

An ordinary optical instrument such as a light microscope does not permit us to see proteins in atomic detail because the distances between atoms are too small. In general, the wavelength required to resolve two objects (recognize the two objects as distinct entities) must

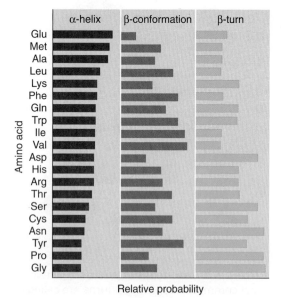

FIGURE 2.27 **Relative probabilities that a given amino acid will occur in the three common types of secondary structure.** (Adapted from D. C. Nelson, and M. M. Cox. 2004. Lehninger Principles of Biochemistry (4th ed.). W. H. Freeman & Company.)

be less than half the distance between the objects. Because distances between atoms linked by covalent bonds are about 0.15 nm, atomic resolution requires the very short wavelengths in x-rays. X-rays do not provide a direct image of protein molecules because currently available lenses and mirrors cannot focus such short wavelengths. X-ray crystallography allows interatomic distances in proteins to be measured by exploiting the fact that the electrons surrounding atoms in a crystalline protein scatter or diffract x-rays. The diffracted x-rays produce a characteristic pattern of spots on a film or detector that is placed behind the protein crystals (FIGURE 2.28). A heavy metal such as uranium is attached to a specific residue without altering protein structure to provide a reference point for data interpretation. Distances between spots and spot intensities provide the necessary information for determining protein structure.

The power of x-ray diffraction crystallography became apparent for all to see in 1957 when John Kendrew used the technique to determine the structure of myoglobin, an oxygen storage protein present in high concentrations in the muscles of diving mammals. A short time later Max Perutz reported the three-dimensional structure of hemoglobin, the oxygen transport protein in red blood cells. Many improvements have been made in x-ray crystallography since the structures of myoglobin and hemoglobin were first reported. Three advances in particular are especially noteworthy: (1) faster computers facilitate data analysis, (2) high-intensity x-ray beams emanating from synchrotrons allow investigators to study

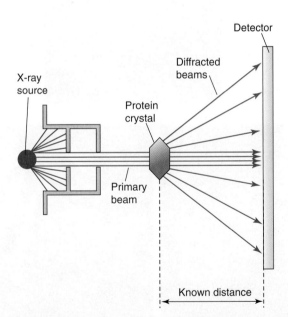

FIGURE 2.28 **Schematic of x-ray crystallography experiment.** When a beam of x-rays (red) hits a crystal, most of the electromagnetic radiation passes right through the crystal. The crystal scatters or diffracts the remaining light in many different directions. The diffracted x-rays produce a characteristic pattern of spots on a film or detector that is placed behind the crystal. Distances between spots and spot intensities provide the necessary information for determining protein structure. (Adapted from C. I. Branden, and J. Tooze. 1999. *Introduction to Protein Structure* (2nd ed.). Garland Science.)

protein crystals that are much smaller than those used in earlier studies of protein structure, and (3) synchrotron radiation at multiple wavelengths eliminates the need to attach heavy metals to specific sites in the protein. These advances have greatly accelerated the rate at which protein crystal structures can be solved. Thanks to these technological advances, much less time is required to solve protein structure problems today.

Recombinant DNA techniques have also played a critical role by allowing investigators to construct cells that produce large quantities of a desired protein. Today, approximately 70,000 protein crystal structures are available at the Research Collaboratory for Structural Bioinformatics Protein Data Bank, and new structures are being added each day. The data are entered as a protein data bank ("pdb") file. A program that serves as a viewer converts the pdb file into a three-dimensional image that can be manipulated on the computer. One excellent viewer, the Discovery Studio Visualizer 3.5® (from Accelrys, Inc., 10188 Telesis Court, Suite 100, San Diego, CA 92121, USA), permits the protein structure to be displayed in different forms and colors. FIGURE 2.29 shows the crystal structure of pancreatic ribonuclease A

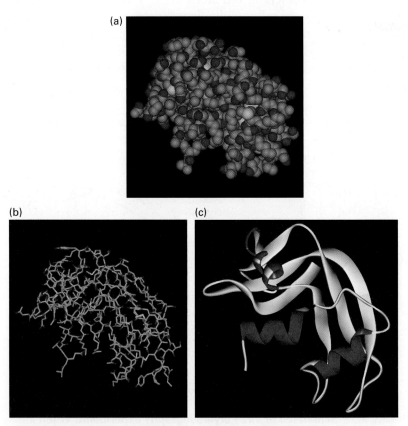

FIGURE 2.29 **Crystal structure of ribonuclease A.** (a) Crystal structure displayed in spacefill form. The colors are in standard CPK (Corey, Pauling, Kultin) color scheme. Carbon is gray, hydrogen white, nitrogen blue, oxygen red, and sulfur yellow. (b) Crystal structure displayed in stick form. The standard CPK color scheme is used. The orientation is as in (a). (c) Crystal structure shown in ribbon form. α-Helices are shown in red and β-conformations in yellow. (Structures from Protein Data Bank ID: 1JVT. L. Vitagliano, et al. 2002. *Proteins* 46:97–104. Prepared by B. E. Tropp.)

(RNase A), an enzyme that digests ribonucleic acids, in a spacefill display, a stick display, and a ribbon display.

Despite its enormous power, x-ray crystallography does have some shortcomings. First, proteins that do not form crystals cannot be examined by this technique. It is usually difficult to obtain membrane proteins in a crystalline form. Most protein structures determined to date are therefore for water-soluble proteins. Second, protein molecules pack close together when they form a crystal. Sometimes this packing causes residues on the protein surface to assume positions that are slightly different from their position in solution. Third, many proteins have regions that are highly disordered, so the structures of these regions cannot be determined by x-ray crystallography.

Initially, there was concern that protein crystal structures might differ from protein structures in aqueous solution. It is now clear, however, that protein crystals contain considerable water, which allows proteins to retain their biological activity. Furthermore, other physical methods, most notably nuclear magnetic resonance spectroscopy, indicate that proteins in solution have the same structures as those determined by x-ray crystallography.

Intrinsically disordered proteins lack an ordered structure under physiological conditions.

Not all proteins fold into ordered structures under physiological conditions. Those that do not, the **intrinsically disordered proteins**, are fairly common in eukaryotes, accounting for about 30% of the protein population. Intrinsically disordered proteins have fewer hydrophobic residues than globular residues. These residues, which help to stabilize globular protein structure by forming a hydrophobic core, are not available to provide the same stability to intrinsically disordered proteins. Intrinsically disordered proteins are also usually richer in polar residues that interact with water and in proline and glycine residues that promote disorder.

Some intrinsically disordered proteins assume stable three-dimensional structures upon interacting with other proteins, nucleic acids, or specific biological molecules. When an ordered complex forms, its structure can be determined by x-ray crystallography. Intrinsically disordered proteins contribute to cell maintenance and viability by participating in molecular recognition, assisting in molecular assembly, playing a role in protein modification, and helping to fold RNA.

The primary structure of a polypeptide determines its tertiary structure.

Heat and certain chemical agents such as acids, detergents, and urea (NH_2CONH_2) cause proteins to lose their biological activity by disrupting the weak noncovalent bonds that stabilize secondary and tertiary structures. Before the weak noncovalent bonds are disrupted, the protein is said to be in its **native state**. After disruption the protein is said to be in a **denatured state**, existing as a mixture of random conformations. It is important to note that *no covalent bonds are* *broken during denaturation.* Hydrophobic segments that are normally buried in the core of water-soluble proteins become exposed after denaturation and bind to hydrophobic segments of other denatured proteins to form water-insoluble aggregates. Protein denaturation is a rather common phenomenon. For example, milk spoils as a result of bacterial growth that produces acidic waste products, which cause the milk to have a sour taste and curdle. The curd is composed of denatured milk proteins that have become water insoluble. An egg's proteins denature when the egg is placed in boiling water.

Some denatured proteins can refold to their native state. Studies of this process, known as **renaturation**, have provided a great deal of information about the folding process. In the mid-1950s Christian Anfinsen and colleagues selected bovine pancreatic RNase A as a model protein for studying *in vitro* renaturation (**FIGURE 2.30**). Eight of the 124 residues in the polypeptide chain are cysteines that pair to form four disulfide bonds. These disulfide bonds help to lock the tertiary structure into place, making it very difficult to denature the enzyme. RNase A can be denatured by simultaneous

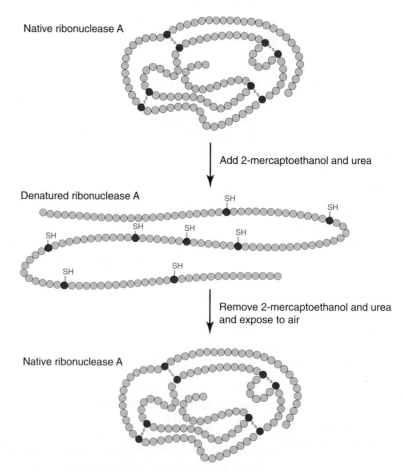

FIGURE 2.30 Denaturation and renaturation of RNase A. Denature pancreatic RNase A with urea in the presence of 2-mercaptoethanol. Renature enzyme by dialyzing denatured RNase to remove the urea and 2-mercaptoethanol and then exposing the polypeptide chain to air to re-form disulfide bonds. (Adapted from L. A. Moran, and K. G. Scrimgeour. 1994. *Biochemistry* (2nd ed.). Prentice Hall.)

treatment with urea and β-mercaptoethanol (CH_2OHCH_2SH). Neither chemical agent by itself is sufficient to denature RNase A. Urea interferes with hydrophobic interactions and disrupts hydrogen bonds, allowing the β-mercaptoethanol to gain access to the disulfide bonds. β-Mercaptoethanol disrupts the disulfide bonds, permitting further urea denaturation. If urea and the mercaptoethanol are slowly removed, perfect renaturation occurs, including formation of the four correct disulfide bonds (disulfide bonds can form spontaneously by oxidation in air). This latter finding is remarkable because there are 10^5 ($7 \times 5 \times 3$) possible ways that eight cysteine residues can combine to form four disulfide bonds. Anfinsen's interpretation of this experiment was that the folding of RNase A is determined exclusively by its amino acid sequence and that the proper disulfide bonds are formed because, during folding, the cysteine residues are correctly placed for joining. Evidence that disulfide bond formation does not direct the folding comes from an experiment in which the β-mercaptoethanol was removed first and oxidation was allowed to occur before removal of the urea, that is, while the RNase A was a random conformation. With this protocol the native molecule was not formed.

The Anfinsen experiment shows that, at least for RNase A, the tertiary structure is determined by the amino acid sequence and is the one with the lowest energy. Similar observations have been made for many other proteins, but not for all. The fact that denatured RNase can fold to form its native structure is remarkable in view of the number of possible conformations. If we assume that each residue has 10 possible conformations available to it, then the 124 residues would have to sample 10^{124} different conformations before achieving the correct one. If the polypeptide chain had to sample each possible conformation, the folding process could not possibly take place during the organism's lifetime. This folding problem, first recognized by Cyrus Levinthal, can only be solved if the polypeptide does not need to sample all possible conformations.

Many models have been proposed in an attempt to explain how the folding problem is solved. One recent proposal, the **zipping and assembly model**, incorporates features from a number of earlier models. According to this model folding begins when small segments along a polypeptide chain fold into secondary structures that are sufficiently stable to survive for a short time. This segmental folding takes place in a pico- or nanosecond time frame with each peptide segment searching for its own conformation independently of other segments. The order of segment folding therefore differs from one polypeptide chain to another. Although not stable enough to retain their conformations on their own, some of the segment structures survive long enough to grow (zip) into more stable structures or coalesce (assemble) with other structures.

In addition to its great theoretical significance, the Anfinsen experiment also had practical applications. At the time of the Anfinsen experiment no one had yet managed to synthesize a polypeptide chain as long as RNase. Some investigators questioned whether it would be worth the effort, because a synthetic polypeptide might not be able to fold into a biologically active form outside of the cell. The Anfinsen experiment showed that polypeptides prepared in the laboratory would have a reasonable chance of folding into biologically active proteins, stimulating efforts to synthesize long polypeptides. Although it is now possible to synthesize polypeptides in the laboratory by standard organic chemical techniques, the more common practice is to isolate or synthesize DNA that codes for the desired protein and then introduce that DNA into a living cell so the cell synthesizes the protein.

Molecular chaperones help proteins to fold inside the cell.

Protein folding is a much more complex process inside the living cell, which has high concentrations of proteins and other macromolecules. Instead of interacting with one another to form the correct tertiary structure, hydrophobic segments in an unfolded polypeptide may interact with hydrophobic segments in other unfolded polypeptide chains to form biologically inactive aggregates. The tendency to form aggregates is increased because stable tertiary structure formation usually requires a complete folding unit (100–300 amino acid residues). However, the complete folding unit does not emerge from the protein synthetic factory, the ribosome, in a single discrete step. Instead, the folding unit slowly emerges from the ribosome as each succeeding amino acid is added to the growing end of the polypeptide chain. Correct folding takes place only after the entire folding unit has emerged from the ribosome. Exposure of hydrophobic segments on the growing polypeptide would make them susceptible to intermolecular aggregation were it not for the presence of specific proteins called **molecular chaperones** that help to stabilize the emerging polypeptide until the entire folding unit has been extruded by the ribosome.

2.9 Myoglobin, Hemoglobin, and the Quaternary Structure Concept

Differences in myoglobin and hemoglobin function are explained by differences in myoglobin and hemoglobin structure.

Myoglobin (an oxygen storage protein) and hemoglobin (an oxygen transport protein) have a special place in the history of molecular biology because they were the first two proteins to have their structures determined. The lessons learned about structure–function relationships from studying these two proteins remain relevant today. We therefore begin our examination of protein structure-function relationships by examining myoglobin and hemoglobin.

Myoglobin, a single polypeptide chain with 153 residues, stores oxygen in muscles. Hemoglobin, a tetramer with two α- and two β-globin chains, transports oxygen in blood. Each α-globin chain has 141 residues, and each β-chain has 146 residues. The folding patterns of the myoglobin and hemoglobin chains are nearly identical (FIGURE 2.31ab). Helical segments within each polypeptide form a pocket containing a large planar ring system with an Fe^{2+} ion at its center, known as heme (Figure 2.31c), which binds oxygen.

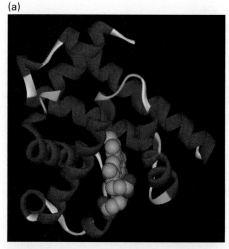

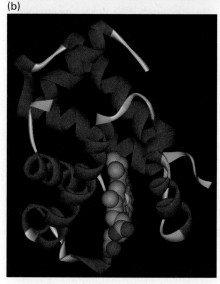

FIGURE 2.31 **Folding pattern of myoglobin and hemoglobin.** (a) Sperm whale myoglobin. (b) Human hemoglobin β-chain. (c) Heme. Polypeptide chains are shown in ribbon display and their heme groups in spacefill display. The accession number for human hemoglobin is 2HHB and that for sperm whale myoglobin is 1VXA. ([a] From Protein Data Bank ID: 1VXA. Yang, F., and Phillipsac, G. N. Jr. 1996. *J Mol Biol* 256:762–774; [b] From Protein Data Bank ID: 2HHB. Fermi, G., et al. 1984. *J Mol Biol* 175:159–174.)

An examination of myoglobin and hemoglobin structure reveals certain general rules that apply to other globular proteins:

1. Globular proteins tend to be compact, with hydrophobic amino acid residues accounting for about 65% of the protein interior.
2. Hydrophilic residues are almost always on the protein surface. Replacing a hydrophilic glutamate residue in position-6 (Glu-6) of the β-globin subunit with a hydrophobic valine residue has a profound influence on hemoglobin structure and function (**BOX 2.2 CLINICAL APPLICATION: SICKLE CELL ANEMIA**).
3. Water is generally excluded from the protein core but fills cavities and crevices when they are present.
4. Hydrophobic interactions, hydrogen bonds, ionic bonds, and van der Waals interactions stabilize tertiary structures (and quaternary structures when they are present as in hemoglobin).
5. Disruption of just a few weak noncovalent bonds can initiate changes that eventually cause the conformation of the entire protein to change. This conformational flexibility is essential for protein function.

BOX 2.2: CLINICAL APPLICATION

Sickle Cell Anemia

In 1904 a West Indian student of African origin who was suffering from anemia, recurrent pains, leg ulcers, jaundice, a low red blood cell count, and an enlarged heart sought the assistance of James Herrick, a Chicago physician. Upon examining the student's blood under the microscope Herrick observed that the red blood cells had a sickle or crescent shape rather than the biconcave appearance normally observed. Herrick hypothesized that the sickle-shaped red blood cells might be the key to understanding the patient's anemia and other symptoms. The disease suffered by the student, which is now known to be an inborn error of metabolism, was given the name sickle cell anemia. Herrick's hypothesis received additional support in 1927 when investigators observed that red blood cells isolated from individuals with sickle cell anemia change from a biconcave to a sickle shape when deprived of oxygen, whereas red blood cells from normal individuals remain biconcave (FIGURE B2.2a).

Linus Pauling and coworkers tried a new approach to the study of sickle cell anemia in 1949. They knew that hemoglobin, the major protein in red blood cells, was responsible for binding oxygen and suspected that individuals with sickle cell anemia have a variant form of hemoglobin. Their hypothesis was tested by comparing the electrophoretic mobilities of normal adult hemoglobin (HbA) and sickle cell hemoglobin (HbS). HbA migrated as though it were slightly more negative than HbS, suggesting that sickle cell anemia is caused by one or more amino acid substitutions in hemoglobin. However, the studies by Pauling and coworkers did not reveal whether the amino acid substitution(s) altered the α-globin chain, β-globin chain, or both, and the studies did not reveal which amino acid(s) was (were) altered.

The problem was solved in an ingenious fashion by Vernon Ingram in 1954. Modern techniques for sequencing polypeptide chains were not yet available, so Ingram could not compare the two forms of hemoglobin amino acid by amino

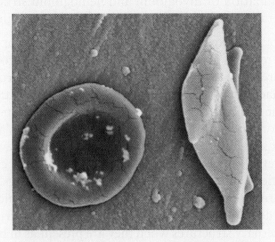

FIGURE B2.2a Biconcave- and sickle-shaped red blood cells. (Photo courtesy of the Sickle Cell Foundation of Georgia/Janice Haney Carr/CDC.)

acid. Instead, he used trypsin to cut HbA and HbS into well-defined fragments, which he spotted onto filter paper and partially separated by electrophoresis. Then he dried the paper, turned it on a right angle, and developed it in a second direction by chromatography. Because each type of protein treated in this fashion yields a unique two-dimensional pattern, the method is called peptide fingerprinting or peptide mapping. With one important exception the "fingerprint" of HbS is identical to that of HbA (FIGURE B2.2b and c).

Amino acid and sequence analyses show the fragment obtained from HbA that does not match up with one from HbS is an octapeptide derived from the N-terminal end of the β-chain. The only difference between the fragments that do not match up is the one derived from HbS contains a valine residue in place of the glutamate residue found in HbA. This substitution was shown to be in the sixth residue from the N-terminal residue of the β-chain (FIGURE B2.2d). The observed substitution is consistent with the observation by Pauling and coworkers that HbA migrates as a slightly more negatively charged protein than does HbS. Ingram's studies showed for the first time that a single amino acid substitution can cause a profound change in protein function. This finding stimulated great interest in the nascent field of molecular biology and reinforced the importance of studying structure–function relationships. Moreover, Ingram's studies introduced peptide mapping, a very powerful technique still used today.

(b) Hemoglobin A (HbA)

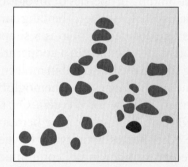

(c) Hemoglobin S (HbS)

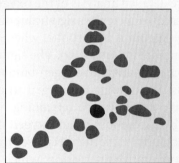

(d) Amino terminal fragment produced from the β-globin chains of HbA and HbS

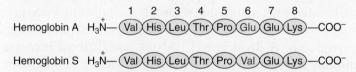

FIGURE B2.2bcd Peptide map of hemoglobin A (HbA) and hemoglobin S (HbS). Schematic diagram shows peptide maps created by trypsin digestion of (b) HbA and (c) HbS. All peptides but one are identical in HbA and HbS. The one peptide that is different, shown in red, has a single amino acid replacement. (d) Glutamate in position-6 of the β-globin chain is replaced BY VALINE. (Adapted from C. Baglioni. 1961. *Biochim Biophys Acta* 48:392–396.)

(continues)

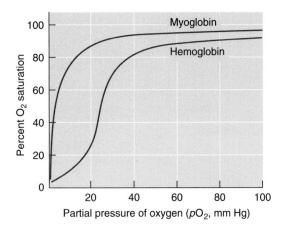

FIGURE 2.32 Oxygen-binding curves for myoglobin and hemoglobin.

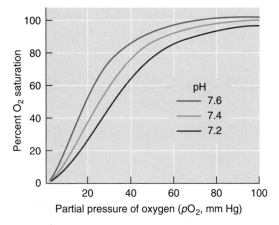

(a) pH effect on oxygen-binding curve

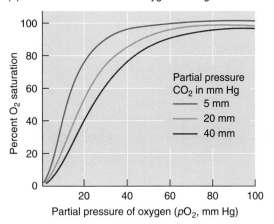

(b) Carbon dioxide effect on oxygen-binding curve

FIGURE 2.33 The Bohr effect. (a) pH effect on oxygen-binding curve. (b) Carbon dioxide effect on oxygen binding curve.

BOX 2.2: CLINICAL APPLICATION

Sickle Cell Anemia (*continued*)

Investigators originally thought that capillaries are blocked because the poorly deformable sickle cells cannot fit through them. More recent studies show that sickle red blood cells, but not normal red blood cells, adhere abnormally to the endothelial cells that line blood vessels. Research activity in the field of sickle cell anemia is now directed toward trying to correct the genetic error and developing an effective therapy to treat the symptoms.

Why have selective pressures not worked to eliminate sickle cell anemia? The rather surprising answer is that selective pressure for another factor actually favors the perpetuation of the sickle cell gene. Individuals who inherit a non-sickling allele from one parent and a sickling allele from the other parent have the sickle cell trait associated with a mild form of anemia. When exposed to *Plasmodium falciparum* or closely related protozoan parasites that cause malaria, they are more resistant to malaria than are individuals with two nonsickling alleles. As might be expected, the greatest incidence of sickle cell anemia and trait occurs in those regions of Africa, the Mediterranean, and India where malaria is most prevalent.

Hemoglobin subunits bind oxygen in a cooperative fashion.

A more detailed analysis of the oxygen-binding properties of myoglobin and hemoglobin reveals significant differences. The oxygen-binding curve for myoglobin is hyperbolic, whereas that for hemoglobin is S-shaped or sigmoidal (**FIGURE 2.32**). The sigmoidal curve indicates a **cooperative interaction** among the oxygen-binding sites in the hemoglobin molecule. That is, the binding of oxygen to one of the hemes in a hemoglobin molecule increases the remaining hemes' affinity for oxygen. On the other hand, the loss of an oxygen molecule from one of the hemes in a hemoglobin molecule decreases the other hemes' affinity for oxygen. This cooperative interaction, which requires some kind of communication among the hemes, permits hemoglobin to bind oxygen in the lungs where oxygen is plentiful and release oxygen near actively respiring cells where oxygen is scarce. Cooperative interaction is impossible in myoglobin because each myoglobin molecule has only one heme group.

Hemoglobin also has other remarkable properties that distinguish it from myoglobin. A drop in pH (**FIGURE 2.33a**) or an increase in carbon dioxide concentration (Figure 2.33b) causes hemoglobin to release oxygen but has no effect on myoglobin (not shown). The pH and carbon dioxide effects are known collectively as the **Bohr effect**, after Christian Bohr, the investigator who first discovered them. The Bohr effect has important physiological consequences. As red blood cells pass through the capillary vessels of rapidly respiring tissues, they are exposed to the carbon dioxide and low pH produced by those tissues. The hemoglobin molecules inside the red blood cells release oxygen and pick up carbon dioxide. When the red blood cells reach the lungs, where oxygen is plentiful, hemoglobin releases carbon dioxide and picks up oxygen.

Hemoglobin has another significant property that is not shared by myoglobin. The small organic phosphate, 2,3-bisphosphoglycerate (BPG), causes hemoglobin to release oxygen (FIGURE 2.34) but has no effect on myoglobin. BPG's influence on the oxygen-binding properties of hemoglobin has important clinical applications (BOX 2.3 CLINICAL APPLICATION: CLINICAL SIGNIFICANCE OF 2,3-BISPHOSPHOGLYCERATE).

The explanation for the cooperative effect and for each of hemoglobin's other special biological properties became clear when investigators compared the structures of the oxygenated and deoxygenated forms of myoglobin and hemoglobin. The conformation of deoxymyoglobin and deoxyhemoglobin polypeptide chains change when the heme group binds oxygen (FIGURE 2.35). The driving force for the conformational change is Fe^{2+} ion movement. In deoxymyoglobin (and each deoxyhemoglobin subunit) the Fe^{2+} ion is about 0.06 nm out of the plane of the heme ring system. In contrast, Fe^{2+} is in the plane of the ring system in oxymyoglobin (and in each oxyhemoglobin subunit). The reason for this position difference, which is barely visible in Figure 2.35, is twofold. First, Fe^{2+} is too large to fit into the center of the heme ring system in deoxymyoglobin (or a deoxyhemoglobin subunit), and, second, electrons in the heme ring system repel electrons in the histidine residue attached to Fe^{2+}. After binding oxygen, Fe^{2+} can move into the plane of the ring system because the effective Fe^{2+} radius decreases and the Fe^{2+} binding energy for oxygen exceeds the repulsive electrostatic forces between the heme ring system and the histidine residue. As Fe^{2+} moves into the plane of the heme ring system, it pulls the attached histidine and its associated helix along with it.

In myoglobin the helix is not locked into place and so is free to move. The situation is different in hemoglobin because the

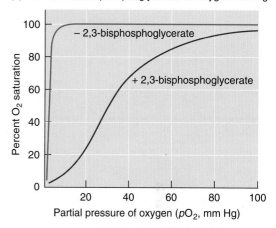

FIGURE 2.34 **Effect of 2,3-bisphosphoglycerate on oxygen binding in hemoglobin.** (a) Structure of BPG. (b) Effect of BPG on hemoglobin oxygen-binding curve.

BOX 2.3: CLINICAL APPLICATION

Clinical Significance of 2,3-Bisphosphoglycerate

The influence of BPG on hemoglobin's oxygen binding curve is an important concern to blood banks. During storage red blood cells break down BPG. As a result the hemoglobin in stored red blood cells exhibits an increasing affinity for oxygen over time. Because of this increased affinity stored red blood cells do not release the needed amount of oxygen to peripheral tissues and cannot be used for transfusions. The solution to this problem is to store blood in the presence of the nucleoside inosine, which the red blood cells metabolize to form BPG.

BPG's influence on hemoglobin's oxygen binding curve also is of concern to athletes training to participate in events at elevated altitudes, where the partial pressure of oxygen is lower than normal. An adaptive mechanism exists for increasing the BPG level of red blood cells when relatively little oxygen reaches the lung. The BPG level of an athlete's red blood cells can be increased by training at the high altitude for a few days. The increase in BPG level will improve the transport of oxygen to muscles and therefore athletic performance.

2.9 Myoglobin, Hemoglobin, and the Quaternary Structure Concept

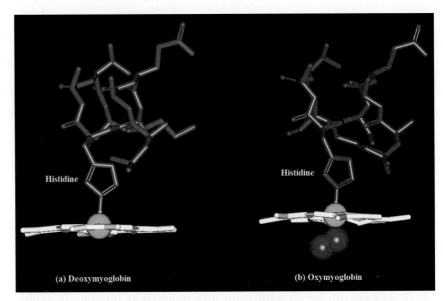

FIGURE 2.35 Movement of Fe^{2+} and its associated helix during the transition from deoxymyoglobin (or deoxyhemoglobin) to oxymyoglobin (or oxyhemoglobin). (a) In deoxymyoglobin (and each deoxyhemoglobin subunit) the Fe^{2+} ion (green sphere) is about 0.06 nm out of the plane of heme's ring system (white). The amino acid residues are shown in blue. (b) The Fe^{2+} moves into the plane of the ring after it binds oxygen (red spheres), pulling its attached histidine with its associated amino acids (red) along with it. ([a] From. Protein Data Bank ID: 4MBN. Takano, T. 1977 *J Mol Biol* 110:569–584; [b] Protein Data Bank ID: 1MBO. From Phillips, S. E. 1980. *J Mol Biol* 142:531–554. Prepared by B. E. Tropp.)

FIGURE 2.36 Human deoxyhemoglobin with 2,3-bisphosphoglycerate bound to the two β-globin chains. The globin chains are in ribbon display with each chain a different color. The heme groups are in stick structure display and the BPG bound to the two β-globin chains in the cavity formed by the four globin chains is in van der Waals display. (From Protein Data Bank ID: 1B86. Richard, V., Dodson, G. G., and Mauguen, Y. 1993. *J Mol Biol* 233:270–274.)

conformation of each chain is stabilized by ionic bonds to other chains. The ionic bonds between hemoglobin chains therefore must be broken before a chain can change from the deoxygenated form to the oxygenated form. Disruption of salt bridges between chains in a hemoglobin molecule not only affects the subunit that binds oxygen but also the subunits with which it interacts. The binding of oxygen to one subunit also loosens restraints on other subunits, making it easier for them to bind oxygen. This cooperative interaction is responsible for the sigmoidal shape of the oxygen-binding curve for hemoglobin.

The Bohr effect can also be explained in terms of hemoglobin's quaternary structure. The number of salt bridges between the polypeptide subunits increases when H^+ or carbon dioxide bind to the polypeptide subunits, stabilizing the structure of the deoxygenated form of hemoglobin. BPG also interacts by stabilizing the structure of deoxygenated hemoglobin; it fits into a cavity between the β-subunits of deoxyhemoglobin, forming an ionic bridge between them (**FIGURE 2.36**). Oxyhemoglobin, which lacks the cavity, cannot bind BPG. Deoxyhemoglobin must release BPG before it can bind oxygen, and oxyhemoglobin must release oxygen before it can bind BPG. Thus, even though the oxygen and BPG binding sites are distinct and distant, binding at one site prevents binding at the other.

Carbon dioxide, H^+, and BPG all lower the ability of hemoglobin to bind oxygen by increasing the number of salt links between the subunits of hemoglobin. Each thus acts at a site distinct from the oxygen-binding site. Molecules that influence protein activity by binding to sites that are distinct from the functional or active sites are called **allosteric effectors**. Binding an allosteric effector to a site on the protein that is some distance from the active site therefore can modify a protein's activity, permitting the protein to "sense" its chemical environment and to act accordingly.

2.10 Immunoglobulin G (IgG) and the Domain Concept

Large polypeptides fold into globular units called domains.

Myoglobin and hemoglobin chains are relatively short and therefore fold into a single compact globular structure. Larger polypeptides tend to fold into two or more compact globular units, called **domains**, which are joined by short lengths of the polypeptide chain. Domains may either behave like completely independent structural units or exhibit varying degrees of structural interaction. Well-defined domains often act as independent folding units and retain at least part of their normal biological activity even if they are split off from the rest of the protein.

The antibody **immunoglobulin G**, or **IgG**, provides an excellent example of a multidomain protein. Vertebrates produce millions of different kinds of IgG molecules. These IgG molecules have two distinct functions. First, each kind of IgG molecule recognizes and attaches itself to a specific chemical grouping (antigen) such as a unique sugar sequence in a bacterial cell wall. Second, the attached IgG serves as a marker, signaling other immune system components to attack and eliminate the foreign invader.

IgG is a tetramer made of two identical **light chains**, each with approximately 214 amino acid residues, and two identical **heavy chains**, each with approximately 446 amino acid residues (**FIGURE 2.37**). Each L-chain is attached to an H-chain by a disulfide bond as well as by weak noncovalent interactions. Disulfide bonds and weak noncovalent interactions also join the two H-chains.

Each L-chain can be divided into two regions of about equal size. The first region, extending from residues 1 to 108, has different sequences in different IgG molecules and is therefore called the variable region of the L-chain (abbreviated V_L). The second region, extending from residue 109 to the end of the chain, has the same sequence in all IgG molecules and is, therefore, called the constant region of the L-chain (abbreviated C_L). The H-chain also has a variable region (V_H) extending from residue 1 to about residue 110. The remainder of the H-chain has three regions of about equal size, each with a constant amino acid sequence. These constant regions, designated C_H1, C_H2, and C_H3, have amino acid sequences that are quite similar, but not identical, to one another.

Each L- and H-region folds into a distinct compact globular structure so that the IgG molecule has a total of 12 domains (Figure 2.37).

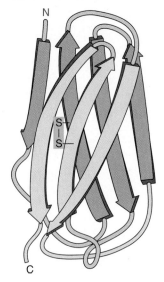

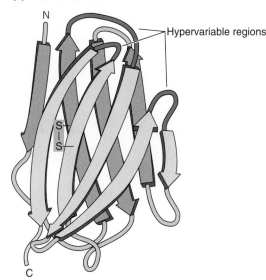

FIGURE 2.38 Folding patterns of the IgG constant and variable domains. (a) Constant region domains from either light or heavy chains fold in the same way, resembling a collapsed barrel with four β-strands on one side and three on the other. (b) Variable region domains of the L- and H-chains also resemble collapsed barrels. However, one side of the barrel has five β-strands rather than three strands. The hypervariable regions, which are shown in red, are loops joining the β-strands in the V_L and V_H domains. (Adapted from Branden, C., and Tooze, J. 1999. *Introduction to Protein Structure* (2nd ed.). Garland Publishing Co.)

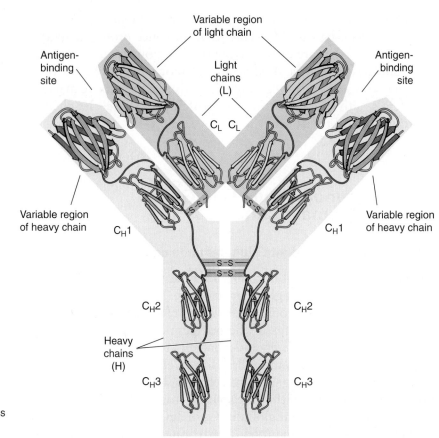

FIGURE 2.37 Schematic of immunoglobulin G (IgG) structure.

C_L and C_H domains have very similar structures, resembling collapsed barrels with four β-strands on one side and three on the other. A short loop joins each β-strand to the one next to it (FIGURE 2.38a). This folding pattern, known as the immunoglobin fold, is present in many other proteins. V_L and V_H domains also resemble collapsed barrels. However, one side of the barrel has five rather than three β-strands (Figure 2.38b). Although the major secondary structure in IgG domains is the β-strand, this is certainly not true for domains in all proteins. Domains are also formed from a mixture of α-helix and β-strands or mainly from α-helix.

Constant and variable region domains have different functions. Constant region domains are responsible for the overall structure of the molecule and for its recognition by other components of the immune system. Variable region domains act as antigen binding sites. Each antigen binding site contains one V_L domain and one V_H domain. Each IgG molecule thus contains two antigen binding sites (Figure 2.37). Three small regions within V_L and another three within V_H display much more variation than the rest of the variable regions. These so-called **hypervariable regions** form loops that extend into the surrounding aqueous environment (Figure 2.38) to make contact with the antigen. The amino acid residues in the hypervariable region determine antibody specificity.

2.11 Enzymes

Enzymes are proteins that catalyze chemical reactions.

Virtually all chemical reactions that take place in the cell are catalyzed by protein molecules called **enzymes**, the subject of this section, or RNA molecules called **ribozymes**. Enzymes catalyze the vast majority of the reactions involved in DNA, RNA, and protein metabolism. A large number of different enzymes are required to catalyze a complex process such as replication, transcription, or translation. To understand these processes we must learn how the enzymes involved in them work. An enzyme often requires one or more cofactors to catalyze a reaction. The cofactor may be a simple metal ion such as Mg^{2+} or a small organic molecule. Organic cofactors, known as **coenzymes**, are usually vitamins or vitamin derivatives. For instance, the cofactor nicotinamide adenine dinucleotide (NAD^+) is a derivative of vitamin B_3 (niacin).

Molecular biologists often use colorimetric and radioactive methods to measure enzyme activity.

The molecule(s) on which an enzyme acts is called its **substrate**(s). Only a small number of substrate molecules, sometimes only one, participate in a single catalyzed reaction. Enzymatic activity is measured by following the rate of substrate breakdown or product formation. Several different physical techniques can be used to monitor an enzyme catalyzed reaction. The two most common techniques used by molecular biologists are colorimetric and radiotracer techniques.

Colorimetric assays are based on the fact that a substrate (or product) absorbs light of a particular wavelength. When a substance that absorbs visible or ultraviolet light is either a substrate or product of an enzyme catalyzed reaction, then substrate breakdown or product formation can be followed using a spectrophotometer. For this reason molecular biologists often work with specially designed substrates that generate products with unique light absorption properties. One such specially designed substrate is o-nitrophenyl-β-D-galactoside (ONPG), which is used to assay β-galactosidase, an enzyme that normally cleaves lactose to form galactose and glucose (FIGURE 2.39a). The bond hydrolyzed, a β-galactoside linkage, is also present in ONPG. β-Galactosidase hydrolyzes the colorless ONPG to form galactose and o-nitrophenoxide, which is intensely yellow (Figure 2.39b). β-Galactosidase activity is therefore readily followed by measuring the concentration of o-nitrophenoxide at a wavelength of 420 nm (blue light).

In a radioactivity assay a radioactive substrate is added to a reaction mixture and the appearance of a radioactive product (or the disappearance of the radioactive substrate) is measured. This type of assay is used to measure the conversion of a radioactive amino acid into a radioactive protein or a radioactive nucleotide into a radioactive nucleic acid. The assay used is based on the following fact: proteins and nucleic acids are insoluble in 0.5 M trichloracetic acid (TCA),

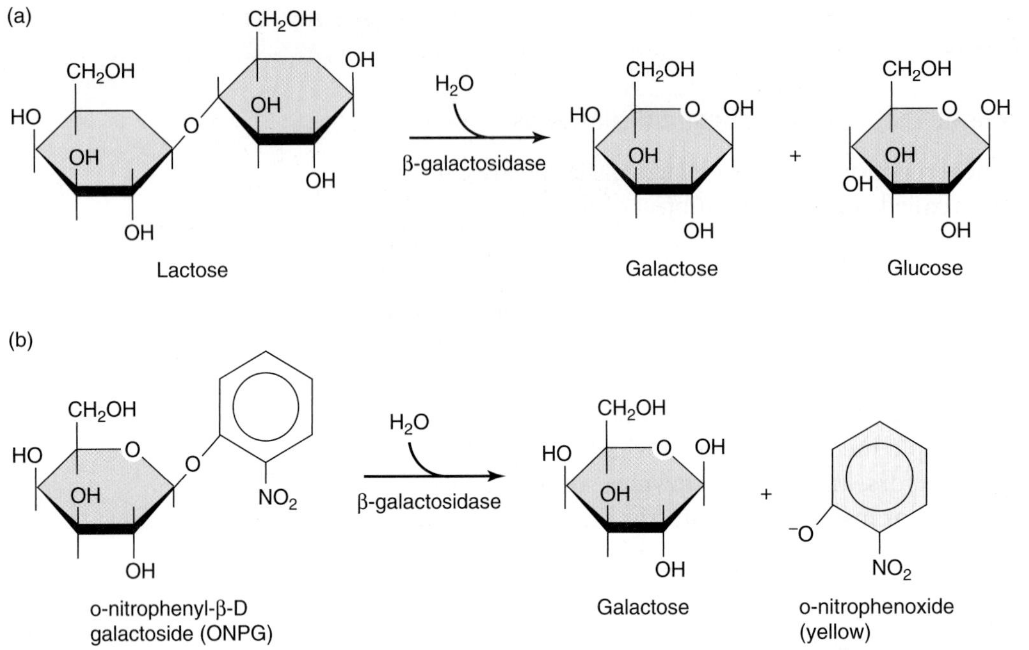

FIGURE 2.39 β-Galactosidase assay. β-Galactosidase converts (a) its natural substrate lactose to galactose and glucose and (b) a synthetic substrate ONPG to galactose and o-nitrophenoxide, which has an intense yellow color.

whereas their precursor amino acids and nucleotides are TCA soluble. We take advantage of these solubility differences to monitor protein and nucleic acid syntheses.

Protein synthesis is monitored by adding a radioactive [^{14}C]amino acid or [^{3}H]amino acid to a reaction mixture that contains the other 19 nonradioactive amino acids, ribosomes, and the appropriate enzymes and factors. After a period of time TCA is added and the mixture is filtered. The acid-soluble radioactive amino acid passes through the filter, whereas acid-insoluble, radioactively labeled protein products are retained on the filter. The radioactivity retained by filter, which is measured using a liquid scintillation counter (**BOX 2.4 IN THE LAB: RADIOACTIVE TRACERS**), indicates the quantity of protein that was synthesized.

DNA synthesis is monitored by using a mixture containing the four deoxyribonucleotide precursors of DNA, one of which is radioactive, the enzyme, and other appropriate components of the mixture. For this example we will assume that [^{14}C]thymidine triphosphate is used. TCA is added to stop the reaction and precipitate newly formed DNA. Then the mixture is filtered. The acid-soluble [^{14}C]thymidine triphosphate passes through the filter, whereas radioactively labeled, acid-insoluble DNA is retained by the filter. Once again the radioactivity retained by filter is determined by using a liquid scintillation counter.

Enzymes lower the energy of activation but do not affect the equilibrium position.

Because enzymes play such a critical role in reactions of interest to molecular biologists, it is important to learn how they work. We begin by drawing a reaction coordinate diagram for an uncatalyzed

BOX 2.4: IN THE LAB

Radioactive Tracers

Radioactive tracer techniques are an outgrowth of nuclear physics and chemistry. Four radioactive isotopes, hydrogen-3 (tritium), carbon-14, phosphorus-32, and sulfur-35, have been especially useful in molecular biology studies. The nuclei of these radioactive isotopes disintegrate to release β-particles. The SI (International System of Units) unit of radioactivity, the Becquerel or Bq, is 1 disintegration per second (dps). Molecular biologists often use two non-SI units, disintegration per minute (dpm) and curies (Ci), when reporting radioactivity (1 Ci = 3.7×10^{10} Bq = 2.2×10^{12} dpm). The specific activity of a radioactive sample is usually given in units such as Bq • μmol^{-1} or μCi • μmol^{-1}.

The energy of a β-particle emitted by a particular isotope is a characteristic of that isotope. β-Particles emitted by phosphorus-32 are so energetic they pass through clothing and skin. Investigators normally work behind a Plexiglas shield and monitor their radiation exposure when working with compounds that contain phosphorus-32. Sulfur-35 releases the next most energetic β-particles followed by carbon-14 and then tritium. Plexiglas shielding is not required when working with compounds that contain sulfur-35, carbon-14, or tritium. Nevertheless, great care must be taken when working with these isotopes. It is essential to read posted rules before working with radioisotopes and to be certain to follow these rules.

Radioisotopes that emit β-particles can be monitored by using a liquid scintillation counter. The radioactive sample is placed in a glass or plastic vial containing a liquid scintillation cocktail. One of the cocktail's components, a fluor, emits a flash of light in response to β-particles. The number of scintillation events is proportional to the number of nuclear disintegrations per second. The liquid scintillation counter detects and records the scintillation events.

reaction in which reactants A and B are converted to products C and D (FIGURE 2.40, plot in red). The extent of reaction, called the reaction coordinate, is plotted on the *x*-axis and the energies of reactants, intermediates, and products are plotted on the *y*-axis. Reactants require enough energy to reach the top of the energy barrier to form a molecular complex, called an **activated complex** or **transition state**, before they can be converted to products. The rate of a reaction is directly related to the fraction of reactant molecules that reach the transition state in a given period of time and depends on the **energy of activation** (E_a), that is, the energy needed to reach the transition state. A catalyst increases the reaction rate by providing an alternative reaction path with a lower energy of activation (Figure 2.40, plot shown in blue). *Catalysts do not change the equilibrium position for a reaction only the rate at which the reaction takes place.*

The catalytic power of enzymes exceeds all artificial catalysts. A typical enzyme accelerates a reaction 10^8- to 10^{10}-fold, though there are enzymes that increase the reaction rate by a factor of 10^{15}. Enzymes

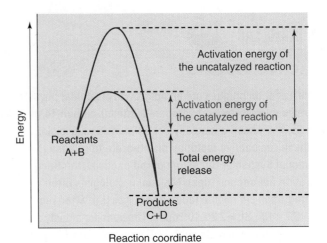

FIGURE 2.40 **Reaction coordinate diagram for an uncatalyzed reaction (red) and an enzyme catalyzed reaction (blue).** The enzyme lowers the activation energy but does not affect the energy released by the reaction. The enzyme therefore increases the reaction rate but does not affect the equilibrium.

are also highly specific in that each catalyzes only a single reaction or set of closely related reactions. Several different hypotheses have been proposed to explain the catalytic properties of enzymes. These hypotheses, which are not mutually exclusive, are as follows:

1. Enzymes lower the energy of activation by stabilizing the transition state. This hypothesis is supported by studies that show IgG molecules, which are formed when animals are injected with a stable analog of the transition state, act as catalysts. Catalytically active antibodies are called **abzymes**.
2. Enzymes lower the energy of activation by putting a strain on a susceptible bond. Distortion of the susceptible bond makes it easier to break the bond.
3. Enzymes lower the energy of activation for reactions involving two or more substrates by holding the substrates near to each other and in the proper orientation.
4. Enzymes lower the energy of activation by forming a transient covalent bond with a reactant molecule that destabilizes some other bond.
5. Enzymes lower the energy of activation by acting as proton donors and acceptors.

All enzyme-catalyzed reactions proceed through an enzyme–substrate complex.

In any enzyme-catalyzed reaction the enzyme, E, always combines with its substrate, S, to form an **enzyme–substrate (ES) complex**, which can then either dissociate to reform substrate or go forward to product, P, and enzyme, E.

$$E + S \underset{k_{-1}}{\overset{k_{+1}}{\rightleftharpoons}} ES \overset{k_{+2}}{\rightarrow} E + P$$

where k_{+1}, k_{-1}, and k_{+2} are rate constants for the reaction. (By convention, kinetic constants for forward reactions k_{+1} and k_{+2} have a (+) symbol in their subscript, and kinetic constants for reverse reactions (k_{-1}) have a (−) symbol in their subscript.) The rate of reaction is directly proportional to the ES concentration. Hence, the **theoretical maximum rate of reaction** (V_{max}), is observed when all the enzyme molecules are present in ES complexes.

The ratio $(k_{-1} + k_{+2})/k_{+1}$, called the **Michaelis constant** (K_m), can provide information about the enzyme's affinity for its substrate. Usually, dissociation of the ES complex is more rapid than conversion of the complex to enzyme and product. When this is the case the value of K_m is a measure of the strength of the ES binding. That is, when $k_{-1} \gg k_{+2}$, K_m approaches k_{-1}/k_{+1}, the dissociation constant for ES complex, and K_m is a measure of an enzyme's affinity for its substrate. *A high K_m value indicates low affinity (and weak binding), and a low K_m value indicates a high affinity (and strong binding).* Strength of binding depends on several conditions, such as temperature, pH, the presence of particular ions, and overall ionic concentration. K_m values vary greatly for different enzymes, ranging from 10^{-6} to 10^{-1} M.

V_{max} and K_m values can be determined from a hyperbolic curve such as that shown in **FIGURE 2.41**, which is generated by plotting the initial rate of reaction (v_o) on the *y*-axis and substrate concentration ([S]) on the *x*-axis. The theoretical maximum velocity (V_{max}) is the limiting velocity obtained as the substrate concentration approaches infinity. The K_m value corresponds to the substrate concentration at which the reaction rate is $V_{max}/2$. It is the same as the concentration at which half of the enzyme molecules in the solution have their active sites occupied by a substrate molecule. Intracellular substrate concentrations tend to range between $0.1\ K_m$ and K_m.

The number of substrate molecules converted to product molecules by an enzyme molecule in a specified time is called the **turnover number**. The turnover number is determined by dividing V_{max} by the molar concentration of the total enzyme present in the reaction mixture.

The specific region on an enzyme responsible for binding the substrate(s) and catalyzing the reaction is called the **active site**. Despite the great variations that exist among enzymes, certain generalizations about the active site are possible. The active site is a cleft or crevice that is usually located on the surface of the protein and represents about 10% to 20% of the enzyme's total volume. The amino acid residues that form the active site come from different regions of the polypeptide chain, sometimes very far apart in the linear sequence. Much of the cleft is lined with hydrophobic side chains, creating an environment that excludes water molecules unless they are reactants. Polar and charged side chains are located at specific sites within the active site, where they contribute to substrate specificity and participate in the catalytic process. Amino acid residues within the active site that directly participate in making and breaking chemical bonds are called catalytic residues.

Two models for enzyme binding have been proposed to explain the extraordinary specificity of enzyme-substrate binding. First, the **lock-and-key model** (**FIGURE 2.42a**) was proposed by Emil Fischer

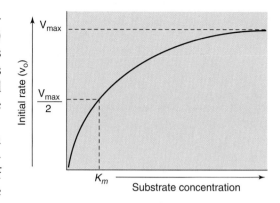

FIGURE 2.41 Calculation of V_{max} and K_m. V_{max} and K_m values can be determined from a hyperbolic curve generated by plotting the initial rates of reaction, v_o, on the *y*-axis and substrate concentration on the *x*-axis. The theoretical maximum velocity, V_{max}, is the limiting velocity obtained as the substrate concentration approaches infinity. The K_m value corresponds to the substrate concentration at which the reaction rate is $V_{max}/2$.

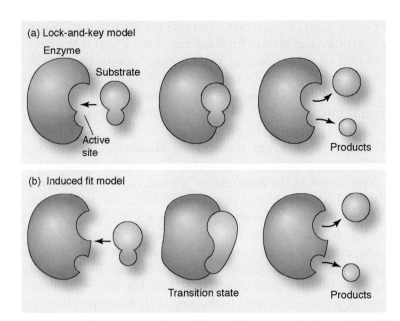

FIGURE 2.42 Enzyme–substrate interaction. There are two models of enzyme binding, (a) lock-and-key and (b) induced fit. In the lock-and-key model the shape of the active site of the enzyme is complementary to the shape of the substrate. In the induced fit model the enzyme changes shape upon binding the substrate and the active site has a shape that is complementary to that of the substrate only after the substrate is bound. (Adapted from Mathews, C. K., et al. 2000. *Biochemistry* (3rd ed.). Prentice Hall.)

in 1894 to explain how an enzyme recognizes its specific substrate. According to this model each substrate has a characteristic geometric shape that fits into a complementary geometric shape in the enzyme. That is, the enzyme behaves like a rigid lock that will only accept a specific substrate key. As more information became available about ES interactions, the lock-and-key model's shortcomings became evident. For example, the model predicts that an enzyme should be able to act on a molecule with the same shape as its substrate but with less bulky substituent groups. Enzymes are usually quite specific for their substrates, however, and rarely act on molecules that are smaller than their substrates.

The second model, the **induced fit model** (Figure 2.42b), proposed by Daniel Koshland in 1958, addressed the known lock-and-key model shortcomings. According to the induced fit model, the enzyme is a flexible structure that changes shape on binding substrate. The induced fit model explains an enzyme's inability to act on small substrate analogs as follows. A molecule with the same shape as the substrate but less bulky substituent groups would not be able to make all the contacts needed to induce the conformational changes. Although the induced fit model is primarily concerned with the change in enzyme shape, it is important to note that the substrate's conformation also usually changes as it binds to the enzyme. In fact, the strain to which the substrate is subjected is often the principal mechanism of catalysis. That is, the substrate is held in an enormously reactive conformation. The induced fit model was proposed before the three-dimensional structures of enzymes were known. X-ray crystallography made it possible to test the model. Studies of enzyme and ES complex structures show that most enzymes do change conformation on binding their substrate.

Molecular details for ES complexes have been worked out for many enzymes.

The first three-dimensional image of an enzyme was of a hen egg-white lysozyme. This enzyme cleaves certain bonds between sugar residues in some of the polysaccharide components of bacterial cell walls and is responsible for maintaining sterility within eggs. FIGURE 2.43 shows egg-white lysozyme bound to a substrate analog. The amino acids that are part of the active site form widely separated clusters along the chain. Only when the chain folds do they come into proximity and form the active site. The true substrate is a hexasaccharide segment that fits into the cleft and is distorted upon binding. The enzyme itself changes shape when the substrate is bound. A variety of interactions (van der Waals, hydrogen, and ionic bonds) stabilizes the binding.

Another enzyme, yeast hexokinase A, which catalyzes the reaction of glucose and adenosine 5′-triphosphate to form glucose-6-phosphate and adenosine 5′-diphosphate, has been studied to examine conformational changes that take place on substrate binding. The enzyme

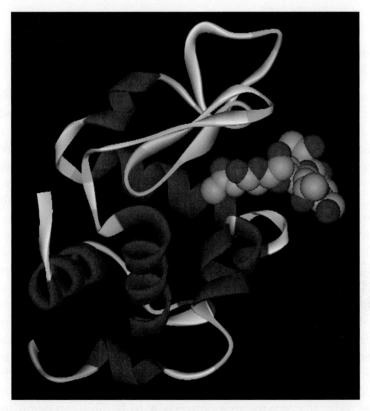

FIGURE 2.43 **Hen egg-white lysozyme in ribbon display.** Helical regions of the protein are shown in red and β-strands in cyan. The bound substrate analog is in van der Waals display. (Structure from Protein Data Bank ID: 1LZB. Maenaka, K., et al., *J. Mol. Biol.* 247[1995]:281–293. Prepared by B. E. Tropp.)

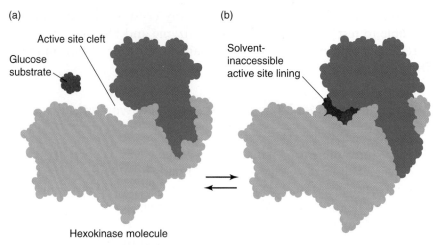

FIGURE 2.44 **Hexokinase, an example of induced fit.** A drawing, roughly to scale, of an idealized hexokinase molecule (a) without and (b) with bound glucose. The two hexokinase domains move together when glucose is bound, creating the catalytic site. The dark blue area in (b) represents solvent inaccessible surface area in the active site cleft that results when the enzyme binds glucose. (Photos courtesy of Thomas A. Steitz, Yale University.)

contains two domains, which move together when glucose binds. This movement creates a catalytic site that is inaccessible to the surrounding water molecules. These changes are shown in the pair of spacefill models in FIGURE 2.44.

Regulatory enzymes control committed steps in biochemical pathways.

Enzyme regulation offers a quick and efficient way to adjust the flow through biochemical pathways, allowing a cell to synthesize products that are in short supply and to stop the synthesis of products that are in abundance. Regulation is usually achieved by control of flow through a few key steps in a biochemical pathway, such as the hypothetical example shown in FIGURE 2.45. In many cases the cell requires only the end products of the pathway (in this example E and G). Pathway intermediates are essential to synthesize E and G but are not otherwise needed by the cell. The pathway has a branchpoint at C. The first reaction after a branchpoint is almost always irreversible, committing the flow of material to that branch. Thus, the first step after a branch is called the **committed step** for that branch.

Enzymes that catalyze the committed steps in a reaction pathway are usually **regulatory enzymes**, which are inhibited by the end product of the branch. E, the end product of the C → D → E branch, inhibits the enzyme that converts C to D. Likewise, G, the end product of the C → F → G branch, inhibits the enzyme that converts C to F. An abundance of both E and G can block the third key regulatory enzyme in the path, the enzyme that converts A to B. This can occur in one of three ways (FIGURE 2.46). C accumulates when both branches are blocked and inhibits the enzyme that converts A to B (Figure 2.46a). Alternatively, E and G act together to block the enzyme that converts

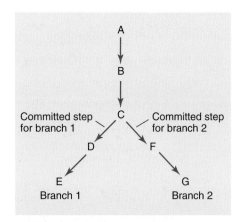

FIGURE 2.45 **Branched pathway.**

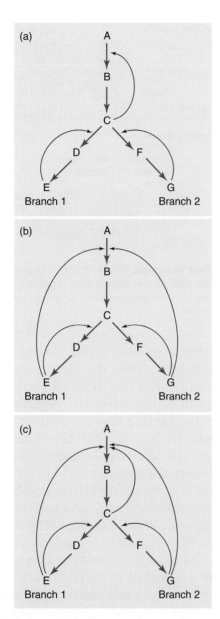

FIGURE 2.46 Alternative methods for feedback inhibition to take place in a pathway with two branch points.

A to B (Figure 2.46b). Sometimes both types of inhibition contribute (Figure 2.46c). The method of control, in which an end product inhibits specific steps in a biochemical pathway, is called **feedback inhibition**.

Regulatory enzymes exhibit sigmoidal kinetics and are stimulated or inhibited by allosteric effectors.

Regulatory enzymes often produce S-shaped or **sigmoidal kinetic curves** when the initial velocity (v_o) is plotted on the y-axis and substrate concentration ([S]) is plotted on the *x*-axis (**FIGURE 2.47**). The reason for the S-shaped curve is exactly the same as that given for the oxygen-binding curve for hemoglobin. The regulatory enzyme is made of subunits that bind substrate in a cooperative fashion.

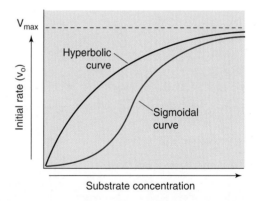

FIGURE 2.47 Sigmoidal kinetics for regulatory enzymes. Plots of substrate concentration versus initial velocity (v_o) generate hyperbolic curves for nonregulatory enzymes but usually generate sigmoidal curves for regulatory enzymes. The sigmoidal curve indicates that substrate binding is cooperative, that is, the binding of a substrate molecule to one enzyme subunit increases the chances that other enzyme subunits will also bind substrate molecules.

2.11 Enzymes

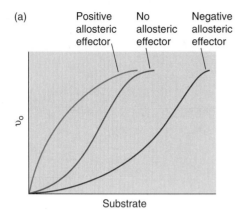

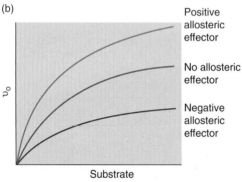

FIGURE 2.48 Effects of positive and negative allosteric effectors. (a) Allosteric effectors can influence concentrations required to give $V_{max}/2$ or (b) allosteric effectors can influence V_{max}.

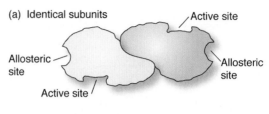

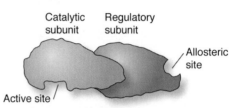

FIGURE 2.49 Regulatory enzymes. Nearly all regulatory enzymes are made up of two or more subunits. The subunits of a regulatory enzyme may be (a) identical to one another or (b) different from one another.

Regulatory enzymes are also inhibited and stimulated by allosteric effectors. Once again, hemoglobin is a useful model. Subunits of an allosteric enzyme can exist in two different conformations, a more active conformation and a less active one. Allosteric effectors act by binding to sites on the enzyme that are distinct from the catalytic site. Positive allosteric effectors stabilize the more active form of the enzyme, whereas negative allosteric effectors stabilize the less active form. Allosteric effectors can be recognized by their effects on enzyme kinetics (FIGURE 2.48). Positive allosteric effectors often act by lowering the substrate concentration required to give half the maximal velocity, whereas negative allosteric effectors often act by increasing this concentration (Figure 2.48a). Alternatively, a positive allosteric effector may increase V_{max} and a negative allosteric effector may lower V_{max} (Figure 2.48b). The activities of many regulatory enzymes are influenced by a combination of positive and negative allosteric effectors.

Regulatory enzymes that are inhibited or stimulated by allosteric effectors are almost always made of two or more subunits (FIGURE 2.49). When the subunits are all identical, each subunit has an active site and at least one allosteric site (Figure 2.49a). When the subunits are different one kind of subunit, the **catalytic subunit**, has the active site and another type of subunit, the **regulatory subunit**, has the allosteric site(s) (Figure 2.49b). Both types of regulatory enzymes play important roles in molecular biology.

Questions and Problems

1. Briefly define, describe, or illustrate each of the following.
   a. α-Amino acid
   b. Globular protein
   c. Fibrous protein
   d. Peptide bond
   e. Oligopeptide
   f. Polypeptide
   g. Gel electrophoresis
   h. Isoelectric pH
   i. Degradation
   j. Overlap method

k. Genome
l. Intron
m. Exon
n. Primary structure
o. Hydrogen bond
p. Ionic bond
q. Hydrophobic interaction
r. van der Waals interactions
s. van der Waals radius
t. Spacefill structure
u. Secondary structure
v. α-Helix
w. β conformation
x. Parallel pleated sheet
y. Antiparallel pleated sheet
z. Turn
aa. Loop
bb. Tertiary structure
cc. Quaternary structure
dd. Supersecondary structure
ee. Intrinsically disordered protein
ff. Zipping and assembly model
gg. Bohr Effect
hh. Cooperative binding
ii. Native state
jj. Denatured state
kk. Renaturation
ll. Molecular chaperone
mm. Domain
nn. Abzyme
oo. Active site
pp. Coenzyme
qq. Ribozyme
rr. Transition state
ss. K_m
tt. V_{max}
uu. Turnover number
vv. Regulatory enzyme
ww. Feedback inhibition
xx. Allosteric effector
yy. Committed step
zz. Sigmoidal kinetic curve

2. Each of the 20 common α-amino acid building blocks of proteins has a unique side chain. Draw the structure of each of the following.
   a. The amino acid that does not have an L-configuration.
   b. The amino acid that lacks a primary amine group.
   c. The amino acid with a methyl group as a side chain.
   d. The amino acid with the most basic side chain.
   e. The three amino acids with aromatic side chains.
   f. The amino acid that forms a disulfide in air.
   g. The two amino acids with amide side chains.
   h. The amino acid with a thioether side chain.
   i. The amino acids with hydroxyl groups in their side chains.
   j. The two amino acids with carboxyl groups in their side chains.

3. Draw the structures for the following three peptides.
   a. Alanylvaline
   b. Phenylalanylglycylcysteine
   c. Valylserylmethionine

4. The C–N bond that links two amino acids has some double-bond character. How does this double-bond character affect peptide conformation?

5. A dipeptide containing alanine and valine was isolated as a degradation product of protein A. A second dipeptide, also containing alanine and valine, was isolated as a degradation product of protein B. The two dipeptides have different physical and chemical properties. How is it possible for dipeptides that each contain alanine and valine to have different chemical and physical properties?

6. A pentapeptide that contains phenylalanine, lysine, alanine, cysteine, and serine can be hydrolyzed by trypsin but not chymotrypsin. Provide an explanation.

7. What will happen to the peptide Ala-Lys-Phe-Gly-Met-Glu when it is treated with the following enzymes and reagents?
   a. trypsin
   b. chymotrypsin
   c. CNBr
   d. phenylisothiocyanate

8. How is it possible for a protein to bind to a cation exchange resin and also to anion exchange resin?

9. Describe a method used to separate two proteins that have very different molecular masses.

10. Hair, a protein that is made of α-keratin, stretches when heated in moist air. Provide a molecular explanation.
11. Polylysine exists as an α-helix at pH 10 and as a random conformation at pH 2. In contrast, polyglutamic acid exists as a random conformation at pH 10 and an α-helix at pH 2. Provide a molecular explanation.
12. Myoglobin contains a heme group and a polypeptide chain. What role does each component play in the oxygen-binding process?
13. Myoglobin and hemoglobin both bind oxygen.
    a. What physiological roles do myoglobin and hemoglobin play in mammals?
    b. List the chemical and structural similarities between myoglobin and hemoglobin.
    c. Draw the oxygen binding curves for myoglobin and hemoglobin. How do these curves differ?
    d. Are the differences between the myoglobin and hemoglobin oxygen-binding curves consistent with the physiological functions of the two proteins? Explain your answer.
    e. Myoglobin has a very similar tertiary structure to the α- and β-globin chains of hemoglobin. In view of these similar tertiary structures, explain the differences in the oxygen-binding curves.
    f. Why do small pH changes influence the oxygen-binding curve of hemoglobin but not myoglobin?
    g. Why does carbon dioxide concentration influence the oxygen-binding curve of hemoglobin but not myoglobin?
14. Investigators attempted to obtain the amino acid sequence of a purified protein. The protein moved as a single band when analyzed by gel electrophoresis and was indeed pure. However, Edman degradation revealed the presence of both alanine and leucine at the amino terminus. How would you explain this observation?
15. The protease papain cuts IgG molecules into three fragments. Each fragment contains two polypeptide chains joined by a disulfide. Two of these three fragments are identical and contain a single antigen-binding site. The third fragment is derived entirely from the heavy chain. Where is the papain cleavage site located? Explain your answer.
16. Reactant molecules must acquire a certain minimum energy, the energy of activation, before they can be converted into products.
    a. How does the transition-state theory explain this energy requirement?
    b. How does the transition-state theory explain the fact that catalysts increase the rate of reaction but have no effect upon the extent of reaction?
17. The lock-and-key and induce fit models explain how an enzyme interacts with its substrate(s). Describe the two models and indicate how they differ.
18. Describe an experiment that you would perform to determine the V_{max} for an enzyme that hydrolyzes ONPG.
19. Describe an experiment you would perform to determine the K_m for an enzyme that hydrolyzes ONPG.
20. An investigator believes a crude extract contains an enzyme that hydrolyzes RNA (an RNase). Describe a method that could be used to test the hypothesis.
21. A reaction pathway is as follows:

$$\begin{array}{c} A \\ \downarrow^{1} \\ B \\ \updownarrow^{2} \\ {}^{3}\swarrow \; C \; \searrow^{4} \\ D \quad\quad G \\ {}^{5}\updownarrow \quad\quad \updownarrow^{7} \\ E \quad\quad H \\ {}^{6}\downarrow \quad\quad \downarrow^{8} \\ F \quad\quad I \end{array}$$

The intermediates in the pathway are indicated by letters and the enzymes that catalyze the steps are indicated by numbers.
   a. Indicate the most likely committed steps in the reaction.
   b. Indicate the most likely regulatory enzymes for the reaction.
   c. Suggest possible intermediates that are likely to be allosteric effectors that cause feedback inhibition.

Suggested Reading

General Overview

Branden, C.-I., and Tooze, J. 1999. *Introduction to Protein Structure* (2nd ed). New York: Garland.

Creighton, T. E. 1993. *Proteins: Structures and Molecular Properties*. New York: W. H. Freeman.

Pauling, L. 1993. How my interest in proteins developed. *Protein Sci* 2:1060–1063.

Petsko, G. A., and Ringe, D. 2003. *Protein Structure and Function*. Sunderland, CT: Sinauer Associates.

Amino Acids and Peptide Bonds

Cozzone, A. J. 2010. Proteins: Fundamental chemical properties. *Encyclopedia of Life Sciences*. pp. 1–10. Hoboken, NJ: John Wiley & Sons.

Deber, C. M., and Brodsky, B. 2001. Proline residues in proteins. *Encyclopedia of Life Sciences*. pp. 1–6. London: Nature.

Dwyer, D. S. 2008. Chemical properties of amino acids. *Wiley Encyclopedia of Chemical Biology*. pp. 1–11. Hoboken, NJ: John Wiley & Sons.

Protein Purification

Deutscher, M. (ed). 1997. *Guide to Protein Purification*. San Diego, CA: Academic Press.

Primary Structure

Smith, J. B. 2001. Peptide sequencing by Edman degradation. *Encyclopedia of Life Sciences*. pp. 1–3. Hoboken, NJ: John Wiley & Sons.

Steen, H., and Mann, M. 2004. The abc's (and xyz's) of peptide sequencing. *Nat Rev Mol Cell Biol* 5:699–711.

Weak Noncovalent Bonds

Chan, H. S. 2002. Amino acid side-chain hydrophobicity. *Encyclopedia of Life Sciences*. pp. 1–7. London: Nature.

Herzfeld, J., and Olbris, D. J. 2002. Hydrophobic effect. *Encyclopedia of Life Sciences*. pp. 1–9. London: Nature.

Hubbard, R. E. 2001. Hydrogen bonds in proteins: role and strength. *Encyclopedia of Life Sciences*. pp. 1–6. London: Nature.

Sharp, K. A. 2001. Water: structure and properties. *Encyclopedia of Life Sciences*. pp. 1–7. London: Nature.

Secondary Structure

Eisenberg, D. 2003. The discovery of the α-helix and β-sheet, the principal structural features of proteins. *Proc Natl Acad Sci USA* 100:11207–11210.

Tertiary Structure

Anfinsen, C. B. 1973. The principles that govern the folding of protein chains. *Science* 181:223–230.

Branden, C.-I., and Tooze, J. 1999. *Introduction to Protein Structure* (2nd ed). New York: Garland.

Whitford, D. 2005. *Proteins: Structure and Function*. West Sussex, UK: John Wiley & Sons.

Myoglobin, Hemoglobin, and the Quaternary Structure Concept

Ho, C., and Lukin, J. A. 2001. Haemoglobin: cooperativity in protein-ligand interactions. *Encyclopedia of Life Sciences*. pp. 1–11. London: Nature.

Ishimori, K. 2002. *Myoglobin. Encyclopedia of Life Sciences*. pp. 1–5. London: Nature.

Jones, S., and Thornton, J. M. 2001. Protein quaternary structure: subunit-subunit interactions. *Encyclopedia of Life Sciences*. pp. 1–7. London: Nature.

Perutz, M. F. 1978. Hemoglobin structure and respiratory transport. *Sci Am* 239:92–125.

Perutz, M. F., Wilkinson, A. J., Paoli, M., and Dodson, G. G. 1998. The stereochemical mechanism of the cooperative effects in hemoglobin revisited. *Ann Rev Biophys Biomol Struct* 27:1–34.

Immunoglobulin Structure and the Domain Concept

Capra, J. D., and Edmundson, A. B. 1977. The antibody combining site. *Sci Am* 236:50–59.

Edelman, G. M. 1970. The structure and function of antibodies. *Sci Am* 223:34–42.

Kumagai, I., and Tsumoto, K. 2001. Antigen-antibody binding. *Encyclopedia of Life Sciences*. pp. 1–7. London: Nature.

Lucas, A. H. 2001. Antibody function. *Encyclopedia of Life Sciences*. pp. 1–8. London: Nature.

Zouali, M. 2001. Antibodies. *Encyclopedia of Life Sciences*. pp. 1–8. London: Nature.

Enzymes

Bugg, T. D. H. 2001. Enzymes: general properties. *Encyclopedia of Life Sciences*. pp. 1–8. London: Nature.

Cleland, W. W. 2001. Enzyme kinetics: steady state. *Encyclopedia of Life Sciences*. pp. 1–5. London: Nature.

Copeland, R. A. 2001. Enzymology methods. *Encyclopedia of Life Sciences*. pp. 1–5. London: Nature.

Fersht, A. (ed). 1998. *Structure and Mechanism in Protein Science: A Guide to Enzyme Catalysis and Protein Folding*. New York: W. H. Freeman.

Ghanem, E., and Raushel, F. M. 2007. Enzymes: The active site. *Encyclopedia of Life Sciences*. pp. 1–7. Hoboken, NJ: John Wiley & Sons.

Gigant, B., and Knossow, M. 2001. Catalytic antibodies. *Encyclopedia of Life Sciences*. pp. 1–7. London: Nature.

Hedstrom, L. 2001. Enzyme specificity and selectivity. *Encyclopedia of Life Sciences*. pp. 1–7. London: Nature.

Koshland, D. K. Jr. 1994. The key-lock theory and the induced fit theory. *Angew Chem Intl Ed Engl* 33:2375–2378.

Martin, B. L. 2007. Regulation by covalent modification. *Encyclopedia of Life Sciences*. pp. 1–7. Hoboken, NJ: John Wiley & Sons.

Mobashery, S., and Kotra, L. P. 2002. Transition state stabilization. *Encyclopedia of Life Sciences*. pp. 1–6. London: Nature.

Nadaraia, S., Yohrling IV, J. V., Jiang, G. C.-T., Flanagan, J. M., and Vrana, K. E. 2007. Enzyme activity: control. *Encyclopedia of Life Sciences*. pp. 1–8. Hoboken, NJ: John Wiley & Sons.

The first part of this chapter builds on B-DNA structure. More specifically, it explores DNA size, fragility, grooves, bending, denaturation, renaturation, and superhelicity. This new information is applied to study enzymes that unwind double-stranded DNA, proteins that stabilize single-stranded DNA, and enzymes that catalyze changes in superhelical structures. Although B-DNA is the predominant DNA conformation inside the cell, other conformations also exist. Some of these conformations and their possible physiological significance are discussed.

The second part of the chapter examines some important aspects of RNA structure. Ribonucleotide building blocks are linked by 5'→3' phosphodiester bonds to form linear polyribonucleotide chains. Cells make many different kinds of RNA chains; each kind has a unique nucleotide sequence (primary structure) and size. Some RNA molecules can perform their functions as unstructured single strands. Many others have distinct secondary and tertiary structures that are required for biological function. Although RNA chains lack the flexibility of polypeptide chains and their component nucleotides lack the variety of functional groups present in amino acid side chains, some RNA molecules fold into structures that bind specific substrates and catalyze chemical reactions. RNA molecules also interact with proteins to form stable ribonucleoprotein complexes.

3.1 DNA Size and Fragility

Most eukaryotic DNA is located in a defined cell nucleus, whereas bacterial and archaeal DNA is present in the cytoplasm.

DNA molecules exist in a wide range of sizes and base compositions in viruses and living organisms. The living world can be divided into three domains, the **eukaryotes**, bacteria, and archaea. Eukaryotic cells have a membrane bound compartment called the cell nucleus that contains most of the cell's DNA. The DNA in the cell nucleus is partitioned into chromosomes. The long DNA molecule in each chromosome winds around protein complexes made of basic proteins called **histones**. The bacteria and archaea, which lack a defined cell nucleus, usually have a single DNA molecule in the cytoplasm. Cells that lack a defined nucleus are called **prokaryotes**. Even though the bacteria and archaea are both classified as prokaryotes because they lack defined nuclei, bacterial and archaeal metabolic machinery differ from one another as much as either differs from eukaryotic metabolic machinery.

DNA molecules vary in size and base composition.

DNA size in base pairs is known for many viral, bacterial, and eukaryotic DNA molecules. DNA molecules isolated from different viruses vary greatly in size, whereas DNA molecules isolated from bacteria tend to exhibit much less size variation. Each kind of chromosome in a eukaryotic nucleus has a DNA with a characteristic size. TABLE 3.1 lists the sizes in base pairs of individual DNA molecules from various

TABLE 3.1 Sizes of Various DNA Molecules	
Source of DNA	Size in Base Pairs (bp)
Plasmid pBR322*	4,361
Simian virus 40 (SV40)	5,200
Phage T7*	39,937
Phage λ*	48,502
F plasmid*	99,159
Vaccinia virus strain WR	194,711
Fowlpox virus	266,145
Mycoplasma genitalium	580,073
Yeast chromosome IV	1,531,929
Escherichia coli	4,639,221
Human chromosome 1	245,522,847

Note: Phages (viruses that infect bacteria) and plasmids marked with an asterisk have *E. coli* as a host. *Mycoplasma genitalium* is the smallest known free-living bacterium. For yeast and humans the molecular mass of the largest DNA molecule in the organism is given.

sources. The length of the duplex DNA molecules can be calculated based on the distance between base pairs of 0.34 nm. The DNA molecules listed in Table 3.1, therefore, range in length from 1.48 μm for plasmid pBR322 DNA to 83,500 μm (8.3 cm!) for human chromosome 1 DNA. In general, more complex organisms require much more DNA than simpler organisms (though the cells of both the toad and the South American lungfish have considerably more DNA than human cells).

DNA molecules are fragile.

DNA molecules are extremely susceptible to breakage by hydrodynamic shear forces (microscopic forces that arise when layers in a liquid move at different velocities) because they are so long and thin (2.0 nm). Ordinary laboratory operations such as pipetting, pouring, and mixing are sufficient to produce shear forces that break DNA. Unbroken DNA molecules shorter than about 300,000 base pair (bp) usually can be isolated from viruses. Unless great care is taken, larger DNA molecules are almost always broken during isolation so the average length of isolated DNA is usually about 40,000 bp. Bacterial DNA, for instance, is fragmented into about 50 to 100 pieces during isolation.

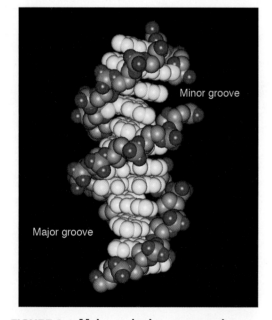

FIGURE 3.1 **Major and minor grooves in B-DNA.** B-DNA is shown as a space-filling structure with the bases in light blue and the rest of the molecule in standard CPK element coloring. (Structure from Protein Data Bank ID: 1BNA Drew, H. R., Wing, R. M., Takano, T., Broka, C., Tanaka, S., Itakura, K., and Dickerson, R. E. 1981. Structure of a B-DNA dodecamer: conformation and dynamics. *Proc Natl Acad Sci USA* 78: 2179–2183. Prepared by B. E. Tropp.)

3.2 Recognition Patterns in the Major and Minor Grooves

Enzymes can recognize specific patterns at the edges of the major and minor grooves.

DNA's length and fragility present a challenge when investigators wish to study DNA *in vitro*. This challenge is technical, however, and does not influence our basic concept of how DNA works. A more fundamental problem results from the fact that base pairs are located

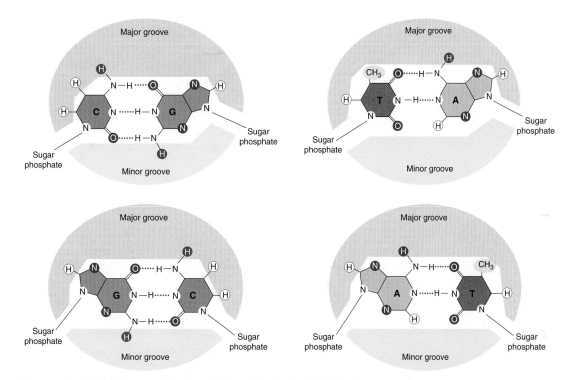

FIGURE 3.2 **Base pair recognition from the edges in the major and minor grooves.** The four types of base pairs are shown. Hydrogen bonds between base pairs are shown as a series of short red lines. Potential hydrogen bond donors are shown in blue and potential hydrogen bond acceptors in red. Nonpolar methyl groups in thymine are yellow, and hydrogen atoms that are attached to carbon atoms and therefore unable to form hydrogen bonds are white. (Modified from C. Branden and J. Tooze. 1999. *Introduction to Protein Structure* (1st ed.). Garland Science. Used with permission of John Tooze, The Rockefeller University.)

within the helix. It therefore was initially difficult to see how enzymes recognize and interact with specific base sequences. One possible solution to the problem is for the two strands that make up the DNA helix to unwind. Although double-stranded DNA does unwind (see below), many enzymes appear to be able to recognize base sequences in the helical structure. Examination of the spacefilling structure shown in FIGURE 3.1 reveals the edges of the major and minor grooves are accessible to enzymes. These grooves arise because the sugar rings lie closer to one side of the base pair than to the other. The grooves' edges are lined with hydrogen bond donors, hydrogen bond acceptors, nonpolar methyl groups, and hydrogen atoms (FIGURE 3.2). Each of the four base pairs projects a unique pattern at the edge of the major groove, but T • A and A • T base pairs project the same pattern at the edge of the minor groove as do C • G and G • C base pairs (FIGURE 3.3). These patterns permit the enzymes to read the sequence from outside the helix.

3.3 DNA Bending

Some base sequences cause DNA to bend.

An immense variety of base sequences has been observed in DNA. Although most sequences do not have any special features that cause them to influence DNA structure, some do. For instance, tracts

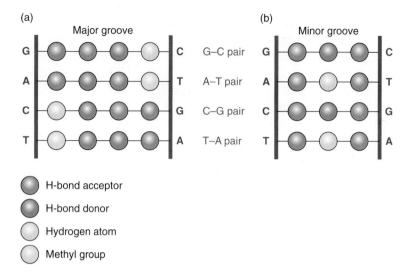

FIGURE 3.3 DNA recognition code. Distinct patterns of hydrogen bond donors, hydrogen bond acceptors, methyl groups, and hydrogen atoms are observed when looking directly at the edges of the base pairs in the major (a) or minor (b) grooves. Each of the four base pairs projects a unique pattern of hydrogen bond donors, hydrogen bond acceptors, methyl groups, and hydrogen atoms at the edge of the major groove. However, the patterns are similar at the edge of the minor groove for T • A and A • T as well as for C • G and G • C. (Modified from C. Branden and J. Tooze. 1999. *Introduction to Protein Structure* (1st ed.). Garland Science. Used with permission of John Tooze, The Rockefeller University.)

consisting of four to six adjacent adenine residues, called **A-tracts**, cause DNA to bend. Other sequences, such as 5′-RGCY-3′, where R is a purine and Y is a pyrimidine, can also cause bending. The local structure of B-DNA thus may differ slightly from the classic linear helix.

3.4 DNA Denaturation and Renaturation

DNA can be denatured.

When the Watson-Crick model was first proposed, many investigators believed the long DNA strands would not be able to unwind and therefore complete strand separation would be impossible. In an attempt to dispel this concern, biophysical chemists tried to show the unwinding process can also take place in the test tube. The approach was to expose DNA to a physical or chemical agent that would disrupt the weak noncovalent interactions (see below) that hold base pairs together without disrupting covalent bonds. Early efforts by Paul Doty and coworkers in the late 1950s showed that DNA solutions undergo a striking drop in viscosity (resistance to flow) when heated. This observation was interpreted to mean the double-helical structure collapses when heated. It seemed probable that the collapsed DNA structure was caused by a conversion of the linear double helix into separate single strands. Several different kinds of experiments helped to establish the two strands do in fact unwind to form separate strands when a DNA solution is heated. For instance, the mass-to-length ratio of DNA before heating is twice that of DNA after heating, and a deoxyribonuclease specific for single-stranded DNA was shown to digest DNA after heating but not before. The transition from the double-helical structure (the native state) to randomly coiled single strands (the denatured state) is called **denaturation**.

The simplest way to detect DNA denaturation is to monitor the ability of DNA in a solution to absorb ultraviolet light at a wavelength of 260 nm, λ_{260}. Purine and pyrimidine bases in nucleic acids absorb 260 nm light strongly. The absorbance at 260 nm, A_{260}, is proportional to concentration. The A_{260} value for double-stranded DNA at a concentration of 50 µg • mL^{-1} is 1.00 unit. Furthermore, the amount of light absorbed by nucleic acids depends on the structure of the molecule. *The more ordered the structure, the less light that is absorbed.* Therefore, double-stranded DNA absorbs less light than the single-stranded chains that form it, and these chains in turn absorb less light than nucleotides released by hydrolysis. For example, three solutions of double-stranded DNA, single-stranded DNA, and free nucleotides, each at 50 µg • mL^{-1}, have the following A_{260} values:

Double-stranded DNA: $A_{260} = 1.00$
Single-stranded DNA: $A_{260} = 1.37$
Free nucleotides: $A_{260} = 1.60$

The increase in A_{260} observed when double-stranded DNA is denatured is termed the **hyperchromic effect**.

If DNA in a solution that is about 0.15 M sodium chloride is slowly heated and the A_{260} is measured at various temperatures, a melting curve such as that shown in FIGURE 3.4 is obtained. The following features of this curve should be noted:

1. The A_{260} remains constant up to temperatures well above those encountered by most living cells in nature.

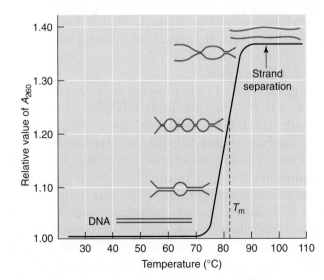

FIGURE 3.4 **DNA melting curve.** A melting curve of DNA showing T_m (the melting temperature) and possible molecular conformations for various degrees of melting.

2. The rise in A_{260} occurs over a range of 6 to 8°C.
3. The maximum A_{260} is about 37% higher than the starting value.

The state of a DNA molecule in different regions of the melting curve is also shown in Figure 3.4. Before the A_{260} rise begins the molecule is fully double stranded. In the rise region noncovalent interactions between base pairs in various segments of the molecule are disrupted; the extent of the disruption increases with temperature. In the initial part of the upper plateau a few noncovalent interactions remain to hold the two strands together until a critical temperature is reached at which the last remaining noncovalent interactions are disrupted and the strands separate completely.

A convenient parameter to characterize a melting transition is the temperature at which the rise in A_{260} is half-complete. This temperature is called the **melting temperature** and is designated T_m.

In the course of studying strand separation, another important fact emerged. If a DNA solution is heated to a temperature at which most but not all noncovalent interactions are disrupted and then cooled to room temperature, A_{260} drops immediately to the initial undenatured value. Additional experiments show the native structure is restored. Therefore, if strand separation is not complete and denaturing conditions are removed, the helix rewinds. Thus, if two separated strands were to come in contact and form even a single base pair at the correct position in the molecule, the native DNA molecule should re-form. We will encounter this phenomenon again when renaturation is described.

Hydrogen bonds stabilize double-stranded DNA.

In 1962 Julius Marmur and Paul Doty isolated DNA from various bacterial species in which the base compositions vary from about 20% G + C to 80% G + C. T_m values from many such DNA molecules are plotted versus percent G + C in **FIGURE 3.5**. Note that T_m increases with increasing percent G + C. This relationship is explained by proposing that hydrogen bonds are at least partially responsible for stabilizing the double-stranded structure. It requires more energy to disrupt the three hydrogen bonds in a G • C base pair than to disrupt the two hydrogen bonds in an A • T base pair.

Measuring T_m values in the presence of a denaturing agent such as urea (NH_2CONH_2) or formamide ($HCONH_2$), which can form hydrogen bonds with DNA bases, supports the role of hydrogen bonds in stabilizing the double-stranded structure. Hydrogen bonds between base pairs have very low energies and so are easily broken. However, hydrogen bonds are also able to rapidly re-form (see below). Denaturing agents shift this equilibrium by forming hydrogen bonds with an unpaired base on one strand and thereby prevent the base from re-forming hydrogen bonds with the complementary unpaired base on the other strand. Thus, denaturing agents maintain the unpaired state at a temperature at which complementary unpaired bases would normally be expected to pair again. Permanent melting of a section of paired bases, therefore, requires less input of thermal energy and T_m is reduced.

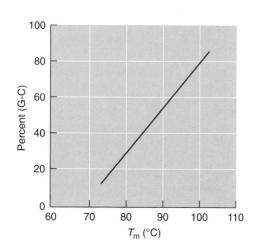

FIGURE 3.5 Effect of G-C content on DNA melting temperature. T_m increases with increasing percent of G + C. The DNA solution contained 0.15 M sodium chloride and 0.015 M sodium citrate.

Base stacking also stabilizes double-stranded DNA.

The planar bases in the double helix are stacked. The forces that stabilize stacking in the double helix include van der Waals forces and hydrophobic effects. We do not know the precise contribution that each of these weak interactions makes to the helical stability because it is difficult to modify the structure of DNA so that just one kind of interaction is altered.

Both base stacking and hydrogen bonds are weak noncovalent interactions and as such are easily disrupted by thermal motion. Stacking is enhanced if the bases are unable to tilt or swing out from a stacked array. Similarly, maximum hydrogen bonding occurs when all bases are pointing in the right direction. Clearly, the two weak interactions reinforce each other. Stacked bases are more easily hydrogen bonded, and, correspondingly, hydrogen-bonded bases, which are oriented by the bonding, stack more easily. If one of the interactions is eliminated, the other is weakened, explaining why T_m drops so markedly after the addition of an agent that destroys either type of interaction.

Ionic strength influences DNA structure.

In addition to the cooperative attractive interactions between adjacent DNA bases and between the two strands, there is an interstrand electrostatic repulsion between the negatively charged phosphates. This strong repulsive force is diminished by placing DNA in a solution that contains a salt such as sodium chloride. The explanation for the salt's effect on the repulsive force is as follows. In the absence of salt the strands repel one another, causing the strands to separate easily. As salt is added positively charged ions such as Na$^+$ form "clouds" of charge around the negatively charged phosphates and effectively shield the phosphates from one another so that the T_m rises. Ultimately, all phosphates are shielded and repulsion ceases; this shielding occurs near the physiological salt concentration of about 0.2 M. However, T_m continues to rise as the sodium chloride concentration increases because purine and pyrimidine solubility decreases, increasing hydrophobic interactions.

The DNA molecule is in a dynamic state.

An important structural feature of the DNA molecule becomes apparent when DNA is examined in the presence of formaldehyde (HCHO). Formaldehyde can react with the NH$_2$ groups of the bases and thus eliminates their ability to hydrogen bond. When formaldehyde is added to double-stranded DNA, the DNA slowly and irreversibly denatures. Because the amino groups must be available to formaldehyde for the reaction to take place, bases must continually unpair and pair (i.e., hydrogen bonds must break and re-form).

A related phenomenon is observed when double-stranded DNA is dissolved in tritiated water ([^{3}H]H$_2$O). There is a rapid exchange between the hydrogen-bonded protons of the bases and the tritium ions in the water. These two observations indicate that DNA is a dynamic structure in which double-stranded regions frequently open to become single-stranded bubbles and then close again. This transient localized melting is called DNA "breathing." Because a G • C base pair has three hydrogen bonds and an A • T base pair has only two, *transient melting occurs more often in regions rich in A • T pairs than in regions rich in G • C pairs.*

Alkali denatures DNA without breaking phosphodiester bonds.

Heat can be used to prepare denatured DNA, which is often an essential step in many experimental protocols. However, high temperature may break phosphodiester bonds so the product of heat denaturation often is a collection of broken single strands. The degradation problem is avoided by using another method to denature DNA. Addition of a base such as sodium hydroxide to the DNA solution removes protons from the ring nitrogen atoms of guanine and thymine, causing the DNA to denature. Because DNA is quite resistant to alkaline hydrolysis, this procedure is the method of choice for denaturing DNA. Acid also causes denaturation but is seldom used for that purpose because acid also causes purine groups to be cleaved from the polynucleotide chain, a process known as **depurination**.

Complementary single strands can anneal to form double-stranded DNA.

A solution of denatured DNA can be treated in such a way that native DNA re-forms. The process is called **DNA renaturation** or **reannealing**, and the re-formed DNA is called **renatured DNA**. The A_{260} of a DNA solution decreases during renaturation, a change known as the **hypochromic effect**. Renaturation has proven to be a valuable tool in molecular biology. It can be used to demonstrate genetic relatedness between different organisms, detect particular species of RNA, determine whether certain sequences occur more than once in the DNA of a particular organism, and locate specific base sequences in a DNA molecule. Two requirements must be met for renaturation to occur:

1. The salt concentration must be high enough so electrostatic repulsion between the phosphates in the two strands is eliminated; usually 0.15 to 0.50 M NaCl is used.
2. The temperature must be high enough to disrupt random intrastrand or interstrand hydrogen bonds. The temperature cannot be too high, however, or stable interstrand base pairing will not occur. The optimal temperature for renaturation is 20 to 25°C below the T_m value.

Renaturation is a slow process compared with denaturation. The rate-limiting step is not the actual rewinding of the helix (which occurs in roughly the same time as unwinding) but the precise collision between complementary strands such that base pairs are formed at the correct positions. Because renaturation is a result only of random motion, it is a concentration-dependent process. At concentrations normally encountered in the laboratory, renaturation takes several hours.

The molecular details of renaturation can be understood by referring to the hypothetical molecule shown in FIGURE 3.6, which contains a sequence that is repeated several times. Assume that each single strand contains 50,000 bases and that the base sequences are complementary. Any short sequence of bases (say, 4–6 bases long) will certainly appear many times in such a molecule and can provide sites for base pairing. Random collision between noncomplementary sequences such as IA and II′ will be ineffective but a collision between IA and IC′ will result in base pairing. This pairing will be short-lived, however, because the bases surrounding these short complementary tracts are not able to pair and stacking stabilization will not occur. At the temperatures used for renaturation these paired regions rapidly become disrupted. As soon as two sequences such as IB and IB′ pair, the adjacent bases will also rapidly pair and the entire double-stranded DNA molecule will "zip up" in a few seconds.

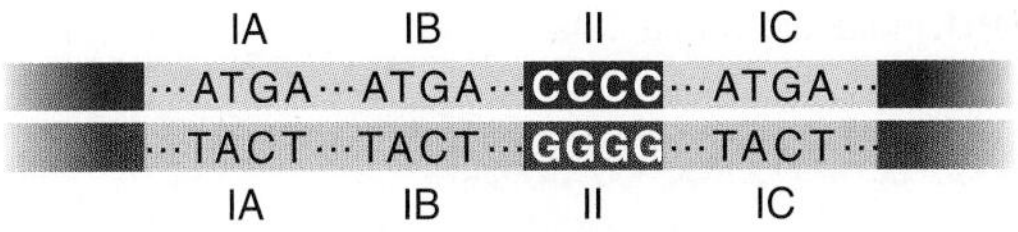

FIGURE 3.6 **Molecular details of renaturation using a hypothetical DNA molecule.** A hypothetical DNA molecule containing a sequence that is repeated several times. The roman numerals on either side of the DNA molecule refer to segments of the DNA molecule that are discussed in the text.

3.5 Helicases

Helicases are motor proteins that use the energy of nucleoside triphosphates to unwind DNA.

Duplex DNA must unwind under physiological conditions during DNA replication. "Molecular motor" enzymes called **helicases** catalyze nucleoside triphosphate–dependent unwinding of double-stranded DNA in cells. Helicases are often part of larger protein complexes, and their activities are influenced by other proteins in the complex.

The observation that a cell-free extract has a DNA-dependent nucleoside triphosphatase activity usually, but not always, indicates a helicase is present. A nucleoside triphosphatase catalyzes the conversion of a nucleoside triphosphate to a nucleoside diphosphate and inorganic phosphate (FIGURE 3.7). A more reliable indication is nucleoside triphosphate–dependent unwinding of double-stranded DNA to produce single strands, which can be detected by their susceptibility to single-strand–specific nucleases. DNA helicases tend to

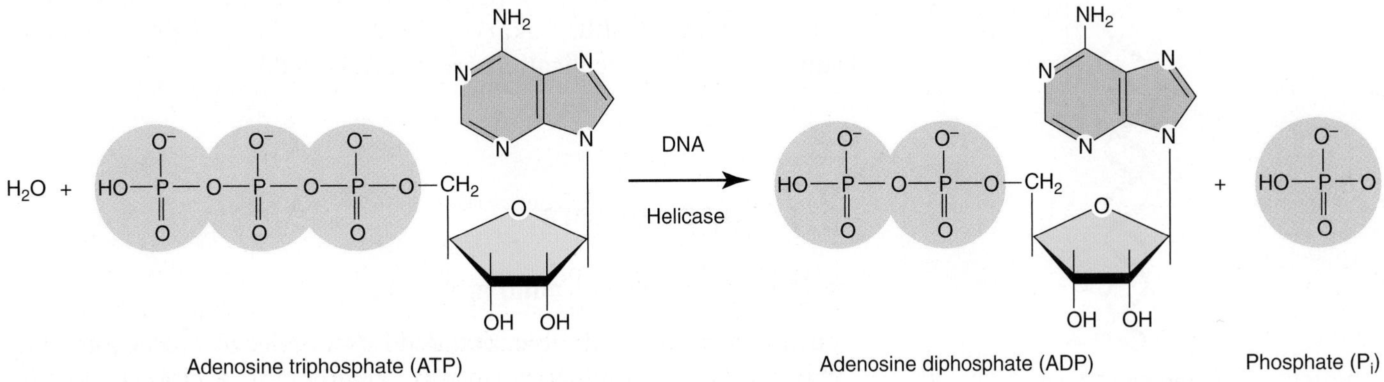

FIGURE 3.7 **DNA-dependent ATP hydrolysis by helicase.**

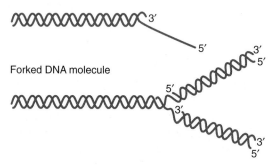

FIGURE 3.8 **DNA structural preferences of different types of helicases.**

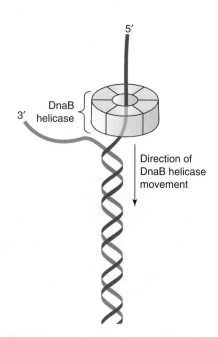

FIGURE 3.9 **Unwinding by bacterial DnaB helicase.**

exhibit structural preferences for their DNA substrates (FIGURE 3.8); some require a forked DNA molecule, others act on DNA with 3′ or 5′ tails, and still others work on blunt-end DNA. Some DNA helicases move along a DNA strand in a 3′→5′ direction, whereas others move along it in a 5′→3′ direction.

At least 14 different DNA helicases have been isolated from the bacteria *Escherichia coli*, 6 from bacterial viruses, 15 from yeast, 8 from plants, and 24 from human cells. An ATP-dependent *E. coli* helicase consisting of six identical subunits is of special interest because it plays an essential role in bacterial DNA replication. This helicase, which is named the **DnaB helicase** for the *dnaB* gene that codes for it, encircles one strand of a forked DNA molecule and moves in a 5′→3′ direction along the encircled strand while excluding the other (FIGURE 3.9).

3.6 Single-Stranded DNA Binding Proteins

Single-stranded DNA binding proteins stabilize single-stranded DNA.

Proteins that bind to single-stranded DNA, **single-stranded DNA binding proteins (SSBs)**, stabilize the transient single-stranded regions formed by the action of helicases on double-stranded DNA. As essential participants in DNA metabolism, SSBs are present in all cells. Some SSBs consist of a single polypeptide, others contain two or more identical polypeptide subunits, and still others are made of different polypeptide subunits. Despite their structural diversity, all SSBs share one important property: they bind more tightly to single-stranded DNA than to double-stranded DNA or RNA.

Bruce Alberts and coworkers isolated the first bacterial SSB, a homotetramer (molecular mass = 75 kDa), from *E. coli* in 1972. Each subunit has one oligonucleotide binding domain. A bacterial cell has about 800 copies of the SSB tetramers. The binding of one *E. coli* SSB tetramer to denatured DNA makes it easier for other SSB tetramers to bind to the DNA (FIGURE 3.10). That is, *E. coli* SSB tetramers bind in a cooperative fashion. SSB binding destabilizes the double helix and causes the melting temperature to be lower. Eukaryotes have a different kind of SBB. Eukaryotic SSB, consisting of three different subunits, is called **replication protein A**, or **RPA**.

3.7 Topoisomers

DNA can exist as a circular molecule.

The ends of a linear double-stranded DNA molecule can be joined by phosphodiester bonds to form a covalently closed, circular double-stranded DNA molecule (FIGURE 3.11). As used here, *circular* means a continuous or unbroken DNA chain rather than a geometric circle. By convention, an intact circular double-stranded DNA molecule is called <u>r</u>eplicating <u>f</u>orm <u>I</u> (**RFI**), and a circular double-stranded DNA molecule with a nick in one strand is called <u>r</u>eplicating <u>f</u>orm <u>II</u> (**RFII**).

Bacterial DNA usually exists as a covalently closed, circular double-stranded DNA molecule.

The existence of covalently closed, circular double-stranded DNA molecules was not noticed for many years because, as mentioned earlier, large DNA molecules usually break during isolation. John Cairns was the first to detect circular DNA molecules in bacteria. The experiment he performed in 1963 was designed to obtain an image of an intact bacterial DNA molecule. He hoped the experiment would reveal whether bacteria have a single large chromosome or many smaller ones. Previous attempts by other investigators to obtain this information failed because DNA is such a fragile molecule. Cairns realized he required a very gentle method to avoid breaking the DNA. He decided to take

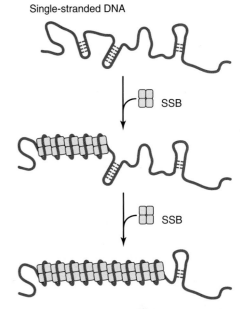

FIGURE 3.10 Cooperative *E. coli* SSB binding to single-stranded DNA.

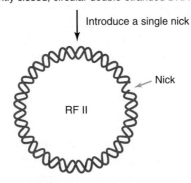

FIGURE 3.11 Closed covalent and nicked circles. The ends of a linear double-stranded DNA molecule can be joined by phosphodiester bonds to form a covalently closed, circular double-stranded DNA molecule. By convention, the intact circular double-stranded DNA molecule is called RFI (replicating form I). RFI is converted to RFII by introducing a single nick (green arrow) in one strand of the covalently closed, circular double-stranded DNA molecule.

advantage of the fact that tritium-labeled DNA emits β-particles, which upon striking a film produce an image of the DNA. This technique of using a radioactively labeled substance to produce an image on a photographic emulsion is called **autoradiography**. Cairns cultured *E. coli* in a medium containing [^{3}H]thymidine for a few hours to label the DNA and then gently released the labeled DNA from the bacteria by treating the cells with a combination of lysozyme (an egg-white enzyme) to digest the bacterial cell wall and detergent to disrupt the cell membrane. After collecting the released DNA on a cellulose ester filter, he coated the dried filter with a photographic emulsion and stored the preparation in the dark for 2 months to allow sufficient time for the β-particles to produce an image. Analysis of the array of dark spots, which appeared after developing the emulsion, revealed that *E. coli* DNA is a double-stranded circular molecule with a contour length of approximately 1 mm, or about 1,000 times longer than the bacteria itself.

Plasmid DNA molecules are used to study the properties of covalently closed, circular double-stranded DNA *in vitro*.

Intact bacterial DNA is too long to be studied conveniently *in vitro*. Fortunately, readily available substitutes allow us to study the properties of covalently closed, circular double-stranded DNA molecules in the laboratory. Many bacteria carry copies of autonomously replicating small circular DNA molecules, called **plasmids**, in addition to their large chromosomal DNA molecule. Plasmids, which range in size from about 0.1% to 5% of the bacterial chromosome, replicate more or less independently of chromosomal DNA replication and hence are transmitted from one generation to the next. Although plasmids are usually dispensable genetic elements, they can have a profound effect on their host bacterial cell. For example, some plasmids make the host bacterial cell resistant to one or more antibiotics, whereas other plasmids allow the host bacterial cell to produce toxins that kill other kinds of microorganisms (**BOX 3.1 CLINICAL APPLICATION: PLASMIDS AND DISEASE**). For now, let's focus on the fact that plasmid DNA is well suited for studying the properties of circularly closed, double-stranded DNA molecules because their small size permits them to remain intact when manipulated in the laboratory.

Covalently closed, circular double-stranded DNA molecules often have superhelical structures.

In a linear duplex a base pair can be distorted without influencing the structure further along the molecule. In contrast, distorting a base pair when the strands form a covalently closed, circular double-stranded DNA influences the rest of the molecule's structure. The mathematical discipline of topology helps to understand this and related phenomena that involve closed covalent circular DNA. A topological property is one that remains unchanged when the object of interest is distorted but not torn or broken. The introduction of a nick into covalently closed, circular double-stranded DNA alters the DNA molecule's topological properties.

BOX 3.1: CLINICAL APPLICATION

Plasmids and Disease

Plasmids confer specific biological properties to the bacteria that harbor them. Two of these properties have important applications to human health. Drug resistance plasmids (R plasmids) have genes that make the bacterial host cells resistant to one or more antibiotics or antibacterial drugs. The first R plasmid was discovered in Japan in the late 1950s when a *Shigella* strain, isolated from a patient suffering from dysentery, was found to be resistant to sulfanilamide, chloramphenicol, streptomycin, and tetracycline. Resistance to just one antibacterial agent might have been due to a rare mutation, but the odds against four independent mutations arising in the same bacterial cell were astronomical. Investigators therefore suspected the *Shigella* strain harbored a plasmid that made it resistant to all four drugs.

Subsequent studies confirmed this suspicion. Some R plasmids act as infective agents that can move from a host cell that harbors it to one that does not. Transmissible R plasmids pose a serious public health threat even when they are present in a nonpathogenic host. Under normal circumstances drug-sensitive nonpathogenic bacteria colonize the skin, mouth, and intestine. When exposed to antibacterial agents these bacteria are eliminated and bacteria that bear R plasmids proliferate because they no longer have competition for the ecological niche. Serious problems arise when nonpathogenic drug-resistant bacteria transmit their R plasmids to invading pathogens, allowing the pathogens to become drug resistant and making clinical treatment difficult, if not impossible.

The transfer of antibiotic resistance genes can take place between bacteria of the same species and also between bacteria with limited phylogenetic relatedness, including transfer between gram-negative and gram-positive species. Unfortunately, the problem extends beyond the unnecessary use of antibiotics in humans. Low doses of certain antibiotics are sometimes added to animal feed to prevent infection and promote growth. Animals given these antibiotics often have bacteria with R plasmids, which can be transferred to bacteria that cause human diseases.

Virulence plasmids have genes that code for toxins that cause many life-threatening illnesses, including dysentery, anthrax, plague, tetanus, and food poisoning. Virulence plasmids are of great concern because they, along with R plasmids, can be exploited to create pathogenic bacteria that can be used in germ warfare. Virulence plasmids are also present in bacteria that infect plants. *Agrobacterium tumafaciens*, a bacterial species that infects fruit trees and other broad leaf plants, contains a tumor-inducing plasmid called the Ti plasmid. Only a portion of the Ti plasmid, called T DNA, is transferred to the plant cell. These T DNA molecules, which are incorporated into plant cell chromosomes, are responsible for inducing the formation of tumors (also called crown galls). T DNA has been modified by recombinant DNA techniques so that it does not cause a tumor. This process, known as disarming, has created a plasmid that can be used for introducing foreign DNA into many plants.

3.7 Topoisomers

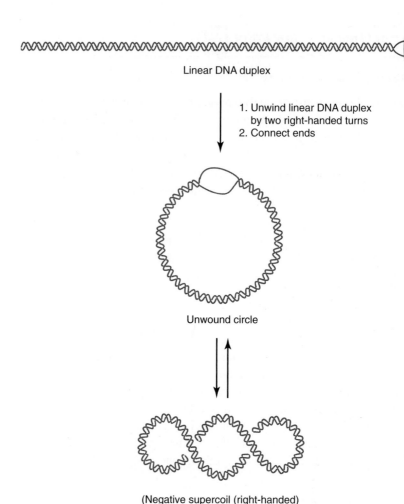

FIGURE 3.12 Formation of a negative supercoil. A linear DNA duplex is unwound twice and its ends are joined to form an unwound covalently closed, circular double-stranded DNA molecule, which then changes shape to form a negative superhelix.

As indicated above, the two ends of a linear DNA helix can be brought together and joined in such a way that each strand is continuous. If, as shown in FIGURE 3.12, one of the ends is untwisted 720 degrees before joining, about 20 bp must unwind (because the linear molecule has 10.5 bp per turn of the helix). However, the DNA molecule has such a strong tendency to maintain a right-handed helical structure with about 10.5 bp per turn that it will deform itself to form a negative **superhelix** (or **supercoil**) with two crossover points or **nodes** (Figure 3.12). A circular DNA molecule with such crossover points is said to **writhe**.

The ends of a linear duplex may also be joined without twisting either strand or after twisting one strand to wind the helix even tighter. When the DNA is not twisted the covalently closed, circular double-stranded DNA forms a **relaxed circle** that has no crossover points and therefore no writhing and so is able to lie unconstrained on a flat surface. When the linear duplex is overwound, the resulting covalently closed, circular double-stranded DNA forms a positive superhelix with writhing in the opposite sense to that observed for the negative superhelix. Two closed covalent circular double-stranded DNA molecules with identical

BOX 3.2: LOOKING DEEPER

Linking Number

One topological property, the linking number (Lk), is particularly useful for describing the topological forms that covalently closed, circular double-stranded DNA molecules assume. The Lk is closely related to the number of times the two sugar phosphate backbones wrap around, or are "linked with," each other. It indicates how often two DNA strands twist about each other to form the helix or how often the helix axis writhes about itself to form the superhelix. As a topological property the Lk for a covalently closed DNA does not change unless a DNA strand breaks.

Lk is the sum of the writhing number, Wr, and twisting number, Tw, where

$$Lk = Wr + Tw$$

The writhing number indicates the number of times the helix axis crosses itself. Each crossover point produces a node such as the two in the negative supercoil shown at the bottom of Figure 3.12. The twisting number indicates the total number of turns of the double-stranded molecule. For a closed covalent, circular double-stranded DNA that lies flat on a surface (a relaxed circle), Tw is the number of times one strand revolves around the other. For example, the twisting number for a relaxed circle with 1050 bp is 100 (1050 bp/10.5 bp). The twisting number is positive for revolutions in right-handed helical regions and negative for a left-handed helix or a left-handed segment.

The linking number, which enables us to distinguish positive from negative supercoiling, cannot be changed without (1) breaking a strand, (2) rotating one strand about the other, and (3) rejoining. A change in the linking number (ΔLk) for a process, therefore, provides information about the mechanism of change. For example, ΔLk tells us something about how enzymes that affect supercoiling do their job. Changes in the linking number are related to changes in the twisting and writhing numbers as follows:

$$\Delta Lk = \Delta Tw + \Delta Wr$$

A decrease in Lk corresponds to some combination of underwinding and negative supercoiling and an increase in Lk reflects some combination of overwinding and positive supercoiling.

---

base pair sequences but different degrees of supercoiling are said to be **topological isomers** or **topoisomers**. FIGURE 3.13 shows the structure of three topoisomers, a relaxed circle, a negative superhelix, and a positive superhelix. Topology provides a quantitative method for characterizing topoisomers (**BOX 3.2 LOOKING DEEPER: LINKING NUMBER**). When a supercoiled, covalently closed, circular double-stranded DNA molecule is treated with a trace quantity of DNase to introduce a single nick into one strand, the resulting nicked molecule uncoils to form a relaxed circle.

Most bacterial DNA molecules are slightly underwound and hence form negative superhelices. The underwinding is not a result of

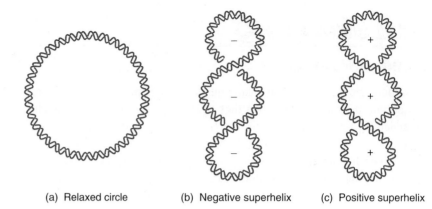

(a) Relaxed circle (b) Negative superhelix (c) Positive superhelix

FIGURE 3.13 Topoisomers. (a) Relaxed circle—the helix axis does not cross itself and the DNA circle lies flat on the plane. (b) Negative superhelix—the front segment of a DNA molecule crosses over the back segment from right to left. (c) Positive superhelix—the front segment of a DNA molecule crosses over the back segment from left to right.

unwinding before end joining but is instead introduced into preexisting circles by an enzyme called **DNA gyrase** (see below). Approximately 75% of the underwinding strain is taken up by writhe. It is important to recall that a DNA molecule is a dynamic structure in which hydrogen bonds break and re-form so that bubbles continuously appear and disappear throughout the supercoiled DNA molecule. At any given instant the fraction of a negatively supercoiled molecule that is single stranded is greater than in a relaxed circle. The sequences in a supercoil that are most likely to be unpaired and form bubbles are those that are more than 90% A • T pairs. A • T rich sequences play important roles in processes such as DNA replication and transcription.

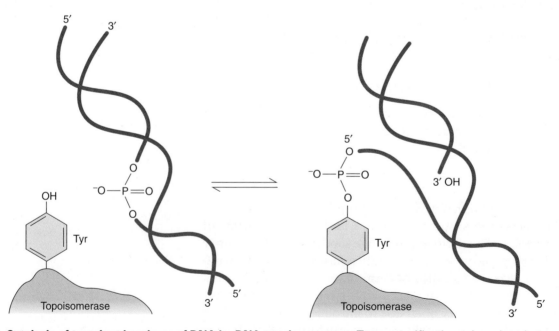

FIGURE 3.14 Catalysis of transient breakage of DNA by DNA topoisomerases. Transesterification takes place between a tyrosol residue on the enzyme and a DNA phosphate group, leading to cleavage of a DNA phosphodiester bond and formation of a covalent enzyme–DNA intermediate. The phosphodiester bond can be re-formed by a reversal of the reaction that is shown.

CHAPTER 3 NUCLEIC ACID STRUCTURE

3.8 Topoisomerases

Topoisomerases catalyze the conversion of one topoisomer into another.

DNA molecules encounter various topological challenges during virtually all stages of their metabolism. Two examples involving bacterial DNA replication will help to illustrate the point. A bacterial DNA molecule must unwind during replication, but because it is a closed covalent circle, unwinding in one region causes overwinding in another region. The resulting torsional strain must be relieved or replication stops. Another topological challenge arises after bacterial DNA synthesis is complete because the two daughter DNA molecules form interlocking rings known as **catenanes**. A mechanism is needed that allows the interlocking rings to separate so the DNA molecules can segregate to daughter cells.

Both of these replication problems as well as a variety of other twisting, writhing, and tangling problems are solved by transient cleavage of the DNA backbone. The enzymes that catalyze this transient cleavage, called **topoisomerases**, convert one topoisomer into another. Topoisomerases are essential for cell viability because they manage DNA topology so that replication, transcription, and other processes involving DNA can take place. Furthermore, topoisomerases are of great practical interest because they are targets for a wide variety of drugs used as antimicrobial and anticancer agents.

Topoisomerases act by cutting DNA molecules and forming transient adducts in which a tyrosine at the active site attaches to the nicked DNA by a phosphodiester bond (FIGURE 3.14). The enzymes are divided into two broad types based on whether they form transient attachments to one or two strands of DNA:

1. *Type I topoisomerases:* Type I topoisomerases form transient attachments to one strand. James Wang detected the first type I topoisomerase, and indeed the first topoisomerase of any kind, in cell free *E. coli* extracts in 1971. An active site tyrosine in the *E. coli* type I topoisomerase forms a transient attachment to the 5′-phosphate end of the cleaved DNA strand (FIGURE 3.15). The enzyme relaxes underwound DNA (negatively supercoiled) by first unwinding a short stretch of double-stranded DNA and then introducing a transient break in one of the strands in the melted region. The unbroken strand is then free to move through the transient break before the nick is resealed (Figure 3.15). The *E. coli* type I topoisomerase's ability to unwind negatively supercoiled DNA decreases as the DNA becomes more relaxed, so the enzyme becomes less and less proficient as the reaction continues.
2. *Type II topoisomerases:* Type II topoisomerases form transient attachments to both DNA strands and require ATP to convert one topoisomer into another. One of the best studied type II topoisomerases is the eukaryotic topoisomerase II. This enzyme works as shown in FIGURE 3.16. The enzyme makes

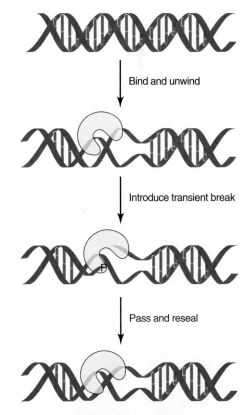

FIGURE 3.15 **Proposed mechanism of relaxation by *E. coli* topoisomerase I.** An active site tyrosine in the *E. coli* type I topoisomerase forms a transient attachment to the 5′-phosphate end of the cleaved DNA strand. The enzyme relaxes underwound DNA (negatively supercoiled) by first melting a short stretch of double-stranded DNA and then introducing a transient break in one of the strands in the melted region. The unbroken strand is then free to move through the transient break before the nick is resealed. (Adapted from Dekker, N. H. et al. 2002. *Proc Natl Acad Sci USA* 99:12126–12131.)

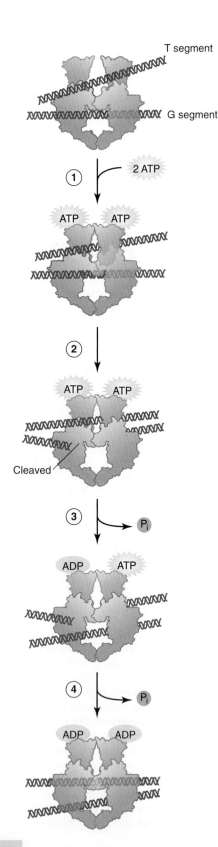

transient cuts in both strands of a double-stranded DNA molecule with the concomitant formation of a phosphomonoester bond between the active-site tyrosine and 5′-end of the cut DNA. Then the enzyme undergoes a conformational change that allows the topoisomerase to pull the two ends of the cut duplex DNA apart to create an opening in the DNA. The DNA region that contains the opening is called the gated (or G-segment) DNA. A second region of duplex DNA from either the same molecule or a different molecule passes through the open DNA gate. This second region of DNA is designated the transported or T-segment. One important bacterial type II topoisomerase, DNA gyrase, has a remarkable catalytic ability. Most type I and type II topoisomerases release topological constraints on DNA and thereby permit negatively supercoiled DNA to be converted into a more relaxed structure. *DNA gyrase has the unique catalytic ability to introduce negative supercoils into covalently closed, circular DNA in the presence of ATP*. DNA gyrase activity is essential for *E. coli* viability because it releases the torsional tension that would otherwise accumulate in front of the point of separation of the two parental DNA strands during DNA replication. Other bacteria also require DNA gyrase for survival. The quinolone antibiotic ciprofloxacin, which inhibits bacterial DNA gyrase, is used to treat a variety of bacterial infections.

3.9 Non-B DNA Conformations

A-DNA is a right-handed double helix with a deep major groove and very shallow minor groove.

B-DNA is the predominant DNA form in living systems but is not the only form. Several non-B DNA conformations also exist in nature, although often just fleetingly. The first non-B conformation to be identified was A-DNA. Rosalind Franklin obtained the x-ray diffraction pattern for A-DNA when she examined DNA fibers at a relative humidity of about 75%. Both A-DNA and B-DNA are right-handed double helices. That is, each spirals in a clockwise direction as an

FIGURE 3.16 Proposed mechanism of action for DNA topoisomerase II. This mechanism involves two DNA segments. Catalysis begins when the topoisomerase binds to two double-stranded DNA segments, which are designated the gate segment or G segment (red) and the transported segment or T segment (dark blue). (Step 1) An ATP molecule binds to the topoisomerase. (Step 2) The G segment is cleaved. (Step 3) The T segment from either the same molecule or a different molecule passes through the open G-segment with the concomitant hydrolysis of one ATP molecule. (Step 4) The G segment is rejoined and the remaining ATP molecule is hydrolyzed. (Adapted from Larsen, A. K., et al. 2003. *Pharmacol Ther* 2:167–181.)

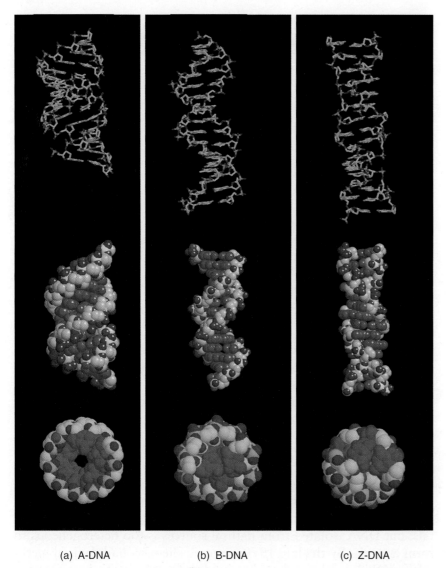

(a) A-DNA   (b) B-DNA   (c) Z-DNA

FIGURE 3.17 DNA conformations. Three conformations of DNA are shown: (a) A-DNA, (b) B-DNA, and (c) Z-DNA. (Top) Structures are in stick display so the orientation of the base pairs with respect to the helix axis is visible. (Middle) Structures are in a spacefill display with the backbone in standard CPK colors and the base pairs in magenta to emphasize the grooves. (Bottom) Structures are once again in a spacefill display but viewed from the top of the DNA molecules. Each DNA molecule contains 12 base pairs. (Top structures from Protein Data Bank 213D. B. Ramakrishnan and M. Sundaralingam, *Biophys. J.* 69 [1995]: 553–558. Prepared by B. E. Tropp; Middle structures from Protein Data Bank IBNA. H. R. Drew, et al., *Proc. Natl. Acad. Sci. USA* 78 [1981]: 2179–2183. Prepared by B. E. Tropp; Bottom structures from Protein Data Bank 2ZNA. A. H.-J. Wang, et al., *Left-handed double helical DNA...* Prepared by B. E. Tropp.)

observer looks down its helical axis of symmetry. Despite this similarity A- and B-DNA differ in many important respects (**FIGURE 3.17a** and **b**):

1. A-DNA has 11 bp per helical turn, whereas B-DNA has 10.5 bp per helical turn.

TABLE 3.2 Comparison of Major Features in A-, B-, and Z-Forms of DNA

Parameter	A-DNA	B-DNA	Z-DNA
Helix sense	Right	Right	Left
Base pairs per turn	11	10.5	12
Axial rise per base pair (nm)	0.26	0.34	0.45
Base pair tilt (°)	20°	−6°	7°
Diameter of helix (nm)	2.3	2.0	1.8

2. The plane of a base pair in A-DNA is tilted 20 degrees away from the perpendicular to the helix axis, whereas the corresponding value for B-DNA is −6 degrees.
3. A-DNA has an axial hole when viewed down its long axis, whereas B-DNA does not.
4. A-DNA has a deep major groove and very shallow minor groove, whereas both the major and minor grooves are about the same depth in B-DNA.

TABLE 3.2 summarizes these differences. Double-stranded DNA seldom, if ever, assumes the A-form in the aqueous environment present in living systems. However, double-stranded RNA (see below) and DNA–RNA hybrids (nucleic acids with one DNA strand and one RNA strand) form right-handed helices that closely approximate A-DNA.

Z-DNA has a left-handed conformation.

Data obtained by using x-ray diffraction analysis of DNA fibers do not provide information about the distances between specific base pairs. Such information only can be obtained by studying DNA crystals, but DNA isolated from natural sources is too heterogeneous to form crystals. By the late 1970s organic chemists had devised methods for DNA synthesis that permitted the synthesis of a homogenous DNA sample that formed a crystal. Alexander Rich and his colleagues took advantage of this advance to prepare crystals of $d(CG)_3$, a self-complementary hexadeoxyribonucleotide:

5′-CGCGCG-3′
3′-GCGCGC-5′

Rich and coworkers examined the x-ray diffraction pattern of the crystalline double-stranded DNA fragment, expecting to see the diffraction pattern for A- or B-DNA. To their great surprise, however, the diffraction pattern had never been seen before, indicating an entirely new DNA conformation. Because the diffraction data showed that the sugar phosphate backbone has a zigzag appearance (FIGURE 3.18), Rich and coworkers called the new form of DNA **Z-DNA**.

Subsequent studies showed that the alternating pyrimidine-purine repeat $d(CA)_3$ and its complement $d(TG)_3$ base pair to form a duplex that also can adopt the Z-conformation. The self-complementary hexadeoxyribonucleotide $d(TA)_3$, however, does not assume a Z-conformation. Even when a duplex can assume the Z-conformation,

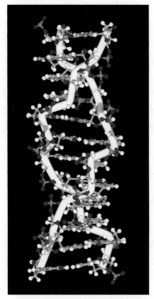 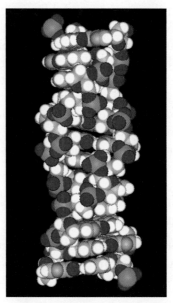

(a) Z-DNA with zig-zag sugar phosphate backbone shown in white

(b) The same Z-DNA with the zigzag sugar phosphate backbone shown in space filling display

FIGURE 3.18 Z-DNA. (a) Z-DNA with its zigzag backbones shown as white tubes. The bases, sugars, and phosphates are shown as ball and stick structures. (b) The same Z-DNA shown in a spacefilling display. (Structure from Protein Data Bank ID: 2ZNA. Wang, A. H. J., et al. *Science*.)

it will not do so unless present in a solution with a high salt concentration. For instance, the 5′-CGCGCG-3′ hexanucleotide requires a sodium chloride concentration greater than 2 M or a magnesium chloride concentration greater than 0.7 M to assume the Z-conformation. The B-conformation is favored at lower salt concentrations. If an alternating d(CG) sequence is contained within a longer DNA tract, such as

5′-TGATCCGCGCGCGAGTCTT-3′
3′-ACTAGGCGCGCGCTCAGAA-5′

the alternating d(CG) sequence can assume the Z-conformation in a 2 M sodium chloride solution, but the rest of the DNA will be in the B-conformation. At least one base pair appears to be disrupted at the Z-DNA–B-DNA junction, forcing at least two bases out of the helix.

Z-DNA is a left-handed helix. The left-handed Z-DNA and right-handed B-DNA are not mirror images but entirely different structures. This difference is immediately obvious when examining the grooves. Z-DNA has a single groove, whereas B-DNA has a major and a minor groove (compare Figures 3.17c and b). The immediate question that arises is whether the left-handed structure has biological significance. Certainly, the intracellular salt concentration does not ever approach 2 M, so *in vivo* salt could not cause the transition. Nevertheless, there is considerable evidence that localized regions of Z-DNA do exist in cells, at least for short periods of time. For instance, antibodies to Z-DNA strongly bind to *Drosophila* salivary gland chromosomes. Additional support comes from the isolation of Z-DNA binding proteins from bacteria, yeast, and animals. Although the functions of these Z-DNA binding proteins have not been completely established, existing evidence suggests they play a role in regulating the expression of a few eukaryotic genes. Z-DNA may play a harmful role in human health. There is mounting evidence that regions with the potential to form Z-DNA are hot spots for DNA double-strand breaks, which can cause chromosomal rearrangements that result in malignant diseases.

Several other kinds of non-B DNA structures appear to exist in nature.

Other kinds of non-B DNA structures have also been observed. Some DNA regions can exist transiently in a cruciform, triplex, or quadruplex structure.

Cruciform Structure
A cruciform structure can only be formed in a region that contains an **inverted repeat sequence**, such as

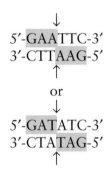

The arrows point to the vertical axis of symmetry: the double-stranded segment to the right of the axis can be superimposed on the one to the left by a 180-degree rotation in the plane of the page. The left-to-right sequence on the top strand hence is repeated right-to-left on the bottom strand. Inverted repeats range in length up to about 50 bp.

Molecular biologists use the term **"palindrome"** when referring to an inverted repeat sequence. However, a lexicologist might take issue with this use of the term because a palindrome is defined as a word (such as "madam") or a phrase (such as "Able was I ere I saw Elba") that reads the same forward or backward on a single line. In theory, DNA molecules that have inverted repeats can exist in two alternate forms, a normal duplex in which base pairs form between the two complementary strands or a **cruciform structure** (FIGURE 3.19) in which base pairs form between complementary regions on the same strand to produce double-stranded branches. Model building and energy calculations show that cruciform structures are somewhat strained compared with normal duplexes. Cruciform structures were originally produced in the laboratory under special conditions, but they also exist in cells.

(a) Inverted repeats

(b) Cruciform structure

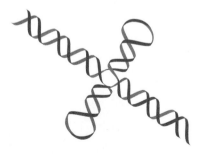

FIGURE 3.19 Cruciform structure. (a) A segment of DNA with inverted repeats. The arrows above and below the structure indicate the inverted repeats. (b) A DNA molecule with an inverted repeat can form a cruciform structure. (Adapted from Bacolla, A., and Wells, R. D. 2004. *J Biol Chem* 279:47411–47414.)

FIGURE 3.20 **A spacefilling model of triplex DNA.** Strands in the Watson-Crick double helix are shown in dark green and purple. The triplet forming oligonucleotide (red) in the major groove is tagged with a psoralen molecule (light green) at its 5′-end. (From Vasquez, K. M., and Glazer, P. M. 2002. Triplex-forming oligonucleotides: principles and applications. *Quart Rev Biophys* 35:89–107.)

Triplex Structures

Under certain conditions the DNA double helix can accommodate a third strand in its major groove to form a triplex structure (FIGURE 3.20). When triplex formation takes place *in vivo*, there is increased likelihood of double-strand breaks and mutations appearing.

Quadruplex Structure

One, two, or four G-rich DNA strands can form a quadruplex (or tetraplex) structure (FIGURE 3.21). The fundamental unit of a quadruplex structure is the quartet, a planar structure containing four guanine groups held together by eight hydrogen bonds. Two or more quartets stack on one another to form a quadruplex. The structure is stabilized by the hydrogen bonds and hydrophobic base stacking. Additional stability is provided by cation–dipole interactions between the eight guanine groups and a metal cation, usually Na^+ or K^+, that sits between two quartets. Intramolecular quadruplexes formed by a single strand are of special interest because the ends of eukaryotic chromosomes have G-rich 3′-overhangs that are rich in guanine. According to one hypothesis quadruplexes formed from these overhangs influence chromosome replication and stability. Quadruplexes may also form in DNA regions that regulate the expression of a few eukaryotic genes.

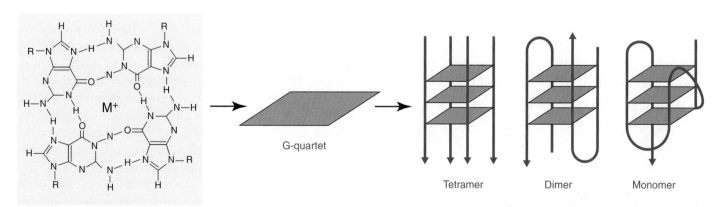

FIGURE 3.21 **Quadruplex structure.** (a) G-quartet. (b) Tetrameric, dimeric, and monomeric G-quadruplexes composed of three G-quartets. (Adapted from Bates, P., Megny, J-L., and Yang, D. 2007. Quartets in G-major. *EMBO Rep* 11:1003–1010.)

3.10 RNA Structure

RNA performs a wide variety of functions in the cell.

When the Watson-Crick model for B-DNA was proposed in 1953, very little was known about RNA function. Some viruses were known to use RNA as their genetic material, but this knowledge did not seem to be relevant to RNA's function(s) in cells because cells use DNA as their hereditary material. The RNA function problem appeared to have been solved when three kinds of RNA molecules were shown to make essential contributions to polypeptide synthesis. Ribosomal RNA molecules combine with proteins to form ribosomes, the ribonucleoprotein complex that serves as the protein synthetic factory. Transfer RNA molecules carry activated amino acids to the ribosomes where messenger RNA specifies the order in which the ribosome adds amino acids to the growing polypeptide chain. Then in the early 1980s Sidney Altman and Thomas Cech, working independently, demonstrated that some RNA molecules act as catalysts called **ribozymes**, a role that molecular biologists had previously assumed to be limited to proteins. The list of RNA functions grows with each passing year. The focus for now is on RNA structure.

RNA secondary structure is dominated by Watson-Crick base pairs.

RNA structure, like protein structure, is divided into primary, secondary, tertiary, and quaternary structures. The primary sequence of an RNA molecule is its base sequence. RNA secondary structure consists of helical regions and various kinds of loops, bulges, and junctions within the helical regions, which are stabilized by Watson-Crick base pairing. The tertiary structure consists of the arrangements of these secondary structures into a three-dimensional structure, which is often compact and stabilized by metal cations. The quaternary structure describes the arrangement of an RNA molecule with respect to other RNA molecules or with protein molecules. It is important to note that some kinds of RNA molecules do not assume a specific three-dimensional structure but instead function as unstructured single strands.

RNA's predominant secondary structure building blocks are helical tracts, which are stabilized by Watson-Crick base pairs and have a similar conformation to A-DNA. A helical tract seldom exceeds 10 successive base pairs before being interrupted by one or more loops or bulges (**FIGURE 3.22**). The smallest bulge results from the presence of a single unpaired base (Figure 3.22a). If the unpaired base is stacked within the helix, the helix bends. If the base is outside the helix, the helix does not bend, but the base can interact with other parts of the same RNA molecule, other RNA molecules, or proteins. Larger bulges result from the presence of additional unpaired bases (Figure 3.22b). An internal loop forms when one or more nucleotides on each RNA strand are unpaired. The smallest loop forms when a single pair of noncomplementary bases, a mismatch pair, interrupts the helical tract (Figure 3.22c). Larger loops form when additional unpaired bases

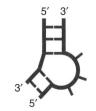

(a) Single nucleotide bulge (b) Three nucleotide bulge

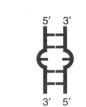
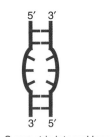

(c) Mismatch pair (d) Symmetric internal loop

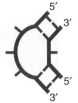

(e) Asymmetric internal loop (f) Hairpin loop

FIGURE 3.22 RNA loop and bulge secondary elements. (Slightly modified from Nowakowski, J., and Tinoco, I. Jr. 1997. RNA structure and stability. *Semin Virol* 8:153–165.)

interrupt the helical tract. A symmetrical internal loop forms when each strand has the same number of opposing unpaired bases (Figure 3.22d), and an asymmetrical internal loop forms when one strand has more unpaired bases than the other (Figure 3.22e). A **hairpin** (or **stem-loop**) **structure** forms when a strand folds back to form a stem that contains Watson-Crick base pairs and a loop (Figure 3.22f). A hairpin loop may be as small as two nucleotides or may be several nucleotides long.

Double helical stems may come together to form a junction (**FIGURE 3.23**). A junction is an important structural element because it helps to establish the overall structure of the RNA molecule. When two helical tracts meet end-to-end at a junction, they form a structure resembling a long helix. This end-to-end interaction, called **coaxial stacking**, helps to stabilize the junction.

RNA tertiary structures are stabilized by interactions between two or more secondary structure elements.

The RNA tertiary structure, which describes the three-dimensional structure of the RNA molecule, results from interactions between two or more secondary structures. Tertiary structures are stabilized by metal cations such as Na^+ and Mg^{2+} that offset charge repulsions that would otherwise prevent the negatively charged sugar phosphate backbone from folding into a condensed structure. Specific structural elements contribute to the tertiary structure. Some of these structural elements are described here.

A **pseudoknot** forms when a base sequence in a hairpin loop pairs with a complementary single-stranded region that is adjacent to the hairpin stem (**FIGURE 3.24**). The two helical tracts stack end-to-end, forming a coaxial stack.

Several kinds of structural interactions can bring distant regions of a large RNA molecule together. Two of these interactions are shown in **FIGURE 3.25**. Kissing hairpins form when unpaired nucleotides in one hairpin base pair with complementary nucleotides in another hairpin (Figure 3.25a). Hairpin loop-bulge contacts form when unpaired nucleotides in a bulge base pair with complementary nucleotides in a hairpin (Figure 3.25b).

Crystal structures have been determined for many RNA molecules. Those for RNA catalysts are especially interesting. The smallest known RNA catalyst, the hammerhead ribozyme (**FIGURE 3.26**), is specified by virus-like RNA molecules that infect plants. The ribozyme's name derives from the shape of its secondary structure.

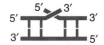
(a) Two-stem junction

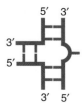
(b) Three-stem junction

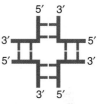
(c) Four-stem junction

FIGURE 3.23 RNA junction secondary elements. (Slightly modified from Nowakowski, J., and Tinoco, I. Jr. 1997. RNA structure and stability. *Semin Virol* 8:153–165.)

3.11 RNA World Hypothesis

The earliest forms of life on earth may have used RNA as both the genetic material and the biological catalysts needed to maintain life.

The discovery that RNA can act as a catalyst has profound implications for the way we view biochemical evolution. Biologists have long speculated about whether the earliest life forms contained DNA or protein. It

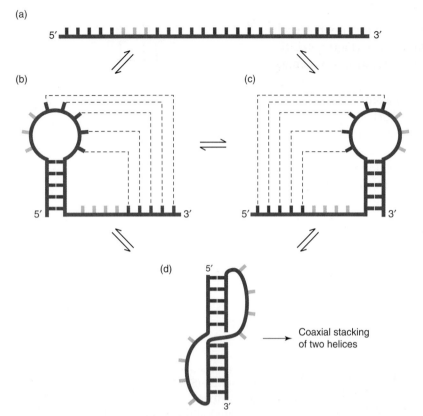

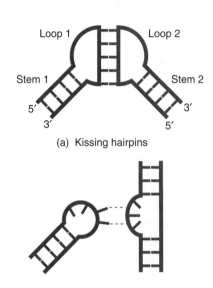

FIGURE 3.24 RNA pseudoknot structure. (a) Unstructured RNA with two pairs of complementary base sequences. One pair is shown in red and the other in blue. (b) The hairpin (stem-loop) structure formed when the complementary sequences shown in red base pair. (c) The hairpin (stem-loop) structure formed when the complementary sequences shown in blue base pair. (d) The pseudoknot that forms when a base sequence in a hairpin loop pairs with a complementary single stranded region that is adjacent to the hairpin stem. The two helical tracts stack end-to-end, forming a coaxial stack. (Slightly modified from Varani, G. 2001. RNA structure. *Encyclopedia of Life Sciences* pp. 1–8. Reproduced with permission from John Wiley & Sons.)

FIGURE 3.25 Interactions that bring distant RNA segments together. (a) Kissing hairpin interaction and (b) hairpin loop-bulge interaction. (Slightly modified from Nowakowski, J., and Tinoco, I. Jr. 1997. RNA structure and stability. *Semin Virol* 8:153–165.)

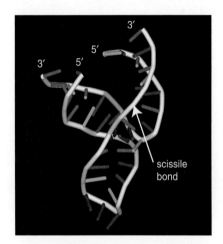

FIGURE 3.26 Crystal structure of the hammerhead ribozyme. The arrow points to the easily cut (scissile) phosphodiester bond. (Structure from Protein Data Bank 379D. J. B. Murray, et al. 1998. *Cell* 92:665–673. Prepared by B. E. Tropp.)

was hard to see how either macromolecule could work in the absence of the other. Protein molecules fold into stable tertiary structures to form catalytic sites but do not transmit genetic information. DNA stores and transmits genetic information, but no naturally occurring DNA has yet been found that can act as a catalyst. The term "naturally occurring" is an important qualifier because DNA molecules have been synthesized in the laboratory that can act as catalysts. Even so, RNA molecules are more versatile because their additional 2′-hydroxyl group permits them to form stable tertiary structures that cannot be achieved by DNA. The discovery that naturally occurring RNA molecules have catalytic properties suggested a solution to the "which came first, the chicken or egg" problem. The earliest progenitors to modern cells probably contained RNA, a molecule that can store genetic information and catalyze biochemical reactions. The earliest life forms thus may have lived in an "RNA world." This hypothesis, first proposed by Walter Gilbert in

1986, suggests that proteins eventually replaced RNA molecules as catalysts for most (but not all) biological reactions because proteins offered more possible sequence and structural alternatives, whereas DNA would eventually replace RNA molecules for genetic storage in most (but not all) biological systems because it is more stable and easy to repair.

Questions and Problems

1. Briefly define, describe, or illustrate each of the following.
   a. Eukaryote
   b. Histones
   c. DNA denaturation
   d. Hyperchromic effect
   e. DNA melting temperature
   f. Depurination
   g. DNA renaturation
   h. Helicase
   i. SSB
   j. Replication protein A
   k. RFI
   l. Superhelix
   m. Catenane
   n. Topoisomer
   o. Topoisomerase
   p. Palindrome
   q. A-DNA
   r. Z-DNA
   s. Cruciform structure
   t. Hairpin structure
   u. Pseudoknot
   v. Ribozyme
   w. Coaxial stacking

2. Briefly explain each of the following.
   a. *E. coli* DNA is about 4.6×10^6 bp long in the living cell, but purified *E. coli* DNA is only about 5×10^4 bp long.
   b. The A_{260} for a linear DNA duplex in solution increases by about 37% when the solution is heated from 25°C to 90°C.
   c. The T_m for a solution containing a linear DNA duplex is lower when the solution contains 0.01 M NaCl than when the solution contains 0.2 M NaCl.
   d. A linear DNA duplex with a 20% G + C content has a lower T_m than a linear DNA duplex with an 80% G + C content.
   e. The optimal temperature for DNA renaturation is about 20 to 25°C below the DNA's melting temperature.
   f. A linear DNA duplex is not sensitive to a DNase that is specific for single-stranded DNA, but the DNA becomes sensitive to hydrolysis if a helicase and ATP are added to the solution.
   g. Bases in a linear DNA duplex become radioactively labeled when the DNA is mixed with [^{3}H]H$_2$O.
   h. The A_{260} of a linear DNA duplex in solution increases when sodium hydroxide is added.
   i. SSB lowers the melting temperature of a linear DNA duplex.
   j. A DNase that removes one nucleotide at a time from the 5′-end of a linear DNA duplex has no effect on a plasmid DNA isolated from *E. coli*.

3. How do SSBs differ from helicases?
4. How does topoisomerase I differ from the DnaB helicase?
5. List four factors that influence the melting temperature of a linear DNA duplex.
6. Compare B-DNA with A-DNA and Z-DNA.
7. DNA gyrase has a catalytic ability that distinguishes it from other type II topoisomerases.
   a. How does DNA gyrase differ from other type II topoisomerases?
   b. Why do antibiotics such as nalidixic acid and novobiocin that inhibit DNA gyrase block bacterial growth?
8. How do type II topoisomerases convert catenanes that are formed after bacterial DNA replication into individual covalently closed circular DNA molecules?
9. How do type I topoisomerases differ from type II topoisomerases?

10. Why does a DNase specific for single-stranded DNA cleave a negatively supercoiled DNA but have no effect on the same closed circular DNA molecule after it has been relaxed by topoisomerase I?
11. A new biological catalyst is discovered that cleaves RNA. Describe some simple tests you could use to determine if the catalyst is an enzyme or a ribozyme.
12. Why do many biologists believe RNA and not DNA or protein was present in the earliest progenitors of modern cells?

Suggested Reading

General

Bloomfield, V. A., Crothers, D. M., and Tinoco, I. Jr. 2000. *Nucleic Acids: Structures, Properties and Functions*. Herndon, VA: University Science Books.

Bowater, R. P. 2005. DNA structure. *Encyclopedia of Life Sciences*. pp. 1–8. Hoboken, NJ: John Wiley & Sons.

Calladine, C. R., and Drew, H. R. 1997. *Understanding DNA* (2nd ed.). San Diego, CA: Academic Press.

Sinden, R. R. 1994. *DNA Structure and Function*. San Diego, CA: Academic Press.

Soukup, G. A. 2001. Nucleic acids: general properties. *Encyclopedia of Life Sciences*. pp. 1–9. Hoboken, NJ: John Wiley & Sons.

DNA Bending

Harvey, S. C., Diakic, M., Griffith, J., et al. 1995. What is the basis of sequence-directed curvature in DNAs containing A tracts? *J Biomol Struct Dyn* 13:301–307.

DNA Denaturation and Renaturation

Marmur, J., and Doty, P. 1961. Thermal renaturation of deoxyribonucleic acids. *J Mol Biol* 3:585–594.

Marmur J., and Doty, P. 1962. Determination of the base composition of deoxyribonucleic acid from its thermal denaturation temperature. *J Mol Biol* 5:109–118.

Helicases

Bailey, S., Eliason, W. K., and Steitz, T. A. 2007. Structure of hexameric DnaB helicase and its complex with a domain of DnaG primase. *Science* 318:459–463.

Egelman, E. 2001. DNA helicases. *Encyclopedia of Life Sciences*. pp. 1–5. London: Nature.

Patel, S. S., and Picha, K. M. 2000. Structure and function of hexameric helicases. *Ann Rev Biochem* 69:651–697.

Singleton, M. R., Dillingham, M. S., and Wigley, D. B. 2007. Structure and mechanism of helicases and nucleic acid translocases. *Ann Rev Biochem* 76:23–50.

Single-Stranded DNA Binding Proteins

Bochkarev, A., and Bochkareva, E. 2004. From RPA to BRCA2: lessons from single stranded DNA binding by the OB-fold. *Curr Opin Struct Biol* 14:36–42.

George, N. P., and Keck, J. L. 2009. Slip sliding on DNA. *Nature* 461:1067–1068.

Iftode, C., Daniely, Y., and Borowiec, J. A. 1999. Replication protein A (RPA): the eukaryotic SSB. *Crit Rev Biochem Molec Biol* 34:141–180.

Lohman, T. M., and Ferrari, M. E. 1994. *Escherichia coli* single-stranded DNA binding protein: multiple DNA-binding modes and cooperativities. *Ann Rev Biochem* 63:527–570.

Raghunathan, S., Kozlov, A., Lohman, T., and Waksman, G. 2000. Structure of the DNA binding domain of *E. coli* SSB bound to ssDNA. *Nat Struct Biol* 7:648–652.

Shamoo, Y. 2002. Single-stranded DNA-binding proteins. *Encyclopedia of Life Sciences*. pp. 1–7. London: Nature.

Wold, M. S. 2001. Eukaryotic replication protein A. *Encyclopedia of Life Sciences*. pp. 1–7. London: Nature.

Topoisomers

Bates, A. D. 2001. Topoisomerases. *Encyclopedia of Life Sciences*. pp. 1–9. London: Nature.

Bates, A. D., and Maxwell, A. 1993. DNA Topology. Washington, DC: IRL Press.

Bauer, W. R., Crick, F. H. C., and White, J. H. 1980. Supercoiled DNA. *Sci Am* 243:100–113.

Bowater, R. P. 2002. Supercoiled DNA structure. *Encyclopedia of Life Sciences*. pp. 1–9. London: Nature.

Hardy, C. D., Crisona, N. J., Stone, M. D., and Cozzarelli, N. R. 2004. Disentangling DNA during replication: a tale of two strands. *Phil Trans Royal Soc Lond B* 359:39–47.

Lebowitz, J. 1990. Through the looking glass: the discovery of supercoiled DNA. *Trends Biochem Sci* 15:202–207.

Lindsley, J. E. 2005. DNA topology: supercoiling and linking. *Encyclopedia of Life Sciences*. pp. 1–7. Hoboken, NJ: John Wiley & Sons.

Mirkin, S. M. 2002. DNA topology: fundamentals. *Encyclopedia of Life Sciences*. pp. 1–11. London: Nature.

Topoisomerases

Changela, A., Perry, K., Taneja, B., and Mondragón, A. 2002. DNA manipulators: caught in the act. *Curr Opin Struct Biol* 13:15–22.

Corbett, K. D., and Berger, J. M. 2004. Structure, molecular mechanisms, and evolutionary relationships in DNA topoisomerases. *Ann Rev Biophys Biomol Struct* 33:95–118.

Nöllmann, M., Crisona, N. J., and Arimondo, P. B. 2007. Thirty years of *Escherichia coli* DNA gyrase: from *in vivo* function to single-molecule mechanism. *Biochimie* 89:490–499.

Schoeffler, A.J., and Berger, J. M. 2008. DNA topoisomerases: harnessing and constraining energy to govern chromosome topology. *Quart Rev Biophys* 41:41–101.

Wang, J. C. 2002. Cellular roles of DNA topoisomerases: a molecular perspective. *Nat Rev Mol Cell Biol* 3:430–440.

Wang, J. C. 2009. A journey in the world of DNA rings and beyond. *Ann Rev Biochem* 78:31–54.

Non-B DNA Conformations

Armitage, B. A. 2007. The rule of four. *Nat Chem Biol* 3:203–204.

Arnott, S. 2006. Historical article: DNA polymorphism and the early history of the double helix. *Trends Biochem Sci* 31:349–354.

Bacolla, A., and Wells, R. D. 2009. Non-B DNA conformations as determinants of mutagenesis and human disease. *Mol Carcinog* 48:273–285.

Bates, P., Megny, J.-L., and Yang, D. 2007. Quartets in G-major. *EMBO Rep* 11:1003–1010.

Burge, S., Parkinson, G. N., Hazel, P., Todd, A. K., and Neidle, S. 2006. Quadruplex DNA: sequence, topology and structure. *Nucleic Acids Res* 34:5402–5415.

Dickerson, R. E. 1992. DNA structures from A to Z. *Methods Enzymol* 211:67–111.

Dickerson, R. E., and Ng, H. L. 2001. DNA structure from A to B. *Proc Natl Acad Sci USA* 98:6986–6988.

Gagna, C. E., Kuo, H., and Lambert, W. C. 1999. Terminal differentiation and left-handed Z-DNA: a review. *Cell Biol Int* 23:1–5.

Herbert, A., and Rich, A. 1999. Left-handed Z-DNA: structure and function. *Genetica* 106:37–47.

Mirkin, S. M. 2008. Discovery of alternative DNA structures: a heroic decade (1979–1989). *Frontiers Biosci* 13:1064–1071.

Potaman, V. N., and Sinden, R. R. 2005. DNA: alternative conformations and biology. *DNA Conformation and Transcription*, T. Ohyama, ed. pp. 1–16. New York: Springer-Verlag.

Rich, A. 2003. The helix: a tale of two puckers. *Nat Struct Biol* 10:247–249.

Rich, A., and Zhang, S. 2003. Timeline: Z-DNA: the long road to biological function. *Nat Rev Genet* 4:566–572.

Rogers, F. A., Lloyd, J. A., and Glazer, P. M. 2005. Triplex-forming oligonucleotides as potential tools for modulation of gene expression. *Curr Med Chem Anti-Cancer Agents* 5:319–326.

Ussery, D. W. 2002. DNA Structure: A-, B- and Z-DNA helix families. *Encyclopedia of Life Sciences*. pp. 1–6. London: Nature.

Vasquez, K. M., and Glazer, P. M. 2002. Triplex-forming oligonucleotides: principles and applications. *Quart Rev Biophys* 35:89–107.

Wells, R. D., Dere, R., Hebert, M. L., Napierala, M., and Son, L. S. 2005. Advances in mechanisms of genetic instability related to hereditary neurological diseases *Nucleic Acids Res* 33:3785–3798.

Wells, R. D. 2007. Non-B DNA conformations, mutagenesis and disease. *Trends Biochem Sci* 32:271–278.

RNA Structure

Batey, R. T., Rambo, R. P., and Doudna, J. A. 1999. Tertiary motifs in RNA structure and folding. *Angew Chem Int Ed* 38:2326–2343.

Bevilacqua, P. C., and Blose, J. M. 2008. Structures, kinetics, thermodynamics, and biological functions of RNA hairpins. *Ann Rev Biophys* 59:79–103.

Brierley, I., Gilbert, R. J. C., and Pennell, S. 2008. RNA pseudoknots and the regulation of protein synthesis. *Biochem Society Trans* 36:684–689.

Carter, R. J., and Holbrook, S. R. 2002. RNA structure: Roles of Me^{2+}. *Encyclopedia of Life Sciences*. pp. 1–7. Hoboken, NJ: John Wiley & Sons.

Cech, T. R. 2002. Ribozymes, the first 20 years. *Biochem Soc Trans* 30:1162–1166.

Cheong, C., and Cheong, H.-K. 2001. RNA structure: tetraloops. *Encyclopedia of Life Sciences*. pp. 1–7. Hoboken, NJ: John Wiley & Sons.

Cruz, J. A., and Westhof, E. 2009. The dynamic landscapes of RNA architecture. *Cell* 136:604–609.

Doudna, J. A. and Cech, T. R. 2002. The chemical repertoire of natural ribozymes. *Nature* 418:222–228.

Draper, D. E. 2004. A guide to ions and RNA structure. *RNA* 10:335–343.

Draper, D. E., Grilley, D., and Soto, A. M. 2005. Ions and RNA folding. *Ann Rev Biophys Biomol Struct* 34:221–243.

Gultyaev, A. P., Pleij, C. W. A., and Westhof, E. 2005. RNA structure: pseudoknots. *Encyclopedia of Life Sciences*. pp. 1–7. Hoboken, NJ: John Wiley & Sons.

Hendrix, D. K., Brenner, S. E., and Holbrook, S. R. 2005. RNA structural motifs: building blocks of a modular biomolecule. *Quart Rev Biophys* 38:221–243.

Hermann, T., and Patel, D. J. 2000. RNA bulges as architectural and recognition motifs. *Structure* 8:R47–R54.

Holbrook, A. R. 2008. Structural principles from large RNAs. *Curr Opin Struct Biol* 15:302–308.

Hou, Y.-M. 2002. Base pairing in RNA: unusual patterns. *Encyclopedia of Life Sciences*. pp. 1–7. Hoboken, NJ: John Wiley & Sons.

Leontis, N. B., and Westhof, E. 2001. Geometric nomenclature and classification of RNA base pairs. *RNA* 7:499–512.

Leontis, N. B., and Westhof, E. 2003. Analysis of RNA motifs. *Curr Opin Struct Biol* 13:300–308.

Leontis, N. B., Lescoute, A., and Westhof, E. 2006. The building blocks and motifs of RNA architecture. *Curr Opin Struct Biol* 16:279–287.

Nowakowski, J., and Tinoco, I. Jr. 1997. RNA structure and stability. *Semin Virol* 8:153–165.

Schroeder, E., Barta, A., and Semrad, K. 2004. Strategies for RNA folding and assembly. *Nat Rev Mol Cell Biol* 5:908–919.

Staple, D. W., and Butcher, S. E. 2005. Pseudoknots: RNA structures with diverse functions. *PLoS Biol* 3:956–959.

Tung, C.-S. 2002. RNA structural motifs. *Encyclopedia of Life Sciences*. pp. 1–4. Hoboken, NJ: John Wiley & Sons.

Varani, G. 2001. RNA structure. *Encyclopedia of Life Sciences*. pp. 1–8. Hoboken, NJ: John Wiley & Sons.

Walter, F., and Westhof, E. 2002. Catalytic RNA. *Encyclopedia of Life Sciences*. pp. 1–12. Hoboken, NJ: John Wiley & Sons.

RNA World Hypothesis

Cech, T. R. 2009. Crawling out of the RNA world. *Cell* 136:599–602.

Gesteland, R. F. (ed.). 2005. *The RNA World* (3rd ed.). New York: Cold Spring Harbor Press.

4

GGGCCTGCAGGATTGCC

Molecular Biology Technology

CHAPTER OUTLINE

4.1 Nucleic Acid Isolation
The method used to isolate DNA depends on the DNA source.

Great care must be taken to protect RNA from degradation during its isolation.

4.2 Physical Techniques Used to Study Macromolecules
Different physical techniques are used to study macromolecules.

Electron microscopy allows us to see nucleic acids and nucleoprotein complexes.

Centrifugation can separate macromolecules and provide information about their size and shape.

Gel electrophoresis separates nucleic acids based on their rate of migration in an electric field.

SDS-PAGE can be used to determine a polypeptide's molecular mass.

4.3 Enzymatic Techniques Used to Manipulate DNA
Nucleases are useful tools in DNA investigations.

Restriction endonucleases cleave within specific nucleotide sequences in DNA.

Restriction endonucleases can be used to construct a restriction map of a DNA molecule.

DNA fragments can be inserted into plasmid DNA vectors.

The Southern blot procedure is used to detect specific DNA fragments.

Northern and Western blotting are used to detect specific RNA and polypeptide molecules, respectively.

DNA polymerase I requires a template-primer.

DNA polymerase I has both 3'→5' and 5'→3' exonuclease activities.

DNA polymerase I can catalyze nick translation.

The polymerase chain reaction (PCR) is used to amplify DNA.

Site-directed mutagenesis can be used to introduce a specific base change within a gene.

The chain termination method for sequencing DNA uses dideoxynucleotides to interrupt DNA synthesis.

DNA sequences can be stitched together by using information obtained from a restriction map.

Shotgun sequencing is used to sequence long DNA molecules.

A new generation of DNA sequencing techniques is now being used for whole genome shotgun sequencing.

The human genome sequence provides considerable new information.

BOX 4.1: IN THE LAB Gel Electrophoresis and Topoisomer Separation

BOX 4.2: IN THE LAB Capillary Gel Electrophoresis

BOX 4.3: IN THE LAB New Generation DNA Sequencers

QUESTIONS AND PROBLEMS

SUGGESTED READING

Image © Alila Sao Mai/ShutterStock, Inc.

Molecular biologists use physical methods and enzyme-catalyzed reactions to characterize and modify nucleic acids. These techniques have become so widely used that you may have encountered some of them in other life science courses. The first part of this chapter describes methods used to isolate and characterize nucleic acids. These methods also apply to proteins and nucleoproteins. The second part describes techniques that use enzymes to manipulate and sequence DNA in the laboratory. Of course, these enzymes have important biological functions in the cell. But for now our attention is directed toward using the enzymes as laboratory tools to cut, sequence, replicate, or otherwise manipulate DNA in the test tube. The biological functions of the enzymes are considered elsewhere.

4.1 Nucleic Acid Isolation

The method used to isolate DNA depends on the DNA source.

DNA isolation, an essential step in many molecular biology experiments, begins with the release of the DNA, often with other biological molecules such as proteins and RNA, from the cells or viruses of interest. Because the biological sources' structures and compositions vary, the particular method used to release the DNA must be tailored to the cellular or viral DNA source. The method of choice depends on how the DNA is enclosed and the percent of total dry weight that is DNA (which varies from about 1% in complex mammalian cells to about 50% in bacterial viruses). Care must be taken to avoid vigorous stirring or other procedures that produce hydrodynamic shear forces that may break DNA. The following procedures are used to isolate viral, bacteria, plasmid, and yeast DNA:

1. *Viral DNA.* An aqueous suspension of virus particles (virions) is gently mixed with phenol, a reagent that is slightly miscible with water. A small amount of phenol enters the aqueous layer, breaking open the protein coat and denaturing individual protein molecules. (Sometimes chloroform and isoamyl alcohol are added to assist the phenol.) Most of the denatured protein either enters the phenol layer or precipitates at the phenol–water interface. The upper aqueous layer, containing the DNA, is carefully separated from the phenol layer and ethanol is added to the aqueous fraction to precipitate DNA. After collection by centrifugation, DNA is dissolved in an aqueous buffer solution.
2. *Bacterial DNA.* The DNA in a bacterial cell is enclosed in a cell envelope consisting of a cell wall and membrane(s). Phenol cannot, by itself, remove this envelope. The envelope can be removed, however, by first treating a bacterial cell suspension with lysozyme to digest the cell wall and then with a detergent such as sodium dodecyl sulfate [SDS; $CH_3(CH_2)_{11}OSO_3^-Na^+$] to disrupt the cell membrane. After

the cell envelope is removed phenol is added with gentle mixing, and the mixture is allowed to stand so the aqueous and phenol phases separate. The upper aqueous phase is collected and treated with ethanol to precipitate DNA and RNA. The precipitate containing DNA and some contaminating RNA is spooled out with a glass rod and excess alcohol removed. Then the spooled DNA is dissolved in a buffered solution containing RNase to digest contaminating RNA. Chloroform is added to the mixture with gentle mixing, and the mixture is allowed to stand so the phases can separate. The upper aqueous phase is collected and treated with ethanol to precipitate DNA, which is spooled out with a glass rod and dissolved in an aqueous buffer solution.

3. *Plasmid DNA.* Plasmid DNA is released from bacteria by first incubating the bacterial suspension with lysozyme to digest the cell wall. Then an SDS–sodium hydroxide solution is added to the mixture. SDS disrupts the cell membrane, and sodium hydroxide hydrolyzes RNA to form a mixture of 2′- and 3′-nucleoside monophosphates. DNA, which lacks 2′-hydroxyl groups, is denatured but not hydrolyzed. The two strands in a linear duplex separate completely after denaturation. Strands derived from the covalently closed, circular double-stranded plasmid DNA, however, remain intertwined after denaturation. When acid is added to neutralize the extract, intertwined plasmid DNA strands rewind to form the intact plasmid. In contrast, the separated bacterial DNA strands combine to form an insoluble aggregate that is removed, along with denatured protein, by centrifugation. The soluble plasmid DNA is precipitated by adding ethanol, collected by centrifugation, and dissolved in an aqueous buffer solution.

4. *Yeast and fungal DNA.* The polysaccharides that make up yeast and fungal cell walls are resistant to lysozyme. Other enzymes (such as cellulase isolated from snails), however, break down these cell walls. Once cell walls have been disrupted, the DNA isolation procedure is similar to that used for bacteria.

Great care must be taken to protect RNA from degradation during its isolation.

One of the greatest concerns when isolating RNA is that the RNA will be degraded by RNases released during the isolation procedure. This problem can be solved by freezing the cells in liquid nitrogen, transferring the frozen cells to a mortar, and grinding the frozen cells with a pestle. Then the ground cells are suspended in an acidic solution containing guanidinium thiocyanate, sodium acetate, phenol, and chloroform. Under these acidic conditions RNA is present in the upper aqueous phase, whereas most of the DNA and proteins are present in either the lower organic phase or the interphase. The aqueous phase containing the RNA is transferred to a clean tube, and ethanol or isopropanol is added. The RNA, which precipitates out of solution, is collected by centrifugation, washed with alcohol, and dissolved in an aqueous buffer solution.

4.2 Physical Techniques Used to Study Macromolecules

Different physical techniques are used to study macromolecules.

Several physical techniques are used to study nucleic acids, proteins, and nucleoproteins. Four of these—electron microscopy, sucrose gradient sedimentation, equilibrium density gradient centrifugation, and electrophoresis—are described briefly so experiments can be understood.

Electron microscopy allows us to see nucleic acids and nucleoprotein complexes.

Three electron microscopy techniques are widely used to study nucleic acids and nucleoprotein complexes:

1. *Metal shadowing:* Metal shadowing provides a three-dimensional image of the surface of a biological sample (FIGURE 4.1). The biological sample is dried on a translucent

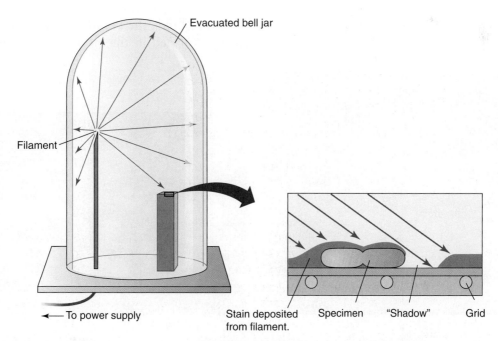

FIGURE 4.1 **Metal shadowing.** This technique allows the observer to view details on the surface of small particles. The sample is dried on a thin film, which is then placed in a vacuum chamber. A filament of a heavy metal such as tungsten is heated and the metal evaporates, forming a thin metal coat on the sample. The observer uses an electron microscope to view the metal coating, which is a replica of the biological sample. (Adapted from Ingraham, J., and Ingraham, C. 2006. *Introduction to Microbiology* (3rd ed.). Academic Internet Publishers.)

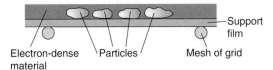

FIGURE 4.2 Negative staining method of visualizing particles by electron microscopy. Four virus particles (and macromolecules) are embedded in a substance that absorbs electrons strongly. As the beam passes through the sample, the fraction of the electrons in the beam that is absorbed depends on the total thickness of the substance; therefore, more electrons will pass through the regions containing each particle and the particle will appear bright against a dark background.

film and then placed in a vacuum chamber, where a filament of a heavy metal such as tungsten is heated. The evaporated metal forms a thin metal coat on the surface of the biological sample. The observer uses an electron microscope to view the metal coating, which is a replica of the biological sample. Metal shadowing is a very useful technique for observing very long molecules when one is primarily interested in their length and linear topology.

2. *Negative staining:* Negative staining provides an alternative means for visualizing macromolecules and virus particles. The macromolecules or virus particles of interest are added to a solution containing a heavy metal salt such as uranyl acetate. The mixture is then deposited as microdroplets onto a supporting grid and allowed to dry. The heavy metal salts form a mold surrounding and covering the macromolecules or virus particles (FIGURE 4.2). The mold surrounding the macromolecules or virus particles is thicker than that covering them. Therefore, when the grid is placed in an electron beam, more electrons pass through the sample than the surrounding region, producing a negative image of the macromolecules or virus particles. FIGURE 4.3 shows proteins and virus particles visualized by using the negative staining technique.

3. *Cryoelectron microscopy:* Cryoelectron microscopy is used to create three-dimensional constructs of protein and nucleoprotein complexes. This technique involves rapidly freezing a droplet of a buffer solution containing the specific protein or nucleoprotein complexes so that a thin amorphous layer of ice forms that holds the complexes in random orientations. A picture is formed in the electron microscope by using a

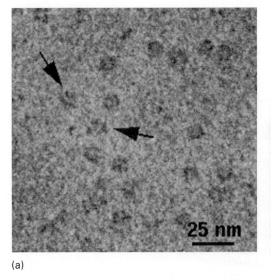
(a)

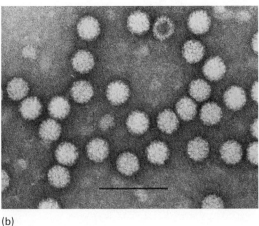
(b)

FIGURE 4.3 Electron micrographs obtained by the negative contrast procedure. (a) Image of negatively stained bacteriophage T7 helicase/primase in the presence of dTDP. (b) Images of negatively stained tomato bushy stunt virus particles. Note the surface details, which show the individual protein molecules that form the protein coat of which each particle is composed. (Part a reproduced with kind permission from Springer Science+Business Media: Ohi, M., et al., *Biol. Proc. Online* 6 [2004]: 23–34. Photo courtesy of Thomas Walz, Harvard Medical School. Part b photo courtesy of Robert G. Milne, Plant Virus Institute, National Research Council, Turin, Italy, with permission from Damion Milne.)

radiation dose that is so low the complexes are not damaged. The resulting micrograph shows hundreds of the protein or nucleoprotein complexes lying in different orientations. Computer-assisted image processing converts the projections into three-dimensional constructs such as those shown for the bacterial ribosome and its subunits in FIGURE 4.4.

Centrifugation can separate macromolecules and provide information about their size and shape.

Several important properties of macromolecules can be determined from the rates at which they move in a centrifugal field. These studies cannot be performed with an ordinary laboratory centrifuge because it cannot produce a sufficient centrifugal force to sediment molecules with enough velocity to overcome randomized diffusion. Modern ultracentrifuges, however, can generate forces as great as 800,000 × gravity, which is more than enough to cause macromolecules to sediment through a solution.

A particle's sedimentation velocity depends on its mass and shape. If two particles have the same shape, the one with the larger mass sediments more rapidly than one with the smaller mass. If two particles both have the same molecular mass, the one with the more compact shape sediments more rapidly under the influence of a centrifugal force. Shape influences sedimentation rate because the more compact a particle is, the less frictional drag it experiences as it sediments through a solution.

The ratio of sedimentation velocity to centrifugal force is called the sedimentation coefficient (s). That is,

$$s = \text{velocity/centrifugal force}$$

The value of s for a particular molecule is often the same in many different solutions, so an s value is frequently considered to be a constant that characterizes a macromolecule. Furthermore, because the s value depends on molecular mass and shape, changes in the s value, as experimental conditions are varied, can be used to monitor changes in molecular aggregation or conformation. For most macromolecules the s value is between 1×10^{-13} and 100×10^{-13} sec. In honor of Theodor Svedberg, the ultracentrifuge's inventor, 10^{-13} seconds is called one Svedberg, or 1S. The s value of a particle formed by the association of two smaller particles cannot be determined by simply adding the s values of the two smaller particles. For example, a 30S bacterial ribosomal subunit combines with a 50S bacterial ribosomal subunit to form a 70S bacterial ribosome (and *not* an 80S bacterial ribosome).

Two different centrifugation techniques are often used in molecular biology laboratories. The first technique is *sucrose gradient centrifugation*. A centrifuge tube is filled with a sucrose solution so the sucrose concentration increases continuously from the top to the bottom of the centrifuge tube (FIGURE 4.5). The protein, nucleic acid, or nucleoprotein solution of interest is carefully layered on top of the sucrose gradient to form a band (or zone). Then the plastic centrifuge tube is placed in a swinging bucket rotor, which is a type of centrifuge head that contains buckets that can swivel on pins. The rotor is placed in the centrifuge. As

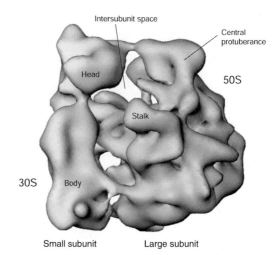

FIGURE 4.4 **Cryoelectron microscopy reconstruction of the *Escherichia coli* ribosome.** (Reproduced from Frank, J., and Agrawal, R. K. 2001. Bacterial ribosomes. *Encyclopedia of Life Sciences*. Reproduced with permission from John Wiley & Sons, Ltd: Chichester. Photo courtesy of Rajendra K. Agrawal, Wadsworth Center.)

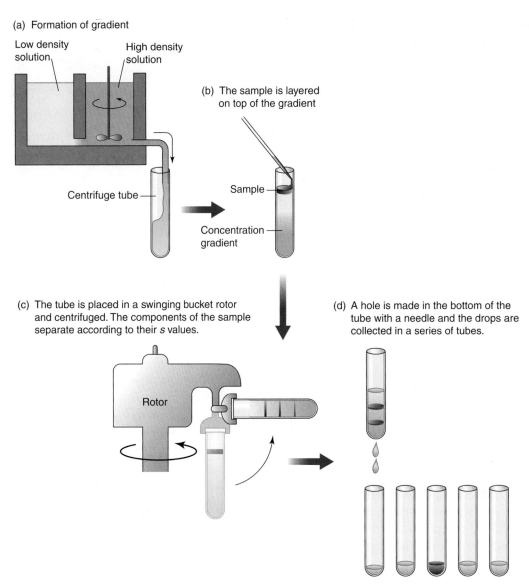

FIGURE 4.5 **Sucrose gradient centrifugation**.

the rotor turns the buckets swivel, causing the centrifuge tube to assume a horizontal position. After centrifugation is complete, the swinging bucket returns to a vertical position and the centrifuge tube inside it is removed. A tiny hole is then punched in the bottom of the centrifuge tube, and drops of the solution are collected in separate tubes. These drops, representing successive layers of solution in the tube, are analyzed to determine the macromolecule's concentrations along the tube.

The second centrifugation technique is *equilibrium density gradient centrifugation*. In this technique, nucleic acids or nucleoproteins are suspended in a CsCl (or Cs_2SO_4) solution at a concentration that is chosen so the solution density is approximately equal to that of the macromolecule. Under the influence of a powerful centrifugal force, Cs^+ and Cl^- (or SO_4^{2-}) ions move toward the bottom of the centrifuge to some extent. They do not accumulate on the bottom of the centrifuge tube, however, because the centrifugal force is not great enough to counteract the tendency for diffusion to maintain a uniform distribution of the ions. After a period of several hours the ions achieve an equilibrium concentration distribution in which there is a nearly linear concentration gradient and, hence, a nearly linear density gradient in the centrifuge tube.

The density is maximal at the bottom of the tube. As the density gradient forms macromolecules begin to migrate. Those in the upper reaches of the tube move toward the bottom, stopping at the position at which their density equals the solution density. Similarly, macromolecules in the lower part of the tube move upward, stopping at the same position. In this way the macromolecules form a narrow band in the tube. If the solution contains macromolecules having different densities, each macromolecule forms a band at the position in the gradient that matches its own density, and thus the macromolecules can be separated.

This technique's resolution is extraordinary. For example, it can separate single-stranded DNA molecules that differ only in the nitrogen isotopes present in their bases. As shown in FIGURE 4.6, single-stranded DNA molecules with the naturally occurring ^{14}N isotope are separated from single-stranded DNA molecules with the heavier ^{15}N isotope. The densities of the ^{14}N and ^{15}N single-stranded DNAs are 1.710 g • cm^{-3} and 1.724 g • cm^{-3}, respectively.

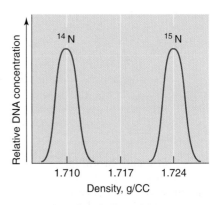

FIGURE 4.6 **Demonstration of DNA strand separation by equilibrium density centrifugation in CsCl.** ^{14}N indicates single-stranded DNA prepared from bacteria cultured in a medium that contained [^{14}N]NH$_4$Cl. ^{15}N indicates single-stranded DNA prepared from bacteria cultured in a medium that contained [^{15}N]NH$_4$Cl.

Gel electrophoresis separates nucleic acids based on their rate of migration in an electric field.

Gel electrophoresis can be used to separate DNA or RNA molecules according to size (FIGURE 4.7). A gel slab, which has small wells at the

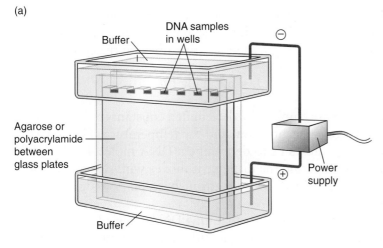

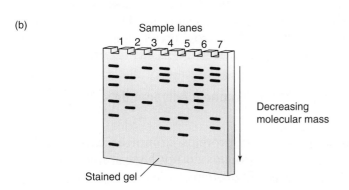

FIGURE 4.7 **Gel electrophoresis.** (a) Apparatus for gel electrophoresis capable of handling several samples simultaneously. An agarose or polyacrylamide suspension is placed in a mold fitted with glass or plastic plates on each side, and an appropriately shaped mold is placed on top of the gel during hardening to make "wells" for the samples. After the gel has hardened the mold on top is removed and samples are placed in the wells. The power supply is connected to the electrophoresis apparatus and the nucleic acids migrate toward the positive electrode (anode). When electrophoresis is complete the power is turned off and the nucleic acids in the gel are made visible by removing the glass or plastic plates and immersing the gel in a solution containing a reagent that binds to or reacts with the separated molecules. (b) The separated components of a sample appear as bands, which may either be colored or fluoresce. The region of a gel in which the components of one sample can move is called a lane. This gel thus has seven lanes.

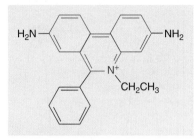

FIGURE 4.8 Ethidium bromide. Ethidium bromide contains a planar phenanthridium ring (shown in red) that inserts between the stacked bases in DNA. DNA absorbs ultraviolet light radiation at 260 nm and transmits it to the bound dye, which re-emits the light at 590 nm in the red-orange region of the spectrum.

top, is prepared on a glass or plastic support. Then the gel and support are immersed in a buffer solution, samples are loaded into the wells, and an electric field is applied across the gel. Negatively charged nucleic acids penetrate and move through the complex network of molecules that comprise the gel and toward the positive electrode or anode. A nucleic acid's migration rate depends on its total charge and shape (that is, its frictional resistance). Its migration rate is also influenced by its molecular mass because surface area, which affects frictional drag, increases with molecular mass. Smaller nucleic acids squeeze through the narrow, tortuous passages in the gel network more easily than larger molecules. Migration rate therefore increases as the molecular mass decreases.

DNA bands can be viewed by staining the gel with ethidium bromide (FIGURE 4.8), which forms a DNA • ethidium bromide complex that fluoresces under ultraviolet light. The sensitivity of detection is sufficiently great that samples with as little as 0.1 µg DNA are easily seen. Because ethidium bromide is carcinogenic, many workers prefer to work with commercially available safer substitutes. Conditions can be adjusted so the rate of migration depends on length in nucleotides for single-stranded nucleic acids or base pairs for double-stranded nucleic acids. Gel electrophoresis can also be used to separate topoisomers (**BOX 4.1 IN THE LAB: GEL ELECTROPHORESIS AND TOPOISOMER SEPARATION**).

Polyacrylamide gels can separate single-stranded DNA molecules that differ in size by just one nucleotide provided the strands are between 5 and 750 nucleotides long. Agarose gels have lower resolving power but separate DNA molecules that range in size from 200 base pair (bp) to about 50,000 bp. Empirical studies show the distance moved by DNA (D) during slab gel electrophoresis depends logarithmically on its length in base pairs, obeying the equation,

$$D = a - b \log bp$$

in which a and b are empirically determined constants that depend on the buffer, the gel concentration, and the temperature. This logarithmic relationship allows us to determine a DNA molecule's length in base pairs by comparing the distance it moves with distances moved by DNA standards of known length in base pairs under the same conditions (FIGURE 4.9).

Although quite effective in separating DNA molecules by size, slab gel electrophoresis is (1) slow (usually requiring hours to complete a run), (2) difficult to automate, and (3) does not easily provide quantitative data. An alternate means for separating DNA in an applied electric field, called **capillary gel electrophoresis**, does not suffer from these disadvantages (**BOX 4.2 IN THE LAB: CAPILLARY ELECTROPHORESIS**).

SDS-PAGE can be used to determine a polypeptide's molecular mass.

The molecular mass of native proteins cannot be determined by gel electrophoresis because proteins do not have a uniform charge across the entire molecule and proteins come in a variety of shapes

BOX 4.1: IN THE LAB

Gel Electrophoresis and Topoisomer Separation

Gel electrophoresis can separate topoisomers (FIGURE B4.1). Because superhelical molecules are compact, they move through the gel more rapidly than relaxed circles with the same mass. Gel electrophoresis therefore serves as a convenient method for detecting topoisomerase activity. The advantages of this method are that it requires very little DNA, is very sensitive, can be used to analyze several samples at one time, requires relatively inexpensive equipment, and is simple to perform.

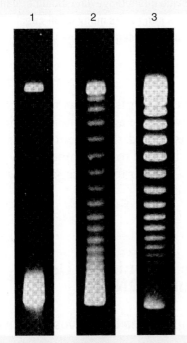

FIGURE B4.1 Agarose gel electrophoresis pattern of covalently closed circular SV40 (simian virus 40) DNA. The DNA was applied to the top of the gel. Lane 1 contains untreated negatively supercoiled native DNA (lower band). Lanes 2 and 3 contain the same DNA that was treated with a type I topoisomerase, which makes a single strand break in only one chain that relaxes negative supercoils by causing successive one unit increases in the linkage number ($\Delta Lk = +1$). The DNA in lanes 2 and 3 were treated for 5 and 30 minutes respectively. (Reproduced from Keller, W. 1975. *Proc Natl Acad Sci USA* 72:2550–2554. Photo courtesy of Walter Keller, University of Basel.)

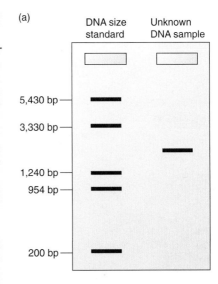

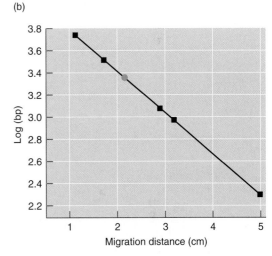

FIGURE 4.9 DNA size determination by gel electrophoresis. (a) A DNA size standard and a DNA of unknown size were placed in the left and right lanes, respectively. (b) A graph showing the migration distance in cm vs. log bp. The points shown in black squares are for the standard DNA and the point shown in a green circle is for the DNA of unknown size. (Modified from illustrations by Michael Blaber, Department of Biomedical Sciences, Florida State University.)

so there is no simple way to predict the migration rate. Proteins could be separated solely on the basis of size if they had a uniform charge and the shape factor could be eliminated. A technique known as **SDS-PAGE** (sodium dodecyl sulfate-polyacrylamide gel electrophoresis) does just that. The detergent SDS and the disulfide bond-breaking agent β-mercaptoethanol ($HSCH_2CH_2OH$) are added to the protein, causing all the polypeptide chains to be denatured, forming rod-shaped structures that are coated with negatively

> **BOX 4.2: IN THE LAB**
>
> **Capillary Electrophoresis**
>
> Separations in capillary electrophoresis are performed in very thin (20–100 μm) capillary tubes made of quartz, glass, or plastic filled with replaceable high-molecular-mass linear polymers. Because the capillary tubes are so thin, heat is rapidly dissipated. This heat dissipation allows separations in electrical fields (typically 100–500 V · cm^{-1}), which are about 10 times higher than those used in slab gel electrophoresis. Capillary electrophoresis thus allows high-resolution separations comparable with those obtained by slab gels in minutes rather than hours. Furthermore, capillary electrophoresis can be automated and provide quantitative data. Commercially available capillary electrophoresis systems can run as many as 96 capillaries at the same time, allowing high-throughput DNA sequence analysis.

charged SDS. The net effect is that, as in the case of DNA, all the proteins migrate toward the anode, the migration rate increases as chain length decreases, and the dependence is logarithmic. SDS-PAGE can therefore be used to determine a polypeptide's molecular mass by comparing its migration rate with the migration rates of polypeptide standards.

4.3 Enzymatic Techniques Used to Manipulate DNA

Nucleases are useful tools in DNA investigations.

Nucleases, enzymes that digest polynucleotides by cleaving the 5'→3' phosphodiester bonds that link neighboring nucleosides, are valuable tools for studying nucleic acids. Those that digest DNA are designated deoxyribonucleases (DNases), whereas those that digest RNA are designated ribonucleases (RNases). Some nucleases digest both kinds of nucleic acids. Some DNases are specific for double-stranded (ds) DNA, others for single-stranded (ss) DNA, and still others act on both kinds of DNA.

Nucleases also differ in the specificity with which they cut the nucleic acid. Nucleases that act within a strand are called **endonucleases**. Some endonucleases are specific and cleave only between particular bases. Nucleases that act only at the end of a nucleic acid, removing a single nucleotide at a time, are called **exonucleases**. Exonucleases that begin cutting at the 5'- or 3'-ends are designated 5'→3' or 3'→5' exonucleases, respectively. Although nucleases have many important biological functions, the present focus is on using them as laboratory tools. Some specific nucleases are described in TABLE 4.1 and others are described when relevant in the text.

TABLE 4.1 Properties of Selected Nucleases

Nuclease	Substrate	Site of Cleavage	Product
Pancreatic RNase	ssRNA in high salt; all RNA in low salt	endonuclease; after C or U	mono- or oligonucleotides with a 3'-P
T1 RNase	ssRNA in high salt; all RNA in low salt	endonuclease; after G	mono- or oligonucleotides with a 3'-P
Pancreatic DNase I	ss or dsDNA	endonuclease	oligonucleotides
Phosphodiesterase venom	RNA or DNA	exonuclease from the 3' end	5'-P mononucleotides
Phosphodiesterase spleen	RNA or DNA	exonuclease from the 5' end	3'-P mononucleotides
Micrococcal nuclease	DNA (or RNA) ss or ds (prefers ssDNA to dsDNA; ssRNA to ds RNA)	endonuclease and exonuclease; at AT or AU rich regions, requires Ca^{2+}	mononucleotides and oligonucleotides with 3'-P ends
S1 nuclease	ssDNA (or ssRNA. S1 nuclease is 5× more active on ssDNA than on ssRNA)	endonuclease	5'-P mononucleotides
Exonuclease I E. coli	ssDNA	3'→5' exonuclease	5'-P mononucleotides plus a terminal dinucleotide
Exonuclease III E. coli	dsDNA	3'→5' exonuclease	5'-P mononucleotides
Exonuclease VII E. coli	ssDNA	3'→5' exonuclease and 5'→3' exonuclease	5'-P mononucleotides

ss = single-stranded; ds = double-stranded.

A specific example, the **S1 endonuclease** isolated from *Aspergillus oryzae*, illustrates how we use nucleases to investigate nucleic acid structure. The S1 endonuclease acts exclusively on single-stranded polynucleotides or on single-stranded regions of double-stranded nucleic acids. It differs from other single-strand–specific enzymes in that the single-stranded region can be as small as one or two bases. Negatively supercoiled DNA contains single-stranded bubbles resulting from localized transient melting. Supercoiled DNA can be cleaved by S1 nuclease because of these regions. In fact, this nuclease can be used to distinguish supercoiled from both nonsupercoiled covalent circles and nicked circular DNA, both of which are resistant to the enzyme. S1 nuclease makes a double-strand break because it acts on both single-stranded branches of the bubble.

Restriction endonucleases cleave within specific nucleotide sequences in DNA.

Molecular biologists recognized they could use nucleases that cut DNA at specific sites to study nucleic acids in the same way protein chemists use trypsin and other specific proteases to cut polypeptide chains. The problem of finding nucleases with sufficient sequence specificity to be useful was solved when investigators discovered a class of nucleases that cleave bacterial virus DNA molecules that enter a bacterial cell, impeding viral replication. These nucleases, called **restriction endonucleases** because they block or restrict viral replication, act only on DNA with specific recognition sequences and only when the recognition sequences are not modified. Host DNA is protected because it has methyl groups attached to specific bases within the recognition sequence.

Four major types of restriction endonucleases are known. We examine just one of these, the type II restriction endonucleases, which are ideally suited for manipulating and studying DNA because they recognize and cleave specific sequences that are 4 to 8 bp long.

Hamilton O. Smith was the first to isolate and characterize a type II restriction endonuclease. The enzyme, isolated from *Haemophilus influenzae* and called HindII, recognizes the set of sequences,

$$5'\cdots\text{GTPy}\downarrow\text{PuAC}\cdots 3'$$
$$3'\cdots\text{CAPu}\uparrow\text{PyTG}\cdots 5'$$

where Py and Pu represent pyrimidine and purine, respectively. The arrows indicate sites at which each strand is cut.

The discovery of HindII motivated investigators to seek other endonucleases that cut within specific nucleotide sequences. The search has been rewarded. More than 2,500 different type II enzymes have been identified, and several of these have been sequenced and biochemically characterized. In many cases two or more different restriction endonucleases recognize the same sequence. Different restriction endonucleases that recognize the same nucleotide sequence and that cleave it in the same position are called **isoschizomers.** Longer recognition sequences are statistically less likely to appear within a DNA molecule than are shorter ones. Restriction endonucleases that recognize 8-bp sequences therefore make many fewer cuts in a DNA molecule than those that recognize 4-bp sequences.

Type II restriction endonuclease recognition sites are inverted repeat sequences (**FIGURE 4.10**). When rotated 180 degrees about the central point in the plane of the page, the recognition site reads exactly as it did before the rotation. Molecular biologists call a sequence with such a dyad axis of symmetry a palindrome. Type II restriction endonucleases cut each strand within the palindrome sequence one time. Some type II enzymes cut each strand at the axis of symmetry to generate **flush** or **blunt ends** (Figure 4.10a). Others make staggered cuts (cuts that are symmetrically placed around the axis of symmetry) to generate **cohesive** or **sticky ends** (Figure 4.10b). In either case phosphodiester bond cleavage generates 3'-hydroxyl and 5'-phosphate ends. **TABLE 4.2** lists some type II restriction endonucleases along with their recognition sites.

Important insights concerning enzyme–DNA interactions have been gained by examining the way restriction endonucleases act on DNA. To be effective in protecting the host cell from viral attack, the host's restriction endonuclease has to cut the viral DNA before its modification enzyme can methylate the DNA. A restriction endonuclease must therefore be able to reach its target site quickly. It does

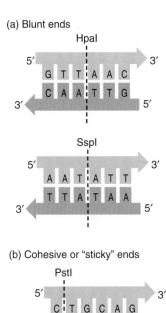

FIGURE 4.10 Restriction endonuclease generated ends. (a) Some restriction endonucleases such as HpaI and SspI cut on the line of symmetry to produce blunt ends. (b) Other restriction endonucleases such as PstI cut on either side of the line of symmetry to produce cohesive ends (also called "sticky" or staggered ends). (Adapted from an illustration by Maria Price Raposa, Carolina Biological Supply Company.)

TABLE 4.2 Sequence Specificity of Some Restriction Endonucleases

Organism of Origin	Restriction Endonuclease	Recognition Sequence
Arthrobacter luteus	AluI	5'...A G↓CT...3'
Anebaena variabilis	Ava I	5'...C↓PyCGPuG...3'[a]
Bacillus amyloliquefaciens H	Bam HI	5'...G↓GATCC...3'
Bacillus globigii	Bgl HI	5'...A↓GATCT...3'
Escherichia coli RY13	Eco RI	5'...G↓AATTC...3'
Escherichia coli J62 pLG74	Eco RV	5'...GAT↓ATC...3'
Haemophilus aegyptius	Hae II	5'...PuGCGC↓Py...3'
Haemophilus aegyptius	Hae III	5'...GG↓CC...3'
Haemophilus haemolyticus	Hha I	5'...GCG↓C...3'
Haemophilus influenzae Rd	Hind II	5'...GTPy↓PuAC...3'
Haemophilus influenzae Rd	Hind III	5'...A↓AGCTT...3'
Haemophilus parainfluenzae	Hpa I	5'...GTT↓AAC...3'
Haemophilus parainfluenzae	Hpa II	5'...C↓CGG...3'
Klebsiella pneumoniae	Kpn I	5'...GGTAC↓C...3'
Moraxella bovis	Mbo I	5'...↓GATC...3'
Nocardia otitidis caviarum	Not I	5'...GC↓GGCCGC...3'
Providencia stuartii	Pst I	5'...CTGCA↓G...3'
Serratia marcescens	Sma I	5'...CCC↓GGG...3'
Streptomyces stanford	Sst I	5'...GAGCT↓C...3'
Xanthomonas malvacearum	XmaI	5'...C↓CCGGG...3'

[a]Py, pyrimidine; Pu, purine.

so by first binding to the foreign DNA in a nonspecific fashion and then scanning the DNA for recognition sequences. This linear diffusion along the DNA molecule allows the restriction endonuclease to find and cleave all recognition sites rapidly and efficiently. The alternative possibility, a three-dimensional diffusion process in which the endonuclease alternately associates with and dissociates from DNA until it finally binds to the recognition sequence, would be too slow to protect the host from viral replication.

Restriction endonucleases can be used to construct a restriction map of a DNA molecule.

Because a particular type II restriction endonuclease recognizes a unique sequence, it can only make a limited number of cuts in an organism's DNA. For example, an endonuclease that recognizes a 6-bp sequence will cut a typical bacterial DNA molecule, which contains roughly 3×10^6 bp, into a few hundred to a few thousand fragments. Smaller DNA molecules, such as phage or plasmid DNA molecules, may have fewer than 10 sites of cutting (frequently, one or two and, often, none). Because of the specificity just mentioned, a particular restriction endonuclease generates a unique family of fragments from a particular DNA molecule. Another restriction endonuclease will generate a different family of fragments from the same DNA molecule.

The family of fragments generated by a single restriction endonuclease can usually be resolved by agarose gel electrophoresis. A fragment's length is determined by comparing its mobility with the mobility of fragments of known length run concurrently. A **restriction map** shows the restriction endonuclease cutting sites along a DNA molecule. One approach to constructing a restriction map is to cut the DNA sample in three ways: with one restriction endonuclease, with a second restriction endonuclease, and with both enzymes (a double digest). The fragments generated are resolved by gel electrophoresis, which also reveals the number of fragments and their lengths in kilobase pair (kbp). The fragmentation pattern and fragment lengths are used to deduce the order of restriction endonuclease cut sites and to assign intervals between them. Because actual lengths are determined by comparing fragment mobilities with the mobilities of standards of known length, all intervals in the map are additive.

A simple example will help to illustrate this approach (FIGURE 4.11). Samples of a 6-kbp linear duplex are cut by BamHI, EcoRV, or both endonucleases into fragments, which are separated by gel electrophoresis as shown in Figure 4.11a. EcoRV cleaves at one site to produce two fragments (2.3 and 3.7 kbp). Inspection of restriction fragments produced by the BamHI and EcoRV mixture reveals that BamHI cuts the 3.7-kbp fragment into two pieces (1.2 and 2.5 kbp) and the 2.3-kbp fragment into two pieces (0.5 and 1.8 kbp). BamHI cleaves at two sites to produce three fragments (4.3 kbp, 1.2 kbp, and 0.5 kbp). A comparison of the fragments produced by BamHI with those

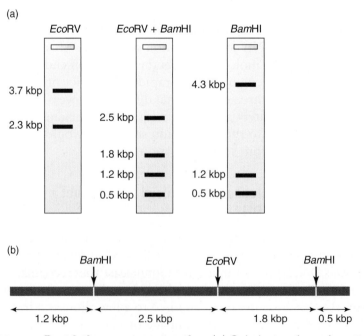

FIGURE 4.11 Restriction map construction. (a) Gel electrophoresis pattern of fragments produced by digesting the same DNA molecule with the restriction endonuclease(s) shown at the top of each lane. Fragment sizes are given in kilobase pairs (kbp). (b) The restriction map deduced from the fragment sizes indicated by gel electrophoresis. Specific restriction site(s) for each enzyme are shown by the vertical arrows.

produced by the double digest indicates that EcoRV cuts the 4.3-kbp fragment into a 2.5-kbp and a 1.8-kbp fragment. This information allows us to determine how the 3.7- and 2.3-kbp fragments produced by EcoRV are joined, yielding the restriction map shown in Figure 4.11b.

Although the principles are simple, the approach does have practical limitations. Some fragments are so small they are difficult to see or they move off the gel. Under these conditions the sum of the fragment sizes will not add up to the size of the original DNA. Two (or more) fragments may be the same size or have such similar sizes the gel does not resolve them. When this occurs one of the bands in the agarose gel will appear to have twice the fluorescence intensity as neighboring bands of similar size in the same gel. Finally, interpretation of results becomes progressively more difficult as the number of bands increase. Results become difficult, if not impossible, to interpret when each restriction endonuclease produces more than seven fragments.

In 1976 Hamilton Smith and Max L. Birnstiel devised a simple method for DNA restriction site mapping. This method reduces the computational complexity by starting with a DNA molecule with a radioactive label at just one end. It exploits the ability of the enzyme polynucleotide kinase to transfer the γ-phosphoryl group from ATP to the free hydroxyl group at the 5′-end of a DNA or RNA molecule. Labeling DNA at one end requires time and effort but can be achieved by the approach shown in FIGURE 4.12. The DNA is treated with phosphatase to remove phosphate groups from the 5′-ends, incubated with polynucleotide kinase and [γ-^{32}P]ATP to add labeled phosphate groups

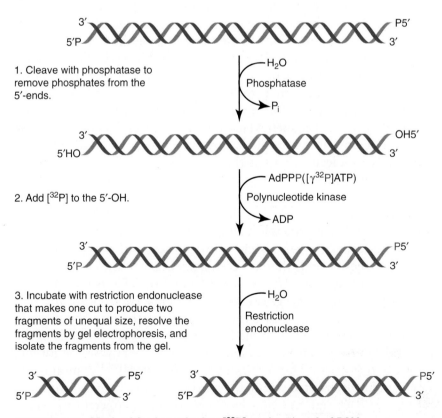

FIGURE 4.12 **Method for introducing [^{32}P] at the 5′-end of DNA.**

4.3 Enzymatic Techniques Used to Manipulate DNA

to the 5′-ends, and then cut with a restriction endonuclease to produce two DNA fragments of unequal size (each with a single labeled end) that are purified by gel electrophoresis.

The restriction map of each labeled fragment can then be determined by placing the labeled fragment in a tube, partially digesting it with a new restriction endonuclease, and then resolving the fragments by gel electrophoresis. A photographic film is then placed on top of the gel and stored in the dark. The β particles released by the ^{32}P reduce the silver ions in the photographic film, producing dark bands over each of the fragments. This autoradiographic detection technique reveals patterns that are easy to interpret, with each lane containing bands of fragments that start at the labeled end and extend to a site to be mapped.

The order of the sites is obtained by reading the bands on the gel from the bottom to the top, i.e., from the smallest fragment (site closest to the radioactively labeled end) to the largest fragment (site furthest from the labeled end). Intervals between cutting sites correspond to differences between fragment sizes. Although simple in principle this method has several shortcomings. As fragments get longer, measurements of chain length become less precise. It is also difficult to find conditions that provide the desired level of cleavage. Furthermore, the rate of cutting is affected by the nucleotides around the cleavage site so that some cutting sites will be cleaved to a greater extent than others.

DNA fragments can be inserted into plasmid DNA vectors.

It is often desirable to generate large amounts of a particular DNA fragment. Although one might guess that replication of a DNA fragment could be accomplished by first introducing the fragment into a cell and then using the cell's replication machinery to do the work of replicating the fragment, this approach rarely works. The reason for the high rate of failure is that DNA molecules must have specific replication sequences before a cell's replication machinery can act on them. Most DNA fragments lack these sequences. There is a simple way to circumvent the problem, however. Insert the DNA fragment into a DNA molecule that has the required replication sequence and then introduce the recombinant DNA molecule into the cell.

Plasmids are excellent carriers or **vectors** for DNA fragments. As illustrated in FIGURE 4.13, the insertion of the DNA fragment into a plasmid is in principle a simple process. A plasmid bearing an antibiotic resistance gene is cut with a restriction endonuclease to produce cohesive ends that are complementary to the cohesive ends of the DNA fragment. *The cut must be at a site that is outside both the antibiotic resistance gene and the sequence required for plasmid DNA replication.* The DNA fragment is inserted into the plasmid DNA by joining the ends of the DNA fragment with the ends of the plasmid DNA. An enzyme called DNA ligase catalyzes the joining or ligation process. The **recombinant plasmid**, which contains the DNA fragment inserted within the plasmid vector, then is introduced into living bacterial cells that have been made competent to accept plasmid DNA.

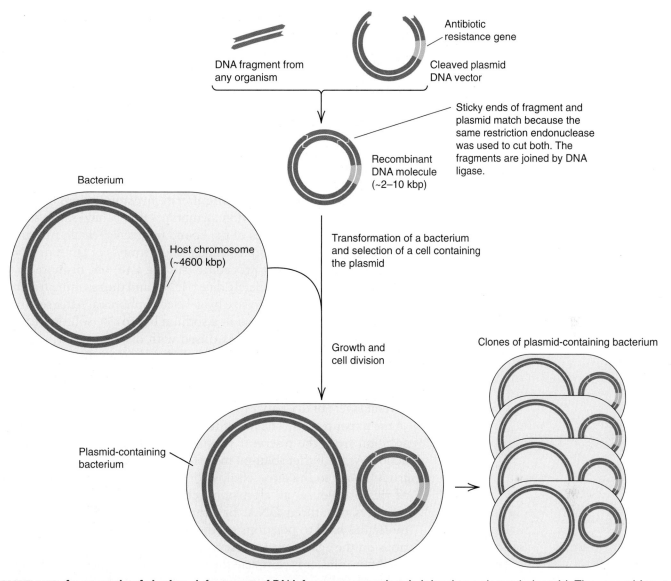

FIGURE 4.13 An example of cloning. A fragment of DNA from any organism is joined to a cleaved plasmid. The recombinant plasmid is then used to transform a bacterial cell, where the recombinant plasmid is replicated and transmitted to the progeny bacteria. The bacterial host chromosome is not drawn to scale. It is typically about 1,000 times larger than the plasmid.

One method for making *Escherichia coli* cells **competent** to take up DNA from the surrounding medium is to incubate the cells with 100 mM calcium chloride solution at 4°C for 30 minutes, centrifuge, resuspend the bacterial pellet in a 75 mM calcium chloride solution containing 15% glycerol, and then freeze 0.2 mL cell samples. The cells are transformed by thawing the samples on ice, adding recombinant DNA, incubating the suspension at 42°C for 90 seconds, and then spreading the suspension on an agar plate containing the selective antibiotic. Each cell containing a recombinant plasmid produces a colony or clone consisting of millions of progeny bacteria with the same recombinant plasmid. For this reason the recombinant plasmid is said to be cloned, and the process is called **cloning**. Other types of autonomously replicating DNA molecules such as viral DNA are also used as cloning vectors.

4.3 Enzymatic Techniques Used to Manipulate DNA

The Southern blot procedure is used to detect specific DNA fragments.

The availability of restriction endonucleases that cleave DNA into unique and fairly small fragments makes it possible to detect small segments within large DNA molecules by taking advantage of the fact that complementary polynucleotide strands (DNA or RNA) can anneal to form double-stranded molecules. In 1975 Edwin M. Southern invented a technique that allows investigators to hybridize a specific polynucleotide (DNA or RNA) to a large number of particular DNA segments without the necessity of purifying individual DNA fragments. This technique, called the **Southern transfer** or **Southern blot procedure** after its inventor, exploits the fact that very thin nitrocellulose or nylon membranes tightly bind single-stranded DNA fragments. The bases of the bound fragments remain free to form hydrogen bonds with complementary single strands of DNA or RNA.

In the Southern blot procedure (FIGURE 4.14) one or more restriction endonucleases completely digest DNA and the resulting fragments are separated according to size by gel electrophoresis. After separation is complete the gel is soaked in a sodium hydroxide solution to denature the DNA. Then the gel is rinsed with distilled water, soaked in buffer solution to adjust the pH, and placed on a sponge that is itself in a reservoir of buffer solution. The top surface of the gel is covered with a nitrocellulose or nylon membrane, which in turn is covered with several layers of dry paper towels, and a heavy weight is placed on top of the paper towels. The paper towels act as a blotter, drawing buffer solution from the reservoir through the various layers by capillary action. As the buffer solution moves up, it carries single-stranded DNA from the gel to the nitrocellulose or nylon membrane. In a variation of this procedure, an electrophoretic process called **electroblotting** transfers denatured DNA. In either case the single-stranded DNA molecules bind to positions on the membrane identical to their positions on the agarose gel, preserving the band pattern.

After transfer is complete the setup is disassembled, and the single-stranded DNA is fixed permanently to the nitrocellulose membrane by baking at 80°C in a vacuum or to the nylon membrane by ultraviolet light–induced cross-linking. Then the membrane is incubated in a solution containing bovine serum albumin, polysucrose, polyvinylpyrolidine, and denatured salmon sperm DNA to eliminate the membrane's inherent ability to bind single-stranded DNA. Denatured DNA that was transferred to the membrane from the electrophoresis gel, however, retains its ability to bind complementary single-stranded DNA or RNA molecules.

DNA sequences of interest bound to the membrane are identified by using a complementary single-stranded DNA or RNA probe with a detectable label. The nitrocellulose or nylon membrane is placed in a buffer solution containing the labeled probe and incubated for several hours at a suitable renaturation temperature to permit the probe to hybridize to its complementary sequence in the DNA bound to the nitrocellulose or nylon membrane. Then the membrane is washed with buffer to remove unbound probe and dried. Autoradiography is used to detect a radioactive probe. DNA fragments complementary to the probe appear as black or stained bands.

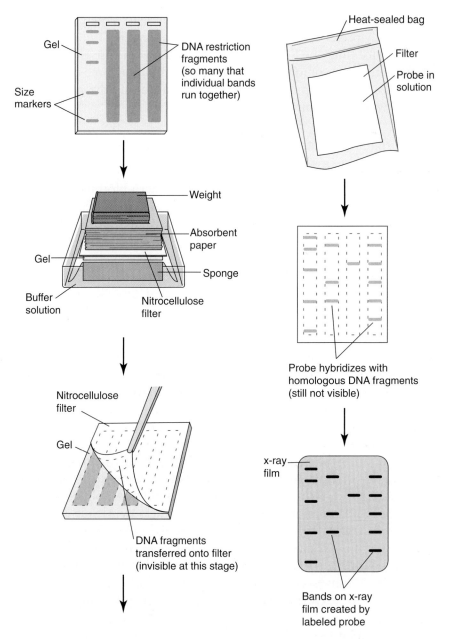

FIGURE 4.14 **Southern blot analysis: an experimental method for identifying a specific DNA fragment in a gel.** DNA is cleaved by one or more restriction endonucleases and the resulting fragments are separated by gel electrophoresis, the resolved DNA fragments are denatured by soaking the gel in a basic solution and denatured fragments are blotted onto a nitrocellulose filter, the nitrocellulose filter is exposed to a probe with a radioactive or other detectable label, and when a radioactive probe is used, the filter is exposed to photographic film and the film is developed.

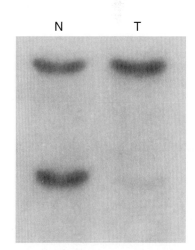

FIGURE 4.15 **Southern blot of genomic DNA from normal tissue (N) and from a tumor (T) from a patient with breast cancer.** Two bands are visible in the normal breast tissue but only the upper band is present in the tumor cells. The faint lower band represents contribution from non-tumor cells in the tumor sample. (Reproduced from Tougas L., et al. 1996. *Clin Invest Med* 19:222–230. Photo courtesy of Serge Jothy, University of Toronto.)

The Southern blot procedure is a very versatile and powerful tool that detects changes in DNA that alter restriction cutting sites or the lengths of segments between restriction cutting sites. It has become an important clinical laboratory tool for identifying genetic problems. For example, the Southern blot in FIGURE 4.15 shows that a band present in normal breast tissue is missing from tumor cells. Today, investigators usually prefer to use a DNA or RNA probe with a fluorescent tag when performing the Southern blot procedure because they can see fluorescent bands of interest without performing the time-consuming autoradiography step and can work without taking the precautions needed when working with radioisotopes.

Northern and Western blotting are used to detect specific RNA and polypeptide molecules, respectively.

In a variant of Southern blotting, called **Northern blotting** (a play on words), RNA molecules are separated by gel electrophoresis and then transferred to nylon or nitrocellulose membranes. Labeled single-stranded RNA or DNA probes can hybridize with the bound RNA. Still another variant is **Western blotting**, in which proteins are separated by gel electrophoresis and then transferred to a nitrocellulose or polyvinyl membrane. Bound protein is detected with labeled or tagged antibody probes.

DNA polymerase I requires a template-primer.

DNA polymerases, enzymes that catalyze DNA synthesis, are among the most important tools in the molecular biologist's toolkit. Several different DNA polymerases are commonly used in the modern molecular biology laboratory. Polymerase selection is determined by the demands of the experiment. For instance, an investigator may decide to use a heat-resistant DNA polymerase isolated from a hyperthermophilic archaeon when the experiment requires the polymerase to retain full activity after exposure to heat. A great deal of our knowledge about how DNA polymerases work derives from studies of DNA polymerase I, an enzyme that was first detected in *E. coli* extracts by Arthur Kornberg and coworkers in 1956. The Roman numeral I distinguishes Kornberg's enzyme from other DNA polymerases that were subsequently discovered to be present in *E. coli*. DNA polymerase I, like all other DNA polymerases, converts deoxyribonucleoside triphosphates into DNA in the presence of preformed DNA according to the following reaction:

$$\left\{ \begin{array}{c} \text{dATP} \\ + \\ \text{dGTP} \\ + \\ \text{dCTP} \\ + \\ \text{dTTP} \end{array} \right\} \xrightleftharpoons[\text{DNA polymerase, Mg}^{2+}]{\text{DNA}} \text{DNA} + \text{pyrophosphate}$$

The enzyme is assayed by adding it to a mixture that also contains the four deoxyribonucleoside triphosphates, one of which is radioactive, a DNA template-primer (see below), and magnesium ions. The reaction is stopped by adding acid, which also precipitates newly formed radioactive DNA. Then the acid-insoluble DNA is separated from the acid-soluble deoxyribonucleoside triphosphates by centrifugation or filtration and its radioactivity is determined in a liquid

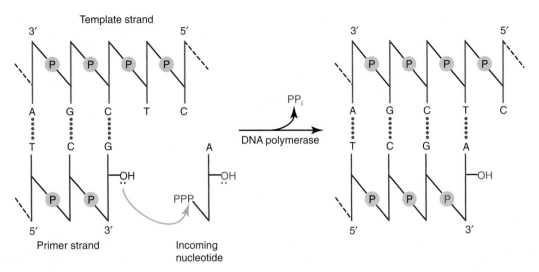

FIGURE 4.16 DNA polymerase I: polymerase function. DNA polymerase I uses the template strand to determine the next nucleotide that it adds to the primer strand.

scintillation counter. All four deoxyribonucleotides must be present for the reaction to take place. The source of the preformed double-stranded DNA does not matter; prokaryotic and eukaryotic DNA work equally well. The two strands of the preformed DNA have different functions at the site of DNA synthesis (FIGURE 4.16). One strand, the **primer strand**, is the site of attachment for incoming deoxyribonucleotides. The other strand, the **template strand**, determines the order of attachment to the primer strand according to Watson-Crick base pairing rules. DNA polymerase I has binding sites for the primer strand, the template strand, and the incoming deoxyribonucleotide. The enzyme catalyzes the nucleophilic displacement of a pyrophosphate group, creating a phosphodiester bond between the 3′-end of the growing primer strand and the 5′-end of the incoming deoxyribonucleotide. Hence, the primer strand grows in a 5′→3′ direction, a characteristic of all known DNA and RNA polymerases. Inorganic pyrophosphate hydrolysis drives the reaction to completion inside the cell.

DNA polymerase I has both 3′→5′ and 5′→3′ exonuclease activities.

Early preparations of *E. coli* DNA polymerase I contained exonuclease activity, which Kornberg and coworkers expected to remove by further enzyme purification. Contrary to their expectation, highly purified DNA polymerase I, consisting of a single polypeptide with 928 amino acid residues (molecular mass = 103 kDa), has both 3′→5′ and 5′→3′ exonuclease activities. The discovery that a single enzyme has the ability to both synthesize and degrade DNA was mystifying. Further studies by Kornberg's group, however, showed the two-exonuclease activities, in fact, do make important contributions to DNA synthesis. The 3′→5′ exonuclease, which catalyzes the sequential hydrolysis of one 5′-deoxynucleotide at a time from the 3′-end, has a proofreading function (FIGURE 4.17).

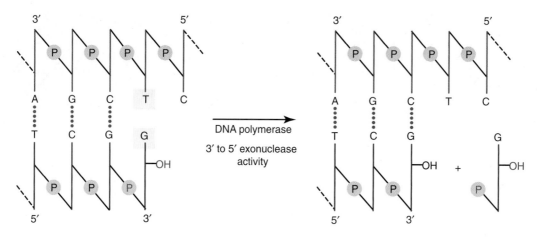

FIGURE 4.17 DNA polymerase I: proofreading function. Although rare, DNA polymerase I does misinsert nucleotides. The 3′→5′ exonuclease activity allows DNA polymerase to remove a misinserted deoxyribonucleotide before the next deoxyribonucleotide is added to the growing primer chain.

Although DNA polymerase I has great specificity, it does make occasional errors either by misinserting a nucleotide or by misaligning the template-primer. Either type of error eventually results in base pair substitution, addition, or deletion. Although errors are rare, occurring at a frequency of about one nucleotide for every 10^4 nucleotides added, failure to correct an error is harmful to the cell. The 3′→5′ exonuclease activity removes the misinsertions before the next deoxyribonucleotide is added to the growing chain. The 5′→3′ exonuclease, which can remove either a mononucleotide or a short oligonucleotide from the 5′-end, plays an editing role in DNA synthesis and, as described below, serves as a useful tool for preparing labeled DNA.

DNA polymerase I is a multifunctional polypeptide (a single polypeptide with more than one catalytic activity) with three domains. Domain 1, which includes the first 325 or so residues at the amino or N-terminus, has 5′→3′ exonuclease activity. Domain 2, consisting of the next 200 or so residues, has 3′→5′ exonuclease activity. Domain 3, which includes the remaining residues at the carboxyl or C-terminus, has polymerase activity. In 1970 Hans Klenow demonstrated that subtilisin (a proteolytic enzyme produced by *Bacillus subtilis*) cleaves DNA polymerase I into two fragments of unequal length (**FIGURE 4.18**). The smaller fragment (molecular mass = 35 kDa) containing domain 1 has 5′→3′ exonuclease activity. The larger one (molecular mass = 68 kDa), known as the **Klenow fragment**, contains domain 2 (3′→5′ exonuclease activity) and domain 3 (polymerase activity).

Thomas Steitz and coworkers determined the three-dimensional structure of the Klenow fragment by x-ray crystallography. The polymerase domain, which is made of three subdomains, resembles a partially opened right hand (**FIGURE 4.19**). The polymerase catalytic site is located in the palm subdomain. As shown in the schematic diagram in **FIGURE 4.20a**, the thumb subdomain appears to contact the minor groove of the primer-template duplex, whereas the finger subdomain binds the uncopied template strand. Three-acidic residues (Asp-705, Asp-882, and Glu-883) form a carboxylate triad (Figure 4.19), which is essential for

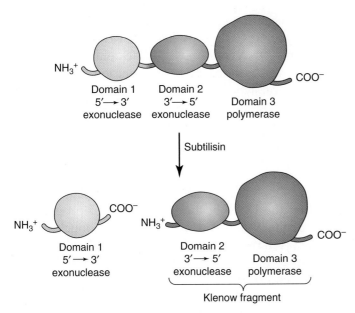

FIGURE 4.18 DNA polymerase I. Generation of a Klenow fragment.

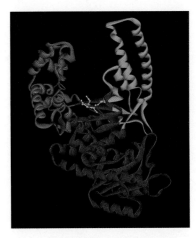

FIGURE 4.19 Ribbon structure of the polymerase domain of the Klenow fragment. The polymerase domain resembles a partially opened right hand. The subdomains corresponding to the thumb, fingers, and palm are colored green, purple, and red, respectively. The carboxylate triad, Asp-705, Asp-882, and Glu-883, at the polymerase active site is shown in yellow. (Structure from the Protein Data Bank ID: 1KFD. Beese, L. S., Friedman, J. M., and Steitz, T. A. 1993. *Biochemistry* 32:14095–14101. Prepared by B. E. Tropp.)

polymerase activity. Two magnesium ions appear to be integral parts of the catalytic site.

Kinetic studies suggest the polymerase stalls when an incorrect nucleotide is attached to the growing end of the primer chain, allowing time for the 3'→5' exonuclease to remove the incorrect nucleotide (Figure 4.20b). X-ray diffraction studies indicate the polymerase and 3'→5' exonuclease catalytic sites are separated by about 3.5 nm. Proofreading can occur by an intra- or

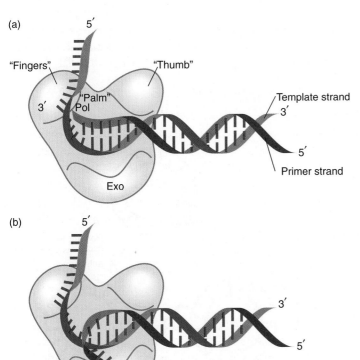

FIGURE 4.20 Schematic of DNA polymerase I in the (a) polymerase mode and (b) proofreading mode. (a) DNA polymerase I in the polymerase mode. The thumb subdomain contacts the minor groove of the template-primer duplex, the finger subdomain binds the uncopied template strand, and the 3'-OH end of the primer strand binds to the polymerase active site. (b) DNA polymerase I in the 3'→5'-exonuclease mode. The thumb subdomain still contacts the minor groove of the primer-template duplex and the finger subdomain still binds the uncopied template strand. However, the 3'-OH end of the primer strand with the misinserted nucleotide binds to the 3'→5' exonuclease catalytic site active site. (Adapted from Joyce, C. M., and Steitz, T. A. 1995. *J Bacteriol* 177:6321–6329.)

4.3 Enzymatic Techniques Used to Manipulate DNA

intermolecular mechanism. In the former case the 3'-end moves from the polymerase catalytic site to the 3'→5' exonuclease catalytic site, whereas in the latter enzyme–DNA complex dissociates and then a different DNA polymerase I removes the mispaired base. A zinc ion and a magnesium ion at the exonuclease catalytic site appear to position a water molecule for a nucleophilic attack on the 5'-phosphate of the mismatched nucleotide.

DNA polymerase I can catalyze nick translation.

As shown in FIGURE 4.21, DNA polymerase I has the remarkable ability to synthesize DNA at a single-strand break (nick). This synthesis requires melting the DNA beyond the nick and progressive strand displacement of the 5'-end. When the 5'→3' exonuclease activity is low or missing (as in the Klenow fragment), DNA synthesis proceeds with strand displacement. When 5'→3' exonuclease activity is present, however, the 5'-end of the displaced strand is digested and the nick moves along the molecule in the direction of synthesis in a process known as **nick translation**. Using radioactive deoxyribonucleoside triphosphates during nick translation converts unlabeled DNA into radioactive DNA with the same nucleotide sequence. Hence, nick translation provides a convenient means for preparing radioactive

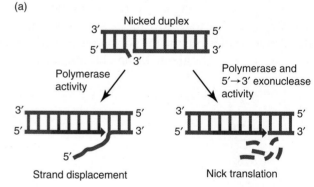
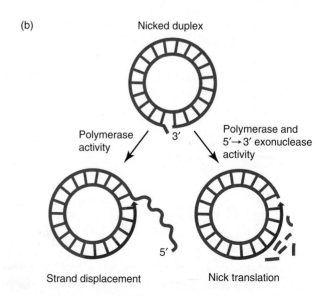

FIGURE 4.21 Strand displacement and nick translation on linear and circular molecules. In nick translation one nucleotide is removed by DNA polymerase I's 5'→3' exonuclease activity for each nucleotide added by the enzyme's polymerase activity. The growing strand is shown in red. (a) Strand displacement and nick translation on nicked linear DNA duplex. (b) Strand displacement and nick translation on nicked circular DNA duplex.

DNA probes for Southern blotting. The same approach can be used to prepare DNA with fluorescent markers. Although nick translation allows us to synthesize DNA with a radioactive or fluorescent tag, it does not lead to the net increase in total DNA that is present.

The polymerase chain reaction (PCR) is used to amplify DNA.

It seems reasonable to suppose that DNA polymerase I should be able to catalyze the synthesis of many copies of a specific double-stranded DNA fragment. It was not until 1983, however, almost three decades after the discovery of DNA polymerase I, that a satisfactory method was finally devised to do so. The method, which was invented by Kary B. Mullis, called the **polymerase chain reaction (PCR)**, is shown in FIGURE 4.22.

The linear duplex to be amplified is heat denatured. Then an oligonucleotide primer (about 20–30 nucleotides in length) is annealed to each of the denatured single strands. Each primer oligonucleotide sequence is selected so it anneals to a sequence in the outer region of the DNA segment to be amplified (the target sequence). The primers are oriented with their 3'-ends directed toward each other so that chain extension copies the region between them. Then DNA polymerase I is added to extend each primer until it reaches the end of the template. The new strands have defined 5'-ends (the 5'-ends of the oligonucleotide primers). This completes cycle 1 of PCR amplification. The same three-step process of (1) denaturing the linear duplex, (2) annealing the primer, and (3) extending DNA primer with DNA polymerase I is repeated in cycle 2 and each of the subsequent cycles. The strands synthesized during cycle 1 (and each succeeding cycle) serve as templates for the next cycle. Repeated DNA polymerase I additions at each cycle can be avoided by using a thermostable DNA polymerase such as the Pfu or Vent DNA polymerases, which are isolated from the hyperthermophilic archaea *Pyrococcus furiosus* and *Thermococcus litoralis* respectively.

PCR machines, known as **thermal cyclers,** which are programmed to shift their temperature up and down during different stages of the cycle, take advantage of thermostable DNA polymerase to allow PCR to be automated. In theory, amplification is exponential so that after n cycles the amplification yield would be 2^n. A 30-cycle amplification therefore would be expected to produce about 10^9 copies of the target segment. This calculation assumes that each cycle proceeds with a 100% efficiency of amplification. Because the efficiency of amplification is estimated to be 60% to 85%, actual amplification yields are lower than the calculated value. Nevertheless, extraordinary amplifications are possible in just a couple of hours because each cycle requires only 4 to 6 minutes.

Among the many important practical applications of PCR are the following: (1) amplification of a segment of a large DNA molecule for subsequent use in genetic engineering, (2) rapid detection of pathogenic bacteria and viruses, (3) detection of inborn errors of metabolism, and (4) detection of tumors. The popular press has described many situations in which PCR has been used in criminal investigations to amplify DNA from saliva, blood, sperm cells, or even a single hair. The amplified DNA is then characterized by restriction endonuclease digestion followed by gel electrophoresis. Amplified DNA can also be sequenced (see below).

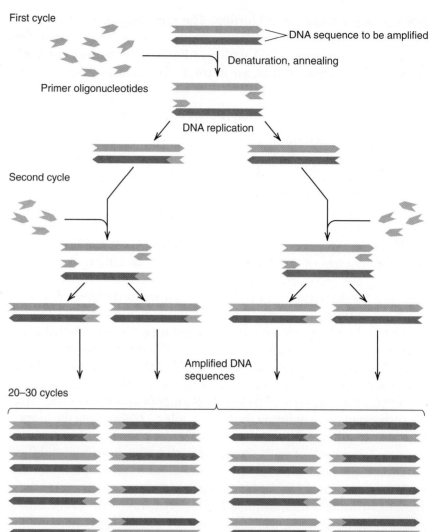

FIGURE 4.22 Schematic for the polymerase chain reaction (PCR), a method for amplifying a specific target DNA sequence. The linear duplex to be amplified (blue) is heat denatured. Then an oligonucleotide primer (orange) is annealed to each of the denatured single strands. DNA polymerase is added to extend each primer. The new strands (red) have defined 5'-ends (the 5'-ends of the oligonucleotide primers). This completes cycle 1 of PCR amplification. The same three-step process of (1) denaturing the linear duplex, (2) annealing the primer, and (3) extending DNA primer with DNA polymerase I is repeated in cycle 2 and each of the subsequent cycles.

Site-directed mutagenesis can be used to introduce a specific base change within a gene.

Once a gene has been cloned and sequenced (see below) it is possible to introduce specific base changes anywhere within the gene. The technique for introducing the specific base changes, called **site-directed mutagenesis**, first conceived by Michael Smith in the mid-1970s, is a very powerful tool for studying how specific amino acid changes influence protein function. FIGURE 4.23 is a simplified schematic for using site-directed mutagenesis to introduce a specific mutation in a gene. A primer with the desired base change(s) is annealed

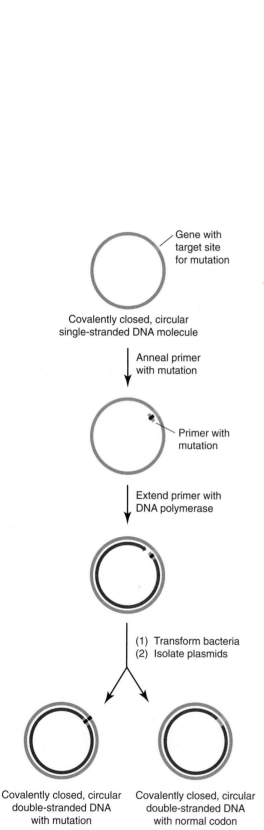

FIGURE 4.23 Schematic for introducing a mutation into a gene by site-directed mutagenesis.

CHAPTER 4 MOLECULAR BIOLOGY TECHNOLOGY

to a closed covalent, circular single-stranded DNA molecule. Then DNA polymerase is added to produce a nicked circular strand. The nicked circle bearing the mutation is introduced into competent bacterial cells, which repair the nick in the mutated plasmid, and the cloned mutated plasmid can be isolated from the transformed bacterial strain.

The chain termination method for sequencing DNA uses dideoxynucleotides to interrupt DNA synthesis.

Frederick Sanger recognized that DNA polymerase's ability to add nucleotides to a primer under the direction of a template could be used to sequence DNA. The Sanger method (also called the **chain termination method**) uses small amounts of dideoxynucleoside triphosphates to cause random termination of primer chain extension. Dideoxynucleoside triphosphates are deoxynucleoside triphosphate analogs that have a hydrogen atom attached to C-3′ in place of a hydroxyl group (**FIGURE 4.24**). Once DNA polymerase has transferred a dideoxynucleotide to the 3′-end of a growing primer chain, further chain extension is impossible because the primer chain must have a 3′-hydroxyl group for the next deoxynucleotide to be attached.

The major features of the chain termination method are summarized in **FIGURE 4.25**. Four sequencing reaction mixtures are prepared. The mixtures differ only in the dideoxynucleoside triphosphate that is added. Each kind of dideoxynucleoside triphosphate is labeled with a different color fluorescent dye. All other components, including the single-stranded DNA molecule to be sequenced, DNA polymerase, the oligonucleotide primer, and the four deoxynucleoside triphosphates, are the same in all four reaction mixtures. The primer oligonucleotide, which determines the particular region within the DNA molecule that will be sequenced, is usually synthesized by chemical methods. Primer chain extension continues in a 5′→3′ direction until the process is terminated by random dideoxynucleotide attachment. As a result of this random chain termination each reaction mixture produces a nested population ladder of extended primer sequences in which all extended chains end with the specific dideoxynucleotide that was added to that reaction mixture. Once synthesis is complete the newly synthesized DNA molecules are denatured and the contents of the four different tubes are combined so the fragments can be separated by polyacrylamide gel electrophoresis or capillary electrophoresis. After fragment separation is complete, sequence information can be read directly from the gel.

A major advance in the dideoxynucleotide method, known as **dideoxy-terminator cycle sequencing**, uses a thermal cycler to amplify reaction products, allowing double-stranded DNA molecules to be sequenced. The reaction mixture, containing an oligonucleotide primer complementary to a segment of only one of the two strands, goes through the rounds of denaturation, annealing, and primer extension described for the PCR. Because the reaction mixture is heated during the denaturation part of the cycle, the DNA polymerase used must

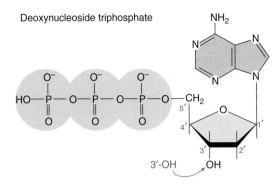
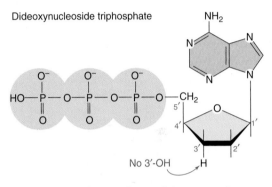

FIGURE 4.24 Comparison of deoxy- and dideoxynucleoside triphosphate structures.

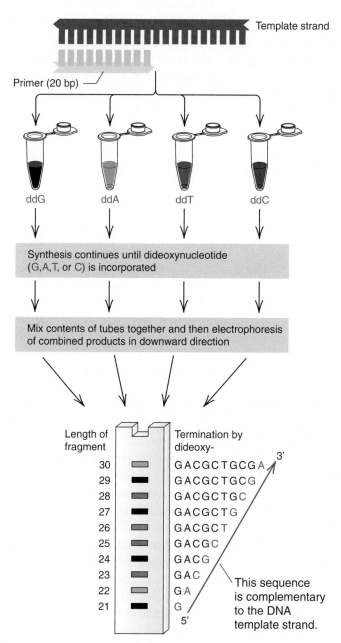

FIGURE 4.25 Chain termination method for sequencing DNA. Lengths of the terminated DNA fragments are shown at the left of the gel. The sequence of the daughter strand is read from the bottom of the gel according to the color of each band as 5'-GACGCTGCGA-3'.

be thermostable. Cycle sequencing amplifies the formation of chain extension products that are terminated by dideoxynucleotides labeled with fluorescent dyes. After the fragments have been separated by size using capillary electrophoresis, a scanner is used to detect the fragments and transfers the information to a computer that stores the nucleotide sequence data. Because the product is amplified, dideoxy-terminator cycle sequencing requires less DNA than the standard technique. Furthermore, both strands of a double-stranded DNA can be sequenced, diminishing the chance of an error.

DNA sequences can be stitched together by using information obtained from a restriction map.

A DNA restriction map can help to sequence long DNA molecules. DNA is cut with the restriction endonuclease(s) used to produce the map. Then the resulting fragments are separated by gel electrophoresis, and each is sequenced by the chain termination method. The restriction map provides the fragment order, which allows the sequences of the small fragments to be stitched together to provide the overall sequence. Although this approach seems fairly straightforward, it has drawbacks. First, the time required (and degree of difficulty) for restriction map construction increases with DNA length. Second, some restriction fragments are too long to permit sequencing without further cutting and others are so short they are lost during purification by gel electrophoresis.

Shotgun sequencing is used to sequence long DNA molecules.

The preferred approach to sequence long DNA molecules is **shotgun sequencing**, a technique devised by Sanger and coworkers in the early 1980s. The approach begins with random DNA cleavage to produce small fragments. Cleavage can be conveniently achieved by subjecting DNA in solution to the shear forces generated by passing the solution through a narrow-gauge syringe. Breakage sites are different for each DNA molecule, producing a heterogeneous population of small fragments. The term "shotgun" refers to the random shredding process. As a result of this random shredding, any given nucleotide in the original large DNA ends up in different small overlapping fragments. Many of the small fragments have 3'- or 5'-overhangs that must be converted to blunt ends before the next sequencing step is possible. This conversion is accomplished by incubating the DNA fragments with the four deoxyribonucleoside triphosphates and a DNA polymerase that fills in 3'-recessed ends and removes protruding 3'-ends to produce blunt ended fragments (FIGURE 4.26). Blunt-end fragments are incubated with polynucleotide kinase and ATP to ensure that 5'-ends are phosphorylated (not shown) and inserted into a cloning vector with a universal primer sequence at its insertion site. A DNA ligase is used that works on blunt ends. Resulting recombinant plasmids are introduced into competent bacteria by transformation. Cloned DNA is isolated, and the short fragments are sequenced using the universal primer. Sequencing reads are stored in a computer that has a sequence assembly software program that searches for overlaps and uses the overlaps to assemble the sequencing reads into a long contiguous sequence.

In 1995 J. Craig Venter, Hamilton Smith, and coworkers used a variant of shotgun sequencing called **whole genome shotgun sequencing** to determine the complete 1.83×10^6 bp genome sequence of *Haemophilus influenzae* Rd, a nonpathogenic variant of a bacterial strain that can cause ear infections and meningitis. This was the first time the genome of a free-living organism had been sequenced. As the term "whole genome shotgun sequencing" suggests, the whole bacterial genome is broken randomly into fragments that are cloned and

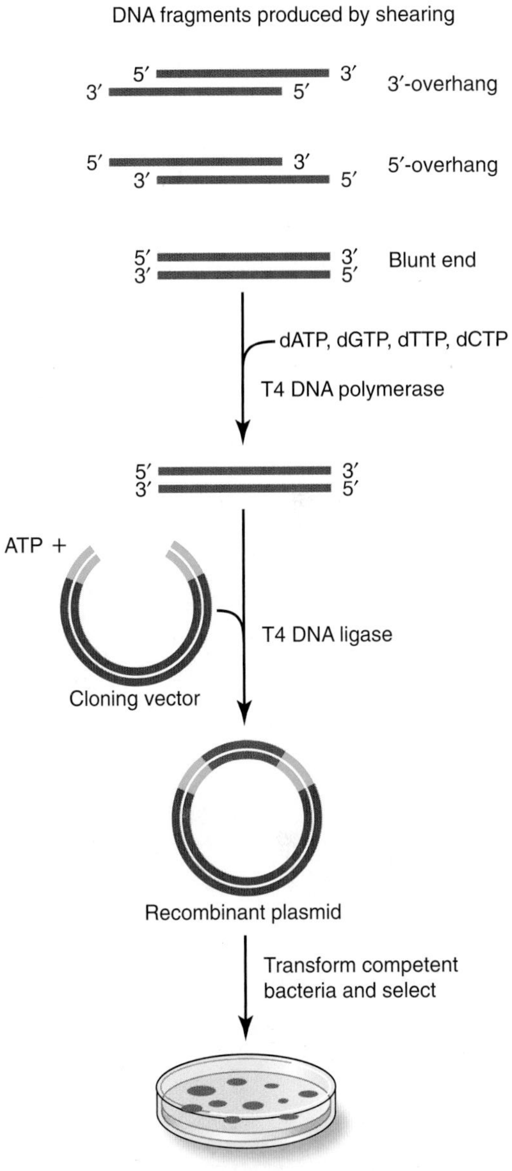

FIGURE 4.26 Recombinant plasmid construction for shotgun sequencing. Fragments formed by mechanical shearing (blue) are incubated with the four deoxyribonucleoside triphosphates and bacteriophage T4 DNA polymerase to fill 3'-recessed ends and remove protruding 3'-ends. Blunt end fragments are incubated with polynucleotide kinase and ATP to ensure that 5'-ends are phosphorylated (not shown) and inserted into a cloning vector (green) with a universal primer sequence (red) at its insertion site. Resulting recombinant plasmids are introduced into competent bacteria by transformation.

4.3 Enzymatic Techniques Used to Manipulate DNA

sequenced. Approximately 20,000 clones were sequenced to ensure that every nucleotide in the genome was included in the final sequence. Based on the sequence information, the *Haemophilus influenzae Rd* genome is predicted to contain 1,743 genes. Five years later Venter and a very large group of coworkers used whole genome shotgun sequencing to determine the sequence of nearly 67% of the approximately 1.8×10^8 bp genome of the fruit fly *Drosophila melanogaster*, including all genetically active sequences.

A new generation of DNA sequencing techniques is now being used for whole genome shotgun sequencing.

The chain termination method, which has been the predominant DNA sequencing technique for the past 30 years, is now being challenged by a new generation of sequencing techniques. This new generation of sequencing techniques allows whole genome shotgun sequencing to be performed much more quickly than was previously possible and at a much lower cost because they are massively parallel (sequencing thousands of DNA strands at one time) and do not require *in vivo* cloning or gel electrophoresis.

The first new generation sequencer was introduced by 454 Life Science, a member of the Roche Group, in 2005. Two years later an improved version, the Genome Sequencer FLX™ (GS FLX), went on the market. This instrument performs about 400,000 parallel sequencing reads with an average read length of 250 bp. It supplies about 10^8 bp of sequence information in 8 hours, enabling a single investigator to sequence an entire bacterial genome in a few days. Not surprisingly, many investigators have started to use this instrument (**BOX 4.3 IN THE LAB: NEW GENERATION DNA SEQUENCERS**).

The human genome sequence provides considerable new information.

The possibility of sequencing the approximately 3 billion base pairs in a single set of human chromosomes (the haploid human genome) was first formally proposed at a meeting sponsored by the U.S. Department of Energy in 1985. The scientific community had a mixed reaction to the proposal. Some investigators argued that sequence information, which could only be obtained through a large-scale effort, was essential for medical diagnosis and treatment. Others argued that the $3 billion needed for the project would be better spent if used to support investigator-initiated projects designed to address specific medical problems. Investigators were also concerned that DNA techniques, sequencing methods, data handling, and data storage then available were not up to the task of sequencing the human genome. Those who supported the project thought that technological advances would produce the necessary tools and that the project should support efforts to make such advances.

The Human Genome Project was officially launched in the United States in 1990 as a $3 billion, 15-year effort sponsored by the National Institutes of Health and Department of Energy. Participating

BOX 4.3: IN THE LAB

New Generation DNA Sequencers

A new generation of DNA sequencers permits rapid and accurate DNA sequence information to be obtained without the need for electrophoresis or *in vivo* cloning. The basic ideas behind these massively parallel sequencing instruments are as follows:

1. Recombinant DNA technology is used to prepare thousands of tiny Sepharose® beads, each with millions of copies of a single-stranded DNA fragment attached to it. The single-stranded DNA fragments attached to any given Sepharose® bead are identical to one another but differ from the fragments attached to other Sepharose® beads. The Sepharose® bead population is large enough to ensure that it contains all the genomic information within an organism. Although the DNA fragments attached to different Sepharose® beads have different sequences, all fragments are constructed to have the same short segment at their 3′-end.
2. The single strands attached to the Sepharose® beads are annealed to a short primer complementary to their 3′-end. Then a slurry containing the Sepharose® beads, DNA polymerase, and enzyme beads with bound ATP sulfurylase and luciferase (see below) is loaded onto a 7 cm × 7.5 cm fiber-optic slide that has 1.6 million wells (FIGURE B4.3a). Each well is so small (44 μm diameter) that only one Sepharose® bead (33 μm diameter) can fit inside. Enzyme beads are quite small and therefore many such beads can fit around the Sepharose® bead with DNA segments. After loading, the fiber-optic slide is placed in the GS FLX sequencer. Nucleotide solutions flow across the fiber-optic slide sequentially in a fixed order (FIGURE B4.3b).

(a) Loading capture beads onto the fiber-optic slide

(b) Schematic showing the location of the fiber-optic slide in the GS FLX instrument

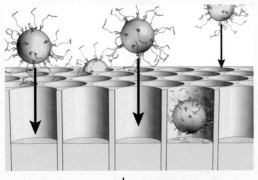
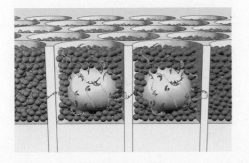
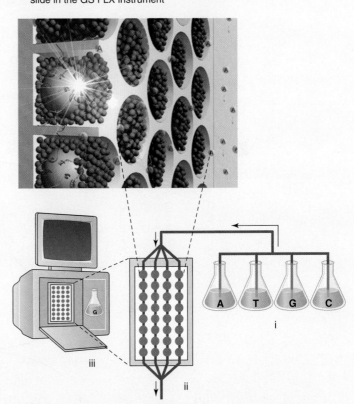

FIGURE B4.3 The fiber-optic slide. (a) Loading beads onto wells. (b) Schematic showing the location of the fiber-optic slide in the GS FLX instrument. The GS FLX instrument has three major subsystems: (i) a flow cell that includes the fiber-optic slide with about one million wells, (ii) a camera to image the fiber-optic slide, and (iii) a computer containing proprietary software needed to collect and analyze the data and assemble the sequence. (Parts a and b [top panel] courtesy of 454 Sequencing. © 2010 Roche Diagnostics. Part b [bottom panel] adapted from 454 Life Sciences, How Is Genome Sequencing Done? Roche Diagnostics, 2010.)

(continues)

BOX 4.3: IN THE LAB

New Generation DNA Sequencers (*continued*)

The sequencing process is best understood by focusing on chemical events that take place in a single well. When the nucleotide that flows into a well is complementary to the next base in the template strand, the nucleotide is added to the primer strand and pyrophosphate is released. If the nucleotide is not complementary, then no reaction takes place. Because the order in which nucleotides flow into a cell is known, sequencing can be performed by correlating pyrophosphate release with this known order of nucleotides that flow into the well.

The problem of finding a quick and accurate way to monitor pyrophosphate release was solved by using three coupled enzyme reactions to monitor pyrophosphate (FIGURE B4.3c). The three reactions are as follows: (1) DNA polymerase releases inorganic

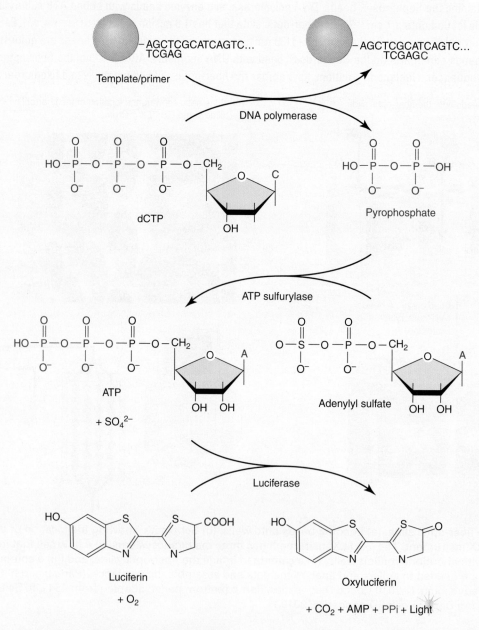

FIGURE B4.3c **Pyrosequencing reactions.**

144   CHAPTER 4 MOLECULAR BIOLOGY TECHNOLOGY

pyrophosphate as it adds each deoxynucleotide to the primer, (2) ATP sulfurylase (ATP: sulfate adenylyltransferase) catalyzes adenylyl group transfer from adenylyl sulfate to the released pyrophosphate to form ATP, and (3) the firefly enzyme luciferase catalyzes an ATP-dependent chemiluminescent reaction in which luciferin conversion to oxyluciferin is accompanied by light emission.

Although the enzyme assay is very fast, reproducible, and quite sensitive, it does have one drawback. dATP is a weak substrate for luciferase, causing background light emission. This difficulty is circumvented by replacing dATP with an analog that is a substrate for DNA polymerase but not for luciferase. The analog, dATPαS [deoxyadenosine 5′-(O-1-thiotriphosphate)], differs from dATP in only one atom; a nonbonding oxygen atom in the α-phosphate group is replaced by a sulfur atom (FIGURE B4.3d).

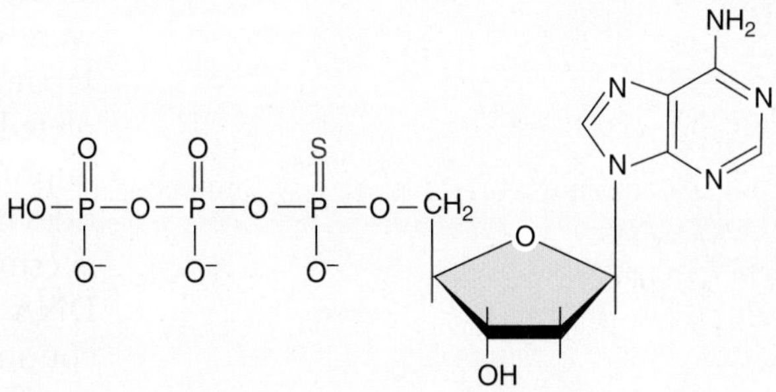

FIGURE B4.3d dATPαS [deoxyadenosine 5′-(O-1-thiotriphosphate)].

The light signal generated in a well each time a nucleotide adds to the primer chain is recorded by a camera. This sequencing technique, called pyrosequencing, has a single read accuracy that is greater than 99.5%. When errors do occur, they usually result from the presence of a homopolymer region in the template DNA. In a limited range (up to about 10 repeated nucleotides), the light signal strength is proportional to the number of nucleotides. The camera can record light emission events as they occur in each of the 1.6 million wells on a single plate and transmit the information to a computer for analysis.

The first application of massively parallel DNA sequencing to human genome sequencing was completed in 2008. Dr. James D. Watson, the co-discoverer of the double helical structure of DNA, provided a blood sample to a small team of investigators, who then determined his DNA sequence in less than 4 months at a cost of $1.5 million. New generation sequencing still has a way to go before it can be used in clinical laboratories. The race is now on to find a technique that will determine an individual's DNA sequence for under $1,000 because DNA sequencing then might become a practical method for diagnosing disease and providing a rational basis for treatment. So-called new-new generation techniques offer considerable promise for reaching the $1,000 goal.

In 2008 Stephan C. Schuster and coworkers sequenced nuclear DNA extracted from hair shafts collected from permafrost remains of the extinct woolly mammoth (*Mammuthus primigenius*). They used hair because it allowed them to remove bacteria and other contaminants without damaging the keratin-encased endogenous DNA. Based on C-14 dating the hair sample was calculated to be about 18,500 years old. Although fragmented and damaged, the ancient DNA could be sequenced using the GS FLX sequencer. Schuster and coworkers determined the sequence for 3.3 billion bases out of an estimated total of about 4 billion bases. Based on their data they estimated that the extinct wooly mammoth and the African elephant are 99.78% identical at the amino acid level. The estimated divergence rate between mammoth and African elephant is half of that between human and chimpanzee.

The DNA of a second extinct species was sequenced at about the same time. Svante Pääbo and coworkers used the GS FLX sequencer to determine the complete sequence of mitochondrial DNA from a bone sample of a Neanderthal individual who lived about 38,000 years ago. Based on the sequence information it appears separation between Neanderthal and present-day human populations took place between 270,000 and 440,000 years ago.

Clearly, DNA sequencing's impact extends well beyond molecular biology. It is providing important new knowledge in all fields of biology and promises to continue doing so at an accelerating rate in the future.

laboratories in the United States, designated genome sequencing centers, were soon joined by sequencing centers in China, France, Germany, Japan, and the United Kingdom. The Human Genome Project completed initial sequencing in 2001. Further efforts closed gaps and improved accuracy. Approximately 2.85 billion of the 3 billion nucleotides in the human genome have been sequenced with an error rate that is estimated to be less than 1 bp in every 10,000 bp. The sequenced DNA includes virtually all active or potentially active genes. Work continues to fill in the remaining gaps in the human genome.

The human genome sequence has provided a great deal of important information. First, the human genome contains about 20,000 to 25,000 protein coding genes, a number that is much smaller than anticipated. Second, DNA that codes for the protein-coding genes accounts for only about 2% of the human genome. Third, gene density varies among chromosomes, with some chromosome being more gene-rich than others. Fourth, several thousand non–protein-coding genes code for RNA molecules. Fifth, gene-dense DNA regions within a chromosome usually have a high G-C content. Sixth, about half the protein-coding genes specify proteins of unknown function. Finally, the human genome sequence is almost identical (99.9%) in all people.

The finished human genome sequence serves as an invaluable reference that is now being used to learn how our genes work, influence metabolic processes, and are linked to diseases. Although the human genome sequence provides considerable valuable information, it is in some ways like a road map that does not show the names of the roads or the towns and cities they pass through. The human genome requires **annotation**. That is, protein-coding genes and their regulatory signals need to be identified, introns and exons indicated, and regions that code for regulatory RNA molecules specified. The annotated human genome has the potential to provide molecular biologists with a factual base that is every bit as important as the periodic table is for chemists.

Questions and Problems

1. Briefly define, describe, or illustrate each of the following.
   a. SDS-Page
   b. Nuclease
   c. Endonuclease
   d. Exonuclease
   e. S1 endonuclease
   f. Restriction endonuclease
   g. Isoschizomers
   h. Blunt ends
   i. Cohesive ends
   j. Restriction map
   k. Recombinant plasmid
   l. Competent bacteria
   m. Southern transfer technique
   n. Northern blotting
   o. Western blotting
   p. DNA polymerase I
   q. Chain termination method
   r. Klenow fragment
   s. Nick translation
   t. Site-directed mutagenesis
   u. Shotgun sequencing
   v. Whole genome shotgun sequencing
   w. Pyrosequencing

2. Why does supercoiled plasmid DNA sediment more rapidly during sucrose gradient centrifugation than a linear duplex with the same number of base pairs?

3. How can you separate linear double-stranded DNA that is uniformly labeled with N-14 from linear double-stranded DNA with the same sequence that is uniformly labeled with N-15?

4. An enzyme has been isolated that migrates as a single band during gel electrophoresis. The same enzyme migrates as two bands of unequal size during SDS-PAGE. How do you explain these results? What other information about the protein can you obtain from SDS-PAGE?

5. How do endonucleases differ from exonucleases?

6. A DNA fragment contains the sequence

   5′-AAGATCGGATCCGAATTCTTTGCA-3′
   3′-TTCTAGCCTAGGCTTAAGAAACGT-5′

   Based on the information presented in Table 4.2, predict the sequence of the fragments that would be produced if the original fragment were incubated with
   a. BamHI
   b. BglII
   c. EcoRI
   d. EcoRV
   e. MboI

7. A 12.0-kb linear double-stranded DNA molecule is cut with BamHI or BglII and then with both enzymes. The results are as follows.

Restriction endonuclease	Restriction fragments (kb)
BamHI	7.5 and 4.5
BglII	6.0, 2.4, and 3.6
BamHI and BglII	6.0, 3.6, 1.5, and 0.9

   Construct a restriction map of the 12.0-kb DNA molecule from the data in the table.

8. EcoRV makes a single cut in each strand of a DNA molecule but produces only one fragment. How is this possible?

9. An investigator isolates a plasmid from *E. coli*. When the plasmid is analyzed by agarose gel electrophoresis, four bands are observed. The plasmid is reported to have a single BamHI restriction site. When the plasmid is incubated with BamHI and analyzed by agarose gel electrophoresis a single band is observed. Explain these results.

10. Describe an approach that you might use to determine the sequence of a cloned 15-kbp DNA fragment.

11. Describe a method you could use to label DNA with 5′-[^{32}P]phosphate at each end. How can you extend this procedure to isolate two fragments that are each labeled with 5′-[^{32}P]phosphate at one end?

12. Based on the fact that DNA polymerase I requires both a template and primer, predict how DNA polymerase I would act on each of the following DNA molecules:

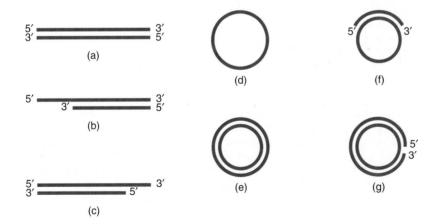

13. A student has isolated plasmids from *E. coli* and would like to make them radioactive. What enzyme(s) and radioactive precursor(s) will be required to label the plasmids? Draw a diagram to illustrate the process.

14. What is a dideoxynucleotide and how is it used to sequence DNA?

15. The gel below was generated when a DNA strand was sequenced using the chain termination method.

The fluorescent dideoxynucleoside triphosphates used had the following colors: ddATP, green; ddTTP, red; ddGTP, blue; and ddCTP, magenta. Determine the nucleotide sequence of the DNA nucleotide chain synthesized from the primer and indicate the 5′ and 3′ ends.

16. The sequence of the entire *E. coli* chromosome is known. How can you use this information and a thermostable DNA polymerase to prepare a DNA fragment that contains a specific 1-kbp chromosomal sequence?

17. You suspect an *E. coli* mutant is missing a DNA segment normally located between two BamHI restriction sites. Describe how the Southern blot procedure could be used to test your hypothesis.

18. Why are fluorescent dideoxynucleoside triphosphates required for chain termination sequencing but not for pyrosequencing?

19. A single-stranded circular DNA has been isolated that contains a gene you would like to alter. How can you change a single nucleotide in the gene?

20. A plasmid has a single BamHI site and a penicillin-resistance gene. The BamHI site is not in the penicillin-resistance gene or a site required for plasmid replication. *E. coli* gene X is located between two BamHI restriction sites, but there are no BamHI restriction sites in the gene. Illustrate how you can use this information to insert gene X into the plasmid and clone gene X.

Suggested Reading

Nucleic Acid Isolation

Bowien, B., and Dürre, P. (eds). 2004. *Nucleic Acids Isolation Methods*. Stevenson Ranch, CA: American Scientific.

Chomczynski, P. 2004. Single-step method of total RNA isolation by guanidine-phenol extraction. *Encyclopedia of Life Sciences*. pp. 1–6. Hoboken, NJ: John Wiley & Sons.

Chomczynski, P., and Mackey, K. 2001. RNA: Methods for preparation. *Encyclopedia of Life Sciences*. pp. 1–2. Hoboken, NJ: John Wiley & Sons.

Rapley, R. (ed). 2000. *The Nucleic Acids Protocol Book*. Totowa, NJ: Humana Press.

Roe, B. A., Crabtree, J. S., and Khan A., 1996. *DNA Isolation and Sequencing*. Hoboken, NJ: John Wiley & Sons.

Physical Techniques Used to Study Macromolecules

Birnie, G. D., and Rickwood, D. (eds). 1978. *Centrifugal Separations in Molecular and Cell Biology*. Tolly, UK: Butterworths.

Bozzola, J. J., and Russell, L. D. 1998. *Electron Microscopy: Principles and Techniques for Biologists* (2nd ed). Sudbury MA: Jones & Bartlett.

Dubochet, J., and Stahlberg, H. 2001. Electron cryomicroscopy. *Encyclopedia of Life Sciences*. pp. 1–5. Hoboken, NJ: John Wiley & Sons.

Graham, J. M., and Rickwood, D. 2001. *Biological Centrifugation*. London, UK: BIOS Scientific Publishers.

Hayat, M. A. 2002. *Principles and Techniques of Electron Microscopy: Biological Applications* (4th ed). Cambridge, UK: Cambridge University Press.

Jones, P. 1996. *Gel Electrophoresis: Nucleic Acids: Essential Techniques*. New York: John Wiley & Sons.

Ludtke, S. J., and Chiu, W. 2001. Electron cryomicroscopy and three-dimensional computer reconstruction of biological molecules. *Encyclopedia of Life Sciences*. pp. 1–5. Hoboken, NJ: John Wiley & Sons.

Price C. A. (ed.) 1982. *Centrifugation in Density Gradients*. San Diego, CA: Academic Press.

Righetti, P. G. 2005. Capillary electrophoresis. *Encyclopedia of Life Sciences*. pp. 1–4. Hoboken, NJ: John Wiley & Sons.

Southern, E. M. 2002. Pulsed-field gel electrophoresis of DNA. *Encyclopedia of Life Sciences*. pp. 1–7. Hoboken, NJ: John Wiley & Sons.

Tao, Y., and Zhang, W. 2000. Recent developments in cryo-electron microscopy reconstruction of single particles. *Curr Opin Struct Biol* 10:616–622.

Unger, V. M. 2001. Electron cryomicroscopy methods. *Curr Opin Struct Biol* 11: 548–554.

Völkl, A. 2002. Ultracentrifugation. *Encyclopedia of Life Sciences*. pp. 1–7. London: Nature.

Westermeier, R. 2005. *Electrophoresis in Practice: A Guide to Methods and Applications of DNA and Protein Separations* (4th ed). New York: John Wiley & Sons.

Enzymatic Techniques Used to Study DNA

Aggarwal, A. K. 1995. Structure and function of restriction endonucleases. *Curr Opin Struct Biol* 5:11–19.

Ansorge, W. J. 2009. Next generation sequencing techniques. *New Biotechnol* 25:195–203.

Bentley, D. R., et al. 2008. Accurate whole human genome sequencing using reversible terminator chemistry. *Nature* 456:53–59.

Bickle, T. A. 1993. The ATP-dependent restriction enzymes. Linn S. M., Lloyd, R. S., and Roberts, R. J. (eds). *Nucleases*. pp. 89–109. Woodbury, NY: Cold Spring Harbor Laboratory Press.

Bickle, T. A., and Krüger, D. H. 1993. Biology of DNA restriction. *Microbiol Rev* 57:434–450.

Bowen, D. J. 2005. *In vitro mutagenesis*. *Encyclopedia of Life Sciences*. pp. 1–6. Hoboken, NJ: John Wiley & Sons.

Chandrasegaran, S. 2001. Restriction enzymes. *Encyclopedia of Life Sciences*. pp. 1–7. London: Nature.

Droege, M., and Hill, B. 2008. The Genome Sequencer FLX™ System—Longer reads, more applications, straight forward bioinformatics and more complete data sets. *J Biotechnol* 136:3–10.

Dunham, I. 2005. Genome sequencing. *Encyclopedia of Life Sciences*. pp. 1–6. Hoboken, NJ: John Wiley & Sons.

Edelheit, O., Hanukoglu, A., and Hanukoglu, A. 2009. Simple and efficient site-directed mutagenesis using two single-primer reactions in parallel to generate mutants for protein structure-function studies. *BMC Biotechnol* 9:61.

França, L. T. C., Carrilho, E., and Kist, T. B. L. 2002. A review of DNA sequencing techniques. *Q Rev Biophys* 35:169–200.

Friedberg, E. C. 2006. The eureka enzyme: the discovery of DNA polymerase. *Nat Rev Mol Cell Biol* 7:143–147.

Green, E. D. 2001. Strategies for the systematic sequencing of complex genomes. *Nat Rev Genet* 2:573–583.

Hardin, S. H. 2001. DNA sequencing. *Encyclopedia of Life Sciences*. pp. 1–6. London: Nature.

Hubbard, T. J. P. 2005. Human genome: draft sequence. *Encyclopedia of Life Sciences*. pp. 1–5. Hoboken, NJ: John Wiley & Sons.

Hutchison III, C. A. 2007. DNA sequencing: bench to bedside and beyond. *Nucleic Acids Res* 35:6227–6237.

Joyce, C. M. 2004. DNA polymerase I, bacterial. *Encycl Biol Chem* 1:720–725.

Kehrer-Sawatzki, H. 2008. Sequencing the human genome: Novel insights into its structure and function. *Encyclopedia of Life Sciences*. pp. 1–9. Hoboken, NJ: John Wiley & Sons.

Lander, E. S., et al. 2001. Initial sequencing and analysis of the human genome. *Nature* 409:860–921.

Mani, M., Kandavelou, K., and Chandrasegaran, S. 2007. Restriction enzymes. *Encyclopedia of Life Sciences*. pp. 1–8. Hoboken, NJ: John Wiley & Sons.

Mardis, E. R. 2008. Next-generation DNA sequencing methods. *Annu Rev Genom Hum Genet* 9:387–402.

Margulies, M., et al. 2005. Genome sequencing in microfabricated high-density picolitre reactors. *Nature* 437:376–380.

Meyers, B. C., Scalabrin, S., and Morgante, M. 2004. Mapping and sequencing complex genomes: Let's get physical! *Nat Rev Genet* 5:578–588.

Miller, W., et al. 2008. Sequencing the nuclear genome of the extinct woolly mammoth. *Nature* 456:387–390.

Mukhopadhyay, R. 2009. DNA sequencers: the next generation. *Anal Chem* 81:1736–1740.

Mullis, K. B. 1990. The unusual origin of the polymerase chain reaction. *Sci Am* 262:56–65.

Nathans, D. 1992. Restriction endonucleases, simian virus 40, and the new genetics. Linsted, J. (ed). *Nobel Lectures, Chemistry 1971–1980*. pp. 498–517. Hackensack, NJ: World Scientific Publishing.

Nyrén, P. 2007. The history of pyrosequencing®. *Methods Mol Biol* 373:1–13.

Pettersson, E., Lundeberg, J., and Ahmadian, A. 2009. Generations of sequencing technologies. *Genomics* 93:105–111.

Roberts R. J., and Halford, S. E. 1993. Type II restriction enzymes. Linn S. M., Lloyd, R. S., and Roberts, R. J. (eds). *Nucleases*. pp. 35–88. Woodbury, NY: Cold Spring Harbor Laboratory Press.

Rothberg, J. M., and Leamon, J. H. 2008. The development and impact of 454 sequencing. *Nat Biotechnol* 26:1117–1124.

Sanger, F. 1992. Determination of nucleotide sequences in DNA. Linsted, J. (ed). *Nobel Lectures, Chemistry 1971–1980*. pp. 431–447. Hackensack, NJ: World Scientific Publishing.

Sanger, F., Nicklen, S., and Coulson, A. R. 1977. DNA sequencing with chain-terminating inhibitors. *Proc Natl Acad Sci USA* 74:5463–5467.

Shendure, J., and Ji, H. 2008. Next-generation DNA sequencing. *Nat Biotechnol* 26:1135–1145.

Smith, H. O. 1992. Nucleotide sequence specificity of restriction endonucleases. Linsted, J. (ed). *Nobel Lectures, Chemistry 1971–1980*. pp. 523–541. Hackensack, NJ: World Scientific Publishing.

Smith, M. 1997. Synthetic DNA and biology. In Malstrom, B. G. (ed). *Nobel Lectures, Physiology or Medicine 1991–1995*. Hackensack, NJ: World Scientific Publishing.

Tűmmler, B., and Mekus, F. 2005. Restriction mapping. *Encyclopedia of Life Sciences*. pp. 1–3. Hoboken, NJ: John Wiley & Sons.

Weber, J. L., and Myers, E. W. 1997. Human whole-genome shotgun sequencing. *Genome Res* 7:401–409.

Wheeler, D. A., et al. 2008. The complete genome of an individual by massively parallel DNA sequencing. *Nature* 452:872–876.

Wilton, S. 2002. Dideoxy sequencing of DNA. *Encyclopedia of Life Sciences*. pp. 1–16. London: Nature.

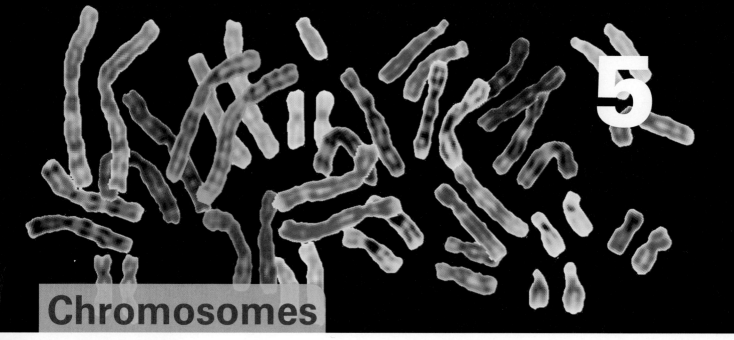

Chromosomes

CHAPTER OUTLINE

5.1 Bacterial Chromatin
Bacterial DNA is located in the nucleoid.

5.2 Introduction to Eukaryotic Chromatin
Eukaryotic chromatin is visible under the light microscope during certain stages of the cell cycle.

5.3 Mitosis
The animal cell life cycle alternates between interphase and mitosis.
Mitosis allows cells to maintain the chromosome number after cell division.

5.4 Meiosis
Meiosis reduces the chromosome number in half.

5.5 Karyotype
Chromosome sites are specified according to nomenclature conventions.
A karyotype shows an individual cell's metaphase chromosomes arranged in pairs and sorted by size.
Fluorescent *in situ* hybridization (FISH) provides a great deal of information about chromosomes.

5.6 The Nucleosome
Five major histone classes interact with DNA in eukaryotic chromatin.
The first level of chromatin organization is the nucleosome.
X-ray crystallography reveals the atomic structure of nucleosome core particles.

The precise nature of the interaction between H1 and the core particle is not known.
We do not know how nucleosomes are arranged in chromatin.
Condensins and topoisomerase II help to stabilize condensed chromosomes.
The scaffold model was proposed to explain higher order chromatin structure.

5.7 The Centromere
The centromere is the site of microtubule attachment.

5.8 The Telomere
The telomere, which is present at either end of a chromosome, is needed for stability.

BOX 5.1: CLINICAL APPLICATION Karyotype and Diagnosis

QUESTIONS AND PROBLEMS

SUGGESTED READING

Image courtesy of Jane Ades/NHGRI.

DNA molecules are extremely long in comparison with the size of living cells and so must be compacted to fit the available space inside a cell. Specific proteins interact with DNA, leading to the formation of a condensed nucleoprotein complex called **chromatin**. Until the mid-1970s biologists' view of chromatin was influenced by the belief that all life on Earth belonged to one of two primary lineages, the eukaryotes (animals, plants, and fungi, which have a defined nucleus) and the prokaryotes (all remaining microscopic organisms that lack a defined cell nucleus). Based on this classification scheme and what was then known about chromatin structure, it seemed likely that prokaryotes would have one type of chromatin structure and eukaryotes would have another. Then in 1977 Carl Woese proposed the prokaryotes actually contain two types of organisms, the bacteria and the archaea.

Once the existence of the three domains of life was established, it seemed possible that each domain would have a characteristic chromatin structure. Experimental studies, however, do not support this possibility. As the following examples illustrate, differences in the chromatin structures of organisms within a single domain are nearly as great as those among organisms belonging to different domains:

1. Variations in chromatin structure exist in bacteria. DNA molecules in most bacterial species that have been studied to date have circular structures like those of the DNA molecules in *Escherichia coli*. DNA molecules in some bacteria, however, such as *Agrobacterium tumefaciens* (a species that infects plants) and *Streptomyces* species, are linear duplexes and in this respect resemble eukaryotic DNA.
2. Most, but not all, eukaryotes contain chromatin formed by interactions between linear duplex DNA and a family of basic proteins called **histones**. However, dinoflagellates, a very large and diverse group of eukaryotic algae, lack histones completely.
3. The archaea exhibit, if anything, even greater variations in chromatin structure than do organisms belonging to the other two domains. Some archaea appear to form nucleoprotein complexes that resemble those present in eukaryotes, whereas others seem to form nucleoprotein complexes that resemble those in bacteria.

As this brief overview shows, it is not possible to examine the chromatin structure from a single kind of organism and expect the information will apply to all organisms within the same domain. We can certainly obtain valuable information about chromatin structure in related organisms, however, and hope this information will eventually permit us to obtain a coherent picture. With this thought in mind, we begin by examining bacterial chromatin structure and then examine chromatin structure in higher animals and plants.

5.1 Bacterial Chromatin

Bacterial DNA is located in the nucleoid.

E. coli DNA, which appears as a closed covalent circle with a total length of about 1,600 μm, must fit into a cylindrical cell with a diameter of about 0.5 μm and a length of about 1 μm. The intracellular DNA is therefore about 1,000-fold more compact than the free DNA. Specific proteins interact with the bacterial DNA to form a highly condensed nucleoprotein complex called the **nucleoid** that occupies about a quarter of the cell's volume (FIGURE 5.1).

Bacterial chromatin can be released from the cell by a gentle cell **lysis** (disruption) technique that avoids DNA breakage or protein denaturation. The released DNA contains a fixed amount of protein and a variable amount of RNA. Most of the RNA is probably nascent RNA (RNA caught in the process of being synthesized) rather than an integral part of the bacterial chromatin.

An electron micrograph of released *E. coli* chromatin reveals multiple loops emerging from a central region, with some loops supercoiled and others relaxed (FIGURE 5.2). Relaxed loops were probably formed as a result of a nick introduced into supercoiled loops by a cellular

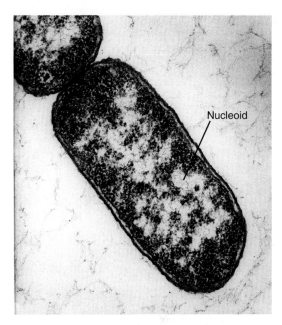

FIGURE 5.1 An electron micrograph of a thin section of *Escherichia coli*. The nucleoid is the light region. (Photo courtesy of the Molecular and Cell Biology Instructional Laboratory Program, University of California, Berkeley.)

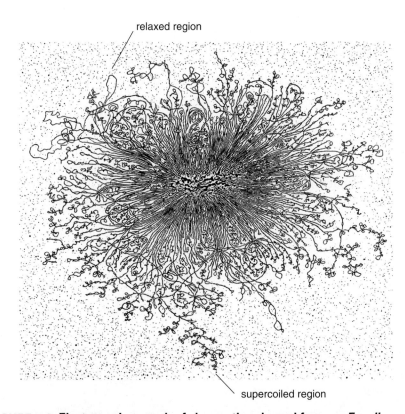

FIGURE 5.2 Electron micrograph of chromatin released from an *E. coli* cell after gentle cell disruption. Multiple loops can be seen emerging from a central region with some loops supercoiled and others relaxed. (Photo courtesy of Bruno Zimm and Ruth Kavenoff. Used with permission of Louis Zimm.)

DNase during the isolation procedure. The fact that supercoiled and relaxed loops are both present indicates that each loop is somehow insulated from the others.

Further support for loop insulation comes from studies in which released *E. coli* chromatin was observed at different times after adding trace quantities of DNase. If a supercoiled DNA molecule receives one nick, the strain of underwinding is immediately removed by free rotation around the opposing sugar-phosphate bond, and all supercoiling is lost. Because a nicked circle is much less compact than a supercoiled molecule of the same molecular mass, the nicked circle sediments much more slowly. One nick thus causes an abrupt decrease (by about 30%) in the sedimentation value. The sedimentation value decreases continuously, however, when DNase introduces one nick at a time into released *E. coli* chromatin. That is, the structure does not change in an all-or-none fashion but proceeds through a large number of intermediate states. This finding indicates that free rotation of the entire DNA molecule does not occur when a single nick is introduced. Electron microscopy studies confirm that as nuclease treatment continues, the number of nonsupercoiled loops increases.

A model of *E. coli* chromatin deduced from sedimentation and electron microscopy studies is shown in FIGURE 5.3. According to this model the bacterial DNA is arranged in supercoiled loops that are fastened to a central protein matrix so each supercoiled loop is topologically independent of all the others. A nick that causes one supercoiled loop to relax would therefore have no effect on other supercoiled loops. The *E. coli* chromosome is estimated to have about 400 such loops, each with an average length of about 10 to 20 kilobase pairs. Biochemical and genetic studies show that supercoiled loops are dynamic structures, which change during cell growth and division. This change allows the entire chromosome to be accessible to the transcription machinery and other enzymes throughout the cell cycle.

Although supercoiling makes an important contribution to bacterial DNA compaction, it is not the only factor. Macromolecular crowding and DNA-binding proteins also contribute to DNA compaction. High

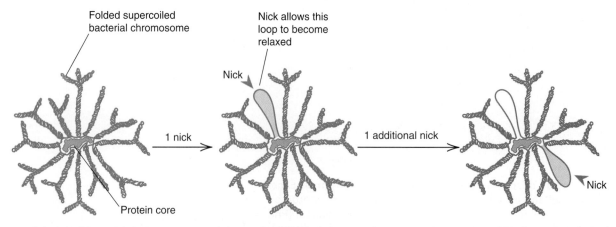

FIGURE 5.3 Model of bacterial chromosome folding. The *E. coli* chromosome has about 400 negatively supercoiled loops attached to a central protein matrix. Each loop (average length of about 10 kilobase pair) is topologically independent of the others.

CHAPTER 5 CHROMOSOMES

intracellular soluble macromolecule concentration limits the available aqueous volume, forcing the DNA to become more compact. Crowding also favors interactions between DNA and the proteins that bind to it. Some of the proteins that help to determine bacterial chromatin architecture are as follows:

1. *The MukB protein*. The MukB protein helps to organize and compact DNA.
2. *H-NS (histone-like nucleoid structuring) protein*. The active form of H-NS appears to be a homodimer (or perhaps an oligomer), which form bridges across DNA segments. Little is known about the way that the protein bridges DNA.
3. *DNA bending proteins*. Three nucleoid associated proteins bend DNA. Two of these, IHF (integration host factor) and HU (heat unstable) protein are closely related heterodimers. The third nucleoid associated protein that bends DNA, FIS (factor for inversion stimulation), is a homodimer.

5.2 Introduction to Eukaryotic Chromatin

Eukaryotic chromatin is visible under the light microscope during certain stages of the cell cycle.

Eukaryotic chromatin can be seen as highly condensed structures called **chromosomes** under a light microscope during certain stages of cell division. As biologists carefully examined chromosomes in many different kinds of cells from various higher animals, two important conclusions emerged. First, reproductive or **germ cells** have a characteristic number of chromosomes (n). This number, called the **haploid number**, is 3, 4, 23, and 30 for germ cells from the mosquito, fruit fly, human, and cattle, respectively. Second, nonreproductive or **somatic cells**, such as those from lung, kidney, and brain, contain two versions of each chromosome (one from each parent) called **homologs**. Because somatic cells have twice the number of chromosomes as germ cells they are said to be **diploid** (2n).

5.3 Mitosis

The animal cell life cycle alternates between interphase and mitosis.

Mitosis is a type of nuclear division that ensures the two daughter cells resulting from cell division each has the same number of chromosomes as the parent cell. A eukaryotic cell spends only a part of its **life cycle** (the time between its formation by parent cell division and its own division to form two daughter cells) in mitosis. The remainder of the time, often approaching 90% of its life cycle, is spent in **interphase**, a stage

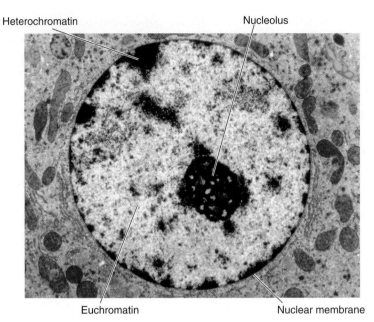

FIGURE 5.4 Electron micrograph of a liver cell nucleus. (© Phototake, Inc./ Alamy Images.)

during which DNA, RNA, protein, and other biological molecules are synthesized.

Two forms of chromatin, a less condensed form called **euchromatin** and a more condensed form called **heterochromatin**, are present in the eukaryotic nucleus during interphase (FIGURE 5.4). Euchromatin, the predominant form, is actively transcribed during interphase. In contrast, heterochromatin, which tends to be located near the nuclear membrane, is not actively transcribed. Nearly all regions of the human genome that remain to be sequenced are in the heterochromatin.

DNA replication and histone synthesis occur during only a part of interphase, the DNA synthetic phase or **S phase**. The S phase is bracketed by two gap phases, G_1 and G_2, so that the stages in the life cycle are in the order $G_1 \rightarrow S \rightarrow G_2 \rightarrow M$ (FIGURE 5.5). The timing of S, G_2, and M tend to be relatively uniform for a given type of somatic cell. However, time spent in G_1, a period of active protein, lipid, and carbohydrate synthesis, is quite variable. Some eukaryotic cells spend almost their entire life cycle in G_1.

Mitosis allows cells to maintain the chromosome number after cell division.

Even though mitosis is a continuous process, it is usually divided into four stages for convenience. Chromosomal changes during these four stages, which occur in the order prophase → metaphase → anaphase → telophase, are depicted in FIGURE 5.6 and summarized below.

1. *Prophase*. Chromatin, which was replicated during the S phase of interphase, condenses to form visibly distinct chromosomes. Each chromosome is divided along its long axis into two identical

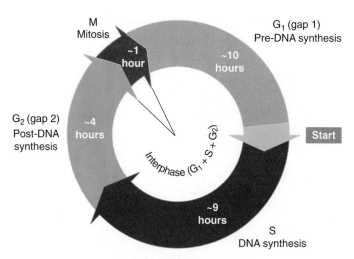

FIGURE 5.5 The cell cycle of a typical mammalian cell growing in tissue culture with a generation time of 24 hours.

subunits called **sister chromatids** that are held together by a protein called **cohesin**. Cohesin is enriched about threefold in a 20- to 50-kb domain flanking a specific chromosomal region known as the **centromere** (see below), relative to its concentration on chromosome arms. As prophase ends the nuclear region involved in ribosomal RNA synthesis, called the **nucleolus**, disappears and the nuclear membrane disassembles to form membrane vesicles. A **mitotic spindle**, consisting of fiber-like bundles of protein molecules called **microtubules**, starts to form.

2. *Metaphase.* The assembly of the mitotic spindle is completed, and the spindle moves to the region previously occupied by the nucleus. Several spindle fibers attach to each chromosome in a region of the centromere called the **kinetochore** (see below). Once attachment is complete the chromosomes move toward the center of the cell until the kinetochores lie in an imaginary plane equidistant from the two spindle poles. Each chromosome must be attached to both poles of the spindle, and the chromosomes must be properly aligned along the imaginary plane equidistant from the spindle poles for mitosis to proceed to the next stage. We can learn a great deal by examining chromosomes during the metaphase (see below).

3. *Anaphase.* Cohesin is cleaved, and cohesion between sister chromatids is dissolved. The two sister chromatids (now considered to be separate chromosomes) move toward opposite spindle poles so that an equal number of identical chromosomes are located at either end of the spindle as anaphase comes to a close. The number of chromosomes in each group is the same as that present in the cell nucleus at the start of interphase.

4. *Telophase.* The spindle disappears, nuclear membranes form around the two groups of chromosomes, and nucleoli re-form. Chromosomes become less and less condensed until they can no longer be seen with a light microscope. The cells divide to produce two identical daughter cells.

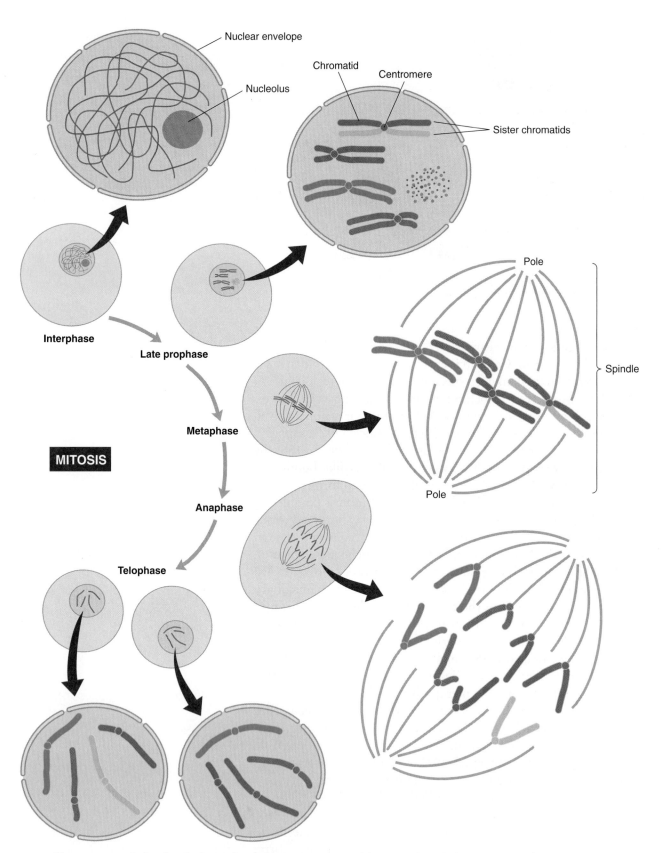

FIGURE 5.6 Chromosome behavior during mitosis in an organism with two pairs of chromosomes (red/rose vs. green/blue). At each stage the smaller inner diagram represents the entire cell, and the larger diagram is an exploded view showing the chromosomes at that stage.

5.4 Meiosis

Meiosis reduces the chromosome number in half.

Eukaryotes use **meiosis** to reduce the chromosome number in half during sexual reproduction. Unlike mitosis, which maintains the chromosome number, meiosis produces gametes (sperm and egg cells) in animals or spores in fungi that have a haploid chromosome number. Meiosis requires two successive nuclear divisions to accomplish this task. The diploid number is reestablished by zygote formation resulting from the union of two haploid cells during sexual reproduction. The first and second meiotic divisions go through prophase, metaphase, anaphase, and telophase stages (FIGURE 5.7).

First Meiotic Division

1. *Prophase I.* Prophase I begins after DNA replication is completed. It is of special interest because chromosome homologs exchange DNA, leading to genetic recombination during this stage. Some of the key changes during prophase I are as follows. The chromatin starts to condense. Beads, called chromomeres, appear at irregular intervals along the length of the chromosome. The number, size, and arrangement of chromomeres are unique for each kind of chromosome. Homologous chromosomes pair. This pairing (**synapsis**) begins at the chromosome tips and continues along the chromosome, establishing specific chromomere–chromomere pairing. Fully paired chromosomes are called **bivalents** (indicating that each pair consists of two types of chromosomes) or **tetrads** (indicating that each homologous pair contains four closely associated chromatids). Paired chromosomes, which are very close together, begin a chromosome exchange process known as crossing over. Condensation reaches a state in which the sister chromatids in each chromosome become visible. Homologous chromosomes start to separate. Complete separation cannot occur at this stage, however, because the homologous chromosomes are joined by cross-connections called **chiasmata** (singular, chiasma = crosspiece), which are produced when nonsister chromatids break and reunite during the crossing-over process. Prophase I is divided into the five substages: leptotene, zygotene, pachytene, diplotene, and diakinesis. Figure 5.7 summarizes the important events that take place during each of these substages.
2. *Metaphase I.* Spindle fibers from one pole make contact with one chromosome in a homologous pair, while spindle fibers from the other pole make contact with the other chromosome in the pair. Each chromosome moves into the metaphase plate (the imaginary plane that is equidistant from each spindle pole).
3. *Anaphase I.* Homologous chromosomes are pulled apart and begin to move to opposite poles. The two members of each homologous pair are separated so that each pole has the haploid number of chromosomes.

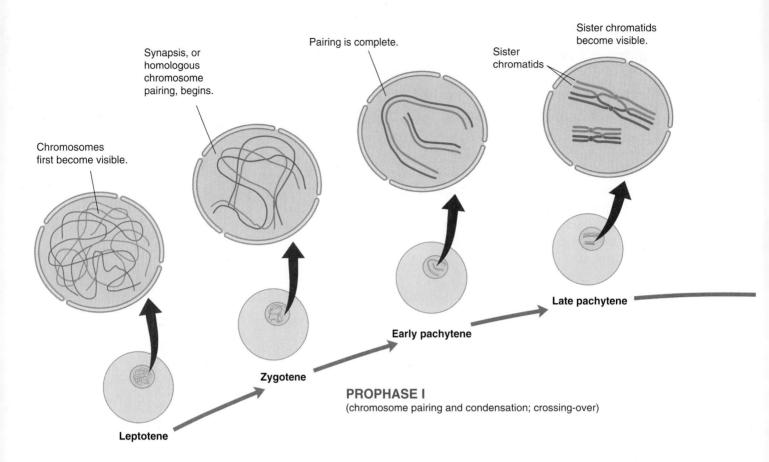
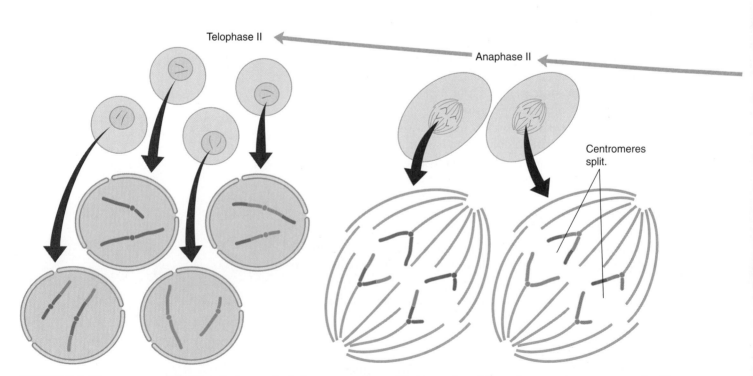

FIGURE 5.7 Chromosome behavior during meiosis in an organism with two pairs of homologous chromosomes (red/rose and green/blue). At each stage the small diagram represents the entire cell and the larger diagram is an expanded view of the chromosomes at that stage.

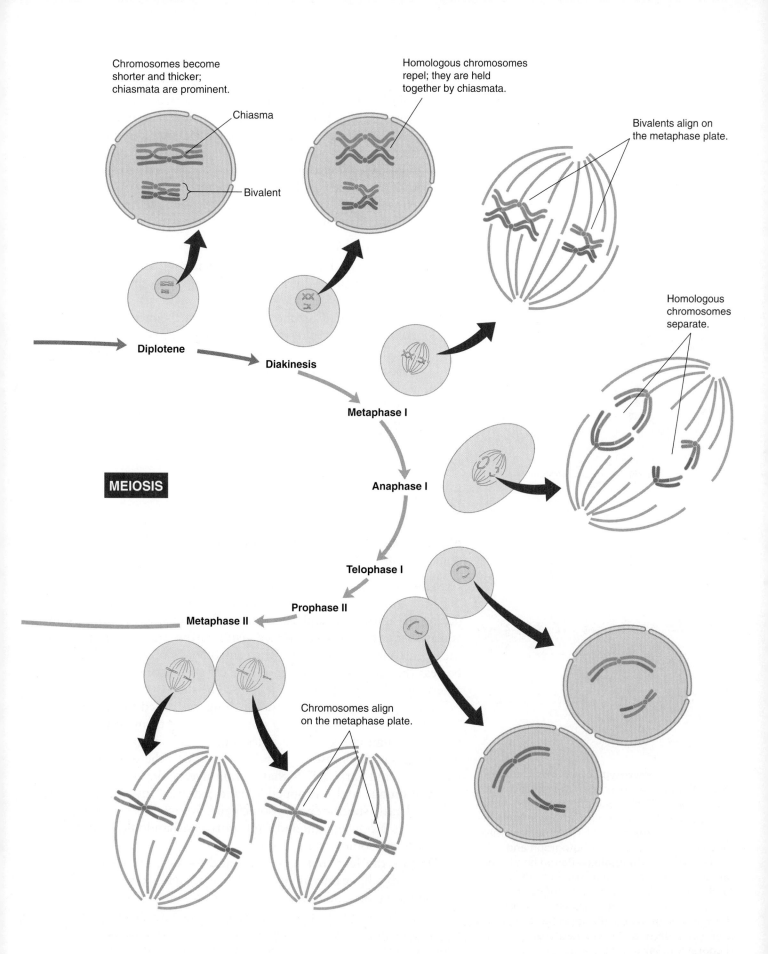

5.4 Meiosis

4. *Telophase I.* The single spindle disassembles and two new spindles form in the region that had been occupied by the spindle poles. In some species the nuclear membrane re-forms, whereas in others the chromosomes enter directly into the second meiotic division. There is a seamless transition from telophase I to prophase II. In fact, the transition is so seamless that prophase II is almost nonexistent in many organisms. No chromosomal replication occurs between the first and second meiotic divisions.

Second Meiotic Division

1. *Prophase II.* The chromosomes remaining at the two poles of the first meiotic division begin to move to the midpoints of the two newly formed spindles.
2. *Metaphase II.* Chromosomes align on the metaphase plate.
3. *Anaphase II.* Cohesins are cleaved, centromeres split, and the chromatids move to opposite poles of the spindle. *Each chromatid is now considered to be a separate chromosome.*
4. *Telophase II.* The chromosomes decondense and nuclear membranes form around the four division products to produce four haploid nuclei.

5.5 Karyotype

Chromosome sites are specified according to nomenclature conventions.

Cytogenetics is the scientific discipline concerned with the study of the physical appearance of chromosomes. Chromosomes are examined during the mitotic metaphase, the stage in which each chromosome consists of two highly condensed sister chromatids held together at their centromere by cohesins. The centromere's position, which may be near the center of the chromosome, off-center, or close to one end, determines the chromatid arm lengths. The shorter of the two arms is called the **p arm** (petite arm) and the longer is called the **q arm**.

Investigators attempted to use size and centromere position to classify human chromosome pairs, but these criteria do not allow unambiguous chromosome identification. With the discovery in the late 1960s and early 1970s that certain dyes stain chromosomes to produce unique and reproducible band patterns, investigators were able to identify specific chromosome pairs. For example, staining with Giemsa dye produces light and dark transverse bands along the length of the chromosome (**FIGURE 5.8a**).

Despite the fact that a typical band looks quite narrow when viewed with a light microscope, it actually extends over more than a million base pairs and dozens of genes. Even when different kinds of chromosomes are the same size and have their centromeres located in the same place, they can still be distinguished by their unique "bar code" patterns. The bands along a chromosome are part of clearly

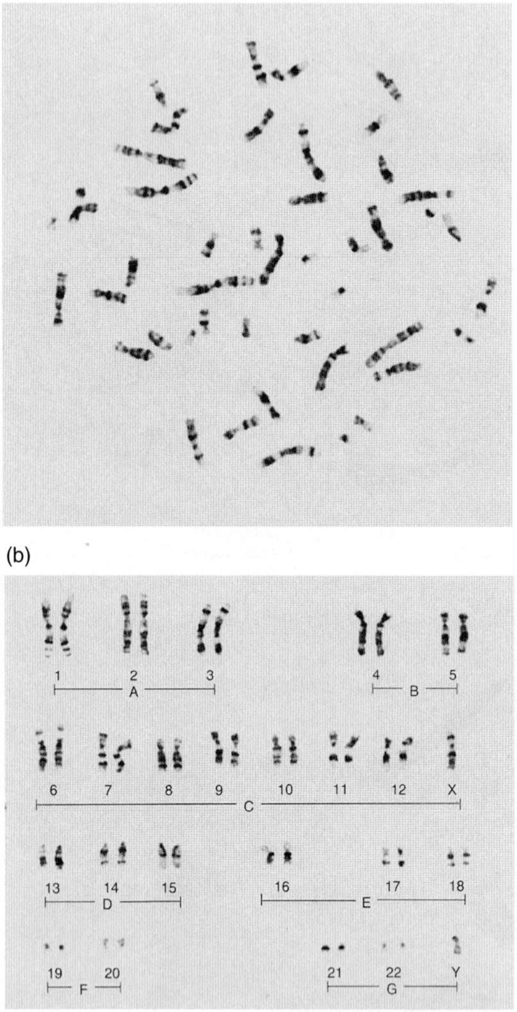

FIGURE 5.8 A karyotype of a normal human male. Blood cells arrested in metaphase were stained with Giemsa and photographed with a microscope. (a) The chromosomes as seen in the cell by microscopy. (b) The chromosomes have been cut out of the photograph and paired with their homologs. Paired homologs are arranged according to size in groups A, B, and so forth. The largest homologous pair is in the upper left. X and Y sex chromosomes have not been paired. (Photos courtesy of Patricia A. Jacobs, Wessex Regional Genetics Laboratory, Salisbury District Hospital.)

delimited regions. Nomenclature conventions for specifying sites in a chromosome, which are illustrated for human chromosome 1 in FIGURE 5.9, are as follows:

1. Each chromosome is assigned a number, with the largest being assigned number 1.
2. The short arm is designated p and the long arm is q.
3. Regions and bands are numbered consecutively from the centromere outward along each chromosome arm.
4. Chromosome number, arm, region number, and band number are written in order.

A karyotype shows an individual cell's metaphase chromosomes arranged in pairs and sorted by size.

A great deal of information can be obtained from digital images of stained chromosomes taken at metaphase. The digital images are cut and pasted with a computer, arranging chromosome pairs by size, shape, and banding pattern to facilitate interpretation. By convention, chromosome pairs are arranged in decreasing order of size to produce an arrangement called a **karyotype**. Figure 5.8b, a human karyotype preparation, reveals the presence of 23 pairs of chromosomes. Twenty-two chromosome pairs, called **autosomes**, are the same in males and females. The remaining pair, the **sex chromosomes**, determines the individual's sex. Male sex chromosomes consist of one X and one Y chromosome, whereas female sex chromosomes consist of the two X chromosomes. A single Y chromosome is sufficient to produce maleness, whereas its absence is required for femaleness. The Y chromosome is smaller than the X chromosome, and the two have different banding patterns. A great deal of information can be obtained by examining karyotype preparations (**BOX 5.1 CLINICAL APPLICATION: KARYOTYPE AND DIAGNOSIS**).

Fluorescent *in situ* hybridization (FISH) provides a great deal of information about chromosomes.

Classical staining techniques do not provide sufficient sensitivity to detect translocations, deletions, or insertions that involve small segments within a chromosome. Investigators have taken advantage of lessons learned from molecular biology to devise a very sensitive technique for detecting even very small chromosomal changes in a sample fixed to a microscope slide.

This technique, called **fluorescent *in situ* hybridization (FISH)**, takes advantage of the fact that a DNA probe with an attached fluorescent dye binds to a specific DNA sequence within a denatured chromosome (FIGURE 5.10). The fluorescent probes can be prepared by nick translation or by the polymerase chain reaction. FISH has a wide variety of applications, which include (1) detecting **aneuploidy** (a condition in which cells have either more or less than the normal diploid number of chromosomes), (2) identifying chromosomal aberrations, and (3) locating genes and other DNA segments on a chromosome. It can also be used to locate DNA segments during interphase when the chromosome is not visible.

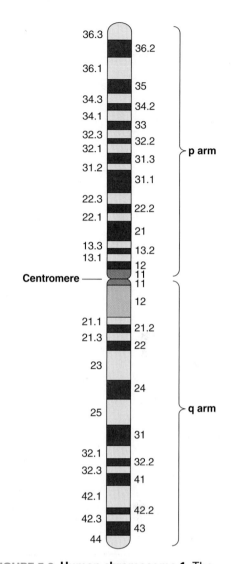

FIGURE 5.9 **Human chromosome 1.** The short arm is designated p and the long arm q. Regions and bands are numbered consecutively from the centromere outward along each chromosome arm. (Modified from HYPERLINK http://www.genome.jp.kegg. Used with permission of Kanehisa Laboratories Bioinformatics Center, Institute for Chemical Research, Kyoto University and the Human Genome Center, Institute for Medical Science, University of Tokyo.)

BOX 5.1: CLINICAL APPLICATION

Karyotype and Diagnosis

Human karyotype preparations provide important clinical information. Although almost any population of dividing cells can be used to obtain the metaphase cells required for such a preparation, blood, bone marrow, fibroblasts, and amniotic fluid are the most common sources of cells to be analyzed. Isolated cells are incubated with a plant or bacterial protein called a mitogen that binds to receptors on the outer surface of the cell membrane and induces mitosis. After 3 or 4 days of incubation, a drug such as colchicine, which disrupts mitotic spindles, is added to arrest dividing cells in metaphase. The population, now enriched in metaphase cells, is stained and analyzed.

Many congenital problems can be identified by chromosomal anomalies. Down syndrome, which occurs in about 1 in every 800 births, will serve as an illustrative example. Approximately 95% of individuals with Down syndrome have three copies of chromosome 21. Down syndrome is associated with mild to severe forms of mental retardation, which is often accompanied by various medical problems, including epilepsy, heart defects, and a marked susceptibility to respiratory infections. Individuals with Down syndrome also appear to age at an accelerated rate and have a high probability of showing the clinical signs and symptoms of Alzheimer-type dementia after age 40.

A condition in which cells have either more or less than the normal diploid number of chromosomes is called aneuploidy. Most cases of simple aneuploidy are caused by errors in chromosomal segregation during meiosis. If pairs of homologous chromosomes do not separate during the first meiotic division or if the centromere joining sister chromatids does not separate during the second meiotic division, the gametes formed have too many or too few chromosomes. Defects in the cohesin pathway are responsible for many errors in chromosomal segregation.

It seems reasonable to suppose that cells with three copies of chromosome 21 overproduce specific proteins that somehow modify fetal development. The challenge is to discover which genes are involved and how their gene products work. The problem is complicated because several genes may be involved. In this regard it is interesting to note that some forms of Down syndrome are caused by chromosomal rearrangements in which a segment of chromosome 21 is attached to another chromosome. In this case the individual with Down syndrome has two copies of chromosome 21 and a part of chromosome 21 attached to another chromosome. Unfortunately, the chromosome 21 segment is still so large it is not yet possible to identify the specific genes responsible for Down syndrome.

A rearrangement of chromosomal material in which part of one chromosome is joined to some other chromosome is called translocation. One of the best-characterized examples of translocation occurs in chronic myelogenous leukemia. Studies of chromosomes in tumor cells reveal reciprocal translocation of material from chromosome 9 to 22 to produce what is known as the Philadelphia chromosome (FIGURE B5.1). This translocation moves a gene (*abl*) from its normal location on chromosome 9 to a new location on chromosome 22, leading to an altered *abl* gene. This altered *abl* gene produces an abnormal protein that activates constitutively (all the time) a number of cell processes that normally are turned on only under special conditions.

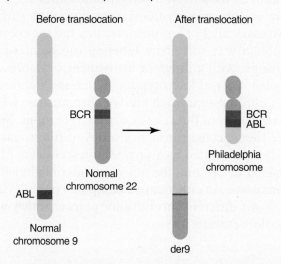

FIGURE B5.1. The Philadelphia chromosome. The Philadelphia chromosome, which is present in individuals suffering from chronic myelogenous leukemia, forms when a piece of chromosome 9 changes place with a piece of chromosome 22. It is the extra-short chromosome formed by the exchange that contains the abnormal *bcr-abl* gene. The other product of the exchange, the extra-long chromosome, is called *der9*.

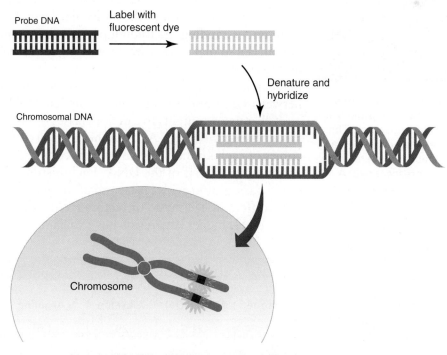

FIGURE 5.10 The fluorescent *in situ* hybridization (FISH) technique. A DNA probe with an attached fluorescent DNA (yellow) binds to a specific DNA sequence within a denatured chromosome. This technique is used to detect aneuploidy, identify chromosomal aberrations, and locate DNA segments or genes. (Adapted from an illustration by Darryl Leja, National Human Genome Research Institute [www.genome.gov].)

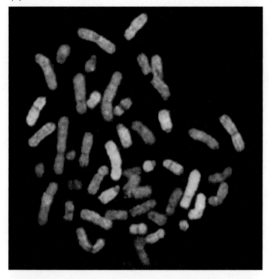

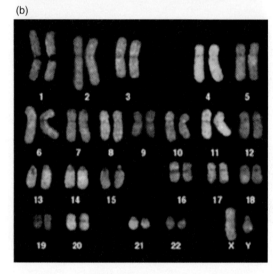

FIGURE 5.11 **Spectral karyotyping of human chromosomes.** (a) Labeled painting probes are hybridized to metaphase chromosomes. The stained chromosomes are viewed through a series of filters, each of which transmits only light emitted by a single fluorescent dye. A computer provides a composite picture that shows each kind of chromosome in a distinct color. (b) The same chromosomes rearranged so that homologous chromosomes are shown as pairs. Somatic chromosomes are shown in decreasing size order with the largest chromosome pair in the upper left. The sex chromosome pair is shown to the right of the smallest chromosome pair in the bottom row. (Photos courtesy of Johannes Wienberg and Thomas Ried, National Institutes of Health.)

A variation of FISH, called chromosome painting, uses a fluorescent dye bound to DNA pieces that bind all along a particular chromosome. The major shortcoming of FISH and chromosome painting is that they cannot be used to study all chromosomes at the same time, because there are not enough fluorescent dyes with sufficient color differences to mark all 23 chromosomes in a unique color.

This problem was solved by labeling the painting probes for each chromosome with a different assortment of fluorescent dyes, a technique called spectral karyotyping. When the fluorescent probes hybridize to a chromosome, each kind of chromosome is labeled with a different assortment of fluorescent dye combinations. Stained chromosomes are then viewed through a series of filters, each of which transmits only light emitted by a single fluorescent dye. Alternatively, an interferometer determines the full spectrum of light emitted by the stained chromosome. In either case a computer provides a composite picture that shows different chromosome pairs as if they were stained in different colors (FIGURE 5.11).

5.6 The Nucleosome

Five major histone classes interact with DNA in eukaryotic chromatin.

Clearly, a chromosome has an elaborate structure and undergoes complex changes during the cell cycle. The task now before molecular and cell biologists is to describe chromosomal structure and explain chromosomal behavior at the molecular level. We are still a long way from understanding the folding process that condenses the DNA in a human cell so that DNA with a total length of over 2 m fits into the cell nucleus with a diameter of about 5 μm. Considerable progress has been made in understanding the first level of organization, the interactions between DNA and histones that compacts DNA by about sevenfold.

Chromatin contains five major classes of histones: H1, H2A, H2B, H3, and H4 (TABLE 5.1). A typical human cell contains about 60 million copies of each kind of histone so their combined mass in chromatin is about equal to that of the DNA. Each histone has a high percentage of the basic amino acids lysine and arginine, but the lysine-to-arginine ratio differs in each type of histone. With the possible exception of histone H4, higher organisms have variants of each histone subtype.

TABLE 5.1 Histone Composition				
Histone	Molecular Mass (kDa)	% Lysine	% Arginine	% Lysine + Arginine
H1	~21.0	29	1.5	30.5
H2A	14.5	11	9.5	20.5
H2B	13.7	16	6.5	22.5
H3	15.3	10	13.5	23.5
H4	11.3	11	14.0	25.0

The positively charged side chains of lysine and arginine residues enable histones to bind to the negatively charged phosphates of the DNA. This electrostatic attraction is an important stabilizing force in chromatin because if chromatin is placed in solutions of high salt concentration (for example, 0.5 M NaCl), which breaks down electrostatic interactions, the chromatin dissociates to yield free histones and free DNA. Moreover, chromatin can be reconstituted by mixing purified histones and DNA in a concentrated salt solution and then gradually lowering the salt concentration by dialysis. Although this result shows that no other components are needed to form chromatin, the rate of chromatin assembly observed under these conditions is much slower than it is in the cell. Protein factors are needed for chromatin formation in the cell.

Reconstitution experiments also have been carried out in which histones from different organisms are mixed. Usually, almost any combination of histones works because, except for H1, the histones from different organisms are very much alike. In fact, H3 amino acid sequences are nearly identical from one organism to the next (sometimes one or two of the amino acids differ). The same is true for H4. For instance, H4 from a cow differs by only two amino acids from H4 from peas—arginine for lysine and isoleucine for valine—which shows the structure of histones has not changed in the billion years since plants and animals diverged.

The first level of chromatin organization is the nucleosome.

When viewed under the electron microscope, uncondensed chromatin from interphase cells resembles beads on a string (FIGURE 5.12). Each bead is a nucleoprotein complex called a **nucleosome** formed by winding DNA around a protein assembly consisting of eight histone molecules. The DNA connecting two nucleosomes is called **linker DNA**.

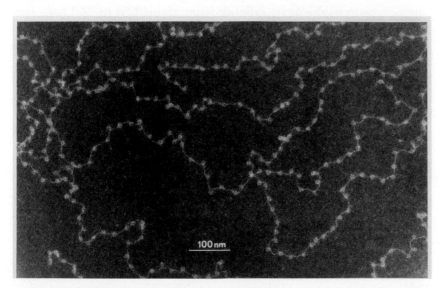

FIGURE 5.12 **Electron micrograph of chromatin.** The beadlike nucleosome particles have diameters of approximately 11 nm. (© Ada & Donald Olins/Biological Photo Service.)

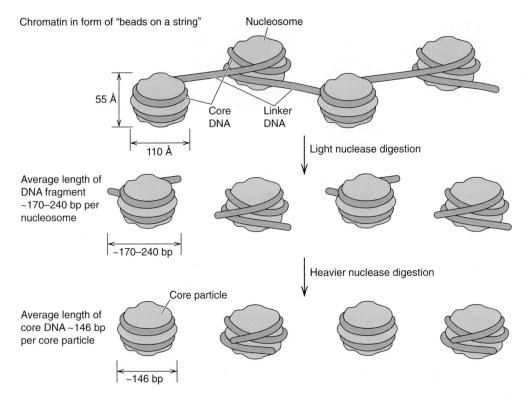

FIGURE 5.13 **"Beads on a string" structure of chromatin.** Brief treatment of chromatin with micrococcal nuclease cleaves DNA in the linker region releasing free nucleosomes that contain 170 to 240 bp of DNA. More extensive treatment results in digestion of all but the 146 bp of DNA in intimate contact with the octameric protein complex.

Micrococcal nuclease can cleave the linker DNA but cannot gain access to DNA that is wrapped around the protein assembly. Consequently, brief digestion with micrococcal nuclease cleaves the chromatin to produce free nucleosomes (FIGURE 5.13).

The length of DNA extracted from these nucleosomes varies from one organism to the next, ranging from about 170 to 240 base pair (bp). This variation results from different sizes of the linker DNA between the nucleosomes. Surprisingly, linker length in the same organism may also vary from one tissue to another (for example, brain vs. liver). The significance of this variation is unknown.

Prolonged nuclease digestion gradually cleaves additional nucleotides until the DNA is about 146 bp long. The structure that remains, the **nucleosome core particle**, consists of an octameric protein complex (two copies each of H2A, H2B, H3, and H4) with a 146-bp DNA fragment wound around it.

X-ray crystallography reveals the atomic structure of nucleosome core particles.

In 1997 Timothy J. Richmond and colleagues determined the structure of the nucleosome core particle at a resolution of 0.28 nm (FIGURE 5.14). This remarkable accomplishment was achieved by constructing nucleosome cores from a 146-bp palindromic DNA molecule with a defined sequence and H2A, H2B, H3, and H4 histones synthesized by *E. coli*.

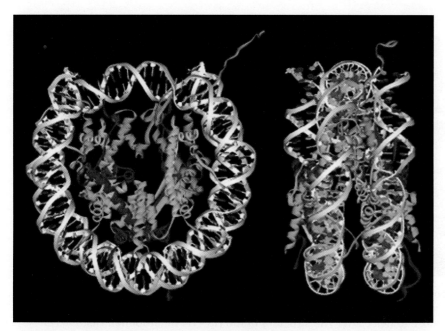

FIGURE 5.14 **Nucleosome core particle.** Ribbon traces for the 146-bp DNA phosphodiester backbone (white) and eight histone protein main chains (blue, H3; yellow, H4; red, H2A; and green, H2B) are shown from two different perspectives. (Structures from Protein Data Bank, ID: 1AOI. Luger, K., et al. 1997. *Nature* 389:251–260. Prepared by B. E. Tropp.)

Although bacteria do not normally synthesize histones, recombinant DNA technology produced bacterial cells that could do so. Bacteria that synthesize histones were created to obtain histone molecules that had not been modified by eukaryotic enzymes that acetylate, methylate, and phosphorylate the polypeptides at various sites. Each kind of histone synthesized by the bacteria was, therefore, a homogeneous protein.

Two H2A, H2B, H3, and H4 histone molecules are required to form a nucleosome core particle. The four histones have a similar folding pattern known as the histone fold, which consists of three α-helices connected by two short loops. Structures for histones 3 and 4 are shown in FIGURES 5.15a and 15b, respectively. The central α-helix is about twice as long as the two flanking α-helices. An H3 subunit and H4 subunit interact to form an H3 • H4 heterodimer in which loop 1 of one subunit is adjacent to loop 2 of the other subunit (Figure 5.15c). The dimer is stabilized by interactions between their antiparallel long α-helices. H2A and H2B subunits interact in a similar way to form H2A • H2B heterodimers. Two H3 • H4 heterodimers pair to form a tetramer through contacts between their H3 subunits. Each H2A • H2B heterodimer binds to the (H3 • H4)$_2$ tetramer through contacts between H2B and H4. The resulting histone octamer has exact twofold symmetry.

Core DNA makes 1.65 turns as it wraps around a helical ramp in the octamer, to generate a left-handed toroidal supercoil with a radius of about 4.25 nm. The DNA's twist value, which varies over the length of the DNA, averages 10.2 bp per turn. This average value is less than

5.6 The Nucleosome

(a) Histone H3   (b) Histone H4   (c) Histone H3•H4 heterodimer

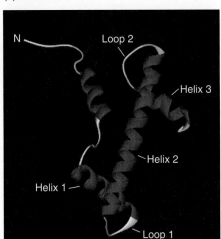
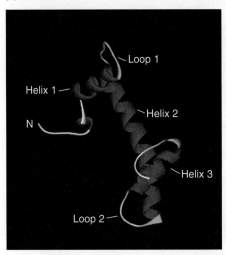
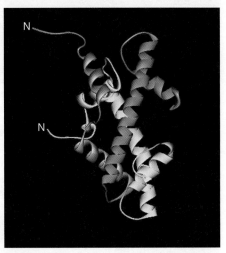

FIGURE 5.15 **The histone fold and histone heterodimer structure.** The histone fold in (a) H3 and (b) H4. The three α-helices (red) that are part of the histone fold are labeled helix 1, helix, 2, and helix 3. Other helices (red) are not labeled. Loop 1 connects helix 1 to helix 2 and loop 2 connects helix 2 to helix 3. (c) The structure of the H3 • H4 heterodimer. Histone H3 is shown in blue and histone H4 in yellow. The orientation of H3 and H4 are identical to their orientations in (a) and (b), respectively. (Structures from Protein Data Bank, ID: 1AOI. Luger, K., et al. 1997. *Nature* 389:251–260. Prepared by B. E. Tropp.)

the 10.5 to 10.6 bp per turn observed for naked DNA in solution. Thus, the DNA in the nucleosome is negatively supercoiled. Each H2A • H2B and H3 • H4 heterodimer binds 27 to 28 bp of DNA, generating bends that cause the DNA to follow a non-uniform path as it wraps around the histone octamer. DNA bending compresses the minor groove. A-T rich sequences accommodate this compression more readily than G-C rich sequences. Therefore, nucleosomes tend to position themselves on DNA so that A-T rich minor grooves contact the histone octamer. Extensive DNA–protein interactions take place through hydrogen bonding between hydrogen atoms in the peptide bonds and oxygen atoms in phosphate groups that form the DNA backbone.

As a general rule, histone octamers do not remain at a specific site on the DNA but are highly dynamic, moving from one position on DNA to another as the cell's requirements change. In a few rare cases, however, nucleosomes are located at specific positions on the DNA. Such specific positioning is often due to the presence of a nonhistone protein that binds to a specific DNA sequence and then either stimulates or blocks nucleosome formation at adjacent sites.

N-terminal histone tails help to regulate chromatin structure and chromatin function and are therefore of considerable interest. Unfortunately, large segments of these tails are disordered and therefore cannot be seen by x-ray crystallography. The visible tail regions appear to be extended structures that interact with the DNA through the minor groove. For instance, the N-terminal tails of H3 and H2B pass through a channel created by the minor grooves in adjacent superhelical turns. Moreover, the N-terminal H4 tail makes many contacts with the face of an H2A • H2B dimer of a neighboring nucleosome core. This interparticle protein–protein contact probably helps to stabilize a higher order nucleosome packing arrangement.

The precise nature of the interaction between H1 and the core particle is not known.

Thus far we have examined the contribution that four of the five histones make to chromatin structure. The fifth histone, histone H1, is present as part of a nucleoprotein particle called the **chromatosome** that is produced by briefly digesting chromatin with micrococcal nuclease. The chromatosome consists of a DNA that is about 166 bp long and wrapped around an octameric histone core and held in place by histone H1. A chromatosome can be converted to a nucleosome core particle by subjecting it to further micrococcal nuclease digestion to remove 10 bp of linker DNA from either end. Chromatosomes have not been prepared in crystalline form. Therefore, the precise nature of the interaction between H1 and the rest of the chromatosome particle is not known.

Eleven different human genes have been identified that code for histone H1 subtypes. All H1 subtypes have a common three-domain structure. A globular central domain is attached to a short N-terminal domain and a long lysine and arginine-rich C-terminal domain. Both the globular and C-terminal domains contribute to the protein's ability to bind to chromatosomal DNA. Based on biochemical, genetic, and physical evidence Tom Misteli and coworkers proposed that H1 interacts with chromatosomal DNA, as shown in FIGURE 5.16. According to this model, H1 seals about 1.6 turns of DNA by binding to an exit/entrance site on the surface of the chromatosome. Considerable evidence indicates that histone H1 contributes to the higher-order folding states of chromatin.

We do not know how nucleosomes are arranged in chromatin.

Despite extensive efforts over more than 30 years, we still do not know how a chain of nucleosomes folds into higher order structures.

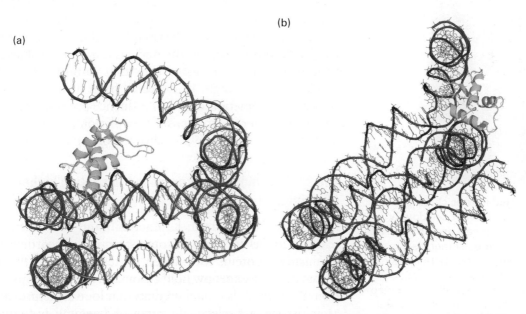

FIGURE 5.16 A molecular model of H1 interaction with the chromatosome. Viewed from the top (a) and the side (b) of the chromatosome. The H1 globular domain is shown in green, nucleosomal DNA is red, and the axis of two-fold symmetry is blue. The other histones are not shown. (Photo courtesy of David T. Brown, University of Mississippi Medical Center.)

5.6 The Nucleosome

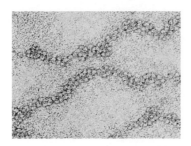

FIGURE 5.17 **The 30-nm chromatin fiber.** Electron micrograph of the 30-nm chromatin fiber in mouse chromosomes. (Courtesy of Barbara A. Hamkalo, University of California, Irvine.)

For some time investigators believed the next level of chromatin organization is a 30-nm fiber, which is visible under the electron microscope after a sample is fixed chemically, dehydrated, and embedded in plastic. FIGURE 5.17 shows the 30-nm chromatin fiber from a mouse chromosome. The existence of the 30-nm chromatin fiber, however, has been questioned by Jacques Dubochet and coworkers, who used cryoelectron microscopy to examine mitotic chromosomes in frozen hydrated sections of human cells that had not been fixed or stained. They did not see any evidence of the 30-nm fiber even though they could see microtubules of comparable size. Their experiments suggested that the mitotic chromosomes viewed in earlier electron microscopy experiments might have been altered during sample preparation. If so, the 30-nm fibers viewed in previous electron microscopy studies might have been artifacts. Even if the condensed chromosome does not have a discrete 30-nm fiber, the 30-nm fiber may be an intermediate in the compaction process.

Condensins and topoisomerase II help to stabilize condensed chromosomes.

In vitro studies show that three proteins, topoisomerase II, condensin I, and condensin II, are required for higher level chromatin condensation in vertebrates. Topoisomerase II disentangles DNA from different chromosomes, permitting each chromosome to fold into a discrete compact structure. Condensin I has a DNA-dependent ATPase activity, which permits it to introduce positive supercoils into closed circular DNA. The contribution that this catalytic activity makes to chromosome structure remains to be determined. Further studies are required to determine whether condensin II also has this activity. Histochemical studies of chromosomes from vertebrate cells reveal that condensin I, condensin II, and topoisomerase II are located along the long chromatid axis (FIGURE 5.18). Condensin I remains in the cytoplasm until the nuclear membrane is disassembled during prophase. In contrast, condensin II is present in the nucleus during interphase. Experiments have been performed using cells that cannot make condensin I or condensin II. As might be expected from their intracellular location, condensin I depletion does not influence prophase chromosome condensation, whereas condensin II depletion causes a significant delay in the initiation of prophase chromosome condensation.

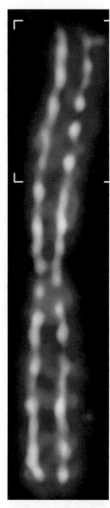

FIGURE 5.18 **Isolated mitotic chromosome from a human cervical tumor cell doubly stained for topoisomerase II and condensin.** The immunostain for topoisomerase II is green and that for condensin is red. DNA is stained blue. (Reproduced from *Dev. Cell*, vol. 4, K. Maeshima and U. K. Laemmli, A two-step scaffolding model for mitotic..., pp. 467–480, copyright 2003, with permission from Elsevier. Photo courtesy of Ulrich K. Laemmli, University of Geneva, Switzerland.)

The scaffold model was proposed to explain higher order chromatin structure.

In the late 1970s Ulrich K. Laemmli and coworkers proposed the **scaffold model** for chromosome structure. According to this model nonhistone proteins form a central scaffold along the long axis of a chromatid and somehow hold chromatin fibers in loops that extend out from the axis. This model predicts that loop structures are maintained after histones are selectively removed. Laemmli and James R. Paulson tested the model by first removing histones from isolated mitotic chromosomes and then examining the histone depleted chromosomes

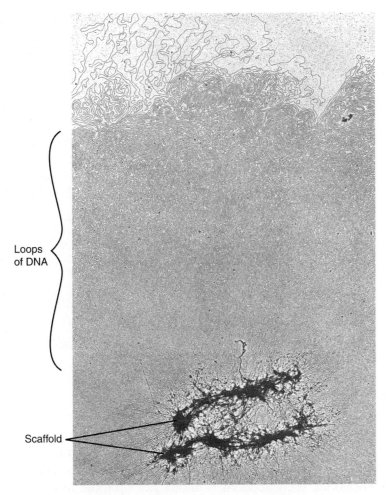

FIGURE 5.19 Scaffold structure and part of surrounding DNA. A structure called a scaffold becomes visible after the histones are depleted by treating the chromosome with 2 M NaCl. DNA loops extend from the scaffold. (Reproduced from Paulson, J. R., and Laemmli, U. K. 1977. The structure of histone..., *Cell* 12:817–828. Copyright 1977, with permission from Elsevier. Photo courtesy of Ulrich K. Laemmli, University of Geneva, Switzerland.)

under the electron microscope. As predicted by the model, the histone-depleted chromosomes have DNA loops attached to a protein scaffold (FIGURE 5.19).

5.7 The Centromere

The centromere is the site of microtubule attachment.

Each chromosome has a distinct morphological region, the **centromere**, which is visible as a thin region in the metaphase chromosome (FIGURE 5.20). The centromere is the chromosomal substructure responsible for a eukaryotic cell's ability to accurately partition sister chromatids between two daughter cells during mitosis and meiosis. It is usually the last attachment site between sister chromatids during

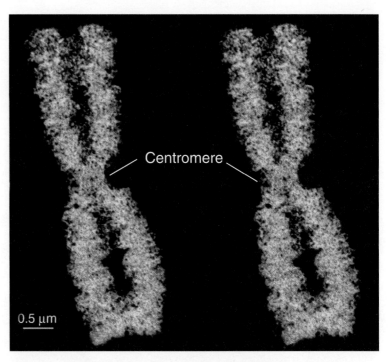

FIGURE 5.20 Chinese hamster ovary cell metaphase chromosome.
(Photo courtesy of Peter Engelhardt, University of Helsinki, Finland.)

mitosis. The connection is usually broken at anaphase by cleaving the cohesins that hold the sister chromatids together.

In budding yeast cells centromere DNA consists of a specific nucleotide sequence that is about 125 bp long. In contrast, human centromere DNA has a hierarchic organization. A 171-bp sequence is repeated with slight nucleotide sequence variations to form a higher repeat, which is in turn repeated with high sequence conservation to form a still higher order repeat, called an α-satellite DNA array, which ranges in size from 200 to 9,000 kb (**FIGURE 5.21**).

The centromere contains a specialized form of chromatin that is assembled by packing the DNA with histones and other proteins. The nonhistone proteins can be divided into two groups, constitutive and passenger proteins. Constitutive proteins remain associated with the centromere throughout the entire cell life cycle, whereas passenger proteins associate with the centromere during one stage of the life cycle but not others. One of the constitutive proteins, **CENP-A** (**centromere protein A**), is a homolog of histone H3 that combines with the other histones to form a new type of nucleosome that is dispersed among normal nucleosomes in the centromere.

Other constitutive and passenger proteins assemble to form the microtubule attachment site called the **kinetochore**. Electron microscopic analysis has shown that a single spindle-protein fiber is attached to the kinetochore in chromosomes present in budding yeast (*Saccharomyces cerevisiae*). Human cell chromosomes, which are much larger than yeast chromosomes, have several spindle fibers attached to their kinetochores. The kinetochore consists of two regions, an inner kinetochore region that interacts with the CENP-A containing

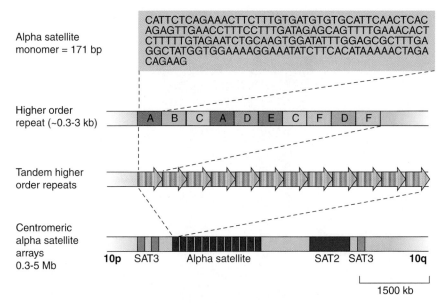

FIGURE 5.21 **DNA organization in human centromeres.** The hierarchic organization of a satellite is illustrated. A 171-bp sequence is repeated with slight nucleotide sequence variations to form a higher repeat, which in turn is repeated with high sequence conservation to form a still higher order repeat called an α-satellite DNA array, which ranges in size from 200 to 9,000 kb. At bottom is a diagram of the centromere region of chromosome 10, illustrating the ~2-Mb α-satellite array with surrounding pericentric satellite arrays (SAT2 and SAT3). (Adapted from Cleveland, D. W., et al. 2003. *Cell* 112:407–421.)

nucleosomes and an outer kinetochore region that is the site of attachment for microtubules that form the mitotic spindle fiber. FIGURE 5.22 shows a model proposed by K. H. Choo and coworkers for human centromere and the kinetochore structures.

5.8 The Telomere

The telomere, which is present at either end of a chromosome, is needed for stability.

Independent studies by two distinguished geneticists, Hermann J. Muller and Barbara McClintock, performed in the late 1930s revealed that the ends of a eukaryotic chromosome are essential for the chromosome's stability. Muller, who coined the term **telomere** (Gr. *telos* = end, *meros* = part) to describe chromosome ends, became aware of their importance as a result of experiments in which x-rays were used to produce deletions in *Drosophila* chromosomes. Muller was able to recover stable chromosomes with internal deletions but could not recover chromosomes with end deletions. Based on these results Muller concluded that telomeres are required for chromosome stability. McClintock reached a similar conclusion when she observed that broken chromosomes in maize fused together, underwent structural changes, or were degraded.

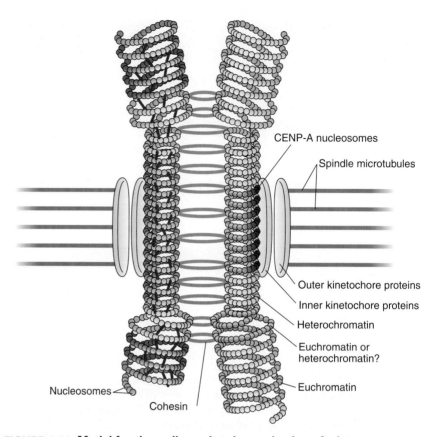

FIGURE 5.22 Model for three-dimensional organization of a human centromere. A chromatin fiber in the centromere may fold into a cylindrical coil. This arrangement would allow the individual blocks of nucleosomes that contain centromere protein A (CENP-A) molecules to form a single domain consisting of a series of repetitive functional units on the metaphase chromosome face on the poleward side, where they can interact with the remaining foundation of kinetochore proteins. On the opposite cylindrical face, intervening histone H3 blocks would also form a single domain. The extent to which this inner domain is heterochromatic or euchromatic is uncertain. This uncertainty has implications for the localization of cohesin (shown in gray), a multi-subunit protein complex that is necessary to maintain chromatid cohesion. CENP-A is shown in red; related centromere proteins are shown in purple or green. Heterochromatin is shown in yellow, euchromatin is shown in blue, and regions of uncertain heterochromatin or euchromatin status are shown in half blue and half yellow. Outer kinetochore proteins are shown in cyan. (Adapted from Amor, D. J., et al. 2004. *Trends Cell Biol* 14:359–368.)

Very little progress was made in understanding telomere structure until 1978 when Elizabeth H. Blackburn and Joseph G. Gall succeeded in isolating telomeres from *Tetrahymena thermophilia*, a ciliated protozoan with two kinds of nuclei, a micronucleus and a macronucleus. The micronucleus, which has five chromosome pairs, serves as the germ line nucleus, transferring genetic information from one generation to the next, but is not involved in transcription. The macronucleus, which is transcriptionally active, is derived from the micronucleus. During a stage of the *T. thermophilia* life cycle, chromosomes in the micronucleus are split into segments that then are amplified to produce thousands of "minichromosomes." The ends of this minichromosome were shown to have the short simple sequence, TTGGGG, repeated over and over.

The studies with *T. thermophilia* minichromosomes motivated investigators to determine whether chromosomes from other organisms also have simple sequence repeats at their ends. Telomeric DNAs from a variety of eukaryotes have now been characterized. In most cases the DNA consists of a simple repeat array in which the repeat unit is usually 5 to 8 bp in length.

Most telomeric DNA sequences have a high G content in the strand that runs 5'→3' toward the end of the chromosome. For example, the repeat array is TTAGGG in humans and other vertebrates and TTTAGGG in the single-celled green algae *Chorella*. The repeat pattern for these simple sequences varies from a few hundred base pairs in yeast and protozoans to a few thousand base pairs in vertebrates. The G-rich strand ends in a 3' single stranded overhang (FIGURE 5.23). That is, the G-rich strand extends beyond the 5'-end of the complementary C-rich strand. The overhang contains from two to three repeat units in simple eukaryotes but approaches 30 repeat units in humans.

Electron microscopy studies performed by Titia de Lange and coworkers in 1999 revealed that the G-rich strand's 3'-overhang folds back to invade the double-stranded region in the mammalian telomere to form a **t-loop** (telomere loop) and a **D-loop** (displacement loop; FIGURE 5.24). Specific proteins are required to establish and maintain telomere structure.

Although we focus here on proteins that protect the mammalian telomere, similar proteins are also present in other kinds of eukaryotes. Two telomere proteins, TRF1 and TRF2 (TTAGGG repeat binding factors 1 and 2), bind to the double-stranded region of the telomere as preformed homodimers. Although both proteins have similar

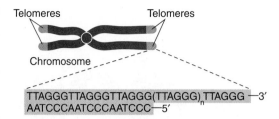

FIGURE 5.23 3'-Telomere overhang. Telomeres, shown in pink, are located at the ends of the sister chromatids. Each telomere has a G-rich 3'-overhang. In simple eukaryotes n usually has a value of 1 or 2, but in humans it may reach about 30.

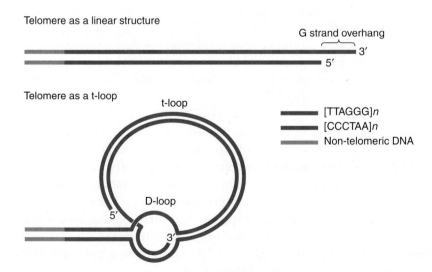

FIGURE 5.24 Structure of telomere t- and D-loops. A t-loop structure forms when the 3'-G strand extension at the end of a chromosome (telomere) invades duplex telomeric repeats, thereby forming a displacement loop (D-loop). (Adapted from Lundblad, V. 2000. *Science* 288:2141–2142.)

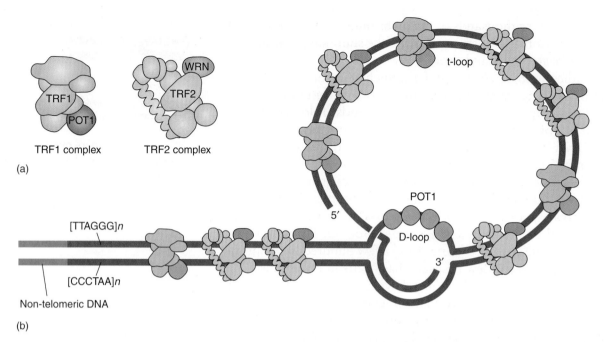

FIGURE 5.25 **Proposed structure of the human telomeric complex.** (a) TRF1 and TRF2 each recruit other proteins to the telomere. The major role of the TRF2 complex appears to be to protect the chromosome ends. The TRF1 complex appears to regulate telomerase-mediated telomere maintenance. (b) The 3′-G strand overhang in human telomeres, an array of TTAGGG repeats that is 100 to 200 nucleotides long, is connected to an array of TTAGGG repeats in the linear duplex region that is 2 to 30 kb long. The telomere DNA can exist as a t-loop in which the 3′-G strand overhang invades the duplex-repeat array to form a displacement or D-loop of TTAGGG repeats. Other configurations are also possible. POT1 (protection of telomeres 1) protein binds to the single-stranded TTAGGG-repeat DNA that results from t-loop formation. Two double-stranded TTAGGG repeat binding factors (TRFs), TRF1 and TRF2, interact with the duplex repeats. (Adapted from de Lange, T. 2004. *Nat Rev Mol Cell Biol* 5:323–329.)

structures, they have different functions. TRF1 is a negative regulator of telomere length. Cells that overproduce TRF1 appear to have shorter telomeres than normal. TRF2 participates in t-loop formation and also appears to be required to cap and protect the chromosome ends. TRF1 and TRF2 each recruit other proteins to the telomere.

The recruited proteins also influence telomere formation and structure. Defects in some of these proteins have been associated with inborn errors of metabolism. For example, one of the proteins recruited by TRF2, the Werner syndrome helicase (WRN), is defective in individuals with Werner syndrome, a genetic disease associated with premature aging. A second type of DNA binding protein, POT1 (protection of telomeres), binds to single-stranded DNA containing a TTAGGGTTAGGG decamer and is probably associated with the single strand in the D-loop. FIGURE 5.25 shows a model of the mammalian telomere with associated proteins. TRF1, TRF2, POT1, and three other proteins form a complex called shelterin that enables cells to distinguish their natural chromosome ends from DNA breaks. The three additional proteins, TINT1 (TRF1 interacting protein), TPP1, and Rap1, are not shown in Figure 5.25.

The cell must maintain its telomeres. Failure to do so leads to a wide assortment of serious problems, ranging from chromosome fusion to programmed cell death. Eukaryotic chromosomes lose DNA from their ends as a result of replication and require a special enzyme called telomerase to replace the lost DNA. Most human somatic cells lack telomerase and so cannot replace the telomere DNA lost during replication.

Failure to replace telomere DNA is probably responsible for the fact that somatic cells go through a finite number of divisions when studied in tissue culture, a phenomenon directly related to the aging process. Germ cells have telomerase and so can divide indefinitely. Most cancer cells also have telomerase. Telomerase is not an essential feature of tumor cells, however, because some tumor cells have an alternate means for maintaining their telomeres. Nevertheless, telomerase is an attractive target for treating many different types of cancers.

Questions and Problems

1. Briefly define, describe, or illustrate each of the following.
   - a. Chromatin
   - b. Histones
   - c. Nucleoid
   - d. Chromosome
   - e. Haploid
   - f. Diploid
   - g. Germ cells
   - h. Somatic cells
   - i. Homologs
   - j. Life cycle
   - k. Interphase
   - l. Euchromatin
   - m. Heterochromatin
   - n. S phase
   - o. G_1 and G_2 phases
   - p. Sister chromatid
   - q. Cohesin
   - r. Centromere
   - s. Nucleolus
   - t. Mitotic spindle
   - u. Microtubules
   - v. Synapsis
   - w. Bivalents
   - x. Chiasmata
   - y. Kinetochore
   - z. p and q arms
   - aa. Karyotype
   - bb. Autosomes
   - cc. Sex chromosomes
   - dd. Fluorescent *in situ* hybridization (FISH)
   - ee. Aneuploidy
   - ff. Nucleosome
   - gg. Linker DNA
   - hh. Nucleosome core particle
   - ii. Chromatosome
   - jj. Centrosome
   - kk. Kinetochore
   - ll. CENP-A
   - mm. t-loop
   - nn. D-loop
   - oo. Telomere

2. What is the bacterial nucleoid?
3. What factors contribute to bacterial chromatin compaction?
4. Describe the evidence that suggests bacterial chromatin is arranged in large loops.
5. Which contains more DNA, a rat sperm cell or a rat liver cell? Explain your answer.
6. During which phase of mitosis do the following events occur?
   - a. Chromatin condenses to form visibly distinct chromosomes.
   - b. The assembly of the mitotic spindle is completed and the spindle moves to the region previously occupied by the nucleus.
   - c. The nuclear region involved in ribosomal RNA synthesis, called the nucleolus, disappears and the nuclear membrane disassembles to form membrane vesicles.
   - d. Cohesin is cleaved and cohesion between sister chromatids is dissolved.
   - e. The spindle disappears, nuclear membranes form around the two groups of chromosomes, and nucleoli re-form.
   - f. Several spindle fibers attach to each chromosome in a region of the centromere called the kinetochore.
   - g. Chromosomes become less and less condensed until they can no longer be seen with a light microscope.

h. Spindle fiber attachment to the kinetochore is complete and the chromosomes move toward the center of the cell until the kinetochores lie in an imaginary plane equidistant from the two spindle poles.

7. Briefly describe or illustrate the process that maintains the chromosome number during cell division. Your answer should include the names of each stage and a description of what happens during each stage.

8. How does meiosis differ from mitosis?

9. A physician suspects that a patient has a deletion in 5q. What type of laboratory test might the physician recommend to test the hypothesis?

10. What happens when eukaryotic chromatin is briefly digested with micrococcal nuclease? What is the significance of the digestion products?

11. What are histones and how do they contribute to chromatin structure?

12. How does the amino acid composition of histones influence the proteins ability to perform their function?

13. How does H1 function differ from the function of the other four kinds of histones?

14. Describe the basic structural feature(s) of the nucleosome. How does a nucleosome differ from a chromatosome?

15. What functions do the centromere and kinetochore play during mitosis? How does the centromere differ from other regions of the chromosome?

16. What is the telomere?

17. What evidence suggests that telomeres are essential for chromosome stability?

18. Briefly describe or illustrate the way that DNA is organized in the telomere.

Suggested Reading

Bacterial Chromatin

Bendich, A. J., and Drlica, K. 2000. Prokaryotic and eukaryotic chromosomes: what's the difference. *Bioessays* 22:481–486.

Case, R. B., Chang, Y.-P., Smith, S. B., et al. 2004. The bacterial condensin MukBEF compacts DNA into a repetitive, stable structure. *Science* 305:222–227.

Cunha, S., Odjik, T., Süleymanoglu, E., and Woldringh, C. L. 2001. Isolation of the *Escherichia coli* nucleoid. *Biochimie* 83:149–154.

Dame, R. T. 2005. The role of nucleoid-associated proteins in the organization and compaction of bacterial chromatin. *Mol Microbiol* 56:858–870.

Dame, R. T., and Goosen, N. 2002. HU: promoting or counteracting DNA compaction? *Fed Eur Biochem Sci* 529:151–156.

de Vries, R. 2010. DNA condensation in bacteria: Interplay between macromolecular crowding and nucleoid proteins. *Biochimie* 92:1715–1721.

Dillon, S. C., and Dorman, C. J. 2010. Bacterial nucleoid-associated proteins, nucleoid structure and gene expression. *Nat Rev Microbiol* 8:185–195.

Hirano, T. 2005. SMC proteins and chromosome mechanics: from bacteria to humans. *Philos Trans R Soc Lond B Biol Sci* 360:507–514.

Hirano, T. 2006. At the heart of the chromosome: SMC proteins in action. *Nat Rev Mol Cell Biol* 7:311–322.

Kellenberger, E. 2006. Bacterial chromosome. *Encyclopedia of Life Sciences*. pp. 1–8. Hoboken, NJ: John Wiley & Sons.

Luijsterburg, M. S., Noom, M. C., Wuite, G. J. L., and Dame, R. T. 2006. The architectural role of nucleoid-associated proteins in the organization of bacterial chromatin: a molecular perspective. *J Struct Biol* 156:262–272.

Rimsky, S. 2004. Structure of the histone-like protein H-NS and its role in regulation and genome superstructure. *Curr Opin Microbiol* 7:109–114.

Rybenkov, V. V. 2009. Towards the architecture of the chromosomal architects. *Nat Struc Mol Biol* 16:104–105.

Saier, M. H. Jr. 2008. The bacterial chromosome. *Crit Rev Biochem Mol Biol* 43:89–134.

Sheratt, D. J. 2003. Bacterial chromosome dynamics. *Science* 301:780–785.

Tendeng, C., and Berlin, P. N. 2003. H-NS in gram-negative bacteria: a family of multifaceted proteins. *Trends Microbiol* 11:511–518.

Thanbichler, M., and Shapiro, L. 2006. Chromosome organization and segregation in bacteria. *J Struct Biol* 156:292–303.

Thanbichler, M., Wang, S. C., and Shapiro, L. 2005. The bacterial nucleoid: a highly organized and dynamic structure. *J Cell Biochem* 96:506–521.

Valens, M., Penaud, S., Rossignol, M., et al. 2004. Macrodomain organization of the *Escherichia coli* chromosome. *EMBO J* 23:4330–4341.

Wu, L. J. 2004. Structure and segregation of the bacterial nucleoid. *Curr Opin Genet Dev* 14:126–132.

Mitosis

Allison, D. C., Nestor, A. L., and Isaka, T. 2005. Chromosomes during cell division. *Encyclopedia of Life Sciences*. pp. 1–11. Hoboken, NJ: John Wiley & Sons.

Appels, R., Morris, R., Gill, B. S., and May C. E. 1998. *Chromosome Biology*. Norwell, MA: Kluwer Academic Publishers.

Criqui, M. C., and Genschik, P. 2002. Mitosis in plants: how far we have come at the molecular level? *Curr Opin Plant Biol* 5:487–493.

Gonczy, P. 2002. Nuclear envelope: torn apart at mitosis. *Curr Biol* 12:R242–R244.

Eissenberg, J. C., and Elgin, S. 2005. Heterochromatin and euchromatin. *Encyclopedia of Life Sciences*. pp. 1–7. Hoboken, NJ: John Wiley & Sons.

Gartenberg, M. 2009. Heterochromatin and the cohesion of sister chromatids. *Chromosome Res* 17:229–238.

Jordan, M. A., and Wilson, L. 1999. The use and action of drugs in analyzing mitosis. *Methods Cell Biol* 61:267–295.

McKee, B. D. 2004. Homologous pairing and chromosome dynamics in meiosis and mitosis. *Biochim Biophys Acta* 1677:165–180.

McNairn, A. J., and Gerton, J. L. 2008. The chromosome glue gets a little stickier. *Trends Genet* 24:382–389.

Mitchison, T. J., and Salmon, E. D. 2001. Mitosis: a history of division. *Nat Cell Biol* 3:E17–E21.

Moore, C. M., and Best, R. G. 2007. Chromosome mechanics. *Encyclopedia of Life Sciences*. pp. 1–10. Hoboken, NJ: John Wiley & Sons.

Nasmyth, K., and Haering, C. H. 2009. Cohesin: its roles and mechanisms. *Annu Rev Genet* 43:525–558.

Onn, I., Heidinger-Pauli, J. M. Guacci, V, Ünal, E., and Koshland, D. E. 2008. Sister chromatid cohesion: a simple concept with a complex reality. *Annu Rev Cell Dev Biol* 24:105–129.

Pines, J., and Rieder, C. L. 2001. Re-staging mitosis: a contemporary view of mitotic progression. *Nat Cell Biol* 3:E3–E6.

Peters, J.-M., and Hauf, S. 2005. Meiosis and mitosis: molecular control of chromosome separation. *Encyclopedia of Life Sciences*. pp. 1–7. Hoboken, NJ: John Wiley & Sons.

Peters, J.-M., Tedeschi, A. and Schmitz, J. 2008. The cohesin complex and its role in chromosome biology. *Genes Dev* 22:3089–3114.

Rieder, C. L., and Khodjakov, A. 2003. Mitosis through the microscope: advances in seeing inside live dividing cells. *Science* 300:91–96.

Russell, P. 1998. Checkpoints on the road to mitosis. *Trends Biochem Sci* 23:399–402.

Stukenberg, P. T. 2003. Mitosis: long-range signals guide microtubules. *Curr Biol* 13:R848–R850.

Watrin, E., and Legagneux, V. 2003. Introduction to chromosome dynamics in mitosis. *Biol Cell* 95:507–513.

Meiosis

John, B. 1990. *Meiosis*. Cambridge, UK: Cambridge University Press.

Kleckner, N. 1996. Meiosis: how could it work? *Proc Natl Acad Sci USA* 93:8167–8174.

Kohli, J., and Hartsuiker, E. 2008. Meiosis. *Encyclopedia of Life Sciences*. pp. 1–8. Hoboken, NJ: John Wiley & Sons.

Lichten M. 2001. Meiotic recombination: breaking the genome to save it. *Curr Biol* 11:R253–R256.

Page, S. L., and Hawley, R. S. 2003. Chromosome choreography: the meiotic ballet. *Science* 301:785–789.

Roeder, G. S. 1997. Meiotic chromosomes: it takes two to tango. *Genes Dev* 11:2600–2621.

Uhlmann, F. 2001. Chromosome cohesion and segregation in mitosis and meiosis. *Curr Opin Cell Biol* 13:754–761.

Karyotype

Bickmore, W. A. 2001. Karyotype analysis and chromosome banding. *Encyclopedia of Life Sciences*. pp. 1–7. London: Nature.

Blennow, E. 2005. Banding techniques. *Encyclopedia of Life Sciences*. pp. 1–5. Hoboken, NJ: John Wiley & Sons.

Craig, J. M., and Bickmore, W. A. 1993. Chromosome bands—flavours to savour. *Bioessays* 15:349–354.

Gosden, J. R. 1994. Chromosome analysis protocols. Walker, J. M. (ed). *Methods in Molecular Biology* (vol. 29). Totawa, NJ: Humana Press.

Moore, C. M., and Best, R. G. 2001. Chromosome preparation and banding. *Encyclopedia of Life Sciences*. pp. 1–7. London: Nature.

Rooney, D. E., and Czepulkowski, B. H. (eds). 1992. *Human Cytogenetics: A Practical Approach, vol. I, Constitutional Analysis* (2nd ed.). Washington, DC: IRL Press.

Rooney, D. E., and Czepulkowski, B. H. (eds). 1994. *Human Cytogenetics: Essential Data*. Hoboken, NJ: John Wiley & Sons.

Sandberg, A. 1990. *The Chromosomes in Human Cancer and Leukemia* (2nd ed.). Frankfurt, Germany: Elsevier.

Schröck, E., du Manoir, S., Veldman, T., et al. 1996. Multicolor spectral karyotyping of human chromosomes. *Science* 273:494–497.

Shaffer, L. G. 2005. Karyotype interpretation. *Encyclopedia of Life Sciences*. pp. 1–7. Hoboken, NJ: John Wiley & Sons.

Speicher, M. R. 2005. Chromosome. *Encyclopedia of Life Sciences*. pp. 1–7. Hoboken, NJ: John Wiley & Sons.

Nucleosome

Brown, T. D., Izard, T., and Misteli, T. 2006. Mapping the interaction surface of linker histone $H1^0$ with the nucleosome of native chromatin in vivo. *Nat Struct Mol Biol* 13:250–255.

Doenecke, D. 2005. Histones: From gene organization to biological roles. *Encyclopedia of Life Sciences*. pp. 1–7. Hoboken, NJ: John Wiley & Sons.

Dutnall, R., and Ramakrishnan, V. 1997. Twists and turns of the nucleosome: tails without ends. *Structure* 5:1255–1259.

Happel, N., and Doenecke, D. 2009. Histone H1 and its isoforms: contribution to chromatin structure and function. *Gene* 431:1–12.

Kornberg, R. D., and Lorch, Y. 1999. Twenty-five years of the nucleosome, fundamental particle of the eukaryote chromosome. *Cell* 98:285–294.

Khorasanizadeh, S. 2004. The nucleosome: from genomic organization to genomic regulation. *Cell* 116:259–272.

Luger, K., Mäder, A. W., Richmond, R. K., et al. 1997. Crystal structure of the nucleosome core particle at 2.8 A resolution. *Nature* 389:251–260.

Luger, K., and Richmond, T. J. 1998. DNA binding within the nucleosome core. *Curr Opin Struct Biol* 8:33–40.

Luger, K., and Richmond, T. J. 1998. The histone tails of the nucleosome. *Curr Opin Genet Dev* 8:140–146.

Luger, K. 2001. Nucleosomes: structure and function. *Encyclopedia of Life Sciences*. pp. 1–8. London: Nature.

Sandman, K., Soares, D., and Reeve, J. N. 2001. Molecular components of the archaeal nucleosome. *Biochimie* 83:277–281.

Widom, J. 1998. Structure, dynamics, and function of chromatin in vitro. *Annu Rev Biophys Biomol Struct* 27:285–327.

Wolfe, A. P. 2001. Nucleosomes: detailed structure and mutation. *Encyclopedia of Life Sciences*. pp. 1–8. London: Nature.

Woodcock, C. L., Skoultchi, A. I., and Fan, Y. 2006. Role of linker histone in chromatin structure and function: stoichiometry and nucleosome repeat length. *Chromosomal Res* 14:17–25.

Wu, J., and Grunstein, M. 2000. 25 years after the nucleosome model: chromatin modifications. *Trends Biochem Sci* 25:619–623.

The Scaffold Model

Belmont, A. S. 2002. Mitotic chromosome scaffold structure: new approaches to an old controversy. *Proc Natl Acad Sci USA* 99:15855–15857.

Belmont, A. S. 2006. Mitotic chromosome structure and condensation. *Curr Opin Cell Biol* 18:632–638.

Hirano, T. 2005. Condensins: organizing and segregating the genome. *Curr Biol* 15:R265–R275.

Losada, A., and Hirano, T. 2005. Dynamic molecular linkers of the genome: the first decade of SMC proteins. *Genes Dev* 19:1269–1287.

Maeshima, K., and Eltsov, M. 2008. Packaging the genome: the structure of mitotic chromosomes. *J Biochem* 143:145–153.

Maeshima, K., and Laemmli, U. K. 2003. A two-step scaffolding model for mitotic chromosome assembly. *Dev Cell* 4:467–480.

The Centromere

Amor, D. J., Kalitsis, P., Sumer, H., and Choo, K. H. 2004. Building the centromere: from foundation proteins to 3D organization. *Trends Cell Biol* 14:359–368.

Black, B. E., and Basset, E. A. 2008. The histone variant CENP-A and centromere specification. *Curr Opin Cell Biol* 20:91–100.

Black, B. E., Foltz, D. R., Chakravarthy, S., et al. 2004. Structural determinants for generating centromeric chromatin. *Nature* 430:578–582.

Cleveland, D. W., Mao, Y., and Sullivan, K. F. 2003. Centromeres and kinetochores: from epigenetics to mitotic checkpoint signaling. *Cell* 112:407–421.

Henikoff, S., and Dalal, Y. 2005. Centromeric chromatin: what makes it unique? *Curr Opin Genet Dev* 15:1–8.

Hill, E., and Williams, R. 2009. Super-coil me: Sizing up centromeric nucleosomes. *J Cell Biol* 186:453–456.

Murphy, T. D., and Karpen, G. H. 1998. Centromeres take flight: alpha satellite and quest for the human centromere. *Cell* 93:317–320.

Przewloka, M. R., and Glover, D. M. 2009. The kinetochore and the centromere: a working long distance relationship. *Annu Rev Genet* 43:439–465.

Sullivan, B. A., and Karpen, G. H. 2004. Centromeric chromatin exhibits a histone modification pattern that is distinct from euchromatin and heterochromatin. *Nat Struct Mol Biol* 11:1076–1083.

The Telomere

Baumann, P., and Cech, T. R. 2001. POT1, the putative telomere end-binding protein in fission yeast and humans. *Science* 292:1171–1174.

Biasco, M. A. 2003. Mammalian telomeres and telomerase: why they matter for cancer and aging. *Eur J Cell Biol* 82:441–446.

Blackburn, E. H. 2001. Telomeres. *Encyclopedia of Life Sciences*. pp. 1–7. London: Nature.

Cech, T. R. 2004. Beginning to understand the end of the chromosome. *Cell* 116:273–279.

Court, R., Chapman, L., Fairall, L., and Rhodes, D. 2004. How the human telomeric proteins TRF1 and TRF2 recognize telomeric DNA: a view from high-resolution crystal structures. *EMBO Rep* 6:39–45.

de Lange, T. 2002. Protection of mammalian telomeres. *Oncogene* 21:532–540.

de Lange, T. 2004. T-loops and the origin of telomeres. *Nat Rev Mol Cell Biol* 5:323–329.

de Lange, T., and DePinho R. A. 1999. Unlimited mileage from telomerase? *Science* 283:947–949.

de Lange, T., and Jacks, T. 1999. For better or worse? Telomerase inhibition and cancer. *Cell* 98:273–275.

Fletcher, T. M. 2003. Telomere higher-order structure and genomic instability. *IUBMB Life* 55:443–449.

Greider, C. 1998. Telomerase activity, cell proliferation, and cancer. *Proc Natl Acad Sci USA* 95:90–92.

Greider, C. 1999. Telomeres do D-loop-T-loop. *Cell* 97:419–422.

Lei, M., Podell, E. R., Baumann, P., and Cech, T. R. 2003. DNA self-recognition in the structure of POT1 bound to telomeric single-stranded DNA. *Nature* 426:198–203.

Lei, M., Podell, E. R., and Cech, T. R. 2004. Structure of human POT1 bound to telomeric single-stranded DNA provides a model for chromosome end-protection. *Nat Struct Mol Biol* 11:1223–1229.

Palm, W., and de Lange, T. 2008. How shelterin protects mammalian telomeres. *Annu Rev Genet* 42:301–304.

Pardue, M.-L., and DeBaryshe, P. G. 1999. Telomeres and telomerase: more than the end of the line. *Chromosoma* 108:73–82.

Pardue, M. L., and DeBaryshe. 2001. Telomeres in cell function: cancer and ageing. *Encyclopedia of Life Sciences*. pp. 1–6. London: Nature.

Rhodes, D., Fairall, L., Simonsson, T., et al. Telomere architecture. *EMBO Rep* 3:1139–1145.

Shay, J. W. 1999. At the end of the millennium, a view of the end. *Nat Genet* 23:382–383.

Wei, C., and Price, C. M. 2003. Protecting the terminus: t-loops and telomere end-binding proteins. *Cell Mol Life Sci* 60:2282–2294.

Genetic Analysis in Molecular Biology

CHAPTER OUTLINE

6.1 *Drosophila melanogaster*
Many fundamental genetic principles were discovered by studying the common fruit fly.

6.2 *Escherichia coli*
E. coli is a gram-negative bacterium.
Bacteria can be cultured in liquid or solid media.
Plating can be used to detect auxotrophs.
Specific notations, conventions, and terminology are used in bacterial genetics.
Cells with altered genes are called mutants.
Some mutants display the mutant phenotype under all conditions, whereas others display it only under certain conditions.
Mutations can be classified on the basis of the changes in the DNA.
Mutants have many uses in molecular biology.
A genetic test known as complementation can be used to determine the number of genes responsible for a phenotype.
E. coli cells can exchange genetic information by conjugation.
The F plasmid can integrate into a bacterial chromosome and carry it into a recipient cell.
Bacterial mating experiments can be used to produce an *E. coli* genetic map.
F′ plasmids contain part of the bacterial chromosome.
Plasmid replication control functions are usually clustered in a region called the basic replicon.

6.3 Budding Yeast (*Saccharomyces cerevisiae*)
Yeasts are unicellular eukaryotes.
Specific notations, conventions, and terminology are used in yeast genetics.
Yeast cells exist in haploid and diploid stages.

6.4 New Tools for Studying Human Genetics
Different humans may exhibit DNA sequence variations within the same DNA region.
Restriction fragment length polymorphisms facilitate genetic analysis in humans and other organisms.
Somatic cell genetics can be used to map genes in higher organisms.

BOX 6.1: LOOKING DEEPER Plasmid Functions Required for Mating

BOX 6.2: IN THE LAB DNA Fingerprint

QUESTIONS AND PROBLEMS

SUGGESTED READING

Image © whatie/ShutterStock, Inc.

The term "gene" (Greek *genos* = birth) was coined by the Danish botanist William Louis Johannsen in 1909 to describe the basic unit of heredity. Early investigators worked to determine the specific locations of genes with respect to one another on the chromosomes of organisms by performing mating experiments. The first studies of this type were performed by Gregor Mendel in the 1860s. Mendel was very fortunate in the physical traits he chose to study in pea plants because the alleles responsible for these traits were on different chromosomes and therefore inherited independently of one another. In 1905, shortly after Mendel's work had been rediscovered, William Bateson, Edith Rebecca Saunders, and Reginald C. Punnett performed genetic crosses with sweet pea plants that showed different physical traits may also be inherited together. That is, they discovered exceptions to Mendel's law of independent assortment. Based on their findings, Bateson and coworkers proposed that certain alleles must somehow be linked, but they were unable to provide a physical explanation for this linkage. A few years later investigators used the fruit fly (*Drosophila melanogaster*) as a model system to show that linked genes are real physical entities. Not surprisingly, as more genetic information was gained about inheritance in fruit fly, there was more incentive to select this organism for further study. Thus, the fruit fly became the favorite model organism for geneticists and remained so for many years.

By the 1950s interest started to shift toward other living systems, particularly the bacteria *Escherichia coli* and some of the viruses that infect it. This shift occurred because bacteria divide very rapidly, permitting genetic experiments to be completed in a day or two. Thanks to the rapidity of bacterial genetic experiments, investigators were able to determine the locations of hundreds of genes on the *E. coli* chromosomes. These gene locations are summarized in the form of a **genetic map**, which is quite helpful when one wishes to clone, move, disrupt, or otherwise manipulate a gene. Although bacterial genetic studies provide important insights into gene structure and function, they do not address important questions that are unique to eukaryotic cells. A species of budding yeast called *Saccharomyces cerevisiae*, which has been the subject of extensive biochemical studies, has proven to be an excellent model system for the study of eukaryotic genes because it grows rapidly and is easy to maintain in the laboratory.

This chapter examines some important aspects of genetic analysis in fruit flies, *E. coli*, and *S. cerevisiae*. Some new techniques used to study genes in multicellular organisms also are introduced. Even though the discussion that follows is limited to a few organisms, the concepts developed are applicable to all organisms.

6.1 *Drosophila melanogaster*

Many fundamental genetic principles were discovered by studying the common fruit fly.

In 1908 Thomas Hunt Morgan attempted to study heredity in animals such as mice and rats but quickly came to realize the animals' slow reproduction rates limited the amount of genetic data that could be

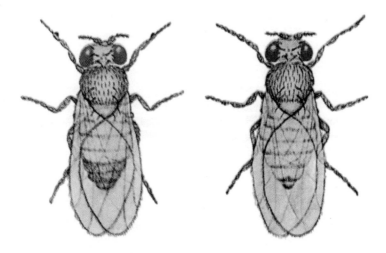

FIGURE 6.1 **Male (left) and female (right) fruit fly.**

collected within a reasonable span of time. Furthermore, maintaining rats or mice required considerable space and expense. Morgan therefore sought an alternative animal system that did not suffer from these disadvantages. His attention was soon directed to a tiny winged-insect, *Drosophila melanogaster*, which is known as the common fruit fly because it feeds on decaying fruit. The fruit fly has the advantages of being fertile all year long, producing a new generation every 12 days, and requiring very little space (about 1,000 flies, each about 2.5 mm long, can be collected in a 1-quart glass milk bottle). Furthermore, the fruit fly has just four pairs of chromosomes: one pair of sex chromosomes and three pairs of autosomes. Male and female flies are easily distinguished based on striping patterns, external genitalia, and the presence of dense bristles on the legs of male flies (FIGURE 6.1).

Morgan's experiments suggested that chromosomes sometimes exchange segments in a process that he called crossing over. He proposed that crossing over takes place during meiosis when homologous chromosomes pair and involves a physical exchange between the chromosomes to produce recombinant chromosomes (FIGURE 6.2). This hypothesis was confirmed in 1931 by studies in *Drosophila* and maize that showed a correlation between crossing over and an exchange between chromosomes by using chromosomal variations that could be seen under a light microscope.

The molecular mechanism of recombination is complex. For present purposes it is sufficient to assume that two chromosomes align with one another, a cut is made in both chromosomes at random but matching points, and the four fragments are then joined together to form two new combinations of genes. This crude model accounts for only some of the features of genetic exchange, but these features are in fact the only ones of concern at this time.

Morgan proposed the frequency of recombinant meiotic products resulting from the genetic exchange would increase as the distance between the genes increases. In 1911 Alfred H. Sturtevant, then an undergraduate student working in Morgan's laboratory, used

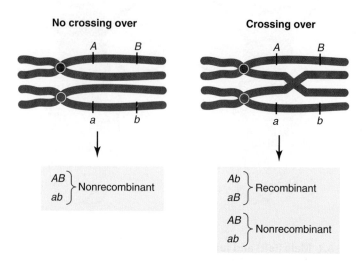

FIGURE 6.2 Genetic recombination between homologous chromosomes. (a) No crossing over between the *A* and *B* genes gives rise to only nonrecombinant gametes. (b) Crossing over between the *A* and *B* genes gives rise to the recombinant gametes Ab and aB and the nonrecombinant gametes AB and ab.

recombination frequencies to construct a linkage map that showed the order and spacing of genes on the *Drosophila* X chromosome. The unit of distance along a genetic map that corresponds to a 1% frequency of recombination is called a centimorgan (cM) in honor of Morgan. In humans, a cM corresponds to an average physical distance of about 1×10^6 base pair (bp).

Most of the genetic mapping studies performed in the first part of the twentieth century were based on the principle that the distance along the chromosome between two recombining genes determines the recombination frequency. As long as the two genes are not too close to one another, the recombination frequency is proportional to distance, because chromosomal crossovers take place at random. If crossing over takes place between two chromosomes with the alleles *Abc* and *aBC* arranged in alphabetical order and equally spaced,

$$\begin{array}{ccc} A & b & c \\ \hline & \times & \\ a & B & C \end{array}$$

there will be twice as many *AC* recombinants as *AB* recombinants, because loci *A* and *C* are twice as far apart as loci *A* and *B*.

Because the recombination frequency is proportional to distance, recombination frequency can be used to determine the arrangement of genes on the chromosome. This can be seen in a simple example in which three genes—*A*, *B*, and *C*—are arranged along a single chromosome in an unknown order. If we assume the recombination frequency between genes *A* and *B* is 1% ($A \times B = 1\%$) and the recombination frequency between genes *B* and *C* is 2% ($B \times C = 2\%$), then the two

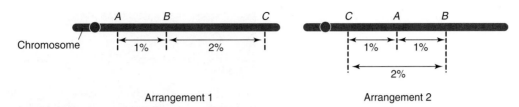

FIGURE 6.3 Mapping. The arrangements for which the recombination frequency between genes *A* and *B* is 1%, and the recombination frequency between genes *B* and *C* is 2%.

arrangements shown in **FIGURE 6.3** are consistent with the data. The correct order can be determined from the recombination frequency between genes *A* and *C*. Let us assume this is 1%. If that is the case, only arrangement 2 is possible. The order is C A B for these genes and the relative separation constitute a genetic map.

Any number of genes can be mapped in this way. For instance, consider a fourth gene, *D*, in the preceding example. If $D \times B = 0.5\%$, *D* must be located 0.5 units either to the left or to the right of *B*. If $A \times D = 1.5\%$, *D* is clearly to the right of *B* and the gene order is C A B D. If $A \times D = 0.5\%$, the gene order would be C A D B.

The analysis just given has been oversimplified because the occurrence of multiple exchanges has been ignored. If two crossover events occur between two genes, then recombination between the two genes is not observed, because the second event cancels the effect of the first. Discussion of this important point can be found in any genetics textbook; however, the effect of multiple exchanges is unimportant for the simple considerations described here. The general approach outlined above is used to map the genes in plants and animals by performing mating experiments.

6.2 *Escherichia coli*

E. coli is a gram-negative bacterium.

Many basic concepts in molecular biology have their origins in studies of the genetics of *E. coli* and the viruses that infect it. Because a basic knowledge of bacterial structure, physiology, and growth is required to fully appreciate these genetic studies, we start with a brief introduction to bacteria. Like all other cells a bacterial cell is enclosed by a membrane, which separates the cytoplasm from the surroundings. A rigid multilayered cell wall made of **peptidoglycan** (a combination of polysaccharides and peptides) surrounds the membrane, protecting the cell from osmotic and mechanical damage and giving the cell its characteristic shape, which usually is spherical, rod-like, or spiral (**FIGURE 6.4**).

Bacteria can be divided into two major groups based on the thickness of their cell walls by using a staining procedure that was devised by the Danish bacteriologist Christian Gram in 1884. In brief, a bacterial

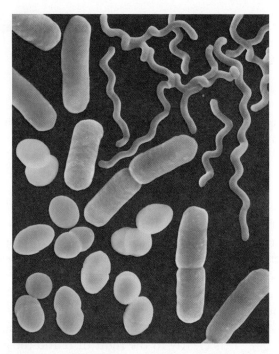

FIGURE 6.4 **Photocomposite of three common types of bacterial morphology—coccus, bacillus, and spirillum.** These spherical, rod, and spiral shaped morphologies are typical of such genera as *Streptococcus* or *Staphylococcus* (green), *Escherichia* or *Bacillus* (orange), and *Leptospira* or *Spirillum* (yellow), respectively. The colors shown here do not represent actual colors of the organisms. (© Phototake, Inc./Alamy Images.)

suspension is placed on a microscope slide and air dried. Then the slide is covered with a crystal violet solution for a short time, washed with water, and exposed to an iodine solution. Both gram-positive and gram-negative bacteria now have a purple color due to the presence of a crystal violet-iodine complex in their cytoplasm. Washing the slide with an acetone-ethanol solution decolorizes the gram-negative bacteria but not the gram-positive bacteria. (This difference occurs because the thick cell walls of gram-positive bacteria prevent dye extraction, whereas the thin cell walls of gram-negative bacteria do not.) Finally, cells are treated with a second dye, safranin, which does not alter the purple color of gram-positive bacteria such as *Bacillus subtilis* but causes gram-negative bacteria such as *E. coli* to stain pink. Thus, the distinct staining properties of gram-positive bacteria (purple) and gram-negative bacteria (pink) are determined by differences in their cell walls.

One other important structural feature distinguishes gram-negative bacteria from gram-positive bacteria. Gram-negative bacteria have an additional membrane called the **outer membrane** that surrounds their cell wall, but gram-positive bacteria do not (FIGURE 6.5). Small water-soluble molecules pass through protein channels called **porins** in the outer membrane. Secreted proteins are too large to pass through the porins, however, and so most are retained in the space between the cell membrane and outer membrane called the **periplasmic space**. Some periplasmic proteins bind nutrients and assist in their transport across the cell membrane, whereas others are digestive enzymes. The outer

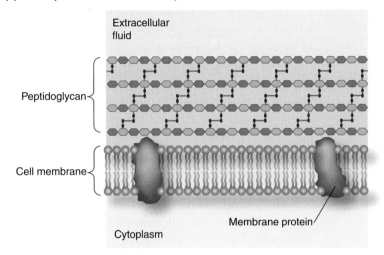

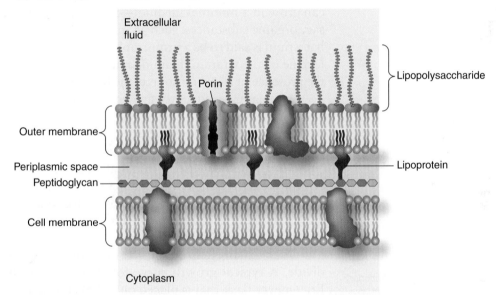

FIGURE 6.5 Bacterial cell envelopes. Bacteria are classified in two groups, gram-positive and gram-negative bacteria. (a) Gram-positive bacteria have a very thick cell wall made of peptidoglycan surrounding the cell or plasma membrane. (b) Gram-negative bacteria have a thinner cell wall made of peptidoglycan surrounding the cell (also called plasma or inner) membrane. A second membrane called the outer membrane surrounds the cell wall. The outer leaflet of the outer membrane contains lipopolysaccharides. Lipoproteins extend from the peptidoglycan to the outer membrane. The space between the inner and outer membranes is called the periplasmic space.

membrane serves as a barrier to detergents and hydrophobic antibiotics, protecting the cell from them. Gram-positive bacteria, which lack an outer membrane, are usually much more sensitive to detergents and hydrophobic antibiotics than are gram-negative bacteria.

Bacteria can be cultured in liquid or solid media.

Bacteria are easy to culture, divide rapidly, and have relatively simple nutritional requirements compared with cells in multicellular organisms. The bacterium that has served as perhaps the most important model system in molecular biology is the gram-negative bacterium *E. coli*, which divides about every 30 to 60 minutes at 37°C under laboratory conditions. A single *E. coli* cell thus multiplies exponentially to become 10^9 bacteria in less than a day.

Bacteria can grow in a liquid growth medium or on a solid surface. A population growing in a liquid medium is called a bacterial culture. If the liquid is a complex extract of biological material, it is called a **broth**. An example is tryptone broth, which contains the milk protein casein hydrolyzed by the digestive enzyme trypsin to yield a mixture of amino acids and small peptides. If the growth medium is a simple chemically defined mixture containing no organic compounds other than a carbon source such as a sugar, it is called a **minimal medium**. A typical minimal medium contains sodium, potassium, magnesium, calcium, ammonium, chloride, phosphate, sulfate, and a few trace metal ions as well as a carbon source such as glucose or glycerol. Bacteria grow more rapidly in a minimal medium containing glucose than in a minimal medium with any other single carbon source. If a bacterium can grow in a minimal medium—that is, if it can synthesize all necessary organic substances, such as amino acids, vitamins, and lipids—the bacterium is said to be a **prototroph**. If a nutrient such as the amino acid leucine or the vitamin thiamine must be added for growth to occur in the presence of a carbon source, the bacterium is termed an **auxotroph**.

Bacteria are frequently grown on solid surfaces. Agar, a gelling agent obtained from seaweed, which is resistant to enzymes from most microorganisms, is universally used to solidify liquid medium. A solid growth medium is called a nutrient agar if a broth medium is gelled or a minimal agar if a minimal medium is gelled. Solid media are typically placed in a shallow glass or plastic flat-bottomed dish with a lid called a Petri dish. In laboratory jargon a Petri dish containing a solid medium is called a plate and the act of depositing bacteria on the agar surface is called plating.

When *E. coli* cells are placed in a liquid medium, they grow and divide. A typical growth curve for an *E. coli* culture growing in an Erlenmeyer flask that is placed on a shaker and maintained at 37°C is shown in **FIGURE 6.6**. After an initial period of slow growth called the **lag phase**, the bacteria begin a period of rapid, exponential growth in which they divide at a fixed time interval called the **doubling time**. The number of cells per milliliter, the **cell density**, doubles repeatedly, giving rise to a logarithmic increase in cell number. This stage of growth of the bacterial culture, called the **log phase**, continues until the *E. coli* achieve a cell density of about 10^9 cells • mL^{-1}. At this cell density the O_2 supply and pH limit the rate of cell growth. *E. coli* growth usually ceases at a cell density of 2 to 3×10^9 cells • mL^{-1} and the bacteria enter the **stationary phase**. Considerably higher cell densities can be achieved by increasing the rate of aeration, maintaining the medium at optimal pH, or both. The terms used in this section are also used in discussing the growth of all microorganisms and frequently of animal cells.

A bacterium growing on an agar surface also divides. Because most bacteria are not very motile on a solid surface, the progeny bacteria remain very near the location of the original bacterium. The number of progeny increases so much that a visible cluster of bacteria appears.

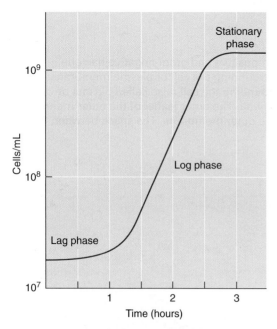

FIGURE 6.6 **Typical bacterial growth curve.**

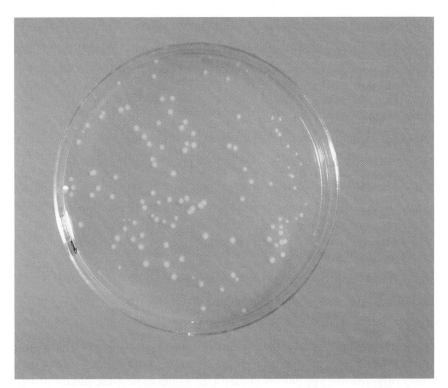

FIGURE 6.7 **A Petri dish with bacterial colonies that have formed on agar.** (Photo courtesy of Dr. Jim Feeley/CDC.)

Such a cluster, which arises from a single cell, is called a colony or **clone** (FIGURE 6.7). Colony formation allows one to determine the number of bacteria in a culture. For instance, if 100 cells are plated, 100 colonies will be visible the next day. If 0.1 mL of a 10^6-fold dilution of a bacterial culture is plated and 200 colonies appear, the cell density in the original culture is $(200/0.1)(10^6) = 2 \times 10^9$ cells/mL. One can also monitor bacterial cell growth by following the culture's turbidity (cloudiness), which results from the presence of the suspended bacteria. Turbidity is easy to follow with an instrument such as a colorimeter or a spectrophotometer that detects light transmission through a suspension.

Investigators sometimes need to isolate a genetically pure colony from a culture that contains a mixture of closely related bacteria that have different genotypes. They accomplish this task by using a technique called "streaking." A sterile toothpick is dipped into the culture or allowed to touch the colony. This toothpick is streaked across the agar surface in a Petri dish. The number of cells deposited at the end of the streak is much lower than at the beginning. A second sterile toothpick is passed across the end of the original streak to form a new streak. Two additional sterile toothpicks are used to make a third and fourth streak. The number of bacteria spread decreases greatly from one streak to another. Then the plate is incubated overnight at 37°C. There are so many bacteria in the first two streaks that no distinct colonies are observed. But the fourth streak (and sometimes the third streak) contains so few bacteria that distinct colonies appear. Because each colony arises from a single bacterium deposited during the streaking process, it represents a genetically homogenous cell population.

6.2 *Escherichia coli*

Plating can be used to detect auxotrophs.

Plating is a convenient way to determine if a bacterium is an auxotroph. Several hundred bacteria are spread on minimal agar and nutrient agar plates, which are stored overnight in a constant-temperature incubator. Several hundred colonies are subsequently observed on the nutrient agar plate because it contains so many substances it can satisfy the requirements of nearly any bacterium. If colonies are also observed on the minimal agar, the bacterium is a prototroph; if no colonies are observed, it is an auxotroph and some required substance is missing from the minimal agar plates. Minimal agar plates are then prepared with various supplements. If the bacterium is a leucine auxotroph, the addition of leucine alone will enable a colony to form. If both leucine and histidine must be added, the bacterium is auxotrophic for both of these substances.

Specific notations, conventions, and terminology are used in bacterial genetics.

Bacterial geneticists use specific notations to indicate genotypic and phenotypic traits. As we will see below, another system of notations and conventions applies to yeast. The phenotype of a bacterial cell that can synthesize the amino acid leucine is denoted Leu^+, whereas the phenotype of a bacterial cell that cannot do so is denoted Leu^-. Note the symbol for the phenotype has three letters, begins with an uppercase letter, and is not italicized. A Leu^- cell lacks a gene needed to synthesize leucine; this gene is denoted *leu*$^-$ (or in some books, *leu*, without the minus sign). Note that the genotype is written in lowercase letters and is italicized. A gene that specifies a particular protein is said to be the **structural gene** for that protein.

A bacterial cell may require several genes to synthesize leucine. These genes are usually denoted *leuA*, *leuB*, and so on, with all letters italicized. A Leu^+ (leucine-synthesizing) cell must have one functional copy of every requisite gene. Thus, its genotype must be *leuA*$^+$, *leuB*$^+$, and so on and would normally be summarized by writing *leu*$^+$ unless it is important for some reason to state the genotype of each gene. A Leu^+ cell might also be diploid for *leu* genes and have a defective gene such as *leuA*$^-$ in one chromosomal set. The two haploid sets are separated in their notation by a diagonal line; the genotype is therefore *leu*$^+$/*leuA*$^-$. Some genes are responsible for resistance and sensitivity to certain extracellular substances such as antibiotics. The genotypes of an ampicillin-resistant cell and an ampicillin-sensitive cell are written *amp*^{R} and *amp*^{S}, respectively; the phenotypes are correspondingly Amp^R and Amp^S.

In summary, for bacteria the following conventions are used:

1. Abbreviations of phenotypes contain three regular typeface letters (the first uppercase) with a superscript "+" or "−" to denote presence or absence of the designated character and superscript "S" and "R" for antibiotic sensitivity and resistance, respectively.
2. The genotype designation is always lowercase, and all components of the symbol are italicized.

CHAPTER 6 GENETIC ANALYSIS IN MOLECULAR BIOLOGY

We also have occasion to designate particular mutants of a gene. Mutants are usually numbered in the order in which they were isolated. The fifty-eighth and seventy-ninth leucine mutants discovered, thus, would be written *leu58* and *leu79*. To denote mutants of a particular gene one would write *leuA58* and *leuB79* if the mutations were in the *leuA* and *leuB* genes, respectively.

The term "**genome**" is useful, although it has evolved to have various meanings. It is correctly defined as the genetic complement (the set of all genes and genetic signals) of a cell or virus. With eukaryotes the term is often used to refer to one complete (haploid) set of chromosomes. In laboratory jargon, when discussing bacteria, phages, or most animal and plant viruses, the term often refers to their single DNA or RNA molecule(s), a classically incorrect but now accepted usage.

Cells with altered genes are called mutants.

We have seen that a gene can be either functional or nonfunctional and that these states are denoted by a superscript (+) or (−) after the gene abbreviation. The functional form of a gene is sometimes called **wild type** because presumably this is the form found in nature. This term is ambiguous though, because often the (−) form is the one that is prevalent. For example, many bacterial species isolated from nature carry the genes for lactose metabolism, yet in many strains the gene is either nonfunctional or missing, that is, *lac⁻*.

We try to use the term "wild type" as little as possible. The precise genetic term "allele" is used to indicate that there are alternative forms of a gene, and sometimes the (+) and (−) forms are called the (+) allele and the (−) allele, respectively. It is very common to call the nonfunctional (−) form of an allele a mutant form. Strictly speaking, a mutant is an organism with a genotype (or, more precisely, a DNA base sequence) that differs from that found in nature. However, it is more convenient (and definitely it is common jargon) to equate the terms "mutant gene" and "nonfunctional gene," and we will use the word "mutant" in that sense. A word of caution is needed, however, because in some instances a mutant has a more active enzyme than its parent.

The process by which the genetic information of an organism is altered in a stable, heritable manner is called mutagenesis. In nature and in the laboratory genetic alteration sometimes arises spontaneously. That is, mutations sometimes arise without any help from the experimenter. This spontaneous process is called **spontaneous mutagenesis**. Mutagenesis can also be induced by exposing cells to chemical agents called **mutagens** or physical agents such as ultraviolet light and x-rays, which cause chemical alterations of the genetic material.

Some mutants display the mutant phenotype under all conditions, whereas others display it only under certain conditions.

Mutants can be classified in several ways. One classification is based on the conditions in which the mutant character is expressed. An absolute defective mutant displays the mutant phenotype under all conditions; that is, if a bacterium requires leucine for growth in all culture media and at all temperatures, it is an absolute defective. A **conditional mutant** does not always show the mutant phenotype. Instead, its behavior depends on physical conditions and sometimes on the presence of other mutations. An important example of a conditional mutant is a **temperature-sensitive (Ts) mutant**, which behaves normally below 30°C (the **permissive temperature**) and as a mutant above 42°C (the **nonpermissive temperature**); intermediate states are usually observed between these temperatures. Note that the gene does not mutate above 42°C; rather, the product of the gene is inactive above 42°C.

Temperature-sensitive mutants have been of great use in the laboratory because they enable one to turn off the activity of a gene product simply by raising the temperature. In many cases the temperature-sensitive defect is reversible, so the activity of the gene product can be regained by lowering the temperature. Temperature-sensitive mutants are especially useful for studying the function(s) of an essential gene. The temperature-sensitive mutant can be maintained at a permissive temperature and then shifted to a nonpermissive temperature for study.

Another widely encountered type of conditional mutant is the **suppressor-sensitive mutant**, which exhibits the mutant phenotype in some strains but not in others. The difference is a consequence of the presence of particular gene products, called **suppressors**. These suppressor gene products either compensate for the defect in the mutant or, in a variety of ways, enable an altered gene to produce a functional gene product. In the jargon of molecular biology, one says that the phenotype of a suppressor-sensitive mutant "depends on the genetic background." A mutation is sometimes designated as suppressor-sensitive by adding the regular typeface letters Am (refers to amber), Oc (for ochre), or Op (for opal) in parentheses. (Amber, ochre, and opal are terms that arose from a private laboratory joke.) The symbols are regular typeface and capitalized because they represent a phenotype. If the mutant has a number, it is also added. Thus, if mutation 35 in the *leu* gene is a suppressor-sensitive Am type, it would be written *leu*35(Am).

Mutations can be classified on the basis of the changes in the DNA.

Another method for classifying mutations is based on the number of changes that have occurred in the genetic material (that is, the number of DNA base pairs that have changed; FIGURE 6.8). If only one base pair change has occurred, the mutant is called a **point mutation**. Sometimes a mutation occurs by means of removal of all or part of a gene; in that case the mutation is a **deletion**. Still another type of mutation, an **insertion**, involves placing additional DNA, ranging in size from a single base pair to a DNA segment that is thousands of base pairs long, within a gene.

A mutant sometimes regains its original phenotype. This occurs by means of chemical changes in the mutant organism's genetic material that either reverse the original mutation or compensates for it. The process of regaining the original phenotype is called reversion, and an organism that has reverted is called a **revertant**.

(a) Functioning gene (nonmutant)

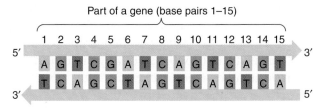

(b) Point mutant

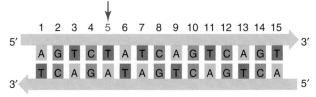

(c) Double mutant

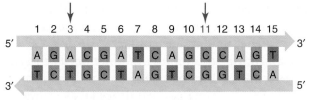

(d) Deletion mutant

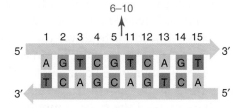

(e) Insertion mutant

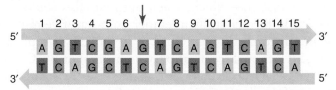

FIGURE 6.8 Schematic of normal DNA molecule and various kinds of mutants. (a) Part of a functioning (nonmutant) gene. (b) Point mutant—base pair 5 is changed from G-C to T-A. (c) Double mutant—base pair 3 is changed from T-A to A-T and base pair 11 is changed from T-A to C-G. (d) Deletion mutant—base pairs 6–10 are deleted. (e) Insertion mutant—a G-C base pair is inserted between base pairs 6 and 7.

Mutants have many uses in molecular biology.

Some of the most significant advances in molecular biology have come about by the use of mutants. Mutants can help us in many ways:

1. *Define a biological function:* Temperature-sensitive mutants are especially useful for defining functions. For example, temperature-sensitive mutants of *E. coli* have been isolated

that fail to synthesize DNA. These mutants fall into over a dozen distinct classes, suggesting there may be at least a dozen different polypeptides required for DNA synthesis.

2. *Elucidate a metabolic pathway:* The metabolic pathway for a specific nutrient can be worked out by monitoring intermediates that accumulate when each gene that codes for an enzyme in the pathway is inactivated by mutation.

3. *Provide information about genetic regulation:* Many mutants have been isolated that alter the amount of a particular protein that is synthesized or the way the amount synthesized responds to external signals. These mutants define regulatory systems. For example, enzymes involved in lactose metabolism in *E. coli* are normally not detectable but appear only after lactose is added to the growth medium. Mutants have been isolated in which these enzymes are always present, whether or not lactose is also present. This finding indicates that some gene is responsible for turning the system of enzyme production on and off, and the regulatory gene product must be responsive to the presence and absence of lactose.

4. *Match an enzyme to a biological function:* The *E. coli* enzyme DNA polymerase I was studied in great detail for many years. Purified DNA polymerase I catalyzes DNA synthesis *in vitro*, so it was believed this polymerase activity was solely responsible for *in vivo* bacterial DNA synthesis. An *E. coli* mutant (*polA⁻*) was isolated, however, in which the activity of DNA polymerase I was reduced 50-fold, yet the mutant bacterium grew and synthesized DNA normally. This observation strongly suggested that DNA polymerase I could not be the only enzyme that catalyzes DNA synthesis in the cell. Indeed, biochemical analysis of cell extracts of the *polA⁻* mutant showed the existence of two other enzymes—DNA polymerase II and DNA polymerase III—which, when purified, also could synthesize DNA. In further study a temperature-sensitive mutation in a gene called *dnaE* was found to be unable to replicate DNA at 42°C, though replication was normal at 30°C. The three enzymes DNA polymerases I, II, and III were isolated from cultures of the *dnaE⁻* (Ts) mutant and each enzyme was assayed. Although DNA polymerases I and II were active at both 30 and 42°C, DNA polymerase III was active at 30°C but not at 42°C. DNA polymerase III was therefore determined to be the product of the *dnaE* gene and an enzyme that is essential for DNA replication.

5. *Determine an antibiotic's target:* The antibiotic rifampicin, which is used to treat various bacterial diseases including tuberculosis, prevents synthesis of RNA. When first discovered it was not known whether rifampicin might act by preventing synthesis of precursor molecules, by binding to DNA and thereby preventing the DNA from being transcribed into RNA, or by binding to RNA polymerase, the enzyme responsible for synthesizing RNA. Mutants were isolated that were resistant to rifampicin. These mutants were of two types: those in which

the bacterial cell envelope was altered such that rifampicin could not enter the cell (an uninformative type of mutant in terms of RNA synthesis), and those in which the RNA polymerase was slightly altered. The finding of the latter mutants proved that the antibiotic acts by binding to RNA polymerase.

6. *Indicate that two different polypeptides interact:* A hypothetical example helps to illustrate how mutants can provide information about protein interactions. Suppose that mutants in two genes *a* and *b*, which are responsible for synthesizing the proteins A and B, fail to carry out some process. Both gene products, thus, are necessary for this process to occur. These gene products may act consecutively or interact to form a single functional unit consisting of both products. Interaction as a single unit is often indicated by reversion studies. When a^- mutant revertants are sought, one sometimes finds (by additional genetic analysis) that the revertants have a mutation in gene *b*. When this type of reversion occurs, one often finds that other b^- mutants (those not formed by reversion of an a^- mutant) revert as a result of (different) mutations in gene *a*. The interpretation of these results is as follows. Proteins A and B are subunits in an active AB protein complex (FIGURE 6.9). An alteration in either the A or B subunit that impedes subunit interaction prevents active AB complex formation. A compensating alteration in the other subunit can then enable the interaction to occur again. Such an interpretation has frequently been found to be correct.

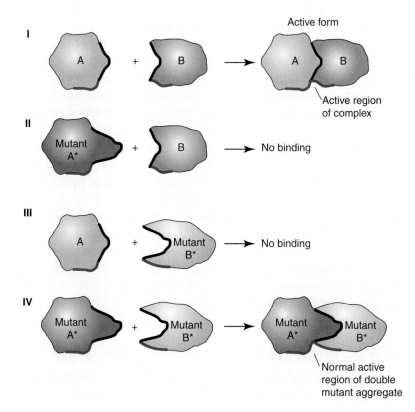

FIGURE 6.9 **Schematic diagram showing how two separately inactive mutant proteins can combine to form a functioning protein complex.** Sites of interaction of proteins A and B are denoted by heavy black lines. Asterisks indicate mutant polypeptides. Components of the AB complex active site are shown in red. Only subunits in I and IV can form a complex with a normal active region.

A genetic test known as complementation can be used to determine the number of genes responsible for a phenotype.

A particular phenotype is frequently the result of the activity of many genes. In the study of any genetic system it is always important to know the number of genes that constitute the system. The genetic test used to evaluate this number, called **complementation**, requires two copies of the genetic unit to be tested are present in the same cell. This requirement can be met in bacteria by constructing a **partial diploid**, that is, a cell containing one complete set of genes and duplicates of some of these genes. (We see how cells of this type are constructed later in this chapter.) A partial diploid is described by writing the genotype of each set of alleles on either side of a diagonal line. For example, $b^+c^+d^+/a^+b^-c^-d^-e^+ \ldots z^+$ indicates that a chromosomal segment containing alleles b, c, and d is present in a cell with a single chromosome containing all of the alleles $a, b, c, \ldots, z$. Usually, only the duplicated genes are indicated, so this partial diploid is designated $b^+c^+d^+/b^-c^-d^-$.

Let's consider a hypothetical bacterium that synthesizes a green pigment from the combined action of genes a, b, and c. The genes code for the enzymes A, B, C, which we assume to act sequentially to form the pigment. If there were a mutation in any of these genes, no pigment would be made. Pigment is made by the partial diploid $a^-b^+c^+/a^+b^-c^+$, however, because the cell contains a set of genes that produce functional proteins A, B, and C. B and C will be made from the $a^-b^+c^+$ chromosome and A and C from the $a^+b^-c^+$ chromosome. In a partial diploid $a_1^-b^+c^+/a_2^-b^+c^+$, in which a_1^- and a_2^- are two different mutations in gene a, no pigment can be made because the bacterium does not contain a functional A protein. The two mutations a^- and b^- of the partial diploid $a^-b^+c^+/a^+b^-c^+$ are said to complement one another because the phenotype of the partial diploid containing them is A^+B^+. In contrast the two mutations a_1^- and a_2^- of the partial diploid $a_1^-b^+c^+/a_2^-b^+c^+$ do not complement one another because the phenotype of the partial diploid containing these mutations is A^-.

Suppose that now a mutation x^-, which also blocks green pigment formation, has been isolated, but the gene in which the mutation has occurred has not been characterized. By constructing a set of partial diploids this gene can be identified by a complementation test. As a start we might test the genes a, b, and c with the partial diploids a^-/x^- (I), b^-/x^- (II), and c^-/x^- (III). If diploids I and II make pigment, the mutation cannot be in genes a or b. If no pigment is made by diploid III, the mutation must be in gene c. If pigment were made in all three diploids, then the important conclusion that mutation x^- is in none of the genes a, b, or c could be drawn. Furthermore, because we have assumed that a, b, and c are each pigment genes, the fact that x is not in any of these genes but prevents pigment formation is evidence that pigment formation requires at least four genes ("at least" because more genes might still be discovered).

E. coli cells can exchange genetic information by conjugation.

Genetic studies of *E. coli* and its viruses provided an essential part of the foundation on which molecular biology is built. Genetics was

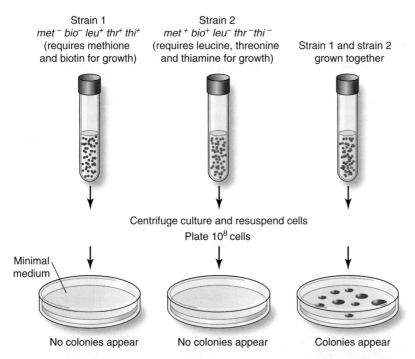

FIGURE 6.10 The Lederberg–Tatum experiment providing the first evidence for bacterial conjugation. The genotype for strain 1 is met^-, bio^-, leu^+, thr^+, thi^+ and that for strain 2 is met^+, bio^+, leu^-, thr^-, thi^-. The symbols bio^- and thi^- indicate requirements for the vitamins biotin and thiamine, respectively. Other abbreviations are standard for amino acids. Colonies are observed on minimal medium plates only after strains 1 and 2 have been permitted to grow together.

already a mature discipline in 1946 when Joshua Lederberg, then a 21-year-old graduate student, and his advisor Edward L. Tatum first showed that one might be able to perform genetic studies in *E. coli*. Their approach, shown in **FIGURE 6.10**, was simple and direct. They selected two strains of *E. coli* with different nutritional requirements. Strain 1 required methionine and biotin for growth, whereas strain 2 required threonine, leucine, and thiamine. No colonies were formed when each strain was plated separately on minimal medium. This result was expected because the probability of two or three mutations all reverting in a single cell is extremely low. An entirely different result was observed when the two strains were mixed and incubated together before plating. In this case a few colonies appeared on the minimal medium, suggesting that genetic material had somehow been exchanged between the two strains.

Subsequent studies by Lederberg and others demonstrated that *E. coli* possesses two mating types: donors (males) and recipients (females). Lederberg and Tatum were fortunate in their choice of strains because not all *E. coli* strains mate. Maleness is determined by a plasmid that is about 100 kilobase pairs long, known as the **fertility factor (F factor)** or **F plasmid**. A male cell, which has the F plasmid, is designated F^+, whereas a female cell, which lacks the F plasmid, is designated F^-. Male cells can transfer the F plasmid to female cells in a process known as mating or **conjugation**. The mating process can be divided into four stages (**FIGURE 6.11**):

1. *Formation of specific donor–recipient pairs through effective contact.* Effective contact begins when the **sex pilus,** a hair-like appendage that projects from the outer surface of the male cell, binds to a receptor on the surface of the female cell

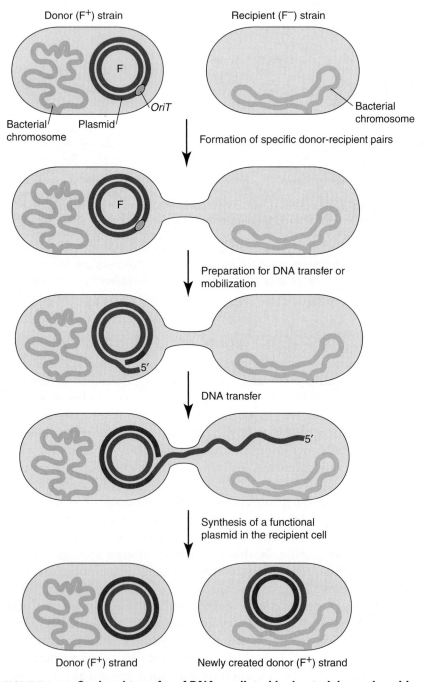

FIGURE 6.11 Conjugal transfer of DNA mediated by bacterial sex plasmids. Transfer of F plasmid during bacterial conjugation. The F plasmid, *oriT*, and bacterial chromosome are not drawn to scale. *E. coli* chromosomal DNA is about 50 times longer than F factor DNA.

CHAPTER 6 GENETIC ANALYSIS IN MOLECULAR BIOLOGY

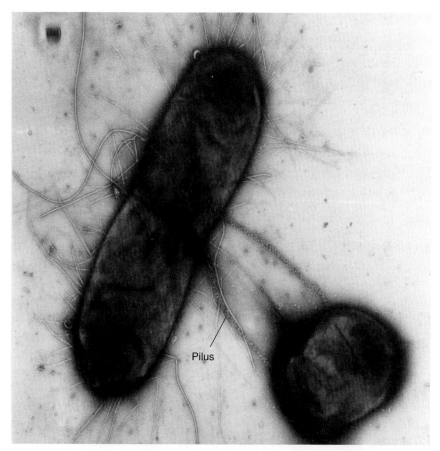

FIGURE 6.12 F pili connecting mating *E. coli* cells. Small bacterial viruses that bind to F pili were added to mating bacteria to make the F pili more visible when viewed by electron microscopy. The virus coated F pili are the top and bottom appendages between the mating bacteria. Other appendages are not coated by the virus particles and are presumed to be common pili but not F pili. (Photo courtesy of Emeritus Professor Ron Skurray, School of Biological Sciences, University of Sydney.)

(**FIGURE 6.12**). Pili are about 2 to 20 μm long and about 8 to 9 nm in diameter.

2. *Preparation for DNA transfer or mobilization.* A single-strand break is introduced in a unique base sequence within the F plasmid called the **transfer origin** or *oriT*.
3. *DNA transfer.* DNA replication in the donor cell takes place through a mechanism known as **rolling circle replication** in which the DNA replication machinery in the donor cell extends the 3′-end of the nicked strand by using the intact complementary strand as a template and displaces the other end of the nicked strand. The nicked DNA is transferred in a 5′→3′ direction from the donor to the recipient cell. Recent studies suggest that the displaced single-stranded DNA moves from the male to the female strain through the sex pilus.
4. *Synthesis of a functional plasmid in the recipient cell.* The single strand that enters the recipient strain serves as a template for the synthesis of a complementary strand. Then the newly formed

BOX 6.1: LOOKING DEEPER

Plasmid Functions Required for Mating

Approximately 40 genes within the F factor are needed for successful mating and DNA transfer to occur. The F factor and other plasmids that have these or similar genes are called self-transmissible plasmids. When a plasmid is missing one or more of these genes, the cell that bears it cannot perform some or all the functions required for mating and DNA transfer. A plasmid with genes for making effective contact but missing those needed for mobilization is called a conjugative plasmid because a cell that bears it can form a conjugation junction with a recipient cell but cannot transfer plasmid DNA to the recipient. Plasmids with genes for DNA mobilization but that are missing genes for making effective contact are called mobilizable plasmids. A cell bearing a mobilizable plasmid can transfer the plasmid to a recipient if it has some other means for making effective contact. For instance, another plasmid within the cell may supply genes needed to form the conjugation junction. Plasmids lacking genes required for making effective contact and for mobilization are called nontransmissible plasmids.

---

double strand cyclizes to form a new F plasmid, converting the recipient into a donor strain. Plasmid DNA replication also occurs in the donor cell so that the donor retains a copy of the F plasmid and its mating type remains unchanged.

Approximately 40 genes within the F factor are needed for successful DNA transfer to occur (**BOX 6.1 LOOKING DEEPER: PLASMID FUNCTIONS REQUIRED FOR MATING**).

The F plasmid can integrate into a bacterial chromosome and carry it into a recipient cell.

The F plasmid also has the ability to become integrated into the bacterial chromosome. When such integration takes place, the F plasmid and the bacterial chromosome become a single, circular DNA molecule and the F plasmid DNA behaves as if it were part of this chromosome, thereby increasing the size of the chromosome. The integration process happens rarely, but it is possible to purify cells that are the progeny of the cell in which integration occurred. Such cells are called **Hfr** (**h**igh **f**requency of **r**ecombination) males.

When an Hfr culture is mixed with an F$^-$ culture, conjugation also occurs as described for the F plasmid, although the material transferred is different from that in an F$^+$ × F$^-$ mating. This time, under the influence of the F insert, DNA replication of the whole chromosome starts within the F insert and a replica of the chromosome is transferred to the F$^-$ cell (**FIGURE 6.13**). Replication starts after the chromosome is nicked within *oriT*. Part of the F sequence is the first DNA to enter the female cell. The remaining part enters only after the entire bacterial chromosome has been transferred. Moreover, because bacteria are very small and in constant motion caused by incessant water molecule bombardment (Brownian motion) and because it takes 100 minutes to transfer an entire chromosome, the mating pair usually breaks apart before transfer is complete. The female, thus, receives both a small functionless fragment of F and a large fragment of the Hfr chromosome, which may contain hundreds of bacterial genes. In an Hfr × F$^-$ mating the mated female therefore almost always remains a female.

In the presence of the new chromosomal fragment the female's recombination system causes genetic exchanges to take place, and a recombinant F$^-$ cell often results. Thus, in a mating between an Hfr *leu*$^+$ culture and an F$^-$*leu*$^-$ culture, recombinant F$^-$*leu*$^+$ cells arise. One can distinguish recombinant *leu*$^+$ females from *leu*$^+$ males by starting out with male and female cells with genetic differences that permit only the female cell to grow on a selective agar medium. A common method is to use antibiotic resistance. For instance, consider an Hfr that is not only Leu$^+$ but also streptomycin-sensitive (Str^S) and a female that is both Leu$^-$ and streptomycin-resistant (Str^R). If the *leu* gene is near the origin of transfer but the *str* gene is not, the mating process is likely to be disrupted before the *str* gene can enter the F$^-$ cell. Then plating the mated cell mixture onto agar containing streptomycin but lacking leucine (1) selectively kills the Hfr Str^S cells and (2) does not lead to growth of the F$^-$ cells unless these also possess the *leu*$^+$

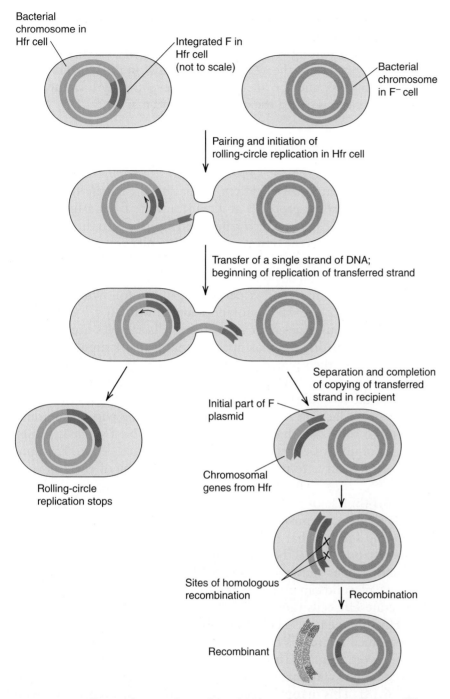

FIGURE 6.13 Stages in transfer and production of recombinants in an Hfr × F⁻ mating. Pairing (conjugation) initiated rolling-circle replication within the F sequence in the Hfr cell results in the transfer of a single strand of DNA. The single strand is converted into double-stranded DNA in the recipient. The mating cells usually break apart before the entire chromosome is transferred. Recombination takes place between the Hfr fragment and the F⁻ chromosome and leads to recombinants containing genes from the Hfr chromosome. Note that only a part of the F factor is transferred. This part of the F factor is not incorporated into the recipient chromosome. The recipient remains F⁻.

allele. Thus, any cell that survives has the genotype leu^+ str^R and is a recombinant female. Only these Leu⁺ Str^R recombinants can form a colony. When a mating is done in this way the transferred allele that is selected by means of the agar conditions (leu^+ in this case) is called a **selective marker** and the allele used to prevent growth of the male (str^R in this case) is called the **counterselective marker**.

Bacterial mating experiments have been used to produce an *E. coli* genetic map.

An important feature of an Hfr × F⁻ mating is that the transfer of the Hfr chromosome proceeds at a constant rate from a fixed point determined by the site at which F has been inserted into the Hfr chromosome. This means the times at which particular genetic loci enter a female are directly related to the positions of these loci on the chromosome. Thus, a map can be obtained from the time of entry of each gene.

Time-of-entry mapping is performed in the following way. An Hfr with an $a^+b^+c^+d^+e^+$ str^S genotype is mated with an $a^-b^-c^-d^-e^-$ str^R female. At various times after mixing the cells samples are removed and agitated violently to break apart all mating pairs simultaneously. Every sample is then plated on five different agar plates containing streptomycin, each of which also contains a different combination of four of the five substances A through E. Thus, colonies that grow on agar plates lacking A are a^+ str^R, those growing without B are b^+ str^R, and so forth. All these data can be plotted on a single graph to give a set of time-of-entry curves, as shown in FIGURE 6.14a. Extrapolation of each curve to the time axis yields the time of entry of each gene $a^+, b^+, \ldots, e^+$. These times can be placed on a map, as shown in Figure 6.14b. Use of a second female that is $b^-e^-f^-g^-h^-$ str^R can then provide the relative positions of three additional genes. These might give a map such as that in Figure 6.14c. Because genes *b* and *e* are common to both maps, the two maps can be combined to form a more complete map, as shown in Figure 6.14d.

The F plasmid can integrate at numerous sites in the chromosome to generate Hfr cells that have different origins of transfer. Each of these Hfr strains can also be used to obtain maps. When separate maps are combined a circle is eventually obtained. For instance, the map obtained from another Hfr might be that shown in Figure 6.14e, which when combined with that in panel (d) yields the circular map shown in Figure 6.14f. This mapping technique has been used repeatedly with hundreds of *E. coli* genes to generate an extraordinarily useful map, one with great significance in the development of molecular genetics. A simplified form of the *E. coli* genetic map is presented in FIGURE 6.15. The discovery in the mid-1950s that the *E. coli* genetic map is circular provided the first suggestion that the *E. coli* chromosome might be a circular DNA molecule, as was later found to be the case.

The *E. coli* chromosome has now been sequenced and the sequence data fit perfectly with the genetic map. Nucleotide sequence information has had a profound effect on molecular biology. Before the sequence was known investigators isolated *E. coli* mutants on the basis

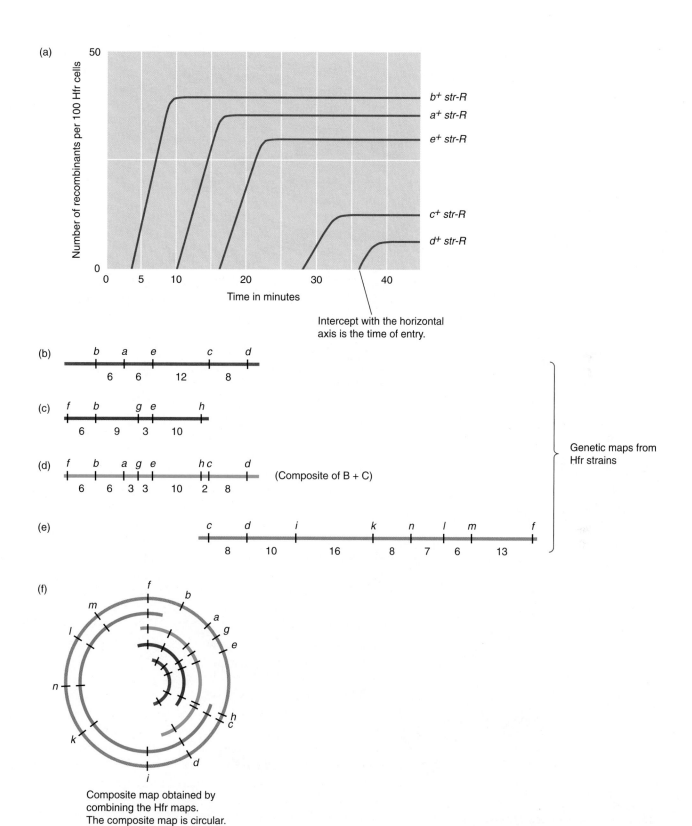

FIGURE 6.14 Time-of-entry mapping. (a) Time-of-entry curves for one Hfr strain. (b) The linear map derived from the data in part a. (c) A linear map obtained with the same Hfr but with a different F⁻ strain containing the alleles b^- e^- f^- g^- h^-. (d) A composite map formed from the maps in parts b and c. (e) A linear map from another Hfr strain. (f) The circular map (gold) obtained by combining the two maps (green and blue) of parts d and e.

6.2 *Escherichia coli*

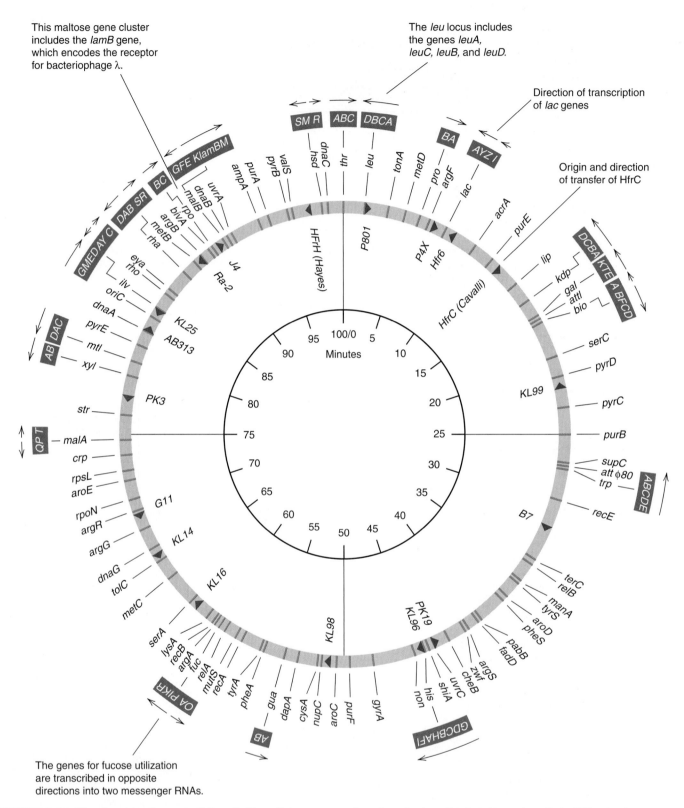

FIGURE 6.15 **Circular genetic map of *E. coli*.** Map distances are given in minutes; the total map length is 100 minutes. For some of the loci that encode functionally related gene products, the map order of the clustered genes is shown, along with the direction of transcription and length of transcript (black arrows). The purple arrowheads show the origin and direction of transfer of a number of Hfr strains. For example, HfrH transfers *thr* very early, followed by *leu* and other genes in a clockwise direction.

of their phenotype and then used genetic techniques to determine the gene's position on the chromosome. Today, we know the positions of all the *E. coli* genes but still do not know the functions of many of them. When mutants with interesting new phenotypes are discovered, the genes can be cloned (or amplified by the polymerase chain reaction) and then sequenced. The gene's position on the genetic map can be determined by comparing its sequence with that of the entire *E. coli* chromosome.

F′ plasmids contain part of the bacterial chromosome.

As already stated, an Hfr cell is produced when the F plasmid stably integrates into the chromosome. The F plasmid can also be excised at a low frequency. When this happens the excised circular DNA is sometimes found to contain genes that were adjacent to F in the chromosome (FIGURE 6.16). A plasmid containing both F genes and chromosomal genes is called an **F′ plasmid**. It is usual to describe an F′ plasmid by stating the genes it is known to possess—for example, F′ *lac pro* contains the genes for lactose utilization and proline synthesis. F′ plasmids can also be transferred from an F′ male to a female.

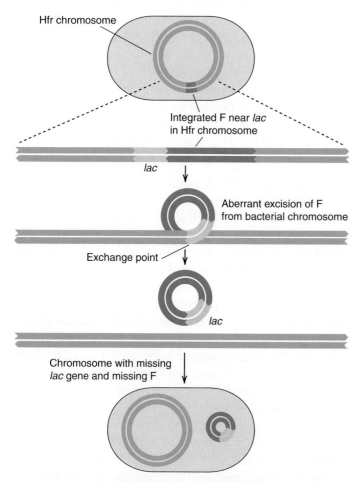

FIGURE 6.16 **Formation of an F′ *lac* plasmid by aberrant excision of F from an Hfr chromosome.** Breakage and reunion are between nonhomologous regions. The bacterial *lac* gene is shown in light blue.

This transfer occurs sufficiently rapidly that the entire F′ is usually transferred before the mating pair breaks apart. Thus, the female recipient is converted to an F′ male in an F′ × F⁻ mating.

F′ plasmids have been used to construct partial diploid bacteria such as those described above for use in the complementation test. A *lac⁺/lac⁻* partial diploid can be constructed by mating an F′*lac⁺ str^S* male with a *lac⁻ str^R* female and selecting for a Lac⁺ Str^R colony. Cells in the colony carry two copies of the *lac* gene: the *lac⁺* allele brought to the female in the F′ and the *lac⁻* allele already present in the female chromosome. To denote this the genotype of the cell is written F′*lac⁺/lac⁻ str^R*, as is usual with partial diploid cells. By convention, genes carried on F′ plasmids are written at the left of the diagonal line.

Plasmid replication control functions are usually clustered in a region called the basic replicon.

When first discovered the F factor seemed to be a unique type of hereditary unit. It soon became evident that the F factor is just one of many extrachromosomal DNA molecules that can replicate in *E. coli* and other organisms. The bacterial host normally supplies most of the enzymes needed for plasmid DNA replication. Plasmid DNA, however, has elements that regulate this replication. These plasmid elements are usually clustered in a 1- to 3-kb region called the **basic replicon**, which is defined experimentally as the smallest piece of plasmid DNA that can replicate with the normal copy number.

Each kind of plasmid has its own unique basic replicon. A basic replicon that functions in one bacterial species usually does not function in other species. For example, most *E. coli* plasmids do not replicate in *Bacillus subtilis*. The basic replicon includes four elements:

1. The origin of replication region (about 50–200 bp long) is the initiation site for DNA synthesis.
2. The initiator region codes for protein and RNA molecules that are required to initiate DNA synthesis.
3. The replication copy control region codes for protein and RNA molecules that regulate the rate of plasmid replication. Some basic replicons are subject to much tighter regulation than others. These regulatory differences explain why some plasmids, the **stringent plasmids**, are maintained at only one or two copies per cell, whereas others, the **relaxed plasmids**, are maintained at multiple copies per cell. Although most relaxed plasmids are maintained at between 10 and 20 plasmids per cell, some are maintained at a few hundred copies per cell. The F factor, which is usually present at one or two copies per cell, is a stringent plasmid.
4. The partitioning system region codes for proteins that determines plasmid distribution at cell division. The partitioning system may be unnecessary when there is a high copy number (many plasmids per cell) because the probability that each daughter cell will receive some plasmids is very high even if the distribution is unequal.

6.3 Budding Yeast (*Saccharomyces cerevisiae*)

Yeasts are unicellular eukaryotes.

Studies using *E. coli* as a model system over the past half-century have provided fundamental insights into the way that information flows from DNA to RNA to protein, thereby placing the discipline of molecular biology on a solid foundation. As a prokaryote, however, *E. coli* is not an effective model system for answering questions that are unique to eukaryotic cells. For instance, *E. coli* studies cannot provide information about nucleosome formation and cannot reveal how RNA is processed in the cell nucleus.

Investigators, searching for a model system that could be used to study fundamental problems such as these in eukaryotes, found that the unicellular yeasts offer many advantages. Like bacteria, yeast cells grow in liquid suspensions in chemically defined media, in complex broths, or on a solid surface to form colonies. Moreover, yeasts grow very rapidly; the doubling time for most yeast strains is only about twice that of *E. coli*. The intracellular organization of a yeast cell is typical of a eukaryotic cell.

Despite these and other advantages, investigators encountered one problem with using yeast as a model system that hindered their studies. The yeast cell membrane is surrounded by a tough thick cell wall made of polysaccharides and polypeptides, making it very difficult to disrupt the cell without causing major damage to the cell organelles. This problem was solved when a snail gut enzyme was shown to degrade the yeast cell wall, allowing yeast cells to be gently disrupted.

S. cerevisiae (commonly known as bakers' yeast), the most thoroughly studied yeast from both a biochemical and genetic point of view, has an ovoid shape and a diameter of about 3 μm. *S. cerevisiae* (and other yeasts) exist in both a haploid and a diploid stage. The haploid cell contains 16 chromosomes, all of which have been sequenced. The yeast nucleus has about 3.5 times more DNA than is present in an *E. coli* cell but about 250-fold less DNA than is present in a human cell nucleus. Yeast therefore probably represents one of the simplest organisms with all of the genetic information required of a eukaryotic cell.

S. cerevisiae cells divide by budding during both their haploid and diploid stages (FIGURE 6.17). The daughter cell emerges as a small bud attached to the mother cell and enlarges until it is almost as big as the mother cell. Chromosomal replication takes place in the mother cell as the bud enlarges and is followed by mitosis. One of the two identical nuclei is transferred to the bud before it pinches off from the mother cell. In contrast to bacteria, which divide to yield twin progeny cells, yeast have a clear mother–daughter relation in the sense that the mother retains a scar on the cell wall at the site of budding. Not all yeasts divide by budding. For instance, *Schizosaccharomyces pombe*, another yeast species that has been extensively studied by molecular biologists, divides

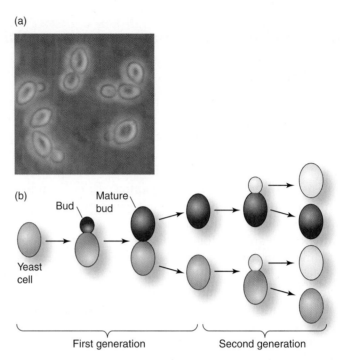

FIGURE 6.17 **Budding yeast.** (a) This light micrograph shows how yeast cells form small protuberances, called "buds," that gradually enlarge until they reach the size of the mother cell and become separated from it, ready to divide again in the same manner. (b) Budding, showing the mother–daughter relation. The first-generation daughter, which is also the second-generation mother, is shown in the darker color. (Photo courtesy of Breck Byers, University of Washington.)

by binary fission. In most other ways, however, *S. pombe* is very similar to bakers' yeast.

An initial step in the study of *S. cerevisiae* has been to determine if biochemical mechanisms worked out for *E. coli* and that apply to many other bacteria are also valid for the simple eukaryote. The results are quite informative. The basic metabolism of yeast and bacteria are quite similar. For instance, both use the same pathways to synthesize purine and pyrimidine nucleotides. However, the overall process for synthesizing macromolecules (DNA, RNA, and protein) in yeast are not those of bacteria but instead those of higher eukaryotes. Yeast has thus become an outstanding model system for studying eukaryotic macromolecular synthesis, gene organization, and regulation of gene expression. So much is now known about yeast genetics and biochemistry that yeast is an attractive choice for fundamental studies in molecular biology. Other unicellular eukaryotes such as the alga *Chlamydomonas* and the protozoan *Tetrahymena* also are used as model systems in eukaryotic molecular biology.

Specific notations, conventions, and terminology are used in yeast genetics.

A nutritional requirement in yeast is usually specified in regular typeface letters followed by a superscript plus or minus sign. Thus, Leu^+

indicates the yeast can make leucine and Leu⁻ indicates it cannot. Three italicized letters are usually (but not always) used to specify a yeast gene. Many gene names are based on a mutant phenotype. For instance, the first three letters in an amino acid's name indicate a gene that is essential for the synthesis of that amino acid. Uppercase letters such as *LEU* indicate the gene is dominant and lowercase letters such as *leu* indicate it is recessive.

Genes are also named for the proteins or RNA molecules they encode. For example, the gene that encodes calmodulin (a regulatory protein that binds calcium ions) is called *CMD1*. The genetic locus (position) is given by a number (rather than a letter) following directly after the gene symbol (for example, *leu2*). A wild-type gene is designated with a superscript plus sign placed just after locus number (*LEU2$^+$*). The allele number is placed after the locus number and separated from it by a hyphen (for example, *leu2-5*). The name of a protein specified by a gene is indicated by writing the gene name in regular typeface as a proper noun (first letter in uppercase) often followed by the letter p. For instance the protein product of *STE3* is Ste3p or Ste3. The rules described for genetic nomenclature in yeast are summarized in TABLE 6.1.

Yeast cells exist in haploid and diploid stages.

One major advantage of yeast as an experimental system is that it has both haploid and diploid stages. There are two haploid mating types: Mat**a** (**a**) and Matα (α), which can mate to form a stable **a**/α diploid. The existence of the mating types means that elegant genetic experiments can be done that complement physicochemical studies and that both the mechanisms of meiosis and the chromosomal interactions responsible for genetic recombination can be studied. When starved for nutrients, an **a**/α diploid cell undergoes meiosis to form four haploid cells called **yeast spores**, which are encased in a sac called an **ascus**. If removed from the ascus and separated from one another, each of the four yeast spores (two **a** cells and two α cells) can grow and

TABLE 6.1 Genetic Nomenclature, Using *LEU2* as an Example

Gene Symbol	Definition
LEU$^+$	All wild type alleles controlling leucine requirement
leu2⁻	Any *leu2* allele conferring a leucine requirement
LEU2$^+$	The wild type allele
leu2-9A	Specific allele or mutation
Leu$^+$	A strain not requiring leucine
Leu⁻	A strain requiring leucine
Leu2p or Leu2	The protein encoded by *LEU2*
LEU2 mRNA	The mRNA transcribed from *LEU2*
Leu2-Δ1	A specific complete or partial deletion of *LEU2*

FIGURE 6.18 **Aspects of the *S. cerevisiae* cell cycle**. The blue and red colors indicate two different mating types (**a** and α). The colors are brought together in diploid cells to indicate that the duplicate nucleus has both **a** and α determinants. (Adapted from Strathern, J. N. 1981. *The Molecular Biology of the Yeast Saccharomyces*, Monograph 11A. Cold Spring Harbor Laboratory Press.)

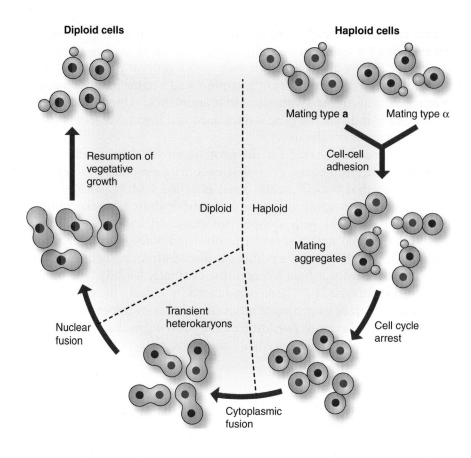

divide. When **a** cells and α cells are mixed, they mate to re-form the diploid state.

The yeast mating process occurs in discrete stages, summarized in FIGURE 6.18. Cells of an opposite mating type agglutinate shortly after they are mixed. Glycoproteins on the cell surface act as glue-like substances, causing the cells to stick together. Many of the cells have buds at the beginning of the mating process. Buds are no longer evident 90 to 120 minutes later, however, because the cell division cycle is arrested at the G_1 stage. Arrested cells start to grow, assuming raindrop-like shapes. When the tips of the cells within a mating pair touch, enzymes remodel the cell walls and membranes so that a barrier no longer divides the two cells. The two haploid nuclei, which are now in the same compartment, fuse to form a diploid nucleus.

6.4 New Tools for Studying Human Genetics

Different humans may exhibit DNA sequence variations within the same DNA region.

Until the early 1980s planned mating experiments such as those performed by Gregor Mendel in the mid-1860s were the primary (and usually the only) method available for the genetic analysis of higher organisms. Certain higher organisms such as the fruit fly and maize served as preferred model systems because they are easy to maintain and produce large numbers of progeny within a relatively short time. Geneticists also performed mating experiments with mice, rats, and other animals even though these experiments were more time-consuming and expensive. Because ethical considerations prohibit human experiments of this type, other methods had to be devised to study human genetics. Although considerable information was obtained by correlating family trees with specific traits such as color blindness or inborn errors of metabolism such as Tay-Sachs disease, genetic studies in humans lagged behind those in many other biological systems.

New tools were needed to investigate human genetics. Recombinant DNA technology came to the rescue. Recombinant DNA tools work because homologous human chromosomes, although very similar, usually differ in approximately 0.2% to 0.5% of their base pairs. Two types of DNA sequence differences or **polymorphisms** are of special interest for genetic studies:

1. *Single nucleotide polymorphism:* A single nucleotide polymorphism, or SNP (pronounced "snip"), occurs when a single nucleotide at a specific DNA site is replaced by some other nucleotide in at least 1% of the total population. Approximately 1 in every 100 to 300 nucleotides along the 3-billion-base human genome is a SNP. SNPs occur in both coding regions (regions that code for proteins and functional RNA molecules) and noncoding regions. Even when present in a protein coding region, SNPs often do not affect cell function because the base change either does not produce an amino acid change or the amino acid change that it does produce does not affect protein function. Approximately 3 million unique human SNPs are now known and listed in various public and private data banks.

2. *Tandem repeat polymorphisms:* Tandem repeat polymorphisms occur when a DNA sequence, ranging from 1 to up to 100 bp, is repeated head to tail a different number of times at a distinct site in homologous chromosomes. These repeats occur in eukaryotic, bacterial, and archaeal chromosomes. FIGURE 6.19 shows DNA molecules that differ in the number of tandem repeats. In this example the number of repeats varies from 1 to 10. There are two types of tandem repeat polymorphisms. The first type, **microsatellites**, have 1- to 6-bp-long repeat sequences that are reiterated 5 to 100 or more times. They are distributed randomly throughout human and other vertebrate genomes and are present in both protein-coding and noncoding regions. The second type, **minisatellites**, have 7 up to about 100 nucleotide repeat sequences that are imperfectly reiterated 5 to 100 times. Human and other vertebrate genomes have many hundreds to thousands of minisatellites, each with its own distinct repeat unit.

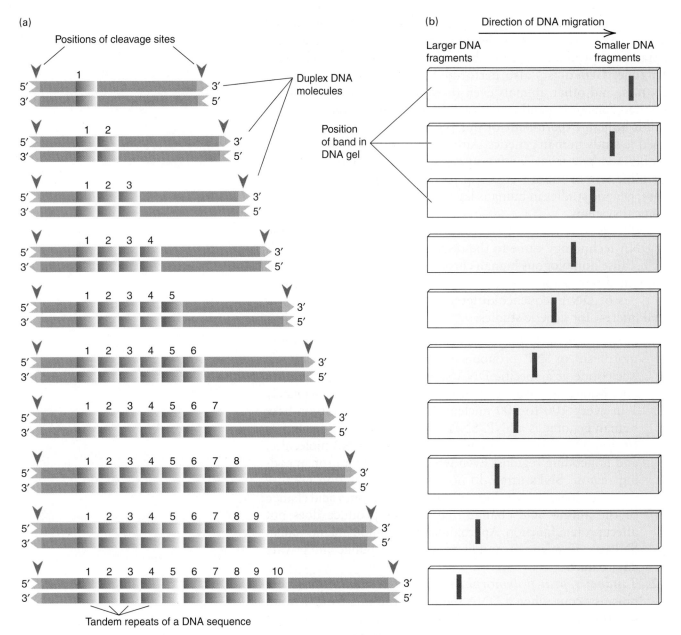

FIGURE 6.19 Tandem repeats of a DNA sequence. A genetic polymorphism in which the alleles in a population differ in the number of copies of a DNA sequence that is repeated in tandem along the chromosome. (a) This example shows alleles in which the repeat number varies from 1 to 10. (b) Cleavage at restriction sites flanking the repeat yields a unique fragment length for each allele that can be seen when the fragments are analyzed by gel electrophoresis. The alleles also can be distinguished by the size of the fragment amplified by polymerase chain reaction using primers that flank the repeat.

Restriction fragment length polymorphisms facilitate genetic analysis in humans and other organisms.

SNPs sometimes create or eliminate restriction endonuclease cleavage sites in one of the two homologous chromosomes. In consequence, restriction endonucleases sometimes cleave corresponding regions of homologous chromosomes into fragments of different sizes. These fragments, called **restriction fragment length polymorphisms (RFLPs)**, are physical traits that can be followed as they are passed from one generation to the next. The gene that codes for the β-globin chain in human hemoglobin provides an important clinical application (FIGURE 6.20). The mutation responsible for sickle cell anemia, which converts Glu-6 to Val-6, also removes a cleavage site for the restriction endonuclease DdeI (Figure 6.20a and b). This change permits the gene for sickle cell anemia to be tracked as it passes from one generation to the next (Figure 6.20c).

In 1980 David Botstein, Ronald W. Davis, and Mark Skolnick proposed that RFLPs could be used as markers to identify genes to

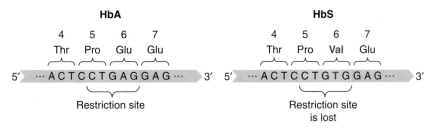

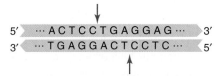

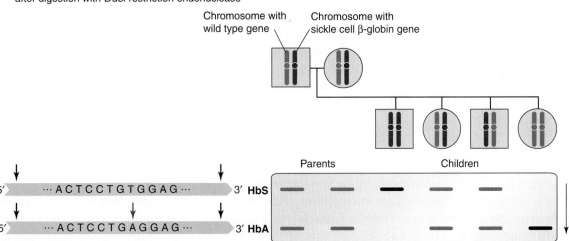

FIGURE 6.20 Use of restriction fragment length polymorphisms to distinguish chromosomes bearing a normal β-globin allele from one with a sickle cell β-globin allele. (a) The nucleotide sequences that code for residues 4 to 7 in the normal β-globin chain (HbA) and the sickle cell β-globin chain (HbS). Note that the restriction endonuclease site (CTGAG), which is present in the normal β-globin allele, is missing in the HbS β-globin allele. (b) Restriction site for the restriction endonuclease DdeI. The red arrows indicate cleavage sites. (c) Southern blot analysis of DNA from parents and children after digestion with the DdeI restriction endonuclease. Chromosomes bearing the normal and sickle cell β-globin alleles are shown in blue and red, respectively. A radioactive probe is used to detect the DNA that codes for the β-globin protein. A DdeI restriction site is missing from the sickle cell β-globin gene. The restriction fragment from DNA with this allele is therefore larger than that for DNA with the normal allele. The electrophoretic bands are darker when an individual has two alleles of one type.

6.4 New Tools for Studying Human Genetics

which they are tightly linked; that is, they proposed that an RFLP can be used to find a gene that remains together with the RFLP after homologous chromosomes segregate into sperm or egg cells during meiosis. When such tight linkage exists, one can find the gene of interest by searching for the RFLP. Because SNPs normally produce only two RFLP variants, RFLPs that result from point mutations are of rather limited use for tracing alleles. Fortunately, tandem repeat polymorphisms do not suffer from this shortcoming. Hence, microsatellites or minisatellites located between two restriction endonuclease sites can help to follow the fate of tightly linked genes in homologous chromosomes. The approach is similar to that described for the β-globin gene. Instead of using a probe that binds to a specific gene sequence, however, the probe is selected for its ability to bind to the microsatellite or minisatellite during Southern blot analysis. Investigators also use tandem repeat polymorphisms to obtain a DNA fingerprint that can identify an individual (**BOX 6.2 IN THE LAB: DNA FINGERPRINT**).

BOX 6.2: IN THE LAB

DNA Fingerprint

Tandem repeat polymorphisms can also be used to obtain a DNA fingerprint that is unique for each individual (with the possible exception of identical twins). Alec Jeffreys and coworkers discovered this application in 1985 while studying the myoglobin gene. As part of their studies, they observed that one of the introns in the myoglobin gene has a minisatellite, which has a different number of repeats in DNA from different people. Jeffreys and his coworkers hoped to use the minisatellite as a marker to locate the myoglobin gene on the chromosome. They therefore prepared a probe for the minisatellite, which they planned to use to identify the myoglobin gene. During one experiment Jeffreys and coworkers digested DNA samples obtained from different individuals with a restriction endonuclease, resolved the fragments by gel electrophoresis, and then used the probe to detect the myoglobin gene. Jeffreys and coworkers expected to see two labeled bands, each corresponding to a different myoglobin allele. To their great surprise they observed that every analyzed sample produced many labeled bands. They quickly realized the probe was binding to a large number of different chromosomal regions from each person. Moreover, each individual's band pattern was unique. The probe could therefore be used to fingerprint the DNA. It soon became clear that probes that detect other minisatellites and microsatellites could be used to obtain unique DNA fingerprints. Although there is some ambiguity when a single kind of probe is used to obtain a DNA fingerprint, this ambiguity becomes negligible when several probes are used.

DNA fingerprinting based on restriction endonuclease sites can provide an unambiguous identification of an individual when one has an ample supply of DNA. For instance, one can determine paternity by taking DNA samples from the father and child. In many situations, however, the DNA that is to be tested is available in only trace quantities. For instance, there may be very little DNA left behind at a crime scene, perhaps just a hair follicle or some saliva. Trace amounts of DNA may also be available to identify a person who dies in a major catastrophe, perhaps just what is present on a hairbrush or toothbrush. Fortunately, polymerase chain reaction amplification allows even trace quantities of DNA to be fingerprinted.

Tandem repeat polymorphisms are also used to diagnose genetic diseases. This diagnostic approach is especially important when the molecular basis of the genetic disease is not known. If a particular length (or version) of a microsatellite or minisatellite is very near to the defective allele, it can also be used to isolate and clone the gene. Once cloned, the gene can be sequenced and the information used to find the protein product. The protein products of the genes that cause muscular dystrophy and cystic fibrosis were found in this way.

Somatic cell genetics can be used to map genes in higher organisms.

A second approach for studying human genetics became possible with the discovery that cultured human and rodent cells fuse to form hybrid cells, which upon continued cultivation progressively lose human chromosomes. The reason human chromosomes are lost is still a mystery. The availability of hybrid cells that contain mostly rodent chromosomes allows investigators to map genes in the few remaining human chromosomes. This approach, called **somatic cell genetics**, provides a method for identifying the chromosome on which a gene of interest is located.

Questions and Problems

1. Briefly define, describe, or illustrate each of the following.
    a. Peptidoglycan
    b. Bacterial outer membrane
    c. Porin
    d. Periplasmic space
    e. Broth
    f. Minimal medium
    g. Prototroph
    h. Autotroph
    i. Lag phase
    j. Doubling time
    k. Cell density
    l. Log phase
    m. Stationary phase
    n. Structural gene
    o. Genome
    p. Wild type
    q. Conditional mutant
    r. Temperature-sensitive mutant
    s. Suppressor-sensitive mutant
    t. Spontaneous mutagenesis
    u. Mutagen
    v. Point mutation
    w. Deletion
    x. Insertion
    y. Revertant
    z. Complementation
    aa. Partial diploid
    bb. Fertility factor
    cc. Bacterial conjugation
    dd. Sex pilus
    ee. *OriT*
    ff. Rolling circle replication
    gg. Hfr
    hh. Selective marker
    ii. Counterselective marker
    jj. F′ plasmid
    kk. Basic replicon
    ll. Stringent plasmid
    mm. Relaxed plasmid
    nn. Yeast spore
    oo. Ascus
    pp. Polymorphisms
    qq. Microsatellite
    rr. Minisatellite
    ss. RFLP

2. The three genes—*A*, *B*, and *C*—are arranged along a single chromosome in an unknown order. The recombination frequency between genes *A* and *B* is 0.5% ($A \times B = 0.5\%$), the recombination frequency between genes *B* and *C* is 1% ($B \times C = 1\%$), and the recombination frequency between genes *A* and *C* is 1.5%. What is the arrangement of genes *A*, *B*, and *C* on the chromosome?

3. Why did early investigators tend to use *Drosophila melanogaster* as their model system rather than mice or rats?

4. Gram-negative bacteria are usually more resistant to detergents than gram-positive bacteria. Explain this difference in sensitivity based on your knowledge of the structural features of the two kinds of bacteria.

5. There are many absolute defective *E. coli* mutants for arginine or uracil synthesis but none for ribosome synthesis. Provide an explanation for this finding. What type of mutant can be isolated to study ribosome synthesis? Explain your answer.

6. Two spontaneous mutants were isolated that require the amino acid methionine for growth. It is relatively easy to obtain spontaneous revertants of mutant 1

but has so far proved impossible to obtain a spontaneous revertant of mutant 2. What can you tell about the nature of the two mutations from this information?

7. Describe a simple experiment that you could use to show that mating takes place in bacteria.

8. Describe the state of the F factor in F$^-$, F$^+$, Hfr, and F′ *E. coli* strains. How can you use an F′ and an F$^-$ strain to construct a partial diploid?

9. Strain Hfr 1 transfers the arg$^+$ marker to an F$^-$ strain last during conjugation. Hfr 1 was streaked on a nutrient agar medium to obtain individual pure colonies. Cells from each of the purified colonies were tested for their ability to mate with the F$^-$ strain. With one exception, cells from all the purified colonies behaved as expected during mating experiments with the F$^-$ strain. Cells from the one exceptional colony transferred the arg$^+$ marker to F$^-$ cells early in mating to form arg$^+$ cells that carry the F factor. Explain the events leading to the formation of the one exceptional colony.

10. A streptomycin-sensitive Hfr strain, which is *arg$^+$*, *his$^+$*, and *trp$^+$*, was mated with a streptomycin-resistant F$^-$ strain, which is *arg$^-$*, *his$^-$*, and *trp$^-$*. Cell samples were removed from the conjugation mixture every 5 minutes, agitated, and spread on a set of three different selective medium plates. All three selective medium plates contained minimal agar, glucose, and streptomycin, and two of the three required amino acids. The only difference between the plates was the missing required amino acid. The following data summarize the number of bacterial colonies observed on the plates containing the specified amino acids at the given times. Use the data given in the table to show the positions of *arg*, *his*, and *trp* on the genetic map.

	Number of Colonies		
Time (min)	Trp and His	Arg and His	Arg and Trp
0	0	0	0
5	0	0	0
10	0	0	0
15	0	25	0
20	0	153	0
25	12	160	0
30	77	158	0
35	120	155	0
40	121	156	0
45	127	154	15
50	125	155	46
55	126	152	52

11. What are RFLPs? How can they be used to distinguish one individual from another? How can they be used to determine the location of a specific gene?

Suggested Reading

Drosophila melanogaster

Kohler, R. E. 1994. *Lords of the Fly: Drosophila Genetics and the Experimental Life* (Vol. 15). Illinois: University of Chicago Press.

Rubin, G. M., and Lewis, E. B. 2000. A brief history of *Drosophila's* contributions to genome research. *Science* 287:2216–2218.

St. Johnston, D. 2002. The art and design of genetic screens: *Drosophila melanogaster*. *Nat Rev Genet* 3:176–188.

Sturtevant, A. H. 1959. Thomas Hunt Morgan biographical memoirs. *Natl Acad Sci USA*. Retrieved from http://books.nap.edu/html/biomems/tmorgan.pdf

E. coli

Beckwith, J., and Silhavy, T. J. 1992. *Power of Bacterial Genetics: A Literature-Based Course.* Woodbury, NY: Cold Spring Harbor Laboratory Press.

Black, J. Q. 2001. *Microbiology* (5th ed). Hoboken, NJ: John Wiley and Sons.

Clewell, D. B. 2005. Antibiotic resistance plasmids in bacteria. *Encyclopedia of Life Sciences.* pp. 1–6. Hoboken, NJ: John Wiley and Sons.

Cupples, C. G. 2001. *Escherichia coli* and development of bacterial genetics. *Encyclopedia of Life Sciences.* pp. 1–6. Hoboken, NJ: John Wiley and Sons.

Dale, J. W., and Park, S. F. 2004. *Molecular Genetics of Bacteria.* Hoboken, NJ: John Wiley and Sons.

de la Cruz, F., Frost, L. S., Meyer, R. J., and Zechner, E. L. 2010. Conjugative DNA metabolism in gram-negative bacteria. *FEMS Microbiol Rev* 34:18–40.

Funnell, B. E., and Phillips, G. J. (eds). 2004. *Plasmid Biology.* Washington, DC: American Society of Microbiology Press.

Klotz, M. G. 2005. Bacterial genetics. *Encyclopedia of Life Sciences.* pp. 1–6. Hoboken, NJ: John Wiley and Sons.

Miller, J. H. 1999. *A Short Course in Bacterial Genetics: A Laboratory Manual and Handbook for* Escherichia coli *and Related Bacteria.* Woodbury, NY: Cold Spring Harbor Laboratory Press.

Mobashery, S., and Azucena, E. F. Jr. 2002. Bacterial antibiotic resistance. *Encyclopedia of Life Sciences.* pp. 1–6. Hoboken, NJ: John Wiley and Sons.

Nikaido, H. 2009. Multidrug resistance in bacteria. *Annu Rev Biochem* 78:119–146.

Novick, R. P. 2001. Plasmids. *Encyclopedia of Life Sciences.* pp. 1–8. Hoboken, NJ: John Wiley and Sons.

Snyder, L., and Champness, W. 2002. *Molecular Genetics of Bacteria.* Washington, DC: American Society of Microbiology Press.

Streips, U. N., and Yashin, R. E. (eds). 2002. *Modern Microbial Genetics* (2nd ed). Hoboken, NJ: John Wiley and Sons.

Summers, D. K. 1996. *The Biology of Plasmids.* Malden, MA: Blackwell Science.

Thomas, C. M., and Summers, D. 2008. Bacterial plasmids. *Encyclopedia of Life Sciences.* pp. 1–9. Hoboken, NJ: John Wiley and Sons.

Tortora, G. J., Funke, B. R., and Case, C. L. 2003. *Microbiology: An Introduction* (8th ed). Upper Saddle River, NJ: Benjamin Cummings.

Trempy, J. 2003. *Fundamental Bacterial Genetics.* Malden, MA: Blackwell Science.

Budding Yeast (*Saccharomyces cerevisiae*)

Abelson, J. N., Simon, M. I., Guthrie, C., and Fink, G. R. (eds). 2004. *Methods in Enzymology,* Volume 194: *Guide to Yeast Genetics and Molecular Biology,* Part A. San Diego: Academic Press.

Adams, A., Gottschling, D. E., Kaiser, C. A., and Stearns, T. (eds). 1997. *Methods in Yeast Genetics.* Woodbury, NY: Cold Spring Harbor Laboratory Press.

Amberg, D. C., Burke, D. J., and Strathern, J. N. 2005. *Methods in Yeast Genetics.* Woodbury, NY: Cold Spring Harbor Laboratory Press.

Burke, D., Dawson, D., and Stearns, T. 2004. *Methods in Yeast Genetics.* Woodbury, NY: Cold Spring Harbor Laboratory Press.

Guthrie, C., and Fink, G. R. 2002. *Methods in Enzymology,* Volume 350: *Guide to Yeast Genetics and Molecular Cell Biology Part B.* San Diego, CA: Academic Press.

Mell, J. C., and Burgess, S. M. 2002. Yeast as a model genetic organism. *Encyclopedia of Life Sciences.* pp. 1–8. Hoboken, NJ: John Wiley and Sons.

Pausch, M. H., Kirsch, D. R., and Silverman, S. J. 2005. *Saccharomyces cerevisiae*: applications. *Encyclopedia of Life Sciences.* pp. 1–7. Hoboken, NJ: John Wiley and Sons.

Phaff, H. J. 2001. Yeasts. *Encyclopedia of Life Sciences.* pp. 1–11. Hoboken, NJ: John Wiley and Sons.

Scheiner-Bobis, G. 2009. Gene expression in yeast. *Encyclopedia of Life Sciences.* pp. 1–5. Hoboken, NJ: John Wiley and Sons.

New Tools for Studying Human Genetics

Butler, J. M. 2005. *Forensic DNA Typing: Biology, Technology, and Genetics Behind STR Markers* (2nd ed). San Diego: Academic Press.

Davis, J. M. 2002. *Basic Cell Culture: A Practical Approach.* Oxford, UK: Oxford University Press.

Debrauwere, H., Gendrel, C. G., Lechat, S., and Dutreix. M, 1997. Differences and similarities between various tandem repeat sequences: minisatellites and microsatellites. *Biochimie* 79:577–586.

Epplen, J. T., and Böhringer, S. 2005. Microsatellites. *Encyclopedia of Life Sciences*. pp. 1–4. Hoboken, NJ: John Wiley and Sons.

Epplen, J. T., and Kunstmann, E. M. 2005. Minisatellites. *Encyclopedia of Life Sciences*. pp. 1–4. Hoboken, NJ: John Wiley and Sons.

Harris, H. 1995. *The Cells of the Body: A History of Somatic Cell Genetics*. Woodbury, NY: Cold Spring Harbor Laboratory Press.

Housman, D. E. 1995. DNA on trial—the molecular basis of DNA fingerprinting. *N Engl J Med* 332:534–535.

Jeffreys, A. J., Wilson, V., and Thein, S. L. 1985. Hypervariable minisatellite regions in human DNA. *Nature* 314:67–73.

Lee, H., and Tirnady, F. 2003. *Blood Evidence: How DNA Is Revolutionizing the Way We Solve Crimes*. Jackson, TN: Perseus Publishing.

Nakamura, Y., Leppert, M., O'Connell, P., et al. 1987. Variable number of tandem repeat (VNTR) markers for human gene mapping. *Science* 235:1616–1622.

Rudin, N., and Inman, K. 2001. *An Introduction to Forensic DNA Analysis* (2nd ed). Boca Raton, FL: CRC Press.

Sedivy, J. M., and Dutriaux, A. 1999. Gene targeting and somatic cell genetics—a rebirth or a coming of age? *Trends Genet* 15:88–90.

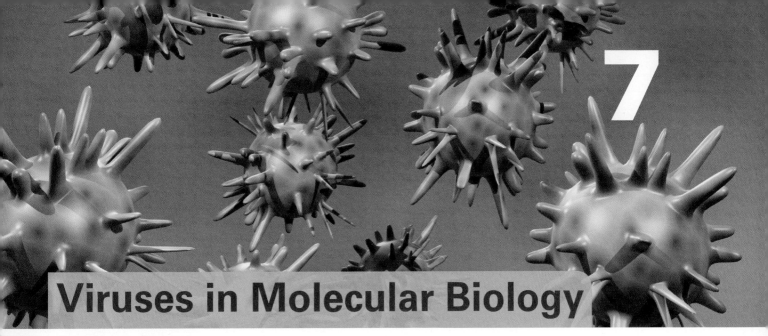

Viruses in Molecular Biology

CHAPTER OUTLINE

7.1 Introduction to the Bacteriophages
Bacteriophages were of interest because they seemed to have the potential to serve as therapeutic agents to treat bacterial diseases.
Investigators belonging to the "Phage Group" were the first to use viruses as model systems to study fundamental questions about gene structure and function.
Bacteriophages come in different sizes and shapes.
Bacteriophages have lytic, lysogenic, and chronic life cycles.
Bacteriophages form plaques on a bacterial lawn.

7.2 Virulent Bacteriophages
The T-even phages are tailed viruses that inject their DNA into bacteria.
Bacterial and phage RNA polymerases help to reel phage T7 DNA into the bacterial cell.
E. coli phage φX174 contains a single-stranded circular DNA molecule.
Some phages have single-stranded RNA as their genetic material.

7.3 Temperate Phages
E. coli phage λ DNA can replicate through a lytic or lysogenic life cycle.
E. coli phage P1 can also replicate through a lytic or lysogenic life cycle.

7.4 Chronic Phages
A chronic phage programs the host cell for continued virion particle release without killing the cell.

7.5 Animal Viruses
Animal viruses can serve as excellent models for addressing fundamental questions in molecular biology.
Polyomaviruses contain circular double-stranded DNA.
Adenoviruses have linear blunt-ended, double-stranded DNA with an inverted repeat at each end.
Retroviruses use reverse transcriptase to make a DNA copy of their RNA genome.

BOX 7.1: LOOKING DEEPER T-Even Phage Assembly

BOX 7.2: IN THE LAB Generalized Transduction

BOX 7.3: IN THE LAB Phage M13 Cloning Vector

BOX 7.4: IN THE LAB Complementary DNA Synthesis

QUESTIONS AND PROBLEMS

SUGGESTED READING

Viruses are obligate parasites that invade cells and then use the cells' metabolic machinery to replicate. By itself a virus particle or **virion** can persist, but it can neither grow nor replicate except within a living host cell. Virions come in a wide variety of sizes and shapes (FIGURE 7.1). Some kinds of virions have membrane envelopes surrounding them, whereas others do not. Virions, which are nearly always too small (10–200 nm) to be seen with a light microscope, can be viewed with an electron microscope.

A virion consists of a nucleic acid, either DNA or RNA, surrounded by a protective protein coat or **capsid**. Depending on the kind of virion, the nucleic acid may be either single- or double-stranded DNA or RNA and either linear or circular. The lengths of viral nucleic acids vary greatly from one type of virus to another, but even the longest viral nucleic acid is much shorter than its host's DNA and, hence, codes for many fewer proteins than the host genome. Viral genes code for proteins that form the capsid as well as for enzymes essential for viral reproduction inside the host. Viruses that have very few genes rely on

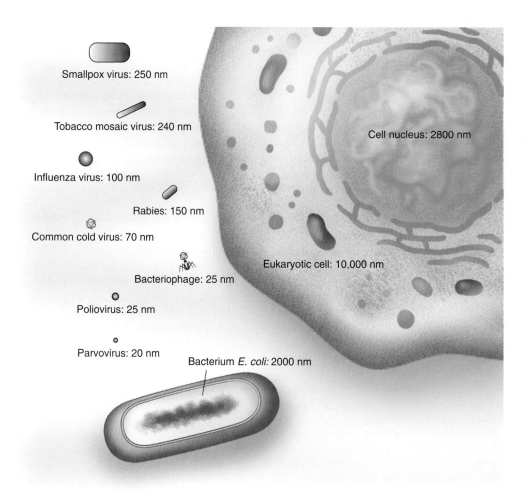

FIGURE 7.1 Size relationships among viruses. The sizes of various viruses relative to a eukaryotic cell, a cell nucleus, and the bacterium *E. coli*. The smallpox virus approximates the smallest prokaryote, the mycoplasmas, in size.

the host to provide nearly all the machinery they require for reproduction, whereas viruses with many genes are less reliant on the host. At a very minimum a virus requires the host cell's ribosomes, protein-synthesizing factors, amino acids, and energy-generating systems to reproduce. A virus is therefore not a living organism but rather a piece of genetic material that can only reproduce in a metabolizing cell.

Viruses are classified according to their host organisms, particle morphology, and nucleic acid composition. Despite their enormous diversity, all viruses must perform certain essential functions. They must be able to deliver their nucleic acid to the inside of a cell, protect their nucleic acid from harmful physical and chemical agents, convert the host cell to a virus factory, and create some means to allow progeny viruses to escape from infected cells. These functions are carried out in a variety of ways by different virus species. Each kind of virus infects a limited number of different kinds of organisms. Some kinds of viruses appear to infect only a single kind of organism, others infect closely related species, and still others infect widely different organisms, even jumping between plant and animal kingdoms.

Many of the fundamental principles that helped to establish molecular biology as a distinct scientific discipline were discovered by studying viruses. Although the discipline of molecular biology has benefited greatly from the study of viruses, benefits have also flowed in the other direction. Virologists take advantage of the principles and techniques discovered by molecular biologists to examine how viruses cause disease and devise treatments for these diseases. In fact, the disciplines of molecular biology and virology are so intertwined that it is virtually impossible to study one discipline without some knowledge of the other. This chapter provides some basic information molecular biologists need to know about viruses. It is not intended to provide a comprehensive view of viruses but rather to examine a few of the virus model systems that have had and continue to have a profound influence on the study of molecular biology.

7.1 Introduction to the Bacteriophages

Bacteriophages were of interest because they seemed to have the potential to serve as therapeutic agents to treat bacterial diseases.

The first clue to the existence of bacterial viruses came in 1896 when the British bacteriologist Ernest Hankin observed that the sewage-ridden Ganges and Jumna Rivers in India contained an unidentified substance that passed through a fine porcelain filter and had the ability to kill *Vibrio cholerae*, the bacterial species that causes the often fatal disease cholera. Hankin recognized that the filterable agent helped to control the level of the pathogenic bacteria in the river by killing the bacteria but did not attempt to determine the nature of the agent. The British physician Frederick Twort encountered a similar filterable agent in 1915 and suggested it might be a virus but did not pursue this idea further.

Two years later the French physician Felix d'Herelle noticed that a bacteria-free filtrate prepared from the stools of French troops who were recovering from dysentery had the ability to kill the bacteria that cause dysentery. Further studies showed the filterable agent caused turbid (cloudy) bacterial cultures to become clear almost as if the agent were eating the bacteria. For this reason d'Herelle called the agent a **bacteriophage** (Gr. *phago*, to eat). This term, and its shortened version **phage**, remains in common use for bacterial viruses to this day. d'Herelle thought bacteriophages might be used to treat bacterial infections. Although initial clinical trials appeared to offer some support for this idea, more extensive clinical studies did not produce reproducible cures.

The failure of bacteriophages to cure bacterial diseases was due in part to the appearance of phage-resistant bacterial mutants and in part to the destruction of the bacteriophages by the patients' immune or digestive systems. With the discovery of antibiotics, interest in using bacteriophages to treat bacterial infections waned, and very little was done to pursue this clinical application for many years. Now that so many bacterial pathogens have become antibiotic-resistant there is renewed interest in the potential of bacteriophages to prevent or treat bacterial diseases. For example, the food industry is currently exploring the use of bacteriophages to disinfect foods, and the U.S. Food and Drug Administration has approved a physician-initiated trial to determine whether a bacteriophage cocktail can effectively destroy *Staphylococcus*, *Pseudomonas*, and *Escherichia coli* in skin wounds and burns.

Investigators belonging to the "Phage Group" were the first to use viruses as model systems to study fundamental questions about gene structure and function.

Even though bacteriophages did not seem to hold great promise as therapeutic agents, they remained fascinating subjects for scientific investigations. By the 1940s a loosely knit, informal network of investigators decided to use bacteriophages to address fundamental questions related to gene structure and function. This network, which came to be known as the Phage Group, owed much to the intellectual leadership of two outstanding members, Max Delbrück and Salvador Luria. They convinced other talented scientists to join them in using a specific bacteriophage that infects *E. coli* as a model system so results from different laboratories could be compared and investigators could build on one another's work. The bacteriophage they selected, bacteriophage T4, was one of seven different types of phages that had been isolated from raw sewage in Brooklyn, New York, several years earlier.

The bacteriophages were named bacteriophage type 1, type 2, and so forth. The names were later shortened to bacteriophage T1, T2, T3, T4, T5, T6, and T7, where "T" is short for type. By chance, bacteriophages T2, T4, and T6 turned out to be very similar to one another and so are called the **T-even phages**. Two of the T-odd bacteriophages, T3 and T7, are also similar to one another. T1 and T5 differ from the other five bacteriophages and also from each other. Many additional phages have now been isolated that are closely related to one or another of the seven original phage types.

Bacteriophages come in different sizes and shapes.

The bacteriophages described in this chapter fall into three structural categories (FIGURE 7.2):

1. *Icosahedral tailless phage:* The highly compact nucleic acid is located within the capsid, which is in the shape of an icosahedron (an almost spherical structure having 20 triangular faces, 30 edges where two faces meet, and 12 corners where five edges meet).
2. *Icosahedral phage with tail:* The icosahedral head containing the compact DNA has a tail attached to it. There are many variations on the basic structure of the tailed phages. For example, either the length and width of the head may be the same or the length may be greater than the width; however, short fat heads are not seen. The tail may be very short (barely visible in electron micrographs) or up to four times the length of the head and it may be flexible or rigid. A complex baseplate may also be present on the tail; when present, it typically has from one to six tail fibers.
3. *Filamentous phage:* The nucleic acid, which has an extended helical form, is embedded in a cylindrically shaped capsid.

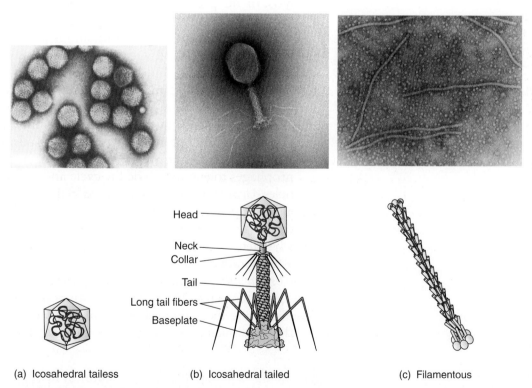

FIGURE 7.2 **The three major morphological classes of phages.** (a) Icosahedral, tailless: bacteriophage ϕX174. (b) Icosahedral, tailed: bacteriophage T4. (c) Filamentous: bacteriophage M13. (Top) Electron micrographs. (Bottom) Schematic diagrams of the phages. The tailed phages do not always have a collar and can have from 0 to 6 tail fibers, the number depending on the phage type. (Part a photo courtesy of Cornelia Büchen-Osmond, ICTVdB, Columbia University. Part b photo courtesy of Robert Duda, University of Pittsburgh. Part c image supplied by the Centre for Bioimaging, Rothamsted Research, Ltd.)

Bacteriophages have lytic, lysogenic, and chronic life cycles.

Bacteriophages, like all other viruses, are not considered to be living organisms because they depend on the host cell's metabolic system and ribosomes for their reproduction. Nevertheless, the term "phage life cycle" is commonly used to describe the various stages of viral infection from the initial contact of the virus with its host cell to the ultimate release of progeny virus particles. Three types of phage life cycles have been observed:

1. *Lytic life cycle:* The bacteriophage's nucleic acid enters the host cell and in a complicated but understandable way converts the bacterium to a phage-synthesizing factory. Within about an hour (the time varying with the phage species) the infected bacterium's cell envelope (cell wall and membrane) is disrupted and 100 or more progeny phages are released. This disruptive release is termed **lysis** and the suspension of newly synthesized phages is called a **phage lysate**. A phage that is capable only of lytic growth, which is designated a **virulent phage**, always kills its host.

2. *Lysogenic life cycle:* Bacteriophages that go through the lysogenic life cycle contain double-stranded DNA, which has two alternate fates upon entering the host cell. It can convert the bacterium to a phage-synthesizing factory, which produces large numbers of progeny phages that are released after cell lysis, or the phage DNA can be integrated into the bacterial chromosome and replicate as part of the bacterial chromosome or in some cases form an autonomously replicating circular DNA molecule. Integrated DNA, which is noninfectious, is called a **prophage**. Under normal growth conditions a prophage remains integrated in the bacterial chromosome. On occasion, however, a prophage is excised from the bacterial chromosome and converts the bacterium into a phage-synthesizing factory, which eventually ruptures to release the progeny phage. Physical or chemical agents that damage DNA induce prophages to enter the lytic life cycle and produce phage particles. Ultraviolet light is often used in the laboratory to induce a prophage to enter the lytic life cycle. Because a bacterial cell carrying a prophage has the potential for cell lysis, it is termed a **lysogen**. A phage that sometimes exists as a prophage is called a **temperate phage**.

3. *Chronic infection life cycle:* Bacteriophages that go through the chronic infection life cycle are continuously released without disrupting the cell envelope (wall and membrane) or killing the host. Such phages are called **chronic phages**.

Bacteriophages form plaques on a bacterial lawn.

Phages multiply much faster than bacteria. When cultured in rich medium *E. coli* doubles in number in about half an hour. In contrast, each bacterial cell infected by a single virulent phage particle such as T4 gives rise to more than 100 progeny in the same time period. Released phage particles can then infect more bacteria. Phages released in the second cycle of infection can infect even more bacteria. Therefore, in 2 hours there are four cycles of infection for both a bacterium and a phage. A single bacterium, however, becomes $2^4 = 16$ bacteria and a single phage becomes $100^4 = 10^8$ phage particles.

Virulent phage particles are easily counted by a technique known as the **plaque assay** in the following way. When 10^8 bacteria are plated on nutrient agar, the resulting colonies appear as a confluent, turbid layer of bacteria called a lawn. To achieve maximum uniformity of the turbidity of the lawn, the bacteria are suspended in a small volume of warm liquid agar that then is poured onto the surface of the solid medium. The liquid agar, which contains 0.7% agar compared with 1.5% in agar plates, forms a soft agar layer on cooling. Known as **top agar**, this soft layer provides a very smooth surface so the bacteria can grow with much uniformity (FIGURE 7.3a).

If phage particles are added along with the bacteria, they attach to the bacteria in the agar. Shortly afterward, each infected bacterial cell lyses and releases about 100 phage particles, each of which attaches to nearby bacteria. These infected bacteria in turn release a burst of phage particles that then can infect other bacteria in the vicinity. These multiple cycles of infection continue, and after several hours the phage has destroyed all of the bacteria at a single localized area in the agar, giving rise to a clear, transparent circular region in the turbid, confluent layer (Figure 7.3b). This clear region is called a **plaque**. Because each phage particle forms one plaque, the number of individual phage particles put on the plate can be determined by counting plaques. One can thus determine the number of infective phage particles present in the original phage suspension. A similar approach can be used when studying temperate or chronic phage.

Temperate phages produce turbid plaques instead of clear plaques. The turbidity arises because temperate phages have a lysogenic as well as lytic life cycle. Most infected bacteria are lysed and killed by the temperate phages, but some cells become lysogens and survive. The growth of these lysogenic cells gives the plaques their turbid appearance. Chronic phages also produce plaques even though there is no cell death. Phage infection slows the rate of bacterial growth, so the plaques are regions of low cell density. Plaque assays are useful not only because they permit us to determine the number of phages in a suspension but also because they allow us to isolate pure phage strains (given that each plaque arises from a single phage). The isolation of pure phage strains is very important for genetic studies and preparing phages to be used in recombinant DNA experiments.

With this general introduction to bacteriophages as a background, we are ready to examine a few specific phages that are instructive model systems in molecular biology. Our focus is on phages that infect *E. coli* because, as a general rule, these phages have been the most thoroughly studied and made the greatest contributions to the development of the field of molecular biology. Although we do not examine many fascinating phages that infect other bacteria, most of the information presented on coliphages is directly applicable to phages that infect other bacteria.

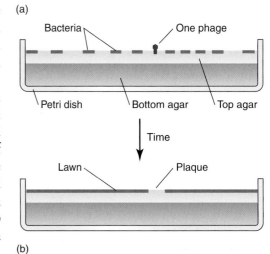

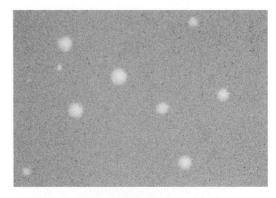

FIGURE 7.3 **Phage plaque.** (a) Schematic drawing of plaque formation. Bacteria grow and form a translucent lawn. There are no bacteria in the vicinity of the plaque, which remains transparent. Neither phage nor bacteria are drawn to scale; both are much smaller. (b) Phage T4 plaques on a bacterial lawn. (Part b © Ken Wagner/Visuals Unlimited.)

7.2 Virulent Bacteriophages

The T-even phages are tailed viruses that inject their DNA into bacteria.

We begin our examination of viruses that infect bacteria by considering the T-even bacteriophages, which are closely related tailed viruses (FIGURE 7.4). The T-even bacteriophage head is an elongated icosahedron, which is connected to a neck surrounded by a collar with attached protein whiskers. The neck is also attached to a tail that ends in a multiprotein baseplate with tail fibers attached to it. The tail consists of an inner rigid core that is surrounded by a contractile sheath.

The DNA in T-even viruses is a linear duplex that is about 600 times longer than the phage head, indicating that the DNA is tightly packed in the phage head. Phage T4 DNA contains adenine, thymine, and guanine and the cytosine derivative 5-hydroxymethylcytosine (FIGURE 7.5a), which base pairs with guanine. The hydroxymethyl group is often modified by glucose group attachment (Figure 7.5b).

T-even phage DNA is terminally redundant and circularly permuted. The term "terminally redundant" refers to the fact that approximately 2% of the viral DNA is repeated at each end of the molecule. As shown in FIGURE 7.6, a terminally redundant linear duplex can form a circle under laboratory conditions when a 3′-exonuclease is used to trim back a single strand at each end. Terminal redundancy is a property of an individual phage T4 DNA molecule. The term "circularly permuted" indicates that terminally redundant ends differ from one phage particle to another (FIGURE 7.7). This difference is observed even within a phage population from a plaque derived from a single phage particle. The gene order is the same in the DNA from all phage particles, but the DNA molecules are circularly permuted. Circular permutation is a property of the phage T4 population.

The phage T4 infection process begins with phage tail fibers contacting specific outer membrane receptor proteins on the bacterial cell surface. Attachment triggers a series of events that ends with the T-even

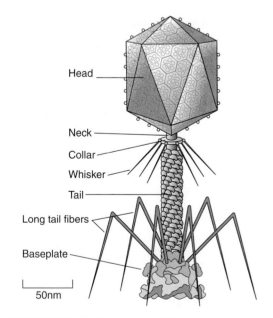

FIGURE 7.4 Basic structural features of bacteriophage T4. (Adapted from Leimana, P. G., et al. 2003. *Cell Mol Life Sci* 60:2356–2370.)

FIGURE 7.5 5-Hydroxymethylcytosine (HMC) and glucosylated 5-hydroxymethylcytosine. (a) If the CH_2OH group (red) in HMC were replaced by hydrogen, the molecule would be cytosine. (b) The glucose group in the glucosylated HMC is shown in blue.

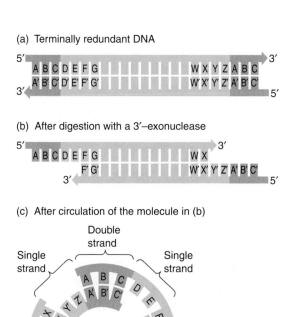

FIGURE 7.6 Terminally redundant linear duplex. Approximately 2% of the viral DNA is repeated at each end of the molecule (shown in red).

phage injecting its DNA into the cytoplasm of the bacterial cell. The T-even phage life cycle is shown in the schematic diagram in **FIGURE 7.8**. **TABLE 7.1** indicates some of the key events that take place during the T-even phage life cycle and the times at which they occur. The main feature to be noticed is the orderly sequence of events.

Shortly after infection the phage DNA codes for proteins that turn off bacterial functions necessary for continued bacterial growth.

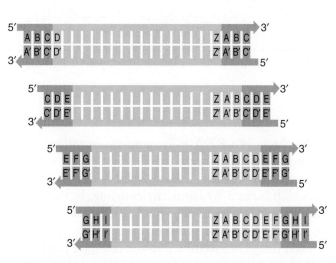

FIGURE 7.7 Circularly permuted collection of terminally redundant DNA molecules. Terminally redundant regions are shown in red.

7.2 Virulent Bacteriophages

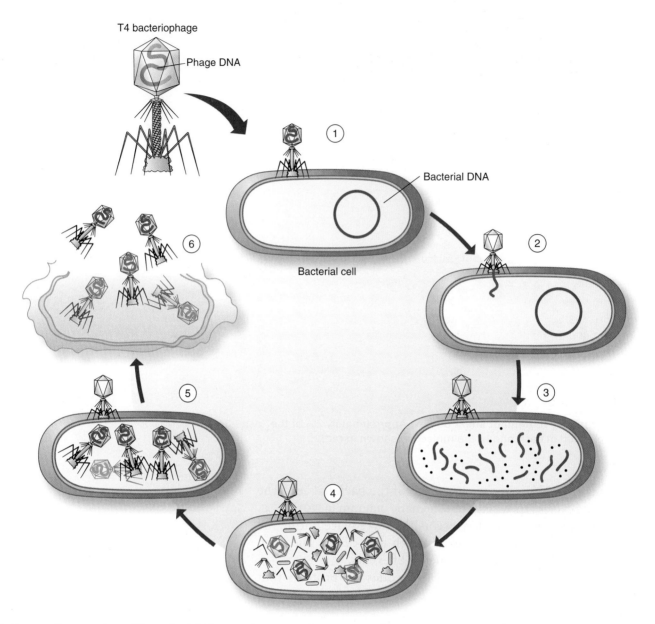

FIGURE 7.8 T-even phage life cycle. (1) T-even phage attaches to a specific host bacterial cell. (2) Phage injects its DNA into the host cell. (3) Phage DNA codes for proteins that digest the bacterial DNA, killing the bacterial cell. (4) Phage takes over the bacterial machinery to make new phage parts. (5) The process ends with the assembly of new phages and (6) the disruption of the bacterial cell wall to release about 100 new phages into the culture medium. (Adapted from Thiel, K. 2004. *Nat Biotechnol* 22:31–36.)

For example, the host's RNA polymerase is modified so host genes are recognized poorly. Furthermore, the first phage messenger RNA (mRNA) formed directs the synthesis of DNases that rapidly degrade host DNA, which lacks 5-hydromethylcytosine. In time, the bacterial DNA is almost totally degraded to nucleotides and is not available as a template for replication or transcription.

The T-even phage is assembled in the cell by a process somewhat similar to that used by car manufacturers. The various parts are first synthesized and then assembled to form the complete virus particle.

TABLE 7.1 T–Even Phage Life Cycle at 37°C	
Time (min)	Change That Takes Place
0	Tail fibers attach to a specific outer membrane protein. Injection of phage DNA probably occurs within seconds of attachment.
1	Host DNA, RNA, and protein syntheses are totally turned off.
2	Synthesis of first mRNA begins.
3	Degradation of bacterial DNA begins.
5	Phage DNA synthesis is initiated.
9	Synthesis of "late" mRNA begins.
12	Completed heads and tails appear.
15	First complete phage particle appears.
22	Lysis of bacteria; release of progeny phage.

Production of complete phage particles can be separated into the following four processes:

1. *DNA replication:* DNA synthesis involves the formation of long DNA molecules called **concatemers**, which consist of many copies of T-even phage DNA molecules linked head-to-tail. Both DNA replication and recombination are required for concatemer formation. The T-even phage replication system is similar to that used by other biological systems.

2. *Head and tail formation:* Phage heads and tails form separately as evidenced by the observation that mutants with defects in head formation make normal tails and mutants with defects in tail formation make normal **procapsids** (head precursors).

3. *DNA packaging:* The enzyme **terminase**, a T-even phage DNA packaging motor, uses the chemical energy stored in ATP to package phage DNA into the procapsid. When the procapsid is filled with DNA so that no more can be packaged inside it, the terminase cuts the DNA. Then the terminase and unpackaged DNA dissociate from the filled procapsid and interact with a new procapsid. Terminase does not cut at a unique base sequence. If it did, T-even phage DNA would not be circularly permuted. Instead, terminase makes cuts at positions determined by the amount of DNA that can fit in a head. The essential point is that the DNA content of a T-even phage particle is greater than the length of DNA required to encode the T-even phage proteins. When cutting a headful from a concatemer, the final segment of DNA packaged is a duplicate of the DNA that is packaged first. This duplication results in terminally redundant T-even phage DNA. The first segment of the second DNA molecule packaged is not the same as the first segment of the first phage. Furthermore, because the second phage must also be terminally redundant, a third phage-DNA molecule must begin with still another segment. The collection of DNA molecules in the phages produced

by a single infected bacterium is therefore a circularly permuted set. Thus, the *headful mechanism* of DNA packaging explains how both terminal redundancy and circular permutation arise.

4. *Virion assembly:* The T-even phage is assembled in the cell by a process in which virion parts are first synthesized and then assembled to form the complete virus particle (**BOX 7.1 LOOKING DEEPER: T-EVEN PHAGE ASSEMBLY**).

The final stage in the T-even phage life cycle, lysis, is an abrupt event. The timing must be tightly regulated because early lysis would yield too few phages and late lysis would diminish the opportunity for an explosive reproduction cycle in new host cells. T-even phages, require two proteins encoded by the phage DNA for lysis. The first, an endolysin, is an enzyme that degrades the cell wall. The second, holin, assembles to form a pore in the cell membrane through which the endolysin moves to reach the cell wall. After endolysin degrades the cell wall, internal osmotic pressure causes cell lysis.

Bacterial and phage RNA polymerases help to reel phage T7 DNA into the bacterial cell.

E. coli phage T7 has an icosahedral head and a short tail (**FIGURE 7.9**). Its linear double-stranded DNA is terminally redundant but not circularly

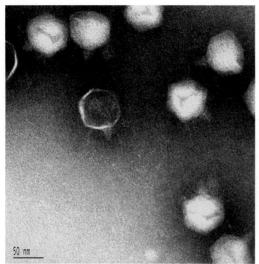

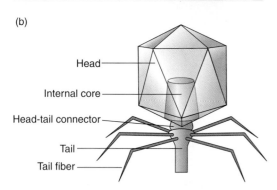

FIGURE 7.9 Phage T7. (a) Electron micrograph of phage T7. The tail fibers are not visible. The bar is 50 nm. (b) Schematic of the T7 virion indicating its protein components. (Part a photo courtesy of Dean Scholl, National Institute of Health; part b adapted from Calendar, R. L. 2005. *The Bacteriophages* [2nd ed]. Oxford University Press.)

BOX 7.1: LOOKING DEEPER

T-Even Phage Assembly

T-even phage assembly has been studied by two techniques, both of which require a large collection of phage mutants unable to produce finished particles. In one method different cultures of cells, each infected with a particular mutant, are lysed and examined by electron microscopy. This procedure shows that heads are made in the absence of tail synthesis and that tails are made by a mutant unable to synthesize heads; head and tail assembly are independent processes.

The second technique is a complementation assay of a type widely used to purify proteins. If two extracts of infected cells, one lacking heads and the other lacking tails, are mixed, phage particles form *in vitro* (**FIGURE B7.1a**). The "headless" extract also can be fractionated and a component can be isolated that allows a "tailless" extract to make tails. In this way a protein in the tail assembly pathway can be isolated and identified.

Assembly studies have shown T-even phage assembly requires two types of components: structural proteins and morphogenetic enzymes. Some structural components assemble spontaneously to form phage structures, whereas others need the help of enzymes. A few host-encoded factors are also needed for head assembly. A schematic for the T-even phage assembly pathway is shown in **FIGURE B7.1b**.

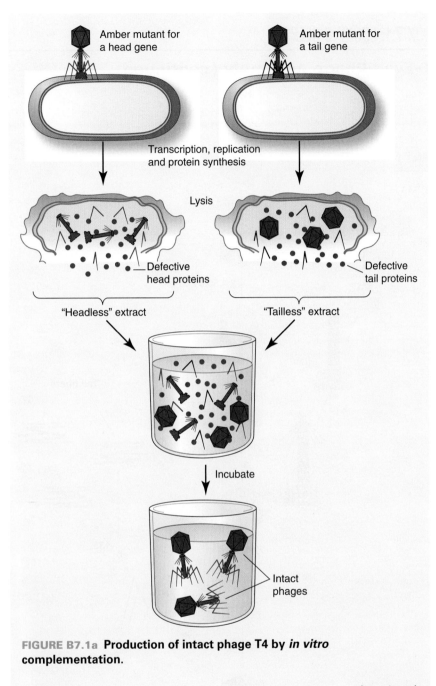

FIGURE B7.1a Production of intact phage T4 by *in vitro* complementation.

(*continues*)

7.2 Virulent Bacteriophages

BOX 7.1: LOOKING DEEPER

T-Even Phage Assembly (*continued*)

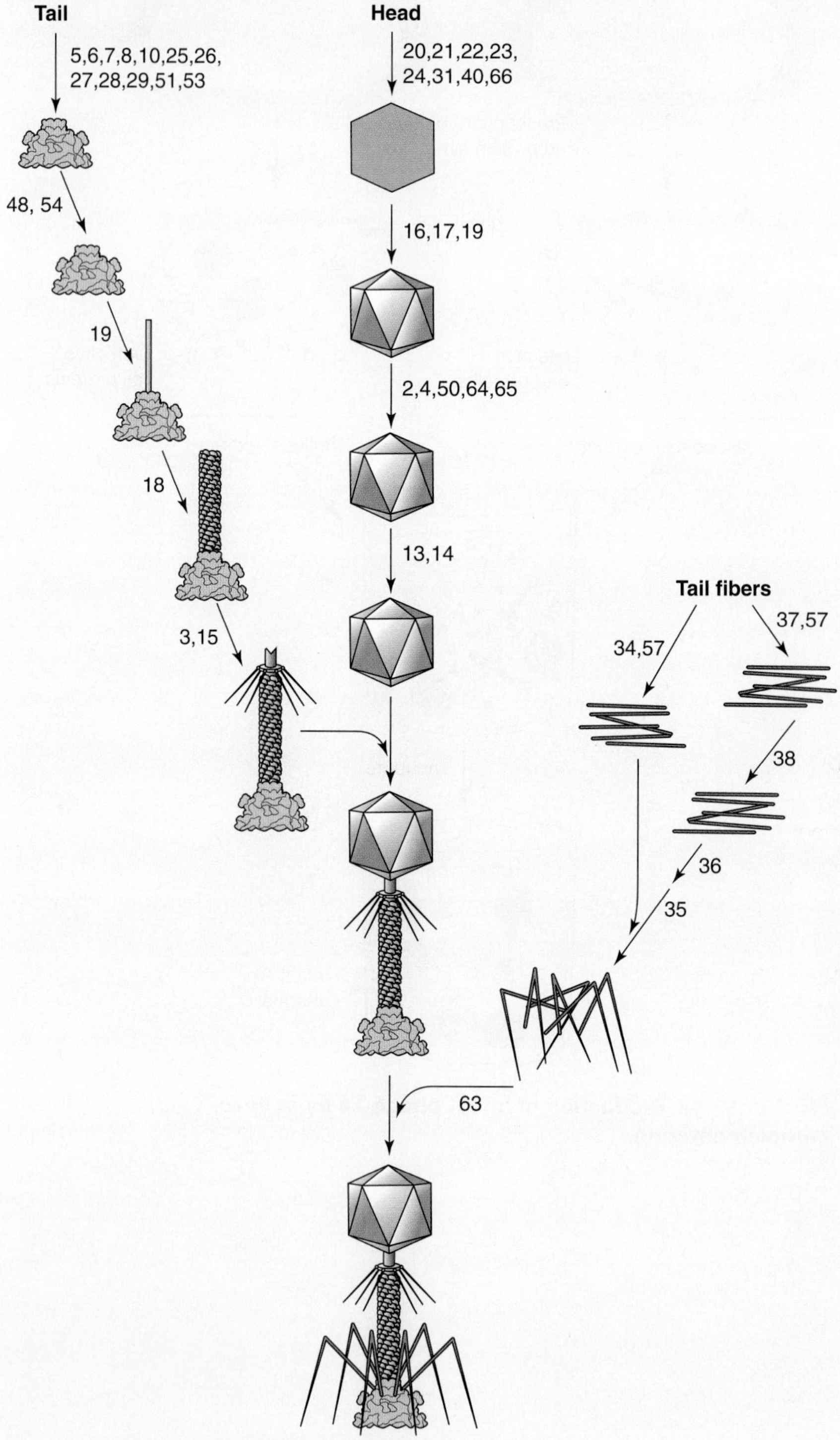

Tail
5,6,7,8,10,25,26,
27,28,29,51,53

48, 54

19

18

3,15

Head
20,21,22,23,
24,31,40,66

16,17,19

2,4,50,64,65

13,14

Tail fibers
34,57   37,57

38

36

35

63

FIGURE B7.1b T-even phage assembly pathway. The numbers designate phage T4 genes.

CHAPTER 7 VIRUSES IN MOLECULAR BIOLOGY

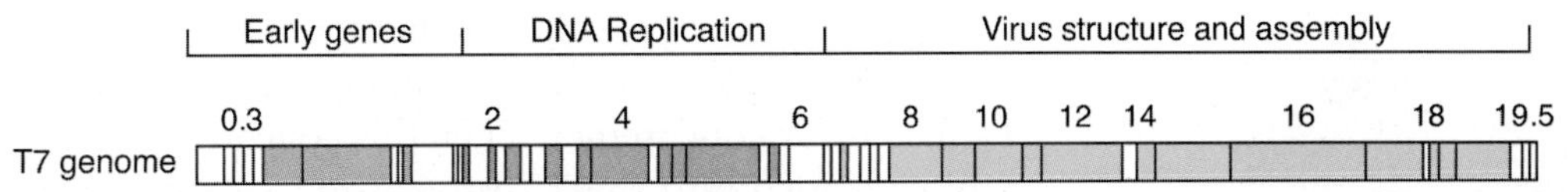

FIGURE 7.10 Phage T7 genetic map. Box colors indicate a simplified assignment to functional classes as shown by the heading. (Modified from Häuser, R., Blasche, S., Dokland, T., Haggård-Ljungquist, E., von Brunn, A., Salas, M., Casjens, S., Molineux, I., and Uetz, P. Bacteriophage protein-protein interactions. *Adv Virus Res* 2012;83:219–298. Courtesy of Peter Uetz, Virginia Commonwealth University.)

permuted. Phage T7 DNA contains the normal bases, adenine, thymine, guanine, and cytosine. A notable feature of the organization of the phage T7 genes is that they are clustered according to function and are arranged to provide a continuous order according to the time of their function (FIGURE 7.10).

TABLE 7.2 indicates some of the key events that take place during the T7 life cycle and the times at which they occur. Phage T7 first makes contact with the bacterial cell through the interaction of its tail fibers with lipopolysaccharides on the bacterial cell surface. The mechanism of phage T7 DNA entry into the bacterial cell is completely different from T-even phage DNA entry. Phage T7, which is tightly bound to the bacterial cell, releases two proteins that form a channel across the bacterial envelope. This channel appears to act as a motor to drive a short phage T7 DNA segment into the bacterial cell. Then *E. coli* RNA polymerase, which is assumed to be stationary, reels a longer phage T7 DNA segment into the bacterial cell. Finally, newly synthesized phage T7 RNA polymerase reels in the remaining DNA.

E. coli RNA polymerase synthesizes RNA during the early stage of infection. The bacterial RNA polymerase is inactivated after synthesizing mRNA that codes for the phage T7 RNA polymerase. This inactivation, which reduces transcription of the bacterial DNA, is an important step in the takeover of the bacterium by the phage. The phage T7 RNA polymerase synthesizes mRNA that codes for viral proteins.

TABLE 7.2 The T7 Life Cycle at 30°

Time (min)	Change That Takes Place
1	Phage attachment.
0–1	Start of phage DNA entry into the cell.
2	Initiation of synthesis of phage mRNA.
4	Turning off of host transcription begins.
8–9	Initiation of phage DNA synthesis.
8–10	Initiation of synthesis of structural proteins; completion of injection of phage DNA.
15	First phage appears.
25	Lysis and release of progeny phage.

Replication of the original phage T7 DNA generates linear duplexes with short single-stranded terminally redundant tails, which anneal to form a linear concatemer. Then this concatemer is cut into unit lengths and packaged into the T7 procapsid. A protein product formed late in infection binds to and inhibits phage T7 RNA polymerase and also acts as an enzyme to digest the cell wall and trigger cell lysis.

E. coli phage ɸX174 contains a single-stranded circular DNA molecule.

E. coli phage ɸX174 has a circular, single-stranded DNA molecule enclosed in an icosahedral head with attached spikes (FIGURE 7.11). The viral DNA, which is 5,386 nucleotides long, codes for only 11 polypeptides (FIGURE 7.12), each of which has been isolated. The number of amino acids contained in these polypeptides exceeds this very small DNA molecule's predicted coding capacity. This finding is explained by the fact that ɸX174 has overlapping genes that are translated in different reading frames. For instance, gene *A* shares nucleotides with genes *B* and *K*.

Phage ɸX174 binds to a lipopolysaccharide receptor on the *E. coli* outer cell membrane. Once attached to the bacterium, the single-stranded DNA and at least one spike protein are ejected through a channel in the spike into the bacterial cell's periplasmic space (the space between the inner and outer membranes). After ɸX174 DNA penetrates the inner membrane, DNA and protein synthesis take place and new virus particles are assembled. A great deal is known about phage ɸX174 infection, but we only examine a few aspects of DNA replication in this section.

The mode of ɸX174 DNA replication is of special interest because the phage has the problem of making an identical copy of a single strand. Clearly, the template cannot yield progeny molecules in a single step. The following terminology is used when discussing the replication of single-stranded DNA molecules. The strand contained in the virus particle has the same sequence as mRNA and it or any strand with the same base sequence is called a (+) strand. A strand having the complementary base sequence is called a (−) strand. The parental single-stranded DNA molecule, the viral or (+) strand, is converted to a covalently closed, double-stranded DNA molecule called replicative form I (RFI). This conversion occurs before transcription begins and hence depends on host enzymes exclusively. The newly formed (−) strand in the supercoiled RFI is transcribed, the resulting mRNA is translated, and a multifunctional protein, the A protein, is made.

As shown in FIGURE 7.13, the A protein makes a single nick in the (+) strand of the supercoiled RFI (between bases 4305 and 4306) and remains covalently linked to the 5′-P terminus. The nicked molecule is called replicative form II (RFII). *E. coli* proteins (synthesized before infection) then cause the parental (+) strand to be displaced from RFII by looped rolling-circle replication. When one round of replication is completed the displaced (+) strand is cleaved from the looped rolling circle and is converted to a single-stranded, closed, covalent, circular DNA molecule, which serves as a template for the synthesis of

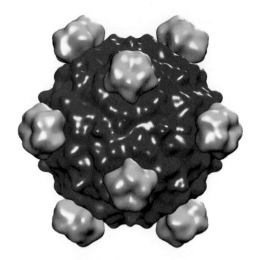

FIGURE 7.11 ɸX174 virion. Capsid and spike proteins are colored blue and yellow, respectively. (Reprinted from Bernal, R., et al. 2004. The Phi-X174 protein J mediates DNA . . . *J Mol Biol* 337:1109–1122, copyright 2004, with permission from Elsevier [http://www.sciencedirect.com/science/journal/00222836]. Photo courtesy of Michael G. Rossmann, Purdue University, and Ricardo A. Bernal, University of Texas, El Paso.)

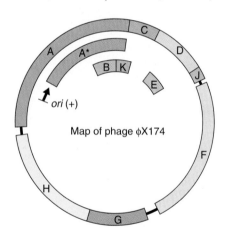

Gene	Function
ori	Origin of plus strand replication
Gene A	RF replication
Gene A*	Shut off host DNA synthesis
Gene B	Capsid morphogenesis
Gene C	DNA maturation
Gene D	Capsid morphogenesis
Gene E	Cell lysis
Gene F	Major coat protein
Gene G	Major spike protein
Gene H	Minor spike protein, adsorption
Gene J	Core protein, DNA condensation
Gene K	Function unknown

FIGURE 7.12 **Map of phage ɸX174.** (Adapted from an illustration by New England BioLabs, Inc.)

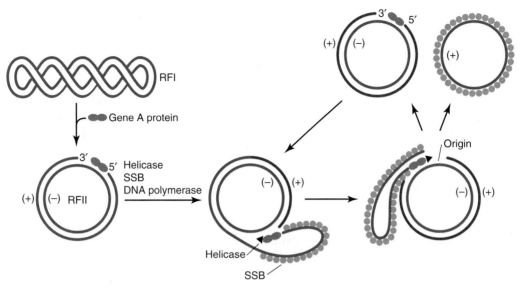

FIGURE 7.13 Looped rolling circle replication of φX174 DNA. The gene A protein nicks the supercoiled DNA and binds to the 5′-terminus of the (+) strand. Rolling circle replication takes place to generate a daughter strand (red) and a displaced (+) single strand that is coated with single-stranded DNA binding protein (SSB) and still covalently linked to the A protein. When the entire (+) strand is displaced, it is cleaved by the joining activity of the A protein. The cycle is ready to begin anew. Note the (−) strand is never cleaved.

another (−) strand; the result is a new RFI. After 30 to 40 RFI copies are formed the phage packaging system begins to capture progeny (+) strands before they can serve as templates for the synthesis of (−) strands. Packaging of φX174 DNA requires seven phage proteins, four of which are present in the finished phage particle. The first completed virion appears about 10 minutes after infection. Cell lysis does not take place until 30 minutes after infection, by which time about 500 phage particles have been synthesized.

Some phages have single-stranded RNA as their genetic material.

In 1960 Norton Zinder and Timothy Loeb set out to discover whether phages exist that infect only F⁺ and Hfr (high frequency of recombination) but not F⁻ *E. coli*. They found several such phages in the sewers of New York City. One of these, the f2 phage, was later shown to have single-stranded RNA rather than DNA as its genetic material. Subsequent studies led to the discovery of additional examples of single-stranded RNA phages such as R17, MS2, and Qβ. The RNA strand in each of these phages acts as mRNA and therefore is designated a (+) strand. These RNA phages are an important source of easily isolated, homogeneous mRNA. As such, they have been particularly valuable for answering questions related to various aspects of protein synthesis.

Phages f2, R17, MS2, and Qβ are tailless icosahedrons (**FIGURE 7.14**). Each contains a single-stranded linear RNA molecule with about 3,600 to 4,300 nucleotides and a great deal of intramolecular hydrogen bonding. The phage RNA serves as both a replication template and an

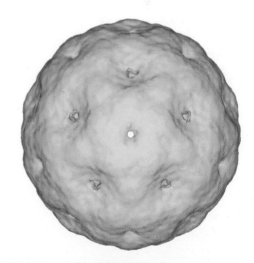

FIGURE 7.14 Cryoelectron microscopy reconstructions of MS2. (Reproduced from Koning, R., et al. 2003. Visualization by cryo-electron microscopy . . . *J Mol Biol* 332:415–422, copyright 2003, with permission from Elsevier. Photo courtesy of Roman Koning, Leiden University Medical Center, The Netherlands.)

7.2 Virulent Bacteriophages

mRNA. The RNA phages bind to the pili that extend from the surface of F⁺ and Hfr cells. After entering the cell the phage RNA acts as an mRNA to code for the synthesis of coat protein and an RNA-dependent RNA polymerase. As shown in FIGURE 7.15, the RNA-dependent RNA polymerase copies the viral (+) strand to generate a (−) strand. While synthesis is proceeding the (−) strand is in contact with the (+) strand only at the polymerization site. For the most part the replicative form is therefore single-stranded. Initiation of several (−) strands occurs before the first (−) strand is complete, and the replicative form is branched. The (−) strands are released and immediately used by the RNA-dependent RNA polymerase to form (+) strands. Some of the (+) strands return to the ribosomes for synthesis of more coat proteins and others are packaged. All progeny contain (+) strands exclusively. A typical burst size is 5,000 to 10,000, which is very large compared with the burst of 100 to a few hundred for DNA phages.

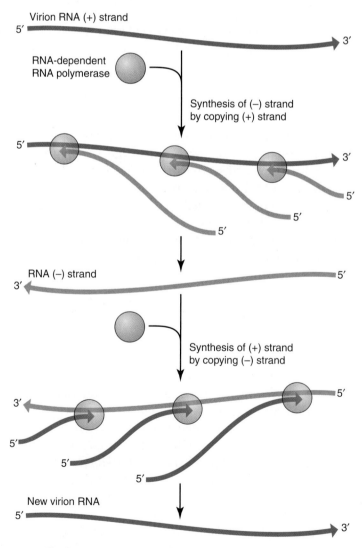

FIGURE 7.15 Replication of phage RNA by RNA-dependent RNA polymerase.

CHAPTER 7 VIRUSES IN MOLECULAR BIOLOGY

7.3 Temperate Phages

E. coli phage λ DNA can replicate through a lytic or lysogenic life cycle.

E. coli phage lambda (λ) has an icosahedral head and a long, noncontractile, flexible tail that ends in a tip structure (**FIGURE 7.16a**). Four nonessential tail fibers extend from the junction of the tail and the tip structure. The virion's DNA, which is located in its head, is a 48,502-base pair (bp) linear duplex with a 12 nucleotide 5′-overhang at each end. Phage λ genes are clustered according to function (Figure 7.16b). For example, the head, tail, replication, and recombination genes form four distinct clusters.

Although the present discussion is limited to phage λ, this phage is just one member of a very large family of similar phages that infect a wide variety of bacteria and are collectively known as the **lambdoid phages**. The tip structure at the end of the phage λ tail plays a critical role in bacterial recognition. It binds to a bacterial protein called LamB on the outer surface of the cell envelope. *E. coli* normally uses its LamB protein to form nonspecific channels that allow small

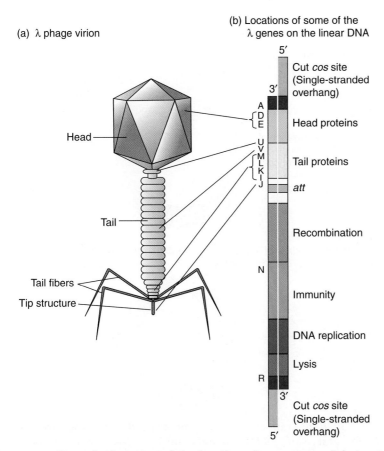

FIGURE 7.16 Bacteriophage λ and the location of some genes for structural proteins. (a) Schematic diagram of bacteriophage λ. (b) Locations for some genes that code for proteins in the virion.

hydrophilic molecules such as the sugar maltose to diffuse through the outer membrane. The phage thus appropriates for its own use a bacterial protein that has an entirely different function in the uninfected cell. The entire phage λ DNA enters the bacterium in less than 5 minutes at 37°C. The mechanism of entry remains to be determined.

After phage λ DNA enters the cell it takes advantage of an unusual structural feature to form a circular structure. The base sequences of the 5′-overhangs, which are known as **cohesive ends** or *cos* **elements**, are complementary to one another (**FIGURE 7.17a**). By forming base pairs between the single-stranded ends, the linear DNA molecule can circularize, yielding a double-stranded circle containing two single-strand breaks (Figure 7.17b). Circularization is easily performed

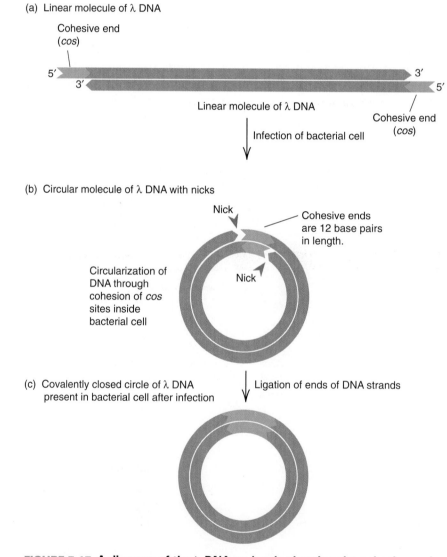

FIGURE 7.17 A diagram of the λ DNA molecule showing the cohesive ends (complementary single-stranded ends). (a) The DNA circularizes by means of base pairing between the cohesive ends. (b and c) The nicked circle that forms is converted into a covalently closed circle by sealing (ligation) of the single-stranded breaks. The length of the cohesive ends is 12 bp in a total molecule of 48,502 bp.

CHAPTER 7 VIRUSES IN MOLECULAR BIOLOGY

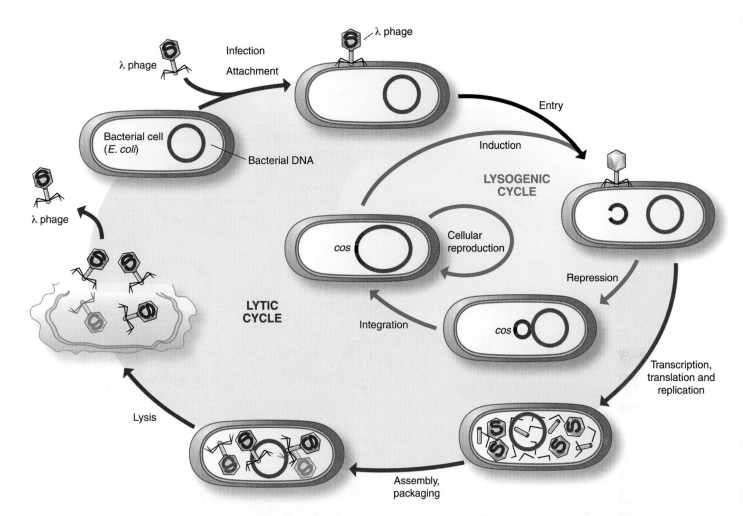

FIGURE 7.18 Life cycle of phage λ, a typical temperate phage. The tip of the phage tail attaches to a receptor on the bacterial cell surface and the phage DNA enters the cell. The empty phage protein shell remains outside the bacterial cell. After λ DNA enters the cell it takes advantage of its complementary 5′-overhangs, which are known as cohesive ends or *cos* elements, to form a circle. In the lytic cycle the DNA is transcribed, translated, and replicated. DNA replication generates long concatemers that are required to package DNA into the phage head. After DNA packaging, phage tails add to the protein shell and new phages are released after cell lysis. In the lysogenic cycle phage development is repressed and the circular phage DNA integrates into the bacterial chromosome. The resulting lysogenic bacteria can replicate indefinitely. They also can be induced to return to the lytic cycle, however, by ultraviolet light (or chemical mutagens) with the excision of phage DNA from the chromosome. (Adapted from Campbell, A. 2003. *Nat Rev Genet* 4:471–477.)

in the laboratory and also occurs in an infected cell within a few minutes after infection. No bases are missing in the newly formed double-stranded region of this circle, so DNA ligase can convert the molecule to a closed covalent circle (Figure 7.17c). The circular, double-stranded DNA can either enter the lytic or lysogenic cycle (**FIGURE 7.18**):

1. *The lytic cycle:* Bacterial DNA gyrase converts the relaxed, closed, covalent, circular DNA to supercoiled DNA. Two types of phage λ DNA replication, **θ replication** and rolling circle replication, take place during the lytic cycle (**FIGURE 7.19**). The term "θ replication" was coined because

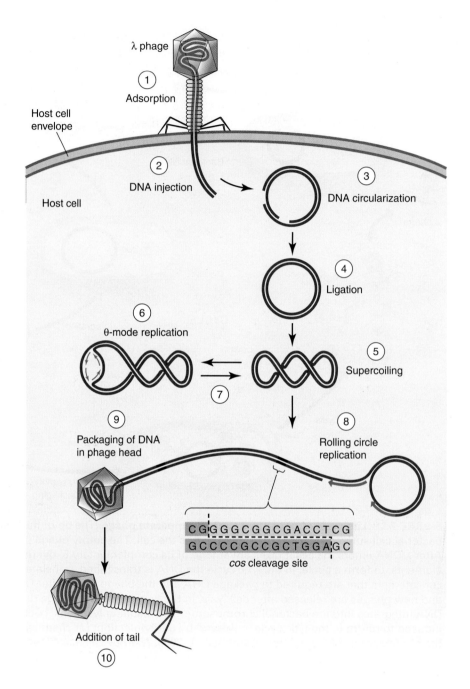

FIGURE 7.19 **DNA replication in the lytic mode of phage λ.** (1) The virion attaches to the host cell. (2) The linear DNA enters the cell. (3) The complementary *cos* elements pair to form a nicked circle. (4) DNA ligase converts the nicked circle to a relaxed covalent circle. (5) DNA gyrase converts the relaxed circular DNA to a supercoil. (6 and 7) Bidirectional or θ (theta) replication produces new DNA. (8) θ replication continues and rolling circle replication starts. In the later stages of infection, replication takes place exclusively through the rolling circle mode. The curved blue arrows indicate the most recently synthesized DNA at the growing points and the arrowheads represent the 3′-ends of the growing DNA chains. (9) The concatemeric DNA produced by the rolling circle mode is specifically cleaved at its *cos* site and is packaged into phage heads. (10) Tails are added to complete the assembly of the mature phage particles, which are each capable of initiating a new round of infection. (Adapted from Voet, D., and Voet, J. G. 2005. *Biochemistry* [3rd ed]. John Wiley & Sons.)

the replication intermediate resembles the Greek letter θ (theta). It is important to note that θ replication of phage λ DNA begins at a specific site on the circular DNA and produces two growing points, which move in opposite directions around the λ DNA circle. θ replication's primary function is to increase the number of templates available for transcription and to provide circular DNA molecules for the next stage of replication, rolling circle replication. As the lytic cycle continues DNA synthesis switches from θ replication to rolling circle replication, which generates long concatemers that are required to package DNA in the phage head. Because the ends of the DNA molecule in the phage λ particle always have single-stranded 5′-ends, long concatemers must be cut at their *cos* sites to generate these termini. This cutting is accomplished by a sequence-specific terminase, which is associated with newly formed empty heads. Terminase-cutting requires two *cos* sites or one *cos* site and a free cohesive end on a single DNA molecule.

2. *The lysogenic cycle:* Infection with phage λ can have a second outcome. Instead of replicating by the lytic mode, the supercoiled phage λ DNA can insert at a specific site of the bacterial chromosome, becoming a prophage. When λ DNA integrates it is inserted at a preferred position in the *E. coli* chromosome. This site, between the genes for galactose utilization (*gal*) and biotin synthesis (*bio*), is called the λ attachment site and designated *att*. The resulting λ lysogen is generally quite stable and can replicate nearly indefinitely without release of phage. If a lysogenic bacterium were to become damaged, however, it would be to the advantage of the prophage to initiate the lytic cycle. This initiation does occur, and the signal to enter the lytic mode is DNA damage caused by ultraviolet light, x-rays, or a chemical agent that disrupts or modifies the DNA. Phage λ DNA excision is usually a very precise process, but on rare occasions (one cell in 10^6 or 10^7 cells) an excision error is made and a small piece of bacterial DNA is excised along with the λ DNA. The recombinant λ DNA is replicated and packaged into a λ particle, which can deliver the bacterial DNA to a new bacterial host in a process known as **specialized transduction**. Only genes adjacent to the prophage are transferred from the lysogen to another bacterial cell.

E. coli phage P1 can also replicate through a lytic or lysogenic life cycle.

The phage P1 particle has a tail that consists of a tube surrounded by a contractile sheath and is connected to an icosahedral head at one end and a baseplate with six kinked tail fibers at the other (FIGURE 7.20). The head contains a linear duplex DNA molecule, which has a terminal redundancy of 10 to 15 kilobase pair (kbp) and is circularly permuted. The virion binds to a lipopolysaccharide on the outer surface of the bacterial envelope. Phage P1 has a fairly broad host range, infecting *E. coli* as well as several other gram-negative bacterial species. Phage

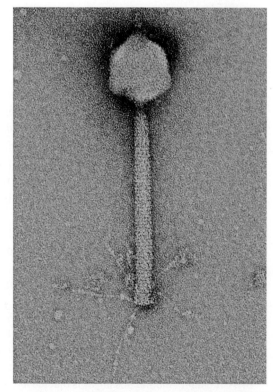

FIGURE 7.20 **Electron micrograph of phage P1.** (Photo courtesy of Michael Wurtz and the Biocenter at the University of Basel.)

P1 has been a very important tool in molecular biology because it allows us to transfer genes from one bacterium to another (**BOX 7.2 IN THE LAB: GENERALIZED TRANSDUCTION**).

Based on the presence of the contractile sheath, P1 DNA is thought to be injected into the bacterial cell. Once inside the cell the linear duplex is converted into a closed covalent circle by a host recombination system that acts on the terminally redundant ends. Bacterial DNA gyrase acts on this relaxed circular DNA, converting it into a supercoil that can enter the lytic or lysogenic cycle. Commitment to the lytic or lysogenic cycle is determined by genetic switches that act through the transcription machinery:

1. *The lytic cycle:* In the lytic cycle DNA synthesis starts with θ replication and then switches to rolling circle replication. At approximately 45 minutes after infection the cells are filled with DNA concatemers, empty phage heads, and phage tails. DNA packaging starts when the phage P1 terminase cleaves a specific nucleotide sequence, termed the *pac* sequence, in the DNA concatemer. Then cut DNA is packaged unidirectionally from the cleaved *pac* end into an empty head until that head is full. When packaging is complete, the DNA is cut at a nonsequence-dependent site. The DNA end remaining outside the head is used in the next P1 sequential packaging event.
2. *The lysogenic cycle:* The circular P1 DNA produced after infection can also enter a passive state in which it behaves like a plasmid

BOX 7.2: IN THE LAB

Generalized Transduction

Phage P1 has been a very important tool in molecular biology because it allows us to transfer genes from one bacterium to another (**FIGURE B7.2**). This process in which a phage particle carries random DNA fragments from one bacterium to another, called generalized transduction, was discovered by Joshua Lederberg and Norton D. Zinder in 1952. The phage particles that carry bacterial genes are designated generalized transducing particles. Phage P1 has two characteristics that make it well suited to function as a generalized transducing particle. First, it codes for a nuclease that slowly degrades bacterial DNA, so when packaging begins the host DNA is present as large fragments. Second, the P1 packaging system is not very fastidious, so the size of the DNA molecule does not have to be exactly the same in all particles. The result is that in the population of particles produced when infected bacteria lyse, rare particles contain a bacterial DNA fragment. Because fragmentation of the host DNA is a random process, these rare particles contain fragments derived from all regions of the host DNA. A sufficiently large population of phage P1 progeny will, therefore, contain at least one particle possessing each host gene. On the average, for any particular gene, roughly one virion particle out of 10^6 viable phages contains a bacterial fragment. These rare particles do not produce P1 phage progeny when they infect bacteria because they contain no P1 DNA. Instead, the bacterial DNA is injected into the host cell.

Let us now examine the events that ensue when one of these rare particles carrying a bacterial DNA fragment attaches to a bacterial cell. Consider a particle that has emerged from an infected wild-type *E. coli* and that contains the gene for leucine (*leu*) synthesis. Such a particle is denoted P1 *leu*⁺. Consider further that the P1 *leu*⁺ particle attaches to a bacterium with a *leu*⁻ genotype and injects its DNA. The bacterium will survive because no phage genes are injected and its nucleases indeed may degrade the injected linear DNA fragment. Another possibility is that the *leu*⁺ segment will be incorporated into the host DNA by genetic recombination, resulting in replacement of the *leu*⁻ allele by a *leu*⁺ allele to produce a *leu*⁺ transductant. In this way the genotype of the recipient cell would be converted from *leu*⁻ to *leu*⁺. Transducing phage particles are detected by their ability to transfer genes.

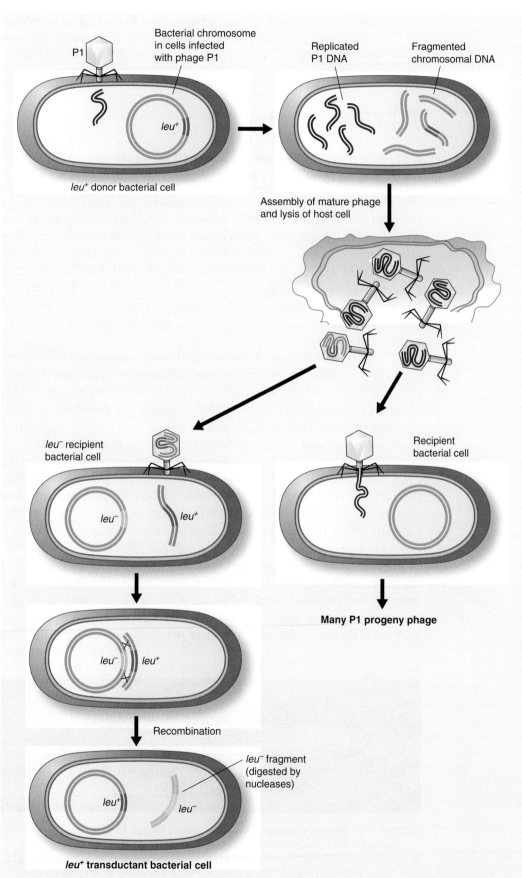

FIGURE B7.2 **Generalized transduction.** Phage P1 infects a *leu⁺* donor, yielding predominantly normal P1 progeny but an occasional particle carries *leu⁺* (or any other small segment of bacterial DNA) instead of phage DNA. When the phage population infects a *leu⁻* bacterial culture, the transducing particle can yield a *leu⁺* transductant. Note that the recombination step requires two exchanges.

7.3 Temperate Phages

and is maintained at one or two copies per cell. The prophage DNA replicates once per bacterial life cycle so that when the bacterium divides, each daughter cell receives a P1 DNA molecule.

7.4 Chronic Phages

A chronic phage programs the host cell for continued virion particle release without killing the cell.

The three best studied chronic coliphages, M13, fd, and f1, are similar filamentous phages with 98% identical DNA sequences. Each is a male-specific phage with a single-stranded, closed covalent circular DNA approximately 6,400 nucleotides long. The DNA molecule is encased in a 930-nm hollow tube (FIGURE 7.21). Unlike the other phages described the filamentous phages neither kill their host cell nor cause it to lyse. Instead, the newly formed virions are continuously released as the cells continue to grow, albeit at a slower rate than uninfected cells.

The hollow tube that surrounds the DNA contains approximately 2,700 copies of the major coat protein, which overlap one another like shingles on the side of a house. Bacterial infection by phages M13, fd, and f1 is a multistep process that begins when the phage particle attaches to the tip of the F pilus. The major coat protein is stripped from the virion and integrated into the bacterial cell membrane. The circular single-stranded DNA, which is released into the bacterial cell's cytoplasm, has an imperfect but stable hairpin. Once in the cytoplasm the DNA is coated with single-stranded DNA binding protein. The single-stranded DNA binding protein, however, does not bind to the hairpin, which then serves as the start site for (–) strand synthesis.

Although some details are different, the filamentous phages use the same fundamental strategy for DNA replication as that described above for φX174. Phage particles are assembled in the bacterial inner membrane. For each completed phage particle a bud forms on the cell surface and a phage is extruded. This process can continue indefinitely without damage to the cell. The first progeny virions appear in the

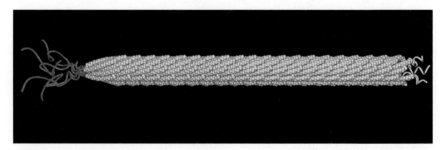

FIGURE 7.21 **Model of phage M13.** The major phage M13 protein is shown in gray and other phage proteins in orange and red. The proteins are not drawn to scale. (Reproduced from Mao, C., et al. 2004. *Science* 303:213–217. Reprinted with permission from AAAS. Photo courtesy of Angela Belcher, Massachusetts Institute of Technology.)

BOX 7.3: IN THE LAB

Phage M13 Cloning Vector

Joachim Messing recognized that phage M13 might be a very useful cloning vector if it were possible to introduce the foreign DNA into the phage DNA without disrupting an essential phage gene. This was a difficult task in the early 1970s because not much was known about phage M13 genes. Messing succeeded in obtaining the desired vector by inserting a cloning sequence with unique restriction endonuclease sites into a region between two M13 genes. The basic approach to constructing an M13 vector with an inserted foreign DNA segment is as follows: (1) Double-stranded M13 vector DNA is isolated from infected cells and cut with a restriction endonuclease. (2) The foreign DNA is inserted between the restriction sites and the DNA molecules joined by DNA ligase. (3) Competent bacteria are transformed with the recombinant DNA. (4) Phage progeny are collected and the single-stranded DNA containing the inserted foreign DNA is isolated. There is no strict limitation to the size of the foreign DNA fragment that can be inserted into the phage DNA because the phage DNA is not inserted into a preformed structure. The single-stranded DNA is well suited for sequence analysis by the chain termination method and for site-directed mutagenesis.

A related cloning vector, a phagemid, can replicate as a double-stranded plasmid in a bacterial cell but also can be induced to produce large quantities of single-stranded recombinant DNA when desired. A phagemid is constructed by introducing plasmid and phage M13 origins of replication into a double-stranded circular DNA molecule. Under normal growth conditions the phagemid replicates as a plasmid in bacteria. It can be induced to replicate as a phage, however, by infecting the bacterial cell with a helper phage, which supplies the gene products required to produce a virion containing a single-stranded recombinant DNA molecule. The phagemid thus offers the convenience of doing recombinant DNA experiments with a plasmid and the advantage of being able to obtain large quantities of the single-stranded recombinant DNA.

---

culture medium about 10 minutes after infection at 37°C. The number of virion particles increases exponentially for the next 40 minutes and then continues at a linear rate. After about 1 hour the medium contains about 1,000 virions per cell. Phage M13 is a convenient cloning vector (**BOX 7.3 IN THE LAB: PHAGE M13 CLONING VECTOR**).

7.5 Animal Viruses

Animal viruses can serve as excellent models for addressing fundamental questions in molecular biology.

Molecular biologists are interested in eukaryotic viruses for much the same reasons we are interested in bacteriophages. Fundamental questions concerning the molecular biology of eukaryotic cells can be answered by studying virus-infected eukaryotic cells. For instance, we gain insights into how eukaryotic cells replicate their own DNA by examining how they replicate viral DNA and how eukaryotic cells synthesize their own mRNA and proteins by studying how they synthesize viral mRNA and proteins. The choice of the specific virus to be used is usually based on the ease of maintaining and propagating the virus in the laboratory and its suitability for answering a specific set of questions. Investigators, however, are also mindful of the fact that the new information they hope to obtain might lead to better methods to treat specific viral diseases.

The three kinds of animal viruses we examine in this section have each contributed to our basic understanding of the molecular biology of eukaryotic cells and each can cause serious illnesses. We focus on these three kinds of animal viruses because you will need to be familiar with them when examining eukaryotic replication, transcription, translation, and related processes. You may wish to consult a textbook on virology to obtain further information about these viruses as well as others that infect animals, plants, and fungi.

Polyomaviruses contain circular double-stranded DNA.

The polyomavirus was detected in mice in 1953. Its name derives from its ability to cause solid tumors at multiple sites in rodents under laboratory conditions. A closely related virus, SV40 (simian virus 40), was discovered in monkey kidney cells in 1960 and later shown to contaminate early preparations of the Salk and Sabin polio vaccines, which were prepared by propagating the polio virus on monkey kidney cells. Fortunately, the contaminating SV40 does not appear to have caused tumors in vaccinated individuals. The mouse polyomavirus and SV40 belong to a family of viruses called *Polyomaviridae*, which have now been recovered from a wide variety of additional organisms, including cows, birds, and humans, and appear to be species-specific.

The *Polyomaviridae* contain a circular double-stranded DNA molecule (about 5.2 kbp), which is located in an icosahedral capsid with a diameter of about 60 nm (FIGURE 7.22). The capsid consists of three proteins: VP1, VP2, and VP3. VP1 is the major capsid protein, accounting for about 80% of the total capsid protein, whereas VP2 and VP3 together account for the remaining 20%. The DNA in the virus particle is wound around nucleosome cores made of histones H2A, H2B, H3, and H4 so the DNA in the resulting minichromosome is highly condensed.

The SV40 genome is divided into a regulatory region, an early region, and a late region (FIGURE 7.23). The regulatory region (about 400 bp) includes the origin of replication and signals that regulate the transcription of the other two regions. The early and late regions are transcribed from different strands and in opposite directions. The early region codes for three proteins, large T antigen, small t antigen, and 17kT protein. The late region codes for the coat proteins VP1, VP2, and VP3.

A simplified version of the SV40 virus replication cycle is shown in FIGURE 7.24. The virion attaches to a glycolipid receptor on the cell surface, is transported into the host cell by endocytosis, and then moves to the cell nucleus through a pathway not described here. Transcription of the early region produces mRNA molecules that are translated to form T antigens. The large T antigen is required to initiate viral DNA replication at the origin of replication. All other proteins needed for DNA replication are supplied by the host cell. Like host cell DNA replication, viral DNA replication takes place during the S phase of the cell cycle. Studies of SV40 DNA replication provide considerable information about the host cell's replication apparatus.

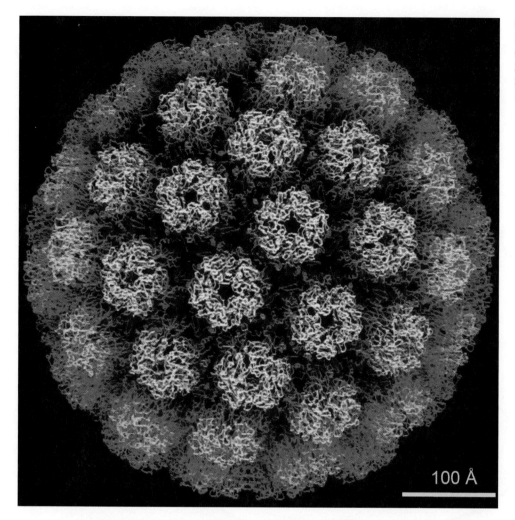

FIGURE 7.22 **V40 virion.** (Reproduced from Liddington R. C., et al. 1991. *Nature* 354: 278–284. Photo courtesy of Stephen C. Harrison, Harvard Medical School.)

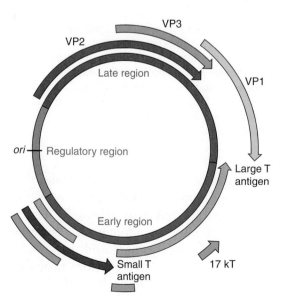

FIGURE 7.23 **Simplified map of SV40 genome.** The SV40 genome is divided into an early region (genes transcribed early in infection), a late region (genes transcribed late in infection), and a regulatory region (genes that control transcription of the early and late regions). The regulatory region also contains the origin of replication (*ori*). The early region codes for the large T antigen, small t antigen, and 17 kT antigen. The large T antigen is the only SV40 product required for viral DNA replication. The late region codes for the coat proteins VP1, VP2, and VP3. The early and late regions are transcribed from different strands and in opposite directions. The SV40 genome, which contains only 5,243 bp, is able to code for so many proteins because of a process called alternative splicing that allows a nucleotide sequence to be incorporated into two or more different mRNA molecules. (Adapted from Sillivan, C. S., and Pipas, J. M. 2002. *Microbiol Mol Biol Rev* 66:179–202.)

7.5 Animal Viruses

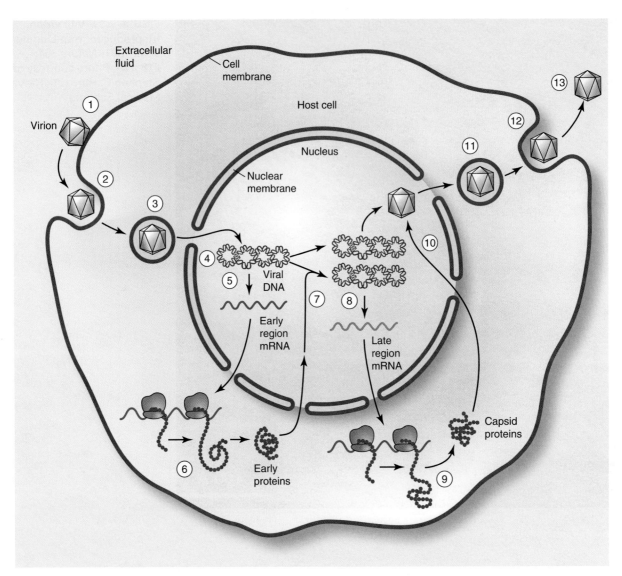

FIGURE 7.24 **Simplified polyomavirus replication cycle.** (1) The virion binds to the cell surface, (2) enters the cell by endocytosis, (3) is transported to the cell nucleus by a route that remains to be determined, and (4) is uncoated. (5) The released viral DNA is transcribed to form early region mRNA molecules, which are (6) translated to form early proteins (T antigens). (7) The viral DNA is also replicated and (8) transcribed to produce late region mRNA molecules, which are (9) translated to form capsid proteins (VP1, VP2, and VP3). (10) Progeny virions are assembled in the nucleus, and (11) the assembled virions move to cytoplasmic vesicles by a process that is still unclear. (12) The vesicles fuse with the cell membrane, (13) releasing the virions. Virions probably also leak out of the nucleus and are released after cell death. (Adapted from Knipe, D. M., et al. 1996. *Fundamental Virology* [3rd ed]. Lippincott Williams & Wilkins.)

Transcription of the late region produces mRNA molecules that are translated to synthesize the capsid proteins (VP1, VP2, and VP3), which assemble with the replicated viral DNA to form progeny virions in the nucleus. The assembled virions move to cytoplasmic vesicles in a process that is the subject of active investigation, and the vesicles fuse with the cell membrane to release the virions. Virions probably also leak out of the nucleus and are released after cell death. In rare cases viral DNA inserts at random positions in a host cell chromosome. Cells with inserted viral DNA are often transformed to tumor cells. The large T antigen interferes with the normal activity of two tumor suppressors, and the small t antigen inhibits a specific protein phosphatase. Both are required for transformation to take place.

Adenoviruses have linear blunt-ended, double-stranded DNA with an inverted repeat at each end.

Adenoviruses, which were isolated in 1953, were so-named because they were first found in cells derived from human adenoids. Adenoviruses cause respiratory and gastrointestinal diseases primarily in children and conjunctivitis in people of all ages. They also have been detected in a wide variety of other warm-blooded animals, including rodents, birds, and various domesticated mammals.

A typical virion has a linear, blunt-ended, double-stranded DNA (35–45 kbp) with an inverted 103-bp repetition at each end. The virus DNA codes for at least 12 different proteins. A simplified version of the genome organization is shown in FIGURE 7.25. Early genes, which are indicated by the prefix "E," are distributed throughout the viral DNA on both strands. In contrast, late genes, which are indicated by the prefix "L," are transcribed from a single DNA strand. The viral DNA is encased in an icosahedral capsid (diameter of 80–110 nm).

A simplified version of the adenovirus replication cycle is shown in schematic form in FIGURE 7.26. Virus particles bind to receptors on the cell surface and are taken into the cell by a process called receptor-mediated endocytosis. The virion particle dissociates in a series of steps, starting with the initial binding to the cell surface receptors. The virion DNA moves to the nucleus, where it is transcribed by the host cell's transcription machinery. Two of the early proteins formed, DNA polymerase and single-stranded DNA binding protein, work together with host cell enzymes to replicate the viral DNA. Then late genes are transcribed and translated to form viral coat proteins, which assemble with the newly formed viral DNA to form progeny virions. The virus blocks protein synthesis during this late stage of infection, resulting in cell lysis and the release of progeny virion particles.

Adenoviruses have not only been important model systems for studying mRNA synthesis, they have also served as vectors for carrying

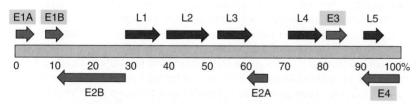

FIGURE 7.25 **Simplified map of the genome of adenovirus.** Early genes, designated by the "E" prefix, are shown in green and late genes, designated by the "L" prefix, are shown in purple. Arrows show the direction of transcription. Genes in boxes are those that can be removed during the production of a replication-defective virus for gene therapy protocols. The *E1A* gene (which encodes the initial viral transcription unit) must be removed to prevent the recombinant virus from replicating. Other genes can be deleted to make more space for the insertion of larger foreign DNA fragments. (Adapted from Wood, K. J., and Fry, J. 1998. *Exp Rev Mol Med* 11:1–20.)

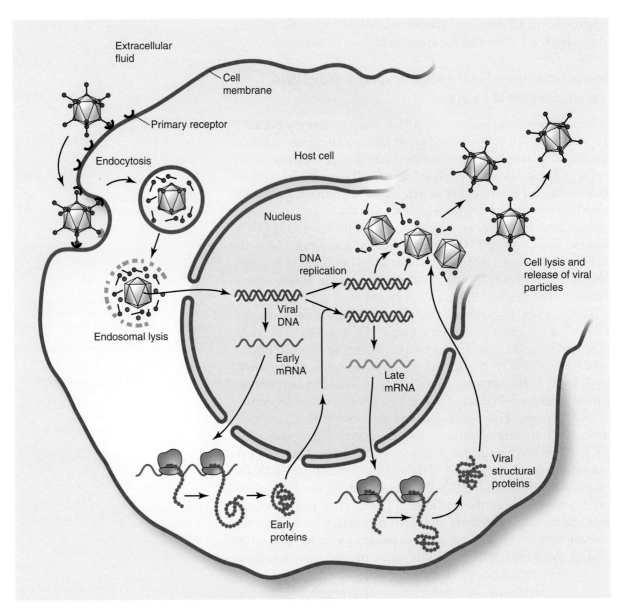

FIGURE 7.26 Simplified adenovirus replication cycle. The fiber attached to a coat protein has a knob (shown as a red sphere) at its end that contacts a receptor on the cell surface. This interaction facilitates the binding of the virus to a secondary receptor on the cell surface. The bound virion enters the cell by receptor-mediated endocytosis and is later released into the cytoplasm by acid-enhanced lysis of the endosomal vesicle. The virion particle dissociates, and the released DNA moves to the nucleus, where it is transcribed by the host cell's transcription machinery. The first virus-encoded protein that is formed, E1A, activates transcription of other early viral transcriptional units. Two of the early proteins formed, DNA polymerase and single-stranded DNA binding protein, work with host cell enzymes to replicate the viral DNA. Then late genes are transcribed and translated to form viral coat proteins, which assemble with newly formed viral DNA to form progeny virions. The virus blocks protein synthesis during the late stage of infection, resulting in cell lysis and the release of progeny virion particles. (Adapted from Mysiak, M. E. 2004. Molecular architecture of the preinitiation complex in adenovirus DNA replication. Master's thesis, Universiteit Utrecht.)

foreign DNA into host cells. Tragically, an early attempt to use a modified adenovirus as a vector in a gene therapy experiment resulted in the death of a young man who was being treated for an inborn error in metabolism. Considerable work is now being done to make the adenovirus vector and other vectors used for human gene therapy experiments safe.

Retroviruses use reverse transcriptase to make a DNA copy of their RNA genome.

The retroviruses are probably the most extensively studied of all animal viruses at the present time. The primary reason for the intense activity in retrovirology is that one member of the retrovirus family, HIV, causes AIDS. Much of the research on retroviruses is driven by the desire to find effective methods to prevent and treat AIDS.

There are many other compelling reasons to study the retroviruses. At the most fundamental level the retrovirus replication cycle requires information flow from RNA to DNA, termed **reverse transcription**, an entirely unexpected pathway when it was first proposed by Howard Temin in the mid-1960s (see below). The prefix "retro" in the family name derives from this "backward" flow of genetic information. The retroviral enzyme that catalyzes RNA-dependent DNA synthesis, better known as reverse transcriptase, is an important tool in molecular biology and recombinant DNA technology (see **BOX 7.4 IN THE LAB: COMPLEMENTARY DNA SYNTHESIS**). Furthermore, genetically altered retroviruses, which are constructed so they are missing genetic information needed for the virus to propagate in host cells, are used as vectors to introduce foreign genes into animal cells and may, with continued improvements, prove to be safe vectors for use in human gene therapy.

Members of the retrovirus family all share certain common features. First, all family members contain two identical RNA molecules (7,000–10,000 kb) called (+) strands because they have the same sequence information as mRNA. These RNA molecules serve as templates for the reverse transcriptase to make DNA but are not required to direct ribosomes to make viral proteins (see below). All retrovirus RNA molecules contain three protein coding regions, which are designated *gag*, *pol*, and *env* (FIGURE 7.27a). The *gag* (group-specific antigen) region codes for a polypeptide, which is called a **polyprotein**, because it is eventually cleaved to produce the following three distinct proteins:

1. *Matrix protein:* The matrix protein is closely associated with the membrane surrounding the virus.
2. *Capsid protein:* The capsid protein forms the core shell, the virion's major internal structure.
3. *Nucleoprotein:* The nucleoprotein is a basic protein that binds to the RNA.

The locations of these three proteins as well as other proteins in the retrovirus particle are shown in Figure 7.27b. One other protein should be mentioned in connection with the *gag* region. In some retroviruses the *gag* region also codes for protease, in others the *pol* region codes for this enzyme, and in still others the protease coding region is between the *pol* and *gag* regions. Protease is essential for viral reproduction because it is required for polyprotein cleavage. In fact, protease inhibitors have proven to be effective drugs for treating individuals with AIDS.

The *pol* (polymerase) region codes for a second polyprotein, which is cleaved to produce reverse transcriptase and integrase (see below). The *env* (envelope) region codes for still another polyprotein,

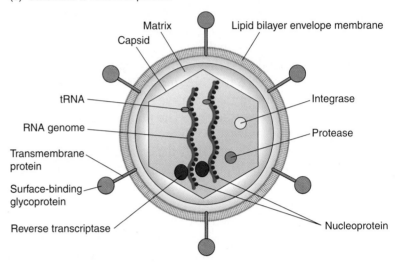

FIGURE 7.27 Retroviral particle and genome structure. (a) Genome organization of a simple retrovirus. (b) Retrovirus particle showing the approximate location of its components. (Adapted from Pedersen, F. S., and Duch, M. 2006. *Encyclopedia of Life Sciences*. John Wiley & Sons.)

which is cleaved to produce a transmembrane protein located in the membrane envelope surrounding the virion, and a surface glycoprotein that binds to the transmembrane protein on the external membrane surface.

Retroviral RNA molecules also share other common features. They all have a modified guanosine at their 5′-end and a poly(A) sequence at their 3′-end that are called a cap and a tail, respectively. Cap and tail structures are typically found in eukaryotic mRNA.

Retroviruses enter the target cell in much the same way other enveloped viruses enter their host cells. The virion attaches to a specific receptor on the cell surface and then the virion membrane fuses with the cell membrane so the core of the particle can enter the cell cytoplasm. Each member of the retrovirus family binds to cell receptors that are specific for it. For example, the HIV virus binds to a receptor called CD4, which is present on the surface of human T-helper cells, immature thymocytes, monocytes, and macrophages. HIV must also bind to a co-receptor on the surface of the target cell before it can enter that cell. Individuals lacking the co-receptor are more resistant to HIV infection than are those who have the co-receptor, suggesting the co-receptor might be a target for new drugs to treat AIDS.

The next steps in retrovirus infection are quite remarkable because they require an RNA template to synthesize DNA. Until the mid-1960s investigators believed all RNA viruses use an RNA-dependent RNA polymerase to replicate their RNA. This method of RNA replication is certainly used by a wide variety of RNA viruses, including RNA phages, polio virus, and tobacco mosaic virus. An entirely different situation was observed for the Rous sarcoma virus, an RNA virus discovered by Peyton Rous in 1911 while investigating the origins of a tumor in chickens. Rous observed that healthy chickens developed tumors when injected with an extract prepared from an excised tumor. Further experiments showed the causative agent must be very small because the extract retained its ability to cause tumors even after being passed through a fine filter that did not let bacteria pass through. Based on this and related experiments, Rous concluded the chicken tumor must be caused by a virus.

Contemporary investigators questioned the relevance of Rous' observations to human cancer because most forms of human cancer are not caused by infectious agents. However, later studies performed by J. Michael Bishop and Harold E. Varmus showed that sequences in Rous sarcoma virus RNA are occasionally replaced by host cell genes that code for protein products that stimulate cell growth and division. These rare defective viruses cannot replicate without the assistance of other viruses. They can, however, transform normal cells into tumor cells.

Howard Temin's hypothesis that DNA synthesis is a required step in Rous sarcoma virus replication was subject to considerable criticism when first proposed in the mid-1960s. The hypothesis gained quick acceptance in 1970, however, when Temin and David Baltimore independently demonstrated that the virion particle contains an RNA-dependent DNA polymerase. Temin proposed the "provirus model" to explain the requirement for DNA synthesis. According to this model, retrovirus replication can be divided into an early and a late stage (FIGURE 7.28):

1. *Early stage:* The virus particle supplies reverse transcriptase and integrase. Reverse transcriptase acts on viral RNA in the cytoplasm to synthesize a complementary DNA strand. An RNase H activity associated with reverse transcriptase degrades the viral RNA in the RNA–DNA hybrid. Then the reverse transcriptase uses the newly formed DNA strand as a template to form double-stranded DNA. The double-stranded DNA moves into the nucleus, where it encounters integrase, which integrates the DNA into a random site in one of the host chromosomes to produce the provirus. The provirus continues to replicate as part of the host chromosome, and if insertion takes place in a germ cell, the provirus will be passed from one generation to the next. Human DNA contains many sequences that appear to have been derived from retroviral DNA.
2. *Late stage:* The late stage of the replication cycle requires host cell enzymes. Cellular transcription machinery transcribes the inserted viral DNA to form mRNA that directs viral protein synthesis and viral RNA that assembles with the viral proteins to form progeny virus particles. Assembly takes place at the cell membrane, and the mature virions are released from the cell by budding. This release process does not result in cell death.

You may wish to refer back to sections in this chapter when you encounter a description of a virus. Information about other viruses and a more detailed examination of those viruses described here can be found in standard virology textbooks.

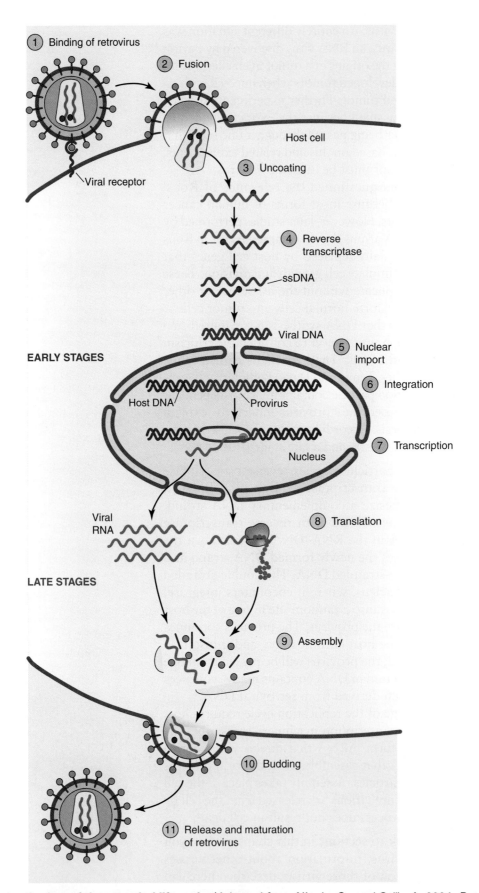

FIGURE 7.28 A schematic view of the retroviral life cycle. (Adapted from Nisole, S., and Saïb, A. 2004. *Retrovirology* 1:9.)

BOX 7.4: IN THE LAB

Complementary DNA Synthesis

Many nucleic acid manipulations start with RNA rather than DNA. When this is the case, it is often convenient to make a DNA copy of the RNA molecule. A reverse transcriptase prepared from avian myeloblastosis virus or Moloney murine leukemia virus is commonly used for this purpose. These reverse transcriptases, like DNA polymerase I, require a template-primer and catalyze deoxynucleotidyl group transfer from deoxynucleoside triphosphates to the 3'-end of the growing primer chain. Reverse transcriptases thus catalyze 5'→3' chain extension to produce a new DNA strand complementary to the template strand. Although the similarities with DNA polymerase I are striking, important differences also exist. For instance, reverse transcriptases can use either an RNA or a DNA chain as a template, lack a 5'→3' proofreading function, and have RNase H activities that digest RNA when it is present in either a DNA–RNA hybrid or a double-stranded RNA molecule. These properties make reverse transcriptase an ideal laboratory tool to copy an RNA template to produce either single- or double-stranded DNA. One of the most common applications is the synthesis of a DNA copy of mRNA, called complementary DNA or cDNA (FIGURE B7.4). Reverse transcriptase can be used to obtain cDNAs for the complete set of RNA molecules produced by the genome at any one time. This collection of transcripts is called the transcriptome. cDNA molecules prepared by using reverse transcriptase are readily sequenced.

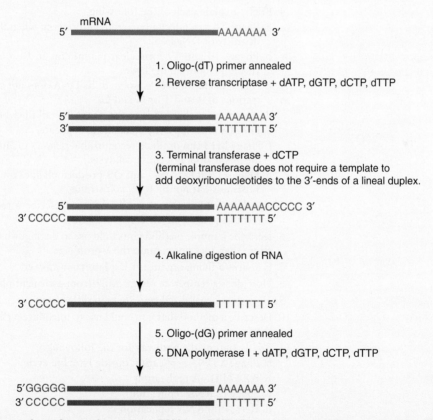

FIGURE B7.4 **Preparation of complementary DNA or cDNA.** Reverse transcriptase produces a cDNA strand (blue) to an mRNA molecule (green). The RNA can then be digested by alkaline hydrolysis as shown here or by RNase digestion and the resulting single-stranded DNA used as a template to make a cDNA (also shown in blue). A primer is required for cDNA synthesis. In this example terminal transferase is used to add oligo-(dC) to the 3'-end of the single stranded DNA so it can anneal to an oligo-(dG) primer.

Questions and Problems

1. Briefly define, describe, or illustrate each of the following.
   a. Virion
   b. Capsid
   c. Bacteriophage
   d. T-even phage
   e. Phage lysate
   f. Virulent phage
   g. Prophage
   h. Lysogen
   i. Temperate phage
   j. Chronic phage
   k. Top agar
   l. Plaque
   m. Concatemer
   n. Terminase
   o. Procapsid
   p. *cos* element
   q. θ replication
   r. Specialized transduction
   s. Generalized transduction
   t. Reverse transcription
   u. Polyprotein

2. Pathogenic strains of *E. coli* cause a bloody diarrhea due to toxins it secretes when it infects human intestinal tracts. Infection with these pathogens may result in kidney failure. If bacteriophages were available that would lyse these pathogenic bacteria, could they be used to treat the disease? Explain your answer.

3. A few microliters of a phage T4 lysate is added to a slightly turbid (cloudy) *E. coli* culture growing in broth. The turbidity of the culture increases for about 30 minutes, but then the culture becomes completely clear with some stringy material floating in it. Explain these observations.

4. Briefly explain each of the following.
   a. Phage T4 can infect a bacterial culture even when the culture contains a large quantity of DNase.
   b. Phages T4 and T7 form clear plaques on an *E. coli* lawn, but phage λ forms turbid plaques on the lawn.
   c. A new bacteriophage is isolated. Its DNA does not contain equimolar concentrations of A and T or G and C.
   d. An antibody that binds to the phage T4 tail fiber blocks viral infection.
   e. Phage λ must have *cos* elements to replicate.
   f. Phage M13 can produce recognizable plaques on an *E. coli* lawn even though the phage does not cause cell lysis.
   g. Phages f2, R17, MS2, and Qβ produce plaques on lawns formed by Hfr and F$^+$ strains but not by F$^-$ *E. coli* strains.

5. You have prepared a phage T7 lysate. Describe a technique you could use to determine phage particle concentration.

6. Describe a simple method you could use to distinguish a λ lysogen from its parent strain that does not contain the λ prophage.

7. Is a virus a living organism? Explain your answer.

8. How does a temperate phage differ from a virulent phage?

9. How does generalized transduction differ from specialized transduction?

10. Describe a method that you could use to introduce a *phe$^+$* allele into an Hfr strain that is *phe$^-$*.

11. Briefly outline the pathway for the following:
    a. Phage λ DNA replication during lytic life cycle
    b. ϕX174 DNA replication
    c. Phage f2 RNA replication
    d. Retroviral RNA replication

12. What role does reverse transcriptase play in retroviral RNA replication?

Suggested Reading

General Overview

Ackermann, H.-W. 2007. Bacteriophages: tailed. *Encyclopedia of Life Sciences.* pp. 1–7. Hoboken, NJ: John Wiley & Sons.

Ackermann, H.-W. 2009. Phage classification and characterization. *Methods Mol Biol* 501:127–140.

Breitbart, M., and Rohwer, F. 2005. Here a virus, there a virus, everywhere the same virus? *Trends Microbiol* 13:278–284.

Cairns, J., Stent, G. S., and Watson, J. (eds). 1966. *Phage and the Origins of Molecular Biology*. Woodbury, NY: Cold Spring Harbor Laboratory Press.

Calendar, R. (ed). 2005. *The Bacteriophages*. Oxford, UK: Oxford University Press.

Callanan, M. J., and Klaenhammer, T. R. 2008. Bacteriophages in industry. *Encyclopedia of Life Sciences*. pp. 1–8. Hoboken, NJ: John Wiley & Sons.

Campbell, A. 2003. The future of bacteriophage biology. *Nat Rev Genet* 4:471–477.

Garcia, P., Martınez, P. B., Obeso, J. M., and Rodriguez, A. 2008. Bacteriophages and their application in food safety. *Lett App Microbiol* 47:479–485.

Guttman, B. S., and Kutter, E. 2002. Bacteriophage genetics. Streips, U. N., and Yasbin, R. E. (eds). *Modern Microbial Genetics* (2nd ed). pp. 85–126. Hoboken, NJ: John Wiley & Sons.

Guttman, B., Raya, R., and Kutter, E. 2005. *Bacteriophages: Biology and Applications*. Boca Raton, FL: CRC Press.

Harper, D. R., and Kutter, E. 2008. Bacteriophage: therapeutic uses. *Encyclopedia of Life Sciences*. pp. 1–7. Hoboken, NJ: John Wiley & Sons.

Mattley, M., and Spencer, J. 2008. Bacteriophage therapy-cooked goose or Phoenix rising? *Curr Opin Biotechnol* 19:608–612.

McGrath, S., Fitzgerald, G. F., and van Sinderen, D. 2004. The impact of bacteriophage genomics. *Curr Opin Biotechnol* 15:95–99.

Minor, P. D. 2007. Viruses. *Encyclopedia of Life Sciences*. pp. 1–10. Hoboken, NJ: John Wiley & Sons.

Sharp, R. 2001. Bacteriophages: biology and history. *J Chem Technol Biotechnol* 76:667–672.

Summers, W. C. 2001. Bacteriophage therapy. *Annu Rev Microbiol* 55:437–451.

Thiel, K. 2004. Old dogma, new tricks—21st century phage therapy. *Nat Biotechnol* 22:31–36.

Wagner, E., and Hewlett, M. 2003. *Basic Virology* (2nd ed). Malden, MA: Blackwell Publishing.

Virulent Bacteriophages

Abedon, S. T. 2000. The murky origin of Snow White and her T-even dwarfs. *Genetics* 155:481–486.

Ackerman, H. W., and Krisch, H. M. 1997. A catalogue of T4-type bacteriophages. *Arch Virol* 142:2329–2342.

Epstein, R. H., Bolle, A., Steinberg, C., et al. 1963. Physiological studies of conditional lethal mutants of bacteriophage T4. *Cold Spring Harb Symp Quant Biol* 28:375–392.

Hayashi, M., Aoyama, A., Richardson, D. L., and Hayashi, M. N. 1988. Biology of the bacteriophage ϕX174. Calendar, R. (ed). *The Bacteriophages*. pp. 1–71. New York: Plenum Press.

Hendrix, R. 2008. Cell architecture comes to phage biology. *Mol Microbiol* 68:1077–1078.

Johnson, J. E., and Chiu, W. 2007. DNA packaging and delivery machines in tailed bacteriophages. *Curr Opin Struct Biol* 17:237–243.

Karam, J. D. (ed.) 1994. *Molecular Biology of Bacteriophage T4*. Washington, DC: American Society for Microbiology.

Kreuzer, K. N. 2000. Recombination-dependent DNA replication in phage T4. *Trends Biochem Sci* 25:165–173.

Leiman, P. G., Kanamarua, S., Mesyanzhinov, V. V., et al. 2003. Structure and morphogenesis of bacteriophage T4. *Cell Mol Life Sci* 60:2356–2370.

Loeb T., and Zinder, N. D. 1961. A bacteriophage containing RNA. *Proc Natl Acad Sci USA* 47:282–289.

Mathews, C. K. 2005. Bacteriophage T4. *Encyclopedia of Life Sciences*. pp. 1–9. Hoboken, NJ: John Wiley & Sons.

Mathews, C. K., Kutter, E. M., Mosig, G., and Berget, P. (eds). 1983. *Bacteriophage T4*. Washington, DC: American Society for Microbiology.

Miller, E. S., Kutter, E., Mosig, G., et al. 2003. Bacteriophage T4 genome. *Microbiol Mol Biol Rev* 67:86–156.

Molineux, I. J. 2001. No syringes please, ejection of phage T7 DNA from the virion is enzyme driven. *Mol Microbiol* 40:1–8.

Ortin, J., and Parra, F. 2006. The structure and function of RNA replication. *Annu Rev Microbiol* 60:305–326.

Rao, V. B., and Feiss, M. 2008. The bacteriophage DNA packaging motor. *Annu Rev Genet* 42:647–681.

Rossman, M. G., Mesyanzhinov, V. V., Arisaka, F., and Leiman, P. G. 2004. The bacteriophage T4 DNA injection machine. *Curr Opin Struct Biol* 14:171–180.

Sinsheimer, R. L. 1959. A single-stranded deoxyribonucleic acid from bacteriophage φX174. *J Mol Biol* 1:43–53.

Van Duin, J. 2005. Bacteriophages with ssRNA. *Encyclopedia of Life Sciences*. pp. 1–6. Hoboken, NJ: John Wiley & Sons.

van Duin, J., and Tsareva, N. A. 2005. Single-stranded RNA phages. R. Calendar (ed). *The Bacteriophages*. Oxford, UK: Oxford University Press.

Temperate Bacteriophage

Bertani, G. 2004. Lysogeny at mid-twentieth century: P1, P2, and other experimental systems. *J Bacteriol* 186:595–600.

Campbell, A. 1994. Comparative molecular biology of lambdoid phages. *Annu Rev Microbiol* 48:193–222.

Casjens, S. R., and Hendrix, R. W. 2001. Bacteriophage lambda and its relatives. *Encyclopedia of Life Sciences*, pp. 1–8. London, UK: Nature.

Catalano, C., Cue, D., and Feiss, M. 1995. Virus DNA packaging: the strategy used by phage lambda. *Mol Microbiol* 16:1075–1086.

Friedman, D. I., and Court, D. L. 2001. Bacteriophage lambda: alive and still doing its thing. *Curr Opin Micriobiol* 4:201–207.

Gottesman, M. 1999. Bacteriophage λ: the untold story. *J Mol Biol* 293:177–180.

Gottesman, M. E., and Weisberg, R. A. 2004. Little lambda, who made thee? *Microbiol Mol Biol Rev* 68:796–813.

Hendrix R., Roberts J., Stahl F., and Weisberg R. (eds). 1983. *Lambda II*. Woodbury, NY: Cold Spring Harbor Laboratory Press.

Lederberg, E. M., and Lederberg, J. 1953. Genetic studies of lysogenicity in *E. coli*. *Genetics* 38:51–64.

Yarmolinsky, M. B. 2004. Bacteriophage P1 in retrospect and in prospect. *J Bacteriol* 186:7025–7028.

Zinder, N. D., and Lederberg, J. 1952. Genetic exchange in Salmonella. *J Bacteriol* 64:679–699.

Chronic Phages

Benhar, I. 2001. Biotechnological applications of phage and cell display. *Biotechnol Adv* 19:1–33.

Marvin, D. A. 1998. Filamentous phage structure, infection, and assembly. *Curr Biol* 8:150–158.

Rodi, D. J., and Makowski, L. 1999. Phage-display—finding a needle in a vast molecular haystack. *Curr Opin Biotechnol* 10:87–93.

Russell, M., and Model, P. 2005. Filamentous phage. R. Calendar (ed). *The Bacteriophages*. Oxford, UK: Oxford University Press.

Sidhu, S. S. 2001. Engineering M13 for phage display. *Biomol Eng* 18:57–63.

Animal Viruses

Atwood, W. J., and Shah, K. V. 2001. Polyomaviruses. *Encyclopedia of Life Sciences*. pp. 1–6. London: Nature.

Barré-Sinoussi, F. 1996. HIV as the cause of AIDS. *Lancet* 348:31–35.

Benjamin, T. L. 2001. Polyoma viruses: old findings and new challenges. *Virology* 289:167–173.

Blattner, W., Gallo, R. C., and Temin, H. M. 1988. HIV causes AIDS. *Science* 241:515–516.

Bukrinsky, M. I. 2001. HIV life cycle and inherited coreceptors. *Encyclopedia of Life Sciences*, pp. 1–6. London: Nature.

Coffin, J. M., Hughes, S. H., and Varmus, H. E. (eds). 1997. *Retroviruses*. Woodbury, NY: Cold Spring Harbor Laboratory Press.

Cole, C. N. 1996. Polyomaviridae: the viruses and their replication. B. N. Fields, P. M. Howley, D. M. Knipe, et al. (eds). *Fundamental Virology* (3rd ed). pp. 917–945. Baltimore: Lippincott, Williams and Wilkins.

Crandall, K. A. 2001. Human immunodeficiency viruses (HIV). *Encyclopedia of Life Sciences*, pp. 1–7. London: Nature.

Flint, S. J. 2001. Adenoviruses. *Encyclopedia of Life Sciences*, pp. 1–14. London: Nature.

Gallo, R. C. 2004. HIV-1: a look back from 20 years. *DNA Cell Biol* 23:191–192.

Gallo, R. C., and Montagnier, L. 2003. The discovery of HIV as the cause of AIDS. *N Engl J Med* 349:2283–2285.

Garcea, R. L., and Imperiale, M. J. 2003. Simian virus 40 infection of humans. *J Virol* 77:5039–5045.

Gazdar, A. F., Butel, J. S., and Carbone, M. 2002. SV40 and human tumors: myth and association or causality? *Nat Rev Cancer* 2:957–964.

Graham, F. L., and Hitt, M. M. 2007. Adenovirus vectors in gene therapy. *Encyclopedia of Life Sciences*. pp. 1–6. Hoboken, NJ: John Wiley & Sons.

Hu, W.-S., and Pathak, V. K. 2000. Design of retroviral vectors and helper cells for gene therapy. *Pharmacol Rev* 52:493–511.

Hunter, E. 1997. Viral entry and receptors. J. M. Coffin, S. H. Hughes, and H. E. Varmus (eds). *Retroviruses*. pp. 71–120. Woodbury, NY: Cold Spring Harbor Laboratory Press.

Lednicky, J. A., and Butel, J. S. 1999. Polyomaviruses and human tumors: a brief review of current concepts and interpretations. *Front Biosci* 4:D153–D164.

Löwer, R., Löwer, J., and Kurth, R. 1996. The viruses in all of us: characteristics and biological significance of human endogenous retrovirus sequences. *Proc Natl Acad Sci USA* 93:5177–5184.

Nelson, P. N., Carnegie, P. R., Martin, J., et al. 2003. Demystified... human endogenous retroviruses. *J Clin Pathol Mol Pathol* 56:11–18.

Nemerow, G. R., Pache, L., Reddy, C., and Stewart, P. L. 2009. Insights into adenovirus host cell interactions from structural studies. *Virology* 384:380–388.

Neu, U., Stehle, T., and Atwood, W. J. 2009. The *Polyomaviridae*: contributions of virus structure to our understanding of virus receptors and infectious entry. *Virology* 384:389–399.

Novembre, F. J. 2001. Simian retroviruses. *Encyclopedia of Life Sciences*, pp. 1–8. London: Nature.

Pedersen, F. S., and Duch, M. 2006. Retroviral replication. *Encyclopedia of Life Sciences*. pp. 1–8. Hoboken, NJ: John Wiley & Sons.

Pedersen, F. S., and Duch, M. 2001. Retroviruses in human gene therapy. *Encyclopedia of Life Sciences*, pp. 1–10. London: Nature.

Pipas, J. M. 2009. SV40: Cell transformation and tumorigenesis. *Virology* 384:294–303.

Russell, W. C. 2009. Adenoviruses: update on structure and function. *J Gen Virol* 90:1–20.

Shenk, T. E. 1996. Adenoviridae: the viruses and their replication. B. N. Fields, P. M. Howley, D. M. Knipe, et al. (eds). *Fundamental Virology* (3rd ed). pp. 979–1016. Baltimore: Lippincott, Williams and Wilkins.

Sullivan, C. S., and Pipas, J. M. 2002. T antigens of simian virus 40: molecular chaperones for viral replication and tumorigenesis. *Microbiol Mol Biol Rev* 66:179–202.

Turner, B. G., and Summers, M. F. 1999. Structural biology of HIV. *J Mol Biol* 285:1–32.

Whittaker, G. R., Kann, M., and Helenius, A. 2000. Viral entry into the nucleus. *Annu Rev Cell Dev Biol* 16:627–651.

8
DNA Replication

CHAPTER OUTLINE

8.1 General Features of DNA Replication
DNA replication is semiconservative.
Bacterial and eukaryotic DNA replication is bidirectional.
The DNA strand that grows in an overall 3'→5' direction is formed by joining short fragments.
DNA ligase connects adjacent Okazaki fragments.
RNA serves as a primer for Okazaki fragment synthesis.

8.2 Bacterial DNA Replication
The bacterial replication machinery has been isolated and examined *in vitro*.
Mutant studies provide important information about the enzymes involved in bacterial DNA replication.
The replicon model proposes that an initiator protein must bind to a DNA sequence called a replicator at the start of replication.
E. coli chromosomal replication begins at *oriC*.
Several enzymes act together at the replication fork.
DNA polymerase III is required for bacterial DNA replication.
DNA polymerase III holoenzyme has three distinct kinds of subassemblies.
The core polymerase has one subunit with 5'→3' polymerase activity and another with 3'→5' exonuclease activity.
The sliding clamp forms a ring around DNA, tethering the remainder of the polymerase holoenzyme to the DNA.
The clamp loader places the sliding clamp around DNA.
The replisome catalyzes coordinated leading and lagging strand DNA synthesis at the replication fork.
E. coli DNA replication terminates when the two growing forks meet in the terminus region, which is located 180 degrees around the circular chromosome from the origin.
The terminus utilization substance binds to *Ter* sites.
Topoisomerase IV and recombinase separate newly formed sister chromosomes.

8.3 Eukaryotic DNA Replication
The SV40 DNA replication system is a good model for *in vitro* eukaryotic DNA replication.
Eukaryotic replication machinery must replicate long linear duplexes with multiple origins of replication.
Eukaryotic chromosomes have many origins of replication.
Yeast origins of replication, called autonomously replicating sequences, determine the site of DNA chain initiation.
Pol δ and Pol ε are primarily responsible for copying the lagging- and leading-strand templates, respectively.
Studies of the *Tetrahymena* and yeast telomeres suggest that a terminal transferase-like enzyme is required for telomere formation.
Telomerase uses an RNA template to add nucleotide repeats to chromosome ends.
Telomerase plays an important role in solving the end-replication problem.

8.4 Replication Coupled Chromatin Synthesis
Chromatin disassembly and reassembly are tightly coupled to DNA replication.

BOX 8.1: LOOKING DEEPER Replication Regulation at the Initiation Stage

BOX 8.2: IN THE LAB Measurement of Processivity

BOX 8.3 LOOKING DEEPER Helicase and Primer Coordination

BOX 8.4: CLINICAL APPLICATION Telomerase in Aging and Cancer

QUESTIONS AND PROBLEMS

SUGGESTED READING

The purpose of DNA replication is to copy a DNA molecule so the two resulting daughter DNA molecules have the same genetic information as the original parent DNA. The process must proceed with great fidelity so information is not lost or changed. When the Watson-Crick model of double-stranded DNA was first proposed in 1953, many investigators assumed just a few enzymes might suffice to catalyze DNA replication. Investigators studying DNA replication, however, soon became aware that the process is considerably more complex than first assumed and requires many different kinds of enzymes.

The complexity of the DNA replication process results in part from the following facts: (1) helicases and energy are required to unwind the double helix, (2) single-strand DNA binding proteins are needed to prevent the single DNA strands produced by the action of helicases from being hydrolyzed by nucleases or forming intrastrand base pairs, (3) DNA ligases are required to join nicked DNA strands, (4) safeguards are necessary to prevent replication errors and eliminate the rare errors that do occur, (5) topoisomerases are required to release the torsional strain that builds up as a circular DNA or long-linear duplex replicates, and (6) special protein factors and enzymes are needed to initiate DNA replication and to terminate it. This chapter first considers a few general features of the replication process and then examines the sequence of events that take place during the initiation, elongation, and termination stages of DNA replication in bacteria and eukaryotes.

8.1 General Features of DNA Replication

DNA replication is semiconservative.

Watson and Crick recognized their DNA model suggests a replication mechanism in which the two parental strands separate, allowing each separated strand to serve as a template for the synthesis of a complementary strand (FIGURE 8.1). According to this replication mechanism, each double-stranded daughter DNA molecule has a conserved DNA strand that is derived from the parental DNA and a newly synthesized strand (FIGURE 8.2a). When this **semiconservative model of replication** was proposed, DNA denaturation was not understood and strand separation was, for a variety of reasons, considered to be impossible. Two alternative models were, therefore, also considered.

The first of these alternative models, the **conservative model of replication,** makes the following two assumptions: (1) the double helix unwinds at the replication site only to the extent needed for the base sequence there to be read by the polymerizing enzyme and (2) the two original strands rewind after replication so that one of the two DNA molecules present after replication contains both original strands (is conserved) and the other DNA molecule is made of two new strands (Figure 8.2b). The second alternative model, the **dispersive model of replication,** shares some of the features of the conservative model but predicts that each strand of the daughter DNA molecules has

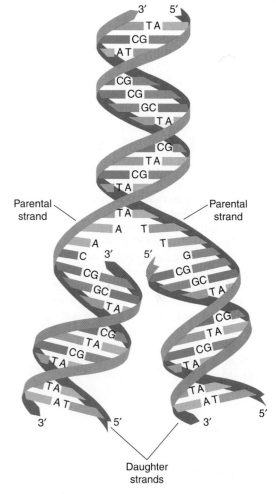

FIGURE 8.1 **Watson-Crick model of DNA replication.** Watson and Crick proposed that DNA replicates by separating the two parental strands and creating two new daughter strands by pairing new nucleotides with template nucleotides using the A:T and G:C pairing rule. Conserved parental strands are in blue and newly synthesized strands in red. The details of DNA replication are considerably more complex than shown here because all new DNA strands are synthesized by adding new nucleotides to the 3′-end of existing primer chains. All new chains are therefore synthesized in a 5′→3′ direction.

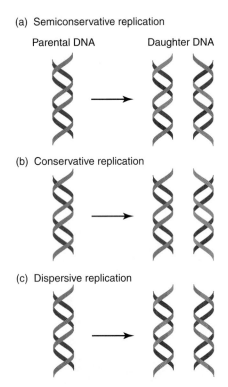

FIGURE 8.2 Three models of replication.
(a) Semiconservative DNA replication predicts that each daughter helix will have one conserved (parental) DNA strand and one newly synthesized DNA strand.
(b) Conservative DNA replication predicts that one daughter DNA molecule will have two conserved (parental) DNA strands, whereas the other daughter DNA molecule will have two new DNA strands. (c) Dispersive DNA replication predicts that each daughter DNA molecule will have interspersed sections of both old and new DNA.

interspersed sections of both old and new DNA (Figure 8.2c). Comparison of the three mechanisms of replication, shown in Figure 8.2, reveals that they make different predictions about the composition of daughter DNA molecules after one or two rounds of replication. It therefore would be possible to establish the correct model if some method could be devised to distinguish between new and old DNA strands.

In 1958, just 5 years after the Watson-Crick model was proposed, Matthew Meselson and Franklin Stahl performed an equilibrium density gradient centrifugation experiment that demonstrated DNA replication is semiconservative. They started by culturing *Escherichia coli* for many generations in a growth medium containing [^{15}N]NH$_4$Cl as the sole source of nitrogen so the purine and pyrimidine bases in DNA were uniformly labeled with ^{15}N. Because ^{15}N is a heavy nitrogen isotope, this medium is often referred to as "heavy" growth medium. After the bases in DNA were uniformly labeled with ^{15}N, the bacteria were transferred to a "light" growth medium containing [^{14}N]NH$_4$Cl. Meselson and Stahl then (1) removed samples from the culture at various times, (2) extracted DNA from the samples, (3) added the extracted DNA to a concentrated cesium chloride solution, and (4) subjected the mixture to high-speed centrifugation for about 1 day to establish a cesium chloride equilibrium density gradient. FIGURE 8.3 shows the results obtained by analyzing DNA extracted from samples at generations 0, 1, 2, 3, and 4 after the bacteria were transferred to the "light" growth medium:

1. *Generation 0:* All bacterial DNA is uniformly labeled with ^{15}N and migrates as a single band in the cesium chloride gradient.
2. *Generation 1:* The semiconservative model predicts that one generation after transfer to the "light" growth medium, each daughter DNA molecule should have a new [^{14}N]DNA (light or L) strand and a parental [^{15}N]DNA (heavy or H) strand, producing a hybrid DNA molecule with a density between that of a DNA molecule with two H-strands and one with two L-strands. This is exactly what is observed at generation 1. The conservative model predicts the appearance of two DNA molecules, one with two H-strands and one with two L-strands. The data at generation 1 therefore clearly rule out the conservative model but do not rule out the dispersive model. Data from the next generation, however, do so.
3. *Generation 2:* The semiconservative model predicts that two generations after transfer to the "light" medium, two DNA molecules should have a hybrid density (one H-strand and one L-DNA strand) and two DNA molecules should be made entirely of L-DNA strands. The data at generation 2 reveal two bands containing equal quantities of DNA. One band has the density predicted for a hybrid DNA and the other band has the density predicted for DNA with two L-strands. In contrast, the dispersive model predicts that all the DNA molecules should contain DNA strands with interspersed segments of H-DNA and L-DNA. Clearly, the data for the DNA extracted from cells at generation 2 are consistent with the semiconservative model of replication and argue against the dispersive model.

4. *Generations 3 and 4:* The semiconservative model is also the only one that is fully consistent with the data obtained three or four generations after transfer to the "light" medium.

A second experiment confirmed the structure of the hybrid DNA found after one generation. In this experiment Meselson and Stahl denatured the hybrid DNA by heating it to 100°C and then centrifuged the denatured DNA in cesium chloride. The heated DNA yielded two bands: one had the density expected for single-stranded L-DNA and the other the density expected for single-stranded H-DNA. This denaturation experiment proves the hybrid duplex does in fact consist of one L-strand and one H-strand. Subsequent studies showed that DNA replication in other organisms also follows the semiconservative model.

Bacterial and eukaryotic DNA replication is bidirectional.

At the time Meselson and Stahl performed their experiment, investigators mistakenly believed naturally occurring *E. coli* DNA molecules were linear duplexes. Autoradiography studies performed by John Cairns in 1963, however, showed *E. coli* DNA is circular. Cairns also obtained autoradiograms of bacterial DNA molecules in the process of replication. Because images of these molecules resembled the Greek letter θ (theta), Cairns called the replicating intermediates **θ-structures**. One of the most famous autoradiograms from the Cairns' collection is shown in FIGURE 8.4, along with Cairns interpretation.

The θ-structure can be explained by **unidirectional** (FIGURE 8.5a) or **bidirectional replication** (Figure 8.5b). In unidirectional replication a single growing point moves around the circular DNA until replication is complete. In bidirectional replication two growing points start at the same site and move in opposite directions until they meet at the opposite side of the circle. A region called the **replication bubble**, which contains newly synthesized DNA, grows as DNA synthesis continues. Cairns' autoradiography experiments do not distinguish between unidirectional and bidirectional replication and so a new approach was required.

Bidirectional replication can be distinguished from unidirectional replication by performing a radioisotope labeling experiment. Cells are incubated with [^{3}H]thymidine with a low specific activity (microcuries per micromole) to lightly label the replication bubble. Then the cells are incubated with [^{3}H]thymidine with a very high specific activity for a short time. Bidirectional replication predicts a lightly labeled bubble with heavily labeled segments at both ends, whereas unidirectional replication predicts a lightly labeled replication bubble with a short heavily labeled segment at just one end (FIGURE 8.6). Experiments in bacteria and eukaryotes show heavily labeled segments at both ends, indicating *DNA replication is bidirectional*.

The DNA strand that grows in an overall 3′→5′ direction is formed by joining short fragments.

In vivo studies of bacterial DNA replication led investigators to conclude that both new DNA strands grow in the same direction as the

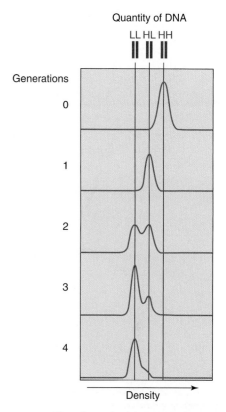

FIGURE 8.3 Meselson-Stahl experiment showing semiconservative DNA replication. *E. coli* were cultured in a growth medium containing [^{15}N]NH$_4$Cl as the sole source of nitrogen for several generations so purine and pyrimidine bases in DNA were uniformly labeled with the ^{15}N heavy isotope. Then the cells were transferred to a new growth medium containing [^{14}N]NH$_4$Cl. Samples were removed at intervals, and their extracted DNA was subjected to equilibrium density centrifugation. During centrifugation each type of DNA molecule moves until it comes to rest at a position in the centrifuge tube at which its density equals the density of the CsCl solution at that position. The column labeled "generations" indicates the number of generations the cells were incubated in the growth medium containing [^{14}N]NH$_4$Cl. The tracings show the DNA density for the indicated generation. H indicates a "heavy" DNA strand labeled with ^{15}N and L indicates a "light" DNA strand labeled with ^{14}N.

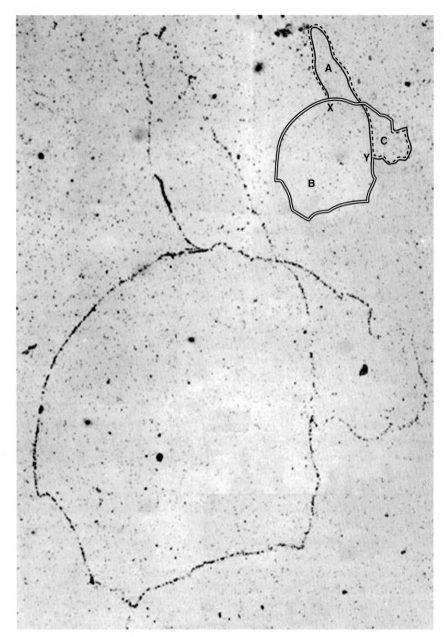

FIGURE 8.4 Demonstration of a θ structure intermediate in *E. coli* chromosome replication. Bottom: *E. coli* cells were cultured in the presence of [³H]thymidine for one entire generation and part of a second. Then the DNA was gently extracted and covered with a photographic gel. Weak β-rays released by the tritium produce dark spots on the photographic gel. Top: Cairns diagrammatic interpretation shows that the structure can be divided into three sections (A, B, and C) that arise at two forks (X and Y). (Reproduced from Cairns, J. 1964. *Cold Spring Harb Symp Quant Biol* 28:44. Copyright 1964, Cold Spring Harbor Laboratory Press. Photo courtesy of John Cairns, University of Oxford.)

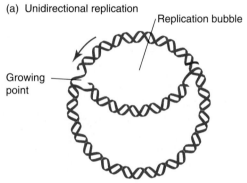

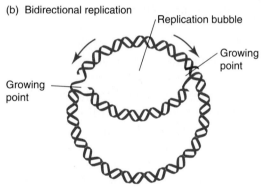

FIGURE 8.5 Two alternate methods of replication. (a) In unidirectional replication a single growing point moves in the direction shown by the red arrow. (b) In bidirectional replication two growing points move in opposite directions, as shown by the red arrows. Parental DNA is shown in blue and newly replicated DNA is shown in red.

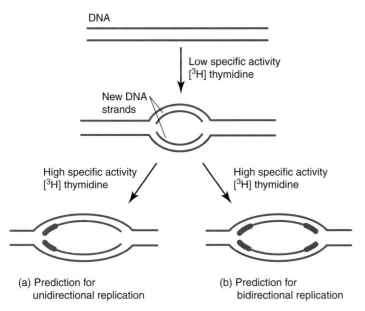

FIGURE 8.6 **Test for bidirectional bacterial DNA replication.** Cells are incubated in a medium containing [³H]thymidine with a low specific activity to lightly label new strands in the replication bubble (thin red lines). Then the cells are diluted into a medium containing [³H]thymidine with a high specific activity and incubated for a short time. (a) Unidirectional replication predicts the two lightly labeled strands in the replication bubble will each have one short highly labeled strand on the same side of the replication bubble (thick red lines). (b) Bidirectional replication predicts the two lightly labeled strands (thin red lines) will have two highly labeled segments (thick red lines): one on each side. Experiments in bacteria and eukaryotes show heavily labeled segments at both ends, indicating DNA replication is bidirectional.

replication fork (the Y-shaped structure where DNA is being synthesized). One new chain must therefore grow in an overall 5'→3' direction, while the other grows in an overall 3'→5' direction. The requirement for overall 3'→5' chain growth presents a serious problem because all known DNA polymerases catalyze 5'→3' chain growth by adding new nucleotides to the 3'-end of the growing chain. Despite extensive efforts to find a DNA polymerase that catalyze 3'→5' chain growth, no such enzyme has ever been found. How then does the replication machinery synthesize the daughter strand that grows 3'→5'?

In 1968 Reiji Okazaki, working in Japan, suggested an ingenious solution to the problem. He proposed the DNA strand that grows in an overall 3'→5' direction is synthesized discontinuously. According to this proposal DNA polymerase first makes small fragments in a 5'→3' direction and then the cell joins successive fragments to make the long DNA strand. Because there was no logical requirement for discontinuous synthesis of the strand growing in an overall 5'→3' direction, Okazaki could not be sure whether its synthesis was continuous or discontinuous. He therefore proposed two alternative models (FIGURE 8.7). According to the first model, the **discontinuous model of replication**, the replication machinery makes both new strands by first synthesizing small DNA fragments in a 5'→3' direction and then joining successive

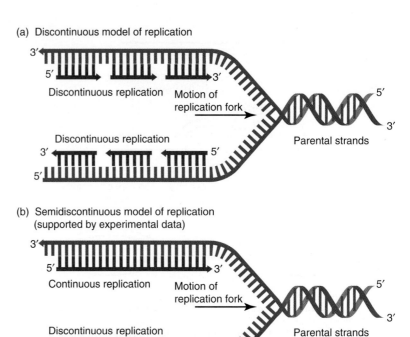

FIGURE 8.7 Models proposed by Reiji Okazaki to explain *in vivo* DNA replication. Parental strands are shown in blue and newly replicated strands are shown in red. The arrowheads indicate the 5′→3′ direction of DNA polymerase synthesis.

fragments. According to the second, the **semidiscontinuous model of replication,** the replication machinery makes the chain growing in an overall 3′→5′ direction by discontinuous synthesis but makes the chain growing in an overall 5′→3′ direction by the continuous addition of nucleotides to its 3′-end. The discontinuous model predicts that all newly formed DNA should exist as small fragments, whereas the semidiscontinuous model predicts that half the newly formed DNA should exist as fragments and the other half as long DNA strands.

In one particularly informative experiment, Okazaki cultured *E. coli* in the presence of [^{3}H]thymidine at 20°C. The low temperature was used to slow down the replication process and thus make it easier to follow changes in DNA as a function of time. Samples were removed at various times. Then the DNA was carefully layered on top of an alkaline sucrose gradient to form a band and analyzed by sucrose gradient centrifugation. The results of this experiment are shown in FIGURE 8.8. After 5 or 10 seconds nearly all newly formed radioactive DNA was present in a slowly sedimenting fraction near the top of the centrifuge tube. The radioactivity in this fraction continued to increase until about 30 seconds and then leveled off. Based on sedimentation rates Okazaki estimated the DNA in the slowly sedimenting fraction was between 1,000 and 2,000 nucleotides long. Under similar conditions denatured *E. coli* DNA is usually 20 to 50 times longer (it would, of course, be longer if the strands were not broken in the course of isolation). A second labeled DNA fraction, barely noticeable in the

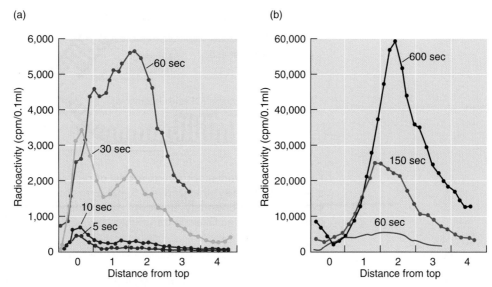

FIGURE 8.8 **Okazaki's pulse labeling experiment.** *E. coli* were cultured in the presence of [³H]thymidine at 20°C for the indicated times and then the DNA was extracted and analyzed on alkaline sucrose gradients. The scales shown on the *y*-axis are different in (a) and (b) because more radioactive label was incorporated into DNA during the longer incubation periods shown in (b). The top of the gradient is on the left and the bottom is on the right. (Adapted from Okazaki, R., et al. 1968. *Proc Natl Acad Sci USA* 59:598–605.)

10-second sample, continued to increase in size in each subsequent sample and by 60 seconds it was the predominant form.

These results appear to fit the discontinuous model of replication (Figure 8.8a), in which all newly synthesized DNA first appear as small fragments that are only later stitched together to form longer strands. However, subsequent studies by other investigators indicated that *E. coli* has a DNA repair system that introduces nicks in newly synthesized DNA. When DNA replication is studied in the absence of this repair system, only about half the newly formed DNA is present in short fragments. Therefore, *DNA replication is semidiscontinuous.* Semidiscontinuous replication has been observed in all organisms and DNA viruses studied to date. The adjacent short fragments (now called **Okazaki fragments** in honor of their discoverer) formed during eukaryotic DNA replication are about 100 to 200 nucleotides long (or about 10% the size of those formed during bacterial DNA replication).

The synthesis of the discontinuously replicating strand lags behind that of the continuously replicating strand. For this reason the discontinuously synthesized strand is called the **lagging strand** and the continuously synthesized strand is called the **leading strand.** As a consequence of semidiscontinuous DNA replication, the template strand for discontinuous DNA replication has a short single-stranded region at the replication fork (FIGURE 8.9).

DNA ligase connects adjacent Okazaki fragments.

The semidiscontinuous model of replication requires an enzyme to join adjacent Okazaki fragments. DNA ligase, an enzyme discovered in 1967 by investigators interested in DNA recombination and repair,

8.1 General Features of DNA Replication

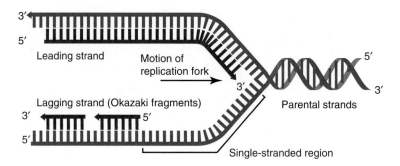

FIGURE 8.9 **Semidiscontinuous model of replication.** Both daughter strands (red) are synthesized in a 5'→3' direction. The leading strand is synthesized continuously, however, and the lagging strand is synthesized discontinuously as Okazaki fragments.

seemed to be an excellent candidate. Okazaki established that DNA ligase is the enzyme required to connect DNA fragments formed during semidiscontinuous replication by demonstrating that *E. coli* mutants with a defective DNA ligase accumulate large quantities of fragments.

DNA ligase connects the phosphate group at the lagging strand's 5'-end to the 3'-hydroxyl group at the end of the newly formed Okazaki fragment (FIGURE 8.10). Some DNA ligases use ATP to supply the energy needed for this reaction, whereas others use NAD^+. Phage T4 DNA ligase and all known eukaryotic DNA ligases are ATP-dependent. Bacterial DNA ligases are NAD^+-dependent. NAD^+ is a somewhat surprising energy source because it usually serves as a cofactor in redox reactions. Bacterial DNA ligases use the energy present in the phosphoanhydride bond that links the nicotinamide mononucleotide (NMN) and adenylate groups.

RNA serves as a primer for Okazaki fragment synthesis.

The semidiscontinuous model of replication requires the repeated initiation of DNA synthesis. Bacterial DNA polymerases, however, cannot

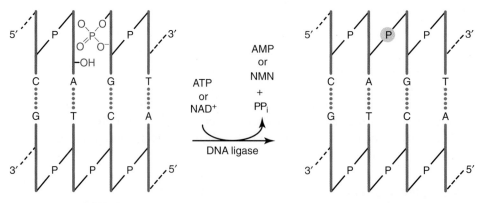

FIGURE 8.10 **DNA ligase activity.** DNA ligase connects the phosphate group at the lagging strand's 5'-end to the 3'-hydroxyl group at the end of the newly formed Okazaki fragment. Some DNA ligases use ATP to supply the energy needed for this reaction while others use NAD^+.

initiate the synthesis of a new DNA strand; they can only add deoxyribonucleotides to the 3′-ends of preexisting primers. Then how does Okazaki fragment synthesis begin? Okazaki answered this question by showing bacteria synthesize RNA oligonucleotides that serve as primers for DNA polymerase (FIGURE 8.11). The ribonucleotides have to be removed and replaced by deoxyribonucleotides before DNA ligase joins adjacent fragments. The bacterial pathway for this process is shown in FIGURE 8.12.

8.2 Bacterial DNA Replication

The bacterial replication machinery has been isolated and examined *in vitro*.

Although the *in vivo* experiments described above provide considerable information about DNA replication, they do not shed much light on the cellular apparatus responsible for DNA replication. The best way to learn about the replication apparatus is to isolate it and then examine how each part works by itself and in concert with the other parts. This approach has been used to study the replication apparatus of viruses, bacteria, eukaryotes, and the archaea. Because the *E. coli* replication apparatus has been the most thoroughly investigated, we begin by examining it.

The *E. coli* replication apparatus is made of many different protein components that must be assembled on DNA before replication

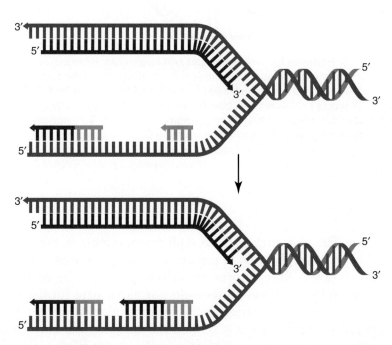

FIGURE 8.11 RNA primers for Okazaki fragments. The synthesis of Okazaki fragments is primed by short RNA segments. RNA primers are shown in green, parental DNA in blue, and new DNA in red.

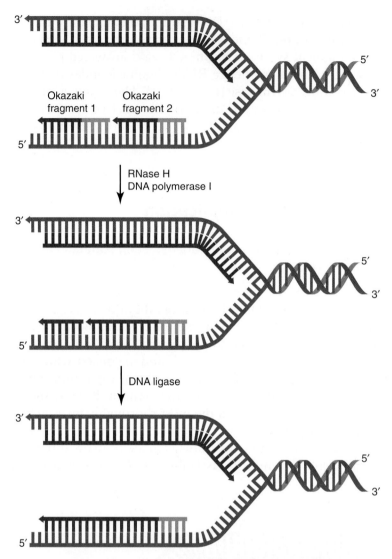

FIGURE 8.12 Pathway for processing Okazaki fragments. Ribonuclease H (RNase H) removes all ribonucleotides at the 5'-terminus of the Okazaki fragment except for the ribonucleotide that is linked to the DNA end. Then DNA polymerase I catalyzes nick translation. Its polymerase activity extends the 3'-end of the newly formed Okazaki fragment and its 5'→3' exonuclease removes the remaining ribonucleotide from the 5'-terminus of the neighboring Okazaki fragment. Finally, DNA ligase joins adjacent Okazaki fragments to form the lagging strand. RNA primers are shown in green, parental DNA in blue, and new DNA in red.

can begin. The first stage of DNA replication, *initiation*, involves the assembly of the replication apparatus at a unique site on the bacterial chromosome. Specific proteins help to assemble the replication apparatus. Some of these proteins do not participate further in the replication process. Initiation is followed by *elongation*, a process in which the leading strand is synthesized continuously and the lagging strand is synthesized discontinuously. A complex replication machine is required for DNA elongation. The final stage of replication, *termination*, begins when the two-replication forks meet about half way around the bacterial DNA molecule.

Mutant studies provide important information about the enzymes involved in bacterial DNA replication.

Molecular biologists have learned a great deal about the DNA replication process by studying mutants with blocks in specific steps in replication. Because a defect in DNA synthesis would be lethal, most replication mutants that have been studied are temperature-sensitive. Early replication mutants were isolated before the functions of the altered genes were known and therefore were named *dnaA*, *dnaB*, *dnaC*, and so forth. The proteins encoded by these genes were named DnaA, DnaB, DnaC, and so forth. Some of the genes identified by mutant analysis were later shown to influence DNA synthesis indirectly rather than directly. For instance, the gene originally named *dnaF* was renamed *nrdA* when it was shown to be the structural gene for a subunit in ribonucleoside diphosphate reductase, the enzyme that converts ribonucleotides to deoxyribonucleotides. As more was learned about the replication process and additional mutants were discovered, investigators named the new genes according to their function. For instance, *gyrA* and *gyrB*, the structural genes for the two subunits in DNA gyrase, were named for their function.

The replicon model proposes that an initiator protein must bind to a DNA sequence called a replicator at the start of replication.

Although genes replicate as part of larger chromosomes, most cannot replicate as independent units. An independently replicating DNA unit, such as a bacterial chromosome or a plasmid, which can maintain a stable presence in a cell, is called a **replicon**.

In the early 1960s François Jacob and Sydney Brenner proposed the **replicon model** to explain how DNA molecules replicate autonomously. The replicon model requires two specific components, an **initiator protein** and a **replicator**. The initiator protein binds to the replicator, a specific set of sequences within the replicon. Because the replicator must be part of the DNA to be replicated, it is said to be *cis*-acting. Once bound to the replicator, the initiator helps to unwind the DNA and recruit components of the replication machinery. The DnaA protein is the initiator for bacterial DNA replication.

The specific site within the replicator at which replication is initiated is called the **origin of replication** (*ori*). Bacterial cells usually require just one origin of replication. Because a bacterial replicator is usually quite short (200–300 base pair [bp]), the terms "replicator" and "origin of replication" tend to be used interchangeably even though technically the origin of replication is just a part of the replicator. Eukaryotic chromosomes, which are much longer than bacterial chromosomes, require many origins of replication (see below).

E. coli chromosomal replication begins at *oriC*.

The *E. coli* origin of replication, ***oriC***, is located at minute 84 of the *E. coli* genetic map. The *E. coli* chromosome's large size and fragility precludes studying *oriC* structure and function in the intact chromosome *in vitro*. Fortunately, *oriC* structure and function can be studied *in vitro* by constructing a plasmid that depends on *oriC* for its replication. The method used to construct this plasmid is shown in FIGURE 8.13. A restriction endonuclease is used to cut the bacterial chromosome into fragments. The same enzyme is also used to cut a plasmid DNA into a fragment containing a drug-resistance gene (ampicillin in this example) but lacking the plasmid origin of replication. After mixing, bacterial and plasmid DNA fragments are joined by DNA ligase, the resulting recombinant plasmid DNA is used to transform bacteria, and transformed bacteria are spread onto growth medium that contains ampicillin. Only cells bearing recombinant plasmid DNA with a drug-resistance gene and *oriC* form colonies under these growth conditions. Intracellular nucleases eventually destroy DNA molecules that cannot replicate autonomously. The minimal *oriC* sequence needed for plasmid replication, which was determined by trimming the ends of the cloned oriC region, is 245 bp long.

The minimal *oriC* sequence has the following specific features that are essential for the initiation of DNA replication (FIGURE 8.14). Its right side has five 9-bp (9-mer) sites (R1–R5) known as **DnaA boxes,** and its left side has a **DNA unwinding element** (DUE) that has three AT-rich 13 bp (13-mer) elements. *E. coli oriC* has one other very notable feature. Based on a completely random nucleotide sequence, one might expect *oriC* to

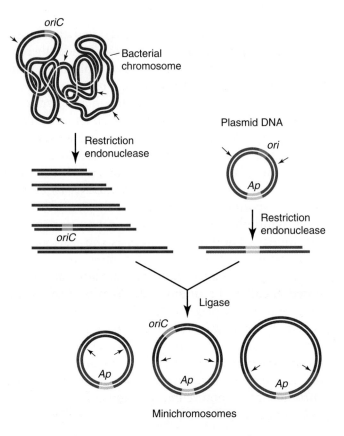

FIGURE 8.13 Isolation of the *E. coli* replication origin *oriC*. Plasmid DNA is digested with a restriction endonuclease to obtain a fragment that lacks the plasmid's origin of replication but retains its drug-resistance gene (ampicillin in this example). The bacterial chromosome is also digested with the restriction endonuclease to obtain a pool of different fragments. The plasmid and bacterial fragments are mixed and then joined by DNA ligase. The resulting recombinant plasmids are used to transform *E. coli*. Only transformants with a recombinant plasmid bearing both the drug-resistance gene and *oriC* will form colonies. (Adapted from Kornberg, A., and Baker, T. A. 1991. *DNA Replication* (2nd ed). W. H. Freeman and Company.)

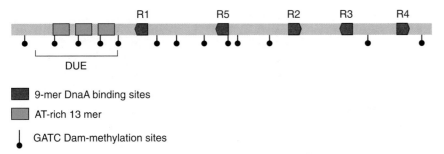

FIGURE 8.14 Minimal *oriC* region required for minichromosome replication. (Adapted from Stepankiw, N., et al. 2009. *Mol Microbiol* 74:467–479.)

have one or two GATC sites. In fact, it has several GATC sites, strongly suggesting these sites have an important function that has been conserved during evolution. These GATC sites help to regulate DNA replication.

Four proteins, DnaA (initiator), DnaB (helicase), DnaC (loader), and DnaG (primase), play critical roles in the DNA replication initiation process. The functions of these proteins, summarized in the simplified schematic shown in FIGURE 8.15, are as follows. DnaA proteins combine with ATP to form DNA • ATP complexes, which bind to the DnaA boxes and form a filament assembly. This filament assembly causes the AT-rich DUE to melt. The DnaC loader and DnaB helicase exist as a $DnaB_6$ • $DnaC_6$ complex with an ATP molecule bound to each DnaC subunit. A DnaB hexamer is loaded onto each of the single strands generated when the DUE melts. DnaC, which plays an essential role in loading the DnaB helicases onto the single strands, dissociates from the complex after helicase loading is completed. The two hexameric DnaB helicases that are loaded onto the nascent replication forks have opposite orientations, allowing them to move away from each other. After this movement DnaG primase binds to the DnaB helicase and catalyzes RNA primer synthesis.

The first pair of primers, which are formed on either side of the initiation site, serve to prime leading strand formation. Primase initiates leading strand synthesis only one time during each DNA replication cycle. Once leading strand synthesis starts, it continues in a $5' \rightarrow 3'$ direction until replication is complete (or the replication machinery encounters a problem such as damaged DNA). Primase must act once in every 1,000 to 2,000 nucleotides during lagging strand synthesis to initiate Okazaki fragment synthesis. Primase's contribution to lagging strand synthesis is examined in the section describing the elongation stage of DNA replication. The initiation stage ends with the addition of the DNA replication machine to the replication fork.

Bacterial DNA replication is regulated at the initiation stage of replication, ensuring the bacterial chromosome is replicated only one time during the cell cycle. DnaA plays an important role in this regulation. The approximately 1,000 DnaA molecules present in each *E. coli* molecule are stable. Intracellular DnaA • ATP levels, however, rise and fall during the cell cycle. Approximately 80% of the DnaA is present in the active DnaA • ATP level just before the initiation of a new round of DNA replication but falls to about 20% after replication is initiated. Three systems regulate the timing and synchrony of replication initiation (**BOX 8.1 LOOKING DEEPER: REPLICATION REGULATION AT THE INITIATION STAGE**).

FIGURE 8.15 Model for loading DnaB onto single strands at DUE. (1) Two DnaB$_6$ • DnaC$_6$ complexes load onto the unwound DUE strands so they end up in opposite orientations. The upper and lower DUE strands are blue and red, respectively. (2) ATP hydrolysis (not shown) leads to DnaC release, freeing each DnaB helicase to move in a 5'→3' direction and thereby extend the unwound region of *oriC* to about 65 nucleotides. (3) DnaG primase binds to DnaB helicases and synthesizes primers (green). (Adapted from Mott, M. et al. 2008. *Cell* 134:623–634.)

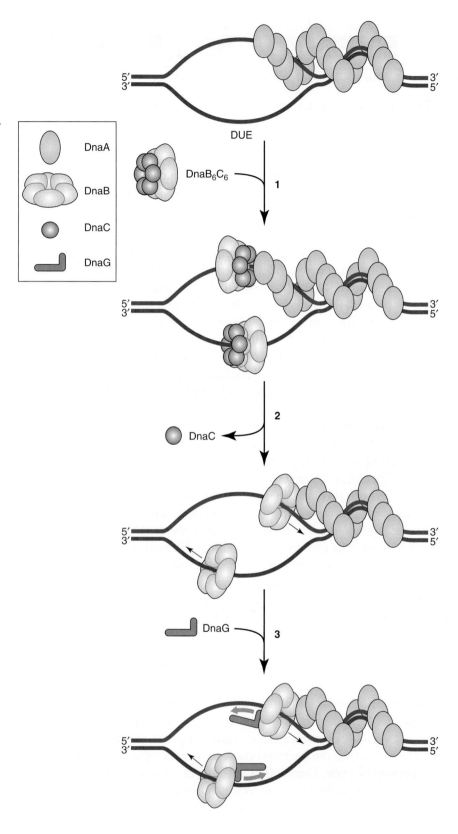

BOX 8.1: LOOKING DEEPER

Replication Regulation at the Initiation Stage

Three systems regulate the timing and synchrony of replication initiation: (1) regulatory inactivation of DnaA, (2) titration of free DnaA molecules, and (3) sequestration by the SeqA protein.

1. *Regulatory inactivation of DnaA (RIDA):* RIDA is a postinitiation regulatory mechanism that blocks further initiations by converting active DnaA • ATP into inactive DnaA • ADP. The two essential components of the RIDA system are the sliding clamp (see text) and a protein called Hda (homologous to DnaA). Hda binds to the sliding clamp and stimulates DnaA to hydrolyze ATP. This hydrolysis causes a rapid decrease in the intracellular DnaA • ATP level, preventing further DNA replication initiations.
2. *DnaA titration:* A DnaA titration system also lowers the free intracellular DnaA concentration. The *E. coli datA* (DnaA titration A) locus, which maps at 94.7 minutes of the chromosomal map, is essential for DnaA titration. The *datA* locus contains five 9-mers with very high affinities for DnaA. In fact, their affinity for DnaA is so high the *datA* locus can titrate eight times more DnaA molecules than the region spanning *oriC* and the neighboring gene *mioC*. Because the *datA* locus is a relatively short distance from *oriC*, it replicates early in the replication cycle to double the number of high affinity DnaA binding sites. The *datA* locus appears to function as a reservoir that collects excess DnaA molecules until the next initiation event. In support of this mechanism, genetic studies show that cells with a *datA* deletion have extra initiation events.
3. *Sequestration:* As mentioned in the text, *oriC* has a much higher frequency of GATC sequences than would be predicted from completely random nucleotide sequences. DNA adenine methyltransferase transfers methyl groups from S-adenosylmethionines to the adenines in these GATC sequences. Because methylation occurs after DNA replication, only the parental strand is methylated in newly synthesized DNA. The hemimethylated *oriC* is sequestered so it cannot participate in a new round of replication. This sequestration requires SeqA protein, which binds to the hemimethylated GATC sites and thereby prevents DnaA • ATP from binding to DnaA binding sites in *oriC*. SeqA therefore acts as a regulatory protein that keeps all origins inactivated for about one-third of a generation. Further study is required to determine how sequestration ends. It is possible, however, that spontaneous dissociation of SeqA and subsequent methylation by DNA adenine methyltransferase may suffice.

Several enzymes act together at the replication fork.

DNA elongation is a complex process that involves cooperative interactions among many different proteins to synthesize two new DNA chains at an astonishing rate. Tania Baker and Stephen Bell illustrated the amazing job *E. coli* replication enzymes perform while synthesizing DNA. They began by changing the scale so the DNA duplex is 1 m in diameter. On this scale the replication fork moves at about 600 km/h (375 mph) and the replication machinery would be about the size of a FedEx® delivery truck. The following is their insightful analogy:

> Replicating the *E. coli* genome would be a 40 min, 400 km trip for two such machines, which would, on average make an error only once every 170 km. The mechanical prowess of this machine is even more impressive given that it synthesizes two chains simultaneously as it moves. Although one strand is synthesized in the same direction as the fork is moving, the other chain (the lagging strand) is synthesized in a piecemeal fashion (as Okazaki fragments) and in the opposite direction of the overall fork movement. As a result, about once a second one delivery person (i.e., polymerase active site) associated with the truck must take a detour, coming off and then rejoining the template DNA strand to synthesize the 0.2 km fragments. (Baker, T. A. and Bell, S. P. 1998. Polymerases and the replisome: machines within machines. *Cell* 92:295–305.)

This analogy underscores the enormous complexity and incredible speed of the DNA replication process. DNA polymerase, the bacterial counterpart of the FedEx® delivery truck, acts as the replication machine. It extends the leading strand by continuous nucleotide attachment to the 3′-hydroxy terminus, while helping to form the lagging strand by extending the primer until the resulting Okazaki fragment reaches the 5′-end of the adjacent fragment.

DNA polymerase III is required for bacterial DNA replication.

Investigators initially assumed DNA polymerase I was the sole polymerase required for bacterial DNA synthesis. The first indication that this assumption was wrong came in 1969 when Paula Delucia and John Cairns isolated an *E. coli* mutant that lacked DNA polymerase I activity but continued to synthesize DNA and grow normally. The possibility that the mutant might have a low level of DNA polymerase I activity that allowed it to synthesize DNA was ruled out when an *E. coli* mutant with a deletion in *polA* (the gene that encodes DNA polymerase I) was shown to also synthesize DNA.

The most likely explanation for the *polA* mutant's ability to synthesize DNA is that some other DNA polymerase is present and that enzyme is responsible for DNA synthesis. In support of this hypothesis two new enzymes—DNA polymerase II and DNA polymerase III—were detected in *polA* mutant extracts when gapped DNA (created by partial hydrolysis of nicked DNA with an exonuclease) was used as a template. The two new polymerases add nucleotides to the 3′-end of the primer strand in the order specified by the template strand. Neither enzyme had been detected in bacterial extracts before because DNA polymerase I is so active it masks their activity.

The next task was to determine what role, if any, the new DNA polymerases play in DNA replication. Once again, a genetic approach helped to provide the answer. Mutants lacking DNA polymerase II synthesize DNA normally, indicating this enzyme is not essential for bacterial DNA replication. In contrast, temperature-sensitive DNA polymerase III mutants replicate DNA at 30°C but not at 42°C, indicating DNA polymerase III is required for bacterial DNA synthesis.

Although genetic studies indicated that DNA polymerase III plays an essential role in bacterial DNA synthesis, the purified enzyme had few of the properties expected of the replication enzyme. For instance, DNA polymerase III could not extend a unique primer completely around a single-stranded circular DNA template even when large quantities of enzyme and substrate were added to reaction mixtures for long periods of time. Furthermore, DNA polymerase III synthesized DNA at about 20 nucleotides • s^{-1}, an exceptionally slow rate when compared with the 1,000 nucleotides • s^{-1} observed in the living cell. The rate is slow because the enzyme dissociates from its DNA template rather frequently.

Enzymes that remain tightly associated with their template through many cycles of nucleotide addition are said to be highly processive. **Processivity** provides a quantitative measure of a polymerase's ability to remain tightly associated with its DNA template during DNA (or RNA) synthesis. A polymerase's processivity equals the average number of nucleotides the enzyme attaches to a growing chain before it dissociates from the DNA template. A polymerase with a low processivity value dissociates from its template after only a few nucleotides are added, catalyzing nonprocessive or **distributive replication**. Distributive replication is inefficient because the polymerase requires considerable time to associate with its template after each dissociation event. The processivity of a DNA polymerase can be determined by a simple laboratory experiment (**BOX 8.2 IN THE LAB: MEASUREMENT OF PROCESSIVITY**).

DNA polymerase III holoenzyme has three distinct kinds of subassemblies.

A possible explanation for DNA polymerase III's low activity is that additional proteins are needed to increase its processivity. Arthur Kornberg and coworkers thought they might be able to detect these additional proteins by using a circular single-stranded phage DNA as a template. Their choice of template was influenced by the knowledge that the circular single-stranded DNA molecules in bacteriophages such as M13 and φX174 are so small (5–6 kilobases long) they lack sufficient genetic information to code for most enzymes needed for DNA replication. The bacteriophages therefore rely on bacterial enzymes to catalyze the first stage of DNA replication, the conversion of circular single-stranded DNA to the double-stranded replication form. It thus seemed reasonable to expect that bacterial extracts could extend a primer annealed to circular single-stranded bacteriophage DNA completely around the template.

BOX 8.2: IN THE LAB

Measurement of Processivity

One common procedure used to determine processivity is as follows. A DNA primer-template substrate is prepared in which the primer is relatively short, of a known and fixed length, and labeled with either a radioactive isotope or a fluorescent marker. DNA polymerase is incubated with the labeled primer-template in the absence of deoxynucleoside triphosphates, allowing the enzyme to bind to the primer-template but not to extend the primer chain. Then deoxynucleoside triphosphates and a large excess of unlabeled DNA primer-template (the "trap") are added. The prebound DNA polymerase extends the labeled primer until the polymerase dissociates from the primer-template. After dissociation the polymerase is free to bind to either labeled substrate or the unlabeled "trap" DNA. It almost always binds to the unlabeled "trap" DNA, which is present in large excess. Because the "trap" prevents DNA polymerase from associating with the labeled substrate again, the number of nucleotides added to the labeled primer corresponds to the product of one round of processive synthesis catalyzed by the polymerase. After extension is complete products are separated by gel electrophoresis, and the distribution of extended labeled primer is visualized by autoradiography or scanning for the fluorescent marker. Even though the "trap" DNA is extended, it cannot be seen because it is not labeled.

This primer extension is exactly what Kornberg and coworkers observed (FIGURE 8.16). Taking advantage of this assay, Kornberg's laboratory used standard protein fractionation procedures to purify the fully functional replicase, DNA polymerase III holoenzyme, which contains three distinct subassemblies. One of these subassemblies, the core polymerase subassembly, is responsible for the DNA polymerase III activity that was originally detected in *polA* mutant extracts.

Because each cell contains only about 10 to 20 copies of DNA polymerase III holoenzyme, considerable effort was required to obtain

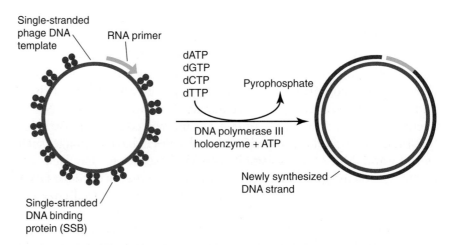

FIGURE 8.16 Action of DNA polymerase III holoenzyme on a primed single-strand DNA binding protein–coated circular single-stranded phage DNA molecule.

sufficient quantities of the enzyme for study. This effort was rewarded when purified holoenzyme was shown to catalyze DNA synthesis at a rate of about 750 nucleotides • s^{-1}. The rapid synthetic rate, a result of the holoenzyme's high processivity, allows the enzyme to add several thousand nucleotides to a primer annealed to a circular single-stranded bacteriophage DNA molecule without dissociating from the DNA. DNA polymerase III holoenzyme thus synthesizes DNA at a much faster rate and with a much higher processivity than DNA polymerase I. One other important difference was noted. DNA polymerase III holoenzyme requires ATP for full activity, whereas DNA polymerase I does not. We return to this ATP requirement when we examine the way the holoenzyme works.

DNA polymerase III holoenzyme has nine distinct polypeptides. The genes that code for these polypeptides have been cloned, allowing investigators to construct bacteria that overproduce the polypeptides. Isolated polypeptide subunits, which are available in ample supply, have been used to constitute DNA polymerase III holoenzyme as well as various subassemblies. The three subassemblies are called the **core polymerase, clamp loader,** and **sliding clamp.** Each subassembly makes a unique contribution to holoenzyme function. We can gain considerable insight into the way the replication machine works by examining subassembly function. We, therefore, begin by examining the core polymerase, the subassembly responsible for adding nucleotides to the 3′-hydroxyl end of growing polynucleotide chains.

The core polymerase has one subunit with 5′→3′ polymerase activity and another with 3′→5′ exonuclease activity.

The subassembly responsible for chain extension, the core polymerase (previously known as DNA polymerase III), consists of an α-subunit, ε-subunit, and θ-subunit:

1. *α-Subunit:* The α-subunit (PolIIIα), encoded by *dnaE*, catalyzes 5′→3′ chain growth and is essential for DNA replication. Thomas Steitz and coworkers used recombinant DNA techniques to insert the gene that codes for the polIIIα subunit from *Thermus aquaticus*, a gram-negative thermophilic bacterial strain, into an *E. coli* plasmid. Then they introduced the recombinant plasmid into *E. coli* and purified the highly expressed polIIIα subunit from the transformed cells. The purified subunit was used to prepare a PolIIIα • DNA • dATP ternary complex. The crystal structure for this complex reveals that PolIIIα has six domains (**FIGURE 8.17**). Three of these, the palm, finger, and thumb domains, are present in all DNA polymerases. The finger domain interacts with the incoming deoxyribonucleoside triphosphate, the palm domain contains the catalytic site, and the thumb domain grips the DNA substrate. The arrangement of the finger, palm, and thumb domains is different from the arrangement of the corresponding domains in DNA polymerase I. Thus, the polIIIα subunit

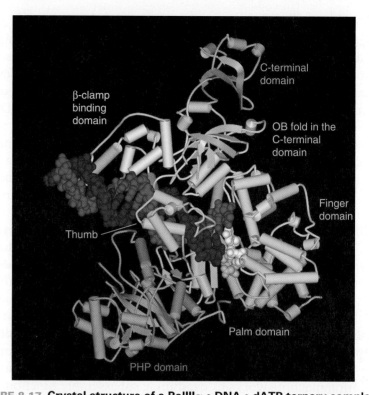

FIGURE 8.17 Crystal structure of a PolIIIα • DNA • dATP ternary complex. The protein structure is shown in a schematic display. The six domains are as indicated. The oligonucleotide binding (OB) fold (light pink) is part of the C-terminal domain (dark pink). Primer and template DNA strands are shown as a red and blue spacefill structures, respectively. The thumb domain is green. The incoming deoxyribonucleoside triphosphate is shown as a white spacefill structure. The three aspartate residues at the catalytic site in the palm domain are shown as gold spacefill structures. (Protein Data Bank ID: 3E0D. Wing, R. A., et al. 2008. *J Mol Biol* 382:859–869.)

and DNA polymerase I belong to different DNA polymerase families. Two of the additional PolIIIα subunit domains interact with the sliding clamp subassembly, a ring-shaped protein that increases core polymerase processivity by tethering the core polymerase to its DNA substrate (see below). The function of the last domain, the so-called PHP domain, remains to be determined.

2. *ε-Subunit:* The ε-subunit, which is encoded by *dnaQ* (also called *mutD*), has 3'→5' exonuclease activity. On those rare occasions when the polymerase makes an error, the exonuclease removes the mispaired nucleotide, thereby providing a proofreading function. Consistent with this function, cells with defective ε-subunits have high mutation rates.

3. *θ-Subunit:* The θ-subunit, which is encoded by *holE*, seems to stimulate the ε-subunit. This function is not essential, however, because DNA polymerase III holoenzyme isolated from *holE* deletion mutants is fully active. Thus, the θ-subunit does not appear to have a unique function at this time.

The α- and ε-subunits form a 1:1 complex in which each subunit's activity is greater than it is in the free subunit. The θ-subunit joins this complex by binding to ε to form a linear α-ε-θ arrangement. Associations among core polymerase subassembly subunits are so tight denaturing agents are required for dissociation. Detailed information about core polymerase subassembly structure awaits crystallographic studies. The core polymerase subassembly does not bind tightly to DNA. We now examine the sliding clamp subassembly, which tethers the core polymerase subassembly to DNA.

The sliding clamp forms a ring around DNA, tethering the remainder of the polymerase holoenzyme to the DNA.

The sliding clamp (also known as the β-dimer or the β-clamp) is made of two identical polypeptide subunits that are encoded by *dnaN*. Michael O'Donnell and coworkers determined the crystal structure of the *E. coli* sliding clamp on DNA (FIGURE 8.18). Two semicircular shaped polypeptide subunits are arranged head-to-tail with a 3.5 nm diameter central opening. Each subunit has three domains (Figure 8.18a). Each domain within a subunit has a different amino acid sequence. Nevertheless, all three domains have the same folding pattern: two β-strands are on the outside and two α-helices on the inside. The arrangement of the α-helices in the sliding clamp allows it to move along the encircled DNA.

The sliding clamp has two distinct faces, a C-terminal face and an N-terminal face (Figure 8.18b). As the names suggest, the former is characterized by protruding C-termini and the latter by protruding N-termini. The core polymerase and the clamp loader bind to the C-terminal face. The sliding clamp therefore increases the holoenzyme's processivity by acting as a tether for the remainder of the holoenzyme. Other proteins also bind to the C-terminal face, including four additional bacterial DNA polymerases, various DNA repair enzymes, and DNA ligase.

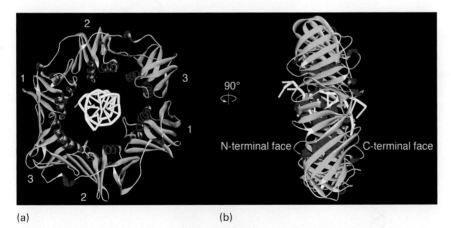

FIGURE 8.18. Structure of an *E. coli* sliding clamp on DNA. The two subunits are shown as solid ribbons. α-Helices are red, β-structures are cyan, and turns are white. The DNA strands are in tube form and are white. (a) Crystal structure of the β dimer (sliding clamp) with bound DNA viewed looking down the DNA axis. Domains in each subunit are numbered 1, 2, and 3 on each subunit. (b) The same sliding clamp with bound DNA viewed from the side with the N- and C-terminal faces labeled. The C-terminal face has been implicated in sliding clamp interactions with other proteins. (Protein Data Bank ID: 3BEP. Georgescu, R. E., et al. 2008. *Cell* 132:43–54.)

The clamp loader places the sliding clamp around DNA.

The clamp loader uses energy provided by ATP to load the sliding clamp onto a DNA template-primer with a 5′-overhang. This energy requirement explains why the DNA polymerase III holoenzyme needs ATP. Much of our knowledge about the clamp loader has been obtained from studies of a minimal clamp loader, which was constructed *in vitro* to contain just five of the seven subunits normally present in the clamp loader. This minimal clamp loader contains three identical γ-subunits, a δ-subunit, and a δ′-subunit. The minimal clamp loader behaves like a machine with moving parts that loads the sliding clamp onto a DNA primer–template complex (FIGURE 8.19). The γ-subunits, which are the only polypeptide subunits that bind and hydrolyze ATP, act as the machine's motor. The δ′-subunit, which appears to be stationary, regulates the δ-subunit's ability to bind the sliding clamp. In the absence of ATP, δ′ prevents δ from binding the sliding clamp and opening it. When ATP binds to the γ-subunits, the δ-subunit pulls away from δ′ and thereby gains the ability to bind the sliding clamp. The clamp loader • sliding clamp complex has a high affinity for a DNA primer–template complex with a 5′-overhang. After the clamp loader • sliding clamp complex interacts with the primed DNA, bound ATP molecules are hydrolyzed, causing the clamp loader • sliding clamp complex to dissociate and allowing the sliding clamp to close around the DNA.

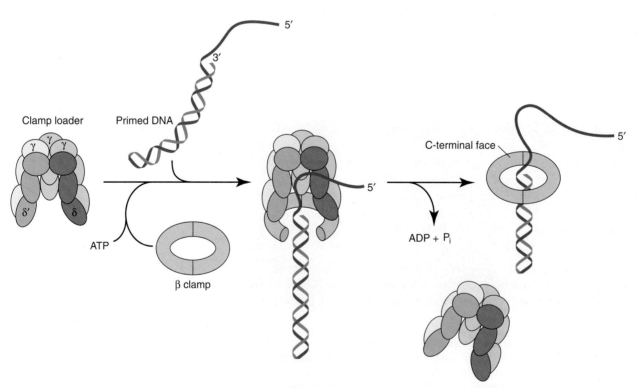

FIGURE 8.19 **The clamp loading cycle.** ATP binds to the clamp loader, allowing the clamp loader and the β clamp to form a tight complex. The clamp loader • sliding clamp complex has a high affinity for a DNA primer-template complex with a 5′-overhang. After the clamp loader • sliding clamp complex interacts with the primed DNA, bound ATP molecules are hydrolyzed, causing the clamp loader • sliding clamp complex to dissociate and allowing the sliding clamp to close around the DNA. (Adapted from Pomerantz, R. T., and O'Donnell, M. 2007. *Trends Microbiol* 15:156–164.)

The bacterial cell has two different kinds of clamp loaders. One exists as a free form and the other as a subassembly of the DNA polymerase III holoenzyme. The free form contains three γ-subunits, whereas the subassembly contains three τ-subunits. The γ- and τ-subunits are encoded by the same gene and therefore very similar. The major difference between the two is that the shorter γ-subunit is missing a carboxyl terminal region that is present in the τ-subunit. This region, which allows the τ-subunit to bind to the core polymerase and the DnaB helicase, is essential for high processivity but not for loading the sliding clamp onto a primer-template. The τ-subunit is essential for DNA replication and must be present for cells to remain viable. In contrast, cells that lack the γ-subunit continue to synthesize DNA and are viable. This finding indicates that the free form of the clamp loader is not essential for replication. Further work is necessary to determine the free form's function(s) in the cell.

The replisome catalyzes coordinated leading and lagging strand DNA synthesis at the replication fork.

By the late 1970s data generated from autoradiography and electron microscopy experiments showed that leading and lagging strand synthesis are coordinated at the replication fork. Investigators found it difficult to understand, however, how the replication machinery could coordinate this synthesis because DNA polymerases at the replication fork must move in opposite directions along the two antiparallel DNA template strands. In 1980 Bruce Alberts proposed the **trombone model of replication** to explain how coordination of replication of leading and lagging strands might take place. As shown in an updated version of this model in FIGURE 8.20, the lagging template strand loops out as each Okazaki fragment is formed. This looping allows the pair of core polymerase subassemblies to coordinate leading and lagging strand synthesis while each moves in a $5' \rightarrow 3'$ direction along its template strand. Recent electron microscopy studies show the trombone structure exists in cells.

When Alberts proposed the trombone model, relatively little was known about DNA polymerase III holoenzyme structure. The discovery that each holoenzyme has three core polymerases, however, fits the trombone model and explains how the DNA replication machinery coordinates leading and lagging strand synthesis at the replication fork. At a minimum this replication machinery or **replisome** consists of DNA polymerase III holoenzyme, helicase, primase, and single-strand DNA binding protein. Fundamental questions still exist concerning how the activities of these enzymes and protein factors are coordinated during replication. For example, how can we explain the fact that helicase moves $5' \rightarrow 3'$ on the lagging strand while primase moves in the opposite direction as it synthesizes the RNA primer (**BOX 8.3 LOOKING DEEPER: HELICASE AND PRIMER COORDINATION**)?

Other enzymes also contribute to DNA replication. The functions of DNA polymerase I, RNase H, and DNA ligase were described earlier. DNA gyrase and topoisomerase I also make an essential contribution to bacterial DNA replication. Because the bacterial chromosome is a covalently closed circular DNA molecule, unwinding at

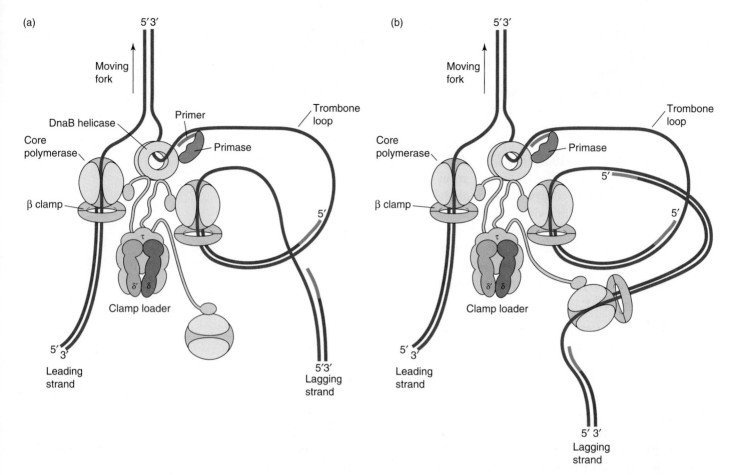

FIGURE 8.20 DNA replisome with three core polymerases. (a) The third core polymerase is off DNA. This situation may take place when only one core polymerase is required for lagging strand synthesis. (b) Two core polymerases are involved in lagging strand synthesis. The trombone loop and replisome components are labeled. For simplicity, only a single primase is shown associated with DnaB helicase and only the δ, δ′, and τ-subunits are shown. Parental DNA is dark blue, newly synthesized DNA is red, and primer is green. (Adapted from McInerney, P. et al. 2007. *Mol Cell* 27:527–538.)

BOX 8.3: LOOKING DEEPER

Helicase and Primer Coordination

DnaB helicase moves in a 5′→3′ direction along the lagging strand as it unwinds double-stranded DNA. Primase, however, which needs to be associated with helicase to function, must move in the opposite direction to synthesize the primer. Three models have been proposed to solve this problem (FIGURE B8.3):

1. *The pausing model:* Helicase movement pauses to allow primase to synthesize the primer and then resumes. This model appears to be consistent with observations for the phage T7 replisome.
2. *The disassembly model:* One or more primase subunits dissociate from the helicase and then stay behind to synthesize a primer, while the helicase and its remaining primase subunits continue to move along the lagging strand. The helicase needs to recruit replacement primase subunits for each cycle of Okazaki fragment formation. This model appears to be consistent with observations for the *E. coli* replisome.

3. *The priming loop model:* The helicase and primase remain associated. The helicase continues to move in a 5'→3' direction along the lagging strand as it unwinds the double-stranded DNA, while the associated primase moves in the opposite direction to synthesize primer. This arrangement results in the formation of a transient single-stranded DNA loop in the lagging-strand template. This loop, designated the priming loop, is subsequently released to become part of the trombone loop when the primer is passed to the core polymerase associated with the lagging strand. This model seems to fit data for the phage T4 replisome but may also apply to bacterial replisomes.

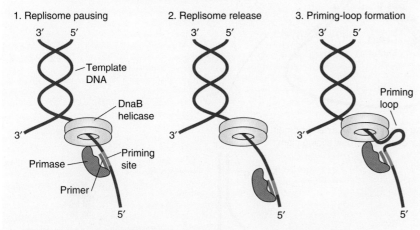

FIGURE B8.3 Three priming mechanisms. (Adapted from N. E. Dixon, *Nature* 462 [2009]: 854–855.)

the replication fork leads to tighter winding ahead of the replication fork. DNA gyrase and topoisomerase I act cooperatively to relieve the resulting torsional strain, allowing the replication fork to continue moving along the bacterial chromosome.

Continuous replication of the leading strand is readily explained by the sliding clamp's ability to serve as a tether for the core polymerase. This ability, however, presents a problem for lagging strand synthesis because the sliding clamp must release the core polymerase after each Okazaki fragment is completed. FIGURE 8.21 shows the core polymerase cycle on the lagging strand.

There are three steps in the lagging strand cycle: (1) core polymerase extends the Okazaki fragment and the clamp loader loads a new sliding clamp, (2) core polymerase dissociates from the sliding clamp upon reaching an adjacent duplex, and (3) core polymerase binds to the new sliding clamp and synthesizes DNA. The trombone loop is reset after each Okazaki fragment has been completed. The core polymerase is not free to diffuse away because it remains bound to the τ-subunit. The replication fork is now ready to begin a new cycle of Okazaki fragment formation.

Some mechanism is required to permit the core polymerase to be released from the sliding clamp when Okazaki fragment synthesis is complete. The *E. coli* replication machinery uses two different release

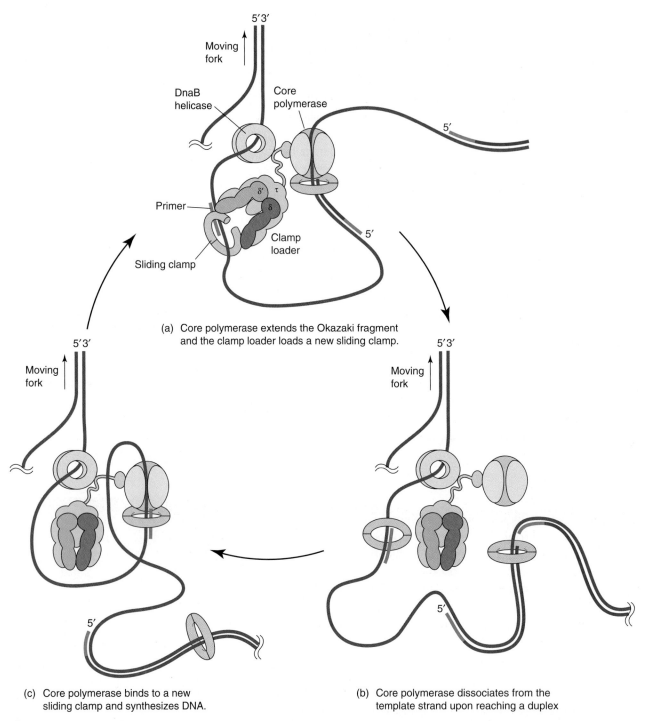

FIGURE 8.21 Core polymerase cycle on the lagging strand. Parental DNA is dark blue, newly synthesized DNA is red, and primers are green. For simplicity, leading strand replication is not shown and only one core polymerase is shown to be involved in lagging strand synthesis. (Adapted from Leu, F.P., et al. 2003. *Mol Cell* 11:315–327.)

pathways. The first is known as "**collision release**" because core polymerase remains bound to the sliding clamp until it "collides" with a duplex. The second is called either "**premature release**" because core polymerase is released before Okazaki fragment formation is complete or "**signaling release**" because the presence of a new primer or newly assembled sliding clamp appears to signal release.

Sliding clamps left behind after completion of Okazaki fragment synthesis bind DNA polymerase I and DNA ligase, which participate in the pathway that joins adjacent fragments (Figure 8.12). Abandoned sliding clamps must eventually be recycled, however, because each cell has about 300 sliding clamps to produce about 10 times that number of Okazaki fragments. Free δ-subunits, which are in excess over other clamp-loading subunits in the cell, are the most likely clamp-unloading agents. The free δ-subunit is very active in unloading sliding clamps from DNA but cannot load clamps onto DNA. The free clamp loader may also unload abandoned sliding clamps.

E. coli DNA replication terminates when the two growing forks meet in the terminus region, which is located 180 degrees around the circular chromosome from the origin.

The two replication forks initiated from *oriC* eventually meet at a termination region on the opposite side of the chromosome, triggering a series of events that lead to the completion of chromosome synthesis and then chromosome separation. The movement of the replication fork within the termination region is arrested or at least forced to pause for a long time at specific **termination sites** (*Ter* sites). *Ter* sites are conserved 11-bp sequences that may be present in two orientations. When present in one orientation they allow the replication machinery to pass through, but when present in the opposite orientation they stop replication. *E. coli* has a total of 10 *Ter* sites that are present in two clusters, each containing 5 *Ter* sites. As shown in **FIGURE 8.22**, *TerC, TerB, TerF, TerG,* and *TerJ* block the progress of a replication fork that moves in a clockwise direction, whereas *TerA, TerD, TerE, TerI,* and *TerH* block the progress in the opposite direction.

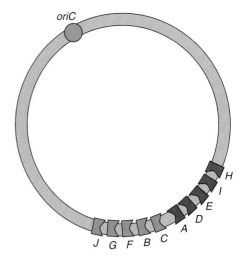

FIGURE 8.22 Locations of the *E. coli* replication terminator (*Ter*) sites. Ten terminators flank the terminus region. Clockwise replication is arrested by the terminators *TerC, TerB, TerF, TerG,* and *TerJ* and counterclockwise replication by terminators and *TerA, TerD, TerE, TerI,* and *TerH.* (Adapted from Mulugu, S., et al. 2001. *Proc Natl Acad Sci USA* 98:9569–9574.)

The terminus utilization substance binds to *Ter* sites.

A protein called the **terminus utilization substance** (**Tus**) binds to the *Ter* sites. The Tus protein binds to *Ter* as a monomer with a very high affinity, ensuring the polar function of the Tus • *Ter* complex (**FIGURE 8.23**). When present in the proper orientation the Tus • *Ter* complex arrests the progress of the replication fork by interfering with DnaB helicase's ability to unwind DNA at the replication fork. The physiological significance of the Tus • *Ter* system remains an open question because null mutants lacking the Tus protein grow normally under a variety of growth conditions. The gram-positive bacteria *Bacillus subtilis* uses an entirely different protein in place of Tus, one that acts as a dimer to bind to termination sites that are approximately 30-bp long. This difference is surprising because *E. coli* and *B. subtilis* replication systems are quite similar in other respects.

Topoisomerase IV and recombinase separate newly formed sister chromosomes.

Two additional enzymes, topoisomerase IV (a type-2 topoisomerase) and recombinase (**FIGURE 8.24**), are needed to allow the newly formed

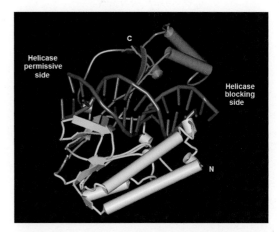

FIGURE 8.23 Crystal structure of the Tus • *Ter* complex of *E. coli* showing the DNA-binding region of β-strands and the helicase blocking end. Notice the protein has a polarity and the side that blocks the DnaB helicase is on the right. The protein is shown in schematic form with the N-terminal domain in yellow and the C-terminal domain in green. The amino and carboxyl ends are indicated by N and C, respectively. The DNA is shown as a blue tube structure. (Protein Data Bank ID: 1ECR. Kamada, K., et al. 1996. *Nature* 383:598.603.)

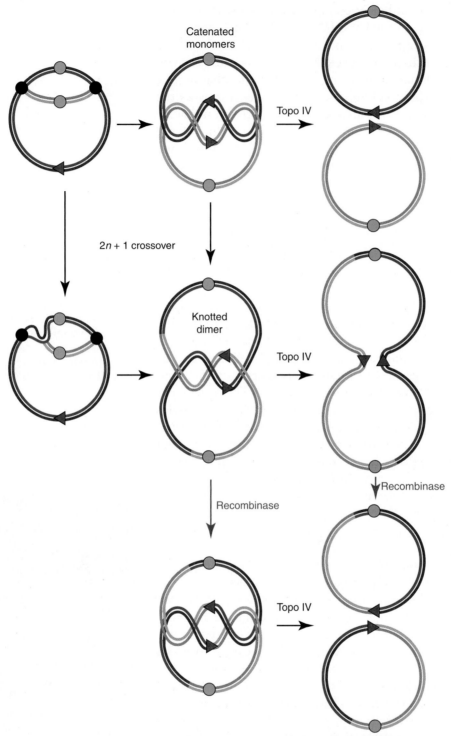

FIGURE 8.24 **Links between replication, recombination, and chromosome segregation.** Topoisomerase IV (Topo IV) is sufficient to produce separated monomers when the newly formed sister chromosomes form catenanes (interlocking rings) but are not covalently joined. When an odd number of crossovers takes place between homologous regions in newly formed sister chromosomes, it generates a knotted dimer in which the sister chromosomes are covalently joined. Then both recombinase and topoisomerase IV (Topo IV) are required to separate monomers. Origins are shown as green circles. Replication forks and associated replication machinery are shown as black circles. The recombination site *dif* that is recognized by recombinase is shown as a red triangle. Newly synthesized strands are shown in lighter shades. (Adapted from Sherratt, D.J., Lau, I.F., and Barre, F.X. 2001. *Curr Opin Microbiol* 4:653–659.)

sister chromosomes to separate. Topoisomerase IV allows the interlinked chromosomes to separate. However, it cannot help when an odd number of recombination events occur during the replication process, causing the two chromosomes to become joined by covalent bonds and form a dimer. Dimerization presents a problem because the two sister chromosomes cannot separate and move to daughter cells. Recombinase acts at a specific site within the termination region called *dif* (*d*eletion-*i*nduced *f*ilamentation), converting the dimer into two separate daughter chromosomes. The site's name derives from the observation that *dif* deletion mutants continue to grow lengthwise, form long filaments, and eventually die because they cannot partition their joined chromosomes to complete cell division.

8.3 Eukaryotic DNA Replication

The SV40 DNA replication system is a good model for *in vitro* eukaryotic DNA replication.

We begin our study of eukaryotic DNA replication by examining SV40 DNA replication because *in vitro* studies of this system provide many important insights into eukaryotic DNA replication. SV40 DNA is a covalently closed circular duplex with a single origin of replication. Its replication, therefore does not require components of the eukaryotic replication machinery that coordinate replication from multiple origins or extend telomeres that are shortened during replication. In 1984 Joachim J. Li and Thomas J. Kelly demonstrated that a soluble extract derived from SV40-infected monkey cells can replicate SV40 DNA. Subsequent studies demonstrated that extracts from uninfected monkey cells also support SV40 DNA replication if they are supplemented with viral encoded large tumor (T) antigen, which functions as both an initiator and helicase. *The host cell provides all other enzymes required for SV40 DNA replication.*

Six identical T antigen molecules combine to form a helicase that interacts with the SV40 origin of replication, which contains about 300 bp. However, only 64 of these base pairs, the so-called core sequence, are essential for DNA replication. Nonessential elements on either side of the core sequence facilitate the initiation process. Further studies are required to determine how the T antigen unwinds DNA and elucidate the steps leading to the assembly of the replication machinery at the replication fork.

The T antigen works together with the eukaryotic cell's single-stranded DNA binding protein (replication protein A) to recruit DNA polymerase α-primase (Pol α), a four-subunit protein complex, to the newly developing replication bubble. As the only eukaryotic enzyme that can synthesize primer, Pol α is an essential participant in initiation. Its crystal structure has not yet been determined. Pol α has two distinct catalytic activities, primase and DNA polymerase, which reside in separate subunits. The other two subunits appear to play important structural roles, but no enzyme activities have as yet been assigned to them. Mammalian Pol α does not have proofreading activity. Primase forms RNA oligomers that are 8 to 12 nucleotides long, which serve as primers for DNA synthesis. Then the weakly processive Pol α DNA polymerase incorporates 15 to 25 deoxyribonucleotides before dissociating from the DNA template. The oligonucleotide synthesized by Pol α thus consists of an RNA primer joined to a short DNA segment (**initiator DNA**). The first pair of initiator DNA molecules, which are formed on either side of the core sequence, serve to prime leading strand formation. Initiator DNA molecules formed by Pol α later in the replication process prime lagging strand synthesis.

Further work in Kelly's and other laboratories led to the identification and characterization of the following additional cellular proteins required for SV40 DNA replication:

1. *DNA polymerase δ (Pol δ):* Pol δ catalyzes both leading and lagging strand DNA synthesis in the SV40 DNA replication system.
2. *PCNA:* The eukaryotic sliding clamp was identified as an antigen in the nuclei of proliferating cells before its function was known and was therefore called PCNA (proliferating cell nuclear antigen). PCNA has a similar quaternary structure to that of the bacterial sliding clamp. This similarity is remarkable because the protein subunits in the eukaryotic and bacterial sliding clamps are not homologous and the eukaryotic clamp has three subunits, whereas the bacterial sliding clamp has two subunits (FIGURE 8.25). Pol δ, FEN1 (see below), and DNA ligase all bind to PCNA.
3. *Replication factor C:* Replication factor C, the eukaryotic clamp loader, couples ATP hydrolysis to the assembly of the PCNA sliding clamp onto a primed template. Replication factor C, which has five subunits that are homologous to the subunits of the *E. coli* clamp loader, works in a similar (but not identical) way to the *E. coli* clamp loader.

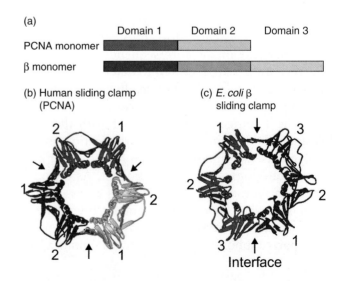

FIGURE 8.25 Human and *E. coli* β clamps. (a) The human sliding PCNA subunit is about two-thirds the size of the *E. coli* β-subunit. The former contains two domains (shown in shades of blue) and the latter three domains (shown in shades of purple). (b) The crystal structure of the human sliding clamp (PCNA) is similar to PCNAs from other eukaryotes. The three subunits are shown in red, yellow, and blue. (c) The crystal structure of the *E. coli* sliding clamp. The two subunits are shown in red and blue. Interfaces between subunits are indicated by arrows and domains within a subunit are numbered (1 or 2 for PCNA and 1, 2, or 3 for the bacterial clamp). (Part a adapted from Bruck, I., and O'Donnell, M. 2001. *Genome Biol* 2:reviews 3001.1–3001.3. Parts b and c reproduced from Bruck, I., and O'Donnell, M. 2001. *Genome Biol* 2:reviews 3001.1–3001.3. Photos courtesy of Michael O'Donnell, Rockefeller University.)

4. *Flap endonuclease (FEN1)*: FEN1 is a 5′-endo/exonuclease that recognizes a double-stranded DNA with a 5′-unannealed flap and cleaves at the base of the flap (FIGURE 8.26). FEN1 acts together with Pol δ and DNA ligase during Okazaki fragment maturation (see below).
5. *DNA ligase*: DNA ligase seals the nick between adjacent Okazaki strands.

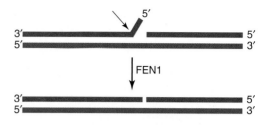

FIGURE 8.26 **Flap endonuclease 1 (FEN1) cleavage.** FEN1 recognizes a double-stranded DNA with a 5′-unannealed flap and makes an endonucleolytic cleavage at the base of the flap.

After extending an initiator DNA on the lagging strand by about 100 to 200 nucleotides, Pol δ runs into the RNA primer of the previous Okazaki fragment. The RNA primer must be removed before DNA ligase joins the two fragments. Continued Pol δ synthesis displaces one or two nucleotides at the 5′-end of the previous Okazaki fragment. Endonucleolytic cleavage by FEN1 removes the flap. Other pathways probably also contribute to RNA primer removal. For example, RNase H may participate in primer removal. FIGURE 8.27 shows how several of the important enzymes and proteins required for SV40 DNA elongation may be organized at the replication fork.

Eukaryotic replication machinery must replicate long linear duplexes with multiple origins of replication.

Major differences between bacterial and eukaryotic DNA replication arise because a bacterial chromosome is a closed covalent circle,

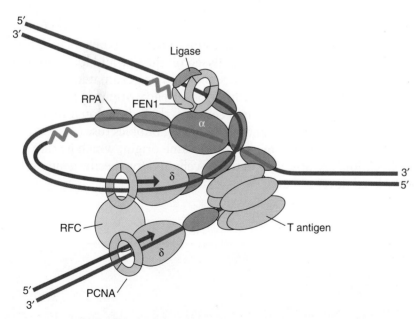

FIGURE 8.27 **Model for the organization of proteins and enzymes at SV40 replication fork.** Primer joined to initiator DNA is shown in green, SV40 DNA that is being replicated in blue, and newly synthesized DNA in red. The abbreviations used are as follows: α, Polα; δ, Polδ; FEN1, flap endonuclease; ligase, DNA ligase; PCNA, proliferating cell nuclear antigen; RPA, replication protein A; and RFC, replication factor C. (Adapted from Stillman, B. 2008. *Mol Cell* 30:259–260.)

whereas a eukaryotic chromosome is a long linear duplex. These differences lead to the following important consequences:

1. A bacterial chromosome has a single origin of replication, whereas each eukaryotic chromosome has several origins of replication. Eukaryotes have regulatory mechanisms that ensure each origin fires once, and only once, during the S phase of the life cycle.
2. Bacteria require only one DNA polymerase, DNA polymerase III holoenzyme, for normal replication fork propagation, whereas eukaryotes require three different enzymes (see below).
3. Bacterial DNA replication terminates when the two replicating forks meet about half way around the bacterial chromosome and then are joined. Linear eukaryotic chromosomes require a special enzyme, telomerase, to form their ends.

We must keep these differences in mind as we examine the different stages of eukaryotic replication: initiation, elongation, and termination. Although there is no single eukaryotic model system in which all aspects of DNA replication have been studied, much of our current knowledge of eukaryotic DNA replication has come from studies with yeast. Where possible, we focus on the yeast replication system, but we must be mindful that DNA replication in other eukaryotes may differ in some details.

Eukaryotic chromosomes have many origins of replication.

The long linear eukaryotic DNA molecules initiate DNA replication at multiple sites. Evidence for this comes from electron micrographs of replicating eukaryotic DNA, which reveal multiple replication bubbles (FIGURE 8.28) and from autoradiograms of tritium-labeled DNA fibers isolated from proliferating cells. Based on the size and location of the replication bubbles, origins of replication are estimated to be from 10 to 300 kilobases apart. Initiation at each origin is restricted to once per S phase during normal cell growth. Bidirectional replication bubble growth continues until neighboring bubbles fuse (FIGURE 8.29). Rare multiple initiation events at the same origin, which lead to gene amplification, are observed in tumor cells and for specific genes during normal cell development.

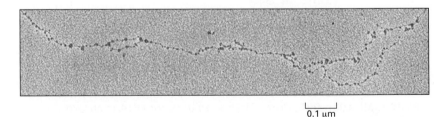

FIGURE 8.28. An electron micrograph of a replicating chromosome in an early embryo of *Drosophila melanogaster*. Each origin is apparent as a replication bubble along the DNA strand. (Courtesy of Victoria E. Foe, Center for Cell Dynamics, University of Washington.)

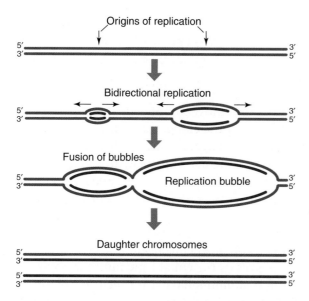

FIGURE 8.29 Schematic showing the fusion of a pair of replication bubbles.

Yeast origins of replication, called autonomously replicating sequences, determine the site of DNA chain initiation.

In 1980 Ronald W. Davis and coworkers discovered that yeast origins of replication, called **autonomously replicating sequence (ARS) elements**, allow plasmids containing selectable genes to replicate autonomously in yeast cells. The experiment shown in **FIGURE 8.30** illustrates one method for demonstrating the existence of ARS elements. Two sets of circular plasmids, one containing an ARS element and *LEU* and the other just containing *LEU*, are introduced into *leu⁻* yeast cells, which are then spread on agar growth medium lacking leucine. A large number of colonies form when cells transformed with ARS plasmids are spread on the agar medium because the plasmids can replicate autonomously. In contrast, very few colonies appear when cells transformed with plasmids lacking ARS are spread on the agar medium and the colonies that do form contain cells in which *LEU* is integrated into the yeast genome. Yeast cells have approximately 300 ARS elements spread over their 16 chromosomes, whereas cells from higher animals probably have thousands. A possible reason why a single origin of replication suffices for *E. coli* chromosome replication while several ARS elements are needed for yeast chromosome replication is that the bacterial replication fork migrates about 30 times faster than a yeast chromosome replication fork.

The initiation stage of DNA replication in eukaryotes is more complex than that in bacteria. A major reason for this complexity is that eukaryotes have to control initiations at many origins, whereas bacteria have to control initiation at just one origin. Eukaryotic cells require many additional proteins to ensure each origin fires once and only once during each cell cycle. If an origin failed to fire once, then daughter cells might have harmful or perhaps lethal deletion

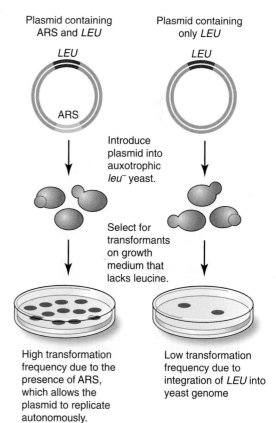

FIGURE 8.30 Method for detecting yeast autonomously replicating sequences (ARS). Two sets of plasmids, one with an ARS element and *LEU* and the other with just *LEU*, are introduced into *leu⁻* yeast cells. Then the transformed yeast cells are spread on agar growth medium lacking leucine. Many colonies appear when cells transformed with ARS plasmids are spread because these plasmids can replicate autonomously. In contrast, very few colonies appear when plasmids lacking ARS are spread and the colonies that do form contain cells in which *LEU* is integrated into the yeast genome.

8.3 Eukaryotic DNA Replication

mutations. If an origin fired more than once, the daughter cells would have multiple copies of a genetic region that could lead to harmful protein or metabolite accumulation or to harmful chromosomal rearrangements. Despite the increased complexity of the initiation stage of replication in eukaryotes, the result is similar: the establishment of a pair of replisomes that move in opposite directions so the replication process can proceed to the elongation stage.

In 1992 Stephen P. Bell and Bruce Stillman isolated a yeast protein that binds to ARS elements. This protein, called the origin recognition complex (ORC), consists of six different protein subunits, each of which has been conserved during evolution. At least five ORC subunits bind ATP and have ATPase activity. The genes that code for each ORC subunit have been cloned from yeast and other eukaryotes, facilitating genetic and biochemical studies. Two lines of evidence suggest that ORC serves as the eukaryotic initiator. First, yeast mutants with altered ORC subunits have a problem initiating chromosomal DNA synthesis and maintaining ARS plasmids. Second, nucleotide substitutions in the ARS, which reduce binding to ORC *in vitro*, also lower plasmid stability *in vivo*.

Yeast ORC, which is bound to DNA throughout the yeast life cycle, recruits a cell division cycle protein during late M/early G_1 phase. This recruited protein, called Cdc6, is yet another protein that binds ATP and has ATPase activity. The gene that codes for Cdc6, *CDC6*, was first identified by using a genetic screen designed to identify mutants that interfere with normal progression through the cell cycle. Yeast mutants that lack a functional Cdc6 do not replicate DNA. Cdc6 is degraded by proteases once initiation is complete, helping to ensure each origin fires just one time during the cell cycle. Two additional proteins are required to form the prereplication complex. These proteins, MCM2-7 (helicase) and Cdt1, combine to form a complex during M phase. MCM2-7 helicase, like bacterial DnaB helicase, has six subunits that each bind ATP and have ATPase activity (FIGURE 8.31).

Pol δ and Pol ε are primarily responsible for copying the lagging- and leading-strand templates, respectively.

Eukaryotes require three different DNA polymerases for normal replication fork propagation. Two of these, Pol α and Pol δ, are also essential for SV40 replication. The third, Pol ε, is not required for SV40 DNA replication and was therefore identified later than the other two. All three polymerases have four subunits in most eukaryotes studied to date. One known exception, yeast Pol δ, has only three subunits. FIGURE 8.32 shows how the subunits are organized in yeast Pol α, Pol δ, and Pol ε. Crystal structures are not yet available for these enzymes or their subunits.

1. *Pol α*: Pol α has limited processivity and lacks 3'→5' exonuclease activity. Its Pri1 subunit has primase activity that forms short RNA primers (8–12 nucleotides). Its largest subunit, Pol1, adds 15 to 25 deoxyribonucleotides to the 3'-end of the

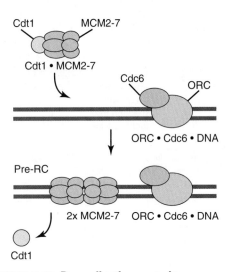

FIGURE 8.31 Prereplication complex formation. An ORC • Cdc6 • DNA complex is joined by a Cdt1 • MCM2 • 7 complex. MCM2-7's quaternary structure changes from a single hexamer before loading to a double hexamer after loading. Although not shown, ATP hydrolysis is required during at least two steps in prereplication complex formation: (1) Cdc6 hydrolyzes ATP while assisting Cdt1 to load MCM2-7 onto DNA. (2) ORC1 hydrolyzes ATP to release MCM2-7 from ORC so that the process of prereplication complex formation can be completed. ORC is shown in blue, Cdc6 in tan, Cdt1 in yellow, and MCM2-7 in purple. (Adapted from Evrin, C., et al. 2009. *Proc Natl Acad Sci USA* 106:20240–20245.)

primer to form initiator DNA. The two other Pol α subunits stabilize and regulate the catalytic subunits.

2. *Pol δ: Pol δ is responsible for lagging strand synthesis.* Its large subunit has both 5'→3' DNA polymerase activity and 3'→5' exonuclease activity. Pol δ has a low processivity value, which becomes quite high when Pol δ is tethered to DNA by PCNA.

3. *Pol ε: Pol ε is primarily responsible for catalyzing leading strand synthesis.* Although Pol ε is not required for SV40 DNA replication, eukaryotes require it to remain viable. Its largest subunit, Pol2, has both 5'→3' DNA polymerase activity and 3'→5' exonuclease activity. Pol ε has considerable binding affinity for single- and double-stranded DNA and may be able to function independently of PCNA. The discovery that Pol ε is required for eukaryotic DNA replication raises the question of why it is not also required for the SV40 DNA replication. The answer probably relates to the fact that SV40 replication uses the T antigen as both the initiator protein and helicase. Pol ε may need to interact with a eukaryotic enzyme or protein factor that is not used during SV40 DNA replication.

FIGURE 8.33 shows how several of the important enzymes and proteins required for eukaryotic DNA elongation may be organized at the replication fork. For simplicity some important proteins that participate in DNA replication are not included. There are two important

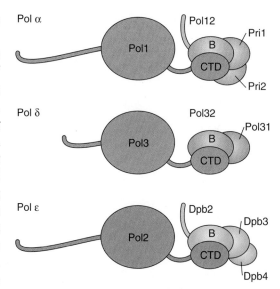

FIGURE 8.32 Subunit organization for yeast Pol α, Pol δ, and Pol ε. Each polymerase has a catalytic subunit, a conserved regulatory (or B) subunit, and other subunits with functions that remain to be determined. The catalytic subunits in Pol α, Pol δ, and Pol ε are called Pol1, Pol3, and Pol2, respectively. The regulatory subunits in Pol α, Pol δ, and Pol ε are called Pol12, Pol32, and Dpb2, respectively. Other subunits in each polymerase are as follows: Pri1 (primase activity) and Pri2 in Pol α; Pol31 in Pol δ; and Dpb3 and Dpb4 in Pol ε. (Modified from Klinge, S., et al., *EMBO J.*, 28 [2009]: 1978–1987.)

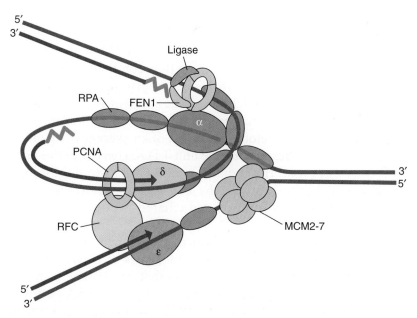

FIGURE 8.33 Simplified model for the organization of proteins and enzymes at the eukaryotic replication fork. Primer joined to initiator DNA is shown in green, DNA that is being replicated in blue, and newly synthesized DNA in red. The abbreviations used are as follows: α, Pol α; δ, Pol δ; ε, Pol ε; PCNA, proliferating cell nuclear antigen; FEN1, flap endonuclease; ligase, DNA ligase; RPA, replication protein A; and RFC, replication factor C. (Adapted from Stillman, B. 2008. *Mol Cell* 30:259–260.)

8.3 Eukaryotic DNA Replication

differences from SV40 DNA replication fork: (1) MCM2-7 serves as the helicase during eukaryotic DNA replication, whereas the T antigen performs this function during SV40 DNA replication, and (2) the eukaryotic DNA replication system uses Pol ε for leading strand synthesis and Pol δ for lagging strand synthesis, whereas the SV40 DNA replication system uses Pol δ for both leading and lagging strand formation. Leading- and lagging-strand synthesis are coordinate replication processes in eukaryotes just as they are in bacteria, but for convenience we consider the two processes separately.

1. Leading-strand synthesis begins after Pol α completes initiator DNA synthesis. Then Pol ε somehow replaces Pol α at the replication fork and continues leading-strand DNA synthesis until two replication bubbles meet.
2. Lagging-strand synthesis involves Okazaki fragment formation and maturation. As the replication machinery moves the replication bubble continues to grow and Pol α binding sites become available on the lagging strand. Pol α synthesizes initiator DNA. Then a switch occurs in which Pol δ replaces Pol α. The steps in Okazaki fragment maturation are the same as those described for SV40 replication.

Studies of the *Tetrahymena* and yeast telomeres suggest that a terminal transferase-like enzyme is required for telomere formation.

Eukaryotic DNA polymerases work by extending a preformed primer and therefore cannot copy the very end of a linear duplex. Molecular biologists needed to learn more about the DNA structure at the end of eukaryotic chromosomes before they could address this "end replication problem." Elizabeth H. Blackburn and Joseph G. Gall isolated telomeres from minichromosomes present in macronuclei of the ciliated protozoan *Tetrahymena thermophilia* in the late 1970s. They showed that these telomeres consist of an average of 50 tandem repeats of the simple hexanucleotide unit TTGGG and that this G-rich strand extends beyond the complementary C-rich strand, forming a 3′-overhang. Additional studies by Blackburn and others revealed that minichromosomes from related ciliated protozoa had similar telomeres.

Further studies were needed to learn whether telomeres are unique structures that are limited to minichromosomes of ciliated protozoa or are evolutionarily conserved structures that are present in chromosomes from other eukaryotes. Blackburn and Jack Szostak performed an experiment to address this issue in 1982. Szostak had been studying plasmid replication and recombination in yeast. Circular plasmids with an origin of replication replicate in yeast. Yeast will not support the replication of the same plasmid, however, if it is present as a linear duplex. If the ends of the linear duplex are homologous to yeast DNA, the plasmid is integrated into a yeast chromosome. If the linear duplex does not share a region of homology with a yeast chromosome, then yeast cells usually degrade the duplex. On rare occasions yeast cells somehow join the ends of the linear duplex to form a circular DNA molecule that can replicate in yeast. Blackburn and Szostak thought yeast cells might be tricked into replicating a linear plasmid with a *Tetrahymena* telomere at each end. Although they considered the possibility to be a long shot, the experiment was worth doing because it was easy to perform and a successful outcome would provide important insights into telomere function. They therefore constructed the desired recombinant plasmid and introduced it into yeast as shown in FIGURE 8.34. Then they isolated plasmids from the transformed yeast cells and subjected them to Southern blot analysis. About half the isolated plasmids migrated as linear duplexes, indicating that the *Tetrahymena* telomeres function in yeast. The fact that telomeres work in evolutionarily distant organisms suggests extraordinary functional conservation.

Szostak and Blackburn realized they could construct an ideal cloning vector for a yeast telomere by removing a *Tetrahymena* telomere

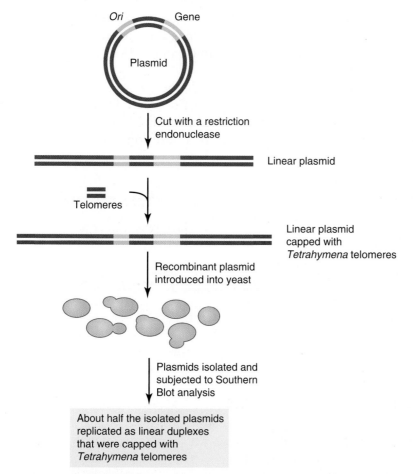

FIGURE 8.34 **Experiment showing that *Tetrahymena* telomeres function in yeast cells.** A circular plasmid with an origin of replication (*ori*) and a selectable marker (gene) was cut with a restriction endonuclease. A *Tetrahymena* telomere was added to each end of the linear plasmid and the recombinant plasmid introduced into yeast. Plasmids were isolated from the transformed cells and subjected to Southern blot analysis. Approximately half the isolated plasmids were linear duplexes with a *Tetrahymena* telomere at each end.

8.3 Eukaryotic DNA Replication

from one end of the recombinant linear plasmid. The steps involved in cloning the yeast telomere, shown in FIGURE 8.35, are as follows:

1. Remove a *Tetrahymena* telomere (t) from one end of the linear plasmid.
2. Digest yeast DNA with a restriction endonuclease to produce a DNA fragment pool with only rare fragments derived from chromosome ends.
3. Mix the DNA fragments with the modified linear plasmid and add DNA ligase to join the two types of fragments.
4. Transform yeast cells with the recombinant plasmids.

Most of the recombinant DNA molecules produced could not replicate in yeast because they were attached to a *Tetrahymena* telomere at one end and a non-telomeric yeast DNA fragment at the other end. The rare recombinant DNA molecule with a *Tetrahymena* telomere (t) at one end and a yeast telomere (y) at the other end, however, was able to replicate in yeast. Sequencing studies showed the yeast telomere has TG_{1-3} repeats. Szostak and Blackburn were able to construct a **yeast artificial chromosome** by attaching yeast telomeres to both ends of a long linear duplex that contained an origin of replication and a yeast centromere.

Szostak and Blackburn also made the remarkable discovery that yeast cells add about 200 bp of yeast telomere DNA to the ends of linear plasmids capped with *Tetrahymena* telomeres. They thought the most likely explanation for this addition was that yeast cells

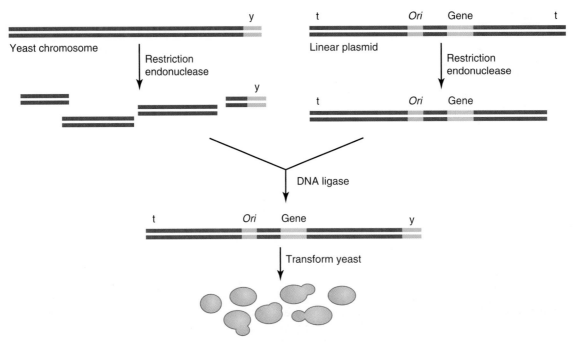

FIGURE 8.35 **Method for cloning yeast telomere.** A recombinant linear plasmid, constructed as shown in Figure 8.34, was digested with a restriction endonuclease to remove a *Tetrahymena* telomere (t) from one end. Yeast DNA (shown in red) was digested with a restriction endonuclease to produce a DNA fragment pool, which was mixed with the modified linear plasmid. DNA ligase was added to the mixture to join yeast fragments to the end of the linear plasmid that lacked a telomere and the recombinant DNA was introduced to yeast. Rare recombinant DNA molecules, which had a *Tetrahymena* telomere (t) at one end and a yeast telomere (y) at the other end, replicated in yeast.

have a terminal transferase-like enzyme that adds nucleotides to the ends of the linear duplex. It was difficult to see, however, how this enzyme could have the required specificity to add nucleotide repeats to the ends.

Telomerase uses an RNA template to add nucleotide repeats to chromosome ends.

Carol Greider joined Blackburn's laboratory as a graduate student in 1984 and set out to find the terminal transferase-like enzyme that had been predicted to add nucleotide repeats to chromosome ends. She succeeded by using an assay mixture that contained synthetic DNA oligonucleotide (TTGGGG)$_4$ as substrate, [^{32}P]dGTP, dTTP, and crude *Tetrahymena* cell extract. She followed the progress of the reaction by using gel electrophoresis to separate labeled DNA. The gel showed that hundreds of d(TTGGGG) repeats were added to the primer. Greider called the enzyme that catalyzes the addition of the nucleotide repeats telomerase. Telomerase plays a role in aging and cancer (**BOX 8.4 CLINICAL APPLICATION: TELOMERASE IN AGING AND CANCER**).

Further studies by Greider and Blackburn revealed that telomerase contains an RNA molecule essential for its function. They isolated the 159-nucleotide RNA subunit from *Tetrahymena* telomerase and discovered it contains an internal 5'-CAACCCCAA-3' sequence, which is complementary to the d(TTGGGG)$_n$ telomeric repeat in *Tetrahymena*. Subsequent studies revealed telomerase RNA molecules from other organisms differ in size, sequence, or both. Nevertheless, each telomerase RNA molecule has a sequence that is complementary to the telomeric DNA of the organism from which the telomerase is isolated. These results suggest that a short sequence within telomerase RNA acts as a template for telomeric DNA synthesis. Proof for this hypothesis comes from experiments that show telomerase activity is lost after RNase digestion but can be restored by removing RNase and adding back intact RNA. When the added RNA has a nucleotide substitution in its template sequence, the telomeric DNA produced has the predicted change. Genetic studies with *Saccharomyces cerevisiae* also indicate that telomerase RNA is essential for telomerase function because the gene that specifies telomerase RNA (*TLC1*) is essential for telomere maintenance. Telomerase therefore acts as an RNA-dependent DNA polymerase or reverse transcriptase that supplies its own template RNA molecule.

FIGURE 8.36 is a schematic for a possible telomerase reaction cycle. At the start of the reaction cycle telomerase binds to the DNA primer with the telomeric RNA template base pairing with the primer's 3'-end. Then the active site of the polymerase adds deoxyribonucleotides onto the primer 3'-end. Once this addition is complete, the enzyme moves to the new 3'-end of the primer strand. Telomerase's ability to carry out a single round of DNA synthesis is termed "nucleotide addition processivity." Realignment of the same enzyme for a second round of addition is termed "repeat addition processivity."

BOX 8.4: CLINICAL APPLICATION

Telomerase in Aging and Cancer

Somatic cells from humans and other multicellular animals divide a variable but limited number of times when cultured and then enter senescence, a state in which they are alive but no longer divide. Some animal cells, most notably germ line cells and most cancer cells, can replicate indefinitely without entering senescence and so are said to be immortal. Experiments performed by Howard Cooke in 1986 indicated that germ line cells have longer telomeres than somatic cells. Cooke speculated that telomerase might not be active in normal somatic human cells. Subsequent studies revealed that somatic cells have very low telomerase activity.

Studies by Andrea G. Bodnar and coworkers have shown that when vectors bearing genes for the telomerase reverse transcriptase subunit are introduced into human fibroblasts or retinal pigment epithelial cells, which have little if any telomerase activity, the cells continue to divide well beyond their normal lifespan without entering senescence and appear normal when viewed under the microscope. Moreover, their telomeres are much longer than those of normal human fibroblasts or retinal pigment epithelial cells. These results establish a causal relationship between telomere shortening and cellular senescence, suggesting the existence of a "mitotic clock" that regulates telomere size. Similar conclusions have been reached by studying yeast cells, which normally have active telomerase and so do not enter a senescent state. Yeast mutants that lack either telomerase RNA or telomerase reverse transcriptase, however, become senescent.

As indicated above, most cancer cells have telomerase activity, which raises the possibility that telomerase may serve as a target for chemotherapeutic agents. There is also reason, however, to suspect that telomerase will not prove to be an effective target. Maria Blasco and coworkers used recombinant DNA techniques to construct a strain of mice that lack the telomerase RNA gene. Remarkably, mice that are homologous for the missing gene appear to be normal and fertile. Continuous inbreeding has produced six generations of the mutant mice. Moreover, somatic cells from mutant mice are readily converted to tumor cells.

One must be cautious in applying the information obtained with mice to humans because mouse telomeres are on average 5 to 10 times longer than human telomeres. The mutant mouse experiments raise the possibility of alternative mechanisms for maintaining telomeres such as genetic recombination. Most organisms use telomerase to maintain their telomeres; however, a few organisms do not require telomerase to solve the end-replication problem.

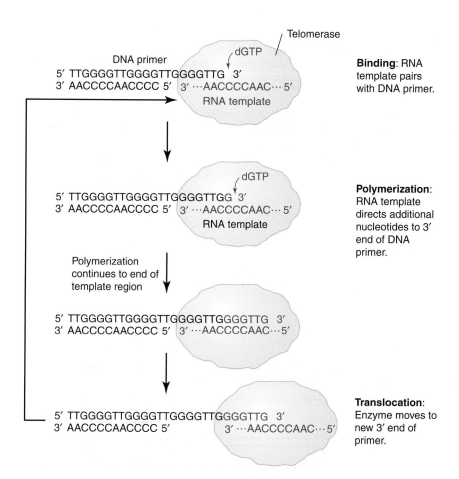

FIGURE 8.36 Telomerase reaction cycle. The reactions shown are those revealed by studying the *Tetrahymena* telomerase. The cycle consists of three steps, binding, polymerization, and translocation, to reposition the RNA template. (Adapted from Chatziantoniou, V.D. 2001. *Pathol Oncol Res* 7:161–170.)

Telomerase plays an important role in solving the end-replication problem.

Telomere structure and telomerase's role in telomere synthesis provide important new insights into the end-replication problem. Semiconservative replication initiates at origins internal to the telomeric repeats and the replication forks move toward the chromosome ends (FIGURE 8.37a). The C-rich strand (denoted by C) is always assembled by lagging-strand synthesis and the G-rich strand (denoted by G) is always assembled by leading-strand synthesis (Figure 8.37b). Before the structure of the telomere was elucidated, investigators thought lagging-strand synthesis might present a problem because removing the RNA primer at the 5'-end of the last Okazaki fragment would produce a shortened new C-rich strand. New C-rich strand synthesis, however, does not necessarily present a problem for the replication machinery because the G-rich stand

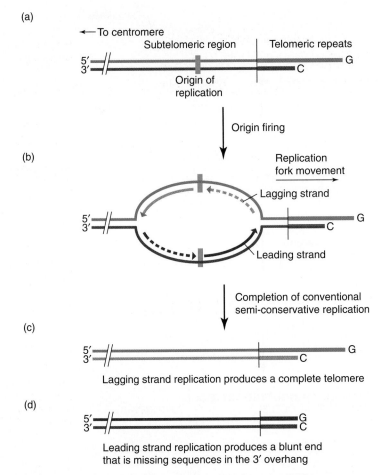

FIGURE 8.37 Leading strand problem. (a) The telomere G-rich strand (G, shown in light blue) ends in a 3'-single strand that overhangs the C-rich strand (C, shown in dark blue). (b) Semiconservative replication initiates at origins internal to the telomeric repeats and the replication forks move toward the chromosome end. (c) Synthesis of lagging strands does not necessarily present a problem for the replication machinery because the RNA primer at the end of the last Okazaki fragment can be removed without the loss of information. (d) Synthesis of leading strands introduces a problem because replication stops at the 5'-end of the C-rich template to produce a blunt end with a resulting loss of sequence information. (Adapted from Chakhparonian, M., and Wellinger, R.J. 2003. *Trends Genet* 19:439–446.)

has a 3'-overhang. The RNA primer at the end of the last Okazaki fragment therefore can be removed without the loss of information (Figure 8.37c).

In contrast, leading-strand replication presents a serious problem because replication of the new G-rich strand will stop at the 5'-end of the C-rich template strand to produce a blunt end with a resulting loss of sequence information (Figure 8.37d). Unless something is done to restore the lost sequence, this processing step will cause the telomere to become shorter. If this shortening process were to continue through several more replication cycles, then the telomere would be

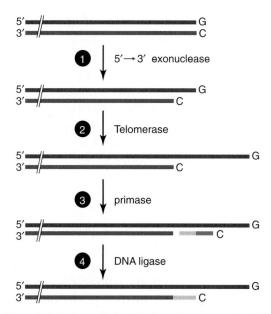

FIGURE 8.38. **Hypothesis for solving the leading strand problem.**

lost entirely and the chromosome would no longer be able to survive as an independent replicating unit.

Although telomerase cannot extend a DNA chain from a blunt end, the pathway shown in FIGURE 8.38 shows how the leading strand can be extended. The steps in this pathway are as follows: (1) a strand-specific 5'→3' exonuclease removes nucleotides from the 5'-end of the C-rich strand to generate a G-rich 3'-overhang, (2) telomerase extends the G-rich 3'-overhang generated by the exonuclease, and (3) the newly synthesized region serves as a template for standard semiconservative replication, restoring the 5'-end of C-rich strand.

8.4 Replication Coupled Chromatin Synthesis

Chromatin disassembly and reassembly are tightly coupled to DNA replication.

The DNA synthetic machinery must be able to make direct contact with its DNA substrate for replication to take place. However, the basic unit of structure in eukaryotic chromatin, the nucleosome, blocks such direct contact. It therefore seems reasonable to propose that cells can disassemble chromatin just before the replication fork and reassemble the chromatin just after the replication fork. Electron microscopy studies using the SV40 minichromosome as a model system are consistent with this hypothesis. These studies show that a stretch of about 300 bp of nucleosome-free (naked) DNA is present just before the replication fork and another stretch of at least 250 bp of naked DNA is present just after the replication fork.

Histone chaperones play a central role in replication coupled chromatin formation by preventing nonspecific interactions between positively charged histones and negatively charged DNA. They thereby permit ordered chromatin disassembly or assembly without themselves becoming a permanent part of nucleosomes. **Chromatin modifiers** also participate in chromatin formation by adding acetyl, methyl, phosphate, or other groups to histones or removing these groups from modified histones. FIGURE 8.39 summarizes how some histone chaperones and chromatin modifiers interact during replication coupled chromatin formation. For convenience, we divide replication coupled chromatin formation into two stepwise processes, chromatin disassembly and chromatin assembly.

Chromatin disassembly begins when the histone chaperone FACT (facilitates chromatin transcription) removes H2A • H2B dimers from nucleosomes. At least one other histone chaperone also appears able to perform this function. Then the MCM2-7 helicase releases the (H3 • H4)$_2$ tetramer that is left behind as it unwinds the double-double stranded DNA. The MCM2-7 helicase probably binds the released tetramer, ensuring it remains close to the replicating fork. The histone chaperone Asf1 (anti-silencing factor) can split an (H3 • H4)$_2$ tetramer into 2 dimers (see below).

Histones from disassembled nucleosomes ("old" histones) and newly synthesized histones ("new" histones) are used to assemble nucleosomes on newly formed daughter duplexes. The stable form for newly synthesized and disassembled H2A and H2B is the H2A • H2B dimer. The stable form of newly synthesized H3 and H4 is the H3 • H4 dimer. There is uncertainty, however, about the stable form of H3 and H4 derived from nucleosome disassembly. The observation that Asf1 can split (H3 • H4)$_2$ tetramers obtained from disassembled nucleosomes into H3 • H4 dimers suggests the dimer is the stable form. Metabolic labeling studies, however, suggest the (H3 • H4)$_2$ tetramer is the stable form because nearly all nucleosomes assembled on newly formed daughter duplexes contain either all newly synthesized H3 and H4 histones or all old H3 and H4 histones. It is difficult to explain this observation if nucleosome disassembly produces H3 • H4 dimers because then a new H3 • H4 dimer would be expected to combine with an old H3 • H4 dimer to form a mixed tetramer. As shown in Figure 8.39, new and old (H3 • H4)$_2$ tetramers appear to distribute randomly on the two new daughter DNA duplexes.

During the first stage of chromatin assembly, the histone chaperones CAF-1 (chromatin assembly factor) and Asf1 appear to work together to deposit an (H3 • H4)$_2$ tetramer (or two H3 • H4 dimers) on newly replicated DNA. Almost all newly formed H3 is acetylated on lysine 56. This acetylation appears to help recruit H3 • H4 dimers to the histone chaperones that are positioned at the replication fork to promote H3 • H4 dimer assembly into DNA. Rtt106, another histone chaperone that binds H3 • H4, also appears to participate in nucleosome assembly, but its specific contribution remains to be determined. During the next step in chromatin assembly, FACT

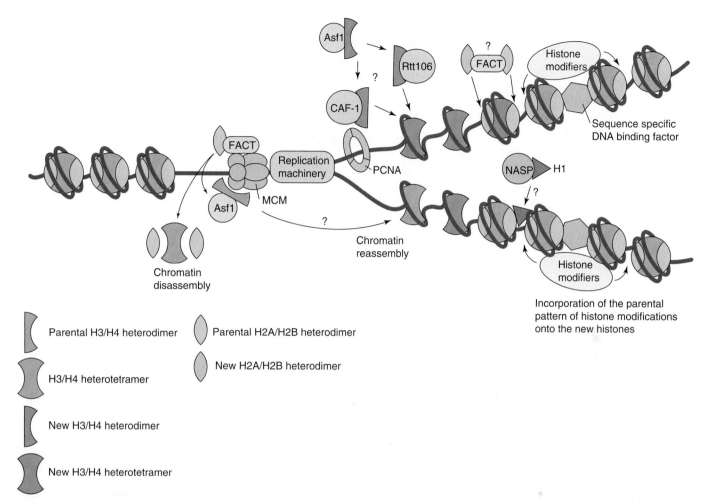

FIGURE 8.39 **Replication coupled chromatin disassembly and assembly.** *Chromatin disassembly:* Chromatin disassembly begins when the histone chaperone FACT (facilitates chromatin transcription) removes H2A • H2B dimers from nucleosomes. Then the MCM2-7 helicase releases the (H3 • H4)$_2$ tetramer that is left behind as it unwinds the double-double stranded DNA. The MCM2-7 helicase probably binds the released tetramer, ensuring it remains close to the replicating fork. *Chromatin assembly*: The histone chaperones CAF-1 (chromatin assembly factor) and Asf1 appear to work together to deposit an (H3 • H4)$_2$ tetramer (or two H3 • H4 dimers) on newly replicated DNA during the first step in nucleosome assembly. Rtt106, another histone chaperone that binds H3 • H4, also appears to participate in nucleosome assembly but its specific contribution remains to be determined. During the next step in nucleosome assembly, FACT appears to deposit H2A • H2B dimers onto (H3 • H4)$_2$ tetramers. Finally, yet another histone chaperone, NASP (nuclear autoantigenic sperm protein), helps deposit histone H1 onto linker DNA between adjacent nucleosomes. Question marks indicate uncertainty about participation in a specific step. (Reprinted from *The Lancet*, Ransom, M., Dennehey, B. K., and Tyler, J. K., *Cell* 140 [2010]: 183–195, with permission from Elsevier.)

deposits H2A • H2B dimers onto (H3 • H4)$_2$ tetramers. Finally, yet another histone chaperone, NASP (nuclear autoantigenic sperm protein), helps deposit histone H1 onto linker DNA between adjacent nucleosomes.

Although *in vitro* studies show histone factors can work together to produce nucleosomes on duplex DNA, the nucleosomes are irregularly spaced. ATP-dependent **chromatin remodelers** such as ACF (ATP-utilizing chromatin assembly and remodeling factor) and CHRAC (chromatin accessibility complex) are required to generate regular physiological nucleosome spacing. Both chromatin remodelers contain a common subunit, ISWI (imitation switch), which can promote nucleosome spacing by itself, although not as efficiently as when it is part of the chromatin remodeler.

Questions and Problems

1. Briefly define, describe, or illustrate each of the following.
   a. Semiconservative model of replication
   b. Conservative model of replication
   c. Dispersive model of replication
   d. Unidirectional replication
   e. Bidirectional replication
   f. Replication bubble
   g. Discontinuous model of replication
   h. Semidiscontinuous model of replication
   i. Replication fork
   j. Okazaki fragment
   k. Leading strand
   l. Lagging strand
   m. Replicon model
   n. Initiator protein
   o. Replicator
   p. cis-acting
   q. Origin of replication
   r. oriC
   s. DnaA boxes
   t. DUE
   u. Processivity
   v. Distributive replication
   w. Core polymerase
   x. Clamp loader
   y. Sliding clamp
   z. Trombone model of replication
   aa. Replisome
   bb. Collision release
   cc. Premature release or signaling release
   dd. Ter site
   ee. Initiator DNA
   ff. Autonomously replicating sequence elements
   gg. Histone chaperones
   hh. Chromatin modifiers
   ii. Chromatin remodelers

2. Briefly describe experiments that show that DNA replication is
   a. Semiconservative
   b. Semidiscontinuous
   c. Bidirectional

3. During replication elongation both new DNA strands grow in the same direction at the replication fork. How is this type of replication possible if all known DNA polymerases catalyze DNA synthesis in a 5'→3' direction?

4. Bacterial DNA synthesis starts at oriC.
   a. How can you isolate oriC so it can be subjected to in vitro studies?
   b. Briefly describe the structural features of oriC that are important for replication initiation.

5. DnaA, DnaB, DnaC, and DnaG make important contributions to the bacterial replication initiation process.
   a. What contribution(s) does each of these proteins make to the initiation process?
   b. Which of these proteins is also necessary for DNA replication elongation?

6. DNA replication was monitored in a temperature-sensitive DnaA mutant. DNA replication takes place normally at 20°C and continues for several minutes after the culture temperature is raised to 42°C. How do you explain the continued DNA synthesis?

7. Fluoroquinolones are antibiotics that inhibit DNA gyrase. What effect would you expect these antibiotics to have on bacterial DNA replication? Explain your answer.

8. A novel drug has recently been demonstrated to inhibit the NAD$^+$-dependent bacterial DNA ligase. What effect would you expect this drug to have on bacterial DNA replication in general and Okazaki fragment formation in specific?

9. DNA replication was monitored in a temperature-sensitive DnaG mutant. DNA replication takes place normally at 20°C but stops immediately after raising the culture temperature to 42°C. How do you explain this rapid inhibition?
10. DNA polymerase III holoenzyme consists of three different kinds of subassemblies. What are these subassemblies called and what is the function of each subassembly?
11. How does the trombone model of replication explain that a single DNA polymerase III holoenzyme can catalyze the synthesis of both the leading and lagging strands of DNA?
12. What functions do *ter* sites and the Tus protein play in the termination of *E. coli* DNA replication?
13. What role does the T antigen play in SV40 DNA replication?
14. What proteins are present at the SV40 DNA replication fork? What is the function of each of these proteins?
15. How does eukaryotic chromosomal DNA replication differ from *E. coli* DNA replication?
16. What functions do Pol α, Pol δ, and Pol ε have in eukaryotic DNA replication?
17. Which eukaryotic enzyme serves the same role as bacterial primase by forming the primer required for lagging strand synthesis?
18. Why is telomerase required for eukaryotic chromosomal DNA synthesis but not for SV40 chromosomal synthesis?
19. What evidence supports the idea that the RNA molecule present in telomerase is essential for telomere synthesis?
20. Describe an experimental method that can be used to clone a yeast telomere?
21. What contributions do histone chaperones, chromatin modifiers, and chromatin remodelers make to eukaryotic chromosome synthesis?

Suggested Reading

Overview

Baker, T. A., and Bell, S. P. 1998. Polymerases and the replisome: machines within machines. *Cell* 92:295–305.

Benkovic, S. J., Valentine, A. M., and Salinas, F. 2001. Replisome-mediated DNA replication. *Annu Rev Biochem* 70:181–208.

Kornberg, A., and Baker, T. 1992. *DNA Replication* (2nd ed). New York: W. H. Freeman.

Johnson, A., and O'Donnell, M. 2005. Cellular DNA replicases: components and dynamics at the replication fork. *Annu Rev Biochem* 74:283–315.

General Features of DNA Replication

Cairns, J. 1963. The chromosome of *Escherichia coli*. *Cold Spring Harb Symp Quant Biol* 18:43–46.

Mesleson, M., and Stahl, F. 1958. The replication of DNA in *Escherichia coli*. *Proc Natl Acad Sci USA* 44:671–682.

Okazaki, R., Okazaki, T., Sakabe, K., et al. 1968. Mechanism of DNA chain growth. I. Possible discontinuity and unusual secondary structures of newly synthesized chains. *Proc Natl Acad Sci USA* 59:598–605.

Bacterial DNA Initiation

Cunningham, E. L., and Burger, J. M. 2005. Unraveling the early steps of prokaryotic replication. *Curr Opin Struct Biol* 15:68–76.

Erzberger, J. P., Pirruccello, M. M., and Berger, J. M. 2002. The structure of bacterial DnaA: implications for general mechanisms underlying DNA replication initiation. *EMBO J* 21:4763–4773.

Kaguni, J. M. 2006. DnaA: controlling the initiation of bacterial DNA replication and more. *Annu Rev Microbiol* 60:351–371.

Katayama, T., Ozaki, S., Keyamura, K., and Fujimitsu, K. 2010. Regulation of the replication cycle: conserved and diverse regulatory systems for DnaA and *oriC*. *Nat Rev Microbiol* 8:163–170.

Leonard, A. C., and Grimwade, J. E. 2010. Regulating DnaA complex assembly: it is time to fill the gaps. *Curr Opin Microbiol* 13:766–772.

Messer, W. 2002. The bacterial replication initiator DnaA. DnaA and *oriC*, the bacterial mode to initiate DNA replication. *FEMS Microbiol Rev* 26:355–374.

Messer, W. 2005. Prokaryotic replication origins: Structure and function in the initiation of DNA replication. *Encyclopedia of Life Sciences*. pp. 1–7. Hoboken, NJ: John Wiley & Sons.

Mott, M. L., and Berger, J. M. 2007. DNA replication initiation: mechanisms and regulation in bacteria. *Nat Rev Microbiol* 5:343–354.

Ozaki, S., and Katayama, T. 2009. DnaA structure, function, and dynamics in the initiation at the chromosomal origin. *Plasmid* 62:71–78.

Robinson, N. P., and Bell, S. D. 2005. Origins of DNA replication in the three domains of life. *FEBS J* 272:3757–3766.

Bacterial DNA Elongation

Bailey, S., Wing, R. A., and Steitz, T. A. 2006. The structure of *T. aquaticus* DNA polymerase III is distinct from eukaryotic replicative DNA polymerase. *Cell* 126:893–904.

Bamabara, R. A., Fay, P. J., and Mallaber, L. M. 1995. Methods of analyzing processivity. *Methods Enzymol* 262:270–280.

Bailey, S., Eliason, W. K., and Steitz, T. A. 2007. The crystal structure of the *Thermus aquaticus* DnaB helicase monomer. *Nucleic Acids Res* 35:4728–4736.

Bloom, L. F., and Goodman, M. F. 2001. Polymerase processivity: measurement and mechanisms. *Encyclopedia of Life Sciences*. pp. 1–6. Hoboken, NJ: John Wiley & Sons.

Bowman, G. D., Goedken, E. R., Kazmirski, S. L., et al. 2005. DNA polymerase clamp loaders and DNA recognition. *FEBS Lett* 579:863–867.

Bruck, I., and O'Donnell, M. 2001. The ring-type polymerase sliding clamp family. *Genome Biol* 2:1–3001.

Coman, M. M., Jin, M., Ceapa, R., et al. 2004. Dual functions, clamp opening and primer-template recognition, define a key clamp loader subunit. *J Mol Biol* 342:1457–1469.

Corn, J. E., and Berger, J. M. 2006. Regulation of bacterial priming and daughter strand synthesis through helicase-primase interactions. *Nucleic Acids Res* 34:4082–4088.

Corn, J. E., Pelton, J. G., and Berger, J. M. 2008. Identification of a DNA primase template tracking site redefines the geometry of primer synthesis. *Nat Struct Mol Biol* 15:163–169.

Davey, M. J., Jeruzalmi, D., Kuriyan, J., and O'Donnell, M. 2002. Motors and switches: AAA+ machines within the replisome. *Nat Rev Mol Cell Biol* 3:1–10.

Dixon, N. E. 2009. Prime-time looping. *Nature* 462:854–855.

Ellison, V., and Stillman, B. 2001. Opening of the clamp: an intimate view of an ATP-driven biological machine. *Cell* 106:655–660.

Georgescu, R. E., Kim, S.-S., Yurieva, O., et al. 2008. Structure of a sliding clamp on DNA. *Cell* 132:43–54.

Hamdan, S. M., and Richardson, C. C. 2009. Motors, switches, and contacts in the replisome. *Annu Rev Biochem* 78:205–243.

Indiani, C., and O'Donnell, M. 2003. Mechanism of the δ wrench in opening the β sliding clamp. *J Biol Chem* 278:40272–40281.

Indiani, C., and O'Donnell, M. 2006. The replication clamp-loading machine at work in the three domains of life. *Nat Rev Mol Cell Biol* 7:751–761.

Jeruzalmi, D., O'Donnell, M., and Kuriyan, J. 2002. Clamp loaders and sliding clamps. *Curr Opin Struct Biol* 12:217–224.

Kuchta, R. D., and Stengel, G. 2010. Mechanisms and evolution of DNA primases. *Biochim Biophys Acta* 1804:1180–1189.

Langston, L. D., and O'Donnell, M. 2006. DNA replication: keep moving and don't mind the gap. *Mol Cell* 23:155–160.

Langston, L. D., Indiani, C., and O'Donnell, M. 2009. Whither the replisome. *Cell Cycle* 8:2686–2691.

López deSaro F., Georgescu, R. E., Leu, F., and O'Donnell, M. 2004. Protein trafficking on sliding clamps. *Philos Trans R Soc Lond Biol Sci* 359:25–30.

Lovett, S. T. 2007. Polymerase switching in DNA replication. *Mol Cell* 27:523–526.

Marians, K. J. 2008. Understanding how the replisome works. *Nat Struct Mol Biol* 15:125–127.

McInerney, P., Johnson, A., Katz, F., and O'Donnell, M. 2007. Characterization of a triple DNA polymerase replisome. *Mol Cell* 27:527–538.

O'Donnell, M. 2006. Replisome architecture and dynamics in *Escherichia coli*. *J Biol Chem* 281:10653–10656.

O'Donnell, M., Jeruzalmi, D., and Kuriyan, J. 2001. Clamp loader structure predicts the architecture of DNA polymerase III holoenzyme and RFC. *Curr Biol* 11:R935–R946.

Pomerantz, R. T., and O'Donnell, M. 2007. Replisome mechanics: insights into a twin DNA polymerase machine. *Trends Microbiol* 15:156–164.

Soultanas, P. 2005. The bacterial helicase-primase interaction: a common structural/functional module. *Structure* 13:839–844.

Tanaka, T., and Masai, H. 2010. Bacterial replication fork: Synthesis of lagging strand. *Encyclopedia of Life Sciences*. pp. 1–5. Hoboken, NJ: John Wiley & Sons

Wing, R. A., Bailey, S., and Steitz, T. A. 2008. Insights into the replisome form the structure of a ternary complex of DNA polymerase III α-subunit. *J Mol Biol* 382:859–869.

Yao, N. Y., and O'Donnell, M. 2008. Replisome structure and conformational dyanamics underlie fork progression past obstacles. *Curr Opin Cell Biol* 21:336–343.

Bacterial Replication Termination

Bastia, D., Zzaman, S, Krings. G., et al. 2008. Replication termination mechanism as revealed by *Tus*-mediated polar arrest of sliding helicase. *Proc Natl Acad Sci USA* 105:12831–12836.

Bussiere, D. E., and Bastia, D. 1999. Termination of DNA replication of bacterial and plasmid chromosomes. *Mol Microbiol* 31:1611–1618.

Duggin, I. G., and Wilce, J. A. 2005. Termination of replication in bacteria. *Encyclopedia of Life Sciences*. pp. 1–7. Hoboken, NJ: John Wiley & Sons.

Kaplan, D. L., and Bastia, D. 2009. Mechanism of polar arrest of a replication fork. *Mol Microbiol* 72:279–285.

Mulcair, M. D., Schaeffer, P. M., Oakley, A. J., et al. 2006. A molecular mousetrap determines polarity of termination of DNA replication in *E. coli*. *Cell* 125:1309–1319.

Mulugu, S., Potnis, A., Shamsuzzaman, et al. 2001. Mechanism of termination of DNA replication of *Escherichia coli* involves helicase-contrahelicase interaction. *Proc Natl Acad Sci USA* 98:9569–9574.

Sherratt, D. J., Lau, I. F., and Barre, F. X. 2001. Chromosome segregation. *Curr Opin Microbiol* 4:653–659.

The SV40 DNA Replication System

Fanning, E., and Zhao, K. 2009. SV40 DNA replication: from the A gene to a nanomachine. *Virology* 384:352–359.

Herendeen, D., and Kelly, T. J. 1996. SV40 DNA replication J. Julian Blow (ed). *Eukaryotic DNA Replication.* Oxford, UK: Oxford University Press.

Li, D., Zhao, R., Lileystrom, W., et al. 2003. Structure of the replicative helicase of the oncoprotein SV40 large tumor antigen. *Nature* 423:512–518.

Eukaryotic Replication Initiation

Aladjem, M. I., and Fanning, E. 2004. The replicon revisited: an old model learns new tricks in metazoan chromosomes. *EMBO Rep* 5:666–691.

Bell, S. P. 2002. The origin recognition complex: from simple origins to complex functions. *Genes Dev* 16:659–672.

Bell, S. P., and Dutta, A. 2002. DNA replication in eukaryotic cells. *Annu Rev Biochem* 71:333–374.

Bielinsly, A.-K., and Gerbi, S. A. 2001. Where it starts: eukaryotic origins of DNA replication. *J Cell Sci* 114:643–651.

Bryant, J. A. 2010. Replication of nuclear DNA. *Progr Botany* 71:25–60.

Costa, A., and Onesti, S. 2009. Structural biology of MCM helicases. *Crit Rev Biochem Mol Biol* 44:326–342.

Cvetic, C., and Walter, J. C. 2005. Eukaryotic origins of DNA replication: could you please be more specific? *Cell Dev Biol* 16:343–353.

Françon, P., and Méchali, M. 2006. DNA replication origins. *Encyclopedia of Life Sciences.* pp. 1–5. Hoboken, NJ: John Wiley & Sons.

Ge, X. Q., and Blow, J. J. 2009. Conserved steps in eukaryotic DNA replication. L. S. Cox (ed). *Molecular Themes in DNA Repair.* Cambridge, UK: Royal Society of Chemistry.

Kawakami, H., and Katayama, T. 2010. DnaA, ORC, and CDC6: similarity beyond the domains of life and diversity. *Biochem Cell Biol* 88:49–62.

Kreigstein, H. J., and Hogness, D. S. 1974. Mechanism of DNA replication in *Drosophila* chromosomes: structure of replication forks and evidence for bidirectionality. *Proc Natl Acad Sci USA* 71:135–139.

Pospiech, H., Grosse, F., and Pisani, F. M. 2010. The initiation step of eukaryotic DNA replication. *Subcell Biochem* 50:79–104.

Remus, D., and Diffley, F. X. 2009. Eukaryotic DNA replication: lock and load, then fire. *Curr Opin Cell Biol* 21:771–777.

Scholefield, G., Veening, J.-W., and Murray, H. 2011. DnaA and ORC: more than DNA replication initiators. *Trends Cell Biol* 21:188–194.

Schwob, E. 2004. Flexibility and governance in eukaryotic DNA replication. *Curr Opin Microbiol* 7:680–690.

Sclafani, R. A., and Holzen, T. M. 2007. Cell cycle regulation of DNA replication. *Annu Rev Genet* 41:237–280.

Toone, W. M., Aerne, B. L., and Morgan, B. A. 1997. Getting started: regulating the initiation of DNA replication in yeast. *Annu Rev Microbiol* 51:125–149.

Weinreich, M. DeBeer, M. A. P., and Fox, C. A. 2003. The activities of eukaryotic replication origins in chromatin. *Biochim Biophys Acta* 1677:142–157.

Wigley, D. B. 2009. ORC proteins: marking the start. *Curr Opin Struct Biol* 19:72–78.

Eukaryotic Replication Elongation

Bowman, G. D., O'Donnell, M., and Kuriyan, J. 2004. Structural analysis of a eukaryotic sliding DNA clamp-clamp loader complex. *Nature* 420:724–730.

Cerritelli, S., and Crouch, R. J. 2009. Ribonuclease H: the enzymes in eukaryotes. *FEBS J* 276:1494–1505.

Ellenberger, T., and Tomkinson, A. E. 2008. Eukaryotic DNA ligases: structural and function insights. *Annu Rev Biochem* 77:313–338.

Garg, P., and Burgers, M. J. 2005. DNA polymerases that propagate the eukaryotic DNA replication fork. *Crit Rev Biochem Mol Biol* 40:115–128.

Gilbert, D. M. 2004. In search of the holy replicator. *Nat Rev Mol Cell Biol* 3:1–8.

Indiani, C., and O'Donnell, M. O. 2006. The replication clamp-loading machine at work in the three domains of life. *Nat Rev Mol Cell Biol* 7:751–761.

Klinge, S., Núñez-Ramirez, R., Llorca, O., and Pellegrini, L. 2009. 3D architecture of DNA Pol α reveals the functional core of multi-subunit replicative polymerases. *EMBO J* 28:1978–1987.

Kunkel, T. A., and Burgers, P. M. 2008. Dividing the workload at a eukaryotic replication fork. *Trends Cell Biol* 18:521–527.

Labib, K., and Gambus, A. 2007. A key role for the GINS complex at DNA replication forks. *Trends Cell Biol* 17:271–278.

Li, J. J., and Kelly, T. J. 1984. Simian virus 40 DNA replication in vitro. *Proc Natl Acad Sci USA* 81:6973–6977.

Liu, H., Kao, H.-I., Bambara, R. A. 2004. Flap endonuclease I: a central component of DNA metabolism. *Annu Rev Biochem* 73:589–615.

Maga, G., and Hübscher, U. 2003. Proliferating cell nuclear antigen (PCNA): a dancer with many partners. *J Cell Sci* 116:3051–3060.

McElhinny, S. A. N., Gordenin, E. A., Stith, C. M., Burgers, P. M. J., and Kunkel, T. A. 2008. Division of labor at the eukaryotic replication fork. *Mol Cell* 30:137–144.

Moldovan, G.-L., Pfander, B., Jentsch, S. 2007. PCNA, the maestro of the replication fork. *Cell* 129:665–679.

Pavlov, Y. I., and Shcherbakova, P. V. 2010. DNA polymerases at the eukaryotic replication fork—20 years later. *Mutat Res* 685:45–53.

Pursell, Z. F., and Kunkel, T. A. 2008. DNA polymerase ε: a polymerase of unusual size (and complexity). *Prog Nucleic Acid Res Mol Biol* 82:101–145.

Stillman, B. 2008. DNA polymerases at the replication fork in eukaryotes. *Mol Cell* 30:259–260.

Waga, S., and Stillman, B. 1998. The DNA replication fork in eukaryotic cells. *Annu Rev Biochem* 67:721–751.

Walther, A. P., and Wold, M. S. 2001. Eukaryotic replication fork. *Encyclopedia of Life Sciences*. pp. 1–8. London: Nature.

Telomere and Telomerase

Autexier, C., and Lue, N. F. 2006. The structure and function of telomerase reverse transcriptase. *Annu Rev Biochem* 75:493–517.

Blackburn, E. H. 2005. Telomeres and telomerase: their mechanisms of action and the effects of altering their functions. *FEBS Lett* 579:859–862.

Blackburn, E. H. 2001. Telomeres. *Encyclopedia of Life Sciences*. pp. 1–7. London: Nature.

Blackburn, E. H., Greider, C. W., and Szostak, J. W. 2006. Telomeres and telomerase: the path from maize, *Tetrahymena* and yeast to human cancer and aging. *Nat Med* 12:1133–1138.

Cech, T. R. 2004. Beginning to understand the end of the chromosome. *Cell* 116:273–279.

Chakhparonian, M., and Wellinger, R. J. 2003. Telomere maintenance and DNA replication: how closely are these two connected? *Trend Genet* 19:439–446.

Chan, S. R. W. L., and Blackburn, E. H. 2004. Telomeres and telomerase. *Philos Trans R Soc Lond B Biol Sci* 359:109–121.

Chatziantoniou, V. D. 2001. Telomerase: biological function and potential role in cancer management. *Pathol Oncol Res* 7:161–170.

de Lange, T. 2006. Lasker laurels for telomerase. *Cell* 126:1017–1020.

Gilson, E., and Géli, V. 2007. How telomeres are replicated. *Nat Rev Mol Cell Biol* 8:825–838.

Gilson, E., and Ségal-Benirdjian, E. 2010. The telomere story or the triumph of an open-minded research. *Biochimie* 92:321–326.

Harrington, L. 2003. Biochemical aspects of telomerase function. *Cancer Lett* 194:139–154.

Lansdorp, P. M. 2005. Major cutbacks at chromosome ends. *Trends Biochem Sci* 30:388–395.

Lingner, J., Cooper, J. P., and Cech, T. R. 1995. Telomerase and DNA end replication: no longer a lagging strand problem? *Science* 269:1533–1534.

Lue, N. F. 2004. Adding to the ends: what makes telomerase processive and how important is it? Bioessays 26:955–962.

Makarov, V. L., Hirose, Y., and Langmore, J. P. 1997. Long G tails at both ends of human chromosomes suggest a C strand degradation mechanism for telomere shortening. *Cell* 88:657–666.

Osterhage, J. L., and Friedman, K. L. 2009. Chromosome end maintenance by telomerase. *J Biol Chem* 284:16061–16065.

Oulette, M. M., and Choi, K. H. 2007. Telomeres and telomerase in ageing and cancer. *Encyclopedia of Life Sciences*. pp. 1–6. Hoboken, NJ: John Wiley & Sons.

Pardue, M.-L., and DeBaryshe, G. 2001. Telomeres in cell function: cancer and ageing. *Encyclopedia of Life Sciences*, pp. 1–6. London: Nature.

Sfeir, A. J., Chai, W., Shay, J. W., and Wright, W. E. 2005. Telomere-end processing: the terminal nucleotides of human chromosomes. *Mol Cell* 18:131–138.

Sekaran, V. G., Soares, J., and Jarstfer, M. B. 2010. Structures of telomerase subunits provide functional insights. *Biochim Biophys Acta* 1804:1190–1201.

Shore, D., and Bianchi, A. 2009. Telomere length regulation: coupling DNA processing to feedback regulation of telomerase. *EMBO J* 28:2309–2322.

Theimer, C. A., and Feigon, J. 2006. Structure and function of telomerase RNA. *Curr Opin Struct Biol* 16:307–318.

Vega, L. R., Mateyak, M. K., and Zakian, V. A. 2003. Getting to the end: telomerase access in yeast and humans. *Nat Rev Mol Cell Biol* 4:948–959.

Wellinger, R. J., Ethier, K., Labrecque, P., and Zakian, V. A. 1996. Evidence for a new step in telomere maintenance. *Cell* 85:423–433.

Replication Coupled Chromatin Synthesis

Corpet, A., and Almouzni, G. 2008. Making copies of chromatin: the challenge of nucleosomal organization and epigenetic information. *Trends Cell Biol* 19:29–41.

Groth, A. 2009. Replicating chromatin: a tale of histones. *Biochem Cell Biol* 87:51–63.

Groth, A. Rocha, W., Verreault, A., and Almouzni, G. 2007. Chromatin challenges during DNA replication and repair. *Cell* 128:721–733.

Jasencakova, Z., and Groth, A. 2010. Restoring chromatin after replication: how new and old histone marks come together. *Semin Cell Dev Biol* 21:231–237.

Krebs, J. E., and Peterson, C. L. 2000. Understanding "active" chromatin: a historical perspective of chromatin remodelling. *Crit Rev Eukaryot Gene Expr* 10:1–12.

Ransom, M., Dennehey, B. K., and Tyler, J. K. 2010. Chaperoning histones during replication and repair. *Cell* 140:183–195.

Tyler, J. K. 2002. Chromatin assembly. *Eur J Biochem* 269:2268–2274.

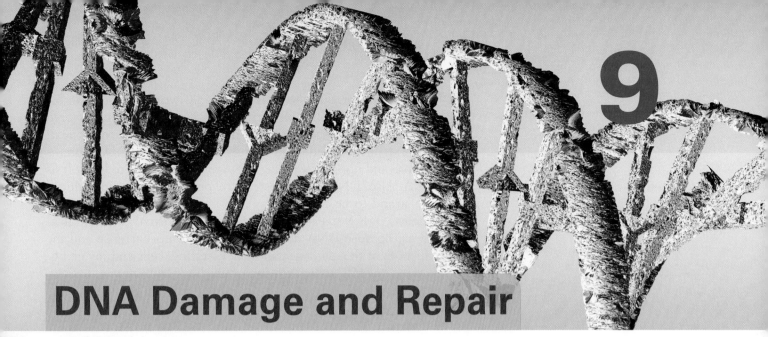

DNA Damage and Repair

CHAPTER OUTLINE

9.1 Radiation Damage
Ultraviolet light causes cyclobutane pyrimidine dimer formation and (6-4) photoproduct formation.

X-rays and gamma rays cause many different types of DNA damage.

9.2 DNA Instability in Water
DNA is damaged by hydrolytic cleavage reactions.

9.3 Oxidative Damage
Reactive oxygen species damage DNA.

9.4 Alkylation Damage by Monoadduct Formation
Alkylating agents damage DNA by transferring alkyl groups to centers of negative charge.

Many environmental agents must be modified by cell metabolism before they can alkylate DNA.

9.5 Chemical Cross-Linking Agents
Chemical cross-linking agents block DNA strand separation.

9.6 Mutagen and Carcinogen Detection
Mutagens can be detected based on their ability to restore mutant gene activity.

9.7 Direct Reversal of Damage
Photolyase reverses damage caused by cyclobutane pyrimidine dimer formation.

O^6-Alkylguanine, O^4-alkylthymine, and phosphotriesters can be repaired by direct alkyl group removal by a suicide enzyme.

9.8 Base Excision Repair
The base excision repair pathway removes and replaces damaged or inappropriate bases.

9.9 Nucleotide Excision Repair
Nucleotide excision repair removes bulky adducts from DNA by excising an oligonucleotide bearing the lesion and replacing it with new DNA.

9.10 Mismatch Repair
The DNA mismatch repair system removes mismatches and short insertions or deletions that are present in DNA.

9.11 SOS Response and Translesion DNA Synthesis
Error-prone DNA polymerases catalyze translesion DNA synthesis.

RecA and LexA regulate the *E. coli* SOS response.

The SOS signal induces the synthesis of DNA polymerases II, IV, and V.

Human cells have at least 14 different template-dependent DNA polymerases.

BOX 9.1: CLINICAL APPLICATION The Interstrand Cross-linking Agents Cisplatin and Psoralen

BOX 9.2: CLINICAL APPLICATION Xeroderma Pigmentosum

QUESTIONS AND PROBLEMS

SUGGESTED READING

Image © Bruce Rolff/ShutterStock, Inc.

One might predict the genetic material would be quite stable under normal physiological conditions so information stored within it is passed accurately from one generation to the next. To some extent this prediction is true. The fact that DNA contains deoxyribose and not ribose makes DNA less susceptible to hydrolysis than RNA. Also, DNA's duplex structure permits damage introduced into one strand to be corrected by information present in the other. Nonetheless, DNA is easily damaged under normal physiological conditions. Because DNA damage is a common occurrence and the consequences of this damage can be devastating, cells employ elaborate enzyme systems, which consume substantial amounts of energy, to restore damaged DNA to its original sequence and structure. This restoration process is called **DNA repair**. Despite the considerable cellular resources devoted to DNA repair, occasional changes in DNA sequence do occur. Although most of these changes are harmful, some must take place for new species to evolve.

Many different kinds of physical and chemical agents damage DNA. Some are **endogenous agents**, which are formed inside the cell by normal metabolic pathways, whereas others are **exogenous agents**, which come from the surrounding environment. Some exogenous agents act directly on DNA, and others must be modified by the cell's enzymes before they can damage DNA. Cells usually can survive DNA damage that cannot be repaired if the replication and transcription machinery can continue to perform their functions using the damaged DNA. Replication through the damaged region, however, may produce mutations. Cells cannot survive if the replication or transcription machinery cannot act on the damaged DNA. Hence, agents that damage DNA can cause mutations, kill cells, or both.

DNA damaging agents that cause mutations, called **mutagens**, also cause cancer in higher animals and so are said to be **carcinogens**. Carcinogenic agents work by generating mutations in two different kinds of genes. The first kind, the **proto-oncogenes**, codes for proteins that promote normal cell growth and division. Mutations that alter proto-oncogenes convert these genes into **oncogenes**, which synthesize a larger quantity of the normal protein or a more active form of the protein. The second kind, the **tumor suppressor genes**, codes for proteins that cause cell division and replication to slow down. Mutations that inactivate tumor suppressor genes remove a brake that helps to regulate cell division.

A single mutation is usually not sufficient to convert a normal cell to a cancer cell. Many mutations, which can accumulate over a long time, however, may be sufficient to do so. Hence, it is very important to keep our exposure to exogenous DNA damaging agents to a minimum. Somewhat paradoxically, individuals with certain types of cancers are intentionally exposed to DNA damaging agents, but with great caution. The reason for this exposure is the hope that the cytotoxic effects of the DNA damaging agents will kill rapidly dividing cancer cells. However, DNA damaging agents are not magic bullets that limit their attack to DNA in cancer cells. They also damage DNA in normal cells, leading to mutations that may eventually cause normal cells to be transformed to cancer cells. Careful risk-to-benefit analysis is required for a chemotherapeutic agent (anticancer drug) that works by causing DNA damage.

Several different pathways have evolved to repair damaged DNA. These repair pathways can be divided into two major groups, those that cause direct reversal of DNA damage and those that excise defective elements and replace them with normal nucleotides. In some instances the cell cannot repair the damage but nevertheless can survive because it has metabolic machinery that permits it to tolerate the damage.

The first part of this chapter examines some of the different types of DNA damage that take place in cells, and the second part examines pathways that cells use to reverse DNA damage, excise and replace damaged elements, or tolerate the damage. Although each repair pathway tends to correct a specific kind of DNA damage, there is considerable overlap in this specificity so that two or more repair pathways can correct the same type of damage.

9.1 Radiation Damage

Ultraviolet light causes cyclobutane pyrimidine dimer formation and (6-4) photoproduct formation.

Cells exposed to high energy electromagnetic radiation, which includes ultraviolet (UV) light (wavelengths 100–400 nm) and two forms of ionizing radiation, x-rays (wavelengths 0.01–99 nm) and gamma rays (wavelengths < 0.01 nm), experience considerable damage to their DNA. We begin by examining the major types of damage caused by UV light, which is divided into three bands: UV-C (100–295 nm), UV-B (296–320 nm), and UV-A (321–400 nm). UV-C includes λ_{260}, the DNA absorption maximum, and so would cause a great deal of DNA damage to exposed organisms if it were able to penetrate the earth's atmosphere. Fortunately, very little UV-C reaches the earth's surface because the ozone in the stratosphere prevents it from doing so. The ozone layer cannot be taken for granted, however, because it can be depleted by the release of chlorofluorohydrocarbons and other industrial chemicals into the atmosphere. Ozone depletion would lead to a much greater incidence of cancer of the skin, the only human tissue subject to direct UV damage. Laboratory studies designed to study the way UV light damages DNA are usually conducted with germicidal lamps that produce UV-C light. UV-B, which accounts for about 10% of the UV radiation reaching the earth's surface, is responsible for most of the DNA damage in skin. UV-A, which accounts for most UV light reaching earth, penetrates the skin more deeply than UV-B but is less efficient at causing DNA damage. The sunlight's tanning effects are largely due to UV-A, but constant exposure to UV-A comes at the steep price of skin aging, wrinkling, and an increased risk of skin cancer.

Two major types of pyrimidine dimers, the **cyclobutane pyrimidine dimers** and the **(6-4) photoproducts** (FIGURE 9.1), account for nearly all UV-induced DNA damage. Cyclobutane pyrimidine dimers, which

FIGURE 9.1 Ultraviolet light promoted cyclobutane pyrimidine dimer and (6-4) photoproduct formation.
(a) Ultraviolet light promotes the formation of cyclobutane pyrimidine dimers by introducing two new bonds between adjacent pyrimidines (in this case a cytosine and a thymine) on the same DNA strand. One bond connects the C-5 atoms and the second connects the C-6 atoms. (b) Ultraviolet light also promotes the formation of (6-4) photoproducts by introducing a bond between the C-6 atom of one pyrimidine and the C-4 atom of an adjacent pyrimidine (in this case a cytosine and a thymine) on the same DNA strand. Although this figure shows the bonds between the pyrimidine rings, it does not show the way the pyrimidine rings are arranged in space. (Adapted from Friedberg, E. C., et al. 2005. *DNA Repair and Mutagenesis* (2nd ed). ASM Press.)

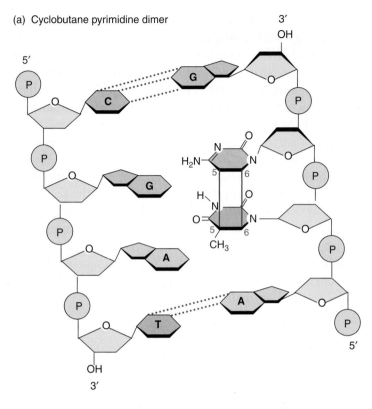

(a) Cyclobutane pyrimidine dimer

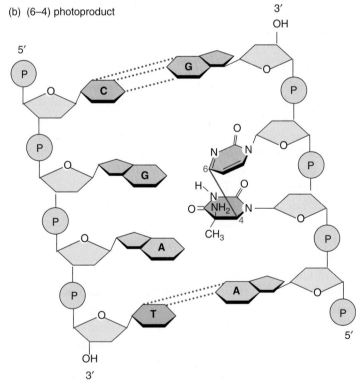

(b) (6–4) photoproduct

account for about 75% of UV-induced damage, result from the formation of two bonds between adjacent pyrimidines. The first bond connects the C-5 atoms on the two rings and the second connects the C-6 atoms. The most common cyclobutane pyrimidine dimer is the thymine–thymine dimer. Cytosine–thymine and cytosine–cytosine dimers also form but at slower rates. B-DNA can accommodate a single dimer, but the helical axis is forced to bend. The (6-4) photoproducts result from the formation of a bond between the C-6 atom of the 3′-pyrimidine (either thymine or cytosine) and the C-4 atom of the 5′-pyrimidine (usually cytosine). The (6-4) photoproduct causes a major distortion in B-DNA because the two pyrimidine rings are perpendicular to each other. If not removed, a pyrimidine dimer or (6-4) photoproduct can interfere with the normal operation of the replication and transcription machinery, resulting in mutations and cell death.

X-rays and gamma rays cause many different types of DNA damage.

Ionizing radiation directly or indirectly generates many different kinds of DNA lesions. Direct damage takes place when DNA or water tightly bound to it absorbs the radiation. Indirect damage takes place when water molecules or other molecules surrounding the DNA absorb the radiation and form reactive species that then damage the DNA. Lesions may be isolated or clustered (many lesions within a few helical turns). One type of clustered lesion, the double-strand break, is generally thought to be the primary reason ionizing radiation is so lethal to cells. Double-strand breaks are also responsible for various chromosomal aberrations such as deletions, duplications, **inversions** (segments breaking away from the chromosome, inverting from end to end, and then reinserting at the original breakage site), and **translocations** (segments breaking away from the chromosome and then reinserting at new sites).

Approximately 65% of the damage to DNA that is caused by x-rays and γ-rays is due to indirect effects, primarily through the transfer of photons to water. The photon transfer activates water, causing the formation of three highly reactive chemical species, $H_2O^{\bullet+}$ (water radical cation), $\bullet OH$ (hydroxide radical), and O_2^- (superoxide), which each attacks and damages biological molecules they encounter. A wide variety of chemical changes take place when that molecule happens to be DNA. Specific changes caused by hydroxide and superoxide radicals are described later in this chapter when we examine oxidative damage.

9.2 DNA Instability in Water

DNA is damaged by hydrolytic cleavage reactions.

DNA has three kinds of bonds with the potential for hydrolytic cleavage: (1) phosphodiester bonds, (2) N-glycosyl bonds, and (3) bonds linking amine groups to the ring structures in cytosine, adenine, and

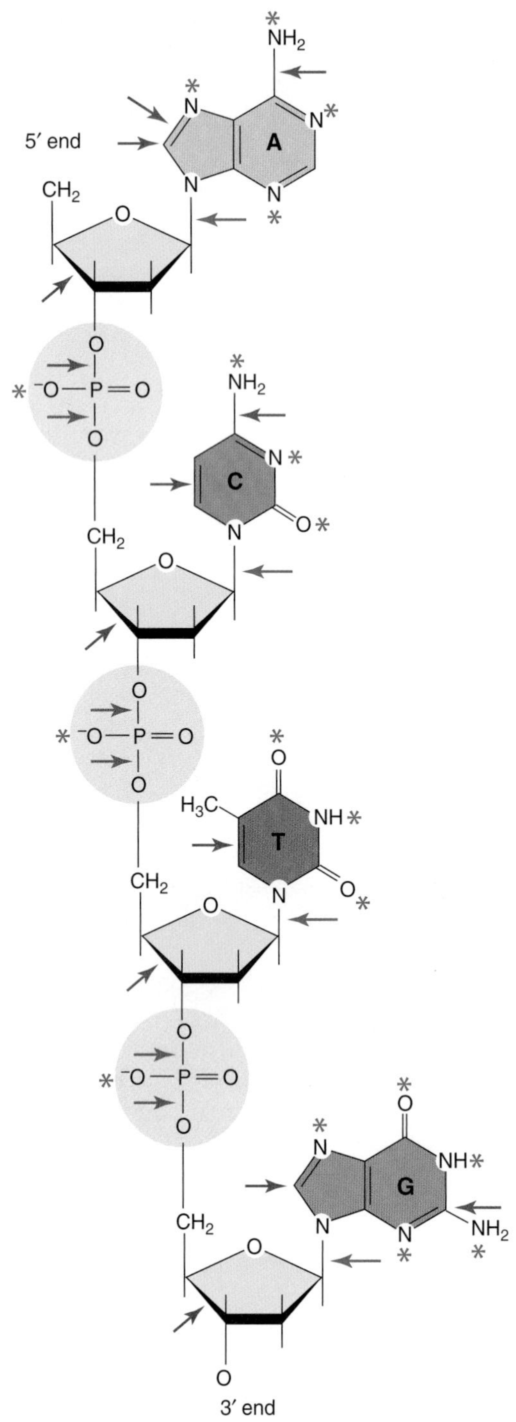

FIGURE 9.2 Sites of chemical damage to DNA. The bases (A, T, G, and C), deoxyribose groups, and phosphodiester bonds are vulnerable to attack by many exogenous and endogenous agents. Brown, green, and blue arrows point to phosphodiester bonds, N-glycosylic bonds, and bonds to exocyclic amine groups, respectively, that can be cleaved by water. Red arrows point to sites of attack by reactive oxygen species on the deoxyribose groups and bases. Green asterisks indicate electron-rich atoms that are attacked by alkylating agents. (Adapted from Doetsch, P. W. 2001. *Encyclopedia of Life Sciences*. John Wiley & Sons.)

guanine (**FIGURE 9.2**). Spontaneous phosphodiester bond cleavage, which introduces a nick into a DNA strand, is a very rare occurrence and probably does not make a significant contribution to DNA damage. N-glycosyl bond cleavage leads to the formation of an **abasic site**, which is also known as an AP (for apurinic and apyrimidinic) site (**FIGURE 9.3**). According to current estimates, about 10,000 purine and 500 pyrimidine bases are lost from DNA in a mammalian cell nucleus each day. These observations are consistent with *in vitro* experiments showing that purine N-glycosyl bonds are more easily hydrolyzed than pyrimidine N-glycosyl bonds. AP site formation sensitizes the neighboring 3'-phosphodiester bond to cleavage. A DNA strand with one or more AP sites makes a poor template because it lacks the information required to direct accurate replication and transcription.

Water-mediated deamination converts cytosine, guanine, and adenine to uracil, xanthine, and hypoxanthine, respectively (**FIGURE 9.4**). Hydrolytic cytosine deamination is estimated to take place about 100 to 500 times a day in a mammalian cell, whereas combined guanine and adenine deaminations are estimated to occur at about 1% or 2% of that value. The conversion of guanine to xanthine may result in mutations or arrested DNA synthesis because xanthine does not form stable base pairs with either cytosine or thymine. The conversion of adenine to hypoxanthine will, if not repaired, cause a T–A base pair to be replaced by a C–G base pair. Likewise, an uncorrected deamination that converts C to U will cause a C–G base pair to be replaced by a T–A base pair. This type of point mutation, in which a pyrimidine–purine base pair is replaced by a different pyrimidine–purine base pair, is called a **transition mutation** (**FIGURE 9.5a**). A second type of point mutation, called a **transversion mutation** (Figure 9.5b), involves replacing a purine–pyrimidine base pair with a pyrimidine–purine base pair (see below).

A few cytosine bases in eukaryotic DNA are converted to the modified base 5-methylcytosine. This modified base is concentrated in so-called **CpG islands**, which are small segments of DNA often present in regulatory elements termed **promoters** that are located just before the transcription unit that they regulate. The term CpG island derives from the fact that the CpG sequence is present in these DNA sections at a much higher frequency than in the rest of the DNA. The frequency of spontaneous deamination of 5-methylcytosine bases in CpG islands is even greater than that for cytosine. The product in this case, however, is thymine and not uracil, resulting in the conversion of a C–G base pair to a T–A base pair.

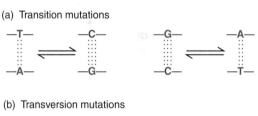

FIGURE 9.3 Hydrolytic cleavage of an N-glycosylic bond.

Nitrous acid (HNO_2), which is formed from nitrites used as preservatives in processed meats such as bacon, sausage, and hot dogs, reacts with the amine groups attached to the ring structures in cytosine, adenine, and guanine, greatly increasing their rate of deamination. Bisulfite (HSO_3^-), an additive that is sometimes present in wine, beer, fruit juices, and dried fruits, also greatly increases the rate of cytosine deamination but does not affect purine or 5-methylcytosine deamination.

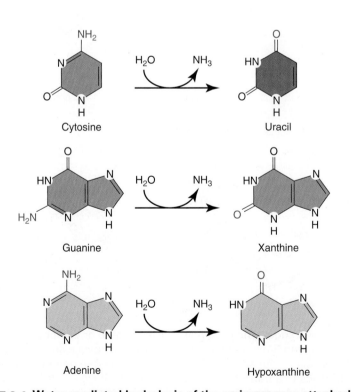

FIGURE 9.4 **Water-mediated hydrolysis of the amino groups attached to cytosine, guanine, and adenine.**

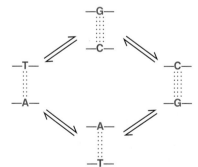

FIGURE 9.5 **Transition and transversion mutations.** (a) In a transition mutation, one pyrimidine-purine base pair replaces another base pair in the same pyrimidine-purine relationship. (b) In a transversion mutation, a pyrimidine-purine base pair is replaced by a purine-pyrimidine base pair (or vice versa). Purine bases are blue and pyrimidine bases are red.

9.2 DNA Instability in Water

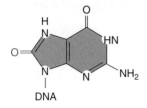

(a) 8-oxoguanine

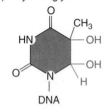

(b) Thymine glycol

FIGURE 9.6 Two of the modified bases produced by oxidative damage to DNA caused by the hydroxyl radical.

9.3 Oxidative Damage

Reactive oxygen species damage DNA.

The hydroxyl radical (•OH) appears to be the main reactive oxygen species that damages cellular DNA. These radicals are produced by ionizing radiation or from hydrogen peroxide (H_2O_2) in the presence of a metal ion such as Fe^{2+}. Relatively little hydrogen peroxide is present in cells because two enzymes, catalase and peroxidase, destroy it. Hydroxyl radicals, whether generated by ionizing radiation or from hydrogen peroxide, are known to cause more than 80 different kinds of base damage. Two of the oxidized base products, 8-oxoguanine (oxoG) and thymine glycol, are shown in FIGURE 9.6. 8-Oxoguanine can base pair with adenine or cytosine (FIGURE 9.7). If uncorrected, the resulting 8-oxoG–A base pair will be replicated to form a T–A base pair, causing a transversion mutation. Thymine glycol inhibits DNA replication and is therefore cytotoxic.

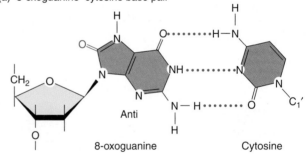

(a) 8-oxoguanine–cytosine base pair

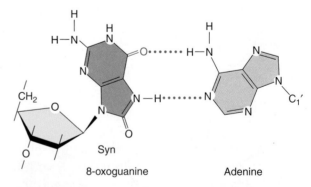

(b) 8-oxoguanine–adenine base pair

FIGURE 9.7 8-Oxoguanine base pairs with cytosine or adenine. (a) In the anti-conformation 8-oxoguanine base pairs with cytosine and (b) in the syn form it base pairs with adenine. If uncorrected, the 8-oxoguanine-adenine base pair leads to a transversion mutation in which a G–C base pair is replaced by a T–A base pair.

9.4 Alkylation Damage by Monoadduct Formation

Alkylating agents damage DNA by transferring alkyl groups to centers of negative charge.

DNA has electron-rich atoms that are readily attacked by electron-seeking chemicals termed **electrophiles** (Gr. *electros* = electron and *philos* = loving) or electrophilic agents. This section examines DNA damage caused by a highly reactive group of electrophiles called **alkylating agents** because they transfer methyl, ethyl, or larger alkyl groups to the electron-rich atoms in DNA. Alkylation takes place at (1) nitrogen and oxygen atoms external to the base ring systems, (2) nitrogen atoms in the base ring systems except those linked to deoxyribose, and (3) nonbridging oxygen atoms in phosphate groups. Many different kinds of naturally occurring and synthetic chemical agents are known to transfer alkyl groups to DNA.

The product formed by attaching a chemical group to DNA is called an **adduct**. If the chemical group attaches to a single site on the DNA, then the product is termed a **monoadduct**. For example, the exposure of DNA to dimethylnitrosamine leads to the production of a monoadduct in which a single methyl group attaches to DNA. The structures of dimethylnitrosamine and a few other simple agents that methylate DNA are shown in FIGURE 9.8. These methylating agents represent only a small fraction of the known alkylating agents, many of which transfer much larger alkyl groups to DNA.

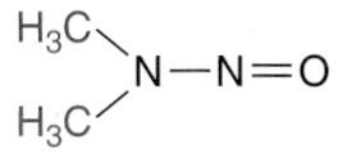

Dimethylnitrosamine

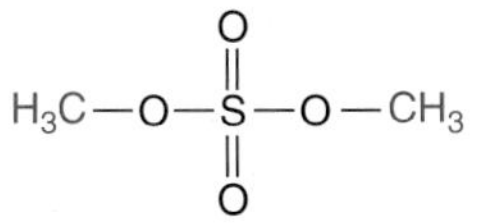

Dimethylsulfate

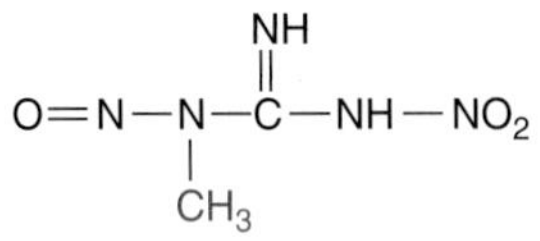

N-Methyl-N′-nitro-N-nitrosoguanidine

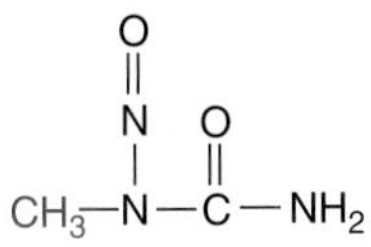

N-Methyl-N-nitrosourea (MNU)

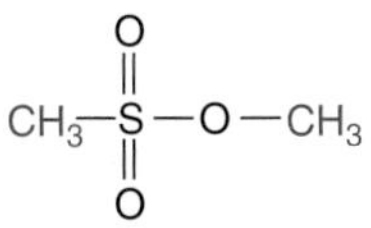

Methylmethane sulfonate

FIGURE 9.8 **Chemical structures of a few simple DNA methylating agents.** Methyl groups that are transferred to DNA are shown in red.

Many environmental agents must be modified by cell metabolism before they can alkylate DNA.

Many environmental agents become active alkylating agents only after they are metabolized in the cell. The first clue to the existence of these compounds came from a discovery made in 1775 by Percival Potts, an English surgeon. Potts noticed that several male patients with cancer of the scrotum had worked as chimney sweeps as young boys. This observation led him to propose a causal relationship between exposure to soot in the chimney and cancer. Potts thus became the first to point out the relationship between a hazardous workplace environment and cancer. We now know that the hazardous environmental material in soot is a mixture of polycyclic aromatic hydrocarbons formed by the incomplete combustion of the burning wood or coal used as fuel. Similar types of polycyclic aromatic hydrocarbons are also present in tobacco smoke and charbroiled meats. The structures of a few of the more than 100 different members of the polycyclic aromatic hydrocarbon family are shown in FIGURE 9.9. The common structural feature in these and all other polycyclic aromatic hydrocarbons is two or more fused aromatic rings. These hydrophobic hydrocarbons are not able to damage DNA unless they are metabolically activated.

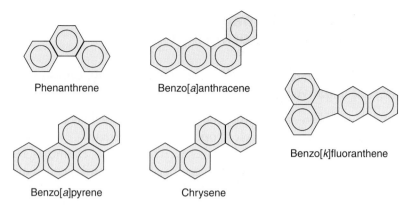

FIGURE 9.9 A few representative members of the polycyclic aromatic hydrocarbon (PAH) family.

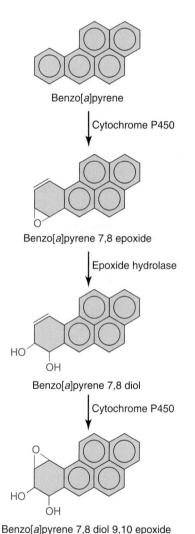

FIGURE 9.10 **Metabolic activation of benzo[a]pyrene, a polycyclic aromatic hydrocarbon.** Benzo[a]pyrene activation requires two kinds of enzymes, cytochrome P450 and epoxide hydrolase. The 9,10 epoxide product attacks DNA to form an adduct that is not shown in this figure.

The pathway for converting one of the polycyclic aromatic hydrocarbons, benzo[a]pyrene, into an active epoxide alkylating agent is shown in FIGURE 9.10. This pathway requires cytochrome P450 enzymes and epoxide hydrolase, which are located in the endoplasmic reticulum. Both enzymes normally make a valuable contribution to the cell's survival and well-being. Cytochrome P450 enzymes catalyze oxidation reactions that convert metabolic precursors to essential biomolecules and also help to detoxify harmful drugs by making them water soluble so they can be excreted in the urine. Cytochrome P450 enzymes are not very specific, however, and so are able to act on polycyclic aromatic hydrocarbons such as benzo[a]pyrene, adding oxygen atoms to form reactive three-membered epoxide rings. The epoxides then alkylate DNA, causing replication errors that result in mutations, which ultimately convert a normal cell into a cancer cell.

The aflatoxins, another class of chemical carcinogens that must be activated before damaging DNA, are produced by *Aspergillus flavus* and *Aspergillus parasiticus*, fungi that grow on peanuts and grains such as rice and corn. Animals feeding on contaminated peanuts or grains containing aflatoxins exhibit markedly increased rates of liver diseases, including liver cancer. Aflatoxin B_1, the most potent toxin produced by *A. flavus*, presents a particularly serious health threat in the United States. Cytochrome P450 converts aflatoxin B_1 into an epoxide derivative that damages DNA (FIGURE 9.11). Under ideal conditions a tripeptide called glutathione will attack the epoxide ring, making the aflatoxin derivative soluble so it can be excreted in the urine. Some of the reactive epoxide derivatives, however, escape attack by glutathione and therefor, are free to attack guanine rings in DNA. The flat aflatoxin ring system inserts between DNA bases (FIGURE 9.12), causing helical distortion that in turn leads to replication errors.

CHAPTER 9 DNA DAMAGE AND REPAIR

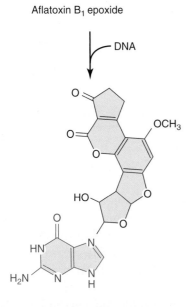

FIGURE 9.11 Metabolic activation of aflatoxin B_1. Aflatoxin B_1, a toxic product of certain fungi that grow on grains, must be metabolically activated before it can add to DNA. Cytochrome P450 is required to introduce an epoxide group that activates aflatoxin B_1, which then forms an adduct to the N-7 position in guanine.

Aflatoxin attached to the N7 position of the guanine base, which is shown in red

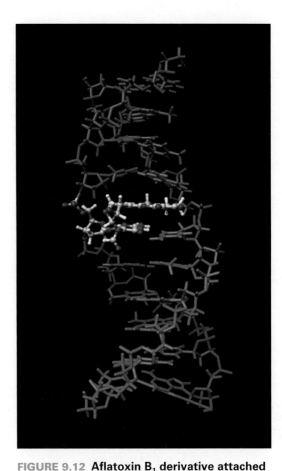

FIGURE 9.12 Aflatoxin B_1 derivative attached to DNA. Activated aflatoxin attaches to a guanine in the DNA helix and inserts its bulky ring system between the DNA bases. This insertion distorts the helix and leads to replication errors when the DNA is replicated. The guanine imidazole ring has opened under the conditions used in this experiment. One DNA strand is colored blue and the other magenta. Atoms derived from aflatoxin and the guanine nucleotide to which it binds are in standard CPK colors. (Photo courtesy of Kyle L. Crown and Michael P. Stone, Vanderbilt University.)

9.5 Chemical Cross-Linking Agents

Chemical cross-linking agents block DNA strand separation.

Each of the alkylating agents described to this point forms a monoadduct. Many alkylating agents, however, have two reactive sites and therefore can form intrastrand (within the same strand) or interstrand (connecting opposite strands) cross-links. Interstrand cross-links are of special interest because they prevent strand separation and, if not corrected, are lethal. One of the simplest cross-linking agents, nitrogen mustard gas (bis[2-chloroethyl]methylamine; FIGURE 9.13a) was originally developed by the military as a chemical warfare agent that attacks the central nervous system. Later studies showed that nitrogen mustard

FIGURE 9.13 **Nitrogen mustard gas, an agent that causes crosslink formation.** (a) Structure of bis(2-chloroethyl) methylamine (nitrogen mustard gas). (b) Nitrogen mustard gas forms an interstrand crosslink through N7 positions of two guanine bases on opposite strands of a DNA double helix. (Part b adapted from Friedberg, E. C., et al. 2005. *DNA Repair and Mutagenesis* [2nd ed]. ASM Press.)

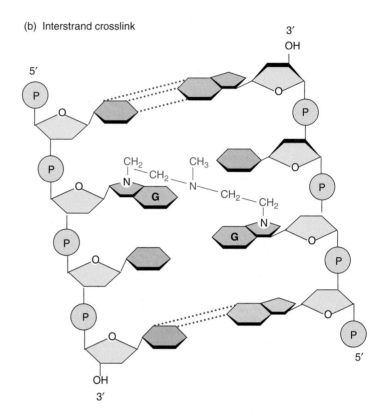

gas also damages DNA by forming interstrand cross-links. It does so by attacking N-7 on two guanines, which are on opposite strands of DNA double helix (Figure 9.13b). Although a very toxic substance, nitrogen mustard gas has found clinical application as a chemotherapeutic agent for treating certain forms of lymphoma and leukemia. Other chemical cross-linking agents also have important clinical applications (**BOX 9.1 CLINICAL APPLICATION: THE INTERSTRAND CROSS-LINKING AGENTS CISPLATIN AND PSORALEN**).

BOX 9.1: CLINICAL APPLICATION

The Interstrand Cross-Linking Agents Cisplatin and Psoralen

Cisplatin (**FIGURE B9.1a**) was discovered by Barnett Rosenberg in the mid-1960s while he was examining the effects electric currents have on *E. coli* growth. Treated bacteria grew to 300 times their normal length because the treatment blocked cell division but no other growth processes. Further studies showed the growth effect was not caused by the electric field but by a compound formed in a reaction between the supposedly unreactive platinum electrodes and components in the bacterial suspension. The compound was later shown to be cisplatin.

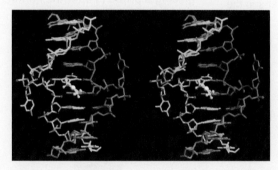

(a) Cisplatin

(b) Covalent structure of a cisplatin cross-link

(c) Stereoview of a cisplatin intrastrand cross-link

(d) Oligonucleotide used to obtain data for the model shown in (c)

```
   1 2 3 4 5 6 7 8 9 10
5' C A T A G C T A T G 3'

3' G T A T C G A T A C 5'
  10' 9' 8' 7' 6' 5' 4' 3' 2' 1'
```

FIGURE B9.1a–d Cisplatin. (a) The chemical structure of cisplatin. (b) The chemical structure of the crosslink formed by interaction of cisplatin with two purine bases on the same or opposite strands of a DNA double helix. (c) A stereoview of cisplatin interstrand crosslink. The platinum ion is shown as a light blue sphere. (d) Oligonucleotide used to obtain data required to solve the structure shown in (c). (Part b adapted from Huang, H., et al. 1995. *Science* 270:1842–1845. Part c reproduced from Huang, H., et al. 1995. *Science* 270:1842–1845. Reprinted with permission from AAAS. Photo courtesy of Paul B. Hopkins, University of Washington.)

Experiments designed to determine if cisplatin also inhibits the division of other kinds of cells revealed that cisplatin blocks the division of tumor cells. After entering a cell by passive diffusion or active transport, cisplatin is converted to a highly reactive, charged complex that forms a crosslink between two purine bases on the same or the opposite strand of a double stranded DNA (**FIGURE B9.1b–d**). Cisplatin's cytotoxic effects appear to be due to interstrand cross-linking, which blocks the replication and transcription machinery.

(*continues*)

BOX 9.1: CLINICAL APPLICATION

The Interstrand Cross-Linking Agents Cisplatin and Psoralen (*continued*)

Although an effective chemotherapeutic agent for treating cancer of the bladder, ovaries, and testicles, cisplatin's clinical application is limited by significant side effects and drug resistance. A great deal of effort is currently being devoted to finding other metallo-organic cytotoxic agents that cause cross-linking in DNA but do not have the same side effects as cisplatin.

Another cross-linking agent, psoralen, is used to treat skin disorders. Psoralen, a naturally occurring substance synthesized by some members of the carrot plant family, must be photoactivated before it can alkylate DNA. The planar psoralen molecule, which consists of a furan ring fused to a heterobicyclic ring system called coumarin (FIGURE B9.1e), inserts between base pairs in the

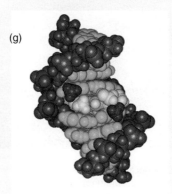

FIGURE B9.1e Psoralen. Psoralen is a planar tricyclic compound, which consists of a furan ring (red) fused to a coumarin ring system (black). Atoms in the ring system are numbered as indicated. Note the positions of atoms in the furan ring are indicated by placing a prime (′) after the number.

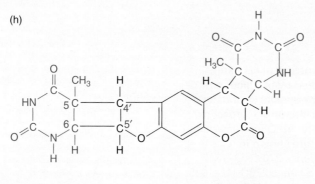

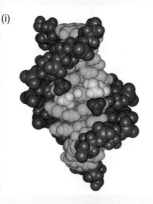

FIGURE B9.1f–i Psoralen photoproducts. (f) When activated by light with a wavelength of 400–450 nm psoralen reacts with DNA to form a monoadduct. The main product is the 4′,5′ monoadduct photoproduct in which psoralen adds to a thymine base (shown in red). (g) Spacefilling model of the structure of monoadduct viewed into the major groove showing the edge of the psoralen derivative between the T–A base pairs. The psoralen derivative is in yellow, and the thymine methyl groups are in red. (h) When activated by light with a wavelength of 320–400 nm the psoralen group in the 4′,5′ monoadduct is activated and can add to a pyrimidine base (shown in blue) on the opposite DNA strand to form a crosslink. (i) Spacefilling model of the structure of crosslink product viewed into the major groove showing the edge of the psoralen derivative between T–A base pairs. The psoralen derivative is in yellow and the thymine methyl groups are in red. (Parts f and g adapted from Bethea, D., et al. 1999. *J Dermatol Sci* 19:78–88. Parts h and i reprinted with permission from Spielmann, H. P., et al. 1995. *Biochemistry* 34:12937–12953. Copyright 1995, American Chemical Society. Photo courtesy of H. Peter Spielmann, University of Kentucky.)

DNA molecule. Upon exposure to light with a wavelength of 400 to 450 nm the furan ring in psoralen becomes activated and adds across the 5,6 double bond in a pyrimidine base (usually thymine) to form the 4′,5′-monoadduct (where 4′ and 5′ refer to positions on the psoralen furan ring), as shown in **FIGURE B9.1f** and g. The planar tricyclic psoralen derivative in the monoadduct is in position to combine with a second pyrimidine base on the opposite DNA strand, but it must be activated by light with a wavelength of 320 to 400 nm before it can do so. The resulting photoproduct (**FIGURE B9.1h** and i) contains a cross-link between pyrimidine bases on opposite strands of the DNA duplex. If not properly repaired, psoralen damage causes mutations and is lethal to cells.

Psoralen plus UV-A is used to treat psoriasis, a chronic inflammatory and proliferative skin cell disease, as well as other skin disorders. This treatment must be administered with great care, however, because, although quite effective, it increases the risk of developing squamous cell skin cancer. Related furocoumarins, which have been isolated from plants or synthesized in the laboratory, are currently under study in the hope of finding a drug with psoralen's beneficial effects but without its harmful side effects.

9.6 Mutagen and Carcinogen Detection

Mutagens can be detected based on their ability to restore mutant gene activity.

The pharmaceutical, food, textile, petroleum, and other industries synthesize tens of thousands of new chemicals each year in the hope of finding marketable products. Most chemicals do not pass the initial screening tests and are not studied further. The few that do pass need to be subjected to a wide variety of additional tests. A major concern is that a chemical may serve the purpose for which it was intended but nevertheless not be marketable because it is a mutagen. Therefore, it is necessary to have a rapid and effective test available to detect chemical mutagens and thereby ensure they do not come to market. Mutagen detection is important because DNA damage in a germ cell can result in birth defects and DNA damage in somatic cells can result in cancer.

Bruce Ames started to think about this safety issue in 1964 while reading the ingredients on a box of potato chips. He wondered how he could tell whether preservatives and other additives might be carcinogenic. His curiosity eventually led to his developing a test, which, after undergoing several modifications and improvements, is still used today to determine whether a chemical substance has a high probability of being a carcinogen.

In its most basic form, the **Ames test** involves determining whether the chemical to be tested causes a histidine-requiring mutant of the gram-negative bacteria *Salmonella typhimurium* that has a base substitution, deletion, or insertion in a *his* gene to revert to the His$^+$ phenotype. The test is performed by adding approximately 10^8 bacteria, the chemical to be tested, and a trace quantity of histidine to molten top agar. Histidine is included to allow cells to divide a few times so the mutagens that require DNA replication have a chance to work but the histidine is at too low a concentration to permit His⁻ bacteria to form visible colonies. After gentle agitation top agar mixtures are poured onto glucose minimal agar plates. Colonies are counted after the plates have been incubated for 48 hours. A significant increase in the reversion frequency above that obtained in the absence of the tested chemical identifies the tested chemical as a mutagen. The reversion frequency depends on the concentration of the substance being tested and, for a known mutagen or carcinogen, correlates roughly with its known effectiveness.

In its original form the Ames test failed to demonstrate the mutagenicity of several potent carcinogens. The reason for this failure is that some substances such as benzo[a]pyrene and aflatoxin B1 must be activated by cytochrome P450 enzymes before they can damage DNA. Because bacteria lack these enzymes, their DNA is not damaged by chemical agents that must be activated. Ames modified his test to correct for this problem by adding rat liver **microsomes** (a cell fraction that contains the endoplasmic reticulum) and required coenzymes to the top agar. The addition of the microsomal fraction, which is rich in cytochrome P450 enzymes, makes the Ames test sensitive to chemicals that must be activated by cytochrome P450 enzymes before they can damage DNA and cause mutagenesis.

Ames increased the sensitivity of the test by making three changes to the tester strains. First, he inactivated the *uvrB* gene, which codes for a protein required for nucleotide excision repair (see below). Second, he introduced a mutation that makes the bacterial outer membrane more permeable to large molecules, increasing the likelihood that large chemical mutagens will be able to reach DNA inside the cell. Third, he introduced plasmids with genes that code for DNA polymerases that catalyze error prone DNA synthesis (see below), which greatly increases the mutation rate. Although other tests also have been developed to detect mutagens and carcinogens, the Ames test remains the most widely used.

The Ames test has now been used with tens of thousands of substances and mixtures (such as industrial chemicals, food additives, pesticides, hair dyes, and cosmetics), and numerous unsuspected substances have been found to stimulate reversion in this test. A high frequency of reversion does not mean the substance is definitely a carcinogen but only that it has a high probability of being so. As a result of these tests many industries have reformulated their products. For example, the cosmetic industry has changed the formulation of many hair dyes and cosmetics to render them nonmutagenic. Ultimate proof of carcinogenicity is determined by testing for tumor formation in laboratory animals. The Ames test and several other microbiological tests are used to reduce the number of substances that have to be tested in animals because to date only a few percent of more than 500 substances known from animal experiments to be carcinogens failed to increase the reversion frequency in the Ames test. A basic assumption behind the Ames test and other tests that use bacteria as test organisms is that a chemical that causes mutations in bacteria will also do so in other organisms. This assumption appears to be well founded because the basic mechanisms of DNA damage are the same in all organisms.

9.7 Direct Reversal of Damage

Photolyase reverses damage caused by cyclobutane pyrimidine dimer formation.

Our examination of DNA repair begins with a look at enzymes that catalyze the direct reversal of DNA damage. The first clue to the existence of an enzyme that catalyzes the direct reversal of DNA damage was reported by Albert Kelner in 1949, 4 years before the Watson-Crick model was proposed. Kelner's research was motivated by his desire to isolate antibiotic-resistant bacterial mutants. To do so he first irradiated bacteria with UV light at doses that killed most of the bacteria and then tested the survivors to isolate the desired mutants. Even though Kelner was very careful to keep the experimental conditions the same, he noticed a great deal of variation in the number of survivors from one experiment to another. After considerable effort Kelner finally discovered the explanation for the puzzling inconsistent results. Cells that were placed in the dark after UV irradiation had a much lower survival rate than those exposed to the light coming through the laboratory window. Exposure to visible light thus reversed UV light's bactericidal effects.

Within several weeks of Kelner's discovery, Renato Dulbecco observed a similar phenomenon while studying UV-irradiated phage T2. Once again, the phenomenon manifested itself as an unexpected lack of reproducibility. Dulbecco prepared multiple medium agar plates, each containing the same numbers of UV-irradiated phages and phage-sensitive bacteria, and placed the plates in a stack. Each of the stacked plates should have had about the same number of plaques. The plaque number decreased dramatically, however, going from top to bottom. Dulbecco explained this observation by proposing that the plates on the top of the pile were exposed to more light from the fluorescent bulb used to illuminate the laboratory than were the plates on the bottom. He tested this hypothesis by exposing some of the plates to fluorescent light while keeping others in the dark. As predicted, the numbers of plaques on plates exposed to the light were a great deal higher than those left in the dark. The bacteria were somehow using the visible light to repair the damaged phage DNA. The chemical basis for this light-dependent phenomenon, which Dulbecco called **photoreactivation,** remained to be elucidated.

Claud S. Rupert and coworkers devised an *in vitro* photoreactivation system in 1957, taking a major step toward determining the chemical mechanism of photoreactivation. Their straightforward approach was to isolate DNA from the gram-negative bacteria *Haemophilus influenzae,* irradiate the bacterial DNA with UV light to inactivate its transforming ability, and then demonstrate that a cell-free

Escherichia coli extract, acting in the presence of visible light, restores transforming activity. Although this study had the potential to open the way for the purification and characterization of the photoreactivation enzyme, investigators still needed to establish the chemical nature of the DNA damage that was repaired.

This problem was solved over the next few years when investigators demonstrated that UV irradiation induces cyclobutane pyrimidine dimer formation. Further studies showed the photoreactivation enzyme reverses UV-induced damage by using the energy provided by blue light (350–450 nm) to drive cyclobutane ring disruption in cyclobutane pyrimidine dimers (**FIGURE 9.14**). With the recognition that the photoreactivation enzyme catalyzes the disruption of carbon–carbon bonds, it was given the more descriptive name of **cyclobutane pyrimidine dimer photolyase**. Bacterial cells that lack this photolyase cannot repair cyclobutane pyrimidine dimer lesions by photoreactivation.

Cyclobutane pyrimidine dimer photolyases are present in a wide variety of organisms including bacteria, the archaea, plants, and animals but not in humans or other placental mammals. The photolyases have two light-absorbing pigments. The first, which is part of the active site, is required to repair cyclobutane pyrimidine dimer damage. The second, which is not required for repair, captures light with wavelengths that would not otherwise be available to the first pigment and transfers the energy to the first pigment. It thereby permits the photolyase to use available light energy much more efficiently than would otherwise be possible.

As described above, UV irradiation also induces the formation of a second type of pyrimidine dimer, the (6-4) photoproduct. Although the name of this lesion derives from the chemical bond linking carbon-6 of one pyrimidine ring to carbon-4 of an adjacent pyrimidine ring on the same strand, additional chemical changes also take place during

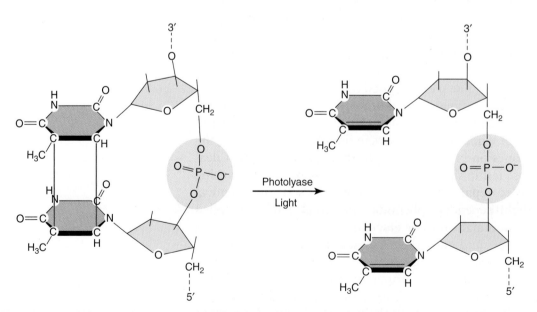

FIGURE 9.14 Photolyase catalyzes a light driven reaction that disrupts the cyclobutane ring in cyclobutane pyrimidine dimers, reversing the damaging effect of UV irradiation.

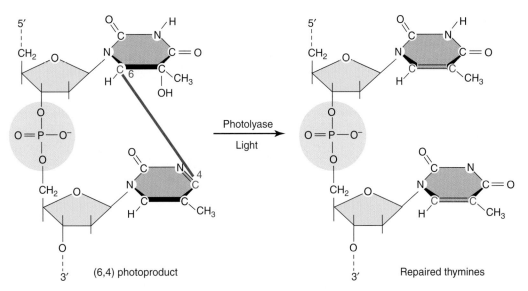

FIGURE 9.15 Reaction catalyzed by the (6-4) photolyase. In addition to cleaving the bond shown in red, the (6-4) photolyase also has to reverse other changes to the DNA caused by the UV damage.

dimer formation. Taking note of the additional chemical changes, investigators thought it would be quite unlikely that a single enzyme would be able to catalyze direct reversal of a (6-4) photoproduct lesion. It therefore came as quite a surprise when Takeshi Todo, Taisei Nomura, and coworkers reported in 1993 that *Drosophila melanogaster* has a photolyase that reverses (6-4) photoproduct lesions. This photolyase, which was designated the **(6-4) photolyase** to distinguish it from the cyclobutane pyrimidine dimer photolyase, catalyzes the reaction shown in **FIGURE 9.15**. Although widely distributed in plants and animals, the (6-4) photolyase has not been detected in bacteria or mammals that have been tested to date. As we see below, organisms can also repair dimer lesions introduced by UV irradiation by excising damaged nucleotides and replacing them with normal nucleotides. This type of excision repair is the major pathway for repairing UV-induced damage to DNA in humans and other organisms that lack both photolyases.

O^6-Alkylguanine, O^4-alkylthymine, and phosphotriesters can be repaired by direct alkyl group removal by a suicide enzyme.

Another means of direct damage reversal is dealkylation. Direct dealkylation reactions have probably been most extensively studied in *E. coli*. Two different O^6-alkylguanine DNA alkyltransferases selectively remove the alkyl groups by transferring them to their own cysteine residues. The first, O^6-alkylguanine DNA alkyltransferase I, can remove methyl and other alkyl groups attached to O-6 in guanine. In addition, the enzyme can remove alkyl groups attached to O-4 in thymine and to phosphotriesters. O^6-alkylguanine DNA alkyltransferase I is a monomer that has a flexible linker connecting its N- and C-terminal domains (**FIGURE 9.16**). Each domain has an active site that performs a specific function. The N-terminal domain transfers an alkyl group from a phosphotriester to one of its own cysteine

9.7 Direct Reversal of Damage

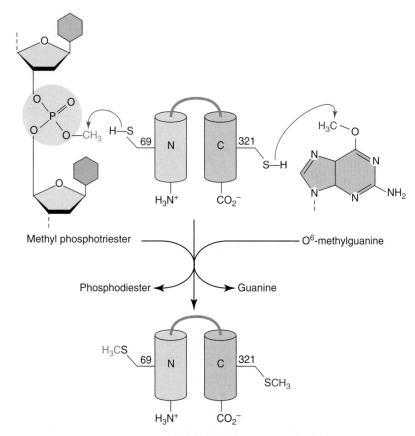

FIGURE 9.16 O^6-alkylguanine and phosphotriester lesions repaired by O^6-alkylguanine DNA alkyltransferase I. O^6-alkylguanine DNA alkyltransferase I contains N- and C-terminal domains that are connected by a flexible linker. The N-terminal domain of O^6-alkylguanine DNA alkyltransferase I transfers an alkyl group (green) from a phosphotriester to one of its own cysteine residues, Cys-69. The C-terminal domain transfers an alkyl group (red) from either O^6-alkylguanine or O^4-alkylthymine (not shown) to one of its own cysteine residues, Cys-321. (Adapted from Myers, L. C., et al. 1992. *Biochemistry* 31:4541–4547.)

residues, Cys-69. The C-terminal domain transfers an alkyl group from either O^6-alkylguanine or O^4-alkylthymine (not shown) to one of its own cysteine residues, Cys-321. Once alkylated, the protein cannot be regenerated and therefore behaves more like an alkyl transfer agent than a classical enzyme. Proteins such as O^6-alkylguanine DNA alkyltransferase I, which lose activity after acting only one time, are called **suicide enzymes**.

O^6-Alkylguanine DNA alkyltransferase I has one additional remarkable function. After methylation at Cys-69, the protein is converted to a transcriptional activator that stimulates the transcription of the gene that codes for it as well as a few other genes that code for proteins that repair DNA damage. This additional activity permits the bacteria to adapt to environments in which they are exposed to alkylating agents by synthesizing more copies of O^6-alkylguanine DNA alkyltransferase I, which then repair the damage. O^6-alkylguanine

DNA alkyltransferase I that is synthesized after all alkylation damage has been repaired will remain unmethylated and in this form block transcription of the same genes that were activated by the methylated protein.

O^6-Alkylguanine DNA alkyltransferase II, the second *E. coli* alkyl transfer protein, has properties that are very similar to those of the C-terminal domain of O^6-alkylguanine DNA alkyltransferase I. It also transfers a single alkyl group from the O-6 position in guanine or the O-4 position in thymine to one of its own cysteine residues and then loses activity and so is also classified as a suicide enzyme. O^6-alkylguanine DNA alkyltransferase II, however, does not appear to be subject to genetic regulation. Its function appears to be to protect bacteria from alkylation damage during the time it takes for the gene that codes for O^6-alkylguanine DNA alkyltransferase I to be fully expressed.

The eukaryotic protein that removes alkyl groups from O^6-methylguanine and O^4-methylthymine is similar to the bacterial O^6-alkylguanine DNA alkyltransferase II. Human alkylguanine DNA alkyltransferase is of considerable clinical interest because many chemotherapeutic agents used to destroy cancer cells are alkylating agents. When alkyltransferase activity in tumor cells is very high, these agents are not effective. On the other hand, when alkyltransferase activity in the surrounding healthy cells is too low, the chemotherapeutic agents will kill these cells. The human protein is also important because it helps to protect against carcinogenic alkylating agents.

9.8 Base Excision Repair

The base excision repair pathway removes and replaces damaged or inappropriate bases.

Many types of DNA damage cannot be repaired by a single enzyme that catalyzes direct damage reversal. Instead, repair requires participation of several different enzymes, each performing a specific task in a multistep pathway. Damage to DNA bases caused by deamination, oxidation, and alkylation is mainly repaired by one such multistep pathway, **base excision repair**. Enzymes in the base excision repair pathway also participate in single-strand break repair. Base excision repair derives its name from the first step in the pathway, N-glycosyl bond cleavage, which excises the damaged or inappropriate base from the DNA to form an abasic site. Because no single enzyme can distinguish the four bases normally present in DNA from the wide variety of altered bases generated by alkylation, deamination, and oxidation, cells must use many different enzymes to perform this function. Some N-glycosylases are monofunctional enzymes with only the ability to excise a damaged base (FIGURE 9.17a). Others also have an AP lyase activity that cleaves the bond between the sugar and the phosphate 3′ to the damaged site (Figure 9.17b). An enzyme with just N-glycosylase activity is designated a DNA glycosylase, whereas one that also has AP lyase activity is designated a DNA glycosylase/lyase.

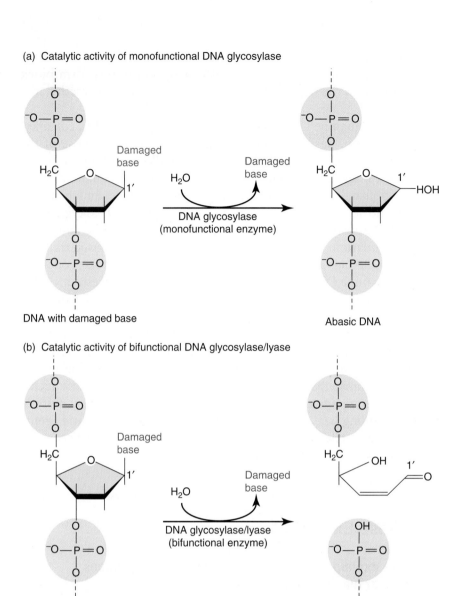

FIGURE 9.17 **Monofunctional and bifunctional DNA glycosylases.** (a) Monofunctional DNA glycosylases excise a damaged base. (b) Bifunctional DNA glycosylases also have an AP lyase activity that cleaves the bond between the sugar and the phosphate group 3′ to the AP site.

The base excision repair pathway, which can be divided into two stages, is similar in bacteria and eukaryotes. For simplicity, only eukaryotic enzymes are indicated in FIGURE 9.18. The first stage, *base excision and chain cleavage* (purple arrows), requires DNA glycosylase (or DNA glycosylase/lyase) and AP endonuclease. DNA glycosylase excises the damaged base to form an AP (apurinic and apyrimidinic) site, and then AP endonuclease hydrolyzes the phosphodiester bond 5′ to the AP site to generate a nick that has a 5′-deoxyribose phosphate (5′-dRP) on one side and a 3′-OH on the other. The second stage, *nucleotide replacement and ligation*, can continue by either of

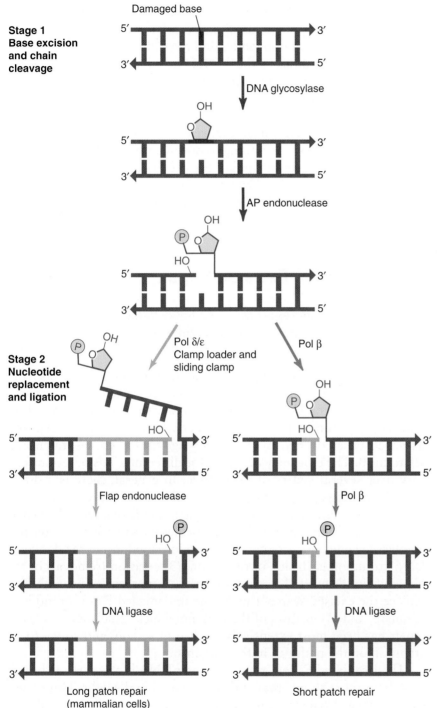

FIGURE 9.18 **Base excision repair in eukaryotes starting with a monofunctional DNA glycosylase.** Stage 1 (purple arrows): A DNA glycosylase excises a damaged or inappropriate base to produce an abasic or AP site and then an AP endonuclease cleaves the DNA backbone 5′ of the AP site to generate a 5′-deoxyribose phosphate end and a free 3′-OH end. Stage 2 (short patch repair, orange arrows): DNA polymerase β (Pol β) adds a nucleotide to the 3′-OH end and removes the 5′-deoxyribose phosphate group. Then DNA ligase joins the ends to form an intact strand. Stage 2 (long patch repair, blue arrows): DNA polymerase δ or ε acting with the clamp loader and sliding clamp adds 2 or more nucleotides to the 3′-hydroxyl group while at the same time displacing the 5′-deoxyribosephosphate group to generate a flap. Flap endonuclease removes the flap and DNA ligase connects the ends. (Adapted from Schaerer, O. D. 2003. *Angew Chem Int Ed Engl* 34:2946–2974.)

two subpathways. The first subpathway, **short patch repair**, replaces only one nucleotide (orange arrows in Figure 9.18). DNA polymerase β (Pol β), a repair enzyme, adds a deoxyribonucleotide to the 3′-OH end on one side of the nick and removes the 5′-deoxyribose phosphate from the 5′-end on the other side of the nick. DNA ligase completes the repair process by joining the adjacent ends. The second subpathway,

9.8 Base Excision Repair

long patch repair, replaces two or more nucleotides (blue arrows in Figure 9.18). DNA polymerase δ or ε catalyzes chain extension with the assistance of the clamp loader and the sliding clamp. As the polymerase adds nucleotides to the 3′-OH end on one side of the nick it displaces the 5′-deoxyribose phosphate end on the other side of the nick. Flap endonuclease (FEN1) cleaves the displaced strand and DNA ligase seals the remaining nick.

The regulatory mechanism that selects short or long patch repair is not well understood. According to one hypothesis selection depends on the 5′-deoxyribose phosphate formed by the AP endonuclease. If this group can be removed by Pol β, then the cell will use short patch repair; if it cannot, then the cell will use long patch repair.

9.9 Nucleotide Excision Repair

Nucleotide excision repair removes bulky adducts from DNA by excising an oligonucleotide bearing the lesion and replacing it with new DNA.

The nucleotide excision repair pathway removes bulky adducts from DNA, correcting for many different types of structurally unrelated DNA damage. For instance, nucleotide excision repair excises UV-induced cyclobutane pyrimidine dimers, (6-4) photoproducts, and damaged bases formed by alkylating agents such as psoralen and aflatoxin B. The efficiency of repair for different kinds of lesions can vary over several orders of magnitude. In general, there is a direct correlation between the amount of helical distortion produced by the lesion and the efficiency with which the lesion is removed. The basic nucleotide excision repair pathway, which is the same in all organisms, involves (1) damage recognition, (2) an incision (cut) in the damaged DNA strand on each side of the lesion, (3) excision (removal) of the oligonucleotide created by the incisions, (4) synthesis of new DNA to replace the excised segment using the undamaged DNA strand as a template, and (5) ligation of the remaining nick. Nucleotide excision repair has been most extensively studied in *E. coli*, and so we focus attention on the pathway in that organism.

UV-irradiated *E. coli* can repair their DNA in the dark, although they do so more slowly than when incubated in the light. This observation suggests bacteria can use some process in addition to photoreactivation to repair UV-induced light damage. Richard Setlow and William Carrier and, working independently, Richard Boyce and Paul Howard-Flanders used a similar approach to investigate this alternative process in 1964. Both groups cultured *E. coli* in the presence of [^{3}H]thymine to label the DNA and then irradiated the cells with UV light to induce thymine cyclobutane dimer formation. Then they (1) incubated the UV-irradiated cells in the dark so photoreactivation could not take place, (2) removed samples after various incubation times and added trichloracetic acid to them,

(3) separated acid-insoluble DNA from acid-soluble oligonucleotides, (4) digested the DNA and oligonucleotides to release intact thymine cyclobutane dimers, and (5) detected the released dimers by chromatography. The experiments revealed that as the incubation time in the dark increases, cyclobutane thymine dimers disappear from the acid-insoluble DNA and appear in the acid-soluble oligonucleotide fraction. These results were correctly interpreted to mean that bacteria can excise an oligonucleotide containing a lesion and replace the excised oligonucleotide with newly synthesized DNA. Subsequent studies showed that eukaryotes and the archaea also have nucleotide excision repair pathways.

Genetic studies revealed that three *E. coli* genes—*uvrA*, *uvrB*, and *uvrC* (for UV radiation)—code for proteins that are essential for damage recognition, incision, and excision. All three genes have been cloned, and the proteins they encode (UvrA, UvrB, and UvrC) have been purified and characterized. Although the three proteins do not combine to form a stable ternary complex, the polypeptides nevertheless are said to be part of a UvrABC damage-specific endonuclease, or UvrABC endonuclease for short. Some investigators prefer the term UvrABC excinuclease to indicate the proteins participate in excision and incision reactions.

The bacterial nucleotide excision pathway is shown in **FIGURE 9.19**. UvrB, a helicase, interacts with both UvrA and UvrC, although not at the same time. UvrB combines with UvrA to form a $UvrA_2 \bullet UvrB_2$ complex, which usually binds to DNA at some distance from the damaged site. Then the UvrB helicase catalyzes ATP-dependent movement of the $UvrA_2 \bullet UvrB_2$ complex along the DNA until the protein complex encounters a bulky adduct or helical distortion. The two UvrA subunits bind to DNA regions on either side of the damaged site but do not make direct contact with the lesion. This binding causes the double helix to bend and unwind. Then the UvrA subunits are released in an ATP-dependent reaction with a concomitant conformational change in UvrB that produces a stable $UvrB_2 \bullet DNA$ complex. Thus, UvrA uses energy provided by ATP to facilitate the formation of a stable $UvrB_2 \bullet DNA$ complex that would not otherwise form and then dissociates from the complex. Next, two UvrC subunits bind to the $UvrB_2 \bullet DNA$ complex and make two cuts, one on each side of the lesion. The first cut of the damaged strand is four nucleotides toward the 3′-end from the lesion and the second is seven nucleotides toward the 5′-end from the lesion. Finally, three additional enzymes complete the repair process: (1) UvrD, a helicase, unwinds the damaged oligonucleotide; (2) DNA polymerase I uses the undamaged strand as a template to fill in the gap; and (3) DNA ligase seals the remaining nick to complete the repair process. Eukaryotes have a similar nucleotide excision repair system, but the enzymes that participate in the eukaryotic pathway are not homologous to the bacterial enzymes. Individuals with a defect in one of the enzymes that participate in nucleotide excision repair suffer from an autosomal recessive disease called xeroderma pigmentosum (**BOX 9.2 CLINICAL APPLICATION: XERODERMA PIGMENTOSUM**).

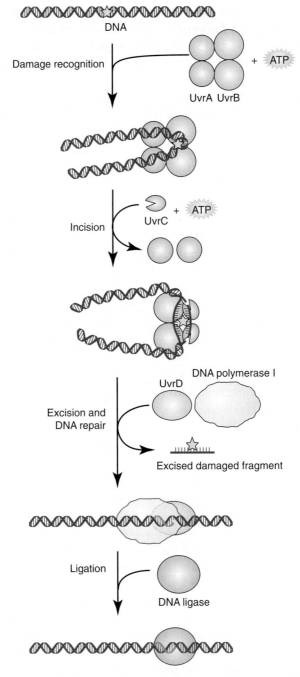

FIGURE 9.19 Bacterial nucleotide excision repair pathway. UvrB interacts with UvrA to form a UvrA$_2$ • UvrB$_2$ complex, which usually binds to DNA at some distance from the damaged site (not shown). Then the complex moves along the DNA until it encounters a bulky adduct (yellow star) or helical distortion. The two UvrA subunits bind on either side of the damaged site without making direct contact with the lesion, causing the double helix to bend and unwind. After this local unwinding, UvrA subunits are released to produce a stable UvrB$_2$ • DNA complex. Next, UvrC subunits bind to the UvrB$_2$ • DNA complex and make a cut on each side of the lesion. Finally, UvrD, a helicase, unwinds the damaged oligonucleotide, DNA polymerase I uses the undamaged strand as a template to fill in the gap, and DNA ligase seals the remaining nick to complete the repair process.

BOX 9.2: CLINICAL APPLICATION

Xeroderma Pigmentosum

The first hint that mammalian cells also carry out nucleotide excision repair came from experiments performed in the early 1960s by Robert Painter and coworkers, who had developed an autoradiographic technique to study [^{3}H]thymidine incorporation into mammalian cell DNA. Cells in the S-stage of the cell cycle incorporate large quantities of [^{3}H]thymidine into DNA, whereas cells in other stages of the cell cycle incorporate little, if any, radioactive label into DNA. Painter and coworkers observed that mammalian cells not in the S stage of the cell cycle gained the ability to incorporate low levels of [^{3}H]thymidine into DNA after UV irradiation and proposed this "unscheduled" DNA synthesis was part of a repair pathway.

James E. Cleaver, a young investigator working in Painter's laboratory, helped to establish the connection between unscheduled DNA synthesis and nucleotide excision repair in 1968. Cleaver was aware of the then recently published work showing bacteria have a nucleotide excision repair pathway and suspected the unscheduled DNA synthesis was due to a mammalian nucleotide excision repair pathway. He hoped to obtain evidence to support his hypothesis by first isolating UV-sensitive mammalian cell lines and then showing they did *not* incorporate low levels of [^{3}H]thymidine into DNA after UV irradiation. When preparing to isolate the desired cell lines he came across a newspaper article describing xeroderma pigmentosum (XP), an autosomal recessive disease. Each parent of an XP child carries an XP mutation but displays no obvious symptoms of the disease. In contrast, the XP child has a 1,000-fold greater likelihood of developing skin cancer, especially in parts of the body exposed to sunlight such as the hands, face, and neck. The incidence of skin cancer is directly related to the sunlight exposure. Without considerable precautions to avoid sunlight, the median age at which skin cancer first appears is about 8 years, roughly 50 years earlier than that for the general population.

The relationship between sunlight exposure and the incidence of skin cancer may be explained by proposing that sunlight causes damage to DNA in skin cells in all individuals but that normal skin cells can repair the damage, whereas XP cells cannot. Cleaver correctly deduced that XP cells could be used in place of the UV-sensitive mutants he had planned to isolate to test whether mammalian cells have a nucleotide excision repair system. As anticipated, XP cells failed to incorporate low levels of [^{3}H]thymidine after they were irradiated with UV light. Cleaver also demonstrated that wild-type human cells have a nucleotide excision repair system, which is missing in XP cells.

Dirk Bootsma and colleagues provided important new insights into the mammalian nucleotide excision repair system in the early 1970s after noticing significant variations in the levels of DNA repair in cells obtained from different XP patients. The existence of these variations suggested the possibility that

(continues)

BOX 9.2: CLINICAL APPLICATION

Xeroderma Pigmentosum (*continued*)

XP may result from a defect in one of several different genes. Bootsma and coworkers verified this possibility by fusing fibroblasts isolated from one patient with those isolated from another patient. When a fibroblast, for example XP-A, from one patient was fused to a fibroblast, for example XP-B, from another patient, the fused cell was not hypersensitive to UV light and was able to perform unscheduled DNA synthesis. These fusion experiments indicated that each cell makes a protein required for nucleotide excision repair that is not made by the other. In other words, the XP-A cell makes a normal XPB protein and the XP-B cell makes a normal XPA protein. Systematic cell fusion experiments using all available XP cell lines eventually revealed that human cells have seven genes, *XPA* to *XPG*, that code for polypeptides that participate in nucleotide excision repair.

Depending on the precise location of the DNA lesion, the nucleotide excision repair system may act on an untranscribed DNA strand or an actively transcribed strand. The pathway for repairing untranscribed DNA is designated nucleotide excision repair. Some investigators refer to it as the global genome excision repair pathway because most of the cell's DNA is not being transcribed at any given time. The pathway for repairing transcribed DNA, the transcription-coupled nucleotide excision repair pathway, was discovered when investigators noticed lesions that block the progress of the transcription machinery are repaired more rapidly in transcriptionally active DNA regions than in transcriptionally silent DNA regions. The transcription-coupled excision repair pathway is specific to the transcribed strand.

Progress in understanding the mammalian nucleotide excision repair system required purification and characterization of the XP proteins. *In vitro* assays were devised to detect these proteins. In one commonly used assay, XP protein activity is followed by first preparing a DNA substrate with a radioactively labeled cyclobutane thymine dimer and then monitoring the excision of the oligonucleotide bearing the thymine dimer. Isolation of XP proteins was greatly facilitated by the availability of XP cell lines with specific XP protein defects. Cell-free extracts from normal cells were subjected to standard protein fractionation techniques and the fractions tested to determine if they could complement the activity of an extract prepared from an XP cell line with a known defect. For instance, the XPC protein was isolated by fractionating cell extracts that complemented the activity of XP-C cell extracts. Investigators were eventually able to purify the mammalian enzymes that participate in *in vitro* nucleotide excision repair. Further study revealed these enzymes catalyze a series of steps similar to those described for the bacterial system.

9.10 Mismatch Repair

The DNA mismatch repair system removes mismatches and short insertions or deletions that are present in DNA.

Mismatch repair corrects rare base pair mismatches and short deletions or insertions that appear in DNA after replication. DNA polymerases introduce about one mispaired nucleotide per 10^5 nucleotides. The 3'→5' proofreading exonuclease increases replication fidelity by 100-fold by removing mispaired nucleotides before the growing chain can be extended further. Although an error frequency of 1 nucleotide in 10^7 may seem extremely low, it would result in a high mutation rate. The second type of error that arises during replication, short insertions and deletions, occurs because repeated-sequence motifs such as $[CA]_n$ or $[A]_n$ in microsatellites sometimes dissociate and then re-anneal incorrectly. As a result of this slippage the newly synthesized strand will have a different number of repeats than the template strand. Introduction of an insertion or deletion into the newly synthesized DNA is likely to produce a mutation if it affects a region that codes for a protein or an essential RNA molecule. Cells with a nonfunctional mismatch repair system have a high rate of mutation due to their inability to efficiently repair base pair mismatches, short insertions, or short deletions that arise during replication.

We begin our examination of mismatch repair by considering the *E. coli* mismatch repair system because this system has been the most extensively studied. Although this system provides valuable lessons for studying mismatch repair in other organisms, it differs from the mismatch repair systems used by gram-positive bacteria and eukaryotes in one important respect. The *E. coli* mismatch repair system can distinguish a newly synthesized strand from a parental strand because only the parental strand has methyl groups attached to sites with the sequence GATC.

E. coli has a deoxyadenosine methylase that transfers methyl groups from S-adenosylmethionine molecules to deoxyadenosines in GATC sequences. The timing of methylation by deoxyadenosine methylase, however, lags behind that of nucleotide addition at the replication fork by about 2 minutes, so the newly synthesized strand is transiently unmethylated. The *E. coli* mismatch repair system exploits this period of transient unmethylation to identify and cut GATC sites in a newly synthesized strand with a mismatch.

Genetic and biochemical studies have demonstrated that three *E. coli* proteins, MutS, MutL, and MutH, are dedicated to mismatch repair. Although these proteins are essential for mismatch repair, they are not sufficient. Several additional enzymes and protein factors also make important contributions. Among these enzymes and protein factors are UvrD (helicase), single-stranded DNA binding protein (SSB), 5'→3' exonucleases, 3'→5' exonucleases, DNA polymerase III holoenzyme, DNA ligase, and deoxyadenosine methylase. The *E. coli* mismatch repair system has been reconstituted *in vitro* from purified proteins, permitting us to determine how each enzyme and protein factor contributes to the process. The results of the *in vitro* studies are summarized in **FIGURE 9.20**.

The process begins when MutS binds to the mismatch (Figure 9.20). MutS recruits MutL in an ATP-dependent fashion. Then the MutS • MutL complex activates MutH, which makes an incision at the nearest unmethylated GATC site, either 5′ or 3′ to the mismatch, in the newly synthesized strand. MutH does not bind or cleave fully methylated GATC sites. UvrD unwinds the DNA, and SSB binds to the resulting single strands. When the incision is 5′ to the mismatch, a 5′→3′ exonuclease hydrolyzes the nicked strand in a 5′→3′ direction. When the incision is 3′ to the mismatch, a 3′→5′ exonuclease hydrolyzes the nicked strand in a 3′→5′ direction. DNA polymerase III holoenzyme

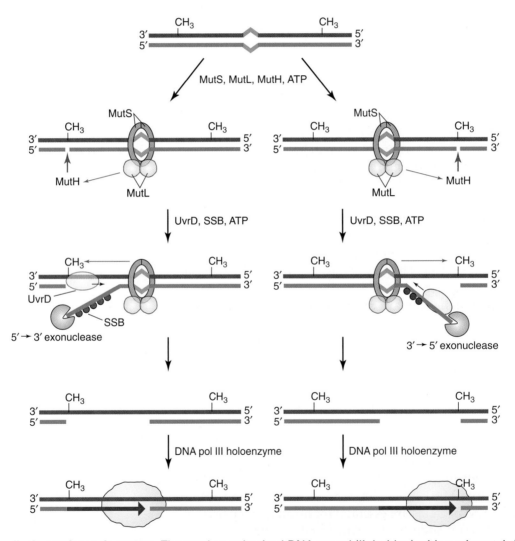

FIGURE 9.20 *E. coli* mismatch repair system. The newly synthesized DNA strand (light blue) with a mismatch (orange) is transiently unmethylated at GATC sites. The parent strand (dark blue) has methylated GATC sites. The mismatch-activated MutS • MutL • ATP complex stimulates the MutH endonuclease to incise the nearest unmethylated GATC sequence (either 5′ or 3′ from the mismatch). UvrD helicase unwinds the DNA and SSB binds to the single strands. If the incision is 5′ to the mismatch (left), then a 5′→3′ exonuclease hydrolyzes the nicked strand in a 5′→3′ direction. If the incision is 3′ to the mismatch (right), then a 3′→5′ exonuclease hydrolyzes the nicked strand in a 3′→5′ direction. DNA polymerase III holoenzyme fills the gap with new DNA (shown in red). DNA ligase seals the remaining nick and deoxyadenosine methylase adds a methyl group to the GATC site (not shown in figure). (Adapted from Lyer, R. R., et al. 2006. *Chem Rev* 106:302–323.)

fills the gap with new DNA (shown in red). DNA ligase seals the remaining nick, and deoxyadenosine methylase adds a methyl group to the GATC site.

All organisms that have a mismatch repair system have MutS and MutL homologs. MutH, however, is present only in *E. coli* and some other gram-negative bacteria. Organisms that lack MutH require some means other than the recognition of the unmethylated GATC site to distinguish between newly synthesized DNA strands and parental strands. The *E. coli* mismatch repair system suggests a possible explanation. MutH is not required if the DNA molecule already has a nick on either the 5′ or 3′ side of the mismatch.

Eukaryotes have proteins that are similar to the bacterial MutS and MutL proteins but lack a homolog to MutH. The eukaryotic MutL has endonuclease activity. Three human MutS homologs, designated MSH2, MSH3, and MSH6, participate in mismatch repair. MSH2 and MSH6 combine to form a heterodimer called MutSα, and MSH2 and MSH3 combine to form a second heterodimer called MutSβ. MutSα initiates mismatch repair at single mismatches and small insertion/deletion loops, whereas MutSβ only initiates mismatch repair at insertion/deletion loops of various sizes. The human mismatch repair system can recognize newly synthesized lagging strand DNA because the mismatched base is on an Okazaki fragment. Further work is required to determine how eukaryotes distinguish between the newly synthesized leading strand and its template strand. Individuals with a nonfunctional mismatch repair system due to a defective MutS or MutL suffer from nonpolyposis colon cancer, an autosomal recessive syndrome that greatly increases their predisposition to develop intestinal cancer.

9.11 SOS Response and Translesion DNA Synthesis

Error-prone DNA polymerases catalyze translesion DNA synthesis.

Replicative DNA polymerases such as the DNA polymerase III holoenzyme in *E. coli* and DNA polymerases δ and ε in eukaryotes occasionally encounter DNA lesions such as abasic (AP) sites or cyclobutane pyrimidine dimers that have escaped DNA repair systems. Such lesions can derail the replication complex, causing the replicative polymerase subassembly to dissociate from the DNA and possibly also from the sliding clamp. Cells survive this event by using a special class of DNA polymerases, the error-prone DNA polymerases, to catalyze DNA synthesis across the lesion. Once this translesion DNA synthesis is complete, replicative DNA polymerases can resume their normal task. Error-prone DNA polymerases catalyze DNA synthesis with a much lower fidelity than the replicative DNA polymerases. Because of their lower fidelity, the error-prone DNA polymerases are able to extend the primer strand even when the template strand is damaged. The advantage of using the error-prone DNA polymerases is that the cell is usually able to survive DNA damage that has not been repaired.

This survival comes at a cost, however, because translesion DNA synthesis introduces mutations. Important insights into error-prone DNA synthesis and the enzymes that catalyze it come from studies of the SOS response in *E. coli*.

RecA and LexA regulate the *E. coli* SOS response.

The term SOS response may seem like a strange name for a complex biological response to DNA damage. Miroslav Radman suggested the international distress signal, SOS, be applied to the signal that activates a multigene response to DNA damage in *E. coli* after UV irradiation or a chemically induced chromosomal change. Although the total number of genes induced by the SOS signal was not known in 1973 when Radman made his suggestion, subsequent studies revealed the SOS signal induces more than 40 *E. coli* SOS genes. Some induced genes such as *uvrA*, *uvrB*, and *uvrD* code for proteins that are used to repair damaged DNA, others code for proteins that participate in recombinational DNA repair, and still others code for error-prone DNA polymerases.

The SOS response is regulated by two proteins, LexA and RecA. The LexA protein is a homodimer that binds to regulatory sites that lie just 5′ of the transcription initiation sites of genes that respond to the SOS signal. The regulatory sites, 20-base pair sequences that vary slightly from one transcription unit to another, are called **operators**. Each operator has a twofold axis of symmetry, permitting one LexA subunit to bind to the left half-site and another to bind to the right half-site.

The important point for understanding how the SOS signal works is that the LexA dimer represses gene transcription when bound to the operator. Moreover, the degree of repression is directly related to the LexA dimer's affinity for the SOS gene operator. Each LexA subunit folds into two domains that are joined by a short connecting peptide. The N-terminal domain binds to an operator half-site and the C-terminal domain has a dimerization element that holds the two LexA subunits in the dimer together. The C-terminal domain also has a latent protease activity that is an important part of the signal response. Pheobe A. Rice and coworkers determined the structure for a LexA dimer bound to DNA (**FIGURE 9.21**).

RecA is a multifunctional protein that is essential for regulation of the SOS response, error-prone DNA synthesis, and, as indicated by its name, homologous recombination. The first step in the SOS response is the appearance of a single-stranded region of DNA, which is formed in response to a DNA lesion that blocks the progress of the replicating fork. RecA combines with the single-stranded DNA region to form a nucleoprotein filament that acts as the SOS response signal. Filament assembly does not take place during active replication because the SSB protein is usually present in high enough concentration to successfully compete with RecA for the relatively short single-stranded regions present at the normal replication fork. RecA nucleoprotein filament assembly, which requires ATP but not its hydrolysis, proceeds in a 5′ to 3′ direction with RecA molecules adding to the 3′ end of

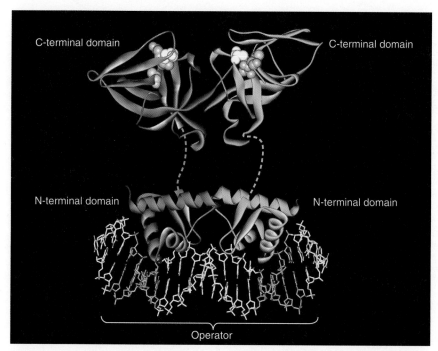

FIGURE 9.21 Model of LexA dimer docked on a *recA* operator site. LexA subunits are shown as blue and green ribbons. The N-terminal domain of each LexA subunit binds to a DNA sequence called the operator. The five residue linker connecting the two domains, which is disordered and so not visible, is shown as a dashed line. The operator DNA has the sequence $CTGTN_8ACAG$ (where N can be any nucleotide). CTGT and ACAG are shown as orange stick figures and the other nucleotides as white stick figures. The C-terminal domain has elements necessary for dimerization. Ser-119 and Lys-156 are part of the active site for a latent protease. Ser-119 is shown as a white spacefill structure and Lys-156 (here mutated to alanine) is shown as a pink spacefill structure. Cleavage takes place between Ala-84 and Gly-85, which are shown as a pink spacefill structures. (Structure from Protein Data Bank 3JSO. Zhang, A. P. P., Pigli, Y. Z., and Rice, P. A. 2010. *Nature* 466:883–886. Prepared by B. E. Tropp.)

the single-stranded DNA segment (**FIGURE 9.22**). RecA has an ATPase activity that can convert bound ATP to ADP. This hydrolysis is accompanied by the conversion of RecA from a conformation with a high affinity for DNA to one with a low affinity for DNA. ATP hydrolysis is required for nucleoprotein filament disassembly, which also takes place in a 5'-to-3' direction.

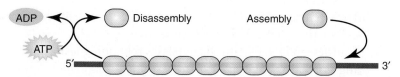

FIGURE 9.22 Assembly and disassembly of RecA nucleofilaments. The assembly and disassembly of RecA nucleofilaments are unidirectional processes, each moving in a 5'-to-3' direction along a single stranded DNA segment. More specifically, RecA adds at the 3'-proximal end and leaves from the 5'-proximal end. Disassembly requires ATP hydrolysis. (Adapted from Schlacher, K., et al. 2006. *Chem Rev* 106:406–419.)

FIGURE 9.23 Regulation of the SOS response. LexA protein (green) binds to the operator, which lies just before the SOS response genes, repressing their transcription. After UV irradiation, RecA (orange) adds to single-stranded regions of DNA to form a nucleoprotein filament. RecA in the nucleoprotein filament acts as a co-protease to activate the latent protease activity in C-terminal domain of LexA. The activated protease splits LexA into two fragments. The N-terminal fragment containing the DNA binding domain loses its high affinity for the SOS response gene operators. As a result the SOS genes are no longer repressed. The product of one SOS response gene, *dinB* (pink), forms DNA polymerase IV. The products of two other SOS response genes, *umuC* and *umuD* (blue), contribute to the formation of DNA polymerase V. Both DNA polymerases IV and V can catalyze translesion synthesis. (Adapted from Schlacher, K., et al. 2006. *Chem Rev* 106:406–419.)

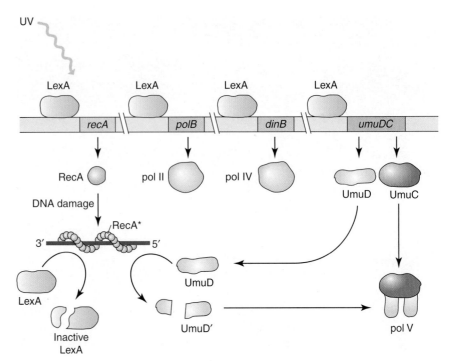

Key steps in the SOS response are shown in **FIGURE 9.23**. RecA in the nucleoprotein filament acts as a co-protease for LexA by activating the latent protease in the LexA C-terminal domain, without directly participating in proteolytic cleavage. Once activated, the LexA protease catalyzes its own cleavage, splitting into an N-terminal fragment and a C-terminal fragment. The N-terminal fragment, which contains the DNA binding domain, no longer binds to the operator with high affinity. Cleavage of LexA activates transcription of the SOS genes, including the *recA* and *lexA* genes. When the *recA* gene is fully induced, approximately 1,000 RecA monomers form a nucleoprotein filament on a segment of DNA that is about as long as an Okazaki fragment. Even a short segment of single-stranded DNA bound by 50 to 60 RecA monomers produces an SOS response signal strong enough to ensure the inactivation of LexA within a couple of minutes. The SOS response can be divided into three stages:

1. Lesions are removed by base excision repair and nucleotide excision repair.
2. RecA participates in recombinational repair, which corrects lesions that remain after the first stage.
3. Translesion DNA synthesis bypasses remaining lesions, permitting normal DNA synthesis to resume.

The LexA dimer cannot accumulate as long as RecA is present in the nucleoprotein filament. The nucleoprotein filament, however, disassembles after DNA repair or translesion DNA synthesis is complete, allowing the LexA dimer to once again accumulate and repress the SOS genes.

The SOS signal induces the synthesis of DNA polymerases II, IV, and V.

DNA damage induces the synthesis of three DNA polymerases that help the cell to deal with DNA damage that escapes the repair systems. The first of these enzymes, DNA polymerase II, the *polB* gene product, is a high fidelity polymerase that contains a 3′→5′ exonuclease activity. Normally present at about 30 to 50 molecules per cell, its level increases about sevenfold in response to the SOS signal. DNA polymerase II appears to play an important role in restarting DNA replication after lesion bypass DNA synthesis is complete. It does so by extending a mismatched primer terminus when DNA polymerase III has difficulty in doing so. DNA polymerase II can also synthesize across an abasic (AP) site and bypass some bulky adducts by primer realignment that takes advantage of repeated nucleotide sequences and introduces insertions or deletions.

The second DNA polymerase that is induced in response to the SOS signal is DNA polymerase IV, a product of the *dinB* (damaged induced) gene. The uninduced level of DNA polymerase IV, about 250 molecules per cell, increases about 10-fold on induction by the SOS signal. DNA polymerase IV can perform either error-free or error-prone translesion DNA synthesis. Its fidelity is determined by the nature of the lesion and the nucleotides that surround the lesion. Remarkably, the DNA polymerase III core subassembly and DNA polymerase IV both bind to the same sliding clamp. A choice must therefore be made between the two polymerases at a template–primer junction. The DNA polymerase III core subassembly is selected during normal and uninterrupted replication. DNA polymerase IV gains control when some impediment stalls the progress of the DNA polymerase III core subassembly. Once the stall is relieved, the DNA polymerase III core subassembly, which remains bound to the sliding clamp, regains control and resumes normal replication. The precise mechanism that leads to switching between the two DNA polymerases remains to be elucidated.

The third DNA polymerase induced in response to the SOS signal is DNA polymerase V, a heterotrimer. The protein subunits in DNA polymerase V are products of the *umuC* and *umuD* (UV mutagenesis) genes. DNA polymerase V is below detectable levels in uninduced bacteria. After induction, UmuC and UmuD protein levels increase to about 200 and 2,400 copies per cell, respectively. The RecA nucleoprotein filament activates a latent protease activity in UmuD in much the same way it activates the latent protease in LexA. Once activated, the protease cleaves UmuD, converting it to UmuD′. Then two molecules of UmuD′ combine with a molecule of UmuC to form DNA polymerase V (UmuD′$_2$C). DNA polymerase V requires the presence of the clamp loader, the sliding clamp, SSB, and RecA for efficient DNA translesion synthesis.

DNA polymerase V is converted to an active UmuD′$_2$C • RecA • ATP complex called the **mutasome**. The UmuC subunit contains the DNA polymerase activity. The mutasome can catalyze translesion synthesis through several different types of DNA lesions, including cyclobutane thymine dimers, (6-4) photoproducts, abasic (AP) sites, and large covalent adducts. Its low fidelity permits it to insert a nucleotide opposite a lesion and then add nucleotides to the resulting 3′-end. Bacteria with a nonfunctional mutasome do not accumulate mutations in response to UV light or to various chemical mutagens. Hence, such bacteria cannot be used in the Ames test. The mutasome not only introduces DNA in sites across from a lesion but also introduces untargeted mutations in sites that appear to be undamaged. Thus, cells require the mutasome to survive damage that is not repaired by DNA repair systems, but this enzyme must be highly regulated so it cannot introduce mutations into undamaged DNA when replicative DNA polymerases are not blocked by DNA damage.

Human cells have at least 14 different template-dependent DNA polymerases.

The discoveries of *E. coli* DNA polymerases IV and V motivated investigators to search the eukaryotic genome data banks to determine if eukaryotes also contained additional DNA polymerases. As a result of this type of search and related biochemical and genetic studies, it has now been established that human cells have at least 14 different template-dependent DNA polymerases. The names and some physiological properties of the eukaryotic DNA polymerases are given in TABLE 9.1.

TABLE 9.1 Proposed Nomenclature for Eukaryotic DNA Polymerase

Greek Name	HUGO name[1]	Class	Proposed Main Function
α (alpha)	POLA	B	DNA replication
β (beta)	POLB	X	Base excision repair
γ (gamma)	POLG	A	Mitochondrial replication
δ (delta)	POLDI	B	DNA replication
ε (epsilon)	POLE	B	DNA replication
ζ (zeta)	POLZ	B	Bypass synthesis
η (eta)	POLH	Y	Bypass synthesis
υ (theta)	POLQ	A	DNA repair
ι (iota)	POLI	Y	Bypass synthesis
κ (kappa)	POLK	Y	Bypass synthesis
λ (lambda)	POLL	X	Base excision repair
μ (mu)	POLM	X	Non-homologous end joining
σ (sigma)	POLS	X	Sister chromatid cohesion
	REVIL	Y	Bypass synthesis
	TDT	X	Antigen receptor diversity

Note: [1]HUGO is an acronym for Human Genome Organization.
Source: Modified from P. M. J. et al., *J Biol. Chem.* 276 (2001): 43487–43490.
© 2001, The American Society for Biochemistry and Molecular Biology.

The original Greek names have been replaced by names suggested by the Human Genome Organization (HUGO). Unfortunately, the HUGO recommendations are not always followed in the literature, and the older Greek names are often used. DNA polymerases are grouped into classes depending on sequence homologies and structure similarities. The *E. coli* DNA polymerases I, II, III, and IV/V are founding members of classes A, B, C, and Y, respectively. Eukaryotic PolB (DNA polymerase β) is the founding member of class X. Although each kind of eukaryotic DNA polymerase appears to have a specific function, some polymerases appear to be able to stand in for others.

Questions and Problems

1. Briefly define, describe, or illustrate each of the following.
   a. DNA repair
   b. Endogenous agent
   c. Exogenous agent
   d. Mutagen
   e. Carcinogen
   f. Proto-oncogene
   g. Oncogene
   h. Tumor suppressor gene
   i. Cyclobutane pyrimidine dimer
   j. (6-4) Photoproduct
   k. Inversion
   l. Translocation
   m. Abasic site
   n. Transition mutation
   o. Transversion mutation
   p. CpG island
   q. Electrophile
   r. Alkylating agent
   s. Monoadduct
   t. Ames test
   u. Microsome
   v. Photoreactivation
   w. Suicide enzyme
   x. Short patch repair
   y. Long patch repair
   z. Mutasome

2. Some people sit out in the sun to obtain a "healthy" tan. Based on your knowledge of UV light's effect on DNA, what do you think about long exposure to sunlight? Be specific.

3. X-rays damage DNA.
   a. How do x-rays cause this damage?
   b. X-ray damage to DNA can be isolated or clustered. Which type of damage is more harmful to the cell? Explain your answer.

4. Many years ago shoe stores had x-ray machines that permitted customers to see how well their new shoes fit. Based on your knowledge of the effects of x-rays on DNA, what do you think of this practice? Can you think of any instances where we may be overexposed to x-rays today?

5. Briefly describe how the deamination of an adenine or cytosine base in DNA can lead to a point mutation. What type of point mutation will be formed in each case?

6. Alkylating agents cause various types of DNA damage. Draw a short oligodeoxynucleotide that contains A, G, C, and T and show the sites at which alkylation may take place.

7. Polycyclic aromatic hydrocarbons do not damage DNA in *E. coli* but do damage DNA in human cells. Explain this observation.

8. Why is it usually easier for a cell to repair damage caused by a monoadduct such as a methyl or ethyl group than to repair damage caused by a chemical cross-linking agent such as nitrogen mustard gas?

9. Briefly describe an experimental approach you would take to determine if a newly discovered organic molecule is a mutagen or carcinogen.

10. *E. coli* has a single enzyme that can remove a methyl group from a phosphotriester, O^6-methylguanine, or O^4-methylthymine. Identify this enzyme and describe how it performs these demethylations. How does this enzyme regulate DNA repair?

11. Show the pathway that a bacterial cell can use to repair an abasic site. What additional enzymes are required to remove a damaged base?

12. Bacteria that are exposed to UV-C light accumulate cyclobutane pyrimidine dimers. The number of dimers drops dramatically after the irradiated bacteria are exposed to visible light. How do you explain these observations?
13. What enzymes do bacteria need to repair a cyclobutane pyrimidine dimer or a (6-4) photodimer in the dark?
14. A mutation in genes for each of the following proteins makes the corresponding mutant more sensitive to UV light than is its wild-type parent. Provide an explanation for the increased sensitivity of each mutant.
    a. DNA polymerase I
    b. UvrA
    c. UvrB
    d. UvrC
    e. UvrD
15. *E. coli* is able to detect a mismatch that forms during DNA replication. How do the cells distinguish the newly formed strand with the error from the parental strand? How do the cells remove the mismatched base and fill in the resulting gap?
16. What roles do the RecA and LexA proteins play in the *E. coli* SOS response?
17. DNA polymerases IV and V are error-prone polymerases. Why do cells require such polymerases?
18. Briefly explain why the sensitivity of the Ames test increases when the test cells have a high level of DNA polymerase V.

Suggested Reading

Historical

Friedberg, E. C. 1997. *Correcting the Blueprint of Life*. Woodbury, NY: Cold Spring Harbor Laboratory Press.

General

Dalhus, B., Laerdahl, J. K., Backe, P. H., and Bjørås, M. 2009. DNA base repair–recognition and initiation of catalysis. *FEMS Microbiol Rev* 33:1044–1078.

Friedberg, E. C., Walker, G. C., Siede, W., et al. 2006. *DNA Repair and Mutagenesis* (2nd ed). Washington, DC: ASM Press.

Sancar, A., Lidsey-Boltz, L. A., Ünsel-Kaçmaz, K., and Linn, S. 2004. Molecular mechanisms of mammalian DNA repair and the DNA damage checkpoints. *Annu Rev Biochem* 73:39–85.

Schärer, O. D. 2003. Chemistry and biology of DNA repair. *Angew Chem Int Ed* 42:2946–2974.

DNA Damage

Bethea, D., Fullmer, B., Syed, S., et al. 1999. Psoralen photobiology and photochemotherapy: 50 years of science and medicine. *J Dermatol Sci* 19:78–88.

Cooke, M. S., Evans, M. D., Dizdaroglu, M., and Lunec, J. 2003. Oxidative DNA damage: mechanisms, mutation, and disease. *FASEB J* 17:1195–1214.

Doetsch, P. W. 2001. DNA damage. *The Encyclopedia of Life Sciences*. pp. 1–7. London: Nature.

Gasparro, F. P., Liao, B., Foley, P. J., et al. 1998. Psoralen photochemotherapy, clinical efficacy, and photomutagenicity: the role of molecular epidemiology in minimizing risks. *Environ Mol Mutagen* 31:105–112.

Goodsell, D. S. 2001. The molecular perspective: cytochrome P450. *Oncologist* 6:205–206.

Guengerich, F. P. 2003. Cytochrome P450 oxidations in the generation of reactive electrophiles: epoxidation and related reactions. *Arch Biochem Biophys* 409:59–71.

Jamieson, E. R., and Lippard, S. J. 1999. Structure, recognition, and processing of cisplatin DNA adducts. *Chem Rev* 99:2467–2498.

Marnett, L. J. 2000. Oxyradicals and DNA damage. *Carcinogenesis* 21:361–370.

McCann, J., Choi, E., Yamasaki, E., and Ames, B. N. 1975. Detection of carcinogens as mutagens in the *Salmonella*/microsome test: assay of 300 chemicals. *Proc Natl Acad Sci USA* 72:5135–5139.

Moore, B. S., Morris, L. P., and Doetsch, P. W. 2009. DNA damage. *Encyclopedia of Life Sciences*. pp. 1–10. Hoboken, NJ: John Wiley & Sons.

Rosenberg, B., Van Camp, L., Trosko, J. E., and Mansour, V. H. 1969. Platinum compounds: a new class of potent antitumor agents. *Nature* 222:385–386.

Spielmann, H. P., Dwyer, T. J., Hearst, H. E., and Wemmer, D. E. 1995. Solution structures of psoralen monoadducted and cross-linked DNA oligomers by NMR spectroscopy and restrained molecular dynamics. *Biochemistry* 34:12937–12953.

Spratt, T. E., and Levy, D. E. 1997. Structure of the hydrogen bonding complex of O^6-methylguanine with cytosine and thymine during DNA replication. *Nucleic Acids Res* 25:3354–3361.

Tsutomu Shimada, T., and Fujii-Kuriyama, Y. 2004. Metabolic activation of polycyclic aromatic hydrocarbons to carcinogens by cytochromes P450 1A1 and 1B1. *Cancer Sci* 95:1–6.

Williams, J. H., Phillips, T. D., Jolly, P. E., et al. 2004. Human aflatoxicosis in developing countries: a review of toxicology, exposure, potential health consequences, and interventions. *Am J Clin Nutr* 80:1106–1122.

Zhang, C. X., and Lippard, S. J. 2003. New metal complexes as potential therapeutics. *Curr Opin Chem Biol* 7:481–489.

Detection of Mutagens

Ames, B. N. 2003. An enthusiasm for metabolism. *J Biol Chem* 278:4369–4380.

Ames, B. N., Durston, W. E., Yamasaki, E., and Lee, F. D. 1973. Carcinogens are mutagens: a simple test system combining liver homogenates for activation and bacteria for detection. *Proc Natl Acad Sci USA* 70:2281–2285.

Mortelmans, K., and Zeiger, E. 2000. The Ames *Salmonella*/microsome mutagenicity assay. *Mutat Res* 455:29–60.

Webrzyn, G., and Czyz, A. 2003. Detection of mutagenic pollution of natural environment using microbiological assays. *J Appl Microbiol* 95:1175–1181.

Damage Reversal

Begley, T. J., and Samson, L. D. 2004. Reversing DNA damage with a directional bias. *Nat Struct Mol Biol* 11:688–690.

Goosen, N., and Moolenaar, G. F. 2008. Repair of UV damage in bacteria. *DNA Repair* 7:353–379.

Kao, Y.-T., Saxena, C., Wang, L., et al. 2005. Direct observation of thymine dimer repair in DNA by photolyase. *Proc Natl Acad Sci USA* 102:16128–16132.

Kelner, A. 1949. Effect of visible light on the recovery of *Streptomyces griseus conidia* from ultraviolet irradiation injury. *Proc Natl Acad Sci USA* 35:73–79.

Mees, A., Klar, T., Gnau, P., et al. 2004. Crystal structure of a photolyase bound to a CPD-like DNA lesion after in situ repair. *Science* 306:1789–1793.

Memisoglu, A., and Samson, L. D. 2001. DNA repair by reversal of damage. *Encyclopedia of Life Sciences*. pp. 1–8. London: Nature.

Mishina, Y., Duguid, E. M., and He, C. 2006. Direct reversal of DNA alkylation damage. *Chem Rev* 106:215–232.

Müller, M., and Carell, T. 2009. Structural biology of DNA photolyases and cryptochromes. *Curr Opin Struct Biol* 19:277–285.

Rupert, C. S., Goodgal, S. H., and Herriott, R. M. 1957. Photoreactivation in vitro of ultraviolet inactivated *Hemophilus influenzae* transforming factor. *J Gen Physiol* 41:451–471.

Sancar, A. 2003. Structure and function of DNA photolyase and cryptochrome blue-light photoreceptors. *Chem Rev* 103:2203–2237.

Sancar, A. 2008. Structure and function of photolyase and *in vivo* enzymology: 50th anniversary. *J Biol Chem* 283:32153–32157.

Schärer, O. D. 2003. Chemistry and biology of DNA repair. *Angewandte Chemie Int Ed* 42:2946–2974.

Base Excision Repair

Barnes, D. E., and Lindahl, T. 2004. Repair and genetic consequences of endogenous DNA base damage in mammalian cells. *Annu Rev Biochem* 38:445–476.

Baute, J., and Depicker, A. 2008. Base excision repair and its role in maintaining genome stability. *Crit Rev Biochem Mol Biol* 43:239–276.

Beard, W. A., and Wilson, S. H. 2006. Structure and mechanism of DNA polymerase. *Chem Rev* 106:361–382.

Hegde, M. L., Hazra, T., and Mitra, S. 2008. Early steps in the DNA base excision/single-strand interruption repair pathway in mammalian cells. *Cell Res* 18:27–47.

Krokan, H. E., Nilsen, H., Skorpen, F., et al. 2000. Base excision repair of DNA in mammalian cells. *FEBS Lett* 476:73–77.

Memisoglu, A., and Samson, L. 2000. Base excision repair in yeast and mammals. *Mutat Res* 451:39–51.

Nilsen, H., and Krokan, H. E. 2001. Base excision repair in a network of defense and tolerance. *Carcinogenesis* 22:987–998.

Robertson, A. B., Klungland, A., Rognes, T., and Leiros, I. 2009. Base excision repair: the long and short of it. *Cell Mol Life Sci* 66:981–993.

Sander, M., and Wilson, S. H. 2002. Base excision repair, AP endonucleases, and DNA glycosylases. *Encyclopedia of Life Sciences*. pp. 1–9. London: Nature.

Seeberg, E., Eide, L., and Bjørås, M. 1995. The base excision pathway. *Trends Biochem Sci* 20:391–397.

Nucleotide Excision and Repair

Batty, D. P., and Wood, R. D. 2000. Damage recognition in nucleotide excision repair of DNA. *Gene* 241:193–204.

Cleaver, J. E. 2001. Xeroderma pigmentosum: the first of the cellular caretakers. *Trends Biochem Sci* 26:398–401.

Cleaver, J. E., Karplus, K., Kashani-Sabet, M., and Limoli, C. L. 2001. Nucleotide excision repair "a legacy of creativity." *Mutat Res* 485:23–36.

Doetsch, P. W. 2001. DNA repair disorders. *Encyclopedia of Life Sciences*. pp. 1–6. London: Nature.

Friedberg, E. C. 2000. Nucleotide excision repair and cancer disposition. *Am J Pathol* 157:693–701.

Friedberg, E. C. 2001. How nucleotide excision repair protects against cancer. *Nat Rev Cancer* 1:22–33.

Gillet, L. C. J., and Schärer, O. D. 2006. Molecular mechanisms of mammalian global genome nucleotide excision repair. *Chem Rev* 106:253–276.

Hutsell, S. Q., and Sancar, A. 2005. Nucleotide excision repair, oxidative damage, DNA sequence polymorphisms, and cancer treatment. *Clin Cancer Res* 11:1355–1357.

Lainé, J. P., and Egly, J.-M. 2006. When transcription and repair meet: a complex system. *Trends Genet* 22:430–436.

Lehmann, A. R. 1995. Nucleotide excision repair and the link with transcription. *Trends Biochem Sci* 20:402–405.

Mitchell, J. R., Hoeijmakers, J. H. J., and Niedernhofer, L. J. 2003. Divide and conquer: nucleotide excision repair battles cancer and ageing. *Curr Opin Cell Biol* 15:232–240.

Pakotiprapha, D., Samuels, M., Shen, K. 2012. Structure and mechanism of the UvrA–UvrB DNA damage sensor. *Nat Struct Mol Biol* 19:291–298.

Park, C.-J., and Choi, B. S. 2006. The protein shuffle: sequential interactions among components of the human nucleotide excision repair pathway. *FEBS J* 273:1600–1608.

Rasmussen, R. E., and Painter R. B. 1964. Evidence for repair of ultra-violet damaged deoxyribonucleic acid in cultured mammalian cells. *Nature* 203:1360–1362.

Robertson, A. B., Klungland, A., Rognes, T., and Leiros, I. 2009. Base excision repair: the long and short of it. *Cell Mol Life Sci* 66:981–993.

Truglio, J. J., Croteau, D. L., Van Houten, B., and Kisker, C. 2006. Prokaryotic nucleotide excision repair: the UvrABC system. *Chem Rev* 106:233–252.

Mismatch Repair

Constantin, N., Dzantiev, L., Kadyrov, F. A., and Modrich, P. 2005. Human mismatch repair reconstitution of a nick-directed bidirectional reaction. *J Biol Chem* 280:39752–39761.

de Wind, N., and Hays, J. B. 2001. Mismatch repair: praying for genome stability. *Curr Biol* 11:R545–R548.

Fishel, R. 1999. Signaling mismatch repair in cancer. *Nat Med* 5:1239–1241.

Fleck, T. M., Kunz, C., and Fleck, O. 2002. DNA mismatch repair and mutation avoidance pathways. *J Cell Physiol* 191:28–41.

Harfe, B. D., and Jinks-Robinson, S. 2000. DNA mismatch repair and genetic instability. *Annu Rev Genet* 34:359–399.

Iyer, R. R., Pluciennick, A., Burdett, V., and Modrich, P. L. 2006. DNA mismatch repair: functions and mechanisms. *Chem Rev* 106:302–323.

Jiricny, J. 2000. Mismatch repair: the praying hands of fidelity. *Curr Biol* 10:R788–R790.

Jiricny, J. 2006. The multifaceted mismatch-repair system. *Nat Rev Mol Cell Biol* 7:335–346.

Jiricny, J. 2006. MutLα: at the cutting edge of mismatch repair. *Cell* 126:239–241.

Kadyrov, F. A., Dzantiev, L., Constantin, N., and Modrich, P. 2006. Endonucleolytic function of MutLα in human mismatch repair. *Cell* 126:297–308.

Kolodner, R. D. 1995. Mismatch repair: mechanisms and relationship to cancer susceptibility. *Trends Biochem Sci* 20:397–401.

Kunkel, T. A., and Erie, D. A. 2005. DNA mismatch repair. *Annu Rev Biochem* 74:681–710.

Li, G.-M. 2008. Mechanisms and functions of DNA mismatch repair. *Cell Res* 18:85–98.

Marti, T. M., Kunz, C., and Fleck, O. 2002. DNA mismatch repair and mutation avoidance pathways. *J Cell Physiol* 191:28–41.

Modrich, P. 2006. Mechanisms in eukaryotic mismatch repair. *J Biol Chem* 281:30305–30309.

Peltomäki, P. 2001. DNA mismatch repair and cancer. *Mutat Res* 488:77–85.

Polosina, Y. Y., and Cupples, C. G. 2009. MutL: conducting the cell's response to mismatched and misaligned DNA. *BioEssays* 32:51–59.

Schofield, M. J., and Hsieh, P. 2003. DNA mismatch repair: molecular mechanisms and biological function. *Annu Rev Microbiol* 57:579–608.

Sixma, T. K. 2001. DNA mismatch repair: MutS structures bound to mismatches. *Curr Opin Struct Biol* 11:47–52.

SOS Response and Translesion DNA Synthesis

Baynton, K., and Fuchs, R. P. P. 2000. Lesions in DNA: hurdles for polymerases. *Trends Biochem Sci* 25:74–79.

Burgers, P. M. J., Koonen, E. V., Bruford, E., et al. 2001. Eukaryotic DNA polymerases: proposal for a revised nomenclature. *J Biol Chem* 276:43487–43490.

Butala, M., Žgur-Bertok, D., and Busby, S. J. W. 2009. The bacterial LexA transcriptional repressor. *Cell Mol Life Sci* 66:82–93.

Chattopadhyaya R., Ghosh, K., and Namboodiri, V. M. 2000. Model of a LexA repressor dimer bound to recA operator. *J Biomol Struct Dyn* 18:181–197.

Chattopadhyaya, R., and Pal, A. 2004. Improved model of a LexA repressor dimer bound to recA operator. *J Biomol Struct Dyn* 5:681–689.

Clark, A. J., and Margulies, A. D. 1965. Isolation and characterization of recombination-deficient mutants of *Escherichia coli* K12. *Proc Natl Acad Sci USA* 53:451–459.

Defals, M., and Devoret, R. 2001. SOS response. *Encyclopedia of Life Sciences*. pp. 1–9. London: Nature.

Friedberg, E. C., Lehmann, A. R., and Fuchs, R. P. P. 2005. Trading places: how do DNA polymerases switch during translesion DNA synthesis? *Mol Cell* 18:499–505.

Friedberg, E. C., Wagner, W., and Radman, M. 2002. Specialized DNA polymerases, cellular survival, and genesis of mutations. *Science* 296:1627–1630.

Goodman, M. F. 2000. Coping with replication "train wrecks" in *Escherichia coli* using Pol V, Pol II, and RecA proteins. *Trends Biochem Sci* 25:189–195.

Goodman, M. F. 2002. Error-prone repair DNA polymerases in prokaryotes and eukaryotes. *Annu Rev Biochem* 71:17–50.

Goodman, M. F., and Tippin, B. 2000. Sloppier copier DNA polymerases involved in genome repair. *Curr Opin Genet Dev* 10:162–168.

Guengerich, F. P. 2006. Interactions of carcinogen-bound DNA with individual RNA polymerases. *Chem Rev* 106:420–452.

Hübscher, U. 2005. DNA polymerases: Eukaryotic. *Encyclopedia of Life Sciences*. pp. 1–4. New York: John Wiley & Sons.

Janion, C. 2001. Some aspects of the SOS response system—a critical survey. *Acta Biochim Polon* 48:559–610.

Jiang, Q., Karata, K., Woodgate, R., et al. 2009. The active form of DNA polymerase V is UmuD$'_2$C–RecA–ATP *Nature* 460:259–263.

Kunkel, T. A., Pavlov, Y., and Bebenek, K. 2003. Functions of human DNA polymerases and suggested by their properties, including fidelity with undamaged DNA templates. *DNA Repair (Amst)* 2:135–149.

Livneh, Z. 2001. DNA damage control by novel DNA polymerases: translesion replication and mutagenesis. *J Biol Chem* 276:25639–25642.

López de Saro, F. J., Georgescu, R. E., Goodman, M. F., and O'Donnell, M. 2003. Competitive processivity-clamp usage by DNA polymerases during DNA replication and repair. *EMBO J* 22:6408–6418.

McIherney, P., and O'Donnell, M. 2004. Functional uncoupling of twin polymerases. *J Biol Chem* 279:21543–21551.

Miller, J. H. 2005. Perspective on mutagenesis and repair: the standard model and alternate modes of mutagenesis. *Crit Rev Biochem Mol Biol* 40:155–179.

Patel, M., Jiang, Q., Woodgate, R. et al. 2010. A new model for SOS-induced mutagenesis: how RecA protein activates DNA polymerase V. *Crit Rev Biochem Mol Biol* 45:171–184.

Plosky, B. S., and Woodgate, R. 2004. Switching from high-fidelity replicases to low-fidelity lesion-bypass polymerases. *Curr Opin Genet Dev* 14:113–119.

Prakash, S., Johnson, R. E., and Prakash, L. 2005. Eukaryotic translesion synthesis DNA polymerases: specificity of structure and function. *Annu Rev Biochem* 74:317–353.

Ramadan, K., Shevelev, I., and Hübscher, U. 2004. The DNA-polymerase-X family: controllers of DNA quality? *Nat Rev Mol Cell Biol* 5:1038–1043.

Rattray, A. J., and Strathern, J. N. 2003. Error-prone DNA polymerases: when making a mistake is the only way to go ahead. *Annu Rev Genet* 37:31–66.

Rosenberg, S. M. 2001. Evolving responsively: adaptive mutation. *Nat Rev Genet* 2:504–515.

Schlacher, K., and Goodman, M. F. 2007. Lessons from 50 years of SOS DNA-damage-induced mutagenesis. *Mol Cell Biol* 8:587–594.

Schlacher, K., Pham, P., Cox, M. M., and Goodman, M. F. 2006. Roles of DNA polymerase V and RecA protein in SOS damage-induced mutation. *Chem Rev* 106:406–419.

Scherbakova, P. V., and Fijalkowska, I. J. 2006. Translesion synthesis DNA polymerases and control of genome stability. *Front Biosci* 11:2496–2517.

Showalter, A. K., Lamarch, B. J., Bakhtina, M., et al. 2006. Mechanistic comparison of high-fidelity and error prone DNA polymerases and ligases involved in DNA repair. *Chem Rev* 106:340–360.

Steitz, T. A., and Yin, Y. W. 2003. Accuracy, lesion bypass, strand displacement and translocation by DNA polymerases. *Philos Trans R Soc (Lond B)* 359:17–23.

Tippin, B., Pham, P., and Goodman, M. F. 2004. Error-prone replication for better or worse. *Trends Microbiol* 12:288–295.

Yang, W. 2003. Damage repair DNA polymerases Y. *Curr Opin Struct Biol* 13:23–30.

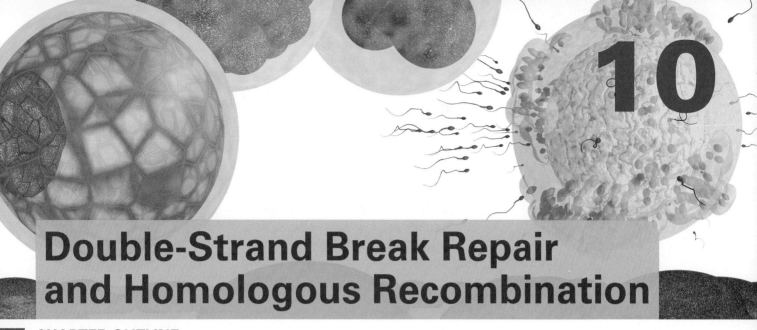

Double-Strand Break Repair and Homologous Recombination

CHAPTER OUTLINE

10.1 Replication Restart After Bacterial Replication Fork Collapse
Cells use homologous recombination to restart replication after a replication fork collapses.
The RecBCD protein processes the broken end of the double-strand break to form a single-stranded 3′-OH tail.
RecA acts as a recombinase to catalyze strand invasion.
The RuvAB protein complex catalyzes branch migration in *E. coli*.
RuvC helps convert the DNA with the Holliday junction into a DNA fork.
PriA helps to load DnaB on the reestablished DNA fork to restart replication.

10.2 Mitotic Recombination
Eukaryotic cells use homologous recombination to repair double-strand breaks during the late S stage or the G_2 stage of the cell cycle.

10.3 Gene Knockouts
Homologous recombination can be used to create gene knockouts in mice.

10.4 Nonhomologous End-Joining
Nonhomologous end-joining connects broken ends that have no homology.

10.5 Meiotic Recombination
Homologous recombination during meiosis can result in gene conversion.
Homologous recombination during meiosis requires some enzymes that are unique to it.

QUESTIONS AND PROBLEMS

SUGGESTED READING

Image © Scivit/ShutterStock, Inc.

In this chapter we examine the repair of **double-strand breaks**, which cannot be repaired by using information present in a complementary strand because both DNA strands are cut at the same (or nearly the same) position on the DNA duplex. Many enzymes that participate in the repair of double-strand breaks are essential so mutations that cause a complete loss of their function are lethal. Even mutations that cause one of these repair enzymes to have a partial loss of function can have harmful consequences. For instance, humans with defects in enzymes or proteins that participate in the repair of double-strand breaks may have a developmental disorder, a compromised immune system, or an increased susceptibility to various forms of cancer.

Cells use two major pathways to repair double-strand breaks. The first is called **homologous recombination** because repair enzymes use the information in one DNA duplex to repair a double-strand break in an identical or nearly identical (homologous) DNA duplex. Homologous recombination usually proceeds with high fidelity so no mutations are introduced into the repaired duplex. The second pathway is called **nonhomologous end-joining** because repair enzymes join broken DNA ends with no homology or microhomology limited to one to four nucleotides. Although nonhomologous end-joining sometimes takes place with high fidelity, at other times it leads to nucleotide additions or deletions at the repair joint. Mutations that arise at the repair joint are almost always preferable to the alternative, loss of large amounts of genetic information—often the entire chromosome.

We begin our examination of double-strand break repair by examining how bacterial cells repair a double-strand break that arises when the bacterial replication machinery encounters a nick in one strand. Information gained from studying this repair pathway is then applied to examining the repair of double-strand breaks in somatic cells during different stages of the cell cycle and to the genetic exchanges that take place during the first meiotic division.

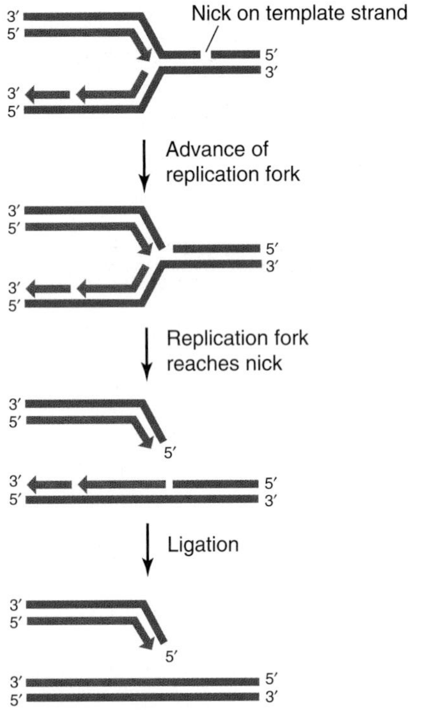

10.1 Replication Restart After Bacterial Replication Fork Collapse

Cells use homologous recombination to restart replication after a replication fork collapses.

Double-strand breaks are generated from nicks in a single strand that arise during normal DNA synthesis and repair. For instance, DNA ligase may not have had a chance to repair a nick formed during base or nucleotide exchange repair. **FIGURE 10.1** shows the formation of a double-strand break when the replication fork encounters a nick on the template for leading strand synthesis. One branch of the fork separates from the rest of the fork in a process known as **fork collapse**. Cells require specific enzymes to restart replication at a collapsed fork. These enzymes have been most extensively investigated in *Escherichia coli*. The pathway used to restart replication after fork collapse involves (1) single-stranded 3′-OH tail formation, (2) strand

FIGURE 10.1 Replication fork collapse. The replication fork reaches a nick in the leading strand template, producing a double-strand break that separates one branch of the fork from the parental DNA. This separation process is called fork collapse. Okazaki fragments in the lagging strand are joined by DNA ligase. A similar type of fork collapse takes place when a nick is present in the lagging strand template.

FIGURE 10.2 **Bacterial pathway to repair a collapsed replication fork and resume replication.** A collapsed replication fork (see Figure 10.1) is repaired as follows: (1) The RecBCD complex cuts back (resects) the blunt end of the separated branch to form a single-stranded 3′-OH tail. (2) RecA binds to the newly created 3′-overhang and acts as a recombinase to catalyze strand invasion into the parental duplex. (3) RuvAB catalyzes branch migration in the direction shown by the large cyan arrow to form a cross-shaped structure between the two duplexes called a Holliday junction (HJ). (4) The resolvase RuvC cleaves the Holiday junction at the site indicated by red arrows. DNA ligase (not shown) joins the strands to generate a DNA fork. (5) PriA loads the DnaB helicase on the DNA fork and then other components of the replication machinery add to reestablish an active replication fork. Further information about each of these steps is presented in the text.

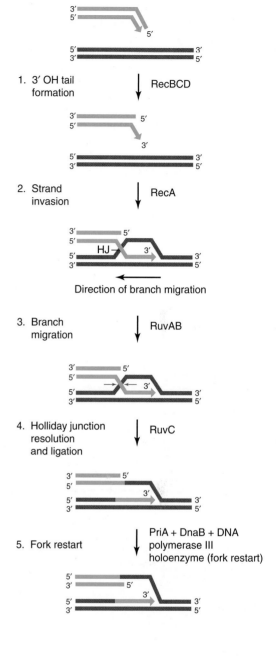

invasion and Holliday junction formation, (3) branch migration, (4) Holliday junction resolution and ligation, and (5) replication fork restart. This process is shown in **FIGURE 10.2**, and the individual steps are described in greater detail below.

The RecBCD protein processes the broken end of the double-strand break to form a single-strand 3′-OH tail.

The first clue to the existence of bacterial enzymes that participate in the repair of double-strand breaks came from studying *E. coli* mutants selected for their inability to perform genetic recombination. The defective genes were named *recA*, *recB*, *recC*, and *recD* to indicate the <u>rec</u>ombination defect. The *recA* gene product is RecA. The polypeptides encoded by the *recB*, *recC*, and *recD* genes form a RecBCD protein complex that binds to the blunt end of the branch released as a result of replication fork collapse. The crystal structure for this enzyme–substrate complex indicates that RecBCD unwinds the blunt end (**FIGURE 10.3**).

Although crystal structures are not yet available for the subsequent steps catalyzed by the RecBCD complex, biochemical and genetic studies provide important insights into the contributions the enzyme complex makes to the reestablishment of the replication fork. RecB and RecD have ATP-dependent helicase activities. RecB moves along one strand in a 3′→5′ direction, while RecD moves along the complementary strand in a 5′→3′ direction so that both helicases move inward from the blunt end (**FIGURE 10.4**). The RecB subunit also has a domain with nuclease activity that is connected to the rest of the subunit by a flexible linker. This nuclease can act on the end of either strand. The RecC subunit interacts with both of the other subunits, increasing their processivity and regulating their rate of movement. Initially, RecD is the faster moving helicase, and the nuclease preferentially hydrolyzes the single strand with the 3′ end. This degradation pattern continues until the RecBCD complex reaches a 5′-GCTGGTGG-3′ sequence, which is known as a ***chi*** (<u>c</u>rossover <u>h</u>otspot <u>i</u>nstigator) **site**. Because the *chi* site is not symmetric, recombination takes place in only one direction. About one *chi* site is present for every 5,000 base pairs. This is a much higher frequency of occurrence than would be predicted on a statistical basis. When a

FIGURE 10.3 Crystal structure showing how the RecBCD complex binds to the blunt end of a DNA substrate analog. The DNA substrate analog has a hairpin structure. Its single-stranded 3'-OH tail interacts with the RecB subunit (orange) and its 5' end interacts with the RecD subunit (green). The RecB subunit has 3'→5' helicase activity and the RecD subunit has 5'→3' helicase activity. The RecB subunit also has nuclease activity that is in a domain that is connected to the rest of the subunit by a flexible linker. The RecC subunit interacts with both of the other subunits, increasing their processivity and regulating their rate of movement. (Structure from Protein Data Bank 3K70. Saikrishnan, K., Griffiths, S. P., Cook, N., Court, R., and Wigley, D. B. 2008. *EMBO J* 27:2222–2229 Prepared by B. E. Tropp.)

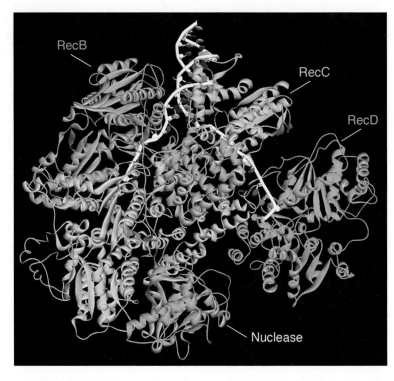

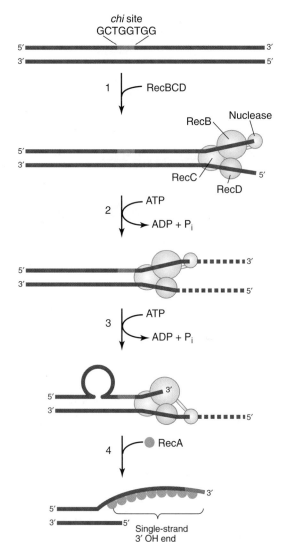

RecBCD complex encounters a *chi* site it undergoes a conformational change that slows the rate at which the RecD helicase moves so that RecB becomes the faster helicase. The conformational change also causes the nuclease to exclusively hydrolyze the single strand with the 5' end. These activity changes lead to the formation of a single-stranded 3'-OH tail. RecBCD also helps to load RecA onto the newly generated single-stranded 3'-OH tail to form a RecA coated nucleoprotein filament and dissociates from the complex.

RecA acts as a recombinase to catalyze strand invasion.

The RecA-coated nucleoprotein filament engages in a search for homology. When the filament encounters a homologous DNA duplex, RecA acts as a recombinase and catalyzes ATP-dependent strand invasion (Figure 10.2, step 2). Additional details for the strand invasion process are shown in **FIGURE 10.5**. A **displacement loop (D-loop)** forms as the

FIGURE 10.4 RecBCD processing double-strand break ends. (1) RecBCD binds to the blunt end of the branch released as a result of replication fork collapse. (2) RecB and RecD are ATP-dependent helicases. Initially, RecD (5'→3' helicase) is the faster moving helicase and the nuclease preferentially hydrolyzes the strand with the 3' end. This degradation pattern continues until the RecBCD complex reaches a *chi* site (orange region). (3) After encountering the *chi* site the RecBCD complex undergoes a conformational change that slows down the RecD helicase so that RecB (3'→5' helicase) becomes the faster helicase. The conformational change also causes the nuclease to exclusively hydrolyze the single strand with the 5' end. (4) These activity changes lead to the formation of a 3'-OH single-strand overhang. The RecBCD loads RecA onto this 3'-OH single-strand overhang and dissociates from the complex.

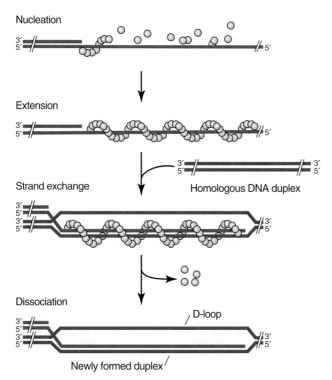

FIGURE 10.5 Strand invasion and displacement loop (D-loop) formation. The RecA coated nucleoprotein filament searches a DNA duplex for homology and, upon encountering it, catalyzes ATP-dependent strand exchange. The net effect is that the invading strand displaces the duplex strand that has the same nucleotide sequence. The displaced strand forms a displacement loop (D-loop). RecA dissociates from the DNA during this strand invasion process. (Adapted from Holthausen, J. T., Wyman, C., and Kanaar, R. 2010. *DNA Repair* 9:1264–1272.)

end of the invading strand displaces a strand in the duplex DNA with an identical (or nearly identical) sequence. RecA dissociates from the DNA with concomitant ATP hydrolysis during this process. The newly formed duplex can be extended by unidirectional branch migration through local denaturation of the original DNA duplex and annealing of the complementary single strand. The outcome of strand invasion is to form a cross-shaped structure between the two duplexes. This structure is called a **Holliday junction** after Robin Holliday, who proposed its existence in 1964 to explain a specific type of genetic exchange observed during meiosis in yeast. The crossed-strand Holliday junction can rotate at the crossover to form an unfolded junction (**FIGURE 10.6**).

The RuvAB protein complex catalyzes branch migration in *E. coli*.

Although branch migration can take place in the absence of a catalyst, the process is very slow. In bacteria this process is catalyzed by the RuvAB protein complex (Figure 10.2, step 3). The two kinds of subunits, RuvA and RuvB, are named for the genes that code for them. These genes, *ruvA* and *ruvB*, were originally identified because mutants with defects in either gene exhibit decreased resistance to UV light. **FIGURE 10.7** shows a schematic of the RuvAB complex as it acts at an

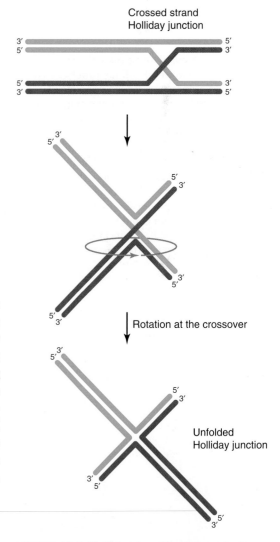

FIGURE 10.6 Holliday junction isomerization. The Holliday junction can undergo a rotation of the DNA strands around the crossed strand junction (indicated by the circular green arrow) to produce the unfolded or open structure shown on the bottom. This rotation process is termed Holliday junction isomerization.

10.1 Replication Restart After Bacterial Replication Fork Collapse

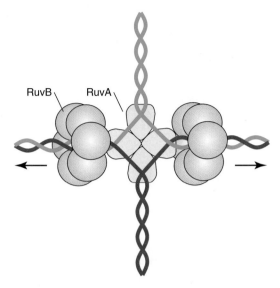

FIGURE 10.7 **RuvAB complex catalyzed branch migration at a Holliday junction.** The RuvA protein complex, which consists of eight RuvA subunits, recruits two ring-shaped RuvB hexamers to the Holliday junction so each RuvB hexamer has a Holliday junction DNA arm within its central cavity. Only the front four RuvA subunits are visible in this perspective. The RuvB hexamers use the energy supplied by ATP to force the DNA arms in the direction shown by the arrows and thereby promote rapid branch migration. (Adapted from Rice, K. P., and Cox, M. M. 2001. *Encyclopedia of Life Sciences*.)

unfolded Holliday junction. An octamer of RuvA subunits interacts with the Holliday junction. Six RuvB subunits form a ring-shaped ATP-dependent DNA pump. The RuvA protein complex recruits two of these ring-shaped hexamers to the Holliday junction so each RuvB hexamer has one of the Holliday junction's DNA arms within its central cavity. The RuvB hexamers use the energy supplied by ATP to force the DNA arms outward and thereby promote rapid branch migration.

RuvC helps convert the DNA with the Holliday junction into a DNA fork.

The cell still needs to convert the DNA structure with a Holliday junction back to a DNA fork. The RuvC protein acting in concert with DNA ligase accomplishes this task (Figure 10.2, step 4). RuvC, a homodimer, has a nuclease activity that is specific for symmetric Holliday junction cleavage. This cleavage process is called **resolution**, and enzymes such as RuvC that have the ability to catalyze resolution are called **resolvases**. The RuvC homodimer binds to and cuts the strands exiting the RuvB ring to generate a DNA fork with two nicks. DNA ligase repairs the two nicks to reestablish the DNA fork.

PriA helps to load DnaB on the reestablished DNA fork to restart replication.

The underlying principle of reestablishing the collapsed replication fork is that replication can be initiated at sites other than replication origins. The initiator protein DnaA is required to load the DnaB helicase at the origin of replication, *oriC*. Another protein, called PriA, is required to load DnaB at sites other than *oriC* (Figure 10.2, step 5). Once DnaB is loaded, other components of the replication machinery assemble to form the active replication fork.

10.2 Mitotic Recombination

Eukaryotic cells use homologous recombination to repair double-strand breaks during the late S stage or the G_2 stage of the cell cycle.

Eukaryotic cells replicate DNA during the S phase of the cell cycle. So when a cell is in the late S stage or the G_2 stage of the cell cycle, each of its chromosomes is divided along its long axis into two identical subunits called sister chromatids. The cell can use information in the duplex DNA of one sister chromatid to repair a double-strand break in the other sister chromatid by using homologous recombination. Although the cell can also use a homologous chromosome to provide the necessary sequence information, this rarely happens during the S or G_2 stages of the cell cycle because homologous chromosomes are much further apart than sister chromatids. Several highly conserved enzymes and protein factors participate in this repair.

Eukaryotes use two related homologous recombination pathways, **synthesis-dependent strand annealing** and **double-strand break repair**, to repair double-strand breaks. The early stages in both pathways are identical and involve double-strand break end processing as well as Rad51 nucleoprotein filament formation and strand invasion.

1. *Double-strand break end processing* (FIGURE 10.8): The broken DNA ends are subject to 5'→3' degradation to generate 3'-OH single-stranded tails. The human and yeast enzymes that catalyze 5'→3' degradation to generate 3'-OH single-stranded tails have different names. For simplicity we use the human enzyme names. The MRN protein complex, which has nuclease activity, binds to each double-strand break end and acts along with a second nuclease, CtIP, to catalyze limited 5'→3' degradation. Then EXO1 nuclease catalyzes more extensive 5'→3' degradation or Bloom helicase unwinds the duplex DNA while DNA2 nuclease catalyzes processive 5'→3' degradation.

2. *Rad51 nucleoprotein filament formation and strand invasion* (FIGURE 10.9): The newly generated single-stranded 3'-OH tails are coated with the eukaryotic single-strand binding protein, replication protein A (RPA), which prevents the tails from forming secondary structures and protects them from degradation (Figure 10.9). The RPA-coated tails induce a series of protein phosphorylation reactions that lead to cell cycle arrest, providing the cell with sufficient time to repair the double-strand break. Then Rad51, a homolog of the bacterial recombinase

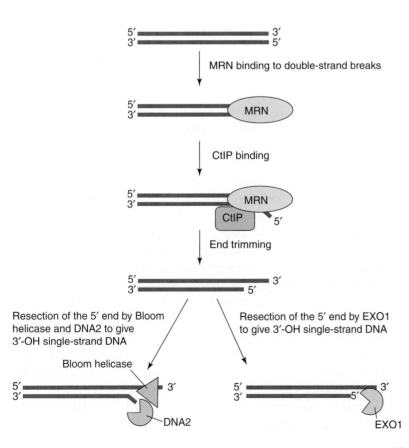

FIGURE 10.8 **Model for double-strand break end processing.** After a double-strand break is formed, the MRN protein complex with nuclease activity binds to each double-strand break end and acts along with a second nuclease, CtIP, to catalyze limited 5'→3' degradation. More extensive 5'→3' degradation takes place via two different pathways. In the first, the EXO1 nuclease catalyzes processive 5'→3' degradation. In the second, Bloom helicase in humans unwinds the duplex DNA while DNA2 nuclease catalyzes processive 5'→3' degradation. (Adapted from Mimitou, E. P., and Symington, L. S. 2009. *DNA Repair* 8:983–995.)

FIGURE 10.9 Rad 51 nucleoprotein filament formation and strand invasion. The newly generated single-stranded 3′-OH tail is coated with the eukaryotic single-strand binding protein, replication protein A (RPA), which prevents the tail from forming secondary structures and protects it from degradation. Rad51, which is homologous to the bacterial recombinase RecA, replaces RPA. The human mediator protein, BRCA2, assists in this replacement process. Once formed, the Rad51 nucleoprotein filament engages in a search for homology in the intact homologous DNA duplex and, upon encountering a homologous DNA duplex, promotes strand invasion. (Adapted from Pardo, B., Gómez-González, B., and Aguilera, A. 2009. *Cell Mol Life Sci* 66:1039–1056.)

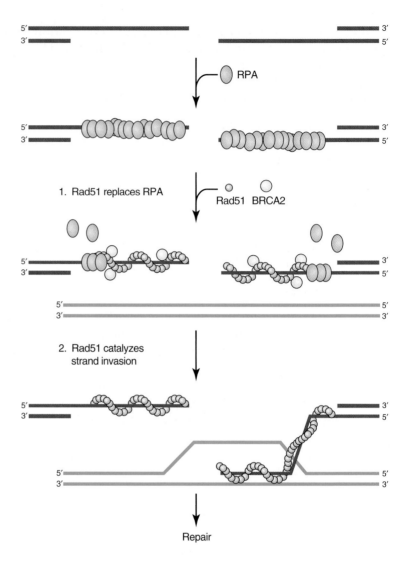

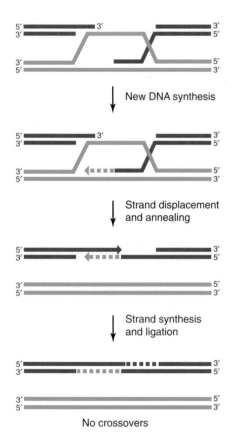

FIGURE 10.10 Synthesis-dependent strand annealing. Recombination is initiated as shown in Figure 10.9. The invading 3′-OH single-stranded tail is extended by DNA synthesis and then displaced. The displaced strand anneals to the complementary strand on the other side of the double-strand break. Remaining gaps are filled in by DNA synthesis and nicks are ligated. No Holliday junction is formed, and the product is always noncrossover. Newly synthesized DNA is represented by dashed lines. The light and dark blue shades of newly synthesized DNA indicate the template strand that directed the synthesis.

RecA, replaces RPA. BRCA2 (named for *BRCA2*, the human breast cancer susceptibility gene 2 that codes for it) functions as a mediator to assist in this replacement process. Mutations in the *BRCA2* gene are associated with familial breast and ovarian cancers. One of the newly formed Rad51 nucleoprotein filaments searches for homology in the intact homologous DNA duplex. Upon encountering a homologous region, Rad51 promotes strand invasion. After D-loop formation, double-strand break repair proceeds through either synthesis dependent strand annealing or double-strand break repair.

3. *Synthesis-dependent strand annealing* (FIGURE 10.10): The invading 3′-OH single-stranded tail is extended by DNA synthesis and then displaced from the D-loop. The displaced strand anneals to the complementary strand on the other side of the double-strand break. Remaining gaps are filled in by DNA synthesis and nicks are ligated.

4. *Double-strand break repair* (FIGURE 10.11): The invading 3′-OH single-stranded tail is once again extended by DNA synthesis.

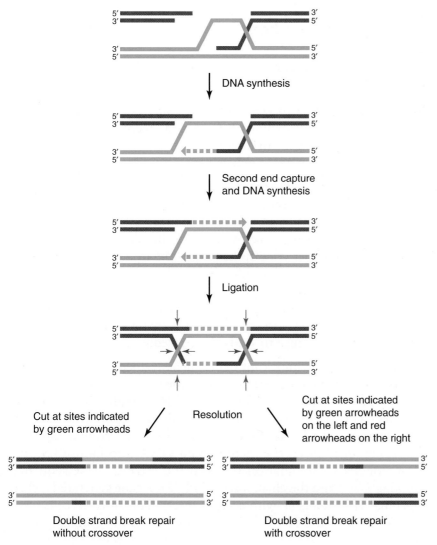

FIGURE 10.11 **Double-strand break repair.** Recombination is initiated as shown in Figure 10.9. Extension of the invading 3'-OH single-stranded tail by DNA synthesis enlarges the D-loop. Once the displaced loop can pair with the other side of the break, the second double-strand break end is captured. DNA synthesis, which fills in the gaps, is followed by ligation to form two Holliday junctions. If both Holliday junctions are cleaved at sites indicated by the green arrowheads, the resulting DNA duplexes retain their original DNA on either side of the repaired double-strand break. The same is true if the duplexes are cleaved at the sites indicated by the red arrowheads (not shown). In contrast, cleavage at the sites indicated by the green arrowheads on the left and the red arrowheads on the right cause a DNA segment exchange on either side of double-stranded DNA break. This type of DNA segment exchange is termed a crossover. Newly synthesized DNA is represented by dashed lines. The light and dark blue shades of newly synthesized DNA indicate the template strand that directed the synthesis.

As the D-loop grows bigger the displaced D-loop strand captures the second end on the other side of the double-strand break. Then DNA synthesis fills in the gaps. Ligation generates two Holliday junctions. A resolvase cleaves the Holliday junctions to produce two intact DNA duplexes. If both Holliday junctions are cleaved at sites indicated by the green arrowheads, the resulting DNA duplexes retain their original DNA on either side of the repaired double-strand break. The same is true if the duplexes are cleaved at the sites indicated by the red arrowheads (not shown). In contrast, cleavage at the sites indicated by the green arrowheads on the left and the red arrowheads on the right cause a DNA segment exchange on either side of double-stranded DNA break. This type of DNA segment exchange is termed a **crossover**. A crossover is also produced if cleavage takes place at the sites indicated by the green arrowheads on the right and the red arrowheads on the left. Crossovers that take place between identical sister chromatids are genetically silent and relatively rare during mitotic homologous recombination.

10.2 Mitotic Recombination

10.3 Gene Knockouts

Homologous recombination can be used to create gene knockouts in mice.

In the 1980s Mario Capecchi and Oliver Smithies devised a gene targeting technique that exploits homologous recombination to delete or disrupt a specific gene of interest in mice. Their technique is based on the following important experimental advances:

1. **Embryonic stem (ES) cells** isolated from a mouse embryo in an early developmental stage can grow stably in culture.
2. Recombinant DNA can be introduced into mouse ES cells.
3. A transformed ES cell with a specific disrupted gene can be selected under conditions that eliminate random insertions.
4. Transformed mouse ES cells develop into germ line or other tissues after injection into an early mouse embryo. The discovery that transformed mouse ES cell's can develop into germ line cells is important because the gene knockout has to be heritable in mice to study the consequences of the knockout.

In the first gene knockout experiments Capecchi and coworkers constructed a recombinant DNA molecule with the selectable antibiotic resistance marker, Neo^R, inserted within a specific mouse *HGPRT* gene segment (FIGURE 10.12). *HGPRT* codes for an enzyme that converts guanine and its analogs into nucleoside-5′-monophosphates. Then they introduced the recombinant DNA into mouse ES cells. Homologous recombination between the *HGPRT* gene sequences flanking Neo^R and the chromosomal *HGPRT* gene generated a chromosomal *HGPRT* gene with a Neo^R insert. Transformed cells could be selected because Neo^R confers resistance to the aminoglycoside antibiotic G418, which kills neomycin-sensitive cells. The *HGPRT* knockout was confirmed by showing the recombinant cells could grow in special medium containing 6-thioguanine, which is a poison to cells that make the active *HGPRT* enzyme. The efficiency of targeted knockout of the *HGPRT* gene to random insertion of the Neo^R gene was about 1:1,000.

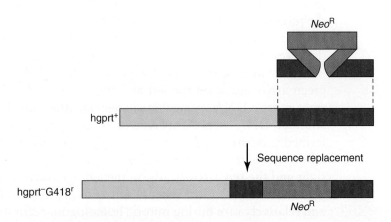

FIGURE 10.12 Disruption of the mouse *HGPRT* gene. A DNA molecule was constructed that contained the selectable antibiotic resistance marker, Neo^R (magenta), in place of a specific segment (dark blue) of the mouse *HGPRT* gene. The recombinant DNA was introduced into mouse embryonic stem (ES) cells. Homologous recombination between *HGPRT* gene sequences flanking the Neo^R insertion and the chromosomal *HGPRT* gene resulted in a replacement of the chromosomal *HGPRT* segment with Neo^R. Transformed ES cells could be selected because Neo^R confers resistance to the aminoglycoside antibiotic G418, which kills neomycin-sensitive cells. (Adapted from Capecchi, M. 2001. *Nat Med* 7:1086–1090.)

Capecchi and coworkers developed a **positive-negative selection technique** to enrich for targeted gene disruption (FIGURE 10.13). The recombinant DNA used in this technique has a Neo^R gene within a segment of the gene of interest and the <u>h</u>erpes <u>s</u>implex <u>v</u>irus gene *HSV-tk* at its end. The *HSV-tk* gene codes for a viral <u>t</u>hymidine <u>k</u>inase, which can convert a thymidine analog called FIAU to a lethal nucleotide analog. The recombinant DNA is introduced into mouse ES cells. If homologous recombination takes place between the sequences that flank Neo^R and the chromosomal gene of interest, then the transformed cell will have the Neo^R marker within the gene of interest but will not have the *HSV-tk* gene in the chromosome (Figure 10.13a). If, on the other hand, the recombinant DNA randomly inserts into the chromosome, then the transformed cell will have both the Neo^R and *HSV-tk* genes somewhere within one of the mouse ES chromosomes (Figure 10.13b). One can select transformants that have targeted gene disruption by culturing the cells in a growth medium that contains the

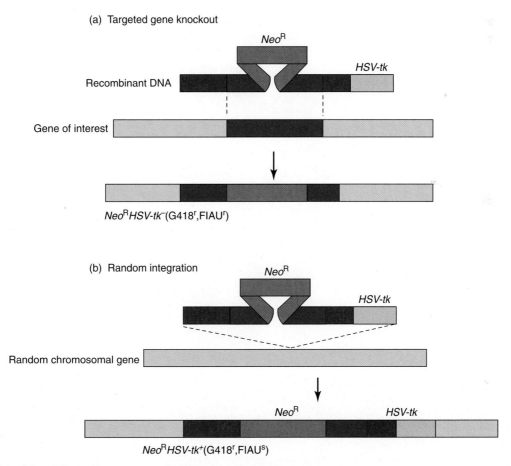

FIGURE 10.13 **Positive-negative selection to select precise targeting events.** The DNA construct has the Neo^R gene (magenta) in place of a segment (dark blue) of the gene of interest and also carries the *HSV-tk* gene (gold) at one end. The Neo^R gene confers resistance to aminoglycoside antibiotic G418, whereas the *HSV-tk* gene confers sensitivity to a thymidine analog (FIAU). (a) Gene targeting events produce transformed mouse ES cells that have the Neo^R gene but not the *HSV-tk* gene. Such transformants are resistant to both G418 and FIAU. (b) Random integration events produce mouse ES cells that have both the Neo^R and *HSV-tk* genes. Such transformants are resistant to the aminoglycoside antibiotic G418 but sensitive to FIAU, allowing for the selection of the rare gene targeting events against a background of random insertion events. (Adapted from Capecchi, M. 2001. *Nat Med* 7:1086–1090.)

10.3 Gene Knockouts

FIGURE 10.14 Gene replacement in mice. (a) Embryonic stem (ES) cells are altered by gene targeting to express an altered version of a gene. The cells are expanded into colonies, and cells from each colony are tested for the correct replacement. (b) The chosen ES cells are injected into an early mouse embryo, the blastocyst, which is then introduced into a pseudo-pregnant recipient mouse. Mice produced from the ES cells have a coat color derived from the injected ES cells, here shown in pink. In a few cases the ES cells will populate the germ line. When these mice are bred with normal mice, all cells in the progeny mice have the altered gene. (Adapted from Alberts, B., et al. 2002. *Molecular Biology of the Cell* [4th ed]. Garland Science.)

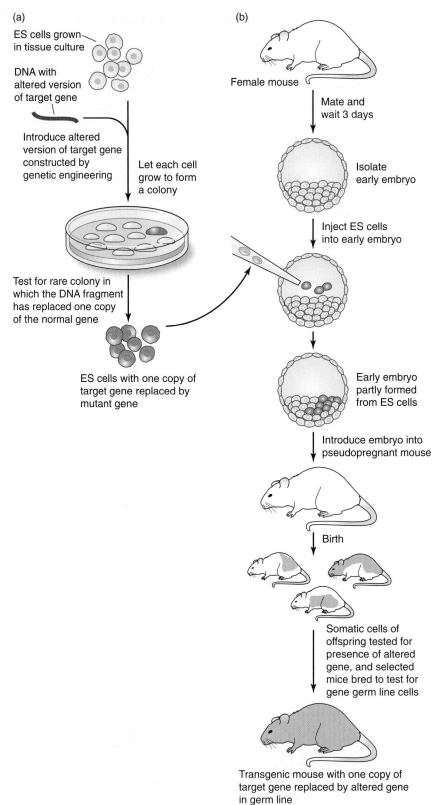

aminoglycoside antibiotic G418 (positive selection) to kill cells that do not have the Neo^R gene (positive selection) and the thymidine analog FIAU to kill cells that have the *HSV-tk* gene (negative selection). This approach allows for the selection of the rare gene targeting events against the background of random insertion events.

After the appropriate gene-targeted knockout has been made in ES cells, the cells are injected into a mouse early embryo. The strain of the mouse used to provide the early embryo differs from that of the injected ES cells by genes that determine mouse coat color so it is possible to identify mice that have the disrupted ES cells (FIGURE 10.14).

Once the early embryos are injected with the ES cells, they are implanted into pseudo-pregnant female mice. The mice that come from the implanted early embryos are called **chimeric mice** because they are a mixture of cells derived from the injected ES cells and cells from the early embryo. Chimeric mice show both coat colors, those from the ES cells and those from the early embryo cells. Breeding of these mice provides the next generation of mice that carry a germ line knockout of the gene of interest. These mice, called **transgenic mice**, are heterozygous for the knockout gene of interest and are used for further studies and matings to generate homozygous knockout mice.

10.4 Nonhomologous End-Joining

Nonhomologous end-joining connects broken ends that have no homology.

Nonhomologous end-joining repairs double-strand breaks during the early S phase and G_1 phase of the cell cycle when a closely associated homologous sister sequence chromatid is unavailable. Ku, an essential protein for nonhomologous end-joining, was first discovered in the cell nucleus as a target of autoantibodies in a patient with an autoimmune disease called scleroderma. The Ku protein, which is named after the patient, is a ring-shaped heterodimer that binds to broken DNA ends at the onset of nonhomologous end-joining (FIGURE 10.15). In mammals the Ku protein recruits the DNA-dependent protein kinase (DNA-PK) to form a new protein complex, which phosphorylates several proteins, including a specialized ligase called DNA ligase IV. If the broken ends can be joined, then DNA ligase IV does so. If the broken ends are "dirty" because they lack either a 5'-phosphate end or a 3'-hydroxy end, they cannot serve as a substrate for DNA ligase. "Dirty" ends are produced by radiation, oxidation, radiomimetics (chemicals that make double-strand breaks), and enzymes such as nucleases and defective topoisomerases. Nucleases and DNA polymerases are required to process the "dirty" ends so DNA ligase IV can join them. This processing usually produces small mutations consisting of a few nucleotides of insertion or deletion at the end-joints. The immune system uses nonhomologous end-joining to make a mature immunoglobulin gene from separated modular parts.

FIGURE 10.15 **Nonhomologous end-joining.** The Ku protein, a ring-shaped heterodimer, binds to broken DNA ends at the start of nonhomologous end-joining. If the broken ends can be joined, then DNA ligase IV does so. If the broken ends are "dirty" because they lack either a 5′-phosphate end or a 3′-hydroxy end, they cannot serve as a substrate for DNA ligase. Nucleases and DNA polymerases are required to process the "dirty" ends so DNA ligase IV can join them. (Adapted from Pardo, B., Gómez-González, B., and Aguilera, A. 2009. *Cell Mol Life Sci* 66:1039–1056.)

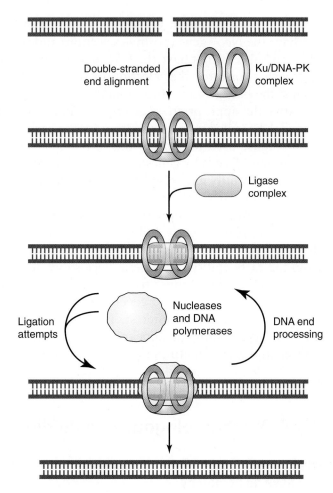

10.5 Meiotic Recombination

Homologous recombination during meiosis can result in gene conversion.

Homologous recombination is responsible for much of the genetic diversity among progeny of common parentage and is also essential for correct segregation of homologous chromosome pairs at the first meiotic division. When the process goes awry, homologous chromosomes are not held together and therefore segregate randomly during the first meiotic division, giving rise to aneuploid gametes that are not functional. Homologous recombination is also responsible for some novel aspects of the recombination that takes place during the first meiotic division.

According to Mendel's law of segregation, alleles should segregate equally into gametes. In most fungi the four spore products of a single meiosis are recovered together in a sac called an **ascus**. If a diploid parent cell has two different alleles for a single gene (*A* and *a*), then Mendel's first law predicts that two spores should have the *A* allele and two should have the *a* allele (FIGURE 10.16). This 2:2 segregation reflects the constancy of the gene, which changes only at a very low frequency due to mutation. In yeast and many other fungi meiotic products, called tetrads, occasionally deviate from the 2:2 segregation

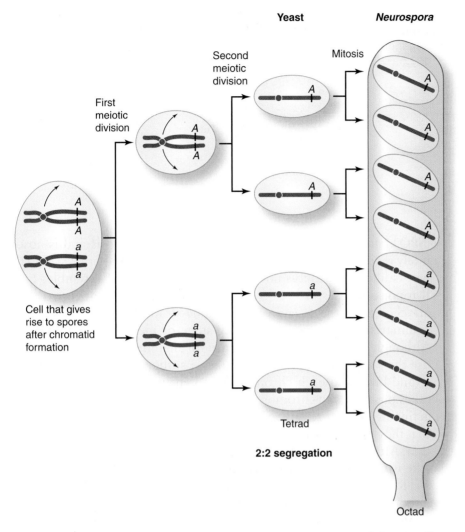

FIGURE 10.16 **Meiosis in yeast and *Neurospora*.** In most fungi meiosis produces four spore products that are present in a sac called an ascus. If a yeast cell has two different alleles for a single gene (*A* and *a*), then Mendel's first law predicts that two spores should have the *A* allele and the other two should have the *a* allele. In the bread mold *Neurospora*, a mitotic division takes place after meiosis so that eight spores. Half of these spores have the *A* allele and the other half have the *a* allele. (Adapted from Griffiths, J. F., et al. 1999. *Modern Genetic Analysis* [2nd ed]. W. H. Freeman and Company.)

and have a 3:1 or 1:3 segregation (FIGURE 10.17). This deviation occurs at a frequency of about 1% for any given gene but can reach up to 50% in some cases. In other fungi such as *Neurospora*, where there is one mitotic division after meiosis before ascus formation, the segregation is 4:4. Then the occasional deviants have 6:2 segregation.

This irregular segregation process that allows sequence information to be transferred from one DNA duplex to a homologous duplex without altering the sequence of the first duplex is called **gene conversion.** Subsequent studies showed that gene conversion also takes place in other diploid organisms and is an important outcome of homologous recombination. Biochemical studies of homologous recombination during meiosis explain how gene conversion takes place.

Homologous recombination during meiosis requires some enzymes that are unique to it.

Meiotic recombination occurs at much higher frequencies than mitotic recombination. This difference is primarily due to an endonuclease

10.5 Meiotic Recombination

FIGURE 10.17 Gene conversions. In yeast and many other fungi meiotic products, called tetrads, occasionally deviate from the 2:2 segregation and have a 3:1 or 1:3 segregation. In *Neurosopora* the occasional deviants have 6:2 or 2:6 segregation.

Yeast			Neurospora		
Normal segregation	Gene conversion		Normal segregation	Gene conversion	
A	A	A	A	A	A
A	A	a	A	A	A
a	A	a	A	A	a
a	a	a	A	A	a
			a	A	a
			a	A	a
			a	a	a
			a	a	a
2:2	3:1	1:3	4:4	6:2	2:6

(SPO11 in mammals) that is expressed only in meiosis and is absolutely required for formation of meiotic double-strand breaks. SPO11 works like DNA topoisomerases, and indeed SPO11 resembles type II topoisomerases. It breaks both strands of DNA through covalent attachment to the 5′ end of the double-strand break. **FIGURE 10.18a** summarizes the steps in the early stage of meiotic recombination. Although the enzymes indicated are for mammals, other eukaryotes use the same pathway. SPO11 makes a sufficient number of double-strand breaks in each chromosome to ensure at least one crossover takes place in each chromosome arm during meiosis. MRN and CtIP, acting in concert, remove SPO11 by endonucleolytic cleavage. Recall that these nucleases also participate in mitotic homologous recombination. After SPO11 is removed EXO1 (or DNA2 and helicase) catalyze extensive 5′→3′ degradation to produce single stranded 3′-OH tails.

Recall that during mitosis the sister chromatid is the preferred donor of nucleotide sequence information for the repair of a double-strand break. In contrast, *during the first meiotic division a nonsister chromatid on a homologous chromosome is the preferred donor of nucleotide sequence information for recombination*. The distribution of crossovers along a chromosome is not spaced randomly, but rather crossovers are separated by large distances. When a crossover occurs at one position on a chromosome, it is highly unlikely that a second crossover occurs nearby. This phenomenon, known as **interference**, limits the number of crossovers per chromosome arm.

As indicated in Figure 10.18a, two recombinases of the RecA family, Rad51 and Dmc1, are active in meiosis. Dmc1, unlike Rad51, is expressed only in meiosis. Mutation of the gene that codes for either protein results in a failure to complete meiosis and a reduction in crossover formation as assayed by physical methods. In yeast, Rad51 overexpression can partially suppress *dmc1* mutants, showing that Rad51 and Dmc1 perform related roles *in vivo*.

Studies of meiotic recombination in yeast indicate two distinct waves of homologous recombination. During the first wave most double-strand breaks are repaired by a synthesis-dependent strand-annealing mechanism (Figure 10.18b), giving rise to meiotic gene conversion but no crossovers. The segment of the repaired DNA duplex responsible for gene conversion is indicated by the DNA duplex segment shown in the gray box. The two strands in this DNA duplex segment are not

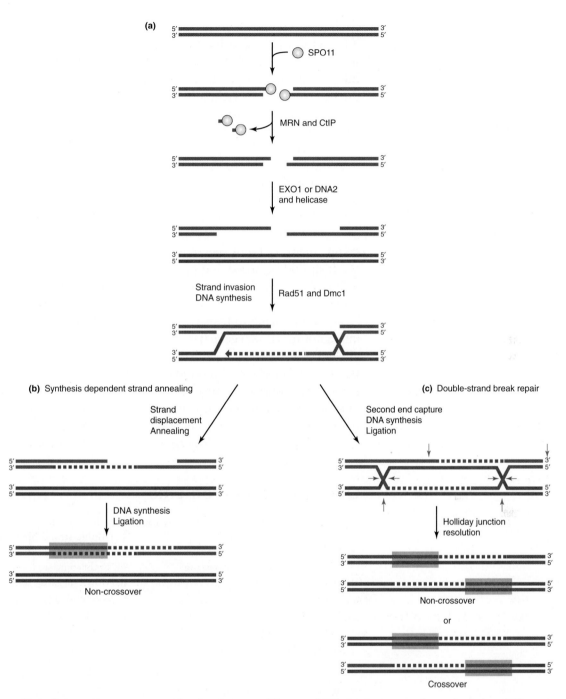

FIGURE 10.18 Meiotic homologous recombination in yeast. The two DNA duplexes are shown in red and blue to emphasize that recombination takes place between chromatids on homologous rather than sister chromosomes. The color of newly synthesized DNA (indicated by dashed lines) indicates the template strand that directed the synthesis. (a) *Single-stranded 3′-OH tail formation and strand invasion:* SPO11 breaks both strands of DNA through covalent attachment to the 5′ end of the double-strand break. MRN and CtIP act in concert to remove SPO11 by endonucleolytic cleavage. After SPO11 removal, EXO1 (or DNA2 and helicase) catalyze extensive 5′→3′ degradation to produce single strand 3′-OH tails. Two recombinases, Rad51 and Dmc1, catalyze strand invasion. (b) *Synthesis-dependent strand-annealing:* The invading single-stranded 3′-OH tail is extended by DNA synthesis and then displaced. The displaced strand anneals to the complementary strand on the other side of the double-strand break. Remaining gaps are filled in by DNA synthesis and nicks are ligated. No Holliday junctions are formed and there is no crossing over. (c) *Double-strand break repair:* The D-loop is enlarged as DNA polymerase extends the single-stranded 3′-OH tail and the second double-strand break end is captured. DNA synthesis fills in the gaps and the resulting nicks are ligated to form two Holliday junctions. Double-strand break repair can lead to crossovers or noncrossovers depending on how the Holliday junction is resolved. Mismatch repair corrects mismatched bases in the heteroduplexes (gray boxes) to generate a fully complementary DNA segment and thus causes gene conversion.

10.5 Meiotic Recombination

completely complementary because each strand has nucleotide sequence information provided by a different homologous chromosome. This segment is said to be a **heteroduplex** because the two strands are not fully complementary. Mismatch repair corrects mismatched bases in the heteroduplex to yield a fully complementary DNA segment. Thus, heteroduplex formation, a natural outcome of synthesis-dependent strand annealing, together with mismatch repair, can produce meiotic gene conversion. During the second wave (Figure 10.18c) the remaining double-strand breaks are repaired by a double-strand break repair mechanism that is preferentially resolved in the crossover mode. Heteroduplexes (gray boxes) present on either side of the repaired double-strand break are responsible for gene conversion. Mismatch repair corrects mismatched bases in the heteroduplexes to generate a fully complementary DNA segment and thus causes gene conversion.

It should be noted that gene conversion can also take place during mitosis but rarely does so because double-strand break repair seldom involves chromatids on homologous chromosomes. The effects of gene conversion during mitosis can be quite harmful if a gene is converted into a nonfunctional gene with a resultant **loss of heterozygosity** for that allele. For instance, if a gene that codes for an active tumor suppressor is converted into a nonfunctional tumor suppressor by gene conversion during mitosis, the cell may be transformed into a cancer cell. This type of gene conversion, which is known to occur on rare occasions, explains why an individual with a single defective tumor suppressor gene is at an increased risk of developing certain types of cancer.

Questions and Problems

1. Briefly define, describe, or illustrate each of the following.
   a. Double-strand break
   b. Homologous recombination
   c. Nonhomologous end-joining
   d. Fork collapse
   e. *Chi* site
   f. D-loop
   g. Holliday junction
   h. Resolution
   i. Resolvase
   j. Synthesis-dependent strand annealing
   k. Double-strand break repair
   l. Crossover
   m. Embryonic stem cell
   n. Positive-negative selection technique
   o. Chimeric mice
   p. Transgenic mice
   q. Gene conversion
   r. Interference
   s. Heteroduplex
   t. Loss of heterozygosity

2. Why is an unrepaired double-strand break usually lethal to a bacterial cell?

3. What happens if the replication fork reaches a single-strand nick before it can be repaired?

4. What contribution does each of the following enzymes or protein factors make to the restart of replication after replication fork collapse?
   a. RecBCD
   b. RecA
   c. RuvAB
   d. RuvC
   e. PriA

5. What contribution does the *chi* site make to the restart of replication after replication fork collapse?

6. Why do eukaryotic cells use homologous recombination between sister chromatids to repair a double-strand break during the G_2 and late S phase of the cell cycle but not during other phases of the cell cycle?

7. What contribution does each of the following enzymes or protein factors make to the repair of a double-strand break in human somatic cells?
   a. MRN and CtIP
   b. EXO1
   c. RPA
   d. Rad51
8. What contribution does the BRCA2 protein make to the repair of a double-strand break?
9. In what ways are synthesis-dependent strand annealing and double-strand break repair similar? In what ways are the two pathways different?
10. Investigators who wish to construct a mouse with a specific gene knockout must be able to distinguish transformed ES cells with a targeted gene knockout from transformed ES cells with random integration events. How do they do so?
11. Why might it be disadvantageous for cells to use nonhomologous end-joining rather than homologous recombination to repair double-strand breaks during the G_1 phase and early S phase of the cell cycle?
12. How do mutations arise during nonhomologous end-joining repair of double-strand breaks?
13. What function does each of the following enzymes or protein factors have in meiotic recombination in mammals?
   a. SPO11
   b. MRN and CtIP
   c. EXO1
   d. DNA2 and helicase
   e. Rad51
   f. Dmc1
14. How do heteroduplexes form during synthesis-dependent strand annealing and double-strand break repair during the first meiotic division? How do heteroduplexes contribute to gene conversion?
15. Double-strand break repair generates both gene conversions and crossovers, but synthesis-dependent strand annealing only generates gene conversions. Explain why this is so.

Suggested Reading

Replication Restart After Bacterial Replication Fork Collapse

Cox, M. M. 2002. The bacterial RecA protein and the recombinational DNA repair of stalled replication forks. *Annu Rev Biochem* 71:71–100.

Cox, M. M. 2002. The nonmutagenic repair of broken replication forks via recombination. *Mutat Res* 510:107–120.

Cox, M. M. 2007. Motoring along with the bacterial RecA protein. *Nat Rev Mol Cell Biol* 8:127–138.

Dillingham, M. S., and Kowalczykowski, S. C. 2008. RecBCD enzyme and the repair of double-stranded DNA breaks. *Microbiol Mol Biol Rev* 72:642–671.

Egelman, E. H. 2009. Implications of the RecA structure. *Biol Rep*. http://F1000.com/Reports/Biology/content/1/7

Heller, R. C., and Marians, K. J. 2006. Replisome assembly and the direct restart of stalled replication forks. *Nat Rev Mol Cell Biol* 7:932–943.

Holthausen, J. T., Wymana, C., and Kanaar, R. 2010. Regulation of DNA strand exchange in homologous recombination. *DNA Repair* 9:1264–1272.

Kreuzer, K. N. 2005. Interplay between DNA replication and recombination in prokaryotes. *Annu Rev Microbiol* 59:43–67.

Lilley, D. M. J., and White, M. F. 2001. The junction-resolving enzymes. *Nat Rev Mol Cell Biol* 2:433–443.

Lovett, S. T. 2003. Connecting replication and recombination. *Mol Cell* 11554–11556.

Lusetti, S. L., and Cox, M. M. 2002. The bacteria RecA protein and the recombinational DNA repair of stalled replication forks. *Annu Rev Biochem* 71:71–100.

Marians, K. J. 2000. Replication and recombination intersect. *Curr Opin Genet Dev* 10:151–156.

Marians, K. J. 2003. Mechanisms of replication fork restart in *Escherichia coli*. *Phil Trans R Soc Lond B* 359:71–77.

Masai, H., Tanaka, T., and Kohda, D. 2010. Stalled replication forks: making ends meet for recognition and stabilization. *Bioessays* 32:687–697.

McGlynn, P., and Lloyd, R. G. 2002. Recombinational repair and restart of damaged replication forks. *Nat Rev Mol Cell Biol* 3:859–870.

Michel, B., Boubakri, H., Baharoglu, Z. et al. 2007. Recombination proteins and rescue of arrested replication forks. *DNA Repair* 6:967–980.

Michel, B., Grompone, G., Florès, and Bidnenko, V. 2004. Multiple pathways process stalled replication forks. *Proc Natl Acad Sci USA* 101:12783–12788.

Niu, H., Raymond, S., and Sung, P. 2009. Multiplicity of DNA end resection machineries in chromosome break repair. *Genes Dev* 23:1481–1486.

Patel, S. S. 2010. One motor driving two translocases. *Nat Struct Mol Biol* 17:1166–1167.

Persky, N. S., and Lovett, S. 2008. Mechanism of recombination: lessons from *E. coli*. *Crit Rev Biochem Mol Biol* 43:347–370.

Rice, K. P., and Cox, M. M. 2001. Recombinational DNA repair in bacteria: postreplication. *Encyclopedia of Life Sciences*. pp. 1–8. Hoboken, NJ: John Wiley & Sons.

Saikrishnan, K., Griffiths, S. P., Cook, N., Court, R., and Wigley, D. B. 2008. DNA binding to RecD: role of the 1B domain in SF1B helicase activity. *EMBO J* 27:2222–2229.

Singleton, M. R., Dillingham, M. S., Gaudier, M. et al. 2004. Crystal structure of RecBCD enzyme reveals a machine for processing DNA breaks. *Nature* 432:187–193.

West, S. C. 1998. RuvA gets x-rayed on Holliday. *Cell* 94:699–701.

Wigley, D. B. 2007. RecBCD: The supercar of DNA repair. *Cell* 131:651–653.

Wu, C. G., Bradford, C., and Lohman, T. M. 2010. *Escherichia coli* RecBC helicase has two translocase activities controlled by a single ATPase motor. *Nat Struct Mol Biol* 17:1210–1217.

Wyman, C. 2011. Mechanistic insight from chaos: how RecA mediates strand exchange. *Structure* 19:1031–1032.

Mitotic Recombination

Helleday, T., Lo, J., van Gent, D. C., and Engelward, B. P. 2007. DNA double-strand break repair: from mechanistic understanding to cancer treatment. *DNA Repair* 6:923–935.

Burgoyne, P. S., Mahadeviah, S. K., and Turner, J. M. A. 2007. The management of DNA double-strand breaks in mitotic G_2, and in mammalian meiosis viewed from a mitotic G_2 perspective. *Bioessays* 29:974–986.

Cann, K. L., and Hicks, G. G. 2007. Regulation of the cellular DNA double-strand break response. *Biochem Cell Biol* 85:663–674.

Game, J. C. 2001. Recombinational DNA repair in eukaryotes. *Encyclopedia of Life Sciences*. pp. 1–7. Hoboken, NJ: John Wiley & Sons.

Heyer, W.-D., Ehmsen, K. T., and Liu, J. 2010. Regulation of homologous recombination in eukaryotes. *Annu Rev Genet* 44:113–139.

Hiom, K. 2010. Coping with DNA double-strand breaks. *DNA Repair* 9:1256–1263.

Kass, E. M., and Jasin, M. 2010. Collaboration and competition between DNA double-strand break repair pathways. *FEBS Lett* 584:3703–3708.

Klein, H. L. 2008. DNA endgames. *Nature* 455:740–741.

Mimitou, E. P., and Symington, L. S. 2009. DNA end resection: many nucleases make light work. *DNA Repair* 8:983–995.

Mimitou, E. P., and Symington, L. S. 2009. Nucleases and helicases take center stage in homologous recombination. *Trends Biochem Sci* 34:264–272.

Mimitou, E. P., and Symington, L. S. 2011. DNA end resection—Unraveling the tail. *DNA Repair* 10:344–348.

Moynahan, M. E., and Jasin, M. 2010. Mitotic homologous recombination maintains genomic stability and suppresses tumorigenesis. *Nat Rev Mol Cell Biol* 11:196–207.

Pardo, B., Gómez-González, B., and Aguilera, A. 2009. DNA double-strand break repair: how to fix a broken relationship. *Cell Mol Life Sci* 66:1039–1056.

Potenski, C. J., and Klein, H. L. 2011. The expanding arena of DNA repair. *Nature* 471:48–49.

San Fiippo, J., Sung, P., and Klein, H. 2008. Mechanism of eukaryotic homologous recombination. *Annu Rev Biochem* 77:229–257.

Sung, P. 2001. Homologous genetic recombination in eukaryotes. *Encyclopedia of Life Sciences*. pp. 1–5. Hoboken, NJ: John Wiley & Sons.

Sung, P., and Klein, H. 2006. Mechanism of homologous recombination: mediators and helicases take on regulatory functions. *Nat Rev Mol Cell Biol* 7:739–750.

Symington, L. S., and Holloman, W. K. 2008. Resolving resolvases: the final act? *Mol Cell* 32:603–604.

Wyman, C., and Kanaar, R. 2006. DNA double-strand break repair: all's well that ends well. *Annu Rev Genet* 40:363–383.

Gene Knockouts

Capecchi, M. R. 2005. Gene targeting in mice: functional analysis of the mammalian genome for the twenty-first century. *Nat Rev Genet* 6:507–512.

Muller, U. 1999. Ten years of gene targeting: targeted mouse mutants, from vector design to phenotype analysis. *Mech Dev* 82:3–21.

Smithies, O., Gregg, R. G., Boggs, S. S., et al. 1985. Insertion of DNA sequences into the human chromosomal beta-globin locus by homologous recombination. *Nature* 317:230–234.

Thomas, K. R., and Capecchi, M. R. 1986. Introduction of homologous DNA sequences into mammalian cells induces mutations in the cognate gene. *Nature* 324:34–38.

Nonhomologous End-Joining

Daley, J. M., Palmbos, P. L., Wu, D., and Wilson, T. E. 2005. Nonhomologous end joining in yeast. *Annu Rev Genet* 39:431–451.

Gellert, M. 2002. V(D)J recombination: RAG proteins, repair factors, and regulation. *Annu Rev Biochem* 71:101–132.

Lieber, M. R. 2008. The mechanism of human nonhomologous DNA end joining. *J Biol Chem* 283:1–5.

Lieber, M. R. 2010. The mechanism of double-strand DNA break repair by the nonhomologous DNA end-joining pathway. *Annu Rev Biochem* 79:181–211.

Lieber, M. R., Ma, Y., Pannicke, U., and Schwarz, K. 2004. The mechanism of vertebrate nonhomologous DNA end joining and its role in V(D)J recombination. *DNA Repair (Amst)* 3:817–826.

Roth, D. B. 2003. Restraining the V(D)J recombinase. *Nat Rev Immunol* 3:656–666.

Meiotic Recombination

Hurles, M. E. 2002. Gene conversion. *Encyclopedia of Life Sciences*. pp. 1–7. Hoboken, NJ: John Wiley & Sons.

Keeney, S. 2001. Mechanism and control of meiotic recombination initiation. *Curr Top Dev Biol* 52:1–53.

Keeney, S., and Neale, M. J. 2006. Initiation of meiotic recombination by formation of DNA double-strand breaks: mechanism and regulation. *Biochem Soc Trans* 34:523–525.

Lichen, M. 2008. Thoroughly modern meiosis. *Nature* 454:421–422.

San Fiippo, J., Sung, P., and Klein, H. 2008. Mechanism of eukaryotic homologous recombination. *Annu Rev Biochem* 77:229–257.

Schwacha, A., and Kleckner, N. 1995. Identification of double Holliday junctions as intermediates in meiotic recombination. *Cell* 83:783–791.

Sun, H., Treco, D., Schultes, N. P., and Szostak, J. W. 1989. Double-strand breaks at an initiation site for meiotic gene conversion. *Nature* 338:87–90.

11 Transposable Elements

CHAPTER OUTLINE

11.1 Bacterial Transposable Elements

Insertion sequences (IS elements) are simple mobile genetic elements.

A composite transposon consists of a pair of IS elements on either side of one or more genes.

Tn5 and Tn10 transposons move via a cut-and-paste transposition mechanism.

The Tn3 transposon uses replicative transposition to move to a new site while maintaining a copy at the original site.

Transposons are useful tools for bacterial studies.

11.2 Eukaryotic Transposable Elements

Some eukaryotic transposable elements move through a DNA-only mechanism and others require an RNA intermediate.

The eukaryotic "hAT" mobile elements move by a DNA only cut-and-paste transposition mechanism.

Antibody and T-cell receptor genes are formed by a mechanism that resembles the Hermes transposition mechanism.

A long dormant transposable element in fish called *Sleeping Beauty* has been restored to full activity by recombinant DNA technology.

LTR retrotransposons copy an RNA intermediate to make cDNA, which then is integrated into the genome by a cut-and-paste mechanism.

Non-LTR elements move using a target-primed reverse transcription mechanism.

Alu is the most common SINE in the human genome.

There is considerable debate concerning what benefits, if any, retrotransposons make to the host cells.

BOX 11.1: IN THE LAB Reporter Genes

BOX 11.2: LOOKING DEEPER Barbara McClintock and the Discovery of "Jumping Genes"

PROBLEMS AND QUESTIONS

SUGGESTED READING

Image © TRINACRIA PHOTO/ShutterStock, Inc.

All organisms have elaborate replication and repair systems to ensure their DNA is stably maintained from one generation to the next. One might therefore expect the genome to be static with few, if any, DNA rearrangements taking place. Contrary to this expectation, virtually all organisms have mobile DNA sequences called **transposable elements** or **transposons** that can move from one location to another in the genome. The movement of these mobile DNA sequences is called **transposition**. The frequency of element transposition in a bacterial or eukaryotic cell tends to be on the order of 10^{-7} to 10^{-2} transposition events per generation. There is considerable selective pressure for both the host organism and the transposable element to evolve to keep the transposition frequency low. A high transposition frequency increases the likelihood of a transposable element disrupting an essential gene and causing the host organism to die. Moreover, the presence of identical transposable elements within the genome increases the probability that homologous recombination between the mobile genetic elements would lead to deletions or harmful chromosomal rearrangements. A transposable element cannot continue to exist if its host dies. Many transposable elements have, therefore, evolved within their host to do as little harm as possible, and some even provide some benefits to the host. For example, some transposable elements in bacteria carry antibiotic resistance genes.

Because transposons are probably most thoroughly studied in bacteria, the first part of this chapter examines some transposons found in bacteria. The lessons learned from studying these transposons can be applied to some but certainly not all eukaryotic transposons. The second part of this chapter examines eukaryotic transposons.

11.1 Bacterial Transposable Elements

Insertion sequences (IS elements) are simple mobile genetic elements.

Bacteria contain simple transposable elements called **insertion sequences** or **IS elements**. James A. Shapiro identified the first such element while studying a mutation in the *Escherichia coli* galactose gene *galK*, which codes for galactokinase. Subsequent studies revealed the mutation in the *galK* gene was caused by a 770-base pair (bp) insertion sequence with imperfect 30-bp **terminal inverted repeats**. The IS element was also shown to have a *Tnp* gene between the terminal inverted repeats that codes for a recombinase called **transposase**. The structure of the IS element identified by Shapiro, now called IS1, is shown in FIGURE 11.1. It is important to note there is a short **direct repeat** (identical sequence in the same orientation) of the host DNA on either side of the inserted transposon. Although our focus for now is on the transposon itself, and not the surrounding host DNA, we will return to the direct repeats below when we examine the transposition mechanism. More than 1,000 different bacterial IS elements, ranging in size from about 750 to 2,500 bp long, have now been identified.

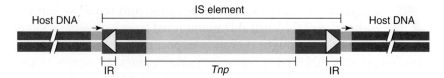

FIGURE 11.1 **Anatomy of an IS element.** The IS element is shown inserted into host DNA (red). There is a short direct repeat (identical sequence in the same orientation) of host DNA on either side of the inserted transposon. These direct repeats are shown in pink with black arrows above them to indicate they have the same orientation. The IS element has an inverted repeat or IR (shown as yellow arrows) at each end. It also has a single gene, *Tnp* (orange), which codes for a recombinase known as transposase. DNA segments between the inverted repeats and *Tnp* are shown in dark blue.

Any given bacterial cell usually has only a few different kinds of IS elements. The IS elements can be assigned to about 20 families based on various characteristics such as the nature of their terminal inverted repeats, which range in size from 10 to 40 bp, and their transposases.

A composite transposon consists of a pair of IS elements on either side of one or more genes.

IS elements are building blocks for a still more complex structure called a **composite transposon,** which forms when an IS element inserts on either side of one or more genes (FIGURE 11.2). When the outermost ends of the two IS elements are used for transposition, they allow the

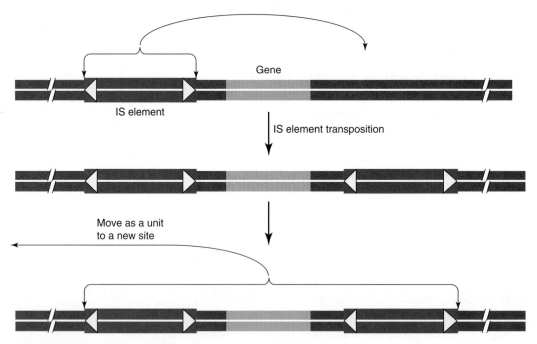

FIGURE 11.2 **Gene mobilization by composite transposons.** IS elements are building blocks for a still more complex structure called a composite transposon, which forms when an IS element inserts on either side of one or more genes. When the outermost ends of the two IS elements are used for transposition, they allow the movement of the entire composite transposon (the two IS elements and the intervening genes). Over time, the transposase gene in one IS element may lose function and need to rely on the transposase gene encoded by the other IS element for mobilization. (Adapted from L. Snyder and W. Champness. *Molecular Genetics of Bacteria*, Second edition. ASM Press, 2003.)

movement of the entire composite transposon (the two IS elements and the intervening DNA). Over time, the transposase gene in one IS element may lose function and need to rely on the transposase gene encoded by the other IS element for mobilization. Composite transposons can prove beneficial to their bacterial hosts if the genes between the IS elements offer some advantage such as antibiotic resistance or the ability to metabolize a carbon source.

FIGURE 11.3 shows the structures of Tn5 and Tn10, two composite transposons in *E. coli* that have been extensively studied. The IS elements on the left and right ends of the Tn5 transposon are designated IS50L and IS50R, respectively. The corresponding IS elements in the Tn10 transposon are designated IS10L and IS10R. Only the IS elements on the right side of each of these composite transposons have a gene that codes for an active transposase. The <u>o</u>utside <u>e</u>nd of each IS element is designated OE and the <u>i</u>nside <u>e</u>nd is designated IE.

Tn*5* and Tn*10* transposons move via a cut-and-paste transposition mechanism.

Transposons Tn5 and Tn10 each codes for a transposase, which physically moves the corresponding transposon from its original position to a new position. Transposition takes place through a **cut-and-paste mechanism**. The Tn5 and Tn10 transposases belong to the DDE family of transposases. The family name derives from the fact that the active enzyme has two aspartic acid (D) residues and one glutamic acid (E) residue at its catalytic site. These acidic residues bind essential divalent metal ion cofactors. The Tn5 transposase is probably the best studied of the bacterial DDE transposases; therefore, we focus our attention on it. The information obtained from studying the Tn5 transposase, however, also applies to the Tn10 transposase and other members of the DDE transposase family.

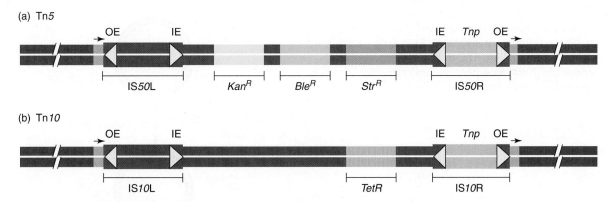

FIGURE 11.3 **Anatomies of composite transposons Tn5 and Tn10.** There is a short direct repeat (identical sequence in the same orientation) of host DNA (red) on either side of the inserted transposons. These direct repeats are shown in pink with black arrows above them to indicate they have the same orientation. (a) Tn5 has IS elements at each end. The elements on the left and right ends are designated IS50L and IS50R, respectively. The outside end of each IS element is designated OE and the inside end is designated IE. IS50R has a functional *Tnp* gene, which codes for an active transposase, whereas IS50L does not. *Kan*^{R}, *Ble*^{R}, and *Str*^{R} indicate genes for resistance to kanamycin, bleomycin, and streptomycin, respectively. (b) Tn10 also has IS elements at each end. The elements on the left and right ends are designated IS10L and IS10R, respectively. Only IS10R has a functional *Tnp* gene that codes for an active transposase. *Tet*^{R} indicates a gene for tetracycline-resistance.

The transposition process catalyzed by Tn5 transposase takes place in several steps (FIGURE 11.4):

1. *Binding* (Figure 11.4, step 1): A Tn5 transposase (orange oval) binds to the outside end (OE) of each IS element. The DNA protein complex that forms contains a Tn5 transposase dimer, which is bound to the inverted repeats at the ends of a bent Tn5 transposon. The details of this binding process must still be worked out. Two transposase molecules may form a dimer, which then binds to the inverted repeats in a bent DNA molecule, or each transposase molecule may bind to an inverted repeat and the DNA may then bend to allow the transposase molecules to form a dimer.

2. *Cutting* (Figure 11.4, step 2): The Tn5 transposase cuts both DNA strands at the outside end (OE) of one IS element and then repeats the cutting process at the outside end of the other IS element. Ivan Rayment and coworkers determined the crystal structure of the Tn5 transposase bound to cut Tn5 transposon DNA ends (FIGURE 11.5). Each transposase subunit interacts

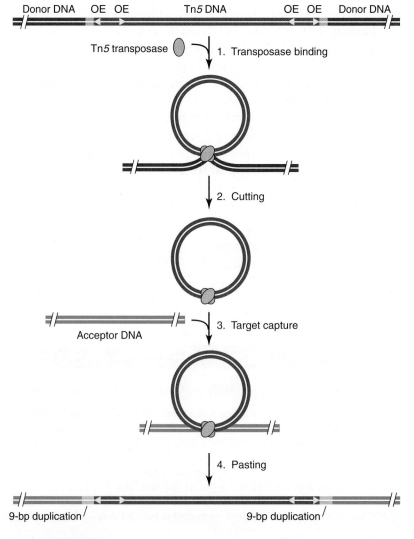

FIGURE 11.4 Steps in hairpin-mediated transposition process of bacterial transposon Tn5. (1) *Binding:* A Tn5 transposase (orange oval) binds to the outside end (OE) of each IS element. The DNA protein complex that forms contains a Tn5 transposase dimer, which is bound to the inverted repeats at the ends of a bent Tn5 transposon. (2) *Cutting:* The Tn5 transposase cuts both DNA strands at the outside end (OE) of one IS element and then repeats the cutting process at the outside end of the other IS element. (3) *Target capture:* The Tn5 transposase • Tn5 DNA complex binds to the target DNA (light blue) to form a target capture complex. (4) *Pasting:* The Tn5 transposase inserts the free Tn5 DNA into the target DNA. (Modified from *Curr. Opin. Struct. Biol.*, vol. 14, M. Steiniger-White, I. Rayment and W. S. Reznikoff, Structure/function insights into Tn5 transposition, pp. 50–57, copyright 2004, with permission from Elsevier [http://www.sciencedirect.com/science/journal/0959440X].)

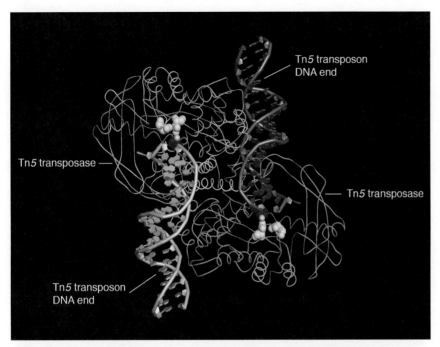

FIGURE 11.5 Crystal structure of transposon Tn5 transposase bound to cut transposon Tn5 DNA ends. Each Tn5 transposase subunit interacts with both Tn5 DNA ends. Cutting cannot take place unless the transposase dimer binds correctly to the DNA, ensuring both ends of the transposon will be cut. One Tn5 DNA end is shown in dark and the other in light blue. One transposase subunit is green and the other orange. Aspartate and glutamate residues in the catalytic site are shown as yellow and red spacefill structures, respectively. Magnesium ions bound to the DDE site are shown as purple spheres. (Protein Data Bank ID: 1MUH. Davies, D. R. et al. 2000. *Science* 289:77–85.)

with both Tn5 DNA ends. Cutting cannot take place unless the transposase dimer binds correctly to the DNA, ensuring both ends of the transposon will be cut. William S. Reznikoff and coworkers proposed a mechanism for the cutting at each end based on the crystal structure and their biochemical studies (**FIGURE 11.6**). The steps in this cutting process are as follows: (a) An activated water molecule cleaves a phosphodiester bond in one DNA strand at an OE on an IS element side of Tn5. (b) The newly generated 3′-OH group attacks the opposite DNA strand at the same OE to form a hairpin structure and free the donor DNA on that side of the Tn5 transposon from the transposon. (c) A second activated water molecule cuts the hairpin intermediate. The Tn5 transposase repeats the same reactions at the OE on the IS element on the other side of the Tn5 transposon, completely freeing the Tn5 transposon from the donor DNA. Tn5 excision produces a double-strand break in the donor DNA that needs to be repaired (see below).

3. *Target capture* (Figure 11.4 step 3): The Tn5 transposase • Tn5 DNA complex binds to the target DNA (light blue) to form a target capture complex. Tn5 preferentially transposes

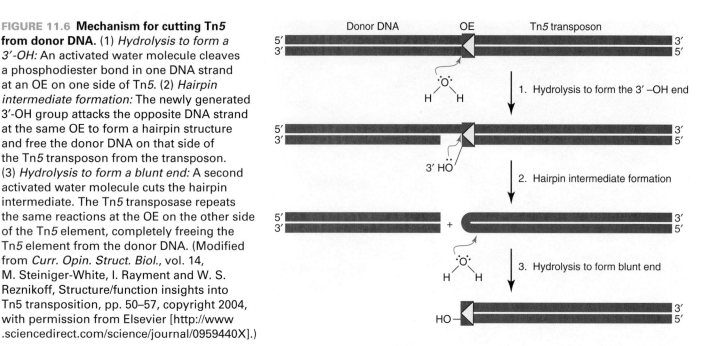

FIGURE 11.6 **Mechanism for cutting Tn5 from donor DNA.** (1) *Hydrolysis to form a 3′-OH:* An activated water molecule cleaves a phosphodiester bond in one DNA strand at an OE on one side of Tn5. (2) *Hairpin intermediate formation:* The newly generated 3′-OH group attacks the opposite DNA strand at the same OE to form a hairpin structure and free the donor DNA on that side of the Tn5 transposon from the transposon. (3) *Hydrolysis to form a blunt end:* A second activated water molecule cuts the hairpin intermediate. The Tn5 transposase repeats the same reactions at the OE on the other side of the Tn5 element, completely freeing the Tn5 element from the donor DNA. (Modified from *Curr. Opin. Struct. Biol.*, vol. 14, M. Steiniger-White, I. Rayment and W. S. Reznikoff, Structure/function insights into Tn5 transposition, pp. 50–57, copyright 2004, with permission from Elsevier [http://www.sciencedirect.com/science/journal/0959440X].)

into sites that have a 5′-AGNTYWRANCT-3′ sequence (where R = purine, Y = pyrimidine, W = adenine or thymine, and N = any nucleotide) or a slight variation of this sequence.

4. *Pasting* (Figure 11.4 step 4): The Tn5 transposase inserts the free Tn5 DNA into the target DNA, generating a 9-bp direct repeat on either side of the Tn5 insert. The mechanism for this pasting process, shown in FIGURE 11.7, is as follows. A 3′-OH group at each end of the transposon (dark blue) attacks a phosphodiester bond on each side of a 9-bp target DNA sequence (pink), causing the transposon to integrate by covalent bond formation between the 3′-OH groups of the transposon ends and the 5′-phosphate groups of the target DNA. DNA repair enzymes then fill in the gap that forms and repair the resulting nick.

Because the Tn5 transposon needs to multiply to be maintained over time, it requires some method to replicate. This requirement probably explains why Tn5 has evolved a molecular mechanism to upregulate transposition in response to the methylation state of DNA after replication. In *E. coli* and related bacteria the enzyme DNA adenine methylase methylates adenine bases in the sequence GATC. Normally, all GATC sites in the chromosome on both DNA strands are methylated on the adenine bases at these sites. After DNA replication, however, the chromosome is transiently hemimethylated. That is, the old strand of DNA contains a methyl group at the GATC site but the newly replicated strand has not yet had a chance to be methylated.

The hemimethylated state of the element increases the frequency of Tn5 transposition in two distinct ways. First, the end sequences are better bound by the transposase when they are hemimethylated. Second, transcription of the transposase gene is sensitive to the hemi-methylation state. The fully methylated state favors transcription of the transposase gene to form a truncated transposase protein that acts

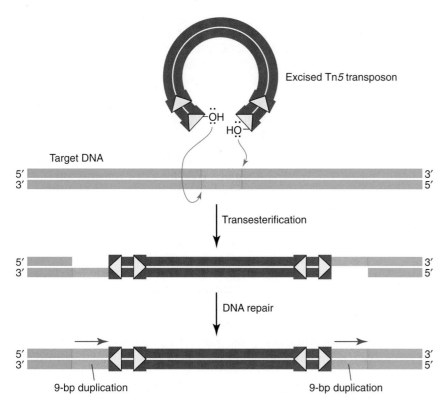

FIGURE 11.7 Mechanism used by transposase to insert transposon Tn5 into target DNA. The Tn5 transposase inserts the Tn5 DNA (dark blue) from the Tn5 DNA • transposase complex into the target DNA (light blue). The 3'-OH group at each end of the transposon attacks a phosphodiester bond on each side of a 9-bp target DNA sequence (pink), causing the transposon to integrate by covalent bond formation between the 3'-OH groups at the transposon ends and the 5'-phosphate groups of the target DNA. DNA repair enzymes then fill the gaps that form and repair the resulting nicks.

as an inhibitor of transposition. Immediately after replication DNA is hemimethylated, favoring the formation of the full-length transposase that catalyzes transposition to new positions in the cell. These findings suggest a mechanism for Tn5 transposon replication.

Before DNA replication the bacterial chromosome has a single fully methylated Tn5 transposon in its chromosome. Immediately after the replication fork passes through the Tn5 DNA, the two newly formed DNA duplexes each has a hemimethylated Tn5 insert, permitting the transposon to move to a site in front of the replication fork (FIGURE 11.8). The double-strand break produced by Tn5 excision can be repaired by homologous recombination using the other newly synthesized DNA duplex to provide the required sequence information. Further DNA replication leads to the presence of four Tn5 transposons within the host cell before cell division. After cell division each daughter cell has one Tn5 transposon at the original chromosomal site and a second Tn5 transposon at a new chromosomal site.

The Tn3 transposon uses replicative transposition to move to a new site while maintaining a copy at the original site.

Some transposons move to a new site while maintaining a copy at their original site. These transposons use a mechanism called replicative transposition that does not involve a hairpin intermediate. The bacterial transposon Tn3 helps to illustrate replicative transposition. As shown in FIGURE 11.9, Tn3, a 4,957-bp long element, has 38-bp terminal inverted repeats flanking *Tnp*, *Res*, and *Amp^R* genes. The *Tnp* and *Res* genes code for a DDE transposase and a recombinase

11.1 Bacterial Transposable Elements

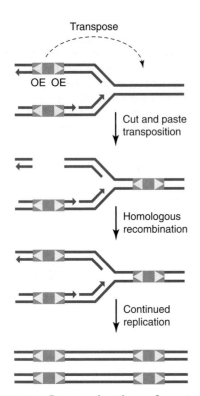

FIGURE 11.8 **Proposed pathway for cut-and-paste transposon replication.** Before DNA replication the bacterial chromosome has a single fully methylated Tn5 transposon in its chromosome (see text). Immediately after the replication fork passes through the Tn5 DNA, the two newly formed DNA duplexes will each have a hemimethylated Tn5 insert, permitting the transposon to move to a site in front of the replication fork. The double-strand break produced by Tn5 excision can be repaired by homologous recombination using the other newly synthesized DNA duplex to provide the required sequence information. Further DNA replication leads to the presence of four Tn5 transposons within the host cell before cell division. After cell division each daughter cell has one Tn5 transposon at the original chromosomal site and a second Tn5 transposon at a new chromosomal site.

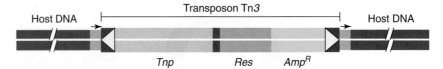

FIGURE 11.9 **Anatomy of Tn3, a noncomposite transposon.** There is a short direct repeat (identical sequence in the same orientation) of host DNA (dark blue) on either side of the inserted transposon. These direct repeats are shown in pink with black arrows above them to indicate they have the same orientation. Transposon Tn3 has inverted terminal repeats flanking *Tnp*, *Res*, and *Amp*^{R} genes. The *Tnp* and *Res* genes code for a DDE transposase and a recombinase called resolvase, respectively. *Amp*^{R} codes for β-lactamase, which confers resistance to β-lactam antibiotics such as ampicillin.

called resolvase (see below), respectively. *Amp*^{R} codes for β-lactamase, conferring resistance to β-lactam antibiotics such as ampicillin. A transposable element such as Tn3, which has inverted repeats rather than IS elements flanking *Tnp* and other genes, is called a **noncomposite transposon**. Some noncomposite transposons such as Tn3 move by replicative transposition, whereas others move by a cut-and-paste mechanism.

The steps in replicative transposition of a Tn3 element from a donor DNA (dark blue) to a target DNA (light blue), which are shown in FIGURE 11.10, are as follows:

1. Tn3 transposase nicks the donor DNA at the 3′-ends of the element (cyan arrows). Then the released free 3′-hydroxyl groups attack a pair of phosphodiester bonds in the target DNA that are 5-bp apart (pink arrows).
2. Bacterial DNA polymerase extends the 3′-ends of the target DNA and DNA ligase repairs the remaining pair of nicks to form a product in which the donor and target DNA molecules are fused together in a structure called a **cointegrate.**
3. Resolvase, the Tn3's *Res* gene product, catalyzes recombination between the two resolution sites (green) to separate the donor and acceptor DNA molecules. The Tn3 transposon in each DNA molecule is flanked by 5-bp direct repeats. The Tn3 element in the donor DNA retains its original direct repeat (cyan). The Tn3 that has moved to the acceptor DNA also has a direct repeat (red) but its sequence differs from that in the donor DNA.

The barrier between moving as a replicative transposon or a cut-and-paste transposon may be very low. For example, the bacterial insertion sequence IS*903* has the unusual ability to transpose by the cut-and-paste mechanism and by replicative transposition. Most products are formed by the cut-and-paste mechanism. A single mutation in the nucleotide immediately flanking the transposon can tip the balance from moving as a cut-and-paste element to moving as a replicative element. This could allow a diversity of recombination reactions each with a different evolutionary outcome.

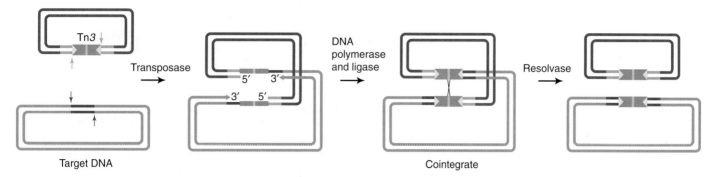

FIGURE 11.10 Replicative transposition pathway used by transposon Tn3. The steps for transposition of a Tn3 element from a donor DNA (dark blue) to a target DNA (light blue) are as follows. (1) Tn3 transposase nicks the donor DNA at the 3'-ends of the element (cyan arrows). Then the released free 3'-hydroxyl groups attack a pair of phosphodiester bonds in the target DNA that are 5 bp apart (red arrows). (2) Bacterial DNA polymerase extends the 3'-ends of the target DNA and DNA ligase repairs the remaining pair of nicks to form a product in which the donor and target DNA molecules are fused together in a structure called a cointegrate. (3) Resolvase catalyzes recombination between the two resolution sites (green segment) to separate the donor and acceptor DNA molecules. The Tn3 transposon in each resulting DNA molecule is flanked by 5-bp direct repeats. The Tn3 element in the donor DNA retains its original direct repeat (cyan). The Tn3 element that moved to the acceptor DNA also has a direct repeat (red) but its sequence differs from that in the donor DNA. (Adapted from http://www.sci.sdsu.edu/~smaloy/MicrobialGenetics/topics/transposons/repl-tpn.html)

Transposons are useful tools for bacterial studies.

Shortly after bacterial transposons were discovered, investigators devised techniques that allowed them to insert transposons into the bacterial chromosome. These techniques involve using a phage (or plasmid vector), which cannot replicate or integrate in the target bacterial host, to introduce a transposon bearing an antibiotic resistance gene into the host. After infection (or transformation) bacteria are spread on a rich agar medium containing the desired antibiotic, and antibiotic-resistant colonies are selected and purified. This approach has one serious drawback: the transposons remain mobile. Subsequent generations will, therefore, have transposons at many new sites. This transposon stability problem complicates the interpretation of genetic and biochemical data.

The mobility problem is solved by constructing **synthetic elements** with terminal inverted repeats that carry a selectable marker such as antibiotic resistance genes but lack the *Tnp* gene that codes for transposase. **FIGURE 11.11** compares the structure of a "natural" transposon and a synthetic element derived from it. The natural transposon, which has a *Tnp* gene, does not require the assistance of any other genetic element for mobility. In contrast, the synthetic element does require assistance of another genetic element that bears the *Tnp* gene for mobility. Thus, the natural transposon is **autonomous** and the synthetic element, sometimes termed a defective transposon, is **nonautonomous**. Defective transposons are introduced into a bacterial cell by a vector, such as a bacteriophage or plasmid, which cannot replicate or integrate in the target host. In a typical experiment a bacteriophage or plasmid vector containing a defective transposon that confers resistance to a specific antibiotic is introduced into bacteria.

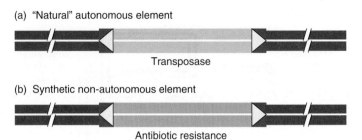

FIGURE 11.11 Synthetic element. (a) The natural autonomous element has a gene that codes for transposase flanked by terminal inverted repeats. (b) The synthetic element derived from the natural autonomous element is nonautonomous because the gene that codes for transposase has been replaced by a gene that confers antibiotic resistance.

The gene specifying the transposase is present in a separate vector that is already present in the bacteria. Bacteria are spread on a solid growth medium containing the antibiotic. Bacteria that are resistant to the transposon-encoded marker can only produce colonies if the element moves from the vector to the host genome. Antibiotic resistant colonies are washed from the plate to create a transposon mutagenized pool. All the bacteria in the pool have a transposition insertion at a different random position in their genome. A sufficiently large pool size is collected to be reasonably sure a transposon has inserted in every gene in the genome somewhere in the pool of cells. Then the inserted transposons are transferred to a second bacterial strain by generalized transduction. The transduced bacteria retain the inserted selectable marker but lack transposase activity. Because defective transposons do not encode their own transposase and the bacteria do not provide this activity, the issue of transposon stability is solved.

Typically, the transduced bacterial pool is screened for a particular phenotype or, where possible, a specific trait is selected. Conditional phenotypes can be screened using techniques where an archive copy of a transposon-containing clone is maintained when testing a variety of lethal conditions. One possible downside to these types of experiments is that transposon insertion mutations are almost exclusively complete null mutations. However, transposon insertion mutations in the 3′-end of an essential gene sometimes occur that do not inactivate the gene, and occasionally these insertions reveal interesting phenotypes.

Large-scale transposon mutagenesis experiments also have been useful for identifying essential genes. If the insertion sites of a sufficient number of transposon insertions can be analyzed in an organism and the genome is sufficiently small, one can begin to get a picture of the essential genes because they are never inactivated in the experiment. This type of experiment involves sequencing a large number of insertions and also requires the transposon insertion to be completely random. In these experiments a gene is assumed to be essential if a transposon insertion can never be identified in that particular gene. Synthetic elements have also been constructed that permit investigators to study the regulation of gene transcription (**BOX 11.1 IN THE LAB: REPORTER GENES**).

BOX 11.1: IN THE LAB

Reporter Genes

Synthetic elements have been constructed that contain an antibiotic resistance marker and a reporter gene that lacks a promoter (transcription initiation signal). **FIGURE B11.1** shows how a synthetic element bearing a modified *lacZ* reporter gene that lacks a promoter (transcription initiation signal) can be used to study gene regulation. A plasmid (or phage) vector, which cannot integrate or replicate in the target bacterial host, is used to introduce the synthetic element into bacteria. After transformation (or infection) bacteria are spread on nutrient agar containing the appropriate antibiotic and X-Gal (5-bromo-4-chloro-3-indolyl-β-D-galactopyranoside). All bacteria that have the synthetic element insert will form colonies. Only bacteria that have the synthetic element inserted into a target gene with an active promoter, however, can produce the *lacZ* gene product β-galactosidase, which hydrolyzes X-Gal. These bacteria, which form a substance that is converted to a blue dye, are easy to recognize as blue colonies. In a common experiment the investigator monitors bacterial colonies under different growth conditions to determine those genes that are induced under a specific set of conditions. Colonies containing cells with genes that are activated in response to various stresses such as DNA-damaging agents, temperature extremes, nutrient starvation, or toxic compounds can easily be detected by their blue color. Then the location of the individual element insert can be determined to discover the genes that are induced.

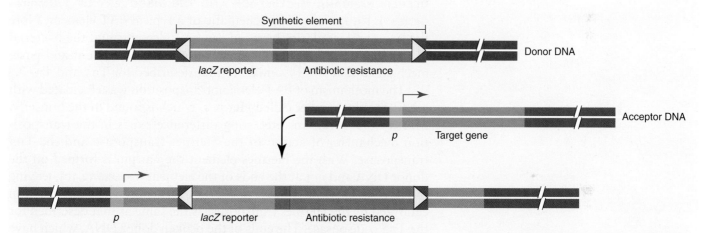

FIGURE B11.1 Synthetic element with *lacZ* reporter gene. The synthetic element used in reporter fusion experiments usually has an antibiotic resistance marker (brown). The synthetic element also encodes the *lacZ* reporter gene (purple) that lacks its own promoter (transcription initiation signal). The *lacZ* gene can be expressed when the synthetic element inserts in the correct orientation into a target gene (orange) after an active target gene promoter (p), which is shown in green. The donor DNA is shown in red and the acceptor DNA in blue.

11.2 Eukaryotic Transposable Elements

Some eukaryotic transposable elements move through a DNA-only mechanism and others require an RNA intermediate.

Eukaryotes have two major classes of transposable elements that are distinguished on the basis of their transposition mechanism. Class I elements require an RNA intermediate to transpose, and class II elements transpose by a DNA-only mechanism similar to that used by the bacterial cut-and-paste transposable elements. Both eukaryotic classes have some autonomous elements and others that are nonautonomous.

We begin by examining the class II elements because of their similarity to the bacterial transposons.

The eukaryotic "hAT" mobile elements move by a DNA only cut-and-paste transposition mechanism.

The "*h*AT" mobile elements are a large family of mobile elements found in a wide variety of eukaryotes, including fungi, plants, invertebrates, and vertebrates. The term *h*AT derives from the first letters in the names of the three founding members of this family, the *hobo* element in *Drosophila*, the *Ac* element in maize, and the Tam3 element in the snapdragon plant. The *Ac* element and *Ds*, a defective element derived from *Ac*, have a very special place in the transposable element field because they were the first transposable elements to be identified. The pioneering experiments performed by Barbara McClintock beginning in the mid-1940s opened the door to future transposon investigation by her as well as others (**BOX 11.2 LOOKING DEEPER: BARBARA McCLINTOCK AND THE DISCOVERY OF "JUMPING GENES"**). **FIGURE 11.12** is a schematic of a typical *h*AT element. Note the structural similarity between the *h*AT elements and the bacterial IS elements. The *h*AT elements move by a DNA only cut-and-paste mechanism that is very similar to that described for Tn5 and Tn10.

The mechanism of *h*AT element transposition was elucidated with a *h*AT family member called Hermes, which is found in the housefly, *Musca domestica*. An interesting difference exists in the transposition mechanism of action of the Hermes transposase and the Tn5 transposase. With the Hermes element the hairpin is formed on the donor DNA and not at the ends of the element (**FIGURE 11.13**), leaving the Hermes element with clean breaks at its ends. The paste reaction catalyzed by the Hermes transposase is the same as that described for the Tn5 transposase. The ends of the broken donor DNA, which have hairpin structures, have to be repaired by homologous recombination or nonhomologous end-joining.

Antibody and T-cell receptor genes are formed by a mechanism that resembles the Hermes transposition mechanism.

The adaptive immune system in vertebrates can generate a seemingly limitless number of antibodies and T-cell receptors against diverse

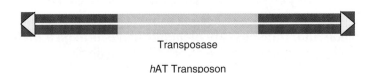

Transposase

*h*AT Transposon

FIGURE 11.12 Schematic of a typical *h*AT mobile transposable element. The "*h*AT" mobile elements are a large family of mobile elements found in a wide variety of eukaryotes, including fungi, plants, invertebrates, and vertebrates. The term *h*AT derives from the first letters in the names of the three founding members of this family, the *hobo* element in *Drosophila*, the *Ac* element in maize, and the Tam3 element in the snapdragon plant.

BOX 11.2: LOOKING DEEPER

Barbara McClintock and the Discovery of "Jumping Genes"

Barbara McClintock discovered genetic transposition in maize in the mid-1940s. Her studies showed that certain chromosomal elements can "jump" in specific genetic crosses. Her first clue to the existence of jumping genetic elements came from the observation that in some maize strains DNA breaks always occur at the same position in chromosome 9. Further studies revealed that two elements are required for the DNA breaks to take place. A break-inducing genetic entity, the dissociation element (*Ds*), has to be present at the breakage site, and a second discrete entity, the activator element (*Ac*), has to be present elsewhere in the genome. McClintock's next major breakthrough came from the discovery that alteration in kernel pigment patterns correlated with *Ds* element movement (FIGURE B11.2a). The gene responsible for pigment color is inactivated when the *Ds* element inserts into it (Figure B11.2b). The *Ac* element is required to supply the recombinase needed for *Ds* movement. If the *Ds* element were again mobilized by *Ac*, an autonomous element that produces transposase, the progeny could revert back to purple during development of the kernel if the site was correctly repaired (Figure B11.2c). Additional studies showed the *Ds* element is an *Ac* element that has lost the ability to make the recombinase, more specifically called a transposase, needed to move the element. The *Ac* element can also move and does not require assistance of any other genetic element to do so. In other words, *Ac* is an autonomous element and *Ds* is a nonautonomous element. Geneticists initially had difficulty in reconciling McClintock's discovery of "jumping genes" with the notion the genes in a chromosome appeared to be arranged in a specific order that does not change. The significance of McClintock's work became apparent when investigators discovered transposable elements in bacteria and then in other organisms.

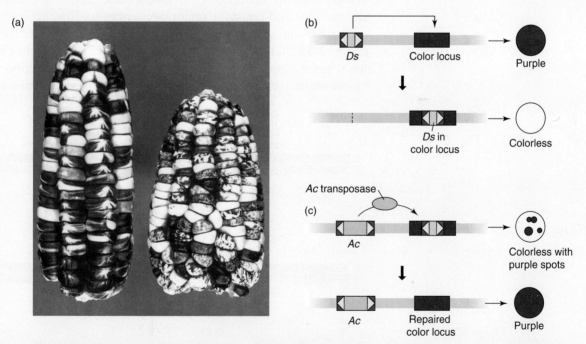

FIGURE B11.2 **Movement of autonomous and nonautonomous transposable elements can alter the pigmentation of maize.** (a) Changes in kernel coloration could be explained by the movement of DNA elements in the chromosome. (b) Maize kernels are purple when the responsible genes are not interrupted by endogenous DNA elements like the nonautonomous transposable element called *Ds*. An insertion element like *Ds* can jump into a locus responsible for purple pigmentation before development of the coat. (c) If the *Ds* element were again mobilized by *Ac*, an autonomous element that produces transposase, the progeny could revert back to purple during development of the kernel if the site was correctly repaired. (Part a reproduced from Jones, R. N. 2005. *Cytogenet Genome Res* 109:90–103, with permission from S. Karger AG, Basel. Photo courtesy of Neil Jones, The University of Wales, Aberystwyth. Parts b and c adapted from Jones, R. N. 2005. *Cytogenet Genome Res* 109:90–103.)

(*continues*)

BOX 11.2: LOOKING DEEPER

Barbara McClintock and the Discovery of "Jumping Genes" (continued)

The original *Ac* element identified by McClintock still serves as an important tool in maize. One property of the *Ac* element that can be viewed as a positive or a negative is its propensity to transpose to regions of the chromosome that are genetically close. Although local hopping can be a frustration to investigators, it also can be a tool for localized mutagenesis of a region. Genetic strategies have been used to isolate *Ac* elements across all 10 maize chromosomes. These elements should prove useful for more focused *Ac* mutagenesis for subregions of the maize genome. *Ac* elements are useful also for generating mutations at a position after they transpose out of a gene. Mutations that occur in the donor site after transposition result from repair of the host chromosome through nonhomologous end-joining. This repair process produces footprints at the site of the original transposon insertion that can be used to obtain a variety of different mutations.

pathogens and antigens. It does this through programmed site-specific recombination of the immunoglobulin gene precursors, generating up to 10^{11} different antibodies. The idea that the immunoglobulin genes are formed from noncontiguous gene segments was first proposed in 1965 by William J. Dreyer and J. Claude Bennett. They suggested (1) the variable and constant regions of the immunoglobulin genes are encoded by different DNA sequences; (2) some process brings the variable and constant region sequences together to form a single transcriptional unit, one for the heavy chain and another for the light chain of an immunoglobulin; and (3) each heavy or light chain gene locus has multiple *V* region sequences but only one *C* region sequence before rearrangement in B cells. In 1976 Susumu Tonegawa showed this

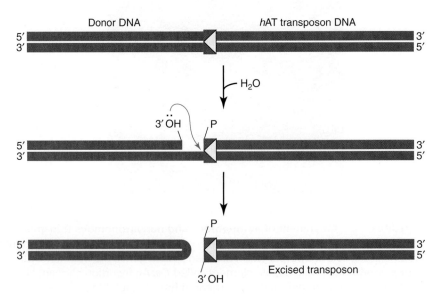

FIGURE 11.13 **Mechanism of Hermes transposition.** The Hermes transposable element, a member of the *h*AT family, is found in the housefly, *Musca domestica*. Its mechanism of transposition is similar to that used by the bacterial transposon Tn*5* with one important difference. With the Hermes element the hairpin is formed on the donor DNA and not at the ends of the element. (Adapted from L. Zhou, et al., *Nature* 432 [2004]: 995–1001.)

suggested organization was correct by demonstrating the variable and constant regions are separated in the DNA from mouse embryos but are adjacent in mature B cells. Tonegawa's work set the stage for the demonstration of programmed somatic DNA rearrangements in B cells.

Subsequent studies revealed that heavy-chain genes are organized as repeated segments of variable (V), diversity (D), and joining (J) segments (FIGURE 11.14). Heavy chains of immunoglobulins are formed by joining V, D, and J segments to the constant region sequence during B-cell development through a recombination mechanism similar to that used in Hermes transposition. Light-chain genes are organized as repeated segments of V and J segments. The light chains of mature immunoglobulins are formed by joining V and J sequences to the constant region sequence during B-cell development. Diversity in the immune system is achieved by different recombination processes. There are three levels of diversity:

1. *Combinatorial diversity:* Specific V, (D), and J segments are selected.
2. *Junctional diversity:* The process that joins the V, (D), and J coding segments is imprecise, resulting in the deletion or addition of nucleotides at the joining junction.
3. *Pairing diversity:* Heavy and light chains combine to make an antigen receptor protein.

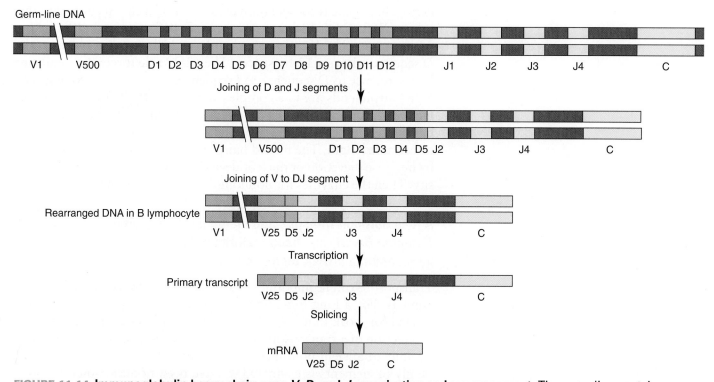

FIGURE 11.14 **Immunoglobulin heavy chain gene *V*, *D*, and *J* organization and rearrangement.** The germ line contains multiple *V*, *D*, and *J* regions. In immunoglobulin heavy chain genes the *D* and *J* regions are joined first. Next, a *V* region is joined to the *DJ* region to form a *VDJ* gene. The C region is joined to the *VDJ* region at the RNA level by splicing to make mRNA encoding the *VDJC* regions. (Adapted from Cooper, G. M., and Hausman, R. E. 2000. *The Cell: A Molecular Approach* (2nd ed). Sinauer Associates.)

11.2 Eukaryotic Transposable Elements

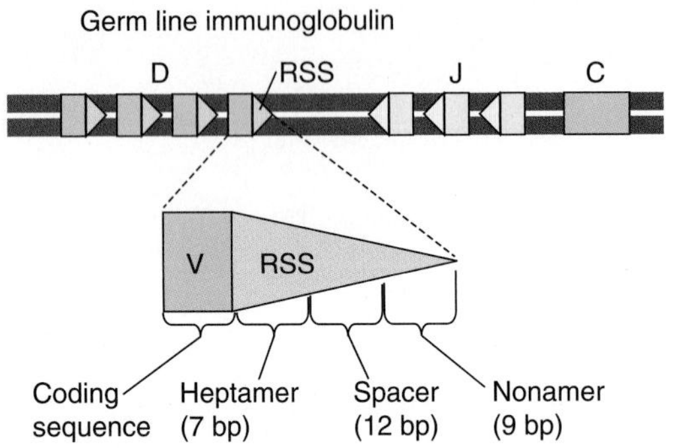

FIGURE 11.15 The recombination signal sequences (RSS) in V(D)J recombination. Each V, D, and J gene coding sequence is flanked by a RSS that contains conserved heptamer and nonamer sequences flanking a spacer of 12 or 23 bp. (Adapted from Roth, D. B. 2003. *Nat Rev Immunol* 3:656–666.)

V(D)J recombination occurs by programmed double-strand breaks. The process of V(D)J recombination must be carefully regulated to guard against joining incorrect double-strand break ends. A similar mechanism is used to ensure proper V(D)J rearrangements take place for both heavy- and light-chain genes. We examine the mechanism as applied to heavy-chain gene rearrangement. The V, D, and J segments have a **recombination signal sequence (RSS)** at each end. The RSSs participate in the recombination reaction that joins segments. RSS consists of a conserved heptamer (7 bp) and nonamer (9 bp) elements, which are separated by a spacer region of 12 or 23 nucleotides (FIGURE 11.15).

The RSSs are the targets of two enzymes that cleave the RSS in a defined manner (FIGURE 11.16). These enzymes are named RAG-1 and RAG-2 for the recombination-activating genes that code for them. The *RAG* genes are expressed only during lymphocyte development, thus limiting the production of dangerous double-strand breaks to a stage where there is a need for the programmed rearrangements. The RAG proteins make a single-strand nick at the junction between the RSS and the coding sequence, forming a free 3′-OH group. The RAG • RSS complex then pairs with another RAG • RSS complex, and the free 3′-OH ends attack the opposite strand through transesterification to form a double-strand break at the RSS–coding sequence junction. During the process of forming the double-strand break, the ends of the coding sequence region become sealed by a *hairpin*. To prevent joining of two V sequences, two D sequences, or two J sequences, heavy-chain gene recombination uses the 12/23 rule. According to this rule an RSS with a 12-bp spacer can only be joined to an RSS with a 23-bp spacer. The V and J sequences have 23-bp spacers, whereas the D sequences have a 12-bp spacer. The 12/23 rule forces the joining to be V sequence to D sequence to J sequence. Once the two coding ends with hairpin ends are in a complex with the RAG proteins and other factors, the hairpins are cleaved by a specific nuclease. Cleavage of the hairpin sequences is not at a specific sequence so the opened ends are of variable length. The nicked hairpin ends may be further processed by nuclease digestion or nucleotide addition to create junctional diversity. Then the enzymes for nonhomologous end-joining connect ends.

A long dormant transposable element in fish called *Sleeping Beauty* has been restored to full activity by recombinant DNA technology.

Transposons can be important tools used to study various multicellular animals. For a long time, however, no useful transposable elements existed for work in mammalian systems. An interesting story for the production of a transposon that works in mammalian cells starts with a family of ancient nonfunctional transposons identified in fish. This family of elements is estimated to have been nonfunctional for over 14 million years. Zsuzanna Izsvák and coworkers determined the sequences of these defective elements and identified several translation termination mutations in the transposase gene. Moreover, they were able to predict some of the missense mutations that occurred over time by comparing many different nonfunctional copies of the element. Izsvák and coworkers realized they might be able to use this information to convert a nonfunctional element to a functional transposon. They used site-directed mutagenesis to make the necessary nucleotide changes required to reestablish transposon autonomy. The restored functional transposon is called **Sleeping Beauty** because the element had been nonfunctional (or asleep) for millions of years. A synthetic element derived from the restored *Sleeping Beauty* transposon is able to transpose in human cells when introduced with another DNA molecule that contains the gene required to synthesize transposase. Related synthetic elements have been shown to transpose in mouse somatic cells, where they serve as a tool for gene inactivation. *Sleeping Beauty* holds great promise as a tool for studying vertebrate genomics and as a vector for gene therapy.

LTR retrotransposons copy an RNA intermediate to make cDNA, which then is integrated into the genome by a cut-and-paste mechanism.

The final group of eukaryotic mobile elements to be discussed here transpose via an RNA intermediate. These retrotransposon elements, which make up about 90% of all transposable elements in the human genome, can be divided into two general classes depending on whether or not they have a **long terminal repeat (LTR)** at each end. Those that have LTRs are called **LTR retrotransposons,** and those that do not are called **non-LTR retrotransposons.**

LTR retrotransposons have a similar genetic makeup to retroviruses. Unlike retroviruses, however, which appear to be limited to vertebrate genomes, LTR retrotransposons are widely distributed in eukaryotes. In fact, the first LTR retrotransposons to be investigated were the Ty1 element in yeast and the *copia* element in *Drosophila*. The human genome contains about 7×10^5 LTR retrotransposon insertions, which account for about 8% of the DNA. Very few of the human LTR retrotransposon insertions have retained the ability to transpose.

A typical LTR retrotransposon has direct repeats that are about 300 to 500 bp long flanking *pol* and *gag* genes (FIGURE 11.17a). The *pol* gene codes for a long polypeptide, which is cleaved later to yield a reverse transcriptase, an integrase, and a protease. The *gag* gene codes for a structural protein known as the group-specific antigen protein, which interacts with RNA to form a virus-like particle. A major difference between LTR retrotransposons and retroviral DNAs is that most retrotransposons lack *env* genes (Figure 11.17b). The *env* gene product, an envelope protein, is essential for producing an infective virion particle. LTR retrotransposons that lack an *env* gene are therefore not infective agents.

A great deal of our information about the mechanisms of LTR retrotransposition comes from studies with yeast LTR retrotransposons (FIGURE 11.18). It is likely, however, that LTR retrotransposons present in other organisms use a similar mechanism. Retrotransposition begins when a cellular RNA polymerase catalyzes transcription starting at a site within the LTR adjacent to the *gag* gene. Then the

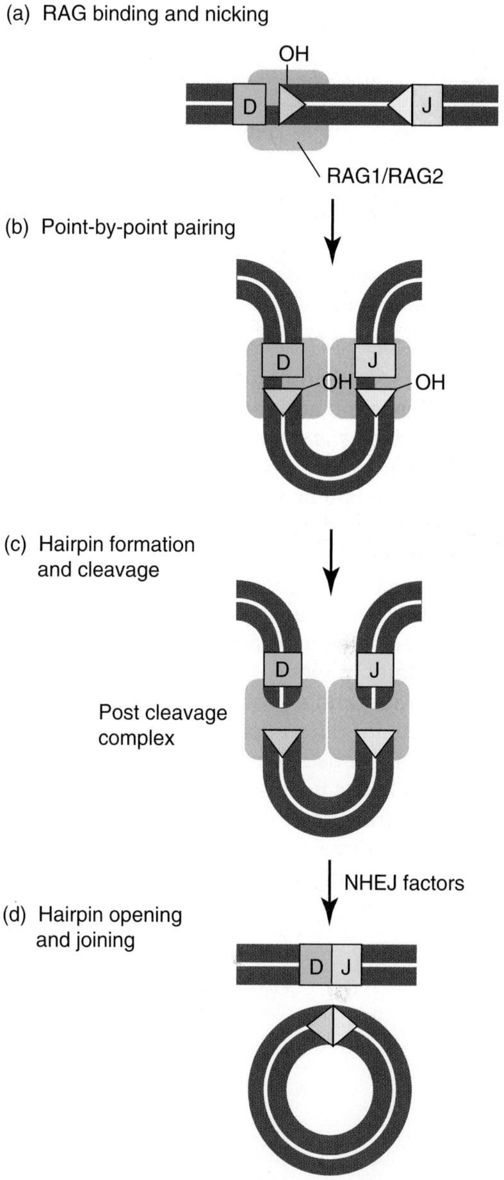

FIGURE 11.16 The V(D)J reaction recombination reaction. (a) RAG1 and RAG2 bind to a recombination signal sequence (RSS) and make a single-strand nick. (b) The RAG1 • RAG2 • RSS complex pairs with another RSS and makes another nick on the opposite strand to make a double-strand break. (c) The coding ends are formed into hairpins by the cleavage and remain associated with RAG1 • RAG2 complex, and the RSS ends in a post-cleavage complex. (d) The hairpins are opened by a nuclease and nonhomologous end-joining (NHEJ) factors to join the D and J ends together and the two signal ends together. (Adapted from Roth, D. B. 2003. *Nat Rev Immunol* 3:656–666.)

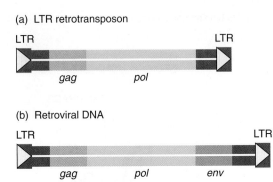

FIGURE 11.17 Comparison of LTR retrotransposon and retroviral DNA.
(a) *LTR retrotransposon:* A typical LTR retrotransposon has direct repeats that are about 300 to 500 bp long (yellow arrows) flanking *gag* and *pol* genes. The *gag* gene codes for a structural protein known as the group specific antigen protein, which interacts with RNA to form a virus-like particle. The *pol* gene codes for a long polypeptide, which is cleaved later to yield a reverse transcriptase, an integrase, and a protease. (b) *Retroviral DNA:* A major difference between LTR retrotransposons and retroviral DNAs is that retroviral DNAs have *env* genes, whereas most LTR transposons do not. The *env* gene product, an envelope protein, is essential for producing an infective virion particle.

RNA moves to the cytoplasm where it is translated by ribosomes to form the long polypeptide and the group-specific antigen protein. The long polypeptide is cut to produce protease, reverse transcriptase, and integrase. Group-specific antigens usually interact with two RNA molecules, reverse transcriptase, and integrase to form a virus-like particle. The RNA provides the sequence information required to make a full-length DNA copy. A tRNA molecule serves as the primer for DNA synthesis, which takes place in several steps that are not shown in the figure. Then the DNA • integrase complex enters the nucleus, where the integrase functions as a transposase and inserts the retrotransposon DNA into the genome by a cut-and-paste mechanism similar to that described for Tn5.

Non-LTR elements move using a target-primed reverse transcription mechanism.

Eukaryotes have two major types of non-LTR retrotransposons, **long interspersed nucleotide elements (LINEs)** and **short interspersed nucleotide elements (SINEs)**. LINEs are usually about 5 to 7 kbp long, whereas SINEs are usually less than 500 kbp long. LINEs and SINEs are the most abundant transposable elements in the human genome, which has more than 1 million insertions of each type scattered throughout. Some LINEs are autonomous elements that encode proteins required for their transposition, but most do not encode the proteins required for transposition and therefore depend on proteins specified by autonomous LINEs for transposition.

The mechanism of LINE movement was derived from studies with a LINE called the R2Bm element, which is present in the silkworm *Bombyx mori*. This element has a 5′-untranslated region (5′-UTR) at one end and a 3′-untranslated region (3′-UTR) at the other end (FIGURE 11.19). The UTRs flank a region that codes for a single polypeptide with reverse transcriptase and endonuclease activities. A cellular RNA polymerase uses R2Bm integrated into the host chromosome to synthesize an RNA molecule, which serves as an intermediate for **target-primed reverse transposition** mechanism. This transposition mechanism does not depend on LTRs because the reverse transcription is primed by the 3′-ends of the target DNA, which are generated by stepwise breaks in the target DNA (FIGURE 11.20).

The steps in target-primed reverse transposition are as follows. (1) A cellular RNA polymerase interacts with the promoter in the 5′-UTR and transcribes the LINE. The newly synthesized RNA is translated to make a polypeptide with endonuclease and reverse transcriptase activity. (2) The endonuclease makes a single nick in one strand in the target DNA. (3) Reverse transcriptase uses the newly synthesized RNA as a template to extend the nicked 3′-end. (4) The endonuclease makes a single nick in the other strand of the target DNA. (5) Reverse transcriptase extends the 3′-end of this new nick, and the newly formed DNA displaces the RNA. (6) DNA ligase repairs the nicks.

The human genome has several different types of LINE inserts. Only one of these, however, LINE-1, seems capable of autonomous transposition. An estimated 1 in 50 humans will experience a new

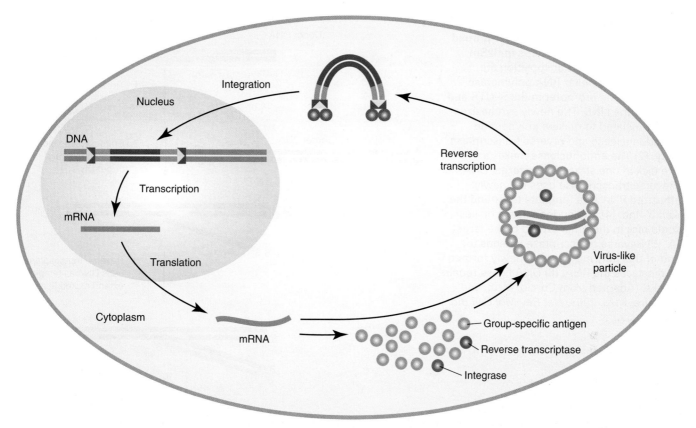

FIGURE 11.18 LTR retrotransposon movement. Retrotransposition begins when a cellular RNA polymerase catalyzes transcription starting at a site within the LTR adjacent to the *gag* gene. The RNA (green) moves to the cytoplasm where it is translated by ribosomes (not shown) to form a long polypeptide (not shown) and the group-specific antigen protein (pink circles). The long polypeptide is cut to produce protease, reverse transcriptase (blue circle), and integrase (red circle). Group-specific antigen usually interacts with two RNA molecules, reverse transcriptase, and integrase to form a virus-like particle. The RNA provides the sequence information required to make a full-length DNA copy. This process takes place in several steps, which are not shown in the figure. The DNA • integrase complex enters the nucleus, where the integrase functions as a transposase and inserts the retrotransposon DNA into the genome by a cut-and-paste mechanism similar to that described for Tn5. (Adapted from Havecker, E. R., Gao, X., and Voytas, D. F. 2004. *Genome Biol* 5:225, Figure 1.)

LINE-1 integration event in their genome, occurring either in the parental germ cells or early in their own embryonic development. LINE-1 structure is shown in FIGURE 11.21. LINE-1s, which constitute approximately 17% of human DNA, are about 6 kbp long and have 5'-UTRs with about 900 bp and 3'-UTRs with about 200 bp, which are followed by polyadenylic acid tails of variable length.

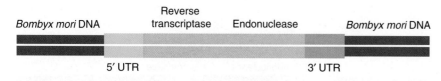

R2Bm element integrated into *Bombyx mori* DNA

FIGURE 11.19 The anatomy of R2Bm, a non-LTR LINE element. R2Bm is a non-LTR LINE element present in the silkworm *Bombyx mori*. It has a 5'-UTR at one end and a 3'-UTR at the other end. The UTRs flank a region that codes for a single polypeptide with reverse transcriptase and endonuclease activities.

11.2 Eukaryotic Transposable Elements

FIGURE 11.20 **R2Bm LINE element insertion into the host chromosome by target-primed reverse transposition.** The steps in R2Bm target-primed reverse transposition are as follows. (1) A cellular RNA polymerase interacts with the promoter in the 5'-UTR and transcribes the LINE. The newly synthesized RNA is translated to make a polypeptide with endonuclease and reverse transcriptase activity. (2) The endonuclease makes a single nick in one strand in the target DNA. (3) Reverse transcriptase uses the newly synthesized RNA as a template to extend the nicked 3'-end. (4) The endonuclease makes a single nick in the other strand of the target DNA. (5) Reverse transcriptase extends the 3'-end of this new nick and the newly formed DNA displaces the RNA. (6) DNA ligase repairs the nicks. (Adapted from Curcio, M. J., and Derbyshire, K. M. 2003. *Nat Rev Mol Cell Biol* 4:9, Figure 5a.)

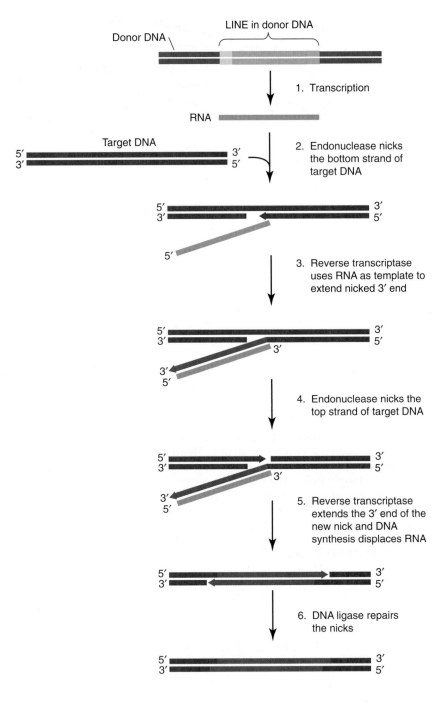

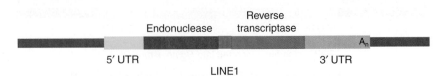

FIGURE 11.21 **The anatomy of human LINE-1.** A human LINE-1 is about 6 kilobases long and has a 5'-UTR with about 900 bp and a 3'-UTR with about 200 bp, which is followed by a polyadenylic acid tail (As) of variable length.

Alu is the most common SINE in the human genome.

SINEs resemble LINEs in having a 5′-UTR at one end and a 3′-UTR at the other end but are much shorter because they do not have coding information for endonuclease or reverse transcriptase. The most common SINE in the human genome is the *Alu* element, which is so named because it has an *Alu* restriction endonuclease site. *Alu* is estimated to make up about 11% of the human genome. The 1.1 million *Alu* elements in the human genome are believed to trace back to a single insertion event that occurred into the *7SL RNA* gene in an ancient primate. The 7SL RNA is part of a nucleoprotein called the signal recognition complex that binds to a specific N-terminal sequence in some proteins as they leave the ribosome and then directs the complex to a protein conducting channel in the endoplasmic reticulum. *Alu*, like other SINEs, is a nonautonomous retroelement. SINE transposition depends on proteins encoded by functional LINEs. Functional LINEs can move more elements than just nonfunctional LINEs and SINEs. It also appears that cellular mRNAs can be processed by the LINE-1 machinery and inserted into new positions in the chromosome. Evidence for this processing and insertion comes from the observation that nonfunctional genes, called **processed pseudogenes,** contain a poly(A) end and appear to reside at preferred LINE-1 insertion sties. These processed pseudogenes comprise an estimated 0.5% of the human genome.

There is considerable debate concerning what benefits, if any, retrotransposons make to the host cells.

When retrotransposon inserts were originally discovered in the human genome, many investigators thought it was "junk DNA" that offered no benefit to the host. Because the retrotransposon DNA seemed to exist only to allow the formation of new retrotransposons, they appeared to be "selfish" DNA. Retrotransposons are known to be responsible for several human diseases, which are caused by retrotransposon insertion into an essential gene. A pair of identical retrotransposon inserts can also provide the homology required for the production of deletions and harmful gene rearrangements through homologous recombination.

Retroelements may also provide benefits to the host genome as well. It is clear that retroelements can use chromosome breaks formed as a result of other processes as targets for insertion. It remains unclear if the retroelements aid in the repair process or merely insert in a break that would normally be fixed through other means, such as nonhomologous end-joining or replication-mediated repair. LINE-1 elements can participate in a gene fusion event with adjacent genes, a process that can contribute to the formation of new hybrid genes. Hundreds of examples have been observed where human proteins contain portions of proteins derived from LINE-1 and *Alu*. It remains to be determined whether these proteins are part of the normal functioning of our cells. LINEs and SINEs also can impact gene expression by altering the expression of genes by shuffling transcription regulatory elements. Changes in multicellular genomes may benefit greatly from the diversity that can be supplied by the movement of LINEs and SINEs and by other processes related to their abundance in genomes.

Questions and Problems

1. Briefly define, describe, or illustrate each of the following.
   a. Transposable element
   b. Transposition
   c. IS element
   d. Terminal inverted repeat
   e. Transposase
   f. Composite transposon
   g. Cut-and-paste mechanism
   h. Direct repeat
   i. Replicative transposition
   j. Noncomposite transposon
   k. Synthetic element
   l. Autonomous element
   m. Nonautonomous element
   n. Recombination signal sequence
   o. *Sleeping Beauty*
   p. Long terminal repeat
   q. LTR retrotransposon
   r. Non-LTR retrotransposon
   s. LINEs
   t. SINEs
   u. Target-primed reverse transposition
   v. Processed pseudogene

2. You suspect a new mutation that appears in *E. coli* is caused by an IS element.
   a. Base analogs and frameshift mutagens do not cause the mutation to revert. Are these observations consistent with your hypothesis that the mutation is due to an IS element? Explain.
   b. Previous studies indicate there is an EcoRI restriction site on either side of the gene that has been mutated but there are no EcoRI sites within the gene. How might you use this information to test your hypothesis that the mutation is caused by an IS element? Explain.
   c. Would it be possible to obtain useful information if the IS element has an EcoRI restriction site? Explain.

3. IS elements, composite transposons, and noncomposite transposons are all transposable elements. Prepare a table that lists the ways in which each of these transposable elements are similar and in which they are different.

4. How might one use recombinant DNA technology to convert an IS element to a composite transposon with a gene that confers resistance to tetracycline? How might one use recombinant DNA technology to convert an IS element to a noncomposite transposon with a gene that confers resistance to ampicillin?

5. The Tn5 transposase functions as a dimer. How does the fact that each Tn5 transposase subunit in the transposase • Tn5 DNA complex makes contact with both ends of the Tn5 DNA contribute to the success of Tn5 transposition?

6. After Tn5 inserts into a target site, it is flanked by 9-bp direct repeats that derived from the target DNA. Is the cut-and-paste mechanism of transposition consistent with this observation? Explain your conclusion.

7. The cut-and-paste mechanism of transposition involves cutting a transposable element from a donor site and pasting it into a target site. How do transposons that use this transposition mechanism replicate? Use a diagram to explain your answer.

8. Transposon Tn3 moves by replicative transposition. Why does this mechanism of transposition require a resolvase?

9. A plasmid vector, which cannot replicate or integrate into the *E. coli* chromosome, bears a synthetic element with an Amp^R gene. Describe how you can use the plasmid vector and an *E. coli* strain with a plasmid that codes for transposase to isolate a mutant with the synthetic element inserted within the *lacZ* gene.

10. You wish to examine the regulation of an *E. coli* gene that codes for a protein with no known enzymatic activity. How would you use a reporter fusion to accomplish your goal?

11. What contribution does each of the following make to retrotransposition by LTR retrotransposons?
    a. Host cell RNA polymerase
    b. Group specific antigen
    c. Reverse transcriptase
    d. Integrase

12. There are three levels of diversity in the immune system.
    a. Describe V(D)J recombination and explain how it contributes to antibody diversity.
    b. V(D)J recombination is one of three processes that contribute to antibody diversity. What are the other two processes and how do they contribute to antibody diversity?
13. LTR retrotransposons are one of the two major types of retrotransposons in eukaryotes.
    a. What are the major structural features of an LTR retrotransposon?
    b. How do LTR retrotransposons move from one site in the genome to another?
14. Non-LTR retrotransposons consist of LINEs and SINEs.
    a. What are the major structural features of LINEs and SINEs?
    b. Some LINEs are autonomous, but all SINEs are nonautonomous. What differences exist between autonomous LINEs and nonautonomous SINEs?
    c. How do autonomous LINEs move from one site in the genome to another?
    d. Although SINEs are nonautonomous, they do occasionally move from one genomic site to another. How is this mobility possible?

Suggested Reading

General Interest

Bushman, F. 2002. *Lateral DNA Transfer, Mechanisms and Consequences*. Woodbury, NY: Cold Spring Harbor Press.

Capy, P., and Deragon, J.-M. 2005. Transposons. *Encyclopedia of Life Sciences*. pp. 1–4. Hoboken, NJ: John Wiley & Sons.

Craig, N. J. 2002. Transposases and integrases. *Encyclopedia of Life Sciences*. pp. 1–7. Hoboken, NJ: John Wiley & Sons.

Craig, N. L., Craigie R., Gellert M., and Lambowitz, A. M. (eds). 2002. *Mobile DNA* (2nd ed). Washington, DC: ASM Press.

Curcio, M. J., and Derbyshire, K. M. 2003. The outs and ins of transposition: from mu to kangaroo. *Nat Rev Mol Cell Biol* 4:865–877.

Hickman, A. B., Chandler, M., and Dyda, F. 2010. Integrating prokaryotes and eukaryotes: DNA transposases in light of structure. *Crit Rev Biochem Mol Biol* 45:50–69.

Nesmelova, I. V., and Hackett, P. B. 2010. DDE transposases: structural similarity and diversity. *Adv Drug Delivery Rev* 62:1187–1195.

Roberts, A. P., Chandler, M., Courvalin, P. et al. 2008. Revised nomenclature for transposable genetic elements. *Plasmid* 60:167–173.

Bacterial Transposable Elements

Ahmed, A. 2009. Alternative mechanisms of Tn5 transposition. *PLoS Genet* 5:1–5.

Braam, L. A. M., and Reznikoff, W. S. 2001. DNA transposition: classes and mechanisms. *Encyclopedia of Life Sciences*. pp. 1–8. Hoboken, NJ: John Wiley & Sons.

Choi, K.-H., and Kim, K.-J. 2009. Applications of transposon-based gene delivery system in bacteria. *J Microbiol Biotechnol* 19:217–228.

Davies, D. R., Goryhin, I. Y., Reznikoff, W. S., and Rayment, I. 2000. Three-dimensional structure of the Tn5 synaptic complex transposition intermediate. *Science* 289:77–85.

Haniford, D. B. 2006. Transposone dynamics and regulation in Tn10 transposition. *Crit Rev Biochem Mol Biol* 41:407–424.

Harshey, R. M. 2001. Transposons: prokaryotic. *Encyclopedia of Life Sciences*. pp. 1–7. Hoboken, NJ: John Wiley & Sons.

Judson, N., and Mekalanos, J. J. 2000. Transposon-based approaches to identify essential bacterial genes. *Trends Microbiol* 8:521–526.

Maloy, S. R. 2007. Use of antibiotic-resistant transposons for mutagenesis. *Methods Enzymol* 421:11–17.

Olurunniji, F. J., and Stark, W. M. 2010. Catalysis of site-specific recombination by Tn3 resolvase. *Biochem Soc Trans* 38:417–421.

Reznikoff, W. S. 2003. Tn5 as a model for understanding DNA transposition. *Mol Microbiol* 47:1199–1206.

Reznikoff, W. S. 2006. Tn5 transposition: a molecular tool for studying protein structure-function. *Biochem Soc Trans* 34(pt 2):320–323.

Reznikoff, W. S. 2008. Transposon Tn5. *Ann Rev Genet* 42:269–286.

Shapiro, J. A. 2009. Letting *Escherichia coli* teach me about genome engineering. *Genetics* 183:1205–1214.

Shuman, H. A., and Silhavy, T. J. 2003. The art and design of genetic screens: *Escherichia coli*. *Nat Rev Genet* 4:419–431.

Siguler, P., Filée, J., and Chandler, M. 2006. Insertion sequences in prokaryotic genomes. *Curr Opin Microbiol* 9:526–531.

Vikváryová, M., and Valková, D. 2004. Transposons—the useful genetic tools. *Biol Bratisl* 59:309–318.

Eukaryotic Transposable Elements

Aronovich, E. L., McIvor, R. S., and Hackett, P. B. 2011. The Sleeping Beauty transposon system: a non-viral vector for gene therapy. *Human Mol Genet* 20:R14–R20.

Batzer, M. A., and Deininger, P. L. 2002. *Alu* repeats and human genomic diversity. *Nat Rev Genet* 3:370–380.

Beauregard, A., Curcio, M. J., and Belfort, M. 2008. The take and give between retrotransposable elements and their hosts. *Annu Rev Genet* 42:587–617.

Beck, C. R., Garcia-Perez, J. L., Badge, R. M., and Moran, J. V. LINE-1 elements in structural variation and disease. *Annu Rev Genomics Human Genet* 12:187–215.

Becker, H.-A., and Lönnig, W.-E. 2001. Transposons: eukaryotic. *Encyclopedia of Life Sciences*. pp. 1–10. Hoboken, NJ: John Wiley & Sons.

Bouuaert, C. C., and Chalmers, R. M. Gene therapy vectors: the prospects and potentials of the cut-and-paste transposons. *Genetica* 138:473–484.

Comfort, N. C. 2003. *The Tangled Field: Barbara McClintock's Search for the Patterns of Genetic Control*. Cambridge, MA: Harvard University Press.

Cordaux, R., and Batzer, M. A. 2008. Evolutionary emergence of genes through retrotransposition. *Encyclopedia of Life Sciences*. pp. 1–7. Hoboken, NJ: John Wiley & Sons.

Eickbush, T. H., and Eickbush, D. G. 2005. Transposable elements: evolution. *Encyclopedia of Life Sciences*. pp. 1–6. Hoboken, NJ: John Wiley & Sons.

Eickbush, T. H., and Jamburuthuoda, V. K. 2008. The diversity of retrotransposons and the properties of their reverse transcriptases. *Virus Res* 134:221–234.

Furano, A. V., and Boissinot, S. 2008. Long interspersed nuclear elements (LINEs): evolution. *Encyclopedia of Life Sciences*. pp. 1–6. Hoboken, NJ: John Wiley & Sons.

Gellert, M. 2005. V(D)J recombination: RAG proteins, repair factors, and regulation. *Annu Rev Biochem* 71:101–132.

Gogvadze, E., and Buzdin, A. 2009. Retroelements and their impact on genome evolution and functioning. *Cell Mol Life Sci* 66:3727–3742.

Goodier, J. L., and Kazazian, H. H. Jr. 2005. Long ineterspersed nuclear elements (LINEs). *Encyclopedia of Life Sciences*. pp. 1–5. Hoboken, NJ: John Wiley & Sons.

Goodier, J. L., and Kazazian, H. H. Jr. 2008. Retrotransposons revisited: the restraint and rehabilitation of parasites. *Cell* 135:23–35.

Han, J. S. 2010. Non-long terminal repeat (non-LTR) retrotransposons: mechanisms, recent developments, and unanswered questions. *Mobile DNA* 1:15.

Havecker, E. R., Gao, X., and Voytas, D. F. 2004. The diversity of LTR retrotransposons. *Genome Biol* 5:225.

Hickman, A. B., Perez, Z. N., Zhou, P., et al. 2005. Molecular architecture of a eukaryotic DNA transposase. *Nat Struct Mol Biol* 12:715–721.

Ivics, Z., Hackett, P. B., Plasterk, R. H., Izsvák, Z. 1997. Molecular reconstruction of *Sleeping Beauty*, a Tc1-like transposon from fish, and its transposition in human cells. *Cell* 91:501–510.

Jones, R. N. 2005. McClintock's controlling elements: the full story. *Cytogenet Genome Res* 109:90–103.

Kazazian, H. H. Jr. 2004. Mobile elements: drivers of genome evolution. *Science* 303:1626–1632.

Keller, E. F., and Mandelbrot, B. B. 1984. *A Feeling for the Organism: The Life and Work of Barbara McClintock*. New York: W. H. Freeman.

Morrish, T. A., Gilbert, N., Myers, J. S., et al. 2002. DNA repair mediated by endonuclease-independent LINE-1 retrotransposition. *Nat Genet* 31:159–165.

Muñoz-López, M., and Garcia-Pèrez, J. L. 2010. DNA transposons: nature and applications in genomics. *Curr Genomics* 11:115–126.

Roy-Engel, A. M., and Belancio, V. P. 2011. Retrotransposons and human disease. *Encyclopedia of Life Sciences*. pp. 1–8. Hoboken, NJ: John Wiley & Sons.

Schmid, C. W., and Rubin, C. M. 2005. Short interspersed elements (SINEs). *Encyclopedia of Life Sciences*. pp. 1–4. Hoboken, NJ: John Wiley & Sons.

Weiner, A. M. 2002. SINEs and LINEs: the art of biting the hand that feeds you. *Curr Opin Cell Biol* 14:343–350.

Zhou, L., Mitra, R., Atkinson, P. W., et al. 2004. Transposition of the *h*AT element links transposable elements and V(D)J recombination. *Nature* 432:995–1001.

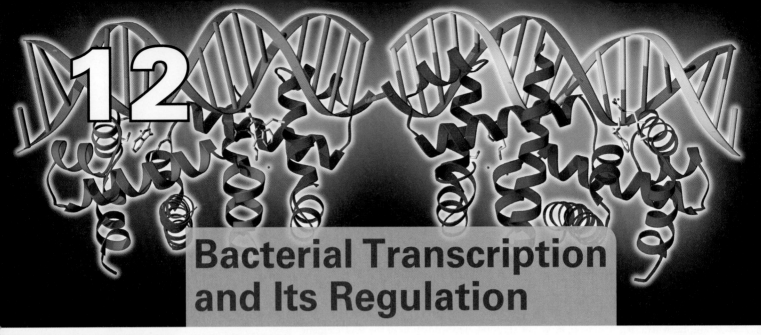

12. Bacterial Transcription and Its Regulation

CHAPTER OUTLINE

12.1 Introduction to the Bacterial RNA Polymerase Catalyzed Reaction.
RNA polymerase requires a DNA template and four nucleoside triphosphates to synthesize RNA.
Bacterial RNA polymerases are large multisubunit proteins.

12.2 Initiation Stage
Bacterial RNA polymerase holoenzyme consists of a core enzyme and sigma factor.
A transcription unit must have an initiation signal called a promoter for accurate and efficient transcription to take place.
Active σ^{70} promoters have −10 and −35 boxes.
Genetic and biochemical studies provide additional information about bacterial promoters.
DNA footprinting shows that σ^{70} RNA polymerase holoenzyme binds to promoter DNA to form a closed and then an open complex.
Bacterial RNA polymerase crystal structures show how the enzyme is organized and provide insights into how it works.
Members of the σ^{70} family have four conserved domains.
The σ^{70} RNA polymerase holoenzyme goes through several rounds of abortive initiation before promoter escape.
RNA polymerase "scrunches" DNA during transcription initiation.

12.3 Transcription Elongation Complex
The transcription elongation complex is a highly processive molecular motor.
Pauses influence the overall transcription elongation rate.
RNA polymerase can detect and remove incorrectly incorporated nucleotides.

12.4 Transcription Termination
Bacterial transcription machinery releases RNA strands at intrinsic and Rho-dependent terminators.

12.5 Messenger RNA
Bacterial mRNA may be monocistronic or polycistronic.
Bacterial mRNA usually has a short lifetime compared with other kinds of bacterial RNA.
Controlling the rate of mRNA synthesis can regulate the flow of genetic information.
Messenger RNA synthesis can be controlled by negative and positive regulation.

12.6 Lactose Operon
The E. coli genes lacZ, lacY, and lacA code for β-galactosidase, lactose permease, and β-galactoside transacetylase, respectively.
The lac structural genes are regulated.
Genetic studies provide information about the regulation of lac mRNA.
The operon model explains the regulation of the lactose system.
Allolactose is the true inducer of the lactose operon.
The Lac repressor binds to the lac operator in vitro.
The lac operon has three lac operators.
The Lac repressor is a dimer of dimers, where each dimer binds to one lac operator sequence.

12.7 Catabolite Repression
E. coli uses glucose in preference to lactose.
The cAMP receptor protein combines with 3′,5′-cyclic adenylate to form a positive regulator or activator.
The cAMP • CRP complex binds to an activator site (AS) upstream from the lac promoter and activates lac operon transcription.
Glucose causes catabolite repression through cAMP modulation and inducer exclusion mechanisms.
cAMP • CRP activates more than 100 operons.

12.8 Tryptophan Operon
The tryptophan (trp) operon is regulated at the levels of transcription initiation, elongation, and termination.

BOX 12.1: LOOKING DEEPER Alternate Sigma Factors

BOX 12.2: CLINICAL APPLICATION Rifamycin

BOX 12.3: LOOKING DEEPER RNA Polymerase Movement

QUESTIONS AND PROBLEMS

SUGGESTED READING

Image © LAGUNA DESIGN/Photo Researchers, Inc.

Cells express their genes by transferring information from DNA to RNA and then from RNA to polypeptides. The former type of information transfer is termed **transcription** because the information remains in nucleotide sequences, whereas the latter type of information transfer is called **translation** because the information is translated from nucleotide sequences into amino acid sequences. This chapter examines the transcription process that takes place in bacteria. Enzymes that copy information present in DNA templates into RNA molecules are called DNA-dependent RNA polymerases, or RNA polymerases for short. RNA polymerases were purified first from bacteria and later from the eukaryotes and archaea. RNA synthesis in all three domains of life takes place in three stages: initiation, elongation, and termination.

Bacterial, eukaryotic, and archaeal RNA polymerases are large multisubunit enzymes that require the assistance of additional factors to recognize specific genes. Initial studies suggested RNA polymerases from organisms belonging to each domain might be unique to that domain. Although the RNA polymerases do in fact differ in several important respects, more recent studies show the catalytic core of the enzymes from all three domains are remarkably similar. Lessons learned from studying an RNA polymerase from an organism belonging to one domain, therefore, can provide useful information for studying RNA polymerases from organisms belonging to the other two domains.

Despite the fact that bacterial, eukaryotic, and archaeal transcription share many important similarities, important differences also exist. For example, details of the bacterial and eukaryotic initiation and termination processes differ, involving different factors and different mechanisms. The mechanisms that regulate bacterial and eukaryotic RNA synthesis also differ, as do the pathways for converting the primary transcripts into mature RNA molecules.

In vitro and *in vivo* studies, mostly involving the *Escherichia coli* RNA polymerase, have provided important insights into how bacterial RNA polymerases work. Nobody has yet succeeded in obtaining a crystal structure for *E. coli* RNA polymerase, but such structures have been obtained for RNA polymerases isolated from two other gram-negative bacteria, the extreme thermophiles *Thermus aquaticus* and *Thermus thermophilus*. The high degree of sequence homology present in bacterial RNA polymerases permits us to apply information obtained from the RNA polymerase from one bacterial strain to RNA polymerases of other bacterial strains. This chapter begins with an introduction to the RNA polymerase catalyzed reaction, then examines the initiation, elongation, and termination stages of bacterial transcription, and concludes by examining some mechanisms that bacteria use to regulate mRNA synthesis.

12.1 Introduction to the Bacterial RNA Polymerase Catalyzed Reaction

RNA polymerase requires a DNA template and four nucleoside triphosphates to synthesize RNA.

Several different groups detected bacterial RNA polymerases at about the same time in the early 1960s. Under most growth conditions *E. coli* has about 1,000 to 2,000 RNA polymerase molecules per cell. The enzyme catalyzes nucleoside monophosphate group transfer from a nucleoside triphosphate (NTP) to the 3′-end of the growing RNA chain (or the first nucleoside triphosphate) so that chain growth proceeds in a 5′→3′ direction (**FIGURE 12.1**). The essential chemical characteristics of RNA synthesis are as follows:

1. Phosphodiester bond formation takes place as the result of a nucleophilic attack of the 3′-hydroxyl group on the growing chain (or first nucleoside triphosphate) on the α-phosphoryl group of the incoming NTP. This reaction is similar to the one that takes place during DNA synthesis. Also, as with DNA synthesis, pyrophosphate hydrolysis drives the reaction to completion.
2. The DNA template sequence determines the RNA sequence. Each base added to the growing 3′-end of the RNA chain is chosen by its ability to pair with a complementary base in the template strand. Thus, the bases C, T, G, and A in a DNA strand cause G, A, C, and U, respectively, to appear in the newly synthesized RNA molecule.
3. All four ribonucleoside triphosphates (adenosine 5′-triphosphate [ATP], guanosine 5′-triphosphate [GTP], cytidine 5′-triphosphate [CTP], and uridine 5′-triphosphate [UTP]) are required for RNA synthesis. When a single nucleotide is omitted RNA synthesis stops at the point where that nucleotide must be added.
4. The RNA chain grows in the 5′→3′ direction; that is, nucleotides are added only to the 3′-OH end of the growing chain. This direction of chain growth is the same as that in DNA synthesis.
5. RNA polymerases, in contrast with DNA polymerases, can initiate chain growth *without* a primer.
6. Only ribonucleoside 5′-triphosphates participate in RNA synthesis. The first base to be laid down in the initiation event is a triphosphate. Its 3′-OH group is the point of attachment for the second nucleotide and its 5′-triphosphate group remains at the 5′-end throughout chain elongation.

Only one of the two DNA strands in a given double-stranded DNA region acts as the template strand, dictating the sequence of the newly synthesized RNA molecule. The complementary DNA strand, the nontemplate strand, has the same base sequence as the RNA molecule (except that U replaces T) and is commonly

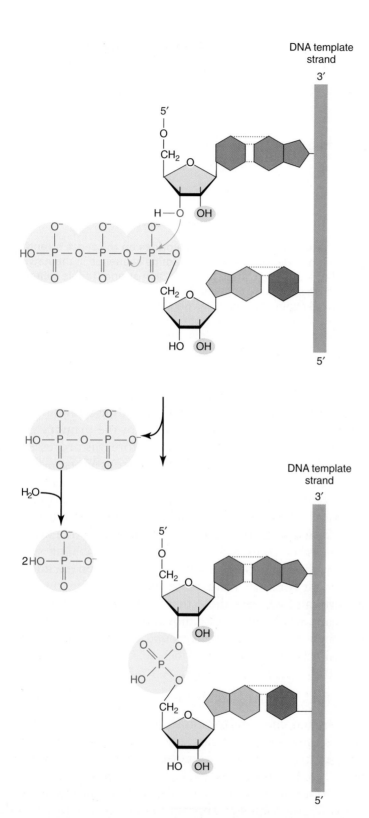

FIGURE 12.1 RNA polymerase catalyzed phosphodiester bond formation.
RNA polymerase catalyzes nucleoside monophosphate group transfer from a nucleoside triphosphate (NTP) to the 3'-end of a growing RNA chain (or the first nucleoside triphosphate) so chain growth proceeds in a 5'→3' direction. (Adapted from Berg, J. M., et al. 2002. *Biochemistry* (5th ed). W. H. Freeman and Company.)

12.1 Introduction to the Bacterial RNA Polymerase Catalyzed Reaction

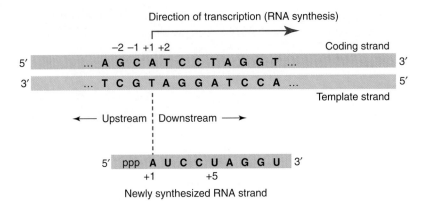

FIGURE 12.2 Rules for numbering nucleotides on the sense strand. The template strand (blue) dictates the nucleotide sequence in the newly formed RNA strand (green). The complementary DNA strand (red) is called the coding or sense strand. By convention the nucleotide at the transcription start site on the coding strand is designated as position +1, the next nucleotide as position +2, and so forth. Sequences that come after the transcription start site (on the 3′-side) are downstream, whereas those that come before it (on the 5′-side) are upstream.

referred to as the **coding** or **sense strand** (FIGURE 12.2). Sequence information is usually given for just the coding strand because this information also provides the sequence of the **primary transcript** (the newly synthesized RNA molecule before it is processed.) Template strand sequence is easily obtained from Watson-Crick base pairing rules.

By convention, the nucleotide at the transcription start site on the coding strand is designated as position +1, the next nucleotide is at position +2, and so forth (Figure 12.2). The nucleotide that immediately precedes the transcription start site on the coding strand is at position −1, the nucleotide before that is at position −2, and so forth. Sequences that come *after* the transcription start site (on the 3′-side) are **downstream**, whereas those that come *before* it (on the 5′-side) are **upstream**. RNA polymerase moves downstream as it transcribes a DNA template. With this background information in mind, let us now examine bacterial RNA polymerase.

Bacterial RNA polymerases are large multisubunit proteins.

Purification and characterization studies show that the fully active form of *E. coli* RNA polymerase, the RNA polymerase holoenzyme, is a large multisubunit protein (molecular mass = 459 kDa) that has a total of six subunits ($\alpha_2\beta\beta'\omega\sigma$) (TABLE 12.1). RNA polymerase holoenzymes from other bacteria have the same multisubunit structure, and each subunit sequence appears to have been conserved during evolution.

RNA polymerase activity is usually assayed using a reaction mixture that contains RNA polymerase (which may be in a crude extract), DNA, Mg^{2+}, three nonradioactive NTPs, and one radioactive NTP labeled either in the base with ^3H or ^{14}C or in the phosphate attached to the ribose with ^{32}P. The reaction is stopped

TABLE 12.1 *E. coli* RNA Polymerase Subunits

Subunit	Number of Subunits in Holoenzyme	Gene	Map Position	Molecular Mass (Da)	Function
alpha (α)	2	rpoA	74.10	36,511	Required for enzyme assembly; interacts with some regulatory proteins
beta (β)	1	rpoB	90.08	150,616	Forms a pincer and is the site of rifampicin action
beta' (β')	1	rpoC	90.16	155,159	Forms a pincer and provides an absolutely conserved –NADFDGD– motif that is essential for catalysis
omega (ω)	1	rpoZ	82.34	10,105	Helps in enzyme assembly but is not required for enzyme activity
sigma (σ)	1	rpoD	69.21	70,263	Directs enzyme to promoters but is not required for phosphodiester bond formation.

by adding trichloroacetic acid, which causes the newly formed RNA but not the NTP precursors to become insoluble, and the precipitate is collected by filtration or centrifugation. The amount of radioactivity in the precipitate is proportional to the amount of RNA synthesized.

The large size of the bacterial holoenzyme compared with typical enzymes raises the question of whether some complex feature of the polymerization reaction requires such a large enzyme. In fact, bacteriophage T7 RNA polymerase, which has only one polypeptide subunit with a molecular mass 99 kDa and a tertiary structure (FIGURE 12.3) reminiscent of the Klenow fragment, synthesizes phage T7 RNA very efficiently. Clearly, the polymerization reaction itself does not require a huge multisubunit bacterial enzyme.

Clues to understanding reasons for the size differences between the *E. coli* and the phage T7 RNA polymerases come from (1) attempting to use the phage T7 enzyme to transcribe *E. coli* DNA and (2) studying the gene organization in phage T7 DNA. First, phage T7 RNA polymerase can transcribe at best a small fraction of the *E. coli* genes. Second, phage T7 genes are arranged in only a few transcription units, so there are only a few RNA polymerase binding sites. In *E. coli*, RNA polymerase must be able to recognize approximately 4,300 genes, which are signaled by about 1,000 different binding sites and also respond to a large number of regulatory proteins that alter the polymerase's ability to recognize a binding site. It seems likely that these multiple requirements necessitate the large multisubunit *E. coli* enzyme.

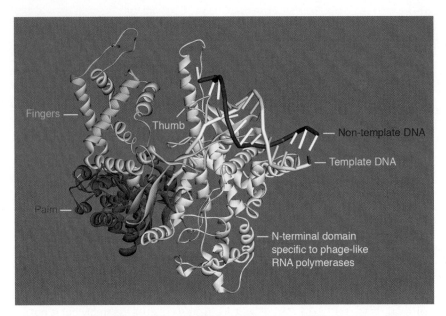

FIGURE 12.3 Structure of the T7 RNA polymerase • DNA complex. The polymerase subdomains are colored red "palm," green "thumb," and orange "fingers." The N-terminal domain (which is specific to all phage-like RNA polymerases) is yellow. Template and nontemplate DNA strands are shown in light and dark blue, respectively. (Structure from Protein Data Bank ID: 1CEZ. Cheetham, G. M., Jeruzalmi, D., and Steitz, T. A. 1999. *Nature* 399:80–83. Prepared by B. E. Tropp.)

12.2 Initiation Stage

Bacterial RNA polymerase holoenzyme consists of a core enzyme and sigma factor.

Early studies with an *E. coli* RNA polymerase holoenzyme missing the ω subunit showed that $\alpha_2\beta\beta'\sigma$ can dissociate under physiological conditions to form a core polymerase ($\alpha_2\beta\beta'$) and sigma (σ) factor as follows:

$$\underset{\text{RNA polymerase holoenzyme}}{\alpha_2\beta\beta'\sigma} \rightleftharpoons \underset{\text{core polymerase}}{\alpha_2\beta\beta'} + \underset{\text{sigma factor}}{\sigma}$$

The equilibrium lies far to the left with a dissociation constant of 10^{-9}. Richard Burgess and colleagues were able to separate the two proteins using phosphocellulose ion exchange chromatography, which shifts the equilibrium because core polymerase binds to the resin but σ factor does not (**FIGURE 12.4**).

The $\alpha_2\beta\beta'$ core polymerase can synthesize RNA using single-stranded DNA or a nicked double-stranded DNA as a template, but it cannot synthesize RNA using an intact double-stranded DNA molecule as a template. Hence, neither the σ factor nor the ω subunit is required for phosphodiester bond formation. The $\alpha_2\beta\beta'$ core polymerase can use nicked DNA double-stranded DNA as a template

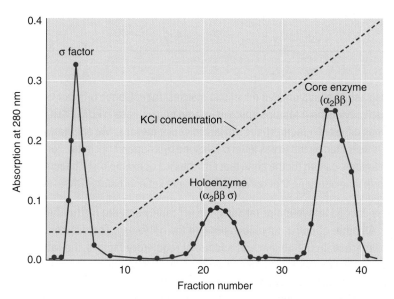

FIGURE 12.4 Chromatographic separation of core polymerase from the sigma (σ) factor. After being placed on a column containing phosphocellulose ion exchange resin, the protein was eluted with aqueous KCl solution at the indicated concentrations. Three protein peaks were observed by monitoring the light absorption at 280 nm. Analysis of the three peaks revealed that most of the holoenzyme dissociated into σ factor and core enzyme ($α_2ββ'$). Some of the original enzyme remained in its associated form ($α_2ββ'σ$), however, and is present in the central peak. The ω subunit was not detected in this study. (Adapted from Burgess, R. R., et al. 1969. *Nature* 221:43–46.)

for RNA synthesis because DNA unwinds at nicks to generate non-physiological initiation sites. The $α_2ββ'$ core polymerase remains intact under physiological conditions. When teased apart under nonphysiological conditions and purified, α, β, and β' each lacks the ability to catalyze RNA synthesis by itself.

A transcription unit must have an initiation signal called a promoter for accurate and efficient transcription to take place.

In contrast to the core polymerase ($α_2ββ'$), the RNA polymerase holoenzyme ($α_2ββ'σ$) can use intact double-stranded DNA as a template for RNA synthesis. Transcription begins when the holoenzyme recognizes and binds to a specific transcription initiation signal called a **promoter** at the beginning of a transcription unit. The existence of promoters was first demonstrated by isolation of a particular class of mutations in *E. coli* that prevent cells from synthesizing enzymes required for lactose metabolism. These mutations, termed promoter mutations, not only result in a lack of gene activity but also cannot be complemented because a promoter can only stimulate transcription of a DNA sequence on the same DNA molecule.

Sigma factor is essential for promoter DNA recognition but does not bind to promoter DNA on its own. *E. coli* has several different kinds of σ factors, each recognizing a unique promoter sequence (**BOX 12.1 LOOKING DEEPER: ALTERNATE SIGMA FACTORS**). The primary

BOX 12.1: LOOKING DEEPER

Alternate Sigma Factors

E. coli has seven different kinds of σ subunits, initially distinguished based on their molecular masses. At first, each σ factor was designated by using the Greek letter σ followed by a superscript specifying the σ factor's approximate molecular mass in kDa (TABLE B12.1). Because this nomenclature system becomes ambiguous when different σ factors have similar molecular masses, two alternative nomenclature systems were introduced. In the first, each σ factor is indicated by the Greek letter σ followed by a superscript letter. For example, *E. coli* σ^{70} is called σ^D in this system. In the second, each σ factor is named for the gene that encodes it. For example, σ^{70} becomes RpoD. Each kind of σ factor combines with core polymerase to form a holoenzyme that recognizes a unique set of promoter DNA sequences.

Based on sequence homologies *E. coli* σ subunits can be divided in two major families, the σ^{70} family and the σ^{54} (σ^N family). The only σ^N family member present in *E. coli* is the σ^N factor. All other *E. coli* σ subunits belong to the σ^{70} family. The different *E. coli* σ subunits compete with one another for the core RNA polymerase. In the absence of additional regulatory factors, the rate of transcription depends solely on promoter strength and the concentration of specific holoenzymes. The σ^{70} RNA polymerase holoenzyme transcribes genes that code for proteins required for normal exponential growth, which account for most of the bacteria's genes. σ^{70} is, therefore, an essential protein. RNA polymerase holoenzymes containing one of the other σ subunits recognize specific sets of genes distributed throughout the chromosome that code for proteins needed for cell survival under some special environmental condition such as heat shock (σ^H) or nitrogen deprivation (σ^N). Table B12.1 summarizes the properties and functions of the different *E. coli* σ subunits. All σ subunits bind to RNA polymerase core enzymes, allowing the holoenzyme to interact with specific promoter elements. Many, possibly all, σ subunits also are targets of accessory ligands that regulate their activity.

The number of alternative σ subunits present in a bacterial cell varies greatly from one bacterial species to another. For instance, *Mycoplasma* sp. has only one kind of σ subunit, whereas *Streptomyces coelicolor* has at least 66 different kinds of σ subunits. Sigma subunits sometimes have novel functions. For instance, four sporulation-specific σ subunits allow *Bacillus subtilis* to respond to certain types of nutrient deprivation by becoming metabolically dormant cells or spores that are surrounded by a protective multilayer envelope. Spore formation is a complex process, requiring many enzymes and proteins that are not synthesized by exponentially growing cells. During spore formation preexisting sigma subunits are destroyed and new sigma subunits are formed, which combine with the core enzyme to form a holoenzyme that recognizes promoters of genes needed for spore formation.

The σ^{54}-RNA polymerase is unique among bacterial RNA polymerase holoenzymes because it cannot initiate transcription without the assistance of an activator protein. The activation process, which is shown schematically in FIGURE B12.1, begins with the binding of activator protein to a site on the bacterial chromosome called an enhancer, which is usually 100 bp or more upstream from the σ^{54} promoter. Interaction between the activator proteins and σ^{54} RNA polymerase holoenzyme is possible because the

TABLE B12.1 *E. coli* Sigma Factors

Sigma Factor	Molecular Mass (kDa)	Gene	Map Position (min)	Functions	Consensus Sequence[a]		
					−35 region	spacer bp	−10 region
σ^{70} or σ^D	70	*rpoD*	69.21	Major sigma factor during exponential growth	TTGACA	16–18 bp	TATAAT
σ^{32} or σ^H	32	*rpoH*	77.55	Transcription heat-shock genes	CTTGAA	13–15 bp	CCCCAT**n**T
σ^{24} or σ^E	24	*rpoE*	58.36	Response to periplasmic stress	GAACTT	16–17 bp	TCTGA
σ^{28} or σ^F	28	*rpoF*	43.09	Expression late flagellar genes	CTAAA	15 bp	GCCCATAA
σ^{18} or σ^{FecI}	18	*FecI*	92.35	Iron-isocitrate transport	GAAAAT	15 bp	TGTCCT
σ^{38} or σ^S	38	*rpoS*	61.75	Expression stationary phase genes	TTGACA	14–18 bp	CTAYACTT
σ^{54} or σ^N	54	*rpoN*	72.05	Nitrogen metabolism genes	−25 CTGGCAC	6 bp	−12 TTGCA

[a]**n** is any nucleotide and Y is a pyrimidine.

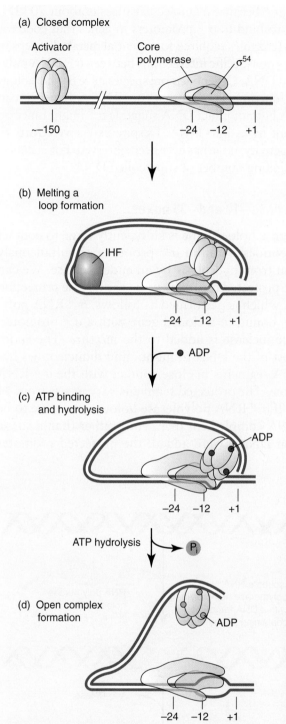

FIGURE B12.1 Schematic illustrating the route to open complex formation by the σ⁵⁴ RNA polymerase holoenzyme. (a) The activator (yellow), a ring-shaped hexamer, binds to enhancer sequences that are about 150 bp upstream of the transcription start site. The σ⁵⁴ RNA polymerase holoenzyme binds to its cognate promoter sequences to form the closed complex. (b) A site of localized melting is generated around position −12. The nucleotide sequence between the two protein-binding sites may produce DNA bending, but an additional protein, such as the integration host factor (IHF), is sometimes required to assist the bending. (c) The addition of nucleoside triphosphates (NTPs, shown as small red circles) is required for a stable interaction between activator and σ⁵⁴-RNA polymerase holoenzyme. (d) After nucleotide hydrolysis the activator • polymerase complex becomes destabilized, resulting in open complex formation. (Adapted from Burrows, P. C. 2003. *Bioessays* 25:1150–1153.)

DNA segment between the two protein-binding sites bends to form a loop. Although the nucleotide sequence between the two protein-binding sites may produce DNA bending, an additional protein such as the integration host factor is sometimes required to assist the bending. Activator proteins are themselves subject to regulation by the binding of a small effector molecule or, more commonly, by the addition of a phosphate group to a specific site on the activator protein. Activation induces an ATPase activity within the activator protein that is essential for unwinding DNA in the promoter region so that transcription can begin.

12.2 Initiation Stage

E. coli σ factor, called σ^{70} because its molecular mass is about 70 kDa, helps core polymerase to bind to σ^{70} promoters in genes that code for housekeeping enzymes (enzymes required for essential metabolic steps in the cell) and destabilizes nonspecific interactions between σ^{70} RNA polymerase holoenzyme and DNA. Other bacterial species have σ^{70} homologs that perform the same function. The fact that core RNA polymerase can transcribe nicked DNA but not intact DNA suggests σ^{70} might function by introducing transient nicks into DNA. Despite extensive efforts to observe such a nicking activity, none has ever been observed. For instance, σ^{70} does not alter the linking number of supercoiled DNA.

Active σ^{70} promoters have −10 and −35 boxes.

The σ^{70} RNA polymerase holoenzyme is sufficiently large to contact many deoxyribonucleotides within a σ^{70} promoter simultaneously and protect this region from hydrolysis by an endonuclease. We can estimate the size of the protected region by using the **DNase protection method** (FIGURE 12.5), which is performed as follows. σ^{70} RNA polymerase holoenzyme is bound to a cloned gene with a σ^{70} promoter, and then a DNA endonuclease is added to the mixture. The endonuclease degrades most of the DNA to mono- and dinucleotides but leaves untouched DNA segments in close contact with the σ^{70} RNA polymerase holoenzyme. The protected segments vary in size from 41 to 44 base pairs (bp). If σ^{70} RNA polymerase holoenzyme were to be added to the total DNA complement of *E. coli* (rather than a single cloned gene) and then DNase were added, the protected promoter

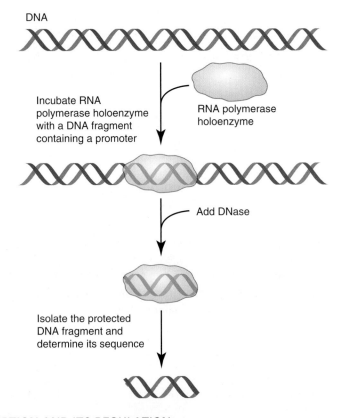

FIGURE 12.5 DNase protection assay. RNA polymerase holoenzyme binds to a DNA fragment containing a promoter. The enzyme protects the promoter from DNase hydrolysis. The protected region is isolated and sequenced. This technique can be used to characterize any sequence-specific protein–DNA interaction.

fragments would consist of about 1,000 different DNA segments, each derived from a particular gene or set of adjacent genes.

In 1975 David Pribnow used the DNase protection method to analyze σ^{70} promoters from different genes that had been protected by the σ^{70} RNA polymerase holoenzyme. The protected fragments were sequenced, as were RNA molecules synthesized from each gene *in vitro*. The 5'-terminus of the RNA molecule revealed the initiation start site on the complementary DNA template strand. Pribnow recognized an important common feature in the protected DNA fragments. A 6-bp sequence centered about 10 bp before (upstream from) the transcription start site is conserved. This hexamer is now known as the **–10 box** for its location or the **Pribnow box** for its discoverer. Examination of 300 *E. coli* σ^{70} promoters has shown the frequency of occurrence of bases in –10 boxes is as follows (the subscript is the frequency):

$$T_{77}A_{76}T_{60}A_{61}A_{56}T_{82}$$

If the sequences were totally unrelated, one would expect each base to occur at each position 25% of the time, but instead the sequences are AT-rich. This AT-rich region melts during the early stage of transcription initiation.

Bacterial σ^{70} promoters usually have a second conserved six-base sequence, centered about 35 bp upstream from the transcription initiation site and, therefore, called the **–35 box**. Examination of 300 *E. coli* σ^{70} promoters revealed the frequency of occurrence of bases in the –35 box is as follows (the subscript is the frequency):

$$T_{69}T_{79}G_{61}A_{56}C_{54}A_{54}$$

An idealized sequence such as that specified for the –10 box (or the –35 box), which indicates the most frequently found base in each position of many actual sequences, is termed a **consensus sequence**. Sequences of actual –10 and –35 boxes usually differ from the consensus sequence. These differences, which are evident in the *E. coli* σ^{70} promoter sequences recognized by σ^{70} RNA polymerase holoenzyme (**FIGURE 12.6**), allow the cell to regulate genes based on the strength of RNA polymerase binding—an important regulatory mechanism.

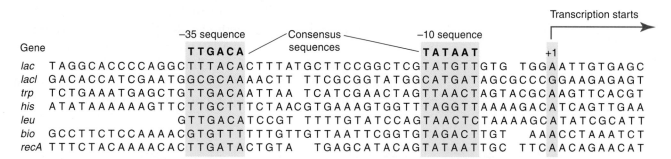

FIGURE 12.6 Base sequences in promoter regions of several *E. coli* genes. The consensus sequences located 10 and 35 nucleotides upstream from the transcription start site (+1) are indicated. Promoters vary tremendously in their ability to promote transcription. Much of the variation in promoter strength results from differences between the promoter elements and the consensus sequences in the –10 and –35 boxes.

FIGURE 12.7 Mutations that alter the *E. coli lac* promoter. Four nucleotide substitutions are shown in the coding strand for the *lac* promoter. The two substitutions shown in red weaken the promoter and are therefore called down mutations. The two shown in green strengthen the promoter and are therefore called up mutations. The −10 box is shaded in yellow and the −35 box in light blue. Many base changes that are known to alter promoter activity are located in or near the −10 box or are clustered around base −35 and thus define an important site. Deletions or additions in the spaces between the −10 and −35 boxes usually decrease promoter activity.

Genetic and biochemical studies provide additional information about bacterial promoters.

Additional information about promoters has been obtained by introducing mutations into promoters that alter transcription initiation. The rationale is that if a base change affects promoter activity, that base must be contained in a critical promoter region. Four such mutations are shown for the *E. coli lac* promoter in FIGURE 12.7. Two of these mutations are in the −10 box and stimulate transcription initiation, whereas the other two are in the −35 box and inhibit transcription initiation. In general, promoters that support active transcription, termed *strong promoters*, have −10 and −35 boxes that are close to the consensus sequences and a spacer between the two boxes that is 17 ± 1 bp. Mutations in the −10 or −35 boxes that cause their sequences to differ from the consensus sequences usually lead to weaker promoter activity as evidenced by lower rates of initiation of RNA synthesis. The σ^{70} promoters in bacterial genes that code for ribosomal RNA have a region called the **up element** (FIGURE 12.8), which is located just upstream of the −35 box that interacts with α subunits in the σ^{70} RNA polymerase holoenzyme. This additional interaction is responsible for an increase in promoter strength that accounts for a high rate of ribosomal RNA synthesis.

DNA footprinting shows that σ^{70} RNA polymerase holoenzyme binds to promoter DNA to form a closed and then an open complex.

We can learn a great deal about the sequence specificity of a DNA binding protein by using a method called the **DNA footprinting technique** (FIGURE 12.9). A particular piece of double-stranded DNA, labeled in one strand at its 5′-terminus with ^{32}P (or in its 3′-terminus with a labeled nucleotide), is mixed with a DNA-binding protein of interest. Then DNase I is added, but so briefly that, on average, each

FIGURE 12.8 DNA promoter elements. Moving downstream, promoter elements are as follows: the UP element, −35 box (−35), and the −10 box (−10).

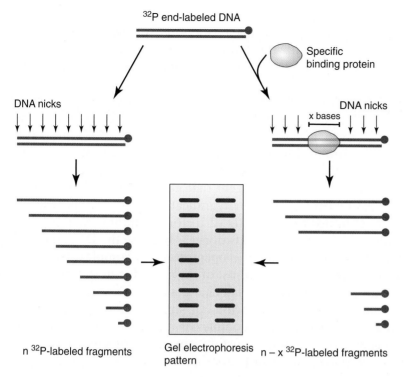

FIGURE 12.9 **DNA footprinting technique.** Double-stranded DNA, labeled in one strand at its 5′-terminus with ^{32}P (shown in red), is treated with DNase I for a very short time so that, on average, each DNA molecule receives no more than one single-strand break. The same experiment is performed in the presence of a protein that binds to specific sites on the DNA. The DNA sites with the bound protein are protected from the action of DNase I. Certain fragments, therefore, will not be present. The missing bands in the gel pattern identify the binding site(s) on the DNA. (Adapted from Stryer, L. 1995. *Biochemistry* (4th ed). W. H. Freeman and Company.)

DNA molecule receives no more than one single-strand break. (This brief exposure to DNase I is in marked contrast to the long exposure used in the DNase protection method.) Nicking occurs at all positions except those protected by the DNA-binding protein. Then the DNA is isolated, denatured, and analyzed by gel electrophoresis.

The radioactive bands observed after gel electrophoresis correspond to a set of molecules with sizes that are determined by the nick positions in relation to the radioactively labeled end. If the DNA contains *n* base pairs and the DNA-binding protein is *not* added, *n* sizes of DNA fragments will be present. If, however, the DNA-binding protein is added and prevents DNase I from gaining access to *x* base pairs, only *n–x* different sizes of DNA fragments will be represented. Two DNA samples are compared in Figure 12.9, one without the DNA-binding protein (to obtain the positions of the *n* bands) and one with the DNA-binding protein (to determine the positions of the missing bands). The missing bands in the gel pattern identify the binding site(s) on the DNA.

DNA footprinting experiments reveal the σ^{70} RNA polymerase holoenzyme • σ^{70} promoter complex changes conformation during the transcription initiation process. These changes are evident

12.2 Initiation Stage

when one compares the DNA footprint pattern obtained from an RNA σ⁷⁰ polymerase holoenzyme • σ⁷⁰ promoter DNA complex formed at 0°C with the same complex formed at 37°C. The low temperature prevents the conformational change from taking place. When the complex forms at 0°C, the protected DNA region extends from positions −55 to −10. Because the DNA remains completely double helical, this complex is designated a **closed promoter complex** (FIGURE 12.10a). When the complex forms at 37°C, the protected DNA region is longer, extending from positions −55 to +20 (Figure 12.10b). Furthermore, positions −12 to +2 become sensitive to KMnO₄, a chemical reagent that attacks single- but not double-stranded DNA. This sensitivity indicates the strands have separated (opened) in positions −12 to +2 to form a bubble that exposes the transcription start site at position +1. Because the DNA has melted, this complex is designated an **open promoter complex**. Note the AT-rich −10 box accounts for most of the melted region, explaining why a mutation in the −10 box that changes an AT base pair to a GC base pair causes a decrease in promoter strength. Because the protected DNA region in the open complex is about 26 nm and the longest holoenzyme dimension is just 15 nm, the protected DNA segment must wrap around the holoenzyme.

Bacterial RNA polymerase crystal structures show how the enzyme is organized and provide insights into how it works.

In 1999 Seth Darst and coworkers determined the crystal structure for the *T. aquaticus (Taq)* core RNA polymerase (FIGURE 12.11). The protein has a total of five subunits that are present in the stoichiometry α₂ββ'ω. Each subunit is homologous with its *E. coli* counterpart. The ω subunit, which is not always present in isolated *E. coli* core RNA polymerase, helps to assemble the core RNA polymerase but is not required for RNA synthesis. As evident from the orientation presented in Figure 12.11, the core polymerase resembles a crab claw. One pincer

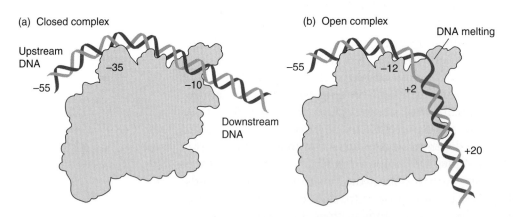

FIGURE 12.10 Open and closed promoter complexes. (a) Closed complex (RNA polymerase holoenzyme bound to promoter DNA at 0°C). The DNase footprint extends from positions −55 to −10. The entire promoter DNA remains as a double helix. (b) Open complex (RNA polymerase holoenzyme bound to promoter DNA at 37°C). The DNase footprint extends from positions −55 to +20 and a bubble opens from positions −12 to +2. The position numbers shown in the figure are at approximate locations. (Adapted from Murakami, K. S., and Darst, S. A. 2003. *Curr Opin Struct Biol* 131:31–39.)

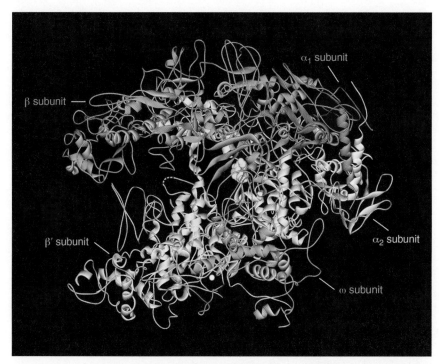

FIGURE 12.11 Structure of the *Thermus aquaticus* core RNA polymerase. The crystal structure of the *T. aquaticus* core RNA polymerase is shown as a ribbon structure. The three aspartates that are part of the absolutely conserved –NADFDGD– motif at the catalytic site are shown as yellow spacefill structures. The magnesium ion bound to these aspartates is shown as a light green sphere. The bridge helix and trigger loop in the β′ subunit are shown in yellow and cyan, respectively. The part of the trigger loop shown as a dashed line is not resolved in the crystal structure. The zinc ion associated with the β′ subunit is shown as a white sphere. (Structure from Protein Data Bank 1HQM. Minakhin, L. et al. 2001. *Proc Natl Acad Sci USA* 98:892–897. Prepared by B. E. Tropp.)

is almost entirely β subunit and the other almost entirely β′ subunit. The pincers function as a DNA clamp. The core RNA polymerase is about 15 nm long (from the tips of the claws to the back) and 11 nm wide. The following are other noteworthy features of the *Taq* core RNA polymerase:

1. The two α subunits combine to form a dimer, which interacts with the β and β′ subunits. The arrangement is not symmetrical, however, because one α subunit, α_1, interacts with the β subunit, whereas the other α subunit, α_2, interacts with the β′ subunit. No residue in either α subunit has access to the internal channel of the core RNA polymerase where catalysis occurs.
2. The β and β′ subunits, which together account for about 60% of the core RNA polymerase mass, interact extensively with each other. The catalytic site is formed by one such interaction in the active site channel. *The β′ subunit provides an absolutely conserved -NADFDGD- motif that is essential for catalysis.* The three aspartates in this motif, along with the magnesium ion bound to them, are required for phosphodiester

12.2 Initiation Stage

bond formation. The incoming NTP carries a second required magnesium ion to the active site.

3. The β′ subunit has a bridge helix and trigger loop adjacent to the catalytic site that play essential roles in the nucleotide addition cycle (see below).

The high-resolution image obtained for the *Taq* core polymerase is thought to closely approximate core RNA polymerase structures from *E. coli* and other bacteria because (1) the subunits in core RNA polymerases are conserved in bacteria; (2) the *Taq* core RNA polymerase crystal structure corresponds in both size and shape to lower resolution electron microscopy images of the *E. coli* protein; (3) secondary structures predicted from *E. coli* β subunit and β′ subunit sequences are consistent with the secondary structures actually observed in the *Taq* RNA polymerase crystal structure; (4) amino acid substitutions that make *E. coli* RNA polymerase resistant to rifamycin, an antibiotic that inhibits the enzyme (**BOX 12.2 CLINICAL APPLICATION: RIFAMYCIN**),

BOX 12.2: CLINICAL APPLICATION

Rifamycin

The rifamycins, a family of antibiotics produced by strains of *Streptomyces*, are used to treat a wide variety of bacterial infections. Unfortunately, many bacterial strains, including some that cause tuberculosis, have become resistant to rifamycin and its derivatives. A semisynthetic derivative of rifamycin, rifampicin, is often used to study bacterial transcription. **FIGURE B12.2a** shows the rifampicin structure, and Figure B12.2b is a three-dimensional structure of the *Thermus aquaticus* core RNA polymerase with bound rifampicin. Rifampicin-resistant mutants map within *rpoB*, the structural gene for the β subunit. Rifampicin binds in a pocket within the β subunit close to the active site and obstructs the growing RNA chain's path, blocking synthesis after the first or second phosphodiester bond is formed. The inhibited enzyme remains bound to the promoter, preventing uninhibited enzyme from initiating transcription.

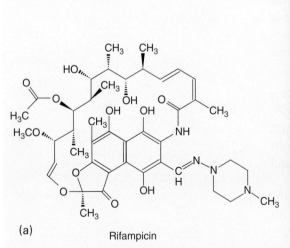
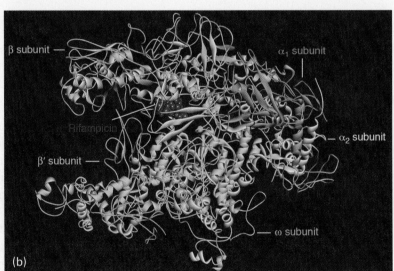

FIGURE B12.2 Rifampicin pocket. (a) Rifampicin structure. (b) Three-dimensional structure of *Thermus aquaticus* core RNA polymerase with bound rifampicin. The rifampicin is shown as a red spacefill structure. (Part b Protein Data Bank: 1I6V Campbell, E. A., et al. 2001. *Cell* 104:901–912. Prepared by B. E. Tropp.)

are scattered along the β chain, but these same amino acids cluster around a pocket in core polymerase; and (5) genetic and biochemical studies show that residues that bind the initiating NTP substrate are scattered throughout the *E. coli* β subunit sequence but are clustered together in the core RNA polymerase structure.

In 2002 Shigeyuki Yokoyama and coworkers determined the crystal structure for the *T. thermophilus* σ^{70} RNA polymerase holoenzyme (FIGURE 12.12). The most important structural difference between the core polymerase and the holoenzyme is that the holoenzyme contains the σ^{70} subunit, which is located almost entirely on the core's surface, with the exception of a segment that is buried within the core molecule.

Although core polymerase and holoenzyme crystal structures provide considerable information about RNA polymerase structure, they provide no direct information about interactions that exist among core polymerase, σ factor, and DNA. Darst and coworkers addressed these and related structural issues by preparing crystals that contain the *Taq* σ^{70} RNA polymerase holoenzyme bound to a synthetic DNA with a fork-junction sequence (the joint between double-stranded DNA and single-stranded DNA in the transcription bubble). The DNA molecule they synthesized for this purpose (FIGURE 12.13a) has a double-stranded −35 box and a mostly open −10 box. The crystal structure of the σ^{70} RNA polymerase holoenzyme • fork-junction

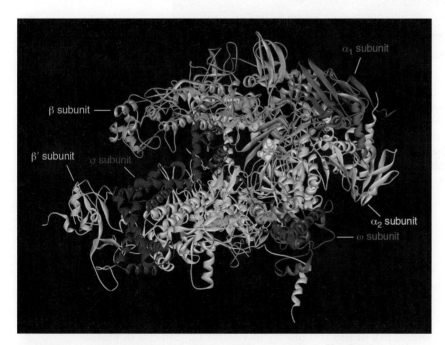

FIGURE 12.12 **Structure of the *Thermus thermophilus* σ^{70} RNA polymerase holoenzyme.** The crystal structure of the *T. thermophilus* σ^{70} RNA polymerase holoenzyme is shown as a ribbon structure. The three aspartates that are part of the absolutely conserved −NADFDGD− motif at the catalytic site are shown as yellow spacefill structures. The magnesium ion bound to these aspartates is shown as a light green sphere. The bridge helix and trigger loop in the β′ subunit are shown in yellow and cyan, respectively. (Structure from Protein Data Bank 1IW7. Vassylyev, D. G., et al. 2002. *Nature* 417:712–719. Prepared by B. E. Tropp.)

(a) Synthetic fork-junction DNA

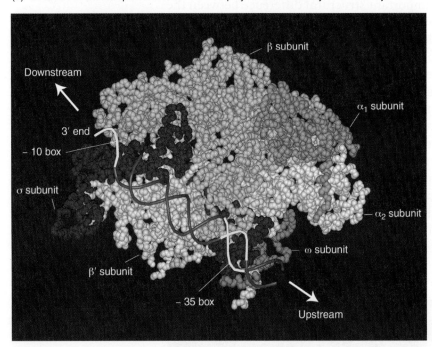

(b) Structure of the complex between the RNA polymerase holoenzyme and fork-junction.

FIGURE 12.13 Structure of *Thermus aquaticus* σ⁷⁰ RNA polymerase holoenzyme bound to fork-junction DNA. (a) The synthetic fork-junction DNA used to prepare the crystal that contained the RNA polymerase holoenzyme • fork-junction DNA complex. (b) Structure of the complex between the RNA polymerase holoenzyme and the synthetic fork-junction DNA. The core RNA polymerase is shown as a spacefill structure. DNA is shown as a blue tube with the −35 and −10 boxes in yellow. The magnesium ion at the active site is shown as a light green sphere. (Part a is modified from K. S. Murakami, et al., *Science* 296 [2002]: 1285–1290. Reprinted with permission from AAAS. Part b structure derived from Protein Data Bank 1L9Z. Murakami, K. S., et al. 2002. *Science* 296:1285–1290. Prepared by B. E. Tropp.)

DNA complex, which appears to resemble the open complex, is shown in Figure 12.13b. The fork-junction DNA lies across one face of the holoenzyme, entirely outside the RNA polymerase active site. All sequence specific −10 and −35 box contacts with the RNA polymerase take place through interactions with the σ^{70} subunit.

Members of the σ^{70} family have four conserved domains.

The σ^{70} subunit has four conserved domains, each of which can be subdivided into smaller highly conserved regions. Numbering for σ^{70} factor domains begins at the N-terminus. Small highly conserved regions are specified by two numbers separated by a decimal. The first number indicates the domain and the second indicates a small highly conserved region within the domain. For example, $\sigma_{2.4}$ specifies the fourth small highly conserved region in the second domain. The σ^{70} factor assumes a

FIGURE 12.14 **Interactions between DNA promoter elements and σ^{70}.** The σ^{70} subunit is shown with the N-terminus on the right and the C-terminus on the left. The red arrow indicates the initiation site and direction of transcription.

different conformation when it binds to core polymerase. One notable change is the disruption of an interaction between negatively charged region 1.1 and domain 4. This interaction prevents the free σ^{70} factor from binding DNA. In essence, region 1.1 acts as a DNA mimic, which competes with promoter DNA for the binding site on domain 4. FIGURE 12.14 shows the specific interactions between σ^{70} when it is part of the σ^{70}-RNA polymerase holoenzyme and the −10 and −35 boxes.

Promoter opening probably begins with the first adenine in the −10 box (the underlined A in TATAAT) flipping out into a hydrophobic pocket in σ^{70} and then DNA melting extends downstream past the transcription start site (+1) to complete the transcription bubble in the open promoter complex. Then the template strand somehow enters the active site through a positively charged tunnel and the β and β′ pincers clamp down on the downstream double-stranded DNA in the major channel. RNA synthesis begins when NTP substrates, which move through a secondary channel, reach the catalytic site.

The σ^{70} RNA polymerase holoenzyme goes through several rounds of abortive initiation before promoter escape.

When an assay mixture contains σ^{70} RNA polymerase holoenzyme, DNA, and only the first two NTP substrates, RNA synthesis comes to a halt and the nascent RNA is released. Although the cessation of RNA synthesis is expected because the full complement of NTPs is not present, the release of nascent RNA is surprising. Even more surprising, nascent RNA ranging from about 8 to 10 nucleotides is also released even when the reaction mixture contains all four NTPs. These experiments indicate that a significant amount of **abortive initiation** takes place during the early stage of the σ^{70} RNA polymerase holoenzyme catalyzed reaction. Escape from abortive cycling to productive RNA synthesis is termed **promoter escape.**

RNA polymerase "scrunches" DNA during transcription initiation.

The observation that abortive initiation leads to the formation of oligonucleotide products that are up to 8 to 10 nucleotides in length suggests the RNA polymerase active center moves relative to the DNA that it acts on. Yet in apparent contrast, DNA footprinting experiments indicate the enzyme does not move during abortive transcription because it protects the same upstream DNA fragment before and after abortive RNA synthesis. Three models have been offered to reconcile these seemingly contradictory observations (FIGURE 12.15). The **scrunching model**

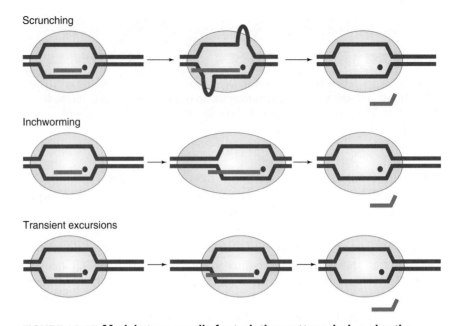

FIGURE 12.15 Models to reconcile footprinting pattern during abortive initiation. The transcription initiation complex is shown with RNA polymerase (gray oval), DNA (blue lines), nascent RNA (green line), and a magnesium ion (red sphere) at the active site. The scrunching model (top) proposes that RNA polymerase holoenzyme unwinds adjacent DNA segments and pulls the unwound DNA into itself during initial transcription. The energy stored in the scrunched DNA is then used for promoter escape. The inchworming model (middle) proposes that the leading edge of RNA polymerase advances during the early stage of transcription initiation to allow the active center to move forward, while the other end of the enzyme remains anchored to the upstream DNA region. The energy stored in the stretched protein is used for promoter escape. The transient excursion model (bottom) proposes transient cycles of forward enzyme motion during abortive RNA synthesis and backward enzyme movement after abortive transcript release with long intervals between cycles. The long interval times between cycles was postulated to explain the footprinting observations. (Modified from Herbert, K. M., et al. 2008. *Annu Rev Biochem* 77:149–176. Copyright 2008 by Annual Reviews, Inc. Reproduced with permission of Annual Reviews, Inc., in the format Textbook via Copyright Clearance Center.)

proposes that DNA is pulled into the RNA polymerase holoenzyme as the DNA unwinds to form the open complex. The energy stored in the scrunched DNA is then used for promoter escape. The **inchworming model** proposes the leading edge of RNA polymerase advances during the early stage of transcription initiation to allow the active center to move forward, while the other end of the enzyme remains anchored to the upstream DNA region. The energy stored in the stretched protein is used for promoter escape. Finally, the **transient excursion model** proposes transient cycles of forward enzyme motion during abortive RNA synthesis and backward enzyme movement after abortive transcript release with long intervals between cycles. The long interval times between cycles was postulated to explain the footprinting observations.

Richard H. Ebright and coworkers performed two types of experiments to distinguish among the models. First, they attached pairs of fluorescent tags to specific sites on the RNA polymerase holoenzyme or DNA and then used these tags to monitor the distances within single molecules of abortively initiating transcription initiation complexes. They observed that the RNA polymerase holoenzyme does not stretch to reach adjacent DNA segments (eliminating the inchworm model) and does not move to reach adjacent DNA segments (eliminating the transient excursions model). Instead, their observations indicated that the RNA polymerase holoenzyme remains stationary and pulls adjacent DNA into itself, supporting the scrunching model. In the second type of experiment Ebright and coworkers demonstrated that RNA polymerase holoenzyme unwinds adjacent DNA segments and pulls the unwound DNA into itself during initial transcription ("scrunching") and then rewinds the unwound DNA when the RNA polymerase leaves the initiation site and begins to move down the DNA ("unscrunching"). Energy stored in the system during the scrunching stage is used during promoter escape to break interactions between the RNA polymerase holoenzyme and the initiation site and to allow RNA polymerase to move forward and catalyze transcription elongation.

12.3 Transcription Elongation Complex

The transcription elongation complex is a highly processive molecular motor.

As the initiation stage comes to an end, the RNA polymerase conformation changes to generate a **transcription elongation complex**, consisting of core RNA polymerase, template DNA, and a growing RNA chain. RNA chain elongation involves a catalytic cycle in which the following occurs:

1. NTPs pass through the secondary channel to reach the catalytic site. This passage may limit the elongation rate because there is only a one in four chance the correct nucleotide will move through the secondary channel and reach the binding site.
2. The 3′-hydroxyl group at the growing end of the RNA strand makes a nucleophilic attack on the α-phosphoryl group of the correct incoming NTP to form a 5′ to 3′ phosphodiester bond.
3. The core RNA polymerase moves one nucleotide downstream. RNA polymerase moves along the DNA template at a rate of about 30 nucleotides • s^{-1}. The incoming NTPs provide sufficient energy to synthesize the phosphodiester bond and drive the RNA core polymerase one nucleotide downstream.

Biochemical studies provide important insights into the transcription elongation complex's structure. The DNase footprint of the transcription elongation complex is about 35 bp shorter than that of the initiating complex. Nevertheless, the transcription elongation complex is the more stable of the two complexes (see below). Approximately 14 bp within the region protected from DNase are melted, forming a transcription bubble (FIGURE 12.16). The first eight nucleotides within this bubble are paired with the RNA chain. The transcription bubble

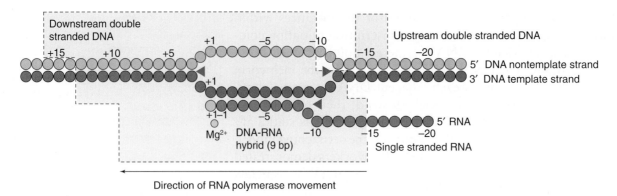

FIGURE 12.16 A schematic representation of nucleic acids within the transcription elongation complex. Transcription is taking place from right to left. The DNA is represented by dark blue circles for the template strand and light blue circles for the nontemplate strand. The RNA chain is represented by dark green circles and the incoming nucleotide is shown in light green. Nucleotide residue positions are numbered relative to the position of the incoming nucleotide substrate, which is designated +1. DNA and RNA segments protected by RNA polymerase during chemical and enzymatic footprinting experiments are indicated by tan shading. Brown triangles indicate that RNA polymerase actively controls the length of the transcription bubble and the RNA-DNA hybrid. (Adapted from Darst, S. A. 2001. *Curr Opin Struct Biol* 11:155–162.)

size appears to remain the same during the elongation process as RNA polymerase moves downstream because double-stranded DNA opens in front of the bubble and reforms behind it.

Conventions for numbering nucleotides in the transcription elongation complex are as follows. The entry position for the incoming nucleotide is +1 and nucleotides downstream from this position (nucleotides in front of moving RNA polymerase) are +2, +3, and so forth. The 3′-terminus of the RNA strand is at position −1, and nucleotides upstream from this position (nucleotides behind moving RNA polymerase) are −2, −3, and so forth. The transcription bubble extends from about −12 to +2 bp. The transcription elongation complex must be stable to synthesize RNA molecules that are 10^4 nucleotides or longer. High processivity is obligatory because a growing RNA chain cannot be extended after the transcription elongation complex dissociates. One indication of transcription elongation complex stability is that it remains intact in 0.5 M potassium chloride. The initiating complex dissociates at this salt concentration.

In 2007 Dimitry G. Vassylyev and coworkers determined the crystal structure of a *T. thermophilus* transcription elongation complex with a core polymerase bound to a synthetic scaffold, which contained 14 bp of downstream DNA, 9 bp of an RNA–DNA hybrid, and seven single-stranded nucleotides of the displaced RNA transcript (**FIGURE 12.17a**). This crystal structure, along with biochemical data, provides sufficient information to build the model for the transcription elongation complex shown in Figure 12.17b. The model shows that three adjacent nucleic acid-binding sites, the downstream duplex-binding site, the RNA–DNA hybrid binding site, and the RNA-binding site, hold the transcription elongation complex together:

1. *Duplex-binding site:* The duplex binding site is a deep cleft formed by the β′ subunit that encircles approximately 9 bp of downstream double-stranded DNA. The DNA makes direct contact with the protein through van der Waals and electrostatic

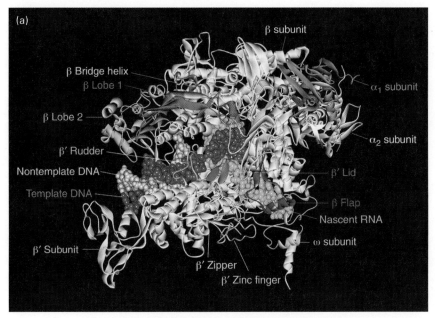

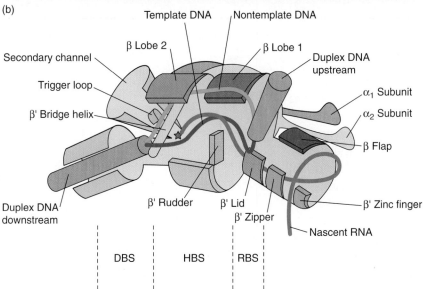

FIGURE 12.17 **The transcription elongation complex.** (a) The crystal structure of a *Thermus thermophilus* transcription elongation complex. The structure is a main channel view of RNA polymerase. Core polymerase is bound to a synthetic nucleic acid scaffold, which contains 14 bp of downstream DNA, 9 bp of an RNA–DNA hybrid, and seven single-stranded nucleotides of displaced RNA transcript. Protein subunits are shown as ribbon structures and nucleic acids as space-fill structures. (b) A schematic model of the transcription elongation complex based on the crystal structure and biochemical studies. The abbreviations DBS, HBS, and RBS at the bottom are for duplex binding site, RNA–DNA hybrid binding site, and RNA binding site, respectively. The red star indicates the catalytic site. (Part a structure from Protein Data Bank 2O5I. Vassylyev, D. G., et al. 2007. *Nature* 448:157–162. Prepared by B. E. Tropp. Part b modified from Nudler, E. 2009. *Annu Rev Biochem* 78:335–361. Copyright 2009 by Annual Reviews, Inc. Reproduced with permission of Annual Reviews, Inc., in the format Textbook via Copyright Clearance Center.)

interactions that must be weak enough to allow the DNA to slide through the cleft. The two DNA strands begin to separate at about position +2 and the DNA makes a sharp 90-degree bend.

2. *RNA–DNA hybrid binding site:* The RNA–DNA hybrid binding site is part of the main channel formed by the β and β′ subunits. It begins slightly after the β′ bridge helix and extends to the β′ lid, surrounding 8 to 9 bp of the RNA–DNA hybrid. Weak interactions between the protein and the RNA–DNA hybrid permit the hybrid to slide through the channel. The β′ rudder may help to stabilize the hybrid through contacts with the template DNA at positions −9, −10, and −11. The β′ lid acts as a wedge that helps the RNA to separate from the DNA and thereby prevent the hybrid from becoming overextended. It may also help stabilize the hybrid through contacts with

the last bases in the hybrid. Furthermore, the lid's flexibility appears to allow the hybrid region to vary from 7 to 10 bp.

3. *RNA binding site:* The RNA binding site, which contains evolutionarily conserved amino acids, is located between the RNA–DNA hybrid binding site and RNA exit channel. It interacts with the single-stranded RNA that has separated from the hybrid. After separation the RNA passes through a narrow pore, formed in part by the β′ lid, enters a wider channel bounded by the β′ zipper and β′ zinc finger on one side and the β flap on the other, and exits the RNA polymerase.

The trigger loop and the β′ bridge helix help to move RNA polymerase forward by one nucleotide during each nucleotide addition cycle (**BOX 12.3 LOOKING DEEPER: RNA POLYMERASE MOVEMENT**).

BOX 12.3 LOOKING DEEPER

RNA Polymerase Movement

The fact that the duplex DNA, DNA–RNA hybrid, and RNA binding sites each surround the nucleic acids ensures transcription elongation is a highly processive process. Moreover, the weak protein–nucleic acid interactions permit lateral enzyme movement along the nucleic acids. Molecular biologists initially thought NTP hydrolysis served as the direct energy source for forward enzyme movement. Recent studies suggest, however, that RNA polymerase movement along the nucleic acids is driven by thermal energy so the transcription elongation complex resembles a Brownian ratchet machine. Because the terms used to describe RNA polymerase movement derive from mechanical systems, it is helpful to consider the stationary pawl and reciprocating pawl mechanical ratchet wheel machines shown in **FIGURE B12.3ab**. In the first machine (Figure B12.3a) a stationary pawl (wedge) blocks reverse wheel motion and thereby prevents random back and forth movement. In the second machine (Figure B12.3b) the pawl oscillates back and forth, pushing the ratchet wheel forward. With this brief background we are ready to examine how RNA polymerase moves forward during transcription elongation.

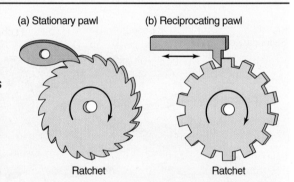

FIGURE B12.3ab Two mechanical ratchet wheel machines. (a) A stationary pawl (wedge) blocks reverse wheel motion and thereby prevents random back and forth movement. (b) A reciprocating pawl oscillates back and forth, pushing the ratchet wheel forward. (Adapted from Bar-Nahum, G., et al. 2005. *Cell* 120:183–193.)

RNA polymerase's active site has two moving parts, the trigger loop and β′ bridge helix (see Figure 12.17b), that contribute to NTP binding and RNA polymerase movement. **FIGURE B12.3c** presents a model proposed by Evgeny Nudler and coworkers that shows the contributions that trigger loop and β′ bridge helix make to the nucleotide addition cycle. The trigger loop is a flexible structure that can assume an unfolded conformation or a folded α-helical conformation. At the start of the cycle (the posttranslocated state) the trigger loop is in the unfolded or open conformation. When the correct nucleotide binds to the active site, the trigger loop assumes the folded α-helical or closed conformation and helps to stabilize nucleotide binding to the active site. The tightly bound NTP acts as a stationary pawl to block lateral RNA polymerase movement. After phosphodiester bond formation pyrophosphate is released and the trigger loop returns to its unfolded state. The elongation complex is now in the pretranslocated state. The trigger loop oscillates between the unfolded conformation and an intermediate form, which presses against the bridge helix. This pressure causes the bridge helix to push against the RNA–DNA hybrid, forcing the elongation complex to enter an active posttranslocated state. Thus, the trigger loop and bridge helix function like a reciprocating pawl, facilitating RNA polymerase movement in the absence of any additional NTP.

FIGURE B12.3c Nucleotide addition cycle. At the start of the cycle (the posttranslocated state) the trigger loop (light blue) is in the unfolded or open conformation. When the correct NTP (shown as green nucleotide with orange phosphates) binds to the active site, the trigger loop assumes a folded α-helical or closed conformation and helps to stabilize NTP binding. The tightly bound NTP acts as a stationary pawl to block lateral RNA polymerase movement. After phosphodiester bond formation, pyrophosphate is released and the trigger loop returns to its unfolded state. The elongation complex is now in the posttranslocated state. The trigger loop oscillates between the unfolded conformation and an intermediate form, which presses against the bridge helix (yellow). This pressure causes the bridge helix to push against the RNA–DNA hybrid, forcing the elongation complex to enter an active posttranslocated state. Thus, the trigger loop and bridge helix function like a reciprocal pawl, facilitating RNA polymerase movement in the absence of NTP. Template DNA is in dark blue stick figures, and RNA is in dark green stick figures. Magenta circles indicate magnesium ions I and II. (Modified from Nudler, E. 2009. *Annu Rev Biochem* 78:335–361. Copyright 2009 by Annual Reviews, Inc. Reproduced with permission of Annual Reviews, Inc., in the format Textbook via Copyright Clearance Center.)

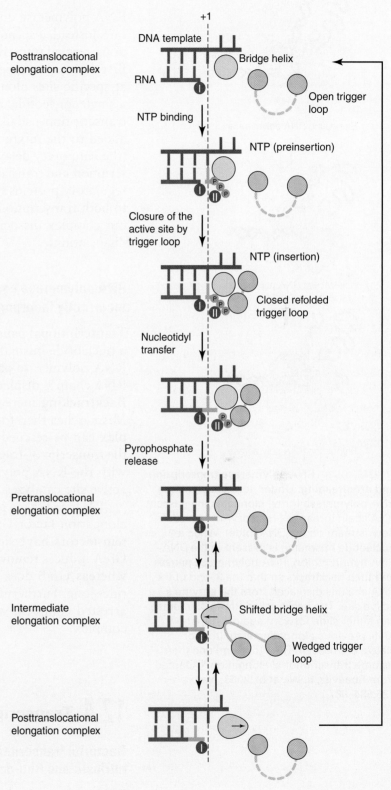

12.3 Transcription Elongation Complex

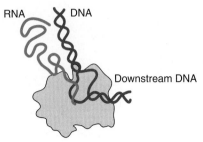
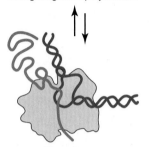
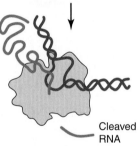

FIGURE 12.18 RNA polymerase transcription and proofreading. Under normal conditions RNA polymerase (gray) elongates the nascent RNA chain (green) as the enzyme moves downstream on the DNA (blue). When a nucleotide mismatch is present in the DNA–RNA hybrid region, RNA polymerase pauses and then backtracks so that the 3'-end of the RNA chain is displaced from the enzyme's active site. The backtracked RNA polymerase can either slide forward again, returning to its previous elongating state (top) or cleave the nascent RNA (bottom) and then resume transcriptional elongation. (Adapted from Shaevitz, J. W., et al. 2003. *Nature* 426:684–687.)

Pauses influence the overall transcription elongation rate.

RNA polymerase does not move along the DNA template at a fixed rate. Instead, it spends proportionally more time at some template positions, termed **transcription pause sites,** than at others. Pause sites were first detected in experiments in which RNA polymerase was stopped at specific sites along the DNA template, usually by deprivation of a single nucleotide, and then allowed to resume synthesis. Although transcription usually resumes right after the missing nucleotide is added to the mixture, pauses were observed at some sites. *Pausing* (the temporary delay in chain elongation) helps to synchronize transcription and translation, slows RNA polymerase movement to allow regulatory proteins to interact with the complex, and probably leads to both **transcription arrest** (the complete halt of transcription without complex dissociation) and **transcription termination** (complex dissociation).

RNA polymerase can detect and remove incorrectly incorporated nucleotides.

Transcriptional pausing is an important step in proofreading. When a nucleotide mismatch is present in the DNA–RNA hybrid region, RNA polymerase pauses and then backtracks so the 3'-end of the RNA chain is displaced from the enzyme's active site (FIGURE 12.18). **Backtracking** increases as RNA–DNA hybrid stability decreases. Mismatches therefore promote backtracking. A backtracked complex can be rescued by internal hydrolytic cleavage and release of the transcript 3'-fragment to generate a new 3'-end that is in register with the RNA polymerase catalytic center. The RNA polymerase active site catalyzes this cleavage reaction but at a very slow rate.

Cleavage takes place much more rapidly when the transcription elongation factor GreA or GreB is present. Although the two protein factors have similar sequences, they have different functions. GreA induces removal of fragments that are 2 to 3 nucleotides long, whereas GreB does the same for fragments as large as 18 nucleotides long. Furthermore, GreA can only prevent the formation of an arrested complex, whereas GreB can rescue a preexisting arrested complex.

12.4 Transcription Termination

Bacterial transcription machinery releases RNA strands at intrinsic and Rho-dependent terminators.

RNA polymerase is a highly processive macromolecular machine. Specific mechanisms are required to release RNA polymerase from the transcription elongation complex. Two transcription termination

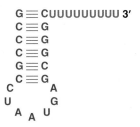

FIGURE 12.19 Transcription termination signal for the intrinsic termination pathway. (Adapted from Mooney, R. A., Artsimovitch, I., and Landick, R. 1998. *J Bacteriol* 180:3265–3275.)

pathways, **intrinsic termination** and **Rho-dependent termination**, contribute about equally to this release in *E. coli*.

The intrinsic termination pathway is so named because it takes advantage of the core RNA polymerase's intrinsic catalytic activity to terminate transcription. Although nucleotide sequences specifying intrinsic terminators are present on DNA, it is actually the nascent RNA (and not DNA) that triggers the transcription termination response. Two sequence motifs on the nascent RNA strand are essential for intrinsic terminator function (FIGURE 12.19). The first motif, a G-C rich inverted repeat, allows the RNA to fold into a stem and loop structure that reaches to within seven to nine nucleotides of the 3'-end of the nascent RNA strand. Mutations that maintain the stable stem structure are usually tolerated. Those that decrease the stem structure's stability tend to reduce or eliminate termination. Multiple mutations within the stem region also may lead to loss of terminator activity even though the stem structure retains its stability. Secondary RNA structure is therefore important in determining intrinsic terminator activity, but nucleotide sequence within the stem also appears to make a contribution. The second motif, a run of 8 to 10 nucleotides that consists mostly of uridines, comes immediately after the stem and loop structure. Replacing the uridines in the second motif with other nucleotides converts an intrinsic terminator into a pause signal. Intrinsic terminators appear to act by first causing the transcription elongation complex to pause and then to release the nascent RNA chain.

The Rho-dependent transcription termination pathway requires a ring-shaped helicase with six identical polypeptide subunits called the Rho factor. Rho factor can be detected *in vitro* as an ATP-dependent helicase that releases RNA from an RNA–DNA hybrid. FIGURE 12.20 shows how Rho factor may contribute to transcription termination. According to this model, Rho factor loads onto a nascent mRNA at a cytosine-rich region that contains 40 or more nucleotides known as the **Rho utilization (*rut*) site**. Then the Rho factor moves in a 5'→3' direction in pursuit of the RNA polymerase. Upon reaching RNA polymerase, Rho somehow allows the RNA polymerase to disengage from the transcription elongation complex.

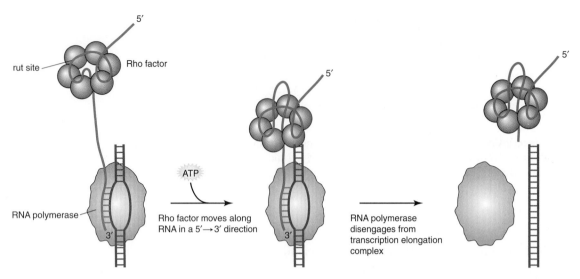

FIGURE 12.20 Rho factor transcription termination. Rho factor hydrolyzes ATP, driving itself along the mRNA in a 5′ to 3′ direction and then somehow causes RNA polymerase to disengage from the transcription elongation complex. (Modified from *Curr. Biol.*, vol. 13, D. L. Kaplan and M. O'Donnell, Rho Factor: Transcription Termination in Four Steps, pp. R714–R716, copyright 2003, with permission from Elsevier [http://www.sciencedirect.com/science/journal/09609822].)

Rho factor is not present in all bacteria and has not been identified in eukaryotes or archaea.

12.5 Messenger RNA

Bacterial mRNA may be monocistronic or polycistronic.

Bacterial RNA polymerase holoenzyme is required to synthesize the three major classes of bacterial RNA: messenger RNA (mRNA), ribosomal RNA, and transfer RNA. Additional enzymes and protein factors regulate the synthesis of these RNA molecules. The present discussion is limited to the regulation of bacterial mRNA synthesis. Our examination of the regulation of mRNA synthesis begins by considering some characteristics of mRNA, the sequence of which is determined by a specific template DNA sequence within the bacterial chromosome. Because transcription and translation proceed in a 5′→3′ direction and both processes take place in the same cellular compartment, the bacterial protein synthetic machinery can start to read the 5′-end of mRNA before the 3′-end is formed. A bacterial cell therefore does not have a chance to alter a nascent mRNA molecule before the protein synthetic machinery begins to translate it. The situation is different in eukaryotes, which carry out transcription in the cell nucleus and translation in the cytoplasm. The eukaryotic cell thus can convert the primary transcript to mature mRNA in the cell nucleus before the mRNA is required to direct protein synthesis in the cytoplasm.

The protein synthetic machinery reads the mRNA nucleotide sequence in groups of three bases or **codons**. Each codon specifies an amino acid or a termination signal. The protein synthetic machinery begins polypeptide synthesis at a start codon located toward the 5′-end of the mRNA and continues synthesis in a 5′→3′ direction until it encounters a termination codon. The segment of mRNA that codes for a polypeptide chain is called an **open reading frame** because the protein synthetic machinery begins reading the segment at a specific start codon and stops reading it at a specific termination codon. A DNA segment corresponding to an open reading frame plus the translational start and stop signals for protein synthesis is called a **cistron,** and an mRNA encoding a single polypeptide is called monocistronic mRNA. Although "cistron" and "gene" are sometimes used interchangeably to describe bacterial DNA segments that specify polypeptides, the term "gene" has a broader meaning because it also includes the promoter region and applies to DNA segments that code for RNA molecules such as transfer RNA and ribosomal RNA that are not translated.

Bacterial mRNA molecules often contain two or more cistrons. In fact, bacterial **polycistronic mRNA** molecules are actually more common than bacterial monocistronic mRNA molecules. Each cistron within a polycistronic mRNA specifies a specific polypeptide chain. Furthermore, cistrons contained in polycistronic mRNA often specify proteins for a single metabolic pathway. For example, one *E. coli* mRNA has three cistrons, each coding for a different protein required for lactose metabolism, and another contains five cistrons, each coding for a different enzyme required for tryptophan synthesis.

Using polycistronic mRNA is a way for a cell to regulate synthesis of related proteins coordinately. With a polycistronic mRNA molecule the synthesis of several related proteins—in similar quantities and at the same time—can be regulated by a single signal. The sizes of bacterial mRNA molecules vary within a broad range. A monocistronic mRNA with a 900 nucleotide open reading frame (300 codons) is needed to specify a polypeptide that is 300 amino acid residues long. Of course, a polycistronic mRNA would be much longer.

In addition to reading frames and start and stop sequences for translation, other regions in mRNA are significant. For example, mRNA translation rarely, if ever, starts exactly at the 5'-end and stops exactly at the 3'-end. Instead, initiation of synthesis of the first polypeptide chain of a polycistronic mRNA may begin hundreds of nucleotides from the 5'-RNA terminus. The untranslated RNA sequence before the coding region is called the **5'-untranslated region.** The untranslated sequence after the coding region is called the **3'-untranslated region.** Polycistronic mRNA molecules also usually contain intercistronic sequences (spacers) that are tens of bases long.

Bacterial mRNA usually has a short lifetime compared with other kinds of bacterial RNA.

An important characteristic of bacterial mRNA is that its lifetime is short compared with other types of bacterial RNA molecules. The half-life of a typical bacterial mRNA molecule is a few minutes. Although mRNA's short lifetime may seem wasteful, it has an important regulatory function. A cell can turn off the synthesis of a protein that is no longer needed by turning off synthesis of the mRNA that encodes the protein. Soon after, none of that particular mRNA remains and synthesis of the protein ceases, allowing bacterial cells to save energy because they are not forced to synthesize proteins they no longer need.

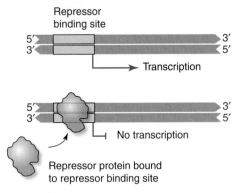

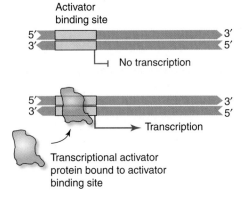

FIGURE 12.21 The distinction between negative and positive regulation. (a) In negative regulation the "default" state of the gene is one in which transcription takes place. The binding of a repressor protein to the DNA molecule prevents transcription. (b) In positive regulation the default state is one in which transcription does not take place. The binding of a transcriptional activator stimulates transcription. A single genetic element may be regulated both positively and negatively; in such a case transcription requires the binding of the transcriptional activator and the absence of the repressor protein.

Controlling the rate of mRNA synthesis can regulate the flow of genetic information.

Bacterial cells can control gene expression (the flow of information from gene to protein) by regulating the rate of gene transcription. We have already encountered one mechanism that is used to regulate bacterial gene transcription. Genes with strong promoters are continuously transcribed at high rates, whereas those with weak promoters are continuously transcribed at low rates. Some other regulatory mechanism is required when gene transcription rates must vary in response to changing physiological conditions. Bacteria achieve this goal by turning on the transcription of specific genes when the gene products are needed and turning off transcription when the gene products are not needed. Actually, there are no known examples of switching a gene's transcription completely off. When transcription is in the "off" state, there always remains a basal level of gene expression. This basal level often amounts to only one or two transcription events per cell generation and thus very little mRNA synthesis. For convenience, when discussing transcription we use the term "off" but what is meant is "very low."

In bacterial systems in which several enzymes act in sequence in a single metabolic pathway, it is often the case that either all these enzymes are present or all are absent. This phenomenon, called **coordinate regulation,** results from control of the synthesis of a single polycistronic mRNA that encodes all the gene products. There are several mechanisms for this type of regulation, as seen below.

Messenger RNA synthesis can be controlled by negative and positive regulation.

The molecular mechanism of mRNA regulation can be divided into two major categories: **negative regulation** and **positive regulation** (FIGURE 12.21). In negative regulation a **repressor** turns off the transcription of one or more genes. In positive regulation an **activator** turns on the transcription of one or more genes. Negative regulation and positive regulation are not mutually exclusive. Many genes respond to both types of regulation. TABLE 12.2 summarizes the properties of negative and positive control.

We now examine the mechanisms that *E. coli* use to regulate the genes responsible for lactose utilization. Studies of this system have provided both the language and the principles required to understand genetic regulation.

TABLE 12.2 mRNA Regulation		
Binding of Regulator to DNA	Positive	Negative
Yes	On	Off
No	Off	On
Note: If a system is both positively and negatively regulated, it is "on" when the positive regulator is bound to the DNA and the negative regulator is not bound to the DNA.		

12.6 Lactose Operon

The *E. coli* genes *lacZ*, *lacY*, and *lacA* code for β-galactosidase, lactose permease, and β-galactoside transacetylase, respectively.

In *E. coli* two proteins are necessary for lactose metabolism. These proteins are the enzyme β-galactosidase, which cleaves lactose to yield galactose and glucose (FIGURE 12.22), and a carrier molecule, lactose permease, which transports lactose (and other galactosides) into the cell. The existence of the two proteins was first shown by a combination of genetic experiments and biochemical analysis.

First, hundreds of Lac⁻ mutants (unable to use lactose as a carbon source) were isolated. By genetic manipulation some of these mutations were moved from the *E. coli* chromosome to an F'*lac* plasmid (a plasmid carrying the genes for lactose utilization), and then partial diploids having the genotypes F'*lac*⁻/*lac*⁺ or F'*lac*⁺/*lac*⁻ were constructed. (The relevant genotype of the plasmid is given to the left of the diagonal line and that of the bacterial chromosome to the right.) It was observed that these diploids always have a Lac⁺ phenotype (that is, they make β-galactosidase), which shows none of the *lac*⁻ mutants makes an inhibitor that blocks *lac* gene function.

Partial diploids were also constructed in which both the chromosome and the F'*lac* plasmid were *lac*⁻. Using different pairs of *lac*⁻ mutants, some pairs were observed to have a Lac⁺ phenotype, whereas others were observed to have a Lac⁻ phenotype. This complementation test showed all the mutants initially isolated fell into one of two groups, which were called *lacZ* and *lacY*. Mutants in the two groups have the property that the partial diploids F'*lacY*⁻*lacZ*⁺/*lacY*⁺*lacZ*⁻ and F'*lacY*⁺*lacZ*⁻/*lacY*⁻*lacZ*⁺ have a Lac⁺ phenotype and the genotypes F'*lacY*⁻*lacZ*⁺/*lacY*⁻*lacZ*⁺ and F'*lacY*⁺*lacZ*⁻/*lacY*⁺*lacZ*⁻ have the Lac⁻ phenotype. The existence of two complementation groups was good evidence of at least two genes in the *lac* system.

The *lacZ* gene is the structural gene for β-galactosidase. This enzyme is readily detected by a simple colorimetric assay that takes advantage of the fact that β-galactosidase catalyzes the hydrolysis of *o*-nitrophenyl-β-galactoside, a lactose analog (FIGURE 12.23). The product *o*-nitrophenoxide, which is yellow, can be detected by light absorption at a wavelength of 420 nm. The function of the *lacY* gene product as lactose permease was strongly suggested by experiments that showed *lacZ*⁺*lacY*⁻ cells cannot transport [^{14}C]lactose into the cell but *lacZ*⁻ *lacY*⁺ cells can.

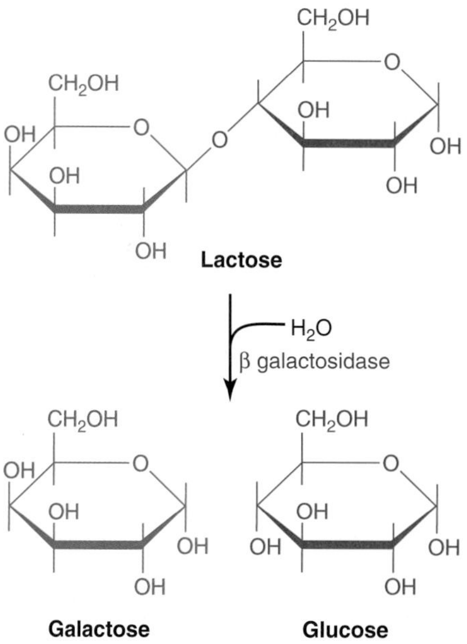

FIGURE 12.22 β-Galactosidase catalyzed lactose hydrolysis.

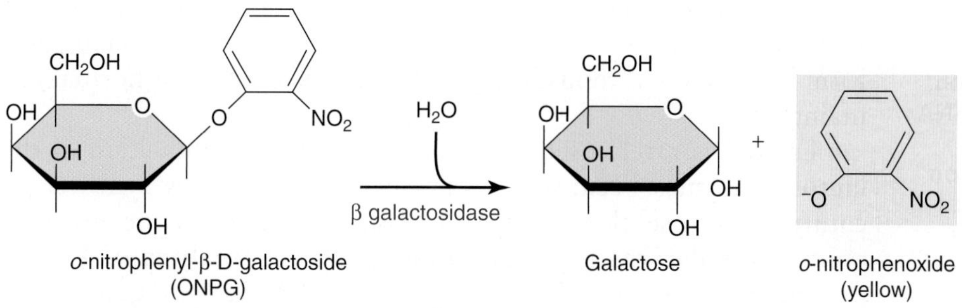

FIGURE 12.23 β-Galactosidase assay using *o*-phenyl-β-D-nitrophenylgalactoside as a substrate.

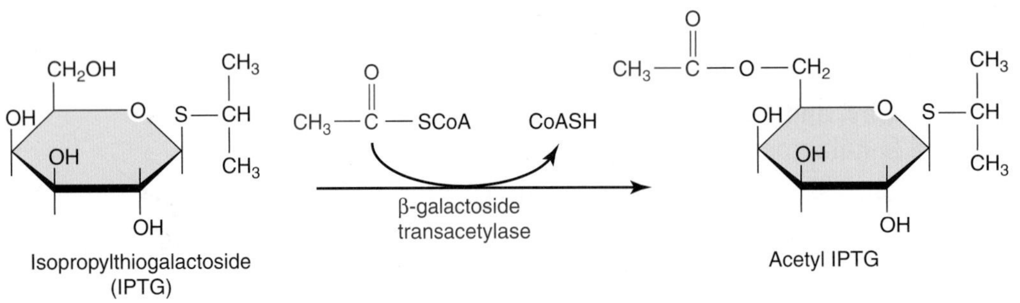

FIGURE 12.24 Reaction catalyzed by β-galactoside transacetylase.

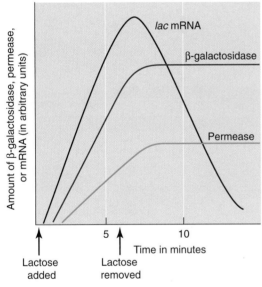

FIGURE 12.25 The on-off nature of the *lac* system. Lac mRNA appears soon after lactose is added; β-galactosidase and permease appear at about the same time but are delayed with respect to mRNA synthesis because of the time required for translation. When lactose is removed no more *lac* mRNA is made, and the amount of *lac* mRNA decreases because of the usual degradation of mRNA. Both β-galactosidase and lactose permease are stable. Their concentrations remain constant even though no more enzymes can be synthesized.

Investigators discovered a third gene, *lacA*, while studying lactose transport. The *lacA* gene product is a β-galactoside transacetylase, which transfers an acetyl group from acetyl-CoA to lactose and its analogs (FIGURE 12.24). *lacA* was not detected at the same time as *lacZ* and *lacY* because its gene product, transacetylase, is not required for lactose catabolism. The precise role of *lacA* is still a matter of conjecture. One hypothesis is that transacetylase detoxifies lactose analogs that would harm cells.

The *lac* structural genes are regulated.

When *E. coli* with a *lac*$^+$ genotype is cultured in a lactose-free medium (also lacking glucose, a point discussed shortly), the intracellular concentrations of β-galactosidase, permease, and transacetylase are exceedingly low—roughly one or two molecules of each protein per bacterial cell. When lactose is added to the growth medium, however, the concentration of these proteins increases in a coordinate fashion to about 10^5 molecules per cell (or about 1% of the total cellular protein). This phenomenon is shown for β-galactosidase and permease in FIGURE 12.25. Furthermore, lactose addition triggers the synthesis of *lac* mRNA as evidenced by studies in which mRNA, labeled with [^{32}P]phosphate at various times after lactose addition, is hybridized to DNA that carries *lac* genes (Figure 12.25).

Enzymes such as β-galactosidase, lactose permease, and transacetylase are said to be **inducible enzymes** because their rate of synthesis increases in response to the addition of a small molecule (lactose) to the medium. Other enzymes, called **repressible enzymes**, exhibit a decreased rate of synthesis in response to the addition of a small molecule in the medium. For instance, the addition of tryptophan to the growth medium causes *E. coli* to greatly decrease the rate at which it produces enzymes needed for tryptophan synthesis. Still other enzymes, called **constitutive enzymes,** are synthesized at fixed rates under all growth conditions. Constitutive enzymes usually perform basic cellular "housekeeping" functions needed for normal cell maintenance.

Lactose is rarely used in experiments to study induction of the lactose enzymes because the β-galactosidase that is synthesized catalyzes lactose cleavage. As a result the lactose concentration continually decreases, which complicates the analysis of kinetic experiments. Instead, two sulfur-containing lactose analogs are used, isopropylthiogalactoside (IPTG) and thiomethylgalactoside, which are effective inducers without being substrates of β-galactosidase (FIGURE 12.26). Inducers having this property are called **gratuitous inducers**.

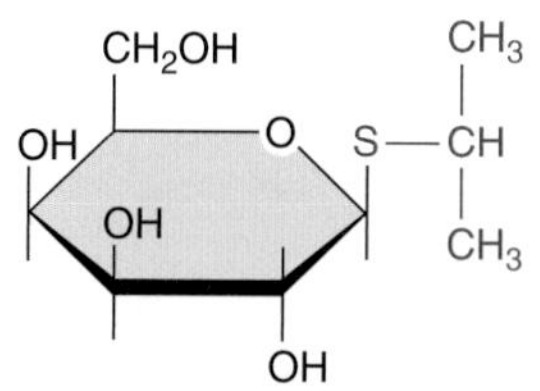

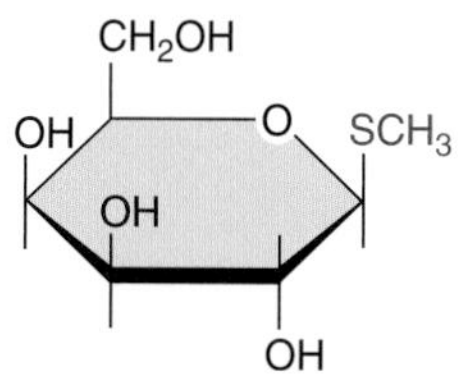

Genetic studies provide information about the regulation of *lac* mRNA.

Two French investigators, Jacques Monod and François Jacob, sometimes working together and sometimes with other investigators, performed a series of genetic and biochemical experiments in the late 1950s that helped to elucidate the mechanism that regulates the *lac* system. They began by isolating constitutive *E. coli* mutants that make *lac* mRNA (and hence β-galactosidase, permease, and transacetylase) both in the presence and absence of an inducer. Then they constructed a variety of partial diploid cells containing constitutive mutants and observed the cell's ability to synthesize β-galactosidase.

These mutations appeared to be of two types, termed *lacI* and *lacO^C* (TABLE 12.3). The *lacI^-* mutations behave like typical minus mutations in most genes and are recessive (entries 3 and 4). Because *lac* mRNA synthesis is off in a *lacI^+* cell and on in a *lacI^-* mutant, the *lacI* gene is apparently a regulatory gene that codes for a product that acts as an inhibitor to keep the *lac* structural genes turned off. A *lacI^-* mutant lacks the inhibitor and thus is constitutive. A *lacI^+*/*lacI^-* partial diploid has one good copy of the *lacI* gene product, so the system is inhibited. Monod and Jacob called the *lacI* gene product the **Lac repressor**. Their original genetic experiments did not indicate whether the Lac repressor is a protein or an RNA molecule.

This question was answered when an *E. coli* mutant with a polypeptide chain termination codon inside *lacI* was isolated and found to synthesize β-galactosidase constitutively. The most likely explanation for the constitutive lactose system is that the mutant strain synthesizes a truncated repressor protein that cannot block transcription of the lac genes. This conclusion was confirmed when the Lac repressor was purified and characterized (see below). Genetic mapping experiments placed the *lacI* gene adjacent to the *lacZ* gene and established the gene order *lacI lacZ lacY lacA*.

A striking property of the *lacO^c* mutations is that in certain cases they are dominant (entries 1, 2, and 6 in Table 12.3). The significance

FIGURE 12.26 Two lactose analogs that are gratuitous inducers.

TABLE 12.3 Characteristics of Partial Diploids Having Various Combinations of *lacI* and *lacO* Alleles

	Genotype	Constitutive or Inducible Synthesis of *lac* mRNA
1.	F'*lacO^c lacZ^+*/*lacO^+ lacZ^+*	Constitutive
2.	F'*lacO^+ lacZ^+*/*lacO^c lacZ^+*	Constitutive
3.	F'*lacI^- lacZ^+*/*lacI^+ lacZ^+*	Inducible
4.	F'*lacI^+ lacZ^+*/*lacI^- lacZ^+*	Inducible
5.	F'*lacO^c lacZ^-*/*lacO^+ lacZ^+*	Inducible
6.	F'*lacO^c lacZ^+*/*lacO^+ lacZ^-*	Constitutive

of the dominance of the *lacO^c* mutations becomes clear from the properties of the partial diploids shown in entries 5 and 6. Both combinations are *Lac⁺*, because there is a functional *lacZ* gene. Entry 5, however, shows β-galactosidase synthesis is inducible even though a *lacO^c* mutation is present. In contrast, entry 6 shows β-galactosidase synthesis is constitutive. The difference between the two combinations in entries 5 and 6 is the *lacO^c* mutation is carried on a DNA molecule that also has a *lacZ⁻* mutation in entry 5, whereas in entry 6 *lacO^c* and *lacZ⁺* are carried on the same DNA molecule. Thus, *lacO^c* causes constitutive synthesis of β-galactosidase only when *lacO^c* and *lacZ⁺* are on the same DNA molecule, that is, when *lacO^c* and *lacZ⁺* are in *cis*.

This conclusion was confirmed when an immunological test capable of detecting a mutant β-galactosidase showed (1) the mutant enzyme is synthesized constitutively in a *lacO^c lacZ⁻/lacO⁺ lacZ⁺* partial diploid (entry 5) and (2) the wild-type enzyme is synthesized only if an inducer is added. This immunological experiment takes advantage of the fact that a purified antibody to β-galactosidase also reacts with the mutant protein as long as the structural differences between wild-type and mutant protein are not too great. A reaction of this type, in which an antibody that is raised in response to one protein is used to detect a closely related protein, is called a cross-reaction, and the closely related protein (mutant β-galactosidase in this experiment) is called **cross-reacting material.** Thus the presence of cross-reacting material, which can be detected by a variety of standard immunological procedures, is indicative of the presence of mutant protein.

Genetic mapping experiments showed all *lacO^c* mutations are located between genes *lacI* and *lacZ*, so the gene order of the five elements of the *lac* system is *lacI lacO lacZ lacY lacA*. Together these experiments lead to the conclusion that *lacO^c* mutations define a regulatory site rather than a gene (because mutations in coding genes should be complementable) and that the *lacO* region determines whether synthesis of the product of the adjacent *lacZ* gene is inducible or constitutive. The *lacO* site is called the **operator.**

The operon model explains the regulation of the lactose system.

Monod and Jacob proposed the **operon model** in 1961 to explain how the *lac* system is regulated. The term "operon" refers to two or more contiguous genes and the genetic elements that regulate their transcription in a coordinate fashion. Promoters had not yet been discovered when Monod and Jacob proposed the operon model but were readily incorporated into the operon model after their discovery. FIGURE 12.27 shows a revised version of the original *lac* operon model that includes the *lac* promoter. The five major features of the model are as follows:

1. The products of the *lacZ*, *lacY*, and *lacA* genes are encoded in a single polycistronic *lac* mRNA molecule.
2. The promoter for this mRNA molecule is immediately adjacent to the *lacO* region. Promoter mutations (p^-) that are completely incapable of making β-galactosidase, permease, and transacetylase have been isolated. The promoter is located between *lacI* and *lacO*.

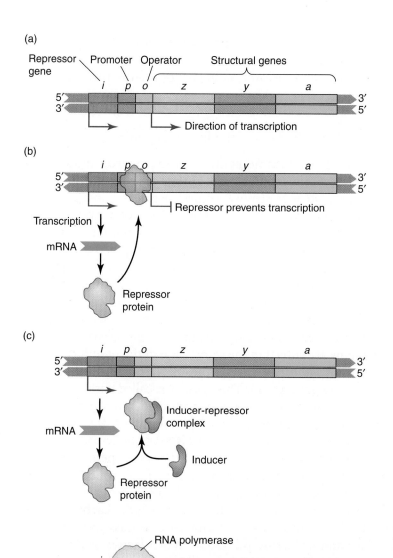

FIGURE 12.27 *Lac* **operon model.** (a) A map of the *lac* operon not drawn to scale. The *p* and the *o* sites are actually much smaller than the other regions and together comprise only 84 bp. (b) A diagram of the *lac* operon in the repressed state. (c) A diagram of the *lac* operon in the induced state. The inducer alters the repressor's conformation so that the repressor can no longer bind to the operator. The abbreviations *l*, *p*, *o*, *z*, *y*, and *a* are used instead of *lacI*, *lacO*, and so on.

3. The operator is a sequence of bases (in the DNA) to which the repressor protein binds.
4. When the repressor protein is bound to the operator, *lac* mRNA transcription cannot take place.
5. Inducers stimulate *lac* mRNA synthesis by binding to the repressor. This binding alters the repressor's conformation so it cannot bind to the operator. In the presence of an inducer, therefore, the operator is unoccupied and the promoter is available for initiation of mRNA synthesis. This state is called **derepression**.

12.6 Lactose Operon

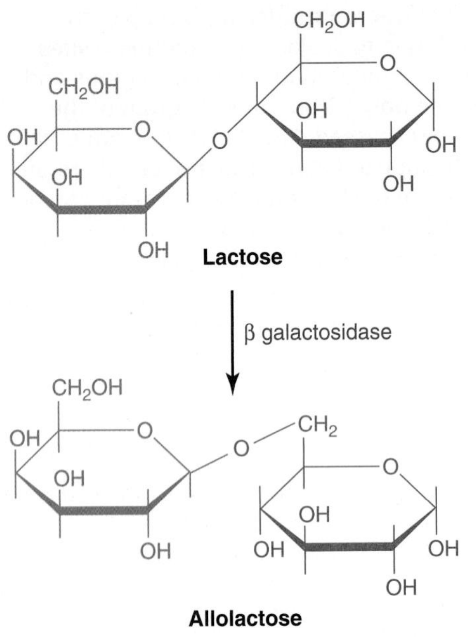

FIGURE 12.28 *Lac* operon model. Synthesis of allolactose, the actual inducer of the *lac* operon.

This simple model explains many features of the *lac* system and of other negatively regulated genetic systems. This explanation is incomplete, however (see below), as the *lac* operon is also subject to positive regulation.

Allolactose is the true inducer of the lactose operon.

Two related problems became evident as the operon model was tested. First, inducers must enter a cell if they are to bind to repressor molecules, yet lactose transport requires permease, and permease synthesis requires induction. We thus must explain how the inducer gets into a cell in the first place. Second, the isolated Lac repressor does not bind lactose (4-O-β-D-galactopyranosyl-D-glucose) but does bind a lactose isomer called allolactose (6-O-β-D-galactopyranosyl-D-glucose). Remarkably, β-galactosidase, the enzyme that catalyzes lactose hydrolysis, also converts a small proportion of lactose to allolactose (FIGURE 12.28). Induction of the synthesis of β-galactosidase by lactose therefore requires β-galactosidase to be present.

Both problems are solved in the same way; in the uninduced state a small amount of *lac* mRNA is synthesized (roughly one mRNA molecule per cell per generation). This synthesis, called **basal synthesis**, occurs because the binding of repressor to the operator is never infinitely strong. Thus, even though the repressor binds tightly to the operator, it occasionally comes off and an RNA polymerase molecule can initiate transcription during the instant that the operator is free.

We can now describe in molecular terms the sequence of events after addition of a small amount of lactose to a growing Lac⁺ culture. Consider bacteria growing in a medium in which the carbon source is glycerol. Each bacterial cell contains one or two molecules of β-galactosidase and of lactose permease. When lactose is added the few permease molecules transport a few lactose molecules into the cell and the few β-galactosidase molecules convert some of these lactose molecules into allolactose. An allolactose molecule then binds to a repressor molecule that is sitting on the operator, and the repressor is inactivated and falls off the operator. Synthesis of *lac* mRNA begins, and these mRNA molecules are translated to produce hundreds of β-galactosidase and permease molecules. The new permease molecules allow lactose molecules to pour into the cell. Most of the lactose molecules are cleaved to yield glucose and galactose, but some are converted to allolactose molecules, which bind to and inactivate all the intracellular repressor molecules. (Repressor is made continuously, though at a very low rate, so there is usually sufficient allolactose to maintain the cell in the derepressed state.) Thus, *lac* mRNA is synthesized at a high rate, and the permease and β-galactosidase concentrations become quite high. The glucose and galactose molecules produced by the cleavage reaction are used as carbon and energy sources.

Ultimately, all the lactose in the growth medium and within the cells is consumed. Then the allolactose concentration within the cell drops so there is not sufficient allolactose to bind to a repressor. The repressor binds to the operator, reestablishing repression and thereby blocking further synthesis of *lac* mRNA. In bacteria most mRNA molecules have a half-life of only a few minutes. Hence, in less than one generation there is little remaining *lac* mRNA, and synthesis of β-galactosidase and permease ceases. These proteins are quite stable but are gradually diluted out as the cells divide. Note that if lactose were added again to the growth medium one generation after the original lactose had been depleted, cleavage of lactose would begin immediately because the cells already have adequate permease and β-galactosidase.

The Lac repressor binds to the *lac* operator *in vitro*.

An important step in proving the principal hypothesis of the operon model was the isolation of the Lac repressor and the demonstration of its expected properties. Walter Gilbert and Benno Müller-Hill succeeded in isolating the Lac repressor from *E. coli* extracts in 1966. Their approach was to fractionate proteins by standard techniques and then assay individual fractions for their ability to bind [^{14}C]IPTG (one of the gratuitous inducers). Binding was detected by equilibrium dialysis, as shown in FIGURE 12.29. Genetic studies support the hypothesis that the IPTG-binding protein is the Lac repressor. For instance, extracts prepared from *lacI*$^-$ mutants have little or no affinity for [^{14}C]IPTG. The Lac repressor is made of four identical subunits that each binds an IPTG molecule. Crude cell extracts bind about 20 to 40 molecules of IPTG per cell, so there are roughly 5 to 10 repressor tetramers per cell.

Lac repressor mRNA is transcribed constitutively from a weak promoter. Mutants have been isolated in which the weak *lacI* promoter is converted to a strong promoter. These mutants, which overproduce the Lac repressor, are noninducible because they cannot accumulate enough intracellular inducer to overcome repression. Repressor-overproducers have been extremely valuable experimentally because high concentrations of repressor (about 1% of the cellular protein) have in turn meant very large amounts of repressor could be purified, providing sufficient amounts for physical study and characterization. The specific binding of repressor to the operator sequence and the inhibition of this binding by an inducer have been demonstrated by using purified repressor. An important procedure for studying repressor-operator binding is the **nitrocellulose filter assay**. Proteins stick to these filters, but DNA does not. If a mixture of repressor and radioactive *lac* DNA is passed through such a filter, radioactivity will be retained on the filter if the protein and the *lac* DNA form a complex. The data shown in TABLE 12.4 were obtained by means of this test. The results indicate that Lac repressor binds to DNA with a normal *lac* operator but fails to bind to DNA with a *lacO*^{c} mutant operator. Furthermore, IPTG prevents

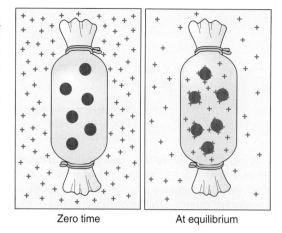

FIGURE 12.29 Equilibrium dialysis. A dialysis bag filled with a cell extract containing repressor (red circles) is placed in a solution containing radioactive IPTG (blue crosses). The radioactive IPTG can bind to the repressor. In the absence of a repressor the concentrations of free IPTG would be the same inside and outside the bag. Because the repressor binds some of the radioactive IPTG, the concentration of radioactive IPTG is greater inside the bag than outside.

TABLE 12.4 Demonstration of Repressor-Operator Binding by the Filter-Binding Assay

Mixture Applied to Filter	^{14}C Bound to the Filter
[^{14}C]*lac* DNA	No
[^{14}C]*lac* DNA + repressor	Yes
[^{14}C]*lac* DNA + repressor + IPTG	No
[^{14}C]*lacO*^{c} DNA + repressor	No

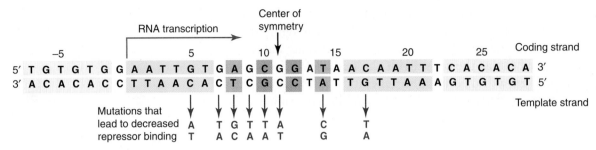

FIGURE 12.30 Thirty-five base pairs protected by the *lac* operator. Footprinting studies reveal that the *lac* repressor protects 35 bp from −7 to +28 so the operator includes both the transcriptional start site at +1 and part of the *lac* promoter's −10 box. Regions of symmetry are shown in the same color. Some of the *lac* operon mutations that decrease the *lac* repressor's affinity for operon DNA are shown in magenta.

the Lac repressor from binding to DNA with a normal *lac* operator. These studies confirm the major predictions of the operon model.

In vitro studies provide additional information about the binding of the Lac repressor to the operator. Footprinting studies reveal repressor protects 35 bp from −7 to +28 so the operator includes both the transcriptional start site and part of the *lac* promoter's −10 box (FIGURE 12.30). The operator has a twofold axis of symmetry that passes through the base pair at position +11. The sequence is not a perfect palindrome, however, because only 28 of the 35 bp have the required symmetry. Mutations in the *lac* operator interfere with the Lac repressor's ability to bind to the operator and lead to constitutive enzyme production. Thus, Lac repressor makes important contacts with bases in the *lac* operator.

The *lac* operon has three *lac* operators.

After the *lac* operon was sequenced it eventually became clear that it has two additional operators. The original operator with its center of symmetry at position +11 is now designated $lacO_1$. Auxiliary operators $lacO_2$ and $lacO_3$ have their centers of symmetry at positions +412 and −82, respectively. Thus, $lacO_3$ is upstream of the *lac* promoter, whereas $lacO_2$ is located in *lacZ* (FIGURE 12.31).

The discoveries of $lacO_2$ and $lacO_3$ led investigators to ask whether these auxiliary operators participate in *lac* operon regulation. In 1990 Müller-Hill and coworkers performed a series of genetic experiments that provided an affirmative answer to this question. Their approach was to alter one or more *lac* operators and then determine the alteration's effect on repression. Point mutations in $lacO_1$ caused a 5- to 50-fold decrease in repression, but some repression was still observed. Destruction of either $lacO_2$ or $lacO_3$ caused a twofold decrease in repression, and destruction of both auxiliary operators caused a 70-fold decrease in repression. The ability to repress the *lac* system was completely lost in cells in which $lacO_1$ and either of the auxiliary operators were destroyed. These results indicate $lacO_1$ plays a major role in repression but the two auxiliary operators also make important contributions. We now examine the structure of the Lac repressor to see how it interacts with the *lac* operators.

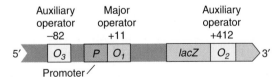

FIGURE 12.31 The three *lac* operators. The *lac* operon has three operators. The major operator, $lacO_1(O_1)$, is centered at position +11 and two auxiliary operators, $lacO_2(O_2)$ and $lacO_3(O_3)$, are centered at positions +412 and −82, respectively.

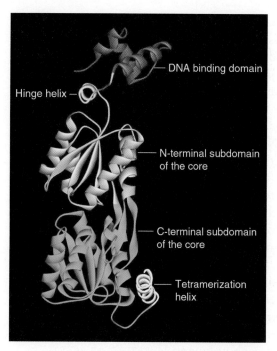

FIGURE 12.32 **The *lac* repressor subunit.** The *lac* repressor monomer has four functional domains. The first, starting from the N-terminal domain or headpiece (residues 1–45), is the DNA binding domain and colored orange. The second, the hinge region (residues 46–62), is colored yellow. The third, the ligand binding domain or core (residues 63–340), which has a distinct N-terminal subdomain (blue) and C-terminal subdomains (green). The fourth, the tetramerization helix (residues 341–357), is colored white. This figure is a composite of two crystal structures. (The structures of the N-terminal domain and hinge regions are from Protein Data Bank ID: 1EFA. Bell, C. E., and Lewis, M. 2000. *Nat Struct Biol* 7:209–214 and the structures of the core domain and tetramerization helix are from Protein Data Bank ID: 1TLF. Friedman, A. M., Fischmann, T. O., and Steitz, T. A. 1995. *Science* 268:1721–1727. Prepared by B. E. Tropp.)

The Lac repressor is a dimer of dimers, where each dimer binds to one *lac* operator sequence.

FIGURE 12.32 shows the structure of a single Lac repressor subunit. As indicated in the figure, the subunit has four distinct functional units:

1. *N-terminal domain* (or *head piece*): The N-terminal domain (residues 1–45) binds *lac* operator DNA. A **helix-turn-helix motif** within the headpiece recognizes and binds appropriate operator DNA (FIGURE 12.33). Helix-turn-helix motifs, which are often present in proteins that bind to DNA, consist of two short α-helical segments containing between seven and nine residues separated by a β-turn (a tight turn involving four

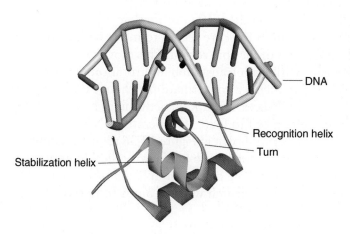

FIGURE 12.33 **A complex of Lac repressor N-terminal domain ("headpiece") with an 11-bp half-operator corresponding to the left half of wild-type *lac* operator.** The recognition and stabilization helices in the helix-turn-helix motif are shown in red and green, respectively. The turn is shown in orange. (Structure from Protein Data Bank 1LCC. Chuprina, V. P., et al. 1993. *J Mol Biol* 234:446–462. Prepared by B. E. Tropp.)

12.6 Lactose Operon

amino acid residues). One of the two α-helices, designated the **recognition helix**, is positioned in the major groove of the DNA so amino acid residues within this helix make specific contacts with bases in the DNA. The other helix, which is designated the **stabilization helix**, helps to stabilize the complex.

2. *Hinge region*: The hinge region (residues 46–62), which joins the N-terminal headpiece to the core domain, is disordered in the absence of DNA but folds into an α-helix upon binding to operator DNA. Two hinge helices in a Lac repressor dimer associate through van der Waals interactions and the paired hinge helices bind to the center of the *lac* operator in the minor groove and bend the DNA (**FIGURE 12.34**).

3. *Ligand binding domain* (or *core domain*): The ligand binding domain (residues 62–340) is divided into an N-terminal subdomain and a C-terminal subdomain. The two kinds of subdomain have similar folding patterns. Weak noncovalent interactions between N-terminal subdomains as well as between C-terminal subdomains make important contributions to Lac repressor dimer formation. IPTG fits in a pocket between the two subdomains in each polypeptide (**FIGURE 12.35a**).

4. *Tetramerization helix*: The tetramerization helix (residues 341–357) associates with corresponding regions of the other three subunits of the Lac repressor to form a four-helix bundle that produces a Lac repressor tetramer with a V-shape (Figure 12.35a). Mutations or deletions that alter the tetramerization helix result in the formation of dimers rather than tetramers. The dimers have partial repressor function. Although the tetramer retains its V-shape in the Lac repressor • DNA complex (Figure 12.35b), its conformation differs from that in the repressor • IPTG complex (Figure 12.35a).

The crystal structures of Lac repressor bound to DNA or IPTG suggest a possible mechanism for the Lac repressor to act as a molecular switch that turns the Lac system off and on. Kathleen Matthews and colleagues suggested the model shown in (**FIGURE 12.36**) to explain how a Lac repressor dimer might work. According to this model, the helix-turn-helix motif in each N-terminal domain makes contact with bases in the major groove of a Lac operator half-site. Furthermore, the hinge helices insert into the minor groove of the central region of the structure, causing the DNA to bend. Binding IPTG to the Lac repressor causes the N-terminal subdomain of the core region to move with respect to the C-terminal subdomain of the core region. This movement is similar to screwing on a cap (with both translation and rotation), closing the IPTG-binding site and inducing a conformational change that disrupts the structures of the hinge helices. Disruption of the hinge helix structure frees the helix-turn-helix motif to leave the operator DNA.

The switching mechanism shown in Figure 12.36 requires a dimeric Lac repressor and therefore does not explain why the Lac repressor has evolved to be a tetramer. The explanation is that the Lac repressor is a more effective genetic switch when it binds to two operators

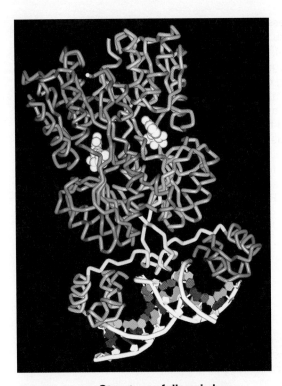

FIGURE 12.34 Structure of dimeric Lac repressor created by recombinant DNA techniques so that the carboxyl terminal domains are missing. The two hinge regions (one on each subunit), which are shown in yellow, fold into α-helices that associate with one another through van der Waals interactions and bind to DNA in the minor groove. Other parts of the repressor subunits are shown in green for one subunit and orange for the other subunit. Each subunit has a bound O-nitrophenyl-β-fucoside (ONPF), a galactoside analog, which stabilizes the Lac repressor in the conformation that binds to the *lac* operator. ONPF is shown as a white spacefill structure. (Structure from Protein Data Bank 1EFA. Bell, C. E., and Lewis, M. 2000. *Nat Struct Biol* 7:209–214. Prepared by B. E. Tropp.)

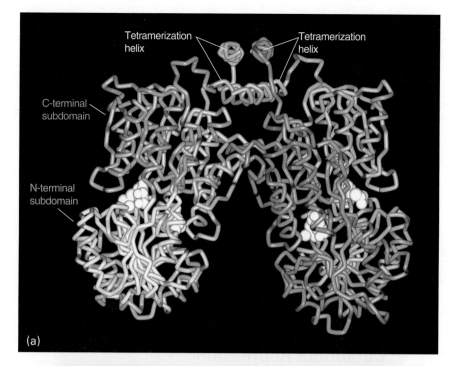

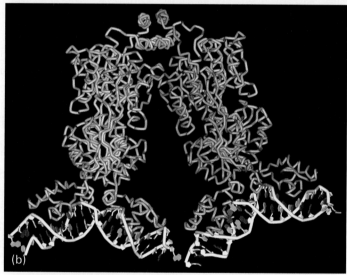

FIGURE 12.35 Tetrameric Lac repressor structure. Each monomer in the Lac repressor tetramer is shown as a different color tube structure (magenta, blue, green, or orange). (a) Repressor • IPTG complex. The N- and C-terminal core subdomains are shown for the magenta subunit. IPTG (white spacefill structure) fits in a pocket between the two subdomains in each polypeptide. DNA-binding domains are not shown because these domains are not well ordered and so not visible in the crystal structure. (b) Repressor • DNA complex. The 21 base-pair DNA duplexes that are bound by each dimer are shown as tubes and rings. The Lac tetramer appears to be a tethered dimer of dimers. (Part a structure from Protein Data Bank 1LBH. Lewis, M., et al. 1996. *Science* 271:1247–1254. Prepared by B. E. Tropp. Part b structure from Protein Data Bank 1LBG. Lewis, M., et al. 1996. *Science* 271:1247–1254. Prepared by B. E. Tropp.)

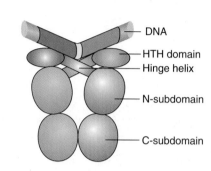

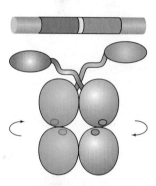

FIGURE 12.36 Model showing the reactions that are elicited when IPTG binds to the Lac repressor. (a) Lac repressor dimer bound to DNA. The Lac repressor makes contact with similarly colored DNA half-sites. The N-terminal helix-turn-helix (HTH) domains (ovals) contact B-form DNA in the major groove. The hinge helices insert into the minor groove at the central region of the structure, causing a significant bend in the DNA. This contact region is indicated by the yellow color in the central segment of operator DNA. (b) When IPTG (blue ovals) binds to the Lac repressor, the N-terminal core subdomain moves with respect to the C-terminal core subdomain in a manner that suggests screwing on a cap (with both rotation and translation), closing the sugar binding site. The conformational change disrupts the hinge-helix structures (indicated by thick wavy lines) and frees the helix-turn-helix domains (ovals), allowing them to dissociate from the DNA. (Adapted from Matthews, K. S., Falcon, C. M., and Swint-Kruse, L. 2000. *Nat Struct Biol* 7:184–187.)

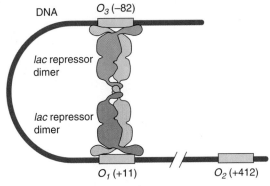

at the same time. Recall that the *lac* operon has two auxiliary operators, *lacO₂* and *lacO₃*, in addition to the major operator, *lacO₁*. DNA looping permits the tetrameric Lac repressor to bind to *lacO₁* and one of the auxiliary operators (FIGURE 12.37). The cell would need one superoperator to achieve the same level of repression that it achieves with two weak cooperating operators.

Experiments designed to explain how the Lac repressor prevents the transcription of the *lac* operon have led to conflicting hypotheses. Gilbert and coworkers originally proposed the Lac repressor prevents RNA polymerase holoenzyme from binding to the promoter. Later studies by other investigators suggested the Lac repressor does not prevent RNA polymerase holoenzyme from binding to the promoter but instead prevents the conversion of a closed *lac* promoter to an open *lac* promoter. Recent thermodynamic studies of repression suggest Gilbert's original proposal may have been correct after all.

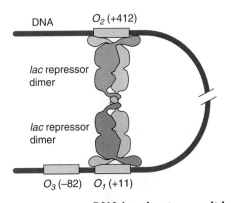

FIGURE 12.37 **DNA looping to permit Lac repressor to bind to two operators.** The *lac* operon DNA is shown in blue, *lac* operators in yellow, and the *lac* repressor dimers in shades of green or blue. The slanted double lines in the DNA indicate that not all DNA is shown. (Adapted from Müller-Hill, B. 1998. *Curr Opin Microbiol* 1:145–151.)

12.7 Catabolite Repression

E. coli uses glucose in preference to lactose.

The function of β-galactosidase in lactose metabolism is to hydrolyze lactose to form glucose and galactose. If the growth medium contains both glucose and lactose, then in the interest of efficiency there is no need for a cell to turn on the *lac* operon. Experiments performed by Monod in the mid-1940s demonstrated that cells behave according to this logic. *E. coli* cells incubated in the presence of glucose and lactose do not start to make β-galactosidase until all the exogenous glucose is consumed (FIGURE 12.38). These findings were later extended to lactose permease and transacetylase, the two other proteins specified by the *lac* operon. Lactose enzymes are not made when glucose is present because no *lac* mRNA is made. Transcription-level inhibition of the lactose enzymes and a variety of other inducible enzymes

FIGURE 12.38 **Effect of changing the carbon source from glucose to lactose.** *E. coli* are cultured in medium that contains glucose and lactose. The cells initially use glucose as the carbon source. At the time indicated by the arrow, glucose is in such short supply that the *E. coli* cell mass stops increasing. The absorbance at 610 nm is a measure of the total cell mass (black curve). A short time after glucose depletion the bacterial cells start to make β-galactosidase and use lactose as the carbon source. Enzyme activity is indicated in arbitrarily defined units (red curve).

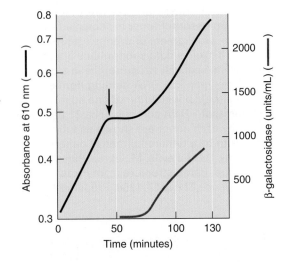

by glucose (or other readily used carbon sources) is called **catabolite repression**.

The cAMP receptor protein combines with 3′,5′-cyclic adenylate to form a positive regulator or activator.

The mechanism by which glucose inhibits β-galactosidase synthesis remained a complete mystery for about 20 years after Monod first observed the phenomenon. Richard S. Makman and Earl W. Sutherland found an important clue to solving the mystery in 1965 when they observed the intracellular concentration of 3′,5′-cyclic adenylate, or cAMP (FIGURE 12.39), drops from about 10^{-4} M to 10^{-7} M when glucose is added to a growing culture of *E. coli*.

Genetic studies confirmed cAMP's involvement in catabolite repression. Two mutant classes were isolated that could not synthesize *lac* enzymes when cultured in a medium containing lactose but no glucose. Class I mutants regained the ability to synthesize *lac* enzymes when cAMP was added to the growth medium, but class II mutants did not.

Subsequent studies showed class I mutants have defects in adenylate cyclase, the enzyme that converts ATP to cAMP (FIGURE 12.40). The structural gene for adenylate cyclase, *cya*, maps at minute 85.98. Adenylate cyclase exists in an active form that is *phosphorylated* (it contains an attached phosphate group) and an inactive form that is dephosphorylated. Class II mutants have defects in a protein that binds cAMP. This protein, called the cAMP receptor protein (CRP) or the catabolite activator protein (CAP), is encoded by the *crp* gene, which maps at minute 75.09. *In vitro* studies have shown that CRP and cAMP form a complex, denoted cAMP • CRP complex, which is needed to activate the *lac* system. The cAMP • CRP requirement is independent of the repression system because *cya* and *crp* mutants cannot make *lac* mRNA even if a *lacI*⁻ or *lacO*ᶜ mutation is present. The cAMP • CRP complex is a positive regulator or activator. In the absence of the cAMP • CRP complex, the *lac* promoter is quite weak because its −10 box differs significantly from the consensus sequence. A mutant *lac* promoter with a −10 box that has the consensus sequence does not require the cAMP • CRP complex for transcription activation.

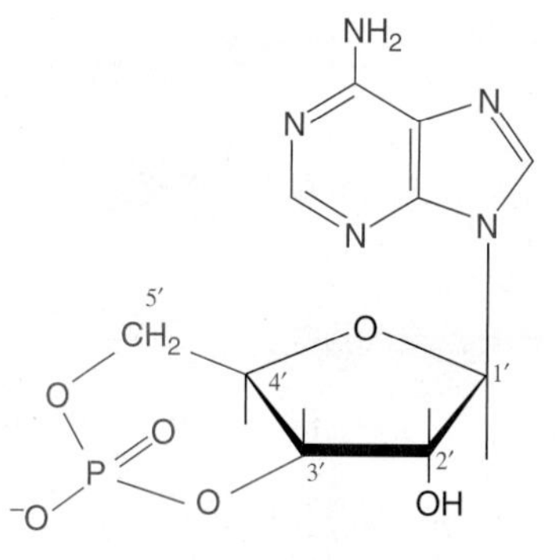

FIGURE 12.39 3′,5′-Cyclic adenylate (cAMP).

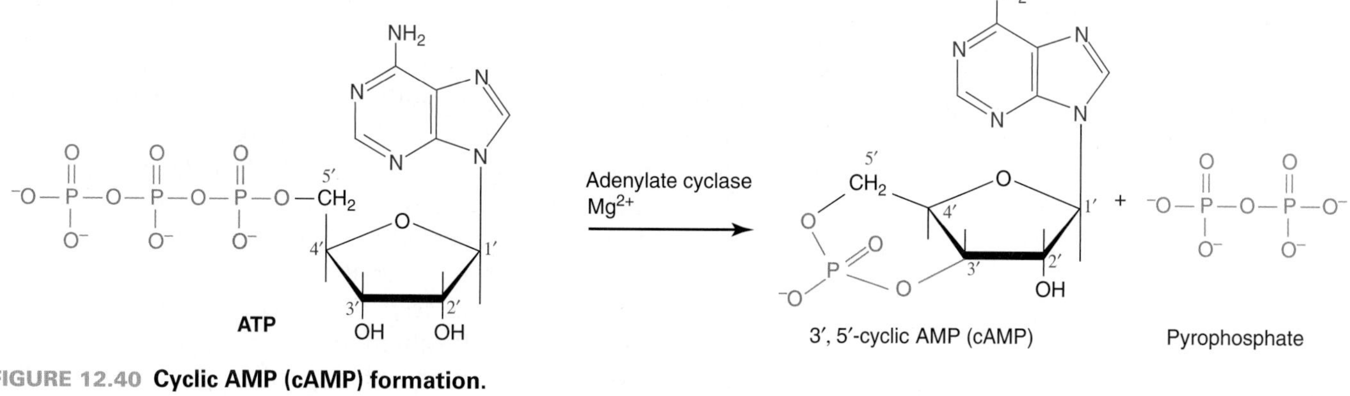

FIGURE 12.40 Cyclic AMP (cAMP) formation.

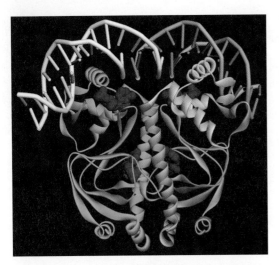

FIGURE 12.41 **The cAMP receptor protein (CRP) dimer bound to DNA and two cAMP molecules per monomer.** The N-terminal and C-terminal domains in one CRP monomer are colored blue and yellow, respectively. The DNA half-site bound by this CRP monomer is gray. The other CRP monomer is shown in green and the DNA bound to it is white. Each monomer is bound to two cAMP molecules shown in red spacefill form. In each CRP one cAMP lies on the helix-turn-helix and close to the DNA and the other lies in the C-terminal domain. (Protein Data Bank: 2CGP Passner, J. M., and Steitz, T. A. 1997. *Proc Natl Acad Sci USA* 94:2843–2847. Prepared by B. E. Tropp.)

The cAMP • CRP complex binds to an activator site (AS) upstream from the *lac* promoter and activates *lac* operon transcription.

Jonathan Passner and Thomas Steitz determined the crystal structure for the cAMP • CRP complex bound to DNA (FIGURE 12.41). CRP consists of two chemically identical polypeptide chains of 209 amino acid residues. Each chain has an N-terminal domain and a C-terminal domain. The N-terminal domain has a pocket for binding cAMP and the C-terminal domain contains a helix-turn-helix motif that binds DNA. In the absence of cAMP, CRP • DNA interactions are nonspecific and weak. The cAMP • CRP complex, however, binds very tightly to a specific DNA sequence designated the **activator site (AS)**. In the *lac* operon the center of AS is 61.5 bp upstream from the transcription start site (that is, between nucleotides −61 and −62). Many other bacterial operons are activated by cAMP • CRP. Each of these operons also contains at least one AS. Comparing AS sequences reveals the following 22-bp consensus sequence with twofold symmetry:

5′-AAATGTGATCTAGATCACATTT-3′
3′-TTTACACTAGATCTAGTGTAAA-5′

The most highly conserved nucleotides in AS are the two TGTGA motifs. Mutations that alter nucleotides in the TGTGA motifs lead to decreased *lac* operon transcription. Each CRP subunit binds to half of the AS. The interactions between the C-terminal domain's helix-turn-helix motif and AS are similar to those described for Lac repressor and operator DNA. In fact, helix-turn-helix interactions with DNA are a recurring theme in DNA-protein recognition.

Binding of the cAMP • CRP complex to the AS consensus sequence is so tight an operon containing this sequence would be permanently switched on. It is therefore not surprising that actual cAMP • CRP activator sites differ from the consensus sequence. Activator sites in different operons compete for the cAMP • CRP complex with the activator preferentially binding to sequences that most closely resemble the consensus sequence.

In addition to specific contacts with AS, the cAMP • CRP complex also makes specific contacts with RNA polymerase holoenzyme. Genetic studies show the contact site in RNA polymerase holoenzyme is located on the α subunit, which consists of an N-terminal domain, a C-terminal domain (αCTD), and a flexible linker region that joins the two domains. A mutant RNA polymerase with truncated αCTDs cannot transcribe the *lac* operon in the presence or absence of cAMP • CRP. However, the same enzyme can transcribe a *lac* operon with a consensus sequence in its promoter's −10 box. Furthermore, chemical cross-linking experiments show CRP and αCTD are adjacent to one another. These observations indicate cAMP • CRP interacts with αCTD. Experiments with RNA polymerases, which have one full-length α subunit and one truncated α subunit lacking αCTD, reveal that transcription activation requires RNA polymerase to have just one intact α subunit. It does not seem to matter if the intact α subunit is the one associated with the β subunit or the one associated with the β′ subunit. The α subunit interacts with a specific region in CRP called

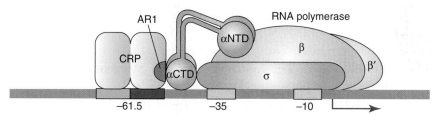

FIGURE 12.42 Transcriptional activation of the *E. coli lac* operon. The symbols are CRP, cAMP receptor protein; σ, σ factor; αCTD, carboxyl terminal domain α subunit; αNTD, amino terminal domain α subunit; β, β subunit; and β', β' subunit. The magenta oval represents activating region 1 (AR1) of CRP. The promoter in the lac operon is a prototype of a class of promoters known as class 1 cAMP-dependent promoters. Transcription activation requires direct protein–protein interaction between AR1 of the downstream subunit of CAP and one copy of αCTD. (Adapted from Busby, S., and Ebright, R. H. 1999. *J Mol Biol* 293:199–213.)

activating region 1 (AR1). The interactions between cAMP • CRP and RNA polymerase holoenzyme are summarized in **FIGURE 12.42**.

Glucose causes catabolite repression through cAMP modulation and inducer exclusion mechanisms.

Based on the information presented above, it seems reasonable to propose that glucose somehow inhibits phosphorylation of adenylate cyclase, thereby preventing cAMP formation. The next challenge was to find the link between glucose metabolism and adenylate kinase phosphorylation. Biochemical and genetic studies indicated the link is a glucose-specific, phosphoenolpyruvate-dependent phosphotransferase system. This system, depicted in **FIGURE 12.43**, uses energy supplied by phosphoenolpyruvate to phosphorylate glucose as it transports the sugar across the inner cell membrane.

The system requires four proteins. Two of these, enzyme I and the histidine-containing protein (HPr), are also components of other sugar transport systems and therefore are unlikely to be direct participants in a glucose-specific phenomenon. The two other proteins, enzyme IIA (EIIA) and enzyme IIBC (EIIBC), are specific for the glucose

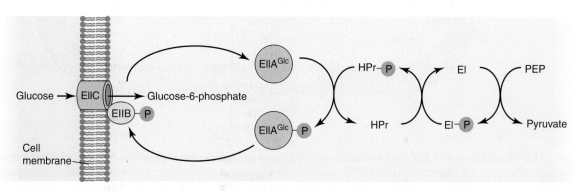

FIGURE 12.43 Glucose-specific, phosphoenolpyruvate-dependent phosphotransferase system (PTS). Phosphate is transferred from phosphoenolpyruvate (PEP) to the exogenous glucose simultaneously with glucose transport. This flow is mediated by enzyme I, HPr, and carbohydrate-specific EIIA and EIIBC. (Adapted from Stülke, J., and Hillen, W. 1999. *Curr Opin Microbiol* 2:195–201.)

transport system and are more likely participants in a glucose-specific phenomenon.

EIIA participates in catabolite repression by two different mechanisms (FIGURE 12.44). The first mechanism is based on the fact that EIIA can transfer a phosphoryl group from HPr-P to either EIIBC or adenylate cyclase. The preferred substrate is EIIBC, which then transfers the phosphoryl group to glucose to form glucose-6-phosphate. When glucose is unavailable EIIBC will be fully phosphorylated and EIIA-P has no other alternative but to transfer its phosphoryl group to adenylate cyclase. Phosphorylation changes the inactive dephosphorylated form of adenylate cyclase to the active phosphorylated form of adenylate cyclase, which then converts ATP to cAMP. Thus, glucose interferes with the conversion of the inactive form of adenylate cyclase to the active form.

Investigators initially thought glucose's influence on adenylate cyclase activity was solely responsible for glucose's effect on *lac* operon transcription. The following observation forced investigators to consider the possibility that glucose may also influence *lac* operon transcription in another way. A *cya*⁻ mutant with an altered CRP that does not require cAMP to activate the *lac* operon is subject to catabolite repression. Because mutant cells synthesize much less β-galactosidase in the presence of glucose, cAMP cannot be the sole modulator of *lac* mRNA synthesis.

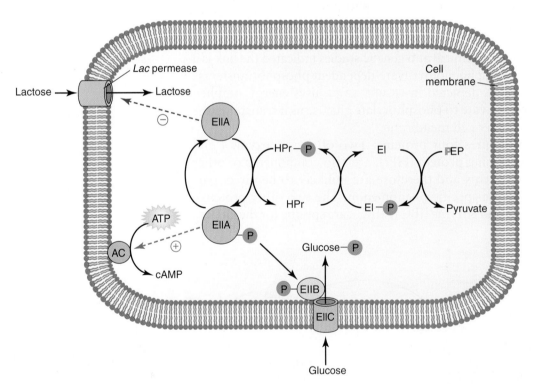

FIGURE 12.44 cAMP-modulated and induced exclusion mechanisms for catabolite repression. EIIA participates in catabolite repression by two different mechanisms. (1) cAMP modulated mechanism, where EIIA can transfer a phosphoryl group from HPr-P to either EIIBC or adenylate cyclase (AC). The preferred substrate is EIIBC, which then transfers the phosphoryl group to glucose to form glucose-6-phosphate. When glucose is unavailable EIIBC will be fully phosphorylated, forcing EIIA-P to transfer its phosphoryl group to adenylate cyclase. Phosphorylation changes the inactive dephosphorylated form of adenylate cyclase to the active phosphorylated form of adenylate cyclase, which then converts ATP to cAMP. (2) Inducer exclusion mechanism where the dephosphorylated form of EIIA binds to lactose permease and inactivates it.

Further study revealed a second mechanism by which glucose influences *lac* operon transcription. This mechanism, called **inducer exclusion,** also involves the glucose transport system. When glucose is present, EIIA-P transfers its phosphate group to the sugar and the dephosphorylated form of EIIA binds to lactose permease and inactivates it. Lactose permease inactivation prevents lactose from entering the cell and being converted to allolactose. In the absence of allolactose the repressor remains bound to the operator and the *lac* operon is turned off. The cAMP modulation and inducer exclusion mechanisms may not be the only ones that contribute to glucose's ability to inhibit *lac* mRNA formation. For example, regulation of CRP synthesis may also be important. Thus, *lac* operon regulation, one of the best studied of all gene regulation systems, remains a very active area of inquiry.

cAMP • CRP activates more than 100 operons.

Enzymes responsible for the catabolism of many other organic molecules, including galactose, arabinose, sorbitol, and glycerol, are synthesized by inducible operons. Each of these operons cannot be induced if glucose is present. These are called **catabolite-sensitive operons.** A network of operons that is under the control of a single regulatory protein such as the cAMP • CRP complex is called a **modulon.**

A simple genetic experiment shows the cAMP • CRP complex participates in the regulation of many operons. A single spontaneous mutation in a gene for the catabolism of a sugar such as lactose, galactose, arabinose, or maltose arises with a frequency of roughly 10^{-6}. A double mutation, lac^-mal^-, would arise at a frequency of 10^{-12}, which for all practical purposes cannot be measured. However, double mutants that are phenotypically Lac⁻Mal⁻ or Gal⁻Ara⁻ do arise at a measurable frequency. These apparent double mutants are not the result of mutations in the two sugar operons but always turn out to be crp^- or cya^-. Furthermore, if a Lac⁻Mal⁻ mutant appears as a result of a single mutation, the protein products of the other catabolite-sensitive operons are also not synthesized. Biochemical experiments with a few of these catabolite-sensitive operons indicate that binding of the cAMP • CRP complex occurs before the promoter region in each of these systems.

12.8 Tryptophan Operon

The tryptophan (*trp*) operon is regulated at the levels of transcription initiation, elongation, and termination.

The tryptophan (*trp*) operon, which we now examine, is regulated at the transcription initiation, elongation, and termination stages. It consists of a promoter, an operator, a leader (*trpL*), an attenuator, and five structural genes, designated *trpE*, *trpD*, *trpC*, *trpB*, and *trpA* (**FIGURE 12.45**). Consistent with the fact that the *trp* operon specifies biosynthetic rather than degradative enzymes, it does not have a

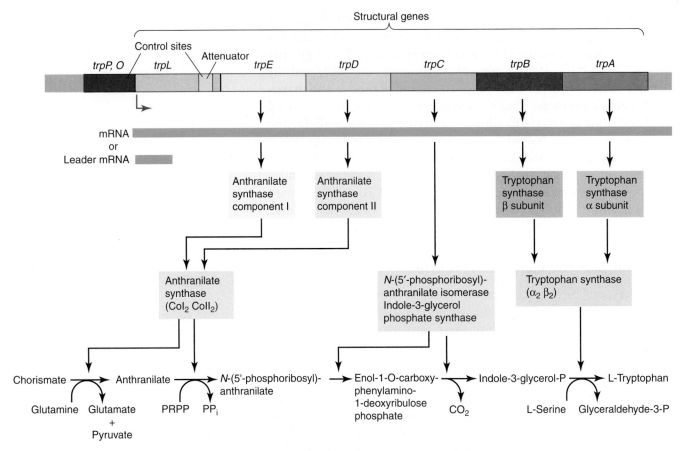

FIGURE 12.45 *E. coli* **tryptophan (*trp*) operon, the enzyme it specifies, and the reactions they catalyze.** PRPP, phosphoribosylpyrophosphate. (Adapted from Yanofsky, C. 1971. *J Am Med Assoc* 218:1026–1035).

cAMP • CRP activation site. The operator is part of a coarse on–off control, and the leader and attenuator allow for finer control. The leader sequence, which codes for a short peptide (see below), is the first region of the *trp* operon to be transcribed. RNA polymerase then moves forward to transcribe the five structural genes in the order *trpE* → *trpA* to form the polycistronic *trp* mRNA. The five polypeptides specified by the structural genes form three enzymes that are essential for tryptophan biosynthesis.

Early studies revealed the *trp* system is turned off when tryptophan is added to an *E. coli* culture, suggesting the operon is repressed rather than induced. Repression, like induction, involves negative regulation of transcription initiation by a regulatory protein. However, the *Trp regulatory protein does not bind to the operator until after forming a complex with tryptophan* (**FIGURE 12.46**). The structural gene for the Trp regulatory protein, *trpR*, is located a considerable distance away from the *trp* operon. This distance does not present a problem because the regulatory protein diffuses throughout the cell. The biologically active form of the regulatory protein is a homodimer. Each subunit contains a helix-turn-helix motif that can bind to a *trp* operator half-site. The *trp* operator and promoter regions have significant overlap. Binding of the TrpR • tryptophan complex and RNA polymerase, therefore, are mutually exclusive.

(a) Transcription occurs

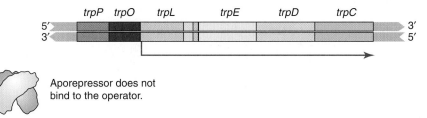

Aporepressor does not bind to the operator.

(b) Transcription is repressed

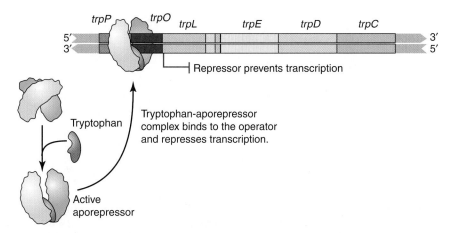

Repressor prevents transcription

Tryptophan

Tryptophan-aporepressor complex binds to the operator and represses transcription.

Active aporepressor

FIGURE 12.46 Regulation of the *E. coli trp* operon. (a) By itself, the Trp aporepressor does not bind to the operator, and transcription occurs. (b) In the presence of sufficient tryptophan the combination of aporepressor and tryptophan forms the active repressor that binds to the operator and transcription is repressed.

The TrpR • tryptophan complex serves as a coarse on–off switch that turns the tryptophan operon off when tryptophan levels are high. A fine control mechanism also exists that allows cells to regulate their tryptophan enzyme concentration according to the tryptophan concentration. An important clue to the existence of this fine control mechanism came from an experiment performed by Charles Yanofsky in 1972 that showed *E. coli* mutants that lack a functional TrpR protein increase *trp* operon transcription after being starved for tryptophan. If the TrpR • tryptophan complex were the only regulatory factor, then transcription of the *trp* operon should not have increased.

Further studies revealed *trp* mRNA has a 162 nucleotide sequence before the first codon in *trpE*, designated the leader, or *trpL*, that plays an essential role in the fine control mechanism. Constitutive mutants exhibit a sixfold increase in tryptophan enzyme synthesis when bases 123 to 150 within the leader are deleted. This 28 base sequence is called the **attenuator**. The attenuator can fold into a stem-and-loop structure with the potential to function as a rho-independent transcription terminator (**FIGURE 12.47**). Evidence that the attenuator actually terminates transcription comes from the observation that wild-type cells cultured in the presence of tryptophan terminate synthesis of most *trp* mRNA molecules when they are 140 nucleotides long, well short of full-length polycistronic *trp* mRNA. Deleting the attenuator

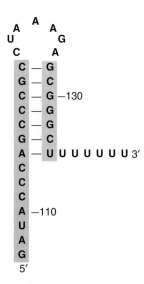

FIGURE 12.47 The terminal region of the *trp* attenuator region. The attenuator can fold into a stem-and-loop structure with the potential to function as a rho-independent transcription terminator. (Adapted from Voet, D., Voet, J. G., and Pratt, C. W. 2000. *Fundamentals of Biochemistry* (1st ed). John Wiley & Sons.)

12.8 Tryptophan Operon

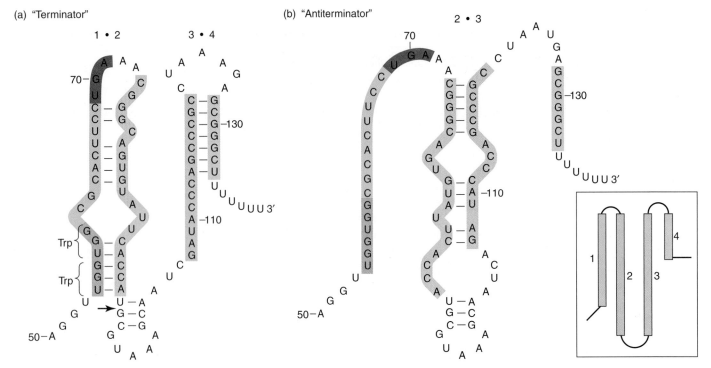

FIGURE 12.48 Alternative secondary structures of *trpL* mRNA. Bases are numbered from the 5′-end of the transcript. Segments that base pair to form intrastrand secondary structure are numbered according to the inset. (a) 1 • 2 and 3 • 4 (terminator) hairpins signal transcriptional pausing and termination, respectively. The arrow indicates the site at which RNA polymerase pauses until it is approached by the moving ribosome. (b) Alternative 2 • 3 (antiterminator) hairpin that signals transcriptional read-through at the attenuator by preventing terminator 3 • 4 formation. (Adapted from Voet, D., Voet, J. G., and Pratt, C. W. 2000. *Fundamentals of Biochemistry* (1st ed). John Wiley & Sons.)

removes the transcription termination site, allowing RNA polymerase to complete *trp* mRNA synthesis.

The *trp* leader has four complementary segments that can interact to form two sets of mutually exclusive hairpin structures (FIGURE 12.48). Segments 1 and 2 can base pair to form hairpin 1 • 2, while segments 3 and 4 base pair to form hairpin 3 • 4 (Figure 12.48a). Alternatively, segments 2 and 3 can base pair to form hairpin 2 • 3 (Figure 12.48b). The purified *trp* leader mRNA forms the 1 • 2 and 3 • 4 hairpins because this secondary structure has the highest degree of hydrogen bonding and therefore the greatest stability. The 3 • 4 hairpin, which contains the attenuator sequence, is a transcription terminator. It follows that RNA polymerase would be able to synthesize full-length *trp* mRNA if conditions were somehow favorable for 2 • 3 hairpin formation rather than 3 • 4 hairpin formation.

The attenuation model, proposed by Yanofsky, explains how cells block 3 • 4 hairpin formation when tryptophan concentrations are low. This model builds on the observation that bacterial ribosomes start to translate mRNA molecules before RNA synthesis is complete. Coupled transcription-translation is possible in bacteria because (1) ribosomes and DNA are in the same cell compartment and (2) mRNA is transcribed in a 5′→3′ direction and translated in the same direction. The attenuation model is based on the fact that

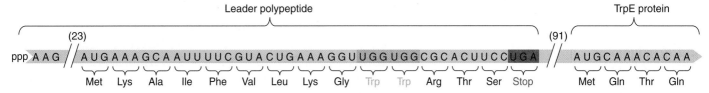

FIGURE 12.49 The sequence of nucleotides in *trp* leader mRNA, showing the leader polypeptide, the two tryptophan codons (orange), and the beginning of the TrpE protein. The numbers 23 and 91 indicate the numbers of nucleotides in the sequence which, for clarity, is not shown.

the leader sequence codes for a peptide that is 14 amino acids long with tryptophan residues at positions 10 and 11 (FIGURE 12.49). The presence of a pair of adjacent tryptophans is unusual because tryptophan normally accounts for about 1% of the amino acid residues in a protein.

RNA polymerase molecules that escape repression begin synthesizing *trp* mRNA. RNA polymerase continues transcribing the *trp* operon until it encounters a pause site located just after segment 2 of the leader sequence. Because the leader sequence is at the 5′-end of *trp* mRNA, it is the first sequence available for translation. Ribosomes begin translating the leader sequence at its AUG start codon. When the moving ribosome reaches the paused RNA polymerase, the paused RNA polymerase resumes transcription. Pausing serves the important role of synchronizing the transcription and translation processes. The subsequent fate of the RNA polymerase depends on the tryptophan concentration (FIGURE 12.50).

1. When the tryptophan concentration is high, the ribosome moves past the tryptophan codons in segment 1 and continues

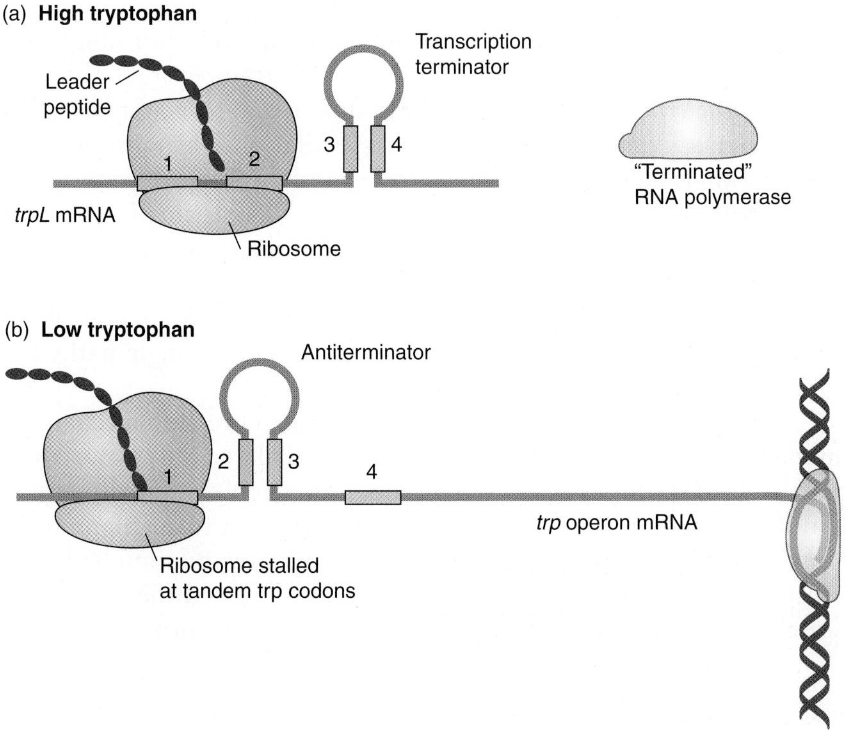

FIGURE 12.50 Attenuation in the *trp* operon. (a) When tryptophan is abundant the ribosome reads through the tandem Trp codons in segment 1 of *trpL* mRNA, reaching segment 2 and preventing formation of the base-paired 2 • 3 hairpin. The 3 • 4 hairpin, an essential component of the transcriptional terminator, can then form, thus aborting transcription. (b) When tryptophan is scarce the ribosome stalls on the tandem Trp codons of segment 1, permitting the formation of the 2 • 3 hairpin. This formation prevents the formation of the 3 • 4 hairpin, allowing RNA polymerase to transcribe through the unformed terminator and continue transcribing the *trp* operon. (Adapted from Voet, D., Voet, J. G., and Pratt, C. W. 2000. *Fundamentals of Biochemistry* (1st ed). John Wiley & Sons.)

12.8 Tryptophan Operon

to translate the leader region until it encounters the stop codon (UGA) between segments 1 and 2 and falls off the nascent mRNA molecule. Once free of the ribosome, segment 1 pairs with segment 2 to form the 1 • 2 hairpin. RNA polymerase continues to transcribe the leader region, synthesizing segment 3 and then segment 4. These two segments pair to form the rho-independent transcription terminator, which causes RNA polymerase to fall off the DNA template, preventing *trpE* transcription.

2. When the tryptophan concentration is very low, the bulky ribosome will pause at the tryptophan codons (UGG) on segment 1, preventing segment 1 from pairing with segment 2 to form the 1 • 2 hairpin. Segment 2 is therefore free to interact with segment 3 to form the 2 • 3 hairpin (the so-called **antiterminator**) as soon as RNA polymerase completes the synthesis of segment 3. Then RNA polymerase can continue to synthesize the complete *trp* mRNA because the 3 • 4 hairpin (the transcription terminator) is not formed.

Thus, if tryptophan is present in excess, transcription termination occurs at the attenuator and little enzyme is synthesized; if tryptophan is absent, transcription termination does not occur and tryptophan enzymes are made. Many operons responsible for amino acid biosynthesis are regulated by attenuators equipped with the base-pairing mechanism for competition as described for the *trp* operon. For example, this method of transcription regulation has been described for the histidine, threonine, leucine, isoleucine-valine, and phenylalanine operons of the bacteria *E. coli*, *Salmonella typhimurium*, and *Serratia marcescens*. Except for the phenylalanine operon, each lacks a repressor-operator system and is regulated solely by attenuation.

Because these operons are regulated adequately (although the range of expression is not as great as that of the *trp* operon), one might ask why the *trp* and *phe* operons have dual regulatory systems. This question has given rise to considerable speculation. The most obvious explanation is that the separate effects expand the range of tryptophan concentration in which regulation occurs. Whereas this is certainly true, there is more to it because the *trp* repressor also regulates the activity of four other operons. Such an operon network, which consists of two or more operons that are (1) regulated by a common regulatory protein and its effector(s) and (2) associated with a single pathway, function, or process, is called a **regulon.**

One hypothesis to explain the origin of the repression mechanism in the *trp* operon is that the *trp* operon was regulated only by attenuation in the distant past. The Trp repressor, probably originally served a different function, to combine with tryptophan to regulate *aroH* (one of the five operons regulated by the Trp repressor complex). The existence of a tryptophan repressor complex allowed the possibility for a region adjacent to the *trp* promoter to change into an operator that can bind it.

Questions and Problems

1. Briefly define, describe, or illustrate each of the following.
   a. Transcription
   b. Translation
   c. Primary transcript
   d. Downstream
   e. Upstream
   f. Promoter
   g. Consensus sequence
   h. −10 and −35 boxes
   i. DNase protection method
   j. Up element
   k. DNA footprinting technique
   l. Open promoter complex
   m. Abortive initiation
   n. Promoter escape
   o. Scrunching model
   p. Transcription elongation complex
   q. Backtracking
   r. Intrinsic termination
   s. Rho-dependent termination
   t. The *rut* site
   u. Open reading frame
   v. Cistron
   w. Polycistronic mRNA
   x. 5′-Untranslated region
   y. 3′-Untranslated region
   z. Coordinate regulation
   aa. Negative regulation
   bb. Positive regulation
   cc. Repressor
   dd. Activator
   ee. Inducible enzyme
   ff. Repressible enzyme
   gg. Constitutive enzyme
   hh. Gratuitous inducer
   ii. Lac repressor
   jj. Cross-reacting material
   kk. Operon model
   ll. Derepression
   mm. Basal synthesis
   nn. Nitrocellulose filter assay
   oo. Helix-turn-helix motif
   pp. Recognition helix
   qq. Stabilization helix
   rr. Catabolite repression
   ss. Activator site
   tt. Inducer exclusion
   uu. Catabolite sensitive operons
   vv. Modulon
   ww. Attenuator
   xx. Antiterminator
   yy. Regulon

2. Describe an experimental method to detect RNA polymerase activity.

3. How are the RNA polymerase and DNA polymerase I catalyzed reactions similar? How are the two enzyme catalyzed reactions different?

4. The bacterial RNA polymerase is a much bigger protein than DNA polymerase I. Is this difference in size required because it is more difficult to form a phosphodiester bond involving ribonucleotides than deoxyribonucleotides? Provide experimental evidence to support your answer.

5. The core RNA polymerase can use nicked DNA as a template but cannot use intact double-stranded DNA as a template.
   a. What is the core RNA polymerase?
   b. Why can the core RNA polymerase use nicked DNA as a template?
   c. What additional factor must be added to the core RNA polymerase to obtain an enzyme that can use intact double-stranded DNA as a template?
   d. Does this additional factor introduce transient nicks to allow the core RNA polymerase to use intact DNA as a template? Provide experimental evidence to support your answer.

6. The β subunit in bacterial RNA polymerase has a highly conserved NADFDGD motif. What is the function of this motif and what happens to RNA polymerase activity if one of the aspartate subunits is changed to alanine?

7. The strength of a bacterial promoter depends on how close the promoter's sequence is to the consensus sequence.
   a. Describe an experimental approach that you might take to increase the strength of a weak promoter.
   b. Would a bacterial strain be able to grow faster if all of its genes had strong promoters? Explain your answer.
   c. Describe a situation when a bacteria benefits from having a gene with a weak promoter.

8. DNA footprinting experiments have provided important insights related to transcription initiation.
   a. Briefly describe the DNA footprinting technique.
   b. How has the DNA footprinting technique helped to advance our understanding of the transcription initiation process?

9. When a σ^{70} RNA polymerase holoenzyme • promoter DNA complex forms at 0°C, the protected DNA is not sensitive to $KMnO_4$ oxidation. The protected DNA becomes sensitive to $KMnO_4$ oxidation, however, when the temperature is raised to 37°C. What is the explanation for this change in sensitivity? What is the significance of this change in sensitivity?

10. Briefly outline the events that take place during transcription initiation.

11. The RNA polymerase transcription elongation complex occasionally makes errors. How does the enzyme correct these errors?

12. Describe the two mechanisms that the bacterial transcription machinery uses to terminate transcription.

13. The rate of β-galactosidase synthesis in *E. coli* is influenced by the bacteria's genotype and its growth conditions. Indicate whether you would expect the rate of β-galactosidase synthesis to increase, decrease, or remain the same after each of the following changes to the growth medium. Explain your answer.
    a. *E. coli* cells are cultured in minimal medium containing glycerol as the sole carbon source. Then lactose is added to the medium.
    b. *E. coli* cells are cultured in minimal medium containing glucose as the sole carbon source. Then lactose is added to the medium.
    c. *E. coli* cells are cultured in minimal medium containing both glucose and lactose as carbon sources. Then cAMP is added to the medium.
    d. An *E. coli lacI⁻* mutant is cultured in minimal medium that contains glycerol as the sole carbon source. Then lactose is added to the medium.
    e. An *E. coli lacI⁻* mutant is cultured in minimal medium that contains glycerol as the sole carbon source. Then glucose is added to the medium.
    f. An *E. coli lacI⁻* mutant is cultured in minimal medium that contains glucose as the sole carbon source. Then cAMP is added to the medium.
    g. An *E. coli crp⁻* mutant is cultured in minimal medium that contains lactose and glucose as the sole carbon sources. Then cAMP is added to the medium.
    h. An *E. coli cya⁻* mutant is cultured in minimal medium that contains lactose as the sole carbon source. Then cAMP is added to the medium.
    i. An *E. coli* O^C mutant is cultured in minimal medium that contains glycerol as the sole carbon source. Then lactose is added to the medium.

14. A spontaneous *E. coli* mutant has lost the ability to use either maltose or galactose as a sole carbon source even when cAMP is added to the medium. Provide a hypothesis to explain these findings and suggest an experimental method to test your hypothesis.

15. How is the regulation of the *lac* and *trp* operons similar? How is the regulation of the two operons different?

16. UGG codes for tryptophan and the codon UGC codes for cysteine. How would the regulation of the *trp* operon be changed by replacing the UGGUGG sequence in *trpL* with UGCUGC? Would the two codon changes influence regulation caused by the Trp regulatory protein? Explain your answer.

Suggested Reading

Historical

Hurwitz, J. 2005. The discovery of RNA polymerase. *J Biol Chem* 280:42477–42488.

Initiation Stage

Artsimovitch, I. 2008. Post-initiation control by the initiation factor sigma. *Mol Microbiol* 68:1–3.

Borukhov, S., and Nudler, E. 2003. RNA polymerase holoenzyme: structure, function, and biological implications. *Curr Opin Microbiol* 6:93–100.

Burgess, R. R., Travers, A. A., Dunn, J. J., and Bautz, E. K. 1969. Factor stimulating transcription by RNA polymerase. *Nature* 221:43–46.

Campbell, E. A., Korzheva, N., Mustaev, A., et al. 2001. Structural mechanism for rifampicin inhibition of bacterial RNA polymerase. *Cell* 104:901–912.

Campbell, E. A., Muzzin, O., Chlenov, M., et al. 2002. Structure of the bacterial RNA polymerase promoter specificity σ subunit. *Mol Cell* 9:527–539.

Campbell, E. A., Westblade, L. F., and Darst, S. A. 2008. Regulation of bacterial RNA polymerase σ factor activity: a structural perspective. *Curr Opin Microbiol* 11:121–127.

Cramer, P. 2002. Multisubunit RNA polymerases. *Curr Opin Struct Biol* 12:89–97.

Cheetham, G. M. T., Jeruzalmi, D., and Steitz, T. A. 1999. Structural basis for initiation of transcription from an RNA polymerase-promoter complex. *Nature* 399:80–83.

Darst, S. A. 2001. Bacterial RNA polymerase. *Curr Opin Struct Biol* 11:155–162.

Erie, D. A. 2002. The many conformational states of RNA polymerase elongation complexes and their roles in the regulation of transcription. *Biochim Biophys Acta* 1577:224–239.

Feklistov, A., and Darst, S. A. 2009. Promoter recognition by bacterial alternative σ factors: the price of high selectivity. *Genes Dev* 23:2371–2375.

Finn, R. D., Orlova, E. V., Gowen, B., et al. 2000. *Escherichia coli* core and holoenzyme structures. *EMBO J* 19:6833–6844.

Gralla, J. D. 2000. Signaling through sigma. *Nat Struct Biol* 7:530–532.

Gruber, T. M., and Gross, C. A. 2003. Multiple sigma subunits and the partitioning of bacterial transcription space. *Annu Rev Microbiol* 57:441–466.

Hellman, J. D. 2005. Sigma factors in gene expression. *Encyclopedia of Life Sciences*, pp. 1–7. Hoboken, NJ: John Wiley & Sons.

Hinton, D. M. 2005. Molecular gymnastics: distortion of an RNA polymerase σ factor. *Trend Microbiol* 13:140–143.

Hsu, L. M. 2002. Open season on RNA polymerase. *Nat Struct Biol* 9:502–504.

Hsu, L. M. 2002. Promoter clearance and escape in prokaryotes. *Biochim Biophys Acta* 1577:191–207.

Hsu, L. M. 2009. Monitoring abortive initiation. *Methods* 47:25–36.

Kapnidis, A. N., Margeat, E., Ho, S. O., et al. 2006. Initial transcription by RNA polymerase proceeds through a DNA-scrunching mechanism. *Science* 314:1144–1147.

Koo, B-M., Rhodium, V. A., Nonaka, G., deHaseth, P. L., and Gross, C. A. 2009. Reduced capacity of alternative σs to melt promoters ensures stringent promoter recognition. *Genes Dev* 23:2426–2436.

Mekler, V., Kortkhonjia, E., Mukhopadhyay, J., et al. 2002. Structural organization of bacterial RNA polymerase holoenzyme and the RNA polymerase-promoter open complex. *Cell* 108:599–614.

Miroslavona, N. S., and Busby, S. J. W. 2006. Investigations of the molecular structure of bacterial promoters. *Biochem Soc Symp* 73:1–10.

Murakami, K., Masuda, S., Campbell, E. A., et al. 2002. Structural basis of transcription initiation: an RNA polymerase holoenzyme-DNA complex. *Science* 296:1285–1290.

Murakami, K. S., and Darst, S. A. 2003. Bacterial RNA polymerases: the wholo story. *Curr Opin Struct Biol* 13:31–39.

Murakami, K. S., Masuda, S., and Darst, S. A. 2002. Structural basis of transcription initiation: RNA polymerase holoenzyme at 4 Å resolution. *Science* 296:1280–1284.

Nickels, B. E., and Hochschild, A. 2004. Regulation of RNA polymerase through the secondary channel. *Cell* 118:281–284.

Polyakov, A., Severinova, E., and Darst, S. A. 1995. Three-dimensional structure of *E. coli* core RNA polymerase: promoter binding and elongation conformations of the enzyme. *Cell* 83:365–373.

Revyakin, A., Liu, C., Ebright, R. H., and Strick, T. R. 2006. Abortive initiation and productive initiation by RNA polymerase involve scrunching. *Science* 314:1139–1143.

Roberts, J. W. 2006. RNA polymerase, a scrunching machine. *Science* 314:1097–1098.

Ross, W., and Gourse, R. L. 2009. Analysis of RNA polymerase-promoter complex formation. *Methods* 47:13–24.

Schmitz, A., and Galas, D. J. 1979. The interaction of RNA polymerase and lac repressor with the lac control region. *Nucleic Acids Res* 6:111–137.

Typas, A., Becker, G., and Hengge, R. 2007. The molecular basis of selective promoter activation by the σ^S subunit of RNA polymerase. *Mol Microbiol* 63: 1296–1306.

Vassylyev, D. G., Sekine, S., Laptenko, O., et al. 2002. Crystal structure of a bacterial RNA polymerase holoenzyme at 2.6 Å resolution. *Nature* 417:712–719.

Wösten, M. M. S. M. 1997. Eubacterial sigma-factors. *FEMS Microbiol Rev* 22:127–150.

Woychik, N. A. M., and Reinberg, D. 2001. RNA polymerases: subunits and functional domains. *Encyclopedia of Life Sciences*, pp. 1–8. London: Nature.

Young, B. A., Gruber, T. M., and Gross, C. A. 2002. Views of transcription initiation. *Cell* 109:417–420.

Zhang, G., Campbell, E. A., Minakhin, L., et al. 1999. Crystal structure of *Thermus aquaticus* core RNA polymerase at 3.3 Å resolution. *Cell* 98:811–824.

Zhang, X., Chaney, M., Wigneshwaraj, S. R., et al. 2002. Mechanochemical ATPases and transcriptional activation. *Mol Microbiol* 45:895–903.

Transcription Elongation Complex

Artsimovitch, I., and Landick, R. 2000. Pausing by bacterial RNA polymerase is mediated by mechanistically distinct classes of signals. *Proc Natl Acad Sci USA* 97:7090–7095.

Bar-Nahum, G. Epstein, V., Ruckenstein, A. E., et al. 2005. A ratchet mechanism of transcription elongation and its control. *Cell* 120:183–193.

Borukhov, S., Lee, J., and Laptenko, O. 2005. Bacterial transcription elongation factors: new insights into molecular mechanism of action. *Mol Microbiol* 55:1315–1324.

Borukhov, S., and Nudler, E. 2007. RNA polymerase: the vehicle of transcription. *Trends Microbiol* 16:126–134.

Bureckner, F., Oritz, J., and Cramer, P. 2009. A movie of the RNA polymerase nucleotide addition cycle. *Curr Opin Struct Biol* 19:294–299.

Fish, R. N., and Kane, C. M. 2002. Promoting elongation with transcript cleavage stimulatory factors. *Biochim Biophys Acta* 1577:287–307.

Gelles, J., and Landick, R. 1998. RNA polymerase as a molecular motor. *Cell* 93:13–16.

Kiereeva, M., Zashlev, M., and Burton, Z. F. 2010. Translocation by multi-subunit RNA polymerases. *Biochim Biophys Acta* 1799:389–401.

Korzheva, N., Mustaev, A., Kozlov, M., et al. 2000. A structural model of transcription elongation. *Science* 289:619–625.

Herbert, K. M., Greenleaf, W. J., and Block, S. M. 2008. Single-molecule studies of RNA polymerase: motoring along. *Annu Rev Biochem* 77:149–176.

Kireeva, M., Kashlev, M., and Burton, Z. F. 2010. Translocation by multi-subunit RNA polymerases. *Biochim Biophys Acta* 1799:389–401.

Landick, R. 2001. RNA polymerase clamps down. *Cell* 105:567–570.

Marr, M. T., and Roberts, J. W. 2000. Function of transcription cleavage factors GreA and GreB at a regulatory pause site. *Mol Cell* 6:1275–1285.

Mooney, R. A., Artisimovitch, I., and Landick, R. 1998. Information processing by RNA polymerase: recognition of regulatory signals during RNA chain elongation. *J Bacteriol* 180:3265–3275.

Neuman, K. C., Abbondanzieri, E. A., Landick, R., et al. 2003. Ubiquitous transcriptional pausing is independent of RNA polymerase backtracking. *Cell* 115: 437–447.

Nudler, E. 2009. RNA polymerase active center: the molecular engine of transcription. *Annu Rev Biochem* 78:335–361.

Opalka, N., Chlenov, M., Chacon, P., et al. 2003. Structure and function of the transcription elongation factor GreB bound to bacterial RNA polymerase. *Cell* 114:335–345.

Proshkin, S., Rahmouni, A. R., Mironov, A., and Nudler, E. 2010. Cooperation between translating ribosomes and RNA polymerase in transcription elongation. *Science* 328:504–508.

Roberts, J. W., Shankar, S., and Filter, J. J. 2008. RNA polymerase elongation factors. *Annu Rev Biochem* 62:211–233.

Shaevitz, J. W., Abbondanzieri, E. A., Landick, R., and Block, S. M. 2003. Backtracking by single RNA polymerase molecules observed at near-base-pair resolution. *Nature* 426:684–687.

Sosunova, E., Sosunova, V., Kozlov, M., et al. 2003. Donation of catalytic residues to RNA polymerase active center by transcription factor Gre. *Proc Natl Acad Sci USA* 100:15469–15474.

Svetlov, V., and Nudler, E. 2009. Macromolecular micromovements: how RNA polymerase translocates. *Curr Opin Struct Biol* 19:1–7.

Sydow, J. F., and Cramer, P. 2009. RNA polymerase fidelity and transcriptional proofreading. *Curr Opin Struct Biol* 19:732–739.

Vassylyev, D. G. 2009. Elongation by RNA polymerase: a race through roadblocks. *Curr Opin Struct Biol* 19:1–10.

Vassylyev, D. G., Vassylyeva, M. N., Perederina, A., Tahirov, T. H., and Artsimovitch, I. 2007. Structural basis for transcription elongation by bacterial RNA polymerase. *Nature* 448:157–162.

Vassylyev, D. G., Vassylyeva, M. N., Zhang, J., Palangat, M., Artsimovitch, I., and Landick, R. 2007. Structural basis for substrate loading in bacterial RNA polymerase. *Nature* 448:163–168.

Zhang, J., Palangar, M., and Landick, R. 2010 Role of the RNA polymerase trigger loop in catalysis and pausing. *Nat Struct Mol Biol* 17:99–104.

Transcription Termination

Burmann, B. M., Schweimer, Luo, X, et al. 2010. A NusE:NusG complex links transcription and translation. *Science* 238:501–504.

Kaplan, D. L., and O'Donnell, M. 2003. Rho factor: transcription termination in four steps. *Curr Biol* 13:R714–R716.

Passman, Z., and von Hippel, P. H. 2000. Regulation of Rho-dependent transcription termination by NusG is specific to the *Escherichia coli* elongation complex. *Biochemistry* 39:5573–5585.

Richardson, J. P. 2001. Transcript elongation and termination. *Encyclopedia of Life Sciences*. pp. 1–7. London: Nature.

Richardson, J. P. 2002. Rho-dependent termination and ATPases in transcript termination. *Biochim Biophys Acta* 1577:251–260.

Skordalakes, E., and Berger, J. M. 2003. Structure of the Rho transcription terminator: mechanism of mRNA recognition and helicase loading. *Cell* 114:135–146.

Messenger RNA

Grunberg-Manago, M. 1999. Messenger RNA stability and its role in control of gene expression in bacteria and phages. *Annu Rev Genet* 33:193–227.

Lactose Operon

Adhya, S. 1996. The *Lac* and *Gal* operons today. E. C. C. Lin and S. Lynch (eds). *Regulation of Gene Expression in* Escherichia coli. Georgetown, TX: R. G. Landes.

Beckwith, J. 1996. The operon: an historical account. F. C. Neidhardt (ed). Escherichia coli *and* Salmonella typhimurium. pp. 1227–1231. Washington, DC: ASM Press.

Bell, C. E., and Lewis, M. 2000. A closer view of the conformation of the Lac repressor bound to operator. *Nat Struct Biol* 7:209–214.

Bell, C. E., and Lewis, M. 2001. The Lac repressor: a second generation of structural and functional studies. *Curr Opin Struct Biol* 11:19–25.

Borukhov, S., and Lee, J. 2005. RNA polymerase structure and function at *lac* operon. *Crit Rev Biol* 328:576–587.

Jacob, F., and Monod, J. 1961. Genetic regulatory mechanisms in the synthesis of proteins. *J Mol Biol* 3:318–356.

Kercher, M. A., Lu, P., and Lewis, M. 1997. *Lac* repressor–operator complex. *Curr Opin Struct Biol* 7:76–85.

Lewis, M. 2005. The *lac* repressor. *Crit Rev Biol* 328:521–548.

Lewis, M., Chang, G., Horton, N. C., et al. 1996. Crystal structure of the lactose operon repressor and its complexes with DNA and inducer. *Science* 271: 1247–1254.

Matthews, K. S., Falcon, C. M., and Swint-Kruse, L. 2000. Relieving repression. *Nat Struct Biol* 7:184–187.

Müller-Hill, B. 1998. Some repressors of bacterial transcription. *Curr Opin Microbiol* 1:145–151.

Ullmann, A. 2009. *Escherichia coli* lactose operon. *Encyclopedia of Life Sciences*. pp. 1–8. Hoboken, NJ: John Wiley and Sons.

Catabolite Repression

Busby, S., and Ebright, R. H. 1999. Transcription activation by catabolite activator protein (CAP). *J Mol Biol* 293:199–213.

Busby, S., and Kolb, A. 1996. The CAP modulon. E. C. C. Lin and S. Lynch (eds). *Regulation of Gene Expression in* Escherichia coli. Georgetown, TX: R. G. Landes Company.

Deutscher, J. 2008. The mechanisms of carbon catabolite repression in bacteria. *Curr Opin Microbiol* 11:87–93.

Deutscher, J., Francke, C., and Postma, P. W. 2006. How phosphotransferase system-related protein phosphorylation regulates carbohydrate metabolism in bacteria. *Microbiol Mol Biol Rev* 70:939–1031.

Harman, J. G. 2001. Allosteric regulation of the cAMP receptor protein. *Biochim Biophys Acta* 1547:1–17.

Kimata, K., Takahashi, H., Inada, T., et al. 1997. cAMP receptor protein plays a crucial role in glucose-lactose diauxie by activating the major glucose transporter gene in *Escherichia coli*. *Proc Natl Acad Sci USA* 94:2914–2919.

Lawson, C. L., Swigon, D., Murakami, K. S., et al. 2004. Catabolite activator protein: DNA binding and transcription activation. *Curr Opin Struct Biol* 14:1–11.

Lengeler, J. W. 1996. The phosphoenolpyruvate-dependent carbohydrate: phosphotransferase system (PTS) and control of carbon source utilization. E. C. C. Lin and S. Lynch (eds). *Regulation of Gene Expression in* Escherichia coli. Georgetown, TX: R. G. Landes.

Passner, J. M., and Steitz, T. A. 1997. The structure of a CAP-DNA complex having two cAMP molecules bound to each monomer. *Proc Natl Acad Sci USA* 94:2843–2847.

Rhodius, V. A., and Busby, S. J. W. 2000. Transcription activation by the *Escherichia coli* cyclic AMP receptor protein: determinants within activating region 3. *J Mol Biol* 299:295–310.

Stülke, J., and Hillen, W. 1999. Carbon catabolite repression in bacteria. *Curr Opin Microbiol* 2:195–201.

Tryptophan Operon

Fisher, R. F., and Yanofsky, C. 1983. Mutations of the β subunit of RNA polymerase alter both transcription pausing and transcription termination in the *trp* operon leader region *in vitro*. *J Biol Chem* 258:8146–8150.

Henkin, T. M., and Yanofsky, C. 2002. Regulation by transcription attenuation in bacteria: how RNA provides instructions for transcription termination/antitermination decisions. *Bioessays* 24:700–707.

Landick R, Turnbough, C. L. Jr, and Yanofsky C. 1996. Transcription attenuation. F. C. Neidhardt, et al. (eds). Escherichia coli *and* Salmonella: *Cellular and Molecular Biology*. pp. 1263–1286. Washington, DC: ASM Press.

Merino, E., Jensen, R. A., and Yanofsky, C. 2008. Evolution of bacterial *trp* operons and their regulation. *Curr Opin Microbiol* 11:78–86.

Merino, E., and Yanofsky, C. 2005. Transcription attenuation: a highly conserved regulatory strategy used by bacteria. *Trends Genet* 21:260–264.

Oxender, D. L., Zurawski, G., and Yanofsky, C. 1979. Attenuation in the *Escherichia coli* tryptophan operon: role of RNA secondary structure involving the tryptophan codon region. *Proc Natl Acad Sci USA* 76:5524–5528.

Yanofsky, C. 2000. Transcription attenuation: once viewed as a novel regulatory strategy. *J Bacteriol* 182:1–8.

Yanofsky, C. 2003. Using studies on tryptophan metabolism to answer basic biological questions. *J Biol Chem* 278:10859–10878.

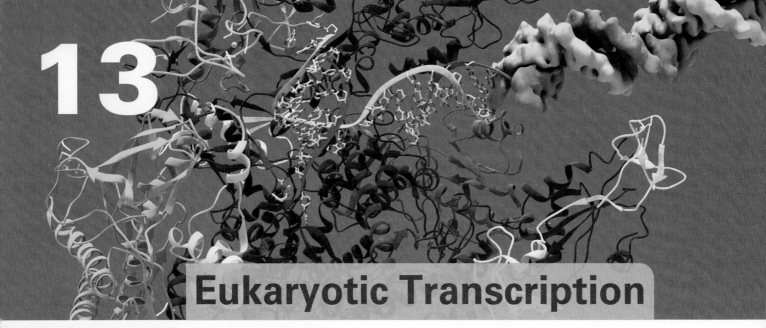

13 Eukaryotic Transcription

CHAPTER OUTLINE

13.1 Introduction to Eukaryotic Nuclear RNA Polymerases
- The eukaryotic cell nucleus has three different kinds of RNA polymerase.
- RNA polymerases I, II, and III can be distinguished by their sensitivities to inhibitors.
- Each nuclear RNA polymerase has some subunits that are unique to it and some it shares with one or both of the two other nuclear RNA polymerases.

13.2 RNA Polymerase II Structure
- Yeast RNA polymerase II crystal structures help explain how the enzyme works.
- The crystal structure has been determined for the complete 12-subunit yeast RNA polymerase II bound to a transcription bubble and product RNA.
- Nuclear RNA polymerases have limited synthetic capacities.

13.3 Core Promoter for Protein-Coding Genes
- The core promoter for protein-coding genes extends from 40 bp upstream of the transcription start site to 40 bp downstream from this site.

13.4 General Transcription Factors: Basal Transcription
- RNA polymerase II requires the assistance of general transcription factors to transcribe naked DNA from specific transcription start sites.
- TFIID or its TBP subunit must bind to a TATA core promoter before other general transcription factors can do so.
- General transcription factors and RNA polymerase interact at the promoter to form a preinitiation complex.

13.5 Transcription Elongation
- The C-terminal domain of the largest RNA polymerase subunit must be phosphorylated for chain elongation to proceed.
- A variety of transcription elongation factors helps to suppress transient pausing during elongation.
- Elongation factor TFIIS reactivates arrested RNA polymerase II.

13.6 Regulatory Promoters, Enhancers, and Silencers
- Linker-scanning mutagenesis reveals the regulatory promoter's presence just upstream from the core promoter.
- Enhancers stimulate transcription and silencers block transcription.
- The upstream activating sequence regulates genes in yeast.

13.7 Introduction to Transcription Activator Proteins for Protein-Coding Genes
- Transcription activator proteins help to recruit the transcription machinery.
- A combinatorial process determines gene activity.
- DNA affinity chromatography can be used to purify transcription activator proteins.
- A transcription activator protein's ability to stimulate gene transcription can be determined by a transfection assay.

13.8 DNA-Binding Domains in Transcription Activator Proteins
- Transcription activator proteins are commonly grouped according to the structures of their DNA-binding domains.

13.9 Activation Domains in Transcription Activator Proteins
- Activation domains tend to be intrinsically disordered.
- Gal4 has DNA-binding, dimerization, and activation domains.
- Other proteins act along with Gal4 to regulate *GAL* genes.
- The activation domain must associate with a DNA-binding domain to stimulate transcription.

13.10 Mediator
- Squelching occurs when transcription activator proteins compete for a limiting transcription machinery component.
- Mediator is required for activated transcription.
- The yeast Mediator complex associates with activators at the UAS in active yeast genes.

Image © Ramon Andrade/Photo Researchers, Inc.

13.11 Epigenetic Modifications

Cells remodel or modify chromatin to make the DNA in chromatin accessible to the transcription machinery.

Histone modification influences transcription of protein-coding genes.

DNA methylation plays an important role in determining whether chromatin will be silenced or actively expressed in vertebrates.

Epigenetics is the study of inherited changes in phenotype caused by changes in chromatin other than changes in DNA sequence.

13.12 RNA Polymerase I Catalyzed Transcription

RNA polymerase I is required to synthesize 5.8S, 18S, and 28S rRNA.

The rRNA transcription unit promoter consists of a core promoter and an upstream promoter element (UPE).

5.8S, 18S, and 23S rRNA syntheses take place in the nucleolus.

RNA polymerase I is a multisubunit enzyme with a structure similar to that of RNA polymerase II.

The upstream binding factor and selectivity factor, working together, recruit RNA polymerase I to the rDNA promoter to form a preinitiation complex.

RNA polymerase I forms a transcription elongation complex, leaving UBF and SL1/TIF-1B behind.

RNA polymerase I transcription termination requires the assistance of a termination factor and a release protein.

13.13 RNA Polymerase III Catalyzed Transcription

RNA polymerase III transcripts are short RNA molecules with a variety of biological functions.

RNA polymerase III transcription units have three different types of promoters.

RNA polymerase III does not appear to require additional factors for transcription elongation or termination.

BOX 13.1: LOOKING DEEPER The Archaeal RNA Polymerase

BOX 13.2: LOOKING DEEPER TFIIB Assists in Open Promoter Formation

BOX 13.3: CLINICAL APPLICATION The AP-1 Family of Transcription Activator Proteins

BOX 13.4: IN THE LAB Yeast Two-Hybrid Assay

BOX 13.5: LOOKING DEEPER Genomic Imprinting

QUESTIONS AND PROBLEMS

SUGGESTED READING

At the most fundamental chemical level, eukaryotic and bacterial RNA syntheses are similar. Both processes begin when two ribonucleoside triphosphates, which are lined up on a DNA template strand, join to form a dinucleoside tetraphosphate with a hydroxyl group at its 3′-end and a triphosphate group at its 5′-end. Both processes continue by extending the dinucleotide chain in a 5′→3′ direction when ribonucleotides are added to the 3′-end of the growing chain in the order specified by the DNA template until a termination signal is reached.

Based on these chemical similarities one might expect bacterial and eukaryotic transcription machinery to have similar catalytic sites. Bacterial and eukaryotic core RNA polymerases do in fact share important structural and functional features. Many important differences also exist between the bacterial and eukaryotic transcription machinery. These differences are as follows:

1. Bacteria use a single RNA polymerase to synthesize ribosomal RNA (rRNA), messenger RNA (mRNA), and transfer RNA (tRNA). In contrast, eukaryotes use a specific dedicated nuclear enzyme to synthesize each kind of RNA.
2. Bacterial RNA polymerase requires the assistance of at most one or two accessory factors to transcribe genes. Eukaryotic RNA polymerases require several such factors.
3. Bacterial RNA polymerase holoenzyme has direct access to its DNA template, whereas the eukaryotic transcription machinery has difficulty in reaching its DNA template because eukaryotic DNA interacts with histones to form nucleosomes,

which in turn form more compact chromatin structures. An important consequence of chromatin structure is that eukaryotic genes tend to be turned off in the absence of regulatory proteins. In contrast, bacterial genes tend to be turned on or require only one or two regulatory proteins such as cAMP receptor protein to become fully active. Additional factors are required to move and modify histone octamers that block access to eukaryotic genes before RNA synthesis can occur. Hence, chromatin structure introduces a level of transcriptional complexity in eukaryotes that does not exist in bacteria.

We begin this chapter by describing early experiments demonstrating eukaryotic cells have three different nuclear RNA polymerases and then explore the contribution each of these polymerases makes to eukaryotic RNA synthesis. The best studied of these enzymes is RNA polymerase II. For this reason, and because RNA polymerase II transcribes protein-coding genes, we examine its structure and function before examining the other two kinds of nuclear RNA polymerases, RNA polymerases I and III.

13.1 Introduction to Eukaryotic Nuclear RNA Polymerases

The eukaryotic cell nucleus has three different kinds of RNA polymerase.

Samuel B. Weiss and Leonard Gladstone provided the first clear evidence for the existence of eukaryotic DNA-dependent RNA polymerase in 1959, about 1 year before the enzyme was detected in bacteria. Their studies showed disrupted rat liver cell nuclei convert radioactive ribonucleoside triphosphates into an acid-insoluble, RNase-sensitive product with a base composition similar to rat liver DNA. Moreover, this conversion required the presence of all four ribonucleoside triphosphates and did not occur in the presence of DNase. Further progress was hindered by the inability to extract the RNA polymerase activity from disrupted nuclei. Several different factors contributed to this failure. One of the most important—the presence of inhibitory proteins—was eventually minimized by adjusting the extraction buffer's ionic strength so RNA polymerase was extracted preferentially.

During the 1960s the research groups of John J. Furth, Jamshed R. Tata, and Pierre M. Chambon each studied distinct RNA polymerase activities, never realizing the true tripartite nature of eukaryotic RNA synthesis. Although there were fleeting suggestions that eukaryotes might have multiple RNA polymerases, this was considered to be quite unlikely because *Escherichia coli* has only a single kind of core RNA polymerase. It was not until the late 1960s that a graduate student, Robert G. Roeder, hypothesized that there were three distinct RNA polymerase activities. He proposed a series of experiments to prove his hypothesis both to himself and his doubting mentor, William J. Rutter.

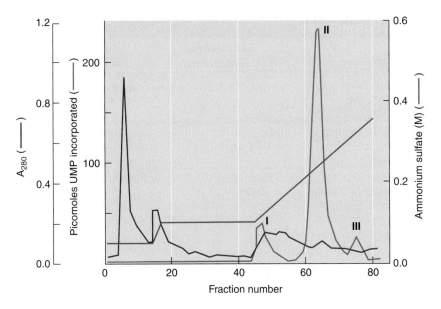

FIGURE 13.1 Separation of RNA polymerase activities by DEAE-Sephadex chromatography. Robert G. Roeder and William J. Rutter separated RNA polymerases from sea urchin embryos by DEAE-Sephadex chromatography. Red, protein measured by absorbance at 280 nm; green, RNA polymerase activity measured by incorporation of radioactively labeled UMP; purple, ammonium sulfate concentration used in elution buffer. (Adapted from Roeder, R. G., and Rutter, W. J. 1970. *Proc Natl Acad Sci USA* 65:675–682.)

A major breakthrough occurred in 1970 when Roeder and Rutter used DEAE-Sephadex chromatography to fractionate proteins in the nuclear extract (FIGURE 13.1). This fractionation technique takes advantage of the fact that negative charges on the protein surface interact with positive charges on the DEAE-Sephadex beads so a protein's ability to bind to the beads tends to increase with its net negative charge. Roeder and Rutter detected three separate peaks of RNA polymerase activity when they applied sea urchin embryo or rat liver nuclear extracts to DEAE-Sephadex and then released the bound protein with a buffer solution containing increasing ammonium sulfate concentrations. The activities, which were named RNA polymerase I, RNA polymerase II, and RNA polymerase III in the order of their release from the column, subsequently were shown to be three distinct enzymes. Once these experiments were completed, it became clear that Tata had been studying RNA polymerase I, Chambon RNA polymerase II, and Furth RNA polymerase III.

Recently, Craig S. Pickard and coworkers discovered two new nuclear RNA polymerases, RNA polymerases IV and V, are present in land plants and some algae. Some of the subunits in the new RNA polymerases are identical to those in RNA polymerase II and others are homologous to those in RNA polymerase II. RNA polymerases IV and V appear to synthesize small interfering RNA, referred to as siRNA, molecules used for siRNA-mediated gene silencing. Because RNA polymerases IV and V have not yet been subject to *in vitro* studies, they are not discussed further.

For convenience we use the term "nuclear RNA polymerases" to distinguish RNA polymerases I, II, and III, the three RNA polymerases present in all eukaryotic cell nuclei from RNA polymerases that are present in the mitochondria of all eukaryotes and in the chloroplasts of plants. The three nuclear RNA polymerases have different physiological roles. Some of these roles are summarized in TABLE 13.1.

13.1 Introduction to Eukaryotic Nuclear RNA Polymerases

TABLE 13.1 Comparing the Three Eukaryotic RNA Polymerases

Enzyme	Location	RNA Products	Sensitivity to α-amanitin	Sensitivity to actinomycin D
RNA polymerase I	Nucleolus	Pre-rRNA (leading to 5.8S, 18S, and 28S rRNA)	Resistant	Very sensitive
RNA polymerase II	Nucleoplasm	Pre-mRNA and some small nuclear RNAs (snRNAs)	50% inhibition at 0.02 μg/mL	Slightly sensitive
RNA polymerase III	Nucleoplasm	tRNA, 5S rRNA, U6 snRNA (spliceosomal RNA), and 7SL RNA (signal recognition particle RNA)	50% inhibition at 20 μg/mL	Slightly sensitive

RNA polymerases I, II, and III can be distinguished by their sensitivities to inhibitors.

One way in which the three RNA polymerases can be distinguished from one another is their sensitivity to α-amanitin (Table 13-1), a toxic octapeptide produced by *Amanita phalloides*, a poison mushroom known as the *death cap* (FIGURE 13.2). When added at a concentration as low as 0.02 μg · mL^{-1}, α-amanitin causes 50% inhibition of elongation by RNA polymerase II. A thousand times greater concentration is required to cause equivalent inhibition of RNA polymerase III. The toxic octapeptide has no effect on RNA polymerase I. The sensitivities of the three RNA polymerases can be exploited to block the synthesis of specific types of RNA in the living cell. Thus, very low α-amanitin concentrations block mRNA synthesis while permitting the continued synthesis of most other kinds of RNA. Higher concentrations also block the synthesis of 5S rRNA, tRNA, and most other small RNA molecules while permitting continued synthesis of other rRNA.

The transcription reactions catalyzed by the three RNA polymerases also exhibit differential sensitivity to actinomycin D. In this case, however, the reaction catalyzed by RNA polymerase I is the

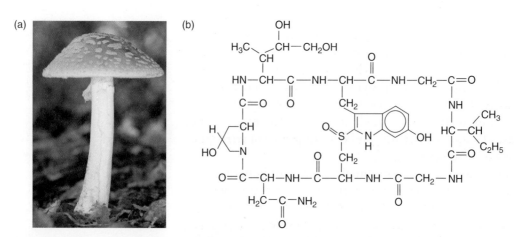

FIGURE 13.2 α-Amanitin. (a) *Amanita phalloides* (also known as the Death Cap). (b) Structural formula of α-amanitin. (Part a © Niels-DK/Alamy Images. Part b adapted from Defendenti, C., et al. 1998. *Forensic Sci Int* 92:59–68.)

most sensitive. In contrast to α-amanitin, which binds to the enzyme, actinomycin D (FIGURE 13.3a) binds to the double-stranded DNA template. The heterocycle ring system in actinomycin D inserts itself between C-G and G-C base pairs, preventing the DNA strand separation required for transcription and replication (Figure 13.3b). RNA polymerase I catalyzed reactions tend to be the most sensitive to actinomycin D because the rRNA genes it transcribes are GC-rich.

Each nuclear RNA polymerase has some subunits that are unique to it and some it shares with one or both of the two other nuclear RNA polymerases.

RNA polymerases I, II, and III in *Saccharomyces cerevisiae* have 14, 12, and 17 polypeptide subunits, respectively, making these enzymes considerably more complex and difficult to study than their bacterial counterpart. Genes that code for each of yeast's 12 RNA polymerase II subunits, designated *RPB1* to *RPB12*, have been identified and cloned. (The *RPB* gene designation is a holdover from an alternate system for RNA polymerase nomenclature in which the letters A, B, and C were used in place of Roman numerals I, II, and III. Genes that code for subunits in RNA polymerases I and III are designated *RPA* and *RPC*, respectively, followed by a number to indicate the specific subunit.) Remarkably, at least 10 of the human RNA polymerase II genes can be substituted for their counterparts in yeast. The significance of this result is that information obtained by studying RNA polymerase II in yeast applies to RNA polymerase II obtained from other eukaryotes, including humans.

Although the large number of subunits present in each of the nuclear RNA polymerases presents serious challenges to our understanding of how each enzyme works, homology studies indicate the situation is simpler than it at first seems. To begin, five subunits in yeast RNA polymerase II (Rpb5, Rpb6, Rpb8, Rpb10, and Rpb12) are also in yeast RNA polymerases I and III. Information gained by studying the functions of each of these five subunits in one nuclear RNA polymerase will undoubtedly provide information about their functions in the other two. Furthermore, this information tells us that one or more of the other subunits must be responsible for the specificity differences that exist among the nuclear RNA polymerases. Homologies not only exist among nuclear RNA polymerase subunits but also extend to bacterial and archaeal RNA polymerases (TABLE 13.2). The archaeal RNA polymerase is especially noteworthy because its subunit number is similar to that of RNA polymerase II. Natural selection does not therefore appear to have favored extra subunits in nuclear RNA polymerases to solve some problem related to the polymerase's presence in the cell nucleus.

Although the archaeal RNA polymerase bears a closer structural relationship to nuclear RNA polymerases than to bacterial RNA polymerase, the two prokaryotic RNA polymerases share one important functional feature: both enzymes synthesize mRNA, rRNA, and tRNA. Homology studies reveal that each bacterial core RNA polymerase subunit (β, β', α, and ω) has at least one homolog in each nuclear RNA polymerase. For example, comparison of the bacterial subunits with those in RNA

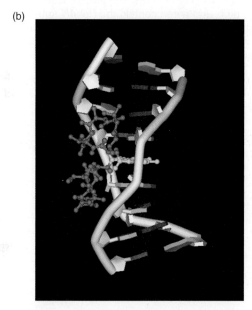

FIGURE 13.3 **Actinomycin D.** (a) The ring system (green) in actinomycin D binds between C-G and G-C base pairs in double-stranded DNA molecules. (b) DNA (5'-D[GpApApGpCpTpTpC]-3') complexed with actinomycin D. The ring system of actinomycin D that intercalates between C-G and G-C base pairs in DNA is shown in yellow. The rest of the actinomycin D molecule is shown in orange. Actinomycin D inhibits transcription by preventing strand separation. (Protein Data Bank ID: 2D55. Shinomiya, M., et al. 1995. *Biochemistry* 34:8481–8491. Prepared by B. E. Tropp.)

13.1 Introduction to Eukaryotic Nuclear RNA Polymerases

Pol I, Pol II, and Pol III are RNA polymerases I, II, and III, respectively. The subunits in each row are either homologous or identical to one another. The five identical subunits in the three eukaryotic nuclear RNA polymerases are shown in red. The two identical subunits in Pol I and Pol III are shaded in blue. When subunits in the same row are not identical, they are homologous. For example, subunits Rpa135, Rpb2, Rpc128, B, and β are homologous. The archaeal RNA polymerase requires two subunits, A and A′, to replace Rpb1. It has been recommended that the standard archaeal nomenclature system (shown in parentheses) be replaced by that shown, which is consistent with the eukaryotic nomenclature system. The area shaded in yellow contains the ten subunits that make up the core of the eukaryotic nuclear RNA polymerases. The areas shaded in blue, pink, and green contain the Rpb4/7 subcomplex, the transcription factor IIF (TF IIF)-like complex, and the RNA polymerase III subcomplex, respectively.

TABLE 13.2 RNA Polymerase Subunits

Eukaryotic			Archaeal		E. coli
Pol I	Pol II	Pol III			
Rpa190	Rpb1	Rpc160	Rpo 1N / Rpo 1C	(A′) / (A″)	β′
Rpa135	Rpb2	Rpc128	Rpo2	(B)	β
Rpc40	Rpb3	Rpc40	Rpo3	(D)	α
Rpc19	Rpb11	Rpc19	Rpo11	(L)	α
Rpb6	Rpb6	Rpb6	Rpo6	(K)	ω
Rpb5	Rpb5	Rpb5	Rpo5	(H)	
Rpb8	Rpb8	Rpb8	Rpo8	(G)	
Rpb10	Rpb10	Rpb10	Rpo10	(N)	
Rpb12	Rpb12	Rpb12	Rpo12	(P)	
Rpa12.2	Rpb9	Rpc11			
Rpa14	Rpb4	Rpc17	Rpo4	(F)	
Rpa43	Rpb7	Rpc25	Rpo7	(E)	
Rpa49		Rpc37			
Rpa34.5		Rpc53			
		Rpc82			
		Rpc34			
		Rpc31			
			Rpo13		
14 subunits	12 subunits	17 subunits	13 subunits		5 subunits

polymerase II reveals β′ is homologous to Rpb1, β to Rpb2, α to Rpb3 and Rpb11, and ω to Rpb6 (Table 13.2). It therefore seems reasonable to conclude that the conserved amino acid sequences have especially important roles in RNA synthesis and that a structural relationship exists between the bacterial and nuclear enzymes. Comparisons of crystal structures described in the next section show this is indeed true.

13.2 RNA Polymerase II Structure

Yeast RNA polymerase II crystal structures help explain how the enzyme works.

The large size and complex structure of RNA polymerase II presented a major challenge for investigators attempting to determine its structure. In 2005 two groups, one led by Roger Kornberg and the other by Patrick Cramer, reported crystal structures for the 12-subunit yeast RNA polymerase II. FIGURE 13.4 shows the structure reported by the Cramer laboratory. As predicted from amino acid sequence homologies, yeast RNA polymerase II bears a striking resemblance to *Thermus aquaticus* RNA polymerase (FIGURE 13.5). The similarities are greatest in the region surrounding the cleft, where the active site is located, and in regions that interact with the nascent RNA. As expected for a region that interacts with nucleic acids, a relatively high proportion of the amino acids that form the cleft in both the bacterial and yeast

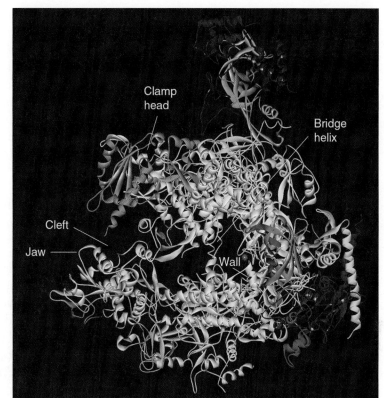

FIGURE 13.4 Ribbon representation of the complete 12-subunit yeast RNA polymerase II structure. Colors for specific subunits are indicated in the color code diagram. Numbers indicate the Rpb subunits. Some specific features of RNA polymerase II are indicated. The carboxyl terminal domain (CTD) of Rpb1 is largely unstructured and so is not visible. The active site magnesium is shown as a magenta sphere. (Structure from Protein Data Bank 1WCM. Armache, K.-J., et al. 2005. *J Biol Chem* 280:7131–7134. Prepared by B. E. Tropp.)

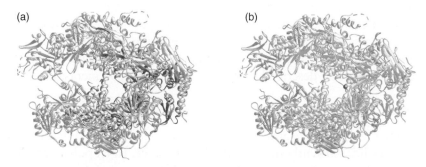

FIGURE 13.5 A conserved RNA polymerase core structure. (a) Blocks of sequence homology between the two largest subunits of bacterial and eukaryotic RNA polymerases are in red. (b) Regions of structural homology between RNA polymerase II and bacterial RNA polymerase, as judged by a corresponding course of the polypeptide backbone, are in green. (Reproduced from Cramer, P., Bushnell, D. A., and Kornberg, R. D. 2001. *Science* 292:1863–1876. Reprinted with permission from AAAS. Photos courtesy of Roger D. Kornberg, Stanford University School of Medicine.)

13.2 RNA Polymerase II Structure

RNA polymerases have positively charged side chains. The inner core region of RNA polymerase II shares many structural features with bacterial RNA polymerases, including the cleft, active site, bridge helix, trigger loop, rudder, lid, and secondary channel.

Crystal structures are not yet available for RNA polymerases I and III. Nevertheless, various lines of chemical and physical evidence indicate all three nuclear RNA polymerases have similar structures. The crystal structure for the archaeal RNA polymerase is similar to that for RNA polymerase II (**BOX 13.1 LOOKING DEEPER: THE ARCHAEAL RNA POLYMERASE**).

The crystal structure has been determined for the complete 12-subunit yeast RNA polymerase II bound to a transcription bubble and product RNA.

Investigators realized they might be able to obtain a "snapshot" of the transcription elongation complex if they could prepare the complex in crystal form. Cramer and coworkers approached the problem by assembling a complex that contained the 12-subunit RNA polymerase II bound to a transcription bubble and product RNA (**FIGURE 13.6a**). The crystal structure and a schematic cut-away view of the 12-subunit RNA polymerase II elongation complex are shown in Figures 13.6b and c, respectively. Protein "jaws" grip the downstream DNA that enters the enzyme. The DNA extends along the cleft toward the catalytic site and starts to unwind at position +3, where +1 corresponds to the position of the template base for the incoming nucleotide. Nucleoside triphosphates reach the active site by passing through a funnel-shaped secondary channel on the underside of the enzyme and then through a pore near the active site. The 3'-end of the growing RNA chain lies above the pore. The active site resembles that present in bacterial RNA polymerase. The DNA–RNA hybrid, consisting of 7 to 8 base pair (bp) that emerges from the active site, is forced to change direction and turn by about 90 degrees with respect to the downstream DNA, because the "wall" at the end of the cleft blocks straight passage through the enzyme. The enzyme helps to separate the DNA–RNA strands. The nucleotide addition cycle for RNA polymerase II is the same as that described for the bacterial RNA polymerases.

Nuclear RNA polymerases have limited synthetic capacities.

Despite their large size and complexity, the nuclear RNA polymerases have very limited synthetic capacity. For instance, RNA polymerase II, acting on its own, can initiate a low level of transcription from random sites when the DNA template has a nick, a single-stranded gap, or a 3'-overhang. It cannot, however, initiate transcription from specific start sites within an intact double-stranded DNA template without the assistance of other proteins. RNA polymerases I and III also require assistance from additional protein factors to catalyze specific transcription. Moreover, each of the nuclear RNA polymerases requires its own specific set of transcription factors to assist it in locating the transcription start site and making RNA. Because the details of the transcription process differ for each of the nuclear RNA polymerases, it is necessary to examine each one separately.

BOX 13.1: LOOKING DEEPER

The Archaeal RNA Polymerase

Archaeal RNA polymerase has a subunit composition and molecular architecture that is very similar to RNA polymerase II. **FIGURE B13.1** shows the crystal structure for the RNA polymerase from *Sulfolobus shibatae*, a hyperthermophilic archaeon that was first identified living in acidic geothermal hot springs. The archaeal RNA polymerase crystal structure bears a remarkable resemblance to the yeast RNA polymerase II crystal structure shown in Figure 13.4. The subunits in the two structures are very similar with three exceptions: (1) the largest archaeal RNA polymerase subunit, Rpb1, is divided into two subunits, Rpo1N and Rpo1C; (2) the archaeal RNA polymerase does not have a subunit equivalent to Rpb9; and (3) the archaeal RNA polymerase has an Rpo13 subunit that is not present in RNA polymerase II. Until recently, each archaeal RNA polymerase subunit was designated by an upper case letter (see Table 13.3). Nicola G. A. Abrescia and coworkers proposed replacing this nomenclature system with one that calls attention to the similarities between archaeal RNA polymerase and RNA polymerase II. According to their proposal, the name of each archaeal RNA polymerase subunit begins with the prefix "Rpo" followed by a number. The number is assigned so that each archaeal RNA polymerase subunit has the same number as its RNA polymerase II counterpart (Table 13.3).

In 2002 Finn Werner and Robert O. J. Weinzierl used recombinant DNA technology to insert genes for RNA polymerase subunits from *Methanococcus jannaschii*, a hyperthermophilic archaeon, into plasmids that replicate in *E. coli*. Then they purified the subunits from the different clones and mixed the subunits together under *in vitro* conditions to assemble a functional archaeal RNA polymerase. The success of this reconstitution experiment allowed Weinzierl and coworkers to examine structure–function relationships for archaeal RNA polymerase in a way not possible with RNA polymerase II because all attempts to assemble RNA polymerase II under *in vitro* conditions have thus far been unsuccessful. Two factors may have contributed to the successful *in vitro* archaeal RNA polymerase assembly. First, the archaeal RNA polymerase is derived from a hyperthermophilic organism, which may lead to greater enzyme stability.

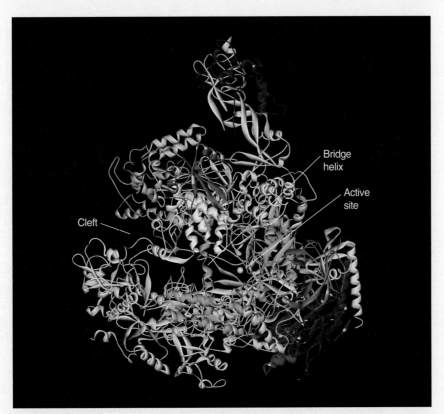
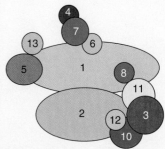

FIGURE B13.1 Crystal structure for the RNA polymerase from *Sulfolobus shibatae*, a hyperthermophilic archaeon. Colors of specific Rpo subunits are indicated in the color code diagram. Rpo1N and Rpo1C are shown in light and dark gray, respectively. (Structure from Protein Data Bank 2WAQ. Korkhin, Y., et al. 2009. *PLoS Biol* 7:e1000102. Prepared by B. E. Tropp.)

(*continues*)

BOX 13.1: LOOKING DEEPER

The Archaeal RNA Polymerase (*continued*)

Second, the two largest RNA polymerase II subunits are each split into two subunits in *M. jannaschii* RNA polymerase, which may make it easier for the enzyme to assemble. Note the Rpo2 subunit from *M. jannaschii* RNA polymerase differs in this respect from the Rpo2 subunit in the *S. shibatae* RNA polymerase. For convenience we refer to *M. jannaschii* RNA polymerase II's Rpo1N and Rpo1C subunits as Rpo1 and its two Rpo2 subunits as Rpo2. Rpo3, Rpo10, Rpo11, and Rpo12 form an assembly platform for Rpo1 and Rpo2. Rpo3 and Rpo11 are homologous to the bacterial α subunits, which form the α dimer assembly platform for the β and β′ subunits. The assembly platform (Rpo3, Rpo10, Rpo11, and Rpo12) and catalytic subunits (Rpo1 and Rpo2) form the minimal subunit complex necessary and sufficient for promoter-directed transcription. The remaining subunits may contribute to archaeal RNA polymerase stability, regulation, or both. Weinzierl and coworkers adapted their reconstitution assembly technique to devise an automated, high-throughput method for producing and characterizing RNA polymerase variants. This technique permits them to create 19 amino acid substitutions at any single site of interest and study the effect that each substitution has on enzyme activity. They can also assemble RNA polymerase molecules with desired mutations in two or more subunits. This important new technique has provided considerable information about the way that the bridge helix and trigger loop interact during RNA synthesis.

The archaeal RNA polymerase offers one additional research advantage when compared with RNA polymerase II. It requires just three general transcription factors (see Section 13.4 General Transcription Factors: Basal Transcription), TBP, TFB (homologous to the eukaryotic TFIIB), and TFE (homologous to the N-terminal region of the largest eukaryotic TFIIE subunit). An understanding of the archaeal transcription system undoubtedly will lead to a better understanding of the eukaryotic transcription system.

13.3 Core Promoter for Protein-Coding Genes

The core promoter for protein-coding genes extends from 40 bp upstream of the transcription start site to 40 bp downstream from this site.

Investigators assumed that eukaryotic promoters, like their bacterial counterparts, would be just upstream from the transcription start site. Once methods became available for transcription start site identification, investigators started to search for RNA polymerase II promoters just upstream of transcription start sites. Initial efforts to characterize the eukaryotic promoter concentrated on highly expressed protein-coding genes, such as those that code for hemoglobin, histone, and ovalbumin, because these genes were the easiest to study at the time. These efforts were rewarded in 1977 when David Hogness and coworkers discovered a consensus sequence, TATAXAX (where X is an A or T), located 25 to 30 bp upstream from the transcription start site in histone genes. As additional highly expressed protein-coding genes from animal cells and viruses became available for study, they too were shown to have this sequence, now called the **TATA box**, 25 to 30 bp upstream from their transcription start sites. The TATA box also was observed just upstream from the transcription start site in protein-coding genes from plants and fungi. Its position in *S. cerevisiae*, however, was observed to vary from 30 to 120 bp upstream from the transcription start site.

The common occurrence of the TATA box suggested it might be essential for the transcription of all protein-coding genes. But as less

(a)

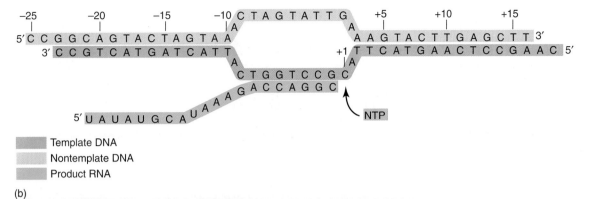

- Template DNA
- Nontemplate DNA
- Product RNA

(b)

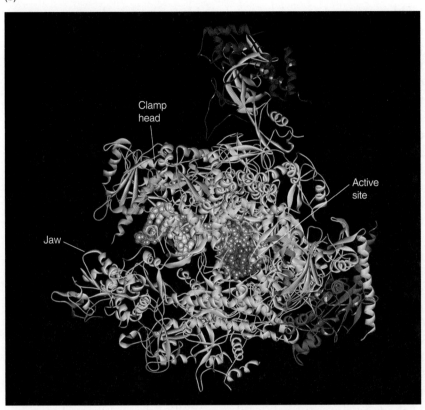

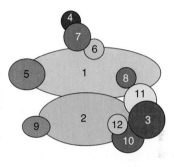

FIGURE 13.6 The yeast RNA polymerase II transcription elongation complex.
(a) Schematic diagram of the transcription bubble and product mRNA. The bubble, formed by a 41-mer DNA duplex with an 11 nucleotide mismatch region, has an RNA 20-mer with eight 3′-terminal nucleotides complementary to the DNA template in the bubble. (b) A ribbon model of the RNA polymerase II subunits and nucleic acids in the transcription elongation complex. Colors of specific subunits are indicated in the color code diagram above. Numbers indicate the Rpb subunits. Nucleic acids are shown as spacefill structures. Template DNA, nontemplate DNA, and product RNA are shown as dark blue, light blue, and green spacefill structures, respectively. The incoming nucleoside triphosphate is shown as an orange spacefill structure and the magnesium ion at the active site is shown as a pink sphere. (c) A schematic cut-away view of the 12 subunit RNA polymerase II transcription elongation complex. The view is the same as that shown in the ribbon model in (b). The DNA template and nontemplate strands are shown in dark blue and light blue, respectively. The nascent RNA chain is shown in green. Dashed lines indicate uncertainty about exact strand position. Nucleotide triphosphates, NTPs, enter the active site by first passing through the secondary channel and then through a pore. The magnesium ion shown as a magenta sphere indicates the active site's location. (Part a adapted from Armache, K. J., Kettenberger, H., and Cramer, P. 2005. *Curr Opin Struct Biol* 15:197–203. Part b structure from Protein Data Bank 1Y77. Kettenberger, H., Armache, K.-J., and Cramer, P. 2004. *Mol Cell* 16:955–965. Prepared by B. E. Tropp. Part c adapted from Hahn, S. 2004. *Nat Struct Mol Biol* 11:394–403.)

(c)

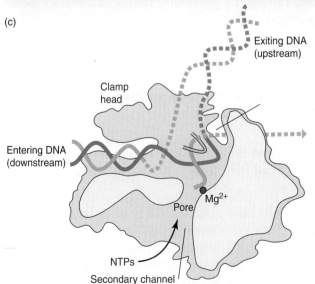

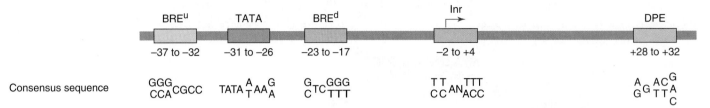

FIGURE 13.7 Core promoter elements that contribute to basal transcription in multicellular animals. (Adapted from Maston, G. A., et al. 2006. *Annu Rev Genom Hum Genet* 7:29–59.)

highly expressed genes became available for study, it became apparent that most promoters *lack* a TATA box. The discovery of TATA-less promoters suggested other short DNA sequences might be able to replace the TATA box function in transcription initiation. Several elements have been identified that either replace the TATA box in transcription initiation or act along with it during transcription initiation. The DNA region that includes these elements, the **core promoter**, extends from about 40 bp upstream of the transcription start site to about 40 bp downstream from this site. FIGURE 13.7 shows the TATA box and some other elements present in core promoters from multicellular animals. The **initiator (Inr) element**, which is probably the most commonly occurring element in the core promoter, flanks the transcription start site. The **downstream promoter element (DPE)** is downstream of Inr. Two other elements, **TFIIB recognition elements (BRE) BRE^u** and **BRE^d**, flank the TATA box. The superscript "u" indicates BRE^u is upstream of the TATA box, whereas the superscript "d" indicates BRE^d is downstream of the TATA box. Approximately 10,000 known human core promoters were surveyed for the TATA, Inr, DPE, and BRE elements. Inr, the most common element, was present in nearly half of the core promoters, whereas DPE and BRE were each present in about a quarter of the core promoters, and TATA boxes were present in about one-eighth of the core promoters. Remarkably, almost a quarter of the analyzed promoters had none of the four elements, raising the possibility that additional elements remain to be discovered.

13.4 General Transcription Factors: Basal Transcription

RNA polymerase II requires the assistance of general transcription factors to transcribe naked DNA from specific transcription start sites.

RNA polymerase II requires the assistance of protein factors to bind to the core promoter. This requirement was first demonstrated by studying specific initiation at the major late promoter of adenovirus DNA, which controls highly expressed genes for structural proteins in the virus particle. Purified RNA polymerase II cannot catalyze specific initiation at this promoter but gains the ability to do so when a soluble cell-free extract from human KB cells (cells derived from an oral epidermoid carcinoma) is added to it. In 1979 Robert G. Roeder and coworkers used classical protein fractionation techniques to isolate protein factors from the KB cell extract that assist RNA polymerase II. These protein factors, or **general transcription factors** as they are now known, were named TFIIA, TFIIB, TFIID, TFIIE, TFIIF, and TFIIH. The first two letters, TF, indicate the protein is a general transcription factor; the Roman numeral II signifies the factor supports RNA polymerase II transcription; and the final letter was assigned based on the protein fractionation scheme rather than on protein function. (The letters C and G are missing because later studies showed the proteins originally assigned these letters are not transcription factors for RNA polymerase II.) Subsequent studies by Roeder and other investigators demonstrated that these general transcription factors are present in all eukaryotes from yeast to humans. TABLE 13.3 summarizes the functions of yeast general transcription factors. Counterparts in other eukaryotes serve the same functions. The general transcription factors are described below in greater detail.

RNA polymerase II, acting together with general transcription factors, comprises the minimum transcription machinery required for correct transcription initiation of a linear duplex with a core promoter that has a TATA box. This minimum transcription machinery determines the transcription start site and the direction of transcription. Because the level of transcription catalyzed by RNA polymerase II

TABLE 13.3 General Transcription Factors		
Factor	**No. of Subunits**	**Functions**
TFIIA	2	Stabilizes TBP and TFIID binding. Blocks the inhibitory effects of TAF1 and other proteins.
TFIIB	1	Stabilizes TFIID-promoter binding. Contributes to transcription start site selection. Helps recruit RNA polymerase II • TFIIF complex to the core promoter.
TFIID (TBP and TAFs)	14	Binds to the TATA box, Inr, and DPE. It can deform promoter DNA and serve as a platform for the assembly of TFIIB.
TFIIE	2	Helps to recruit TFIIH to the core promoter and is required for promoter melting.
TFIIF	3	Binds RNA polymerase II and is involved in recruiting the polymerase to the pre-initiation complex. Required to recruit EFIIE and EFIIH to the pre-initiation complex.
TFIIH	10	Functions in transcription and DNA repair. It has kinase and helicase activities and is essential for open complex formation.

13.4 General Transcription Factors: Basal Transcription

together with the general transcription factors is much lower than that observed in the cell, it is called **basal transcription**. For this reason some investigators prefer to use the term basal transcription factor instead of general transcription factor. RNA polymerase II and general (basal) transcription factors assemble at the transcription start site to form a **preinitiation complex**.

TFIID or its TBP subunit must bind to a TATA core promoter before other general transcription factors can do so.

A major breakthrough in our understanding of how general transcription factors interact with RNA polymerase II promoters occurred in 1988 when Stephen Buratowski and coworkers showed a yeast protein can substitute for mammalian TFIID in a reconstituted mammalian transcription system. This demonstration led to the purification of the yeast protein called the TATA-binding protein (TBP) that specifically binds to the TATA box. The yeast gene encoding TBP was cloned and sequenced, providing the information needed to locate homologous genes in other organisms.

Yeast TBP has a 60-residue N-terminal region and a 180-residue C-terminal region. Mutants that lack the N-terminal region are viable, indicating this region is not required for transcription. As might be expected for a nonessential region, the N-terminal region varies in size and sequence from one type of organism to another. In contrast, the C-terminal region is both essential for transcription and highly conserved. The overall folding pattern of the C-terminal region is the same for all TBP molecules that have been studied to date. The crystal structure for TBP from the flowering plant *Arabidopsis thaliana* resembles a molecular saddle with two stirrups (FIGURE 13.8). The C-terminal region has two identically folded domains, each containing a β-sheet and a long and a short α-helix. This high degree of internal symmetry suggests the C-terminal region of TBP probably evolved by the duplication of an ancestral gene. If so, this duplication must have occurred before the archaea and eukaryotes diverged because the C-terminus of archaeal TBP is approximately 40% identical to that of eukaryotic TBP and the two kinds of TBP have very similar structures.

The crystal structure of the C-terminal region of *A. thaliana* TBP bound to the TATA box also has been determined (FIGURE 13.9). Protein–DNA interactions occur through an induced fit mechanism involving conformational changes in both TBP and the TATA box. Amino acids on the underside of the saddle are very hydrophobic and interact with the hydrophobic groups in the DNA minor groove. This interaction is noteworthy because proteins nearly always interact with the major groove. A pair of phenylalanine side chains insert after the first T-A base pair in the TATA box, causing the minor groove to widen (Figure 13.9). A kink in the DNA forms as a result of the abrupt transition to a partially unwound right-handed double helix. A second pair of phenylalanine side chains insert before the last A-T base pair of the TATA box, producing a second kink and an equally abrupt return to B-form DNA. The net effect is that the DNA bends by about 80 degrees. The two "stirrups" assist in this bending.

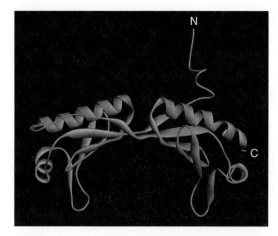

FIGURE 13.8 **The three-dimensional structure of the C-terminal region of TBP from the flowering plant *Arabidopsis thaliana*.** The C-terminal region of the TBP molecule, which resembles a molecular saddle with two stirrups, has two identically folded domains. Each domain contains a β-sheet and a long and a short α-helix. (Structure from Protein Data Bank 1VOK. Nikolov, D. B. and Burley, S. 1994. *Nat Struct Biol* 1:621–637. Prepared by B. E. Tropp.)

FIGURE 13.9 **The three-dimensional structure of the C-terminal region of TBP from *Arabidopsis thaliana* bound to the TATA box of the adenovirus major late promoter (AdMLP).** Protein–DNA interactions occur through an induced fit mechanism involving conformational changes in both TBP and the TATA box. A pair of phenylalanine side chains (red stick structures: left side) insert after the first T-A base pair in the TATA box, causing the minor groove to widen. A kink in the DNA forms as a result of the abrupt transition to a partially unwound right-handed double helix. A second pair of phenylalanine side chains (red stick structures: right side) insert before the last A-T base pair of the TATA box, producing a second kink and an equally abrupt return to B-form DNA. The net effect is that the DNA bends by about 80 degrees. The two "stirrups" assist in this bending. (Structure from Protein Data Bank 1QNE. Patikoglou, G. A. et al. 1999. *Genes Dev* 13:3217–3230. Prepared by B. E. Tropp.)

The convex (upper) side of TBP binds additional protein factors that are required for transcription.

Further study showed that TBP is the TATA binding subunit in TFIID. The additional proteins present in TFIID, called TBP-associated factors (TAFs), are required to transcribe genes that lack a TATA box as well as for the high levels of transcription that occur within the cell. TFIID in all organisms from yeast to humans contain a core set of 13 TAFs that are evolutionarily highly conserved. These TAFs are designated TAF1 to TAF13. TFIID is horseshoe shaped with its TBP subunit just above the central cavity of the horseshoe (FIGURE 13.10). Although crystal structures for the TFIID • core promoter complex are not yet available, the TBP subunit in TFIID probably binds to the TATA box pretty much as described for free TBP.

TFIID also participates in the transcription of protein-coding genes that lack a TATA box. In this case, however, it is the TBP associated factors or TAFs (and not TBP) that bind to regulatory elements in the core promoter. FIGURE 13.11 is a schematic showing the interactions of human TAFs with Inr and DPE in a promoter that lacks the TATA box. As depicted, TAF2 interacts with Inr, while TAF6 and TAF9 interact with DPE. This binding ensures TBP will be positioned so the upper (convex) part of its saddle is available to interact with other proteins required for transcription. The TFIID conformation may vary depending on the nature of the contacts made with the core promoter, exposing different regions of TBP and the TAFs. Even more

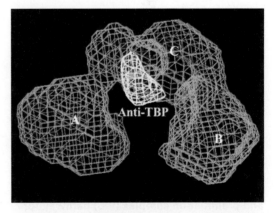

FIGURE 13.10 **Mapping of TBP on human TFIID.** The blue mesh shows the horseshoe shape of the TFIID complex, which contains thirteen TBP associated factors (TAFs). The three lobes in the horseshoe are indicated by the letters A, B, and C. The yellow mesh shows the location of antibody that binds to TBP. (Reproduced from Andell, F. III, et al. 1999. *Science* 286:2153–2156. Reprinted with permission from AAAS. Photo courtesy of Robert Tijan, University of California, Berkeley.)

13.4 General Transcription Factors: Basal Transcription

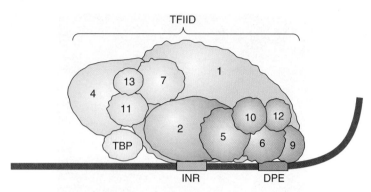

FIGURE 13.11 Schematic composite showing the interactions of various TFIID subunits with the core promoter. Several TFIID subunits have been implicated in binding to the core promoter, including TBP (yellow), which binds to the TATA box when it is present, TAF2 (blue), which interacts with the initiator (Inr), and TAF6 (orange) and TAF9 (orange), which interact with the downstream promoter element (DPE). Many, but not all, of the 13 highly conserved TAFs are shown in this figure. (Adapted from Näär, A. M., Lemon, B. D., and Tijian, R. 2001. *Annu Rev Biochem* 70:475–501.)

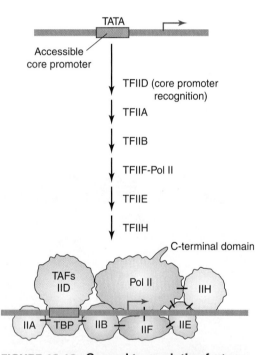

FIGURE 13.12 General transcription factors and preinitiation complex (PIC) pathway for RNA polymerase II promoters with a TATA containing core promoter. The preinitiation complex is assembled on DNA in a multistep process in which TFIID (D) binds first, followed by TFIIA (A), TFIIB (B), a preformed complex of RNA polymerase II (Pol II), and TFIIF (F), TFIIE (E), and TFIIH (H). Some of the known interactions among the general transcription factors and RNA polymerase II are indicated by lines. The red arrow indicates the transcription initiation site. (Adapted from Roeder, R. G. 2003. *Nat Med* 9:1239–1244.)

variation is possible because cells appear to have more than one kind of TFIID. This variation results in part from the fact that eukaryotes have TBP-like proteins that can replace TBP and in part from the fact that noncore TAFs and TAF-like factors may also be present.

One additional point needs to be made before we can examine the way other general transcription factors and RNA polymerase II become part of the preinitiation complex. TBP • TATA box (or TFIID • TATA box) complex formation and nucleosome formation are mutually exclusive processes; that is, TBP (or TFIID) cannot bind to DNA regions that are part of a nucleosome and histones cannot interact with DNA that is part of a TBP • TATA box (or TFIID • TATA box) complex. As we see below, special protein factors and enzymes are required to move or remove nucleosomes so transcription can start. We steer clear of this problem for now by considering naked double-stranded DNA templates.

General transcription factors and RNA polymerase interact at the promoter to form a preinitiation complex.

When the core promoter has a TATA box, the other general transcription factors add to the TBP • TATA box (or TFIID • TATA box) complex in the order shown in FIGURE 13.12. This order may vary at core promoters that do not have a TATA box. The contributions that TFIIA, TFIIB, TFIIF, TFIIE, and TFIIH make to transcription initiation are as follows:

1. *TFIIA:* Human TFIIA, a heterotrimer, is needed for basal transcription when TFIID is used to assemble the preinitiation complex but not when TBP is used. TFIIA's stimulation in the presence of TFIID appears to be related to TFIIA's ability to block the inhibitory effects of TAF1 and other proteins. This finding suggests that TFIIA should be classified as a regulatory

protein rather than as a general transcription factor. Because TFIIA is not required to reconstitute a minimum transcription system, it is not considered further.

2. *TFIIB*: TFIIB helps to convert the closed promoter to an open promoter. Human TFIIB is a single polypeptide. Homologs have been identified in a wide variety of other organisms including plants, yeast, and the archaea. In 1995 Stephen K. Burley and coworkers determined the crystal structure of a ternary complex of human TFIIB bound to TBP from *A. thaliana*, which was in turn bound to the TATA box of the adenovirus late major promoter (FIGURE 13.13). Patrick Cramer and coworkers determined the crystal structure for a yeast RNA polymerase II • TFIIB complex, which helps to explain how TFIIB contributes to converting the closed promoter to an open promoter (**BOX 13.2 LOOKING DEEPER: TFIIB ASSISTS IN OPEN PROMOTER FORMATION**). When the DNA being transcribed has a negatively supercoiled core promoter with a TATA box, TBP and TFIIB alone are sufficient to permit RNA polymerase II to catalyze basal transcription. When the core promoter is not negatively supercoiled, RNA polymerase II also requires TFIIF, TFIIE, and TFIIH for basal transcription.

3. *TFIIF*: TFIIF is a heterotetramer of the $\alpha_2\beta_2$ type. The TFIIF$_\alpha$ subunit possesses a serine/threonine kinase activity, allowing it to phosphorylate a serine and threonine residue on its own chain. This autophosphorylation may play some role in transcription regulation. The preformed RNA polymerase II • TFIIF complex binds to the RNA polymerase II • TFIID • TFIIB complex.

4. *TFIIE*: Human TFIIE, a heterotetramer of the $\alpha_2\beta_2$ type, has a dumbbell shape. It seems to interact with the RNA polymerase II "jaws" at the downstream end of the active center cleft, placing it in a position to interact with promoter DNA about 25 bp downstream of the transcription start site. Its major functions appear to be to recruit TFIIH to the RNA polymerase II complex and then regulate TFIIH's helicase and kinase activities (see below).

5. *TFIIH*: Human TFIIH has 10 polypeptide subunits and a molecular mass of about 500 kDa. Two of the subunits, designated XPB and XPD, have $3' \rightarrow 5'$ and $5' \rightarrow 3'$ helicase activities, respectively. The XP designation derives from the fact that individuals who lack a functional XPB or XPD have an inborn error in metabolism known as xeroderma pigmentosum, which results from an inability to repair certain types of DNA damage. The helicases help unwind the DNA in the region over the active center cleft. TFIIH also has a third enzymatic activity, cyclin-dependent protein kinase. A **cyclin** is a protein that is expressed at different levels throughout the cell cycle. When a threshold level of the cyclin is reached, it interacts with a specific protein kinase such as cyclin-dependent protein kinase. This interaction stimulates the protein kinase to phosphorylate one or more specific proteins, enabling the phosphorylated proteins to perform functions required for cell division.

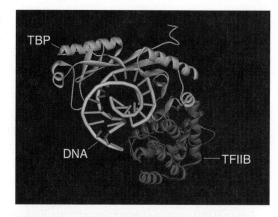

FIGURE 13.13 **Crystal structure of a ternary complex of the human TFIIB bound the C-terminal region of TBP from *Arabidopsis thaliana*, which is in turn bound to TATA box of the adenovirus late major promoter.** (Structure from Protein Data Bank 1VOL. Nikolov, D. B., et al. 1995. *Nature* 377:119–128. Prepared by B. E. Tropp.)

BOX 13.2: LOOKING DEEPER

TFIIB Assists in Open Promoter Formation

In 2009 Patrick Cramer and coworkers attempted to determine the crystal structure for the yeast RNA polymerase II • TBP • TFIIB • DNA complex but were unable to do so because TBP and DNA dissociated from the complex during crystal formation. Despite this setback, their effort proved to be quite worthwhile because the crystals they did obtain contained the RNA polymerase II • TFIIB complex. FIGURE B13.2a shows the crystal structure of this complex with RNA polymerase II subunits in ribbon form and TFIIB in spacefill form. Note the RNA polymerase II in this figure has been rotated around a vertical axis by 180 degrees with respect to the RNA polymerase II shown in Figure 13.4 so the enzyme is pointing in the opposite direction. This rotation permits us to see important interactions between TFIIB and RNA polymerase II. FIGURE B13.2b shows TFIIB in the same orientation as in Figure B13.2a but in ribbon form with the RNA polymerase subunits omitted. Moving from the N- to the C-terminus, TFIIB structural features are as follows: (1) B-ribbon with a bound zinc ion, (2) B-reader, (3) B linker, and (4) B-core.

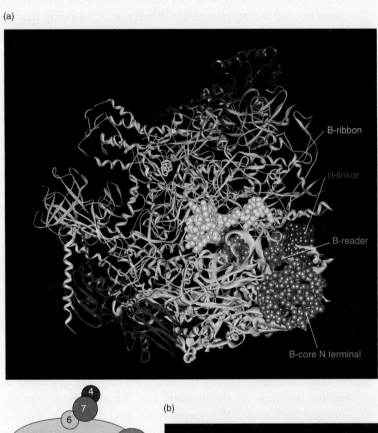

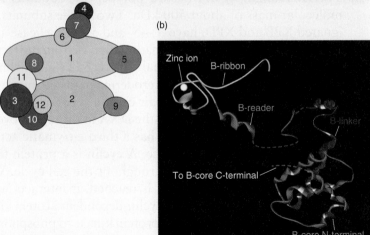

FIGURE B13.2a AND b **Interaction between TFIIB and yeast RNA polymerase II.** (a) RNA polymerase • TFIIB crystal structure. RNA polymerase is shown in ribbon form. The subunits are as indicated by the color code of the insert below. Numbers in the insert indicate Rpb subunits. TFIIB is shown in spacefill form with its specific regions labeled. (b) Ribbon form of TFIIB in the same orientation as in (a) but with the RNA polymerase subunits omitted. The colors of TFIIB structural features are the same in (a) and (b). Dashed lines indicate uncertainty about exact strand position. (Structures from Protein Data Bank 3K1F. Kostrewa, D., et al. 2009. *Nature* 462:323–330. Prepared by B. E. Tropp.)

The RNA polymerase II • TFIIB crystal structure reveals the following: (1) the B-ribbon contacts a docking site on Rpb1 near the RNA exit site; (2) the B-reader is located near the active site, where it is proposed to help read the DNA sequence during transcription start site selection; (3) the B-linker is located near the polymerase's rudder; and (4) the B-core is located above polymerase's wall. Chemical crosslinking and mutant studies are consistent with the crystal structure.

Based on the TFIIB • RNA polymerase II crystal structure, the TBP • TFIIB • TATA box crystal structure, and biochemical data, Cramer and coworkers propose the model shown in FIGURES B13.2c and d for the transcription initiation complex and its conversion to the transcription elongation complex. The basic steps in this model are as follows:

1. *Closed complex formation:* The TATA box is positioned above the cleft in RNA polymerase II near the B-core.
2. *Open complex formation:* The B-linker assists the bound DNA to melt about 20 bp downstream of the TATA box to initiate transcription bubble formation. The released template strand slides into the RNA polymerase II cleft, filling the template tunnel, and the downstream double-stranded DNA enters the downstream cleft. The B-reader helps stabilize the transcription bubble near the active site.
3. *DNA start site scanning:* The template strand makes its way through the template tunnel beside the active site and is scanned for an Inr element with the assistance of the B-reader.
4. *RNA chain initiation:* RNA synthesis begins at the open promoter when the first two nucleotides line up on the template strand in the newly formed transcription bubble and RNA polymerase II catalyzes phosphodiester bond formation.
5. *Abortive transcription:* Continued ribonucleotide addition leads to the formation of short transcripts. Many of these short transcripts are released, possibly because the B-reader loop blocks chain growth.
6. *Promoter escape:* Chain extension beyond seven nucleotides triggers TFIIB release and the formation of the transcription elongation complex.

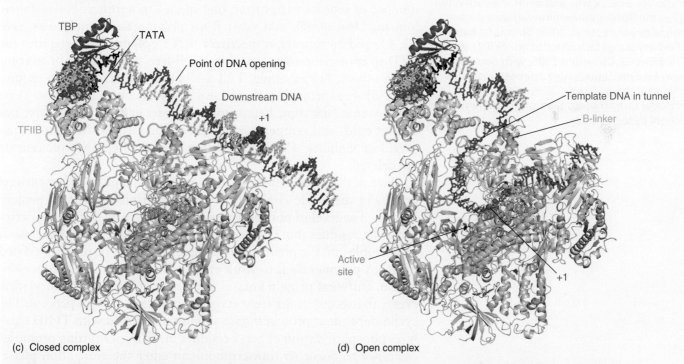

FIGURE B13.2c AND d Models of closed and open complexes. (a) Closed complex. DNA template and nontemplate strands are in blue and cyan, respectively. TFIIB is shown as a green ribbon and TBP as a magenta ribbon. RNA polymerase subunits are shown as gray ribbon structures. The TATA box is in black and magnesium ion at the active site is shown as a magenta ball. (b) Open complex. The nucleotide in the template strand corresponding to position +1 is shown in space-fill form. (Adapted from Kostrewa, D., et al. 2009. *Nature* 462:323–330. Photos courtesy of Patrick Cramer, Ludwig-Maximilians-Universität München.)

13.4 General Transcription Factors: Basal Transcription

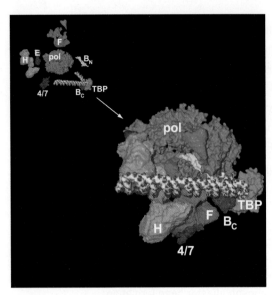

FIGURE 13.14 RNA polymerase II transcription initiation complex. This model was assembled by combining results from x-ray diffraction and electron microscopy. Bc and Bn are the C and N-termini of EFIIB. E, F, and H are EFIIE, EFIIF, and EFIIH, respectively. 4/7 is the Rpb4/7 heterodimer. (Reproduced from Boeger, H., et al. 2005. Structural basis of eukaryotic gene transcription. *FEBS Lett* 579:899–903. Copyright 2005, with permission from Elsevier [http://www.sciencedirect.com/science/journal/00145793]. Photo courtesy of Roger D. Kornberg, Stanford University School of Medicine.)

Combining results from x-ray diffraction and electron microscopy, Roger Kornberg and coworkers proposed a model for the preinitiation complex that includes RNA polymerase II and the general transcription factors (FIGURE 13.14).

13.5 Transcription Elongation

The C-terminal domain of the largest RNA polymerase subunit must be phosphorylated for chain elongation to proceed.

The carboxyl terminal domain, CTD, of the largest subunit in RNA polymerase II, Rpb1, has an important role in the transition from the initiation complex to the elongation complex. This domain has a highly unusual amino acid sequence, consisting of tandem repeats of the heptapeptide Tyr-Ser-Pro-Thr-Ser-Pro-Ser. RNA polymerases I and III do not have comparable domains. The CTD is unstructured and so not visible in RNA polymerase II crystal structures. Although the basic repeating unit in the CTD is the same in all eukaryotes, the number of repeats varies from one species to another. For instance, human, *Drosophila*, and yeast RNA polymerase II molecules have 52, 43, and 26 repeats, respectively. RNA polymerase II requires the CTD to transcribe some genes in a cell-free system but not to transcribe others. For example, TATA-less promoters appear to require CTD, whereas promoters that have a TATA box usually do not. Yeast mutants that lack some heptapeptide repeats in Rpb1 can survive but exhibit cold- and temperature-sensitive phenotypes. Mutants, however, lose viability when more than half of the heptapeptide repeats are deleted.

Five of the seven residues within each heptapeptide have hydroxyl groups in their side chains that can be phosphorylated by protein kinases. The level of phosphorylation varies greatly during transcription. The residues must be dephosphorylated for RNA polymerase II to assemble into the preinitiation complex but must be phosphorylated for RNA polymerase II to work efficiently during transcription elongation. Different protein kinases are responsible for phosphorylation events that occur at different stages of the transcription process. The cyclin-dependent protein kinase subunit associated with TFIIH catalyzes phosphorylation of Ser-5 residues in CTD, permitting promoter clearance to occur so transcription can enter the elongation phase. Ser-2 and Ser-7 residues are phosphorylated during the elongation stage. RNA polymerase II must be dephosphorylated after each round of transcription is completed before the enzyme can reassemble into an initiation complex to begin the next round of transcription. Specific protein phosphatases catalyze the dephosphorylation. A great deal remains to be learned about kinase and phosphatase specificity and regulation. More also needs to be learned about the influence that different combinations of phosphorylated residues in the CTD have on transcription.

A variety of transcription elongation factors helps to suppress transient pausing during elongation.

RNA polymerase II does not move along the template DNA strand in a continuous manner. Instead, the enzyme oscillates between forward and backward movements. Reverse movement of RNA polymerase II along its DNA template, called "backtracking," was described for bacterial RNA polymerase. Backtracking by RNA polymerase II leads to transcriptional pausing and arrest. Backtracking is usually less extensive in the paused state (2–4 nucleotides) than in the arrested state (7–14 nucleotides). Hence, paused RNA polymerase II can return to the transcription state without the assistance of other proteins, whereas arrested RNA polymerase II cannot.

The average rate of nucleotide addition, which is approximately 1,500 nucleotides per minute, appears to be limited by the fraction of time RNA polymerase II spends in the paused state rather than to the average rate of nucleotide addition. A variety of different transcription factors has been observed to increase the rate of transcription elongation by decreasing the time spent in the paused state. Ronald C. Conaway and coworkers propose that one of these, elongin, stimulates transcription by stabilizing the active conformation of RNA polymerase II, increasing the rate at which the inactive form is converted back to the active form, or both methods. Other transcription elongation factors, such as the members of the ELL family, may act in a similar way.

Elongation factor TFIIS reactivates arrested RNA polymerase II.

Arrested RNA polymerase II requires the assistance of an additional transcription factor, TFIIS (also called SII), before it can resume transcription. TFIIS is the eukaryotic counterpart to bacterial GreA and GreB proteins.

RNA polymerase II occasionally makes a mistake and attaches the wrong nucleotide to the growing RNA chain. Such an error produces a mismatched base pair within the transcription bubble that causes a distortion within the DNA–RNA hybrid and destabilizes the elongation complex. RNA polymerase II senses the distortion and backtracks (FIGURE 13.15). TFIIS interacts with the backtracked complex, causing conformational changes in RNA polymerase II. As a result of these conformational changes, the RNA polymerase II active site is converted from a nucleotidyl transferase to a nuclease. RNA cleavage creates a new 3'-end at the active site that allows transcription to resume.

13.6 Regulatory Promoters, Enhancers, and Silencers

Linker-scanning mutagenesis reveals the regulatory promoter's presence just upstream from the core promoter.

The basal RNA polymerase II transcription machinery is very inefficient and by itself accounts for little, if any, of the mRNA synthesized in the cell. Investigators therefore suspected that additional regulatory

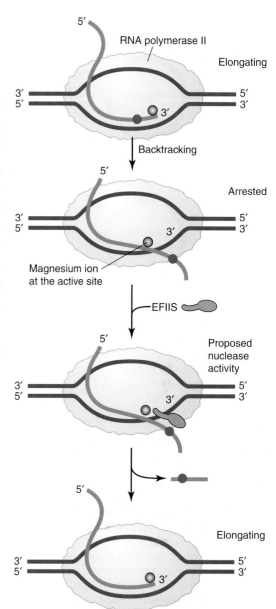

FIGURE 13.15 **Backtracking and TFIIS action.** RNA polymerase II attaches the wrong nucleotide (blue circle) to the growing RNA chain, producing a base pair mismatch. RNA polymerase somehow senses the mismatch and moves backward forcing the 3'-end of the nascent RNA chain to disengage from the active site so a short segment containing the incorrectly incorporated nucleotide is forced out through the secondary channel and into the funnel. This extrusion causes RNA synthesis to be arrested. The transcription elongation factor (EFIIS) (orange amoeboid shape) binds to RNA polymerase II, inducing the enzyme to cleave the RNA stretch that includes the incorrect nucleotide. Cleavage of the RNA segment containing the incorrect nucleotide creates a new 3'-RNA end at the active site so that chain extension can continue. (Adapted from Conaway, J. W., et al. 2000. *Trends Biochem Sci* 25:375–380.)

mechanisms must play a role in the transcription of protein-coding genes. Based on their knowledge of bacterial gene regulation, investigators anticipated eukaryotic regulatory elements for protein-coding genes would be located just upstream from the transcription initiation site. They therefore attempted to detect these regulatory elements by first introducing mutations in this region and then determining the mutations' effects on gene expression. Deleting several base pairs at the same time might seem to offer a rapid method for locating regulatory elements within a gene, but it has a serious shortcoming. Deletion mutations not only remove the nucleotides of interest, they also alter the spacing between flanking DNA sequences. The loss of gene activity that results from deleting a DNA segment thus might be due to the fact that an essential segment was removed, the spacing between flanking sequences was changed, or both.

In 1982 Stephen McKnight and Robert Kingsbury devised a new technique that eliminated the spacing problem, facilitating the search for regulatory elements. This technique, called **linker-scanning mutagenesis**, involves systematic replacement of short DNA segments (usually 3–10 bp) in a region of interest with a DNA linker containing a random sequence of exactly the same size. Although a linker mutation, like a deletion mutation, changes a short DNA segment, it has the advantage of preserving the spacing between nucleotide sequences on either side of the altered segment. Retention of spacing is very important because it allows us to distinguish between effects due to sequence alterations and those due to space changes between flanking sequences.

Linker-scanning mutagenesis was first used to search for the promoter of the thymidine kinase gene in the herpes simplex virus (an icosahedral, enveloped DNA virus responsible for cold sores and genital herpes). Linker mutations were introduced just upstream from the transcription initiation site of a cloned thymidine kinase gene that had been inserted into a plasmid. After mutagenesis, the plasmid was microinjected into oocytes of the frog *Xenopus laevis* to permit the *Xenopus* transcription machinery to transcribe the thymidine kinase gene (FIGURE 13.16).

The experimental results, which are summarized in the schematic diagram in FIGURE 13.17, show thymidine kinase gene transcription is blocked by mutations in four essential regions that are just upstream from the transcription initiation site. Two of these essential regions have the same sequence motif. Other mutations upstream from the transcription initiation site have no effect on transcription. The first essential region, the TATA box, is part of the core promoter. One of the other essential regions contains a **CCAAT box**, whereas the two other essential regions each have a **GC box**. These names derive from the fact that CCAAT and GC boxes have GGCCAATCT and GGGCGG consensus sequences, respectively. These two consensus sequences were later found to be present in many other viral and eukaryotic protein-coding genes and are usually located between 50 and 200 bp upstream from the transcription initiation site. This region, which is called either the **regulatory promoter** because of its function or the **proximal promoter** because of its location, is just upstream from the core promoter (FIGURE 13.18).

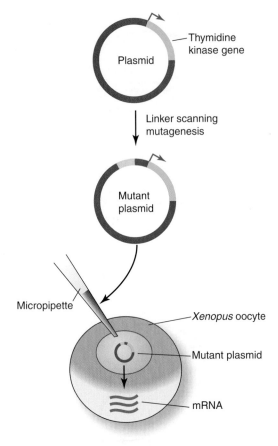

FIGURE 13.16 Linker scanning mutagenesis technique for studying the SV40 thymidine kinase gene. Replace a DNA segment upstream from the transcription initiation site (*arrow*) of a thymidine kinase gene (green), which has been inserted in a plasmid, with a linker DNA (yellow) of equal size. Then microinject mutagenized plasmid DNA into *Xenopus* oocyte to determine whether or not the oocyte transcription machinery can transcribe the thymidine kinase gene to produce mRNA. Structures are not drawn to scale.

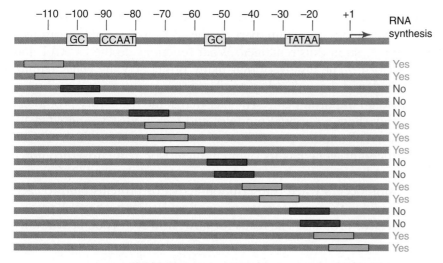

FIGURE 13.17 **Linker scanning mutagenesis studies of the herpes simplex virus (HSV) thymidine kinase gene.** The region of the HSV thymidine kinase gene, which is just upstream from the transcription initiation site, is organized as shown in the top line. The numbers above the line specify the distance in base pairs from the transcription initiation site (+1). The GC, CCAAT, and TATA regulatory elements are shown in yellow boxes. The mutants created by linker-scanning are shown in the series of lines below. Each rectangle represents a position in which a linker replaced a 6–10 nucleotide segment. Mutant DNAs were injected into *Xenopus* oocytes, where the cell's enzymes transcribe the thymidine kinase gene. Green rectangles indicate mutations that produce normal amounts of RNA. Red rectangles indicate mutations that cause reduced levels of RNA synthesis because of a substantial decrease in promoter activity. All mutations that changed sequences in the GC, CCAAT, and TATA regulatory elements decreased promoter activity. (Adapted from Watson, J. D., et al. 1992. *Recombinant DNA* (2nd ed). W. H. Freeman and Company.)

Investigators considered the possibility that the CCAAT box is the eukaryotic counterpart of the bacterial −35 box but were forced to reject the idea for the following reasons. First, many protein-coding genes lack the CCAAT box, and in one extreme situation, *Drosophila*, no regulatory promoter contains this box. Second, when present, the precise position of the CCAAT box within the regulatory promoter varies from one kind of protein-coding gene to another. Finally, regulatory promoters often

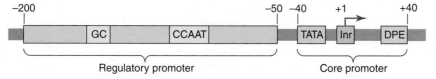

FIGURE 13.18 **The regulatory promoter.** The regulatory promoter (also called the proximal promoter) contains sequence motifs known as promoter proximal elements such as the CCAAT and GC boxes that bind specific transcription activator proteins. The regulatory promoter usually extends from about −200 to −50 with respect to the transcription start site (+1). The core promoter, which consists of a TATA box (TATA), initiator (Inr), and downstream promoter element (DPE), is also shown. The figure is not drawn to scale and does not represent a specific gene.

have more than one CCAAT box. The GC box was also ruled out as a possible eukaryotic counterpart for the −35 box for similar reasons.

The question thus remained: What functions do the CCAAT and GC perform? The discovery that many protein-coding genes require additional modules within their regulatory promoters for full expression provided an important clue. The fact that some regulatory promoters have only these additional modules and lack both GC and CCAAT boxes was still another clue. What emerges is a picture in which (1) each protein-coding gene has its own characteristic regulatory promoter that is made of some unique combination of modules, and (2) full gene expression occurs when transcription activator proteins bind to each module within the regulatory promoter. Regulatory sites are also present for repressor proteins. The focus for now, however, is on activation.

Enhancers stimulate transcription and silencers block transcription.

George Khoury and coworkers discovered a remarkable new regulatory element in 1981 while studying transcription in simian virus 40 (SV40). This virus has a circular double-stranded DNA containing about 5.2 kilobase pair (kbp). SV40 is particularly well suited for transcription studies because it has only two transcription units, the early and late transcription units, which are transcribed in opposite directions from a common control region (FIGURE 13.19). The early transcription unit is activated soon after the virus coat is removed in the cell nucleus and codes for the large tumor (large T) antigen required for DNA replication. The early transcription unit's core promoter contains a TATA box, and its regulatory promoter contains six GC boxes. Khoury and coworkers detected two identical 72-bp sequences just upstream from the regulatory promoter (Figure 13.19). Removing either 72-bp sequence causes a slight inhibition of the early transcription unit, but removing both sequences causes a 100-fold inhibition.

Experiments were performed to determine how the 72-bp sequence influences the transcription of a mammalian β-globin gene when the

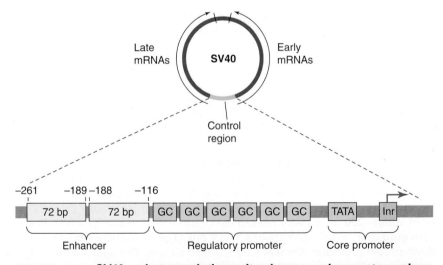

FIGURE 13.19 **SV40 early transcription unit enhancer and promoter regions.**

72-bp sequence is inserted at different positions with respect to the gene's transcription initiation site. This objective was accomplished by inserting the 72-bp sequence into various sites within a plasmid bearing the β-globin gene and then using the recombinant plasmids to transfect cells that do not normally synthesize β-globin. Remarkably, the 72-bp sequence stimulated β-globin mRNA synthesis from the correct initiation site by more than 100-fold when it was inserted a great distance (1 kbp or more) upstream or downstream from the transcription initiation site, and even when it was inserted within the transcription unit. Furthermore, stimulation was the same when the 72-bp sequence was inserted in the forward or backward orientation. The 72-bp sequence's properties were so remarkable at first it seemed possible that it might be a unique genetic element present only in SV40 and perhaps a few related viruses. Additional investigations, however, revealed that animal and plant cells also have regulatory sequences that stimulate transcription in the same way as the 72-bp sequence. These regulatory sequences, which were called **enhancers** because of their ability to stimulate transcription, have two unique features that distinguish them from other regulatory sequences:

1. Enhancers stimulate transcription from the correct transcription initiation site even when they are located several thousand base pairs upstream or downstream from that site.
2. Enhancers stimulate transcription when inserted in either orientation.

Many different enhancers have now been characterized. They range in size from about 50 bp to 1.5 kbp and, like regulatory promoters, consist of a cluster of modules. In fact, many of these modules, including the CCAAT and GC boxes, are the very same ones present in regulatory promoters. According to data gathered by the Encyclopedia of DNA Elements (ENCODE) Project, an international collaboration of research laboratories supported by the National Human Genome Research Institute, any specific human cell type may have about 50,000 functional enhancers. Because there are many cell types, the human genome may have as many as 10^5 to 10^6 enhancers. Each type of enhancer has a characteristic combination of modules. A major structural difference between enhancers and regulatory promoters is the modules appear to be closer together in enhancers. A cell must have transcription activator proteins capable of binding to the modules within an enhancer or a regulatory promoter for full gene expression to occur. This requirement provides a means for gene regulation by controlling the level of functional transcription activator proteins within the cell nucleus.

Eukaryotes also have a negative regulatory element called the **silencer**. Silencers are sequence-specific elements that repress transcription of a target gene. Most silencers function independently of distance and orientation, but some silencers are position dependent. For example, one class of *Drosophila* silencers must be within about 100 bp of the target gene to repress transcription. Silencers are binding sites for negative transcription factors, repressors that act by establishing a repressive chromatin structure (see below), preventing a nearby activator from binding to its DNA-binding site, or in a few cases blocking preinitiation complex formation.

The upstream activating sequence regulates genes in yeast.

Most protein-coding genes in yeast have a single regulatory region, called an **upstream activating sequence (UAS)**, which is located within a few hundred base pairs of the transcription initiation site. Like enhancers in higher organisms, the UAS works at variable distances from the transcription initiation site and in either the forward or reverse orientation. All UAS elements studied to date, however, differ from enhancers in one important respect: UAS elements do not function when located downstream from the transcription initiation site.

13.7 Introduction to Transcription Activator Proteins for Protein-Coding Genes

Transcription activator proteins help to recruit the transcription machinery.

Each of the thousands of protein-coding genes within a eukaryotic cell competes for the limited transcription machinery that is available. The basal transcription machinery requires the assistance of a special class of transcription factors called **transcription activator proteins** to locate protein-coding genes that will be transcribed. Each transcription activator protein has at least two independently folding domains, a **DNA-binding domain** and an **activation domain** (FIGURE 13.20a). The DNA-binding domain makes sequence-specific contacts with modules in a gene's regulatory promoter or enhancer (Figure 13.20b). For instance, DNA-binding domains of Sp1 (selective promoter factor 1) and the C/EBP (CAAT box and enhancer binding protein) transcription activator proteins bind the GC box and the CAAT box, respectively.

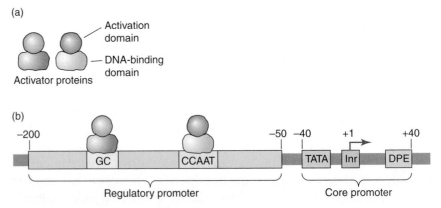

FIGURE 13.20 **A cartoon showing how promoter-selective DNA transcription factors called activator proteins are organized and interact with sequence motifs in the regulatory promoter.** (a) The promoter selective DNA-binding transcription factor called an activator has a DNA-binding domain and an activation domain. (b) The activators' DNA-binding domains bind to specific sequence motifs in the regulatory promoter, leaving the activation domains free to act together to bind transcription factors and thereby recruit the transcription machinery to the gene.

Several different kinds of folding patterns have been observed in the DNA-binding domains, allowing us to assign most transcription factors to a structurally defined family. The activation domain recruits components of the transcription machinery to the gene and then interacts with various components of the transcription machinery to stimulate transcription (see below). We know very little about activation domain structure at this time.

Many transcription activator proteins also have additional structural features. Some of the most important of these are as follows:

1. *Nuclear localization signal:* The nuclear localization signal, a short basic sequence containing arginine and lysine residues, allows transcription activator proteins to move through the nuclear pore complex from the cytoplasm to the nucleoplasm.
2. *Nuclear export signal:* The nuclear export signal permits the transcription activator proteins to move through the nuclear pore complex from the nucleoplasm to the cytoplasm.
3. *Dimerization domain:* The dimerization domain allows one transcription activator protein to pair with an identical transcription activator protein to form a homodimer or with a different transcription activator protein to form a heterodimer.
4. *Ligand binding domain:* Some transcription activators have domains that can bind specific small molecules called ligands.

A combinatorial process determines gene activity.

Initially, investigators thought regulatory promoters and enhancers were completely different types of regulatory elements. The distinction between the two has become blurred, however, because the same modules are often present in both regulatory promoters and the enhancers. This means the same activator protein may bind to both regulatory regions. Furthermore, DNA can form loops that bring activator proteins bound to regulatory promoters near activator proteins bound to enhancers (FIGURE 13.21), so the activation domains can work cooperatively to recruit the transcription machinery to the gene.

Because each gene has a characteristic set of modules in its regulatory promoter and enhancer(s), it binds a unique combination of activator proteins. Gene activity is determined in large part by how well the activation domains work together to recruit the transcription machinery to the gene. Hence, gene activation results from

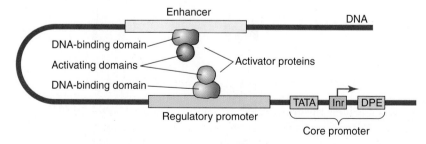

FIGURE 13.21 DNA looping. The DNA between the enhancer and the regulatory promoter loops, allowing the activation domains of activator proteins bound to regulatory promoters to be near one another.

combinatorial control, which allows a cell to use relatively few polypeptides to regulate many protein-coding genes in response to diverse signals.

DNA affinity chromatography can be used to purify transcription activator proteins.

Transcription activator proteins must be purified before they can be characterized. It was difficult to purify the transcription activator proteins, however, because they are present at very low intracellular concentrations. In 1986 James T. Kadonaga and Robert Tjian solved the problem by using **DNA affinity chromatography** to purify transcription activator proteins from crude nuclear extracts or partially purified preparations (FIGURE 13.22). They prepared DNA affinity beads by synthesizing DNA fragments with tandem repeats of the

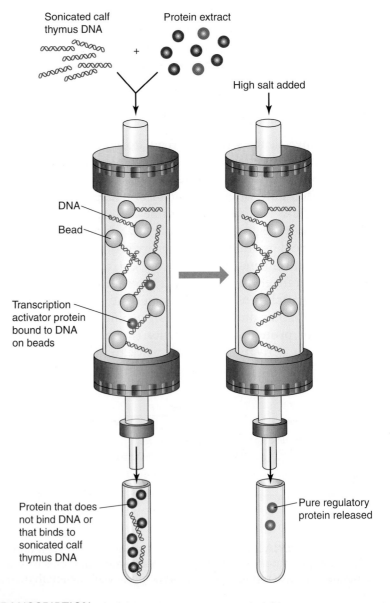

FIGURE 13.22 Schematic of DNA affinity chromatography. Sonicated calf thymus DNA is mixed with a protein extract that contains a few transcription activator proteins (green) and many other proteins (red). The activator proteins do not bind the sonicated calf thymus DNA, but some of the other proteins do bind to it. The protein–DNA mixture is passed through a column containing DNA affinity beads. The beads (yellow) have attached DNA fragments (blue) that have tandem repeats of the sequence motif that binds the transcription activator protein. The transcription activator protein binds specifically to the DNA attached to the beads. The other proteins pass through the affinity column. Transcription activator protein is eluted from the column with buffer that contains a high salt concentration. (Adapted from an illustration from the Natural Toxins Research Center, Texas A&M University.)

sequence motif that binds the transcription activator protein of interest and then linked these DNA fragments to inert polysaccharide beads. The crude nuclear extract or partially purified protein solution containing the transcription activator protein was mixed with sonicated calf thymus competitor DNA, and then the protein–DNA mixture was passed through a column filled with the DNA affinity beads. Proteins that bound nonspecifically to DNA passed through the column after binding to the sonicated thymus DNA. Activator proteins that bound specifically to the DNA affinity beads were eluted from the column by washing the column with a high salt buffer solution. Fractions were assayed by testing eluted proteins for their ability to bind to the sequence motif. Although the amount of a transcription activator protein obtained by DNA affinity chromatography was quite low, it was sufficient for partial or complete amino acid sequence determination. Sequence information then was used to predict possible nucleotide sequences for a segment of the activator protein gene so a probe could be synthesized to search for the activator protein gene.

Several transcription activator protein genes have been cloned in this fashion. Once a transcription activator protein gene has been cloned from one organism, it is usually fairly straightforward to find its counterpart in other organisms by launching a search for homologous sequences. The rapid pace of gene identification and sequence determination in many different eukaryotes including humans has facilitated this process.

A transcription activator protein's ability to stimulate gene transcription can be determined by a transfection assay.

A cloned transcription activator protein gene can be used to test the transcription activator protein's ability to stimulate transcription in a cell-based assay (FIGURE 13.23). Two recombinant plasmids are required to perform this assay. The first recombinant plasmid bears the gene that specifies the transcription factor, and the second bears a reporter gene and the control element recognized by the transcription factor. Several different coding sequences have been used to construct reporter genes. Included among these are β-galactosidase from *E. coli*, luciferase from firefly, and green fluorescent protein from jellyfish. Assays for β-galactosidase and luciferase are sensitive over a wide range of enzyme concentrations and easy to perform. β-Galactosidase activity is detected in a spectrophotometer by taking advantage of the fact that the enzyme converts the artificial substrate *o*-nitrophenylgalactoside to *o*-nitrophenol, a yellow substance that absorbs light at 450 nm. Luciferase activity is monitored in a luminometer or liquid-scintillation counter by exploiting the enzyme's ability to catalyze the ATP-dependent oxidative decarboxylation of luciferin with the release of light. The green fluorescent protein has the advantage of not requiring any additional substrates. It emits a green light when irradiated with ultraviolet light or blue light, allowing direct measurement of its concentration in living cells.

A host cell that lacks both the reporter gene and the transcription activator protein gene is transfected with one or both recombinant plasmids. If the cloned gene in fact does code for the transcription

FIGURE 13.23 **Cell-based assay for the transcription activator protein.** Two recombinant plasmids are required. The first contains the gene for the transcription activator protein, and the second contains a reporter gene and a transcription activator binding site. If the transcription activator protein is synthesized and able to bind to the transcription activator binding site, then the reporter gene will be transcribed to produce mRNA that can be translated to produce a protein such as green fluorescent protein or β-galactosidase that is easy to detect. (Adapted from Lodish, H., et al. 2000. *Molecular Cell Biology* (4th ed). W. H. Freeman and Company.)

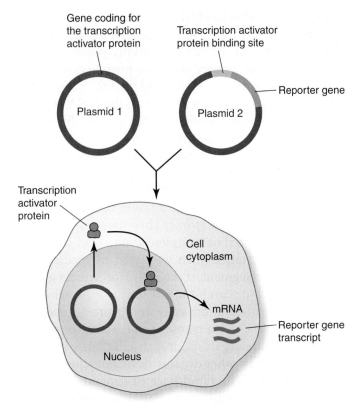

activator protein, expression in the host cell will produce the transcription activator protein, which in turn will stimulate the synthesis of the reporter protein. No stimulation will be observed if only one of the two recombinant plasmids is used. Activator protein can also be tested *in vitro* with the same recombinant reporter gene and required general transcription factors, RNA polymerase II, and other required components of the transcription machinery.

13.8 DNA-Binding Domains in Transcription Activator Proteins

Transcription activator proteins are commonly grouped according to the structures of their DNA-binding domains.

Transcription activator proteins can be classified by the folding patterns of their DNA-binding domains. Although this classification approach is based on structural similarities in just one domain, transcription activator proteins with similar DNA-binding domains often have similar biological functions. With this relationship in mind, we now examine four representative families of transcription activator proteins. The characteristic polypeptide folding pattern of the DNA-binding domain in each of these families is as follows: **helix-turn-helix, Cys_4 zinc finger, basic region leucine zipper**, and the Zn_2Cys_6 **binuclear cluster**.

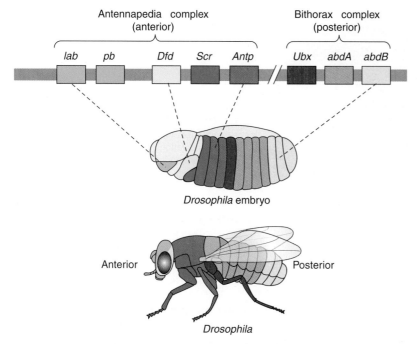

FIGURE 13.24 Arrangement of *Drosophila HOM* genes. The *HOM* genes are located in two small clusters on chromosome 3, designated the *Antennapedia* complex (five genes) and the *bithorax* complex (three genes). *Antennapedia* genes regulate the head and anterior thoracic segments, whereas *bithorax* genes regulate the posterior thoracic and abdominal segments. Gene order on the chromosome parallels body region order along the anterior to posterior axis of the embryo. (Adapted from Raven, P. H., and Johnson, G. B. 2001. *Biology* (6th ed). McGraw-Hill Higher Education.)

Helix-Turn-Helix DNA-Binding Domain

The first clue to the existence of the helix-turn-helix transcription activator proteins came from studies of eight developmental genes in the fruit fly *Drosophila*, the so-called **homeotic (*HOM*) genes**, which "inform" embryonic cells they will become part of the fly's head, thorax, or abdomen; that is, *HOM* genes assign distinct positional identities to cells in different regions along the fly's anterior-posterior axis. As shown in FIGURE 13.24, the *HOM* genes are located in two small clusters on chromosome 3, designated the *Antennapedia* complex (five genes) and the *bithorax* complex (three genes). *Antennapedia* genes regulate the head and anterior thoracic segments, whereas *bithorax* genes regulate the posterior thoracic and abdominal segments. Remarkably, gene order on the chromosome parallels body region order along the anterior to posterior axis of the embryo. Although usually lethal, *HOM* gene mutations can produce incredible phenotypic changes in surviving offspring. For instance, *Antennapedia* gene mutations cause legs to grow on the head where antennae should be located (FIGURE 13.25).

Independent studies by Walter Gehring and Matthew Scott in 1984 showed that homeotic genes in *Drosophila* have a common DNA sequence within them that is about 180 bases long. Subsequent studies by these and other workers showed this conserved sequence, which is designated the **homeobox**, is also present in vertebrate, plant, and fungi homeotic genes as well as in nonhomeotic genes such as the yeast

13.8 DNA-Binding Domains in Transcription Activator Proteins

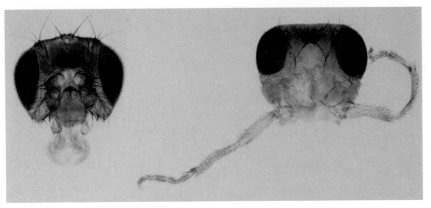

FIGURE 13.25 **Head of a wild-type *Drosophila melanogaster* (left) compared with an *Antennapedia* mutant in which the antennae are replaced by a pair of middle legs (right).** (Photo courtesy of Walter J. Gehring, Biozentrum, University of Basel.)

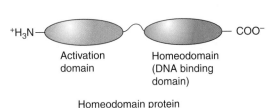

FIGURE 13.26 **Homeodomain protein.**

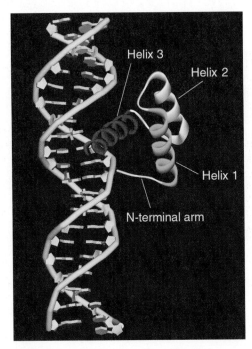

FIGURE 13.27 **Structure of engrailed homeodomain DNA complex.** The protein is presented in ribbon form (colored according to structure) and DNA (blue) in tube form. Helices 1 and 2 are joined by a short loop. Helices 2 and 3 are connected by a turn, forming a helix-turn-helix structural motif that makes sequence specific contacts with DNA. Helix 3 is the recognition helix; its hydrophilic face fits into the major groove of a DNA segment containing a 5'-ATTA-3' (5'-TAAT-3') sequence motif. The hydrophobic face of helix 3 packs against helices 1 and 2 but is not apparent in this perspective. The homeodomain's N-terminal arm helps fasten the protein to DNA by fitting into the minor groove. (Structure from Protein Data Bank 3HDD. Fraenkel, E., et al. 1998. *J Mol Biol* 284:351–361. Prepared by B. E. Tropp.)

mating type genes. Because yeast mating type genes were previously demonstrated to be transcription activator proteins, it seemed reasonable to propose that homeobox proteins might also have the same function. Subsequent studies supported this proposal. The homeobox protein folds into two domains (FIGURE 13.26). The N-terminus forms the activation domain and the C-terminus, consisting of 60-amino acid residues, forms the DNA-binding domain or **homeodomain**.

High-resolution crystal structures are not yet available for intact homeobox proteins (or for any other intact eukaryotic transcription activator protein) because of difficulties in preparing suitable crystals. Peptide fragments containing homeodomains do fold into stable structures, however, which form crystals suitable for x-ray crystallography. Carl O. Pabo and coworkers obtained the first high-resolution crystal structure for a homeodomain in 1994. The peptide fragment they studied, derived from the *Drosophila* homeobox protein engrailed, has a compact structure containing three α-helices (FIGURE 13.27). Helices 1 and 2 are joined by a short loop. *Helices 2 and 3 are connected by a turn, forming a helix-turn-helix structural motif that makes sequence-specific contacts with DNA.* Helix 3 is the recognition helix; its hydrophilic face fits into the major groove of a DNA segment containing a 5'-ATTA-3' (5'-TAAT-3') sequence motif. Under *in vitro* conditions different homeobox proteins appear to have similar affinities for DNA fragments with a 5'-ATTA-3' sequence motif. The same proteins, however, show definite preferences for specific promoters *in vivo*. The greater degree of specificity that is observed *in vivo* appears to be due to the ability of additional proteins to influence binding specificity. A monomeric homeodomain protein can bind to DNA in a sequence-specific manner, but it often attains even greater specificity by interacting with other proteins.

Homeotic genes in vertebrates are called ***Hox*** genes (a contraction of the term homeobox). Mammalian cells have 38 *Hox* genes, which are located in four gene clusters, designated *HoxA*, *HoxB*, *HoxC*, and *HoxD*. Each gene cluster is about 100 kbp long and is located on a separate chromosome. Genes within each cluster are homologous to those within the two *Drosophila* gene clusters (FIGURE 13.28). Moreover,

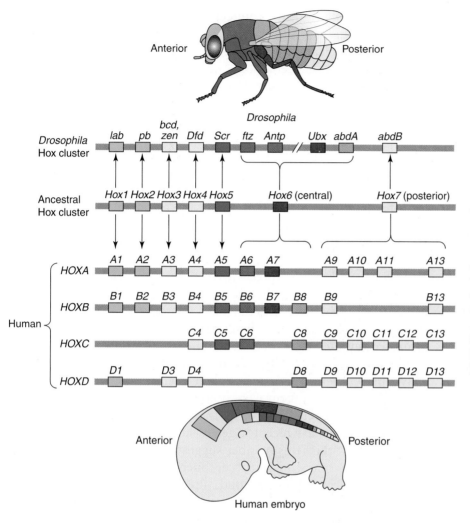

FIGURE 13.28 **Comparison of homeotic gene arrangements in *Drosophila* and humans.** Similar genes, which are arranged in the same order, control the development of the anterior and posterior parts of the bodies of flies and humans. These homeotic genes are located on a single chromosome in the fly (top row of colored squares) and on four separate chromosomes in mammals (lower rows of squares). Studies show an ancestor of all bilateral animals may have had seven *Hox* genes. A hypothetical ancestral *Hox* cluster is shown in the middle, with arrows indicating the predicted origins of fly and mammalian *Hox* genes. The genes are color coded to match the body regions they specify along the anterior-posterior embryo axis. (Modified from Veraksa, A., Del Campo, M., and McGinnis, W. 2000. *Mol Genet Metab* 69:85–100. Copyright 2000, with permission from Elsevier [http://www.sciencedirect.com/science/journal/10967192].)

gene arrangements within each cluster parallel the order of the body regions they specify along the anterior-posterior embryo axis. *Hox* gene mutations cause structural deletions and transformations that are analogous to those caused by *HOM* gene mutations in fruit flies. For instance, *Hox A-3* deletion mutations in mice cause a complicated set of deformities, including an incomplete heart and the absence of thymus and parathyroid glands. As might be expected, most mutants with such major anatomical and physiological abnormalities die at birth. In fact, most homeotic gene mutations result in nonviable organisms.

Cys$_4$ Zinc Finger

Transcription activator proteins belonging to the **nuclear receptor superfamily** have DNA-binding domains with Cys$_4$ zinc finger motifs. Nuclear receptors allow cells in higher organisms to respond to a variety of external and internal chemical signals by increasing or decreasing the transcription of specific genes. Nuclear receptors were initially detected as nuclear proteins that bind various small molecules (ligands) such as steroid hormones, thyroid hormone, vitamin D, or retinoic acid. Investigators use this binding to detect nuclear receptors as they are being purified from cellular extracts by protein fractionation techniques.

Purified nuclear receptor•ligand complexes bind to regulatory promoters and enhancers. Some nuclear receptors such as the thyroid receptor (TR) appear to be located in the cell nucleus in both the free receptor form and in the receptor•ligand complex form. Although both forms of the TR bind to DNA, the free receptor inhibits gene expression, whereas the receptor•thyroxine complex stimulates gene expression. Other nuclear receptors, such as the glucocorticoid receptor, bind in the free form to a cytoplasmic protein complex. When a glucocorticoid steroid binds to a nuclear receptor, it induces a conformational change that causes the glucocorticoid receptor to be released from the cytoplasmic protein complex and then move to the cell nucleus, where it stimulates transcription of specific genes.

Once nuclear receptor genes were cloned and sequenced in the mid-1980s, it became apparent they have a conserved 70-residue sequence that contains eight cysteine residues, suggesting the possibility of two zinc-binding sites that each contain four cysteine residues. A search of the gene data bank for the conserved 70-amino acid sequence revealed that previously unidentified polypeptides have a similar sequence, signifying they too might be nuclear receptors. These polypeptides were designated **orphan nuclear receptors** because the ligands that bind to them were unknown. Based on gene sequences from the human genome project, the best current estimate is a total of 49 human nuclear receptor genes. However, many more nuclear receptors exist than this number suggests for two reasons. First, the active form for many nuclear receptors is a heterodimer, allowing for combinatorial variations. Second, alternate splicing of the primary transcript for a nuclear receptor often produces two or more different mRNA molecules and therefore two or more nuclear receptor isoforms.

Nuclear receptors have a common structural design, shown in FIGURE 13.29. Starting at the amino terminus and moving toward the carboxyl end, the major features are (1) a poorly conserved activation function domain (AF-1), (2) a highly conserved DNA-binding domain (DBD), (3) a variable hinge, (4) a conserved ligand binding domain (the hormone binding site), and (5) a second activation domain (AF-2).

Investigators have used recombinant DNA technology to prepare peptide fragments corresponding to the DNA-binding domains of

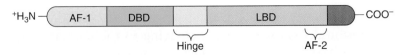

FIGURE 13.29 **Schematic representation of a nuclear receptor.** A typical nuclear receptor contains several functional domains. The variable amino terminal region contains an activation function domain (AF-1). The conserved DNA-binding domain (DBD) recognizes specific DNA sequences. A variable hinge region connects the DNA-binding domain to the ligand-binding domain (LBD). A second activation function domain (AF-2) is located at the end of the ligand-binding domain.

(a)

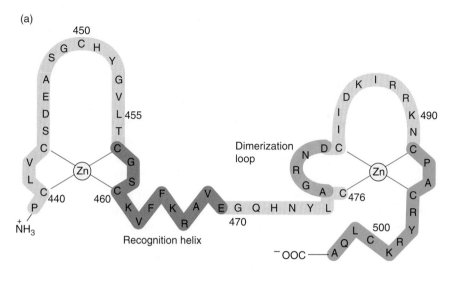

(b)

FIGURE 13.30 The glucocorticoid receptor. (a) The amino acid sequence of the zinc containing DNA-binding domain of the glucocorticoid receptor. The two zinc atoms in this domain are shown in light blue. Each zinc atom binds to four cysteine residues. One of the zinc atoms stabilizes the recognition helix (red), which makes sequence-specific contact with the DNA. The other zinc atom stabilizes a loop involved in formation of the dimeric receptor molecule and the support helix (green). (b) The glucocorticoid receptor binds to a DNA region known as the glucocorticoid response element, GRE. The GRE has two palindromic half sites (light and dark blue) that are separated by a three base pair spacer (NNN, where N is any nucleotide). (Part a modified from Branden, C., and Tooze, J. 1999. *Introduction to Protein Structure* (2nd ed). Garland Science. Used with permission of John Tooze, The Rockefeller University. Part b adapted from Branden, C., and Tooze, J. 1999. *Introduction to Protein Structure* (2nd ed). Garland Science.)

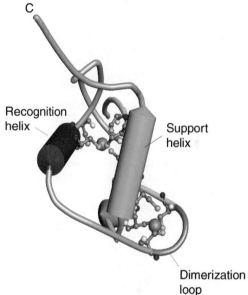

FIGURE 13.31 Crystal structure of the glucocorticoid receptor DNA-binding domain. The glucocorticoid receptor DNA-binding domain (shown in schematic display style) has two Cys$_4$ zinc fingers, each consisting of an irregularly looped string of amino acids followed by an α-helix. The two α-helices cross one another near their midpoints so that the zinc fingers are intertwined. Hydrophobic interactions between conserved residues on the interacting helical faces stabilize the compact globular core. The first α-helix (red) binds to bases in DNA and is therefore called the recognition helix. The second α-helix (green) functions as a support to hold the recognition helix in place. The loop colored purple contacts its counterpart on a second glucocorticoid receptor DNA-binding domain, stabilizing the homodimer structure. Zinc atoms are shown as magenta spheres and the cysteine residues as ball and stick structures with the sulfur atom in yellow. The carboxyl terminus is indicated by the letter C. (Structure from Protein Data Bank 1R4R. Luisi, B. F. 1991. *Nature* 352:497–505. Prepared by B. E. Tropp.)

different nuclear receptors. FIGURE 13.30 shows the sequence of a fragment containing the DNA-binding domain of the glucocorticoid receptor and the DNA sequence to which the domain binds. Notice the DNA-binding region has two Cys$_4$ zinc finger motifs. Steroid receptors such as the glucocorticoid receptor and the estrogen receptor bind to DNA as homodimers. The DNA-binding unit therefore consists of two DNA-binding domains, one from each steroid receptor.

The crystal structure of the glucocorticoid receptor DNA-binding domain shows the arrangement of the two zinc fingers, each consisting of an irregularly looped string of amino acids followed by an α-helix (FIGURE 13.31). The two α-helices cross one another near their midpoints so the zinc fingers are intertwined. Hydrophobic interactions between conserved residues on the interacting helical faces stabilize the compact globular core.

FIGURE 13.32 shows the crystal structure for a glucocorticoid receptor DNA-binding domain dimer interacting with DNA. The two

13.8 DNA-Binding Domains in Transcription Activator Proteins

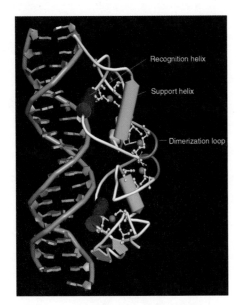

FIGURE 13.32 Glucocorticoid receptor DNA-binding domain bound to DNA. The glucocorticoid receptor (shown in schematic form) binds to DNA (shown in blue tube form). The glucocorticoid receptor is a homodimer that is stabilized by interactions between the dimerization loops (purple) in each monomer. Each subunit also contributes a DNA-binding domain that contains a pair of intertwined Cys_4 zinc fingers. The cysteine residues are shown as yellow ball and stick structures and the zinc atoms as magenta spheres. The two helices in a DNA-binding domain perform different structures. The α-helix (red) in the first zinc finger functions as a recognition helix to make specific contacts with bases in the major groove. The α-helix (green) in the second zinc finger functions as a support to hold the recognition helix in place. (Structure from Protein Data Bank 1R4R. Luisi, B. F. 1991. *Nature* 352:497–505. Prepared by B. E. Tropp.)

DNA-binding domains are arranged head-to-head so the overall structure is symmetric. A dimerization loop formed by the residues between the first two cysteine residues in the second zinc finger (Cys-476 and Cys-492) in one DNA-binding domain interacts with the dimerization loop in the other DNA-binding domain to stabilize the homodimer. The two DNA-binding domains in the homodimer contact bases in the major groove on the same side of the DNA molecule. The intertwined Cys_4 zinc fingers in each DNA-binding domain act as a unit but have different functions. The α-helix in the first zinc finger is a recognition helix that makes specific contacts with bases in the major groove; the α-helix in the second acts as a support to hold the recognition helix in place. The homodimer binds to a DNA sequence called the **hormone response element**. The glucocorticoid and estrogen response elements are imperfect palindromes consisting of inverted 6-bp repeats separated by a 3-bp spacer (FIGURE 13.33). This spacer must be 3 bp, but the nucleotides within it do not appear to matter. Each DNA-binding domain interacts with one 6-bp repeat or half-site. Consensus sequences for glucocorticoid and estrogen receptor half-sites are 5′-AGAACA-3′ and 5′-AGGTCA-3′, respectively.

Active forms of many other nuclear receptors are heterodimers. One important family consists of receptors that form heterodimers with the retinoid X receptor (RXR), which is the receptor for *9-cis* retinoic acid. Dimerization partners include TR, the vitamin D receptor (VDR), the retinoic acid receptor (RAR), and the peroxisome proliferator-activated receptor (PPAR). The DNA-binding domain of RXR only interacts with its heterodimeric partner in the presence of the hormone response element. When hormone response elements are present, DNA-binding domains produce the same dimerization and selectivity patterns as full-length receptors. Heterodimer formation is induced by specific hormone response elements that contain direct repeats with characteristic spacer regions between half-sites. The half-site for all the DNA-binding domains that belong to the RXR family of heterodimers is ATTTCA. The number of base pairs in the spacer region determines the specificity of binding. For instance, adding 1 bp to the spacer in RXR • VDR response element converts it to an RXR • TR response element. Rules that govern specificity according to spacer size, known as the *1–5 rule*, are illustrated in FIGURE 13.34.

The number of combinatorial possibilities is increased because the heterodimer can switch polarities. For instance, RXR • RAR has one polarity when it binds to a response element with a 1-bp spacer and the opposite polarity when it binds to a response element with a 5-bp spacer. Hormone response changes as a result of the polarity switch. Protein–protein interactions between RXR and each of its partners help to ensure correct spacing recognition. Further stabilization is achieved through sequence-specific contacts with bases in the major groove at each of the half-sites. The RXR DNA-binding domain's ability to accommodate so many different dimerization partners is probably because it has many combinatorial binding sites on its surface. FIGURE 13.35 shows the structures for RXR • TR and RXR • RAR DNA-binding domain heterodimers bound to DNA.

Glucocorticoid response element

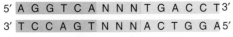

Estrogen response element

FIGURE 13.33 Glucocorticoid and estrogen response elements. Both the glucocorticoid and estrogen response elements are palindromic repeats, consisting of two 6-bp half sites separated by three nucleotides, N (where N can be any nucleotide).

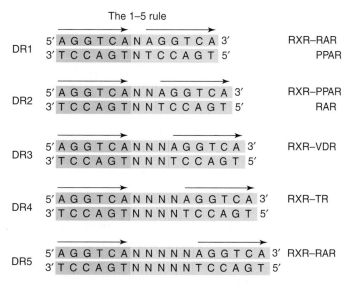

FIGURE 13.34 The 1–5 rule of DNA direct repeat binding by RXR and its nuclear receptor partners. The base pair size of the spacing between the AGGTCA sequences can vary from one to five. (Adapted from Rastinejad, F. 2001. *Curr Opin Struct Biol.* 11:33–38.)

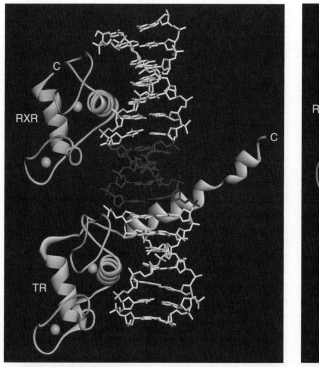

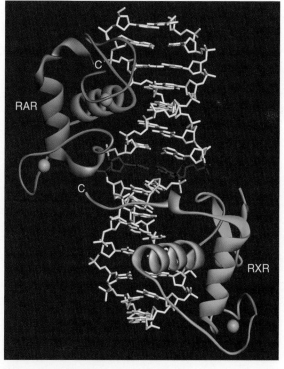

FIGURE 13.35 Structures of DNA-binding complexes involving RXR and their dimer interfaces. DNA is shown as a stick figure. Nucleotides belonging to the spacing element between direct repeats (DR) are shown in red. All other nucleotides are shown in white. Zinc ions are shown as magenta spheres. RXR (retinoid X receptor) is shown in green, TR (thyroid hormone receptor) in yellow, and RAR (retinoic acid receptor) in blue. In each case the protein–protein contacts are formed directly over the minor groove of the spacing, with several protein–DNA phosphate contacts stabilizing the assembly. (a) The RXR • TR DNA-binding domain heterodimeric complex with four nucleotide spacing and (b) the RXR • RAR DNA-binding heterodimeric complex with one nucleotide spacing. (Part a structure from Protein Data Bank 2NLL. Rastinejad, F. et al. 1995. *Nature* 375:203–211. Prepared by B. E. Tropp. Part b structure from Protein Data Bank 1DSZ. Rastinejad, F. et al. 2000. *EMBO J* 19:1045–1054. Prepared by B. E. Tropp.)

13.8 DNA-Binding Domains in Transcription Activator Proteins

Basic Region Leucine Zipper Proteins

The basic region leucine zipper (bzip) proteins have fibrous DNA-binding domains. Members of this family, which bind to DNA as homo- or heterodimers, were initially identified because their polypeptide chains have α-helical segments that are about 30 residues long with a leucine in every seventh position. Because a typical α-helix has 3.6 residues per turn, the presence of a leucine at every seventh position means the leucine side chains are on one face of the helix. A peptide fragment was synthesized corresponding to the leucine repeat region in a yeast transcription activator protein, which is called Gcn4 because it coordinately increases the transcription of several genes involved in the general control of nitrogen metabolism. The synthetic Gcn4 fragments paired in a parallel orientation to form dimers in solution. The crystal structure revealed the two polypeptide fragments coil about each other with a slight left-handed supertwist to produce a **coiled coil** in which the smoothly bent α-helices make tight contacts over the length of the dimer (FIGURE 13.36). The observed helical repeat in the coiled coil is 3.5 residues per turn (0.1 residues per turn less than that in a free α-helix), so that leucine side chains extend out from the same face of the helix after every two turns, making side-to-side contacts in every other turn.

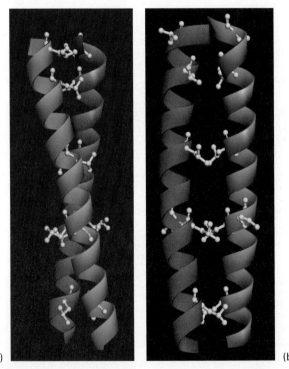

(a) (b)

FIGURE 13.36 **Crystal structure of Gcn4 leucine zipper, a two-stranded parallel coiled coil.** Backbones of the two peptide chains that form the Gcn leucine zipper are shown as red and blue helices. The side chains of the leucine residues are shown as yellow ball and stick structures. The perspective in (a) was chosen to stress the coiled coil, whereas that shown in (b) was chosen to stress the interactions among the leucine side chains. (Structures from Protein Data Bank 2ZTA. O'Shea, E. K., et al. 1991. *Science* 254:539–544. Prepared by B. E. Tropp.)

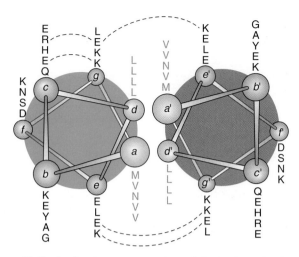

FIGURE 13.37 Helical wheel representation of Gcn4 domain that forms the coiled coil. Each strand of a coiled-coil protein may be viewed as repeated seven residue sequences of the form (a-b-c-d-e-f-g)$_n$, where a, b, c, d, e, f, and g represent consecutive residues on one strand of the coiled coil. Residues on the other strand are indicated by placing a prime (′) after the letter. The first and fourth positions (a and d) are usually hydrophobic amino acids. Residues that form ion pairs are connected with dashed lines. (Adapted from O'Shea, E. K., et al. 1991. *Science* 254:539–544.)

We can gain some additional insight into how side chains in a coiled coil structure interact by looking down the helix axis of each polypeptide. **FIGURE 13.37** shows how this approach applies to the 30-amino acid residues in the Gcn4 leucine zipper. We begin by drawing two **helical wheels**, one for each peptide in the dimer. Each helical wheel contains seven positions, a–g (or a′–g′), corresponding to the seven amino acids required to complete two helical turns. The first seven residues (MKQLEDK) in the 30-residue Gcn4 leucine zipper are placed in order at positions a–g (or a′–g′). Then the next seven residues are placed in order above the first seven, and so forth. Hydrophobic residues tend to be located on the same helical face at positions a and d (or a′ and d′). Positive and negative residues tend to alternate along the helix in positions e and g (or e′ and g′) so attractive intra- and interstrand ion pairs can stabilize the dimer.

Although the leucine zipper structure shows how Gcn4 dimerizes, it does not show how the dimer interacts with DNA. A short sequence of basic amino acids that is located just N-terminal to the leucine zipper motif is required for DNA recognition. This <u>b</u>asic amino acid sequence together with the leucine <u>zip</u>per forms the **bzip DNA-binding domain**. As shown in **FIGURE 13.38**, Gcn4 also has an activation domain toward the center of the polypeptide chain.

Gcn4 appears to resemble forceps in the way that it binds to DNA as is evident from the crystal structure of Gcn4 bound to DNA (**FIGURE 13.39a**). The basic DNA-binding region in each polypeptide chain folds into a helical structure as it fits into the major groove of half-sites on either side of the semi-palindromic DNA (Figure 13.39b). Residues within the DNA-binding region act as hydrogen bond donors to specific bases as well as to unesterified oxygen

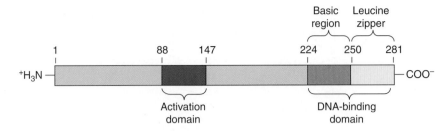

FIGURE 13.38 **Gcn4 transcription activator protein.**

atoms in the phosphodiester backbone. Many other transcription activator proteins are bzip proteins **(BOX 13.3 CLINICAL APPLICATION: THE AP-1 FAMILY OF TRANSCRIPTION ACTIVATOR PROTEINS).**

Zn_2Cys_6 Binuclear Cluster

The Zn_2Cys_6 binuclear cluster family of transcription activator proteins in yeast and other fungi derives its name from the fact that its DNA-binding domain has a folding group with two zinc ions that are

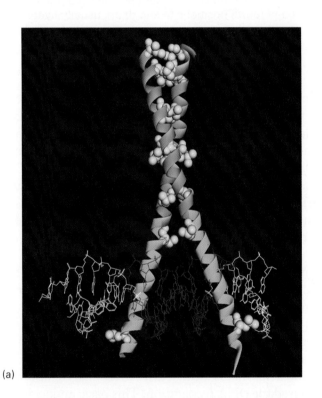

(a)

(b) 5´ T T C C T A T G A C T C A T C C A G T T 3´
3´ A A G G A T A C T G A G T A G G T C A A 5´

FIGURE 13.39 **The structure of a complex between the DNA-binding domain of Gcn4 and a fragment of DNA.** (a) The DNA-binding region, which is basic (residues 226–249), is shown in green. The leucine zipper region (residues 250–281) is shown in blue. Leucine side chains are shown as yellow space filling structures. (b) The DNA used to obtain the crystal structure in (a). Each polypeptide chain forms a helical structure that fits into the major groove of a half-site at either end of the quasi-palindrome that is shown in red. (Part a structure from Protein Data Bank 1YSA. Ellenberger, T. E., et al. 1992. *Cell* 71:1223–1237. Prepared by B. E. Tropp.)

BOX 13.3: CLINICAL APPLICATION

The AP-1 Family of Transcription Activator Proteins

AP-1 proteins are a family of transcription activator proteins that regulate many different genes, including some that control cellular proliferation. The founding member of the AP-1 family was identified as a transcription activator protein that binds to enhancer regions of simian virus 40 (SV40) DNA. The search for the polypeptides that form AP-1 was facilitated by information obtained from earlier studies of retroviruses. In particular, FBJ murine osteosarcoma virus was known to have the *v-fos* gene, which induces bone cancer, and avian sarcoma virus-17 was known to have the *v-jun* (*ju-nana* is the Japanese word for 17) gene, which induces fibrosarcoma in chickens. Genes such as *v-fos* and *v-jun* that transform normal cells into cancer cells are oncogenes, and their protein products are oncoproteins. Subsequent studies showed these viral genes have cellular counterparts called *c-fos* and *c-jun*. In fact, the viral genes originated in the cell. Sequence studies revealed each polypeptide has a bzip region that is quite similar to that in Gcn4 (FIGURE B13.3a), suggesting they too might be bzip transcription activator proteins. Further study revealed that c-Fos and c-Jun combine to form the founding member of the AP-1 family of transcription activator proteins; that is, AP-1 is a c-Fos • c-Jun heterodimer. Additional cellular homologs of each kind of polypeptide were discovered later.

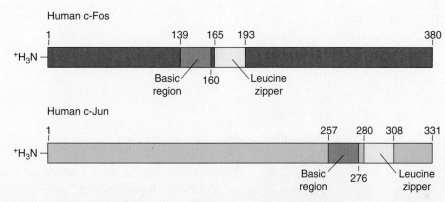

FIGURE B13.3a **Schematic representations of the functional domains of the c-Jun and c-Fos proteins.** The basic region is shown in green and the leucine zipper in yellow.

c-Jun can bind to DNA as a homodimer or as part of a c-Jun • c-Fos heterodimer. c-Fos does not bind to DNA as a homodimer. The reason some polypeptide combinations lead to dimer formation and others do not is apparent from their helical wheel structures (FIGURE B13.3bcd). The side chains that are immediately outside the hydrophobic core (positions e and g) promote dimer formation in Jun • Fos because of attractive charge interactions. Jun • Jun dimer formation is not as favorable as Jun • Fos formation because fewer attractive interactions are possible among the side chains. Fos • Fos dimer does not form because the large number of negatively charged glutamate (E) side chains in positions e and g are mutually repulsive.

The Fos • Jun heterodimer can bind to a DNA recognition element with a TGA(C/G)T(C/A)A sequence motif in two orientations (FIGURE B13.3e). Different AP-1 homo- and heterodimers bind to different DNA recognition elements, allowing for greater diversity of promoter and enhancer specificity. When a *c-fos* or *c-jun* gene is mutated or inappropriately expressed, its product can transform a normal cell into a cancer cell. Normal cellular genes such as *c-fos* and *c-jun* that have the potential to cause cancer when mutated or inappropriately expressed are said to be proto-oncogenes.

(continues)

BOX 13.3: CLINICAL APPLICATION

The AP-1 Family of Transcription Activator Proteins (*continued*)

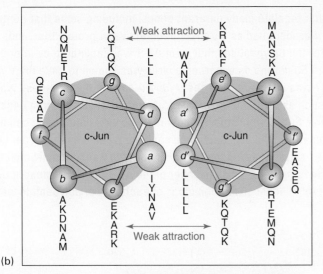

(b)

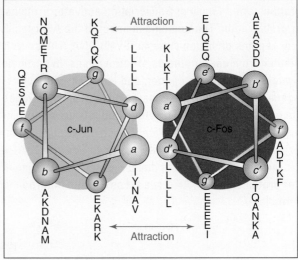

(c)

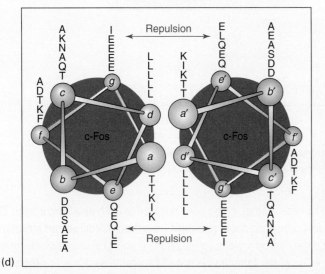

(d)

FIGURE B13.3b, c, AND d Wheel structures for c-Fos and c-Jun. (b) The c-Jun • c-Jun homodimer can form because attractive electrostatic interactions exist between side chains in positions e and g. (c) The c-Jun • c-Fos heterodimer formation is even more favorable than c-Jun • c-Jun homodimer formation because the attractive interactions between side chains in positions e and g are even greater. (d) The c-Fos • c-Fos homodimer does not form because of charge repulsion between the negatively charged side chains in positions e and g.

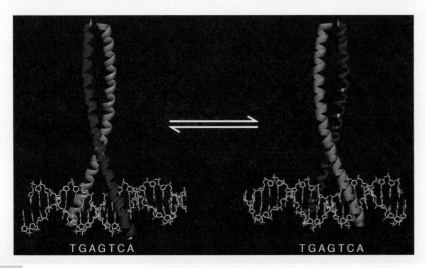

FIGURE B13.3e Fos • Jun heterodimer bound to an asymmetric AP1 site (TGAGTCA) in two orientations. Jun and Fos bzip domains are colored blue and red, respectively. The AP1 recognition element base sequence is indicated below the structures. The two structures represent opposite orientations of Fos • Jun binding to the same DNA sequence and reflect heterodimer rotation by about a half turn around the dimer axis. (Structure from Protein Data Bank 1FOS. Glover, J. N., and Harrison, S. C. 1995. *Nature* 373:257–261. Prepared by B. E. Tropp.)

coordinated to six cysteine residues. The best-studied member of this family is Gal4, a transcription activator protein that regulates genes coding for galactose metabolism in *Saccharomyces cerevisiae*. The active form of Gal4, which consists of a pair of identical 881-residue polypeptide subunits, binds to *Gal4* sites on DNA. Gal4's DNA-binding domain is at its amino terminus, extending from residues 1 to 64. A peptide fragment bearing this amino terminal sequence is a monomer in solution but forms a homodimer upon binding to DNA. The crystal structure for the complex formed when the amino terminal peptide fragment binds to DNA reveals three structural modules (FIGURE 13.40). Residues 1 to 40 form the Zn_2C_6 binuclear cluster, residues 51 to 64 fold into an α-helix that serves as a weak dimerization region, and residues 41 to 50 form a linker that connects the Zn_2C_6 binuclear cluster to the weak dimerization region. The weak dimerization region in one subunit interacts with the weak dimerization region in the other subunit to form a coiled coil that helps to stabilize the dimer.

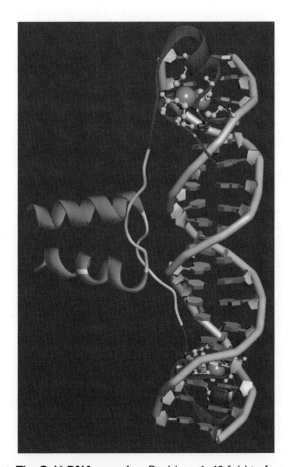

FIGURE 13.40 **The Gal4-DNA complex.** Residues 1–40 fold to form the Zn_2C_6 binuclear cluster (red) that interacts with DNA. Residues 51–64 form a weak dimerization region (magenta). The Zn_2C_6 binuclear cluster and weak dimerization region are connected by a linker (light gray) consisting of residues 41–50. Zinc atoms are shown as magenta spheres and cysteine residues as yellow ball and stick structures. (Structure from Protein Data Bank 1D66. Marmorstein, R., et al. 1992. *Nature* 356:408–414. Prepared by B. E. Tropp.)

13.8 DNA-Binding Domains in Transcription Activator Proteins

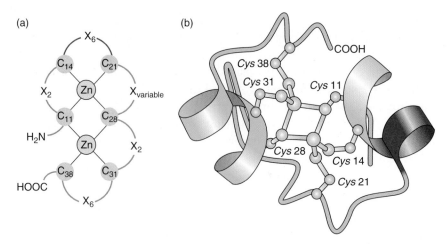

FIGURE 13.41 **DNA-binding Zn_2C_6 binuclear cluster region of the Gal4 subunit.** (a) Schematic diagram showing that the Zn_2C_6 binuclear cluster contains two zinc atoms each bound to four cysteine residues, two of which bridge the zinc atoms. The diagram also shows the number of amino acids in the loop regions between the cysteine ligands. (b) Three-dimensional representation of the Zn_2C_6 binuclear cluster. Folding with the Zn_2C_6 binuclear cluster creates two short α-helices. The region shown in red is involved in the sequence-specific DNA interactions. The Zn_2C_6 binuclear cluster stabilizes the structure to give the proper fold for DNA binding. (Modified from Branden, C., and Tooze, J. 1999. *Introduction to Protein Structure* (1st ed). Garland Science. Used with permission of John Tooze, The Rockefeller University.)

The Gal4 Zn_2C_6 binuclear cluster is shown in FIGURE 13.41. Two of the cysteines bind to both zinc ions, forming a pair of bridges between the zinc ions (Figure 13.41a). Folding within the Zn_2C_6 binuclear cluster creates two short α helices (Figure 13.41b). The C-terminus of the first helix (shown in red in Figure 13.41b) makes specific contacts with bases in the major groove of the *Gal4* site, which is a 17-bp semi-palindrome with highly conserved CCG inverted repeats at either end. Each Zn_2C_6 binuclear cluster in the homodimer makes specific contacts with one of the CCG triplets at the end of the *Gal4* site. The linker and dimerization region also contribute to binding by contacting the phosphate backbone in the 11-bp spacer region. Although the Gal4 DNA-binding domain tolerates base pair substitutions in the 11-bp spacer, it has very little tolerance for deletions or insertions.

Other transcription activator proteins belonging to the Zn_2C_6 binuclear cluster family also bind to semi-palindromes with CCG triplets at either end but exhibit varying specificities for the spacer lengths between the triplets. For instance, the yeast transcription activator protein Ppr1 (pyrimidine pathway regulator 1), which regulates genes involved in pyrimidine nucleotide metabolism, binds to a 12-bp semi-palindrome with a CCG triplet at either end. Binding, however, will only take place if the spacer is 6-bp long.

Richard J. Reece and Mark Ptashne took advantage of the differences in binding specificity between Gal4 and Ppr1 to demonstrate that the linker region (and not the Zn_2C_6 binuclear cluster) determines the spacer length that will be recognized. They did so by studying the specificity of chimeric DNA-binding domains, which they constructed by using recombinant DNA techniques (FIGURE 13.42). They observed Chimera 1 that contains the Ppr1 Zn_2C_6 binuclear cluster, the Gal4 linker, and the Gal4 dimerization domain binds to the Gal4 DNA site, whereas Chimera 2 that contains the Ppr1 Zn_2C_6 binuclear cluster, the Ppr1 linker, and the Gal4 dimerization domain binds to the Ppr1 DNA site. Thus, the highly conserved Zn_2C_6 binuclear cluster recognizes the CCG triplet at either end of the semi-palindrome and the linker region selects among different semi-palindrome sites based on their spacer size.

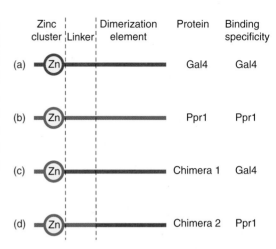

FIGURE 13.42 **Domain swapping experiment with Gal4 and the related transcription activator protein Ppr1.** (a) The Gal4 transcription activator protein binds to the Gal DNA site. (b) The Ppr1 transcription activator protein binds to the Ppr1 DNA site. (c) Chimera 1, which contains the Ppr1 zinc cluster, the Gal4 linker, and the Gal4 dimerization domain, binds to the Gal4 DNA site. (d) Chimera 2, which contains the Ppr1 zinc cluster, the Ppr1 linker, and the Gal4 dimerization domain, binds to the Ppr1 DNA site. Thus, the linker region determines binding specificity. (Modified from Branden, C., and Tooze, J. 1999. *Introduction to Protein Structure* (1st ed). Garland Science. Used with permission of John Tooze, The Rockefeller University.)

13.9 Activation Domains in Transcription Activator Proteins

Activation domains tend to be intrinsically disordered.

Although a great deal is now known about structure–function relationships in DNA-binding domains from a wide variety of transcription activator proteins, much less is known about the activation domains in these same transcription activator proteins. This disparity in knowledge reflects difficulties in studying activation domains. In contrast to DNA-binding domains, which usually fold to form specific three-dimensional structures, activation domains tend to be intrinsically disordered but may assume a defined structure when they interact with other components of the transcription machinery. Furthermore, activation domains do not appear to share common structural features such as conserved amino acid sequences. Activation domain function has been just as difficult to study as activation domain structure. Although activation domains work by binding to specific proteins in the transcription machinery, it is quite difficult to determine the physiological target. In fact, many activation domains appear to have several possible targets, and it is quite possible that many of these targets are physiologically important. DNA-bound transcriptional activator proteins may work by activation by recruitment, activation by conformational change, or by a combination of the two mechanisms:

1. *Activation by recruitment mechanism:* The activation domain interacts with one or more components of the transcription machinery and stabilizes the binding of the component(s) to the template DNA.
2. *Activation by conformational change mechanism:* The activation domain somehow induces a conformational change in one or more components of the transcription machinery that are bound to it and thereby stimulates RNA polymerase II to initiate transcription.

Gal4 has DNA-binding, dimerization, and activation domains.

Despite the formidable difficulties in elucidating the function of the activation domain, a general picture has started to emerge. Because

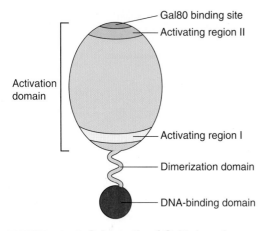

FIGURE 13.43 Schematic of Gal4 domain structure. Gal4 has a DNA-binding domain, dimerization domain, and activation domain. The DNA-binding domain (red) consists of a Zn_2C_6 binuclear cluster and a weak dimerization region (shown in greater detail in Figure 13.40). The dimerization domain, which greatly increases dimer stability, is shown in light blue. The activation domain (green) includes activating region I (yellow) and activating region II (magenta), which function as weak and strong activating regions, respectively. Activating region II contains a Gal80 binding region, which is shown in dark blue. (Adapted from Ptashne, M., and Gann, A. 2002. *Genes & Signals*. Cold Spring Harbor Laboratory Press.)

much of this picture comes from studying transcription activator proteins in yeast, we begin there. Thus far, our examination of Gal4 has been limited to the structure and function of the segment spanning residues 1 to 64, which folds to form the Zn_2C_6 binuclear cluster and the dimerization region that comprise the DNA-binding domain. Two other domains make essential contributions to Gal4 function (FIGURE 13.43). The segment spanning residues 66 to 94 folds to form the dimerization domain and that spanning residues 95 to 881 forms the activation domain. The dimerization domain greatly increases dimer stability. The activation domain includes activating regions I and II, which function as weak and strong activating regions, respectively. It also contains a binding site for a protein called Gal80 that prevents the activation domain from interacting with other transcription factors (see below).

Other proteins act along with Gal4 to regulate *GAL* genes.

Gal4 requires the assistance of several other proteins to switch *GAL* genes on and off in response to the carbon source in the growth medium. Although the discussion that follows focuses on the regulation of *GAL1* (the gene that encodes galactokinase), the lessons learned also apply to the other *GAL* genes. *GAL1* is switched on when galactose is added to growth medium in the absence of glucose and is switched off when the galactose is consumed or glucose is added. Galactose therefore induces *GAL1* gene expression and glucose represses it. Two types of proteins bind to regulatory sites upstream from the *Gal1* transcription start site. The first of these, Gal4, which was described above, binds to four sites within the upstream activating sequence. This sequence designated UAS_{GAL} spans about 118 bp and is located about 275 bp upstream from the transcription initiation site (FIGURE 13.44). Gal4 can stimulate transcription if just one of the four 17-bp sites is present. The second protein, designated the Mig1 protein (multicopy inhibitor of *GAL* gene expression), has a DNA-binding domain that makes sequence-specific contacts with the so-called *Mig1* site, which is located between UAS_{GAL} and the transcription initiation site.

In the absence of both galactose and glucose, Gal4 binds to the four sites in UAS_{GAL} (FIGURE 13.45a). Gal4 binding fails, however, to switch on the transcription of the *GAL1* gene. The reason for this

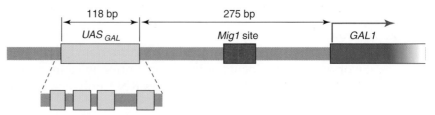

FIGURE 13.44 The *GAL1* gene and the region just upstream. The four Gal4 binding sites within UAS_{GAL} are depicted in the enlarged section. The Mig1 binding site is located between UAS_{GAL} and *GAL1*. UAS_{GAL} spans about 118 bp and each Gal4 site is 17 bp. (Adapted from Ptashne, M., and Gann, A. 2002. *Genes & Signals*. Cold Spring Harbor Laboratory Press.)

(a) Absence of galactose

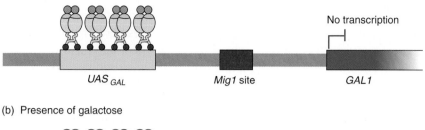

(b) Presence of galactose

FIGURE 13.45 **Effect of galactose as an inducer on *Gal1* gene transcription.** (a) In the absence of galactose, Gal80 dimers (orange spheres) bind to the Gal80 binding site, preventing the activation domain (green and magenta) from interacting with components of the transcription machinery and blocking transcription. (b) In the presence of galactose, Gal3 (not shown) binds to Gal80, causing Gal80 dimers to dissociate and be released from Gal4. The Gal4 activation domain (green and magenta) is now free to interact with components of the transcription machinery so that Gal1 can be transcribed. Color coding for Gal4 is the same as in Figure 13.43 (activating region II magenta, dimerization domain cyan, and Zn_2C_6 binuclear cluster red). For simplicity, the Gal80 binding region and activating region I are not shown.

failure is that Gal80, a homodimer, binds to the Gal80 binding site in the Gal4 activation domain, preventing this domain from interacting with other transcription components. Adding galactose to the growth medium relieves the inhibition caused by Gal80 and permits *GAL1* transcription to take place (Figure 13.45b). Gal3, a cytoplasmic protein homologous to Gal1 but lacking galactokinase activity, is required to relieve the inhibition caused by the Gal80 homodimer. Gal3 binds to galactose and ATP to form a Gal3 • galactose • ATP complex, which somehow promotes the release of Gal80 from Gal4.

GAL1 transcription stops when glucose is added to a growth medium that contains galactose. Glucose acts through a protein kinase called Snf1 kinase (<u>s</u>ucrose <u>n</u>on-<u>f</u>ermenting kinase) to regulate Mig1 function. When glucose is absent, Snf1 kinase phosphorylates Mig1. The phosphorylated form of Mig1 is a cytoplasmic protein and so cannot bind to the *Mig1* site in the nucleus. Glucose inactivates Snf1 kinase so it can no longer phosphorylate Mig1. The dephosphorylated form of Mig1 moves to the nucleus and binds to the *Mig1* site (FIGURE 13.46). Then additional transcription factors bind to Mig1 and block *GAL1* transcription.

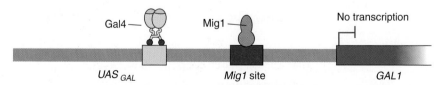

FIGURE 13.46 **The effect of glucose on *Gal1* transcription.** In the presence of glucose the dephosphorylated form of Mig1 binds to the *Mig1* site, a necessary step for *Gal1* gene transcription inhibition.

13.9 Activation Domains in Transcription Activator Proteins

The activation domain must associate with a DNA-binding domain to stimulate transcription.

Starting in the mid-1980s Ptashne and coworkers performed a series of experiments with Gal4 that helped to explain how activation domains work. These experiments typically involved constructing recombinant plasmids bearing genes that code for modified Gal4 transcription activator proteins. Modifications included deleting various segments from the 881-residue Gal4 polypeptide chain or creating hybrid proteins by joining Gal4 peptide fragments to peptide fragments from other proteins. The basic experimental objective—to determine whether the recombinant proteins function as transcription activator proteins—was achieved by first transforming a yeast strain that lacked Gal4 with a plasmid bearing a gene coding for a modified Gal4 protein and a plasmid bearing a reporter gene and then measuring the transformant's ability to produce a reporter gene product such as β-galactosidase. Results from one set of these experiments are summarized in FIGURE 13.47. The normal Gal4 transcription activator protein stimulated transcription of a reporter gene with a UAS_{GAL} site upstream from its transcription initiation site (Figure 13.47a). However, a peptide fragment containing just the first 100 residues in the Gal4 chain (a region that includes the DNA-binding domain, the dimerization domain, and a sequence that directs Gal4 to the cell nucleus) did not activate gene transcription even though it did bind to the UAS_{GAL} site (Figure 13.47b). The complementary fragment (residues 100–881) also did not activate reporter gene transcription (Figure 13.47c). The fact that the activation domain could not stimulate transcription on its own indicates that the conformational change model is not sufficient to explain how the Gal4 activation domain stimulates transcription. It does not rule out,

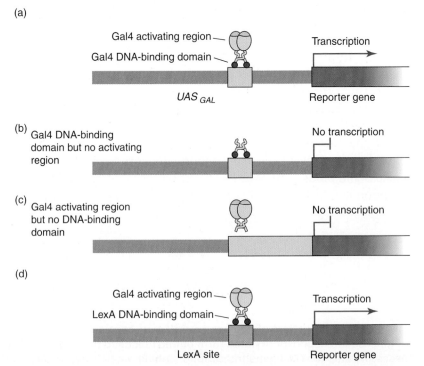

FIGURE 13.47 **Gal4 domain function.** (a) The Gal4 transcription activator stimulates transcription from a reporter gene with a UAS_{GAL} site upstream from the transcription initiation site. (b) A peptide fragment (residues 1–100) containing the Gal4 DNA-binding domain and the dimerization domain does not stimulate transcription from a reporter gene with a UAS_{GAL} site upstream from the transcription initiation site. (c) A complementary Gal4 fragment (residues 100–881) containing the activation regions cannot stimulate transcription from a reporter gene with a UAS_{GAL} site upstream from the transcription initiation site. (d) A LexA-Gal4 hybrid protein containing the DNA-binding domain from the bacterial repressor protein LexA attached to the Gal4 activation region (residues 100–881) stimulates transcription from a reporter gene with a *LexA* site upstream from the transcription initiation site.

however, the possibility that the conformational change model makes some contribution to transcription activation.

Linking the complementary fragment (residues 100–881) to a different DNA-binding domain produced a hybrid protein that did function as a transcription activator protein. For example, attaching the complementary fragment to the DNA-binding domain of an *E. coli* repressor known as LexA produced a LexA-Gal4 hybrid protein that could initiate transcription from a reporter gene with a LexA site upstream from the transcription initiation site (Figure 13.47d). This type of experiment is known as a **domain swap** because the DNA-binding domain from one protein replaces that of another. The fact that the LexA-Gal4 hybrid protein activated transcription is consistent with the activation by recruitment model. The experiments shown in Figure 13.47 indicate that a transcription activator protein requires both a DNA-binding domain and a transcription activation domain to work but that an assortment of DNA-binding domains, even one from bacteria, can replace that from Gal4.

Ptashne and Ma performed deletion analysis experiments that revealed the Gal4 activation domain contains two discrete activating regions. Activating region I extends from residues 148 to 196. Activating region II extends from residues 768 to 881 and also includes the Gal80 repressor-binding site. Attaching either activating region to a DNA-binding domain creates a transcription activator protein; therefore, large portions of Gal4 are not required for gene activation. Each of the activating regions had a much greater than expected number of the negatively charged amino acid residues aspartate and glutamate, causing these regions to have net negative charges. Genetic studies of activating region I showed activating region strength is related to negative charge. Activating region I became weaker when mutations reduced negative charge by replacing aspartate or glutamate residues with uncharged residues. Conversely, activating region I became stronger when mutations increased negative charge by replacing one of the few lysine or arginine residues with uncharged residues. Attempts to demonstrate activating regions from other transcription activator proteins also have net negative charges were successful in some cases but failed in many others.

Net negative charge is not the only factor that determines Gal4 activating region strength. Gal4 activating regions I and II also have hydrophobic residues that appear to contribute to their function. The arrangement of the glutamate and aspartate residues in the hydrophobic background is probably important for activity because scrambling the order of residues within activating region I or II produces nonfunctional peptides. There does not appear to be a requirement for some specific sequence, however, because about 1% of randomly generated *E. coli* peptide fragments can function as activating regions when attached to a DNA-binding domain. One hypothesis is that negatively charged side chains in a hydrophobic background somehow create a sticky peptide segment that can bind to other transcription machinery components and help recruit the components to the preinitiation complex. Natural selection may have favored other types of sticky peptide segments for activating regions that do not have net negative charges.

Ma and Ptashne realized the relative ease with which they were able to construct active hybrid transcription factors suggested there are few, if any, stereospecific restrictions on how the activating region must be attached to the DNA-binding domain. This realization led them to devise an experiment to test whether the DNA-binding domain and the activation region can be on separate polypeptide chains that interact to form a protein complex. Their approach was to modify two polypeptides that interact with each other through noncovalent bonds so the first polypeptide had the DNA-binding domain and the second had the activating region. They constructed the polypeptide with the DNA-binding domain by deleting residues 148 to 850 from Gal4 and the polypeptide with the activating region by inserting an acidic bacterial peptide into the Gal80 repressor. Then they tested the ability of yeast cells that make various combinations of the two polypeptides to transcribe UAS_{GAL}-lacZ fusion genes in their chromosomes. The results of these experiments are summarized in FIGURE 13.48.

Cells that make the Gal4 derivative synthesize low levels of β-galactosidase (Figure 13.48a). This result is expected because the Gal4 derivative still has part of activating region II (residues 851–881). Cells that also make wild-type Gal80 synthesize only trace quantities of β-galactosidase (Figure 13.48b). This result is also expected because the Gal80 protein binds to the Gal4 derivative and represses transcription. Replacing wild-type Gal80 with a Gal80 that has a bacterial peptide insert, which was previously shown to function as an activating region, gives rise to high levels of β-galactosidase synthesis (Figure 13.48c).

These experiments show that a polypeptide with a DNA-binding domain and a second polypeptide with an activating region can combine through noncovalent interactions to form a functional transcriptional activator protein. This experiment is therefore also consistent

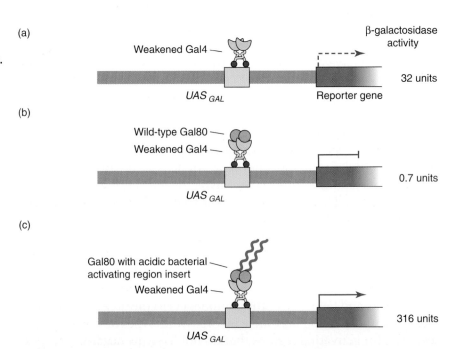

FIGURE 13.48 Experiment testing whether the DNA-binding domain and the activating region can be on separate polypeptide chains. (a) Cells that synthesize a weakened Gal4 in which part of activating region II (residues 851–881, magenta) is attached directly to the dimerization domain (cyan) synthesize a low level of β-galactosidase from a UAS_{GAL}-LacZ fusion gene. (b) Cells that also make wild-type Gal80 (orange) synthesize only trace quantities of β-galactosidase because Gal80 binds to the weakened Gal4 and represses transcription. (c) Cells that synthesize the weakened Gal4 and a modified Gal80 that has a bacterial peptide insert (green) that can function as an activating region synthesize high levels of β-galactosidase.

with the activation by recruitment model. The discovery that the DNA-binding domain and the activation region can be on separate polypeptide chains that combine to form a protein complex that functions as a transcription activator protein led to the development of a very valuable laboratory technique called the yeast two-hybrid assay that permits us to detect polypeptides that interact through noncovalent interactions (**BOX 13.4 IN THE LAB: YEAST TWO-HYBRID ASSAY**).

13.10 Mediator

Squelching occurs when transcription activator proteins compete for a limiting transcription machinery component.

Investigators initially thought eukaryotic transcription activator proteins stimulate gene transcription by making direct contact with one or more components of the basal transcription machinery. Although a considerable body of evidence in fact shows transcription activator proteins do contact basal transcription machinery components such as TBP, TAFs, and TFIIB, these contacts are often not sufficient for activated transcription. An entirely new protein complex, designated Mediator, is required for transcription activator and repressor proteins to work.

The first hint of Mediator's existence came from studies performed in yeast and mammalian cells in 1988 that showed a high intracellular concentration of one transcription activator protein inhibits other transcription activator proteins from stimulating gene transcription. Grace Gill and Mark Ptashne proposed this inhibitory phenomenon, which they called **squelching**, occurs when transcription activator proteins compete among themselves for some limiting target (FIGURE 13.49). TBP, TFIIB, and RNA polymerase II each seemed to be a reasonable candidate for the limiting target because each can bind to transcription activator proteins. The problem, however, was to devise some method for determining which, if any, of these components of the transcription machinery was the limiting target. Roger Kornberg and coworkers thought one method for solving the problem would be to reproduce the squelching phenomenon in a cell-free system. After achieving this goal by adding transcription activator proteins to a yeast nuclear extract, they demonstrated squelching persisted even after excess quantities of RNA polymerase II or any given general transcription factor was added to the extract. These experiments indicated that some still unknown factor was the limiting target responsible for squelching. As we see in the next section, the search for this target led to the discovery of **Mediator**.

Mediator is required for activated transcription.

The next major step in the discovery of Mediator occurred when Kornberg and coworkers noticed a protein fraction, which they had prepared while trying to isolate transcription factors from yeast, appeared to enable activated transcription. They suspected this fraction

BOX 13.4 : IN THE LAB

Yeast Two-Hybrid Assay

The yeast two-hybrid assay permits us to detect polypeptides that interact through noncovalent interactions.

In 1989 Stanley Fields and Ok-kyu Song devised an ingenious technique known as the two-hybrid assay that takes advantage of the modular nature of transcription activator proteins to probe protein–protein interactions (FIGURE B13.4). The assay begins with the construction of two plasmids, each directing the expression of a different hybrid protein. The first hybrid protein has some protein, X, fused to a DNA-binding domain and the second has some other protein, Y_1 or Y_2, fused to a strong activation domain (AD). The two plasmids are introduced into a yeast cell with a reporter gene with a regulatory site that is a target for the first hybrid protein's DNA-binding domain.

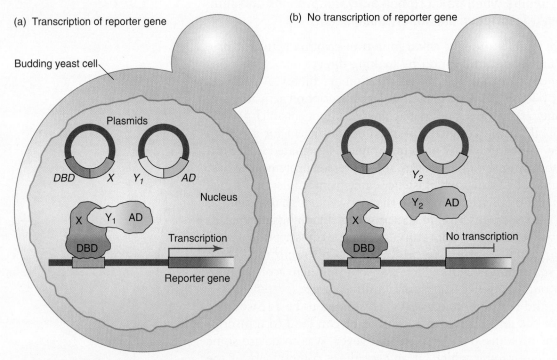

FIGURE B13.4 **How the two-hybrid system works.** (a) The bait protein, protein X, is fused to a DNA-binding domain (DBD). Protein Y_1 is fused to a transcription activation domain (AD). Both hybrid proteins are expressed from plasmids in a yeast cell. Proteins X and Y_1 interact, leading to the activation of a reporter gene, which codes for a protein that permits the yeast to grow on a defined medium. (b) Proteins X and Y_2 do not interact. The reporter gene therefore is not expressed and the yeast cannot grow. (Adapted from Uetz, P. H., Hughes, R. E., and Fields, S. 1998. *Focus* 20:62–64.)

To see how this assay works, let's assume that proteins X and Y_1 interact, whereas proteins X and Y_2 do not. Then a cell with hybrid proteins X-DBD and Y_1-AD will have a functional transcription activator that stimulates reporter gene transcription (Figure B13.4a). In contrast, a cell with hybrid proteins X-DBD and Y_2-AD will not have a functional transcription activator and therefore will be unable to transcribe the reporter gene (Figure B13.4b).

One important application of the two-hybrid assay is to search for proteins that interact with a specific protein of interest. The assay uses the specific protein of interest as "bait" to catch proteins that interact with it ("prey"). The first step is to construct yeast so they have a plasmid directing the synthesis of a hybrid protein with a DNA-binding domain fused to the bait protein (X in Figure B13.4) and a selectable reporter gene such as *LEU2* or *HIS3* with an upstream site for the DNA-binding domain. The next step is to transform the yeast with a plasmid library that has been constructed so each transformed cell will express a hybrid protein with the same activation domain linked to different prey proteins (proteins comparable with Y_1 and Y_2 proteins in Figure B13.4). Only transformants that receive hybrid prey proteins that interact with the hybrid bait protein can grow on minimal medium lacking leucine or histidine. Plasmids can be isolated from these transformants and sequenced and the sequence information used to search the protein data bank to identify the prey protein.

Despite numerous false-positives and negatives, the two-hybrid assay has been quite successful in identifying proteins that interact with one another. The two-hybrid assay also has been useful for identifying mutant proteins with amino acid substitutions that cause stronger or weaker binding to some other target protein.

contained a component of the transcription machinery that somehow interacted with transcription activator proteins and the basal transcription machinery so it could stimulate transcription or serve as a target for squelching. Because the putative transcription component seemed to serve as a go-between for transcription activator proteins and the basal transcription machinery, it was named Mediator.

Kornberg's group set out to isolate Mediator from the protein fraction using an assay system consisting of purified general transcription factors, RNA polymerase II, the transcription activator protein Gcn4, and two different DNA templates (one with Gcn4 binding sites and the other with Gal4 binding sites). Transcription from the two templates produced radioactive RNA molecules of different sizes that were resolved by gel electrophoresis, making it possible to distinguish activated transcription due to Gcn4 sites from basal transcription due to Gal4 sites. Taking advantage of this assay system, Kornberg and

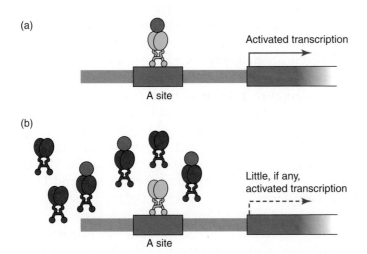

FIGURE 13.49 **Squelching.** (a) Transcription activator A (light green) binds to the target protein (blue), permitting activated transcription to occur. (b) Transcription activator B (red) binds to the target protein (blue) so that very little of the target protein is available to bind to transcription activator A (light green). Squelching results from the competition between the transcription activators for the target protein.

13.10 Mediator

coworkers succeeded in isolating Mediator, a protein complex, which contains 21 different proteins.

Genes that code for each of the 21 subunits in the core Mediator from yeast have been identified. Several of these genes were initially identified through the use of genetic screens to detect mutations that affect transcription. For instance, in 1989 Michael Nonet and Richard Young identified five genes, each coding for a different Mediator polypeptide subunit, while studying a yeast mutant with a truncated CTD in Rpb1. The truncated CTD causes the mutant to have cold- and temperature-sensitive phenotypes. Nonet and Young sought to identify proteins involved in CTD function by searching for suppressor mutations that would restore the mutant's ability to grow in the cold. Several such suppressor mutations were found, leading to the identification of several new genes, which were named *SRB1*, *SRB2*, and so forth to reflect the fact that the genes were first detected in a genetic screen for suppressors of RNA polymerase B (an alternate name for RNA polymerase II). Additional genes for Mediator subunits were identified through the use of genetic screens that were designed to study some specific aspect of transcription. The original names given to these genes reflected the genetic screens that were used or the transcriptional property studied. For example, genes that code for two of the Mediator subunits, *RGR* and *Gal11*, were named for their involvement in resistance to glucose repression and galactose metabolism, respectively. The remaining Mediator genes had not been identified at the time Kornberg's group isolated Mediator and so were named *MED1*, *MED2*, and so forth.

Studies from many different laboratories show Mediator is a fundamental part of the RNA polymerase II transcription machinery in all eukaryotes. Moreover, the Mediator subunits from all organisms seem to be homologous with those in yeast and to have similar shapes. Because this homology has only been recognized recently, the literature contains a rather confusing nomenclature system. An example will help to illustrate the problem. Several mammalian Mediator complexes have been isolated, and each has been given a name to reflect its function or the method used to isolate it. Two of the best-studied mammalian Mediators thus are named thyroid hormone receptor-associated protein (TRAP) and vitamin D receptor-interacting protein (DRIP) despite the fact that these two Mediators are very similar and perhaps identical. A unifying nomenclature system was recently proposed in which nearly all Mediator subunits are named MED followed by a number designation. For instance, Rgr in yeast and TRAP170 in humans are designated MED14, whereas Gal11 in yeast and Arc105 in humans are named MED15. This nomenclature system acknowledges that Mediator was first discovered in yeast. The original yeast MED subunits therefore retain their names (MED1–11). The remaining yeast MED subunits are given names starting with MED12 in order of decreasing molecular mass.

Biochemical, genetic, and electron microscopy studies provide insights into yeast Mediator subunit organization and the interactions between Mediator and RNA polymerase II. Michel Werner and

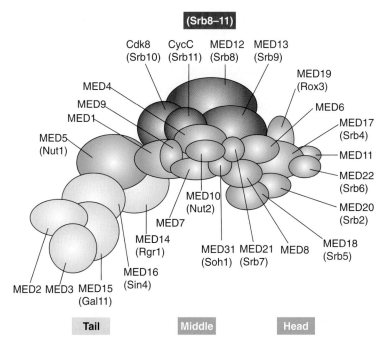

FIGURE 13.50 **Topological organization of yeast Mediator.** This model is based in large part on results from a pair-wise two-hybrid experiment used to investigate protein–protein contacts between budding yeast Mediator subunits as well as the relative size of the Mediator subunits. The core mediator is organized into three subcomplexes, termed the head (blue), middle (green), and tail (yellow) modules. The Srb8-11 module (red) is present in many, but not all, yeast Mediator complexes. Where applicable, older names for the Mediator subunits are given in parentheses below the new names. (Modified from Guglielmi, B., et al. 2004. A high resolution protein interaction map of the yeast Mediator complex. *Nucleic Acids Res* 32:5379–5391. Reprinted by permission of Oxford University Press. Used with permission of Michel Werner, Service de Biologie Intégrative et Génétique Moléculaire CEA-Saclay and Nynke van Berkum and Frank Holstege, University Medical Center Utrecht.)

coworkers used pairwise two-hybrid analysis to investigate protein–protein contacts between budding yeast Mediator subunits. Based on these and related studies as well as the relative sizes of the Mediator subunits, they constructed the topological map of yeast Mediator shown in FIGURE 13.50.

Yeast Mediator appears to be organized into three subcomplexes: the head, middle, and tail. The head subcomplex seems to interact with the largest RNA polymerase II subunit (Rpb1) at its CTD and serve as a signal processor that directly influences RNA polymerase II activity. Cells that lack this subcomplex are nonviable. The middle subcomplex, which contains the MED9/10 module, also appears to interact with CTD. It is thought to function in regulatory signal transfer after binding to the transcription activator protein. The tail subcomplex includes most of the Mediator subunits originally identified by genetic screens for transcription mutants; it is thought to sense signals from gene-specific transcription activator proteins such as Gcn4. Cells that lack the tail domain are viable.

The yeast Mediator complex associates with activators at the UAS in active yeast genes.

The discovery that the approximately 1-MDa Mediator complex is essential for activated transcription in eukaryotes was quite surprising because so many other proteins were already known to participate in this process. Mediator complexes were isolated from various eukaryotes with bound activator protein or RNA polymerase II, suggesting Mediator functions as the communication link from activator protein to RNA polymerase II. Further study was required to determine how Mediator interacts with the preinitiation complex.

13.10 Mediator

Kornberg and coworkers used a technique known as **chromatin immunoprecipitation (ChIP)** to determine how Mediator interacts with promoters. The steps involved in this technique (illustrated in FIGURE 13.51) are as follows. (1) Cells are treated with formaldehyde, a cell-permeable molecule that causes protein–protein and protein–DNA crosslinking. (2) Treated cells are lysed and their isolated chromatin is sheared by ultrasound to produce small fragments of about 100 to 1,000 bp. (3) Antibody is added to precipitate fragments that have the protein of interest bound to them. (4) The crosslinks are removed by incubating the fragments in buffer at low pH. (5) Proteins are removed and the specific primers are added to the free DNA to amplify regions of interest.

The ChIP experiments performed by Kornberg and coworkers showed that Mediator associates with UAS sites under activating conditions and not with the core promoter. Consistent with this finding is that temperature-sensitive TFIIB mutants continued to bind Mediator at the UAS sites even when the temperature was increased, so RNA polymerase could no longer bind to the DNA. Finally, cells were constructed that had a recombinant gene that was missing the TATA box but still had UAS sites. ChIP experiments demonstrated Mediator was able to bind to the UAS sites even though the general transcription factors and RNA polymerase II were no longer bound to the DNA. These ChIP experiments indicate that Mediator probably binds to the activation domain of an activator protein that is already bound to a UAS site. The picture that emerges from these studies is that Mediator functions as a bridge between gene-specific activator proteins and other components of the RNA polymerase transcription machinery at the regulatory sequence (FIGURE 13.52).

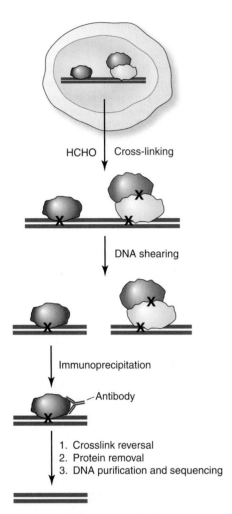

FIGURE 13.51 **The chromatin immunoprecipitation (ChIP) technique.** Cells are cultured under desired experimental conditions. Then a cross-linking agent, usually formaldehyde, is added to trap protein–protein and protein–DNA interactions. After cell lysis, DNA is subject to mechanical shearing and then immunoprecipitated using an antibody or combination of antibodies specific for the protein of interest. In the example shown here, the protein of interest is colored red and the other proteins are colored blue. After immunoprecipitation, the cross-links are reversed at low pH. Following protein removal, DNA is purified. The purified DNA may be studied after amplification by the polymerase chain reaction (PCR). (Modified from Sikder, D., and Kodadek, T. 2005. Genomic studies of transcription factor-DNA interactions. *Curr Opin Chem Biol* 9:38–45. Copyright 2005, with permission from Elsevier [http://www.sciencedirect.com/science/journal/13675931].)

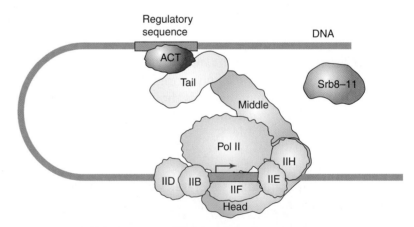

FIGURE 13.52 **Mediator interaction at the promoter.** Mediator (head, middle, and tail) serves as a bridge between the general RNA polymerase II transcription machinery and gene-specific activators (ACT, red). Contacts with RNA polymerase II are localized to the head and middle regions of Mediator. Activator interactions occur mainly in the tail regions of Mediator. The Srb8-11 module is involved in negative regulation of transcription. Only Mediator that lacks the Srb8-11 module can associate with RNA polymerase II. (Adapted from Björklund, S., and Gustafsson, C. M. 2005. *Trends Biochem Sci* 30:240–244.)

In higher eukaryotes, where most genes are regulated by a combination of activator proteins (and repressor proteins), the Mediator probably integrates the signals before transmitting the appropriate signal to RNA polymerase II and other components of the transcription machinery. We are still a long way from knowing just how Mediator and its components function.

13.11 Epigenetic Modifications

Cells remodel or modify chromatin to make the DNA in chromatin accessible to the transcription machinery.

Eukaryotic DNA molecules interact with histone and nonhistone proteins to form chromatin. The basic building block of chromatin, the nucleosome, consists of 1.65 turns of DNA wrapped around an octameric protein complex containing two copies each of the core histones H2A, H2B, H3, and H4. N-terminal tails of the four core histones stick out from the nucleosomes to contact DNA, other histones, and nonhistone proteins. These interactions help to stabilize the nucleosomal fiber as it folds into chromatin. Nucleosomes tend to block the transcription machinery and thereby inhibit gene expression. Chromatin remodeling complexes and modifying enzymes help to make the DNA available to the transcription machinery so the genes can be expressed.

Eukaryotic cells use ATP-dependent chromatin remodeling complexes to reposition nucleosomes, eject nucleosomes, unwrap nucleosomes, and exchange or eject histone dimers (FIGURE 13.53). One polypeptide subunit in each chromatin remodeling complex has ATPase activity. The organization of this subunit is the basis for grouping remodeling complexes into the SWI/SNF, ISWI, CHD, and INO80 families. The following discussion is limited to the yeast SWI/SNF family. Higher animals and plants have chromatin remodeling complexes that are similar in action to those in yeast.

The first clue to the existence of the SWI/SNF family of chromatin remodeling complexes came from genetic studies in yeast. Genetic screens for mating type <u>swi</u>tch mutants (*swi*) and <u>s</u>ucrose <u>n</u>on<u>f</u>ermentation mutants (*snf*), two apparently unrelated processes in yeast, led to the same gene, which is now called *SWI2/SNF2*. Additional studies established a connection between the *SWI2/SNF2* gene product and chromatin remodeling. One such study involved examining the effect that the *SWI2/SNF2* gene product has on chromatin surrounding the promoter for *SUC*, the gene that encodes invertase (the enzyme that hydrolyzes sucrose to fructose and glucose). This chromatin region is resistant to micrococcal nuclease cleavage in wild-type yeast cells that have not been induced for invertase production but becomes sensitive to micrococcal nuclease after the cells are subjected to inducing conditions. In contrast, this same chromatin region resists nuclease cleavage both before and after *SWI2/SNF2* mutants are subjected to inducing conditions.

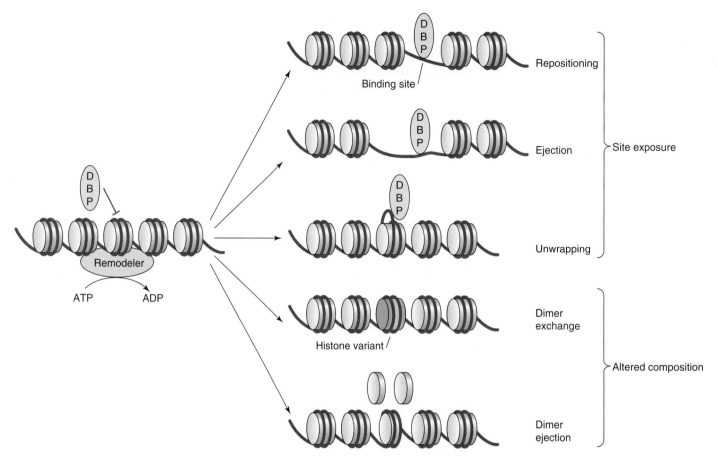

FIGURE 13.53 **Biochemical activities of ATP-dependent chromatin remodeling complexes.** The ATP-dependent chromatin remodeler (green) can reposition nucleosomes, eject nucleosomes, unwrap nucleosomes, exchange dimers, and eject dimers. The net effect is that a DNA site, which was originally covered, becomes available to bind by a DNA-binding protein (DBP) or the nucleosome composition is altered by replacing an H2A • H2B dimer with a histone variant or by ejecting an H2A • H2B dimer. (Reproduced from Clapier, C. R., and Cairns, B. R. 2009. *Annu Rev Microbiol* 78:273–304. Copyright 2009 by Annual Reviews, Inc. Reproduced with permission of Annual Reviews, Inc., in the format Textbook via Copyright Clearance Center.)

The interpretation of these experiments is that wild-type cells reorganize nucleosomes in the chromatin around the *SUC* promoter in response to inducing conditions, but *SWI2/SNF2* mutants do not. The *SWI2/SNF2* gene product was later isolated from yeast as part of a multisubunit Swi/Snf protein complex that uses the energy from ATP hydrolysis to remodel chromatin. The Swi/Snf complex is present at relatively low concentrations within yeast cells (approximately 100 copies per cell). Consistent with this low intracellular concentration, only about 5% of yeast genes require the Swi/Snf complex for activation during unsynchronized growth. Furthermore, mutants that lack this complex are viable. Because chromatin structure has a generally repressive effect on transcription, these findings suggest yeast must have other complexes that remodel chromatin.

In 1996 Roger Kornberg and coworkers isolated another protein complex from yeast that can remodel the structure of chromatin and named it the RSC complex. Some of the protein subunits in the RSC complex are homologous to subunits in the Swi/Snf complex, whereas others are not. The two complexes are assigned to the same chromatin remodeling family because their ATPase subunits are similar. This family is called the SWI/SNF family after its founding member. RSC is essential for cell viability, and its intracellular concentration is about 10-fold greater than that of the SWI/SNF complex. The ATPase subunit in RSC, Sth1, can act on its own to cause DNA to unwrap from nucleosomes but does so at a much slower rate than when it is part of the RSC complex.

Histone modification influences transcription of protein-coding genes.

Pioneering investigations by Vincent Allfrey and Alfred E. Mirsky in 1964 were the first to call attention to the fact that active genes have acetylated histones. Subsequent studies revealed acetylation takes place only at specific lysine residues located in the amino terminal tails of histones. These tails are not required for nucleosomal structural integrity but do contribute to the folding of nucleosome arrays into chromatin. The current hypothesis is that histone tail acetylation prevents interactions that lead to more compact chromatin structures repressive to transcription. Eukaryotes have both nuclear and cytoplasmic histone acetyltransferase (HAT) activities. Cytoplasmic HATs appear to modify histones before the histone's assembly into nucleosomes on newly replicated DNA, whereas nuclear HATs modify histones that are already part of the chromatin structure to make DNA accessible to the transcription machinery. We therefore focus on the nuclear HATs.

C. David Allis and coworkers identified and isolated the first nuclear HAT from *Tetrahymena* in 1996. This protein's sequence turned out to be remarkably similar to that predicted for the product of the yeast *GCN5* gene, which was first identified in a screen for mutants that cannot grow under limiting amino acids conditions. The sequence similarity suggested that Gcn5 is also a HAT. This hypothesis was confirmed when purified Gcn5 was shown to acetylate free histones at specific lysines. However, Gcn5 cannot acetylate histones that are part of a nucleosome. A biochemical search for proteins that contain Gcn5 and do acetylate histones in nucleosomes resulted in the isolation of the Ada and SAGA protein complexes that can do so. Eukaryotic cells also have histone deacetylases that remove acetyl groups from histone N-terminal tails. Deacetylation helps to stabilize the compact chromatin structure, causing the transcription of the affected DNA to be repressed. At least six different protein complexes identified in yeast have histone deacetylase activity. The two major deacetylase complexes, HDA and HDB, have been purified. Corresponding catalytic subunits in each complex share a considerable degree of sequence homology. Higher animals and plants have similar histone deacetylases.

Histones are subject to several other kinds of covalent modification that influence gene activity (FIGURE 13.54). In fact, it is rather remarkable that although histones are among the most highly conserved proteins in nature, they are also subject to more different types of posttranslational modifications than most other proteins. For example, histone kinases add phosphate groups to specific serine or threonine

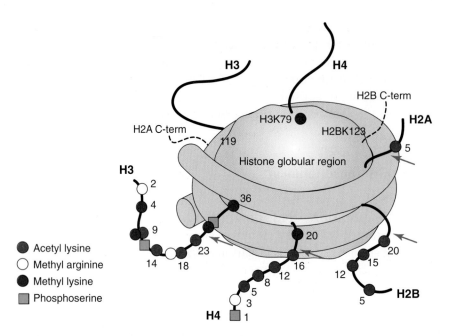

FIGURE 13.54 **Histone modifications on the nucleosome core particle.** The graphic represents a schematic of a nucleosome core particle. For simplicity modifications are shown on only one of the two copies of histones H3 and H4 and only one N-terminal tail is shown for histones H2A and H2B. The C-terminal tails of one histone H2A molecule and one histone H2B molecule are also shown (dashed lines). Colored symbols indicate histone sites that can be modified after a histone is synthesized. Residue numbers are given for each modification. Lysine-9 (K9) in histone H3 can be either acetylated or methylated. Sites marked by green arrows are susceptible to cutting by trypsin in intact nucleosomes. The schematic summarizes data from many different organisms and so any given organism may lack particular modifications. (Modified from Turner, B. M. 2002. Cellular memory and the histone code. *Cell* 111:285–291. Copyright 2002, with permission from Elsevier [http://www.sciencedirect.com/science/journal/00928674].)

residues and histone methyl transferases add methyl groups to specific lysine and arginine residues. A single lysine residue may have up to three attached methyl groups, and a single arginine residue may have one or two attached methyl groups. Enzymes and specific protein factors interact with chromatin based on the positions of the methylated residues and the number of methyl groups attached to them. Protein phosphatases remove phosphate groups, and demethylases remove methyl groups. Some enzymes that modify histones also modify other proteins, making it difficult to determine biological function from mutant studies. Histone-modifying enzymes introduce marks on the histones that are recognized by specific domains or subunits in chromatin remodelers. For example, chromatin remodeling complexes with a bromodomain bind histones that have an acetyl-lysine and those with a chromodomain bind to histones with a methyl-lysine.

The function of remodeling and modifying complexes does not end with the formation of a preinitiation complex poised to initiate transcription. RNA polymerase II also must contend with chromatin structure as it moves along a gene. The presence of nucleosomes in a DNA coding region slows down transcription. Eukaryotic cells use at least two pathways to deal with this problem. (1) Chromatin-remodeling complexes respond to the advancing transcription machinery by displacing histones from DNA. Displaced histones bind to chaperones and reassemble to form nucleosomes after the

DNA region has been transcribed. (2) The transcription machinery appears to be able to move through a nucleosome after only one H2A • H2B heterodimer has been removed. The displaced H2A • H2B heterodimer binds to a histone chaperone, which adds the heterodimer back to the nucleosome after transcription. Further investigations are required to establish the relative contribution each pathway makes to transcription elongation in chromatin and to determine if other pathways also are needed to permit the transcription elongation in chromatin.

DNA methylation plays an important role in determining whether chromatin will be silenced or actively expressed in vertebrates.

Many core promoters in vertebrates are located in CpG islands (CG-rich genomic regions that contain a high frequency of CG dinucleotides). Methylation of cytosine in these CpG islands leads to gene silencing by two mechanisms because (1) methyl-CpG prevents a transcription factor from binding to its cognate site or (2) methyl-CpG acts as a signal for histone modification. Although a few genes are silenced by the first mechanism, most are silenced by the second. Adrian Bird and coworkers took an important step in working out the details of methyl-CpG signaling in 1992 when they identified two nuclear proteins that bind to methyl-CpG. Other proteins were later discovered that also bind to methyl-CpG. For simplicity, we focus attention on MeCP2, one of the two methyl-CpG-binding proteins originally identified by Bird and coworkers. MeCP2 is a single polypeptide with a methyl-CpG binding domain at its N-terminal domain and a transcription repression domain. The transcription repression domain interacts with SIN3A, a transcriptional co-repressor, which then serves as a bridge to histone deacetylase. The resulting histone deacetylase complex silences the gene by increasing chromatin compaction.

Epigenetics is the study of inherited changes in phenotype caused by changes in chromatin other than changes in DNA sequence.

Histone modification patterns within specific chromosome regions appear to be heritable. A heritable code involving covalently modified amino acid residues represents a significant departure from the classical notion that all the information for determining phenotype is stored in the DNA base sequence. In essence, the heritable code hypothesis proposes that daughter cells not only have the same DNA base sequence as the parent cell, but that each chromatin region within the daughter cell has the same DNA methylation pattern and combination of covalently modified histone residues as was present in the parent cell. Because DNA methylation patterns and modified histone molecules determine which DNA segments will be active and which will be silent, this is a form of inheritance that is not directly determined by DNA sequence. We use the term **epigenetics** to describe inherited changes in phenotype that are caused by changes in the chromosome other than changes in DNA sequence.

Different cells and tissues in multicellular organisms acquire different programs for gene expression during development. These programs appear to be largely determined by epigenetic modifications resulting from DNA methylation and histone modifications. Thus, each cell type has unique epigenetic features that depend on its genotype, developmental history, and environmental influences. Epigenetic features tend to become fixed after cell differentiation. For instance, muscle cells retain epigenetic features of muscle cells after cell division. Some cells undergo major epigenetic reprogramming during normal development (**BOX 13.5 LOOKING DEEPER: GENOMIC IMPRINTING**).

13.12 RNA Polymerase I Catalyzed Transcription

RNA polymerase I is required to synthesize 5.8S, 18S, and 28S rRNA.

RNA polymerase I is required to synthesize three of the four different RNA molecules present in the eukaryotic ribosome. These three rRNAs are named 5.8S, 18S, and 28S rRNAs. Their names indicate the characteristic sedimentation coefficient of each kind of rRNA. The fourth rRNA, 5S rRNA, is synthesized by RNA polymerase III (see Section 13.13 RNA Polymerase III Catalyzed Transcription). Studies performed by James Darnell and coworkers beginning in the early 1960s showed the 5.8S, 18S, and 28S rRNAs are synthesized as part of a precursor rRNA (pre-rRNA) that then is processed to form the final rRNA products. Eukaryotic cells probably derive considerable advantage from having the three rRNA molecules made from a single precursor rRNA (pre-rRNA) molecule because this arrangement seems to help coordinate rRNA synthesis and ribosome assembly. It is therefore rather surprising the fourth type of rRNA, 5S rRNA, is not included within pre-rRNA molecules. The selective pressure for separate 5S rRNA genes in higher animals and plants remains to be discovered.

Pre-rRNAs have been identified in many different higher animals and appear to have the general characteristics shown in **FIGURE 13.55** and listed below:

1. rRNA coding sequences are in the order 18S rRNA – 5.8S rRNA – 28S rRNA.
2. The 18S and 5.8S rRNA coding sequences and 5.8S and 28S rRNA coding sequences are separated by **internal transcribed spacer 1** and **internal transcribed spacer 2**, respectively.
3. A **5′-external transcribed sequence** is located upstream from the 18S rRNA coding sequence, and a **3′-external transcribed sequence** is located downstream from the 28S rRNA coding sequence.
4. Internal transcribed spacer sequences and external transcribed sequences are excised when pre-rRNA is processed to form mature rRNA molecules.

BOX 13.5: LOOKING DEEPER

Genomic Imprinting

Mammalian sexual reproduction produces diploid offspring with one copy of each allele inherited from each parent. Classical genetics predicts that maternal and paternal alleles should function equally well. Although this prediction holds true for nearly all genes, there are about 100 known exceptions in mice and a similar number in other mammals. Asymmetrically expressed genes tend to be organized in clusters located on only a few chromosomes. In some cases the maternal allele is transcribed and the paternal allele is silent, whereas in others the reverse is true. This type of asymmetric allelic expression occurs because the allele inherited from each parent has a characteristic methylation pattern, which is generally stable in differentiated somatic cells but reprogrammed (demethylated and remethylated) during the development of germ cells and preimplantation embryos.

The important point is that the methylation pattern, or genomic imprinting, allows the transcription machinery to distinguish homologous chromosomal regions based on parental origin. The best-characterized example of genomic imprinting is probably a genetic mouse cluster that contains the paternally expressed gene *Igf2* (insulin-like growth factor 2) and the maternally expressed gene *H19* (FIGURE B13.5). Both genes are regulated by (1) an *ICR* (imprinting center region) located between them and (2) shared enhancers downstream of *H19*. The *ICR* is methylated in the paternal cluster and unmethylated in the maternal cluster. A protein called CTCF binds to the unmethylated maternal *ICR*, blocking transcription activators bound to the enhancers from stimulating transcription at *Igf2* but permitting them to stimulate transcription at *H19*.

The situation is quite different for the paternal cluster. The methylated *ICR* does not allow CTCF to bind to it and the methylated *H19* promoter prevents the transcription machinery from acting on *H19*. The transcription activators bound to the enhancers, however, can stimulate *Igf2* transcription.

The discovery of genome imprinting solves a very puzzling mammalian reproduction problem. Despite considerable effort, investigators have not been able to obtain a viable organism from egg cells that contain only paternal chromosomes or only maternal chromosomes. Genomic imprinting studies indicate that diploid eggs with chromosomes derived from just one parent cannot develop normally because some asymmetrically expressed genes are essential for normal development. Consistent with this hypothesis, several inborn errors of metabolism have been shown to be caused by genes regulated by genomic imprinting. Inheritance of these metabolic disorders depends on whether the altered allele is in a paternal or maternal chromosome.

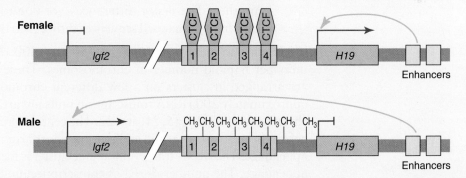

FIGURE B13.5 Transcription regulation by imprinting at mouse *H19* and *Igf2* genes. *H19* and *Igf2* are regulated by (1) an *ICR* (imprinting center region) located between them (shown in green) and (2) shared enhancers downstream of *H19*. The *ICR* is methylated in the paternal cluster and unmethylated in the maternal cluster. A protein called CTCF binds to the unmethylated maternal *ICR*, blocking transcription activators bound to the enhancers from stimulating transcription at *Igf2* but permitting them to stimulate transcription at *H19*. The situation is quite different for the paternal cluster. The methylated *ICR* does not allow CTCF to bind to it, and the methylated *H19* promoter prevents the transcription machinery from acting on *H19*. The transcription activators bound to the enhancers, however, can stimulate *Igf2* transcription. (Modified from Ideraabdullah, F. Y., Vigneau, S., and Bartolomei, M. S. 2008. Genomic imprinting mechanisms in mammals. *Mutat Res* 647:77–85. Copyright 2008, with permission from Elsevier [http://www.sciencedirect.com/science/journal/00275107].)

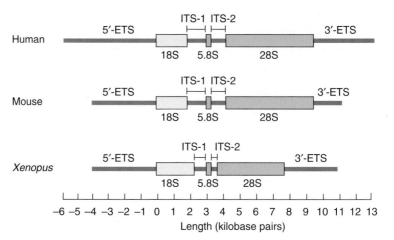

FIGURE 13.55 **Comparison of human, mouse, and *Xenopus* (frog) pre-rRNAs.** Pre-rRNA contains sequences that are converted to 18S, 5.8S, and 28S rRNA molecules. An internal transcribed spacer (ITS-1) separates the 18S and 5.8S rRNA sequences, and a second internal transcribed spacer (ITS-2) separates the 5.8S and 28S rRNA sequences. The pre-rRNA molecules also have a 5'-external transcribed spacer (5'-ETS) upstream from the 18S rRNA and a 3'-external transcribed spacer (3'-ETS) downstream from the 28S rRNA. The scale at the bottom of the figure indicates length in kilobase pairs. (Adapted from Triezenberg, S. J., et al. 1982. *J Biol Chem* 257:7826–7833.)

5. The rRNA sequences tend to be highly conserved, whereas excised sequences do not. This difference in sequence conservation is expected because there is strong selective pressure to conserve rRNA sequences essential for ribosome function but no comparable selective pressure to conserve excised sequences.
6. The lengths of pre-rRNA molecules vary from one species to another. These size differences are due to differences in the lengths of the excised sequences rather than the rRNA sequences.

Higher animals have between 150 and 300 rRNA transcription units per haploid number of chromosomes. These transcription units are arranged in clusters on a few different chromosomes. In humans, approximately 200 rRNA transcription units are arranged in clusters on chromosomes 13, 14, 15, 21, and 22. The centromere in each of these five chromosomes is so near the end that the short arm resembles a dot-like appendage. The rRNA transcription unit cluster is located on this dot-like appendage. The number of rRNA transcription units is much higher in plants, averaging 3,700 copies per haploid number of chromosomes, and much lower in yeast (some yeast strains have as few as 40 rRNA transcription units per cell). rRNA transcription units are amplified in some animals, most notably amphibians, during specific stages of development. For example, the frog oocyte has about 500,000 rRNA transcription unit copies, representing a remarkable 1,000-fold amplification above the number present in the typical frog somatic cell. This large number of rRNA copies is required to meet the demand for new ribosome synthesis.

Because of their arrangement and high rate of transcription, rRNA transcription units can be visualized in spread chromatin preparations using electron microscopy (FIGURE 13.56). The transcripts are so densely

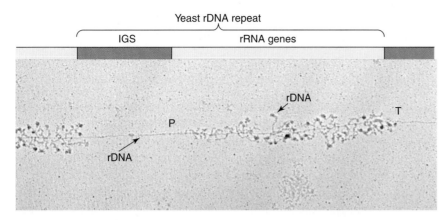

FIGURE 13.56 Electron microscopic image of a yeast nuclear chromatin spread. The repetitive nature of the rDNA is illustrated in an electron microscopic image of a yeast nuclear chromatin spread. Progressively longer rRNAs (stained for associated proteins) emanate from the many RNA polymerase I complexes as they move along the rRNA transcription unit (shown in yellow at the top), beginning at the promoter (P) and finishing at the terminator (T). IGS is an abbreviation for intergenic spacer, the region between two rRNA transcription units. (Reproduced from Russell, J. and Zomerdijk, J. C. B. M. 2005. *Trends Biochem Sci* 30:87–96. Copyright 2005, with permission from Elsevier [http://www.sciencedirect.com/science/journal/09680004]. Photo courtesy of Ann L. Beyer, University of Virginia.)

packed (about 100 transcripts per rRNA transcription unit) they stick out perpendicularly from the DNA, giving each active transcription unit the appearance of a "Christmas tree." The tip of each "tree" is at the transcription initiation site and the broad end is at the transcription termination site.

The rRNA transcription unit promoter consists of a core promoter and an upstream promoter element (UPE).

Individual rRNA transcription units from different organisms have been sequenced. Each rRNA transcription unit has its own promoter and transcription terminator (FIGURE 13.57). The promoter for the rRNA transcription unit consists of two elements: a **core promoter**, which is essential for transcription, and an **upstream promoter element (UPE)**, which is not. The human core promoter begins about 50 bp before the transcription initiation site (+1) and extends about 20 bp beyond it. An adenine-thymine (AT)-rich region within the core promoter that extends from about −10 to +10, called the **rRNA initiator (rInr)**, is unique among rRNA promoter sequences because it is conserved from one species to another. The high AT content appears to permit the closed promoter to change to an open promoter, a prerequisite for transcription initiation. Although it resembles the TATA box of RNA polymerase II promoters, rInr is not a binding site for TBP. The human UPE, which spans a region from −107 to −186, stimulates *in vitro* and *in vivo* transcription initiation by as much as 15- and 100-fold, respectively. Promoters from other species have a similar overall structure but have little sequence homology with the human promoter. Because promoter sequences are species specific, nuclear extracts from

FIGURE 13.57 The rRNA transcription unit and intergenic spacer. Regions that code for 18S, 5.8S, and 28S rRNA are shown in light green, cyan, and green, respectively. The rRNA transcription unit promoter has two elements, a core promoter element (Core), which is brown, and an upstream promoter element (UPE), which is light blue. The core promoter element is essential for transcription, but the upstream promoter element is not. An adenine-thymine (AT)-rich region within the core promoter, the rRNA initiator (rInr), is shown in pink. The high AT content appears to permit the closed promoter to change to an open promoter, an essential prerequisite for transcription initiation. The bent red arrow indicates the transcription start site. The terminator (Term) is indicated in red. (Adapted from Paule, M. R., and White, R. J. 2000. *Nucleic Acids Res* 28:1283–1298.)

organisms of one species can only transcribe rRNA transcription units from other organisms of that same species or a closely related species.

5.8S, 18S, and 28S rRNA syntheses take place in the nucleolus.

Active rRNA transcription units and the RNA polymerase I transcription machinery are located within a well-defined transcription and processing "factory" within the nucleus called the **nucleolus**. Although sometimes referred to as an organelle, the nucleolus is not a true organelle because it is not surrounded by a membrane. The nucleolus is a dynamic structure that is dismantled during mitosis and reassembled at interphase. Its appearance provides clues to certain pathological conditions. For instance, cancer cells often have abnormally large nucleoli.

RNA polymerase I is a multisubunit enzyme with a structure similar to that of RNA polymerase II.

RNA polymerase I appears to contain 11 subunits in mouse and 14 subunits in yeast. This difference may reflect the fact that certain subunits dissociate more easily from the mouse than the yeast enzyme. The yeast genes that code for RNA polymerase I subunits have been cloned. Five of the yeast RNA polymerase I subunits are identical with those in the two other nuclear RNA polymerases, whereas two other subunits are identical with subunits in RNA polymerase III but not RNA polymerase II. The two largest RNA polymerase I subunits, although unique to it, are homologous to the two largest subunits in RNA polymerase II. A high resolution image has not yet been reported for RNA polymerase I. The structure for yeast RNA polymerase I, which was determined by electron microscopy, however, has the same size and shape as that reported for the yeast RNA polymerase II crystal structure, supporting the idea that the two RNA polymerases are quite similar. The only essential function of RNA polymerase I is to transcribe rRNA transcription units.

An RNA polymerase associated factor present in animals called transcription initiation factor 1A (TIF-1A) is tightly associated with the RNA polymerase I core enzyme. TIF-1A appears to be a key factor in growth-dependent regulation of pre-rRNA transcription in response to conditions such as nutrient starvation, aging, and viral infection that down-regulate pre-rRNA transcription. TIF-1A has several phosphorylation sites. Hypophosphorylated TIF-1A does not bind to core RNA polymerase I and does not stimulate transcription. The fact that TIF-1A must be phosphorylated to stimulate transcription suggests a mechanism for regulating pre-rRNA formation. Under favorable growth conditions when rRNA synthesis is needed, cells produce a set of signals that stimulate a protein kinase that activates TIF-1A by adding phosphate groups to it. Under adverse growth conditions when rRNA synthesis is not needed, cells produce a different set of signals that activate a protein phosphatase that inactivates TIF-1A by removing phosphate groups. When cell growth stops, the fraction of RNA polymerase I molecules with bound TIF-1A decreases, although the concentration of total TIF-1A remains unchanged. This observation is consistent with the hypothesis that RNA polymerase I • TIF-1A complex dissociation causes rRNA transcription unit shut-off.

The upstream binding factor and selectivity factor, working together, recruit RNA polymerase I to the rDNA promoter to form a preinitiation complex.

RNA polymerase I also requires two auxiliary transcription factors to make rRNA. The first of these, the upstream binding factor (UBF), is a homodimer. Two homodimers appear to participate in initiation complex formation: one binds to the core promoter and the other to the UPE. The second auxiliary factor is called selectivity factor 1 (SL1) in humans and transcription initiation factor IB (TIF-1B) in mouse. We will follow the common practice of using the term SL1/TIF-1B for the mammalian transcription factor. SL1/TIF-1B contains TBP and five TBP-associated factors and appears to bind to the core promoter. It is species specific, working with rRNA transcription units from the same or closely related species but not with those from other species. For instance, human SL1 works with human and monkey rRNA transcription units but not with mouse or yeast rRNA transcription units. This finding explains why nuclear extracts prepared from an organism belonging to one species will transcribe rRNA transcription units from other organisms of the same or closely related species but not from more distant species. UBF and SL1/TIF-IB bind to the promoter and act synergistically to recruit RNA polymerase I • TIF-1A to the rDNA promoter to form a preinitiation complex (FIGURE 13.58).

RNA polymerase I forms a transcription elongation complex, leaving UBF and SL1/TIF-1B behind.

Once the preinitiation complex has assembled, the RNA polymerase I machinery is ready to initiate transcription. DNA flanking the transcription initiation site melts to create a small bubble that allows nucleotides to align with the template strand, and the first phosphodiester bond is formed. The transcription bubble grows longer as RNA polymerase I moves along the DNA so the transcription bubble eventually becomes 12 to 23 bp long. RNA polymerase I moves down the template strand, leaving UBF and SL1/TIF-1B behind. At least one additional factor, elongation factor SII, associates with the transcription elongation complex. Elongation factor SII therefore participates in transcription elongation by both RNA polymerases I and II. The elongation rate for human RNA polymerase I has been estimated to

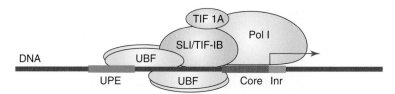

FIGURE 13.58 **Formation of the RNA polymerase I preinitiation complex.** UBF and SL1/TIF-IB bind to the promoter and act synergistically to recruit RNA polymerase I • TIF1A to the rDNA promoter to form a preinitiation complex. (Adapted from Drygin, D., et al. 2010. *Annu Rev Pharmacol Toxicol* 50:133, Figure 1.)

be about 95 nucleotides • s^{-1}. Transcription elongation continues until the RNA polymerase I machinery reaches the transcription terminator. Then transcription is terminated, and the transcript and RNA polymerase I are released. The released polymerase is free to reinitiate transcription from a previously activated and engaged promoter with prebound UBF and SL1/TIF-1B. The continued presence of UBF and SL1/TIF-1B at the promoter allows RNA polymerase I to initiate a new round of transcription much more rapidly than would otherwise be possible. Initiation has been estimated to occur at a minimum of once every 5 seconds *in vivo*.

RNA polymerase I transcription termination requires the assistance of a termination factor and a release protein.

RNA polymerase I transcription terminates at the 3′-end of the gene at specific sequences. In mammals, transcription termination requires the assistance of transcription termination factor 1 (TTF-1) and an RNA polymerase I transcription release factor (PTRF). The termination signal or terminator contains two elements. FIGURE 13.59 shows the arrangement of these two elements in the mouse terminator. The arrangement is similar, but not identical, in other eukaryotes. The upstream T-rich element codes for the last 10 to 12 nucleotides in the terminated transcript. The downstream terminator protein binding element (AGGTCGACCAGA/TT/ANTCCG) is called the **Sal box** because it contains GTCGAC, the Sal I restriction endonuclease cleavage site. The Sal box must be present in the orientation shown for termination to take place. When the RNA polymerase I transcription elongation complex encounters TTF-1 bound to the Sal box, the complex pauses and PTRF interacts with both RNA polymerase I and TTF-1. After pre-RNA is released, an exonuclease trims the 3′-end of the transcript to produce the mature 3′-end. Finally, RNA polymerase I is released. The 45S rRNA is then processed in a multistep pathway requiring several enzymes and factors to form the mature 5.8S, 18S, and 23S rRNAs.

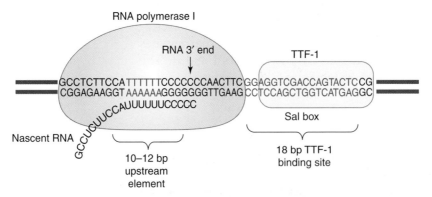

FIGURE 13.59 Mouse transcription terminator. The mouse transcription terminator contains two elements, a T-rich terminator protein-binding site (red) and a Sal box (blue). The terminator protein, transcription terminator 1 (TTF-1), causes the transcription elongation complex to pause. A second transcription factor, the RNA polymerase I transcription release factor (PTRF), which interacts with RNA polymerase I and TTF-1, is required to release the transcript and RNA polymerase I. (Modified from Reeder, R. H., and Lang, W. H. 1997. *Trends Biochem Sci* 22:473–477. Copyright 1997, with permission from Elsevier [http://www.sciencedirect.com/science/journal/09680004].)

13.13 RNA Polymerase III Catalyzed Transcription

RNA polymerase III transcripts are short RNA molecules with a variety of biological functions.

RNA polymerase III transcription units code for a variety of short RNA molecules (usually less than 400 nucleotides long), which are required for protein synthesis or other essential cellular processes. The present discussion is limited to the synthesis of just two of these small RNA molecules, the 5S rRNA in ribosomes and tRNAs.

1. *5S rRNA:* Approximately 120 nucleotides long, 5S rRNA is an essential part of the large ribosome subunit. There are approximately 300 to 500 active 5S rRNA transcription units per haploid number of chromosomes in the human cell. Most of these transcription units are located on chromosome 1.
2. *tRNA:* A typical eukaryotic cell has about 30 to 100 different kinds of tRNA, which vary in length from about 70 to 90 nucleotides long. The tRNAs serve as adaptors during protein synthesis to translate the nucleotide sequences in mRNA to amino acid sequences in the polypeptide product.

RNA polymerase III transcription units have three different types of promoters.

Eukaryotes have three different kinds of promoters in their RNA polymerase III transcription units, referred to as types 1 to 3 (FIGURE 13.60). The architectural structures of these promoters are as follows:

1. *Type 1 promoter:* The type 1 promoter (Figure 13.60a), which was initially characterized by deletion analysis of the *Xenopus laevis* (frog) 5S rRNA transcription unit, is located within rather than before the transcription unit. It contains three short elements, the A box (+50 to +64), the intermediate element or IE (+67 to +72), and a C-box (+80 to +97), which are necessary and sufficient for transcription by frog nuclear extracts. The region within the transcription unit that contains these three promoter elements is called the internal control region. Nucleotides between the three elements are spacers; their sequence does not influence transcription efficiency. Mutations within the three control elements may cause base changes in the 5S rRNA because the control elements are transcribed. The 5S rRNA promoters in other animals have the same architecture as the frog.
2. *Type 2 promoter:* The type 2 promoter (Figure 13.60b), which is present in most tRNA transcription units, is divided into two conserved elements of about 10 bp each, the A box and the B box. There is considerable selective pressure to conserve these two boxes because they also determine important structural features of the tRNA molecule.

Internal promoters

(a) Type 1 promoter (5S rRNA genes)

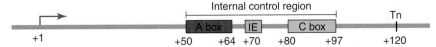

(b) Type 2 promoter (transfer RNA genes)

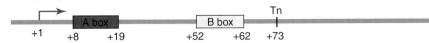

External promoters

(c) Type 3 promoter (mammalian U6 snRNA gene)

FIGURE 13.60 **Organization of the three general types of promoters used by RNA polymerase III.** Transcription initiation and termination sites are indicated by +1 and Tn, respectively. (a) Type 1 promoter (5S RNA genes): the A box, intermediate element (IE), and C-box are colored red, blue, and light green, respectively. (b) Type 2 promoter (tRNA genes): the A and B boxes are colored red and yellow, respectively. (c) Type 3 promoter: the distal sequence element (DSE), proximal sequence element (PSE), and TATA box are colored green, blue, and orange, respectively. (Modified from Paule, M. R., and White, R. J. 2000. Transcription by RNA polymerases I and III. *Nucleic Acids Res* 28:1283–1298. Reprinted by permission of Oxford University Press.)

3. *Type 3 promoter:* The type 3 promoter (Figure 13.60c) is entirely upstream from the transcription initiation site. Because type 3 is not present in transcription units that code for 5S rRNA or tRNA synthesis, it is not described further.

In summary, the RNA polymerase III machinery recognizes three different types of promoters. Two of these are downstream from the transcription initiation site and so also determine the transcript sequence, whereas the third is upstream from this site and so has no effect on the transcript sequence. We now examine the transcription factors required to recruit RNA polymerase III to type 1 and 2 promoters:

1. *Transcription factor interactions at type 1 promoters:* Three transcription factors—TFIIIA, TFIIIB, and TFIIIC—are required to transcribe genes containing a type 1 promoter (FIGURE 13.61a). TFIIIA binds to the internal control region. Then TFIIIC binds to the TFIIIA • DNA complex. The recruitment of TFIIIB, which contains TBP and TBP-associated factors, depends on protein–protein interactions with DNA-bound TFIIIC. TFIIIB recruits RNA polymerase III to the promoter. TFIIIA and TFIIIC do not appear to be directly involved in recruiting RNA polymerase III.
2. *Transcription factor interactions at type 2 promoters:* Only two transcription factors, TFIIIC and TFIIIB, are required to transcribe genes containing a type 2 promoter (Figure 13.61b).

TFIIIC binds to the B and C boxes and then recruits TFIIIB, which in turn recruits RNA polymerase III.

The final component of the RNA polymerase machinery to be recruited to each kind of promoter is RNA polymerase III. Yeast RNA polymerase III contains 17 subunits, making it the largest of the three nuclear RNA polymerases. Ten of the 17 subunits are unique to RNA polymerase III. Two of the remaining subunits are also present in RNA polymerase I, and the remaining five subunits are present in all three nuclear RNA polymerases. The two largest yeast subunits unique to RNA polymerase III are homologous to Rpb1 and Rpb2 in RNA polymerase II. Human RNA polymerase III has also been purified, and with one exception its subunits appear to have yeast subunit counterparts. The one subunit present in yeast but not human RNA polymerase III may be present in the human enzyme but not yet detected because of its small size. Although a crystal structure has not yet been obtained for RNA polymerase III, it seems likely that its core structure will closely resemble the RNA polymerase II structure.

RNA polymerase III does not appear to require additional factors for transcription elongation or termination.

Once the preinitiation complex is assembled at the promoter, the players are in position to begin transcription elongation. The RNA polymerase III transcription machinery melts the DNA flanking the transcription initiation site. TFIIIB is an active participant in this melting event. The first two nucleotides line up on the now exposed template strand and transcription elongation begins. Almost all the RNA polymerase III molecules escape from the promoter without significant pausing or arrest. Once a crystal structure becomes available for RNA polymerase III, it will be interesting to compare it with that of RNA polymerase II to see if there are obvious structural differences that explain why the two polymerases differ in their abilities to escape the promoter. The RNA polymerase III transcription elongation complex moves at about the same rate as the RNA polymerase II transcription elongation complex.

The bound transcription initiation factors might be expected to block transcription elongation through transcription units with type 1 or 2 promoters. This blocking does not seem to occur, however. Moreover, multiple passages of RNA polymerase through a transcription unit do not remove the assembled transcription initiation factors from the transcription unit. One possibility is that RNA polymerase III transiently displaces one transcription factor as it moves through the transcription unit, but protein–protein interactions with other transcription initiation factors permit the displaced factor to remain associated with the promoter.

RNA polymerase III can efficiently terminate transcription in the absence of other factors under *in vitro* conditions. Simple clusters of four or more U residues usually suffice as a termination signal. Further studies are required to determine whether specific termination factors are required *in vivo*.

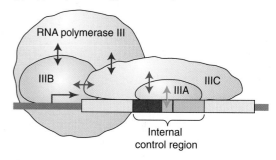

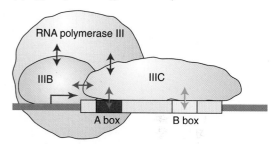

FIGURE 13.61 Initiation complexes formed on RNA polymerase III type 1 and 2 promoters. Interactions at the (a) type 1 promoter and (b) type 2 promoter. Green arrows symbolize interactions of DNA-binding proteins with promoter elements, blue arrows protein–protein contacts among various transcription factors, and purple arrows contacts between RNA polymerase III and transcription factors. (Adapted from Schramm, L., and Hernandez, N. 2002. *Genes Dev* 16:2593–2620.)

Questions and Problems

1. Briefly define, describe, or illustrate each of the following.
    a. TATA box
    b. Core promoter (protein-coding gene)
    c. Initiator (protein-coding gene)
    d. DPE
    e. BRE^u and BRE^d
    f. General transcription factor
    g. Basal transcription
    h. Preinitiation complex
    i. Cyclin
    j. Linker-scanning mutagenesis
    k. CCAAT box
    l. GC box
    m. Regulatory or proximal promoter
    n. Enhancer
    o. Silencer
    p. Upstream activating sequence
    q. Transcription activator protein
    r. DNA-binding domain
    s. Activation domain
    t. Combinatorial control
    u. DNA affinity chromatography
    v. Helix-turn-helix
    w. Cys_4 zinc finger
    x. Basic region leucine zipper
    y. Zn_2Cys_6 binuclear cluster
    z. Homeotic genes
    aa. Homeobox
    bb. Homeodomain
    cc. *Hox* genes
    dd. Nuclear receptor superfamily
    ee. Orphan nuclear receptors
    ff. Hormone response element
    gg. Helical wheels
    hh. bzip DNA-binding domain
    ii. Domain swap
    jj. Squelching
    kk. Mediator
    ll. Chromatin immunoprecipitation
    mm. Epigenetics
    nn. Internal transcribed spacer
    oo. External transcribed spacer
    pp. Core promoter (rRNA gene)
    qq. Upstream promoter element (rRNA gene)
    rr. rRNA Inr
    ss. Nucleolus
    tt. Sal box

2. Investigators have discovered that yeast cells synthesize an RNA molecule that is not mRNA, rRNA, tRNA, or any other known RNA species. They wish to know whether RNA polymerase I, II, or III is responsible for its synthesis. Suggest a simple experimental approach that would help to determine which of the polymerases is responsible for the synthesis of the new type of RNA.

3. In contrast to bacteria, eukaryotes require a distinct RNA polymerase to synthesize rRNAs, mRNAs, and tRNAs. Is this requirement for a distinct RNA polymerase a result of the fact that eukaryotes have a defined nucleus and bacteria do not? Provide evidence to support your answer.

4. Prepare a table showing some of the ways in which bacterial RNA polymerase and eukaryotic RNA polymerase II are similar and some of the ways in which they differ.

5. The core promoter of protein-coding genes plays an important role in their basal transcription.
    a. What evidence led to the discovery of the TATA box?
    b. What is meant by a TATA-less promoter?
    c. How did the discovery of TATA-less promoters alter our concept of promoter structure?
    d. Draw the structure of a core promoter showing some of its important features.
    e. Do the core promoters of all protein-coding genes have all of these features?

6. The TATA-binding protein, TBP, is part of general transcription factor TFIID.
    a. Describe TBP's structure and explain how it interacts with a core promoter that has a TATA box.
    b. Can TBP bind to a core promoter that lacks a TATA box?
    c. What are the other polypeptide subunits present in TFIID?
    d. How does TFIID interact with core promoters that lack TATA boxes?
    e. Is TFIID more likely to bind to protein-coding gene in a naked DNA molecule or a DNA molecule in chromatin? Explain your answer?

7. The general transcription factors and RNA polymerase II interact with the core promoter and with one another to form a preinitiation complex.
   a. Which general transcription factors are essential to form the preinitiation complex when the DNA is not negatively supercoiled? Indicate the order in which these factors combine to form the preinitiation complex.
   b. What function does TFIIB serve?
   c. When a core promoter with a TATA box is part of a negatively supercoiled DNA, RNA polymerase II requires only TBP and TFIIB to catalyze basal transcription. Why is TFIIH required when the DNA is not negatively supercoiled?
   d. Why is the transcription catalyzed by RNA polymerase II in the presence of the general transcription factors called basal transcription?

8. The carboxyl terminal domain, CTD, of the largest subunit in RNA polymerase II, Rpb1, has an important role in the transition from the initiation complex to the elongation complex.
   a. What feature(s) of the CTD is (are) important for this transition?
   b. Do RNA polymerases I or RNA polymerase III also have the same feature(s)?
   c. What contribution does TFIIH make to the transition of RNA polymerase II from the initiation to the elongation stage?

9. RNA polymerase II occasionally attaches the wrong nucleotide to the growing RNA chain.
   a. How does the transcription machinery correct this problem?
   b. What eukaryotic protein factor assists the transcription machinery in this correction process?
   c. How does this correction process compare with that used by bacterial cells?

10. Transcription of the thymidine kinase gene in the herpes simplex virus requires two different sequence motifs that are just upstream from the core promoter.
    a. Describe the experimental technique used to discover the two different sequence motifs.
    b. What is the function of the two sequence motifs?
    c. Do all eukaryotic protein-coding genes have these two sequence motifs?
    d. Is it possible for a protein-coding gene to be fully active without either of the two sequence motifs? Explain your answer.
    e. Does either of the two sequence motifs serve the same function as the bacterial −35 box? Explain your answer.

11. Enhancers and silencers help to regulate the transcription of eukaryotic protein-coding genes.
    a. How do enhancers and regulatory promoters differ? In what ways are they similar?
    b. How do enhancers and silencers differ? In what ways are they similar?
    c. Compare the yeast UAS with an enhancer.

12. Virtually all transcription activator proteins have a DNA-binding domain and an activation domain. Many transcription activator proteins have additional features that are essential for their function. List four of these features and describe their function.

13. You have determined that a transcription activator protein binds to a specific DNA sequence. How can you use this information to purify the transcription activator protein from a crude extract?

14. You have cloned a transcription activator protein gene. How can you use this cloned gene to test the transcription activator protein's ability to stimulate transcription in a cell-based assay?

15. What is the function of homeotic (*HOM*) genes in *Drosophila*?

16. What are the consequences for the fly if a homeotic gene is altered?

17. Homeotic genes in *Drosophila* have a common DNA sequence, the homeobox, which is about 180 bases long. Describe the structure and function of the polypeptide chain encoded by the homeobox.

18. Nuclear receptors have a common structural design.
    a. Draw a schematic structure that shows the common structural design.
    b. Only one of these structural features is highly conserved. Identify this structural feature and describe its important characteristics.

19. Active forms of several nuclear receptors are heterodimers. One important family consists of receptors that form heterodimers with the retinoid X receptor (RXR). The half-site for all of the DNA-binding domains that belong to this family of heterodimers is ATTTCA. How is it possible for hormone response elements that share the same DNA half-site to bind so many different heterodimers with great specificity?

20. What is the major structural feature of the basic region leucine zipper (bzip) proteins? How do these proteins interact with DNA?

21. What is a helical wheel drawing and what type of information can it provide?

22. Gal4, a transcription activator protein, regulates genes coding for galactose metabolism in *Saccharomyces cerevisiae*.
    a. Briefly describe the Gal4 DNA-binding domain?
    b. Both the Gal4 and Ppr1 DNA-binding domains interact with semi-palindromes with CCG triplets at either end but exhibit varying specificities. How is such variation possible? Describe an experiment that supports your answer.

23. DNA-bound transcriptional activator proteins appear to use two mechanisms to influence gene transcription. Briefly describe these two mechanisms. Are they mutually exclusive? Explain your answer.

24. Gal4 requires the assistance of several other proteins to switch *GAL* genes on and off in response to the carbon source in the growth medium.
    a. List the additional proteins that influence *Gal* gene expression.
    b. Draw a figure showing how proteins listed above influence *GAL1* transcription in the presence of galactose.
    c. Draw a figure showing how proteins listed above influence *GAL1* transcription when glucose is added to the medium that contains galactose.

25. Do the DNA-binding domain and the activation domain need to be part of the same molecule to influence transcription? Describe an experiment that supports your answer. How can we take advantage of the information obtained from this experiment to study protein–protein interactions?

26. What is squelching and how did its discovery lead to the detection of Mediator?

27. What function does Mediator play in the transcription of protein-coding genes?

28. Are each of the components in Mediator required to transcribe all protein-coding genes? Provide evidence to support your answer.

29. An investigator has discovered a new yeast mutant in RNA polymerase II that grows very poorly in the cold. *SRB* mutants suppress this cold sensitivity. What is the likely nature of the mutation in RNA polymerase II?

30. Draw a schematic diagram showing how Mediator interacts with other components of the transcription machinery required to transcribe protein-coding genes.

31. Is it possible for two organisms with identical DNA sequences to have different phenotypes? Explain your answer.

32. How do ATP-dependent chromatin remodeling complexes influence transcription?

33. RSC, a member of the SWI/SNF family, is essential for cell viability. What is RSC's function?

34. List the kinds of chemical modifications to histones that can alter the transcription of protein-coding genes.

35. How do HATs influence transcription of a protein-coding gene? How do histone deacetylases influence transcription of a protein-coding gene?

36. What roles do UBF and SL1/TIF-1B play in the transcription of rRNA genes?

37. RNA polymerase I is species specific. What is the basis of this specificity?

38. RNA polymerase III transcribes genes with three different kinds of promoters.
    a. Describe the three different kinds of promoters.
    b. Why do mutations in type I and type II promoters produce altered transcripts?
    c. If type I and type II promoters are located within the gene, then how can RNA polymerase III transcribe the part of the gene that has bound transcription factors?

Suggested Reading

Introduction to Eukaryotic Nuclear RNA Polymerases

Roeder, R. 2003. The eukaryotic transcriptional machinery: complexities and mechanisms unforeseen. *Nat Med* 9:1239–1244.

Roeder, R. G., and Rutter, W. J. 1970. Specific nucleolar and nucleoplasmic RNA polymerases. *Proc Natl Acad Sci USA* 65:675–682.

RNA Polymerase II Structure

Armache, K.-J., Kettenberger, H., and Cramer, P. 2003. Architecture of initation-competent 12-subunit RNA polymerase II. *Proc Natl Acad Sci USA* 100:6964–6968.

Armache, K.-J., Mitterweger, S., Meinhart, A., and Cramer, P. 2005. Structures of complete RNA polymerase II and its subcomplex, Rpb4/7. *J Biol Chem* 280:7131–7134.

Bell, S. D., and Jackson, S. P. 2000. Charting a course through RNA polymerase. *Nat Struct Biol* 7:703–705.

Boeger, H., Bushnell, D. A., Davis, R., et al. 2005. Structural basis of eukaryotic gene transcription. *FEBS Lett* 579:899–903.

Bushnell, D. A., and Kornberg, R. D. 2003. Complete, 12-subunit RNA polymerase II at 4.1-Å resolution: implications for the initiation of transcription. *Proc Natl Acad Sci USA* 100:6969–6973.

Cramer, P. 2002. Multisubunit RNA polymerases. *Curr Opin Struct Biol* 12:89–97.

Cramer, P. 2004. Structure and function of RNA polymerase II. *Adv Protein Chem* 67:1–31.

Cramer, P. 2004. RNA polymerase II structure: from core to functional complexes. *Curr Opin Genet Dev* 14:218–226.

Cramer, P. 2010. Towards molecular biology systems biology of gene transcription and regulation. *Biol Chem* 391:731–735.

Cramer, P., Armache, K.-J., Baumli, S., et al. 2008. Structure of eukaryotic RNA polymerases. *Annu Rev Biochem* 37:337–352.

Cramer, P., Bushnell, D. A., Fu, J., et al. 2000. Architecture of RNA polymerase II and implications for the transcription mechanism. *Science* 288:640–649.

Cramer, P., Bushnell, D. A., and Kornberg, R. D. 2001. Structural basis of transcription: RNA polymerase II at 2.8 Ångstrom resolution. *Science* 292:1863–1876.

Gnatt, A. L., Cramer, P., Fu, J., et al. 2001. Structural basis of transcription: an RNA polymerase II elongation complex at 3.3 Å resolution. *Science* 292:1876–1882.

Hahn, S. 2004. Structure and mechanism of the RNA polymerase II transcription machinery. *Nat Struct Mol Biol* 11:394–403.

Hirata, A., Klein, B., and Murakami, K. S. 2008. The x-ray crystal structure of RNA polymerase from archaea. *Nature* 451:851–854.

Hirata, A., and Murakami, K. S. 2009. Archaeal RNA polymerases. *Curr Opin Struct Biol* 19:1–8.

Korkhin, Y., Unligil, U. M., Littlefield, O., et al. 2009. Evolution of complex RNA polymerases: the complete archaeal RNA polymerase structure. *PLoS Biol* 7:1–10.

Kornberg, R. 2001. The eukaryotic gene transcription machinery. *Biol Chem* 382:1103–1107.

Kornberg, R. D. 2007. The molecular basis of eukaryotic transcription. *Proc Natl Acad Sci USA* 104:12955–12961.

Kostrewa, D., Zeller, M. E., Armache, K.-J., et al. 2009. RNA polymerase II-TFIIB structure and mechanism of transcription initiation. *Nature* 462:323–330.

Nikolov, B. D., and Burley, S. K. 1997. RNA polymerase II transcription initiation: a structural view. *Proc Natl Acad Sci USA* 94:15–22.

Wang, D., Bushnell, D. A., Westover, K. D., et al. 2006. Structural basis of transcription: role of the trigger loop in substrate specificity and catalysis. *Cell* 127:941–954.

Westover, K. D., Bushnell, D. A., and Kornberg, R. D. 2004. Structural basis of transcription separation of RNA from DNA by RNA polymerase II. *Science* 303:1014–1016.

Woychik, N. A., and Hampsey, M. 2002. The RNA polymerase II machinery: structure illuminates function. *Cell* 108: 453–463.

Core Promoter for Protein-Coding Genes

Butler, J. E. F., and Kadonaga, J. T. 2002. The RNA polymerase II core promoter: a key component in the regulation of gene expression. *Genes Dev* 16:2583–2592.

Juven-Gershon, T., Hsu, J.-Y., Theisen, J. W. M., and Kadonaga, J. T. 2008. The RNA polymerase II core promoter—the gateway to transcription. *Curr Opin Cell Biol* 20:253–259.

Juven-Gershon, T., and Kadonaga, J. T. 2010. Regulation of gene expression via the core promoter and the basal transcriptional machinery. *Dev Biol* 339:225–229.

Maston, G. A., Evans, S. K., and Green, M. R. 2006. Transcriptional regulatory elements in the human genome. *Annu Rev Genomics Human Genet* 7:29–59.

Sandelin, A. L., Carninci, P., Lenhard, B., et al. 2007. Mammalian RNA polymerase II core promoters: Insights from genome-wide studies. *Nat Rev Genet* 8:424–436.

General Transcription Factors: Basal Transcription

Asturias, F. J. 2004. RNA polymerase II structure, and organization of the preinitiation complex. *Curr Opin Struct Biol* 14:121–129.

Asturias, F. J. 2009. TFIID: a closer look highlights its complexity. *Structure* 17:1423–1424.

Asturias, F. J., and Craighead, J. L. 2003. RNA polymerase II at initiation. *Proc Natl Acad Sci USA* 100:6893–6895.

Auble, D. T. The dynamic personality of TATA-binding protein. *Trends Biochem Sci* 34:49–52.

Buratowski, S., Hahn, S., Sharp P. A., and Guarente, L. 1988. Function of a yeast TATA element-binding protein in a mammalian transcription system. *Nature* 334:37–42.

Bushnell, D. A., Westover, K. D., Davis, R. E., and Kornberg, R. D. 2004. Structural basis of transcription: an RNA polymerase II–TFIIB cocrystal at 4.5 Angstroms. *Science* 303:983–988.

Cler, E., Papai, G., Schultz, P., and Davidson, I. 2009. Recent advances in understanding the structure and function of general transcription factor TFIID. *Cell Mol Life Sci* 66:2123–2134.

Coulombe, B. 1999. DNA wrapping in transcription initiation by RNA polymerase II. *Biochem Cell Biol* 77:257–264.

Green, M. R. 2000. TBP-associated factors (TAFIIs): multiple, selective transcriptional mediators in common complexes. *Trends Biochem Sci* 25:59–63.

Hahn, S. 2009. New beginnings for transcription. *Nature* 462:292–293.

Hampsey, M., and Reinberg, D. 2001. RNA polymerase II holoenzyme and transcription factors. *Encyclopedia of Life Sciences*. pp. 1–7. London: Nature.

Hochheimer, A., and Tjian, R. 2003. Diversified transcription initiation complexes expand promoter selectivity and tissue-specific gene expression. *Genes Dev* 17:1309–1320.

Juo, Z. S., Chiu, T. K., Leiberman, P. M., et al. 1996. How proteins recognize the TATA box. *J Mol Biol* 261:239–254.

Kostrewa, D., Zeller, M. E., Armache, K.-J., et al. 2009. RNA polymerase II–TFIIB structure and mechanism of transcription initiation. *Nature* 462:323–330.

Krishnamurthy, S., and Hampsey, M. 2009. Eukaryotic transcription initiation. *Curr Biol* 19:R153–R156.

Liu, X., Bushnell, D. A., Wang, D., et al. 2010. Structure of an RNA polymerase II-TFIIB complex and the transcription initiation mechanism. *Science* 237:206–209.

Müller, F., and Tora, L. 2004. The multicoloured world of promoter recognition complexes. *EMBO J* 23:2–8.

Nikolov, D. B., Chen, H., Halay, E. D., et al. 1995. Crystal structure of a TFIIB-TBP-TATA-element ternary complex. *Nature* 377:119–128.

Nogales, E. 2000. Recent structural insights into transcriptional preinitiation complexes. *J Cell Sci* 113:4391–4397.

Reese, J. C. 2003. Basal transcription factors. *Curr Opin Genet Dev* 13:114–118.

Schultz, P., Fribourg, S., Poterszman, A., et al. 2000. Molecular structure of human TFIIH. *Cell* 102:599–607.

Sikorski, T. W., and Buratwoski, S. 2009. The basal initiation machinery: beyond the general transcription factors. *Curr Opin Cell Biol* 21:344–351.

Smale, S. T. 1997. Transcription initiation form TATA-less promoters within eukaryotic protein-coding genes. *Biochim Biophys Acta* 1351:73–88.

Smale, S. T., and Kadonaga, J. T. 2003. The RNA polymerase II core promoter. *Annu Rev Biochem* 72:449–479.

Tora, L., and Timmers, H. T. M. 2010. The TATA box regulates TATA-binding protein (TBP) dynamics *in vivo*. *Trends Biochem Sci* 35:309–314.

Warren, A. 2002. Eukaryotic transcription factors. *Curr Opin Struct Biol* 12:107–114.

Zurita, M., and Merino, C. 2003. The transcriptional complexity of the TFIIH complex. *Trends Genet* 19:578–584.

Transcription Elongation

Armache, K.-J., Kettenberger, H., and Cramer, P. 2005. The dynamic machinery of mRNA elongation. *Curr Opin Struct Biol* 15:197–203.

Gnatt, A. 2002. Elongation by RNA polymerase II: structure-function relationship. *Biochim Biophys Acta* 1577:175–190.

Grohmann, D., Hirtreiter, A., and Werner, F. 2009. Molecular mechanisms of archaeal RNA polymerase. *Biochem Soc Trans* 37:12–17.

Hartzog, G. A. 2003. Transcription elongation by RNA polymerase II. *Curr Opin Genet Dev* 13:119–126.

Kamenski, T., Heilmeir, S., Meihhart, A., and Cramer, P. 2004. Structure and mechanism of RNA polymerase II CTD phosphatases. *Mol Cell* 15:399–407.

Kobor, M. S., and Greenblatt, J. 2002. Regulation of transcription elongation by phosphorylation. *Biochim Biophys Acta* 1577:261–275.

Kohoutek, J. 2009. P-TEFb-the final frontier. *Cell Div* 4:19.

Meinhart, A., Kamenski, T., Hoeppner, S., et al. 2005. A structural perspective of CTD function. *Genes Dev* 19:1401–1415.

Palancade, B., and Bensaude, O. 2003. Investigating RNA polymerase II carboxyl-terminal domain (CTD) phosphorylation. *Eur J Biochem* 270:3859–3870.

Price, D. H. 2002. P-TEFb, a cyclin-dependent kinase controlling elongation by RNA polymerase II. *Mol Cell Biol* 20:2629–2634.

Price, D. H. 2004. RNA polymerase II elongation control in eukaryotes. W. J. Lennarz and W. D. Lane (Eds), *Encyclopedia of Biological Chemistry*. pp. 766–769. New York: Academic Press.

Selth, L. A., Sigurdsson, S., and Svejstrup, J. Q. 2010. Transcript elongation by RNA polymerase II. *Annu Rev Biochem* 79:271–293.

Shilatifard, A. 2004. Transcriptional elongation control by RNA polymerase II: a new frontier. *Biochim Biophys Acta* 1677:79–86.

Shilatifard, A., Conaway, R. C., and Conaway, J. W. 2003. The RNA polymerase II elongation complex. *Annu Rev Biochem* 72:693–715.

Sims, R. J., III, Belotserkovskaya, R., and Reinberg, D. 2004. Elongation by RNA polymerase II: the short and long of it. *Genes Dev* 18:2437–2468.

Svejstrup, J. Q. 2007. Contending with transcriptional arrest during RNAPII transcript elongation. *Trends Biochem Sci* 32:165–171.

Wade, J. T., and Struhl, K. 2008. The transition from transcriptional initiation to elongation. *Curr Opin Genet Dev* 18:130–136.

Wang, D., Bushnell, D. A., Huang, X., et al. 2009. Structural basis of transcription: backtracked RNA polymerase II at 3.4 Angstrom resolution. *Science* 324:1203–1206.

Regulatory Promoters, Enhancers, and Silencers

Blackwood, E. M., and Kadonaga, J. T. 1998. Going the distance: a current view of enhancer action. *Science* 281:60–63.

Kadonaga, J. T., and Tjian, R. 1986. Affinity purification of sequence-specific DNA binding proteins. *Proc Natl Acad Sci USA* 83:5889–5893.

Kozian, D. H., and Kirschbaum, B. J. 1999. Comparative gene-expression analysis. *Trends Biotechnol* 17:73–78.

Lekstrom-Himes, J., and Xanthopoulos, K. G. 1998. Biological role of the CCAAT/enhancer-binding protein family of transcription factors. *J Biol Chem* 273:28545–28548.

Maston, G. A., Evans, S. K., and Green, M. R. 2006. Transcription regulatory elements in human genome. *Annu Rev Genomics Hum Genet* 7:29–59.

Szutorisz, H., Dillon, N., and Tora, L. 2005. The role of enhancers as centres for general transcription factor recruitment. *Trends Biochem Sci* 30:593–599.

Introduction to Transcription Activator Proteins for Protein-Coding Genes

Barberis, A., and Petrascheck, M. 2003. Transcription activation in eukaryotic cells. *Encyclopedia of Life Sciences.* pp. 1–7. London: Nature.

Lawler, J. F., Hyland, E. M., and Boeke, J. D. 2008. Genetic engineering: reporter genes. *Encyclopedia of Life Sciences.* pp. 1–8. Hoboken, NJ: John Wiley & Sons.

Ma, J. 2005. Crossing the line between activation and repression. *Trends Genet* 21:54–59.

Reece, R. J., and Platt, A. 1997. Signaling activation and repression of RNA polymerase II transcription in yeast. *Bioessays* 19:1001–1010.

Warren, A. J. 2002. Eukaryotic transcription factors. *Curr Opin Struct Biol* 12:107–114.

DNA-Binding Domains in Transcription Activator Proteins

Altucci, L., and Gronemeyer, H. 2001. Nuclear receptors in cell life and death. *Trends Endocrinol Metab* 17:460–468.

Bourguet, W., Germain, P., and Gronemeyer, H. 2000. Nuclear receptor ligand-binding domains: three-dimensional structures, molecular interactions and pharmacological implications. *Trends Pharmacol Sci* 21:381–388.

Chambon, P. 2004. How I became one of the fathers of a superfamily. *Nat Med* 10:1027–1031.

Chariot, A., Gielen, J., Mreville, M.-P., and Bours, V. 1999. The homeodomain-containing proteins. *Biochem Pharmacol* 58:1851–1857.

Chinenov, Y., and Kerppola, T. K. 2001. Close encounters of many kinds: Fos-Jun interactions that mediate transcription regulatory specificity. *Oncogene* 20:2438–2452.

Ellenberger, T. E., Brandl, C. J., Struhl, K., and Harrison, S. C. 1992. The GCN4 basic region leucine zipper binds DNA as a dimer of uninterrupted alpha helices: crystal structure of the protein-DNA complex. *Cell* 71:1223–1237.

Fraenkel, E., Rould, M. A., Chambers, K. A., and Pabo, C. O. 1998. Engrailed homeodomain-DNA complex at 2.2 Å resolution: a detailed view of the interface and comparison with other engrailed structures. *J Mol Biol* 284:351–361.

Glass, C. K., and Rosenfeld, M. G. 2000. The coregulator exchange in transcriptional functions of nuclear receptors. *Genes Dev* 14:121–141.

Hard, T., Kellenback, E., Boelens, R., et al. 1990. Solution structure of the glucocorticoid receptor DNA-binding domain. *Science* 249:157–160.

Hinnebusch, A. G., and Natarajan, K. 2002. Gcn4p, a master regulator of gene expression, is controlled at multiple levels of diverse signals of starvation and stress. *Eukaryot Cell* 1:22–32.

Kappen, C. 2000. The homeodomain: an ancient and evolutionary motif in animals and plants. *Comput Chem* 24:95–103.

Krishna, S. S., Majumdar, I., and Grishin, N. V. 2003. Structural classification of zinc fingers. *Nucleic Acid Res* 31:532–550.

Laity, J. H., Lee, B. M., and Wright, P. E. 2001. Zinc finger proteins: new insights into structural and functional diversity. *Curr Opin Struct Biol* 11:39–46.

Landschulz, W. H., Johnson, P. F., and McKnight, S. L. L. 1988. The leucine zipper: a hypothetical structure common to a new class of DNA binding proteins. *Science* 240:1759–1764.

Laughon, A., and Scott, M. P. 1984. Sequence of a *Drosophila* segmentation gene: protein structure homology with DNA-binding proteins. *Nature* 310:25–31.

Lee, K. C., and Kraus, W. L. 2001. Nuclear receptors, coactivators, and chromatin: new approaches, new insights. *Trends Endocrinol Metab* 12:191–197.

Lewis, E. B. 1978. A gene complex controlling segmentation in *Drosophila*. *Nature* 276:565–570.

Luisi, B. F., Xu, W. X., Otwinowski, Z., et al. 1991. Crystallographic analysis of the interaction of the glucocorticoid receptor with DNA. *Nature* 352:497–479.

Luscombe, N. M., Austin, S. E., Berman, H. M., and Thornton, J. M. 2000. An overview of protein-DNA complexes. *Genome Biol* 1:1–37.

Marmorstein, R., Carey, M., Ptashne, M., and Harrison, S. C. 1992. DNA recognition by GAL4: structure of a protein-DNA complex. *Nature* 356:379–380.

McGinnis, W., Garber, R. L., Wirz, J., et al. 1984. A homologous protein-coding sequence in *Drosophila* homeotic genes and its conservation in other metazoans. *Cell* 37:403–408.

McGinnis, W., Levine, M. S., Hafen, E., et al. 1984. A conserved sequence in homeotic genes of the *Drosophila antennapedia* and bithorax complexes. *Nature* 308:428–433.

McKenna, N. J., and O'Malley, B. W. 2002. Combinatorial control of gene expression by nuclear receptors and coregulators. *Cell* 108:465–474.

Natarajan, K., Meyer, M. R., Jackson, B. M., et al. 2001. Transcriptional profiling shows that Gcn4p is a master regulator of gene expression during amino acid starvation in yeast. *Mol Cell Biol* 21:4347–4368.

O'Shea, E. K., Klemm, J. D., Kim, P. S., and Alber, T. 1991. X-ray structure of the GCN4 leucine zipper, a two-stranded, parallel coiled coil. *Science* 254:539–544.

Patikoglou, G., and Burley, S. K. 1997. Eukaryotic transcription factor-DNA complexes. *Annu Rev Biophys Biomol Struct* 26:289–325.

Rastinejad, F. 2001. Retinoid X receptor and its partners in the nuclear receptor family. *Curr Opin Struct Biol* 11:33–38.

Reece, R. J., and Ptashne, M. 1993. Determinants of binding-site specificity among yeast C_6 zinc cluster proteins. *Science* 261:909–911.

Ribeiro, R. C. J., Kushner, P. J., and Baxter, J. D. 1995. The nuclear hormone receptor gene superfamily. *Annu Rev Med* 46:443–453.

Robyr, D., Wolffe, A. P., and Wahli, W. 2000. Nuclear hormone receptor coregulators in action: diversity for shared tasks. *Mol Endocrinol* 14:329–347.

Rosenfeld, M. G., and Glass, C. K. 2001. Coregulator codes of transcriptional regulation by nuclear receptors. *J Biol Chem* 276:36865–36868.

van Dam, H., and Castellazzi, M. 2001. Distinct roles of Jun:Fos and Jun:ATF dimers in oncogenesis. *Oncogene* 20:2453–2464.

Veraksa, A., Del Campo, M., and McGinnis, W. 2000. Developmental patterning genes and their conserved functions: from model organisms to humans. *Mol Genet Metab* 69:85–100.

Vogt, P. K. 2001. Jun, the oncoprotein. *Oncogene* 20:2365–2377.

Xie, W., and Evans, R. M. 2001. Orphan nuclear receptors: the exotics of xenobiotics. *J Biol Chem* 276:37739–37742.

Yen, P. M. 2001. Physiological and molecular basis of thyroid hormone action. *Physiol Rev* 81:1097–1142.

Activation Domains in Transcription Activator Proteins

Brent, R., and Finley, R. L. Jr. 1997. Understanding gene and allele function with two-hybrid methods. *Annu Rev Biochem* 31:663–704.

Breunig, K. D. 2000. Regulation of transcription activation by Gal4p. *Food Technol Biotechnol* 38:287–293.

Ma, J., and Ptashne, M. 1987. Deletion analysis of GAL4 defines two transcriptional activating segments. *Cell* 48:847–853.

Pilauri, V., Bewley, M., Diep, C., and Hopper, J. 2005. Gal80 dimerization and the yeast *GAL* gene switch. *Genetics* 169:1903–1914.

Struhl, K. 1995. Yeast transcriptional regulatory mechanisms. *Annu Rev Genet* 29:651–674.

Uetz, P. H., Hughes, R. E., and Fields, S. 1998. The two-hybrid system: finding likely partners for lonely proteins. *Focus* 20:62–64.

Mediator

Biddick, R., and Young, E. T. 2005. Yeast Mediator and its role in transcriptional regulation. *C R Biol* 328:773–782.

Björklund, S., Buzaite, O., and Hallberg, M. 2001. The yeast Mediator. *Mol Cell* 11:129–136.

Björklund, S., and Gustafsson, C. M. 2005. The yeast Mediator complex and its regulation. *Trends Biochem Sci* 30:240–244.

Blazek, E., Mittler, G., and Meisterernst, M. 2005. The Mediator of RNA polymerase II. *Chromosoma* 113:399–408.

Bourbon, H.-M., Aguilera, A., Ansari, A. Z., et al. 2004. A unified nomenclature for protein subunits of mediator complexes linking transcriptional regulators to RNA polymerase II. *Mol Cell* 14:553–557.

Bryant, G. O., and Ptashne, M. 2003. Independent recruitment *in vivo* by Gal4 of two complexes required for transcription. *Mol Cell* 11:1301–1309.

Cai, G., Imasaki, T., Takagi, Y., and Asturias, F. J. 2009. Mediator structural conservation and implications for the regulation mechanism. *Structure* 17:559–567.

Cai, G., Imasaki, T., Yamada, K., et al. 2010. Mediator head module structure and functional interactions. *Nat Struct Mol Biol* 17:273–279.

Casamassimi, A., and Napoli, C. 2007. Mediator complexes and eukaryotic transcription regulation: an overview. *Biochimie* 89:1439–1446.

Chadick, J. Z., and Asturias, F. J. 2005. Structure of eukaryotic Mediator complexes. *Trends Biochem Sci* 30:264–271.

Cheng, J. X., Gandolfi, M., and Ptashne, M. 2004. Activation of the *Gal1* gene of yeast by pairs of "non-classical" activators. *Curr Biol* 14:1675–1679.

Conaway, R. C., Sato, S., Tomomori-Sato, C., et al. 2005. The mammalian Mediator complex and its role in transcriptional regulation. *Trends Biochem Sci* 30:250–255.

Courey, A. J., and Jia, S. 2001. Transcriptional repression: the long and short of it. *Genes Dev* 15:2786–2796.

Guglielmi, B., van Berkum, N. L., Klapholz, B., et al. 2004. A high resolution protein interaction map of the yeast Mediator complex. *Nucleic Acids Res* 32:5379–5391.

Kim, Y.-J., and Lis, J. T. 2005. Interactions between subunits of *Drosophila* Mediator and activator proteins. *Trends Biochem Sci* 30:245–249.

Kornberg, R. D. 2005. Mediator and the mechanism of transcriptional activation. *Trends Biochem Sci* 30:235–239.

Kuras, L., Borggrefe, T., and Kornberg, R. D. 2003. Association of the Mediator complex with enhancers of active genes. *Proc Natl Acad Sci USA* 100:13887–13991.

Lewis, B. A., and Reinberg, D. 2003. The mediator coactivator complex: functional and physical roles in transcription regulation. *J Cell Sci* 116:3667–3675.

Malik, S., and Roeder, R. G. 2005. Dynamic regulation of Pol II transcription by the mammalian Mediator complex. *Trends Biochem Sci* 30:256–263.

Malik, S., and Roeder, R. G. 2010 The metazoan Mediator co-activator complex as an integrative hub for transcriptional regulation. *Nat Rev Genet* 11:761–772.

Meyers, L. C., Gustafsson, C. M., Hayashibara, K. C., et al. 1999. Mediator protein interactions that selectively abolish activated transcription. *Proc Natl Acad Sci USA* 96:67–72.

Meyers, L. C., and Kornberg, R. D. 2000. Mediator of transcriptional regulation. *Annu Rev Biochem* 69:729–749.

Meyers, L. C., Leuther, K., Bushnell, D. A., et al. 1997. Yeast RNA polymerase II transcription reconstituted with purified proteins. *Methods* 12:212–218.

Rachez, C., and Freedman, L. P. 2001. Mediator complexes and transcription. *Curr Opin Cell Biol* 13:274–280.

Taatjes, D. J. 2010. The human Mediator complex: a versatile, genome-wide regulator of transcription. *Trends Biochem Sci* 35:315–322.

Takagi, Y., Calero, G., Komori, H., et al. 2006. Head module control of mediator interactions. *Mol Cell* 23:355–364.

Takagi, Y., and Kornberg, R. D. 2006. Mediator as a general transcription factor. *J Biol Chem* 281:80–89.

Epigenetic Modifications

Ballestar, E., and Esteller, M. 2005. Methylated DNA-binding proteins. *Encyclopedia of Life Sciences*. pp. 1–5. Hoboken, NJ: John Wiley & Sons.

Becker, P. B., and Hörz, W. 2002. ATP-dependent nucleosome remodeling. *Annu Rev Biochem* 71:247–273.

Berger, S. L. 2002. Histone modifications in transcriptional regulation. *Curr Opin Genet Dev* 12:142–148.

Berger, S. L., Kouzarides, T., Shiekhattar, R., and Shilatifard, A. 2009. An operational definition of epigenetics. *Genes Dev* 23:781–783.

Bodanović, O., and Veenstra, G. J. C. 2009. DNA methylation and methyl-CpG proteins: developmental requirements and function. *Chromosoma* 118:549–565.

Brown, C. E., Lechner, T., Howe, L., and Workman, J. L. 2000. The many HATs of transcription coactivators. *Trends Biochem Sci* 25:15–19.

Cairns, B. R. 2009. The logic of chromatin architecture and remodelling at promoters. *Nature* 461:193–198.

Campose, E. I., and Reinberg, D. 2009. Histones: annotating chromatin. *Annu Rev Genet* 43:559–599.

Chaban, Y., Ezeokonkwo, C., Chung, W.-H., et al. 2008. Structure of a RSC-nucleosome complex and insights into chromatin modeling. *Nat Struct Mol Biol* 15:1272–1277.

Clapier, C. R., and Cairns, B. R. 2009. The biology of chromatin remodeling complexes. *Annu Rev Biochem* 78:273–304.

Constância, M., Murrell, A., and Reik, W. 2005. Genomic imprinting at the transcriptional level. *Encyclopedia of Life Sciences*. pp. 1–7. Hoboken, NJ: John Wiley & Sons.

Cosgrove, M. S., Boeke, J. D., and Wolberger, C. 2004. Regulated nucleosome mobility and the histone code. *Nat Struct Mol Biol* 11:1037–1043.

Dickins, B. J. A., and Kelsey, G. 2008. Evolution of imprinting: imprinted gene function in human disease. *Encyclopedia of Life Sciences*. pp. 1–11. Hoboken, NJ: John Wiley & Sons.

Ferguson-Smith, A. C., and Surani, M. A. 2001. Imprinting and the epigenetic asymmetry between parental genomes. *Science* 293:1086–1089.

Fry, C. J., and Peterson, C. L. 2002. Unlocking the gates of gene expression. *Science* 295:1847–1848.

Fyodorov, D. V., and Kadonaga, J. T. 2001. The many faces of chromatin remodeling: switching beyond transcription. *Cell* 106:523–525.

Gamble, M. J., and Freedman, L. P. 2002. A coactivator code for transcription. *Trends Biochem Sci* 27:165–167.

Gartenberg, M. R. 2000. The Sir proteins of *Saccharomyces cerevisiae*: mediators of transcriptional silencing and much more. *Curr Opin Microbiol* 3:132–137.

Gregory, P. D. 2001. Transcription and chromatin converge, lessons from yeast genetics. *Curr Opin Genet Dev* 11:142–147.

Ideraabdullah, F. Y., Vigneau, S., and Bartolomei, M. S. 2008. Genomic imprinting mechanisms in mammals. *Mutation Res* 647:77–85.

Iizuka, M., and Smith, M. M. 2003. Functional consequences of histone modifications. *Curr Opin Genet Dev* 13:154–160.

Imbalzano, A. N., and Xiao, H. 2004. Functional properties of ATP-dependent chromatin remodeling enzymes. *Adv Protein Chem* 67:157–179.

Jenuwein, T., and Allis, C. D. 2001. Translating the histone code. *Science* 293:1074–1080.

Jiang, C., and Pugh, F. 2009. Nucleosome positioning and gene regulation: advances through genomics. *Nat Rev Genet* 10:161–172.

Jiang, Y., Bressler, J., and Beaudet, A. L. 2004. Epigenetics and human disease. *Annu Rev Genomics Hum Genet* 5:479–510.

Koerner, M. V., and Barlow, D. P. 2010. Genomic imprinting—an epigenetic gene-regulatory model. *Curr Opin Genet Dev* 20:164–170.

Kouzarides, T. 2007. Chromatin modifications and their function. *Cell* 128:693–705.

Kuzmichev, A., and Reinberg, D. 2001. Role of histone deacetylase complexes in the regulation of chromatin metabolism. *Curr Top Microbiol Immunol* 254:35–58.

Li, B., Cary, M., and Workman, J. L. 2007. The role of chromatin during transcription. *Cell* 128:707–719.

Lorch, Y., Maier-Davis, B., and Kornberg, R. D. 2010. Mechanism of chromatin remodeling. *Proc Natl Acad Sci USA* 107:3458–3462.

Lusser, A., and Kadonaga, J. T. 2003. Chromatin remodeling by ATP-dependent molecular machines. *Bioessays* 25:1192–1200.

Margueron, R., and Reinberg, D. 2010. Chromatin structure and the inheritance of epigenetic information. *Nat Rev Genet* 11:285–296.

Marmorstein, R., and Trievel, R. C. 2009. Histone modifying enzymes: structures, mechanisms, and specificities. *Biochim Biophys Acta* 1789:58–68.

Morgan, H. D., Santos, F., Green, K., et al. 2005. Epigenetic reprogramming in mammals. *Hum Mol Genet* 14:R47–R58.

Munshi, A., Shafi, G., Aliya, N., and Jyothy, A. 2009. Histone modifications dictate specific biological readouts. *J Genet Genomics* 36:75–88.

Narlikar, G. J., Fan, H.-Y., and Kingston, R. E. 2002. Cooperation between complexes that regulate chromatin structure and transcription. *Cell* 108:475–487.

Ng, H. H., and Bird, A. 2000. Histone deacetylases: silencers for hire. *Trends Biochem Sci* 25:121–126.

Pérez-Martin, J. 1999. Chromatin and transcription in *Saccharomyces cerevisiae*. *FEMS Microbiol Rev* 23:503–523.

Peterson, C. L. 2002. Chromatin remodeling enzymes: taming the machines. *EMBO Rep* 31:319–322.

Reid, G., Gallais, R., and Métivier, R. 2009. Marking time: the dynamic role of chromatin and covalent modification in transcription. *Int J Biochem Cell Biol* 41:155–163.

Richards, E. J., and Elgin, S. C. R. 2002. Epigenetic codes for heterochromatin formation and silencing: rounding up the usual suspects. *Cell* 108:489–500.

Rutledge, C. E., Lees-Murdock, D. J., and Walsh, C. P. 2010. DNA methylation in development. *Encyclopedia of Life Sciences*. pp. 1–8. Hoboken, NJ: John Wiley & Sons.

Sasaki, H., and Matsui, Y. 2008. Epigenetic events in mammalian germ-cell development: reprogramming and beyond. *Nat Rev Genet* 9:129–140.

Saha, A., Wittmeyer, J., and Cairns, B. R. 2005. Chromatin remodeling through directional DNA translocation from an internal nucleosome site. *Nat Struct Mol Biol* 12:747–755.

Schnitzler, G. R. 2008. Control of nucleosome positions by DNA sequence and remodeling. *Cell Biochem Biophys* 51:67–80.

Segal, E., and Widom, J. 2009. What controls nucleosome positions? *Trends Genet* 25:335–343.

Selth, L. A., Sigurdsson, S., and Svejstrup, J. Q. 2010. Transcription elongation by RNA polymerase II. *Annu Rev Biochem* 79:271–293.

Sims, R. J., 3rd, and Reiberg, D. 2008. Is there a code embedded in proteins that is based on post-translational modifications? *Nat Rev Mol Cell Biol* 9:1–6.

Spotswood, H. T., and Turner, B. M. 2002. An increasingly complex code. *J Clin Invest* 110:577–582.

Sterner, D. E., and Berger, S. L. 2000. Acetylation of histones and transcription-related factors. *Microbiol Mol Biol Rev* 64:435–459.

Struhl, K. 1998. Histone acetylation and transcriptional regulatory mechanisms. *Genes Dev* 12:599–606.

Struhl, K. 1999. Fundamentally different logic of gene regulation in eukaryotes and prokaryotes. *Cell* 98:1–4.

Surani, M. Z. 2001. Imprinting (mammals). *Encyclopedia of Life Sciences*. pp. 1–6. Hoboken, NJ: John Wiley & Sons.

Svejstrup, J. Q. 2004. The RNA polymerase II transcription cycle: cycling through chromatin. *Biochim Biophys Acta* 1677:64–73.

Torok, M. S., and Grant, P. A. 2004. Histone acetyltransferase proteins contribute to transcriptional processes at multiple levels. *Adv Protein Chem* 67:181–199.

Turner, B. M. 2002. Cellular memory and the histone code. *Cell* 111:285–291.

Venters, B. J., and Pugh, B. F. 2009. How eukaryotic genes are transcribed. *Crit Rev Biochem Mol Biol* 44:117–141.

Vignali, M., Hassan, A. H., Neely, K. E., and Workman, J. L. 2000. ATP-dependent chromatin-remodeling complexes. *Mol Cell Biol* 20:1899–1910.

Wolffe, A. P., and Guschin, D. 2000. Chromatin structural features and targets that regulate transcription. *J Struct Biol* 129:102–122.

Wood, A. J., and Oakey, R. J. 2006. Genomic imprinting in mammals: emerging themes and established theories. *PLoS Genet* 2:1677–1685.

Zhang, Y., and Reinberg, D. 2001. Transcription regulation by histone methylation: interplay between different covalent modifications of the core histone tails. *Genes Dev* 15:2343–2360.

RNA Polymerase I Catalyzed Transcription

Caburet, S., Conti, C., Schurra, C., et al. 2005. Human ribosomal RNA gene arrays display a broad range of palindromic structures. *Genome Res* 15:1079–1085.

Comai, L. 2004. Mechanism of RNA polymerase I transcription. *Adv Protein Chem* 67:123–155.

De Carlo, S., Carles, C., Riva, M., and Schultz, P. 2003. Cryo-negative staining reveals conformational flexibility within yeast RNA polymerase I. *J Mol Biol* 329:891–902.

Drygin, D., Rice, W. G., and Grummt, I. 2010. The RNA polymerase I transcription machinery: an emerging target for the treatment of cancer. *Annu Rev Pharmacol Toxicol* 50:131–156.

Haag, J. R., and Pikaard, C. S. 2007. RNA polymerase I: a multifunctional molecular machine. *Cell* 1224–1225.

Hanada, K., Song, C. Z., Yamamoto, K., et al. 1996. RNA polymerase I associated factor 53 binds to the nucleolar transcription factor UBF and functions in specific rDNA transcription. *EMBO J* 15:2217–2226.

Kuhn, C.-D., Geiger, S. R., Baumli, S., et al. 2007. Functional architecture of RNA polymerase I. *Cell* 131:1260–1272.

Moss, T., Langlois, F., Gagnon-Kugler, T., and Stefanovsky, V. 2007. A housekeeper with power of attorney: the rRNA genes in ribosome biogenesis. *Cell Mol Life Sci* 64:29–49.

Moss, T., and Stefanovsky, V. Y. 2002. At the center of eukaryotic life. *Cell* 109:545–548.

Moss, T., Stefanovsky, V., Langlois, F., and Gagnon-Kugler, T. 2006. A new paradigm for the regulation of the mammalian ribosomal RNA genes. *Biochem Soc Trans* 34:1079–1081.

Paule, M. R. (ed.). 1999. *Transcription of Ribosomal RNA Genes by Eukaryotic RNA Polymerase I*. New York: Springer.

Paule, M. R., and White, R. J. 2000. Transcription by RNA polymerases I and III. *Nucleic Acid Res* 28:1283–1298.

Reeder, R. H., and Lang, W. H. 1997. Terminating transcription in eukaryotes: lessons learned from RNA polymerase I. *Trends Biochem Sci* 22:473–477.

Russell, J., and Zomerdijk, J. C. B. M. 2005. RNA polymerase-I-directed rDNA transcription, life and works. *Trends Biochem Sci* 30:87–96.

Sanij, E., and Hannan, R. D. 2009. The role of UBF in regulating the structure and dynamics of transcriptionally active rDNA chromatin. *Epigenetics* 4:374–382.

Sanij, E., Poortingo, G., Sharkey, K., et al. 2008. UBF levels determine the number of active ribosomal RNA genes in mammals. *J Cell Biol* 183:1258–1274.

Thiry, M., and Lafontaine, D. L. J. 2005. Birth of a nucleolus: the evolution of nucleolar compartments. *Trends Cell Biol* 15:194–199.

Werner, M., Thuriaux, P., and Soutourina, J. 2009. Structure-function analysis of RNA polymerases I and III. *Curr Opin Struct Biol* 19:740–745.

White, R. J. 2005. RNA polymerases I and III, growth control and cancer. *Nat Rev Mol Cell Biol* 6:69–78.

RNA Polymerase III Catalyzed Transcription

Canella, D., Praz, V., Reina, J. H., et al. 2010. Defining the RNA polymerase III transcriptome: genome-wide localization of the RNA polymerase III transcription machinery in human cells. *Genome Res* 20:710–721.

Cramer, P. 2006. Recent structural studies of RNA polymerase II and III. *Biochem Soc Trans* 34:1058–1061.

Haeusler, R. A., and Engelke, D. R. 2006. Spatial organization of transcription by RNA polymerase III. *Nucleic Acid Res* 34:4826–4836.

Harismendy, O., Gendrel, C. G., Soularue, P., et al. 2003. Genome-wide location of yeast RNA polymerase III transcription machinery. *EMBO J* 22:4738–4747.

Moqtaderi, Z., Wang, J., and Raha, D. 2010. Genomic binding profiles of functionally distinct RNA polymerase III transcription complexes in human cells. *Nat Struct Mol Biol* 17:635–640

Oler, A. J., Alla, R. K., Roberts, D. N., et al 2010. Human RNA polymerase III transcriptomes and relationships to Pol II promoter chromatin and enhancer-binding factors. *Nat Struct Mol Biol* 17:620–628.

Paule, M. R., and White, R. J. 2000. Transcription by RNA polymerases I and III. *Nucleic Acid Res* 28:1283–1298.

Schramm, L., Hernandez, N. 2002. Recruitment of RNA polymerase III to its target promoters. *Gene Dev* 16:2593–2620.

Werner, M., Thuriaux, P., and Soutourina, J. 2009. Structure-function analysis of RNA polymerases I and III. *Curr Opin Struct Biol* 19:740–745.

White, R. J. 1998. RNA *Polymerase III Transcription* (2nd ed). New York: Springer-Verlag.

White, R. J. 2005. RNA polymerases I and III, growth control and cancer. *Nat Rev Mol Cell Biol* 6:69–78.

Wolin, S. L., and Matera, A. G. 1999. The trials and travels of tRNA. *Genes Dev* 13:1–10.

RNA Polymerase II: Cotranscriptional and Posttranscriptional Processes

CHAPTER OUTLINE

14.1 Pre-mRNA
Eukaryotic cells synthesize large heterogeneous nuclear RNA molecules.

mRNA and hnRNA both have poly(A) tails at their 3'-ends.

14.2 Cap Formation
mRNA molecules have 7-methylguanosine caps at their 5'-ends.

5'-m^{7}G caps are attached to nascent pre-mRNA chains when the chains are 20 to 30 nucleotides long.

All eukaryotes use the same basic pathway to form 5'-m^{7}G caps.

CTD must be phosphorylated on Ser-5 to target a transcript for capping.

14.3 Split Genes
Viral studies revealed that some mRNA molecules are formed by splicing pre-mRNA.

Amino acid coding regions within eukaryotic genes may be interrupted by noncoding regions.

Exons tend to be conserved during evolution, whereas introns usually are not conserved.

A single pre-mRNA can be processed to produce two or more different mRNA molecules.

Combinations of the various splicing patterns within individual genes lead to the formation of multiple mRNAs.

Pre-mRNA requires specific sequences for precise splicing to occur.

Two splicing intermediates resemble lariats.

Splicing consists of two coordinated transesterification reactions.

14.4 Spliceosomes
Aberrant antibodies, which are produced by individuals with certain autoimmune diseases, bind to small nuclear ribonucleoprotein particles (snRNPs).

snRNPs assemble to form a spliceosome, the splicing machine that excises introns.

RNA and protein may both contribute to the spliceosome's catalytic site.

Cells use a variety of mechanisms to regulate splice site selection.

Splicing begins as a cotranscriptional process and continues as a posttranscriptional process.

mRNA splicing and export are coupled processes.

14.5 Cleavage/Polyadenylation and Transcription Termination
Poly(A) tail synthesis and transcription termination are coupled, cotranscriptional processes.

Transcription units often have two or more alternate polyadenylation sites.

Transcription termination takes place downstream from the poly(A) site.

RNA polymerase II transcription termination appears to involve allosteric changes and a 5'→3' exonuclease.

14.6 RNA Editing
RNA editing permits a cell to recode genetic information.

14.7 The Gene Reconsidered
The human proteome contains a much greater variety of proteins than would be predicted from the human genome.

Cotranscriptional and posttranscriptional processes force us to reconsider our concept of the gene.

BOX 14.1: LOOKING DEEPER Self-Splicing RNA

QUESTIONS AND PROBLEMS

SUGGESTED READING

Image from Protein Data Bank ID 2033. D. G. Sashital, V. Venditti, C. G. Angers, G. Cornilescu, and S. E. Butcher. *RNA* 13 [2007]: 328–238. Prepared by B. E. Tropp.

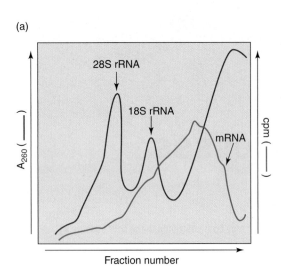

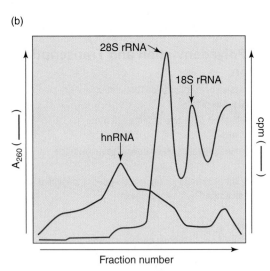

FIGURE 14.1 **Size distribution of HeLa cell mRNA and heterogeneous nuclear RNA (hnRNA) molecules.** Ribosomal RNA profiles, which were determined by measuring λ_{260} nm absorption, are shown in red. (a) HeLa cells were cultured in the presence of [^{32}P] phosphate. RNA isolated from polysomes (two or more ribosomes translating a single mRNA) was analyzed by sucrose gradient centrifugation. The radioactive mRNA profile is shown in green. (b) HeLa cells were cultured in the presence of [^{32}P]phosphate for 5 minutes. Then hnRNA was extracted from cell nuclei and analyzed by sucrose gradient centrifugation. The radioactive hnRNA profile is shown in blue. (Adapted from Darnell, Jr., J. E. 2002. *Nat Med.* 8:1068–1071.)

olecular biologists initially thought eukaryotes would synthesize messenger RNA (mRNA) in much the same way as bacteria. They therefore expected RNA polymerase II would synthesize fully functional mRNA that just needed to be transported from the cell nucleus to the cytoplasm, where it would direct ribosomes to synthesize proteins. It came as a surprise to find the transcript formed by the RNA polymerase II machinery is not mature mRNA but rather a **precursor mRNA (pre-mRNA)** that must be extensively modified before it can function as mRNA. The eukaryotic cell processes the pre-mRNA transcript as it is being synthesized by (1) attaching a guanosine cap to the 5′-end of the nascent chain, (2) excising specific sequences from within the transcript, (3) adding a poly(A) tail to the 3′-end of the mRNA, and (4) terminating transcription. This chapter examines the various stages in cotranscriptional processing as well as reactions that replace one base in a transcript by another (a process termed RNA editing).

14.1 Pre-mRNA

Eukaryotic cells synthesize large heterogeneous nuclear RNA molecules.

The first indication that the RNA polymerase II machinery synthesizes pre-mRNA rather than mRNA came from experiments performed by James Darnell and coworkers in the mid-1960s. In one particularly informative experiment they added [^{3}H]uridine to cultured HeLa cells (human cervical tumor cells) for 5 minutes and then extracted labeled RNA from cell nuclei and analyzed the RNA by sucrose gradient centrifugation. They expected the rapidly labeled nuclear RNA to consist of a population of chains ranging from about 500 to 3,000 nucleotides long, the size distribution of HeLa cell mRNA (FIGURE 14.1a) but instead observed the newly synthesized RNA molecules were about 10 times longer (Figure 14.1b). This rapidly labeled nuclear RNA was dubbed **heterogeneous nuclear RNA (hnRNA)** because it was present in the nucleus and consisted of RNA molecules with a broad size distribution.

Further study indicated the base composition of HeLa cell hnRNA is 43% G+C, just like HeLa cell mRNA. Moreover, HeLa cell hnRNA did not accumulate in the nucleus, signifying it is quite unstable. Based on these observations Darnell proposed that hnRNA (or at least a large percentage of RNA molecules within the hnRNA fraction) are pre-mRNA molecules, which must somehow be trimmed to form shorter mRNA molecules.

mRNA and hnRNA both have poly(A) tails at their 3′-ends.

The next major advance in studying eukaryotic mRNA synthesis took place in the early 1970s as a result of efforts to characterize HeLa cell mRNA by digesting it with ribonucleases. The mRNA was extracted from cytoplasmic complexes known as **polyribosomes**, or **polysomes**, each consisting of two or more ribosomes translating a single mRNA. Two different endonucleases, RNase T_1 (a product of the fungus *Aspergillus oryzae*) and pancreatic RNase, were used to digest the mRNA. RNase T_1 and pancreatic RNase cleave phosphodiester bonds on the 3′-sides of guanylate residues and pyrimidine nucleotides, respectively (FIGURE 14.2). Assuming a more or less random nucleotide distribution within HeLa cell mRNA, one would expect to find the digests would contain a mixture of short oligonucleotides along with some mononucleotides. Although digesting HeLa cell mRNA with either T_1 RNase or pancreatic RNase did indeed produce the expected digestion products, it also produced an entirely unexpected polynucleotide containing 150 to 200 adenylate groups. Further analysis showed these poly(A) segments are attached to the 3′-end of mRNA. With the one notable exception of histone mRNAs, all eukaryotic mRNAs characterized to date have such 3′-poly(A) tails. Yeast poly(A) tails are usually between 50 and 70 nucleotides long, considerably shorter than mammalian poly(A) tails. Subsequent studies showed a large proportion of the rapidly labeled hnRNA molecules also have poly(A) tails, supporting the idea that these hnRNA molecules are converted to mRNA.

Poly(A) tails have a practical laboratory application: their base pairing properties can be used to separate mRNA and pre-mRNA from other kinds of RNA (FIGURE 14.3). This separation is accomplished by passing a mixture of RNA molecules, dissolved in a buffer solution containing a high salt concentration, through a column packed with cellulose fibers linked to oligo(dT). mRNA molecules stick to the column because of base pairing between their poly(A) tails and oligo(dT). The same is true for pre-mRNA molecules. Metal cations from the salt help to stabilize this base pairing by diminishing charge repulsion between phosphate groups. Other kinds of RNA molecules that lack poly(A) tails pass right through the column. Then mRNA molecules (or pre-mRNA molecules) are eluted from the column by washing with a low salt buffer. Importantly, not all the RNA molecules in the hnRNA fraction bind to oligo(dT). Nearly all those that do not bind are destined to become some other kind of cellular RNA, such as ribosomal (rRNA) or transfer RNA (tRNA).

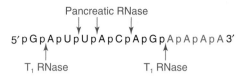

FIGURE 14.2 RNA hydrolysis catalyzed by T_1 RNase and pancreatic RNase. T_1 RNase cleaves on the 3′-side of guanylate and pancreatic RNase cleaves on the 3′-side of pyrimidine nucleotides. Notice that the four adenylates on the 3′-side of the molecule (shown in green) remain intact.

14.2 Cap Formation

mRNA molecules have 7-methylguanosine caps at their 5′-ends.

Because tRNA and rRNA were known to have minor nucleotides formed by adding methyl (–CH$_3$) groups to bases or ribose moieties, it was logical to ask whether eukaryotic mRNA molecules also have methyl groups attached to them. This question could not be answered

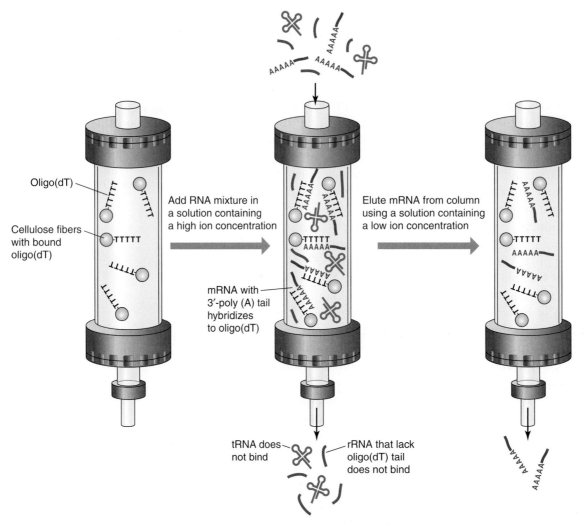

FIGURE 14.3 Purification of mRNA and pre-mRNA from a mixture of RNA molecules by oligo(dT) cellulose chromatography. A mixture of RNA molecules, dissolved in a buffer solution containing a high salt concentration, is passed through a column packed with cellulose fibers linked to oligo(dT). mRNA molecules (red) stick to the column because of base pairing between the poly(A) tails and oligo(dT). The same is true for pre-mRNA molecules. Metal cations from the salt help to stabilize this base pairing by diminishing charge repulsion between phosphate groups. Other kinds of RNA molecules (green and blue) that lack poly(A) tails do not bind to the oligo(dT) and are eluted when the column is washed with buffer containing high salt. The mRNA and pre-mRNA are eluted from the column by washing with a low salt buffer.

unequivocally before oligo(dT) cellulose chromatography became available because there was no way to determine if the methyl groups detected were attached to mRNA or contaminating tRNA and rRNA molecules. The availability of a method to prepare eukaryotic mRNA free of tRNA and rRNA allowed investigators to begin looking for methylated nucleotides in mRNA.

In 1974 Robert Perry and coworkers were among the first to do so. Their approach was to culture mouse L cells (a fibroblast-like cell) in a medium containing [^{3}H-methyl]methionine, a precursor of the methyl group donor S-adenosylmethionine (FIGURE 14.4). They found purified mRNA contains two to three methyl groups per 1,000 nucleotides and the methylated nucleotides were clustered at the 5′-end of the mRNA. Exhaustive digestion of mRNA with

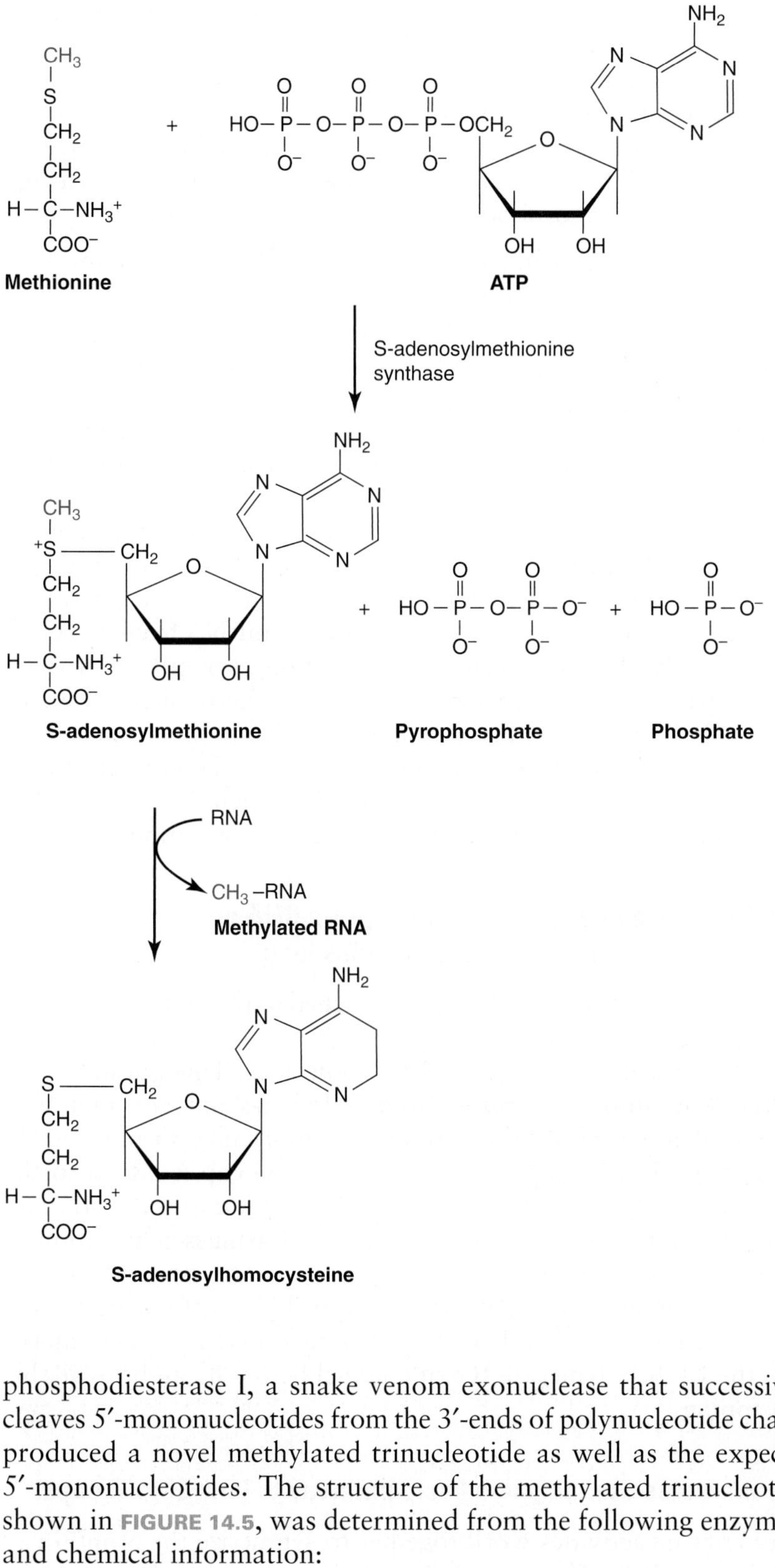

FIGURE 14.4 Conversion of methionine to S-adenosylmethionine. Methionine reacts with ATP to form S-adenosylmethionine, which then can transfer its methyl group to a base in RNA (or DNA) or a ribose in RNA.

phosphodiesterase I, a snake venom exonuclease that successively cleaves 5′-mononucleotides from the 3′-ends of polynucleotide chains, produced a novel methylated trinucleotide as well as the expected 5′-mononucleotides. The structure of the methylated trinucleotide, shown in **FIGURE 14.5**, was determined from the following enzymatic and chemical information:

1. The trinucleotide had a normal 3′-end but lacked a terminal 5′-OH or 5′-phosphate group.
2. Two nucleotides within the trinucleotide had methyl groups attached to them; one methyl group was attached to a guanosine

14.2 Cap Formation

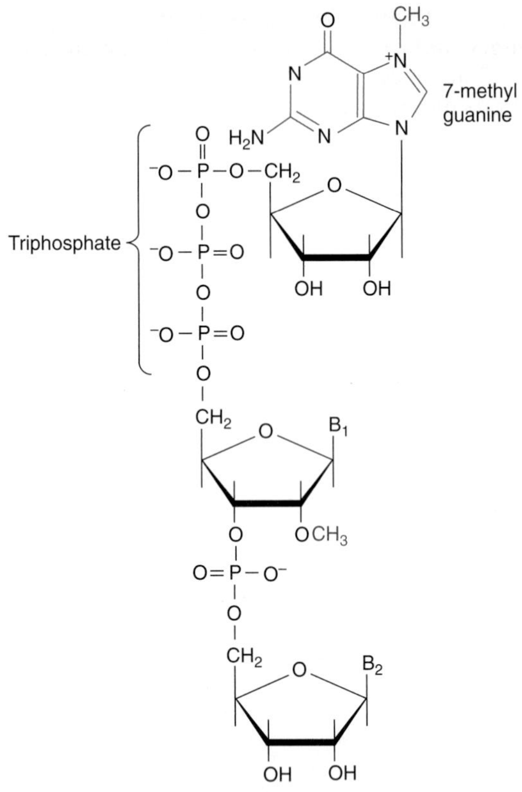

FIGURE 14.5 mRNA 5′-cap structure. The mRNA 5′-cap shown, which is known as cap 1, may be written as m⁷GpppNmpNp, where m is a methyl group, N is any nucleotide, and p is a phosphate group.

at N-7 and the other to C-2′ of the ribose group in the first or initiating nucleotide in mRNA.

3. The N^7-methylguanosine (m^7G) was attached to the first or initiating nucleotide in mRNA through an inverted 5′→5′ triphosphate bridge to form a cap at the 5′-end of mRNA. The cap could be removed by pyrophosphatase catalyzed cleavage but not by phosphodiesterase catalyzed cleavage.

Virtually all eukaryotic mRNAs have **5′-m^7G caps**, as do mRNAs made by many animal viruses. Three different types of cap structures are present in eukaryotic mRNAs. Cap 1, the structure shown in Figure 14.5, is present in mRNAs from multicellular organisms. The other two kinds of caps are called cap 0 and cap 2. The cap 0 structure, which is a feature of yeast mRNAs, has one less methyl group than cap 1. Its single methyl group is attached to guanine at position N-7. The cap 2 structure, which is present in some vertebrate mRNAs, has one more methyl group than the cap 1 structure. The extra methyl group is attached to C-2′ of the second nucleotide in the mRNA. Despite their different methylation patterns, cap 0, cap 1, and cap 2 appear to have similar functions. They protect mRNA from digestion by 5′→3′ exonucleases, influence subsequent stages of mRNA processing, and participate in translation. A protein heterodimer known as the cap binding complex binds to a 5′-m^7G cap in the nucleus. Cap binding complex participates in subsequent stages of cotranscriptional processing, helps mRNA to be transported through the nuclear pore, and is an important factor in the early stage of translation.

5′-m^7G caps are attached to nascent pre-mRNA chains when the chains are 20 to 30 nucleotides long.

Once 5′-m^7G caps were shown to be attached to the first nucleotide in mRNA, it seemed reasonable to ask if 5′-m^7G caps are also attached to the first nucleotide in pre-mRNA molecules. This question was answered by subjecting hnRNA to oligo(dT) column chromatography and demonstrating that pre-mRNA molecules so purified do indeed have 5′-m^7G caps. Because the 5′-m^7G cap is added to the first or initiating nucleotide, it seemed likely the cap is added at an early stage of transcription. In 1993 Eric B. Rasmussen and John T. Lis showed the 5′-m^7G cap forms when nascent transcripts are 20 to 30 nucleotides long. Cap formation probably cannot occur any earlier because the 5′-end of the growing transcript must emerge from the RNA polymerase II exit channel before it can be available for capping.

All eukaryotes use the same basic pathway to form 5′-m^7G caps.

Three enzyme activities work together to synthesize the 5′-m^7G cap (FIGURE 14.6). RNA 5′-triphosphatase cleaves the γ-phosphate from the 5′-triphosphate end of the initiating nucleotide as it emerges from the RNA polymerase II exit channel. Then guanylyltransferase catalyzes guanylyl group transfer from guanosine triphosphate to the newly

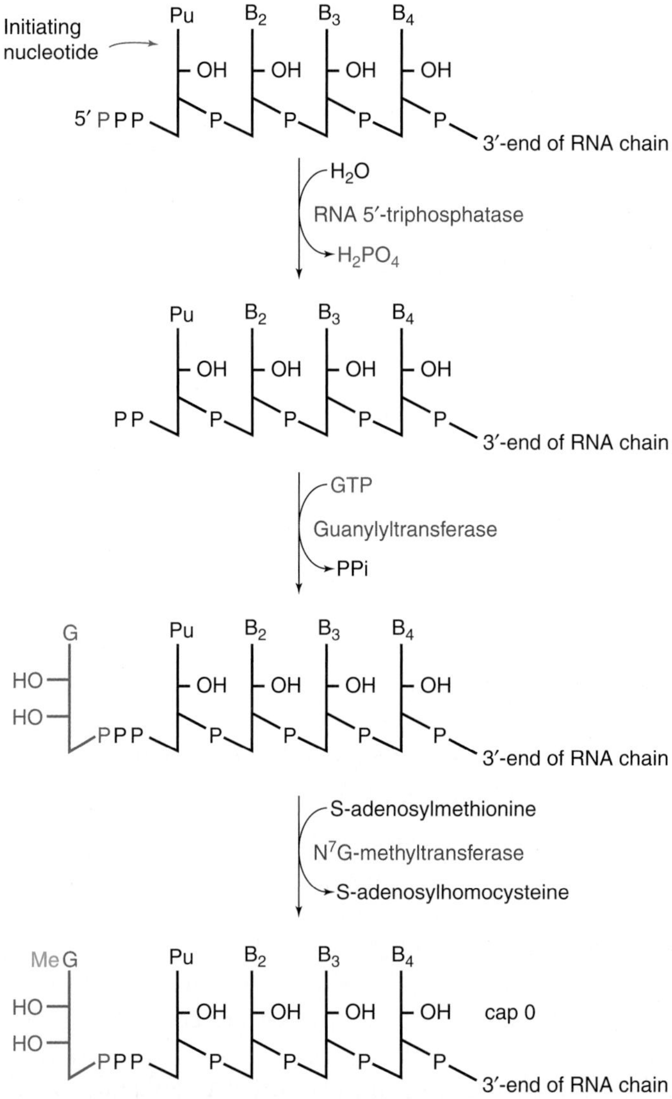

FIGURE 14.6 Capping pathway. The process takes place in three steps: (1) RNA 5′-triphosphatase cleaves the γ-phosphate from the 5′-triphosphate end of the initiating nucleotide as it emerges from the RNA polymerase II exit channel; (2) guanylyltransferase catalyzes guanylyl (GMP) group transfer from guanosine triphosphate (GTP) to the newly created 5′-diphosphate end to form the GpppN cap (where N is usually a purine nucleotide); and (3) methyltransferase transfers a methyl group from S-adenosylmethionine to the N-7 position of the cap guanine.

created 5′-diphosphate end to form the GpppN cap (where N is usually a purine nucleotide). In yeast the two enzyme activities are on separate polypeptide chains that interact to form a heterodimer. In mammals both activities are part of a single bifunctional polypeptide. Methyltransferase, the third enzyme activity required for capping, transfers a methyl group from S-adenosylmethionine to the N-7 position of the cap guanine. In vertebrates an additional methylation reaction converts cap 0 to cap 1 in the nucleus, and then another methylation reaction converts cap 1 to cap 2 in the cytoplasm.

CTD must be phosphorylated on Ser-5 to target a transcript for capping.

When protein-coding genes are modified so they can be transcribed by RNA polymerase I or RNA polymerase III, transcripts are no longer capped. This observation suggests RNA polymerase II has some unique feature that directs the capping machinery to its own transcripts. The carboxyl terminal domain (CTD) of the largest subunit in RNA polymerase II seems a likely candidate because there is no comparable structure in either RNA polymerases I or III. Moreover, the crystal structure of yeast RNA polymerase II reveals CTD is adjacent to the RNA exit channel and therefore is located in just the right position to interact with the transcript as it emerges from polymerase. Biochemical studies show Ser-5 (underlined) in the repeating Tyr-Ser-Pro-Thr-Ser-Pro-Ser heptapeptides in CTD must be phosphorylated for cap formation to take place. Once capping is complete a specific protein phosphatase removes the phosphate group from the Ser-5.

14.3 Split Genes

Viral studies revealed that some mRNA molecules are formed by splicing pre-mRNA.

The observation that both pre-mRNA and mRNA molecules have 5'-m^{7}G caps and 3'-poly(A) tails seemed to rule out the possibility that pre-mRNA is converted to mRNA by trimming one or both ends of the pre-mRNA molecule because such trimming would of necessity remove the 5'-m^{7}G cap, the 3'-poly(A) tail, or both. However, a model in which both ends of pre-mRNA are conserved presented an even greater conceptual challenge because the only way to conserve both ends while shortening the pre-mRNA molecule would be to remove one or more segments from within pre-mRNA and then join the flanking sequences. Investigators were understandably reluctant to accept such an unprecedented splicing mechanism. Then in 1977 two independent research groups, one led by Phillip A. Sharp and the other by Richard Roberts, provided convincing visual evidence that splicing occurs by demonstrating eukaryotic genes have long DNA sequences within them that are missing from mature mRNA.

The Sharp and Roberts laboratories studied viral mRNA formed by HeLa cells infected with adenovirus, a DNA virus that infects the upper respiratory tract. The adenovirus life cycle can be divided into an early and late phase; the latter begins at the onset of viral DNA synthesis. Adenovirus-infected HeLa cells are well suited for studying mRNA formation because they produce an abundant supply of eight different mRNA molecules during the late phase. These late mRNAs can be extracted from polysomes and then separated from one another to provide a single kind of late mRNA. Viral late mRNA formation is an excellent model for host cell mRNA formation because the host cell RNA polymerase II machinery synthesizes the viral mRNA and host cell enzymes add 5'-m^{7}G caps and 3'-poly(A) tails to the viral mRNA.

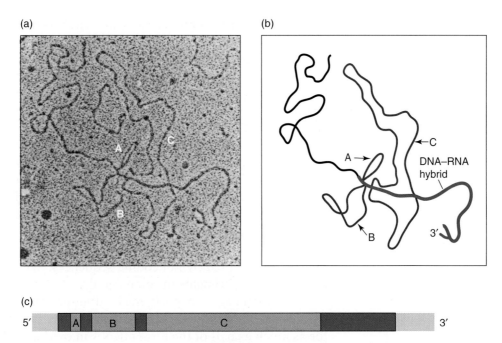

FIGURE 14.7 **An adenovirus gene coding for a viral coat protein is interrupted by segments not present in mRNA.** (a) Electron micrograph of a DNA–RNA hybrid formed by binding an mRNA for viral coat protein with its DNA template. Intervening segments (introns) excised during mRNA maturation are labeled A, B, and C. (b) Schematic of the electron micrograph. The DNA–RNA hybrid region, containing segments retained in the mature mRNA molecule (the exons), is shown in blue. The intervening segments (the introns) excised during mRNA maturation are shown in red and labeled A, B, and C. (c) Representation of the split viral gene for the coat protein. Once again, exons are shown in blue and introns in red. (Part a reproduced from Berget, S. M., Moore, C., and Sharp, P. A. 1977. *Proc Natl Acad Sci USA* 74:3171–3175. Photo courtesy of Phillip A. Sharp, Massachusetts Institute of Technology. Part b modified from Berget, S. M., Moore, C., and Sharp, P. A. 1977. *Proc Natl Acad Sci USA* 74:3171–3175. Part c adapted from Berget, S. M., Moore, C., and Sharp, P. A. 1977. *Proc Natl Acad Sci USA* 74:3171–3175.)

The Sharp and Roberts groups both prepared DNA–RNA hybrids consisting of late mRNA annealed to viral DNA, expecting to find the electron micrographs would reveal which part of the viral genome had produced the mature mRNA molecule. The mRNA did not line up on the DNA as expected. Some electron micrographs showed large loops of unhybridized DNA, clearly demonstrating internal gene sequences were somehow skipped in making the final mRNA.

FIGURE 14.7, a remarkable electron micrograph from the Sharp laboratory, shows a DNA–RNA hybrid formed by annealing a late mRNA for a viral coat protein to denatured viral DNA. The mRNA molecule hybridizes with four DNA regions, which are separated by three long DNA loops. Each loop corresponds to a region of viral DNA for which there is no complementary mRNA sequence. These results suggest nucleotide sequences are removed from within the newly synthesized RNA, the primary transcript, and the flanking sequences are joined to form mRNA molecules.

Amino acid coding regions within eukaryotic genes may be interrupted by noncoding regions.

Soon after the discovery that adenovirus late mRNA consists of transcripts made from noncontiguous viral DNA sequences, molecular biologists showed many eukaryotic pre-mRNAs have coding sequences that are interrupted by noncoding sequences, which are missing from the mature mRNA. Studies performed by Pierre Chambon, beginning in 1977, are particularly instructive in this regard and so we begin by examining them. Chambon's initial objective was to learn how female sex hormones such as estrogen regulate the expression of the gene for ovalbumin, the major egg-white protein produced by laying hen oviduct cells. He suspected the ovalbumin gene is modified in some way during hen oviduct cell differentiation and set out to find evidence for these modifications by comparing the ovalbumin gene in oviduct cells with the ovalbumin gene in a cell that does not synthesize ovalbumin. Although Chambon's hypothesis proved to be incorrect for the ovalbumin gene, the experiments he performed to test the hypothesis revealed the ovalbumin gene contains coding sequences that are interrupted by noncoding sequences. We therefore consider Chambon's experiments in some detail.

Chambon began by purifying ovalbumin mRNA from laying hen oviduct cells, a task made easier by the fact that this mRNA accounts for as much as half of the total mRNA in oviduct cells. Then he used reverse transcriptase to make a single-stranded complementary DNA (cDNA) to the ovalbumin mRNA and DNA polymerase to convert the single-stranded cDNA to double-stranded cDNA (FIGURE 14.8). The double-stranded ovalbumin cDNA was inserted into a plasmid and the resulting recombinant plasmid cloned in *Escherichia coli* to provide an ample supply of DNA for sequence analysis. But the ovalbumin mRNA sequence deduced from the cDNA was unremarkable. The 1,872-nucleotide mRNA contained an 1,158-nucleotide coding sequence (consistent with the 386 amino acids in ovalbumin) that was preceded by a 64-nucleotide untranslated leader sequence at its 5'-end and followed by a 650-nucleotide untranslated terminal sequence at its 3'-end.

Because Chambon's objective was to examine ovalbumin gene structure during different cell differentiation stages, his next step was to use the then new Southern blot procedure to compare ovalbumin genes from laying hen oviduct cells and erythrocytes (red blood cells that do not synthesize ovalbumin). This procedure involves digesting cell DNA with a restriction endonuclease, separating the resulting DNA fragments by electrophoresis, and then detecting the fragment(s) of interest with a radioactive or fluorescent probe. Chambon digested total cell DNA with EcoRI, a restriction endonuclease that does not cleave ovalbumin cDNA, and prepared a radioactive probe by subjecting ovalbumin cDNA to nick translation. Based on the absence of EcoRI sites in ovalbumin cDNA, Chambon expected the restriction endonuclease would cleave on either side of the ovalbumin gene so the radioactive probe would anneal to a single large DNA fragment. Contrary to expectation, he observed four radioactive bands (Eco a–d) in the autoradiogram of the Southern blot when either oviduct or erythrocyte DNA was studied (Figure 14.8). The appearance of four bands was puzzling. How could the ovalbumin gene have EcoRI restriction sites when ovalbumin mRNA does not? The gene splicing model proposed for late mRNA synthesis in adenovirus suggested

FIGURE 14.8 Experiment demonstrating the split hen ovalbumin gene. Pierre Chambon and coworkers prepared double-stranded complementary DNA (cDNA) to the 1,872-nucleotide ovalbumin mRNA by first using reverse transcriptase to make single-stranded cDNA and then using DNA polymerase to convert the single-stranded cDNA to double-stranded cDNA. The double-stranded cDNA was inserted into a plasmid and the resulting recombinant plasmid was cloned in *E. coli*, purified, linearized, labeled with radioactive isotope, and denatured. The labeled cDNA served as a probe to examine the ovalbumin gene from chicken erythrocytes or oviduct cells using the Southern blot procedure. Chromosomal DNA from erythrocytes or oviduct cells was cleaved with EcoRI into fragments ranging in size from 1 to 15 kilobase pair long. Gel electrophoresis separated the fragments by size (with smaller fragments moving faster). The DNA fragments were denatured and the resulting single-stranded DNA was transferred and fixed to a nitrocellulose filter. Then the radioactive probe was added to detect sequences that were complementary to it (and therefore ovalbumin mRNA). Autoradiography revealed the probe anneals to four bands (Eco a–d). Moreover, these bands were in the same position whether the chromosomal DNA had been derived from the chicken erythrocytes (lanes 1 and 2) or from the oviduct cells (lanes 3 and 4). This experiment demonstrates the intact gene has restriction sites and therefore nucleotide sequences that are not present in mature mRNA. (Adapted from Chambon, P. 1981. *Sci Am* 244:60–71. Autoradiogram photo courtesy of Pierre Chambon, Institute of Genetics and Molecular and Cellular Biology, College of France.)

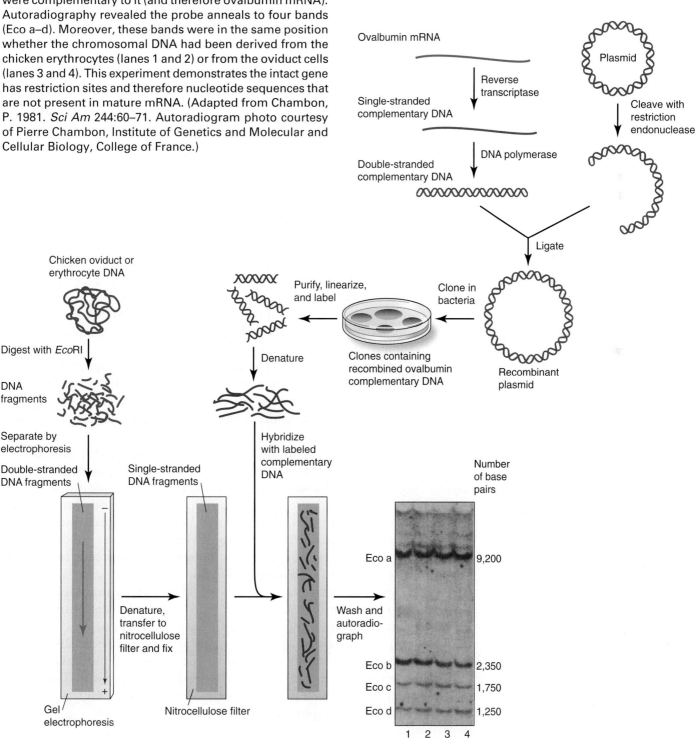

14.3 Split Genes

the hypothesis that the EcoRI sites are present on DNA sequences excised when pre-mRNA is converted to mRNA. According to this **split** or **discontinuous gene hypothesis**, regions that code for amino acids within the ovalbumin gene are interrupted by intragenic regions that do not code for amino acids. Walter Gilbert later coined the term **exon** for an expressed region that codes for amino acids and the term **intron** for a noncoding intervening sequence.

Chambon and coworkers tested the split gene hypothesis by annealing ovalbumin mRNA to single-stranded DNA from a cloned ovalbumin gene. When viewed by electron microscopy, the resulting hybrids had seven loops (introns A–G), each corresponding to a DNA region with no complementary ovalbumin mRNA sequence (**FIGURE 14.9**). The eight DNA sequences that did anneal to the ovalbumin mRNA (exons L, 1–7) were arranged in the same order in both the gene and the mRNA. Two different models were proposed to explain this colinearity:

1. *Model 1:* RNA polymerase II machinery transcribes the amino acid coding sequences and somehow skips the noncoding sequences.
2. *Model 2:* RNA polymerase II machinery transcribes the entire gene to form a long pre-mRNA that is then modified by splicing.

These two models were tested by first annealing hnRNA (rather than mRNA) to single-stranded ovalbumin cDNA and then using electron microscopy to view the hybrid structures. The skipping mechanism predicts that hybrid structures formed by hnRNA should be the same as those formed by mRNA. The splicing mechanism predicts that hnRNA should contain pre-mRNA molecules at different stages of splicing and that some of the hybrids should therefore be missing one or more loops. Electron micrographs revealed hybrids missing one or more loops and in a few cases were missing all the loops, supporting the splicing mechanism. Thus, the experimental results strongly support model 2.

Chambon was by no means the only molecular biologist to encounter a discontinuous gene. Other molecular biologists working at about the same time demonstrated that additional highly expressed eukaryotic genes, including those that code for β-globin, collagen, and insulin, are also split. As more genes became available for study, it became clear that most vertebrate genes are discontinuous. A few vertebrate genes, most notably those that code for histones and interferons, lack introns. We don't know why natural selection favored vertebrate histone and interferon genes that lack introns. Most genes in higher plants (80–85%) are also discontinuous. In contrast to vertebrates and plants fewer than 5% of yeast genes are split, but these discontinuous genes account for about 26% of the expressed transcripts. Thus, most yeast genes do not have introns, but those few that do have them are highly expressed. There is no satisfactory explanation at this time for why more highly expressed yeast genes are split.

(a)

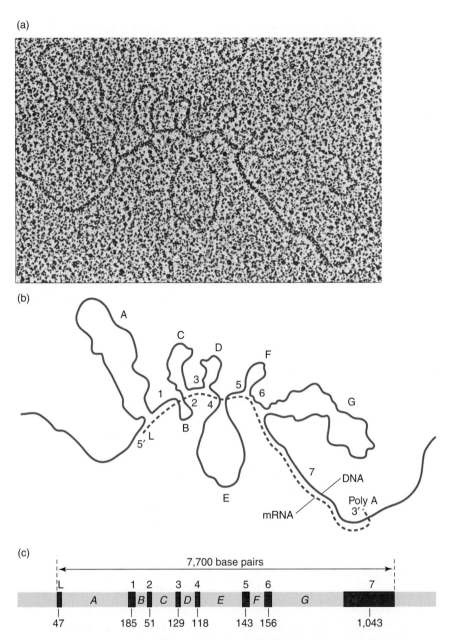

(b)

(c)

FIGURE 14.9 **The organization of the ovalbumin gene as demonstrated by electron microscopy.** (a) Electron micrograph of ovalbumin mRNA hybridized to a single strand of DNA that includes the ovalbumin gene. (b) A schematic of the electron micrograph. Segments of the DNA (blue line) and mRNA (red) that are complementary to each other form a DNA–RNA hybrid. These eight expressed segments (L and 1–7) are termed exons. The seven DNA segments, which loop out from the hybrid (A–G) because they have no complementary sequences in the RNA with which to anneal, are termed intervening sequences or introns. The 5'-m⁷G cap is not shown, but the 3'-poly(A) tail is. (c) A schematic representation of the gene showing the seven introns (gray) and eight exons (red) as well as the number of base pairs in each exon. Intron sizes vary from 251 bp to about 1,600 bp. (Part a reproduced from Chambon, P. 1981. *Sci Am* 244:60–71. Used with permission of Pierre Chambon, Institute of Genetics and Molecular and Cellular Biology, College of France. Parts b and c modified from Chambon, P. 1981. *Sci Am* 244:60–71. Used with permission of Pierre Chambon, Institute of Genetics and Molecular and Cellular Biology, College of France.)

Exons tend to be conserved during evolution, whereas introns usually are not conserved.

The Human Genome Project has provided a great deal of information about introns and exons. TABLE 14.1 summarizes some of this information. Many introns and exons have been identified by aligning cDNA (DNA prepared by using reverse transcriptase to copy mRNA) with genomic DNA. Sequences present in genomic DNA but not cDNA are introns, whereas sequences present in both genomic DNA and cDNA are exons. cDNA fragments (usually 200–500 nucleotides long) prepared by copying one or both ends of an mRNA are often used in place of full-length cDNAs. These fragments are known as **expressed sequence tags** because they (1) represent a snapshot of the genes that are expressed in specific tissues, specific developmental stages, or both and (2) can be used as hybridization probes to tag complementary chromosomal DNA sequences. Special computer software helps to collect, organize, and analyze the enormous amount of data required to identify introns and exons. This convergence of biotechnology and computer technology helped create the exciting new field of **bioinformatics**.

The number of introns varies greatly from one gene to another. We already mentioned most histone genes do not have any introns. At the other extreme the *titin* gene, which codes for a major protein in striated muscles, has 363 introns. Intron size distributions vary considerably from one organism to another (FIGURE 14.10). Human introns tend to be longer than either *Caenorhabditis elegans* (worm) or *Drosophila melanogaster* (fly) introns. The preferred minimum lengths are 47 base pair (bp) for worm, 59 bp for fly, and 87 bp for human. Much longer introns, however, are present in all three organisms. Human intron size distribution is very broad indeed, ranging from about 60 to more than 30,000 bp (Figure 14.10a). The average human intron is 3,365 bp long. The longest known intron, almost 500,000 bp long, is present in the human *NRXN3* gene that codes for neurexin, a protein that functions as a cell adhesion molecule in the nervous system. Because the RNA polymerase II transcription rate is about 1,000 to 2,000 nucleotides per minute, transcription of this one intron requires at least 4 to 8 hours. Introns in higher

TABLE 14.1 Characteristics of Human Genes							
	Internal Exon[a] Size	Exon Number	Intron Size	3'-Untranslated Region	5'-Untranslated Region	Coding Sequence[b]	Gene Size
Median	122 bp	7	1,023 bp	400 bp	240 bp	1,100 bp	14,000 bp
Mean	145 bp	8.8	3,365 bp	770 bp	300 bp	1,340 bp	27,000 bp
Sample size	43,317 exons	3,501 genes	27,238 introns	689 transcripts from chromosome 22	463 transcripts from chromosome 22	1,804 sequence entries	1,804 sequence entries

[a]An internal exon is an exon that has an intron on either side of it.
[b]The coding sequence is the mRNA sequence that codes for the polypeptide.
Source: Adapted from E. S. Lander, et al., *Nature* 409 (2001):860–921.

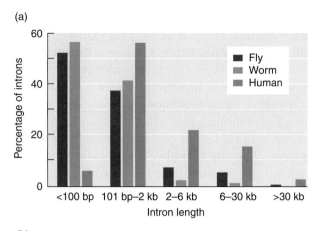
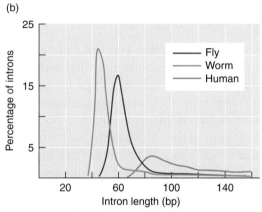

FIGURE 14.10 Size distributions of introns in sequenced genomes.
(a) Introns and (b) short introns (enlarged from a). (Adapted from Lander, E. S., et al. 2001. *Nature* 409:860–921.)

plants tend to be shorter than those in vertebrates, ranging in size from about 60 to 10,000 bp long.

Exons tend to be more uniform in size than introns (**FIGURE 14.11**). The average human gene has about nine exons. Most internal exons in human, worm, and fly are between 50 and 200 bp long (Figure 14.11). The average human exon (145 bp long) is considerably shorter than

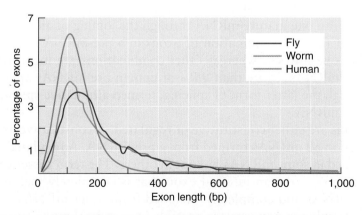

FIGURE 14.11 Size distributions of exons in sequenced genomes.
(Adapted from Lander, E. S., et al. 2001. *Nature* 409:860–921.)

the average human intron. The same holds true for exons and introns in other higher animals and plants. Because exons tend to be so much smaller than introns, most genes are more intron than exon. For instance, more than 98% of the nucleotides in the human dihydrofolate reductase gene are in its introns. There is considerable selective pressure to conserve exon sequence because exon mutations would alter proteins. There is much less selective pressure to maintain intron size or sequence because intron mutations usually do not affect polypeptide sequence. The one notable exception—intron mutations that interfere with splicing—often have devastating effects because they produce mRNA with the wrong coding information. The fact that exon sequences tend to be conserved provides an important means for detecting them. When the sequences of the same gene from two or more organisms are known, identical or very similar sequences are likely to be exons.

Genes in different organisms that evolved from a common ancestral gene through the course of evolution, called **orthologs**, often are very different in length even though they code for similar and sometimes identical polypeptides. Ortholog size differences result from variations in intron rather than exon lengths. Because the coding sequence length is the sum of the exon lengths, one would predict average coding sequence lengths should be similar in different organisms. Consistent with this prediction, average coding sequence lengths in human, worm, and fly are 1,340, 1,311, and 1,497 bp, respectively.

Individual transcription unit sizes vary greatly. Many human transcription units are more than 100 kilobase pair (kbp) long and some are considerably longer. The longest known transcription unit, the dystrophin gene (2.6×10^6 kbp), first came to the attention of molecular biologists because a mutated form causes muscular dystrophy. Coding sequence size distribution tends to be somewhat more uniform than transcription unit size distribution. Some human transcription units, however, have very long coding sequences. The longest known coding sequence (80,760 bp) is produced by the *titin* gene.

A single pre-mRNA can be processed to produce two or more different mRNA molecules.

Splicing may occur so (1) each and every exon in a pre-mRNA is incorporated into one mature mRNA through the joining of all successive exons, or (2) one combination of exons in a pre-mRNA is incorporated in one mRNA, whereas other combinations are incorporated in other mRNAs. More than 95% of human pre-mRNAs appear to go through the latter form of splicing, termed **alternative splicing**, to produce mRNA.

Exons are either constitutive or regulated. A **constitutive exon** is included in all mRNAs formed from a pre-mRNA. A **regulated exon** is included in some mRNAs but not in others. The term **cassette exon** is used to describe an internal exon that is completely included in some mRNAs and completely excluded from other mRNAs. Many alternate splicing patterns are possible for a typical multi-exon pre-mRNA. A cassette exon located between two constitutive exons may

(a) A cassette exon (blue) can be skipped or included

(b) Only one exon in an array of optional exons (blue and green) may be included in mature mRNA

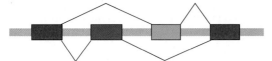

(c) An intron (yellow) may be retained or excluded

(d) An exon may have alternative splice sites at its 3'-end

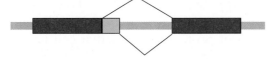

(e) An exon may have alternative splice sites at its 5'-end

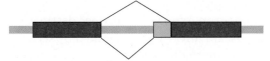

FIGURE 14.12 Patterns of alternate splicing found in nature. (a–e) Red boxes correspond to constitutive exons, blue and green boxes to cassette (or optional) exons, cyan boxes to segments within constitutive exons, and gray regions to introns. The thin lines above and below the pre-mRNA indicate splicing patterns. (Adapted from Graveley, B. R. 2001. *Trends Genet* 17:100–107.)

either be included or skipped when a pre-mRNA is spliced to form mRNA (FIGURE 14.12a). Inclusion of one cassette exon may prevent the inclusion of other cassette exons. As shown in Figure 14.12b, such mutually exclusive splicing occurs when an array of cassette exons is located between two constitutive exons and only one of the cassette exons can be incorporated into mature mRNA. An intron between two constitutive exons may be included in some mRNAs but not in others (Figure 14.12c). A single DNA sequence thus may act as either an intron (when it is excised) or an exon (when it is retained). Length variant exons have alternative splice sites at their 3'-end (Figure 14.12d) or their 5'-end (Figure 14.12e). When mRNAs formed by alternative splicing of a given pre-mRNA are translated, the proteins formed usually share common functions but differ within one or more domains. Sometimes, alternative splicing introduces a termination codon in mRNA, causing a premature end to protein synthesis. Alternative splicing may also alter the reading frame so the same RNA sequence is present in two different reading frames.

Combinations of the various splicing patterns within individual genes lead to the formation of multiple mRNAs.

We illustrate how alternative splicing can generate a diverse collection of proteins from a single gene by considering the *WT1* and *Dscam* genes. The *WT1* gene (FIGURE 14.13) is so named because mutant forms cause Wilms tumor, the most common kidney tumor in children. The *WT1* gene, which is about 50 kb long, codes for multiple variant forms of a 52 to 56 kDa protein by alternative splicing. Each variant form, or **isoform**, has a proline- and glutamine-rich domain at its N-terminus and a DNA-binding domain containing four zinc finger motifs at its C-terminus. Some isoforms have a lysine-threonine-serine (KTS) sequence inserted between zinc fingers 3 and 4, whereas others do not. WT1 isoforms lacking the KTS sequence appear to function as transcription factors, whereas those with the KTS sequence appear to associate with splicing factors.

The *Drosophila Dscam* gene, so named because it codes for protein isoforms that appear to be *Drosophila* homologs of the <u>D</u>owns <u>s</u>yndrome <u>c</u>ell <u>a</u>dhesion <u>m</u>olecule (FIGURE 14.14), illustrates the amazingly diverse array of protein isoforms that can be generated from just one gene through alternative splicing. The developing fly requires the diverse array of Dscam isoforms to assist in the complex task of accurately and reproducibly directing each of the fly's approximately 250,000 growing neuron axons to its proper destination. Exons 4, 6, 9, and 17 in each Dscam mRNA are generated by selecting one variant exon from an array of mutually exclusive exon variants (Figure 14.14a). Exon 4 is selected from an array of 12 variants, exon 6 from an array of 48 variants, exon 9 from an array of 33 variants, and exon 17 from an array of 2 variants. Selection of the exon 4 variant is developmentally regulated, consistent with the notion that at least some of the potential diversity in Dscam isoforms serves to wire neurons correctly. Exons 4, 6, and 9 encode 10 immunoglobulin (Ig) repeats to form an extracellular domain that serves as an axon guidance receptor, whereas exon 17 encodes a transmembrane domain (Figure 14.14b). Because

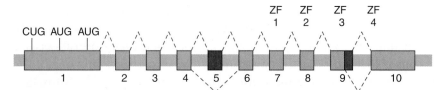

FIGURE 14.13 Alternative splicing of the *WT1* (Wilm's tumor) gene. The *WT1* gene generates up to 24 isoforms as a result of three alternative translation initiation codons in exon 1 (labeled CUG, AUG, and AUG), alternative splicing of cassette exon 5 (red), alternative 5' splice sites in exon 9, and other posttranscriptional modifications. Each isoform has a proline- and glutamine-rich domain at its amino terminus and a DNA-binding domain containing four zinc fingers (ZF) motifs at its carboxyl terminus. Some alternative splices insert three amino acids (KTS) between zinc fingers 3 and 4, whereas others do not. WT1 isoforms lacking KTS appear to function as transcription factors, whereas those that have KTS appear to be associated with splicing factors. (Adapted from Roberts, G. C., and Smith, C. W. J. 2002. *Curr Opin Chem Biol* 6:375–383.)

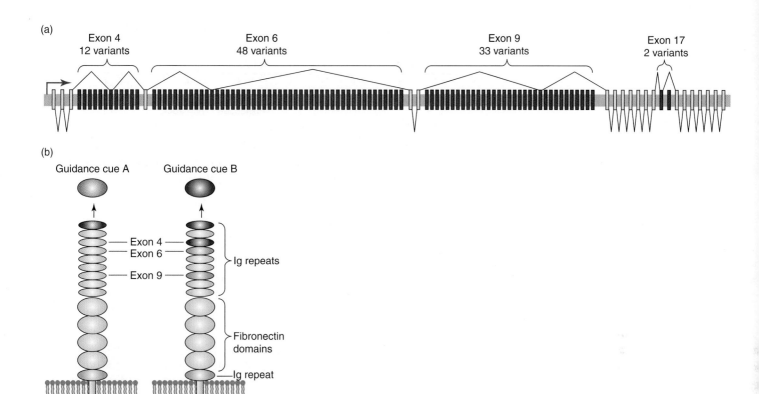

FIGURE 14.14 **Alternative splicing of the *Drosophila Dscam* gene.** The constitutive splicing events are indicated below the gene and alternative splicing events are depicted above the gene. (a) The organization of the *Dscam* gene. The constitutive exons are yellow and the alternative splicing exons are purple. The *Dscam* gene contains four sites of alternative splicing at exons 4, 6, 9, and 17. There are 12 variants of exon 4, 48 variants of exon 6, 33 variants of exon 9, and 2 variants of exon 17. Only one variant exon from each position is included in the Dscam mRNAs. Alternative exons 4, 6, and 9 encode alternative versions of immunoglobulin repeats. (b) Functional consequences of Dscam alternative splicing. The Dscam protein functions as an axon guidance receptor. It is thought that each Dscam variant will interact with a unique set of axon cues. The form of Dscam shown on the left will interact with guidance cue A. The form of Dscam shown on the right contains different sequences encoded by exons 4, 6, and 9 and thus interacts with guidance cue B, rather than guidance cue A. Neurons that express the form of Dscam shown on the right will be attracted in a different direction than neurons expressing the form shown on the left. (Modified from Graveley, B. R. 2001. *Trends Genet* 17:100–107. Copyright 2001, with permission from Elsevier [http://www.sciencedirect.com/science/journal/01689525].)

each of the four exons appears to be generated independently of the other three, the total number of possible Dscam isoforms is 38,016 ($12 \times 48 \times 33 \times 2$), or two- to threefold greater than the total number of transcription units in *Drosophila*.

Pre-mRNA requires specific sequences for precise splicing to occur.

Cells must excise introns with great precision because failure to do so would change the mRNA coding sequence with devastating consequences for the cell. The cell's splicing machinery must therefore unambiguously identify the exon/intron boundary at the start of an intron, termed the **5′-splice site**, and intron/exon boundary at the end of an intron, termed the **3′-splice site**. The yeast splicing machinery requires

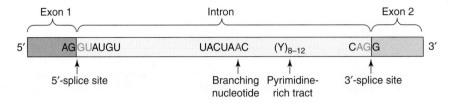

FIGURE 14.15 Sequences in yeast pre-mRNA that help to define the 5′- and 3′-splice sites. Y denotes a pyrimidine nucleotide. The branching nucleotide, A, is shown in red. The GU dinucleotide at the 5′-end of the intron and the AG nucleotide at the 3′-end of the intron are shown in green.

three nearly invariant short nucleotide sequences (FIGURE 14.15), the **5′-splice sequence** (AG/GUAUGU), the **3′-splice sequence** (CAG/G), and the **branchpoint sequence** (UACUAAC), to identify and excise an intron (a slash indicates a splice site). The red "A" in Figure 14.15 identifies the **branching nucleotide** that participates in the chemistry of pre-mRNA splicing (see below). As the names imply, the three short sequences flank the 5′-splice site, the 3′-splice site, and the branching nucleotide, respectively. Many yeast introns also have a stretch of 8 to 12 pyrimidines (mostly uracil), termed a **polypyrimidine tract**, just upstream from the 3′-splice site.

Most animal and plant introns resemble yeast introns in having GU at their 5′-end and AG at their 3′-end. This is certainly true of human introns, where about 98% of the thousands of different confirmed human introns have the GU-AG pattern. Mutations that alter these nucleotides cause inherited metabolic diseases in humans. For instance, certain individuals synthesize defective hemoglobin molecules because a GU sequence at the beginning of one of the β-globin introns has been altered. This alteration, which blocks splicing at the normal splice site and leads to splicing at abnormal sites, causes the production of a nonfunctional β-globin chain. Animal and plant splicing machinery also require 5′-splice sequences, 3′-splice sequences, and branchpoint sequences to identify and splice introns, but the three short sequences are not as highly conserved as in yeast. In humans and other mammals the 5′-splice consensus sequence is AG/GURAGU, the 3′-splice consensus sequence is NYAG/G, and the branchpoint consensus sequence is YNCURAY (where R denotes a purine; Y, a pyrimidine; and N, any nucleotide).

The arrangement of these three short consensus sequences with respect to a typical mammalian GU-AG intron is shown in FIGURE 14.16a. The branching nucleotide is usually 11 to 40 bases upstream from the 3′-splice site. Most GU-AG introns also have a polypyrimidine tract just upstream from the 3′-splice consensus sequence, extending 10 or more nucleotides back into the intron. The same splicing machinery that processes GU-AG introns also appears to process variant introns with GC at their 5′-end. These so-called GC-AG introns account for 0.76% of confirmed human introns. Another 0.10% of human introns begin with AU and end with AC. Consensus sequences associated with these so-called AU-AC introns are shown in Figure 14.16b.

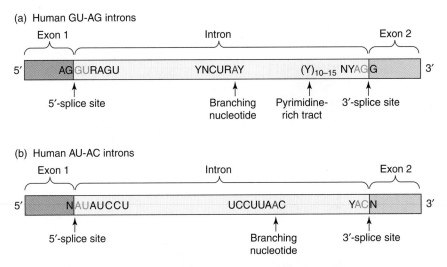

FIGURE 14.16 **Consensus sequences that define human introns.** Consensus sequences that define human (a) GU-AG introns and (b) AU-AC introns. Y is any pyrimidine nucleotide, R is any purine nucleotide, and N is any nucleotide. The branching nucleotide is shown in red and the two terminal dinucleotides in the intron are shown in green.

Two splicing intermediates resemble lariats.

Investigators hoped to find clues to the splicing mechanism by isolating splicing intermediates and then determining the structure of these intermediates. A major breakthrough occurred in the early 1980s with the discovery that HeLa cell extracts, supplemented with ATP and magnesium ions, excise introns from added pre-mRNA. Two research teams, one led by Phillip A. Sharp and the other by Tom Maniatis, took advantage of this *in vitro* splicing system to search for splicing intermediates. Both groups used the following approach: they (1) added ^{32}P-labeled pre-mRNA with a single exon-intron-exon array to the *in vitro* splicing system, (2) separated RNA products formed after various incubation times by polyacrylamide gel electrophoresis, and (3) detected radioactive bands by autoradiography. FIGURE 14.17 shows an informative autoradiogram from an experiment performed by the Maniatis group. Proposed structures for splicing intermediates and products are shown to the right of the autoradiogram. Two of the observed bands, those for the 130- and the 339-nucleotide products (indicated by the red arrows), are of special interest because they are splicing intermediates that contain the intron. Evidence supporting this statement is as follows: (1) each band appears only when ATP and magnesium ions are added to the splicing system, (2) neither band appears when the intron has some nucleotide other than guanylate at its 5'-end, and (3) an intron-specific DNA probe anneals to the RNA in each band. Further studies showed the 339-nucleotide RNA also contained the exon on the 3'-side of the intron.

Electrophoresis studies suggested both the 130- and 339-nucleotide RNA products have unusual structures. A nucleic acid's chain

14.3 Split Genes

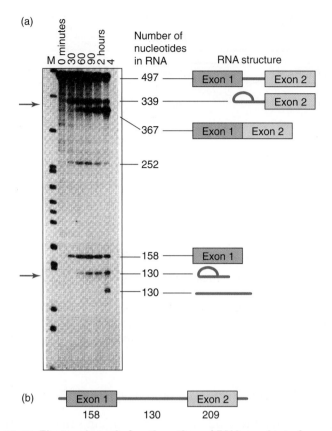

FIGURE 14.17 **Electrophoretic fractionation of RNA products formed by a cell-free HeLa cell splicing system.** (a) [^{32}P]Phosphate-labeled pre-mRNA was added to a cell-free HeLa cell splicing system and incubated in the presence of ATP and magnesium ions. RNA products were removed at various times, fractionated by polyacrylamide gel electrophoresis, and then detected by autoradiography. The time of sample incubation is given at the top of the lane for that sample. Proposed structures for RNA products are shown to the right of the autoradiogram. (b) The structure of the starting pre-mRNA. Red arrows to the left indicate two bands of special interest (see text). (Parts a and b adapted from Ruskin, B. et al. 1984. *Cell* 38:317–331. Photo courtesy of Michael R. Green, University of Massachusetts Medical School.)

length can be determined by comparing its migration distance with migration distances of standards on the same gel. When the chain lengths of the 130- and 339-nucleotide RNA products were determined in this way using a 5% polyacrylamide gel, the calculated chain lengths were significantly greater than the actual values. Moreover, the discrepancies were even greater with a 10% polyacrylamide gel. Similar discrepancies had previously been reported for circular RNA molecules, suggesting the 130- and 339-nucleotide RNA products might be circular. Because rare 2′→5′ branches were known to occur in nuclear RNA molecules with poly(A) tails, it seemed possible the 130- and 339-nucleotide RNA products might have lariat-like structures. Maniatis and coworkers set out to find experimental proof for this model, which predicts the 130- and 339-nucleotide RNA products each has a single 2′→5′ phosphodiester bond like the one shown in

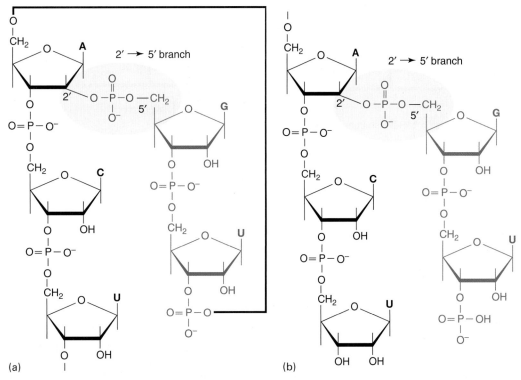

FIGURE 14.18 Lariat structure. (a) Lariat structure showing the 2'→5' branch. (b) Oligonucleotide isolated after lariat structure is digested with RNase.

FIGURE 14.18a. They confirmed this prediction by digesting the RNA molecules with an RNase that cleaves 3'→5' but not 2'→5' phosphodiester bonds, separating the oligonucleotides thus generated by chromatography, and then showing one of the oligonucleotides indeed does have a 2'→5' phosphodiester bond. This oligonucleotide, which is the same for both the 130- and 339-nucleotide RNA products, has an adenylate branching nucleotide linked to guanylate by a 2'→5' phosphodiester bond and to cytidylate by a 3'→5' phosphodiester bond (Figure 14.18b).

The lariat model also predicts that the 2'→5' branch should block reverse transcriptase as it extends a short DNA primer annealed to the 3'-end of the intron (FIGURE 14.19a). Maniatis and coworkers confirmed this prediction by demonstrating a primer annealed to the 3'-end of the intron could be extended up to the 2'→5' branchpoint site but no further. The lariat model also predicts that cleaving the 130- or 339-nucleotide RNA product on the 3'-side of the branchpoint should produce an RNA circle and a linear RNA fragment. Maniatis and coworkers annealed a short DNA fragment to a segment of the intron on the 3'-side of the branchpoint of the 130-nucleotide fragment and then cleaved the DNA–RNA hybrid region with RNase H (Figure 14.19b). As predicted, this cleavage generated a circular RNA and a short linear fragment. An identical circular RNA was produced when the experiment was repeated with the 339-nucleotide fragment, but the linear fragment was considerably longer.

14.3 Split Genes

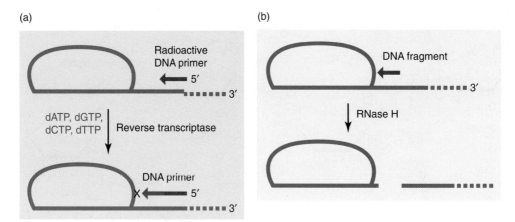

FIGURE 14.19 **Experiments that support the lariat model.** (a) Primer extension. Reverse transcriptase cannot extend a short primer annealed to the 3'-end of the intron beyond the branchpoint in the lariat structure of either the 130 nucleotide or the 339 nucleotide RNA product. (b) A short DNA fragment was annealed to the intron region adjacent to the branchpoint and the hybrid was digested with RNase H, which is specific for DNA–RNA hybrids. Two products were formed, a loop and a linear fragment. The loops were identical when either the 130-nucleotide or 339-nucleotide product was used. However, the linear fragment released was longer when the 339-nucleotide fragment was used.

Splicing consists of two coordinated transesterification reactions.

Kinetic studies using autoradiograms like that in Figure 14.16 show band intensities change with time, providing data that can be used to determine precursor–product relationships. The splicing mechanism that fits the structural and kinetic data best is shown in FIGURE 14.20.

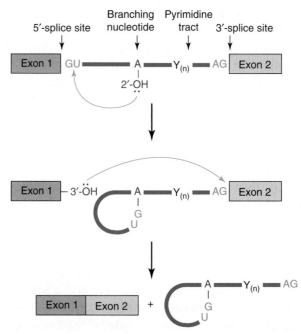

FIGURE 14.20 **The two coordinated transesterification steps in splicing.** First step: The 2'-hydroxyl on the branching nucleotide (A, shown in red) initiates a nucleophilic attack on the 5'-splice site to produce a lariat structure connected to exon 2. Second step: The 3'-hydroxyl generated at the 3'-end of exon 1 during the first transesterification step initiates a nucleophilic attack on the phosphodiester bond at the 3'-splice site. This transesterification reaction joins the exons and releases the intron in a lariat form.

According to this mechanism splicing requires two coordinated transesterification steps.

1. The 2′-hydroxyl of an adenosine within the branchpoint sequence, the branching nucleotide, initiates a nucleophilic attack on the phosphodiester bond at the 5′ exon/intron boundary at the 5′ splice site. The resulting transesterification reaction generates a free 5′-exon (exon 1) and a lariat intermediate with a 2′→5′ branch structure that is still attached to the 3′-exon (exon 2).

2. The 3′-OH of the free exon (exon 1) initiates a nucleophilic attack on the phosphodiester bond at the 3′ intron/exon boundary (the 3′-splice site). The resulting transesterification reaction releases the lariat intron and joins exon 1 and exon 2. Lariat structures do not accumulate *in vivo* because cells have a phosphodiesterase that cleaves 2′→5′ phosphodiester bonds and nucleases that degrade the resulting linear RNA strand. The formation of this linear RNA strand explains why a second 130 nucleotide appears after 4 hours in the autoradiogram shown in Figure 14.17a.

14.4 Spliceosomes

Aberrant antibodies, which are produced by individuals with certain autoimmune diseases, bind to small nuclear ribonucleoprotein particles (snRNPs).

Although it might seem that only a few specific endonucleases and a ligase might suffice to catalyze splicing reactions, eukaryotic cells use a very sophisticated piece of machinery to carry out the steps involved in splicing. As often happens in science, the first clue to the nature of this machinery came from studies in a seemingly unrelated field. People suffering from autoimmune diseases known as mixed connective tissue disease and systemic lupus erythematosus make aberrant antibodies that attack components of their own cells. One such antibody, called the anti-Smith or anti-Sm antibody, binds to small ribonucleoprotein particles in the cell nucleus. Joan Steitz and coworkers suspected these small ribonucleoproteins might help to process pre-mRNA and set about isolating them to study their function(s). The isolation method they devised was to first tag the ribonucleoprotein particles in a human cell extract with antibodies from the serum of a lupus patient and then pull the tagged ribonucleoproteins out of the extract with an insoluble *Staphylococcus aureus* cell wall preparation that binds to antibodies. When Steitz and coworkers analyzed RNA molecules that had been extracted from the ribonucleoprotein particles by polyacrylamide gel electrophoresis, they observed discrete bands corresponding to RNA molecules with chain lengths between 100 and 200 nucleotides long. Because each of the small nuclear RNA molecules is uridine-rich, they are called U1 snRNA, U2 snRNA, U4 snRNA, U5 snRNA, and U6 snRNA. U1, U2, U4, and U5 snRNAs are each synthesized by RNA polymerase II and have a 2,2,7 trimethyl guanosine cap at their 5′-ends (FIGURE 14.21a). U6 snRNA is synthesized by RNA polymerase III and has a γ-methyl phosphate cap (Figure 14.21b).

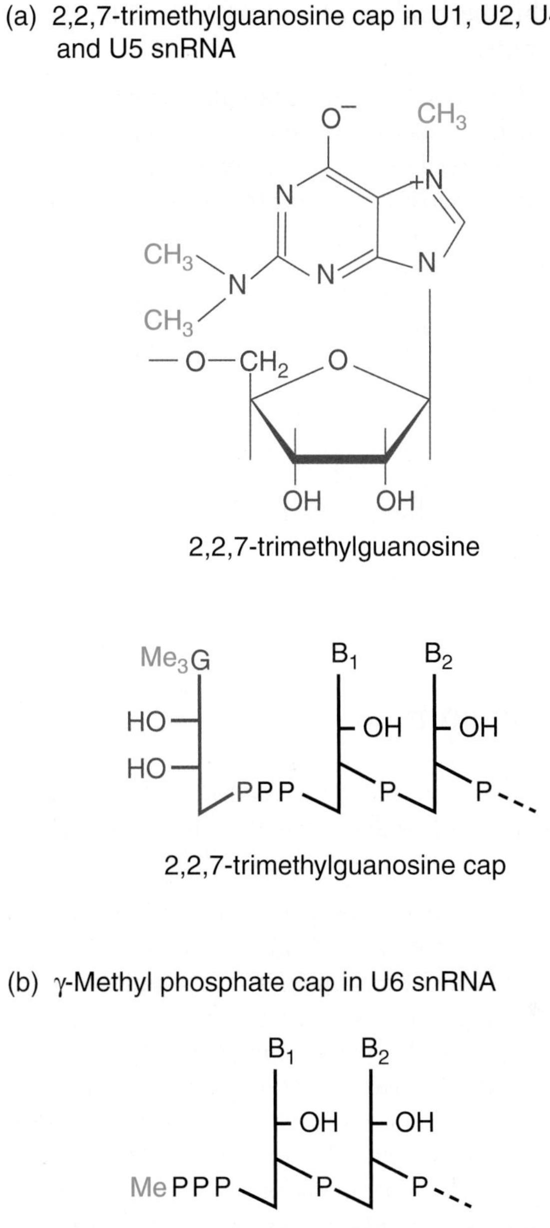

(a) 2,2,7-trimethylguanosine cap in U1, U2, U4, and U5 snRNA

2,2,7-trimethylguanosine

2,2,7-trimethylguanosine cap

(b) γ-Methyl phosphate cap in U6 snRNA

FIGURE 14.21 **Cap structures present in snRNA molecules.** (a) 2,2,7-Trimethyl-guanosine cap in U1, U2, U4, and U5 snRNA shown as a standard chemical structure (top) and as part of a stick structure (bottom) and (b) γ-methyl phosphate cap in U6 snRNA.

Each snRNA is present in its own small nuclear ribonucleoprotein particle (snRNP). The snRNPs are named U1 snRNP, U2 snRNP, U4 snRNP, U5 snRNP, and U6 snRNP after their snRNA component. U4, U5, and U6 snRNPs combine to form a U4/U6 • U5 tri-snRNP particle, which is stabilized by extensive base pairing between U4 and U6 snRNAs and protein–protein interactions between U5 snRNP and the U4/U6 di-snRNP. Although all eukaryotes have the same five kinds of ribonucleoproteins, snRNP concentrations vary from organism to organism. For instance, mammalian cells have approximately 10^5 to 10^6 copies of each snRNP, whereas yeast cells have about 100 to 200 copies of each kind.

snRNPs assemble to form a spliceosome, the splicing machine that excises introns.

When Steitz and coworkers first isolated U snRNPs in the late 1970s, they had no way to study the function of these particles. Once *in vitro* splicing systems became available, however, investigators were quickly able to show U1, U2, U4, U5, and U6 snRNPs are essential splicing machinery components. Some key findings that demonstrated snRNPs are involved in splicing are as follows:

1. Antibodies specific for U1 snRNP block *in vitro* splicing.
2. HeLa cell nuclear extracts lose their ability to splice pre-mRNA when U snRNPs are removed.
3. Splicing does not occur when the first eight nucleotides are removed from the 5'-end of U1 snRNA.
4. U2 snRNP selectively binds to the branching site and U5 snRNP binds to pre-mRNA just upstream of the 3'-splice site.
5. U4 and U6 snRNPs are essential splicing components.

Further experiments showed the five U snRNPs assemble to form a large ribonucleoprotein splicing machine, the **spliceosome**. The total molecular mass of spliceosome, which also contains non-snRNP protein factors, is about 4.8×10^6 Da.

RNA and protein may both contribute to the spliceosome's catalytic site.

In vitro studies show several non-snRNP proteins participate in the spliceosome assembly pathway. One of these, splicing factor 1 (SF1), also called branchpoint binding protein, binds to the branchpoint sequence. Another, U2AF (U2 snRNP auxiliary factor), contains a large subunit, U2AF^{65}, that binds to the polypyrimidine tract and a small subunit, U2AF^{35}, that binds to the AG dinucleotide at the 3'-splice site. FIGURE 14.22 shows the contributions these non-snRNPs and the snRNPs make to the spliceosome assembly pathway. The first complex detected, the E (early) complex, has SF1 and U2AF bound to their respective target sequences and U1 snRNP bound to the 5'-splice site through base pairing between U1 snRNA and the splice site sequence. The E complex is transformed to A complex in the presence of ATP when U2 snRNP binds to the branchpoint sequence through base pairing interactions between U2 snRNA and the branchpoint sequence. A complex is converted to B complex when U4/U6 • U5 tri-snRNP binds near to the 3'-splice site. B complex undergoes a complicated series of rearrangements to form the catalytic or C complex, which catalyzes the first transesterification. Although U1 and U4 snRNPs appear to dissociate as B complex is converted to C complex, they probably remain weakly bound to C complex. Additional conformational changes are undoubtedly required for the spliceosome to catalyze the second transesterification.

Because U2, U5, and U6 snRNP are tightly associated with pre-mRNA at the time of the first transesterification reaction, they may participate in the catalytic process. U6 snRNA, which is highly

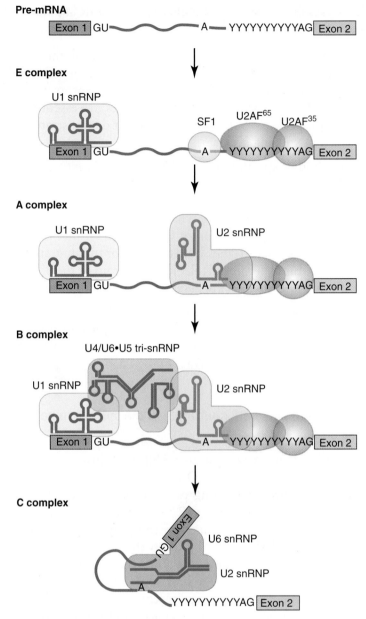

FIGURE 14.22 Spliceosome assembly. During the first step in stepwise spliceosome assembly, U1 snRNP binds to the 5'-splice site. The branchpoint binding protein (SF1) binds to the branchpoint, and U2 auxiliary factors (U2AF65 and U2AF35) bind to the pyrimidine tract and 3'-splice site AG, respectively. This complex, which commits the pre-mRNA to the splicing pathway, is called the early or E complex. Next, the E complex is converted to A complex in the presence of ATP when U2 snRNP binds at the branchpoint. Then U4/U6 • U5 tri-snRNP enters the spliceosome. This entry is followed by a massive rearrangement in which U6 snRNP replaces U1 snRNP at the splice site, U6 and U2 interact, U5 bridges the splice sites, and U1 and U4 become destabilized. This catalytically active rearranged spliceosome is called the C complex. For simplicity, U5 is not shown in the C complex. (Modified from Hertel, K. J., and Graveley, B. R. 2005. *Trends Biochem Sci* 30:115–119. Copyright 2005, with permission from Elsevier [http://www.sciencedirect.com/science/journal/09680004].)

conserved through evolution and very sensitive to chemical changes, is thought to have a catalytic function. There is ample precedent for an RNA molecule serving as a catalyst. Some introns that can catalyze their own excision, the so-called group II introns (**BOX 14.1 LOOKING DEEPER: SELF-SPLICING RNA**), have a folding pattern similar to that of U6 snRNA.

Recent studies suggest a conserved protein called Prp8 (pre-mRNA processing protein), which is an integral part of U5 snRNP, may also contribute to the catalytic site. Studies of Prp8 structure and function have been hampered by the inability to obtain the intact protein in soluble form, probably in large part because the protein is so large (2,413 residues in yeast). Investigators have made some progress in understanding Prp8's function by isolating and studying subdomains. A subdomain near the C-terminus that is about 150 residues long is of particular interest because chemical crosslinking studies show its residues interact with the 5'-splice site. Furthermore, model studies suggest two highly conserved aspartate residues in this subdomain may act together with RNA metal coordination sites to form the catalytic site. If so, the spliceosome is an "RNP-zyme."

Cells use a variety of mechanisms to regulate splice site selection.

Spliceosomes must not only be able to remove introns, but they must do so in a manner that precisely selects correct splice site pairs. Splicing precision presents a much more serious challenge to multicellular organisms than to yeast because (1) multicellular organisms tend to have very long introns with many pseudo-splice sites, whereas yeast tend to have short introns with few, if any, pseudo-splice sites; (2) multicellular organisms tend to have relatively short, degenerate splice sites, whereas yeast have highly conserved splice sites; and (3) many pre-mRNA molecules from higher organisms are subject to alternative splicing, whereas few, if any, yeast pre-mRNAs are subject to alternative splicing.

Multicellular organisms use regulatory proteins to assist in splice-site recognition. These proteins work by first binding to specific sequences in exons or introns and then stimulating or repressing exon recognition. A sequence within an exon or intron that stimulates splice-site selection is called an **exonic splicing enhancer (ESE)** or **intronic splicing enhancer**, respectively. The best studied ESEs are purine-rich sequences with the consensus sequence $(GAR)_n$, where R is any purine. Because ESEs are embedded in coding sequences, a base substitution in an ESE may be quite harmful as a result of its effect on alternative splicing even though the substitution appears to be harmless in terms of its effect on polypeptide sequence.

A family of **splicing regulatory proteins (SR proteins)** appears to be particularly important for ESE recognition. Ten different types of SR proteins have been identified in human cells but none in yeast. SR proteins have modular structures; each has one or two **RNA recognition motifs (RRMs)** at its N-terminus and an arginine/serine-rich

BOX 14.1: LOOKING DEEPER

Self-Splicing RNA

Thomas R. Cech came across a very surprising phenomenon in 1978 while studying rRNA synthesis in the ciliated protozoan *Tetrahymena thermophila* (FIGURE B14.1a). Earlier studies by others showed the cell's macronucleus has about 10,000 autonomously replicating rRNA genes, each producing identical pre-rRNAs at the rate of one copy per second. Based on this information, Cech thought he might be able to isolate *T. thermophila* rRNA genes together with the enzymes and regulatory factors needed to transcribe them. Cech and his students soon discovered the rRNA gene, which contains sequences for the mature 17S, 5.8S, and 26S rRNA, has an approximately 400-nucleotide-long intron within the 26S rRNA sequence (FIGURE B14.1b). The presence of this intron was not a surprise because similar introns were known to be present in rRNA genes from related organisms. However, as we will see shortly, this intron turned out to have very special properties that caused Cech to change his research direction and eventually led him to discover a surprising new RNA function.

Cech and coworkers began their study of *T. thermophila* rRNA gene transcription by incubating nuclear extracts, containing rRNA genes and the transcription machinery, with radioactive nucleoside triphosphates and salts required for transcription. They also included α-amanitin to block formation of mRNA, tRNA, and other small RNA molecules so they could more easily follow the radioactive transcripts that were formed.

Two of the radioactive bands that appeared after gel electrophoresis separation were especially interesting. The first contained a product about the size expected for mature 26S rRNA, and the second contained a low-molecular-weight product of about the size expected for the intron. The presence of these RNA products suggested the *in vitro* transcription system can precisely excise the intron from the newly transcribed pre-rRNA. Because only one other example of *in vitro* intron excision was known at the time (a pre-tRNA from yeast), Cech thought it would be worthwhile to study the splicing process in greater detail. He, therefore, set out to isolate the splicing machinery from *T. thermophila* extracts.

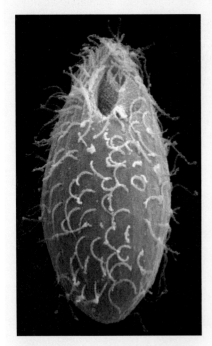

Cech and coworkers needed an unspliced pre-rRNA substrate to assay for the splicing machinery. They were able to extract the desired radioactive substrate from a transcription reaction mixture to which polyamines had been added to block splicing. The radioactive pre-rRNA was added to tubes containing salts, unlabeled nucleotides, and *T. thermophila* extracts. The extract, however, was not included in one of the control tubes to demonstrate it was required for splicing. As expected, pre-rRNA splicing took place in tubes containing the complete reaction mixture. To everyone's amazement, however, splicing took place just as well in the control tube without the extract. Because all biochemical reactions then known were catalyzed by enzymes, Cech and coworkers initially thought extract had accidentally been introduced into the control tube. They, therefore, repeated the experiment, being very careful not to add the *T. thermophila* extract to the tube that was not supposed to contain it. Once again, splicing occurred in the absence of added extract.

When Cech and coworkers performed their experiments, all biochemical reactions were thought to be catalyzed by enzymes. Therefore, Cech and coworkers considered the possibility that the RNA they used as a substrate had actually been spliced inside the cell, but that the mature 26S rRNA and the excised intron somehow remained bound together until mixed with salts and nucleotides. Cech and coworkers decided to examine the splicing reaction that

FIGURE B14.1a **Scanning electron micrograph of a *Tetrahymena* cell.** (Photo courtesy of E. Marlo Nelsen and used with permission of Joseph Frankel, University of Iowa.)

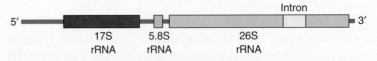

FIGURE B14.1b ***Tetrahymena thermophila* pre-rRNA.**

occurs in the absence of *T. thermophila* extracts more carefully. They soon discovered magnesium ions and GTP (but not ATP, UTP, or CTP) must be present for the low-molecular-weight RNA product to be released. Moreover, the phosphoanhydride bonds in GTP were not required to release the low-molecular-weight RNA (the presumed intron) because guanosine could replace GTP. In a related discovery, Cech and coworkers demonstrated the sequence of the low-molecular-weight RNA product was identical to that predicted for the intron in the rRNA gene in all respects but one. The single difference was the low-molecular-weight RNA product had an extra G at its 5'-end. Cech performed a simple labeling experiment to demonstrate that the extra G came from GTP. He added [^{32}P]GTP to a solution containing pre-rRNA and salts and showed the GTP was covalently attached to the 5'-end of the low-molecular-weight product. Cech and coworkers proposed the splicing mechanism shown in FIGURE B14.1c to explain their experimental results. This mechanism assumes the intron has a binding site for guanosine (or guanine nucleotides) that is analogous to the substrate-binding site in an enzyme. The 3'-hydroxyl of a guanosine (or guanine nucleotide) bound to this site initiates a nucleophilic attack on the 5'-splice site, forming a 3'→5' phosphodiester bond at the 5'-end of the intron and releasing the upstream exon. Magnesium ions appear to be required for this reaction to take place. Once transesterification is complete, the intron goes through a conformational change in which the original guanosine (or GTP) at the G-binding site is replaced by the guanosine at the 3'-end of the intron. The cleaved upstream exon remains base-paired to the intron during this conformational change. In the new conformation, the 3'-end of the upstream exon is in position to attack the 3'-splice site. This transesterification reaction ligates the exons and excises the intron.

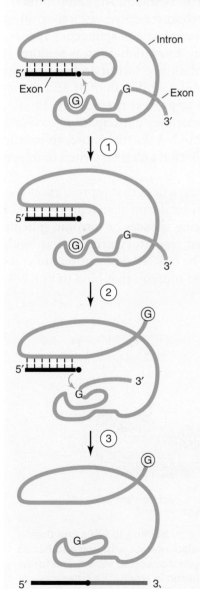

Because the two transesterification reactions proposed in the splicing mechanism have high energies of activation, it seemed unlikely they would take place at any significant rate in the absence of a catalyst. As enzymes were the only known biological catalysts at that time, Cech and coworkers tried to find evidence for enzyme participation. The most likely enzyme source appeared to be a protein that remained bound to the purified pre-rRNA substrate. Cech and coworkers, therefore, tried various approaches to remove or inactivate proteins bound to the pre-rRNA substrate. Splicing continued to take place, however, even after the pre-rRNA was heated in the presence of sodium dodecyl sulfate to denature proteins or treated with proteases to degrade proteins. Cech and coworkers were then forced to consider the unprecedented possibility that the pre-rRNA was itself the catalyst for the reaction.

Because there was absolutely no precedent for an RNA catalyst, Cech and coworkers needed to devise an experiment that would unambiguously demonstrate pre-rRNA can act as a catalyst. One solution would have been to synthesize the pre-rRNA by standard organic synthetic techniques and to then demonstrate intron excision when the synthetic pre-rRNA was incubated in the presence of guanosine (or GTP) and magnesium ions. Unfortunately, this approach was not practical because the pre-rRNA was much too long to be synthesized in an organic chemistry laboratory. So they did the next best thing. They used recombinant DNA technology to transfer a segment of the rRNA gene containing the intron and flanking sequences to a bacterial plasmid and then used bacterial RNA polymerase to synthesize the desired RNA substrate. Remarkably, this substrate, which never came in contact with any *T. thermophila* proteins, could catalyze its own splicing reaction. Cech and coworkers, therefore, unambiguously demonstrated the RNA in the intron catalyzes a chemical reaction. This first example of an RNA catalyst or ribozyme was soon followed by others.

FIGURE B14.1c **Pathway for pre-rRNA intron self-splicing.** Exons are shown in gray and black and the intron in green. Step 1: An intron-bound guanosine or guanosine triphosphate (GTP) (circled) cleaves the 5'-splice site and in the process becomes covalently attached to the 5'-end of the intron. Step 2: A conformational change takes place in which the G at the 3'-end of the intron replaces the original G in the G-binding site. Step 3: The cleaved 5'-exon, which is still base paired to the intron, cleaves the 3'-splice site, allowing the exons to be ligated and the intron to be excised. (Adapted from Doudna, J. A., and Cech, T. R. 2002. *Nature* 418:222–228.)

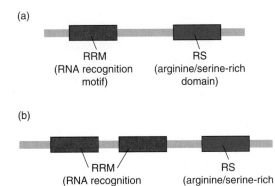

FIGURE 14.23 **SR proteins.** SR proteins have (a) one or (b) two RNA recognition motifs (RRMs) at their N-termini. The RRM binds to specific RNA sequences. SR proteins also have an arginine/serine-rich domain (RS domain) at their C-termini. The RS-domain is required for protein–protein interactions with other SR proteins as well as with SR-related proteins.

domain (RS domain) at its C-terminus (FIGURE 14.23). The RRM is a sequence-specific RNA binding site, whereas the RS domain participates in protein–protein interactions. Different SR proteins may contact one another directly through their RS domains or they may interact through a protein intermediate. In either case at least some of the serine residues in the RS domain must be phosphorylated for interactions to occur, providing the cell with a method to regulate contacts between proteins. Some SR proteins appear to be required for cell survival, whereas others appear to be interchangeable.

SR proteins stimulate recognition of weak splice sites by binding to ESEs and then helping to recruit U1 snRNP to the 5′-splice site and U2AF to the 3′-splice site. SR proteins also interact with one another as well as other proteins to form protein bridges that extend across the intron that is to be excised or across exons (FIGURE 14.24). The former interaction, termed **intron definition**, tends to be used for splice site recognition when the intron is small. The latter interaction, **exon definition**, provides a method for splice site recognition when introns are very long. In yeast, processing of split genes appears to rely primarily on intron definition. Exon definition is seldom, if ever, used because it requires more than one intron and most yeast genes with introns have only one. Multicellular animals and plants use both exon and intron definition to process pre-mRNA. In fact, both exon and intron definition sometimes participate in excising different introns from the same pre-mRNA. Furthermore, splice site recognition may begin with exon definition and then switch to intron definition.

Pre-mRNA molecules also have sequences that repress exon recognition. A sequence within an exon or intron that represses splice site selection is called an **exonic splicing silencer** or **intronic splicing silencer**, respectively. The best studied regulatory protein that binds to silencer sequences is hnRNP A1. The hnRNP designation indicates it is one of a large number of unrelated proteins that bind to hnRNA.

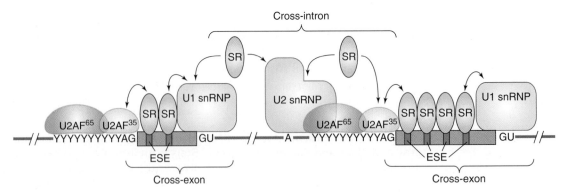

FIGURE 14.24 **Splice site recognition.** SR proteins stimulate recognition of weak splice sites by binding to ESEs and then helping to recruit U1 snRNP to the 5′-splice site and U2AF to the 3′-splice site. SR proteins also interact with one another as well as other proteins to form protein bridges that extend across the intron that is to be excised or across exons. The former interaction, termed intron definition, tends to be used for splice site recognition when the intron is small. The latter interaction, exon definition, provides a method for splice site recognition when introns are very long. (Adapted from Maniatis, T., and Tasic, B. 2002. *Nature* 418:236–243.)

The hnRNP A1 protein, which has two RRMs and a glycine-rich domain, binds to both exonic and intronic splicing silencers. Among the mechanisms proposed to explain how hnRNP A1 represses exon recognition are the following: it blocks SR proteins from binding to adjacent ESEs, interferes with protein–protein interactions required for exon definition, or blocks spliceosome assembly.

Splicing begins as a cotranscriptional process and continues as a posttranscriptional process.

Another factor that determines which exons will be included or excluded is that splicing begins while the nascent pre-mRNA is still being transcribed. Direct evidence for cotranscriptional splicing comes from electron micrographs such as that in FIGURE 14.25, which shows a chromosome as it is being transcribed. Nascent pre-mRNA transcripts are visible as strands of increasing length extending from the DNA template. Progressive formation and loss of various size intron loops near the 5′-end of nascent pre-mRNA strands visible in this micrograph indicate the transcripts are undergoing cotranscriptional splicing. Splicing continues even after transcription is complete. In yeast, pre-mRNA splicing is predominantly posttranscriptional. The fraction of introns removed during transcription versus after transcription is not known in higher plants or animals.

Support for cotranscriptional regulation of splicing comes from experiments showing promoter structure regulates splicing. Changing a promoter can influence splice site selection even when the promoter change has no effect on the promoter's strength or the site of transcription initiation. In an attempt to explain this phenomenon, Yutaka Hirose and James L. Manley proposed the following: (1) the specific nature of the initiation complex assembled on a particular promoter may influence the extent or pattern of phosphorylation of the CTD of the largest RNA polymerase subunit, and this in turn may regulate CTD's ability to participate in splicing, or (2) the promoter may recruit specific factors to the transcription elongation complex that may in turn influence splicing. CTD's location near the transcription exit channel places it in position to transfer snRNPs, SR proteins, and other splice site factors on splice sites and branching sequences as they emerge from the exit channel. CTD therefore may work by stimulating spliceosome assembly, facilitating spliceosome interaction with splice sites, serving as a platform for SR proteins or other splicing factors, or some combination of these and other mechanisms.

Considerable evidence supports the hypothesis that CTD participates in cotranscriptional splicing. *In vitro* experiments show phosphorylated CTD stimulates splicing and anti-CTD antibodies inhibit splicing. Cells with a truncated CTD have a lower splicing efficiency than those with a normal CTD. CTD appears to influence splicing at least in part by stimulating 5′-m^{7}G cap formation. Capping, however, does not completely explain CTD's role in splicing as indicated by the following two observations: (1) a yeast mutant with temperature-sensitive guanylyltransferase synthesizes spliced mRNAs

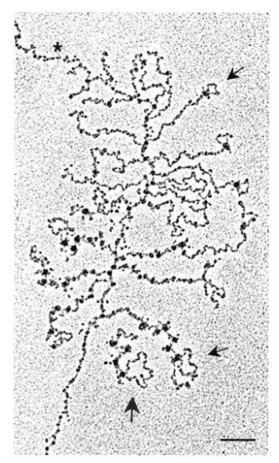

FIGURE 14.25 **Electron micrograph directly visualizing cotranscriptional splicing.** The gene shown is from *Drosophila* and is ~6 kb long. DNA enters at the upper left on the micrograph and exits at the lower left. Transcription initiates at the position marked with an asterisk; nascent RNA transcripts appear as fibrils of increasing length that extend from the DNA template. The transcripts are undergoing cotranscriptional splicing, as indicated by the progressive formation and loss of intron loops of various sizes near the 5′-ends of the transcripts (arrows). The red arrow indicates a transcript near the 3′-end of the gene that is no longer attached to the DNA template and might have been caught in the act of termination and release. Bar = 200 nm. (Caption reprinted and photo reproduced from *Trends Biochem. Sci.*, vol. 25, N. Proudfoot, Connecting transcription to messenger ..., pp. 290–293, copyright 2000, with permission from Elsevier [http://www.sciencedirect.com/science/journal/09680004]. Photo courtesy of Ann L. Beyer, University of Virginia.)

at the restrictive temperature, although much less efficiently than wild-type cells, and (2) frog oocytes splice uncapped pre-mRNAs that are microinjected into them, although less efficiently than capped pre-mRNAs.

mRNA splicing and export are coupled processes.

Export of fully processed mRNA to the cytoplasm is essential for cell survival because it is required for protein synthesis. In contrast, export of introns, partially spliced transcripts, or mutant transcripts to the cytoplasm might decrease the cell's survival chances by damaging the translation machinery or leading to the synthesis of harmful protein products. Eukaryotic cells therefore require some mechanism(s) to distinguish between fully processed mRNA molecules to be exported to the cytoplasm and other RNA polymerase II products that must be retained in the nucleus and degraded. One export mechanism is to couple splicing and export. The first evidence for this coupling came from studies showing mRNAs produced by splicing are more efficiently exported than identical mRNAs transcribed from an intronless DNA. This effect of splicing on export was explained by the observation that spliced mRNAs (but not intronless DNA transcripts) interact with specific proteins to form a distinct mRNA • protein complex that moves through the nuclear pore.

Cells appear to use a family of proteins, termed nonshuttling hnRNP proteins, to prevent excised introns and partially spliced mRNA from leaving the nucleus. These proteins, which bind to intron sequences and package them into hnRNP complexes, have nuclear retention signals that prevent the bound RNA from being exported. Although nonshuttling hnRNP proteins have a low degree of sequence specificity, they do not interact with the relatively short exons because spliceosomes block their way. Spliceosomes also contribute to the retention of partially spliced pre-mRNA within the nucleus. Retained introns are cleaved by a $2' \rightarrow 5'$ phosphodiesterase and then degraded.

14.5 Cleavage/Polyadenylation and Transcription Termination

Poly(A) tail synthesis and transcription termination are coupled, cotranscriptional processes.

Because RNA polymerase II is often required to synthesize very long transcripts, it must be a highly processive enzyme that does not terminate prematurely. Nevertheless, it must terminate at the end of the gene so it does not continue transcribing into the neighboring gene and interfere with the normal transcription of that gene. Failure to terminate would also stop the release of RNA polymerase II, preventing the enzyme from carrying out further transcription. RNA polymerase II transcription termination is a complex process, which involves the following steps:

1. *Pre-mRNA cleavage:* Pre-mRNA is cleaved at a site called the **cleavage/polyadenylation site** or the **poly(A) site**.
2. *Poly(A) addition:* A poly(A) tail is added to the newly generated 3′-end of the RNA molecule.
3. *Transcription termination:* Transcription is terminated downstream from the cleavage/polyadenylation site.

The transcription termination machinery requires recognition elements to help it locate the poly(A) site. These elements differ in both position and sequence in higher animals and yeast. Recognition elements are more highly conserved in higher animals. For simplicity the present discussion is limited to transcription termination in mammals. The poly(A) site in mammals is located between a hexanucleotide (AAUAAA) called the **polyadenylation signal** and a less conserved **U-rich element** (FIGURE 14.26). The polyadenylation signal was originally identified as a highly conserved sequence that is 10 to 30 nucleotides upstream of the cleavage site. Comprehensive mutagenesis studies indicate this hexanucleotide is required for both cleavage and poly(A) addition. Mammalian cleavage/polyadenylation machinery can usually tolerate single base substitutions within the AAUAAA sequence (most commonly A→U at the second position) but it cannot tolerate more extensive substitutions. The U-rich element is located ≤ 30 nucleotides downstream of the poly(A) site. Although the U-rich element is quite variable in both sequence and composition, it usually has one or more stretches of five consecutive U residues (often interrupted by single G residues). A third recognition element, an **auxiliary upstream element**, is often present at a variable distance upstream of the cleavage site. It is usually U-rich or contains $(UGUA)_n$ or $(UAUA)_n$. When present, the auxiliary upstream element enhances the efficiency of cleavage and polyadenylation. The cleavage site is determined by the distance between the polyadenylation signal and U-rich element. About 70% of the cleavage sites in vertebrate pre-mRNAs have an A on the 5′-side of the cleavage site so the first A in the poly(A) tail derives from the pre-mRNA.

The mammalian machinery required to perform two relatively simple tasks, endonucleolytic cleavage and poly(A) addition, requires approximately 85 different proteins. Such complex machinery is

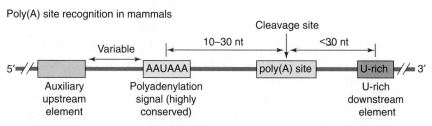

FIGURE 14.26 Schematic representation of poly(A) site recognition in mammals. (Adapted from Gilmartin, G. M. 2005. *Genes Dev* 19:2517–2521.)

14.5 Cleavage/Polyadenylation and Transcription Termination

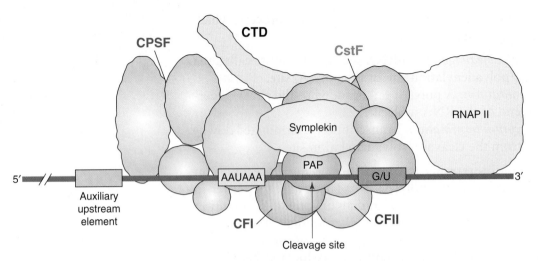

FIGURE 14.27 **Schematic representation of the mammalian polyadenylation machinery.** The abbreviations used are as follows: RNAPII, RNA polymerase II; CTD, carboxyl terminal domain of the largest subunit in RNA polymerase II; CPSF, cleavage/polyadenylation specificity factor; CstF, cleavage stimulation factor; CFI and CFII, cleavage factors I and II, respectively; and PAP, poly(A) polymerase. Poly(A) binding protein II is not shown in this figure. (Adapted from Calvo, O., and Manley, J. L. 2003. *Genes Dev* 17:1321–1327.)

required because cleavage and poly(A) addition are tightly coupled cotranscriptional processes that must take place with a high degree of precision. Fortunately, cleavage and poly(A) addition can be examined separately *in vitro*, simplifying the study of each process.

FIGURE 14.27 a is a schematic showing how several components of the cleavage/polyadenylation machinery are organized at the poly(A) site. The contributions these components make to the cleavage and polyadenylation processes are as follows.

1. *Cleavage/polyadenylation specificity factor:* Cleavage/polyadenylation specificity factor (CPSF), a complex containing five different polypeptide subunits, which binds to the polyadenylation signal AAUAAA, is required for both cleavage and poly(A) addition. One of the five polypeptide subunits is Zn^{2+}-dependent endonuclease, which cleaves pre-mRNA at the poly(A) site. Another subunit binds to AAUAAA with a low degree of specificity. The functions of the other three subunits remain to be determined.
2. *Cleavage stimulation factor:* Cleavage stimulation factor (CstF), a complex containing four subunits, binds to the U-rich sequence and is required for cleavage at the poly(A) site. CstF also interacts with CPSF. This interaction is probably very important because CPSF and CstF each bind much more tightly to pre-mRNA when the other is present.
3. *Cleavage factor I and cleavage factor II:* Cleavage factor I and cleavage factor II are required for cleavage, but little is known about their function.
4. *Poly(A) polymerase:* Poly(A) polymerase adds adenylates to the 3′-end of RNAs and is also required for cleavage. Poly(A) polymerase is recruited to the poly(A) site by CPSF that is already bound to pre-mRNA. Poly(A) polymerase and CPSF

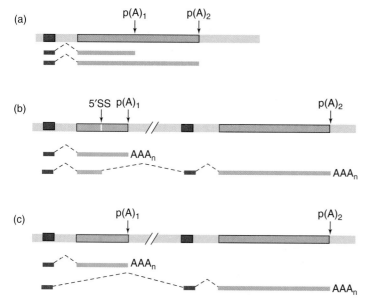

FIGURE 14.28 The 3′-regions of transcription units leading to alternative poly(A) site selection. Three kinds of exon arrangements can produce alternative poly(A) site selection. (a) Some transcription units have multiple poly(A) sites within the terminal exon. Only two tandem poly(A) sites—p(A)$_1$ and p(A)$_2$—are shown here for simplicity but more may be present. (b) Other transcription units have an exon that can serve as an internal or terminal exon depending on physiological conditions. (c) Still other transcription units have two or more alternative 3′-terminal exons. The pre-mRNA is shown at the top of each example with the exons shown in red, blue, and orange boxes and introns as thick light blue lines. The spliced mRNA is shown at the bottom with dashed lines corresponding to introns removed during splicing. The abbreviation 5′SS indicates a splice site. (Adapted from Edwalds-Gilbert, G. K., Veraldi, L., and Milcarek, C. 1997. *Nucleic Acids Res* 25:2547–2561.)

suffice for poly(A) addition to a pre-cleaved RNA substrate, but the rate of reaction is slow.

5. *Symplekin:* Symplekin (derived from the Greek word meaning to tie together) is part of a larger complex that also includes CstF and CPSF. It has been proposed that symplekin helps to assemble or stabilize the CstF complex and thereby helps to hold the complete cleavage/polyadenylation machinery together.
6. *Carboxyl terminal domain of Rpb1:* CTD of Rpb1, the largest subunit in RNA polymerase II, binds to both CPSF and CstF and must be present along with all of the other cleavage/polyadenylation factors for 3′-cleavage to take place.

Transcription units often have two or more alternate polyadenylation sites.

Many transcription units have two or more alternate polyadenylation sites, which may be arranged in three different ways (FIGURE 14.28). Some transcription units have multiple poly(A) sites within the terminal exon (Figure 14.28a). Other transcription units have an exon that can function as an internal or external exon depending on physiological conditions (Figure 14.28b). Still other transcription units have two or more alternative 3′-terminal exons (Figure 14.28c). Further insights can be gained by examining two well-studied systems, the mammalian transcription units for calcitonin and the heavy immunoglobulin chain, IgM.

Calcitonin Transcription Unit

The calcitonin transcription unit has six exons and is alternatively processed in a tissue-specific fashion to produce two proteins (FIGURE 14.29a). In thyroid and most other tissues the first four exons are spliced together and polyadenylated at exon 4 so translation produces calcitonin. In nerve cells the first three exons are joined to exons 5 and

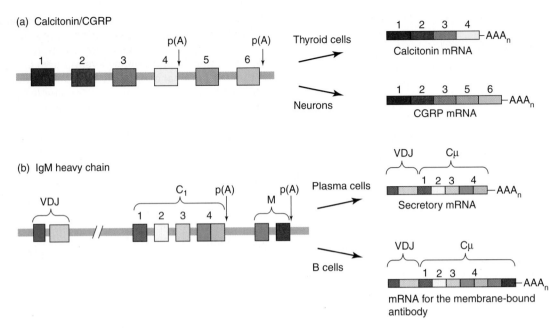

FIGURE 14.29 Alternative polyadenylation of calcitonin/calcitonin gene-related peptide (CGRP) and immunoglobulin M (IgM) transcripts. (a) Organization of exons and introns in calcitonin/calcitonin gene-related peptide pre-mRNA and the structure of primary mRNA produced in thyroid cells or neurons. (b) Alternative processing choices in the IgM heavy chain (mµ) precursor. (Adapted from Zhao, J., Hyman, L., and Moore, C. 1999. *Microbiol Mol Biol Rev* 63:405–445.)

6 and polyadenylated at exon 6 so translation produces a neuropeptide called calcitonin gene-related peptide.

IgM Transcription Unit

The IgM transcription unit codes for two kinds of IgM. B cells synthesize membrane-bound IgM, whereas plasma cells synthesize soluble IgM that can be secreted. The IgM transcription unit has two poly(A) sites (Figure 14.29b). B cells remove the upstream poly(A) site at exon 4 and use a downstream poly(A) site to produce mRNA that codes for membrane-bound IgM. Plasma cells retain the upstream poly(A) to produce mRNA that codes for the shorter, soluble IgM.

Transcription termination takes place downstream from the poly(A) site.

The transcription elongation complex continues to synthesize RNA after passing the poly(A) site. One of the clearest demonstrations that transcription termination takes place after the poly(A) site comes from a technique known as the **nuclear run-on method**. The steps in this method and its application to transcription termination are illustrated in FIGURE 14.30. The basic idea behind this method is the transcription elongation complex continues to synthesize a nascent transcript under *in vitro* conditions after the nucleus is removed from the cell. Although transcription elongation complexes in different nuclei will be at different stages of transcription, only one stage is shown in Figure 14.30 for clarity.

The first step in the nuclear run-on method is to isolate nuclei from cells that are actively transcribing some specific transcription unit. (Yeast cells are made permeable to ribonucleoside triphosphates instead of isolating their nuclei.) Next, the isolated nuclei are suspended in a reaction mixture containing radioactive ribonucleoside triphosphates and incubated for a sufficient time to allow RNA polymerase II to extend the nascent transcript by about 100 nucleotides. The radioactive RNA is purified after the reaction is stopped and hybridized to DNA fragments bound to filters. Each DNA fragment corresponds to a different region within and 3' to the transcription unit under study. Bound radioactive RNA is visualized by autoradiography. DNA fragments upstream of the termination site (fragments A–D in FIGURE 14.30) give significant signals, whereas those downstream from the termination site (fragment E in Figure 14.30) give only background signals. Although the nuclear run-on method is to all intents and purposes an *in vitro* procedure, it does reveal apparent termination regions. A growing number of transcription units transcribed by RNA polymerase II have been analyzed using the nuclear run-on method. The following conclusions have been reached based on these analyses:

1. Transcription termination requires a poly(A) signal.
2. Transcription termination requires factors for 3'-cleavage but does not require factors for polyadenylation or the pre-mRNA to be cleaved.
3. Transcription terminates beyond the 3'-end of mRNA, usually 200 to 2,000 bp downstream from the poly(A) site.
4. Some transcripts terminate efficiently at a single site, whereas others terminate inefficiently over an extended region.

RNA polymerase II transcription termination appears to involve allosteric changes and a 5'→3' exonuclease.

RNA polymerase II must pass the poly(A) site before transcription termination can take place, thereby ensuring transcription termination does not take place until RNA polymerase II has reached the end of the transcription unit. Two models have been proposed to explain how the poly(A) site can stimulate RNA polymerase II transcription termination. The first, known alternatively as the "allosteric" or "antiterminator" model, proposes RNA polymerase II loses a positive elongation/antitermination factor, gains a negative elongation/termination factor, or both as it transcribes past the poly(A) site, causing the polymerase to assume a less processive conformational form that can be more easily released from the DNA template. The second model, the "torpedo" model, proposes that poly(A) site cleavage provides an entry site for a 5'→3' exonuclease, which attacks the newly created uncapped 5'-phosphate end and moves down the nascent RNA toward RNA polymerase II like a guided torpedo until it reaches RNA polymerase II and somehow disrupts the DNA–RNA duplex, thereby triggering transcription termination. The recent discovery of a 5'→3' exonuclease, called Xrn2 in mammals, supports the torpedo model. The yeast 5'→3' exonuclease is called Rat1. These

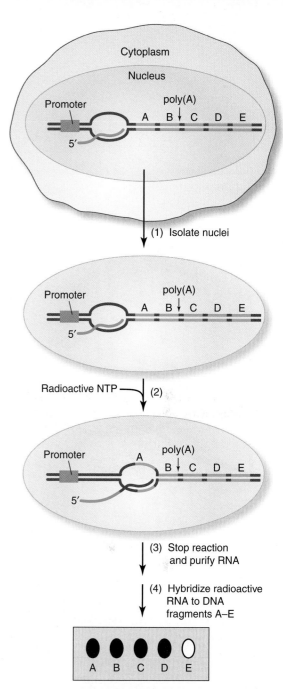

FIGURE 14.30 **Schematic diagram of the nuclear run-on experiment.** Different segments along the DNA template are labeled A–E. The 3'-cleavage site is labeled poly(A). The experimental steps are as follows. (1) Isolate nuclei. (2) Add the isolated nuclei to a reaction mixture containing radioactive nucleoside triphosphates and incubate for a short time so about 100 nucleotides add to the growing chain. (3) Stop the reaction and purify the radioactive (red) RNA molecules. The transcript will have advanced to different positions along the template DNA. (4) Hybridize the purified RNA to DNA fragments corresponding to different regions (A–E) within and 3' to the transcription unit. DNA fragments upstream of the termination site (fragments A–D), give significant signals, whereas those downstream from the termination site (fragment E) give only background signals.

exonucleases have been shown to be necessary but not sufficient for transcription termination.

Experimental evidence supports and refutes each model. For instance, the antiterminator model explains transcription termination at genes in which cleavage at the poly(A) site follows, rather than precedes, transcription termination. The torpedo model does not explain this phenomenon. On the other hand, the torpedo model explains why Xrn2 and Rat1 are necessary for transcription termination, but the antiterminator model does not. Fortunately, the two models are not mutually exclusive. A hybrid model has been proposed that incorporates features of both models and appears to fit the experimental data (FIGURE 14.31). According to this hybrid model the elongation

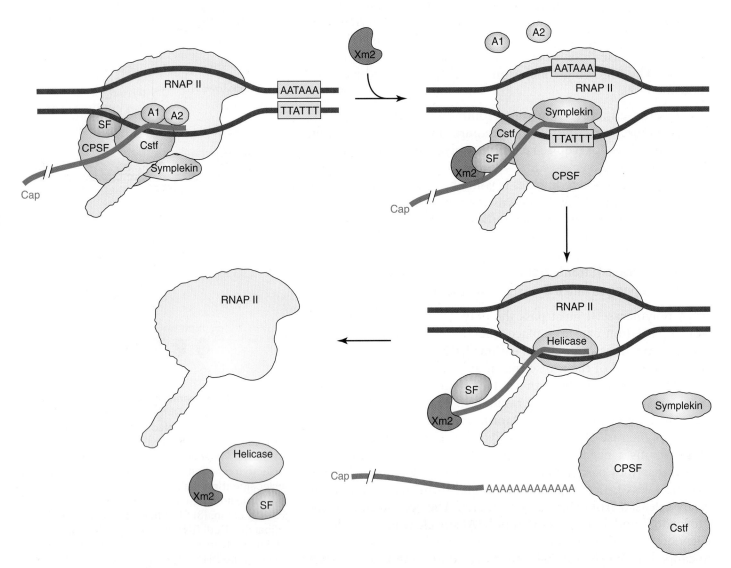

FIGURE 14.31 **Model for termination by RNA polymerase II.** The elongation complex changes conformation upon recognizing the pol(A) site. This conformational change may result from the loss of anti-terminators (A1 and A2), the gain of termination factors (not shown), or both. After the pre-mRNA is cleaved by the cleavage/polyadenylation machinery, a splicing factor (SF) recruits Xrn2. Then Xrn2 digests downstream RNA until it reaches the RNA polymerase. Finally, Xrn2, perhaps with the assistance of a helicase releases RNA polymerase II from the DNA. (Adapted from Richard, P., and Manley, J. L. 2009. *Genes Dev* 23:1247–1269.)

complex changes conformation upon recognizing the poly(A) site. This conformational change may result from the loss of antitermination factors (A1 and A2), the gain of termination factors (not shown), or both. After the pre-mRNA is cleaved by the cleavage/polyadenylation machinery, a splicing factor (SF) helps recruit Xrn2. Xrn2 digests downstream RNA until it reaches the RNA polymerase and then, perhaps with the assistance of a helicase, releases RNA polymerase II from the DNA to complete the termination process.

14.6 RNA Editing

RNA editing permits a cell to recode genetic information.

Although transcriptional processing provides eukaryotic cells with the rather remarkable ability to use a single transcription unit to synthesize a variety of protein isoforms, eukaryotes also use another method to alter RNA before it is translated. They can change an RNA molecule so its sequence is no longer the one predicted from the DNA coding sequence. The process is termed **RNA editing** when the altered RNA molecule could in principle have been encoded by the DNA sequence. The best characterized examples of RNA editing in higher eukaryotes involve deamination reactions in which cytidine is converted to uridine or adenosine is converted to inosine (**FIGURE 14.32**). The translation

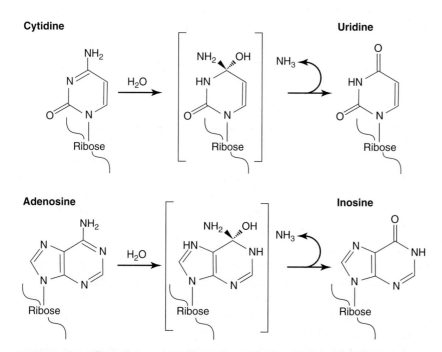

FIGURE 14.32 Reaction mechanism of cytidine and adenosine deaminases acting on RNA. Hydrolytic deamination of cytidine and adenosine leads to uridine and inosine, respectively. Presumably, ammonia is released during the reactions. (Caption reprinted and figure modified from Gerber, A. P., and Keller, W. 2001. *Trends Biochem Sci* 26:376–384. Copyright 2001, with permission from Elsevier http://www.sciencedirect.com/science/journal/09680004.)

machinery reads an inosine in a codon as though it were guanosine. Specific enzymes and associated RNA molecules are required to catalyze each type of editing reaction.

14.7 The Gene Reconsidered

The human proteome contains a much greater variety of proteins than would be predicted from the human genome.

Transcriptional processing and RNA editing have important consequences for the eukaryotic **proteome** (the complete set of proteins expressed during an organism's lifetime). The best current estimate is the human genome contains approximately 20,000 to 25,000 transcription units that code for proteins. Alternative splicing, which affects at least 95% of human pre-mRNAs, probably results in the production of a few hundred thousand different kinds of proteins in the human proteome. RNA editing, alternative transcription initiation sites, and alternative transcription termination sites also contribute to protein diversity. Furthermore, many proteins are modified after they are formed to produce additional variants. The most common types of posttranslational modifications are cleavage, phosphorylation, acetylation, glycosylation, and methylation. Knowing a transcription unit's DNA sequence thus is a starting point for understanding how information is transferred from DNA to protein, but it is not sufficient by itself to tell us the exact nature of the proteins present in a given cell at some specific time.

Cotranscriptional and posttranscriptional processes force us to reconsider our concept of the gene.

Alternative processing raises many fundamental issues, perhaps none more important than the very concept of the gene itself. Before the discovery of alternative pre-mRNA processing a eukaryotic protein-coding gene could be defined as a hereditary unit that codes for a specific polypeptide. This one gene–one polypeptide hypothesis, however, does not work for alternatively processed transcription units that can code for two or more protein isoforms. We must therefore find some other way to define a gene. Although an exon has the minimal amount of information that is expressed as a discrete unit, we cannot define a gene as an exon because exons do not contain all the information we normally associate with genes. Perhaps the best definition we can devise at this time is that a gene is a linear collection of exons that are incorporated into a specific mRNA. The term "gene," however, continues to be used in a more general sense when referring to a transcription unit or polypeptide coding region. Alternative processing also presents challenges for genetic engineering. Some alternatively processed transcription units are known to produce isoforms that have antagonistic effects. Transforming a cell with an alternatively spliced gene may therefore result in the production of a harmful rather than a beneficial protein.

Questions and Problems

1. Briefly define, describe, or illustrate each of the following.
   - a. pre-mRNA
   - b. hnRNA
   - c. Polysome
   - d. 5′-m⁷G cap
   - e. Split gene hypothesis
   - f. Exon
   - g. Intron
   - h. Expressed sequence tags
   - i. Bioinformatics
   - j. Ortholog
   - k. Alternative splicing
   - l. Constitutive exon
   - m. Regulated exon
   - n. Cassette exon
   - o. Isoform
   - p. 5′-Splice site
   - q. 3′-Splice site
   - r. 5′-Splice sequence
   - s. 3′-Splice sequence
   - t. Branchpoint sequence
   - u. Polypyrimidine tract
   - v. Branching nucleotide
   - w. snRNP
   - x. Spliceosome
   - y. Exonic splicing enhancer
   - z. Intronic splicing enhancer
   - aa. SR protein
   - bb. RNA recognition motifs
   - cc. Intron definition
   - dd. Exon definition
   - ee. Exonic splicing silencer
   - ff. Intronic splicing silencer
   - gg. Poly(A) site
   - hh. Polyadenylation signal
   - ii. U-rich element
   - jj. Auxiliary upstream element
   - kk. Nuclear run-on method
   - ll. RNA editing
   - mm. Proteome

2. Prepare a table that summarizes the important similarities and differences that exist between mRNA and pre-mRNA molecules.

3. Describe a method that can be used to separate pre-mRNA molecules from other RNA molecules present in a nuclear extract.

4. An investigator has isolated a new RNA virus that infects mice and suspects the viral RNA has a 3′ poly(A) tail. Suggest a method you could use to determine whether the viral RNA does indeed have a poly(A) tail.

5. Animal cell mRNA molecules have 5′-m⁷G caps.
   - a. What is the source of the methyl group?
   - b. Show the pathway for cap formation.
   - c. At what stage of mRNA formation is the cap added to the RNA molecule?
   - d. Recombinant DNA technology permits us to replace a promoter for the RNA polymerase II transcription machinery with a promoter for the RNA polymerase I or RNA polymerase III transcription machinery. When this type of replacement is made the transcript is no longer capped. Explain this finding based on your knowledge of RNA polymerase II structure and function.

6. Briefly explain each of the following.
   - a. Two EcoRI restriction sites are present in a cloned mouse gene but not in the mouse mRNA formed from that gene.
   - b. A purified histone mRNA dissolved in a buffer solution with a high salt concentration does not bind to cellulose fibers linked to oligo(dT).
   - c. A single mouse transcription unit codes for five distinct polypeptides.
   - d. A *Drosophila* and human enzyme are nearly the same size, have very similar amino acid sequences, and catalyze the same reaction. The human transcription unit, however, is much longer than the *Drosophila* transcription unit.
   - e. A mouse mutant was isolated that produces a β-globin polypeptide that is longer than normal. DNA sequence analysis reveals a single nucleotide substitution in one of the introns but no other alterations.

7. Introns are excised as lariat structures. Describe the experimental evidence that led to this conclusion.

8. Splicing requires two coordinated transesterification steps. Draw a diagram showing these steps.

9. Draw a diagram showing the steps involved in spliceosome formation.

10. Describe the experimental evidence that suggests U6 snRNA helps to catalyze splicing.
11. Describe the experimental evidence that suggests some splicing takes place during transcription.
12. The transcription termination machinery requires recognition elements to help it locate the cleavage/polyadenylation site. Draw a figure that shows these recognition elements.
13. List the components of the mammalian machinery required to cleave the pre-mRNA and add a poly(A) to the newly generated 3′-end. What is the function of each of the components?
14. Describe the experimental evidence that shows the transcription elongation complex continues to synthesize RNA after passing the poly(A) site.

Suggested Reading

General

Bentley, D. 1999. Coupling RNA polymerase II transcription with pre-mRNA processing. *Curr Opin Cell Biol* 11:347–351.

Cramer, P., Srebrow, A., Kadener, S., et al. 2001. Coordination between transcription and pre-mRNA processing. *FEBS Lett* 498:179–182.

Howe, K. J. 2002. RNA polymerase II conducts a symphony of pre-mRNA processing activities. *Biochem Biophys Acta* 1577:308–324.

Maniatis, T., and Reed, R. 2002. An extensive network of coupling among gene expression machines. *Nature* 416:499–506.

Neuberger, K. M. 2002. On the importance of being co-transcriptional. *J Cell Sci* 115:3865–3871.

Proudfoot, N. 2000. Connecting transcription to messenger RNA processing. *Trends Biochem Sci* 25:290–293.

Proudfoot, N. J., Furger, A., and Dye, M. J. 2002. Integrating mRNA processing with transcription. *Cell* 108:501–512.

Pre-mRNA

Darnell, J. E., Jelinek, W. R., and Molloy, G. R. 1973. Biogenesis of mRNA: genetic regulation in mammalian cells. *Science* 181:1215–1221.

Darnell, J. E. Jr. 2002. The surprises of mammalian molecular cell biology. *Nat Med* 8:1068–1071.

Cap Formation

Cho, E. J., Takagi, T., Moore, C. R., and Buratowski, S. 1997. mRNA capping enzyme is recruited to the transcription complex by phosphorylation of the RNA polymerase carboxy-terminal domain. *Genes Dev* 11:3310–3326.

Cougot, N., van Dijk, E., Babajko, S., and Séraphin, B. 2004. Cap-tabolism. *Trends Biochem Sci* 29:436–444.

Cowling, V. H. 2010. Regulation of mRNA cap methylation. *Biochem J* 425:295–302.

Dunyak, D. S., Everdeen, D. S., Albanese, J. G., and Quinn, C. L. 2002. Deletion of individual mRNA capping genes is unexpectedly not lethal to *Candida albicans* and results in modified mRNA cap structures. *Eukaryot Cell* 1:1010–1020.

Fabrega, C., Shen, V., Shuman, S., and Lima, C. D. 2003. Structure of an mRNA capping enzyme bound to the phosphorylated carboxy-terminal domain of RNA polymerase II. *Mol Cell* 11:1549–1561.

Fong, N., and Bentley, D. L. 2001. Capping, splicing, and 3′ processing are independently stimulated by RNA polymerase II: different functions for different segments of the CTD. *Genes Dev* 15:1783–1795.

Furuichi, Y., and Shatkin, A. J. 2005. Caps on eukaryotic mRNAs. *Encyclopedia of Life Sciences*. pp. 1–9. Hoboken, NJ: John Wiley & Sons.

Ho, C. K., and Shuman, S. 1999. Distinct roles for CTD Ser-2 and Ser-5 phosphorylation in the recruitment and allosteric activation of mammalian mRNA capping enzyme. *Mol Cell* 3:405–411.

Kobor, M. S., and Greenblatt, J. 2002. Regulation of transcription elongation by phosphorylation. *Biochim Biophys Acta* 1577:261–275.

Shatkin, A. J., and Manley, J. L. 2000. The ends of the affair: capping and polyadenylation. *Nat Struct Biol* 7:838–842.

Shuman, S. 1997. Origins of mRNA identity: capping enzymes bind to the phosphorylated C-terminal domain of RNA polymerase II. *Proc Natl Acad Sci USA* 94:12758–12760.

Shuman, S. 2001. Structure, mechanism, and evolution of the mRNA capping apparatus. *Progr Nucl Acid Res Mol Biol* 66:3–40.

Split Genes

Berget, S. M. 1995. Exon recognition in vertebrate splicing. *J Biol Chem* 270:2411–2414.

Berget, S. M., Moore, C., and Sharp, P. A. 1977. Spliced segments at the 5′ terminus of adenovirus 2 late mRNA. *Proc Natl Acad Sci USA* 74:3171–3175.

Black, D. L. 1995. Finding splice sites within a wilderness of RNA. *RNA* 1:763–771.

Black, D. L. 2000. Protein diversity from alternative splicing: a challenge for bioinformatics and post-genome biology. *Cell* 103:367–370.

Black, D. L. 2003. Mechanisms of alternative pre-messenger RNA splicing. *Annu Rev Biochem* 72:291–336.

Cáceres, J. F., and Kornblihtt, A. 2002. Alternative splicing: multiple control mechanisms and involvement in human disease. *Trends Genet* 18:186–193.

Chambon, P. 1981. Split genes. *Sci Am* 244:60–71.

Chow, L. T., Gelinas, R. E., Broker, T. R., and Roberts, R. J. 1977. An amazing sequence arrangement at the 5′-ends of adenovirus 2 messenger RNA. *Cell* 12:1–8.

Faustino, N. A., and Cooper, T. A. 2003. Pre-mRNA splicing and human disease. *Genes Dev* 17:419–437.

Gravely, B. R. 2001. Alternative splicing: increasing diversity in the proteomic world. *Trends Genet* 17:100–107.

Hartmann, B., and Valcárcel, J. 2009. Decrypting the genome's alternative messages. *Curr Opin Cell Biol* 21:377–386.

Kriventseva, E. V., Koch, I., Apweiler, R., et al. 2003. Increase of functional diversity by alternative splicing. *Trends Genet* 19:124–128.

Lander, E. S., Linton, L. M., Birren, B., et al. 2002. Initial sequencing and analysis of the human genome *Nature* 409:860–921.

Le Hir, H., Nott, A., and Moore, M. J. 2003. How introns influence and enhance eukaryotic gene expression. *Trends Biochem Sci* 28:215–220.

Lorković, Z. J., Kirk, D. A. W., Lambermaon, M. H. L., and Filipowicz, W. 2000. Pre-mRNA splicing in higher plants. *Trends Plant Sci* 5:160–167.

Lou, H., and Gagel, R. F. 1998. Alternative RNA processing—its role in regulating expression of calcitonin/calcitonin gene-related peptide. *J Endocrinol* 156:401–405.

Maniatis, T., and Tasic, B. 2002. Alternative pre-mRNA splicing and proteome expansion in metazoans. *Nature* 418:236–243.

Modrek, B., and Lee, C. 2002. A genomic view of alternative splicing. *Nat Genet* 30:13–19.

Modrek, B., Resch, A., Grasso, C., and Lee, C. 2001. Genome-wide detection of alternative splicing in expressed sequences of human genes. *Nucleic Acids Res* 29:2850–2859.

Moore, M. J. 2000. Intron recognition comes of age. *Nat Struct Biol* 7:14–16.

Mount, S. M. 2002. Messenger RNA splicing signals. *Encyclopedia of Life Sciences*. pp. 1–7. London: Nature Publishing Group.

Padgett, R. A., Konarska, M. M., Grabowski, P. J., et al. 1984. Lariat RNA's as intermediates and products in splicing messenger RNA precursors. *Science* 325: 898–903.

Roberts, G. C., and Smith, C. W. J. 2002. Alternative splicing: combinatorial output from the genome. *Curr Opin Chem Biol* 6:375–383.

Rosonina, E., and Blencowe, B. J. 2002. Gene expression: the close coupling of transcription and splicing. *Curr Biol* 12:R319–R321.

Sharp, P. A. 1993. Split genes and RNA splicing. N. Ringertz (ed), *Nobel Lectures in Physiology or Medicine 1991–1995*, Hackensack, NJ: World Scientific Publishing.

Smith, C. W. J., and Valcárcel, J. 2000. Alternative pre-mRNA splicing: the logic of combinatorial control. *Trends Biochem Sci* 25:381–387.

Stetefeld, J., and Ruegg, M. A. 2005. Structural and functional diversity generated by alternative mRNA splicing. *Trends Biochem Sci* 30:515–521.

Valcárcel, J., and Smith, C. W. J. 2005. Alternative splicing: cell-type-specific and developmental control. *Encyclopedia of Life Sciences*. pp. 1–9. Hoboken, NJ: John Wiley & Sons.

Wang, J., and Manley, J. L. 1997. Regulation of pre-mRNA splicing in metazoa. *Curr Opin Genet Dev* 7:205–211.

Spliceosomes

Abelson, J. 2008. Is the spliceosome a ribonucleoprotein enzyme? *Nat Struct Mol Biol* 15:1235–1237.

Aguliera, A. 2005. Cotranscriptional mRNP assembly: from the DNA to the nuclear pore. *Curr Opin Cell Biol* 17:1–9.

Beggs, J. D. 2001. Spliceosomal machinery. *Encyclopedia of Life Sciences*. pp. 1–9. London: Nature Publishing Group.

Blencow, B. J. 2000. Exonic splicing enhancers: mechanism of action, diversity and role in human genetic diseases. *Trends Biochem Sci* 25:106–110.

Brow, D. A. 2002. Allosteric cascade of spliceosome activation. *Annu Rev Genet* 36:333–360.

Burge, C. B., Tushchl, T., and Sharp, P. A. 1999. Splicing of precursors to mRNAs by the spliceosomes. R. F. Gesteland, T. R. Cech, and J. F. Atkins (eds). *The RNA World* (2nd ed). Woodbury, NY: Cold Spring Harbor Laboratory Press.

Burge, C. B., Tushchl, T., and Sharp, P. A. 2005. Splicing of precursors to mRNA by the spliceosomes. R. F. Gesteland, J. F. Atkins, and T. R. Cech (eds). *The RNA World* (3rd ed). Woodbury, NY: Cold Spring Harbor Laboratory Press.

Chen, M. C., and Manley, J. L. 2009. Mechanisms of alternative splicing regulation: insights from molecular and genomic approaches. *Nat Rev Mol Cell Biol* 10:741–754.

Collins, C. A., and Guthrie, C. 2000. The question remains: is the spliceosome a ribozyme? *Nat Struct Biol* 7:850–854.

Dreyfuss, G., Kim, V. N., and Kataoka, N. 2002. Messenger-RNA-binding proteins and the messages they carry. *Nat Rev Mol Cell Biol* 3:195–205.

Fu, X. D. 1995. The superfamily of arginine/serine splicing factors. *RNA* 1:663–680.

Goldtrohm, A. C., Greenleaf, A. L., and Garcia-Blanco, M. A. 2001. Co-transcriptional splicing of pre-messenger RNAs: considerations for the mechanism of alternative splicing. *Gene* 277:31–47.

Görnemann, J., Kotovic, K. M., Hujer, K., and Neugebauer, K. M. 2005. Cotranscriptional spliceosome assembly occurs in a stepwise fashion and requires the cap binding complex. *Mol Cell* 19:53–63.

Grabowski, P. J., Padgett, R. A., and Sharp, P. A. 1984. Messenger RNA splicing in vitro: an excised intervening sequence and a potential intermediate. *Cell* 37:415–427.

Grainger, R. J., and Beggs, J. D. 2005. Prp8 protein: at the heart of the spliceosome. *RNA* 11:533–537.

Gravely, B. R. 2000. Sorting out the complexity of SR protein functions. *RNA* 6:1197–1211.

Graveley, B. R. 2004. A protein interaction domain contacts RNA in the prespliceosome. *Mol Cell* 13:302–304.

Hastings, M. L., and Krainer, A. R. 2001. Pre-mRNA splicing in the new millennium. *Curr Opin Cell Biol* 13:302–309.

Hertel, K. J., and Graveley, B. R. 2005. RS domains contact the pre-mRNA throughout spliceosome assembly. *Trends Biochem Sci* 30:115–118.

Hoskins, A. A., Friedman, L. J., Gallagher, S. S., et al. 2011. Ordered and dynamic assembly of single spliceosomes. *Science* 331:1289–1295.

House, A. E., and Lynch, K. W. 2008. Regulation of alternative splicing: more than just the ABCs. *J Biol Chem* 283:1217–1221.

Jurica, M. S. 2008. Detailed close-ups and the big picture of spliceosomes. *Curr Opin Struct Biol* 18:315–320.

Kent, O. A., and MacMillan, A. M. 2002. Early organization of pre-mRNA during spliceosome assembly. *Nat Struct Biol* 9:576–581.

Makarov, E. M., Makarova, O. V., Urlaub, H., et al. 2002. Small nuclear ribonucleoprotein remodeling during catalytic activation of the spliceosome. *Science* 298:2205–2208.

Newman, A. J. 2008. RNA splicing interactions in mRNA splicing. *Encyclopedia of Life Sciences*. pp. 1–7. Hoboken, NJ: John Wiley & Sons.

Nilsen, T. W. 2000. The case for an RNA enzyme. *Nature* 408:782–783.

Nilsen, T. W. 2004. Too hot to splice. *Nat Struct Mol Biol* 11:208–209.

Padgett, R. A. 2005. mRNA splicing: role of snRNAs. *Encyclopedia of Life Sciences*. pp. 1–7. Hoboken, NJ: John Wiley & Sons.

Reed, R. 2000. Mechanisms of fidelity in pre-mRNA splicing. *Curr Opin Cell Biol* 12:340–345.

Reed, R. 2003. Coupling transcription, splicing, and mRNA export. *Curr Opin Cell Biol* 15:326–331.

Reed, R., and Hurt, E. 2002. A conserved mRNA export machinery coupled to pre-mRNA splicing. *Cell* 108:523–531.

Reed, R., and Magni, K. 2001. A new view of mRNA export: separating the wheat from the chaff. *Nat Cell Biol* 3:E201–E204.

Ritchie, D. B., Schellenberg, M. J., Gesner, E. M., et al. 2008. Structural elucidation of a PRP8 core domain from the heart of the spliceosome. *Nat Struct Mol Biol* 15:1199–1205.

Ruskin, B., Krainer, A. R., Maniatis, T., and Green, M. R. 1984. Excision of an intact intron as a novel lariat structure during pre-mRNA splicing *in vitro*. *Cell* 38:317–331.

Staley, J. P. 2002. Hanging on to the branch. *Nat Struct Biol* 9:5–7.

Tacke, R., and Manley, J. L. 1999. Determinants of SR protein specificity. *Curr Opin Cell Biol* 11:358–362.

Tardiff, D. F., Lacadie, S. A., and Rosbash, M. 2006. A genome-wide analysis indicates that yeast pre-mRNA splicing is predominantly posttranscriptional. *Mol Cell* 24:917–929.

Valadkhan, S. 2005. snRNAs as the catalyst of pre-mRNA splicing. *Curr Opin Chem Biol* 9:603–608.

Valadkhan, S., Mohammadi, A., Wachtel, C., and Manley, J. L. 2008. Protein-free spliceosomal snRNAs catalyze a reaction that resembles the first step of splicing. *RNA* 13:2300–2311.

Yang, V. W., Lerner, M. R., Steitz, J. A., and Flint, S. J. 1981. A small nuclear ribonucleoprotein is required for splicing of adenoviral early RNA sequences. *Proc Natl Acad Sci USA* 78:1371–1375.

Yu, Y.-T., Scharl, E. C., Smith, C. M., and Steitz, J. A. 1998. The growing world of small nuclear ribonucleoproteins. R. Gesteland, T. Cech, and J. Atkins (eds). *The RNA World* (2nd ed, pp. 487–524). Woodbury, NY: Cold Spring Harbor Laboratory Press.

Cleavage/Polyadenylation and Transcription Termination

Aranda, A., and Proudfoot, N. 2001. Transcriptional termination factors for RNA polymerase II in yeast. *Mol Cell* 7:1003–1011.

Bates, N., and Hurst, H. 2002. Nuclear run-on assays. *Encyclopedia of Life Sciences*. pp. 1–13. London: Nature Publishing Group.

Birse, C. E., Minivielle-Sebastia, L., Lee, B. A., et al. 1998. Coupling termination of transcription to messenger RNA maturation in yeast. *Science* 280:298–301.

Buratowski, S. 2005. Connections between mRNA 3′ end processing and transcription termination. *Curr Opin Cell Biol* 17:257–261.

Colgan, D. F., and Manley, J. L., 1997. Mechanism and regulation of mRNA polyadenylation. *Genes Dev* 11:2755–2766.

Darnell, J. E., Philipson, L., Wall, R., and Adesnik, M. 1971. Polyadenylic acid sequences: role in conversion of nuclear RNA into messenger RNA. *Science* 174:507–510.

Dye, M. J., and Proudfoot, N. J. 1999. Terminal exon definition occurs cotranslationally and promotes termination of RNA polymerase II. *Mol Cell* 3:371–378.

Dye, M. J., and Proudfoot. N. J. 2001. Multiple transport cleavage precedes polymerase release in termination by RNA polymerase II. *Cell* 105:669–681.

Edwalds-Gilbert, G., Veraldi, K. L., and Milcarek, C. 1997. Alternative poly(A) site selection in complex transcription units: means to an end? *Nucleic Acids Res* 25:2547–2561.

Gilmartin, G. M. 2005. Eukaryotic mRNA 3′-processing: a common means to different ends. *Genes Dev* 19:2517–2521.

Hirose, Y., and Manley, J. L. 1998. RNA polymerase II is an essential mRNA polyadenylation factor. *Nature* 395:93–96.

Kolev, N. G., and Steitz, J. A. 2005. Symplekin and multiple other polyadenylation factors participate in 3′-end maturation of histone mRNAs. *Genes Dev* 19:2583–2592.

Kuehner, J. N., Pearson, E. L., and Moore, C. 2011. Unravelling the means to an end: RNA polymerase II transcription termination. *Mol Cell Biol* 12:1–12.

Luo, W., and Bentley, D. 2004. A ribonucleolytic rat torpedoes RNA polymerase II. *Cell* 119:911–914.

Mandel, C. R., Bai, Y., and Tong, L. 2008. Protein factors in pre-mRNA 3′-end processing. *Cell Mol Life Sci* 65:1099–1122.

Minivielle-Sebastia, L., and Keller, W. 1999. mRNA polyadenylation and its coupling to other RNA processing reactions and to transcription. *Curr Opin Cell Biol* 11:352–357.

Osheim, Y. N., Proudfoot, N. J., and Beyer, A. L. 1999. EM visualization of transcription by RNA polymerase II: downstream termination requires a poly(A) signal but not transcript cleavage. *Mol Cell* 3:379–387.

Osheim, Y. N., Sikes, M. L., and Beyer, A. L. 2002. EM visualization of PolII in *Drosophila*: most genes terminate without prior 3′-end cleavage of nascent transcripts. *Chromosoma* 111:1–12.

Proudfoot, N. 1996. Ending the message is not so simple. *Cell* 87:779–781.

Proudfoot, N. 2004. New perspectives on connecting messenger RNA 3′ end formation to transcription. *Curr Opin Cell Biol* 16:272–278.

Proudfoot, N., and O'Sullivan, J. 2002. Polyadenylation: a tail of two complexes. *Curr Biol* 12:R855–R857.

Richard, P., and Manley, J. L. 2009. Transcription termination by nuclear RNA polymerases. *Genes Dev* 23:1247–1269.

Rondon, A. G., Mishco, H. E., and Proudfoot, N. J. 2008. Terminating transcription in yeast: whether to be a "nerd" or a "rat." *Nat Struct Mol Biol* 15: 775–776.

Rosonina, E., Kaneko, S., and Manley, J. L. 2006. Terminating the transcript: breaking up is hard to do. *Genes Dev* 20:1050–1056.

Ryan, K., Calvo, O., and Manley, J. L. 2004. Evidence that polyadenylation factor CPSF-73 is the mRNA 3′ processing endonuclease. *RNA* 10:565–573.

Ryan, K., Murthy, K. G. K., Kaneko, S., and Manley, J. L. 2002. Requirements of the RNA polymerase II C-terminal domain for reconstituting pre-mRNA 3′-cleavage. *Mol Cell Biol* 22:1684–1692.

Shatkin, A. J., and Manley, J. L. 2000. The ends of the affair: capping and polyadenylation. *Nat Struct Biol* 7:838–842.

Tian, B. 2008. Alternative polyadenylation in the human genome: evolution. *Encyclopedia of Life Sciences*. pp. 1–7. Hoboken, NJ: John Wiley & Sons.

Tran, D. P., Kim, S. J., Park, N. J., et al. 2001. Mechanism of poly(A) signal transduction to RNA polymerase II *in vitro*. *Mol Cell Biol* 21:7495–7508.

Yonaha, M., and Proudfoot, N. J. 2000. Transcriptional termination and coupled polyadenylation *in vitro*. *EMBO J* 19:3770–3777.

Wahle, E., and Rüegsegger, U. 1999. 3′-End processing of pre-mRNA in eukaryotes. *FEMS Microbiol Rev* 23:277–295.

Wilusz, J. E., and Spector, D. L. 2010. An unexpected ending: noncanonical 3′ end processing mechanisms. *RNA* 16:259–266.

Zhang, H., Lee, J. Y., and Tian, B. 2005. Biased alternative polyadenylation in human tissues. *Genome Biol* 6:R100.

Zhao, J., Hyman, L., and Moore, C. 1999. Formation of mRNA 3′-ends in eukaryotes: mechanism, regulation, and interrelationships with other steps in mRNA synthesis. *Microbiol Mol Biol Rev* 63:405–445.

RNA Editing

Blanc, V., and Davidson, N. O. 2003. C-to-U editing: mechanisms leading to genetic diversity. *J Biol Chem* 276:1395–1398.

Estévez, A. M., and Simpson, L. 1999. Uridine insertion/deletion RNA editing in trypanosome mitochondria—a review. *Gene* 240:247–260.

Gerber, A. P., and Keller, W. 2001. RNA editing by base deamination: more enzymes, more targets, new mysteries. *Trends Biochem Sci* 26:376–384.

Maas, S., and Rich, A. 2000. Changing genetic information through RNA editing. *Bioessays* 22:790–802.

Maas, S., Rich, A., and Nishikura, K. 2003. A-to-I editing: recent news and residual mysteries. *J Biol Chem* 278:1391–1394.

Sowden, M. P., and Smith, H. C. 2002. RNA editing. *Encyclopedia of Life Sciences*. pp. 1–7. London: Nature Publishing Group.

15

Small Silencing RNAs

CHAPTER OUTLINE

15.1 RNA Interference (RNAi) Triggered by Exogenous Double-Stranded RNA
The roundworm *Caenorhabditis elegans* is an attractive organism for molecular biology studies.
RNAi was discovered in *C. elegans*.
In vitro studies helped to elucidate the RNAi pathway.
Dicer cleaves long double-stranded RNA into fragments of discrete size.
RISC loading complex is required for siRISC formation.
RNAi blocks virus replication and prevents transposon activation.

15.2 Transitive RNAi
In some organisms RNAi that starts at one site spreads throughout the entire organism.
SID-1, an integral membrane protein in *C. elegans*, assists in the systemic spreading of the silencing signal.
ERI-1, a 3'→5' exonuclease in *C. elegans*, appears to be a negative regulator of RNAi.

15.3 RNAi as an Investigational Tool
RNAi is a powerful tool for investigating functional genomics.

15.4 MicroRNA Pathway
The miRNA pathway blocks mRNA translation or causes mRNA degradation.

15.5 Piwi Interacting RNAs (piRNAs)
piRNAs help to maintain germ line stability in animals.

BOX 15.1: LOOKING DEEPER Dicer from *Giardia intestinalis*

QUESTIONS AND PROBLEMS

SUGGESTED READING

Image © Heiti Paves/ShutterStock, Inc.

As the last millennium drew to a close it appeared that virtually all regulatory factors that influence eukaryotic gene expression were proteins. Then in 1998 Andrew Z. Fire, Craig C. Mello, and coworkers discovered that long double-stranded RNAs can cause sequence-specific gene silencing in a roundworm called *Caenorhabditis elegans*. At its most basic level the gene silencing process that Fire and Mello discovered involves the conversion of a long double-stranded RNA into a small single-stranded RNA of defined size that is incorporated into a ribonucleoprotein complex. The small single-stranded RNA guides the complex to a target messenger RNA (mRNA) and an RNase in the complex cleaves the mRNA. Subsequent studies revealed the existence of other eukaryotic pathways to generate small ribonucleoprotein complexes that regulate gene expression. The discovery that small RNA molecules can regulate gene expression has forced us to revise our views on the mechanisms that regulate gene expression, led to the development of new techniques to study gene function, and promises to radically change the way that clinicians diagnose and treat cancer and various other genetic diseases. We begin by examining the pioneering work by Fire, Mello, and colleagues and then examine related phenomena.

15.1 RNA Interference (RNAi) Triggered by Exogenous Double-Stranded RNA

The roundworm *Caenorhabditis elegans* is an attractive organism for molecular biology studies.

Sydney Brenner first suggested the roundworm might be an excellent model system for studying the molecular biology of a multicellular animal in 1963. He outlined the worm's virtues as a "multicellular organism which has a short life cycle, can be easily cultivated, and is small enough to be handled in large numbers, like a micro-organism." Furthermore, Brenner thought the worm should be amenable to genetic analysis. He also pointed out that the worm has relatively few cells so it should be possible to determine the pattern of cell division and cell fates during the worm's development.

Because many of the experiments to be described in this chapter were performed using the roundworm *C. elegans*, it is necessary to consider some of the worm's important biological characteristics before proceeding further. The adult worm, at about 1 mm long, is just visible to the naked eye. Each adult worm exists as either a male or hermaphrodite (produces sperm and eggs). Male worms have exactly 1,031 somatic cells and hermaphrodites have about 1,000 cells. FIGURE 15.1 shows some of the worm's important anatomical features. The worm feeds on *E. coli* and other bacteria, ingesting and grinding the bacteria in its pharynx (mouth piece) and absorbing nutrients from the resulting extract in its intestine. The worm is estimated to have about 19,000 genes, about 13% of which are organized into operons,

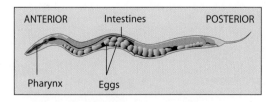

FIGURE 15.1 **A schematic showing some *C. elegans* body parts.**

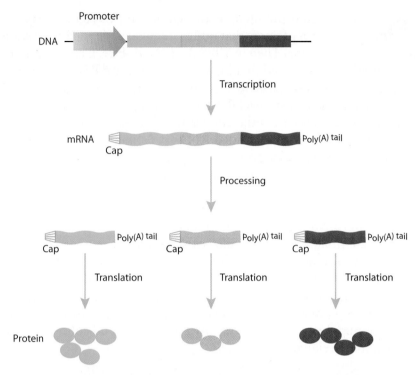

FIGURE 15.2 Transcription and translation from a *C. elegans* operon. Genes in a single operon are transcribed starting at a single promoter to form a pre-mRNA. This pre-mRNA is processed to form mature mRNAs that are each translated into a unique polypeptide.

each containing between two and eight genes. RNA polymerase II transcribes each operon from a single promoter to produce a primary transcript, which then is processed to form mature mRNAs that each codes for a single polypeptide during translation (**FIGURE 15.2**). The following additional features make *C. elegans* an attractive model system for the study of molecular biology:

1. Approximately 50% to 60% of the worm genes are homologous to human genes.
2. The worm is transparent throughout its life, permitting investigators to follow cell development.
3. The worm develops rapidly at 25°C so a single egg can mature into an adult worm in about 3 days.
4. A single worm can produce about 200 to 400 progeny.
5. Worms can be stored in a frozen state at −80°C.
6. Worms are accessible to genetic manipulation.

RNAi was discovered in *C. elegans*.

When Fire, Mello, and coworkers began their studies, molecular biologists knew they could block specific gene expression in *C. elegans* by injecting the worm with sense or antisense RNA and the inhibition would persist well into the next generation. This persistence seemed inconsistent with the fact that cells rapidly degrade single-stranded RNA molecules such as mRNA. Furthermore, it was difficult to understand how a sense RNA strand could block gene expression. Fire, Mello, and coworkers therefore suspected the single-stranded RNAs that investigators used to inject worms were probably contaminated with some biologically active double-stranded RNA molecules. They decided to test their hypothesis by using purified single- and double-stranded RNAs to study the expression of the nonessential fibrous muscle protein gene, *unc-22*, in *C. elegans*. Mutant *unc-22* worms twitch, causing <u>unc</u>oordinated movement that is easy to follow. Fire, Mellow, and coworkers prepared sense and antisense RNA strands corresponding to the *unc-22* gene and annealed the strands to prepare a double-stranded RNA. Then they injected worms in their reproductive organ with the sense strand, the antisense strand, or the double-stranded RNA (FIGURE 15.3). Quantitative studies showed the double-stranded RNA to be about 100-fold more effective in causing uncoordinated movement than either the sense or antisense strands. Double-stranded RNA molecules corresponding to other genes did not affect worm movement, indicating the phenomenon is sequence specific. Mello coined the term **RNA interference (RNAi)** for gene silencing that is caused by double-stranded RNA. RNAi lasts for a very long time and requires only a few molecules per cell, suggesting the double-stranded RNA or a downstream product somehow acts in a catalytic fashion.

Injecting individual worms is a time-consuming task that requires considerable skill. Fortunately, two simpler methods can be used to introduce double-stranded RNA into worms. The first takes advantage of the fact that worms feed on *E. coli*. The *E. coli* strain used to feed the worm is constructed so it contains a plasmid with the same promoter on either side of the gene of interest (FIGURE 15.4). The bacteria, therefore, synthesize both sense and antisense RNAs, which anneal to

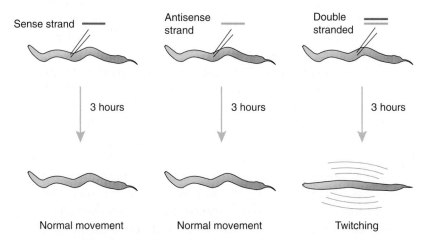

FIGURE 15.3 **RNAi**. *C. elegans* were injected with sense RNA, antisense RNA, and double-stranded RNA corresponding to the *unc-22* gene. The movement of the injected worms was monitored. After 3 hours very few worms injected with the sense or antisense strands exhibited uncoordinated movement due to twitching. In contrast, nearly all the worms injected with double-stranded RNA exhibited uncoordinated movement due to twitching. (Adapted from Daneholt, B. Advanced information: The 2006 Nobel Prize in Physiology or Medicine. http://nobelprize.org/nobel_prizes/medicine/laureates/2006/adv.html)

FIGURE 15.4 **Plasmids that code for double-stranded RNA.** A bacterial strain has a plasmid (dark blue) with a DNA insert (light blue) that codes for RNA corresponding to a target worm gene. The DNA has the same promoter (magenta) on either side so the insert is transcribed in both directions. The red arrows indicate the direction of transcription.

15.1 RNA Interference (RNAi) Triggered by Exogenous Double-Stranded RNA

form double-stranded RNA corresponding to the gene of interest. In one of the earliest feeding experiments recombinant worms constructed to synthesize green fluorescent protein in almost all their somatic cells were allowed to feed on *E. coli* that synthesized double-stranded RNA molecules corresponding to the green fluorescent protein gene, *gfp*. Approximately 12% of the worms studied in this experiment exhibited sequence specific *gfp* gene silencing in all cells but nerve cells. Although it is more convenient to introduce double-stranded RNA into worms by the feeding method, this method is considerably less efficient than injecting double-stranded RNA directly into the worm. The second method is even simpler: worms are simply soaked in a solution that contains the double-stranded RNA of interest. The soaking technique, although not very efficient, can be quite useful for high throughput worm studies.

Fire and coworkers proposed the following three hypotheses to explain how double-stranded RNA might silence worm genes:

1. Double-stranded RNA alters the target DNA.
2. Double-stranded RNA inhibits gene transcription.
3. Double-stranded RNA somehow destabilizes mRNA.

They designed experiments to test each hypothesis. They started by demonstrating that double-stranded RNA does not modify the target DNA, ruling out the first hypothesis. Then they tested the second hypothesis by examining whether a double-stranded RNA corresponding to the first gene in an operon inhibits the expression of downstream genes in that same operon. They observed that the downstream genes are fully expressed. This observation rules out the transcription inhibition hypothesis, which predicts that a double-stranded RNA will silence downstream genes. The transcription inhibition hypothesis is also ruled out because double-stranded RNAs corresponding to promoters and introns do not silence genes. Fire, Mello, and coworkers obtained strong support for the mRNA destabilization hypothesis by demonstrating double-stranded RNA causes a substantial reduction in intracellular target mRNA levels.

Based on these findings Fire and coworkers proposed that *RNAi is a posttranscriptional process in which double-stranded RNA somehow makes the corresponding mRNA unstable*. Some method was needed to learn how double-stranded RNA makes mRNA unstable. Fire, Mello, and coworkers thought they might be able to obtain helpful information by isolating mutant worms that are unable to carry out post-transcriptional gene silencing. They exposed worms to chemical mutagens and let the worms grow for two generations to permit induced mutations to be homozygous. Then they allowed the mutant worms to feed on *E. coli*, which were constructed to express a double-stranded RNA that targeted an essential worm gene. The double-stranded RNA killed worms with an intact RNAi response but had no effect on worms with a defective RNAi response.

Two of the isolated RNA interference deficient (*rde*) mutants are of special interest because they provide important clues to how RNAi works. One of the mutants had an altered *RDE-4* gene. This gene normally codes for a double-stranded RNA binding protein. The other mutant had an altered *RDE-1* gene. This gene normally codes for RDE-1, a member of the **Argonaute protein family**. The family name, which reflects that fact that the founding member was shown to influence plant leaf morphology, has nothing to do with the structure or function of any protein family member. The number of Argonaute genes varies greatly from one eukaryotic species to another. For example, humans have 8 Argonaute genes (4 Ago and 4 Piwi), *Drosophila* have 5 Argonaute genes (2 Ago and 3 Piwi), and *C. elegans* have at least 27 Argonaute genes (5 Ago, 3 Piwi, 18 worm-specific, and 1 that has not yet been assigned). RDE-1 is one of the 5 Ago proteins in *C. elegans*.

In vitro studies helped to elucidate the RNAi pathway.

In 1999 Philip A. Sharp and coworkers initiated a research effort to identify the enzymes involved in the RNAi pathway and characterize the reactions these enzymes catalyze. They decided to use *Drosophila* embryo extracts as their *in vitro* system. Their first step was to determine if the extracts could synthesize luciferase when luciferase mRNA was added to the extracts. They selected luciferase mRNA as the reporter mRNA because *Drosophila* does not normally make luciferase and a very sensitive assay was available to detect how much, if any, enzyme was formed. The luciferase assay exploits the enzyme's ability to catalyze the oxidation of its substrate luciferin to produce light. The equation for this bioluminescent reaction is as follows:

$$\text{Luciferin} + \text{ATP} + O_2 \xrightarrow[\text{Mg}^{2+}]{\text{Luciferase}} \text{Oxyluciferin} + CO_2 + \text{AMP} + \text{PPi} + \text{Light}$$

Sharp and coworkers used two different kinds of luciferase mRNAs in their experiments. The first was prepared by transcribing a cloned firefly luciferase gene and the second by transcribing a cloned sea pansy luciferase gene. The *Drosophila* embryo extracts translated the firefly mRNA to make active firefly luciferase and the sea pansy mRNA to make active sea pansy luciferase.

In their next set of experiments Sharp and coworkers exploited the fact that firefly and sea pansy luciferase genes have different nucleotide sequences and code for luciferases with different luciferin substrate requirements. These experiments involved incubating the *Drosophila* embryo extract with double-stranded RNA corresponding to the firefly luciferase gene for 10 minutes and then adding firefly and sea pansy luciferase mRNAs to the extract. After a 1-hour additional incubation, the mixture was assayed to determine how much, if any, luciferase was synthesized. As anticipated, the *Drosophila* embryo extract synthesized sea pansy luciferase but not firefly luciferase. Sharp and coworkers repeated the experiment, but this time added double-stranded RNA corresponding to the sea pansy luciferase gene. Now the *Drosophila* embryo extract synthesized firefly luciferase but no sea pansy luciferase. These experiments, which demonstrated that RNAi can be studied *in vitro*, set the stage for the isolation and characterization of enzymes and factors that participate in the RNAi pathway.

The following year Gregory J. Hannon and coworkers used a cell-free system prepared from *Drosophila* S2 cells (a cell line derived from a primary culture of late-stage embryonic stem cells) to dissect the RNAi pathway. Their initial goal was to determine whether a double-stranded RNA corresponding to the *E. coli lacZ* gene would cause S2 cell extracts to catalyze sequence specific β-galactosidase mRNA degradation. They began by transfecting S2 cells with double-stranded RNA corresponding to the *E. coli lacZ* gene and then prepared a cell-free extract from the transfected cells to which they added β-galactosidase mRNA. The extract degraded the β-galactosidase mRNA but no other mRNA. Hannon and coworkers proposed that the extract contained a sequence-specific RNAse complex, which they called an **RNA-induced silencing complex (RISC)**. We refer to this complex as siRISC to distinguish it from another form of RISC encountered later in this chapter. In the next phase of their work, Hannon and coworkers used standard protein fractionation techniques to partially purify siRISC. Chemical analysis showed that siRISC contains a 22-nucleotide RNA with a sequence complementary to that in β-galactosidase mRNA. This small RNA guides the siRISC complex to its target mRNA. Subsequent studies revealed that siRISC contains an Argonaute protein, which is essential for its function.

In 2001 Hannon and coworkers performed a second series of *in vitro* experiments, which showed how cells convert the long double-stranded RNA into the small single strand of RNA that guides the siRISC complex to the target mRNA. They began by demonstrating the presence of a new RNAse activity that participates in the conversion of long double-stranded RNA into small double-stranded RNA molecules called **small interfering RNAs (siRNAs)**. Hannon and coworkers separated the new RNase from the siRISC by subjecting the S2 cell extract to high speed centrifugation. The new RNase remained in the supernatant, while the siRISC ended up in the pellet. Hannon and coworkers demonstrated that the new RNase, which they named **Dicer**, belongs to the RNase III family. Subsequent studies revealed that Dicer catalyzes sequence-independent double-stranded RNA cleavage to produce 22-nucleotide fragments. The enzyme does not act on single-stranded RNA. Cells with low levels of Dicer activity cannot carry out double-stranded RNA-dependent gene silencing, supporting the hypothesis that Dicer is required for RNAi.

Based on the experiments described above and more recent experiments, the key steps in the *C. elegans* RNAi pathway are shown in FIGURE 15.5. For simplicity the protein names used are those for *C. elegans* system. Other multicellular animals, however, use the same general pathway. The steps in this pathway are as follows: (1) Dicer combines with RDE-4, a double-stranded RNA binding protein, to form a complex that cleaves long double-stranded RNA into siRNA duplexes. (2) One of the two siRNA strands in the duplex becomes part of siRISC, which also contains Argonaute (RDE-1) and other proteins. (3) The single-stranded RNA guides siRISC to mRNA, and then the Argonaute protein (RDE-1) cleaves the mRNA. Other eukaryotes use the same basic pathway, but some details may differ. With this pathway in mind, let's take a closer look at Dicer and siRISC.

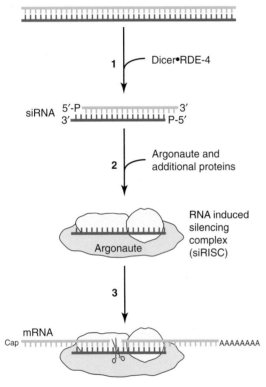

FIGURE 15.5 **The RNAi pathway in *C. elegans*.** (1) Dicer, an enzyme with RNase III activity, combines with RDE-4 (a double-stranded RNA binding protein) to form a complex that cleaves long double-stranded RNA into small double-stranded RNA molecules of defined length called small interfering RNAs (siRNAs). (2) One of the two strands in siRNA combines with an Argonaute protein (RDE-1) and other proteins to form a nucleoprotein complex called the RNA induced silencing complex (siRISC). (3) The single-stranded RNA guides siRISC to mRNA, where the Argonaute protein (RDE-1), cleaves the mRNA. Other eukaryotes use the same basic pathway but some details may differ. (Adapted from Faehnle, C. R., and Joshua-Tor, L. 2007. *Curr Opin Chem Biol* 11:569–577.)

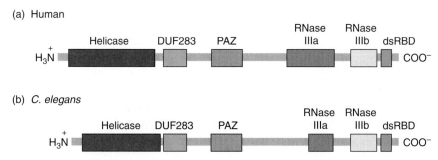

FIGURE 15.6 Representation of Dicer domain structures. (a) Human dicer and (b) *C. elegans* dicer. The domains shown are as follows: ATP/helicase, DUF283 (domain of unknown function), PAZ (Piwi, Argonaute, Zwille), RNase IIIa, RNase IIIb, and dsRBD (double stranded RNA binding domain). (Adapted from Jaskiewicz, L., and Filipowicz, W. 2008. *Curr Top Microbiol Immunol* 320:77–97.)

Dicer cleaves long double-stranded RNA into fragments of discrete size.

Dicer proteins, like other members of the RNase III family, make staggered cuts in double-stranded RNA substrates to form characteristic cleavage fragments that are easy to recognize because each end of the fragment has a two nucleotide 3'-overhang and a 5'-phosphate group (Figure 15.5). Dicers from humans, *C. elegans*, and other multicellular eukaryotes have multidomain structures (**FIGURE 15.6a** and **b**). Starting from the N-terminus these domains are ATP/helicase, DUF283 (domain of unknown function), PAZ (Piwi, Argonaute, Zwille), RNase IIIa, RNase IIIb, and dsRBD (double-stranded RNA binding domain). A crystal structure is not yet available for an intact Dicer from a multicellular eukaryote. Jennifer Doudna and coworkers, however, have solved the crystal structure for an intact Dicer from *Giardia intestinalis*, a common intestinal parasite that infects humans. This structure provides important insights into how Dicer works (**BOX 15.1 LOOKING DEEPER: DICER FROM *GIARDIA INTESTINALIS***).

BOX 15.1: LOOKING DEEPER

Dicer from *Giardia intestinalis*

At present, the only crystal structure for intact Dicer is that for the *G. intestinalis* Dicer. The domain structure for *G. intestinalis* Dicer (**FIGURE B15.1a**) is similar to that for human Dicer (Figure 15.6a) and *C. elegans* Dicer (Figure 15.6b). All three have PAZ, RNase IIIa, and RNase IIIb domains, but *G. intestinalis* Dicer has an alternative form of DUF283 and is missing the helicase and dsRBD domains. Based on the similarities in the domain structures it is likely the *G. intestinalis* crystal structure also provides useful information concerning human and *C. elegans* Dicer structure and function. As shown in **FIGURE B15.1b**, *G. intestinalis* Dicer has an elongated shape resembling a hatchet with the blade at the top and the handle at the bottom. The RNase IIIa and IIIb domains are adjacent to one another in the region corresponding to the blade. Each RNase domain has a pair of divalent metal ions at the active site. The metal ions are manganese ions in the crystal structure but are probably magnesium ions in the cell.

(*continues*)

BOX 15.1: LOOKING DEEPER

Dicer from *Giardia intestinalis* (continued)

The distance between the metal ion pairs is 17.5 Å, which corresponds to the width of the major groove of double-stranded RNA. The *G. intestinalis* Dicer crystal structure suggests the method that Dicer uses to specify RNA fragment length (FIGURE B15.1c). The distance between the PAZ site that binds the 3′-overhang and the RNase IIIa active site is about 65 Å, which corresponds to about 25 base pairs in double-stranded RNA. Thus, Dicer acts as a molecular ruler that measures and cleaves about 25 nucleotides from the end of double-stranded RNA.

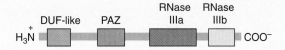

FIGURE B15.1a **Domain structure for Dicer from *Giardia intestinalis*.**

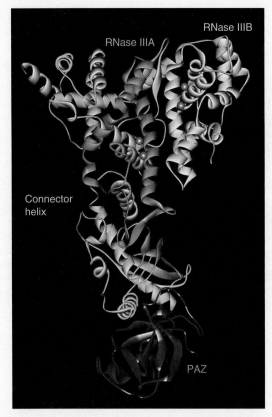

FIGURE B15.1b **Crystal structure of Dicer from *G. intestinalis*.** Each RNase domain has a pair of divalent metal ions at the active site. These metal ions are manganese ions (purple spheres) in the crystal structure but are probably magnesium ions in the cell. The distance between the metal ion pairs is 17.5 Å, which corresponds to the width of the major groove of double-stranded RNA. The distance between the PAZ site that binds the 3′-overhang and the RNase IIIa active site is about 65 Å, which corresponds to about 25 base pairs in double-stranded RNA. (Structure from Protein Data Bank 2FFL. Macrae, I. J., et al. 2006. *Science* 311:195–198. Prepared by B. E. Tropp.)

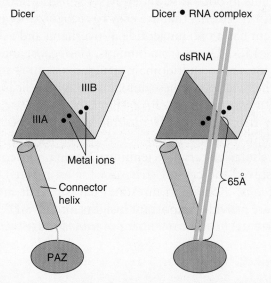

FIGURE B15.1c **Cartoon of *G. intestinalis* Dicer cleaving double-stranded RNA.** The PAZ domain binds one end of the double-stranded RNA (green). The distance between the PAZ binding site and the RNase IIIa active site is about 65 Å, which corresponds to about 25 base pair in double-stranded RNA. Thus, Dicer serves as a molecular ruler that measures and cleaves about 25 nucleotides from the end of double-stranded RNA. (Adapted from Cook, A., and Conti, E. 2006. *Nat Struct Mol Biol* 13:190–192.)

RISC loading complex is required for siRISC formation.

The loading of guide RNA onto an Argonaute protein starts with a siRNA duplex formed by Dicer-mediated cleavage. The two siRNA strands must be separated and one strand, the **passenger strand**, must be discarded, while the other strand, the **guide strand**, becomes part of siRISC. *Drosophila* has two Dicers. One of these, Dicer-2 (Dcr-2), makes an essential contribution to siRISC assembly. Although Dcr-2 is the enzyme that converts long double-stranded RNA into siRNA duplexes, it has an additional role in siRISC formation. Dcr-2 and its partner R2D2, a double-stranded RNA binding protein, combine with siRNA to form a **RISC loading complex** (FIGURE 15.7). The siRISC loading complex contains additional uncharacterized proteins not shown in Figure 15.7. R2D2 binds to the end of the siRNA duplex that has the more stable base pairs and Dcr-2 interacts with the other end. siRISC loading complex can distinguish siRNAs from other small double-stranded RNAs. One way it does so is by discriminating against small double-stranded RNAs that have mismatches corresponding to positions 7 to 11 on the strand destined to be the guide strand.

Argonaute 2 (Ago 2), the *Drosophila* homolog of RDE-1, and perhaps other still uncharacterized proteins (not shown in Figure 15.7) add to the siRISC loading complex to form a **pre-RISC complex**. Then Ago 2 slices the passenger strand, and the resulting passenger strand fragments dissociate from the pre-RISC complex. This dissociation converts the inactive complex into active siRISC, which contains the guide strand. The change from inactive pre-RISC complex to the fully active siRISC requires ATP. Although the pathway shown in Figure 15.7 describes the *Drosophila* system, it also applies to other multicellular animals. Further work is needed to determine why this conversion requires ATP.

RNAi blocks virus replication and prevents transposon activation.

RNAi appears to have arisen in an early eukaryotic ancestor and is conserved in animals, plants, and most fungi. One notable exception, *Saccharomyces cerevisiae*, lacks recognizable homologs of Dicer and Argonaute, but other budding yeast species have active RNAi pathways. Genetic analysis indicates that 10 or more genes are probably needed to synthesize proteins required for RNAi. Eukaryotes, however, seldom encounter highly concentrated double-stranded RNA of identical sequence to one of their protein coding genes under normal physiological conditions. It therefore seems reasonable to ask why the RNAi pathway is highly conserved in eukaryotes as diverse as plants, flies, worms, and humans. One possible explanation is the RNAi pathway provides immunity to viruses that produce double-stranded RNA at some time during their life cycle. Dicer recognizes this double-stranded RNA and cleaves it to produce siRNA, which is converted to the guide RNA that directs siRISC to degrade the viral mRNA necessary for successful viral infection. This explanation for RNAi pathway conservation probably applies to organisms that lack an interferon system to protect them from invading viruses. There

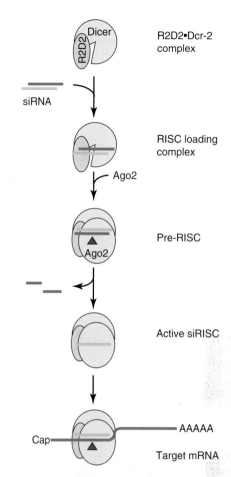

FIGURE 15.7 siRISC assembly in *Drosophila*. Dicer 2 (Dcr-2) and its partner R2D2, a double-stranded RNA binding protein, combine with siRNA to form a RISC loading complex. The Argonaute protein Ago 2 adds to the RISC loading complex to form a pre-RISC complex. Then Ago 2 slices the passenger strand (indicated by a red triangle), and the resulting passenger strand fragments dissociate from the pre-RISC complex, converting the inactive complex into a single-stranded guide containing active siRISC, which binds to a target mRNA. The red triangle indicates the RNA cleavage site. The RISC loading complex and siRISC have additional proteins that have not been characterized and therefore are not shown in this figure. Although the pathway shown is for *Drosophila*, other multicellular animals use the same pathway but some details may differ.

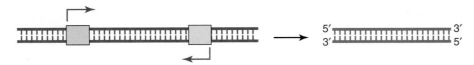

FIGURE 15.8 **Transcription from transposon.** (Adapted from T. W. Nilsen, *Nat. Struct. Mol. Biol.* 15 [2008]: 546–548.)

is probably less selective pressure to conserve the RNAi pathway in humans and other vertebrates, however, which have an active interferon system (see below). Another possible explanation for RNAi pathway conservation comes from studying *C. elegans* mutants that do not synthesize proteins required for the siRNA pathway. These studies show that RNAi blocks the spread of transposons within the genome and thereby prevents the production of abnormal gene products or abnormal gene expression. The double-stranded RNA precursor to siRNA appears to derive from the transcription starting at terminal inverted repeats at the ends of the transposons (FIGURE 15.8).

15.2 Transitive RNAi

In some organisms, RNAi that starts at one site spreads throughout the entire organism.

RNAi in *C. elegans* can spread throughout the worm, even when triggered by a minute quantity of double-stranded RNA injected at a single site. This phenomenon, called **transitive RNAi**, provides useful insights into the amplification mechanism. Transitive RNAi refers to a silencing signal's ability to move along a particular gene. Fire and coworkers discovered this phenomenon while studying worms with an intact *unc-22* gene as well as an *unc-22* gene segment fused to a *gfp* gene. Half the worms studied had the *unc-22* gene segment fused upstream of *gfp* and the other half had it fused downstream of *gfp* (FIGURE 15.9). As expected, all worms injected with double-stranded RNA that targets *gfp* lost the ability to fluoresce. However, the worms that had *unc-22* fused upstream of *gfp* also showed a surprising loss of coordination, indicating the double-stranded RNA that targets *gfp* can somehow also silence *unc-22*.

Based on the observation that silencing is amplified, it seems reasonable to propose that siRNAs formed by Dicer, the primary siRNAs, serve as primers for the synthesis of secondary siRNAs, which in turn lead to the production of additional siRISC. This mechanism for increased siRISC production was ruled out by studies that showed secondary siRNAs have triphosphates at their 5′-ends. This type of 5′-end is most easily explained by proposing that an RNA-dependent RNA polymerase catalyzes unprimed RNA synthesis. The secondary siRNAs with 5′-triphosphate ends, which are 22 or 23 nucleotides in length and always antisense, account for the vast majority of small RNAs in the worm. Their presence indicates that a great deal of amplification takes place. The amplified siRNAs do not associate with

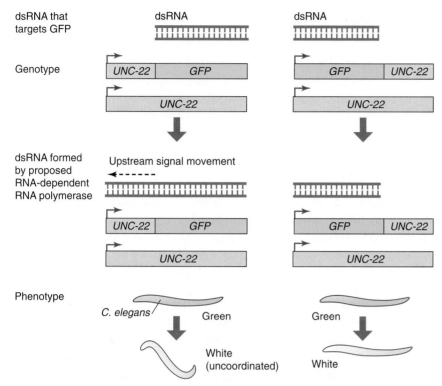

FIGURE 15.9 Transitive RNAi phenomenon. In transitive RNAi in *C. elegans* silencing moves upstream on a specific mRNA target. This phenomenon has been demonstrated by fusing the *gfp* gene to a segment of *unc-22* in a worm that also has the intact *unc-22* gene. Targeting *gfp* abolishes fluorescence but also creates an unexpected phenotype; worm movement becomes uncoordinated when *unc-22* is upstream of *gfp* (left) but not when it is downstream (right). (Adapted from Hannon, G. J. 2002. *Nature* 418:244–251.)

the Argonaute protein RDE-1 but instead associate with one of two secondary Argonaute proteins, SAGO-1 and SAGO-2. These secondary Argonaute proteins lack endonuclease activity and must therefore silence genes by some mechanism other than mRNA cleavage. Further work is required to elucidate the silencing mechanism. siRNA amplification helps to explain how RNA silencing spreads throughout the worm and is passed from one generation to the next. Based on DNA sequence analysis *C. elegans* is predicted to have four different RNA polymerases: RRF-1, RRF-2, RRF-3, and EGO-1. RRF-1 is essential for secondary siRNA production in the adult worm, and EGO-1 may be required in germ line tissues. *C. elegans* RNA-dependent RNA polymerases have not yet been purified or characterized.

Plants also have RNA-dependent RNA polymerase that permits local and systemic RNAi spreading. The plant RNA-dependent RNA polymerase uses the targeted RNA as a template to amplify the silencing signal to form secondary siRNAs both upstream and downstream of the primary siRNA. RNA-dependent RNA polymerases have not been detected in vertebrates or insects, which both appear to lack transitive RNAi. This lack of transitive RNAi can be turned to advantage because it allows us to target individual alternatively spliced mRNA isoforms produced from a single transcription unit.

SID-1, an integral membrane protein in *C. elegans*, assists in the systemic spreading of the silencing signal.

Craig P. Hunter and coworkers isolated *C. elegans* mutants that can silence genes in injected cells but cannot spread RNAi throughout their body. The first systematic RNA interference defective gene they isolated (*sid-1*) codes for an integral membrane protein that is present in all non-neuronal tissues and permits double-stranded RNA to move into cells. In this regard it is interesting to note that RNAi seldom spreads to neurons. Subsequent studies revealed that other organisms, including humans, have SID-1 homologs but *Drosophila* do not. Hunter and coworkers constructed *Drosophila* cells that express *C. elegans* SID-1. The double-stranded RNA concentration required to produce a specific RNAi response is about 100,000-fold lower for these cells than it is for their parent cells that do not make SID-1. *Drosophila* cells that express SID-1 internalize double-stranded RNA very rapidly in a largely energy-independent manner. When *Drosophila* cells coexpress mutant and wild-type SID-1, the mutant SID-1 interferes with wild-type function, suggesting SID-1 works as a multimer. SID-1 does not appear to be required to transport double-stranded RNA out of cells. *C. elegans* use a second membrane protein, SID-2, to take up ingested double-stranded RNA from the intestinal lumen.

Silencing spreads by a different mechanism in plants. Local spreading between neighboring cells takes place via the plasmodesmata, membrane-lined pores that connect the cells. Systemic spreading takes place via the phloem, which normally carries nutrients.

ERI-1, a 3'→5' exonuclease in *C. elegans*, appears to be a negative regulator of RNAi.

Gary Ruvkun and coworkers exploited the fact that RNAi is relatively inefficient in *C. elegans* nerve cells to devise a genetic screen for mutant worms with nerve cells that exhibit enhanced sensitivity to double-stranded RNA interference. Of the 19 candidate mutants, the two with the most enhanced sensitivity to double-stranded RNA both had a mutation in a gene that Ruvkun and coworkers named *eri-1*. After exposure to double-stranded RNA, *eri-1* mutant worms accumulate higher siRNA levels than their wild-type parents. The product of the *eri-1* gene, ERI-1, is a 3'→5' exonuclease found mainly in the cytoplasm. As expected, it is very highly expressed in nerve cells. Based on these studies Ruvkin and coworkers propose that ERI-1 is a negative regulator of RNAi that may act to limit the duration of RNAi or to influence RNAi cell specificity.

15.3 RNAi as an Investigational Tool

RNAi is a powerful tool for investigating functional genomics.

RNAi is a very powerful investigational tool. In the early days of molecular biology mutant phenotypes were used to identify genes, but today the problem is often reversed. Thanks to the great advances in DNA sequencing, thousands of different genes have been identified for which there are no known functions. Until recently, efforts to assign functions for genes that were discovered by large-scale sequencing (**functional genomics**) depended on using genetic recombination to knockout the specific gene of interest. This approach works reasonably well for unicellular organisms such as yeast but presents a problem when studying higher organisms because the methods needed to knockout a specific gene are difficult to perform and require a great deal of time. Moreover, gene knockout does not work for essential genes because unless special precautions are taken the organism cannot survive without the gene. RNAi provides an alternative method for studying gene function when sequence information is available. The functional investigation of a gene that starts with the gene sequence rather than a mutant phenotype is termed **reverse genetics**.

Julie Ahringer and coworkers exploited RNAi to find phenotypes for *C. elegans*' genes identified from sequencing data. They took advantage of the fact that *C. elegans* feeds on *E. coli* to introduce double-stranded RNA into the worms. The double-stranded RNA produced by the bacteria silenced the target worm gene, permitting the investigator to find the phenotype resulting from the inactivation of that gene. Gene inactivation through the use of siRNA is called **gene knockdown**. Ahringer and coworkers constructed 16,757 bacterial strains, each expressing a double-stranded RNA for a different worm gene, and identified mutant phenotypes for 1,722 of the genes studied. Two-thirds of these genes did not have a phenotype previously associated with them. Marc Vidal and coworkers created a second RNAi feeding library for *C. elegans*. The two libraries together account for about 94% of the 19,427 known worm genes. Once the function of a worm gene is known, the knowledge can be used to make an educated guess about the homologous human gene's function.

The specific method for introducing a double-stranded RNA into an organism in a gene knockdown experiment must be tailored to the organism; for example, double-stranded RNAs of more than 30-nucleotides do not trigger the RNAi pathway in humans but instead induce a sequence nonspecific **interferon response**. Interferon, a small protein that helps cells to resist viral infection, activates 2'–5' oligoadenylate synthase, which then converts ATP to 2'–5' oligoadenylate. The 2'–5' oligoadenylate in turn stimulates RNase L to cleave several RNA species, including ribosomal RNA. This cleavage causes the nonspecific inhibition of translation. One method for circumventing the long double-stranded RNA problem is to construct plasmid-based expression vectors that use RNA polymerase III promoters to synthesize short RNA species that do not trigger the interferon response. In the few short years since it was developed, the RNAi approach has helped to determine the function of many genes for which no function had previously been established.

15.4 MicroRNA Pathway

The miRNA pathway blocks mRNA translation or causes mRNA degradation.

A second type of regulatory RNA, called **microRNAs (miRNAs)**, was discovered by investigators interested in *C. elegans* development. Some information about earlier studies on *C. elegans* development performed by John Sulston and H. Robert Horvitz is required to understand the experiments that led to the discovery of miRNAs. In the 1970s Sulston and Horvitz established the pattern of cell division and cell fates that take place as *C. elegans* develops from a single cell to a mature organism. Then they used this information to study underlying mechanisms that regulate patterns of cell division and cell fates by isolating and characterizing mutants with abnormal larval cell lineages. Some of the mutants they studied had altered developmental timing. One such timing mutant, *lin-14* (lineage abnormal 14), was of special interest. Too little *lin-14* gene product caused worms to skip larval stage 1 developmental programs and proceed to larval stage 2 programs. In contrast, too much *lin-14* gene product caused worms to repeat larval stage 1 developmental programs instead of advancing to larval stage 2 programs. These mutant studies indicated that the LIN-14 protein plays an important role in regulating worm development.

In 1993 Victor Ambros and coworkers detected another mutation that altered early *C. elegans* larval development. This mutant, which they named *lin-4*, did not make the normal transition from the first to second larval stage. They initially thought the *lin-4* transcript was translated to make another protein required for normal larval development. Further experiments, however, led them to the remarkable discovery that the *lin-4* transcript is not translated but instead functions as a precursor to a 22-nucleotide RNA that functions as a negative regulator for *lin-14* expression. Sequence studies showed the 22-nucleotide *lin-4* RNA product contains imperfect homology to seven elements in the 3′-untranslated region (3′-UTR) of *lin-14* mRNA. FIGURE 15.10 shows how *lin-4* RNA interacts with two of the *lin-14* mRNA elements.

The *lin-4* gene remained an interesting scientific curiosity until 2000 when Gary Ruvkun and coworkers showed that *C. elegans* requires a second 22-nucleotide RNA, this time encoded by the *let-7* (lethal-7) gene, to proceed from the late larval stage to the adult stage. Further experiments showed the 22-nucleotide *let-7* RNA (1) inhibits the expression of *lin-41* and *lin-42*, two protein coding genes and (2) contains imperfect homologies to specific regions in the 3′-UTRs of *lin-41* and *lin-42* mRNA. Ruvkin and coworkers then demonstrated that *let-7* is conserved and expressed in a wide variety of animals, including flies, mice, chickens, and humans. With this demonstration the study of the 22-nucleotide RNAs moved to the center stage of molecular biology as an important fundamental research problem. Because of their common function in regulating the timing and development of transitions, *lin-4* and *let-7* RNAs

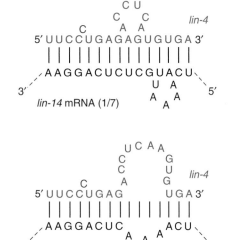

FIGURE 15.10 Examples of proposed interactions between the *C. elegans* lin-4 miRNA and a target mRNA. The 22-nucleotide RNA derived from the *lin-4* transcript is proposed to interact by base pairing with seven sites in the 3′-untranslated region (UTR) of *lineage-14* (*lin-14*). This figure shows the first two of the seven sites in the 3′-UTR of *lin-14* mRNA. The C residue shown in red is predicted to be bulged in four of seven *lin-14* interactions, including the two shown here. The C residue is mutated to U in a strong loss of function *lin-4* mutant. (Adapted from Eddy, S. R. 2001. *Nat Rev Genet* 2:919–929.)

were initially called small temporal RNAs, but this name was later changed to microRNAs (miRNAs).

Thousands of miRNAs have now been identified in animals and plants. According to current estimates miRNA genes account for about 3% of the human genome. A single miRNA may regulate hundreds of distinct target mRNAs, so it is theoretically possible that at least one-third of all human genes are regulated by miRNAs.

Many miRNAs are formed by processing primary miRNA transcripts (pri-miRNAs), which are synthesized by RNA polymerase II. These pri-miRNA transcripts have 5'-m^{7}G caps and 3' poly(A) tails. The pathway for converting pri-miRNA to miRNA is shown in FIGURE 15.11. The early steps in this pathway, which take place in the cell nucleus, depend on whether or not the miRNA sequence is present in an intron. If the miRNA sequence (dark green) is present in an intron, then a spliceosome releases the intron and a lariat debranching enzyme converts the intron to the pre-miRNA hairpin. If the miRNA sequence is *not* present in an intron, then the first processing step is carried out by the **Microprocessor complex**, which has an RNase III and double-stranded RNA binding protein. Humans with a rare inborn error of metabolism called DiGeorge syndrome have a defective double-stranded RNA binding protein. Characteristic symptoms of this syndrome include facial deformities (a cleft palate or lip), heart defects, and immunodeficiency. The Microprocessor complex measures the distance from the single-stranded RNA–double-stranded RNA junction in pri-miRNA before cleaving the pri-mRNA to release an approximately 70-nucleotide pre-miRNA hairpin. The pre-miRNA hairpins derived by the lariat debranching enzyme or by Microprocessor complex cleavage are transported through nuclear pores to the cytoplasm.

After entering the cytoplasm the pre-miRNA hairpins are cleaved on their terminal loop ends by a protein complex that contains Dicer-1 (Dcr-1) and another double-stranded RNA binding protein to produce approximately 22-nucleotide miRNA–miRNA* duplexes (Figure 15.11). The miRNA is usually loaded onto an Argonaute protein to form miRISC. Other proteins in miRISC, which remain to be characterized, are not shown in Figure 15.11. The miRNA guides the miRISC to a target mRNA as directed by base complementary.

miRISC has alternate modes of action that depend on its interaction with the target mRNA (FIGURE 15.12). When there is complete complementarity between the 22-nucleotide guide RNA and a region of the target mRNA, miRISC cleaves the target mRNA. When there is partial complementarity between the guide RNA and regions in the 3'-UTR, miRISC binds to the target mRNA's 3'-UTR and blocks translation by interfering with the action of translation initiation or elongation factors.

The plant miRNA synthetic pathway differs from the animal pathway in several important ways. Plants have a single RNase III, which cleaves both pri-miRNA and pre-miRNA in the nucleus. Then a methyltransferase adds a methyl group to the 3'-ends of the miRNA duplex, stabilizing the molecule. The methylated duplex is transported to the cytoplasm, where it is incorporated into RISC. Plant miRNAs

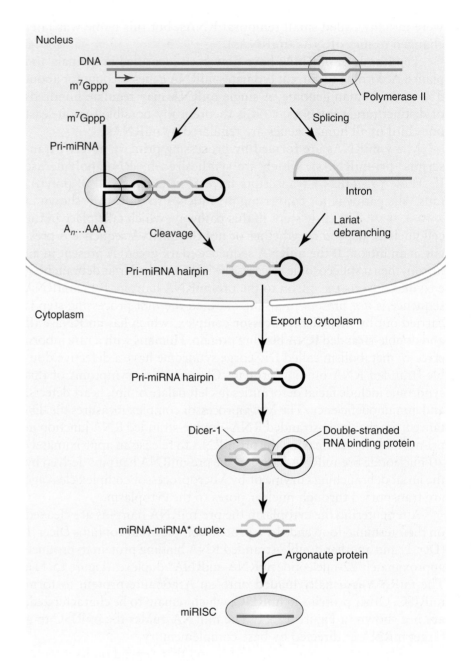

FIGURE 15.11 **Pathway for miRNA formation and action in *Drosophila*.** The early steps in this pathway, which take place in the cell nucleus, depend on whether or not the miRNA sequence is present in an intron. If the miRNA sequence (dark green) is present in an intron, then a spliceosome releases the intron and a lariat debranching enzyme converts the intron to the pre-miRNA hairpin. If the miRNA sequence is *not* present in an intron, then the first processing step is carried out by the Microprocessor complex, which has an RNase III and double-stranded RNA binding protein. The pre-miRNAs formed by the actions of the lariat debranching enzyme and the Microprocessor complex are transported through nuclear pores to the cytoplasm. Once in the cytoplasm the pre-miRNAs are cleaved on their terminal loop ends by a protein complex that contains Dicer-1 (Dcr-1) and a double-stranded RNA binding protein to produce approximately 22-nucleotide miRNA–miRNA* duplexes. The miRNA is usually loaded onto an Argonaute protein to form miRISC. Other proteins in miRISC, which remain to be characterized, are not shown. (Adapted from Ghildiyal, M., and Zamore, P. D. 2009. *Nat Rev Genet* 10:94–108.)

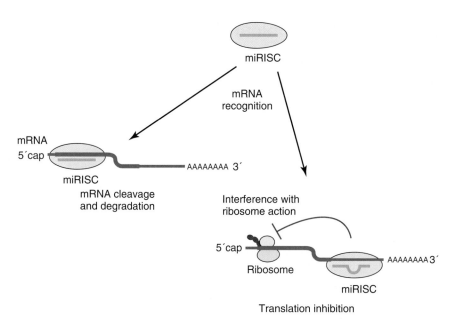

FIGURE 15.12 **Two modes of miRISC action.** When there is complete homology between the guide RNA and a region of the target mRNA, miRISC cleaves the target mRNA. When there is incomplete homology between the 22-nucleotide guide RNA and regions in the 3'-UTR, miRISC binds to the target mRNA's 3'-UTR and blocks translation.

usually are completely complementary to their target mRNA. Furthermore, plant miRNAs tend to bind to sequences in the mRNA's 5'-UTR or its protein-coding sequence rather than to its 3'-UTR. One other important difference between plant and animal miRISCs is that the plant ribonucleoprotein complex usually cleaves its target mRNA.

15.5 Piwi Interacting RNAs (piRNAs)

piRNAs help to maintain the germ line stability in animals.

The **Piwi interacting RNAs (piRNAs)** function in germ cells. They are about 25 to 30 nucleotides long and have 2'-O-methyl groups at their 3'-ends. More than 1.5 million distinct piRNAs have been identified in *Drosophila*. These fly piRNAs, which map to a few hundred clusters, are formed from long, single-stranded pre-RNAs. Further studies are required to elucidate the pathway for piRNA synthesis and to identify and characterize the enzymes involved in this pathway. As its name suggests, piRNAs bind to members of the Piwi subfamily of Argonaute proteins. Piwi function has been most thoroughly studied in *Drosophila*, where they appear to be confined to the nucleoplasm of germ cells and adjacent somatic cells. Piwi proteins help to ensure germ line stability by suppressing potentially harmful transposon mobility. Piwi probably performs the same function in vertebrates. Plant cells do not have Piwi proteins and have adapted a different RNAi-based strategy to control transposon activity.

Questions and Problems

1. Briefly define, describe, or illustrate each of the following.
   a. RNAi
   b. Argonaute protein family
   c. siRISC
   d. siRNA
   e. Dicer
   f. Passenger strand
   g. Guide strand
   h. RISC loading complex
   i. pre-RISC complex
   j. Transitive RNAi
   k. Functional genomics
   l. Reverse genetics
   m. Gene knockdown
   n. Interferon response
   o. miRNAs
   p. Microprocessor complex
   q. piRNA

2. What features do the roundworm *C. elegans* offer that makes it a useful model organism for molecular biology investigations?

3. Describe three methods that can be used to deliver long double-stranded RNA molecules into *C. elegans*.

4. Describe a genetic approach that can be taken to isolate *C. elegans* mutants that block RNAi.

5. What type of mutants are you likely to find when you search for RNAi mutants?

6. Investigators initially thought RNAi might be due to double-stranded RNA's ability to alter the target DNA, inhibit gene transcription, or somehow destabilize mRNA. Describe experiments they performed to distinguish among the three hypotheses. Which hypothesis best fits the data?

7. *Drosophila* embryo extracts were used to show that the RNAi pathway could be observed under *in vitro* conditions.
   a. Describe the experiments performed using luciferase mRNA as a reporter mRNA to demonstrate that RNAi can take place *in vitro*.
   b. Why were these experiments significant?

8. Describe an *in vitro* experiment that shows how long double-stranded RNA affects the target mRNA.

9. Draw the pathway leading from long double-stranded RNA to siRISC.
   a. Describe the roles Dicer and Argonaute protein play in this pathway.
   b. Describe *in vitro* studies that support the pathway.
   c. Describe genetic studies that support the pathway.

10. RNAi appears to have arisen in an early eukaryotic ancestor and is conserved in animals, plants, and most fungi. What type of selective pressures may account for this conservation?

11. RNAi in *C. elegans* can spread throughout the worm, even when triggered by a minute quantity of double-stranded RNA injected at a single site.
    a. What mechanism explains how the injection of a minute quantity of double-stranded RNA can have such a profound and lasting effect on the expression of a specific target gene?
    b. What evidence indicates that siRNAs are not primers for the synthesis of secondary siRNAs?
    c. Amplified secondary siRNAs associate with one of two secondary Argonaute proteins, SAGO-1 and SAGO-2. What is the major functional difference between the SAGO proteins and the Argonaute protein present in siRISC? What important consequence follows from this difference?

12. Describe how investigators can use RNAi to determine the function of a DNA sequence in *C. elegans* for which there is no known function.

13. What is the difference between gene knockout and gene knockdown? What advantages does gene knockdown offer over gene knockout for studying problems in reverse genetics?

14. Describe the experiments that resulted in the discovery of miRNA.

15. An investigator studying the development of *C. elegans* discovers a mutation that blocks a specific stage of development.
    a. Does this mutation have to be present in a specific protein-coding gene? Explain.
    b. If experiments demonstrate the mutation is not in a protein coding gene, what is another likely possibility?
16. Draw a diagram showing the animal pathway leading from the primary transcript formed by RNA polymerase II in the nucleus to the formation of miRISC in the cytoplasm.
17. How does the plant pathway for the synthesis of miRISC differ from the animal pathway?

Suggested Reading

Historical

Ruvkun, G., Wightman, B., and Ha, I. 2004. The 20 years it took to recognize the importance of tiny RNAs. *Cell* S116:s93–s96.

General Overview

Miller, J. 2011. *RNA Interference and Model Organisms*. Sudbury, MA: Jones and Bartlett Publishers.

Paddison, P. J., and Vogt, P. K. (eds). 2008. *RNA Interference*. Berlin and Heidelberg: Springer Verlag.

RNA Interference (RNAi) Triggered by Exogenous Double-Stranded RNA

Ban, J., Shaw, G., Tropea, J. E., et al. 2008. A stepwise model for double-stranded RNA processing by ribonuclease III. *Mol Microbiol* 67:143–154.

Bernstein, E., Caudy, A. A., Hammond, S. M., and Hannon, G. J. 2001. Role for a bidentate ribonuclease in the initiation step of RNA interference. *Nature* 409:363–366.

Boisvert, M.-E., and Simard, M. J. 2008. RNAi pathway in *C. elegans*: the argonautes and collaborators. *Curr Top Microbiol Immunol* 320:21–36.

Caplen, N. J., Parrish, S., Imani, F., et al. 2001. Specific inhibition of gene expression by small double-stranded RNAs in invertebrate and vertebrate systems. *Proc Natl Acad Sci USA* 98:9742–9747.

Elbashir, S. M., Harborth, J., Lendeckel, W., et al. 2001. Duplexes of 21-nucleotide RNAs mediate RNA interference in cultured mammalian cells. *Nature* 411:494–498.

Elbashir, S. M., Lendeckel, W., and Tuschl, T. 2001. RNA interference is mediated by 21- and 22-nucleotide RNAs. *Genes Dev* 15:188–200.

Ender, C., and Meister, G. 2010. Argonaute proteins at a glance. *J Cell Sci* 123:1819–1823.

Fire, A. Z. 2007. Gene silencing by double-stranded RNA. *Angew Chem Int Ed* 46:6966–6984.

Fire, A., Xu, S., Montgomery, M. K., et al. 1998. Potent and specific genetic interference by double-stranded RNA in *Caenorhabditis elegans*. *Nature* 391:806–811.

Fischer, S. E. J. 2010. Small RNA-mediated gene silencing pathways in *C. elegans*. *Int J Biochem Cell Biol* 42:1306–1315.

Ghildiyal, M., and Zamore, P. D. 2009. Small silencing RNAs: an expanding universe. *Nat Rev Genet* 10:94–108.

Hamilton, A. J., and Baulcombe, D. C. 1999. A species of small antisense RNA in posttranscriptional gene silencing in plants. *Science* 286:950–952.

Hammond, S. M., Bernstein, E., Beach, D., and Hannon, G. J. 2000. An RNA-directed nuclease mediates post-transcriptional gene silencing in *Drosophila* cells. *Nature* 404:293–296.

Hannon, G. J. 2002. RNA interference. *Nature* 418:244–251.

Höck, J., and Meister, G. 2008. The Argonaute protein family. *Genome Biol* 9:210.1–201.8.

Hutvagner, G., and Simard, M. 2008. Argonaute proteins: key players in RNA silencing. *Nat Rev Mol Cell Biol* 9:22–32.

Jaskiewicz, L., and Filipowicz, W. 2008. Role of Dicer in posttranscriptional RNA silencing. *Curr Top Microbiol Immunol* 320:77–97.

Jinek, M., and Doudna, J. A. 2009. A three-dimensional view of the molecular machinery of RNA interference. *Nature* 457:405–412.

Kawamata, T., and Tomari, Y. 2010. Making RISC. *Trends Biochem Sci* 35:368–376.

Liu, J., Carmell, M. A., Rivas, F. V., et al. 2004. Argonaute2 is the catalytic engine of mammalian RNAi. *Science* 305:1437–1441.

Liu, Q., and Paroo, Z. 2010. Biochemical principles of small RNA pathways. *Annu Rev Biochem* 79:295–319.

MacRae, I. J., and Doudna, J. A. 2007. Ribonuclease revisited: structural insights into ribonuclease III family enzymes. *Curr Opin Struct Biol* 17:1–8.

MacRae, I. J., Ma, E., Zhou, M., et al. 2008. In vitro reconstitution of the human RISC-loading complex. *Proc Natl Acad Sci USA* 105:512–517.

MacRae, I. J., Zhou, K., and Doudna, J. A. 2007. Structural determinants of RNA recognition and cleavage by Dicer. *Nat Struct Mol Biol* 14:934–940.

MacRae, I. J., Zhou, K., Li, F., et al. 2006. Structural basis for double-stranded RNA processing by Dicer. *Science* 131:195–198.

Mello, C. C. 2007. Return to the RNAi world: Rethinking gene expression and evolution. *Angew Chem Int Ed* 46:6985–6994.

Moazed, D. 2009. Small RNAs in transcriptional gene silencing and genome defence. *Nature* 457:413–420.

Montgomery, M. K., Xu, S., and Fire, A. 1998. RNA as a target of double-stranded RNA-mediated genetic interference in *Caenorhabditis elegans*. *Proc Natl Acad Sci USA* 95:15502–15507.

Naqvi, A. R., Islam, M. N., Choudhury, N. R., et al. 2009. The fascinating world of RNA interference. *Int J Biol Sci* 5:97–117.

Nowotny, M., and Yang, W. 2009. Structural and functional modules in RNA interference. *Curr Opin Struct Biol* 19:286–293.

Pare, J. M., and Hobman, T. C. 2007. Dicer: structure, function, and role in RNA-dependent gene-silencing pathways. J. M. Pare and T. C. Hobman (eds), *Industrial Enzymes* (pp. 421–438). Heidelberg: Springer Netherlands.

Parker, J. S. 2010. How to slice: snapshots of Argonaute in action. *Silence* 1:3.

Parker, J. S., and Barford, D. 2006. Argonaute: a scaffold for the function of short regulatory RNAs. *Trends Biochem Sci* 31:622–630.

Pratt, A. J., and MacRae, I. J. 2009. The RNA-induced silencing complex: a versatile gene-silencing machine. *J Biol Chem* 284:17897–17901.

Rana, T. M. 2007. Illuminating the silence: understanding the structure and function of small RNAs. *Nat Rev Mol Cell Biol* 8:23–36.

Sashital, D. G., and Doudna, J. A. 2010. Structural insights into RNA interference. *Curr Opin Struct Biol* 20:90–97.

Siomi, H., and Siomi, M. C. 2009. On the road to reading the RNA-interference code. *Nature* 457:396–404.

Tabara, H., Sarkissian, M., Kelly, W. G., et al. 1999. The *rde-1* gene, RNA interference, and transposon silencing in *C. elegans*. *Cell* 99:123–132.

Tolia, N. H., and Joshua-Tor, L. 2007. Slicer and argonautes. *Nat Chem Biol* 3:36–43.

Tuschl, T., Zamore, P. D., Lehmann, R., et al. 1999. Targeted mRNA degradation by double-stranded RNA in vitro. *Genes Dev* 13:1391–1397.

Van den Berg, A., Mols, J., and Han, J. 2008. RISC-target interaction: cleavage and translational suppression. *Biochim Biophys Acta* 1779:668–677.

Wang, Y., Juranek, S., Li, H., et al. 2008. Structure of an Argonaute silencing complex with a seed-containing guide DNA and target complex. *Nature* 456:921–926.

Wang, Y., Sheng, G., Juranek, S., et al. 2008. Structure of the guide-strand-containing Argonaute silencing complex. *Nature* 456:209–213.

Zamore, P. D., Tuschl, T., Sharp, P. A., and Bartel, D. P. 2000. RNAi: Double-stranded RNA directs the ATP-dependent cleavage of mRNA at 21 to 23 nucleotide intervals. *Cell* 101:25–33.

Transitive RNAi

Ahlquist, P. 2002. RNA-dependent RNA polymerases, viruses, and RNA silencing. *Science* 296:1270–1273.

Feinberg, E. H., and Hunter, C. P. 2003. Transport of dsRNA into cells by the transmembrane protein SID-1. *Science* 301:1545–1547.

Gent, J. I., Lamm, A. T., Pavalec, D. M., et al. 2010. Distinct phases of siRNA synthesis in an endogenous RNAi pathway in *C. elegans* soma. *Mol Cell* 37:679–689.

Jose, A. M., and Hunter, C. P. 2007. Transport of sequence-specific RNA interference information between cells. *Annu Rev Genet* 41:305–330.

Kennedy, S., Wang, D., and Ruvkun, G. 2004. A conserved siRNA-degrading RNase negatively regulates RNA interference in *C. elegans*. *Nature* 427:645–649.

Matranga, C., and Zamore, P. D. 2007. Small silencing RNAs. *Curr Biol* 17: R789–R793.

Pak, J., and Fire, A. 2007. Distinct populations of primary and secondary effectors during RNAi in *C. elegans*. *Science* 315:241–244.

Sijen, T., Fleenor, J., Simmer, F., et al. 2001. On the role of RNA amplification in dsRNA-triggered gene silencing. *Cell* 107:465–476.

Sijen, T., Steiner, F. A., Thijssen, K. L., and Plasterk, R. H. A. 2007. Secondary siRNAs result from unprimed RNA synthesis and form a distinct class. *Science* 315:244–247.

Talsky, K. B., and Collins, K. 2010. Initiation by a eukaryotic RNA-dependent RNA polymerase requires looping of the template end and is influenced by the template-tailing activity associated with uridyltransferase. *J Biol Chem* 285:27614–27623.

RNAi as an Investigational Tool

Ahringer, J., ed. 2006. Reverse genetics. *WormBook*. The *C. elegans Research Community*. Retrieved December 17, 2010, from http://www.wormbook.org

Kamath, R. S., and Ahringer, J. 2003. Genome-wide RNAi screening in *Caenorhabditis elegans*. *Methods* 30:313–321.

Kamath, R. S., Martinez-Campos, M., Zipperlen, P., Fraser, A. G., and Ahringer, J. 2001. Effectiveness of specific RNA-mediated interference through ingested double-stranded RNA in *Caenorhabditis elegans*. *Genome Biol* 2:1–10.

Rual, J. F., Ceron, J., Koreth, J., et al. 2004. Toward improving *Caenorhabditis elegans* phenome mapping with an ORFeome-based RNAi library. *Genome Res* 14:2162–2168.

MicroRNA Pathway

Bartel, D. P. 2009. MicroRNAs: target recognition and regulatory functions. *Cell* 136:215–233.

Breving, K., and Esquela-Kerscher, A. 2010. The complexities of microRNA regulation: mirandering around the rules. *Int J Biochem Cell Biol* 42:1316–1329.

Brodersen, P., and Voinnet, O. 2009. Revisiting the principles of microRNA target recognition and mode of action. *Nat Rev Mol Cell Biol* 10:141–148.

Bushati, N., and Cohen, S. M. 2007. Micro RNA functions. *Annu Rev Dev Biol* 23:175–205.

Carthew, R. W., and Sontheimer, E. J. 2009. Origins and mechanisms of miRNAs and siRNAs. *Cell* 136:642–655.

Davis, B. N., and Hata, A. 2009. Regulation of microRNA biogenesis: A miRiad of mechanisms. *Cell Commun Signal* 7:18.

Eulalio, A., Huntzinger, E., and Izaurralde, E. 2008. Getting to the root of miRNA-mediated gene silencing. *Cell* 132:9–14.

Faller, M., and Guo, F. 2008. MicroRNA biogenesis: there's more than one way to skin a cat. *Biochim Biophys Acta* 1779:663–667.

Filipowicz, W., Bhattacharyya, S. N., and Sonnenberg, N. 2008. Mechanisms of post-transcriptional regulation by micro-RNAs: are the answers in sight? *Nat Rev Genet* 9:102–114.

Garofalo, M., and Croce, C. M. 2011. microRNAs: Master regulators as potential therapeutics in cancer. *Annu Rev Pharmacol Toxicol* 51:25–43.

Grishok, A., Pasquinelli, A. E., Conte, D., et al. 2001. Genes and mechanisms related to RNA interference regulate expression of small temporal RNAs that control *C. elegans* developmental timing. *Cell* 108:23–34.

Horvitz, H. E. 2002. Worms, life, and death. *Biosci Rep* 23:239–269.

Inui, M., Martello, G., and Piccolo, S. 2010. MicroRNA control of signal transduction. *Nat Rev Cell Mol Biol* 11:252–263.

Kaufman, E. J., and Miska, E. A. 2010. The microRNAs of *Caenorhabditis elegans*. *Semin Cell Dev Biol* 21:728–737.

Kim, V. N., Han, J., and Siomi, M. C. 2009. Biogenesis of small RNAs in animals. *Nat Rev Cell Mol Biol* 10:126–139.

Lee, R. S., Feinbaum, R. L., and Ambros, V. 1993. The *C. elegans* heterochronic gene *lin-4* encodes small RNAs with antisense complementarity to *lin-14*. *Cell* 75:843–854.

Reinhart, B. J., Slack, F. J., Basson, M., et al. 2000. The 21-nucleotide *let-7* RNA regulates developmental timing in *Caenorhabditis elegans*. *Nature* 403:901–906.

Stefani, G., and Slack, F. J. 2008. Small non-coding RNAs in animal development. *Nat Rev Cell Mol Biol* 9:219–229.

Tang, G. 2005. siRNA and miRNA: an insight into RISCs. *Trends Biochem Sci* 30:106–114.

Voinnet, O. 2009. Origin, biogenesis, and activity of plant microRNAs. *Cell* 136:669–687.

Wrightman, B., Ha, I., and Rukun, G. 1993. Posttranscriptional regulation of the heterochronic gene *lin-14* by *lin-4* mediates temporal pattern formation in *C. elegans*. *Cell* 75:855–862.

Piwi Interacting RNAs (piRNAs)

Aravin, A., Hannon, G. J., and Brennecke, J. 2007. The Piwi-piRNA pathway provides an adaptive defense in the transposon arms race. *Science* 318:761–764.

Brennecke, J., Aravin, A. A., Stark, A., et al 2007. Discrete small RNA-generating loci as master regulators of transposon activity in *Drosophila*. *Cell* 128:1089–1103.

Faehnle, C. R., and Joshua-Tor, L. 2007. Argonautes confront new small RNAs. *Curr Opin Chem Biol* 11:569–577.

Farazi, T. A., Juranek, S. A., and Tuschl, T. 2008. The growing catalog of small RNAs and their association with distinct Argonaute/Piwi family members. *Develop* 135:1201–1214.

Senti, K-A., and Brennecke, J. 2010. The piRNA pathway: a fly's perspective on the guardian of the genome. *Trends Genet* 26:499–509.

Siomi, M. C., Manneu, T., and Siomi, H. 2010. How does the royal family of Tudor rule the PIWI-interacting RNA pathway? *Genes Dev* 24:636–646.

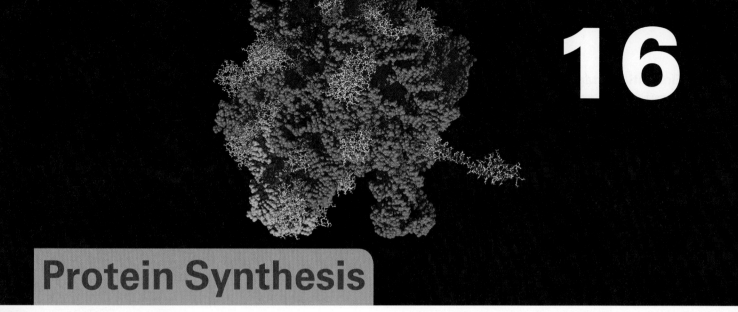

16 Protein Synthesis

CHAPTER OUTLINE

16.1 Introduction to the Ribosome
Protein synthesis takes place on ribosomes.

Bacterial ribosomes are made of a large subunit with a 23S and 5S RNA and a small subunit with 16S RNA.

A eukaryotic ribosome also has a small and a large ribonucleoprotein subunit.

Eukaryotic ribosomes exist free in the cytoplasm or attached to the endoplasmic reticulum.

16.2 Transfer RNA
An amino acid must be attached to a transfer RNA before it can be incorporated into a protein.

All tRNA molecules have CCA_{OH} at their 3'-ends.

An amino acid attaches to tRNA through an ester bond between the amino acid's carboxyl group and the 2'- or 3'-hydroxyl group on adenosine.

Yeast tRNA^{Ala} was the first naturally occurring nucleic acid to be sequenced.

tRNAs have cloverleaf secondary structures.

tRNA molecules fold into L-shaped three-dimensional structures.

16.3 Aminoacyl-tRNA Synthetases
Some aminoacyl-tRNA synthetases have proofreading functions.

Ile-tRNA synthetase has a proofreading function.

Ile-tRNA synthetase can hydrolyze valyl-tRNA^{Ile} and valyl-AMP.

Each aminoacyl-tRNA synthetase can distinguish its cognate tRNAs from all other tRNAs.

Selenocysteine and pyrrolysine are building blocks for polypeptides.

16.4 mRNA and the Genetic Code
mRNA programs ribosomes to synthesize proteins.

Three adjacent bases in mRNA that specify an amino acid are called a codon.

The discovery that poly(U) directs the synthesis of poly(Phe) was the first step in solving the genetic code.

Protein synthesis begins at the amino terminus and ends at the carboxyl terminus.

mRNA is read in a 5' to 3' direction.

Trinucleotides promote the binding of specific aminoacyl-tRNA molecules to ribosomes.

Synthetic messengers with strictly defined base sequences confirmed the genetic code.

Three codons, UAA, UAG, and UGA, are polypeptide chain termination signals.

The genetic code is nonoverlapping, commaless, almost universal, highly degenerate, and unambiguous.

The coding specificity of an aminoacyl-tRNA is determined by the tRNA and not the amino acid.

Some aminoacyl-tRNA molecules bind to more than one codon because there is some play or wobble in the third base of a codon.

16.5 Ribosome Structure
Bacterial 30S (small) subunits and 50S (large) subunits each have unique structures and functions.

Bacterial ribosome structure has been determined at atomic resolution.

16.6 Four Stages of Protein Synthesis
Protein synthesis can be divided into four stages.

16.7 Initiation Stage
Each bacterial mRNA open reading frame has its own start codon.

Bacteria have an initiator methionine tRNA and an elongator methionine tRNA.

The 30S subunit is an obligatory intermediate in polypeptide chain initiation.

Initiation factors participate in the formation of 30S and 70S initiation complexes.

The Shine-Dalgarno sequence in mRNA interacts with the anti-Shine-Dalgarno sequence in the 16S rRNA.

Eukaryotic initiator tRNA is charged with a methionine that is *not* formylated.

Eukaryotic translation initiation proceeds through a scanning mechanism.

Translation initiation factor phosphorylation regulates protein synthesis in eukaryotes.

The translation initiation pathway in archaea appears to be a mixture of the eukaryotic and bacterial pathways.

Image © Martin McCarthy/iStockphoto.

16.8 Elongation Stage

Polypeptide chain elongation requires elongation factors.

The elongation factors act through a repeating cycle.

An EF-Tu • GTP • aminoacyl-tRNA ternary complex carries the aminoacyl-tRNA to the ribosome.

Specific nucleotides in 16S rRNA are essential for sensing the codon–anticodon helix.

EF-Ts is a GDP-GTP exchange protein.

The ribosome is a ribozyme.

The hybrid-states translocation model offers a mechanism for moving tRNA molecules through the ribosome.

16.9 Termination Stage

Bacteria have three protein release factors.

Mutant tRNA molecules can suppress mutations that create termination codons within a reading frame.

16.10 Recycling Stage

The ribosome release factor is required for the bacterial ribosomal complex to disassemble.

16.11 Nascent Polypeptide Processing and Folding

Ribosomes have associated enzymes that process nascent polypeptides and chaperones that help to fold the nascent polypeptides.

16.12 Signal Sequence

The signal sequence plays an important role in directing newly synthesized proteins to specific cellular destinations.

BOX 16.1: LOOKING DEEPER Structural Basis for Posttransfer Editing

BOX 16.2: CLINICAL APPLICATION A Neurodegenerative Disease Caused by an Editing Defect in Alanyl-tRNA Synthetase

BOX 16.3: LOOKING DEEPER Bacterial Pathway for Selenocysteyl-tRNA^{Sec} Formation

BOX 16.4: LOOKING DEEPER Structures of Eukaryotic and Archaeal Ribosomes

BOX 16.5: LOOKING DEEPER Riboswitches

BOX 16.6: LOOKING DEEPER Translocation Mechanism

QUESTIONS AND PROBLEMS

SUGGESTED READING

Protein synthetic studies began even before DNA's structure was known, but these studies proceeded in a rather haphazard way because investigators lacked an intellectual framework to guide their work. The discovery that DNA consists of two polynucleotide strands arranged as a double helix allowed investigators to think about structure–function relationships for the first time.

Francis Crick did just that in 1957 when he proposed the **sequence hypothesis** and the **central dogma**. The sequence hypothesis postulates that the linear sequence of nucleotides in a DNA (or RNA) chain determines the linear sequence of amino acids in a polypeptide chain. Central dogma, which despite its name is a hypothesis, postulates that genetic information flows from DNA to RNA to protein but not from proteins to nucleic acids. Together, these two hypotheses helped provide the theoretical underpinning needed to make advances in the study of protein synthesis.

This chapter begins with a brief introduction to the ribosome's role as the protein synthetic factory and then examines information flow from messenger RNA (mRNA) to polypeptides. The final part of the chapter examines ribosome structure and function to understand how the ribosomes are able to act as universal translators. The primary focus is on bacterial ribosomes, which have been more thoroughly investigated than eukaryotic ribosomes. However, much of the information gained by studying bacterial ribosomes also applies to eukaryotic ribosomes because the two kinds of ribosomes have similar structures and work in the same way.

16.1 Introduction to the Ribosome

Protein synthesis takes place on ribosomes.

RNA's involvement in polypeptide synthesis, which is taken for granted today, was not firmly established until the 1950s. The earliest clues that RNA plays some role in information flow between genes and proteins came from cytochemical and cell fractionation studies performed in the late 1930s and early 1940s. The cytochemical studies showed the following:

1. Most of a eukaryotic cell's DNA and RNA are in its nucleus and cytoplasm, respectively.
2. Cells that are actively engaged in protein synthesis have high levels of cytoplasmic RNA.
3. Cytoplasmic RNA tends to be concentrated in small spherical particles.

Cell fractionation studies revealed a correlation between active protein synthesis and high cytoplasmic RNA levels. Furthermore, pellets obtained by high speed centrifugation of cytoplasmic extracts contained the RNA-rich spherical particles observed in the cytochemical studies. Because these RNA-rich particles contained tightly bound proteins, they are more properly described as ribonucleoproteins. By the early 1950s electron microscopy techniques had improved to the point where investigators could visualize the spherical ribonucleoprotein particles, which appeared to be about 250 Å in diameter. These ribonucleoprotein particles, or ribosomes, as they are now called, function as protein synthetic machines in all cells.

Bacterial ribosomes are made of a large subunit with a 23S and 5S RNA and a small subunit with 16S RNA.

A typical bacterial ribosome has a diameter of about 25 nm, a molecular mass of about 2.5×10^6 Da, and a sedimentation coefficient of 70S. It contains about twice the amount of RNA by mass as protein. Each bacterial cell requires about 70,000 ribosomes to produce the proteins it needs. Isolated bacterial ribosomes dissociate into large and small subunits when divalent cations are removed from the surrounding medium. Thus, a bacterial ribosome with a sedimentation coefficient of 70S dissociates into a 30S and a 50S subunit (FIGURE 16.1). Each *Escherichia coli* 30S subunit contains one 16S RNA (1541 nucleotides) and 21 polypeptide molecules (designated S1–S21). Each 50S subunit contains a 23S RNA (2904 nucleotides), a 5S RNA (120 nucleotides), and 36 polypeptide molecules (designated L1–L36). Bacterial structure and function are examined later in the chapter.

A eukaryotic ribosome also has a small and a large ribonucleoprotein subunit.

All eukaryotic cells have ribosomes in their cytoplasm and mitochondria. Plants also have ribosomes in their chloroplasts. Mitochondrial and chloroplast ribosomes differ from one another as well as from

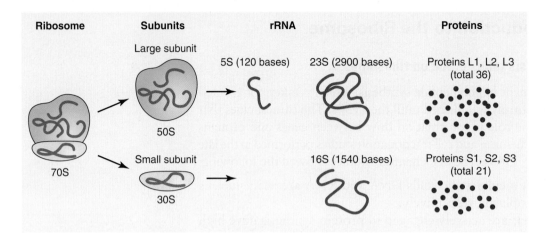

FIGURE 16.1 The bacterial ribosome and its components. (Adapted from an illustration by Kenneth G. Wilson, Department of Botany, Miami University.)

cytoplasmic ribosomes. Nevertheless, the three kinds of eukaryotic ribosomes and the bacterial ribosome work in much the same way. For simplicity we limit the present discussion to the cytoplasmic ribosome, which has a molecular mass of about 4×10^6 Da, a sedimentation coefficient of 80S, and consists of one small and one large subunit (FIGURE 16.2). The small (or 40S) subunit, which is fairly constant in size ($\sim 1.5 \times 10^6$ Da) in ribosomes from all eukaryotes, contains an 18S ribosomal RNA (rRNA) (~1,900 nucleotides) and about 33 different polypeptides. The large (or 60S) subunit varies in size from one species to another. For instance, large subunits in plants and mammals have molecular masses of 2.5×10^6 Da and 3.0×10^6 Da, respectively. Part of

FIGURE 16.2 Components of the eukaryotic 80S ribosome. The 80S ribosome is made of a large subunit that contains 5S rRNA, 5.8S rRNA, 28S rRNA, and about 50 different polypeptides and a small subunit that contains one 18S rRNA and about 33 different polypeptides. The large rRNA in animals, plants, and fungi vary in size. The 28S rRNA shown in this figure is for mammalian ribosomes.

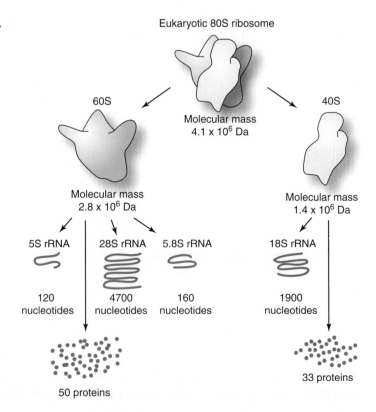

CHAPTER 16 PROTEIN SYNTHESIS

the reason for this variation is the largest of the three rRNA molecules in the large subunit varies in size. This rRNA is about 5,000 nucleotides long (sedimentation coefficient 28S) in mammals but only 3,400 nucleotides long (sedimentation coefficient 25S) in yeast. The two other rRNA components in the large subunit are the 5.8S rRNA (~160 nucleotides) and the 5S rRNA (~120 nucleotides). Large subunits from mammals also have about 50 polypeptide subunits, whereas those from lower eukaryotes have somewhat fewer polypeptides.

Eukaryotic ribosomes exist free in the cytoplasm or attached to the endoplasmic reticulum.

Some eukaryotic ribosomes appear to be free in the cytoplasm, whereas others are attached to the outer surface of a continuous intracellular tubular membrane network called the **endoplasmic reticulum** (FIGURE 16.3). Regions of the endoplasmic reticulum that are studded with ribosomes appear rough or grainy in electron micrographs and are therefore known as the **rough endoplasmic reticulum**. Regions that lack ribosomes have a smooth appearance and are therefore called the **smooth endoplasmic reticulum**. The endoplasmic reticulum cannot be isolated as a continuous membrane network because methods that disrupt the cell membrane also rupture the endoplasmic reticulum. Pieces of endoplasmic reticulum, however, can be isolated by differential centrifugation of a crude cell homogenate (FIGURE 16.4). The crude cell extract is centrifuged at about $600 \times g$ for 5 minutes to remove intact cells and nuclei, and then the postnuclear supernatant is centrifuged at about $15,000 \times g$ for 1 hour to remove mitochondria, lysosomes, and peroxisomes. Further centrifugation at $100,000 \times g$ for 2 hours produces a pellet called the **microsomal fraction** that contains pieces of the endoplasmic reticulum and free ribosomes.

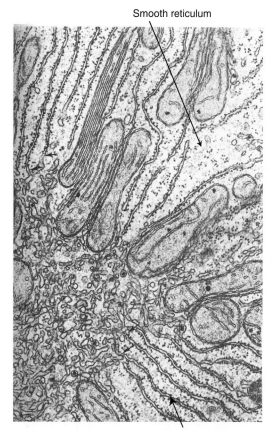

FIGURE 16.3 An electron micrograph of a liver cell showing RER (rough endoplasmic reticulum) and patches of SER (smooth endoplasmic reticulum). (© 1973, Rockefeller University Press. Originally published in *The Journal of Cell Biology.* 56: 746–761. Photo courtesy of Ewald R. Weibel, Universität Bern.)

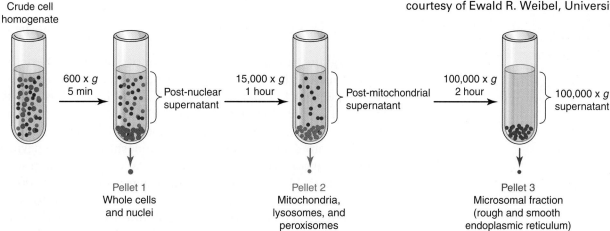

FIGURE 16.4 **Preparation of microsomal and $100,000 \times g$ supernatant fractions.** Components required for cell-free protein synthesis are present in the microsomal and $100,000 \times g$ supernatant fractions. These fractions are isolated by differential centrifugation of the crude cell extract. A low speed centrifugation to remove whole cells and nuclei from the crude cell extract is followed by an intermediate speed centrifugation to remove mitochondria, lysosomes, and peroxisomes. Then the postmitochondrial supernatant is centrifuged at high speeds to produce a pellet (microsomal fraction) and a supernatant ($100,000 \times g$ supernatant). The microsomal fraction contains the cell's rough and smooth endoplasmic reticulum, whereas the $100,000 \times g$ supernatant contains the cell's soluble cytoplasmic components.

16.1 Introduction to the Ribosome

In 1953 Paul Zamecnik and coworkers provided the first convincing evidence that the microsomal fraction makes an important contribution to protein synthesis. Their approach was to inject rats with radioactive amino acids, kill groups of rats at intervals, remove the livers, disrupt the fresh livers, fractionate disrupted liver cells, and determine the amount of radioactive protein that was present in the different cellular fractions. They observed that radioactive proteins first appeared in ribosomes associated with the microsomal fraction, suggesting ribosomes are protein synthetic factories.

Building on these *in vivo* studies, Zamecnik and coworkers worked to develop a cell-free system that would permit them to characterize components of the protein synthetic system and learn how the components work. Their efforts were rewarded by the discovery that a suspension containing rat liver ribosomes, the $100,000 \times g$ supernatant, adenosine triphosphate (ATP), and guanosine triphosphate (GTP) converts acid-soluble [^{14}C]amino acids into acid-insoluble radioactive proteins.

16.2 Transfer RNA

An amino acid must be attached to a transfer RNA before it can be incorporated into a protein.

The next major advance in the study of protein synthesis was made by Mahlon Hoagland while working in Zamecnik's laboratory in 1953. Hoagland's initial objective was to learn how energy is provided for protein synthesis. He thought amino acids must be activated before they can combine to form proteins and suspected this activation was through adenylyl (AMP) group transfer from ATP to α-carboxyl groups on the amino acids.

$$\text{ATP} + \text{amino acid} \rightleftarrows \text{aminoacyl-AMP} + \text{pyrophosphate}$$

Hoagland sought evidence to support this hypothesis by testing various cell fractions to see which, if any, could catalyze amino acid–dependent [^{32}P]pyrophosphate–ATP exchange. His efforts were rewarded by the discovery that the $100,000 \times g$ rat liver supernatant catalyzes vigorous amino acid–dependent [^{32}P]pyrophosphate incorporation into ATP, suggesting the cytoplasm contains enzymes that activate amino acids by reversible adenylylation of the amino acids' carboxyl groups. Hoagland's studies thus suggested that the $100,000 \times g$ supernatant can activate an amino acid by converting it to an aminoacyl-adenylate. Hoagland's preliminary efforts to fractionate proteins in the $100,000 \times g$ supernatant indicated that each amino acid is probably activated by an enzyme that is specific for it. This conclusion was later verified when amino acid activating enzymes were purified and each purified enzyme was shown to be specific for a single amino acid.

Hoagland's experiments tell us how amino acids are activated, but they tell us nothing about how genetic information is transferred from a polynucleotide chain to a polypeptide chain. Zamecnik and coworkers discovered an important piece to the genetic information puzzle when they observed the 100,000 × g supernatant catalyzes the ATP-dependent attachment of [^{14}C]amino acids to soluble RNA molecules in the supernatant. Although Zamecnik knew the 100,000 × g supernatant contains soluble RNA, he, like other investigators at the time, mistakenly assumed this RNA was "junk" produced when ribosomes are degraded. The experiments performed by Zamecnik and coworkers showed that far from being junk, soluble RNA molecules are natural acceptors for activated aminoacyl groups and therefore likely to be essential protein synthetic machinery components.

Zamecnik, Hoagland, and coworkers then performed decisive experiments showing that an amino acid must be attached to soluble RNA before it can be incorporated into proteins. They began by taking advantage of their previous studies to synthesize [^{14}C]aminoacyl-soluble RNA under *in vitro* conditions. Then they added purified [^{14}C] aminoacyl-soluble RNA to a protein synthetic mixture containing 100,000 × g supernatant, ribosomes, ATP, and GTP and monitored radioactive incorporation into protein. As expected, the radioactive label was incorporated into protein. Moreover, different radioactive amino acids were incorporated in an additive fashion. But Zamecnik, Hoagland, and coworkers still needed to rule out the possibility that [^{14}C]amino acids were first released from soluble RNA and then incorporated into protein. They did so by demonstrating that radioactive incorporation is not affected by adding unlabeled amino acids to the reaction mixture. If [^{14}C]amino acids were released from soluble RNA before incorporation into protein, then the unlabeled amino acids would have lowered their specific activity (microcuries per micromole) and caused a marked decrease in radioactive amino acid incorporation. Thus, amino acids attached to soluble RNA are directly transferred to growing polypeptide chains on ribosomes. The term "soluble RNA," indicating a physical property, was eventually replaced by **transfer RNA (tRNA)**, indicating a function.

Building on their pioneering studies, Zamecnik, Hoagland, and coworkers showed the activating enzymes, which they originally detected in the 100,000 × g supernatant, are also responsible for aminoacyl-tRNA synthesis. For this reason these activating enzymes are now called aminoacyl-tRNA synthetases or aminoacyl-tRNA ligases. Although each aminoacyl-tRNA synthetase is specific for the amino acid and tRNA it acts on, all the synthetases catalyze the same two-step reaction pathway (FIGURE 16.5). First, the synthetase catalyzes adenylyl (AMP) group transfer from ATP to the amino acid's α-carboxyl group to form aminoacyl-adenylate and pyrophosphate (Figure 16.3a). Then the synthetase catalyzes aminoacyl group transfer from aminoacyl-adenylate to tRNA to form aminoacyl-tRNA and AMP (Figure 16.3b). The net effect of the two transfer reactions is to convert ATP, amino acid, and tRNA into aminoacyl-tRNA, AMP, and pyrophosphate (Figure 16.3c). tRNA molecules with bound amino acids are said to be charged, whereas those lacking amino acids are said to be uncharged. Although the equilibrium constant for the net reaction is close to 1, the reaction proceeds to completion in the cell because pyrophosphate is hydrolyzed.

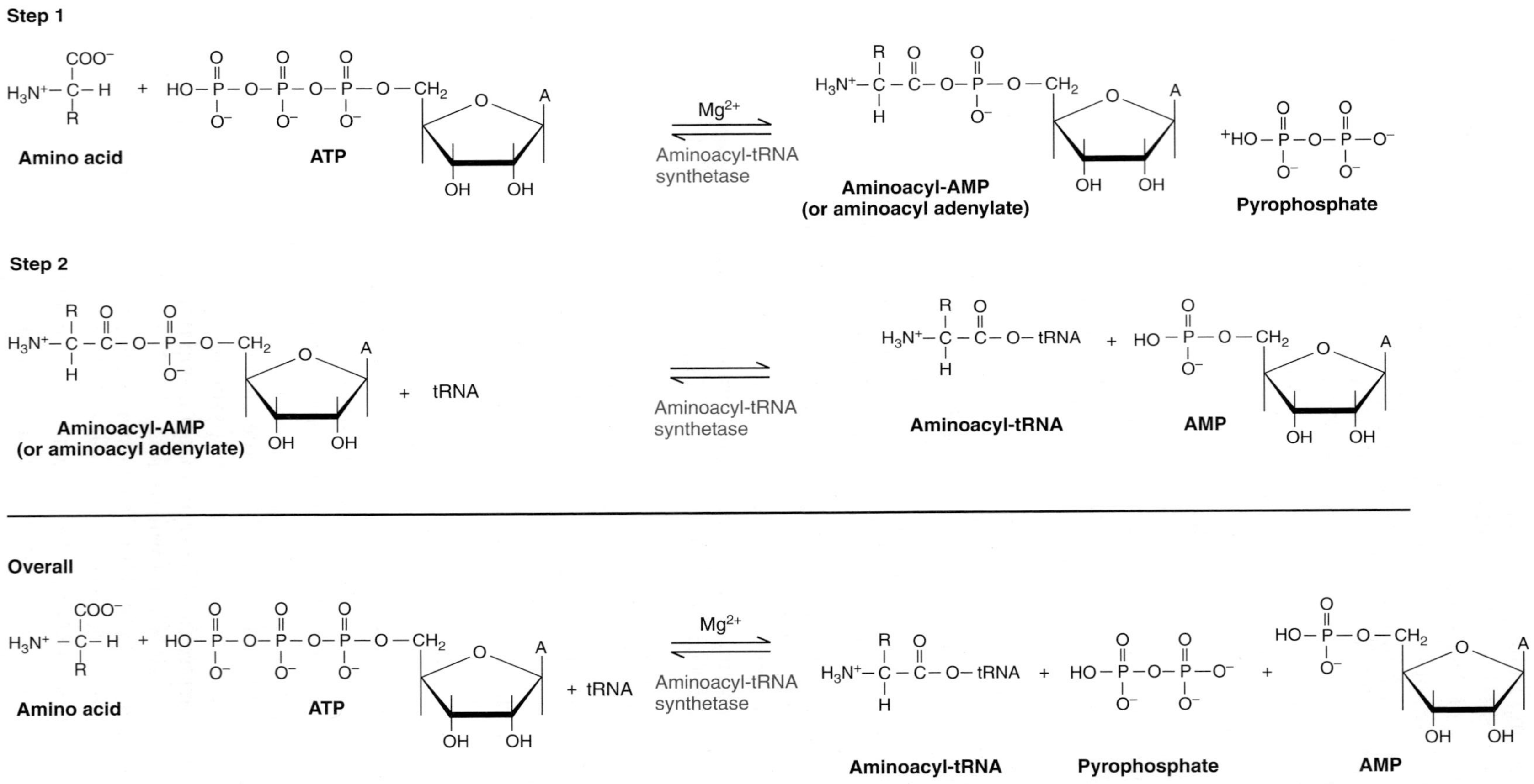

Step 1

Amino acid + ATP ⇌ (Aminoacyl-tRNA synthetase, Mg^{2+}) Aminoacyl-AMP (or aminoacyl adenylate) + Pyrophosphate

Step 2

Aminoacyl-AMP (or aminoacyl adenylate) + tRNA ⇌ (Aminoacyl-tRNA synthetase) Aminoacyl-tRNA + AMP

Overall

Amino acid + ATP + tRNA ⇌ (Aminoacyl-tRNA synthetase, Mg^{2+}) Aminoacyl-tRNA + AMP + Pyrophosphate

FIGURE 16.5 Reactions catalyzed by aminoacyl-tRNA synthetases.

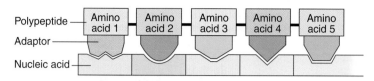

FIGURE 16.6 **The Adaptor Hypothesis.** (Adapted from Voet, D., and Voet, J. G. 2005. *Biochemistry* (3rd ed). John Wiley & Sons.)

At the start of their studies Zamecnik, Hoagland, and coworkers viewed unraveling the protein synthetic pathway as a biochemical problem. By 1956, however, their studies on tRNA had advanced to a stage where they were obliged to start thinking about protein synthesis in terms of the flow of genetic information. Because mRNA had not yet been discovered, the relationship between tRNA and information flow was not obvious. Remarkably, Francis Crick had predicted the existence of a molecule resembling tRNA in an informal note he sent in 1955 to a small group of molecular biologists known as the "RNA Tie Club" (so-named because each club member had a necktie decorated with a specific amino acid or nucleotide). One of the club's major goals was to solve the coding problem. Crick's note called attention to the fact that free amino acids cannot directly line up on a nucleic acid template before polypeptide formation. He therefore proposed that each amino acid must first combine with a short specific adaptor RNA, which then can line up on an RNA template by forming specific base pairs with it (FIGURE 16.6).

Crick's adaptor hypothesis makes two important predictions: (1) there must be at least 20 different adaptor RNA molecules, one for each amino acid, and (2) an amino acid must attach to its cognate (correct) adaptor before it can be incorporated into protein. Zamecnik and Hoagland were not aware of Crick's informal note. However, when James Watson told them about Crick's adaptor hypothesis in late 1956, they immediately realized that tRNA might be Crick's adaptor.

All tRNA molecules have CCA$_{OH}$ at their 3′-ends.

tRNA was such an obviously important molecule that many different laboratories became interested in it. One of the first clues to tRNA structure was the observation that the 100,000 × *g* supernatant contains an exonuclease that cleaves a few exposed nucleotides from the 3′-end of tRNA, rendering the tRNA unable to accept amino acids (FIGURE 16.7a). Further studies showed the 100,000 × *g* supernatant also contains an enzyme that can restore the digested tRNA molecule's ability to accept amino acids when ATP and CTP are added to the mixture (Figure 16.7b). More careful examination of this restoration led to the discovery that all tRNA molecules have CCA$_{OH}$ at their 3′-ends. The key experiments were as follows:

1. [^{14}C]CTP can transfer its CMP group to the 3′-end of the digested tRNA in the absence of ATP.
2. CTP must be present for [^{14}C]ATP to transfer its AMP group.
3. The molar ratio of CMP:AMP transfer is 2:1.

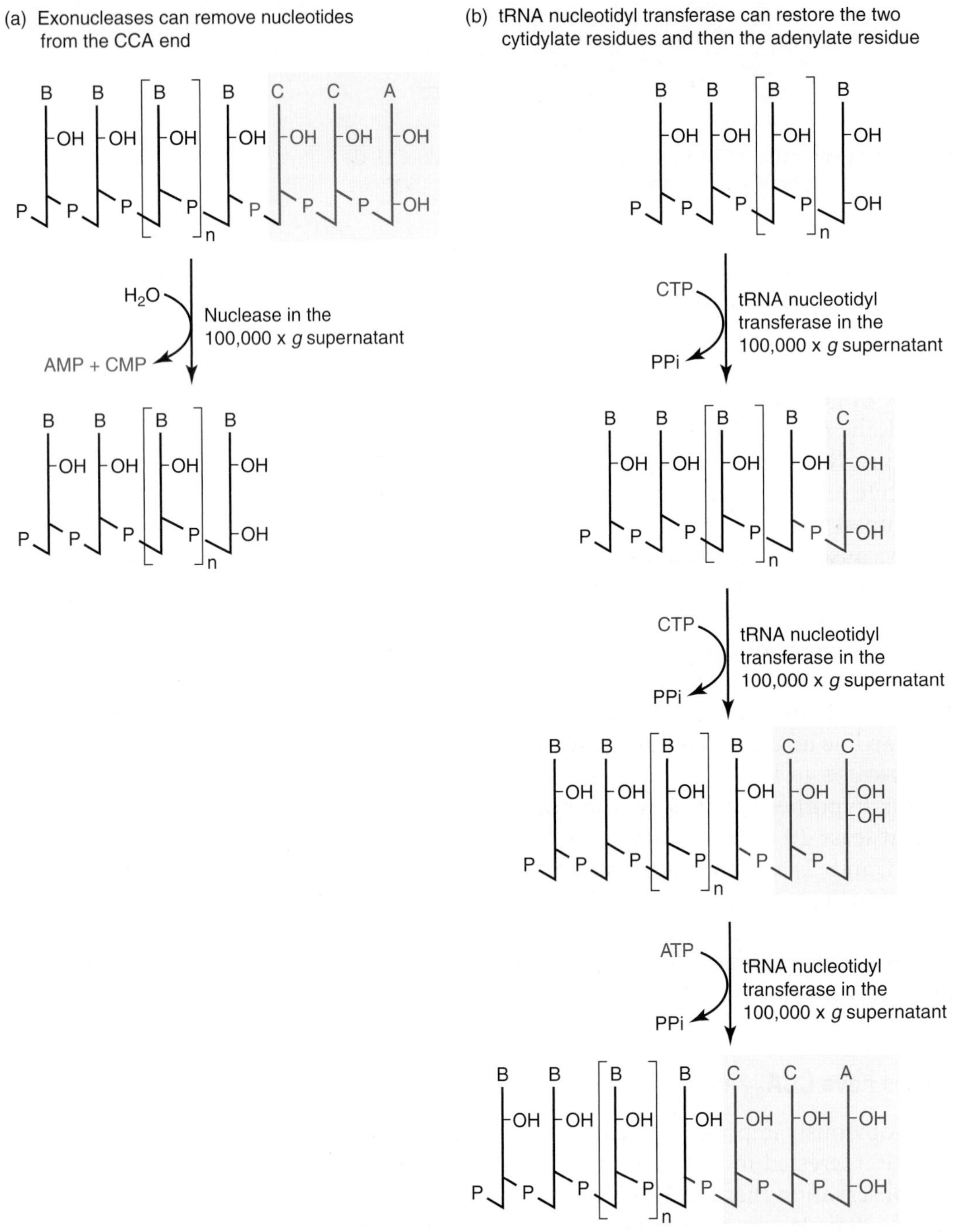

FIGURE 16.7 Reactions at the 3′-end of tRNA. The sequence at the 3′-end of all tRNA molecules is CCA. (a) Exonucleases can inactivate tRNA by removing the CCA end. (b) The enzyme known as tRNA nucleotidyl transferase restores activity by first restoring the two cytidylate residues and then the adenylate residue.

CHAPTER 16 PROTEIN SYNTHESIS

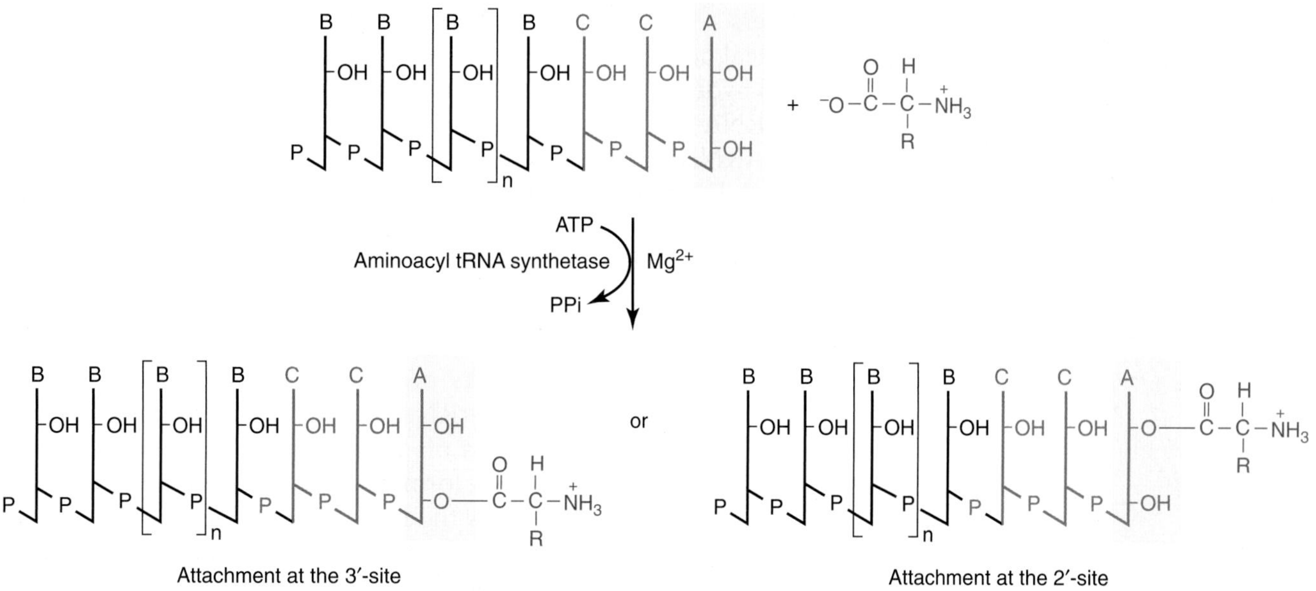

FIGURE 16.8 Site of aminoacyl group attachment to tRNA. Aminoacyl-tRNA synthetases attach aminoacyl groups to the 2′- or 3′-hydroxyl groups of the 3′-terminal adenosine group.

Later studies showed a single enzyme called tRNA nucleotidyl transferase catalyzes all three nucleotide transfer reactions. This enzyme is essential in eukaryotes because most eukaryotic tRNA genes do not code for the CCA$_{OH}$ end. Although bacterial tRNA genes usually do code for the CCA$_{OH}$ end, bacteria require tRNA nucleotidyl transferase to repair tRNA molecules that have lost their CCA$_{OH}$ end.

An amino acid attaches to tRNA through an ester bond between the amino acid's carboxyl group and the 2′- or 3′-hydroxyl group on adenosine.

Paul Berg and coworkers explored the nature of aminoacyl group attachment to bacterial tRNA. They first demonstrated that periodate (IO_4^-), a reagent that specifically cleaves the bond between carbon atoms attached to the 2′- and 3′-hydroxyl groups in the 3′-terminal adenosine, completely inactivates tRNA. Then they showed tRNA loses its ability to accept amino acids much more rapidly when digested with a 3′→5′ exonuclease than when digested with a 5′→3′ exonuclease.

Based on these results it seemed likely the aminoacyl group is attached to the 3′-terminal adenosine in tRNA by an ester bond between the α-carboxyl on the aminoacyl group and the 2′- or 3′-hydroxyl group of the adenosine (**FIGURE 16.8**). If the aminoacyl group is so attached, then pancreatic RNase digestion of aminoacyl-tRNA should release aminoacyl-adenosine (**FIGURE 16.9**). As predicted, RNase digestion does release aminoacyl-adenosine. Moreover, periodate has no effect on the released aminoacyl-adenosine, indicating the amino acid is attached to the 2′- or 3′-hydroxyl group of adenosine. Some aminoacyl-tRNA synthetases add the aminoacyl group to the 2′-hydroxyl, whereas

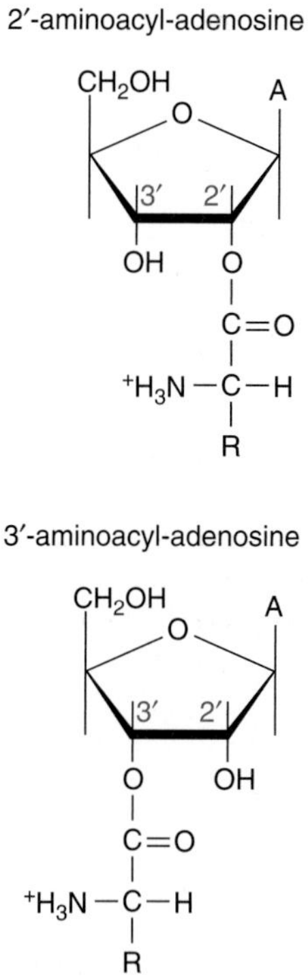

FIGURE 16.9 Aminoacyl-adenosine.

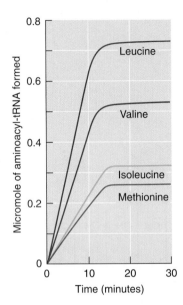

FIGURE 16.10 Kinetics of aminoacyl-tRNA formation. Each reaction mixture contains ATP, magnesium ions, the radioactive amino acid indicated, and the purified *E. coli* aminoacyl tRNA synthetase specific for that amino acid. (Adapted from Berg, P. et al. 1961. *J Biol Chem* 236:1726–1734.)

others add it to the 3′-hydroxyl. The original attachment site is not critical because an aminoacyl group can migrate rapidly between the two hydroxyl groups. The ester bond attaching the aminoacyl group to tRNA is rapidly hydrolyzed under alkaline conditions. For example, Val-tRNA has a half-life of about 5 minutes at pH 8.6 at 37°C. This instability permits aminoacyl groups to be removed under mild alkaline conditions that leave tRNA intact.

Crick's adaptor hypothesis predicts that each amino acid has at least one tRNA specific for it. Two different kinds of experiments confirmed this prediction:

1. Berg and coworkers performed *in vitro* experiments that showed aminoacyl-tRNA formation proceeds linearly with time and then levels off (FIGURE 16.10). Although the extent of aminoacyl-tRNA formation varied from one amino acid to another, tRNA was the limiting factor for each amino acid. When two or more amino acids were included in the same reaction mixture, the total amount of aminoacyl-tRNA formed equaled the sum of the amounts formed when each amino acid was present by itself. Thus, the presence of one amino acid did not affect the extent of incorporation of other amino acids. These experiments are consistent with Crick's prediction that each amino acid attaches to tRNA(s) specific for it.

2. Robert Holley and coworkers purified specific tRNA molecules from a mixture of tRNAs isolated from baker's yeast (*Saccharomyces cerevisiae*), which was chosen because it is available in large amounts at low cost. They purified specific tRNAs from the crude mix using a form of liquid-liquid chromatography known as countercurrent distribution, which separates components from a mixture by repeated distribution between two immiscible liquid phases moving past one another in opposite directions. As predicted by Crick's adaptor hypothesis, each tRNA species accepted only one kind of amino acid. tRNA specificity is indicated by writing the amino acid's three-letter abbreviation as a superscript. For instance, tRNA molecules that accept leucine, valine, and phenylalanine are designated $tRNA^{Leu}$, $tRNA^{Val}$, and $tRNA^{Phe}$, respectively. Some amino acids can attach to more than one tRNA species. Different tRNAs that accept the same amino acid, called **isoacceptors**, are differentiated by adding a number as a subscript. For example, isoacceptors for valine are named $tRNA^{Val}_1$ and $tRNA^{Val}_2$.

DNA sequence studies for entire bacterial genomes show the number of tRNA genes varies from a low of 33 in *Mycoplasma genitalium* (the bacteria thought to have the smallest genome) to a high of 88 in *Bacillus subtilis* (a gram-positive bacteria). Because some tRNA molecules are encoded by more than one tRNA gene, the number of different tRNA molecules is often less than the number of tRNA genes. For instance, *E. coli* has 84 tRNA genes but only 45 different tRNA molecules. Eukaryotes have a much larger number of genes for cytoplasmic tRNA. For example, *S. cerevisiae* has about 270 tRNA genes, whereas humans have about twice that number.

Yeast tRNA^Ala was the first naturally occurring nucleic acid to be sequenced.

Holley and coworkers started the ambitious and difficult task of determining the nucleotide sequence of a tRNA molecule in 1958, a project that took 7 years to complete. Most of the time was spent developing techniques to purify a single species of yeast tRNA^Ala. More than 600 kg of commercial baker's yeast was processed to obtain 200 g of crude tRNA from which 1 g of pure tRNA^Ala was finally isolated. Samples of the purified tRNA were cut with specific nucleases and the resulting fragments isolated and sequenced. Once this arduous task was completed, Holley and coworkers determined the order of fragments within the tRNA molecule by matching overlaps in much the same way that protein chemists determine the sequence of large polypeptides.

Today, it is usually more convenient to sequence the tRNA gene rather than the tRNA. When necessary, however, tRNA can be isolated and sequenced in a matter of weeks rather than years. One of the major reasons for this improvement is a change in the methods used to detect nucleotides. The spectrophotometric methods used by Holley have been replaced by much more sensitive radiolabeling techniques. Consequently, microgram quantities of pure tRNA are sufficient to obtain a complete nucleotide sequence.

tRNAs have cloverleaf secondary structures.

Holley and coworkers examined the nucleotide sequence of tRNA^Ala, hoping to find clues to polyribonucleotide chain folding. They assumed intramolecular base pair formation was the driving force for the folding process. Self-complementary sequences that could form hairpin structures were of special interest. Several alternative folding patterns, each with about the same number of intramolecular base pairs, could be envisioned. The correct folding pattern, however, could not be determined from the sequence of a single tRNA species. As sequence information for other tRNA molecules became available, it became apparent that the only secondary structure common to all was one in which the polynucleotide chain folds into a structure that resembles a three-leaf clover (FIGURE 16.11). Several hundred different tRNA species have now been sequenced from a wide variety of organisms. All cytoplasmic tRNAs are approximately 75 to 95 nucleotides long and have a cloverleaf secondary structure. As shown in Figure 16.11, 15 nucleotides within tRNAs are highly conserved.

In addition to the four major ribonucleosides, each tRNA also contains modified nucleosides. All modifications are introduced after transcription. These modifications include adding one or more methyl groups, reducing a double bond, changing the attachment site to a pyrimidine ring, replacing an oxygen atom with a sulfur atom, and adding a large substituent. Some nucleosides have more than one type of modification. Structures of several of the more than 100 modified nucleosides identified in tRNA are shown in FIGURE 16.12. Three of the uridine derivatives shown—ribothymidine (rT), pseudouridine (ψ),

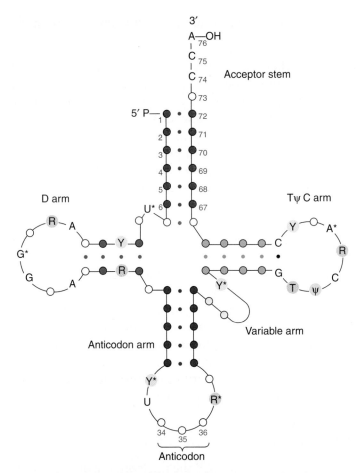

FIGURE 16.11 **The cloverleaf secondary structure.** Invariant bases are indicated by the first letter of their name. Invariant pyrimidines and purines are indicated by R and Y, respectively. The symbol ψ represents pseudouridine. Bases involved in Watson-Crick base pairs are represented by filled circles connected by dots. Other bases are represented by open circles. The D- and variable arms contain different numbers of nucleotides in the various tRNAs. The TψC-, anticodon-, and D-arms each contain a double-stranded stem structure and a loop structure. (Adapted from Voet, D., and Voet, J. G. 2005. *Biochemistry* (3rd ed). John Wiley & Sons.)

and dihydrouridine—are present in virtually all tRNAs. Ribothymidine is formed by methyl group transfer from S-adenosylmethionine to a specific uridine. Pseudouridine is formed by uracil ring rotation so C-5 rather than N-1 is linked to ribose. Dihydrouridine is formed by adding hydrogen atoms across the double bond between carbons 5 and 6 in uracil. Other modified nucleosides are present only at specific sites on a limited number of tRNAs. Many modified nucleosides help the tRNA to perform its function as an adaptor, in some cases assisting an aminoacyl-tRNA synthetase to distinguish between similar tRNA molecules and in others helping the tRNA to interact with mRNA.

The tRNA cloverleaf structure has five arms (Figure 16.11), each named for its function or unique chemical characteristic.

1. The **acceptor stem**, the site of amino acid attachment to tRNA, is formed by base pairing between nucleotides at the 5′-end and nucleotides near the 3′-end of the tRNA molecule. Nearly

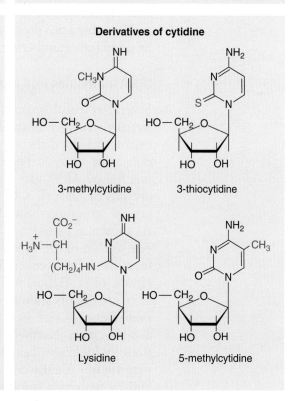

FIGURE 16.12 **Some of the modified nucleosides in tRNA.**

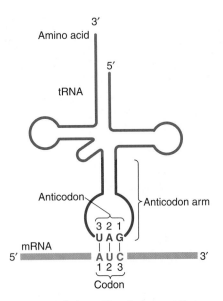

FIGURE 16.13 Interaction between the anticodon on tRNA and the codon on mRNA. Base pairing between the anticodon on a tRNA (magenta) and the codon on mRNA (green) is antiparallel. (Adapted from Nelson, D. C., and Cox, M. M. 2000. *Lehninger Principles of Biochemistry* (3rd ed). W. H. Freeman & Company.)

all acceptor stems are stabilized by seven base pairs. The four nucleotides at the 3'-end, including 3'-CCA$_{OH}$, are not base paired.

2. The **anticodon arm** is a stem-loop structure that lies directly across from the acceptor stem and contains the **anticodon** (a trinucleotide) that base pairs with a **codon** (also a trinucleotide) in mRNA. Base pairing between the complementary trinucleotides takes place in an antiparallel fashion (FIGURE 16.13).
3. The **D-arm** is a stem-loop structure named for one or more dihydrouridine groups that are almost always present within it. The number of nucleotides in the D-arm varies from one tRNA to another.
4. The **TψC arm** is a stem-loop structure named for the TψC sequence present within the loop.
5. The **variable arm**, which lies between the anticodon and TψC stems, contains 4 to 21 residues. Variations here are largely responsible for size differences among the tRNA molecules.

The acceptor stem, anticodon arm, and TψC arm are so conserved in length their residues are usually indicated by a standard numbering system. The first nucleotide at the 5'-terminus is residue 1 and the last residue at the 3'-terminus is residue 76 (Figure 16.11). Adenosine 76 (A76) (in the 5'-CCA$_{OH}$ terminus) is the amino acid attachment site and nucleotides 34 to 36 form the anticodon. Nucleotides in the D- and variable arms present in some tRNAs but not others are indicated by a separate numbering system.

tRNA molecules fold into L-shaped three-dimensional structures.

Several research groups initiated programs to determine tRNA's tertiary structure in the late 1960s. Progress was quite slow at first because of the difficulty in obtaining crystals satisfactory for x-ray diffraction analysis. Finally, in 1974 two independent research groups, one led by Alexander Rich and the other by Aaron Klug, obtained crystals of yeast tRNA^{Phe} suitable for x-ray diffraction analysis. The crystal structure showed that the cloverleaf folds into an L-shape. As shown in FIGURE 16.14, the D- and anticodon arms stack to form one section of the L, while the acceptor stem and TψC arm stack to form the other section. The tertiary structure for tRNA is shown in FIGURE 16.15. Helical regions in the acceptor, anticodon, D-, and TψC arms are usually stabilized by Watson-Crick base pairing as well as by non–Watson-Crick base pairing such as the G:U pair in the tRNA^{Phe} acceptor arm. Nonhelical regions of the tRNA molecule are stabilized by hydrogen bonding interactions between two or three bases that are not usually considered to be complementary to one another and by hydrogen bonding interactions that involve the 2'-hydroxyl group in ribose. 2'-Hydroxyl interactions are especially interesting because they cannot occur in DNA molecules. Different tRNAs have a similar folding pattern, ensuring various components of the protein synthetic machinery can recognize the tRNA after an amino acid has been attached to it. However, tRNAs also have unique features that are recognized by cognate aminoacyl-tRNA synthetases.

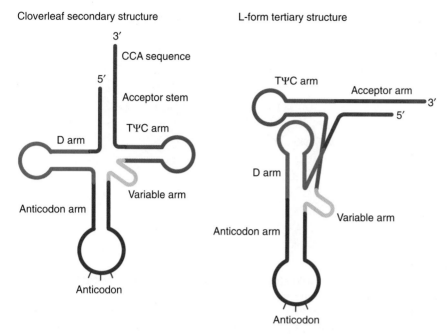

FIGURE 16.14 Folding of cloverleaf structure of tRNA^Phe to produce the L-form.

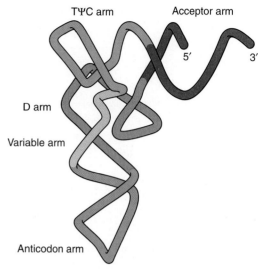

FIGURE 16.15 Three-dimensional structure of tRNA. (Modified from Arnez, J. G., and Moras, D. 1997. *Trends Biochem Sci* 22:211–216. Copyright 1997, with permission from Elsevier [http://www.sciencedirect.com/science/journal/09680004].)

16.3 Aminoacyl-tRNA Synthetases

Some aminoacyl-tRNA synthetases have proofreading functions.

Each aminoacyl-tRNA synthetase must be able to select a single amino acid from a mixture of many others. As early as 1957 Linus Pauling pointed out that a synthetase should make many errors when trying to distinguish between two amino acids such as valine and isoleucine, which have very similar side chains. The error rate can be calculated by comparing the enzyme's binding energy for each amino acid. The calculated error rate for two amino acids that differ in just one $-CH_2-$ group is 1 in 200, based on a binding energy difference of 3.5 kcal mol^{-1}. The fact that this calculated error rate is considerably higher than the experimentally observed error rate, approximately 1 in 10^4 to 1 in 10^5, suggests that some aminoacyl-tRNA synthetases have a method for correcting mistakes. Subsequent studies revealed that several aminoacyl-tRNA synthetases have proofreading activity. We now examine the best studied of these enzymes, the Ile-tRNA synthetase.

Ile-tRNA synthetase has a proofreading function.

The first clear evidence that *E. coli* Ile-tRNA synthetase has a proofreading or editing function was provided by experiments performed by Ann Norris Baldwin and Paul Berg in 1966. Ile-tRNA synthetase catalyzes valyl-AMP formation at approximately 0.5% the rate it catalyzes isoleucyl-AMP formation. Moreover, the enzyme forms a stable complex with aminoacyl-AMP provided tRNA^Ile is not present. Baldwin and Berg took advantage of this stability to isolate each

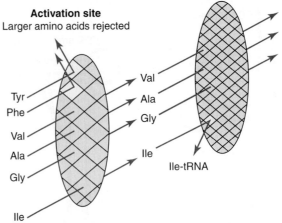

FIGURE 16.16 **The double-sieve mechanism as applied to Ile-tRNA synthetase.** Ile-tRNA synthetase has an aminoacylation site that serves as a "coarse sieve" to exclude amino acids larger than the cognate L-isoleucine and an editing site that serves as a "fine sieve" to eliminate amino acids smaller than isoleucine such as valine. (Modified from Fersht, A. 1998. *Science* 280:541. Reprinted with permission from AAAS.)

enzyme • aminoacyl-AMP complex by molecular sieve chromatography and then tested the effect that tRNA^{Ile} had on each complex. Ile-tRNA^{Ile} was formed when tRNA^{Ile} was added to the enzyme • isoleucyl-AMP complex, completing the aminoacylation reaction. An entirely different kind of result was obtained when tRNA^{Ile} was added to the enzyme • valyl-AMP complex. Instead of transferring the valine to tRNA^{Ile}, the synthetase hydrolyzed the valyl-AMP to produce valine and AMP, thereby demonstrating the synthetase has a proofreading or editing function that prevents misacylation.

In 1977 Alan Fersht proposed the **"double-sieve" model** shown in FIGURE 16.16 to explain the high fidelity of the aminoacyl-tRNA formation. According to this model Ile-tRNA synthetase has an aminoacylation site that serves as a "coarse sieve" to exclude amino acids larger than the cognate L-isoleucine and an editing site that serves as a "fine sieve" to eliminate amino acids smaller than isoleucine such as valine.

Ile-tRNA synthetase can hydrolyze valyl-tRNA^{Ile} and valyl-AMP.

In principle, editing by Ile-tRNA synthetase might take place before (pretransfer) or after (posttransfer) the valyl group is transferred to tRNA^{Ile} (FIGURE 16.17). To test Ile-tRNA synthetase's ability to hydrolyze Val-tRNA^{Ile}, one would need some method to synthesize the substrate Val-tRNA^{Ile}. Emmet Eldred and Paul Schimmel solved this problem in the early 1970s by taking advantage of the fact that under special incubation conditions an aminoacyl-tRNA synthetase from one species sometimes misacylates tRNA molecules from another species. More specifically, they used a yeast extract to attach

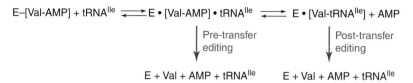

FIGURE 16.17 Pre- and posttransfer editing. The symbol E denotes isoleucyl-tRNA synthetase. (Modified from Nomanbhoy, T. K., Hendrickson, T. L., and Schimmel, P. 1999. *Mol Cell* 4:519–528. Copyright 1999, with permission from Elsevier [http://www.sciencedirect.com/science/journal/10972765].)

valine to purified *E. coli* tRNA^{Ile} and then isolated the misacylated tRNA. Valine was rapidly released when this misacylated tRNA was mixed with *E. coli* Ile-tRNA synthetase, demonstrating the enzyme's ability to perform posttransfer editing. The structural basis for posttransfer editing has been elucidated for the bacterial Ile-tRNA synthetase (**BOX 16.1 LOOKING DEEPER: STRUCTURAL BASIS FOR POSTTRANSFER EDITING**).

The experiments performed by Eldred and Schimmel did not rule out the possibility that Ile-tRNA synthetase also has a pretransfer editing activity. It is difficult to show that pretransfer editing takes place because this process requires tRNA^{Ile} to stimulate the editing function and one would have to eliminate the possibility of Val-tRNA^{Ile} formation followed by hydrolysis. Schimmel and Stephen P. Hale solved the problem by replacing tRNA^{Ile} with a short DNA molecule that binds to the Ile-tRNA synthetase • valyl-AMP complex. They selected the short DNA molecule from a random sequence pool based on its ability to bind to the complex. A DNA or RNA sequence that is selected from a random sequence pool because it can bind to a specific protein or protein complex is called an **aptamer**. The Schimmel and Hale DNA aptamer stimulated Ile-tRNA synthetase to hydrolyze valyl-AMP but did not stimulate the enzyme to hydrolyze isoleucyl-AMP. Because the DNA aptamer could not accept amino acids, valyl-AMP hydrolysis must be due to pretransfer editing.

Failure of proper editing can be harmful to an organism. For example, an alanyl-tRNA synthetase with an editing defect is responsible for a neurodegenerative disease in mice (**BOX 16.2 CLINICAL APPLICATION: A NEURODEGENERATIVE DISEASE CAUSED BY AN EDITING DEFECT IN ALANYL-tRNA SYNTHETASE**).

Each aminoacyl-tRNA synthetase can distinguish its cognate tRNAs from all other tRNAs.

Each aminoacyl-tRNA synthetase must be able to select a family of isoacceptor tRNA molecules from a mixture of many other tRNA molecules. A synthetase does so by searching for a nucleotide, a base pair, a short nucleotide sequence, or some combination of these features that distinguishes its cognate tRNAs from all other tRNAs. One method for identifying specific features of a tRNA recognized by the cognate synthetase is to systematically introduce mutations into the

BOX 16.1: LOOKING DEEPER

Structural Basis for Posttransfer Editing

The structural basis for posttransfer editing has been elucidated for the bacterial Ile-tRNA synthetase. This enzyme has a characteristic ATP binding domain called the Rossmann-fold, named for Michael Rossmann, the investigator who discovered the fold. The Rossmann-fold domain of this enzyme has a polypeptide inserted within it (FIGURE B16.1a). This insert, which is approximately 200 residues long, is called connective polypeptide 1. Three different kinds of experiments show that connective polypeptide 1 forms the editing domain. First, mutations that alter residues in connective polypeptide 1 block the editing function. Second, the connective polypeptide 1 fragment hydrolyzes Val-tRNA^{Ile} without the assistance of any other part of the synthetase. Third, structural studies performed by Shigeyuki Yokoyama and coworkers in 1998 show that valine binds to both the Rossmann-fold and the connective polypeptide 1 domain in Ile-tRNA synthetase from the gram-negative bacteria *Thermus thermophilus*, whereas isoleucine binds to only the former. The double-sieve model predicts this binding difference between the two amino acids. The Rossman-fold domain (the aminoacylation site) acts as the coarse sieve that accommodates both amino acids, whereas connective polypeptide 1 (the editing site) acts as the fine sieve that accommodates only valine.

Ile-tRNA synthetase's aminoacylation and editing sites are more than 25 Å apart, raising the issue of how the 3'-end of a misacylated tRNA moves from the aminoacylation to the editing site. Structural studies performed by Thomas Steitz and his coworkers show that when Ile-tRNA synthetase forms a complex with tRNA^{Ile} and mupirocin (an antibiotic inhibitor that binds to the aminoacylation site), the 3'-terminal of tRNA^{Ile} is located in the connective polypeptide 1 domain (Figure B16.1a). This observation suggests that when valine attaches to tRNA^{Ile}, the tRNA's acceptor stem flips from the aminoacylation site to the editing site while the rest of the RNA molecule remains in place (FIGURE B16.1b).

Connective polypeptide 1 also catalyzes the hydrolysis of valyl-AMP that is mistakenly formed at Ile-tRNA synthetase's aminoacylation site. Many, but probably not all, of the connective polypeptide 1 residues that participate in editing Val-tRNA^{Ile} also participate in valyl-AMP hydrolysis. Ile-tRNA synthetase conformation changes when the enzyme binds tRNA^{Ile} (and presumably the DNA aptamer), allowing valyl-AMP to move through a newly created channel connecting the Rossmann-fold domain to the connective polypeptide 1 domain.

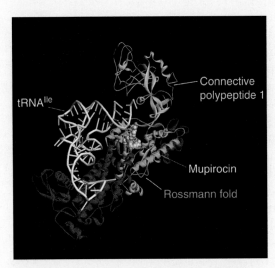

FIGURE B16.1a Structure of the isoleucyl-tRNA synthetase complex with bound tRNA^{Ile} and mupirocin, an inhibitor of the enzyme. The Rossmann fold is shown as a cyan ribbon, connective polypeptide 1 as a yellow ribbon, and other parts of the synthetase as a red ribbon. The tRNA is shown as a white tube and mupirocin (an antibiotic inhibitor that binds to the aminoacylation site) as a white spacefill structure. (Structure from Protein Data Bank 1FFY. Silvian, L. F., Wang, J., and Steitz, T. A. 1999. *Science* 285:1074–1077. Prepared by B. E. Tropp.)

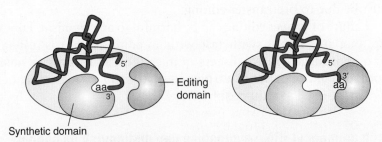

FIGURE B16.1b Structural basis of posttransfer editing. (a) When isoleucyl-tRNA synthetase is in the activation mode of aminoacyl transfer the acceptor strand of the tRNA adopts a hairpin conformation to pack against residues of the Rossmann fold in the synthetic domain. (b) When the enzyme is in the editing mode the tRNA adopts an extended stack conformation, which places the amino acid in the connective polypeptide 1 or editing domain (pocket). (Adapted from Silvian, L. F., Wang, J., and Steitz, T. A. 1999. *Science* 285:1074–1077.)

BOX 16.2: CLINICAL APPLICATION

A Neurodegenerative Disease Caused by an Editing Defect in Alanyl-tRNA Synthetase

Recent studies show a defect in the editing activity of an alanyl-tRNA synthetase causes a neurodegenerative disease in mice. This disease, caused by the autosomal recessive *sti* (sticky) gene, is characterized by tremors that progress to ataxia (lack of muscle coordination during voluntary movements). Histological and biochemical studies reveal that specific cells in the cerebellum of affected mice undergo apoptosis (programmed cell death).

Genetic studies indicate the *sti* gene codes for alanyl-tRNA synthetase. The mutation alters a single residue in the alanyl-tRNA synthetase. More specifically, Ala-374 in the wild-type synthetase is replaced by Glu-374 in the mutant synthetase. This substitution has no effect on aminoacylation activity. Wild-type and mutant alanyl-tRNA synthetases mistakenly attach serine to tRNA^{Ala} at about the same frequency. The same is true for mistaken glycine attachment. In contrast, the mutation does affect the synthetase's editing activity. Studies using Ser-tRNA^{Ala} as substrate indicate that the wild-type synthetase's editing activity is about twofold greater than the mutant synthetase's editing activity. As a result of this decreased editing activity, *sti/sti* mice synthesize altered polypeptide chains in which some alanine residues have been replaced by serine or glycine residues. These altered polypeptides are proposed to have difficulty in folding and therefore to form harmful aggregates. In support of this hypothesis, biochemical and histochemical studies indicate that cells with the mutant alanyl-tRNA synthetase do have increased levels of misfolded proteins.

tRNA and then determine how the mutations affect aminoacylation. Features revealed by *in vitro* and *in vivo* experiments are termed **identity elements** and **recognition elements**, respectively. Because most of the discussion that follows does not include experimental details, we use these two terms interchangeably. The major identity elements cluster in the first few base pairs of the acceptor arm and the anticodon trinucleotide. Synthetases may also recognize unique features in the variable or D-arms or a modified nucleotide. Some identity elements allow positive contacts with the cognate synthetase, whereas others prevent contacts with noncognate synthetases. Two examples help to illustrate the interactions that take place between an aminoacyl-tRNA synthetase and the identity element(s) on its cognate tRNA.

1. E. coli *tRNA^{Ala}*: Schimmel and coworkers performed a series of experiments beginning in 1988 designed to identify the features that *E. coli* Ala-tRNA synthetase recognizes in its cognate tRNA. After ruling out the 15 invariant nucleotides in tRNA^{Ala} as possible identity elements, Schimmel and coworkers used a combination of biochemical and genetic techniques to alter 36 of the remaining 61 nucleotides.

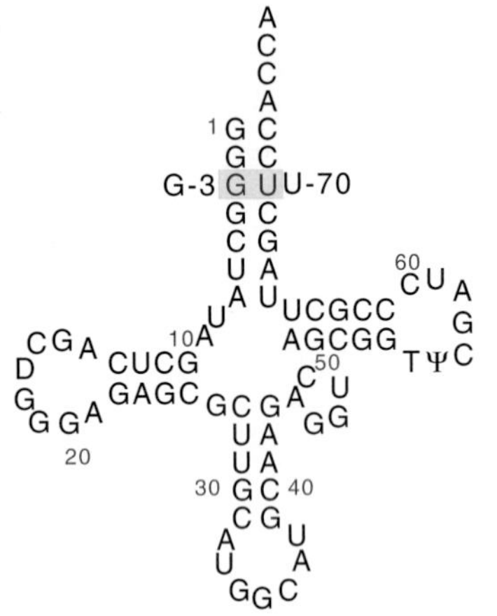

FIGURE 16.18 *E. coli* tRNA^Ala **identity element.** The G3:U70 identity element is shown in red. (Modified with permission from Hou, Y. M., and Schimmel, P. 1989. *Biochemistry* 28: 6800–6804. Copyright 1989 American Chemical Society.)

Two alterations were especially noteworthy. Replacing guanine at position-3 (G3) or uracil at position 70 (U70) with other bases caused tRNA^Ala to lose its ability to accept alanine (FIGURE 16.18). Aminoacylation by Ala-tRNA synthetase was not affected by other substitutions in the acceptor arm or by substitutions in the anticodon-, D-, or TψC arms. These results suggest that the nonstandard G3:U70 base pair is an identity element for tRNA^Ala. To test this possibility the nonstandard G3:U70 base pair was introduced into *E. coli* tRNA^Cys in place of the C3:G70 base pair that is normally present in tRNA^Cys. As a result of this base pair switch, the modified tRNA^Cys became an alanine acceptor that could no longer accept cysteine (FIGURE 16.19). This change in acceptor ability supports the hypothesis that the nonstandard G3:U70 base pair is an identity element for tRNA^Ala and suggests C3:G70 is an identity element for tRNA^Cys. Remarkably, the modified tRNA^Cys still differs from normal tRNA^Ala in 40 bases.

Schimmel and coworkers next tried to determine whether just a part of tRNA^Ala could serve as an alanine acceptor. To that end they synthesized the minihelix^Ala, microhelix^Ala, duplex^Ala, and tetraloop^Ala structures shown in FIGURE 16.20. The *E. coli* tRNA synthetase was able to use each of these RNAs as an alanyl-acceptor, showing the enzyme does not require the anticodon arm or any of the other excised regions to recognize tRNA^Ala. The aminoacyl-tRNA synthetases for histidine, glycine, methionine, valine, isoleucine, serine, glutamine, aspartic acid, cysteine, and tyrosine were later shown to attach amino acids to minihelices derived from the acceptor arm of their cognate tRNA. In contrast to Ala-tRNA synthetase, most of these other synthetases also recognize identity elements in the anticodon. We select the substrate

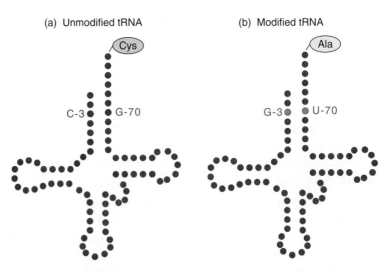

FIGURE 16.19 Effect of base substitution on acceptor activity. Replacing the C3:G70 base pair in tRNACys with a G3:U70 base pair converts the tRNA from one that accepts cysteine to one that accepts alanine.

CHAPTER 16 PROTEIN SYNTHESIS

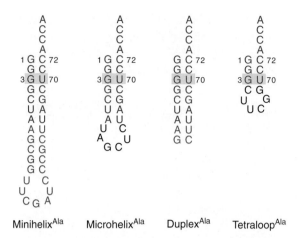

FIGURE 16.20 Structures derived from the acceptor-TΨC arms of *E. coli* tRNA^Ala that accept alanine from *E. coli* tRNA synthetase. Nucleotides present in the acceptor and TΨC arms are shown in red and blue, respectively. The nonstandard G3:U70 base pair is in a yellow box. (Adapted from Beuning, P. J., and Musier-Forsyth, K. 2000. *Biopolymers* 52:1–28.)

for one of these enzymes, tRNA^{Gln}, as our second illustrative example.

2. *E. coli* tRNA^{Gln}: Initial efforts to determine the identity elements in tRNA^{Gln} were similar to those described above for *E. coli* tRNA^{Ala}. *E. coli* tRNA^{Gln} was altered by a combination of biochemical and genetic methods, and then the modified tRNAs were tested for their abilities to accept glutamine. As summarized in FIGURE 16.21, these studies showed that tRNA^{Gln} has identity elements in its acceptor stem, anticodon arm, and D-arm. The crystal structure of the Gln-tRNA synthetase • tRNA^{Gln} complex confirmed the identity elements revealed by studying modified tRNA do indeed make contact with the enzyme (FIGURE 16.22). Similar structural confirmation of the identity element in tRNA^{Ala} is not yet possible because the crystal structure of the enzyme • tRNA complex has not been determined.

Selenocysteine and pyrrolysine are building blocks for polypeptides.

By the mid-1980s there was general agreement that the protein synthetic machinery uses just 20 amino acid building blocks to make proteins. Many modified amino acids were known to be present in proteins, but these were formed by modifying amino acid residues after the polypeptide chain was formed. It was therefore a great surprise when selenocysteine (FIGURE 16.23a) was identified as the twenty-first amino acid. This rare amino acid, an analog of cysteine in which the element selenium replaces sulfur, is present in a few proteins, mostly oxidoreductases, from eukaryotes, bacteria, and archaea (BOX 16.3 **LOOKING DEEPER: BACTERIAL PATHWAY FOR SELENOCYSTEYL-tRNA^{Sec} FORMATION**).

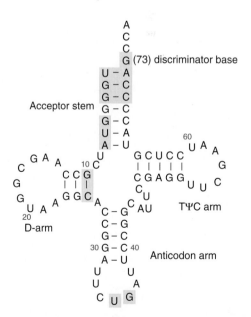

FIGURE 16.21 Secondary structure of tRNA^{Gln} in cloverleaf form. Nucleotides involved in recognition by glutaminyl-tRNA synthetase are highlighted in yellow. (Adapted from Ibba, M., Hong, K.-W., and Söll, D. 1996. *Genes Cells* 1:421–427.)

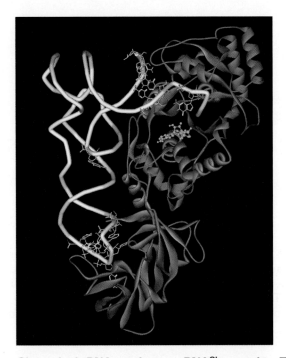

FIGURE 16.22 **Glutaminyl-tRNA synthetase tRNA^Gln complex.** The synthetase is shown as a blue ribbon structure and the tRNA as a white tube structure. Identity elements on the tRNA in contact with glutaminyl-tRNA synthetase are shown as yellow stick structures. The bound ATP is shown as a green ball-and-stick structure. (Structure from Protein Data Bank 1QRS. Arnez, J. G., and Steitz, T. A. 1996. *Biochemistry* 35:14725–14733. Prepared by B. E. Tropp.)

A second unusual amino acid building block, pyrrolysine (Figure 16.23b), was independently discovered by Joseph A. Krzycki and coworkers and Michael K. Chan and coworkers in 2002. At this time pyrrolysine appears to be present in only a few archaeal and bacterial species.

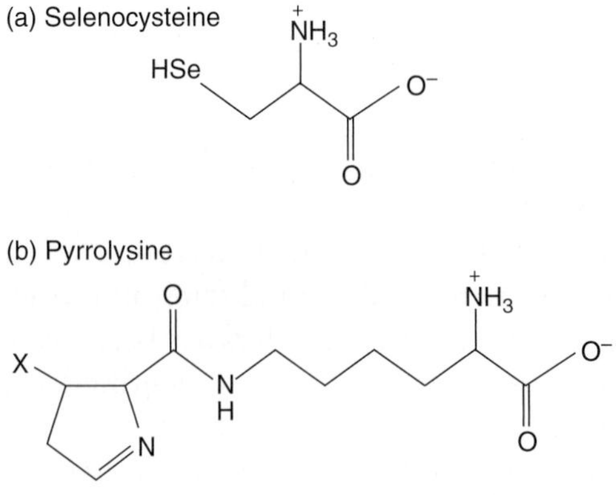

FIGURE 16.23 **The structures of (a) selenocysteine and (b) pyrrolysine.** X indicates a methyl, ammonium, or hydroxyl group. (Part b adapted from M. Ibba and D. Söll, *Curr. Biol.* 12 [2002]: R464–R466.)

CHAPTER 16 PROTEIN SYNTHESIS

BOX 16.3: LOOKING DEEPER

Bacterial Pathway for Selenocysteyl-tRNA^{Sec} Formation

The bacterial pathway for selenocysteyl-tRNA^{Sec} formation, which was elucidated largely through the efforts of August Böck and coworkers beginning in 1986, is shown in FIGURE B16.3. The three steps in this pathway are as follows: (1) selenophosphate synthetase catalyzes the synthesis of selenophosphate from ATP and selenide, (2) Ser-tRNA synthetase attaches a seryl group to tRNA^{Sec}, and (3) selenocysteine synthase catalyzes a pyridoxal phosphate-dependent conversion of selenophosphate and Ser-tRNA^{Sec} to selenocysteyl-tRNA^{Sec}. The tRNA^{Sec} species contains up to 100 nucleotides, making it the longest known tRNA. It also has other unusual features. For example, it has one more nucleotide base pair in its acceptor arm than most other tRNAs and an extended D-arm. Because of these unique features selenocysteyl-tRNA^{Sec} requires its own special translation elongation factor.

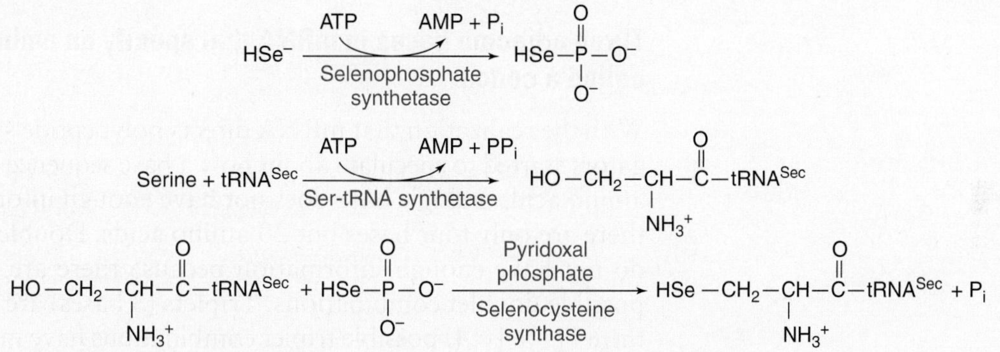

FIGURE B16.3 Bacterial pathway for selenocysteyl-tRNA^{Sec} formation.

16.4 mRNA and the Genetic Code

mRNA programs ribosomes to synthesize proteins.

Although molecular biologists knew a great deal about protein synthesis by the end of the 1950s, they still did not know how genetic information was transferred from the nucleotide sequences in DNA to amino acid sequences in polypeptides. The first clues to the possibility that cells might have a special kind of RNA that programs ribosomes to synthesize specific polypeptides came from studies with phage T2-infected *E. coli*. In 1953 Alfred Hershey reported that infected bacteria rapidly incorporate [^{32}P]phosphate into a small fraction of the cellular RNA molecules and that the labeled RNA is then rapidly degraded. Three years later Elliot Volkin and Lazarus Astrachan made a similar observation but then went on to demonstrate that the rapidly turning over RNA had the same base composition as the viral DNA and not the bacterial DNA. The significance of these experiments was not fully appreciated at the time because investigators thought the genetic instructions were permanently embedded in a ribosome at the time of its synthesis so each ribosome would only be able to make one specific polypeptide.

In 1959 Arthur Pardee, François Jacob, and Jacques Monod performed a series of genetic experiments (later called the PaJaMo experiments) showing that ribosomes are rapidly programmed to synthesize new proteins and do not contain permanently embedded instructions. The PaJaMo experiments began with the transfer of a *lacZ* gene (the structural gene for β-galactosidase) from a donor *E. coli* Hfr (*lacI⁺ lacZ⁺ Sm^S*) strain that could synthesize β-galactosidase to a recipient *E. coli* F⁻ (*lacI⁻ lacZ⁻ Sm^R*) that could not do so. Streptomycin, an antibiotic that blocks protein synthesis in Sm^S cells but not Sm^R cells, was added to the growth medium. The recipient strain started to synthesize β-galactosidase at maximal rates shortly after receiving the *lacZ* gene. Because the recipient strain did not have adequate time to produce new ribosomes dedicated to β-galactosidase synthesis, the PaJaMo experiments indicated that the *lacZ* gene serves as a template for the synthesis of an mRNA molecule, which then directs preexisting ribosomes to synthesize lactose enzymes.

Three adjacent bases in mRNA that specify an amino acid are called a codon.

With the realization that mRNA directs polypeptide synthesis, investigators started to speculate about how a base sequence might specify an amino acid. A single base does not have enough information because there are only four bases but 20 amino acids. Doublets (2 bases) also do not have enough information because there are only 16 (4 × 4) possible doublet combinations. Triplets (3 bases) are feasible because the 64 (4 × 4 × 4) possible triplet combinations have more than enough information to specify 20 amino acids. In fact, there are so many possible triplets one would predict that some amino acids are specified by more than one triplet (a "degenerate code"), some triplets do not specify an amino acid, or both.

Genetic experiments performed by Francis Crick, Sydney Brenner, and coworkers in 1961 showed that amino acids are specified by three nucleotide code words, for which Brenner coined the term "codons." Their experiments were based on the ability of proflavin, an acridine derivative (FIGURE 16.24), to cause a base pair insertion or deletion in DNA. A mutation that adds or deletes a base pair shifts the reading frame and is therefore termed a **frameshift mutation** (FIGURE 16.25). Crick and coworkers began their investigation of the genetic code using

FIGURE 16.24 Proflavin (acridine-3,6-diamine).

Direction of reading →

ABC ABC ABC **ABC** ABC ABC **ABC**···   Normal reading frame in wild type gene

ABC XAB CAB **CAB** CAB CAB **CAB** C···   Inserting **X** shifts reading frame to the right

ABC ABC ABC **ABC** BCA BCA **BC**···   Deleting **A** shifts reading frame to the left

ABC XAB CAB **CAB** CBC ABC **ABC**···   The reading frame is restored in a recombinant gene that contains the insertion and deletion mutations. However, the region between the two mutations has codons that specify amino acids that are different from those in the wild type gene (missense codons).
└──── Missense ────┘

FIGURE 16.25 Effect of insertion and deletion mutations on the reading frame.

proflavin to introduce a mutation into the phage T4 *rIIB* gene. The mutant phage was easy to distinguish from its wild-type parent because of the plaques it formed on a bacterial lawn. Wild-type bacteriophages form small plaques on both *E. coli* B and *E. coli* K12(λ) lawns, whereas *rIIB* mutants form large plaques on *E. coli* B lawns and none at all on *E. coli* K12(λ) lawns. The first *rIIB* mutation isolated, called FC0, was arbitrarily designated (+), as if it had a base pair insertion. (They might also have designated it [−] as if it had a base pair deletion. Although later sequencing studies showed the FC0 mutation is in fact due to an insertion, the conclusions drawn from their experiments would have been the same even if they had guessed wrong.)

If FC0 is a (+) insertion, then it should be possible to isolate pseudo-revertants with a (−) mutation that restores the reading frame. Of course, the region between the (+) and (−) mutations will be altered and so must be tolerant of amino acid substitutions. Crick and coworkers isolated several spontaneous pseudorevertants and separated the (−) mutation from the original (+) mutation by genetic recombination with the wild-type phage. As expected, recombinant phage with a (−) mutation in the *rIIB* gene did not form plaques on *E. coli* K12(λ) lawns. Phage with the (−) *rIIB* mutations were then used in the same way as phage with the original FC0 mutation to find new (+) mutations. Crick and coworkers were therefore able to obtain a large number of (+) and (−) mutants. Further genetic recombination experiments using these (+) and (−) mutations yielded the following results:

1. A cross between two (+) mutants does not produce phage with the wild-type phenotype.
2. A cross between two (−) mutants does not produce phage with the wild-type phenotype.
3. A cross between one (+) and one (−) does produce phage with the wild-type phenotype.

These results were interpreted in the following way. A double mutant of the type (+)(+) or (−)(−) has two additional base pairs or lacks two base pairs, respectively. In both cases reading of the code would be shifted two bases out of phase and a functional protein would not be made. In a (+)(−) double mutant, however, the advanced reading frame after the (+) site would be incorrect, but the correct reading frame would be restored at the (−) site (Figure 16.25). Between the two mutant sites the amino acid sequence would not be that of the wild-type phage but nonetheless the (+)(−) and (−)(+) phage strains would be functional, because both mutant sites were in a noncritical (tolerant) region of the gene. These interpretations proved correct. The point is that as long as the reading frame is restored before the intolerant region is reached, a functional (though not always perfect) protein can be produced.

Because a double (+)(+) mutant never had the wild-type phenotype, the code could not be a two-letter code. If it were a two-letter code, the reading frame would be restored in this combination. The critical experiment to test for a triplet code was the construction of a triple mutant. As expected, the triple mutants (+)(+)(+) and (−)(−)(−) had the wild-type phenotype, whereas the mixed triples, (+)(+)(−) and (+)(−)(−), were still mutant. These genetic experiments provided strong evidence that a codon contains three nucleotides (or a multiple of three nucleotides).

The discovery that poly(U) directs the synthesis of poly(Phe) was the first step in solving the genetic code.

In 1961 Marshall W. Nirenberg and Johann Heinrich Matthaei performed an experiment that led to the eventual solution of the genetic code. Their initial objective was to determine if bacterial or viral RNA could direct *E. coli* ribosomes to make proteins. They started by preparing a 30,000 × g *E. coli* supernatant fraction that others had shown contained ribosomes, tRNA, aminoacyl-tRNA synthetases, and all the translation factors needed for protein synthesis. A typical reaction mixture contained fresh 30,000 × g supernatant, ATP, GTP, Mg^{2+}, 19 unlabeled amino acids, and an [^{14}C]amino acid. The reaction was stopped by adding acid, which also caused protein precipitation. The radioactive protein was collected as a pellet by centrifugation and washed several times before being analyzed for its radioactive content.

Nirenberg and Matthaei observed a significant level of amino acid incorporation in the absence of any external source of mRNA. Adding bacterial or viral RNA seemed to cause a slight increase in radioactive amino acid incorporation. Although the increased amino acid incorporation was too small to be definitive, it suggested the bacterial and viral RNA might be able to direct ribosomes to synthesize proteins. Nirenberg and Matthaei thought the amino acid incorporation observed in the absence of added bacterial or viral RNA was probably due to mRNA already in the 30,000 × g bacterial supernatant fraction. They therefore decided to deplete the bacterial extract of this preexisting mRNA and prevent the synthesis of new mRNA.

Studies by others had shown that *E. coli* extracts stop making proteins shortly after DNase I is added to the extracts. Protein synthesis comes to a halt because mRNA has a short lifetime and cannot be replaced without a DNA template. Nirenberg and Matthaei, therefore, added DNase I to the bacterial extracts to destroy any DNA present. They also forced the bacterial extract to degrade preexisting mRNA and complete the synthesis of nascent polypeptide chains by adding unlabeled amino acids to the extract and incubating the mixture for 40 minutes until amino acid incorporation had nearly ceased. Then the extracts were dialyzed to remove unlabeled amino acids and stored frozen. Thawed extracts retained the ability to support protein synthesis but required an exogenous source of mRNA to do so. The fact that the protein synthetic system lost little, if any, activity after a freeze–thaw cycle meant it was no longer necessary to prepare a fresh bacterial extract before each protein synthesis experiment. Nirenberg and Matthaei made one additional change in the assay procedure. Instead of collecting the protein precipitate by centrifugation, they switched to a faster filtration method. Using their modified assay, Nirenberg and Matthaei demonstrated that bacterial and viral RNA direct ribosomes to make proteins.

Nirenberg and Matthaei thought a synthetic homopolymer such as polyuridylate, poly(U), might also direct ribosomes to make polypeptides. Fortunately, there was a simple enzymatic procedure for making poly(U). Marianne Grunberg-Manago and Severo Ochoa had discovered the required enzyme, polynucleotide phosphorylase, in 1955 and showed it reversibly converts nucleoside diphosphates (NDPs) into polynucleotides, as shown by the following equation:

$$(RNA)_n + NDP \rightleftarrows (RNA)_{n+1} + P_i$$

Incubation with a single NDP produces a homopolymer, and incubation with two or more different NDPs produces a random copolymer (the different nucleotides are randomly distributed along the chain). When first discovered, polynucleotide phosphorylase was thought to synthesize RNA. Further study, however, revealed the enzyme is unsuited for this task because it does not require a template. We now know the enzyme's function is to degrade rather than to synthesize RNA. Although polynucleotide phosphorylase failed to live up to its original promise as an enzyme that synthesizes RNA, it played an important part in solving the genetic code.

Nirenberg and Matthaei tested poly(U)'s ability to direct ribosomes to synthesize polypeptides by adding it to 20 different tubes, each containing the same protein synthetic system but a different [^{14}C]amino acid. They observed amino acid incorporation into a polypeptide in the tube that contained [^{14}C]phenylalanine but in no other tube. Analysis revealed the radioactive polypeptide to be polyphenylalanine. The meaning of this experiment was quite clear. Poly(U) directs ribosomes to make polyphenylalanine. Later experiments showed polyadenylate [poly(A)] directs ribosomes to make polylysine and polycytidylate [poly(C)] directs ribosomes to make polyproline. Polyguanylate [poly(G)] fails to act as a synthetic mRNA because it forms a triple-stranded helix that cannot be translated. If we take the genetic experiments of Crick, Brenner, and coworkers into account, then UUU, AAA, and CCC are the codons for phenylalanine, lysine, and proline, respectively.

Realizing additional coding information could be obtained using random copolymers to direct protein synthesis, the Nirenberg and Ochoa laboratories raced to solve the genetic code. A specific example, a random copolymer containing A and C in a 5:1 ratio, will help to illustrate the rationale and limitations of their initial approach. The AAA triplet frequency in this random copolymer is 0.58 (5/6 × 5/6 × 5/6), whereas the CCC triplet frequency is 4.6×10^{-3} (1/6 × 1/6 × 1/6). Other triplet frequencies (AAC, ACA, CAA, ACC, CAC, and CCA) can be determined in the same way. The random copolymer should direct ribosomes to synthesize polypeptides with amino acid frequencies that are the same as the codon frequencies. One can therefore identify possible codons by comparing actual amino acid frequencies with calculated codon frequencies.

This information is of limited value when a codon contains two different nucleotides, however, because it does not provide the nucleotide sequence within the codon. For instance, we may learn the codon for histidine consists of 2 C + 1 A, but we have no way of knowing whether the actual codon is CCA, CAC, or ACC. Despite these limitations Nirenberg and Ochoa were able to obtain a great deal of useful information about the genetic code. One experiment performed by Ochoa and coworkers is particularly noteworthy because it showed ribosomes read mRNA in a 5'→3' direction. Because this experiment depended on knowing the direction of polypeptide chain growth, we first describe an ingenious experiment performed by Howard Dintzis in 1961 that showed polypeptide chains grow in an amino-to-carboxyl direction.

Protein synthesis begins at the amino terminus and ends at the carboxyl terminus.

Dintzis' approach, summarized in **FIGURE 16.26**, was based on five assumptions. (1) Polypeptide chain formation begins at one end of the polypeptide chain and continues to the other. (2) At any given instant

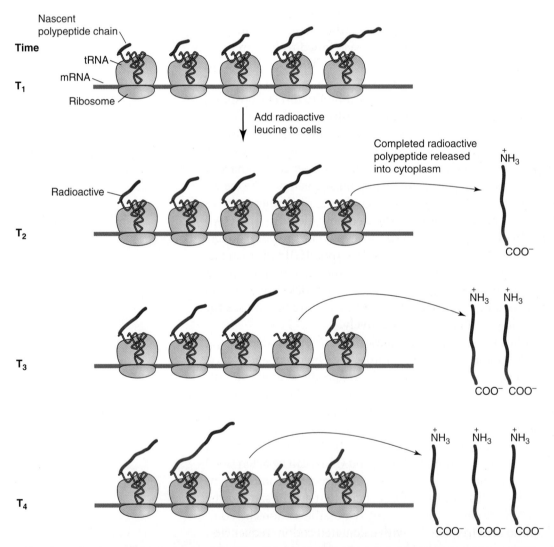

FIGURE 16.26 Direction of polypeptide chain growth. Time T_1—Radioactive leucine is added to rabbit reticulocytes. The ribosomes are at different stages of polypeptide synthesis. Time T_2—Some of the ribosomes complete polypeptide synthesis and release the radioactive polypeptide into the cytoplasm. The radioactive label (shown in red) is present in the carboxyl terminal segment. Time T_3—More of the ribosomes complete polypeptide chain synthesis and release the radioactive polypeptide into the cytoplasm. The radioactive polypeptide is detected in the carboxyl terminal segment as well as the segment next to it. Time T_4—Still more of the ribosomes complete polypeptide chain synthesis and release the radioactive polypeptide into the cytoplasm. The radioactive polypeptide is detected in the carboxyl terminal segment as well as the two adjacent segments. This experiment indicates the carboxyl terminus is the last part of the polypeptide chain that is synthesized. Hence, polypeptide chain synthesis proceeds in an amino-to-carboxyl direction.

the ribosomes in a cell are at different stages of polypeptide synthesis. (3) Ribosomes with longer nascent polypeptide chains will complete and release their polypeptide chains before ribosomes with shorter nascent chains. (4) If cells were incubated with radioactive amino acids for a very short period of time, only those ribosomes with the longest nascent chains would complete synthesis and release labeled polypeptides into the cytoplasm. (5) Only the end of the polypeptide chain that was synthesized last would be radioactively labeled.

Dintzis realized most cells synthesize many different proteins, making it extremely difficult to isolate and characterize a single labeled protein product. He therefore decided to work with immature red blood cells, called reticulocytes, which synthesize two polypeptides, the α- and β-hemoglobin subunits, in large quantities. The reticulocytes were isolated from rabbits that had been made anemic by daily injections of phenylhydrazine so they would have a high rate of red blood cell replacement.

Dintzis added radioactive leucine to a suspension of intact rabbit reticulocytes, which were incubated at 15°C to slow down the rate of protein synthesis. Samples were removed at intervals, cells disrupted, and extracts centrifuged to separate completed soluble proteins from nascent polypeptides still bound to ribosomes. Completed α- and β-globin chains were purified by carboxymethyl cellulose column chromatography and then digested with trypsin. Each of the resulting fragment mixtures was spotted on a piece of filter paper and partially separated by electrophoresis. Then the filter papers were dried, turned at a right angle, and developed in a second direction by chromatography. This two-dimensional separation technique, known as **fingerprinting** or **peptide mapping**, resolved the fragments derived from each chain, permitting the radioactive content of each fragment to be determined. When cells were incubated with radioactive leucine for a very short time, most of the radioactivity appeared in the carboxyl terminal fragment of the completed globin chains (Figure 16.26). This result clearly indicated the carboxyl terminal fragment is the last to be synthesized. When cells were incubated for a slightly longer time, radioactivity was detected in the carboxyl terminal fragment as well as the one next to it. At still longer times a radioactive gradient was observed in which the carboxyl terminal fragment was the most radioactive and the amino terminal fragment was the least radioactive. These observations support the idea that polypeptide synthesis begins at the amino terminal and proceeds by stepwise addition of amino acids to the carboxyl terminal.

mRNA is read in a 5′ to 3′ direction.

Taking advantage of the information supplied by Dintzis, Ochoa and coworkers demonstrated mRNA is read in a 5′ to 3′ direction. They began by preparing an oligonucleotide with the sequence ApApAp(Ap)$_n$ApApC by (1) synthesizing a random copolymer with a 25:1 A-to-C ratio, (2) cutting the copolymer with RNase to generate oligonucleotides with Cp groups at their 3′-ends, and (3) removing the 3′-phosphate groups with phosphatase (FIGURE 16.27). They added the ApApAp(Ap)$_n$ApApC oligonucleotide to a bacterial protein synthetic system that lacked nuclease activity. The system synthesized

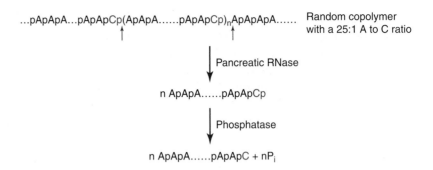

FIGURE 16.27 Preparation of ApApAp . . . ApApC oligonucleotides. A random copolymer, which is synthesized using polynucleotide phosphorylase, contains A to C in a 25:1 ratio. Pancreatic RNase cleaves this random copolymer to generate oligonucleotides ending in Cp (arrows show the sites cut) and phosphatase is used to remove the 3'-P. (Adapted from Salas, M. et al. 1965. *J Biol Chem* 240:3988–3995.)

an oligopeptide with the sequence ^+H_3N-Lys-(Lys)$_{n/3}$-Asn-COO$^-$. Formation of a chain made up almost exclusively of lysine residues was expected because AAA by then was known to be the codon for lysine. The presence of asparagine at the carboxyl end of the polypeptide chain was of great theoretical significance. Because polypeptide synthesis proceeds in the amino-to-carboxyl direction, asparagine's presence at the carboxyl end indicated that AAC at the 3'-end of the polyribonucleotide was the last codon to be read. Thus translation, like transcription, takes place in a 5'→3' direction. Because both processes take place in the same direction, bacteria can begin to translate an mRNA before its transcription is complete. Eukaryotes cannot do so because transcription and translation occur in separate biological compartments.

Trinucleotides promote the binding of specific aminoacyl-tRNA molecules to ribosomes.

In 1964 Nirenberg and Philip Leder devised an ingenious method for assigning codon sequences. Instead of requiring protein synthesis, their method depends on a trinucleotide's ability to promote the binding of an aminoacyl-tRNA to the ribosome. The assay is a relatively simple one (FIGURE 16.28). A mixture consisting of ribosomes, a trinucleotide, and an [^{14}C]aminoacyl-tRNA is incubated for a short time and then passed through a nitrocellulose filter. These filters permit free aminoacyl-tRNA molecules to pass through them but retain ribosomes and any molecules that are bound to the ribosomes (Figure 16.28a and b).

Nirenberg and Leder tested the method's reliability by determining whether trinucleotides containing known codons would promote the binding of the correct aminoacyl-tRNA and no other (Figure 16.28c). As expected, pUpUpU, pApApA, and pCpCpC promoted the specific binding of Phe-tRNA, Lys-tRNA, and Pro-tRNA, respectively. Then they prepared trinucleotides corresponding to the remaining 61 codons and tested each for the ability to bind one of the 20 [^{14}C]aminoacyl-tRNA species. The binding studies permitted about 50 codon assignments to be made. Some trinucleotides did not promote the binding of any aminoacyl-tRNA molecules and others gave ambiguous results. Another method for determining codon assignments was still needed.

Synthetic messengers with strictly defined base sequences confirmed the genetic code.

In the early 1960s H. Gorbind Khorana and his colleagues used a combination of classical organic and enzyme catalyzed synthesis to

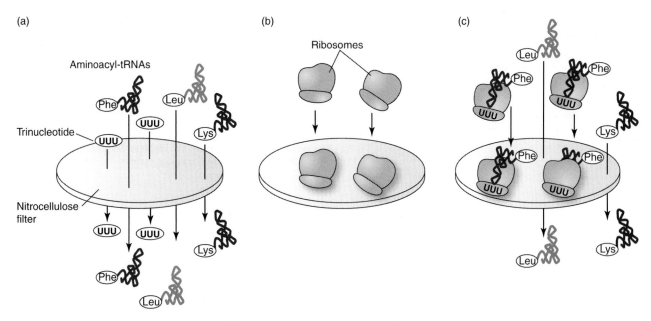

FIGURE 16.28 Trinucleotide binding assay. (a) Both UUU and aminoacyl-tRNA molecules pass through a nitrocellulose filter. (b) Ribosomes stick to the filter. (c) Phe-tRNA and UUU form a complex with ribosomes, which then sticks to the nitrocellulose filter. All other aminoacyl-tRNA molecules pass through the filter.

prepare a number of polyribonucleotides with repeating di-, tri-, and tetranucleotide sequences (FIGURE 16.29). When added to bacterial protein synthetic systems, these polyribonucleotides directed ribosomes to synthesize polypeptides with repeating amino acid sequences. RNA molecules with repeating dinucleotide sequences directed the synthesis of polypeptide molecules with a repeating dipeptide sequence. For example, the dinucleotide sequence poly UC (UCUCUCUCUC . . .) directed the synthesis of a polypeptide with a repeating serine–leucine sequence (Ser-Leu-Ser-Leu . . .; Figure 16.29a).

These results support the hypothesis that the protein synthetic system reads the bases in groups of three (UCU CUC UCU CUC). However, they do not indicate which of the two codons (UCU or CUC) corresponds to serine and which to leucine because polypeptide synthesis could start at an internal codon. Correct assignments were made by considering results obtained from random copolymer experiments, trinucleotide binding assays, and by using synthetic mRNA molecules with repeating tri- and tetranucleotide sequences.

RNA molecules with repeating trinucleotide sequences direct ribosomes to synthesize three different homopolymers. For example, a polyribonucleotide with a repeating AUC sequence (AUCAUCAUC . . .) directs the synthesis of polyisoleucine, polyserine, and polyhistidine molecules (Figure 16.29b). Because natural mRNA directs the synthesis of only one polypeptide, these results indicated that something was missing from most synthetic mRNAs. The missing element, an initiation signal that sets the reading frame by determining the translation start site, is described below. Natural mRNA molecules have the initiation signal but most synthetic RNA molecules do not.

RNA molecules with repeating tetranucleotide sequences direct the synthesis of polypeptide chains with repeating tetrapeptide sequences

16.4 mRNA and the Genetic Code

FIGURE 16.29 **Coding properties of polyribonucleotides with known repeating sequences.**

(a) Coding properties of repeating dinucleotide. Sequences direct the synthesis of polypeptides with repeating dipeptide sequence

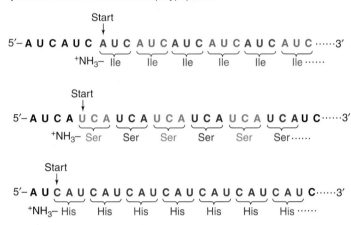

(b) Coding properties of repeating trinucleotide. Sequences direct the synthesis of three different homopolypeptides

(c) Coding properties of repeating tetranucleotides. Sequences direct the synthesis of polypeptides with repeating tetrapeptides

(Figure 16.29c). For example, a polyribonucleotide with a repeating UAUC sequence (UAUCUAUCUAUC . . .) directs the synthesis of a polypeptide with a repeating Tyr-Leu-Ser-Ile sequence.

Three codons, UAA, UAG, and UGA, are polypeptide chain termination signals.

Khorana's studies also identified three chain termination signals. The existence of one of these signals became evident when a polyribonucleotide with a repeating GAUA sequence (GAUAGAUAGAUA . . .) was studied. Instead of directing the synthesis of a polypeptide, this polyribonucleotide directs the synthesis of a tripeptide with the sequence Ile-Asp-Arg. As indicated in FIGURE 16.30, the UAG codon acts as a chain termination signal. Two other codons, UAA and UGA, also were shown to be termination codons. The finding of three stop codons fit nicely with the available genetic evidence. Working independently, the laboratories of Seymour Benzer and Sydney Brenner had demonstrated that some gene mutations cause the gene product to be much shorter than normal. They explained the production of the truncated polypeptides by proposing the mutations change sense codons into stop codons. Mutations that change a sense codon into a stop codon

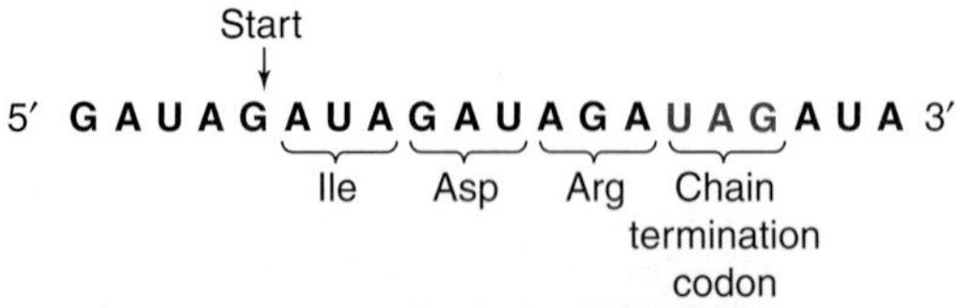

```
        Start
          ↓
5'  G A U A G A U A G A U A G A U A G A U A  3'
         Ile   Asp   Arg   Chain
                           termination
                           codon
```

FIGURE 16.30 **UAG termination codon.**

are termed **nonsense mutations** because they change a triplet from one that specifies an amino acid to one that does not. Three different types of nonsense mutation were shown to exist, and geneticists named them amber, ochre, and opal. These names, which have no physiological significance, were coined in a rather lighthearted spirit. The amber codon is UAG, the ochre codon is UAA, and the opal codon is UGA.

The genetic code is nonoverlapping, commaless, almost universal, highly degenerate, and unambiguous.

All 64 codons have now been assigned. Sixty-one codons determine amino acids, and 3 are translation termination signals. Codon assignments are shown in TABLE 16.1. These assignments have been confirmed

TABLE 16.1 The Genetic Code

First Position (5'-end)	Second Position				Third Position (3'-end)
	U	C	A	G	
U	UUC Phe	UCU Ser	UAU Tyr	UGU Cys	U
	UCC Phe	UCC Ser	UAC Tyr	UGC Cys	C
	UUA Leu	UCA Ser	UAA Stop	UGA Stop	A
	UUG Leu	UCG Ser	UAG Stop	UGG Trp	G
C	CUU Leu	CCU Pro	CAU His	CGU Arg	U
	CUC Leu	CCC Pro	CAC His	CGC Arg	C
	CUA Leu	CCA Pro	CAA Gln	CGA Arg	A
	CUG Leu	CCG Pro	CAG Gln	CGG Arg	G
A	AUU Ile	ACU Thr	AAU Asn	AGU Ser	U
	AUC Ile	ACC Thr	AAC Asn	AGC Ser	C
	AUA Ile	ACA Thr	AAA Lys	AGA Arg	A
	AUG Met / Start	ACG Thr	AAG Lys	AGG Arg	G
G	GUU Val	GCU Ala	GAU Asp	GGU Gly	U
	GUC Val	GCC Ala	GAC Asp	GGC Gly	C
	GUA Val	GCA Ala	GAA Glu	GGA Gly	A
	GUG Val	GCG Ala	GAG Glu	GGG Gly	G

Start Codon
Stop Codon
Nonpolar Side Chain
Uncharged Polar Side Chain
Charged Polar Side Chain

FIGURE 16.31 Difference between an overlapping and a nonoverlapping code. (Top) In an overlapping code a nucleotide (A, B, and C) can be used to specify more than one amino acid (aa). (Bottom) In a nonoverlapping code each nucleotide is part of a single codon that specifies an amino acid.

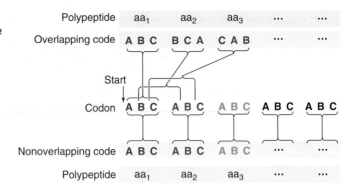

by comparing nucleotide sequences in genes to amino acid sequences in proteins. (DNA sequencing was not possible when the genetic code was solved in the mid-1960s.) Note that amino acids with similar properties tend to have similar codons.

Five general statements can be made about the genetic code.

1. *The genetic code is nonoverlapping.* Each base is part of one and only one codon. The difference between a nonoverlapping and an overlapping code is illustrated in FIGURE 16.31.
2. *The genetic code is commaless.* There are no intervening bases between adjacent codons.
3. *The genetic code is almost universal.* Although coding assignments were originally made in *E. coli*, subsequent studies showed that codons have the same meaning for other prokaryotes and for the cytoplasmic protein synthetic systems of eukaryotes. In consequence, mRNA from one species can be correctly translated by the protein synthetic machinery of another species. Exceptions to the canonical genetic code do exist. For example, the termination codon UGA codes for tryptophan in mitochondria.
4. *The genetic code is highly degenerate.* Most amino acids are specified by two or more codons. Leucine, arginine, and serine are each specified by six different codons. Only two amino acids, methionine and tryptophan, are specified by a single codon. Codons specifying the same amino acid are called synonyms. For example, CCU, CCC, CCA, and CCG are synonyms for proline. As this example illustrates, synonyms usually vary at the third position. Such degeneracy probably helps to protect cells from mutations.
5. *The genetic code is unambiguous under normal physiological conditions.* That is, each codon specifies only one amino acid.

The coding specificity of an aminoacyl-tRNA is determined by the tRNA and not the amino acid.

Although it seemed likely that an aminoacyl-tRNA's specificity for its codon results from anticodon–codon interactions, the studies with synthetic messengers did not rule out the alternate possibility that a codon might somehow recognize the aminoacyl group attached to the tRNA. François Chapeville and coworkers ruled out the latter possibility in

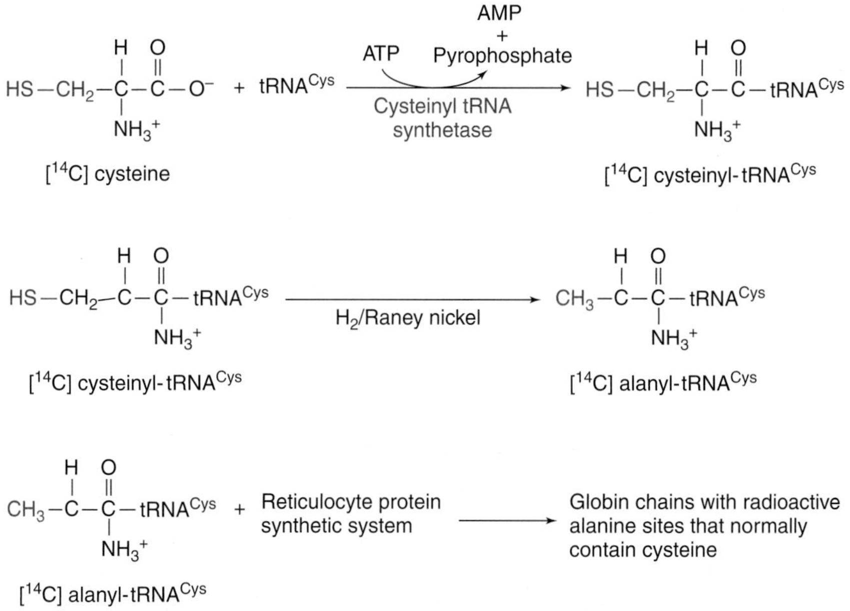

FIGURE 16.32 Experiment showing codon recognizes tRNA and not amino acid.

an experiment performed in 1962. They began by using a combination of enzymatic and organic synthetic techniques to prepare a misacylated tRNA (**FIGURE 16.32**). [^{14}C]Cysteine was attached to tRNA^{Cys} by cysteinyl-tRNA synthetase. The resulting [^{14}C]Cys-tRNA^{Cys} was then converted into [^{14}C]Ala-tRNA^{Cys}. (This was accomplished without disturbing the rest of the molecule by reducing [^{14}C]Cys-tRNA^{Cys} with molecular hydrogen in the presence of a Raney nickel catalyst.) The [^{14}C]Ala-tRNA^{Cys} was added to a protein synthetic system prepared from mammalian reticulocytes. Newly synthesized radioactive hemoglobin molecules were analyzed to determine whether radioactive alanine was incorporated into positions normally occupied by alanine or by cysteine residues. The results were quite clear. Radioactive alanine residues were present in positions that are normally occupied by cysteine residues. The experiment was therefore fully consistent with the adaptor hypothesis, showing the codon recognizes the tRNA molecule and not the aminoacyl group.

Some aminoacyl-tRNA molecules bind to more than one codon because there is some play or wobble in the third base of a codon.

Some tRNA molecules can bind to two or three codons, differing only in the third base. For example, the three codons recognized by tRNA^{Ala} (GCU, GCC, and GCA) all begin with the same two nucleotides but differ in the third nucleotide. tRNA molecules that bind to three different codons all have inosinate in the first position of the anticodon.

Francis Crick proposed the **wobble hypothesis** in 1966 to explain how a single tRNA molecule can bind to two or three different codons. The essential features of this hypothesis are as follows:

1. The codon and anticodon form antiparallel base pairs.
2. The first two bases in the codon form standard Watson-Crick base pairs with the last two bases in the anticodon.

16.4 mRNA and the Genetic Code

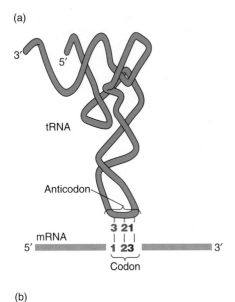

FIGURE 16.33 **Pairing between codon and anticodon.** (a) The first position in the anticodon (red) can form a nonstandard base pair with the third position in the codon (purple). (b) The table shows all possible wobble pairing. For instance, a guanine (G) in the anticodon can pair with either a uracil (U) or cytosine (C) in the codon and an isosine (I) in the anticodon can pair with a uracil (U), cytosine (C), or adenine (A) in the codon.

3. There is a certain amount of play or wobble in base pairing between the first base of the anticodon and the third base of the codon that permits nonstandard base pairing of the type indicated in **FIGURE 16.33**.

The base pairing relationships predicted by the wobble hypothesis have been confirmed by experimental studies.

16.5 Ribosome Structure

Bacterial 30S (small) subunits and 50S (large) subunits each have unique structures and functions.

Thus far, our examination of protein synthesis has been largely concerned with the genetic code with particular emphasis on (1) aminoacyl-tRNA formation and (2) codon assignments. This examination has shown us that all cells use ribosomes as protein synthetic machines that somehow translate the information provided by specific codon sequences on mRNA molecules into specific amino acid sequences in newly synthesized polypeptides. The ribosome is a dynamic machine with many moving parts that work together so the ribosome can translate about 10 to 20 codons per second with an error rate that is usually less than 0.03%. Although bacterial, eukaryotic, and archaeal ribosomes differ in structural detail, they share fundamental structure–function relationships. All ribosomes contain a small and a large subunit that must act together for normal ribosome function. We begin by examining the structure of the bacterial ribosomes because more is known about their structures than is known about the structures of eukaryotic or archaeal ribosomes. Moreover, the information gained by studying bacterial ribosome structure and function provides helpful lessons for studying eukaryotic and archaeal ribosome structure and function.

The bacterial 30S (small) subunit performs the ribosome's decoding function, discriminating between proper and improper codon–anticodon interactions. Approximately one-third of the 30S subunit's mass is in a region designated the head and the remaining two-thirds in a region designated the body (**FIGURE 16.34a**). The upper part of the body has a broad shelf-like protrusion called the platform. The site of codon–anticodon interactions is located in a cleft, which lies between the head and platform. The bacterial 50S (large) subunit contains the ribosome's peptidyl transferase activity, which catalyzes peptide bond formation. It has a flat side that faces the 30S subunit and a convex side that faces the solvent. When viewed flat face on, the 50S subunit has three protuberances that give it a crown-like appearance (Figure 16.34b). Its L1 protein is located in a mushroom-shaped protuberance on the left, its 5S rRNA is in the central protuberance, and its two L7/L12 protein dimers form a flexible stalk on the right. The flat face has a deep groove known as the interface canyon, which runs across the entire width of the subunit and is the site of tRNA binding. The interface canyon is part of the intersubunit space that is visible

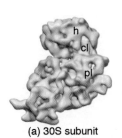
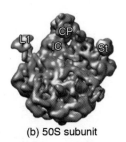
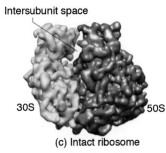

FIGURE 16.34 Cryoelectron microscopy reconstruction of the *Salmonella typhimurium* (close relative to *E. coli*) ribosome. The 70S ribosome structure was computationally separated into 30S (small) subunit and 50S (large) subunit in order to show the topography of the intersubunit space. (a) 30S subunit: landmarks in include the h, head; pl, platform; and cl, cleft. (b) 50S subunit: landmarks include L1, L1 protein; St, L7/L12 stalk; IC, interface canyon (red line with arrow at each end); CP, central protuberance. (c) The intact 70S ribosome at 10Å resolution. (Photo courtesy of Danny Nam Ho and Joachim Frank, HHMI, Columbia University.)

in some orientations of the intact ribosome (Figure 16.34c) and is of special interest because tRNA molecules fit into this space (see below).

Bacterial ribosome structure has been determined at atomic resolution.

Because of its large size and complex structure, the bacterial ribosome presented a major challenge for molecular biologists who wished to visualize it in atomic detail. Ada Yonath and coworkers made an important breakthrough in determining the bacterial ribosome structure in 1980 when they succeeded in growing crystals of *Bacillus steareothermophilus* 50S subunits. Although these crystals only provided a low resolution structure, their formation motivated investigators to seek new and better methods for growing ribosomal crystals that could provide a high resolution structure. The resulting efforts were rewarded in 1999 when the laboratories of Yonath and Venkatraman Ramakrishnan, working independently, solved the crystal structure for the 30S subunit from the thermophilic bacteria *Thermus thermophilus*, and Thomas Steitz, Peter Moore, and coworkers solved the crystal structure for the 50S subunit from the extremely halophilic ("salt-loving") archaeon *Haloarcula marismortui*. Harry F. Noller and coworkers solved the crystal structure for the intact *T. thermophilus* 70S ribosome 2 years later.

The absence of a structure for the *E. coli* ribosome was a cause for mild concern because the *E. coli* ribosome was the model system studied by most molecular biologists and biochemists. Therefore, investigators who were interested in structure–function relationships were forced to make the reasonable assumption that the *E. coli* ribosome would have a very similar structure to that of *T. thermophilus*. This assumption was shown to be correct in 2005 when Jamie H. D. Cate and coworkers solved the structure of the *E. coli* ribosome at a 3.5 Å resolution (FIGURE 16.35). As expected, the size and shape of the high-resolution *E. coli* ribosome structure is fully consistent with the three-dimensional cryoelectron microscopy construct (compare Figures 16.34 and 16.35).

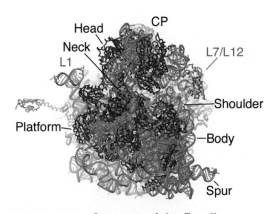

FIGURE 16.35 Structure of the *E. coli* 70S ribosome. The structure is viewed from the solvent side of the 30S subunit. The rRNA and proteins in the 30S subunit are colored light and dark blue, respectively. The 23S rRNA, 5S rRNA, and proteins in the 50S subunit are colored gray, purple, and magenta, respectively. Structural features visible in the 30S subunit include the head, neck, platform, body, shoulder, and spur. Structural features visible in the 50S subunit include L1 (protein L1/rRNA arm) and the central protuberance (CP). The approximate location of the proteins L7/L12, not observed in the structure, is in gray. (Reproduced from Schuwirth, B. S. et al. 2005. *Science* 310:827–834. Reprinted with permission from AAAS. Photo courtesy of Jamie H. Doudna Cate, University of California, Berkeley.)

16.5 Ribosome Structure

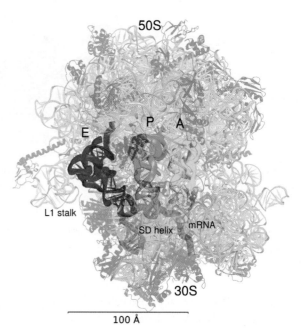

FIGURE 16.36 A model of the structure of the *Thermus thermophilus* 70S ribosome with tRNAs bound to the A, P, and E sites. The tRNA molecules at the A-, P-, and E-sites are colored yellow, orange, and red, respectively. The 16S rRNA in the small subunit is cyan, the 23S and 5S rRNAs in the large subunit are gray and light blue, respectively. Protein subunits in the small and large subunits are blue and magenta, respectively. The mRNA is green. (Reproduced from Korostelev, A., Ermolenko, D. N., and Noller, H. F. 2008. *Curr Opin Chem Biol* 12:674–683. Copyright 2008, with permission from Elsevier [http://www.sciencedirect.com/science/journal/13675931]. Photo courtesy of Andrei Korostelev, University of Massachusetts.)

A model of the *T. thermophilus* 70S ribosome based on a combination of three crystal structures reveals three tRNA-binding sites and the location of an mRNA fragment (FIGURE 16.36). Earlier biochemical studies predicted the existence of the three tRNA binding sites, which were named the **A-site** (aminoacyl-tRNA binding site), the **P-site** (peptidyl-tRNA binding site), and the **E-site** (exit site). The functions of these sites are described below. We know much less about the structures of eukaryotic and archaeal ribosomes (**BOX 16.4 LOOKING DEEPER: STRUCTURES OF EUKARYOTIC AND ARCHAEAL RIBOSOMES**).

16.6 Four Stages of Protein Synthesis

Protein synthesis can be divided into four stages.

With this introduction to ribosomal structure as background, we are now ready to examine protein synthesis, which can be divided into four stages:

1. *The initiation stage:* The reading frame is set by binding initiator tRNAs to ribosomes at start codons with the assistance of **translation initiation factors** (**IFs**). Biochemical studies

BOX 16.4: LOOKING DEEPER

Structures of Eukaryotic and Archaeal Ribosomes

Archaeal and eukaryotic ribosome structures appear to be similar to the bacterial ribosome structure. The best-studied eukaryotic ribosome is that from yeast. Its large (60S) subunit contains three RNA molecules: 25S rRNA (3,392 nucleotides), 5.8S rRNA (158 nucleotides), and 5S rRNA (121 nucleotides). The mammalian counterpart of 25S rRNA has a sedimentation coefficient of 28S and is at least 1,600 nucleotides longer. The 5.8S rRNA is homologous to the 5′-region of the bacterial 23S rRNA. The 60S yeast subunit contains 46 polypeptides. RNA accounts for about 60% of the large subunit's mass and protein for the remaining 40%. The small (40S) subunit has a single 18S rRNA (1,798 nucleotides) and 33 polypeptides. RNA accounts for about 55% of the small subunit's mass and protein for the remaining 45%. All identified yeast ribosomal proteins have homologous counterparts in the mammalian ribosome.

A comparison of eukaryotic rRNA molecules with their bacterial and archaeal counterparts reveals a remarkable degree of primary and secondary structure conservation. The conserved rRNA regions are those required to perform essential ribosomal functions such as the decoding aminoacyl-tRNA and forming peptide bonds. One major reason that eukaryotic rRNA molecules are longer than their bacterial counterparts is additional nucleotide sequences called expansion segments are inserted at specific sites within the conserved rRNA core. Marat Yusupov and coworkers have determined the crystal structure of the yeast ribosome at a resolution of 4.15 Å (FIGURE B16.4).

Although the yeast ribosome is considerably larger than the bacterial ribosome, the two have the same basic architecture. Recognizable bacterial ribosome landmarks including the head, platform, body, and central protuberance (CP) are also present in the yeast ribosome. The rRNA expansion segments are located primarily on solvent exposed subunit surfaces. If eukaryotic ribosome structure–function studies follow the same path as bacterial ribosome structure–function studies, we can look forward to learning a great deal about how the eukaryotic ribosome works in the near future.

Investigators determined the structure of the large ribosomal subunit from the extremely halophilic archaeon *Haloarcula marismortui*. It is remarkably similar to its bacterial counterpart. For example, it has A-, P-, and E-sites in the same positions as the 50S bacterial ribosomal subunit. Crystal structures are not yet available for an intact archaeal ribosome.

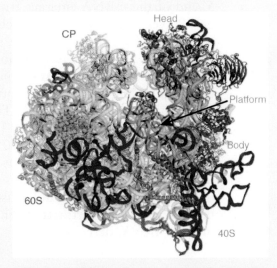

FIGURE B16.4 **The crystal structure of the yeast ribosome.** Proteins are colored dark blue in the 40S subunit and dark yellow in the 60S subunit. Ribosomal RNAs are colored light blue in the 40S subunit and pale yellow in 60S subunit. The rRNA expansion segments, which are located primarily on solvent-exposed subunit surfaces, are shown in red. CP indicates the central protuberance. (Reproduced from Ben-Shem, A. et al. 2010. *Science* 330:1203–1209. Reprinted with permission from AAAS. Photo courtesy of Marat Yusupov, Institut de Génétique et de Biologie Moléculaire et Cellulaire.)

performed in the late 1960s and early 1970s showed that special protein factors are needed to initiate polypeptide chain synthesis. As more information became available about bacterial, eukaryotic, and more recently archaeal polypeptide chain initiation pathways, it became apparent that although the three pathways share some important features, they also differ in many important ways. In fact, the differences among the translation initiation pathways are so great it is best to examine each pathway separately.

2. *The elongation stage:* Amino acids are added to the growing polypeptide chain as codons on mRNA are matched with anticodons on tRNAs with the assistance of **translation elongation factors** (**EFs**).
3. *The chain termination stage:* Termination codons are recognized by **release factors** (**RFs**) and completed polypeptide chains are released from the ribosome.
4. *The recycling stage:* Ribosomal subunits dissociate from each other under the influence of a **ribosomal recycling factor**.

Each protein synthesis stage requires enzymes and protein factors that are unique to it.

16.7 Initiation Stage

Each bacterial mRNA open reading frame has its own start codon.

Bacterial transcription and translation take place in the same cellular compartment, permitting translation to begin shortly after the 5′-end of the mRNA emerges from RNA polymerase's exit channel. Polycistronic mRNAs, which are quite common in bacteria, usually have an initiation codon at the start of each open reading frame and one or more termination codons at the end of each open reading frame. The translation machinery must, therefore, be able to recognize and initiate polypeptide chain synthesis at several different sites in the same mRNA. Translation always starts at a specific initiator codon. In bacteria AUG is the initiation codon about 90% of the time, GUG about 10%, and UUG about 1%. In very rare cases another codon such as AUU may function as an initiation codon. Precise recognition is essential because a phasing error of just one nucleotide would cause translation of the mRNA to be in the wrong reading frame. Each initiation codon binds the same initiator tRNA, described next.

Bacteria have an initiator methionine tRNA and an elongator methionine tRNA.

Kjeld Marcker and Frederick Sanger discovered the existence of the bacterial initiator tRNA while studying methionyl-tRNA synthetase in 1964. They incubated a soluble *E. coli* extract with [^{35}S] methionine, ATP, and tRNA, expecting to synthesize Met-tRNA.

Although they were successful in forming the expected product, they also detected a second product that contained a methionine derivative attached to the tRNA. Further study revealed the second product, N-formylmethionine (fMet), is synthesized by the two-step pathway shown in FIGURE 16.37. Methionyl-tRNA synthetase attaches methionine to tRNA to produce methionyl-tRNA and then methionyl-tRNA formyltransferase transfers a formyl group from N^{10}-formyltetrahydrofolate to methionyl-tRNA to form fMet-tRNA. Mitochondria and chloroplasts synthesize N-formylmethionyl-tRNA by a similar pathway.

Sanger and coworkers recognized the formyl group would prevent fMet from being incorporated into a polypeptide chain at any position other than the amino terminus. Subsequent studies showed that bacterial extracts do in fact incorporate fMet into the amino end of a growing polypeptide. The discovery that bacterial polypeptide synthesis begins with N-formylmethionine was puzzling because fMet is rarely, if ever, found at the amino terminus of a bacterial protein. The N-formyl group's absence was explained when a new enzyme, peptide deformylase, was discovered that cleaves formyl groups from N-termini. Peptide deformylase activity explains why E. coli polypeptides do not have formyl groups at their N-termini but does not explain why the N-terminus is methionine in only about 40% of bacterial polypeptides. The loss of methionine in the other 60% of the polypeptides was explained by the presence of still another enzyme, methionine aminopeptidase, which cleaves the N-terminal methionine from the polypeptides.

Further studies by Sanger and coworkers revealed that E. coli has two methionine tRNA isoacceptors. The first of these, $tRNA^{fMet}$, initiates polypeptide chain synthesis and the second, $tRNA^{Met}$, elongates polypeptide chains (FIGURE 16.38). Initiator tRNA accounts for about 70% of total cellular methionine acceptor activity and elongator tRNA for the remaining 30%. A single methionyl-tRNA synthetase charges both tRNA isoacceptors, recognizing the same identity elements (A73 and the anticodon) in each. Formyltransferase adds a formyl group only to the methionine attached to $tRNA^{fMet}$. Nucleotide substitution studies reveal that $tRNA^{fMet}$ has two major determinants that allow the formyltransferase to distinguish it from $tRNA^{Met}$. First, a mismatch between nucleotides at positions 1 and 72 in $tRNA^{fMet}$ extends the single-stranded stretch at the 3′-end of the acceptor arm to five nucleotides, allowing the formyltransferase catalytic site to gain access to the methionine. Second, an A11:U24 base pair in the D-arm of $tRNA^{fMet}$ interacts with the formyltransferase's C-terminal domain. The discovery of fMet-$tRNA^{fMet}$ proved to be of great importance because it allowed investigators to study the polypeptide chain initiation pathway.

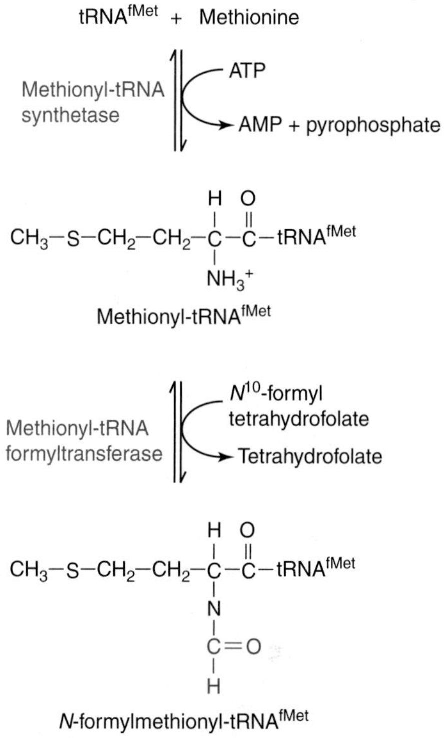

FIGURE 16.37 **N-Formylmethionyl-tRNA^{fMet} formation.**

The 30S subunit is an obligatory intermediate in polypeptide chain initiation.

In 1968 Matthew Meselson and coworkers performed an experiment to follow the distribution of isotopic labels among ribosomes and

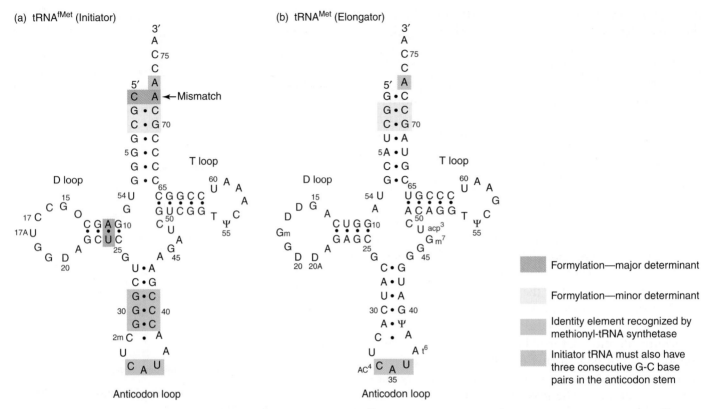

FIGURE 16.38 *E. coli* **tRNA isoacceptors for methionine.** (a) tRNA^{fMet} initiates polypeptide chain synthesis and (b) tRNA^{Met} elongates polypeptide chains. (Adapted from Sperling-Petersen, H. U., Laursen, B. S., and Mortensen, K. K. 2001. *Encyclopedia of Life Sciences.* John Wiley & Sons.)

ribosomal subunits after the transfer of a growing *E. coli* culture from a heavy- to a light-isotopes medium. This experiment showed (1) ribosomes frequently undergo subunit exchange during growth and (2) 30S and 50S subunits remain intact during growth. Based on these results they proposed that 70S ribosomes dissociate into 50S and 30S subunits after completing polypeptide chain synthesis. As investigators started to think about advantages cells might derive from ribosomal dissociation, they considered the possibility that the 30S or 50S subunit might participate in polypeptide chain initiation. This hypothesis received support from *in vitro* experiments that showed (1) poly(AUG) stimulates fMet-tRNA^{fMet} (but not Met-tRNA^{Met}) to bind to the 30S subunit and (2) 50S subunits block this binding. Based on these and related experiments, Masayasu Nomura proposed that fMet-tRNA^{fMet} and mRNA bind to the 30S ribosomal subunit to form a **30S initiation complex** and then the 50S subunit combines with the 30S initiation complex to form a **70S initiation complex.**

Nomura devised a clever experiment to test his model. First he cultured *E. coli* in a growth medium containing heavy isotopes (^2H_2O and $^{15}NH_4^+$) and isolated the "heavy" 70S ribosomes. Then he incubated the "heavy" 70S ribosomes with a synthetic random poly AUG, [^{3}H]fMet-tRNA^{fMet}, and translation initiation factors in the presence of excess "light" 50S ribosomal subunits (prepared from cells cultured in medium containing normal isotopes). Nomura's model predicts the three-step pathway shown in FIGURE 16.39.

- *Step 1.* Heavy 70S ribosomes dissociate to form heavy 50S and 30S subunits.
- *Step 2.* Heavy 30S subunits form 30S initiation complexes containing [^{3}H]fMet-tRNA^{fMet} and mRNA.
- *Step 3.* Light 50S subunits combine with 30S initiation complexes to form hybrid 70S initiation complexes.

Nomura separated heavy from hybrid 70S ribosomes by cesium chloride equilibrium density gradient centrifugation. As predicted by the model, [^{3}H]fMet-tRNA^{fMet} was present in the 70S hybrid fraction. However, the possibility still existed that the heavy 70S ribosomes rapidly exchanged their heavy 50S subunits for light 50S subunits before binding fMet-tRNA^{fMet}, and the newly formed hybrid 70S ribosomes and not the 30S subunits were responsible for binding [^{3}H]fMet-tRNA^{fMet}. Nomura was able to eliminate this possibility by including [^{14}C]Val-tRNA^{Val} in the incubation mixture and showing it binds to heavy 70S ribosomes and not to hybrid 70S ribosomes.

Initiation factors participate in the formation of 30S and 70S initiation complexes.

In 1966 Severo Ochoa and coworkers showed that ribosomes isolated from *E. coli* contain three translation initiation factors, IF1, IF2, and IF3, that stimulate the formation of the 30S and 70S initiation complexes. They extracted the initiation factors from the ribosomes

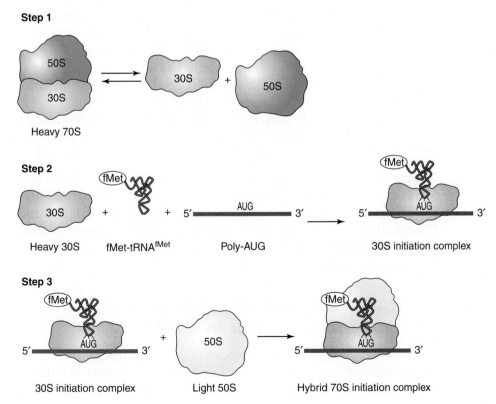

FIGURE 16.39 Outline of Nomura's experiment to test whether the 30S initiation complex is an intermediate in polypeptide chain initiation.

with a 1 M ammonium sulfate solution and used standard protein fractionation techniques to purify the three initiation factors.

IF1, IF2, and IF3 are required for polypeptide chain initiation at physiological magnesium ion concentrations, estimated to be about 5 mM, but not at higher concentrations. High magnesium ion concentrations stabilize the 70S ribosome and permit mRNA and aminoacyl-tRNA molecules to bind directly to the 70S ribosome in the absence of the initiation factors so the normal physiological pathway for polypeptide chain initiation is bypassed. Nirenberg and Matthaei were able to observe poly(U)-directed polyphenylalanine synthesis because their system contained a high magnesium ion concentration.

FIGURE 16.40 summarizes our current understanding of the pathway leading to the formation of the 70S initiation complex under normal physiological conditions:

- *Initiation factor 3:* IF3 binds to the 30S subunit after 70S ribosome dissociation and thereby shifts the equilibrium to favor the dissociated subunits. The crystal structure for intact IF3 has not yet been determined.
- *Initiation factor 1:* IF1, the smallest of the three bacterial initiation factors, is the next initiation factor to bind to the 30S subunit. IF1 is a highly conserved protein that binds at the A-site of the 30S subunit. The IF1 surface that interacts with the 30S subunit is rich in basic amino acids, providing favorable electrostatic interactions between IF1 and the negatively charged 16S rRNA phosphate backbone. IF1 stimulates the activities of IF2 and IF3 by promoting their more efficient binding to the 30S subunit.
- *Initiation factor 2:* IF2, the largest of the three initiation factors, has an N-terminal domain with no known function, a middle domain that binds and hydrolyzes GTP, and a C-terminal domain that interacts with fMet-tRNA^{fMet} (but not Met-tRNA^{Met}). The structure of the intact protein has not been determined at atomic resolution. Experiments designed to determine the order in which IF2 • GTP interacts with fMet-tRNA^{fMet} and the 30S subunit provide conflicting results. Some experiments indicate that IF2 • GTP binds to the 30S subunit before the subunit binds to fMet-tRNA^{fMet} and others indicate the reverse binding order.

The 50S subunit joins the 30S initiation complex to form a 70S initiation complex. GTP hydrolysis takes place immediately after the 50S subunit joins the 30S initiation complex and then IF1, IF2, and IF3 are released. The resulting 70S initiation complex is now ready to begin the elongation cycle by binding the aminoacyl-tRNA specified by the second codon at the A-site.

The Shine-Dalgarno sequence in mRNA interacts with the anti-Shine-Dalgarno sequence in the 16S rRNA.

The bacterial protein synthetic machinery must be able to distinguish the initiator AUG codon from other AUGs that code for internal methionines or from AUGs that occur in alternate reading frames.

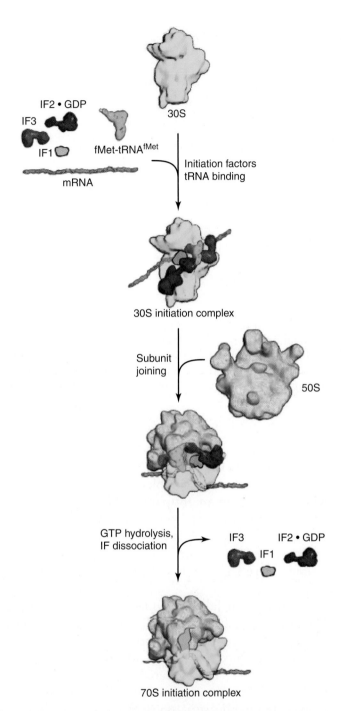

FIGURE 16.40 Translation initiation pathway in bacteria. Components are positioned on the ribosome according to currently available experimental information. (Adapted from Schmeing, T. M., and Ramakrishnan, V. 2009. *Nature* 461:1234–1242.)

In 1974 John Shine and Lynn Dalgarno made two important observations that helped to explain how the bacterial protein synthetic machinery performs this task. First, they noticed most cistrons have a purine-rich consensus sequence (5′-UAAGGAGGU-3′) located about five to seven nucleotides upstream from the initiator codon, with the highlighted nucleotides being the most frequently present

16.7 Initiation Stage

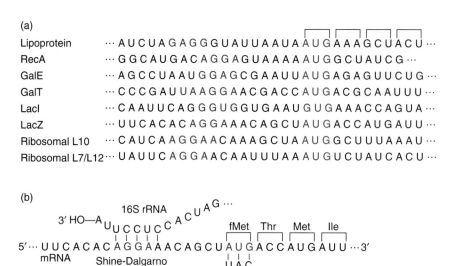

FIGURE 16.41 Shine-Dalgarno sequence. (a) Shine-Dalgarno sequences (ribosome-binding sites), which are located toward the 5′-end of mRNA, are shown for several *E. coli* mRNAs. The initiation (start) codons (blue) are immediately downstream. The optimal spacing between Shine-Dalgarno sequences and initiation sequences is 7–9 nucleotides. (b) Base pairing between the Shine-Dalgarno sequence and the complementary sequence in the 16S rRNA helps to establish the initiation codon and the correct reading frame for translation. (Adapted from Horton, R. H. et al. 2002. *Principles of Biochemistry* (3rd ed). Prentice Hall.)

(FIGURE 16.41a). Second, they noticed the 16S rRNA has a complementary pyrimidine-rich sequence (5′-ACCUCCUUA-3′) near its 3′-end. They therefore proposed that the complementary sequences on mRNA and 16S rRNA base pair in an antiparallel fashion (Figure 16.41b), helping the translation machinery to identify the initiator codon. The purine-rich sequence in mRNA is now called the **Shine-Dalgarno sequence** or ribosome binding site and the complementary sequence near the 3′-end of 16S rRNA is called the **anti-Shine-Dalgarno** sequence.

Joan Steitz and Karen Jakes provided experimental support for the Shine-Dalgarno hypothesis in 1974. They began by preparing a 70S initiation complex containing a short [^{32}P]mRNA fragment with a Shine-Dalgarno sequence. Next they added a ribonuclease called colicin E3 to this complex. They selected colicin E3 because it makes a single cut in 16S rRNA about 50 nucleotides from the 3′-end and so would be expected to release a 16S rRNA fragment with its Shine-Dalgarno sequence still base paired to the [^{32}P]mRNA (FIGURE 16.42). Finally, they added detergent to the mixture to extract proteins under conditions that would not disrupt base pairs. When the RNA was analyzed by gel electrophoresis under nondenaturing conditions, Steitz and Jakes observed a novel radioactive species. As predicted by the Shine-Dalgarno hypothesis, this species contained a [^{32}P]mRNA fragment • 16S rRNA fragment complex, which was not observed when the RNA was heated in urea before loading on the gel. Perhaps the most convincing evidence in support of the Shine-Dalgarno hypothesis is the actual visualization of the Shine-Dalgarno helix in the crystal structure (FIGURE 16.43).

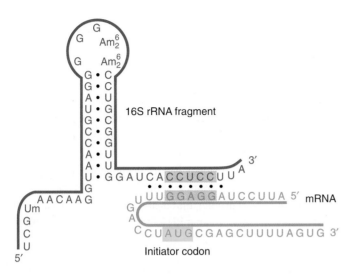

FIGURE 16.42 Binding of mRNA to a complementary sequence in 16S rRNA. The AUG start codon is shadowed in yellow, the Shine-Dalgarno sequence in pink, and the anti-Shine-Dalgarno sequence in tan.

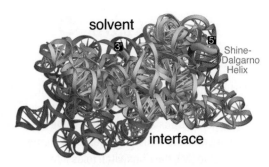

FIGURE 16.43 Shine-Dalgarno sequence as viewed from the top of the head of the 30S subunit. The 16S rRNA is colored cyan and the mRNA is colored yellow. Ribosomal proteins have not been included to simplify the view. The Shine-Dalgarno helix is colored magenta. 5′ and 3′ correspond to the 5′ and 3′ ends of the mRNA. The interface surface faces the 50S subunit and the solvent surface faces the surrounding solvent. (Reproduced from Culver, G. M. 2001. *Structure* 9:751–758. Copyright 2001, with permission from Elsevier [http://www.sciencedirect.com/science/journal/09692126]. Photo courtesy of Gloria M. Culver, University of Rochester.)

Although the Shine-Dalgarno consensus sequence in *E. coli* is 5′-AGGAGGU-3′, that in most normal cistrons contains just three or four of these bases. Longer Shine-Dalgarno sequences are required only if the mRNA secondary structure limits ribosomal access to the start codon. Optimal spacing between the Shine-Dalgarno sequence and the initiation codon is 5 nucleotides, but the Shine-Dalgarno sequence still works if it is up to 13 nucleotides from the initiation codon. The Shine-Dalgarno sequence appears to be dispensable when the AUG codon is exactly at the 5′-end of the mRNA, but the weaker initiation codons, GUG and UUG, do not support initiation under these conditions. The secondary structure of mRNA also influences initiation codon selection by masking AUGs so they cannot serve as initiation codons. The availability of Shine-Dalgarno sequences in some bacterial mRNA molecules is influenced by the presence of small metabolic intermediates (**BOX 16.5 LOOKING DEEPER: RIBOSWITCHES**).

Eukaryotic initiator tRNA is charged with a methionine that is *not* formylated.

Like the bacterial pathway, the eukaryotic translation initiation pathway uses one methionine tRNA for polypeptide chain initiation and another for polypeptide chain elongation. Also as in bacteria, the same methionyl-tRNA synthetase charges both the initiator tRNA (tRNA$_i$, where i is for initiator) and the elongator tRNA (tRNA$_m^{Met}$). There is, however, an important difference between bacteria and eukaryotes. Met-tRNA$_i$ is not formylated, forcing the eukaryotic protein synthetic machinery in the cytoplasm to rely on other features to distinguish initiator tRNA from elongator tRNAs. The major sequence feature that prevents vertebrate initiator tRNA from acting as an elongator tRNA is two specific base pairs, A50:U64 and U51:A63, in the initiator

BOX 16.5: LOOKING DEEPER

Riboswitches

Certain small metabolic intermediates influence translation initiation of bacterial mRNA molecules by stabilizing or destabilizing the way that a specific region, the riboswitch, in the 5'-UTR region folds. **FIGURE B16.5** shows one mechanism of riboswitch-mediated regulation of translation initiation. In the absence of metabolite the riboswitch folds into a structure that permits the Shine-Dalgarno sequence in the mRNA to interact with the anti-Shine-Dalgarno sequence on the 16S rRNA (Figure B16.5a). In the presence of metabolite (M), the riboswitch folds into a structure in which the Shine-Dalgarno sequence is part of a helical structure and so cannot interact with the anti-Shine-Dalgarno sequence (Figure B16.5b). The remarkable feature of the riboswitch is that specificity for the metabolite is determined entirely by the RNA molecule and does not involve protein; that is, the RNA folds into a structure that binds a specific metabolite and no others. For instance, the riboswitch that is part of an mRNA that codes for an enzyme required for thiamine (vitamin B_1) synthesis binds thiamine pyrophosphate.

Genes controlled by riboswitches often code for proteins that participate in the formation or transport of the metabolite that is sensed by the riboswitch.

Other metabolites shown to regulate translation initiation by their interaction with a riboswitch include vitamin B_{12}, flavin mononucleotide, S-adenosylmethionine, guanine, lysine, and glycine. Riboswitches that influence translation initiation have been demonstrated in bacteria and the archaea. In many cases the binding of a metabolite to a riboswitch also triggers premature transcription termination because the resultant RNA folding produces a transcription terminator. Plants and fungi also appear to have riboswitches, but thus far those that have been studied appear to influence splicing rather than translation initiation. In view of the widespread distribution of riboswitches in nature, it would be surprising if riboswitches were not eventually shown to contribute to gene regulation in animals.

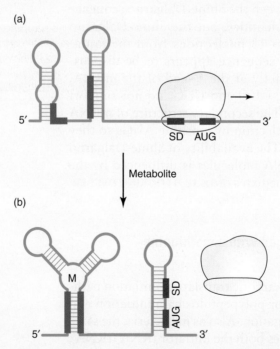

FIGURE B16.5 Riboswitch-mediated control of translation initiation. (a) Translation takes place because the Shine-Dalgarno (SD) sequence in mRNA can form base pairs with the complementary anti-Shine-Dalgarno sequence in the 16S rRNA. (b) Translation is blocked because a specific metabolite (M) stabilizes the riboswitch structure (shown as a hypothetical 3-stem structure) in a folding pattern that prevents SD from forming base pairs with the anti-Shine-Dalgarno sequence. Large and small ribosome subunits are colored blue and yellow, respectively. (Adapted from Nudler, E., and Mironov, A. S. 2004. *Trends Biochem Sci* 29:11–17.)

tRNA (shown in the orange box in **FIGURE 16.44**). Mutating the two base pairs converts the vertebrate initiator tRNA into an elongator tRNA. Other unique sequence and structural features present in eukaryotic initiator tRNA but not in elongator tRNAs are as follows: (1) an A1:U72 base pair at the end of the acceptor stem (mutations that alter the A1:U72 base pair cause a significant decrease in initiator function), (2) a sequence of three consecutive G:C base pairs in the anticodon stem, and (3) A54 and A60 in place of the T54 and Y60 (Y = pyrimidine) present in nearly all elongator tRNAs.

Eukaryotic translation initiation proceeds through a scanning mechanism.

The eukaryotic translation initiation pathway is far more complex than the bacterial pathway, requiring many more initiation factors and steps. Differences between the two pathways arise from the fact that transcription and translation take place in the same compartment in bacteria but in different compartments in eukaryotes. Because eukaryotic mRNA synthesis is completed in the nucleus, the cell has an opportunity to process the mRNA by adding a cap, removing introns, and attaching a poly(A) tail before the mRNA becomes available to ribosomes in the cytoplasm for translation. The mature eukaryotic mRNA that finally emerges in the cytoplasm after transport through the nuclear pores is a structured RNA molecule coated with proteins. In contrast to bacterial mRNA, eukaryotic mRNA does not have a Shine-Dalgarno sequence and usually has just one cistron.

The major pathway for translation initiation in eukaryotes begins with the recruitment of the small ribosomal subunit bearing specific initiation factors to the 5'-end of the mRNA. Once recruited, the small ribosomal subunit scans the mRNA in a 5' to 3' direction searching for the first AUG codon that is in the proper context. In mammals the most efficient AUG initiation codons are embedded within the sequence ACCAUGG (the initiation codon is highlighted in gray), but other sequences also work. Plants and yeast have very different sequences around their initiation codon. The mammalian ribosomal subunit may bypass the first AUG it encounters if a pyrimidine is at position −3 or +4 (the adenosine in AUG is at position +1). When a correct match is found, initiation factors are released and the large subunit joins the small subunit to form an initiation complex in which the initiator tRNA is paired with the initiation codon AUG at the P-site.

This scanning model, first proposed by Marilyn Kozak in 1978, is supported by considerable experimental evidence. There are three especially significant observations. (1) In contrast to bacterial ribosomes, eukaryotic ribosomes cannot bind to, or initiate translation on, short covalently circularized polyribonucleotide molecules that are about 60 nucleotides long because the molecules lack a 5'-end. (2) A stable secondary structure in the region between the 5'-end and the initiation codon (the **5'-untranslated region [5'-UTR]**) lowers translation initiation efficiency by impeding the small ribosomal subunit's movement along mRNA. (3) A stable hairpin structure about

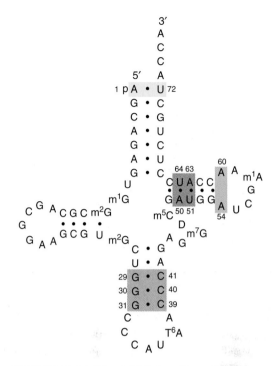

FIGURE 16.44 Cloverleaf structure of the vertebrate initiator tRNA. The unique features present in initiator tRNA are an A1:U72 base pair at the end of the acceptor stem (yellow), three consecutive G:C base pairs in the anticodon stem (purple), and A54 and A60 in the TψC loop (green). Mutations that alter A50:U60 and U51:A63 (orange) convert the initiator tRNA into an elongator tRNA. (Adapted from Drabkin, H. J., Estrella, M., and Rajbhandary, U. L. 1998. *Mol Cell Biol* 18:1459–1466.)

16.7 Initiation Stage

12 nucleotides downstream from the AUG initiation codon enhances initiation efficiency by forcing the small ribosome subunit to pause at the initiation codon. Translation initiation efficiency is also influenced by the length of the 5′-UTR. Systematically shortening the 5′-UTR from 77 to 3 nucleotides was shown to cause a progressive decrease in translation efficiency.

Some mRNAs have internal ribosomal entry sites, usually in their 5′-UTRs, which allow ribosomes to assemble at an internal site. These entry sites were first detected while studying the translation of an RNA molecule isolated from poliovirus. The poliovirus codes for a protease that inactivates an essential initiation factor required for the scanning mechanism. It is, therefore not surprising that cellular mRNAs with internal ribosomal entry sites were initially identified in cells infected with the poliovirus. Although a fascinating story, the translation of these mRNAs, which represent only a small subset of total cellular mRNA, is not described further.

Eukaryotes have at least 12 different initiator factors, which collectively contain at least 23 different polypeptides. Each **eukaryotic translation initiation factor** (eIF) is designated by a number often followed by a letter. For example, the eukaryotic translation initiation factor that binds to GTP and then Met-tRNA$_i$ is called eIF2. The eukaryotic translation initiation pathway is considerably more complex than its bacterial counterpart. The key stages in the eukaryotic translation initiation process are shown in FIGURE 16.45 and described below. This pathway shown does not indicate the contribution(s) that each eukaryotic initiation factor makes to 80S initiation complex formation.

1. *Ternary complex formation*: eIF2 combines with GTP to form an eIF2 • GTP complex, which then binds Met-tRNA$_i$ to generate the eIF2 • GTP • Met-tRNA$_i$ ternary complex.
2. *Ternary complex binding to 40S subunit*: The ternary complex, assisted by translation initiation factors, interacts with the 40S ribosomal subunit to form a complex.
3. *mRNA activation*: Eukaryotic mRNA passes through the nuclear pore into the cytoplasm as a structured molecule that is coated with polypeptides. Translation initiation factors assist in activating the mRNA by removing secondary structures and many of the polypeptides that coat the mRNA.
4. *mRNA entry into complex formed in stage 1*: Translation initiation factors assist the entry of mRNA into the complex formed in stage 1.
5. *5′→3′ Scanning to detect the initiation codon*: The complex formed in stage 2 moves along the mRNA in a 5′ to 3′ direction. Translation initiation factors assist in this ATP-dependent movement, which continues until the AUG initiation codon on the mRNA is aligned with the anticodon on the initiator tRNA.
6. *80S initiation complex formation*: Translation initiation factors facilitate the joining of the 60S subunit to the complex containing the 40S subunit to form the 80S initiation complex.

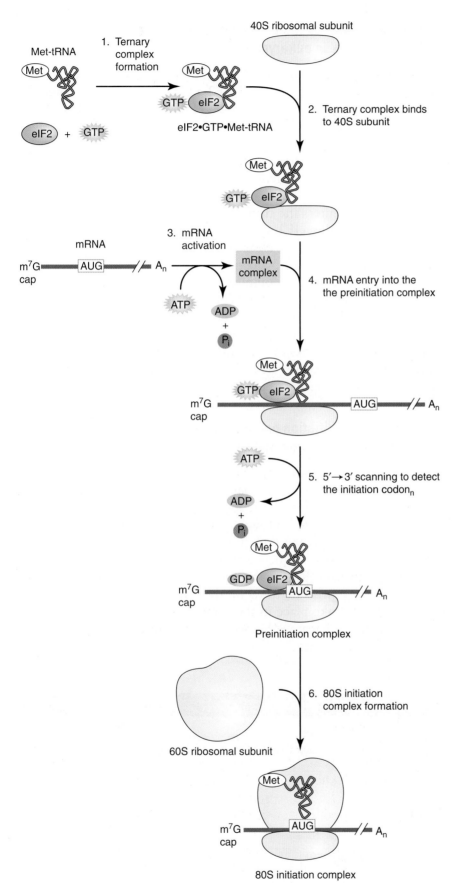

FIGURE 16.45 **Eukaryotic translation initiation pathway.** This simplified figure does not indicate the contribution(s) that each eukaryotic initiation factor makes to 80S initiation complex formation. See the text for a description of the steps.

16.7 Initiation Stage

Translation initiation factor phosphorylation regulates protein synthesis in eukaryotes.

eIF2 • GDP, which is released during 80S initiation complex formation, must be converted back to eIF2 • GTP before a new round of translation initiation can take place. However, eIF2 binds GDP so tightly that a GTP exchange factor, eIF2B, is required to catalyze the exchange reaction. One method eukaryotes use to inhibit total protein synthesis is to attach a phosphate group to eIF2α, one of three subunits in eIF2. Four different protein kinases, each regulated by a different signal, phosphorylate eIF2α (TABLE 16.2). Phosphorylated eIF2, eIF2(αP), enters the translation initiation pathway as part of the ternary complex eIF2(αP) • GTP • Met-tRNA$_i$ and is released as eIF2(αP) • GDP. The released complex binds to eIF2B • GTP, the GTP-GDP exchange protein. The resulting eIF2(αP) • GDP • eIF2B • GTP quaternary complex is a dead-end complex that cannot be converted to eIF2(αP) • GTP. Because the molar concentration ratio of eIF2-to-eIF2B is about 10:1, converting even a small proportion of eIF2 to eIF2(αP) inactivates most, if not all, eIF2B. This inactivation blocks further translation initiation because active eIF2B is needed to regenerate the eIF2 • GTP required for translation initiation to take place.

The translation initiation pathway in archaea appears to be a mixture of the eukaryotic and bacterial pathways.

The translation initiation pathway in the archaea appears to have some features of the bacterial pathway and others of the eukaryotic pathway. As in eukaryotes, the methionine esterified to the initiator tRNA is not formylated. However, like bacterial mRNAs, archaeal mRNAs do not have caps at their 5′-ends, are often polycistronic, and have Shine-Dalgarno sequences before their initiation codons. The similarity between mRNA in bacteria and the archaea suggests there may be selective pressure to retain these features when transcription and translation take place in the same compartment. The initiation factors in the archaea are homologous to those in eukaryotes, but there are fewer of them.

TABLE 16.2 eIf2α Kinases	
Kinase	Activation Signal
Heme controlled repressor (HCR)	Activated when the intracellular heme level falls.
Protein kinase R (PKR)	Activated by double-stranded RNA molecules produced in virus-infected cells.
GCN2	Activated in yeast when medium is deprived of amino acids and in tissue culture in the absence of serum.
PKR-like ER kinase (PERK)	Activated when unfolded polypeptide chains accumulate in the endoplasmic reticulum.

16.8 Elongation Stage

Polypeptide chain elongation requires elongation factors.

Once the translation initiation pathway has established the correct reading frame, the protein synthetic machinery is ready to enter the next stage of protein synthesis, polypeptide chain elongation. A major breakthrough in elucidating the polypeptide elongation pathway came in the mid-1960s when Fritz Lipmann and coworkers showed three translation elongation factors isolated from *E. coli* extracts participate in poly(U)-directed polyphenylalanine synthesis. The bacterial elongation factors are named EF-Tu, EF-Ts, and EF-G. The terms Tu and Ts indicate that the *E. coli* proteins are temperature unstable and temperature stable, respectively. The term G indicates the elongation factor has GTPase activity. A new nomenclature system has been suggested for bacterial elongation factors in which EF-Tu, EF-Ts, and EF-G are called EF1A, EF1B, and EF2, respectively. Here we use the old nomenclature because most molecular biologists continue to use the old system. EF-Tu is the most highly expressed protein in *E. coli* and accounts for about 5% to 10% of total cellular protein. Its intracellular concentration is about 10 times greater than that of ribosomes or EF-G. Eukaryotes have similar elongation factors. The eukaryotic counterparts of EF-Tu, EF-Ts, and EF-G are named eEF1A, eEF1B, and eEF2, respectively.

The elongation factors act through a repeating cycle.

The bacterial polypeptide elongation pathway is cyclic, using the same reaction sequence to add each new amino acid to the growing polypeptide chain (FIGURE 16.46). The eukaryotic pathway is quite similar except that the bacterial elongation factors EF-Tu, EF-Ts, and EF-G are replaced by the eukaryotic elongation factors eEF1A, eEF1B, and eEF2, respectively.

An EF-Tu • GTP • aminoacyl-tRNA ternary complex carries the aminoacyl-tRNA to the ribosome.

EF-Tu • GTP combines with aminoacyl-tRNA to form the ternary complex EF-Tu • GTP • aminoacyl-tRNA, which delivers the aminoacyl-tRNA to the ribosome. The tRNA, which is loosely bound to the ribosome at this stage, is said to be in an A/T state because its anticodon interacts with mRNA at the A-site, while its acceptor end remains bound to the translation elongation factor. The A-site interaction permits the ribosome to test the codon–anticodon match before it incorporates the new amino acid into the growing polypeptide chain. If the match between the codon and anticodon is incorrect, the aminoacyl-tRNA dissociates from the ribosome without stimulating GTP hydrolysis. But if the match is correct, the ribosome changes its conformation, stabilizing the interaction between the tRNA and ribosome and activating EF-Tu's latent GTPase (Figure 16.46, step 2). EF-Tu

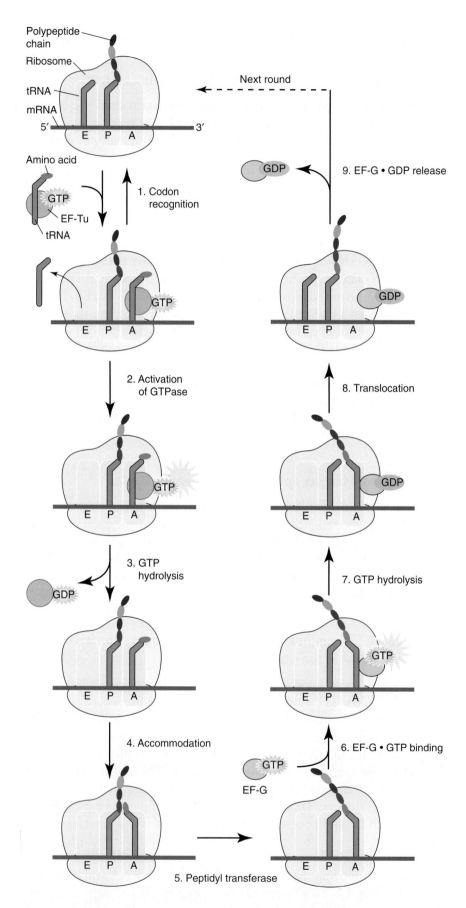

FIGURE 16.46 Polypeptide chain elongation pathway in bacteria. The eukaryotic polypeptide elongation pathway is the same except that eEF1A and eEF2 replace eF-Tu and EF-G, respectively. (Adapted from Ramakrishnan, V. 2002. *Cell* 108:557–572.)

goes through a major conformational change after hydrolyzing GTP, which causes it to release the aminoacyl end of the A-site tRNA and dissociate from the ribosome as EF-Tu • GDP (Figure 16.46, step 3). The now free aminoacyl end of the tRNA moves into the peptidyl transferase center of the large ribosomal subunit, a process that is called **accommodation** (Figure 16.46, step 4). An additional elongation factor, EF-P, is required when f-Met-tRNA^{fMet} is at the P-site. Eukaryotes have a related factor.

Specific nucleotides in 16S rRNA are essential for sensing the codon–anticodon helix.

Almost from the time the genetic code was first deciphered, molecular biologists have tried to determine how the 30S subunit decodes mRNA with such high fidelity. The binding energy for a correct codon–anticodon interaction is only about 2 to 3 kcal mol^{-1} more favorable than it is for a codon–anticodon interaction that has a single nucleotide mismatch. This binding energy difference predicts an error rate about 10- to 100-fold greater than that actually observed. It seemed likely that the ribosome plays some role in stabilizing the correct codon–anticodon interactions. This hypothesis received support from biochemical experiments with bacterial ribosomes that showed N1 methylation of highly conserved adenines at positions 1492 (A1492) and 1493 (A1493) of the 16S rRNA impaired A-site tRNA binding. Similar impairment also was observed when these adenines were changed to guanine or cytosine. Although these experiments indicated that A1492 and A1493 help the ribosome to recognize the shape of the codon–anticodon helix at the A-site, they did not show how they do so.

Ramakrishnan and coworkers turned to x-ray crystallography to solve the problem. They began by soaking an oligonucleotide containing the tRNA^{Phe} anticodon stem-loop and a U$_6$ hexanucleotide into crystals of the *T. thermophilus* 30S ribosomal subunit. The x-ray diffraction data showed that a correct codon–anticodon match causes A1492 and A1493 to flip out of the loop in which they are normally located and the highly conserved guanine at position 530 (G530) to change conformation. In their new conformations A1493 and A1492 interact with the first and second base pairs of the codon–anticodon helix, respectively, while G530 interacts with both the second position of the anticodon and the third position of the codon (FIGURE 16.47). These conformational changes allow the ribosome to closely inspect the first two base pairs of the codon–anticodon helix so it can discriminate between Watson-Crick base pairs and mismatches. The environment of the third or "wobble" position appears to be able to accommodate an uncommon base pair such as GU.

EF-Ts is a GDP-GTP exchange protein.

The EF-Tu • GDP released during the elongation cycle must exchange its GDP for GTP before it can participate in the next elongation cycle. However, it requires the assistance of EF-Ts to do so because EF-Tu's affinity for GDP is about 40 times greater than its affinity for GTP. EF-Ts, a

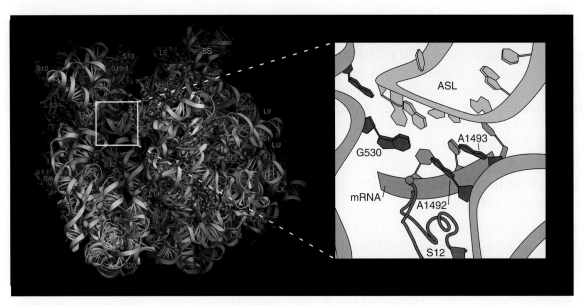

FIGURE 16.47 **Decoding mRNA.** The anticodon stem-loop (ASL) of A-site tRNA (gold) is in the interface cavity between the 30S subunit (left) and the 50S subunit (right). (Inset) A magnification showing the mRNA codon (purple) and cognate tRNA (gold) in the A-site of the 30S subunit. A1492 and A1493 (red) sense Watson-Crick base pairing in the first two bases of the codon–anticodon double helix. G530 (red) and the S12 polypeptide (brown) in the 30S subunit both contact A1492. (Photo reproduced from Yusupov, M. M. et al. 2001. *Science* 292:883–896. Reprinted with permission from AAAS. Photo courtesy of Harry Noller, University of California, Santa Cruz. Illustration modified from Dahlberg, A. E. 2001. *Science* 292:868–869. Reprinted with permission from AAAS.)

guanine nucleotide exchange factor, works as shown in FIGURE 16.48. A highly conserved segment in the N-terminal region of EF-Ts invades the G domain of EF-Tu. This invasion triggers a conformational change in the nucleotide binding pocket, causing EF-Tu to release the GDP that is bound to it. EF-Ts remains bound to EF-Tu, forming an EF-Tu • EF-Ts complex. High intracellular GTP concentrations rapidly convert this EF-Tu • EF-Ts complex into EF-Tu • GTP and EF-Ts.

The eukaryotic elongation factor, eEF1B, has a more complex structure than its bacterial counterpart but catalyzes the same nucleotide exchange reaction. Unlike EF-Ts, which consists of a single polypeptide, human eEF1B contains at least four polypeptide subunits.

The ribosome is a ribozyme.

With the aminoacyl-tRNA bound to the A-site and peptidyl-tRNA (or fMet-tRNA^{fMet}) bound to the P-site, the ribosome is ready for peptide bond formation, which is catalyzed by the peptidyl transferase activity in the 50S ribosomal subunit (Figure 16.46, step 5). Peptidyl group transfer from the peptidyl-tRNA at the P-site to the aminoacyl-tRNA at the A-site produces a deacylated tRNA at the P-site and a peptidyl-tRNA at the A-site.

The initial method for assaying peptidyl transferase activity required an intact ribosome, mRNA, a P-site peptidyl-tRNA, and an A-site aminoacyl-tRNA, making it difficult to determine whether peptidyl transferase or some other factor required for peptide bond formation was being monitored. A more direct assay for peptidyl transferase activity

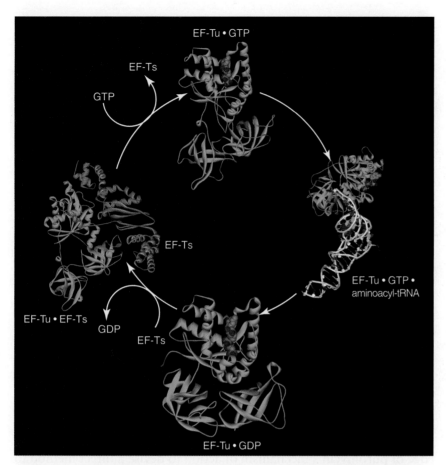

FIGURE 16.48 **Elongation factor Ts (EF-Ts) function in GDP-GTP exchange.** EF-Ts is required for EF-Tu to exchange its GDP for GTP. A highly conserved segment in the N-terminal region of EF-Ts invades the guanine nucleotide binding domain of EF-Tu, triggering a conformational change in the nucleotide binding pocket. This invasion causes EF-Tu to release the GDP that is bound to it. EF-Ts remains bound to EF-Tu, forming an EF-Tu • EF-Ts complex. High intracellular GTP concentrations rapidly convert this EF-Tu • EF-Ts complex into EF-Tu • GTP and EF-Ts. EF-Tu and EF-Ts are shown as green and orange ribbons, respectively. The tRNA molecule is shown as a white ribbon and guanine nucleotides as space-filling structures using standard CPK colors. (EF-Tu • GTP Protein Data Bank ID 1EXM. Part a structure from Protein Data Bank 1TUI. Polekhina, G. et al. 1996. *Structure* 4:1141–1151. Prepared by B. E. Tropp. Part b structure from Protein Data Bank 1TTT. Nissen, P. et al. 1995. *Science* 270:1464–1472. Prepared by B. E. Tropp. Part c structure from Protein Data Bank 1EXM. Hilgenfeld, R., Mesters, J. R., and Hogg, T. 2000. *The Ribosome: Structure, Function, Antibiotics, and Cellular Interaction*, pp. 347–357. ASM Press. Prepared by B. E. Tropp. Part d structure from Protein Data Bank 1EFU. Kawashima, T. et al. 1996. *Nature* 379:511–518. Prepared by B. E. Tropp.)

was required. The antibiotic puromycin, which is produced by the gram-positive bacteria *Streptomyces alboniger*, helped to solve the problem. Puromycin is remarkably similar in structure to the tyrosyl-adenosine group at the 3′-end of Tyr-tRNA (FIGURE 16.49). Because of this close resemblance, peptidyl transferase can use puromycin as a substrate in place of aminoacyl-tRNA at the A-site. As a consequence of this relaxed substrate recognition, the peptidyl transferase transfers a peptidyl group from a peptidyl-tRNA at the P-site to puromycin, causing premature

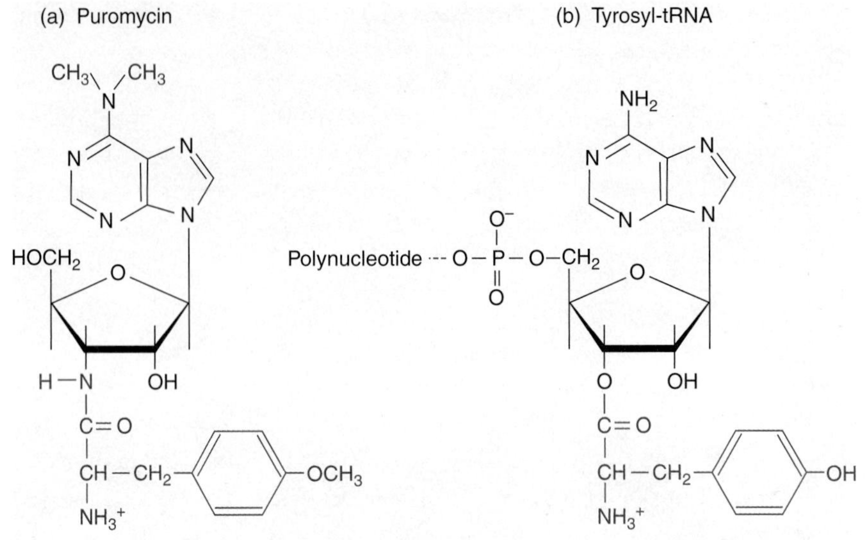

FIGURE 16.49 **Structures of (a) puromycin and (b) the 3′-end of tyrosyl-tRNA.** Major structural features present in puromycin but not in the 3′-end of tyrosyl-tRNA are shown in red. Features common to the tyrosine analog present in puromycin and tyrosine are shown in blue.

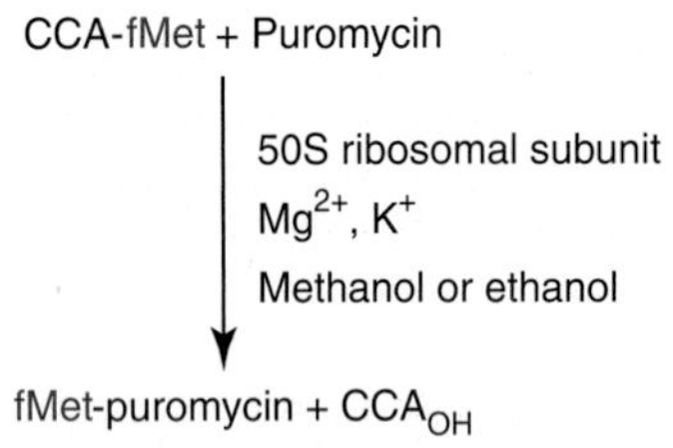

FIGURE 16.50 **Fragment assay for peptidyl transferase activity.**

polypeptide chain termination and peptidyl-puromycin release. Puromycin thus can serve as a model substrate for the peptide bond-forming reaction and is well suited for the study of this reaction.

In the mid-1960s Robin Monro and coworkers used their knowledge of puromycin's mechanism of action to devise a direct method to measure peptidyl transferase activity called the **fragment assay**. This assay measures fMet-puromycin formation when a mixture containing 50S ribosomal subunits, a 3′-fragment of peptidyl-tRNA such as CCA-fMet, and puromycin is incubated in a buffer solution containing 30% to 50% methanol or ethanol (**FIGURE 16.50**). The alcohol appears to enhance the affinity of low-molecular-mass P-site substrates for the 50S ribosomal subunit. The fragment reaction does not require the 30S ribosomal subunit or mRNA. It also does not require ATP or GTP, indicating the peptidyl transferase reaction does not require an energy source other than that already present in the ester bond that links the peptide group to tRNA.

Initially, investigators thought it likely that proteins present in the 50S subunit catalyze the peptidyl transfer reaction. Francis Crick challenged this idea in 1968 because he found it difficult to understand how the first ribosome could have been a protein when the ribosome's job is to make proteins. He therefore proposed that the first ribosome was made entirely of RNA. The discovery that some RNA molecules act as catalysts prompted molecular biologists to seriously consider Crick's hypothesis that RNA forms the catalytic site. This hypothesis received experimental support in 1992 when Harry F. Noller and coworkers showed that ribosomes from the thermophilic bacteria *Thermus aquaticus* completely lost their peptidyl transferase activity when their RNA was degraded by incubation with ribonuclease T1. In contrast, the ribosomes retained more than 80% of their peptidyl transferase activity after their proteins were removed by protease degradation followed by phenol extraction. The peptidyl transferase activity that remained after protein removal appeared to be physiologically significant because it retained its sensitivity to chloramphenicol and carbomycin, two antibiotics known to inhibit peptidyl transferase. Although Noller's experiments suggest that peptidyl transferase activity resides in the ribosomal RNA, they did not completely rule out participation by ribosomal proteins because the treated ribosomes still retained about 10% of their polypeptides.

Before the crystal structure of the 50S ribosomal subunit was known, various methods were used to characterize the peptidyl transferase. Cryoelectron microscopy constructs showed that the CCA ends of A- and P-site tRNAs are close together in the middle of the 50S ribosomal subunit, suggesting peptidyl transferase is located there (**FIGURE 16.51**). Genetic experiments identified specific nucleotides required for peptidyl transferase activity. For example, mutations that alter conserved nucleotides in the 23S rRNA inhibit or alter peptidyl transferase activity, suggesting these nucleotides contribute to the catalytic center. Many of these same conserved nucleotides are protected from small chemical probes when tRNA molecules are bound to the A- and P-sites on the ribosome. Perhaps the most convincing evidence for the fact that RNA and not protein has peptidyl transferase comes from studying atomic structures. The nearest ribosomal protein to the peptidyl transferase center is about 18 Å from this site, indicating the peptidyl transfer reaction is catalyzed by RNA and not protein. It seems likely that the adenosine at the 3'-end of the tRNA, which is linked to the peptidyl group, makes an essential contribution to the catalytic process. More specifically, biochemical studies indicate that the 2'-OH on this adenosine functions as a general acid-base during the catalytic process (**FIGURE 16.52**). The ribosome appears to contribute to the catalytic process by correctly orienting the two substrate molecules and stabilizing the transition state.

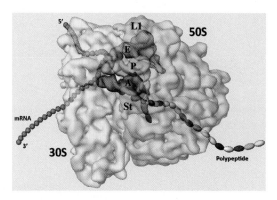

FIGURE 16.51 Cryoelectron micrograph construct of the *E. coli* ribosome prior to peptidyl transfer. The small and large subunits are shown in yellow and blue, respectively. The mRNA is shown in orange. The A- and P-site tRNAs are shown in purple and green, respectively. The E-site is shown in gold. The path of the nascent polypeptide chain through the exit tunnel in the large subunit is shown by the blue, green, and purple ovoid structures. St indicates the stalk and L1 indicates the L1 protein. The CCA ends of A- and P-site tRNAs are close together in the middle of the 50S ribosomal subunit, suggesting that peptidyl transferase is also located there. (Modified from Frank, J. 2001. *BioEssays*. John Wiley & Sons. Photo courtesy of Joachim Frank, HHMI/HRI, Wadsworth Center.)

FIGURE 16.52 tRNA contribution to peptidyl transfer reaction. The terminal adenosine of the peptidyl-tRNA at the P site makes an important contribution to peptidyl transfer. Its 2'-OH (red) acts as a general base and a general acid to facilitate peptidyl transfer. (Adapted from Simonović, M., and Steitz, T. 2009. *Biochim Biophys Acta* 1789:619, Figure 6C.)

16.8 Elongation Stage

The hybrid-states translocation model offers a mechanism for moving tRNA molecules through the ribosome.

Once the peptidyl transferase catalyzed reaction is complete, the P-site tRNA is deacylated and the A-site tRNA has a peptide chain with one additional amino acid residue. Before the next elongation cycle can take place, the peptidyl-tRNA has to move from the A-site to the P-site and the deacylated tRNA has to move from the P-site to the E-site so it can be released from the ribosome (Figure 16.46, steps 6–9). This highly coordinated movement, known as **translocation**, has to be precise so the reading frame of the mRNA is preserved.

Noller proposed the **hybrid-states translocation model**, shown in FIGURE 16.53, to explain the movements of mRNA and tRNAs through the ribosome. This model is based on chemical footprinting experiments performed by Noller and Danesh Moazed in 1989 that monitored the progress of tRNA through the ribosomes. They first demonstrated that N-acetyl-Phe-tRNA binds to the P-site by verifying its full reactivity with puromycin. Based on this experiment they assigned the bases protected by N-acetyl-Phe-tRNA in the 16S and 23S rRNA molecules to the 30S and 50S P-sites, respectively. Then they performed a second footprinting experiment after letting the complex react with puromycin. Puromycin did not affect the 16S rRNA footprint but had a profound effect on the 23S rRNA footprint. The CCA_{OH} end of the now deacylated tRNA moved so it no longer

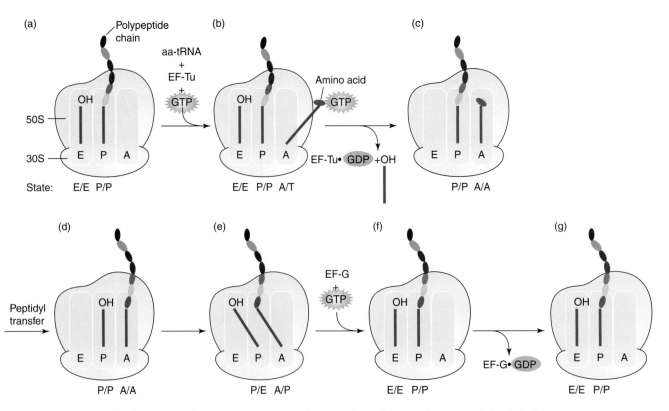

FIGURE 16.53 Schematic representation of our current understanding of the main steps of the hybrid-states translocational cycle. The tRNA molecules are shown as green lines, and the mRNA is not shown. (Adapted from Noller, H. F. et al. 2002. *FEBS Lett* 514:11–16.)

protected P-site bases on the 23S rRNA but instead protected E-site bases. The tRNA therefore appeared to be in a hybrid state with its anticodon end still bound to the 30S P-site but its CCA_{OH} end bound to 50S E-site. This hybrid state is represented as P/E, where the letters before and after the slash indicate the 30S and 50S subunit sites, respectively. Movement from the P/P state to the P/E state takes place spontaneously and requires neither GTP nor an elongation factor.

Noller and Moazed performed a second experiment in which they again bound N-acetyl-Phe-tRNA in the P-site but this time introduced aminoacyl-tRNA (and not puromycin) to the A-site. Footprinting performed after the peptidyl transferase reaction showed the two tRNAs had moved from their A/A and P/P states to A/P and P/E hybrid states. Later footprinting studies showed that EF-G • GTP converts the A/P and P/E hybrid states to P/P and E/E states, respectively. GTP is hydrolyzed during the translocation process (**BOX 16.6 LOOKING DEEPER: TRANSLOCATION MECHANISM**).

16.9 Termination Stage

Bacteria have three protein release factors.

Polypeptide chain termination takes place when a termination (nonsense) codon enters the A-site on the 30S ribosomal subunit. Early studies suggested that the three bases of the termination codon (UAA, UAG, or UGA) provide all the information required for termination. It now appears, however, that bases on either side of the termination codon influence the strength of the stop signal. For instance, UGA is a much stronger stop signal in *E. coli* when followed by a U than when followed by a C. The next two bases downstream as well as bases just upstream from the termination codon also influence stop signal strength but to a lesser extent.

Polypeptide chain termination requires release factors (RFs). Bacteria have three different release factors. RF1 and RF2 interact with termination codons, and RF3 stimulates the rate of peptide release when either RF1 or RF2 is also present but it does not act on its own. RF1 and RF2 both recognize the termination codon UAA. In addition, RF1 recognizes the termination codons UAG and RF2 recognizes the termination codon UGA.

Investigators proposed the following two mechanisms to explain how RF1 and RF2 work: (1) RF1 and RF2 are hydrolases that become activated upon binding to the ribosome, and (2) RF1 and RF2 stimulate a ribosomal hydrolase activity. Biochemical studies show it is the ribosome and not RF1 or RF2 that hydrolyzes the ester bond linking the completed peptide to the tRNA at the P-site. RF1 and RF2 act by triggering a conformational change in the ribosome's peptidyl transferase center, which permits water to attack the ester bond in the peptidyl-tRNA at the P-site. Bacterial RF3 is a nonessential protein that facilitates the dissociation of RF1 and RF2 from the ribosome complex after peptide release. Eukaryotes have two cytoplasmic release factors. One of these, eRF1, is similar to the bacterial RF1 and RF2

BOX 16.6: LOOKING DEEPER

Translocation Mechanism

The translocation mechanism (FIGURE B16.6) involves the following steps. Translocation begins after the polypeptide chain transfers from the tRNA at the P-site to the tRNA at the A-site. The resulting pre-translocation ribosome fluctuates between a classical and a hybrid state. In the classical state the peptidyl tRNA and deacylated tRNAs are in A/A and P/P sites, respectively. In the hybrid state the peptidyl tRNA and deacylated tRNAs are in A/P and P/E sites, respectively. The small subunit rotates around the large subunit in a counterclockwise direction as the ribosome changes from the classical to the hybrid state. EF-G • GTP binds to the ribosome when it is in the A/P and P/E hybrid state, stabilizing this state and inducing rapid GTP hydrolysis. This hydrolysis triggers a conformational change in EF-G, which in turn detaches the mRNA-tRNA complex from the decoding center. As a result of this "unlocking" the mRNA-tRNA complex can move up by one codon on the small subunit while the tRNAs retain their positions on the large subunit. Translocation is completed by a slight backward rotation of the small subunit as EF-G • GDP is released from the ribosome. The ribosome is now returned to the "locking" state, which prevents EF-G • GTP from binding to the ribosome but permits an aminoacyl-tRNA • EF-Tu • GTP complex to bind to the T/A site. Subsequent entry of the aminoacyl-tRNA into the A/A site during accommodation has been proposed to cause the uncharged tRNA to be released from the E/E site.

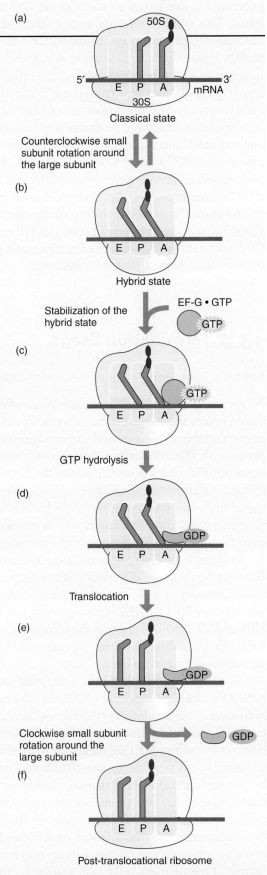

FIGURE B16.6 **A model for EF-G catalyzed translocation.** The pre-translocation ribosome fluctuates between a classical and hybrid state. (a) In the classical state the peptidyl tRNA and deacylated tRNAs are in A/A and P/P sites, respectively. (b) In the hybrid state the peptidyl tRNA and deacylated tRNAs are in A/P and P/E sites, respectively. The small subunit rotates around the large subunit in a counterclockwise direction as the ribosome changes from the classical to the hybrid state. (c) EF-G • GTP binds to the hybrid state ribosome, stabilizing the hybrid state and inducing rapid GTP hydrolysis. This hydrolysis triggers a conformational change in EF-G, which in turn detaches the mRNA-tRNA complex from the decoding center. (d) As a result of this "unlocking" the mRNA-tRNA complex can move up by one codon on the small subunit, whereas the tRNAs retain their positions on the large subunit. (e) Translocation is completed by the backward rotation of the small subunit as EF-G • GDP is released from the ribosome. (Adapted from Agirrezabala, X., and Frank, J. 2009. *Q Rev Biophys* 42:159–200.)

but recognizes all three termination codons. The other, eRF3, acts cooperatively with eRF1 to guarantee efficient termination codon recognition and rapid polypeptide release.

Mutant tRNA molecules can suppress mutations that create termination codons within a reading frame.

A simple mutational event can convert a sense codon into a termination codon. For example, a single base change converts UCG (a sense codon for serine) into UAG (a termination codon). Such nonsense mutations cause the protein synthetic machinery to produce nonfunctional polypeptide fragments. The cell's ability to synthesize a fully functional polypeptide can be restored by a second mutational event. As expected, many secondary mutations convert the termination codon to one that specifies an acceptable amino acid substitute and a few convert it back to the original codon. Some secondary mutations suppress the termination codon by altering the protein synthetic machinery rather than the affected gene. This suppression is often accomplished by changing the anticodon of a minor tRNA molecule so it becomes able to base pair with the termination codon. For example, a single base change in a minor $tRNA^{Ser}$ species converts its CGA anticodon into a CUA anticodon. The cell survives this change because it has a second $tRNA^{Ser}$ species that still recognizes the sense UCG codon for serine. The mutant $tRNA^{Ser}$ suppresses the nonsense mutation by translating UAG as a serine codon. Suppressor tRNA molecules might be expected to cause serious problems by translating normal termination codons and producing abnormally large polypeptide products. This problem is less serious than expected because (1) as a derivative of a minor tRNA molecule, the intracellular suppressor tRNA molecule's concentration is low, and (2) the protein synthetic machinery recognizes normal termination codons in the context of nucleotide sequences.

16.10 Recycling Stage

The ribosome release factor is required for the bacterial ribosomal complex to disassemble.

At the end of the termination stage in bacteria, the ribosome complex is still associated with mRNA and a deacylated tRNA, most probably with its acceptor end in the E-site of the 50S subunit and its anticodon end in the P-site of the 30S subunit. During the recycling stage the ribosomal subunits dissociate from this complex, freeing them to participate in a new round of polypeptide synthesis. In bacteria this process requires a new translation factor known as the ribosome release factor. EF-G • GTP and IF3 somehow assist the ribosome release factor in disassembling the posttermination complex. The details of this disassembly process remain to be determined.

Virtually nothing is known about the recycling stage in eukaryotes. Thus far, no eukaryotic counterpart to the bacterial ribosome release factor has been discovered. One possibility is that one or more eukaryotic ribosomal proteins assist in recycling so a soluble ribosome release factor is not required. After ribosome disassembly the small ribosomal subunit is free to interact with the initiation factors to start a new round of polypeptide synthesis.

16.11 Nascent Polypeptide Processing and Folding

Ribosomes have associated enzymes that process nascent polypeptides and chaperones that help to fold the nascent polypeptides.

The nascent polypeptide passes through a peptide exit tunnel that extends from the peptide transferase center to the ribosome surface (FIGURE 16.54). The exit tunnel in the bacterial ribosome is about 80 to 100 Å long and about 10 Å in diameter at its narrowest point (about 30 Å from the peptidyl transferase center) but widens to about twice that diameter at the rim of the exit point. The exit tunnel can accommodate an α helix with about 60 residues or an extended peptide with about half that number of residues. The space within the exit tunnel may permit the nascent polypeptide chain to assume an α-helical conformation but is too confining to permit more extensive folding. As the nascent peptide chain emerges from the tunnel, it can interact with enzymes that catalyze cotranslational modifications, chaperones that assist in folding and prevent misfolding, and the signal recognition particle that facilitates transport across the cell membrane (see below).

Peptide deformylase and methionine aminopeptidase bind at the rim of the bacterial ribosome's exit pore. The deformylase cleaves the N-terminal formyl group from the nascent polypeptide as it emerges from the exit tunnel, and then the aminopeptidase recognizes about 60% of the different nascent polypeptides and removes their N-terminal methionines. Nascent eukaryotic polypeptides do not have an N-terminal formyl group but do begin with methionine. Ribosome-bound methionine aminopeptidases remove the N-terminal methionine.

In bacteria, cotranslational protein folding is assisted by a 48-kDa chaperone called the **trigger factor**, which binds at the bacterial ribosome's exit pore (Figure 16.54). *E. coli* trigger factor is constitutively expressed, resulting in about two or three trigger factor molecules per ribosome. The trigger factor protein transiently associates with the 50S subunit. Its residence time on the ribosome depends on whether the ribosome has a nascent protein in the exit tunnel. If a nascent protein is not present, the average residence time is about 11 to 15 seconds but this time increases severalfold when a nascent protein is present. The trigger factor binds to hydrophobic patches as they emerge from the ribosome and sometimes remain associated with the segment even after polypeptide chain completion. Moreover, a single nascent polypeptide chain or free polypeptide may have two or more trigger factors associated with it.

A chaperone also associates with the large subunit of the eukaryotic ribosome. This chaperone consists of three different subunits

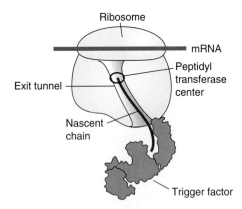

FIGURE 16.54 Schematic of ribosome and associated trigger factor. (Adapted from Hoffmann, A., et al. 2010. *Biochim Biophys Acta* 1803:652, Figure 1.)

that differ in both sequence and structure from the trigger factor. In yeast, deletion of any single subunit results in slow growth and cold sensitivity. Protein folding is a very active area of research because it merits attention as the final step in gene expression. Moreover, several neurological disorders including Alzheimer's, Huntington's, and Parkinson's diseases may be due to the accumulation of toxic proteins, which result from misfolding.

16.12 Signal Sequence

The signal sequence plays an important role in directing newly synthesized proteins to specific cellular destinations.

We now consider another aspect of ribosome function: the ribosome's interaction with the endoplasmic reticulum in eukaryotes and the cell membrane in prokaryotes. Soon after biologists started to investigate ribosomes, they observed that some ribosomes appear to exist free in the cytoplasm, whereas others are bound to the endoplasmic reticulum. The free ribosomes were shown to synthesize cytoplasmic and mitochondrial proteins, whereas the membrane-bound ribosomes synthesized integral membrane proteins, lysosomal proteins, and proteins that were secreted.

The reason some ribosomes bind to the endoplasmic reticulum was not clear at first. One possibility was that cells have two distinct kinds of ribosomes: those that bind to the endoplasmic reticulum and those that do not. Studies by Günter Blobel and David D. Sabatini in 1971 indicated that free and bound ribosomes appeared to be the same. There must therefore be some other explanation for why some ribosomes are bound to the endoplasmic reticulum and others are not.

In 1971 Blobel and Sabatini proposed the **signal hypothesis** to explain how cells determine whether a protein will be synthesized on a free or membrane-bound ribosome. According to the signal hypothesis, (1) free and membrane bound ribosomes are identical, (2) protein synthesis always begins on free ribosomes, and (3) nascent secretory, transmembrane, or lysosomal proteins have sequences of 20 to 30 amino acids at their amino terminus that act as signals to bind the ribosomes to the endoplasmic reticulum. Ribosomes only bind to the endoplasmic reticulum when synthesizing proteins with these signal sequences.

Although signal sequences vary from one protein to another, certain common features can be recognized. Signal sequences can be divided into three parts: a short, positively charged N-terminal region; a central region containing 7 to 13 hydrophobic amino acid residues; and a more polar C-terminal region that includes a cleavage site. Only polypeptides that have a signal sequence can be inserted into the endoplasmic reticulum membrane or transferred across it.

The amino end of a nascent polypeptide with a signal sequence requires the assistance of three components to pass into the lumen of the endoplasmic reticulum or to be integrated into the membrane. These components, which are present in all organisms, are the signal recognition particle (SRP), the SRP receptor (SR), and a protein conducting channel or translocon. In prokaryotes the cell membrane serves the same purpose as the endoplasmic reticulum in eukaryotes. We focus our attention on the mammalian SRP cycle because it was the first one to be discovered. The basic properties of the mammalian components required for the SRP cycle are as follows:

1. *Signal recognition particle:* SRP consists of six proteins and a 7S RNA. One of the protein subunits has a guanine nucleotide binding site and the signal sequence binding site is located within it.
2. *SRP receptor:* SR contains two GTP-binding polypeptide subunits and two subunits that are part of the membrane.
3. *Translocon:* The protein conducting channel, which is made of three polypeptides, acts as a passive conduit for polypeptides.

The three basic steps involved in the mammalian SRP cycle (**FIGURE 16.55**) are as follows: (1) SRP binds tightly to the signal sequence on the nascent polypeptide chain as the signal sequence

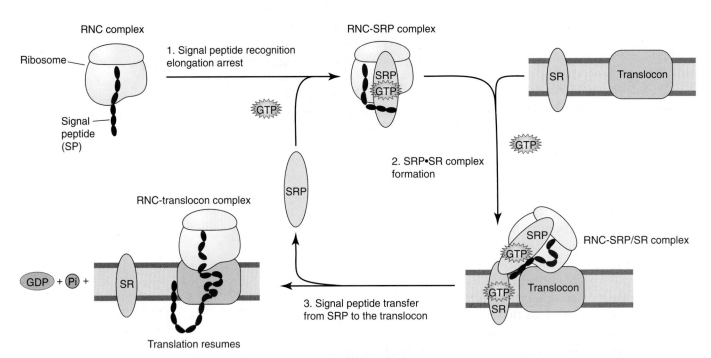

FIGURE 16.55 **Schematic representation of the mammalian signal recognition particle (SRP) cycle.** The three basic steps involved in the mammalian SRP cycle are as follows. (1) SRP binds tightly to the signal sequence on the nascent polypeptide chain as the signal sequence emerges from the ribosomal polypeptide exit tunnel. SRP goes through a conformational change as it binds to the ribosome-nascent chain complex (RNC). As a result of this conformational change, SRP binds to both the peptide exit tunnel and elongation factor-binding site on the ribosome. This binding prevents eEF1A and eEF2 from binding to the ribosome and causes translation to be arrested. (2) SRP carries the ribosome–nascent chain complex to the endoplasmic reticulum, where SRP and the membrane-bound SR interact. At this stage both SRP and SR have bound GTPs. The ribosome docks to the translocon and the signal peptide moves into the protein conducting channel. (3) The GTPs bound to the SRP • SR complex are hydrolyzed, causing SRP • SR to dissociate and releasing SRP from the ribosome. SRP release frees the ribosome to resume polypeptide chain elongation. The growing polypeptide chain moves through the protein conducting channel driven by the energy provided by additional GTP hydrolysis during the translocation stage of the elongation cycle. Upon entering the lumen of the endoplasmic reticulum the signal sequence encounters a specific peptidase called the signal peptidase that cleaves it from the remainder of the polypeptide chain. (Adapted from Luirink, J., and Sinning, I. 2004. *Biochim Biophys Acta* 1694:17–35.)

emerges from the ribosomal polypeptide exit tunnel. SRP goes through a conformational change as it binds to the **ribosome–nascent chain complex**. As a result of this conformational change SRP binds to both the peptide exit tunnel and elongation factor-binding site on the ribosome. This binding prevents eEF1A and eEF2 from binding to the ribosome and causes translation to be arrested. (2) SRP carries the ribosome–nascent chain complex to the endoplasmic reticulum, where SRP and the membrane-bound SR interact. At this stage, both SRP and SR have bound GTPs. The ribosome docks to the translocon and the signal peptide moves into the protein conducting channel. (3) The GTPs bound to the SRP • SR complex are hydrolyzed, causing SRP • SR to dissociate and releasing SRP from the ribosome. SRP release frees the ribosome to resume polypeptide chain elongation. The growing polypeptide chain moves through the protein conducting channel driven by the energy provided by additional GTP hydrolysis during the translocation stage of the elongation cycle. Upon entering the lumen of the endoplasmic reticulum the signal sequence encounters a specific peptidase called the signal peptidase that cleaves it from the remainder of the polypeptide chain. Although Figure 16.55 shows mammalian components, the process is very similar in all organisms.

Secretory and lysosomal proteins pass completely through the endoplasmic reticulum membrane and are directed to their ultimate destination by biochemical modifications such as glycosylation that take place in the lumen of the endoplasmic reticulum and in the Golgi apparatus. Integral membrane proteins have one or more hydrophobic sequences that allow them to move laterally across the translocon and into the membrane, causing the polypeptide chain to embed itself within the membrane.

Questions and Problems

1. Briefly define, describe, or illustrate each of the following.
    a. Sequence hypothesis
    b. Central dogma
    c. Smooth endoplasmic reticulum
    d. Rough endoplasmic reticulum
    e. Microsomal fraction
    f. Transfer RNA
    g. Isoacceptors
    h. Acceptor stem
    i. Codon
    j. Anticodon
    k. Anticodon arm
    l. D-arm
    m. TψC arm
    n. Variable arm
    o. "Double sieve" model
    p. Aptamer
    q. Identity elements
    r. Frameshift mutation
    s. Fingerprinting (peptide mapping)
    t. Nonsense mutations
    u. Wobble hypothesis
    v. A-site
    w. P-site
    x. E-site
    y. Initiation factors
    z. Elongation factors
    aa. Release factors
    bb. Ribosome recycling factor
    cc. 30S Initiation complex
    dd. 70S Initiation complex
    ee. Shine-Dalgarno sequence
    ff. Anti-Shine-Dalgarno sequence
    gg. Accommodation
    hh. Fragment assay
    ii. Translocation
    jj. Hybrid-states translocation model
    kk. Trigger factor
    ll. Signal hypothesis
    mm. ribosome–nascent chain complex

2. *In vitro* studies show that reactions catalyzed by aminoacyl-tRNA synthetases have equilibrium constants of about 1. What drives the reactions to completion in cells?

3. Describe an assay you would use to measure the activity of alanyl-tRNA synthetase?

4. Briefly describe two experiments showing that each aminoacyl-tRNA synthetase is specific for a single amino acid.

5. Describe Francis Crick's adaptor hypothesis. How does tRNA fit into this hypothesis?

6. Describe an experiment that demonstrates aminoacyl-tRNA is an intermediate in protein synthesis. How does this experiment rule out the possibility that tRNA only stores amino acids until they are needed for protein synthesis?

7. Describe an experiment that shows tRNA molecules have a CCA_{OH} at their 3'-ends.

8. Why does periodate inactivate tRNA but have no effect on aminoacyl-tRNA?

9. How would you remove the aminoacyl group from a aminoacyl-tRNA molecule without otherwise altering the tRNA molecule?

10. Describe two approaches for determining identity elements in tRNA molecules.

11. Describe the assay system that Marshall W. Nirenberg and Johann Heinrich Matthaei used to demonstrate that polyU codes for polyphenylalanine.

12. Describe the theoretical basis for using random copolymers to assign codons. What is the fundamental weakness of this approach?

13. How did investigators establish that polypeptide synthesis starts at the amino terminus and continues to the carboxyl terminus?

14. How did investigators determine that mRNA is read in a 5' to 3' direction? What is the significance of the fact that both transcription and translation take place in a 5' to 3' direction?

15. What is the trinucleotide binding assay and how did it contribute to the elucidation of the genetic code?

16. Predict the polypeptide that would be formed when an *in vitro* protein synthetic system uses each of the following mRNA templates.
    a. poly(UG)
    b. poly(UA)
    c. poly(UAC)
    d. poly(AAU)
    e. poly(UCAG)
    f. poly(CUG)

17. Describe the kind of mutation(s) that would convert UUG, the codon for tryptophan, into a nonsense or termination codon.

18. Suppose a meteor from some distant planet landed on earth and carried a virus. Would you be concerned about viral pathogenicity if you knew that life on the distant planet used a different genetic code from that used by organisms on earth? Explain your answer.

19. The genetic code is nonoverlapping, commaless, almost universal, highly degenerate, and unambiguous.
    a. What do each of these terms mean?
    b. Describe the experimental evidence that supports a nonoverlapping, commaless, almost universal, highly degenerate, and unambiguous genetic code.
    c. What is the wobble hypothesis and how does it fit with the fact that genetic code is highly degenerate?

20. What are the major structural features of the 70S bacterial ribosome?

21. What are the four stages of polypeptide synthesis?

22. Bacteria have two different kinds of tRNA^{Met}.
    a. What roles do these tRNA species play in polypeptide synthesis?
    b. Draw the pathway for fMet-tRNA^{fMet} formation.

23. fMet is the first amino acid added to all newly synthesized *E. coli* polypeptides. The formyl group, however, is seldom if ever present in *E. coli* proteins and most *E. coli* proteins do not have a methionine at their amino termini. How do you explain these observations?

24. Describe the experimental evidence showing that the Shine-Dalgarno sequence in mRNA and the anti-Shine-Dalgarno sequence in 16S rRNA interact during the initiation stage of bacterial polypeptide synthesis.

25. How does polypeptide chain initiation in eukaryotes differ from that in prokaryotes?
26. Describe or illustrate the initiation stage in bacterial polypeptide chain synthesis. Discuss the roles of the various components that participate in this process.
27. Describe or illustrate the elongation stage in bacterial polypeptide synthesis. Discuss the role of each of the protein factors that participate in this process.
28. Describe the experimental evidence that supports the notion that it is the RNA and not the proteins in the 50S ribosomal subunit that is essential for the peptidyl transferase reaction.
29. Describe an *in vitro* assay used to measure peptidyl transferase activity in the 50S subunit in the absence of aminoacyl-tRNA or peptidyl-tRNA.
30. How does puromycin block protein synthesis? Based on puromycin's mechanism of action, would you expect puromycin to be a useful antimicrobial agent to treat individuals suffering from a bacterial infection? Explain your answer.
31. Several different components participate in the mammalian SRP cycle. Describe the role of each of the following components in this process.
    a. signal sequence
    b. endoplasmic reticulum
    c. signal recognition particle
    d. SRP receptor
    e. translocon
    f. signal peptidase

Suggested Reading

Historical Overview

Clark, B. F. C. 2001. The crystallization and structural determination of tRNA. *Trends Biochem Sci* 26:511–514.

Hoagland, M. 1996. Biochemistry or molecular biology? The discovery of "soluble RNA." *Trends Biochem Sci* 21:77–80.

Kresge, N., Simoni, R. D., and Hill, R. L. 2006. The purification and sequencing of alanine transfer ribonucleic acid: the work of Robert W. Holley. *J Biol Chem* 281:e7–e9.

Maden, B. E. H. 2003. Historical review: peptidyl transfer, the Monro era. *Trends Biochem Sci* 28:619–624.

Nirenberg, M. 2004. Historical review: deciphering the genetic code—a personal account. *Trends Biochem Sci* 29:46–54.

Pederson, T. 2005. 50 years ago protein synthesis met molecular biology: the discoveries of amino acid activation and transfer RNA. *FASEB J* 19:1583–1584.

Zamecnik, P. 2005. From protein synthesis to genetic insertion. *Annu Rev Biochem* 74:1–28.

tRNA Structure

Apgar, J., Holley, R. W., and Merrill, S. H. 1962. Purification of the alanine-, valine-, histidine-, and tyrosine-acceptor ribonucleic acids from yeast. *J Biol Chem* 237:796–802.

Berg, P., Bergmann, F. H., Ofengand, E. J., and Dieckmann, M. 1961. The enzymic synthesis of amino acyl derivatives of ribonucleic acid. I. The mechanism of leucyl-, valyl-, isoleucyl-, and methionyl ribonucleic acid formation. *J Biol Chem* 236:1726–1734.

Clark, B. F. C. 2001. The crystallization and structural determination of tRNA. *Trends Biochem Sci* 26:511–514.

Goldman, E. 2008. Transfer RNA. *Encyclopedia of Life Sciences*. pp. 1–9. Hoboken, NJ: John Wiley & Sons.

Hoagland, M. B., Stephenson, M. L., Scott, J. F., et al. 1958. A soluble ribonucleic acid intermediate in protein synthesis. *J Biol Chem* 231:241–257.

Holley, R. W., Apgar, J., Everett, G. A., et al. 1965. Structure of a ribonucleic acid. *Science* 147:1462–1465.

Holley, R. W., Everett, G. A., Madison, J. T., and Zamir, A. 1965. Nucleotide sequences in the yeast alanine transfer ribonucleic acid. *J Biol Chem* 240:2122–2128.

Hopper, A. K., and Phizicky, E. M. 2003. tRNA transfers to the limelight. *Genes Dev* 17:162–180.

Kim, S. H., Quigley, G., Suddath, F. L., et al. 1972. The three-dimensional structure of yeast phenylalanine transfer RNA: shape of the molecule at 5.5-Angstrom resolution. *Proc Natl Acad Sci USA* 69:3746–3750.

Rich, A., and Kim, S. H. 1978. The three-dimensional structure of transfer RNA. *Sci Am* 238:52–62.

Robertus, J. D., Ladner, J. E., Finch, J. T., et al. 1974. Structure of yeast phenylalanine tRNA at 3 Å resolution. *Nature* 250:546–551.

Westhof, E., and Auffinger, P. 2001. tRNA structure. *Encyclopedia of Life Sciences*. pp. 1–11. London, UK: Nature Publishing Group.

Aminoacyl-tRNA Synthetases

Ahel, I., Korencic, D., Ibba, M., and Söll, D. 2003. Trans-editing of mischarged tRNAs. *Proc Natl Acad Sci USA* 100:15422–15427.

Antonellis, A., and Green, E. D. 2008. The role of aminoacyl-tRNA synthetases in genetic diseases. *Annu Rev Genomics Hum Genet* 9:87–107.

Arnez, J. G., and Moras, D. 1997. Structural and functional considerations of the aminoacylation reaction. *Trends Biochem Sci* 22:211–216.

Baldwin, A. N., and Berg, P. 1966. Transfer ribonucleic acid-induced hydrolysis of valyladenylate bound to isoleucyl ribonucleic acid synthetase. *J Biol Chem* 241:839–845.

Beebe, K., de Pouplana, L. R., and Schimmel, P. 2003. Elucidation of tRNA-dependent editing by a class II tRNA synthetase and significance for cell viability. *EMBO J* 22:668–675.

Beebe, K., Merriman, E., and Schimmel, P. 2003. Structure-specific tRNA determinants by editing mischarged amino acid. *J Biol Chem* 46:45056–45061.

Beuning, P. J., and Musier-Forsyth, K. 2000. Transfer RNA recognition by aminoacyl-tRNA synthetases. *Biopolymers* 52:1–28.

Beuning, P. J., and Musier-Forsyth, K. 2000. Hydrolytic editing by a class II aminoacyl-tRNA synthetase. *Proc Natl Acad Sci USA* 97:8916–8920.

Böck, A., Thanbichler, R., Rother, M., and Resch, A. 2004. Selenocysteine. M. Ibba, C. S. Francklyn, and S. Cusack (eds), *Aminoacyl-tRNA Synthetases*. Austin, TX: Landes Bioscience.

Commans, S., and Bock, A. 1999. Selenocysteine inserting tRNAs: an overview. *FEMS Microbiol Rev* 23:335–351.

de Pouplana, L. R., and Schimmel, P. 2000. A view into the origin of life: aminoacyl-tRNA synthetases. *Cell Mol Life Sci* 57:865–870.

Eldred, E.W., and Schimmel, P. R. 1972. Rapid deacylation by isoleucyl transfer ribonucleic acid synthetase of isoleucine-specific transfer ribonucleic acid aminoacylated with valine. *J Biol Chem* 247:2961–2964.

Fersht, A. 1998. Protein structure: sieves in sequence. *Science* 280:541.

Francklyn, C. 2008. DNA polymerase and aminoacyl-tRNA synthetases: shared mechanisms for ensuring the fidelity of gene expression. *Biochemistry* 45:11695–11703.

Fukai, S., Nureki, O., Sekine, S. I., et al. 2000. Structural basis for double-sieve discrimination of L-valine from L-isoleucine and L-threonine by the complex tRNA^{Val} and valyl-tRNA synthetase. *Cell* 103:793–803.

Geslain, R., and de Pouplana, L. R. 2004. Regulation of RNA function by aminoacylation and editing? *Trends Genet* 20:604–610.

Giege, R. 2008. Toward a more complete view of tRNA biology. *Nat Struct Mol Biol* 15:1007–1014.

Hale, S. P., and Schimmel, P. 1996. Protein synthesis editing by a DNA aptamer. *Proc Natl Acad Sci USA* 7:2755–2758.

Hatfield, D. L., and Gladyshev, V. N. 2002. How selenium has altered our understanding of the genetic code. *Mol Cell Biol* 22:3565–3576.

Hou, Y. M., and Schimmel, P. 1989. Evidence that a major determinant for the identity of a transfer RNA is conserved in evolution. *Biochemistry* 28:6800–6804.

Ibba, M., Becker, H. D., Stathopoulos, C., et al. 2000. The adaptor hypothesis revisited. *Trends Biochem Sci* 25:311–316.

Ibba, M., Curnow, A. W., and Söll, D. 1997. Aminoacyl-tRNA synthesis: divergent routes to a common goal. *Trends Biochem Sci* 22:39–42.

Ibba, M., Hong, K.-W., and Söll, D. 1996. Glutaminyl-tRNA synthetase: from genetics to molecular recognition. *Genes Cells* 1:421–427.

Ibba, M., and Söll, D. 1999. Quality control mechanisms during translation. *Science* 286:1893–1897.

Ibba, M., and Söll, D. 2000. Aminoacyl-tRNA synthetases. *Annu Rev Biochem* 69:617–650.

Ibba, M., and Söll, D. 2002. Genetic code: introducing pyrrolysine. *Curr Biol* 12:R464–R466.

Ibba, M., and Söll, D. 2004. Aminoacyl-tRNAs: setting the limits of the genetic code. *Genes Dev* 18:731–738.

Jakubowski, H. 2005. Transfer RNA synthetase editing of amino acids. *Encyclopedia of Life Sciences*. pp. 1–10. Hoboken, NJ: John Wiley & Sons.

Linecum, T. L. Jr., Tukalo, M., Yaremchuk, A., et al. 2003. Structural and mechanistic basis of pre- and posttransfer editing by leucyl-tRNA synthetase. *Mol Cell* 11:951–963.

Ling, J., Reynolds, N., and Ibba, M. 2009. Aminoacyl-tRNA synthesis and translational quality control. *Annu Rev Microbiol* 63:61–78.

Lukashenko, N. P. 2010. Expanding genetic code: amino acids 21 and 22 selenocysteine and pyrrolysine. *Russian J Genet* 46:899–916.

Nakama, T., Nueki, O., and Yokoyama, S. 2001. Structural basis for the recognition of isoleucyl-adenylate and an antibiotic, mupirocin, by isoleucyl-tRNA synthetase. *J Biol Chem* 276:47387–47393.

Nureki, O., Vassylyev, D. G., Tateno, M., et al. 1998. Enzyme structure with two catalytic sites for double-sieve selection of substrate. *Science* 280:576–582.

RajBhandary, U. L. 1997. Once there were twenty. *Proc Natl Acad Sci USA* 94:11761–11763.

Ryckelynck, M., Giegé, R., and Frugier, M. 2005. tRNAs and tRNA mimics as cornerstones of aminoacyl-tRNA synthetase regulations. *Biochimie* 87:835–845.

Schimmel, P. 2008. An editing activity that prevents mistranslation and connection to disease. *J Biol Chem* 283:28777–28782.

Schimmel, P. 2008. Development of tRNA synthetases and connection to genetic code and disease. *Protein Sci* 17:1643–1652.

Silvian, L. F., Wang, J., and Steitz, T. A. 1999. Insights into editing from an Ile-tRNA synthetase structure with tRNA^{Ile} and Mupirocin. *Science* 285:1074–1077.

mRNA and the Genetic Code

Agris, P. F. 2004. Decoding the genome: a modified view. *Nucleic Acids Res* 32:223–238.

Brenner, S., Stretton, A. O. W., and Kaplan, S. 1965. Genetic code: the "nonsense" triplets for chain termination and their suppression. *Nature* 206:994–998.

Chapeville, F., Lipmann, F., von Ehrenstein, G., et al. 1962. On the role of soluble ribonucleic acid in coding for amino acids. *Proc Natl Acad Sci USA* 48:1086–1092.

Crick, F. H. C. 1966. Codon-anticodon pairing: the wobble hypothesis. *J Mol Biol* 19:548–555.

Crick, F. H. C., Barnett, L., Brenner, S., and Watts-Tobin, R. J. 1961. General nature of the genetic code for proteins. *Nature* 192:1227–1232.

Davis, B. K. 2004. Expansion of the genetic code in yeast: making life more complex. *Bioessays* 26:111–115.

Dintzis, H. M. 1961. Assembly of the peptide chains of hemoglobin. *Proc Natl Acad Sci USA* 47:247–261.

Khorona, H. G. 1968. Synthesis in the study of nucleic acids. The Fourth Jubilee Lecture. *Biochem J* 109:709–725.

Nirenberg, M. W., and Leder, P. 1964. RNA codewords and protein synthesis: the effect of trinucleotides upon the binding of sRNA to ribosomes. *Science* 145:1399–1407.

Nirenberg, M. W., and Matthaei, J. H. 1961. The dependence of cell-free protein synthesis in *E. coli* upon naturally occurring or synthetic polyribonucleotides. *Proc Natl Acad Sci USA* 47:1588–1602.

Pardee, A., Jacob, F., and Monod, J. 1959. The genetic control and cytoplasmic expression of "inducibility" in the synthesis of beta-galactosidase by *E. coli*. *J Mol Biol* 1:165–178.

Salas, M., Smith, M. A., Stanley, W. M. Jr., et al. 1965. Direction of reading of the genetic message. *J Biol Chem* 240:3988–3995.

Volkin, E., and Astrachan, L. 1956. Intracellular distribution of labeled ribonucleic acid after phage infection of *Escherichia coli*. *Virology* 2:433–437.

Watanabe, K., and Suzuki, T. 2008. Universal genetic code and its natural variations. *Encyclopedia of Life Sciences*. pp. 1–8. Hoboken, NJ: John Wiley & Sons.

Ribosome Structure: Overview

Green, R., and Noller, H. F. 1997. Ribosomes and translation. *Annu Rev Biochem* 66:679–716.

Kapp, L. D., and Lorsch, J. L. 2004. The molecular mechanics of eukaryotic translation. *Annu Rev Biochem* 73:657–704.

Schmeing, T. M., and Ramakrishnan, V. 2009. What recent ribosome structures have revealed about the mechanisms of translation. *Nature* 461:1234–1242.

Wilson, D. N., Blaha, G., Connell, S. R., et al. 2002. Protein synthesis at atomic resolution: mechanistics of translation in the light of highly resolved structures for the ribosomes. *Curr Protein Pept Sci* 3:1–53.

Ribosome Structure

Al-Kardaghi, S., and Liljas, O. K. 2000. A decade of progress in understanding the structural basis of protein synthesis. *Progr Biophys Mol Biol* 73:167–193.

Bashan, A., and Yonath, A. 2008. Correlating ribosome function with high-resolution structures. *Trends Microbiol* 16:326–335.

Berk, V., and Cate, J. H. D. 2007. Insights into protein biosynthesis from structures of bacterial ribosomes. *Curr Opin Struct Biol* 17:302–309.

Culver, G. M. 2001. Meanderings of the mRNA through the ribosome. *Structure* 9:751–758.

Demeshkina, N., Jenner, L., Yusupova, G., and Yusupov, M. 2010. Interactions of the ribosome with mRNA and tRNA. *Curr Opin Struct Biol* 20:1–10.

Frank, J. 2000. The ribosome—a macromolecular machine par excellence. *Chem Biol* 7:R133–R141.

Frank, J. 2001. Structure of the 80S ribosome from Saccharomyces cerevisiae—tRNA-ribosome and subunit-subunit interactions. *Cell* 107:373–386.

Frank, J. 2003. Toward an understanding of the structural basis of translation. *Genome Biol* 4:237.

Frank, J., Agrawal, R. K., and Verschoor, A. 2001. Ribosome shape and structure. *Encyclopedia of Life Sciences*. pp. 1–6. London: Nature.

Korostelev, A., Ermolenko, D. N., and Noller, H. F. 2008. Structural dynamics of the ribosome. *Curr Opin Chem Biol* 12:1–10.

Korostelev, A., and Noller, H. F. 2007. The ribosome in focus: new structures bring new insights. *Trends Biochem Sci* 32:434–441.

Korostelev, A., Trakhanov, S., Laurberg, M., et al. 2007. Interactions and dynamics of the Shine–Dalgarno helix in the 70S ribosome. *Proc Natl Acad Sci USA* 104: 16840–16843.

Korostelev, A., Trakhanov, S., Laurberg, M., and Noller, H. F. 2006. Crystal structure of a 70S ribosome-tRNA complex reveals functional interactions and rearrangements. *Cell* 126:1065–1077.

Moore, P. B. 2001. The ribosome at atomic resolution. *Biochemistry* 40:3143–3250.

Moore, P. B. 2009. The ribosome returned. *J Biol* 8:1–8.

Moore, P. B., and Steitz, T. A. 2002. The involvement of RNA in ribosome function. *Nature* 418:229–235.

Noller, H. F., Hoang, L., and Frerick, K. 2005. The 30S ribosomal P site: a function of 16S rRNA. *FEBS Lett* 579:855–858.

Puglisi, J. D. 2009. Resolving the elegant architecture of the ribosome. *Mol Cell* 36:720–723.

Ramakrishnan, V. 2002. Ribosome structure and the mechanism of translation. *Cell* 108:557–572.

Ramakrishnan, V. 2008. What we have learned from ribosome structures. *Biochem Soc Trans* 36:567–574.

Ramakrishnan, V. 2010. Unravelling the structure of the ribosome (Nobel Lecture). *Angew Chem Int Ed* 49:4355–4380.

Ramakrishnan, V., and Moore, P. B. 2001. Atomic structures at last: the ribosome in 2000. *Curr Opin Struct Biol* 11:144–154.

Schuwirth, B. S., Borovinskaya, M. A., Hau, C. W., et al. 2005. Structures of the bacterial ribosome at 3.5Å resolution. *Science* 310:827–834.

Selmer, M., Dunham, C. M., Murphy, F. F., et al. 2006. Structure of the 70S ribosome complexed with mRNA and tRNA. *Science* 313:1935–1942.

Spahn, C. M. T., Beckmann, R., Eswar, N., et al. 2002. Ribosome as a molecular machine. *FEBS Lett* 514:2–10.

Steitz, T. A. 2008. A structural understanding of the dynamic ribosome machine. *Nat Rev Mol Cell Biol* 9:242–253.

Steitz, T. A. 2010. From the structure and function of the ribosome to new antibiotics (Nobel Lecture). *Angew Chem Int Ed* 49:4381–4398.

Taylor, D. J., Devkota, B., Huang, A. D., et al. 2009. Comprehensive molecular structure of the eukaryotic ribosome. *Structure* 17:1591–1604.

Wimberly, B. T., Brodersen, D. E., Clemons, W. M. Jr., et al. 2000. Structure of the 30S ribosomal subunit. *Nature* 407:327–339.

Yonath, A. 2010. Polar bears, antibiotics, and the evolving ribosome (Nobel Lecture). *Angew Chem Int Ed* 49:4340–4354.

Yusupov, M. M., Yusupova, G. Z., Baucom, A., et al. 2001. Crystal structure of the ribosome at 5.5 Å resolution. *Science* 292:883–896.

Initiation Stage

Boelens, R., and Gualerzi, C. O. 2002. Structure and function of bacterial initiation factors. *Curr Protein Pept Sci* 3:107–119.

Carr-Schmid, A., and Kinzy, T. G. 2001. Messenger RNA: interaction with ribosomes. *Encyclopedia of Life Sciences*. London: *Nature*. pp. 1–6.

Culver, G. M. 2001. Meanderings of the mRNA through the ribosome. *Structure* 9:751–758.

Dever, T. E. 2002. Gene-specific regulation by general translation factors. *Cell* 108:545–556.

Drabkin, H. J., Estrella, M., and Rajbhandary, U. L. 1998. Initiator-elongator discrimination in vertebrate tRNAs for protein synthesis. *Mol Cell Biol* 18:1459–1466.

Grunberg-Manago, M., Suder, S. M., and Joseph, S. 2007. Protein synthesis initiation in bacteria. *Encyclopedia of Life Sciences*. pp. 1–10. Hoboken, NJ: John Wiley & Sons.

Guthrie, C., and Nomura, M. 1968. Initiation of protein synthesis: a critical test of the 30S subunit model. *Nature* 219:232–235.

Hellen, C. U. T., and Pestova, T. V. 2006. Translation initiation: Molecular mechanisms in eukaryotes. *Encyclopedia of Life Sciences*. pp. 1–5. Hoboken, NJ: John Wiley & Sons.

Hellen, C. U. T., and Sarnow, P. 2001. Internal ribosome entry sites in eukaryotic mRNA molecules. *Genes Dev* 15:1593–1612.

Jackson, R. J., Hellen, C. E., and Pestova, T. V. 2010. The mechanism of eukaryotic translation initiation and principles of its regulation. *Nat Rev Mol Cell Biol* 10:113–127.

Kaempfer, R. O. R., Meselson, M., and Raskas, H. J. 1968. Cyclic dissociation into stable subunits and re-formation of ribosomes during bacterial growth. *J Mol Biol* 31: 277–289.

Kapp, L. D., and Lorsch, J. R. 2004. The molecular mechanics of eukaryotic translation. *Annu Rev Biochem* 73:657–704.

Kozak, M. 1999. Initiation of translation in prokaryotes and eukaryotes. *Gene* 234:187–208.

Kozak, M. 2001. New ways of initiating translation in eukaryotes? *Mol Cell Biol* 21:1899–1907.

Kozak, M. 2002. Pushing the limits of the scanning mechanism for initiation of translation. *Gene* 299:1–34.

Laursen, B. S., Sørensen, H. P., Mortensen, K. K., and Sperling-Petersen, H. U. 2005. Initiation of protein synthesis in bacteria. *Microbiol Mol Biol Rev* 69:101–123.

Mandal, M., and Breaker, R. R. 2004. Gene regulation by riboswitches. *Nat Rev Mol Cell Biol* 5:451–463.

Mangroo, D., Wu, X., and Rajbhandary, U. L. 1995. *Escherichia coli* initiator tRNA: structure–function relationships and interactions with the translational machinery. *Biochem Cell Biol* 73:1023–1031.

Marcker, K., and Sanger, F. 1964. N-formyl-methionyl-sRNA. *J Mol Biol* 8: 835–840.

Marintchev, A., and Wagner, G. 2004. Translation initiation: structures, mechanisms, and evolution. *Q Rev Biophys* 37:197–284.

Marshall, R. A., Aitken, C. E., and Puglisi, J. D. 2009. GTP hydrolysis by IF2 guides progression of the ribosome into elongation. *Mol Cell* 35:37–47.

Matthews, M. B. 2002. Lost in translation. *Trends Biochem Sci* 27:267–269.

Meinnel, T. M., Mechulam, Y., and Blanquet, S. 1993. Methionine as translation start signal: a review of the enzymes of the pathway in *Escherichia coli*. *Biochimie* 75:1061–1075.

Merrick, W. C. 2010. Eukaryotic protein synthesis: still a mystery. *J Biol Chem* 285:21197–21201.

Merrick, W. C. 2003. Initiation of protein biosynthesis in eukaryotes. *BAMBEd* 31:378–385.

Myasnikov, A. G., Simonetti, A., Marzi, S., and Klaholz, B. P. 2009. Structure-function insights into prokaryotic and eukaryotic translation initiation. *Curr Opin Struct Biol* 19:300–309.

Nudler, E. 2006. Flipping riboswitches. *Cell* 126:19–22.

Nudler, E., and Mironov, A. S. 2004. The riboswitch control of bacterial metabolism. *Trends Biochem Sci* 29:11–17.

Ochoa, S. 1968. Translation of the genetic message. *Naturwissenschafter* 55:505–514.

Pestova, T. V., and Hellen, C. U. T. 2001. The structure and function of initiation factors in eukaryotic protein synthesis. *Mol Life Sci* 57:651–674.

Pestova, T. V., Kolupaeva, V. G., Lomakin, I. B., et al. 2001. Molecular mechanisms of translation initiation in eukaryotes. *Proc Natl Acad Sci USA* 98:7029–7036.

Preiss, T., and Hentze, M. W. 2003. Starting the protein synthesis machine: eukaryotic translation initiation. *Bioessays* 25:1201–1211.

Proud, C. G. 2001. Regulation of mRNA translation. *Essays Biochem* 37:97–108.

Rajbhandary, U. L. 1994. Minireview: initiator transfer RNAs. *J Bacteriol* 176: 547–552.

Rodnina, M. V., and Wintermeyer, W. 2009. Recent mechanistic insights into eukaryotic ribosomes. *Curr Opin Cell Biol* 21:435–443.

Roll-Mecak, A., Shin, B. S., Dever, T. E., and Burley, S. K. 2001. Engaging the ribosome: universal ifs of translation. *Trends Biochem Sci* 26:705–709.

Sachs, A. B., and Varani, G. 2000. Eukaryotic translation initiation: there are (at least) two sides to every story. *Nat Struct Biol* 7:356–361.

Simonetti, A., Marzi, S., Jenner, L., et al. 2009. A structural view of translation initiation in bacteria. *Cell Mol Life Sci* 66:423–436.

Simonetti, A., Marzi, S., Myasnikov, A. G., et al. 2008. Structure of the 30S translation initiation complex. *Nature* 455:416–420.

Sonenberg, N., and Hinnebusch, A. G. 2009. Regulation of translation initiation in eukaryotes: mechanisms and biological targets. *Cell* 136:731–745.

Sperling-Petersen, H. U., Laursen, B. S., and Mortensen, K. K. 2002. Initiator tRNAs in prokaryotes and eukaryotes. *Encyclopedia of Life Sciences*. pp. 1–7. London: Nature.

Steitz, J. A., and Jakes, K. 1975. How ribosomes select initiator regions in mRNA: base pair formation between the 3′ terminus of 16S rRNA and the mRNA during initiation of protein synthesis in *Escherichia coli*. *Proc Natl Acad Sci USA* 72:4734–4738.

Vitreschak, A. G., Rodionov, D. A., Mironov, A. A., and Gelfand, M. S. 2004. Riboswitches: the oldest mechanism for the regulation of gene expression? *Trend Genet* 20:44–50.

Yuan, Z., Trias, J., and White, R. J. 2001. Deformylase as a novel antibacterial target. *Drug Disc Today* 6:954–961.

Elongation Stage

Agirrezabala, X., and Frank, J. 2009. Elongation in translation as a dynamic interaction among the ribosome, tRNA, and elongation factors EF-G and EF-Tu. *Q Rev Biophys* 43:159–200.

Agrawal, R. K., Spahn, C. M. T., Penczek, P., et al. 2000. Visualization of tRNA movements on the *Escherichia coli* 70S ribosome during the elongation cycle. *J Cell Biol* 150:447–459.

Andersen, G. R., Nissen, P., and Nyborg, J. 2003. Elongation factors in protein biosynthesis. *Trends Biochem Sci* 28:434–441.

Bashan, A., Agmon, I., Zarivich, R., et al. Structural basis of the ribosomal machinery for peptide bond formation, translocation, and nascent chain progression. *Mol Cell* 11:91–102.

Blaha, G., Stanley, R. E., and Steitz, T. A. 2009. Formation of the first peptide bond: the structure of EF-P bound to the 70S ribosome. *Science* 325:966–970.

Brodersen, D. E., and Ramakrishnan, V. 2003. Shape can be seductive. *Nat Struct Biol* 10:78–80.

Cabrita, L. D., Dobson, C. M., and Christodoulou, J. 2010. Protein folding on the ribosome. *Curr Opin Struct Biol* 20:33–45.

Connell, S. R., Takemoto, C., Wilson, D. N., et al. 2007. Structural basis for interaction of the ribosome with the switch regions of GTP-bond elongation factors. *Mol Cell* 25:751–764.

Czworkowski, J. 2001. Elongation factors, bacterial. *Encyclopedia of Life Sciences*. pp. 1–6. London: Nature.

Dahlberg, A. E. 2001. Ribosome structure: the ribosome in action. *Science* 292:868–869.

Ehrenberg, M. 2010. Translocation in slow motion. *Nature* 466:325–326.

Etchells, S. A., and Hartl, F. U. 2004. The dynamic tunnel. *Nat Struct Biol* 11:382–391.

Frank, J. 1998. How the ribosome works. *Am Sci* 86:428–439.

Frank, J. 2003. Electron microscopy of functional ribosome complexes. *Biopolymers* 68:223–233.

Frank, J., and Agrawal, R. K. 2001. Ratchet-like movements between the two ribosomal subunits: their implications in elongation factor recognition and tRNA translocation. *Cold Spring Harb Symp Quant Biol* 66:67–75.

Frank, J., Gao, H., Segupta, H., et al. 2007. The process of mRNA-tRNA translocation. *Proc Natl Acad Sci USA* 104:19671–19678.

Gao, Y.-G., Selmer, M., Dunham, C. M., et al. 2009. The structure of the ribosome with elongation factor G trapped in the posttranslocational state. *Science* 326:694–699.

Gilbert, R. J. C., Fucini, P., Connell, S., et al. 2004. Three dimensional structures of translating ribosomes by cryo-EM. *Mol Cell* 14:57–66.

Hoffmann, A., Bukau, B., and Kramer, G. 2010. Structure and function of the molecular chaperone trigger factor. *Biochim Biophys Acta* 1803:650–661.

Horwich, A. 2004. Sight at the end of the tunnel. *Nature* 431:520–522.

Jenni, S., and Ban, N. 2003. The chemistry of protein synthesis and voyage through the ribosomal tunnel. *Curr Opin Struct Biol* 13:212–219.

Krab, I. M., and Parmeggiani, A. 2002. Mechanisms of EF-Tu, a pioneer GTPase. *Prog Nucleic Acid Res Mol Biol* 71:513–551.

Kramer, G., Boehringer, D., Ban, N., and Bukau, B. 2009. The ribosome as a platform of co-translational processing, folding and targeting of newly synthesized proteins. *Nat Struct Mol Biol* 16:589–597.

Joseph, S. 2003. After the ribosome structure: how does translocation work. *RNA* 9:160–164.

Liljas, A. 2009. Leaps in translational elongation. *Science* 326:677–678.

Lucas-Lenard, J., and Lipmann, F. 1966. Separation of three microbial amino acid polymerization factors. *Proc Natl Acad Sci USA* 55:1562–1566.

Moazed, D., and Noller, H. F. 1989. Intermediate states in the movement of transfer RNA in the ribosome. *Nature* 342:142–148.

Monro, R. E. 1967. Catalysis of peptide bond formation by 50S ribosomal subunits from *Escherichia coli*. *J Mol Biol* 26:147–160.

Moore, P. B., and Steitz, T. A. 2003. After the ribosome structures: how does peptidyl transferase work? *RNA* 9:155–159.

Nierhaus, K. H., and Stiezl, U. 2001. Peptidyl transfer on the ribosome. *Encyclopedia of Life Sciences*. pp. 1–8. London: Nature.

Noller, H. F., Hoffarth, V., and Zimniak, L. 1992. Unusual resistance peptidyl transferase in protein extraction procedures. *Science* 256:1416–1419.

Noller, H. F., Yusopov, M. M., Yusopova, G. Z., et al. 2002. Translocation of tRNA during protein synthesis *FEBS Lett* 514:11–16.

Ogle, J. M., Carter, A. P., and Ramakrishnan, V. 2003. Insights into the decoding mechanism from recent ribosome structures. *Trends Biochem Sci* 28:259–266.

Pech, M., and Nierhuas, K. H. 2008. Ribosomal peptide-bond formation. *Chem Biol* 15:417–419.

Puglisi, J. D., Blanchard, S. C., and Green, R. 2000. Approaching translation at atomic resolution. *Nat Struct Biol* 7:855–861.

Rodnina, M. V., Daviter, T., Gromadski, K., and Wintermeyer, W. 2002. Structural dynamics of ribosomal RNA during decoding of the ribosome. *Biochimie* 84:745–754.

Schmeing, T. M., Huang, K. S., Kitchen, D. E., et al. 2005. Structural insights into the roles of water and the 2′ hydroxyl of the P site tRNA in the peptidyl transferase reaction. *Mol Cell* 20:437–448.

Schmeing, T. M., Huang, K. S., Strobel, S. A., and Steitz, T. A. 2005. An induced-fit mechanism to promote peptide bond formation and exclude hydrolysis of peptidyl-tRNA. *Nature* 438:520–524.

Schmeing, T. M., Voorhees, R. M., Kelley, A., et al. 2009. The crystal structure of the ribosome bound to EF-Tu and aminoacyl-tRNA. *Science* 326:688–694.

Simonović, M., and Steitz, T. A. 2009. A structural view on the mechanism of the ribosome-catalyzed peptide bond formation. *Biochim Biophys Acta* 1789: 612–623.

Sprinzl, M., Brock, S., Huang, Y., and Miovnik, P. 2000. Regulation of GTPases in bacterial translation machinery. *Biol Chem* 381:367–375.

Steitz, T. A., and Moore, P. B. 2000. RNA, the first macromolecular catalyst: the ribosome is a ribozyme. *Trends Biochem Sci* 28:411–418.

Valle, M., Sengupta, J., Swami, N. K., et al. 2002. Cryo-EM reveals an active role for aminoacyl-tRNA in the accommodation process. *EMBO J* 21:3557–3567.

Valle, M., Zavialov, A., Li, W., et al. 2003. Incorporation of aminoacyl-tRNA into the ribosome as seen by cyro-electron microscopy. *Nat Struct Biol* 10:899–906.

Valle, M., Zavialov, A., Sengupta, J., et al. 2003. Locking and unlocking of ribosomal motions. *Cell* 114:123–134.

Voorhes, R. M., Weixlbaumer, A., Loakes, D., et al. 2009. Insights into substrate stabilization from snapshots of the peptidyl transferase center of the intact 70S ribosome. *Nat Struct Mol Biol* 16:528–533.

Voss, N. R., Gerstein, M., Steitz, T. A., and Moore, P. B. 2006. The geometry of the ribosomal polypeptide exit tunnel. *J Mol Biol* 360:893–906.

Weinger, J. S., Kitchen, D., Scaringe, S. A., et al. 2004. Solid phase synthesis and binding affinity of peptidyl transferase transition state mimics containing 2′-OH at P-site position A76. *Nucleic Acids Res* 32:1502–1511.

Weinger, J. S., Parnall, K. M., Dorner, S., et al. 2004. Substrate-assisted catalysis of peptide bond formation by the ribosome. *Nat Struct Mol Biol* 11:1101–1106.

Termination Stage

Brodersen, D. E., and Ramakrishnan, V. 2003. Shape can be seductive. *Nat Struct Biol* 10:78–80.

Cappechi, M. R, 1967. Polypeptide chain termination in vitro: isolation of a release factor. *Proc Natl Acad Sci USA* 58:1144–1151.

Caskey, C. T., Tompkins, R., Scolnick, E., et al. 1968. Sequential translation of trinucleotide codons for the initiation and termination of protein synthesis. *Science* 162:135–138.

Ehrenberg, M., and Tenson, T. 2002. A new beginning of the end of translation. *Nat Struct Biol* 9:85–87.

Herrington, M. B. 2003. Nonsense mutations and suppression. *Encyclopedia of Life Sciences*. pp. 1–6. London: Nature.

Ito, K., Uno, M., and Nakamura, Y. 2000. A tripeptide "anticodon" deciphers stop codons in messenger RNA. *Nature* 403:680–684.

Kisselev, R. L., and Buckingham, R. H. 2000. Translation termination comes of age. *Trends Biochem Sci* 25:561–566.

Kisselev, L., Ehrenberg, M., and Frolova, L. 2003. Termination of translation: interplay of mRNA, rRNAs and release factors? *EMBO J* 22:175–182.

Kjeldgaard, M. 2003. The unfolding story of polypeptide release factors. *Mol Cell* 11:8–10.

Korostelev, A., Asahara, J., Lancaster, L., et al. 2008. Crystal structure of a translation termination complex formed with release factor RF2. *Proc Natl Acad Sci USA* 105:19684–19689.

Laurberg, M., Asahara, H., Korostelev, A., et al. 2008. Structural basis for translation termination on the 70S ribosome. *Nature* 454:852–857.

Nakamura, Y., and Ito, K. 2003. Making sense of mimic in translation termination. *Trends Biochem Sci* 28:99–105.

Rawat, U. B. S., Zavialov, A. V., Sengupta, J., et al. 2003. A cryo-electron microscopic study of ribosome-bound termination factor RF2. *Nature* 421:87–90.

Scolnick, E., Tompkins, R., Caskey, T., and Nirenberg, M. 1968. Release factors differing in specificity for terminator codons. *Proc Natl Acad Sci USA* 61: 768–774.

Tate, W. P., Poole, E. S., and Mannering, S. A. 2001. Protein synthesis termination. *Encyclopedia of Life Sciences*. pp. 1–6. London: Nature.

Weixlbaumer, A., Jin, H., Neubauer, C., et al. 2008. Insights into translational termination from the structure of RF2 bound to the ribosome. *Science* 352: 953–956.

Recycling Stage

Agrawal, R. K., Sharma, M. R., Kiel, M. C., et al. 2004. Visualization of ribosome-recycling factor on the *Escherichia coli* 70S ribosome: functional implications. *Proc Natl Acad Sci USA* 101:8900–8905.

Hirokawa, G., Demeshkina, N., Iwakura, N., et al. 2006. The ribosome-recycling step: consensus or controversy. *Trends Biochem Sci* 31:143–149.

Kim, K. K., Min, K., and Suh, S. W. 2000. Crystal structure of the ribosome recycling factor from *Escherichia coli*. *EMBO J* 19:2362–2370.

Signal Sequence

Adler, N. N., and Johnson, A. E. 2004. Cotranslational membrane protein biogenesis at the endoplasmic reticulum. *J Biol Chem* 279:22787–22790.

Blobel, G., and Dobberstein, B. 1975. Transfer of proteins across membranes. I. Presence of proteolytically processed and unprocessed nascent light immunoglobulin light chains on membrane-bound ribosomes of murine myeloma. *J Cell Biol* 67:835–851.

Halic, M., Becker, T., Pool, M. R., Spahn, C. M. T., et al. 2004. Structure of the signal recognition particle interacting with the elongation-arrested ribosome. *Nature* 427:808–814.

Halic, M., and Beckmann, R. 2005. The signal recognition particle and its interactions during protein targeting. *Curr Opin Struct Biol* 15:116–125.

Herkovits, A. A., Bochkareva, E. S., and Bibi, E. 2000. New prospects in studying the bacterial signal recognition particle pathway. *Mol Microbiol* 38:927–939.

Luirink, J., and Sinning, I. 2004. SRP-mediated protein targeting: structure and function revisited. *Biochim Biophys Acta* 1694:17–35.

Ryan, M. T., and Pfanner, N. 2001. Protein translocation across membranes. *Encyclopedia of Life Sciences*. pp. 1–10. London: Nature.

Wild, K., Rosendal, K. R., and Sinning, I. 2004. A structural step into the SRP cycle. *Mol Microbiol* 53:357–363.

Wild, K., Weichenrieder, O., Strub, K., et al. 2002. Towards the structure of the mammalian signal recognition particle. *Curr Opin Struct Biol* 12:72–81.

Glossary

A tracts Tracts consisting of four to six adjacent adenines that cause DNA to bend.

Abasic site (also called the **AP site**) A location in DNA that does not have either a purine or pyrimidine base, usually due to DNA damage resulting in N-glycosyl bond cleavage.

Abortive initiation The synthesis and release of oligoribonucleotides that takes place before transcription progresses to the elongation stage.

Absolute defective mutant A mutant that displays the mutant phenotype under all conditions.

Abzyme Catalytically active antibody.

Ac **element** A transposable element in maize.

Acceptor stem The site of amino acid attachment to tRNA that is formed by base pairing between nucleotides at the 5′-end and nucleotides near the 3′-end of the tRNA molecule.

Accommodation The step in translation elongation during which the free aminoacyl end of aminoacyl-tRNA is released from EF-Tu and moves to the peptidyl transferase center of the large ribosomal subunit.

ACF (see **ATP-utilizing chromatin assembly and remodeling factor**)

Acquired immunodeficiency syndrome (AIDS) A disease caused by the retrovirus HIV in which the immune system's ability to fight infection is severely reduced by destruction of T lymphocytes.

Actinomycin D An antibiotic (produced by *Streptomyces*), which has a heterocyclic ring system that inserts between C-G and G-C base pairs in DNA and inhibits transcription.

Activated complex (also called **transition state**) The intermediate formed when reactants gain sufficient energy to reach the top of the energy barrier, before being converted to products.

Activating region 1 (AR1) The cAMP receptor protein (CRP) region that interacts with the α subunit of bacterial RNA polymerase.

Activation domain The domain in eukaryotic transcription activator proteins that recruits components of the transcription machinery to the gene and interacts with various components of the transcription machinery to stimulate transcription.

Activator A protein that increases the transcription of one or more genes.

Active site The specific region on an enzyme responsible for binding the substrate(s) and catalyzing the reaction.

Ada complex A protein complex that stimulates transcription at least in part by catalyzing histone acetylation.

Adduct The product formed by attaching a chemical group to DNA.

Adenine A purine base derivative that is one of the four bases present in DNA and RNA molecules.

Adenylate cyclase An enzyme that converts ATP to cyclic AMP and pyrophosphate.

A-DNA A double helix DNA structure that appears as the relative humidity falls to about 75%. It is not normally present in DNA under physiological conditions.

Affinity chromatography A chromatographic technique that takes advantage of the fact that many proteins can bind specific ligands.

Aflatoxins A class of chemical carcinogen produced by some members of the genus *Aspergillus* that must be activated before they can form DNA adducts.

Alanine An amino acid with a methyl side chain, represented by Ala or A.

Alkylating agents A highly reactive group of electrophiles that transfer methyl, ethyl, or larger alkyl groups to the electron-rich atoms in DNA.

O^6-Alkylguanine DNA alkyltransferase I An *E. coli* enzyme that can transfer methyl and other alkyl groups attached to O-6 in guanine or to O-4 in thymine to one of its own cysteine residues. In addition, the enzyme can transfer an alkyl group in a phosphotriester.

O^6-Alkylguanine DNA alkyltransferase II An *E. coli* alkyl transfer protein that transfers a single alkyl group from the O-6 position in guanine or the O-4 position in thymine to one of its own cysteine residues.

Allele One of several alternative forms of a gene at a particular locus on a chromosome.

Allosteric effector A molecule that influences protein activity by binding to a site that is distinct from the active (or functional) sites.

Alpha (α) amanitin A toxin produced by the mushroom *Amanita phalloides* that inhibits transcription by RNA polymerase II.

Alpha (α) amino acid An amino acid in which the central carbon atom is attached to an amino ($-NH_3^+$) group, a carboxylate ($-COO^-$) group, a hydrogen atom, and a side chain ($-R$).

Alpha (α) helix A polypeptide secondary structure in which a linear sequence of amino acids folds into a right-handed helix, with hydrogen bonds inside the helix.

Alpha (α) satellite DNA array A high order repeat found in human centromeric DNA.

Alternative splicing A process that produces two or more distinct mRNA molecules from the same pre-mRNA as a result of selecting different exons for inclusion in the mRNA.

Alu The most common type of short interspersed nucleotide element in the human genome.

Ames test A diagnostic procedure used to detect potential carcinogens by the agent's ability to cause mutations in bacterial cells.

Amino acid residues Amino acids once they are linked together in a peptide. Often called residues for short.

Amino acids Small organic molecules containing both amine and carboxyl functional groups that are the basic building blocks for polypeptides.

Aminoacyl-tRNA synthetases (also called **aminoacyl-tRNA ligases**) A family of enzymes that uses energy stored in ATP to attach amino acids to the 2'- or 3'-OH position in the 3' adenosine of tRNA.

Aneuploidy A condition in which eukaryotic cells have either more or less than the normal diploid number of chromosomes.

Anticodon A trinucleotide in tRNA that base pairs with the codon in mRNA in an antiparallel fashion.

Anticodon arm A stem-loop structure in tRNA that lies directly across from the acceptor stem and contains the anticodon (a trinucleotide) that base pairs with a codon (also a trinucleotide) in mRNA.

Antigen A foreign substance that is capable of eliciting an immune response.

Antiparallel DNA The two strands in double-stranded DNA run in opposite directions so that one strand runs $3' \rightarrow 5'$ in one direction, while the other strand runs $5' \rightarrow 3'$ in the same direction.

Antiparallel pleated sheet A β sheet that is organized so that successive polypeptide chains are oriented in opposite directions.

Anti-Shine-Dalgarno sequence The complementary sequence to the Shine-Dalgarno sequence, located near the 3'-end of 16S rRNA.

Antiterminator A secondary structure in trpL mRNA that allows transcription to proceed through the trp operon.

AP endonuclease An enzyme in the base excision repair pathway that hydrolyzes the phosphodiester bond 5' to the AP site to generate a nick.

AP site (see **Abasic site**)

Aptamer A DNA or RNA sequence that is selected from a random sequence pool because it can bind to a specific protein or protein complex.

Arginine A basic amino acid with a guanidino group in its side chain, represented by Arg or R.

Argonaute protein family A family of proteins that participate in RNA silencing mediated by small non-coding RNAs.

ARS elements (see **Autonomously replicating sequence elements**)

Ascus A yeast sac that contains four spores.

A-site The aminoacyl-tRNA binding site in the ribosome. A tRNA in the A-site accepts a peptidyl group from a peptidyl-tRNA in the P-site.

Asparagine An amide derivative of aspartate with a polar group in its side chain, represented by Asn or N.

Aspartic acid An acidic amino acid with a carboxyl group as part of its side chain, represented by Asp or D.

ATP-utilizing chromatin assembly and remodeling factor (also called **ACF**) An ATP-dependent chromatin remodeling machine that generates regular physiological nucleosome spacing.

Attenuator A control element in *trpL*, the first open reading frame of the *trp* operon, that is responsible for transcriptional regulation.

Autonomous elements Transposons that encode both the transposase and the end sequences required to mobilize the element.

Autonomously replicating sequence (ARS) elements Eukaryotic origins of replication that determine the sites of DNA chain initiation. They are also required for the replication of linear plasmids in yeast cells.

Autoradiography Technique in which a radioactive molecule emits β rays that reduce silver grains on photographic film to produce an image.

Autosomes Chromosomes that are not involved in sex determination such as the 22 chromosome pairs that are the same in human males and females.

Auxiliary upstream element A recognition element, which is often present at a variable distance upstream of the poly(A) cleavage site, that enhances the efficiency of cleavage and polyadenylation.

Auxotroph A mutant that requires a specific nutritional supplement that is not required by the wild-type parent.

B cell (also called B lymphocyte) A type of white blood cell that produces immunoglobulins and mediates humoral immunity.

Backtracking A proofreading mechanism used by RNA polymerase in which the polymerase moves in the reverse direction so the 3′-end of the RNA chain is displaced from the enzyme's active site.

Bacterial culture A population of bacteria growing in a liquid medium.

Bacterial lawn A confluent, turbid layer of bacteria growing on an agar medium.

Bacterial replication initiation The first stage of DNA replication in which the replication apparatus is assembled at a unique site on the bacterial chromosome.

Bacteriophage (see **phage**)

Basal synthesis The low level of transcription of genes in an operon that takes place because repressor occasionally dissociates from the operator.

Basal transcription The low level of correct transcription initiation in eukaryotes at a core promoter in the presence of RNA polymerase II and the general transcription factors.

Base excision repair A multistep pathway that repairs damage to DNA bases caused by deamination, oxidation, and alkylation.

Basic region leucine zipper A basic amino acid sequence together with the leucine zipper that forms the bzip DNA binding domain in a family of transcription activator proteins.

Basic replicon A region of DNA that is defined experimentally as the smallest piece of plasmid DNA that can replicate with the normal copy number.

B-DNA A double helix DNA structure that is stable at a relative humidity of about 92%. The Watson-Crick model is based on data obtained from these crystals.

Beta (β) galactosidase The enzyme that cleaves lactose to yield galactose and glucose.

Beta (β) conformation A protein secondary structure in which the polypeptide chain is extended and has a zigzag conformation.

Beta (β) pleated sheet A secondary protein structure in which two or more polypeptide chain regions, which are arranged in a zigzag structure, interact by hydrogen bonds. The β sheet is designated as parallel if the segments have the same amino to carboxyl orientation and as antiparallel if interacting chains have opposite orientations.

Beta (β) turn The most common kind of protein turn, which consists of four residues, that allows the polypeptide chain to reverse direction.

Bidirectional replication Mechanism used to replicate double-stranded circular DNA in which two growing points start at the same site and move in opposite directions until they meet at the opposite side of the circle.

Bioinformatics A scientific discipline that uses mathematics and computer technology to solve biological problems.

Bivalents (also called **tetrads**) Fully paired homologous chromosomes in which each chromosome contains two sister chromatids.

Blunt DNA ends (also called **flush DNA ends**) End of a DNA fragment that has no single-strand overhang. Blunt ends are produced by a restriction endonuclease that cuts both strands of a double-stranded DNA at a palindrome sequence's center of symmetry.

Body of 30S ribosomal subunit Region of the ribosome containing about two-thirds of the 30S subunit's mass.

Bohr effect A drop in pH or an increase in carbon dioxide concentration causes oxyhemoglobin to release oxygen.

–10 Box (also called **Pribnow box**) A hexamer sequence in the bacterial promoter that is centered about 10 bp upstream from the transcription initiation site.

–35 Box A hexamer sequence in the bacterial promoter that is centered about 35 bp upstream from the transcription initiation site.

Branch migration The process of exchange of base-pairing partners at a Holliday junction formed during homologous recombination.

Branchpoint sequence A consensus sequence that contains the adenylate to which the free 5'-end of the intron joins during splicing.

Branching nucleotide The adenylate in the branchpoint sequence that participates in the chemistry of pre-mRNA splicing.

BRCA2 (see **Human breast cancer susceptibility gene 2**)

BRE (see **TFIIB recognition element**)

Bridge helix An α-helix in the β' subunit of bacterial RNA polymerase or the Rpb1 subunit of eukaryotic RNA polymerase II that oscillates between a straight and bent form to allow the polymerase to move along the DNA template during transcription.

Broth A liquid growth medium that is a complex extract of biological material.

bzip DNA-binding domain Domain comprising a short sequence of basic amino acids that is located just N-terminal to the leucine zipper motif, which is required for DNA recognition.

CAF-1 (see **Chromatin assembly factor–1**)

cAMP receptor protein (also called **CRP**) A protein that binds cAMP and then activates *lac* operon expression.

cAMP • CRP complex Complex between cAMP and CRP needed to activate the *lac* operon and other operons that are subject to catabolite repression.

Cap 0 A 7-methylguanosine in 5'-5' triphosphate linkage with the first nucleotide in mRNA (or pre-mRNA).

Cap 1 A cap 0 structure that also has a methyl group attached to the 2'-O position on the first nucleotide in mRNA (or pre-mRNA).

Cap 2 A cap 0 structure that also has methyl groups attached to the 2'-O positions on the first and second nucleotides in mRNA (or pre-mRNA).

Cap binding complex A protein complex that binds to a 5'-m^{7}G cap on eukaryotic mRNA.

Capillary gel electrophoresis Gel electrophoresis in which separations are performed in very thin capillary tubes.

Capsid The protein coat that surrounds the nucleic acid in a virion.

Carbomycin An antibiotic that blocks bacterial protein synthesis by inhibiting peptidyl transferase activity.

Carcinogens Mutagens that cause cancer.

Cassette exon An internal exon that is completely included in some mRNAs and completely excluded from other mRNAs.

Catabolite repression Transcription-level inhibition by glucose (or other readily used carbon sources) of the *lac* operon and a variety of other operons that code for catabolic enzymes.

Catabolite-sensitive operons Operons that are stimulated by the cAMP • CRP complex.

Catalytic subunit The subunit of a regulatory enzyme that contains the active site.

Catenane A structure consisting of two or more interlocking circular DNA molecules.

CCA nucleotidyl transferase The enzyme responsible for adding CCA to 3'-end of tRNA.

CCAAT box A motif present in some eukaryotic regulatory promoters and enhancers that is recognized by the C/EBP activator protein.

Cdc6 A cell division cycle protein required for replication.

CDK (see **cyclin-dependent protein kinase**)

cDNA (see **complementary DNA**)

Cdt1 A protein required by yeast to recruit MCM2-7 (helicase) to the origin of replication and to initiate the S phase.

Cell density The number of cells per milliliter in a liquid growth medium.

CENP-A (also called **centromere protein A**) A constitutive protein in the centromere that combines with histones to form a new type of nucleosome that is dispersed among normal nucleosomes in the centromere.

Central dogma Hypothesis that proposes genetic information flows from DNA to DNA (DNA replication), from DNA to RNA (transcription), and from RNA to polypeptide (translation).

Centromere A constricted region of a chromosome that includes the site of attachment to the mitotic or meiotic spindle.

Centromere protein A (see **CENP-A**)

Chain termination method A technique that uses dideoribonucleoside triphosphates to terminate *in vitro* DNA replication and thereby produce a nested population ladder of extended primer sequences, which are analyzed to obtain the DNA sequence.

Chargaff's rules (1) Double-stranded DNA has equimolar adenine and thymine concentrations as well as equimolar guanine and cytosine concentrations, and (2) DNA composition varies from one genus to another.

Charged tRNA A tRNA molecule with its attached amino acid.

***Chi* sites** Short consensus sequences (5'-GCTG-GTGG-3') that are recombination hot spots in the *E. coli* chromosome.

Chiasma A cross-connection that is produced when nonsister chromatids break and reunite during the crossing-over process.

Chimeric mice Genetically mosaic mice that are derived by injecting embryonic stem cells from mice of one genotype into an early embryo of mice of another genotype.

ChIP (see **Chromatin immunoprecipitation**)

Chloramphenicol An antibiotic originally derived from a species of *Streptomyces* that inhibits bacterial peptidyl transferase.

Chromatin accessibility complex (also called **CHRAC**) An ATP-dependent chromatin remodeling machine that generates regular physiological nucleosome spacing.

Chromatin assembly factor-1 (also called **CAF-1**) A histone chaperone that deposits histone (H3 • H4)$_2$ tetramer (or two H3 • H4 dimers) onto newly replicated daughter DNA duplexes.

Chromatin immunoprecipitation (also called **ChIP**) A chemical crosslinking technique used to detect proteins that bind to DNA or to one another inside the cell.

Chromatin A nucleoprotein complex consisting of DNA and proteins, which is located in the nucleus of a eukaryotic cell and the nucleoid of a prokaryotic cell.

Chromatin modifiers Enzymes that participate in chromatin formation by adding acetyl, methyl, phosphate, or other groups to histones or removing these groups from modified histones.

Chromatin remodelers Enzymes that alter chromatin by repositioning nucleosomes, ejecting nucleosomes, unwrapping nucleosomes, or exchanging (or ejecting) histone dimers.

Chromatin remodeling A process in which chromatin is altered by repositioning nucleosomes, ejecting nucleosomes, unwrapping nucleosomes, or exchanging (or ejecting) histone dimers.

Chromatosome A nucleoprotein particle that is produced by briefly digesting chromatin with micrococcal nuclease. It consists of a DNA that is about 166 bp long wrapped around an octameric histone core and held in place by histone H1.

Chromomere A small beadlike structure that appears at irregular intervals along the length of the chromosome as chromatin starts to condense during mitosis or meiosis.

Chromosome A DNA–protein complex with many genes on the DNA. Most eukaryotes have several different linear chromosomes. Bacteria usually have a single circular chromosome.

Chromosome painting A variation of fluorescent *in situ* hybridization in which fluorescent dyes are bound to DNA pieces that bind all along a particular chromosome.

Chronic phage A phage that is continuously released from an infected bacterial cell without disrupting the cell envelope or killing the host.

Circularly permuted phage DNA The terminally repeated genes at the ends of certain phage DNA such as the T-even phage differ from one phage particle in the population to another.

***Cis*-acting element** A DNA element that must be on the DNA that it is regulating. For example, the *lac* operator is a *cis*-acting element.

Cisplatin A platinum-based chemical that combines with DNA to form intra- and interstrand cross-links.

***Cis*-splicing** Splicing process in which exons incorporated into mRNA all come from the same pre-mRNA molecule.

Cistron A DNA segment corresponding to an open reading frame plus the translational start and stop signals for protein synthesis.

Clamp loader An ATP-dependent DNA polymerase subassembly that loads the sliding clamp on DNA.

Cleavage/polyadenylation site [also known as **poly(A) site**] Site at which pre-mRNA is cleaved during RNA polymerase II transcription termination.

Cleft of the 30S subunit The site of codon–anticodon interactions, which lies between the head and platform.

Clone All the cells in a colony derive from a single ancestor. The term is often used when the single ancestor carried a specific recombinant plasmid that is present in all the cells in the colony.

Cloning The use of recombinant DNA technology to produce many exact copies of a single gene (or other DNA sequence) by introducing a recombinant plasmid or virus bearing the gene (or DNA sequence) into a cell, which then replicates to produce multiple copies of the gene (or DNA sequence).

Closed promoter complex A promoter complex formed early in transcription initiation stage in which the DNA is still completely double helical.

Coaxial stacking Two RNA helical tracts stacked end-to-end.

Coding strand (also called **sense strand**) The nontemplate DNA strand that has the same nucleotide sequence as the transcribed RNA (except that T replaces U).

Codon A particular sequence of three adjacent nucleotides that code for a specific amino acid.

Coenzyme A nonproteinaceous organic molecule, most commonly a vitamin or vitamin derivative, that is required for enzyme activity.

Cohesin A protein that holds sister chromatids together during the early stages of mitosis but is cleaved during anaphase.

Cohesive end (also called **sticky or staggered end**) A single-strand overhang at the end of a double-stranded DNA fragment (often produced by a staggered cut on the DNA by a type II restriction endonuclease), which can base pair with a complementary single-strand overhang of another DNA fragment.

Cohesive ends (*cos* elements) The base sequences of the 5′-overhangs of linear phage λ DNA.

Coiled coil A protein fold in which a pair of α-helices coil around one another.

Cointegrate A structure formed during replicative transposition in which the donor and target DNA molecules are fused together.

Col plasmids Plasmids that program bacteria to produce antibacterial proteins called colicins.

Collision release A model of lagging strand DNA synthesis that proposes core polymerase remains bound to the sliding clamp until it "collides" with a duplex.

Column chromatography A method used to separate components in a mixture. The mixture is loaded onto a column containing a stationary phase made of beads. Then a liquid solution is passed through the column so the components of the mixture are separated by repeated partitioning between the mobile solution and the immobile solid matrix.

Combinatorial control A eukaryotic mechanism in which gene activity is regulated by a combination of activator (and repressor) proteins, which determine whether the transcription machinery is recruited to the gene.

Combinatorial diversity in antibody formation The first level of antibody diversity, which occurs through V(D)J segment rearrangements.

Committed step in a metabolic pathway The first step after a branch in a biochemical pathway that commits the substrate to proceed down that branch and only that branch.

Competent bacteria A bacterial cell with the ability to take up DNA from the surrounding medium.

Complementary DNA (also called **cDNA**) DNA synthesized by using RNA as a template for reverse transcription.

Complementation A genetic test used to evaluate the number of genes and regulatory elements that constitute a genetic system.

Composite transposons Transposons in which two insertion sequence elements flank additional genes.

Concatemer A long DNA molecule with many copies of a viral DNA molecule linked head-to-tail.

Conditional mutant A mutant that has the mutant phenotype under some conditions but not under others. For instance, the mutant may have the phenotype at high temperatures but not at low temperatures.

Conjugation The process in which F^+ (or Hfr) male bacterial cells transfer the F plasmid (or bacterial chromosome) to F^- cells.

Conjugative plasmid A plasmid with genes that allow a donor bacterial cell to make effective contact with a recipient cell but that lacks genes needed to mobilize the DNA for transfer.

Connective polypeptide 1 (also called **CP1**) A polypeptide inserted into the Rossmann-fold domain of bacterial Ile-tRNA synthetase.

Consensus sequence An idealized sequence that indicates the most frequently found base in each position of many actual sequences.

Conservative model of DNA replication Model of DNA replication in which both strands in one daughter DNA molecule are parental DNA and both strands of the other daughter DNA are new DNA.

Constitutive enzymes Enzymes that are synthesized at fixed rates under all growth conditions.

Constitutive exon An exon that is included in all mRNAs formed from pre-mRNA molecules with the same sequence.

Cooperative interaction The binding of two or more identical molecules to the same macromolecule such that the binding of one molecule makes the binding of a second molecule more likely.

Coordinate regulation Regulation of two or more cistrons by a single regulatory element or system so the polypeptides encoded by the cistrons form in response to the same signal.

Core polymerase The DNA polymerase III holoenzyme subassembly (previously known as DNA polymerase III) that is responsible for the DNA polymerase activity.

Core promoter A DNA region surrounding a eukaryotic transcription start site that usually includes at least one of the following elements—a TATA box, initiator element, downstream promoter element, or TFTIIB recognition element.

Coumarin A heterobicyclic ring system contained in the psoralen molecule.

Counterselective marker The allele used to prevent growth of male bacteria after a mating experiment.

Covalently closed circle A double-stranded DNA molecule in which the ends are joined together to form a single continuous molecule.

CpG islands Regions of a gene that contain high concentrations of cytosine and guanine, often located in promoters.

CP1 (see **connective polypeptide 1**)

Crossover A reciprocal exchange of material between paired homologous chromosomes that occurs during prophase I of meiosis and is responsible for genetic recombination.

Cross-reacting material A protein that is recognized by an antibody that was raised in response to a closely related protein.

Cross-reaction A reaction in which an antibody that is raised in response to one protein is used to detect a closely related protein.

CRP (see **cAMP receptor protein**)

Cruciform structure A structure that is occasionally observed when double-stranded DNA contains an inverted repeat, which allow base pairs to form between complementary regions on the same strand and produce double-stranded branches.

Cryoelectron microscopy An electron microscopy technique that uses computer assisted analysis to reconstruct the structures of complex biological molecules from electron micrographs of rapidly frozen samples.

Cut-and-paste transposition mechanism The reaction pathway that results in the element excising from one DNA site and inserting into a new DNA site.

3′,5′-Cyclic adenylate (cAMP) Second messenger formed from ATP by adenylate cyclase.

Cyclin A protein that is expressed at different levels throughout the cell cycle.

Cyclin-dependent protein kinase (also called **CDK**) A family of kinases, which are inactive unless bound to a cyclin molecule, that are involved in cell cycle control.

Cyclobutane pyrimidine dimer A form of DNA damage caused by ultraviolet irradiation, in which two new bonds form between adjacent pyrimidines on the same DNA strand.

Cyclobutane pyrimidine dimer photolyase The photoreactivation enzyme that reverses cyclobutane pyrimidine dimer lesions.

Cys_4 zinc finger DNA binding domain in proteins belonging to the nuclear receptor superfamily.

Cysteine An amino acid with a sulfhydryl (–SH) group in its side chain, represented by Cys or C.

Cystine A dimer produced when an aqueous cysteine solution is exposed to air, which oxidizes the sulfhydryl groups, causing them to form a disulfide bond. Cystine also arises when cysteine residues in proteins are oxidized.

Cytidine deaminase An RNA-editing enzyme that converts certain cytidines in RNAs into uridines through deamination.

Cytosine A pyrimidine base derivative that is one of the four bases present in DNA and RNA molecules.

D-arm An arm of tRNA that contains a modified form of uridine called dihydrouridine.

DDE transposase A family of transposases so-named because their members have two aspartic acid (D) residues and one glutamic acid (E) residue at their catalytic sites.

Degradative plasmids Plasmids that program host cells to make enzymes that degrade various organic molecules such as hydrocarbons.

Deletion mutation A mutation in which one or more base pairs have been deleted.

Denaturation The loss of secondary structure in DNA, RNA, or protein caused by disrupting weak noncovalent bonds.

Denatured protein The state of a protein after weak noncovalent bonds are disrupted so the protein is no longer in its native conformation and as a result loses biological activity.

Deoxyadenosine methylase An enzyme that transfers methyl groups from S-adenosylmethionine molecules to deoxyadenosines in GATC sequences.

Depurination The process in which purine groups are cleaved from the polynucleotide chain

Derepression A process in which the transcription of one or more genes is "turned on" by removing a repressor from an operator.

Diakinesis The fifth stage of meiosis division I, in which chromosomes reach maximum condensation, spindle formation begins, and the nuclear membrane disaggregates.

Dicer A member of the RNase III enzyme family that binds double-stranded RNA and cuts it into short interfering RNA molecules.

Dideoxy-terminator cycle sequencing A technique that uses a uses a thermal cycler to amplify reaction products formed in the presence of dideoxynucleotides, allowing double-stranded DNA molecules to be sequenced.

***dif* site** A specific site within the bacterial replication termination region where recombinase acts to separate daughter chromosome dimers.

Diploid The state of having two versions of each chromosome, double the number found in a haploid cell.

Diplotene The fourth stage of meiosis division I, in which homologous chromosomes begin to separate.

Direct repeat A pair of identical sequences on a nucleic acid that are in the same orientation.

Discontinuous model of replication A DNA replication model that predicts discontinuous synthesis of both new strands.

Dispersive model of DNA replication Model of DNA replication in which each strand of the daughter DNA molecules has interspersed sections of both old and new DNA.

Distributive replication DNA polymerase dissociates from its template after only a few nucleotides are added.

Displacement loop (also called **D-loop**) The loop formed when a single-stranded DNA displaces one of the strands within a region of double-stranded DNA.

D-loop (see **Displacement loop**)

DnaA boxes Sites in the bacterial replicator that bind DnaA • ATP complexes.

DNA affinity chromatography A protein purification technique based on binding specificity between DNA attached to beads and the protein of interest.

DnaB helicase An ATP-dependent *E. coli* helicase that encircles one strand of a forked DNA molecule and moves in a $5' \rightarrow 3'$ direction along the encircled strand while excluding the other.

DNA denaturation The transition from the double helical structure (the native state) to randomly coiled single strands (the denatured state).

DNA reannealing (see **DNA renaturation**).

DNA renaturation (also known as **DNA reannealing**) The re-formation of double-stranded DNA from denatured DNA.

DNA footprinting technique A technique in which a protein is added to DNA that has been labeled at its 5'- or 3'-end and the complex is incubated with DNase (or a chemical that cleaves DNA) for such a short period of time that on average each DNA molecule receives no more than one single-strand break. Then the DNA is isolated, subjected to gel electrophoresis, and nick positions determined with the respect to the labeled end.

DNA glycosylase An enzyme that catalyzes N-glycosyl bond cleavage to remove an incorrect or altered base from DNA.

DNA glycosylase/lyase An enzyme that has glycosylase and AP lyase activity.

DNA gyrase A bacterial enzyme that uses energy provided by ATP to introduce negative supercoils in circular DNA.

DNA polymerase I A multifunctional bacterial polypeptide with DNA polymerase, 3' to 5' exonuclease and 5' to 3' exonuclease activities.

DNA polymerase II A bacterial DNA polymerase that is induced in response to an SOS signal and contains a 3' to 5' exonuclease activity.

DNA polymerase III holoenzyme The replication machinery responsible for bacterial DNA replication.

DNA polymerase III ε-subunit The subunit in core polymerase (previously known as DNA polymerase III) that has 3' to 5' exonuclease activity, allowing for proofreading of newly replicated DNA.

DNA polymerase IV A bacterial DNA polymerase, which is induced in response to the SOS signal, that can perform either error-free or error-prone translesion DNA synthesis.

DNA polymerase β (also called **Pol β**) A eukaryotic DNA polymerase that uses an undamaged DNA strand to fill in gaps left during DNA repair.

DNA polymerase δ (also called **Pol δ**) A eukaryotic DNA polymerase that is primarily responsible for catalyzing lagging strand synthesis.

DNA polymerase ε (also called **Pol ε**) A eukaryotic DNA polymerase that is primarily responsible for catalyzing leading strand synthesis.

DNA polymerase V A bacterial DNA polymerase, which is induced in response to the SOS signal, that performs error-prone translesion DNA synthesis.

DNA polymerase/α-primase (Pol α) A eukaryotic enzyme with primase and DNA polymerase activities, which are required for replication initiation.

DNA-binding domain An activator protein domain that makes sequence-specific contacts with control elements in a gene's regulatory promoter or enhancer.

DNase protection method A technique that depends on a protein's ability to protect a specific DNA region from extended exposure to DNase.

DNA unwinding element (also called **DUE**) A region within the bacterial replicator, which contains three AT-rich 13-bp elements and melts when DnaA • ATP complexes bind to DnaA boxes.

Domain A region within a long polypeptide chain that folds independently of other regions within the chain and usually has a specific biological function.

Domain swap An experiment in which the domain from one protein is replaced by a corresponding domain from another protein.

Dominant allele An allele that produces its characteristic phenotype whether it is paired with an identical allele or different allele.

Double-strand break Cleavage of both strands of a double-stranded DNA at the same (or nearly the same) site.

Double-sieve model proposes that Ile-tRNA synthetase has an aminoacylation site that serves as a "coarse sieve" to exclude amino acids larger than the cognate L-isoleucine and an editing site that serves as a "fine sieve" to eliminate amino acids smaller than isoleucine such as valine.

Double-strand break repair Model for how double-strand breaks are repaired.

Doubling time The time required for a culture's cell number to double.

Down syndrome A congenital human condition that is caused by an extra copy of chromosome 21 (or an extra copy of a segment of chromosome 21).

Downstream Sequences that come after the transcription start site or more generally 3′ to the point of reference on the sense strand.

Downstream promoter element (also called **DPE**) Sequence motif present in many but not all eukaryotic core promoters that is located downstream from the transcription start site.

Drug resistance (R) plasmids Plasmids that have genes that make host cells resistant to one or more antibiotics or antibacterial drugs.

DUE (see **DNA unwinding element**)

eIF (see eukaryotic translation initiation factor)

Electroblotting A technique in which an electric field is used to transfer a nucleic acid or protein from a gel to a nitrocellulose membrane or some other kind of membrane.

Electrophiles Electron-seeking chemicals that attack electron-rich atoms in DNA.

Elongin A protein factor that stimulates transcription by stabilizing the active conformation of RNA polymerase II, increasing the rate at which the inactive form is converted back to the active form, or both.

Embryonic stem cell (also called **ES cell**) A cell that is isolated from a mouse embryo in an early developmental stage and can grow stably in culture.

Endogenous DNA damaging agents Chemicals formed inside the cell by normal metabolic pathways that damage DNA.

Endonucleases Nucleases that cleave phosphodiester bonds within a nucleic acid chain.

Endoplasmic reticulum A continuous intracellular tubular membrane network involved in the synthesis of lipids, membrane proteins, and secretory proteins.

Energy of activation The energy needed for reactants to reach the transition state.

Enhancer A eukaryotic regulatory sequence that stimulates transcription from the correct transcription initiation site even when it is located several thousand base pair upstream or downstream from the transcription initiation site or inserted in either orientation.

Entropy A measure of disorder, which increases as disorder increases.

Enzymes Proteins that catalyze specific reactions.

Enzyme–substrate complex (also called **ES complex**) The structure formed when an enzyme combines with its substrate.

Epigenetics Inherited changes in phenotype that are caused by changes in the chromosome other than changes in DNA sequence.

Equilibrium density gradient centrifugation A centrifugation technique that separates particles according to their density.

Error-prone DNA polymerases A special class of DNA polymerases that catalyze DNA synthesis across a DNA lesion.

ES complex (see **enzyme-substrate complex**)

ES cell (see **embryonic stem cell**)

ESE (see **Exonic splicing enhancer**)

E-site The ribosomal site from which the tRNA exits.

Established cell line A population derived from a cell that has gained the ability to grow and divide indefinitely.

Ethidium bromide A carcinogenic chemical agent that binds tightly to DNA. Investigators exploit its fluorescence properties to detect DNA after electrophoresis.

Euchromatin The less highly condensed form of eukaryotic chromatin present in the interphase nucleus that is actively transcribed.

Eukaryotes Cells with a defined nucleus.

Eukaryotic translation initiation factor (also called **eIF**) A factor required to initiate translation in eukaryotes. It is designated by a number that is often followed by a letter.

Excinuclease (see **UvrABC endonuclease**)

Exogenous DNA damaging agents Physical and chemical agents that come from the surrounding environment and damage DNA.

Exon An expressed region within a transcription unit that codes for amino acids.

Exon definition Formation of protein bridges across exons.

Exonic splicing enhancer (also called **ESE**) A sequence within an exon that stimulates splice-site selection.

Exonic splicing silencer A sequence within an exon that represses splice-site selection.

Exonuclease A nuclease that cleaves one nucleotide at a time starting at one end of a polynucleotide chain.

Expressed sequence tags cDNA fragments prepared by copying one or both ends of an mRNA.

3′-External transcribed sequence Sequence located downstream from the 28S rRNA coding sequence.

5′-External transcribed sequence Sequence located upstream from the 18S rRNA coding sequence.

F plasmid (also called fertility factor) A bacterial plasmid that is responsible for determining maleness.

F′ plasmid A plasmid that is formed when an F plasmid integrated into a bacterial chromosome is imperfectly excised so the released plasmid contains both F plasmid genes and bacterial genes that had been adjacent to the integrated F plasmid.

Feedback inhibition Enzyme regulatory mechanism in which an end product of a biochemical pathway inhibits a committed step in that pathway.

Fertility factor (see **F plasmid**)

Fibrous proteins Proteins with thread-like shapes, which tend to have structural or mechanical functions and to be water insoluble.

Fingerprinting (also called **peptide mapping**) A method of protein characterization in which a protein is digested with a protease such as trypsin and the resulting peptide fragments are spotted onto filter paper and partially separated by electrophoresis. Then the dried paper is turned 90 degrees and developed in a second direction by chromatography to yield a unique two-dimensional peptide map.

Fluorescent *in situ* hybridization (also called **FISH**) A technique that uses fluorescent probes to detect and localize specific DNA sequences on chromosomes.

Flush DNA ends (see **Blunt DNA ends**)

fMet-tRNA A bacterial initiator tRNA with a bound N-formylmethionine.

Fork collapse An event that takes place when the replication fork encounters a nick on a template strand and one branch of the fork separates from the rest of the fork.

Fragment assay An *in vitro* analysis that measures fMet-puromycin formation when a mixture containing 50S ribosomal subunits, a 3′-fragment of peptidyl-tRNA such as CCA-fMet, and puromycin is incubated in a buffer solution containing 30% to 50% methanol or ethanol.

Frameshift mutation A mutation caused by a deletion or insertion that is not a multiple of three base pairs and therefore changes the reading frame.

Functional genomics Efforts to assign functions for genes that were discovered by large scale sequencing.

Furanose A simple sugar that contains a five-member, furan-based ring structure.

G_1 phase (Gap 1 phase) A period in the cell cycle during interphase that precedes the S phase in which protein, lipid, and carbohydrate synthesis take place.

G_2 phase (Gap 2 phase) A period in the cell cycle during interphase that follows the S phase.

Gal4 A member of the Zn_2Cys_6 binuclear cluster family, which is a transcription activator protein that regulates genes coding for galactose metabolism in *Saccharomyces cerevisiae*.

GC box A sequence motif, which is present in some regulatory promoters and enhancers, that is recognized by the transcription factor Sp1.

Gcn4 A leucine zipper transcription activator protein that coordinately increases the transcription of several genes involved in nitrogen metabolism.

Gel electrophoresis A laboratory separation technique that exploits the fact that a macromolecule such as a protein or nucleic acid moves through an agarose or polyacrylamide gel under the influence of an electric field at a characteristic rate that depends on the molecule's size, shape, and charge.

Gel filtration (also called **molecular exclusion chromatography**) A column chromatography method that separates molecules by size.

Gene The segment of DNA that specifies a polypeptide or RNA chain. Alternatively, a eukaryotic gene is a linear collection of exons that are incorporated into a specific mRNA.

Gene conversion *In vivo* alteration of one strand of a heteroduplex DNA to make it complementary with the other strand at any position where there were mispaired bases.

Gene expression The transcription and translation processes through which genetic information is converted to a product.

Gene knockdown Gene inactivation through the use of siRNA.

Gene knockout A selectable marker is inserted into the coding region of a gene, thereby disrupting its activity.

Gene replacement A knockout in which all or part of the coding sequence is deleted and replaced with a selectable marker.

General transcription factors Protein factors required for basal transcription that enable RNA polymerase II to transcribe naked DNA from specific transcription initiation sites.

Generalized transduction The process in which a phage particle carries random DNA fragments from one bacterium to another.

Genetic engineering (also known as **Recombinant DNA technology**) A group of techniques used to prepare DNA fragments from different sources and then to join these fragments to form an artificial DNA construct, which can be introduced into a living cell where it either replicates autonomously or becomes integrated into a chromosome.

Genetic map A map of a chromosome that shows the positions of the known genes, markers, or both relative to each other.

Genetic recombination The process in which DNA molecules break and the resulting fragments join in new combinations.

Genome All the genetic material in the chromosomes of a specific organism.

Genotype The genetic makeup of an organism.

Germ cells Cells that will give rise to gametes, which have a characteristic number of chromosomes, the haploid number (n).

Globular proteins Proteins with spherical or nearly spherical shapes such as enzymes, transport proteins, and antibodies, which tend to be water-soluble.

Glucocorticoid receptor A transcription activator that binds a glucocorticoid steroid such as cortisol in the cytoplasm and then moves to the cell nucleus, where it stimulates transcription of specific genes.

Glutamic acid An acidic α-amino acid with a carboxyl group as part of its side chain, represented by Glu or E.

Glutamine An amide derivative of glutamate with a polar amide group in its side chain, represented by Gln or Q.

Glutathione A tripeptide antioxidant with many functions, one of which is to help protect cells from the carcinogen aflatoxin.

Glycine An amino acid with a side chain consisting of a single hydrogen atom, represented by Gly or G.

Gratuitous inducer A substance that is an effective inducer of an inducible operon but that is not degraded by the enzymes encoded by the operon.

Group II introns Introns capable of self-splicing.

Guanine A purine base derivative that is one of the four bases present in DNA and RNA molecules.

Guide strand The strand in the siRNA duplex that is assembled into siRISC and guides the complex to the target mRNA.

Hairpin structures (also called stem-loop structures) Double-stranded structure that forms when inverted repeats cause a polynucleotide chain to fold back on itself, forming intrastrand hydrogen bonds.

Haploid cell A cell that has a single set of chromosomes corresponding to one member of each chromosome pair present in a diploid cell.

Haploid number The characteristic number of chromosomes present in a germ cell.

HAT (see **Histone acetyltransferase**)

HAT medium A growth medium containing hypoxanthine, aminopterin, and thymidine.

HbA Normal adult hemoglobin.

HbS Sickle-cell hemoglobin.

HDAC (see **Histone deacetylases**)

Head of the 30S subunit Region of the 30S ribosomal subunit above the cleft that contains about one third of the subunit's mass.

Headful mechanism A mechanism for determining phage DNA length based on the ability to fill the phage head.

Headpiece (see **N-terminal domain of Lac repressor**)

Heavy chain in IgG The larger of two types of subunits in an antibody tetramer.

Helical wheel A planar projection of an amino acid sequence in a polypeptide that is based on the assumption that the polypeptide folds into an α-helix.

Helicase Motor proteins that use the energy of nucleoside triphosphates to unwind DNA.

Helix-turn-helix motif A DNA-binding motif that interacts with specific DNA sequences.

Heme A large heterocyclic planar ring system within a hemoglobin molecule that has Fe^{2+} at its center.

Heterochromatin A highly condensed form of chromatin in the cell nucleus that is not actively transcribed.

Heterodimer protein A protein complex that contains two different polypeptides.

Heteroduplex DNA A DNA double helix containing one or more mismatches formed by annealing single strands from different sources.

Heterogeneous nuclear RNA (also called hnRNA) A term originally used to describe a population of rapidly labeled nuclear RNA molecules with a bread size distribution that is now known to consist mainly of pre-mRNA.

Heterokaryon A cell containing two separate nuclei, which is formed experimentally fusing of two genetically different cells.

Heterotrimer A protein complex that contains three different polypeptides.

Hfr males (also called high frequency of recombination males) Bacterial cells with an F plasmid insert in their chromosomes.

Hinge region of Lac repressor Region of the Lac repressor that joins the N-terminal headpiece to the core domain

Histidine A basic amino acid with an imidazole group in its side chain, represented by His or H.

Histone acetyltransferase (also called HAT) Enzymes that modify histones by addition of acetyl groups.

Histone chaperones Protein factors that associate with histones to facilitate nucleosome assembly and disassembly without themselves becoming a permanent part of nucleosomes.

Histone deacetylases (also called HDAC) Enzymes that remove acetyl groups from histones and may be associated with repressors of transcription.

Histones Proteins rich in lysine and arginine that interact with DNA to form nucleosomes.

HIV (see **Human immunodeficiency virus**)

hnRNA (see **Heterogeneous nuclear RNA**)

Holliday junction An intermediate structure formed during double-strand break repair and homologous recombination that has a crossed-strand DNA structure with a mobile junction.

Homeobox A conserved DNA sequence within homeotic genes that is about 180 bases long and specifies the 60-amino acid residue helix-turn-helix DNA binding domain.

Homeodomain The C-terminal domain of a homeobox protein that contains about 60-amino acid residues, which fold into a helix-turn-helix motif that binds DNA.

Homeotic (*HOM*) genes Genes that assign positional identities to cells during embryonic development.

Homodimer A protein complex that contains two identical polypeptides.

Homolog A term used to describe a pair of chromosomes, genes, or proteins that have similar sequences.

Homologous recombination Recombination involving a reciprocal exchange of DNA sequences with a high degree of similarity.

Homotrimer A protein complex that contains three identical polypeptides.

Hormone response element An imperfect palindromic DNA sequence to which the receptor homodimer binds.

Host restriction and modification A process that permits a bacterium to identify and degrade foreign DNA.

***Hox* genes** Homeotic genes found in vertebrate animals.

Human breast cancer susceptibility gene 2 (also called **BRCA2**) The gene associated with familial breast and ovarian cancers.

Human immunodeficiency virus (also called **HIV**) A retrovirus that infects human T helper cells, immature thymocytes, monocytes, and macrophages and causes acquired immunodeficiency syndrome.

Hybridization A technique in which complementary single-stranded nucleic acids are permitted to interact to form DNA–DNA or DNA–RNA double strands.

Hybrid-states translocation model Model used to explain the movements of mRNA and tRNAs through the ribosome.

Hydrogen bond A bond resulting from electrostatic attraction between a partially negative electronegative oxygen, nitrogen, or fluorine atom and a partially positive hydrogen atom linked to a second oxygen, nitrogen, or fluorine atom.

Hydrophilic residue An amino acid with a charged or polar side chain that is water soluble.

Hydrophobic interactions Weak noncovalent interactions that result because it is entropically more favorable for water molecules to form a single large cage around a cluster of hydrophobic molecules than it is for the water molecules to form individual cages around each hydrophobic molecule.

Hydrophobicity The tendency of nonpolar molecules or groups to aggregate in water.

Hydroxymethylcytosine A modified form of cytosine that base pairs with guanine in phage T4 DNA.

Hyperchromic effect Increased λ_{260} light absorption of single-stranded DNA compared with double-stranded DNA.

Hypervariable region A region within the variable region of an antibody that contributes to antigen-binding specificity.

Hypochromic effect Decreased λ_{260} light absorption of a double-stranded DNA compared with single-stranded DNA.

ICR (see **Internal control region**)

Identity elements Features present in tRNA that are required for recognition by the cognate aminoacyl-tRNA synthetase under *in vitro* conditions.

Immunoglobulin G (IgG) A type of antibody that circulates in the blood and helps to render harmful foreign particles harmless.

Inchworming model A model that proposes the leading edge of RNA polymerase advances during the early stage of transcription initiation to allow the active center to move forward, whereas the other end of the enzyme remains anchored to the upstream DNA region.

Induced fit model Substrate-binding mechanism in which the enzyme changes shape upon binding the substrate and the active site assumes a shape that is complementary to that of the substrate only after the substrate is bound.

Inducer exclusion A model that explains glucose's inhibition of lactose catabolism by proposing the dephosphorylated EIIA that forms in the presence of glucose binds to lactose permease and inactivates it.

Inducible enzyme An enzyme that experiences an increased rate of synthesis when a small molecule (an inducer) is added to the growth medium.

30S Initiation complex Complex containing the small bacterial ribosomal subunit, translation initiation factors, and fMet-tRNA that forms just before interaction with the large ribosome subunit.

70S Initiation complex A complex formed when the 50S subunit combines with the 30S initiation complex that contains fMet-tRNA bound to AUG at the P-site.

Initiator (Inr) element Sequence motif that is part of many but not all eukaryotic core promoters and when present brackets the transcription start site.

Initiator DNA A short DNA sequence attached to RNA, which is synthesized by Polα, and serves as a primer for eukaryotic DNA synthesis.

Initiator protein A protein that binds to the replicator and helps to unwind the DNA and recruit components of the replication machinery.

Inr (see **Initiator element**)

Insertion mutation A mutation in which one or more nucleotide pairs are added to a gene.

Insertion sequence (also called an **IS element**) Mobile element that codes for a transposase, has an inverted terminal repeat at each end, and ranges in size from about 750 to 2,500 bp long.

Intercalation Insertion of a molecule such as ethidium bromide between adjacent base pairs, causing the DNA helix to unwind.

Interference A phenomenon that limits the number of crossovers per chromosome arm.

Interferon response Interferon activates 2′-5′ oligoadenylate synthase, which then converts ATP to 2′-5′ oligoadenylate. The 2′-5′ oligoadenylate in turn stimulates RNase L to cleave several RNA species, including ribosomal RNA. This cleavage causes the nonspecific inhibition of translation.

Intergenic spacer Spacers that separate rRNA coding regions.

Internal control region (also called **ICR**) The region within an RNA polymerase III type I promoter that contains the A box, intermediate element, and C box.

Internal ribosomal entry sites Sites that allow ribosomes to assemble at an internal site.

Internal transcribed spacer 1 Spacer that separates the 18S and 5.8S rRNA sequences.

Internal transcribed spacer 2 Spacer that separates the 5.8S and 28S rRNA sequences.

Interphase The interval in the cell life cycle that consists of G_1, S_1, and G_2 stages.

Intrinsic termination Termination of transcription by bacterial RNA polymerase without the help of any additional factors.

Intrinsically disordered protein A protein that does not fold into an ordered structure.

Intron The intervening sequence of DNA that is transcribed but later removed from the transcript by splicing.

Intron definition Formation of protein bridges across an intron that is to be removed by splicing.

Intronic splicing enhancer A sequence within an intron that stimulates splice-site selection.

Intronic splicing silencer A sequence within an intron that represses splice-site selection.

Inversion A process in which a segment breaks away from the chromosome, inverts from end to end, and then reinserts at the original breakage site.

Inverted repeat sequence Two copies of the same DNA sequence, which are in the opposite orientation on the same DNA molecule.

Ion-exchange chromatography A technique used to separate proteins or other charged molecules in solution that is based on the molecules' affinities for a negatively charged stationary phase (cation exchanger) or a positively charged stationary phase (anion exchanger).

Ionic bond A noncovalent bond resulting from the attraction between positively and negatively charged ionic groups.

IS element (see **Insertion sequence**)

Isoacceptor Different tRNAs that accept the same amino acid.

Isoelectric pH The pH at which a protein has no net charge and thus will not migrate in an electric field.

Isoforms Variant forms of a polypeptide resulting from alternative pre-mRNA splicing.

Isoleucine An amino acid with a hydrocarbon side chain, represented by Ile or L.

Isoschizomers Different restriction endonucleases that recognize the same nucleotide sequence and that cleave it in the same position.

Junctional diversity The second level of diversity that occurs during V(D)J segment rearrangements.

Karyotype A display in which chromosome pairs are arranged in order of decreasing size.

Kinetochore A structure associated with the centromere that attaches to mitotic or meiotic spindles.

Klenow fragment The large fragment produced when E. coli DNA polymerase I is cleaved by subtilisin, exhibiting 3′ to 5′ exonuclease activity and polymerase activity.

K_m (see **Michaelis constant**)

Lac repressor A protein that binds to the *lac* operator and blocks the transcription of the *lac* genes.

Lactose permease A transport protein that is required for the entry of lactose (and other β-galactosides) into the bacterial cell.

Lag phase The initial period of slow bacterial growth after a culture is started.

Lagging strand The strand of DNA that must grow overall in the 3′ to 5′ direction and is synthesized discontinuously in the form of short fragments (5′ to 3′) that are later connected covalently.

Lambdoid phages A very large family of temperate phages that infect a wide variety of bacteria.

Leading strand The strand of DNA that is synthesized continuously in a 5′ to 3′ direction.

Leptotene The first stage of meiosis division I in which the chromosomes begin to condense.

Leucine An amino acid with a hydrocarbon side chain, represented by Leu or L.

Leucine zipper A leucine-rich dimerization region that stabilizes the coiled-coil interaction required for dimer formation in one family of transcription activator proteins.

Life cycle The time between a cell's formation by parent cell division and its own division to form two daughter cells.

Ligand A small organic molecule that binds specifically and noncovalently to another molecule such as a protein.

Ligand binding domain A protein domain to which small molecules bind.

Light chains The smaller of two types of subunits in an antibody tetramer, the N-terminus of which forms part of the antigen recognition site.

LINEs (see **Long interspersed nucleotide elements**)

Linker DNA DNA connecting two nucleosomes.

Linker-scanning mutagenesis A technique used to search for regulatory elements by replacing short DNA segments in a region of interest with a DNA linker containing a random sequence of exactly the same size.

Linking number (Lk) The number of times the two strands of a closed DNA duplex cross over each other.

Liquid growth medium A liquid substance in which bacterial cultures can grow.

Lock-and-key model Substrate-binding mechanism in which the shape of the active site of the enzyme is complementary to the shape of the substrate.

Log phase The stage in the bacterial growth curve when cells grow exponentially.

Long interspersed nucleotide elements (also called **LINEs**) Non-LTR retrotransposons present in mammalian genomes that are usually about 5–7 kbp long.

Long patch repair A branch of the base excision repair pathway in which 2 to 8 nucleotides are replaced.

Long terminal repeat (also called **LTR**) A long DNA sequence that is repeated at both ends of a retroviral DNA or an LTR retrotransposon.

Loop A protein secondary structure consisting of five or more residues, which lacks a definite structure and allows the polypeptide chain to change direction.

Loss of heterozygosity A result of gene conversion during mitosis that can change a functional allele into a nonfunctional allele.

LTR (see **Long terminal repeat**)

LTR retrotransposon A retrotransposon element that has a long-terminal repeat at each end.

Lysine A basic amino acid with an amine group in its side chain, represented by Lys or K.

Lysis Disruption of the plasma membrane of a cell.

Lysogenic cell A bacterial cell carrying a prophage, which has the potential for cell lysis.

Lysozyme An enzyme that cleaves polysaccharide chains in bacterial cell walls.

Macromolecule A very large molecule, also known as a polymer, such as a nucleic acid or protein that is formed by joining many smaller building blocks or monomers.

Major groove The wide groove that winds around the B-DNA helix.

Mass spectrometry A technique in which molecules are converted to gaseous ions that then are sorted according to their mass-to-charge ratios, which cause the ions to have different trajectories in an electric and magnetic field under high vacuum.

MCM2-7 complex (see **Minichromosome maintenance 2-7 complex**)

Mediator A multisubunit protein complex in eukaryotes that acts as a bridge between activator proteins, repressor proteins, and the basal RNA polymerase transcription machinery.

Meiosis A special form of diploid cell division that takes place during germ cell formation in higher animals and plants (or spore formation in yeast). It involves two nuclear divisions in rapid succession to produce four germ cells (or spores), each having a haploid number of chromosomes.

Melting temperature of DNA (also called T_m) The temperature at which half the nucleotides in a double-stranded DNA are in the native form and half are in a denatured form. Experimentally, T_m is the temperature at which λ_{260} absorbance is at the midpoint of that for native and denatured DNA.

Messenger RNA (also called **mRNA**) The RNA molecule specified by a gene that programs a ribosome to synthesize a specific polypeptide.

Metal shadowing A technique used in electron microscopy that provides a three-dimensional image of the surface of a biological sample by coating the surface with an evaporated heavy metal such as tungsten.

Methionine An amino acid with a side chain containing a thioether group, represented by Met or M.

Methionine aminopeptidase An enzyme that cleaves the N-terminal methionine from a polypeptide.

Methionyl-tRNA formyltransferase Enzyme that transfers a formyl group from N^{10}-formyltetrahydrofolate to methionyl-tRNA^{fMet} to form fMet-tRNA^{fMet}.

5′-m⁷G cap N^7-methylguanosine attached to the first or initiating nucleotide in pre-mRNA or mRNA through an inverted 5′→5′ triphosphate bridge to form a cap at the 5′-end of mRNA.

Michaelis constant (also called K_m) The substrate concentration at which one-half of the total enzyme population exists in the form of an enzyme–substrate complex. It is experimentally determined as the substrate concentration at one-half the maximum velocity for an enzyme-catalyzed reaction.

Microprocessor complex A nuclear protein complex, which contains an RNase III and a double-stranded RNA binding protein, that is required for pre-miRNA hairpin formation.

MicroRNA (also called **miRNA**) Very short RNAs present in miRISC that help to regulate gene expression.

Microsatellites Repeat sequences that are 1 to 6 bp long and reiterated 5 to 100 or more times throughout the vertebrate genome.

Microsome A pellet produced by high speed centrifugation that contains pieces of the endoplasmic reticulum and free ribosomes.

Microtubules Thin, hollow filaments made of tubulin that are part of the mitotic and meiotic spindle.

Minimal clamp loader The smallest complex that can load sliding clamps onto DNA.

Minimal medium A simple chemically defined growth medium containing inorganic salts but no organic compounds other than a carbon source such as a sugar.

Minichromosome maintenance 2-7 complex (also called **MCM2-7 complex**) An ATP-dependent eukaryotic helicase that is essential for unwinding DNA during replication.

Minisatellites Repeat sequences that are 7 bp to about 100 bp long and imperfectly reiterated 5 to 100 times.

Minor groove The narrow groove that winds around the B-DNA helix.

miRNA (see **MicroRNA**)

Mitosis Nuclear division in a eukaryote cell, wherein the number of chromosomes is retained during cell division.

Mitotic spindle An array of microtubules that forms in a mitotic eukaryotic cell and moves the duplicated chromosomes to opposite poles of the cell.

Mobilizable plasmid A plasmid with genes required to mobilize DNA transfer but that lack the genes required for a donor cell to make effective contact with a recipient cell.

Model A hypothesis or tentative explanation of the way a system works.

Modulon A network of operons that is under the control of a single regulatory protein.

Molecular biology A discipline in the life sciences that uses tools and concepts provided by the physical sciences to study structure–function relationships among biological molecules such as nucleic acids and proteins that participate in the transmission of genetic information from one generation to the next and the expression of that genetic information within a living system.

Molecular chaperones Specific proteins that help to stabilize an emerging polypeptide until the entire folding unit has been extruded by the ribosome.

Molecular exclusion chromatography (see **Gel filtration**)

Monoadduct The product formed when a chemical group attaches to a single site on a DNA molecule.

Morphogenetic enzymes Enzymes involved in the assembly of phage particles.

mRNA (see **messenger RNA**)

Mutagen A DNA damaging agent that causes mutations.

Mutagenesis The process of formation of a mutant organism.

Mutasome A UmuD′$_2$C • RecA • ATP complex that is capable of translesion synthesis.

NAP-1 (see **Nucleosome assembly protein 1**)

Native state The conformation of a biologically active protein or nucleic acid.

Negative regulation Molecular regulation mechanism in which a repressor turns off the transcription of one or more genes.

Negative regulator A regulatory protein that lowers the rate of transcription.

Negative staining A technique used in electron microscopy in which a negative image of the specimen is created.

N-formylmethionine (also called **fMet**) The methionine derivative involved in initiation of bacterial translation.

N-glycosylic bond The bond in a nucleoside that links C-1 on the furanose ring to a specific nitrogen atom on the base (N-1 in pyrimidines and N-9 in purines).

Nick A single cut in a DNA strand.

Nick translation A process in which *E. coli* DNA polymerase I acts on a nicked DNA duplex so that its 5′ to 3′ exonuclease digests the 5′-end of the nick, while its polymerase activity extends the 3′-end of the nick.

Nicked circle (also called **RFII**) A circular double-stranded DNA molecule with a single nick on one strand.

Nitrocellulose filter assay An analytic technique based on the fact that ribonucleoproteins and proteins stick to nitrocellulose but nucleic acids do not.

Node The site at which the axis of a DNA molecule crosses itself in a supercoiled DNA.

Nonautonomous element A transposon that has the end sequence recognized by a transposase but does not encode the transposase.

Noncomposite transposon A transposable element such as Tn3, which has inverted repeats rather than IS elements flanking *Tnp* and other genes.

Nonhomologous end-joining A DNA repair mechanism used to fix double-strand breaks where little or no homology is present.

Non-LTR retrotransposon A retrotransposon that does not have a long terminal repeat at each end.

Nonpermissive temperature The temperature at which a temperature-sensitive mutant displays the mutant phenotype.

Nonsense mutation A mutation that changes a sense codon to a translation termination codon.

Nonshuttling hnRNP proteins A family of proteins that prevent excised introns and partially spliced mRNA from leaving the nucleus.

Nontransmissable plasmids Plasmids lacking genes required for making effective contact and for mobilization.

Northern blot procedure A technique in which an RNA molecule is separated from other RNA molecules by gel electrophoresis, transferred to a nitrocellulose membrane (or some other membrane), fixed to the membrane, and then detected based on its ability to bind a labeled or tagged probe.

N-terminal domain of Lac repressor (also called **headpiece**) The DNA-binding domain of the Lac repressor.

Nuclear export signal A signal on some proteins that permits the proteins to move through the nuclear pore complex from the nucleoplasm to the cytoplasm.

Nuclear localization signal A short basic sequence containing arginine and lysine residues that allows proteins that have it to move through the nuclear pore complex from the cytoplasm to the nucleoplasm.

Nuclear receptor superfamily Family of receptors that allow cells in higher organisms to respond to a variety of external and internal chemical signals by increasing or decreasing the transcription of specific genes.

Nuclear run-on method Method used to show that transcription termination of protein coding transcription units takes place after the poly(A) site.

Nucleic acid A macromolecule consisting of a series of nitrogenous bases connected to deoxyribose or ribose molecules, which in turn are linked by phosphodiester bonds.

Nuclein The term used by Friedrich Miescher to describe the fraction of white blood cells containing an acidic material with unusually high phosphorus content.

Nucleoid A highly condensed nucleoprotein complex in a bacterial cell that contains the genome and is not enclosed by a nuclear membrane.

Nucleolus A discrete region of the nucleus where ribosomes are produced.

Nucleosome A compact nucleoprotein structure that is the basic structural subunit of chromatin.

Nucleosome assembly protein 1 (also called **NAP-1**) A chaperone protein that deposits histone H2A • H2B dimers on H3 • H4 tetramers to form complete nucleosomes.

Nucleosome core particle The nucleoprotein complex remaining after chromatin is extensively digested with DNase. It consists of an octameric protein complex (two copies each of histones H2A, H2B, H3, and H4) and has a DNA fragment that is about 146 bp long.

Nucleotide excision repair DNA repair pathway that removes bulky adducts from DNA by excising an oligonucleotide bearing the lesion and replacing it with new DNA.

Octameric protein complex The protein complex within the nucleosome core particle that contains two copies each of histones H2A, H2B, H3, and H4.

Okazaki fragments Short stretches of about 1,000 to 2,000 bases in bacteria and 100 to 200 bases in eukaryotes, which are produced during discontinuous DNA replication, that are later joined into a covalently intact strand by DNA ligase.

Oligopeptide Peptides with fewer than 50 amino acid residues.

Oncogene Mutated proto-oncogene, which synthesizes a larger quantity of the normal gene product or a more active form of the gene product.

Open promoter complex A complex formed during the transcription initiation stage in which the DNA strands have separated to form a bubble that exposes the transcription start site.

Open reading frame A sequence of DNA consisting of triplets that specifies a polypeptide starting with an initiation codon and ending with a termination codon.

Operator A regulatory region in prokaryotic DNA that binds a repressor to inhibit the transcription of one or more adjacent structural genes.

Operon model A set of two or more adjacent structural bacterial genes that are transcribed as a single polycistronic mRNA, the promoter that serves as a signal for transcription initiation, and an operator that regulates transcription.

ORC (see **Origin recognition complex**)

oriC A specific site within the *E. coli* replicator at which DNA replication is initiated. Because a bacterial replicator is usually quite short, the terms "replicator" and "origin of replication" tend to be used interchangeably.

oriT (see **Transfer origin**)

Origin of replication (*ori*) The specific site within the replicator at which replication is initiated.

Origin recognition complex (also called **ORC**) A multisubunit protein complex that functions as an initiator protein in eukaryotes.

Orphan nuclear receptors Nuclear receptors that bind to unknown ligands.

Orthologs Genes in different organisms that evolved from a common ancestral gene through the course of evolution.

Outer membrane Additional membrane that surrounds the cell wall in gram-negative (but not gram-positive) bacteria.

Overlap method A sequencing method that depends on having the sequences of two different fragment collections of a single polypeptide. The order of the fragments within the original polypeptide chain is determined by searching the two sets of fragments for overlapping sequences.

Pachytene The third stage of meiosis division I, in which pairing of chromosomes is completed and crossing over begins.

Pairing diversity The third level of antibody diversity, which occurs through the combination of one heavy chain and one light chain to make an antigen receptor protein.

Palindrome Nucleotide sequence on one strand of DNA that is identical to its complementary strand read backward.

Parallel pleated sheet A β-Sheet organized so all polypeptide chains have the same amino to carboxyl direction.

p arm The shorter of the two chromatid arms.

Partial diploid A cell containing one complete set of genes and duplicates of some of these genes.

Passenger strand The strand in the siRNA duplex that is *not* assembled into the siRISC.

PCNA (also called **proliferating cell nuclear antigen**) Sliding clamp subassembly for eukaryotic DNA replication.

PCR (see **Polymerase chain reaction**)

Peptide A short chain of amino acids linked by peptide bonds.

Peptide bond An amide bond that links amino acids.

Peptide deformylase An enzyme that cleaves formyl groups from N-terminal methionine in bacterial proteins.

Peptide mapping (see **Fingerprinting**)

Peptidoglycan A combination of polysaccharides and peptides that make up the cell wall of bacteria.

Periplasmic space The space between the cell membrane and outer membrane in gram-negative bacteria.

Permissive temperature The temperature below which a temperature-sensitive mutant behaves normally.

Petri dish A shallow glass or plastic flat-bottomed dish with a lid in which solid growth medium is usually placed.

Phage (also called **bacteriophage**) A virus that replicates in bacteria.

Phage lambda (λ) A temperate phage that uses a conservative site-specific recombinase to integrate into the host genome.

Phage lysate The suspension of newly synthesized phages released from the host cell during the lytic cycle.

Phage T4 A member of the T-even phage family that injects its DNA into the bacterial host and kills it.

Phagemid A cloning vector that can replicate as a double-stranded plasmid in a bacterial cell or be induced to replicate as a phage to produce large quantities of single-stranded recombinant DNA.

Phenotype The physical traits of an organism.

Phenylalanine An amino acid with an aromatic side chain, represented by Phe or F.

Phosphodiester bonds The bonds in DNA or RNA that link neighboring nucleosides to a common phosphate group.

Phosphoenolpyruvate-dependent phosphotransferase system (also called **PTS**) A system that uses energy supplied by phosphoenolpyruvate to phosphorylate a sugar as it transports the sugar across the inner cell membrane.

Phosphomonoester The ester bond that links a sugar to a single phosphate group.

(6-4) Photolyase A photolyase that reverses (6-4) photoproduct lesions.

(6-4) Photoproducts Product formed by ultraviolet damage to DNA that involves the formation of a bond between the C-6 atom of a 3′-pyrimidine (either thymine or cytosine) and the C-4 atom of the 5′-pyrimidine (usually cytosine).

Piwi interacting RNA (piRNA) Small RNAs that bind to members of the Piwi subfamily of Argonaute proteins.

Plaque A clear area that appears in a bacterial lawn in the presence of a phage.

Plaque assay A technique used to count phages by observing clear spaces caused by phages in a bacterial lawn.

Plasmid An autonomously replicating small circular DNA molecule.

Plate A laboratory term referring to a Petri dish containing a solid medium.

Platform of the 30S subunit A broad shelf-like protrusion on the upper part of the body in the 30S subunit of the ribosome.

Plating The act of depositing bacteria on the agar surface of a solid growth medium.

Point mutant A mutant in which only one base pair change has occurred.

polA The structural gene for *E. coli* DNA polymerase I.

Pol β (see **DNA Polymerase β**)

Pol δ (see **DNA Polymerase δ**)

Pol ε (see **DNA polymerase ε**)

Poly(A) site (see **Cleavage/polyadenylation site**)

Polyadenylation signal A highly conserved sequence 10 to 30 nucleotides before the cleavage site that is required for both cleavage and poly(A) addition.

Polycistronic mRNA Bacterial mRNA molecules that contain two or more cistrons.

Polycyclic aromatic hydrocarbons Chemical compounds that consist of fused aromatic rings that can damage DNA after they are metabolically activated.

Polymerase chain reaction (also called **PCR**) A technique used to amplify regions of DNA using sequence-specific primers and multiple cycles of DNA replication.

Polymorphism The presence of two or more different alleles of a specific gene with a frequency of at least 1% within the population.

Polypeptide An amino acid chain that contains 50 or more amino acid residues.

Polypeptide folding motifs (also called **supersecondary structures**) Certain combinations of secondary structures present in many different proteins, such as hairpin turn, β-α-β unit, helix-turn-helix, and Greek key.

Polyprotein A polypeptide that is cleaved to form two or more functional proteins.

Polypyrimidine tract A stretch of 8 to 12 pyrimidines (mostly uracil) just upstream from the 3′-splice site in many introns.

Polyribosome (also called **Polysome**) A complex in which two or more ribosomes are simultaneously translating the same mRNA.

Porins Protein channels in the outer membrane of gram-negative bacteria that allow small water-soluble molecules to pass through.

Positive regulation Molecular regulation mechanism in which an activator interacts with a promoter, stimulating transcription.

Positive regulator (see **Activator**)

Positive-negative selection technique A selection technique that facilitates the isolation of a targeted gene disruption.

Posttranscriptional processing The molecular modification of RNA after transcription.

POT1 A DNA binding protein that binds to single-stranded DNA containing TTAGGGTTAGGG and is probably associated with the single strand in the D-loop.

Precursor mRNA (also called **pre-mRNA**) The nuclear transcript that is processed by modification and splicing to yield an mRNA.

Preinitiation complex The assembly of transcription factors at the promoter before RNA polymerase binds.

Premature release (also called **Signaling release**) A model that proposes core polymerase is released before Okazaki fragment formation is complete.

pre-mRNA (see **Precursor mRNA**)

Pre-RISC complex A complex formed when Argonaute 2 (Ago 2) and perhaps other still uncharacterized proteins add to the siRISC loading complex.

Pre-tRNA The nuclear transcript that is processed by modification and splicing to yield a tRNA.

Pribnow box (see **−10 box**)

Primary culture A cell culture in which each cell presumably has the same characteristics as the cells in the original tissue and is following the program of limited cell division set up when the tissue formed.

Primary structure The sequence of amino acids in a polypeptide.

Primary transcript The newly synthesized RNA molecule before it is processed.

Primer strand The preformed strand of DNA that is the site of attachment for incoming deoxynucleotides.

Procapsid A protein shell of a virus that lacks nucleic acid.

Processed pseudogene An inactive gene copy that lacks introns, which originates through reverse transcription of mRNA and insertion of a duplex copy into the genome.

Processivity Provides a quantitative measure of a polymerase's ability to remain tightly associated with its DNA template during DNA (or RNA) synthesis.

Prokaryotes Cells that lack a defined nucleus.

Proline An amino acid with a side chain that is part of a five-member ring that includes the α-amino group, represented by Pro or P.

Proliferating cell nuclear antigen (see **PCNA**)

Promoter A DNA sequence at the beginning of a gene that functions as a signal to initiate transcription.

Promoter escape RNA polymerase becomes free of the promoter so it can move downstream to elongate the nascent RNA chain.

Proofreading A mechanism for correcting errors in nucleic acid synthesis that involves removing individual units after they have been added to the chain. The term is also used to describe hydrolysis of a misacylated tRNA or the hydrolysis of an aminoacyl-AMP that was formed in error.

Prophage Phage DNA that resides in a bacterial host in a passive state so it does not produce progeny phage particles or express many viral functions. The phage DNA may be integrated into the bacterial chromosome or exist as an independent DNA molecule that replicates in the cytoplasm.

Protein A polypeptide folded into a biologically active conformation.

Protein conducting channel (see **Translocon**)

Proteome The complete set of proteins that is expressed by an entire genome.

Proto-oncogene A gene that codes for a product that promotes normal cell growth and division but has the potential to be converted to an oncogene.

Protoplast The spherical form of a bacterial cell treated with lysozyme in which the cell wall has been completely removed.

Prototroph A bacterium that can grow in a minimal medium.

Proximal promoter (see **Regulatory promoter**).

Pseudogene A nonfunctional gene that resembles a functional gene.

Pseudoknot An RNA that forms when a base sequence in a hairpin loop pairs with a complementary single-stranded region adjacent to the hairpin stem.

P-site The peptidyl-tRNA binding site in the ribosome. A peptidyl group bound to a tRNA at the P-site is transferred to the aminoacyl-tRNA at the A-site.

Psoralen A naturally occurring substance synthesized by some members of the carrot plant family, which must be photoactivated before it can alkylate DNA.

PTRF (see **RNA polymerase I transcription release factor**)

PTS (see **Phosphoenolpyruvate-dependent phosphotransferase system**)

Pulse-chase experiment An experimental technique in which cells are first incubated for a short period of time with a radioactive molecule (the pulse) and then for additional time with the nonradioactive molecule (the chase).

Purine A heterocycle with a five-member imidazole ring fused to a six-member pyrimidine ring that is the parent compound for both adenine and guanine.

Puromycin An antibiotic that terminates protein synthesis by mimicking a tRNA at the A-site and linking to the nascent protein chain.

Pyrimidine A six-atom nitrogen-containing heterocycle that is the parent compound for uracil, thymine, and cytosine.

Pyrosequencing A DNA sequencing method based on the fact that each deoxyribonucleotide addition to a primer is accompanied by pyrophosphate release that can be monitored with a chemiluminescent reaction.

Pyrrolysine A rare amino acid, which has only been observed in archaeal cells that generate methane.

q arm The longer of the two chromatid arms.

Quaternary structure The way polypeptide chains are arranged in space with respect to one another in proteins with more than one polypeptide chain.

R2D2 An RNA binding protein that is a component of the RISC loading complex.

Radiomimetics Chemicals that make double-strand breaks.

RAG-1 and RAG-2 Enzymes involved in $V(D)J$ recombination, which make a single-strand nick at the junction between the recombination signal sequence and the coding sequence to form a free 3'-OH group.

Recessive allele An allele that produces its characteristic phenotype only when paired with an identical allele.

Recognition elements Features present in tRNA that are required for recognition by the cognate aminoacyl-tRNA synthetase.

Recognition helix The α-helix in a helix-turn-helix polypeptide folding motif that makes specific contacts with DNA.

Recombinant DNA technology (see **Genetic engineering**)

Recombinant plasmid A modified plasmid created in the laboratory that contains a foreign DNA insert.

Recombinase An enzyme that acts at *dif*, converting a bacterial chromosome dimer into two separate daughter chromosomes.

Recombination signal sequence (also called **RSS**) Sequences that flank the V, D, and J segments that participate in the recombination reaction to join the segments together.

Regulated exon An exon that is included in some mRNAs and excluded from other mRNAs.

Regulatory enzyme An enzyme that catalyzes a committed step in a biochemical pathway that regulates the activity of the pathway.

Regulatory promoter (also called **Proximal promoter**) A DNA sequence that is usually located between 50 and 200 bp upstream from the transcription initiation site in a eukaryotic transcription unit that binds activators and repressors to regulate gene transcription.

Regulatory subunit The subunit of a regulatory enzyme with two different kinds of subunits that contains the allosteric site(s).

Regulon An operon network consisting of two or more operons that is regulated by a common regulatory protein and its effector(s) and associated with a single pathway, function, or process.

Relaxed circles Circular DNA that has no superhelical turns and is unconstrained when it lies flat on a surface.

Relaxed plasmids Plasmids that are maintained at multiple copies per cell.

Release factors (also called **RFs**) Factors that recognize termination codons.

Renaturation The return of a denatured protein or nucleic acid to its native state.

Renatured DNA Double-stranded DNA that has reformed from single strands that were produced by denaturation.

Repair of DNA The return of damaged DNA to its normal sequence and structure.

Replicating form I (also called **RFI**) An intact circular double-stranded DNA molecule.

Replicating from II (also called **RFII**) A circular double-stranded DNA molecule with a single nick in one strand.

Replication bubble A region that contains newly synthesized DNA, which grows as DNA synthesis continues.

Replication factor C (also called **RFC**) The eukaryotic clamp loader that facilitates the loading of PCNA (sliding clamp) onto DNA during replication and repair.

Replication fork The Y-shaped structure formed by DNA strand separation where DNA synthesis takes place.

Replication protein A (also called **RPA**) A single-stranded DNA binding protein present in eukaryotes that is involved in DNA replication, DNA repair, and recombination.

Replication termination in bacteria The final stage of DNA replication, which begins when the two-replication forks meet about half way around the DNA molecule.

Replicative transposition A transposon moves to a new site while maintaining a copy at its original site.

Replicator A specific set of nucleotide sequences within the DNA molecule that is recognized by the initiator protein.

Replicon An independently replicating DNA molecule such as a viral, bacterial, or eukaryotic chromosome or a plasmid that can maintain a stable presence in a cell.

Replicon model A model proposed to explain how DNA molecules replicate autonomously.

Replisome The replication machinery consisting of DNA polymerase III holoenzyme, helicase, primase, and single-stranded DNA binding protein.

Reporter fusion A synthetic element containing a selectable genetic marker and a reporter gene that only will be expressed if it inserts into a gene that is already being expressed in the bacterial cell.

Repressible enzymes Enzymes that exhibit a decreased rate of synthesis in response to the addition of a small molecule in the medium.

Repressor A protein that binds to an operator and blocks the transcription of one or more genes.

Resolution The cleavage of the Holliday junction structure to generate two separate nicked duplexes.

Resolvase An enzyme such as RuvC that has the ability to catalyze the cleavage of the Holliday junction structure to generate two separate nicked duplexes.

Restriction endonucleases Enzymes that recognize specific short palindromic sequences of DNA and cleave the duplex.

Restriction fragment length polymorphisms (also called **RFLPs**) Variations in DNA fragment lengths observed when DNAs from different individuals are cut with the same restriction endonuclease. Fragment length differences, which are usually due to genetic mutations, can serve as genetic markers.

Restriction map A linear array of sites on DNA cleaved by various restriction enzymes.

Retrotransposition Transposition via an RNA form in which DNA is transcribed into RNA, then reverse-transcribed into DNA, which is inserted at a new site in the genome.

Reverse genetics The functional investigation of a gene that starts with the gene sequence rather than a mutant phenotype.

Reverse transcriptase An enzyme that uses a template of single-stranded RNA to generate a double-stranded DNA copy.

Reversion The process of regaining an original phenotype after a mutation has occurred.

Revertant A mutant that has undergone reversion to the wild type.

RFI (see **Replication form I**)

RFII (see **Replication form II**)

RFC (see **Replication factor C**)

RFLPs (see **Restriction fragment length polymorphisms**)

RFs (see **Release factors**)

Rho factor A protein that assists E. coli RNA polymerase to terminate transcription at Rho-dependent terminators.

Rho utilization site (also called ***Rut* site**) Cytosine-rich region at which Rho factor loads onto nascent mRNAs.

Rho-dependent termination Termination of transcription by bacterial RNA polymerase in the presence of Rho factor.

Ribonuclease (RNase) An enzyme that cleaves RNA.

Ribosomal recycling factor Factor that influences ribosomal subunits to dissociate from each other after polypeptide chain synthesis is complete.

Ribosome A large ribonucleoprotein complex consisting of a large and a small subunit that functions as a protein synthetic factory to make specific polypeptides as directed by mRNA.

Ribosome–nascent chain complex Ribosomal complex in the process of polypeptide chain formation.

Riboswitch A 5′-untranslated region in mRNA that changes conformation upon binding a small ligand and thereby alters the mRNA's ability to be translated.

Ribozyme An RNA that has catalytic activity.

Rifampicin A semisynthetic derivative of rifamycin that inhibits phosphodiester bond formation during the initiation stage of transcription and has no effect on later stages.

Rifamycins A family of antibiotics produced by strains of *Streptomyces* that inhibit formation of the first or second phosphodiester bonds.

rInr (see **rRNA initiator**)

RISC (see **RNA-induced silencing complex**)

RISC loading complex Protein complex formed when an siRNA duplex combines with a protein complex containing the DCR2 • R2D2 heterodimer and other proteins.

RNA core polymerase A bacterial protein complex containing the RNA polymerase α, β, and β′ subunits in the stoichiometry $\alpha_2\beta\beta'$.

RNA editing A change of sequence at the level of RNA after transcription.

RNA-induced silencing complex (also called **RISC**) A complex that catalyzes target mRNA cleavage.

RNA interference (also called **RNAi**) Gene silencing that is caused by double-stranded RNA.

RNA polymerase An enzyme that copies the information present in a DNA template into RNA.

RNA polymerase holoenzyme A large multisubunit protein that is the fully active form of *E. coli* RNA polymerase.

RNA polymerase I The eukaryotic enzyme responsible for pre-rRNA synthesis.

RNA polymerase I transcription release factor (also called **PTRF**) Protein required in RNA polymerase I transcription termination.

RNA polymerase II The enzyme responsible for eukaryotic mRNA formation.

RNA polymerase III The eukaryotic enzyme responsible for making precursors of 5S rRNA and tRNA.

RNA polymerase lid A loop in the β′ subunit that acts as a wedge to help the RNA to separate from the DNA and thereby prevent the hybrid from becoming overextended.

RNA recognition motif (also called **RRM**) An RNA binding domain.

RNAi (see **RNA interference**)

RNase H An enzyme that digests RNA in a DNA–RNA hybrid.

Rolling circle replication A mechanism of DNA replication in which a nick is introduced in a double-stranded circular DNA to produce a 3′-end that is extended by DNA polymerase, which uses the intact circular strand as a template. The 5′-end of the nicked strand is displaced as the DNA polymerase extends the 3′-end.

Rough endoplasmic reticulum Regions of the endoplasmic reticulum that are studded with ribosomes, which appear rough or grainy in electron micrographs.

Rous sarcoma virus A retrovirus that causes tumor formation in chickens.

RPA (see **Replication protein A**)

RRM (see **RNA recognition motif**)

rRNA initiator (also called **rInr**) An adenine-thymine (AT)-rich region within the core promoter that extends from about −10 to +10 with respect to the transcription initiation site.

RSC complex A protein complex that remodels the structure of chromatin.

RS-domain An arginine/serine-rich domain found at the C-terminus of SR proteins.

RSS (see **Recombination signal sequence**)

Rudder A loop in the β′ subunit that interacts with DNA to prevent it from re-annealing with the RNA during transcription.

12/23 Rule An IgG heavy chain recombination rule that dictates a recombination signal sequence with a 12-bp spacer can only be joined to a recombination signal sequence with a 23-bp spacer.

***Rut* site** (see **Rho utilization site**)

S phase of the eukaryotic life cycle DNA replication and histone synthesis occur during this part of interphase.

S1 endonuclease An endonuclease that acts exclusively on single-stranded polynucleotides or on single-stranded regions of double-stranded molecules.

SAGA complex A protein complex that stimulates transcription at least in part by catalyzing histone acetylation.

Scaffold model A model in which chromatin is predicted to bind to a protein scaffold to form loops.

Scrunching model A transcription initiation model that proposes that DNA is pulled into the RNA polymerase holoenzyme as the DNA unwinds to form the open complex.

SDS-PAGE (see **Sodium dodecyl sulfate-polyacrylamide gel electrophoresis**)

Secondary structure The folding pattern within a segment of a polypeptide chain containing neighboring residues.

Sedimentation coefficient(s) The ratio of sedimentation velocity to centrifugal force.

Selective marker The transferred allele that is selected by means of the agar conditions when plating cells.

Selectivity factor 1 (also called **SL1**) An auxiliary factor required by human RNA polymerase I to synthesize rRNA.

Selenocysteine (Sec) A rare amino acid analog of cysteine in which selenium replaces sulfur.

Self-transmissable plasmid The F factor and other plasmids with genes that code for proteins required for successful mating.

Semiconservative model of DNA replication Model of replication in which each double-stranded daughter DNA molecule will have a conserved DNA strand, which is derived from the parental DNA, and a newly synthesized strand.

Semidiscontinuous model of replication A replication model that predicts discontinuous synthesis of the chain growing in an overall 3′ to 5′ direction but continuous synthesis of the chain growing in an overall 5′ to 3′ direction.

Sense strand (see **Coding strand**)

SeqA protein A protein that binds to the hemimethylated GATC sites within *oriC* so that *oriC* cannot participate in a new round of replication.

Sequence hypothesis The idea that the linear sequence of nucleotides in a DNA (or RNA) chain determines the linear sequence of amino acids in a polypeptide chain.

Serine An amino acid with a hydroxylmethyl (–CH$_2$OH) group as its side chain, represented by Ser or S.

Sex chromosomes The pair of chromosomes that determines the sex of the individual.

Sex pilus A hair-like appendage that projects from the surface of the male bacterial cell and binds to a receptor on the surface of the female cell.

SF1 (see **Splicing factor 1**)

Shelterin A protein complex that enables cells to distinguish their natural chromosome ends from DNA breaks.

Shine-Dalgarno sequence The purine rich consensus sequence AGGAGG centered about 5 to 7 nucleotides before the AUG initiation codon on bacterial mRNA that acts as a ribosome binding site.

Short interspersed nucleotide elements (also called **SINEs**) A non-LTR retrotransposon that is usually less than 500 kbp long.

Short patch repair A branch of the base excision repair pathway used to replace only a single nucleotide.

Shotgun sequencing A technique that uses random DNA cleavage to produce fragments that are inserted into plasmids and then sequenced. The sequencing reads are stored in a computer that has a sequence assembly software program, which searches for overlaps and uses the overlaps to assemble the sequencing reads into a long contiguous sequence.

Sickle cell anemia A genetic disorder in which red blood cells change from a biconcave to a sickle shape when deprived of oxygen due to a substitution of one amino acid in the β globin chain.

Sickle-cell trait The presence of a single allele for sickle cell anemia that results in some resistance to the malarial parasite but does not cause sickle cell anemia.

Sigma (σ) factor A component of bacterial RNA polymerase holoenzyme required for specific binding to promoter sites on DNA.

Sigmoidal kinetic curve An S-shaped curve often formed by regulatory enzymes when the initial velocity (v_o) of a reaction is plotted on the y-axis and substrate concentration ([S]) is plotted on the x-axis.

Signal hypothesis The hypothesis that explains how cells determine whether a protein will be synthesized on a free or membrane-bound ribosome.

Signal peptidase An enzyme within the membrane of the endoplasmic reticulum or bacterial cell membrane that removes the signal sequences from proteins as the proteins are translocated.

Signal recognition particle (also called **SRP**) A ribonucleoprotein complex that recognizes signal sequences during translation and guides the ribosome to the translocation channel.

Signaling release (see **Premature release**)

Silencer Sequence-specific element in eukaryotic DNA that represses transcription of a target gene and usually functions independently of distance and orientation.

Simple retroviruses Retroviruses with RNA that contains just the *gag*, *pol*, and *env* genes.

SINEs (see **Short interspersed nucleotide elements**)

Single nucleotide polymorphism (also called **SNP**) A polymorphism caused by a change in a single nucleotide.

Single-strand DNA binding proteins (also called **SSBs**) Proteins that bind to and stabilize single-stranded DNA.

siRNA (see **Small interfering RNA**)

Sister chromatids Two identical subunits of a divided chromosome that are held together at the centromere.

Site-directed mutagenesis An *in vitro* technique that introduces a mutation at a specific site within DNA.

Sleeping Beauty A transposon that originated from nonfunctional transposons found in fish, which was reactivated using site-directed mutagenesis.

Sliding clamp The subassembly forms a ring around DNA, tethering the remainder of the DNA polymerase holoenzyme to the DNA.

SL1 (see **Selectivity factor 1**)

SL1/TIF-1B A general term used for the mammalian auxiliary transcription factor for RNA polymerase I that is called SL1 in humans and TIF-1B in mouse.

Sm polypeptides A family of closely related polypeptides found in snRNPs.

Small interfering RNA (also called siRNA) Small pieces of double-stranded RNA created by Dicer • RDE-4 during RNAi.

Small nuclear RNA (also called snRNA) A type of small RNA confined to the nucleus, several kinds of which are involved in splicing or other RNA processing reactions.

Smooth endoplasmic reticulum Regions of the endoplasmic reticulum that lack ribosomes and have a smooth appearance.

Snf1 kinase A protein kinase that operates in the glucose repression pathway.

SNP (see **Single nucleotide polymorphism**)

snRNA (see **Small nuclear RNA**)

Sodium dodecyl sulfate-polyacrylamide gel electrophoresis (also called SDS-PAGE) Gel electrophoresis in which proteins that have been denatured by sodium dodecyl sulfate are separated solely on the basis of size.

Somatic cells All cells that are not germ cells or gametes, which each contain two versions of each chromosome.

Somatic cell genetics The study of genes using hybrids formed by fusing somatic cells of different species.

SOS response A process in which DNA damage in *E. coli* caused by UV irradiation or a chemical agent acts as a signal to induce a multigene repair response.

Southern blot procedure (also called Southern transfer) A technique in which a DNA fragment that is generated by restriction endonuclease digestion is separated from other fragments by gel electrophoresis, denatured, transferred to a nitrocellulose membrane (or some other membrane), fixed to the membrane, and then detected based on its ability to bind a labeled or tagged probe.

Spacefill structure A molecular representation in which atoms are shown as van der Waals spheres.

Specialized transduction Imprecise phage λ excision so that a small piece of bacterial DNA is excised along with the λ DNA. Then the recombinant λ DNA is replicated and packaged into a λ particle, which can deliver the bacterial DNA to a new bacterial host.

Spheroplast The spherical form of a cell treated with lysozyme if some cell wall remains.

3′-Splice sequence Short nucleotide sequence on the 3′-end of an intron that identifies where the intron ends.

5′-Splice sequence Short nucleotide sequence on the 5′ end of an intron that identifies where the intron begins.

3′-Splice site The intron/exon boundary at the end of an intron.

5′-Splice site The exon/intron boundary at the start of an intron.

Spliceosome A large ribonucleoprotein splicing machine on which nuclear mRNA precursors are spliced.

Splicing factor 1 (also called SF1) A non-snRNP protein that participates in the spliceosome assembly pathway by binding to the branch point sequence.

Splicing regulatory proteins (also called SR proteins) A family of proteins that appears to be important for exonic splicing enhancer recognition.

Split (discontinuous) gene hypothesis Hypothesis that regions that code for amino acids within a gene are interrupted by intervening sequences (introns) that do not code for amino acids.

Spo11 protein An endonuclease that produces double-strand DNA breaks during meiosis.

Spontaneous mutagenesis The occurrence of a mutation without the application of a mutagen.

Squelching Inhibition that occurs when transcription activator proteins compete among themselves for Mediator.

SR (see **SRP receptor**)

SR proteins (see **Splicing regulatory proteins**)

SRP (see **Signal recognition particle**)

SRP receptor (also called SR) Receptor to which the SRP ribosome nascent polypeptide chain complex binds during the SRP cycle.

SSBs (see **Single-strand DNA binding proteins**)

Stabilization helix The helix in the helix-turn-helix polypeptide folding motif that helps stabilize the Lac repressor–DNA complex.

Staggered end (see **Cohesive end**)

Stationary phase The stage in a bacterial growth curve in which cells stop dividing because of a lack of oxygen, a nutrient, or an unfavorable pH.

Stem-loop structures (see **Hairpin structures**)

Steroid receptors Receptors that bind steroids and interact with DNA as homodimers.

Sticky end (see **Cohesive end**)

Strand invasion A process in which a recombinase enables a broken single-stranded DNA to search for and find homologous sequences in an intact duplex DNA and then displace the like strand in the duplex and forms Watson-Crick base pairs with its complementary strand.

Stringent plasmid A plasmid that is maintained at only one or two copies per cell.

Structural gene A gene that specifies a particular polypeptide.

Substrate The molecule on which an enzyme acts.

Sucrose gradient centrifugation A technique in which a sample is layered on top of a solution in a centrifuge tube that contains sucrose, the concentration of which increases continuously from the top to the bottom of the tube. Proteins or nucleic acids in the sample are separated by placing the centrifuge tube in a swinging bucket rotor and centrifuging at high speeds.

Suicide enzyme A protein that loses enzymatic activity after acting only one time.

Superhelix (also called **Supercoil**) A circular DNA molecule in which the axis of the DNA crosses itself or writhes.

Supersecondary structures (see **Polypeptide folding motifs**)

Suppressors Gene products that can compensate for the defect in a mutant or enable an altered gene to produce a functional gene product.

Suppressor-sensitive mutants A conditional mutant that exhibits the mutant phenotype in some strains but not in others, depending on the presence of suppressors.

Svedberg (S) A unit of sedimentation coefficient corresponding to 10^{-13} seconds.

Swi/Snf complex A chromatin remodeling complex in yeast that uses hydrolysis of ATP to change the organization of nucleosomes.

Synapsis The association of two pairs of sister chromatids that occurs at the start of meiosis.

Synthesis-dependent strand annealing A eukaryotic homologous recombination pathway involving double-strand break end processing, nucleoprotein filament formation, strand invasion, and strand extension. The displaced strand anneals to the complementary strand on the other side of the double-strand break and the remaining gaps are filled in and nicks ligated.

Synthetic element A transposon that is engineered not to encode the transposase but to carry a selectable genetic marker that is flanked by the *cis*-acting transposon ends.

TAFs (see **TBP-associated factors**)

Tandem repeat polymorphism Polymorphisms that occur when a DNA sequence, ranging from 1 to up to 100 bp, is repeated head to tail a different number of times at a distinct site in homologous chromosomes.

Target-primed reverse transposition A mechanism used for non-LTR transposon transposition in which 3′-ends of the target DNA, generated by stepwise breaks, prime reverse transcription.

TATA box A conserved AT-rich sequence in many but not all eukaryotic core promoters that is located about 25 to 30 bp upstream of the transcription initiation site.

TATA-binding protein A polypeptide subunit of the general transcription factor TFIID that specifically binds to the TATA box.

TBP-associated factors (also called **TAFs**) Proteins in TFIID that are required to transcribe genes that lack a TATA box as well as for the high levels of transcription that occur within the cell.

Telomerase A reverse transcriptase that contains its own RNA template, which it uses to create repeating units of one strand at the telomere by adding nucleotides to the 3′ overhang.

Telomere The natural end of a eukaryotic chromosome, consisting of a simple repeating sequence that helps to stabilize the chromosome.

Temperate phage A phage that can exist as a prophage.

Temperature-sensitive mutant A mutant that behaves normally at a permissive temperature (usually 30°C) and as a mutant at nonpermissive temperatures (usually 42°C).

Template strand The preformed strand of DNA that determines the order of attachment to the primer strand according to base pairing rules.

Terminally redundant A sequence of bases is repeated at both ends of a phage DNA molecule.

Terminal inverted repeats The sequence of nucleotides at the end of one strand of a linear duplex is repeated in the opposite physical direction at the end of the other strand.

Terminase A phage DNA packaging motor that uses the chemical energy stored in ATP to package phage DNA into the procapsid.

Termination sites (also called **Ter sites**) Conserved 11-bp sequences in *E. coli* chromosome that arrest or pause the movement of the replication fork within the termination region.

Terminus utilization substance (also called **Tus**) A protein that binds to *Ter* sites and creates a complex that arrests the progress of the replication fork by interfering with DnaB helicase's ability to unwind DNA at the replication fork.

Tertiary structure A protein's entire three-dimensional structure, including spatial arrangements among different segments and among residues within the different segments.

Tetrads (see **Bivalents**)

Tetramerization helix Region of the Lac repressor that associates with corresponding regions of the other three subunits of the structure to form a four-helix bundle, producing a Lac repressor tetramer with a V-shape.

T-even phages Bacteriophages T2, T4, and T6, which are very similar to one another.

TFIIA General transcription factor that stabilizes TBP and TFIID binding and blocks inhibition by an inhibitory TAF subunit.

TFIIB General transcription factor that binds TBP, RNA polymerase II, and promoter DNA.

TFIIB recognition element (also called **BRE**) A specific recognition element in the core promoter to which TFIIB binds.

TFIID General transcription factor that binds TATA element and deforms promoter DNA. It also binds Inr and DPE promoter elements.

TFIIE General trascription factor that helps to recruit TFIIH to the core promoter and is required for promoter melting.

TFIIF General transcription factor with serine/threonine kinase activity that combines with RNA polymerase II to form a RNA polymerase II • TFIIF complex.

TFIIH A general transcription factor that has XPB (5′ to 3′) and XPD (3′ to 5′) ATP-dependent helicases.

Theoretical maximum rate of reaction (V_{max}) The rate of an enzyme-catalyzed reaction when all enzyme molecules are present in enzyme–substrate complexes.

Theta (θ) replication A type of mechanism used to replicate double-stranded circular DNA in which replication intermediates resemble the Greek letter theta.

Theta (θ) structure Circular bacterial DNA molecule in the process of replication that resemble the Greek letter theta.

Theta (θ) subunit A subunit in core polymerase (previously known as DNA polymerase III) that seems to stimulate the ε-subunit but does not seem to have a unique function at this time.

Threonine An amino acid with a hydroxyl (–OH) group in its side chain, represented by Thr or T.

Thymine A pyrimidine derivative that is one of the four bases present in DNA molecules.

Thyroid receptor (also called **TR**) Thyroxine binding nuclear receptor that appears to be located in the cell nucleus in both the free receptor form and in the receptor–ligand complex form.

TIFIA (see **Transcription initiation factor IA**)

TIFIB (see **Transcription initiation factor IB**)

t-loop The "telomere loop" formed when a G-rich strand's 3′-overhang folds back to invade the double-stranded region in the mammalian telomere.

T_m (see **melting temperature of DNA**)

Top agar A soft agar layer that provides a very smooth surface so the bacteria can grow with much uniformity.

Topoisomerase An enzyme that catalyzes the conversion of one topoisomer into another.

Topoisomerase IV A bacterial enzyme that allows the interlinked or catenated chromosomes to separate.

Topoisomers (see **topological isomers**)

Topological isomers (also called **topoisomers**) Molecules of DNA that are identical except for a difference in linking number.

Topological properties Properties of circular DNA that do not change when the molecules are distorted, but do change when the polynucleotide backbone is cut.

TψC arm An arm of tRNA that contains a TψC sequence.

TR (see **thyroid receptor**)

Transcription The biosynthetic pathway that uses the information stored in DNA sequences to direct RNA synthesis.

Transcription activator proteins Transcription factors that have DNA-binding domains that recognize specific DNA sequence motifs or modules and activation domains that bind to one or more components of the transcription machinery.

Transcription arrest The complete halt of transcription without complex dissociation.

Transcription elongation complex A complex consisting of RNA polymerase, template DNA, and a growing RNA chain that forms at the end of the initiation stage.

Transcription initiation factor IA (also called **TIFIA**) A type of RNA polymerase associated factor, tightly associated with the RNA polymerase I core enzyme, which is involved in growth-dependent regulation of pre-rRNA transcription.

Transcription initiation factor 1B (also called **TIF-1B**) An auxiliary factor required by mouse RNA polymerase I to synthesize rRNA.

Transcription pause site A template position at which RNA polymerase is temporarily delayed before it resumes transcription.

Transcription termination The complete halt of transcription with complex dissociation.

Transcription termination factor 1 (also called **TTF-1**) Protein assists in termination of transcription catalyzed by RNA polymerase I.

Transductant A cell that has gained a new character as a result of transduction.

Transfection Introduction of a foreign DNA into a eukaryotic host cell.

Transfer origin (also called *oriT*) A unique base sequence in the F factor where a single-strand break is introduced to begin the process of DNA transfer from male to female cell in conjugation.

Transfer RNA (also called **tRNA**) A small RNA molecule that carries a specific amino acid building block to the ribosome for incorporation into the growing polypeptide chain. A three-nucleotide sequence in tRNA (the anticodon) pairs with a three-nucleotide sequence in the mRNA (the codon) that is bound to the ribosome.

Transforming principle DNA that is taken up by an organism such as a bacterial cell and then alters the genetic makeup of the recipient cell.

Transgenic mice Mice that carry foreign genetic material that was introduced during an early stage of embryonic development.

Transition state (see **activated complex**)

Transposable element (also called **Transposon**) A mobile DNA sequence that can move from one location to another in the genome.

Transposase An enzyme that catalyzes a transposition event.

Transposon (see **Transposable element**).

Transient excursion model A model of transcription initiation that proposes transient cycles of forward RNA polymerase motion during abortive RNA synthesis and backward enzyme movement after abortive transcript release with long intervals between cycles.

Transition mutation Mutation in which a pyrimidine on one strand is replaced by a different pyrimidine and a purine on the other strand is replaced by a different purine.

Transition state (see **Activated complex**)

Transitive RNAi A process referring to the ability of the silencing signal to move along a particular gene.

Translation The biosynthetic pathway that uses the information stored in mRNA sequences to direct polypeptide chain formation.

Translation elongation factors Protein factors required during polypeptide chain elongation.

Translation initiation factors Protein factors required at the initiation stage of protein synthesis.

Translesion DNA synthesis Repair of a lesion in DNA using error-prone DNA polymerases to synthesize DNA across the lesion.

Translocation The movement of the ribosome one codon along mRNA after the addition of each amino acid to the polypeptide chain.

Translocon (also called **Protein conducting channel**) Membrane proteins that provide a channel for displacement of polypeptide segments across a lipid bilayer membrane.

Transposition A process in which a discrete DNA entity can move between DNA sites that lack homology in a reaction catalyzed by a transposase.

Transversion mutation A mutation in which a pyrimidine on one strand is replaced with a purine and a purine on the other strand is replaced with a pyrimidine.

TRF1 (also called **TTAGGG repeat binding factor 1**) A protein that binds to the double-stranded region of the telomere as a preformed homodimer and appears to be a negative regulator of telomere length.

TRF2 (also called **TTAGGG repeat binding factor 2**) A protein that binds to the double-stranded region of the telomere as a preformed homodimer and that appears to participate in t-loop formation and be required to cap and protect the chromosome ends.

Trigger factor A protein that binds to hydrophobic patches as they emerge from the ribosome and sometimes remain associated with the segment even after polypeptide chain completion.

tRNA (see **Transfer RNA**)

tRNA nucleotidyl transferase The enzyme that adds a CCA to the terminus in those precursors of tRNA molecules that lack a terminal CCA.

Trombone model of replication Model of replication that explains how coordination of replication of leading and lagging strands takes place.

trp* leader** (also called ***TrpL) The 141 nucleotide sequence before the first codon in *trpE* in the *trp* operon that plays an essential role in the fine control mechanism.

Tryptophan An amino acid with an aromatic side chain, represented by Trp or W.

TTAGGG repeat binding factor (see **TRF**)

TTF-1 (see **Transcription termination factor 1**)

Tumor suppressor genes Genes that code for products that cause cell division and replication to slow down.

Turnover number The number of substrate molecules converted to product molecules by an enzyme molecule in a specified time.

Tus (see **Terminus utilization substance**)

Twisting number (Tw) The number of times that two DNA strands twist about each other to form the helix.

Tyrosine An aromatic amino acid with a hydroxyl (–OH) group in its aromatic side chain, represented by Tyr or Y.

U2AF (also called U2 snRNP auxiliary factor) A splicing factor that contains a large subunit that binds to the polypyrimidine tract and a small subunit that binds to the AG dinucleotide at the 3′-splice site.

UAS (see **Upstream activating sequence**)

UBF (see **Upstream binding factor**)

Uncharged tRNA tRNA molecules without bound amino acids.

Unidirectional replication Replication in which a single growing point moves around the circular DNA until replication is complete.

3′-Untranslated region (3′-UTR) The section of untranslated RNA after the coding region.

5′-Untranslated region (5′-UTR) The section of untranslated RNA before the coding region.

Up element A sequence located just upstream of the –35 box in *E. coli* ribosomal promoters that increases the promoter's strength, accounting for the high rate of bacterial ribosomal RNA synthesis.

UPE (see **upstream promoter element**)

Upstream Sequences that come before the transcription start site or more generally 5′ to the point of reference on the sense strand.

Upstream activating sequence (also called **UAS**) A regulatory region for protein-coding genes in yeast, located within a few hundred base pairs upstream of the transcription initiation site.

Upstream binding factor (also called **UBF**) A transcription factor that binds to the upstream promoter element in rRNA genes and promotes RNA polymerase I binding and transcription, in synergy with SL1/TIF-1B.

Upstream promoter element (also called **UPE**) Element located upstream from the core promoter in rRNA genes that is not essential for transcription.

Uracil A pyrimidine base derivative that is one of the four bases present in RNA molecules.

U-rich element An element involved in cleavage/polyadenylation site selection in mammals on the 3′-side of the site.

UvrABC endonuclease (also called **Excinuclease**) Bacterial enzymes that participate in nucleotide excision repair.

UvrD A helicase required to excise damaged oligonucleotides.

Valine An amino acid with an isopropyl side chain, represented by Val or V.

van der Waals interactions Weak electrostatic interactions between two polar groups, a polar group and a nonpolar group, or two nonpolar groups.

van der Waals radius Half the distance between two nonbonded atoms, when the repulsive and attractive forces between the atoms are equal.

Variable arm An arm of tRNA that lies between the anticodon and TψC arms.

Vector The DNA of a virus or plasmid used to transmit genetic material to a cell or organism.

Virion A virus particle.

Virulence plasmids Plasmids that convert certain bacteria into pathogens.

Virulent phage A phage that is capable only of lytic growth and therefore always kills its host.

Western blot A technique in which proteins are separated by gel electrophoresis, transferred to a membrane, and detected with labeled or tagged antibody probes.

Whole genome shotgun sequencing A variant of shotgun sequencing in which a cell's whole genome is broken randomly into fragments that are cloned and sequenced.

Wild type The allele form isolated from nature.

Wobble hypothesis The hypothesis that accounts for the ability of a tRNA to recognize more than one codon by unusual pairing with the third base of a codon.

Writhe Each instance in which the helix axis of a supercoiled DNA crosses itself.

Writhing number (Wr) The number of times a duplex axis crosses over itself.

Xeroderma pigmentosum (XP) An autosomal recessive disease in which one of the enzymes in the nucleotide excision repair pathway is defective.

X-ray crystallography A technique that allows us to determine the three-dimensional arrangement of atoms in a molecule by exploiting the fact that electrons surrounding atoms in a crystalline protein scatter or diffract x-rays.

Yeast artificial chromosome A linear DNA duplex that functions as a cloning vector, which is constructed by recombinant DNA techniques to contain a yeast origin of replication, a yeast centromere, and a yeast telomere at each end.

Yeast spore A haploid yeast cell resulting from the meiosis of a diploid yeast cell.

Z-DNA A left-handed double helical conformation of DNA in which the double helix winds to the left in a zig-zag pattern.

Zinc finger A DNA-binding motif that characterizes a class of transcription factor.

Zipping and assembly model A protein folding model that proposes polypeptide folding begins when small segments along the polypeptide chain fold into secondary structures and some of these folded segments survive long enough to grow (zip) into more stable structures or coalesce (assemble) with other structures.

Zn_2Cys_6 binuclear cluster An arrangement of zinc ions and cysteine residues used in some fungi to stabilize the DNA-binding domain.

Zygotene The second stage of meiosis division I, in which the homologous chromosomes pair.

Index

A

Abasic (AP) sites, 322, 323f
Abortive initiation, 425
Acceptor stem, of tRNA, 636, 638, 639f
Ac element, 393f, 393b–394b
N-Acetyl-Phe-tRNA, 684, 685
ACF (ATP-utilizing chromatin assembly and remodeling factor) chromatin remodeler, 309–310
Actinomycin D, RNA polymerase sensitivity to, 470–471, 471f
Activated complexes, 67
Activation domains, in eukaryotic transcription activator proteins, 492f, 492–493, 511–517
　association with DNA-binding domains, 514f, 514–517, 516f
　of Gal4, 511–512, 512f
　GAL gene regulation and, 512f, 512–513, 513f
　intrinsically disordered nature of, 511
Activators, 436
Active site, of an enzyme, 69
Adaptor hypothesis, 634
Adaptor RNS, 631
Adducts, 325
Adenine, 8t
　in DNA, 6, 6f
Adenoviruses, 253f, 253–254, 254f
A-DNA, 15, 98–100, 99f, 100t
Affinity chromatography, 35, 35f
　transcription activator protein purification by, 494f, 494–495
Aflatoxins, DNA damage due to, 326, 327f
Agar, 192
Aging, telomeres and, 304b
Ago 2 (Augonaute 2), 609
Agrobacterium tumefaciens
　chromatin structure in, 152
　Ti plasmid of, 93b
Ahringer, Julie, 613
Alanine, 27, 28f
　abbreviations for, 29t
Alanyl-tRNA synthetase, editing defect in, neurodegenerative disease caused by, 643b
Alberts, Bruce, 90, 287

Alkali, DNA denaturation by, without breaking phosphodiester bonds, 88
Alkaptonuria, 4
Alkylating agents
　cross-link formation and, 327–328, 328b–331b, 328f–330f
　DNA damage caused by, by monoadduct formation, 325–327
　alkyl group transfer and, 325, 325f
　modification of environmental agents to become alkylating agents and, 325–326, 326f, 327f
Alkyl groups, transfer to centers of negative charge, DNA damage caused by, 325, 325f
O^6-Alkylguanine, repair by direct alkyl group removal by suicide enzymes, 335–337, 336f
Alleles, dominant and recessive, 3
Allfrey, Vincent, 525
Allis, C. David, 525
Allolactose, as lactose operon inducer, 442f, 442–443
Allosteric changes, transcription termination of RNA polymerase II, 589–591, 590f
Allosteric effectors, 74
Allosteric model, 589, 590
α-helix, 46–49, 47f
α subunit, of core polymerase, 283–284
Alternative splicing, 566–567, 567f
Altman, Sidney, 104
Altmann, Richard, 3
Alu element, 401
Amanita phalloides, 470, 470f
α-Amanitin, RNA polymerase sensitivity to, 470, 470f
Ambros, Victor, 614
Ames, Bruce, 331, 332
Ames test, 331–333
Amino acid(s), 25
　α-, 25–30
　　peptide bonds linking, 29f, 29–30, 30f
　　structure of, 25–28, 26f–28f
　　abbreviations representing, 28–29, 29t
　　attachment to tRNA, protein synthesis and, 628–629, 630f, 631, 631f, 633f, 633–634, 634f

　basic, 26, 26f
　coding region interruption by noncoding regions and, splicing and, 559–560, 561f, 562, 563f
　side chains of
　　hydrophilic, 48–49
　　hydrophobic, 44–45, 48–49
Amino acid residues, 30
Aminoacyl-tRNA
　binding of, to multiple codons, 659–660, 660f
　coding specificity of, 658–659, 659f
　formation of, 633f, 633–634, 634f
　protein synthesis and, coding specificity of, 658–659, 659f
　trinucleotide promotion of molecules to ribosomes and, 654, 655f
　tRNA molecule movement through ribosome and, 685
Aminoacyl-tRNA ligases, 629, 630f
Aminoacyl-tRNA synthetases, 629, 630f
　cognate distinction by, 641, 643–645, 644f–646f
　proofreading functions of, 639–640, 640f
　selenocysteine and pyrrolysine as building blocks for polypeptides and, 645–646, 646f
　valyl-tRNA and valyl-AMP hydrolysis by, 640–641, 641f
Amp^R gene, 387, 388
Anaphase, of mitosis, 157
Anaphase I, meiotic, 159
Anaphase II, meiotic, 162
Anemia, sickle cell, 58b–60b, 58f, 59f
Aneuploidy, 163, 164b
Anfinsen, Christian, 54
Anion-exchange chromatography, 33, 33f
Annotation, of human genome, 146
Antennapedia complex genes, 497, 497f, 498f
Antibiotic resistance, 93b
　transposons bearing, 389–390, 390f, 391b
Antibiotics, determining target of, 198–199
Antibodies, eukaryotic, formation of, 392, 394–396, 395f–397f
Anticodon arm, of tRNA, 638
Antiparallel pleated sheets, 49
Anti-Shine-Dalgarno sequence, 668–671, 670, 670f, 671f, 672b, 672f
Anti-silencing factor (Asf1) histone chaperone, 308
Antiterminator, 458
Antiterminator model, 589, 590

Page numbers followed by *b*, *f*, or *t* indicate material in boxes, figures, or tables, respectively.

AP-1 family of transcription activator proteins, 507b, 507f, 508f
AP (abasic) sites, 322, 323f
Aptamer, 640, 641
Arabidopsis thaliana, TBP crystal structure in, 480–481, 481f
Archaea
  chromatin structure in, 152
  ribosomes of, structure of, 663b
  RNA polymerases of, 407, 471, 475f, 475b–476b
  translation initiation pathway of, 676
Arc105, 520
Arginine, 26, 26f
  abbreviations for, 29t
Argonaute protein family, 605, 617
Argonaute protein RDE-1, 611
aroH operon, 458
ARS (autonomously replicating sequence) elements, 297, 297f
Ascus(i), 213–214
  gene conversion and, 372–373, 373f
Asf1 (anti-silencing factor) histone chaperone, 308
Asparagine, 27, 27f
  abbreviations for, 29t
Aspartic acid (aspartate), 27, 27f
  abbreviations for, 29t
Aspergillus flavus, aflatoxins produced by, 326, 327f
Aspergillus oryzae, S1 endonuclease of, 123
Aspergillus parasiticus, aflatoxins produced by, 326, 327f
ATP-utilizing chromatin assembly and remodeling factor (ACF) chromatin remodeler, 309–310
A-tracts, 84
Attenuation model, 456–457, 457f
Attenuator, 455f, 455–456
Augonaute 2 (Ago 2), 609
Autonomously replicating sequence (ARS) elements, 297, 297f
Autonomous transposons, 389
Autoradiography, 92
Autosomes, 163
Auxiliary upstream elements, 585
Auxotrophs, 4, 192
  detection of, by plating, 194
Avery, Oswald, 12–13

B
Bacillus steareothermophilus, 50S subunits of, 661
Bacillus subtilis
  DNA replication in, 291
  σ subunits of, 414b
  tRNA genes in, 634
Backtracking, in transcription, 432, 487
Bacteria. *See also specific bacteria*
  chromatin of, 153f, 153–155, 154f
  complementation and, 200
  culture of, 191–193, 192f, 193f
  DNA of
    coding (sense) strand of, 410, 410f
    isolation of, 113–114
    replication of. *See* Bacterial DNA replication
  DNA polymerases of
    DNA polymerase III, in bacterial DNA replication, 280–285
    in replicative transposition, 388
  F plasmids of, 204, 204b, 205f, 206, 209f, 209–210
  gram-negative and gram-positive, 189–191, 190f, 191f
  mutation of, 195–199, 197f
  notations, conventions, and terminology used with, 194–195
  plasmid replication in, 210
  plating to detect, 194
  release factors of, 685, 687
  replication fork collapse in, homologous recombination replication restart after, 360f, 360–364, 361f
  ribosomal complex dissembly in, 687–688
  ribosomes of, structure of, 625, 625f
  RNA polymerase holoenzyme of, structure of, 410–411, 411t, 412–413, 413f
  RNA polymerases of, 407
    crystal structures of, 420–424, 421f, 423f, 424f
    detection and removal of incorrectly incorporated nucleotides and, 432, 432f
    reaction catalyzed by, 408–412
    scrunching of, 425–427, 426f
    size of, 411
    structure of, 410–411, 411t, 412f
  S and R cells of, 11–12, 12f, 13f
  selenocysteyl-tRNA formation in, 647b, 647f
  transcription in. *See* Bacterial transcription
  transposable elements of. *See* Transposable elements (transposons), bacterial
Bacterial DNA replication, 273–293
  bidirectionality of, 267, 268f, 269f
  DNA polymerase III in, 280–285
    subassemblies of, 281–283, 282f
    subunits of, 283–285, 284f
  elongation stage of, 274–275
  enzymes acting at replication fork in, 280
  enzymes involved in, 275
  initiation stage of, 274
    E. coli replication origin and, 276–277, 276f–278f
    regulation of, 279b
  in vitro studies of, 273–275
  mutant studies of, 275
  primer coordination in, 288–289, 289f
  recombinase and, 291, 292f, 293
  replicon model of, 275
  replisome and, 287, 288f, 289–291, 290f
  sliding clamp and, 285f, 285–287
    placement around DNA, 286f, 286–287
  termination stage of, 275
    replication forks and, 291, 291f
    terminus utilization substance and, 291, 291f
    topoisomerase IV and, 291, 292f, 293
  trombone model of, 287, 288f
Bacterial transcription, 406–458
  catabolite repression in, 448–453
    cAMP•CRP complex binding to activator sites and, 450f, 450–451, 451f
    cAMP•CRP operon activation and, 453
    catabolite regulator protein and, 449, 449f
    E. coli use of glucose in preference to lactose and, 448f, 448–449
    glucose caused catabolite repression and, 451f, 451–453, 452f
  elongation complex and, 427–432
    overall rate of, 432
    RNA polymerase detection and removal of incorrectly incorporated nucleotides and, 432, 432f
    structure of, 427–430, 428f, 429f
  eukaryotic transcription compared with, 467–468
  initiation stage of, 412–427
    promoters and, 413, 416f–418f, 417–418
    RNA polymerase crystal structures and, 420–424, 421f, 423f, 424f
    RNA polymerase holoenzyme binding to promoter and, 418–420, 419f, 420f
    RNA polymerase holoenzyme structure and, 412–413, 413f
    RNA polymerase scrunching during, 425–427, 426f
    σ^{70} family conserved domains and, 424–425, 425f
    σ^{70} polymerase holoenzyme escape and, 425
  lactose operon and, 437–448
    allolactose as inducer of, 442f, 442–443
    lac operators of, 444, 444f
    Lac repressor and, 443f, 443t, 443–444, 444f–448f, 445–446
    lacZ, *lacY*, and *lacA* genes and, 437f, 437–438, 438f
    operon model and, 440–442, 441f
    regulation of *lac* mRNA and, 439t, 439–440
    regulation of *lac* structural genes and, 438f, 438–439, 439f
  messenger RNA and, 434–436
    lifetime of, 435
    monocistronic and polycistronic, 434–435
    rate of synthesis of, 436
    regulation of, 436, 436f, 436t
  RNA polymerase catalyzed reaction and, 408–412
    requirements for, 408, 409f, 410, 410f
    RNA polymerase structure and, 410–411, 411t, 412f
  termination stage of
    intrinsic, 433
    Rho-dependent, 433–434, 434f
    RNA strand release at intrinsic and Rho-dependent terminators and, 432–434, 433f, 434f
  tryptophan operon and, 453–458, 454f–457f
Bacteriophage(s), 225–249. *See also specific bacteriophages*
  chronic, 228, 248–249
  plaques produced by, 229
  filamentous, 227, 227f
  life cycles of, 228
  as model systems, 226
  naming of, 226
  plaques formed by, 228–229, 229f
  sizes and shapes of, 227, 227f
  T4, infection process of, 231–232, 232f
  tailed, icosahedral, 227, 227f
  tailless, icosahedral, 227, 227f
  temperate, 228, 241–248
    λ, 241–245, 241f–244f
    P1, 245f, 245–246, 246b–247b, 248
    plaques produced by, 229
  virulent, 228
    infection process of, 230f–232f, 230–234, 233t
    ΦX174, 238f, 238–239, 239f
    plaque assay for counting particles of, 229, 229f
    with single-stranded RNA, 239f, 239–240, 240f
    T7, 234, 234f, 237f, 237t, 237–238
    T-even, 226, 230f–232f, 230–234, 233t, 234b, 235f–236f
    T-odd, 226
Bacteriophage f2, 239, 239f
Bacteriophage λ
  DNA replication in, 241f–242f, 241–245
  DNA size in, 82t
Bacteriophage M13, 238f, 248
  as cloning vector, 249b
Bacteriophage MS2, 239, 239f

Bacteriophage P1, life cycle of, 245f, 245–246, 246b–247b, 248
Bacteriophage ΦX174, DNA molecule of, 238f, 238–239, 239f
Bacteriophage Qβ, 239, 239f
Bacteriophage R17, 239, 239f
Bacteriophage T4, *rIIB* gene of, 648, 648f
Bacteriophage T7, RNA polymerase of, 411, 412f
Baker, Tania, 280
Bakers' yeast. See *Saccharomyces cerevisiae*
Baldwin, Ann Norris, 639
Baltimore, David, 257
Basal synthesis, 442
Basal transcription, 480
Basal transcription factors, in eukaryotic transcription, 478–486
   preinitiation complex formation and, 482f, 482–483, 483f, 486, 486f
   RNA polymerase II requirement for, 478–480, 479t
   TFIID or TBP binding to TATA core promoter and, 480–482, 480f–482f
Base excision repair, 337–340, 338f, 339f
Base pairs, RNA secondary structure and, 104f, 104–105, 105f
Base(s), of DNA, 6f, 6–7, 7f, 14–15
Basic amino acids, 26, 26f
Basic Local Alignment Search Tool (BLAST) program, 41
Basic region leucine zipper (bzip) proteins, of eukaryotic transcription activator proteins, 504–506, 504f–506f
Basic replicon, 210
Bateson, William, 186
B-DNA, 15, 16, 81–100, 99f, 100f
   DNA bending and, 83–84
   DNA denaturation and, 84–86, 85f
   DNA renaturation and, 88–89, 89f
   DNA size and fragility and, 81–82
   helicases and, 89f, 89–90, 90f
   recognition patterns in major and minor grooves and, 82–83, 82f–84f
   single-stranded DNA binding proteins and, 90, 91f
   topoisomerases and, 97–98
   topoisomers and, 95–96
Beadle, George, 4
Becquerel (Bq), 67b
Bell, Stephen P., 280, 298
Bennett, J. Claude, 394
Benzer, Seymour, 656
Benzo[a]pyrene, metabolic activation of, 326, 326f
Berg, Paul, 633, 634, 639
β-conformation, of amino acids, 48f, 49
β-Galactosidase, activity of, monitoring of, 495
Bioinformatics, 564
Birnstiel, Max L., 127
Bishop, J. Michael, 257
2,3-Bisphosphoglycerate, oxygen binding by hemoglobin and, 61b
Bisulfite, cytosine deamination and, 323
bithorax complex genes, 497, 497f
Bivalents, 159
Blackburn, Elizabeth H., 176, 300–303, 303
Blasco, Maria, 304b
BLAST (Basic Local Alignment Search Tool) program, 41
Blobel, Günter, 689
Blunt ends, 124, 124f
Böck, August, 647b
Bodnar, Andrea G., 304b
Bohr effect, 60f, 60–61, 61f

Bombyx mori, R2Bm element of, 398, 399f
Bonds
   noncovalent, weak, in proteins, 43–46
     hydrogen, 43–44, 44f, 49
     hydrophobic interaction and, 44–45, 45f
     ionic, 44
     van der Waals interaction and, 45t, 45–46, 46f
   peptide, 29–31
     amino acid linkage by, 29f, 29–30, 30f
     peptide chain directionality and, 30, 31f
Bootsma, Dirk, 343b–344b
Botstein, David, 217–218
Boyce, Richard, 340
–35 box, 417
Bq (Becquerel), 67b
Branching nucleotides, 570, 570f
Branch migration, RuvAB protein catalysis of, 363–364, 364f
Branchpoint sequence, 570, 570f
BRCA2 gene, 366
BRE (TFIIB recognition elements), 478
BRE[d], 478
Brenner, Sydney, 275, 601, 648, 651, 656
BRE[u], 478
Broth, 192
Buratowski, Stephen, 480
Burgess, Richard, 412
bzip DNA-binding domain, 505, 506f
bzip (basic region leucine zipper) proteins, of eukaryotic transcription activator proteins, 504–506, 504f–506f

C

Caenorhabditis elegans
   RNA interference discovery in, 602–605, 603f
   RNAi pathway of, 606, 606f
CAF-1 (chromatin assembly factor) histone chaperone, 308
Cairns, John, 91–92, 267, 280
Calcitonin transcription unit, 587–588, 588f
cAMP
   catabolic repression through modulation of, 451f, 451–453, 452f
   catabolite regulator protein combination with, 449, 449f
cAMP•CRP complex, 449, 450f, 450–451, 451f
   bacterial
     activation of, 453
     binding to activator sites, 450f, 450–451, 451f
   operon activation by, 453
Cancer, telomeres and, 304b
Capecchi, Mario, 368, 369
Cap formation, RNA polymerase II and, 553–558
   cap attachment to nascent pre-mRNA chains and, 556
   CTD phosphorylation and, 558
   at 5'-ends, 553–556, 555f, 556f
   eukaryotic pathway for, 556–557, 557f
Capillary gel electrophoresis, 120, 122b
Capsid, 224
Capsid protein, in retroviral RNA, 255
Carbon, van der Waals radius of, 45t
Carboxyl terminal domains (CTDs), 486, 558, 583–584
   phosphorylation of, cap formation and, 558
   of Rpb1, 587
Carcinogens, 318
Carrier, William, 340
Cassette exons, 566–567, 567f
Catabolic site, of spliceosomes, 577, 578f, 579
Catabolite regulator protein, bacterial, 449, 449f
Catabolite repression, bacterial, 448–453
   cAMP•CRP complex binding to activator sites and, 450f, 450–451, 451f

cAMP•CRP operon activation and, 453
   catabolite regulator protein and, 449, 449f
   E. coli use of glucose in preference to lactose and, 448f, 448–449
   glucose caused catabolite repression and, 451f, 451–453, 452f
Catabolite-sensitive operons, 453
Catalytic subunits, 74
Cate, Jamie H. D., 661
Catenanes, 96
CCAAT box, 488, 489–490
CCA[OH], at ends of tRNA, 631, 632f, 633
cDNA (complementary DNA)
   LTR retrotransposons and, cut-and-paste mechanism and, 397–398, 398f, 399f
   synthesis of, 259, 259f
Cech, Thomas R., 104, 580f, 580b–581b, 581f
Cell density, 192
CENP-A (centromere protein A), 174
Central dogma, 19–20, 624
Centromere(s), 157
   as microtubule attachment site, 173–175, 174f–176f
Centromere protein A (CENP-A), 174
c-fos gene, 507b, 507f, 508f
CGA anticodon, 687
Chain termination method, 139f, 139–140, 140f, 142
Chambon, Pierre M., 468, 560, 562
Chan, Michael K., 646
Chaperones, 688–689
Chapeville, François, 658–659
Chargaff, Erwin, 14–15
Chargaff's rules, 15
Chemical cross-linking agents, DNA damage by, 327–331
Chiasmata, 159
Chimeric mice, 371
ChIP (chromatin immunoprecipitation), 522, 522f
Choo, K. H., 175
CHRAC (chromatin assembly complex) chromatin remodeler, 309–310
Chromatids, sister, 157
Chromatin, 152, 167f
   bacterial, 153f, 153–155, 154f
   biochemical activities of, 523, 524f
   condensation of, 172, 172f
   epigenetic modifications of, eukaryotic transcription and, 523–525, 524f
   eukaryotic, 155
     histone classes in, 166t, 166–167
   nucleosomes in. See Nucleosome(s)
   remodeling of, 526–527
   scaffold model for structure of, 172–173, 173f
   silencing versus active expression of, DNA methylation and, 527
   synthesis of, coupled with DNA replication, 307–310, 309f
Chromatin assembly complex (CHRAC) chromatin remodeler, 309–310
Chromatin assembly factor (CAF-1) histone chaperone
Chromatin immunoprecipitation (ChIP), 522, 522f
Chromatin modifiers, 308
Chromatin remodelers, 309–310
Chromatin remodeling complexes, 523–525, 524f
Chromatography
   affinity, 35, 35f
     transcription activator protein purification by, 494f, 494–495
   column, 32, 32f
   ion-exchange, 33, 33f
   protein fractionation by, 31–32, 32f, 33–35
Chromomeres, 159

Chromosomal anomalies, congenital disorder identification by, 164b–165b, 165f
Chromosomal rearrangements, 164b
Chromosomal translocation, 164b–165b, 165f
Chromosome(s), 151–179
  centromeres as microtubule attachment site and, 173–175, 174f–176f
  chromatin and. *See* Chromatin
  karyotype and, 162–166
    chromosome type nomenclature and, 162f, 162–163, 163f
    diagnosis and, 164b–165b, 165f
    information provided by, 163, 165f, 166, 166f
    metaphase chromosomes shown by, 163
  in meiosis, 159–162, 160f–161f
    first division of, 159, 162
    second division of, 162
  Mendel's laws of inheritance and, 3
  in mitosis, 155–158, 156f, 157f, 159f
  nucleosomes and. *See* Nucleosome(s)
  nucleotide repeat addition to ends of, 303, 305f
  sex, 163
  telomeres of, 175–179, 177f, 178f
Chromosome painting, 166
Chronic infection life cycle, of bacteriophages, 228
Chronic phages, 228, 248–249
  plaques produced by, 229
cis-acting replicator, 275
Cisplatin, interstrand cross-linking and, 328b–330b, 329f, 330f
Cistrons, 435
c-jun gene, 507b, 507f, 508f
Clamp loader, 283, 286f, 286–287
Cleavage factors I and II, 585
Cleavage/polyadenylation sites. *See* Poly(A) sites
Cleavage/polyadenylation specificity factor (CPSF), 585
Cleavage stimulation factor (CstF), 585
Cleaver, James E., 343b
Clones, 193
Cloning, 129
Cloning vectors, bacteriophage M13 as, 249b
Closed promoter complex, 420
Coaxial stacking, 105
Coding strand, of bacterial DNA, 410, 410f
Codon(s), 20, 434–435
  in mRNA, protein synthesis and, 648f, 648–650
  start, in protein synthesis, 664
Codon-anticodon helix sensing, protein synthesis and, 679, 680f
Coenzymes, 65
Cohesin, 157
Cohesive ends, 124, 124f, 242, 242f
Coiled coils, 504f, 504–505
Cointegration, 388
"Collision release," 290
Colorimetric assays, of enzyme catalyzed reactions, 65, 66f
Column chromatography, 32, 32f
Combinatorial control, 494
Combinatorial diversity, 395–396, 396f
Committed step, 72
Competence, of *Escherichia coli*, to take up DNA, 129
Complementary DNA (cDNA)
  LTR retrotransposons and, cut-and-paste mechanism and, 397–398, 398f, 399f
  synthesis of, 259b, 259f
Complementation, 200
Complementation assay, for T-even phage assembly study, 234b
Composite transposons, 382f, 382–383, 383f
Conaway, Ronald C., 487

Concatemers, 233
Condensins, chromatin condensation and, 172, 172f
Conditional mutants, 196
Congenital disorders, identification by chromosomal anomalies, 164b–165b, 165f
Conjugation, in *E. coli*, 200–204, 201f–203f, 204b
Consensus sequences, 417, 417f
Conservative model of replication, 265
Constant region domains, of immunoglobulin G, 64
Constitutive enzymes, 438
Cooke, Howard, 304b
Cooperative interaction, 60
Coordinate regulation, 436
copia element, 397
Core polymerase, 283
Core promoter
  for protein-coding genes, in eukaryotic transcription, 476, 478, 478f
  in rRNA transcription unit, 531, 531f
  TATA, TFIID or TBP binding to, 480–482, 480f–482f
Corey, Robert R., 29, 47, 49
cos elements, 242, 242f
Counterselective markers, 206
CpG islands, 322, 527
CPSF (cleavage/polyadenylation specificity factor), 585
Cramer, Patrick, 472, 474, 484b, 485b
Crick, Francis H. C., 265, 624, 631, 634, 648, 649, 651, 682
  central dogma and, 19
  DNA structure and, 15–19, 16f–18f
Cross-linking agents, interstrand, 327–328, 328f, 328b–330f, 328b–331b
Crossovers, 367
CRP protein, 449
Cruciform structures, 102, 102f
Cryoelectron microscopy, 116–117, 117f
CstF (cleavage stimulation factor), 585
CTDs. *See* Carboxyl terminal domains (CTDs)
C-terminal region, of TBP, 480
CtIP, 365
CUA anticodon, 687
Culture, of bacteria, 191–193, 192f, 193f
Cut-and-paste transposition mechanisms, 383–387, 384f–388f
  LTR retrotransposons and, 397–398, 398f, 399f
Cyclins, 483
Cyclobutane pyrimidine dimer(s), formation of phytolase reversal of damage caused by, 333–335, 334f, 335f
  ultraviolet light-caused, DNA damage due to, 319, 320f, 321
Cyclobutane pyrimidine dimer phytolase, 334
Cysteine, abbreviations for, 29t
Cys$_4$ zinc finger DNA-binding domains, of eukaryotic transcription activator proteins, 499–502, 500f–503f
Cytogenetics, 162
Cytosine, 8t
  in DNA, 6, 6f
Cytosine deamination, hydrolytic, 322

D

Dalgarno, Lynn, 669
D-arm, of tRNA, 638
Darnell, James, 528, 552
Darst, Seth, 420
Davis, Ronald W., 217–218
DEAE-Sephadex chromatography, eukaryotic RNA polymerase types and, 469, 469f
Deamination, water-mediated, 322, 323f
de Lange, Titia, 177

Delbrück, Max, 226
Deletions, 196, 197f
Delucia, Paula, 280
Denatured state, of proteins, 53–54, 54f
Deoxyhemoglobin, 62
Deoxyribonucleases (DNases), 122
Deoxyribonucleic acid. *See* DNA
Deoxyribonucleosides, 6, 6f, 7
Deoxyribonucleotides, 7
Depurination, of DNA, 88
Derepression, 441
d'Herelle, Felix, 226
Dicer, 606, 607, 607f
  from *Drosophila melanogaster*, 609, 609f
  from *Giardia intestinalis*, 607b–608b, 608f
Dideoxy-terminator cycle sequencing, 139–140
DiGeorge syndrome, 615
Dimerization domain, of transcription activating proteins, 493
Dimethylnitrosamine, 325, 325f
Dimethylsulfate, 325f
Dintzis, Howard, 652–653
Diploid number, 155
Direct repeats, in insertion sequences, 381, 382f
Disarming, 93b
Disassembly model, primase and, 288b, 289f
Discontinuous gene hypothesis, 562
Discontinuous model of replication, 269–270, 270f
Discovery Studio Visualizer 3.5, 52f, 52–53
Dispersive model of replication, 265–266
Displacement loops (D-loops), 177, 177f, 178f
  in homologous recombination, 362–363
Distributive replication, 281
D-loops (displacement loops), 177, 177f, 178f
  in homologous recombination, 362–363
Dmc1, in homologous recombination, 374
DNA
  bacterial
    coding (sense) strand of, 410, 410f
    isolation of, 113–114
    replication of. *See* Bacterial DNA replication
  of bacteriophage ΦX174, 238f, 238–239, 239f
  base composition of, 81–82, 82t
  bases of, 6f, 6–7, 7f, 14–15
    stacking of, 87
  bending of, 83–84
  complementary, synthesis of, 259b, 259f
  damage to. *See* DNA damage; DNA repair
  denaturation of, 84–86, 85f
    by alkali, without breaking phosphodiester bonds, 88
  depurination of, 88
  discovery as hereditary material, 11–15
    chemical experiments supporting, 14–15
    transformation experiments and, 11–14, 12f, 13f
  discovery of, 3
  double-stranded
    annealing of complementary single strands to form, 88–89, 89f
    in vitro study of, 92, 93b
    linking number and, 95b
    stabilization of, 86f, 86–87
    superhelical, 92, 94f, 94–96, 95b, 96f
    unwinding of, 89f, 89–90, 90f
  enzymatic techniques for investigating, 122–146
    chain termination method, 139f, 139–140, 140f, 142
    DNA polymerase I and, 132–137, 133f–136f
    human genome sequence and, 142, 146
    new generation DNA sequencers and, 143b–145b, 143f–145f
    Northern blotting, 132
    nucleases and, 122–123, 123t
    plasmid DNA vectors and, 128–129, 129f

polymerase chain reaction, 137, 138f
restriction endonucleases and, 123–128, 124f, 125t, 126f, 127f
restriction maps and, 141
shotgun sequencing, 141f, 141–142
site-directed mutagenesis and, 138f, 138–139
Southern blotting, 130–132, 131f
Western blotting, 132
eukaryotic, 81
fragility of, 82
fungal, isolation of, 114
histones of, 81
initiator, 294
isolation of, 113–115
looping of, 493, 493f
major groove in, 17
recognition patterns in, 82–83, 82f–84f
methylation of, eukaryotic transcription and, 527
minor groove in, 17
recognition patterns in, 82–83, 82f–84f
plasmid, isolation of, 114
prokaryotic, 81
renaturation (reannealing) of, 88–89, 89f
repair of. See DNA repair
replicating forms I and II of, 91
replication of. See Bacterial DNA replication; DNA replication; Eukaryotic DNA replication
restriction mapping of, 125–128, 126f, 127f
RNA differentiated from, 5, 5f
single-stranded, stabilization by binding proteins, 90, 91f
size of, 81–82, 82t
structure of. See DNA structure
of SV40 virus, 250
synthesis of, translesion, SOS response and, 347–348
T DNA, 93b
viral, isolation of, 113
yeast, isolation of, 114
DnaA
DNA replication initiation and, 277, 278f
regulatory inactivation of, 279b
sequestration of, 279b
titration of, 279b
DnaA boxes, 276–277
DNA affinity chromatography, 35, 35f
transcription activator protein purification by, 494f, 494–495
DnaB
DNA replication initiation and, 277, 278f
reloading on DNA forks, 364
DnaB helicase, 90
primer coordination and, 288b–289b, 289f
DNA-binding domains, 492
bzip, 505, 506f
of eukaryotic transcription activator proteins, 496–511
activation domain association with, 514f, 514–517, 516f
basic region leucine zipper proteins, 504–506, 504f–506f
Cys_4 zinc finger, 499–502, 500f–503f
helix-turn-helix, 497–499, 497f–499f
structures of, 496–511
Zn_2Cys_6 binuclear cluster, 506, 509–511, 509f–511f
of glucocorticoid receptor, 501f, 501–502, 502f
DnaC, DNA replication initiation and, 277, 278f
DNA damage, 318–333
alkylation, by monoadduct formation, 325–327
alkyl group transfer and, 325, 325f
modification of environmental agents to become alkylating agents and, 325–326, 326f, 327f

by chemical cross-linking agents, 327–331
DNA strand separation blockage and, 327–328, 328f
by hydrolytic cleavage reactions, 321–323, 322f, 323f
mutagen detection and, 331–333
oxidative, 324, 324f
radiation-induced, 319–321
ionizing radiation and, 321
ultraviolet light and, 319, 320f, 321
repair of. See DNA repair
DNA fingerprinting, 218b
DNA footprinting, 418–419, 419f
σ^{70} RNA polymerase holoenzyme during transcription initiation and, 419–420, 420f
DNA forks
Holliday junction conversion into, 364
reloading of DnaB on, 364
DnaG, DNA replication initiation and, 277, 278f
DNA gyrase, 96
DNA ligase
in eukaryotic DNA replication, 295
joining of Okazaki fragments by, 271–272, 272f
DNA ligase IV, in nonhomologous end-joining, 369, 369f, 371
DNA packaging, in bacteriophages, 233–234
DNA polymerase(s)
bacterial, in replicative transposition, 388
catalysis of translesion DNA synthesis by, 347–348
eukaryotic, nomenclature for, 352t, 352–353
SOS response and, 352t, 352–353
DNA polymerase I
catalysis of nick translation by, 136f, 136–137
of E. coli, 198
exonuclease activities of, 133–136, 134f, 135f
template-primer required by, 132–133, 133f
DNA polymerase II, synthesis of, SOS signal induction of, 351
DNA polymerase III, in bacterial DNA replication, 280–285
subassemblies of, 281–283, 282f
subunits of, 283–285, 284f
DNA polymerase III holoenzyme
subassemblies of, 281–283, 282f
subunits of, 283–285, 284f
tethering to DNA, 285, 285f
DNA polymerase IV, synthesis of, SOS signal induction of, 351
DNA polymerase V, synthesis of, SOS signal induction of, 351–352
DNA repair, 318, 319, 333–353
base excision repair and, 337–340, 338f, 339f
direct reversal of damage and, 333–337
alkyl group removal by suicide enzymes and, 335–337, 336f
phytolase and, 333–335, 334f, 335f
of double-strand breaks
by homologous recombination. See Homologous recombination
by nonhomologous end-joining, 360, 371, 372f
long patch repair and, 340
mismatch repair and, 345–347, 346f
nucleotide excision repair and, 340–341, 342f
SOS response to damage and, 347–353
DNA polymerase synthesis induced by, 351–352
human DNA polymerases and, 352t, 352–353
RecA and LexA regulation of, 348–350, 349f, 350f
translesion DNA synthesis and, 347–348

DNA replication, 264–310
bacterial. See Bacterial DNA replication
in bacteriophages, 233
phage lambda, 243, 244f, 245
bidirectionality of, 267, 268f, 269f
chromatin synthesis coupled with, 307–310, 309f
conservative model of, 265
dispersive model of, 265–266
eukaryotic. See Eukaryotic DNA replication
growth of DNA strands and, 267, 269–271, 270f–272f
joining of Okazaki fragments in, 271–272, 272f
RNA as primer for Okazaki fragment synthesis and, 273, 273f, 274f
semiconservative model of, 265–267, 265f–267f
semidiscontinuous model of, 270, 270f
template for, 18, 18f
unidirectional, 267, 268f
DNA restriction maps, stitching together of DNA sequences by, 141
DNase protection method, 416f, 416–417
DNA sequencing
of extinct species, 145b
information provided by, 142, 146
new generation DNA sequencers and, 143b–145b, 143f–145f
DNases (deoxyribonucleases), 122
DNA structure, 8, 9f, 10f, 11
A-form, 15, 98–100, 99f, 100t
B-form, 15, 16, 81–100, 99f, 100f
DNA bending and, 83–84
DNA denaturation and, 84–86, 85f
DNA renaturation and, 88–89, 89f
DNA size and fragility and, 81–82
helicases and, 89f, 89–90, 90f
recognition patterns in major and minor grooves and, 82–83, 82f–84f
single-stranded DNA binding proteins and, 90, 91f
topoisomerases and, 97–98
topoisomers and, 95–96
cruciform, 102, 102f
double helix, 15–19, 16f–18f
double-stranded, circular, 90–91, 91f
dynamic nature of, 87–88
ionic strength and, 87
quadruplex, 103, 103f
stick figure representations of, 9f, 11
tetranucleotide, 14–15
triplex, 103, 103f
Z-form, 100t, 100–102, 101f
DNA unwinding element (DUE), 276, 277, 278f
Domains, of immunoglobulin G, 63–64, 64f
Domain swaps, 515
Dominance, law of, 3
Dominant alleles, 3
Donohue Jerry, 16
Doty, Paul, 84, 86
The Double Helix (Watson), 16
Double helix of DNA, 15–19, 16f–18f
Double mutations, 197f
"Double-sieve" model, 640, 640f
Double-strand breaks
repair by homologous recombination. See Homologous recombination
repair by nonhomologous end-joining, 360, 371, 372f
repair of, 365–367, 366f, 367f
Doubling time, 192
Doudna, Jennifer, 607
Downstream nucleotide sequences, 410
Downstream promoter element (DPE), 478
Down syndrome, 164b

DPE (downstream promoter element), 478
Dreyer, William J., 394
DRIP Mediator, 520
Drosophila melanogaster, 186–189
   Dicer of, 609, 609f
   Dscam gene of, 568f, 568–569, 569f
   genome sequencing of, 142
   homeotic genes of, 497–499, 497f–499f
   LTR retrotransposons of, 397
   male and female, 187, 187f
   Morgan's work on, 186–187
   phytolase of, 335, 335f
   SID-1 expression in, 612
   Sturtevant's work on, 187–188
   Z-DNA binding to salivary chromosomes of, 101–102
Drug resistance plasmids, 93b
Dscam gene, 568f, 568–569, 569f
Dubochet, James, 172
DUE (DNA unwinding element), 276, 277, 278f
Dulbecco, Renato, 333
Duplex-binding site, of transcription elongation complex, 428–429

E

E_a (energy of activation), 67
   lowering of, by enzymes, 66–68, 68f
Ebright, Richard H., 426–427
Editing, RNA, 591f, 591–592
Edman, Pehr, 36
Edman degradation, 36–39, 37f, 38f, 40f
eEF1A, 691
eEF2, 691
EF(s). *See* Elongation factors (EFs)
EF-G, 677
EF-G•GTP, 687
EF-Ts, 677
   protein synthesis and, elongation stage of, 679–680, 681f
EF-Tu, 677
EF-Tu•GDP, 679–680, 681f
EF-Tu•GTP aminoacyl-tRNA ternary complex, 677, 679
EG-G•GTP, 686
EGO-1, 611
eIF (eukaryotic translation initiation factor), 674
eIF2•GDP, 676
80S subunit, in protein synthesis, 626, 626f
EIIA, 453
Eldred, Ennet, 640, 641
Electroblotting, 130
Electron microscopy, 115–117
   cryoelectron, 116–117, 117f
   metal shadowing and, 115f, 115–116
   negative staining and, 116, 116f
Electrophiles, 325
Electrophoresis, for protein purification, 35–36, 36f
Elongation complex, bacterial, 427–432
   overall rate of, 432
   RNA polymerase detection and removal of incorrectly incorporated nucleotides and, 432, 432f
   structure of, 427–430, 428f, 429f
Elongation factors (EFs), 664
   in protein synthesis, 677, 678f
Elongation stage
   in eukaryotic transcription, 486–487
      phosphorylation of C-terminal domain of largest RNA polymerase subunit and, 486
      reactivation of arrested RNA polymerase II and, 487
      transcription elongation factors and, 487
   of protein synthesis, 664, 677–685
      codon-anticodon helix sensing and, 679, 680f

EF-Ts and, 679–680, 681f
EF-Tu•GTP•aminoacyl-tRNA ternary complex and, 677, 679
elongation factors and, 677, 678f
hybrid-states translocation model and, 684f, 684–685
ribosomes and, 680–683, 682f, 683f
Embryonic stem (ES) cells, homologous recombination and, 368–369, 368f–370f, 371
Encyclopedia of DNA Elements (ENCODE) Project, 491
Endogenous agents, DNA damage due to, 318
Endonucleases, 122
   in homologous recombination, 373–374, 375(s)
Endoplasmic reticulum, 627, 627f
End-replication problem, telomeres and, 305–307, 306f, 307f
Energy of activation (E_a), 67
   lowering of, by enzymes, 66–68, 68f
Enhancers, in eukaryotic transcription, 490f, 490–491
env gene, LTR retrotransposons and, 398
env protein coding region, retroviral, 255, 256, 256f
Enzyme(s), 65–74. *See also specific enzymes*
   acting at replication fork, 280
   active site of, 69
   activity of, measurement of, 65–66, 66f, 67b
   bacterial DNA replication and, 275
   constitutive, 438
   DNA investigations and. *See* DNA, enzymatic techniques for investigating
   in DNA replication, 287, 289
   energy of activation and, 66–68, 68f
   enzyme-substrate complexes and, 68f, 68–71, 70f
      molecular details for, 71f, 71–72, 72f
   genes related to, 4
   inducible, 438
   regulatory
      control of committed steps in biochemical pathways by, 72f, 72–73, 73f
      sigmoidal kinetics and allosteric effectors and, 73f, 73–74, 74f
   repressible, 438
   required for homologous recombination in meiosis, 373–374, 375f, 376
   substrates of, 65
   suicide, direct alkyl group removal by, 335–337, 336f
Enzyme-substrate (ES) complexes, 68f, 68–71, 70f
   molecular details for, 71f, 71–72, 72f
Epigenetics, 527–528
ε subunit, of core polymerase, 284
Equilibrium density gradient centrifugation, 118–119, 119f
eri-1 gene, 612
ES (embryonic stem) cells, homologous recombination and, 368–369, 368f–370f, 371
Escherichia coli, 189–210
   bacteriophages of. *See also specific bacteriophages*
      as model system, 226
   β-galactosidase formation by, 448f, 448–449
   chromatin of, 153f, 153–154, 154f
      structure of, 152
   conjugation in, 200–204, 201f–203f, 204b
   culture of, 192, 192f
   DNA of
      mismatch repair of, 345–347, 346f
      structure of, 153, 153f
   DNA polymerase I of, 198
   DNA size in, 82t
   genetic mapping of, 206, 207f, 208f, 209

genetic regulation in, 198
lac promoter of, 418, 418f
making competent to take up DNA, 129
methionine tRNA isoacceptors of, 665, 666f
mutants failing to synthesize DNA, 197–198
origin of replication of, 276–277, 276f–278f
replication fork collapse in, homologous recombination replication restart after, 360f, 360–364, 361f
RNA polymerase of, 407, 411
phage T7 and, 237
70S subunit of, 661, 661f
σ subunits of, 414b, 414t
single-stranded DNA binding protein isolated from, 90, 91f
$tRNA^{Ala}$ of, 643–645, 644f, 645f
$tRNA^{Gln}$ of, 645, 645f, 646f
use of glucose in preference to lactose and, catabolite repression and, 448f, 448–449
ES complexes. *See* Enzyme-substrate (ES) complexes
ESE (exonic splicing enhancer), 580
Euchromatin, 156, 156f
Eukaryotes
   cap formation pathway in, 556–557, 557f
   chromatin of, 155
      histone classes in, 166t, 166–167
   DNA of, 81
   DNA polymerase of, nomenclature for, 352t, 352–353
   DNA replication in. *See* Eukaryotic DNA replication
   initiation stage of protein synthesis in
      methionine initiator tRNA and, 671, 673, 673f
      translation initiation factor phosphorylation and, 676, 676t
   methionine initiator tRNA of, 671, 673, 673f
   pre-mRNA synthesis in, 552, 552f
   ribosomes of
      free and attached, 627f, 627–628
      ribonucleoprotein subunit of, 625–627, 626f
      structure of, 625–628, 626f, 627f, 663b, 663f
   RNA polymerases of, 407
   transcription in. *See* Eukaryotic transcription; Transcription activator proteins, eukaryotic
   translation initiation factor phosphorylation in, 676, 676t
   transposable elements of. *See* Transposable elements (transposons), eukaryotic
Eukaryotic DNA replication, 293–307
   bidirectionality of, 267, 268f, 269f
   nucleotide repeat addition to chromosome ends and, 303, 305f
   origins of replication and, 295–296, 296f, 297f
      in yeast, 297f, 297–298, 298f
   Pol α in, 298–299, 299f
   Pol δ in, 299, 299f, 300
   Pol ε in, 299, 299f, 300
   SV40 system as model for, 293–295, 294f, 295f
   telomerase and
      in aging and cancer, 304b
      end-replication problem and, 305–307, 306f, 307f
   telomere formation and, 300–303, 301f, 302f
Eukaryotic transcription, 466–537
   bacterial transcription compared with, 467–468
   core promoter for protein-coding genes and, 476, 478, 478f
   elongation and, 486–487
      phosphorylation of C-terminal domain of largest RNA polymerase subunit and, 486

reactivation of arrested RNA polymerase II and, 487
transcription elongation factors and, 487
enhancers and, 490f, 490–491
epigenetic modifications and, 523–528
of chromatin, 523–525, 524f
DNA methylation and, 527
of histones, 525–527, 526f
general transcription factors and, 478–486
preinitiation complex formation and, 482f, 482–483, 483f, 486, 486f
RNA polymerase II requirement for, 478–480, 479t
TFIID or TBP binding to TATA core promoter and, 480–482, 480f–482f
Mediator and, 517, 519–523
association with activators at UAS in active yeast genes, 521–523, 522f
requirement for activated transcription, 517, 519–521, 521f
squelching and, 517
regulatory promoters and, 487–490, 488f, 489f
RNA polymerase I catalyzed, 528, 530–534
preinitiation complex formation and, 533, 533f
RNA polymerase I structure and, 532
rRNA synthesis and, 528, 530f, 530–531, 531f, 532
rRNA transcription unit promoter and, 531–532, 532f
termination of, 534, 534f
transcription elongation complex formation and, 533–534
RNA polymerase III catalyzed, 535–537
promoter types and, 535–537, 536f
RNA polymerase III transcript structure and, 535
RNA polymerases and, 468–476
crystal structures of, in yeast, 472, 473f, 474
distinguishing by sensitivities to inhibitors, 470f, 470–471, 471f
subunits of, 471–472, 472t, 477f
synthetic capacities of, 474
types of, 468–469, 469f, 470t
transcription activator proteins and. See Transcription activator proteins, eukaryotic
upstream activating sequence and, 491–492
Eukaryotic translation initiation factor (eIF), 674
EXO1, 365
Exogenous agents, DNA damage due to, 318
Exon(s), 39, 41f, 562
cassette, 566–567, 567f
regulated, 566–567, 567f
size and number of, 455f, 464–466, 564t
splicing and, 564t, 564–566, 565f
Exon definition, 582
Exonic splicing enhancer (ESE), 580
Exonic splicing silencer, 582–583
Exonucleases, 122
transcription termination of RNA polymerase II, 589–591, 590f
Expressed sequence tags, 564
3′-External transcribed sequence, 528
5′-External transcribed sequence, 528
Extinct species, DNA sequencing of, 145b

F

FACT (facilitates chromatin transcription) histone chaperone, 308–309
FBJ murine osteosarcoma virus, 507b
FC0 mutation, 648, 648f
FEN1 (flap endonuclease), in eukaryotic DNA replication, 295, 295f
Ferscht, Alan, 640

Fertility factor. See F plasmid(s)
F factor. See F plasmid(s)
Fibrous proteins, 25
Fields, Stanley, 518b
50s subunit, in protein synthesis, 625, 626f
Fingerprinting, 653
Fire, Andrew Z., 601, 602, 603, 604
Fischer, Emil, 69–70
FISH (fluorescence in situ hybridization), 163, 165f
F′ lac plasmids, 209f, 209–210, 437
Flap endonuclease (FEN1), in eukaryotic DNA replication, 295, 295f
Fluorescence in situ hybridization (FISH), 163, 165f
Flush ends, 124, 124f
fMet (N-formylmethionine), 665, 665f
Fork collapse, bacterial, homologous recombination replication restart after, 360f, 360–364, 361f
N-Formylmethionine (fMet), 665, 665f
40S subunit, in protein synthesis, 626
Fowlpox virus, DNA size in, 82t
F plasmid(s), 82t, 201–202
functions required for mating, 204b
integration into bacterial chromosome, 204, 205f, 206
F′ plasmids, 209f, 209–210
Fragment assay, 682, 682f
Frameshift mutations, 648, 648f
Franklin, Rosalind, 15, 16, 98
Fruit fly. See Drosophila melanogaster
Functional genomics, RNAi as investigational tool for, 612–613
Functional groups, of amino acids, 26, 26f
Fungi, gene conversion and, 372–373, 373f
Furth, John J., 468

G

gag gene, LTR retrotransposons and, 397–398, 398f
gag protein coding region, retroviral, 255, 256f
Gal4
activation domains and, 514f, 514–517, 516f
DNA-binding, dimerization, and activation domains of, 511–512, 512f
DNA-binding by, 509f, 509–511, 511f
GAL genes
Gal11, 520
GAL1, transcription of, 512–513, 513f
regulation of, 512f, 512–513, 513f
Gall, Joseph G., 176, 300
Gal11 Mediator, 520
Gal3 protein, 513
Gal4Zn$_2$C$_6$ binuclear cluster, 510, 510f
γ-rays, DNA damage due to, 321
Garrod, Archibald, 4
GATC sequences, mismatch repair and, 345–347, 346f
GC box, 488, 490
Gcn5, 525
GCN5 gene, 525
Gcn4 protein, 504–506, 504f–506f
Gehring, Walter, 497
Gel electrophoresis, 35–36, 36f, 119f–121f, 119–122
capillary, 120, 122b
sodium dodecyl sulfate-polyacrylamide, 121–122
topoisomer separation and, 121, 121f
Gel filtration, 33, 34f
Gene(s). See also specific genes
antibiotic resistance, transposons bearing, 389–390, 390f, 391b
enzymes related to, 4
"jumping," 393f, 393b–394b

Mendel's laws of inheritance and, 3
reporter, 391b, 391f
split. See Splicing
structural, 194
wild type, 195
Gene activity, eukaryotic transcription activator proteins and, 493f, 493–494
Gene concept, 592
Gene conversion, resulting from homologous recombination in meiosis, 372–373, 373f, 374f
Gene expression, asymmetric, 529b, 529f
Gene knockdown, 613
Gene knockouts, created by homologous recombination, 368–369, 368f–370f, 371
Generalized transduction, in bacteriophage P1, 246b, 247f
General transcription factors, in eukaryotic transcription, 478–486
preinitiation complex formation and, 482f, 482–483, 483f, 486, 486f
RNA polymerase II requirement for, 478–480, 479t
TFIID or TBP binding to TATA core promoter and, 480–482, 480f–482f
Gene therapy, adenoviral vector for, 254
Genetic code, 657t, 657–658, 658f
poly(U) direction of poly(Phe) synthesis and, 650–652
synthetic messengers confirming, 654–656, 656f
Genetic engineering, applications of, 20–22
Genetic mapping, 186
of Escherichia coli, 206, 207f, 208f, 209
somatic cell genetics for, 219
Genome, 195
human
annotation of, 146
DNA sequencing of, information provided by, 142, 146
Human Genome Project and, 142, 146, 564
long interspersed nucleotide elements in, 398–399, 400f, 401
short interspersed nucleotide elements in, 401
of SV 40 virus, 250, 251f
Genome Sequencer FLX (GS FLX), 142
Genomic imprinting, 529b, 529f
Genotype, 4
typographic conventions for, 194
Germ cells, 155
Germ line therapy, 21–22
gfp gene, 610
Giardia intestinalis, Dicer from, 607b–608b, 608f
Gilbert, Walter, 106–107, 443, 448, 562
Gill, Grace, 517
Gladstone, Leonard, 468
Globular proteins, 25
Glucocorticoid receptor, DNA-binding domain of, 501f, 501–502, 502f
Glucose
catabolite repression caused by, bacterial transcription and, 451f, 451–453, 452f
E. coli use in preference to lactose, catabolite repression and, 448f, 448–449
Glucose-specific, phosphoenolpyruvate-dependent phosphotransferase system, 451f, 451–453, 452f
Glutamic acid (glutamate), 27, 27f
abbreviations for, 29t
Glutamine, 27, 27f
abbreviations for, 29t
Glycine, 27, 28f
abbreviations for, 29t
G$_1$ phase, 156

INDEX 739

G₂ phase, 156
GpppN cap, 557
Gram, Christian, 189
Gram-negative bacteria, 190–191, 191f
Gram-positive bacteria, 190–191, 191f
Gram staining, 189–190
Gratuitous inducers, 439, 439f
Green fluorescent protein, 495
Greider, Carol, 303
Griffith, Fred, 11
Grunberg-Managa, Marianne, 651
GS FLX (Genome Sequencer FLX), 142
Guanine, 8t
  in DNA, 6, 6f
Guide strand, 609

H

Haemophilus influenzae
  genome sequencing of, 141–142
  HindII of, 124
H2A•H2B dimers, 308–309, 527
Hairpin, 396
Hairpin structure, of RNA, 105
Hale, Stephen P., 640, 641
Haloarcula marismortui, 661
Hankin, Ernest, 225
Hannon, Gregory J., 606
Haploid number, 155
HAT (histone acetyltransferase) activities, 525
*h*AT mobile elements, transposition mechanism of, 392, 392f, 394f
Haworth, Walter N., 5
Haworth structures, 5f, 5–6
Hda1 deacetylase complex, 525
Head formation, in bacteriophages, 233
Head piece, of Lac repressor subunit, 445f, 445–446
Heavy chains, of immunoglobulin G, 63–64, 64f
HeLa cells
  mRNA formation study using, 552–553
  splicing intermediate studies using, 571
Helical wheels, 505
Helicases, 89f, 89–90, 90f
Helix, double, of DNA, 15–19, 16f–18f
Helix-turn-helix DNA-binding domains, of eukaryotic transcription activator proteins, 497–499, 497f–499f
Helix-turn-helix motif, 445f, 445–446
Hemoglobin
  binding to oxygen, 60f–62f, 60–63, 61b
  Bohr effect and, 60f, 60–61, 61f
  HbS and HbA, 58b–59b, 59f
  sickle cell anemia and, 58f, 58b–60b, 59f
  structure of, 56–57, 57f
Hermes element, transposition mechanism of, 392, 394f
Herrick, James, 58b
Hershey, Alfred, 647
Heterochromatin, 156, 156f
Heteroduplex segments, 376
Heterogeneous nuclear RNA (hnRNA), 552
  poly(A) tails in, 553, 553f
Heterozygosity, loss of, 376
Hfr (high frequency of recombination) males, 204, 205f
H19 gene, 529b, 529f
HGPRT gene, homologous recombination and, 368, 368f
High frequency of recombination (Hfr) males, 204, 205f
Hinge region, of Lac repressor subunit, 446, 446f
Hirose, Yutaka, 583
Histidine, 26, 26f
  abbreviations for, 29t
Histone(s), 81, 152
  epigenetic modifications of, eukaryotic transcription and, 525–527, 526f
  in eukaryotic chromatin, 166t, 166–167
Histone acetyltransferase (HAT) activities, 525
Histone chaperones, 308
HIV, 255, 256
hnRNA (heterogeneous nuclear RNA), 552
  poly(A) tails in, 553, 553f
hnRNP A1, 582–583
Hoagland, Mahlon, 628, 629, 631
Hogness, David, 476
Holley, Robert, 634, 635
Holliday, Robin, 363
Holliday junctions, 363, 363f, 364
  conversion into DNA forks, 364
Holoarcula marismortui, 663
Homeobox, 497–498, 498f
Homeodomain, 498
Homeotic *(HOM)* genes, 497–499, 497f–499f
  of *Drosophila melanogaster*, 497–499, 497f–499f
  mammalian, 498–499, 499f
Homolog(s), 155
Homologous recombination, 359–371, 372–376
  gene knockouts created by, 368–369, 368f–370f, 371
  meiotic, 372–376
    enzymes required for, 373–374, 375f, 376
    gene conversion resulting from, 372–373, 373f, 374f
  mitotic, during late S stage or G₂ stage of cell cycle, 364–367, 365f–367f
  replication restart after bacterial replication fork collapse and, 360f, 360–364, 361f
    PriA protein in, 364
    RecA in, 362–363, 363f
    RecBCD protein in, 361–362, 362f
    RuvAB protein in, 363–364, 364f
    RuvC protein in, 364
  waves of, 374, 376
Hormone response element, 502, 502f
Horvitz, H. Robert, 614
Hotchkiss, Rollin, 14
Housefly, Hermes element of, 392, 394f
Howard-Flanders, Paul, 340
Hox genes, 498–499, 499f
HSV-tk genes, in positive-negative selection technique, 369, 369f, 371
HUGO (Human Genome Organization), DNA polymerase nomenclature and, 353
Human chromosome 1, DNA size in, 82t
Human genome
  annotation of, 146
  DNA sequencing of, information provided by, 142, 146
  long interspersed nucleotide elements in, 398–399, 400f, 401
  short interspersed nucleotide elements in, 401
Human Genome Organization (HUGO), DNA polymerase nomenclature and, 353
Human Genome Project, 142, 146, 564
Human proteome, 592
Hunter, Craig P., 612
Hybrid-states translocation model, 684, 684f, 684–685, 685
Hydrogen, van der Waals radius of, 45t
Hydrogen bonds, in proteins, 43–44, 44f, 49
Hydrolytic cleavage reactions, DNA damage by, 321–323, 322f, 323f
Hydrophilic side chains, 48–49
Hydrophobic interaction, in proteins, 44–45, 45f

Hydrophobicity, 44–45, 45f
Hydrophobic side chains, 44–45, 48–49
Hydroxyl radical, DNA damage due to, 324, 324f
Hyperchromic effect, 85
Hypervariable regions, of immunoglobulin G, 64
Hypochromic effect, 88

I

ICR (imprinting center region), 529b
Identity elements, 643
IF(s) (translation initiation factors), 662, 664
IF3, 687
Igf2 gene, 529b, 529f
IgG (immunoglobulin G), domain concept and, 63–64, 64f
IgM transcription unit, 588
Ile-tRNA synthetase
  posttransfer editing and, 642b, 642f
  proofreading function of, 639–640, 640f
Imitation switch (ISWI) chromatin remodeling complex, 310
Immune system, diversity in, recombination processes and, 395–396
Immunoglobulin fold, 64
Immunoglobulin G (IgG), domain concept and, 63–64, 64f
Immunoglobulin M (IgM) transcription unit, 588
Imprinting center region (ICR), 529b
Inchworming model, 426f, 426–427
Independent assortment, law of, 3
Induced fit model, of enzyme-substrate binding, 70f, 70–71, 72f
Inducer(s), gratuitous, 439, 439f
Inducer exclusion, 453
Inducible enzymes, 438
Ingram, Vernon, 58b–59b
Initiation, abortive, 425
Initiation factor(s) (IFs), 662, 664
  transcription, 532, 533, 533f
  translation, 662, 664
    phosphorylation of, in eukaryotes, 676, 676t
    30S and 70S initiation complex formation and, 667–668, 669f
Initiation factor 1 (IF1), 70S initiation complex formation and, 668, 669f
Initiation factor 2 (IF2), 70S initiation complex formation and, 668, 669f
Initiation factor 3 (IF3), 70S initiation complex formation and, 668, 669f
Initiation stage
  of bacterial DNA replication, 274
    E. coli replication origin and, 276–277, 276f–278f
    regulation of, 279b
  of bacterial transcription
    RNA polymerase crystal structures and, 420–424, 421f, 423f, 424f
    RNA polymerase holoenzyme binding to promoter and, 418–420, 419f, 420f
    RNA polymerase holoenzyme structure and, 412–413, 413f
    RNA polymerase scrunching during, 425–427, 426f
    σ⁷⁰ family conserved domains and, 424–425, 425f
    σ⁷⁰ polymerase holoenzyme escape and, 425
  of protein synthesis, 662, 664–676
    archaeal translation initiation pathway and, 676
    eukaryotic, translation initiation factor phosphorylation and, 676, 676t
    eukaryotic methionine initiator tRNA and, 671, 673, 673f

initiator and elongator methionine tRNAs
and, 664–665, 665f, 666f
scanning mechanism and, 673–674, 675f
70S subunit in, 667–668, 669f
Shine-Dalgarno sequence and anti-Shine-
Dalgarno sequence and, 668–671,
670f, 671f
start codons and, 664
30S subunit in, 665–668, 667f, 669f
Initiator DNA, 294
Initiator (Inr) element, 478
Initiator protein, 275
Inr (iitiator) element, 478
Insertion(s), 196, 197f
Insertion sequences, bacterial, 381–382, 382f
Interference, in homologous recombination, 374
Interferon response, 613
Internal transcribed spacer 1, 528
Internal transcribed spacer 2, 528
Internet, molecular biology information on, 22
Interphase, 155–156
Interstrand cross-linking agents, 327–328,
328f–330f, 328b–331b
Intrinsically disordered proteins, 53
Intron(s), 39, 562, 564t, 564–565, 569
size and number of, 564, 564t
splicing and, 564t, 564–566, 565f
Intron definition, 582
Intronic splicing enhancer, 580
Intronic splicing silencer, 582–583
Inversions, ionizing radiation inducing, 321
Inverted repeat sequences, 102
Ion-exchange chromatography, 33, 33f
Ionic bonds, in proteins, 44
Ionizing radiation
DNA damage caused by, 321
DNA damage due to, 321
IPTG (isopropylthiogalactoside), 439, 439f
IS elements, bacterial, 381–383, 382f
in composite transposons, 382f, 382–383, 383f
Isoelectric pH, 36
Isoforms, 568
Isoleucine, 27, 28f
abbreviations for, 29t
Isopropylthiogalactoside (IPTG), 439, 439f
Isoschizomers, 124
Isotopes, radioactive, 67b
ISWI (imitation switch) chromatin remodeling
complex, 310
Izsvák, Zsuzsanna, 396

J
Jacob, François, 275, 439, 440, 647–648
Jakes, Karen, 670
Jeffreys, Alec, 218b
Johannsen, William Louis, 186
"Jumping" genes, 393f, 393b–394b
Junction(s), in RNA, 105, 105f
Junctional diversity, 395

K
Kadonaga, James T., 494
Karyotype, 162–166
chromosome site nomenclature and, 162f,
162–163, 163f
diagnosis and, 164b–165b, 165f
information provided by, 163, 165f, 166, 166f
metaphase chromosomes shown by, 163
Kelly, Thomas J., 293
Kelner, Albert, 333
Kendrew, John, 51
Khorana, H. Gorbind, 654–655, 656
Khoury, George, 490

Kim, Peter S., 50
Kinetochores, 157, 174–175
Kingsbury, Roger, 488
Kissing hairpins, 105, 106f
Klenow, Hans, 134
Klenow fragment, 134–135, 135f
Klug, Aaron, 638
K_m (Michaelis constant), 69, 69f
Kornberg, Arthur, 132, 133, 281, 282
Kornberg, Roger, 472, 486, 517, 519–520, 522, 524
Koshland, Daniel, 70
Kozak, Marilyn, 673
Krzycki, Joseph A., 646
Ku protein, in nonhomologous end-joining, 369,
369f, 371, 372f

L
lacA gene, bacterial transcription and, 437f,
437–438, 438f, 440
lacI gene, bacterial transcription and, 440
lacO gene, bacterial transcription and, 440, 444
lacO^c mutations, 439t, 439–440
lac operators, bacterial, 444, 444f
lac operon transcription, 452–453
Lac repressor, 439
bacterial transcription and, 443f, 443t, 443–444,
444f–448f, 445–446
lac structural genes, bacterial, regulation of, 438f,
438–439, 439f
Lactose operon, bacterial, 437–448
allolactose as inducer of, 442f, 442–443
lac operators of, 444, 444f
Lac repressor and, 443f, 443t, 443–444,
444f–448f, 445–446
lacZ, lacY, and lacA genes and, 437f,
437–438, 438f
operon model and, 440–442, 441f
regulation of lac mRNA and, 439t, 439–440
regulation of lac structural genes and, 438f,
438–439, 439f
lacY gene, bacterial transcription and, 437f,
437–438, 438f, 440
lacZ gene
bacterial transcription and, 437f, 437–438,
438f, 440
in PaJaMo experiments, 648
lacZ reporter gene, 391b, 391f
Laemmli, Ulrich K., 172–173
Lag phase, 192
Lambdoid phages, 241
Lariat model, 573, 574f
Law of dominance, 3
Law of independent assortment, 3
Law of segregation, 3
5'-Leader (5'-untranslated region), 435, 673–674
Leder, Philip, 654
Lederberg, Joshua, 201, 246b
let-7 gene, 614
Leucine, 27, 28f
abbreviations for, 29t
Levinthal, Cyrus, 55
LexA, regulation of SOS response by, 348–350,
349f, 350f
Li, Joachim J., 293
Life cycle, 155
Ligand binding domain
of Lac repressor subunit, 446, 447f
of transcription activating proteins, 493
Light chains, of immunoglobulin G, 63–64, 64f
LINE(s) (long interspersed nucleotide elements),
398–399, 399f, 400f
LINE-1 element, 398–399, 400f, 401
lin-4 gene, 614–615

lin-7 gene, 614–615
lin-14 gene, 614
lin-41 gene, 614
lin-42 gene, 614
Linker DNA, 167–168, 168f
Linker-scanning mutagenesis, 488f, 488–489, 489f
Linking number (Lk), 95b
Lipmann, Fritz, 677
Liquid scintillation counters, 67b
Lis, John T., 556
Lk (linking number), 95b
Lock-and-key model, of enzyme-substrate binding,
69–70, 70f
Loeb, Timothy, 239
Log phase, 192
Long interspersed nucleotide elements (LINEs),
398–399, 399f, 400f
Long patch repair, 340
Long terminal repeats (LTRs), 397
Loops, in proteins, 49
Loss of heterozygosity, 376
LTR(s) (long terminal repeats), 397
LTR retrotransposons, cut-and-paste mechanisms
and, 397–398, 398f, 399f
Luciferase, activity of, monitoring of, 495
Luria, Salvador, 226
Lysine, 26, 26f
abbreviations for, 29t
Lysis, 153, 228
Lysogenic cycle of phages, 228
of bacteriophage P1, 245–246, 248
of phage lambda, 245
Lysogens, 228
Lytic cycle of phages, 228
of bacteriophage P1, 245–246
of phage lambda, 243, 244f, 245

M
Ma, Jun, 515, 516
MacLeod, Colin, 12–13
Macromolecules, physical techniques for studying,
115–122
electron microscopy, 115–117, 115f–117f
equilibrium density gradient centrifugation,
117–119, 118f, 119f
gel electrophoresis, 119f–121f, 119–122,
121b, 122b
Major groove, in DNA, 17
recognition patterns in, 82–83, 82f–84f
Makman, Richard S., 449
Malaria, resistance to, 60
Mammals, homeotic genes of, 498–499, 499f
Mammuthus primigenius, DNA sequencing of, 145b
Maniatis, Tom, 571, 572–573
Manley, James L., 583
Marcker, Kjend, 664
Marmur, Julius, 86
Mass spectrometer sequence technique, for amino
acid sequence determination, 39, 40f
Matrix protein, in retroviral RNA, 255
Matthaei, Johann Heinrich, 650–651, 668
Matthews, Kathleen, 446
McCarty, Maclyn, 12–13
McClintock, Barbara, 175, 392, 393b–394b
McKnight, Stephen, 488
MCM2-7 helicase, 308
MeCP2, 527
MED1 gene, 520
MED2 gene, 520
Mediator, 517, 519–523
requirement for activated transcription, 517,
519–521, 521f
squelching and, 517

MED14, 520
MED15, 520
Meiosis, 159–162, 160f–161f
  first division of, 159, 162
    homologous recombination in, 372–376
      enzymes required for, 373–374, 375f, 376
      gene conversion resulting from, 372–373, 373f, 374f
  second division of, 162
Mello, Craig C., 601, 602, 603
Melting temperature (T_m), 86
Mendel, Gregor, 3, 186, 214–215
Meselson, Matthew, 266–267, 665–666
Messenger RNA. *See* mRNA (messenger RNA)
Messing, Joachim, 249b
Metal shadowing, 115f, 115–116
Metaphase, mitotic, 157
Metaphase I, meiotic, 159
Metaphase II, meiotic, 162
Methanococcus jannaschii, RNA polymerase from, 475b–476b
Methionine, 27–28, 28f
  abbreviations for, 29t
Methionine aminopeptidase, 688
Methionine conversion, to s-adenosylmethionine, 554–556, 555f
Methylation, of alleles, 529b, 529f
7-Methylguanosine caps, in mRNA, 553–556, 555f, 556f
Methylmethane sulfonate, 325f
N-Methyl-N′-nitrosoguanidine, 325f
N-Methyl-N-nitrosourea (MNU), 325f
5′-m⁷G caps, 556–557, 557f
Mice
  chimeric, 371
  homologous recombination in, 368–369, 368f–370f, 371
  transgenic, 371
Michaelis constant (K_m), 69, 69f
Microprocessor complex, 615
MicroRNAs (miRNAs), blockage of mRNA translation and degradation and, 614f, 614–615, 616f, 617, 617f
Microsatellites, 215
Microsomal fraction, 627
Microsomes, 332
Microtubules, 157
  centromeres as attachment site of, 173–175, 174f–176f
Miescher, Friedrich, 3
Mig1 protein, 512, 512f, 513, 513f
Minimal medium, 192
Minisatellites, 215
Minor, Daniel L., Jr., 50
Minor groove, in DNA, 17
  recognition patterns in, 82–83, 82f–84f
miRNAs (microRNAs), blockage of mRNA translation and degradation and, 614f, 614–615, 616f, 617, 617f
Mirsky, Alfred E., 525
Mismatch repair, 345–347, 346f
Misteli, Tom, 171
Mitosis, 155–158
  animal life cycle alternation between interphase and, 155–156, 156f, 157f
  chromosome number after cell division and, 157–158, 159f
  homologous recombination in, during late S stage or G₂ stage of cell cycle, 364–367, 365f–367f
Mitotic spindles, 157
MNU (N-methyl-N-nitrosourea), 325f
Moazed, Danesh, 684, 685

Model, 16
Modulons, 453
Molecular biology
  intellectual foundation of, 2–3
  Internet resources on, 22
  origin of term, 2
  theoretical framework for, 19–20
Molecular chaperones, 56
Monoadducts, 325, 325f
Monod, Jacques, 439, 440, 448, 647–648
Monro, Robin, 682
Moore, Peter, 661
Morgan, Thomas Hunt, 186–187
Mouse
  chimeric, 371
  homologous recombination in, 368–369, 368f–370f, 371
  transgenic, 371
mRNA (messenger RNA), 19, 19f, 20f
  bacterial, 434–436
    lac, genetic studies of regulation of, 439t, 439–440
    lac, regulation of, 439t, 439–440
    lifetime of, 435
    monocistronic and polycistronic, 434–435
    polycistronic, 435
    rate of synthesis of, 436
    regulation of, 436, 436f, 436t
    synthesis of, regulation of, 436, 436f, 436t
  degradation of, miRNA pathway causing, 614f, 614–615, 616f, 617, 617f
  poly(A) tails in, 553, 553f, 554f
  translation of, blockage by miRNA pathway, 614f, 614–615, 616f, 617, 617f
  trp, synthesis of, 457
MSH2 protein, mismatch repair and, 345
MSH3 protein, mismatch repair and, 345
MSH6 protein, mismatch repair and, 345
Muller, Hermann J., 175
Müller-Hill, Benno, 443, 444
Mullis, Kary B., 137
Musca domestica, Hermes element of, 392, 394f
Mutagen(s), 195, 318
Mutagenesis
  site-directed, 138f, 138–139
  spontaneous, 195
  of transposons, 390
Mutants, 195–196
  classification of, 195–196
  conditional, 196
  suppressor-sensitive, 196
  typographic conventions for, 195
  uses in molecular biology, 197–199
Mutasome, 351
Mutations
  double, 197f
  nonsense, 657
  number required to convert normal cells to cancer cells, 318
  point, 196
  transition, 322, 323f
  transversion, 322, 323f
MutH protein, mismatch repair and, 345–347, 346f
MutL protein, mismatch repair and, 345–347, 346f
MutSα heterodimer, 345
MutSβ heterodimer, 345
MutS protein, mismatch repair and, 345–347, 346f
Mycoplasma genitalium
  DNA size in, 82t
  tRNA genes in, 634
Myoglobin
  Bohr effect and, 60f, 60–61, 61f
  structure of, 56–57, 57f

N

Nascent polypeptide processing and folding, 688f, 688–689
NASP (nuclear autoantigenic sperm protein) histone chaperone, 309
Native state, of proteins, 53
Negative regulation, 436, 436f
Negative staining, electron microscopy and, 116, 116f
*Neo*ᴿ gene, homologous recombination and, 368f, 368–369, 369f, 371
Neurospora, gene conversion and, 373, 374f
N-glycosylic bonds, 6–7
Nick translation, DNA polymerase I catalysis of, 136f, 136–137
Nirenberg, Marshall W., 20, 650–651, 654, 668
Nitrocellulose filter assay, 443
Nitrogen, van der Waals radius of, 45t
Nitrous acid, amino group deamination due to, 323
Nodes, of supercoils, 94
Noller, Harry F., 661, 682–683, 684
Nomura, Taisei, 335, 666
Nonautonomous transposons, 389
Noncomposite transposons, 388
Noncovalent bonds, weak, in proteins, 43–46
  hydrogen, 43–44, 44f, 49
  hydrophobic interaction and, 44–45, 45f
  ionic, 44
  van der Waals interaction and, 45t, 45–46, 46f
Nonet, Michael, 520
Nonhomologous end-joining, double-strand break repair by, 360, 371, 372f
Non-LTR retrotransposons, 397
  target-primed reverse transcription mechanisms and, 398–399, 399f, 400f
Nonpermissive temperature, 196
Nonsense mutations, 657
Northern blotting, 132
N-terminal domain, of Lac repressor subunit, 445f, 445–446
N-terminal region, of TBP, 480–481, 481f
NTP. *See* Nucleoside triphosphate (NTP)
Nuclear autoantigenic sperm protein (NASP) histone chaperone, 309
Nuclear export signal, of transcription activating proteins, 493
Nuclear localization signal, of transcription activating proteins, 493
Nuclear receptors, orphan, 500
Nuclear receptor superfamily proteins, 499–502, 500f–503f
Nuclear run-on method, 588–589, 589f
Nucleases
  in DNA investigations, 122–123, 123t
  properties of, 123t
Nucleic acids, 5–11. *See also* DNA; RNA
  discovery of, 3
  isolation of, 113–115
  sequences of, polypeptide sequence determination from, 39, 41, 41f
  sugars in, 4f, 5f, 5–6
Nuclein, 3
Nucleoids, bacterial chromatin in, 153f, 153–155, 154f
Nucleolus, 157
  rRNA synthesis in, 532
Nucleoprotein, in retroviral RNA, 255
Nucleosides, 8t
  formation of, 6f, 6–7, 7f
  nomenclature for, 7
  in tRNA, structures of, 635–636, 637f
Nucleoside triphosphate (NTP)
  bacterial RNA synthesis and, 408
  transcription elongation complex and, 427

Nucleosome(s), 166–173
  arrangement in chromatin, 171–172, 172f
  atomic structure of core particles of, 168–170, 169f, 170f
  reposition of, by ATP-dependent chromatin remodeling, 523, 524f
Nucleosome core particles, 168
  atomic structure of, 168–170, 169f, 179f
  interaction between H1 and, 171, 171f
Nucleotide(s)
  branching, 570, 570f
  formation of, 7f, 7–8, 8t
  in transcription elongation complex, numbering of, 428
Nucleotide excision repair, 340–341, 342f
Nucleotide repeats, addition to chromosome ends, 303, 305f
Nudler, Evgeny, 430b

O

Ochoa, Severo, 651, 653, 667
O'Donnell, Michael, 285
Okazaki fragments
  adjacent, joining of, 271–273, 273f
  formation of, 271, 273
  in trombone model of replication, 287
Okazaki, Reiji, 269, 270, 272
Oligopeptides, 30
Oncogenes, 318
Oncoproteins, 507b
One gene–one enzyme hypothesis, 4
One gene–one polypeptide hypothesis, 4
Open promoter complex, 420
Open reading frames, 435
Operon(s). See also specific operons
  cAMP•CRP activation of, 453
Operon model, 440–442, 441f
  bacterial transcription and, 440–442, 441f
ORC (origin of recognition complex), 298
ori. See Origins of replication (ori)
Origin of recognition complex (ORC), 298
Origins of replication (ori), 275
  of E. coli, 276–277, 276f–278f
  in eukaryotic DNA replication, 295–298, 296f, 297f
    site of DNA chain initiation and, 297f, 297–298, 298f
oriT (transfer origin), in F plasmid, 203
Orphan nuclear receptors, 500
Orthologs, 566
Outer membrane, gram staining and, 190–191, 191f
Ovalbumin, gene for, 560, 561f, 562, 563f
Overlap method, for amino acid sequence determination, 38, 38f
Oxidative DNA damage, 324, 324f
8-Oxoguanine (oxoG), 324, 324f
Oxygen
  hemoglobin binding to, 60f–62f, 60–63, 61b
  van der Waals radius of, 45t
Oxyhemoglobin, 62

P

Pääbo, Svante, 145b
Pabo, Carl O., 498
Painter, Robert, 343b
Pairing diversity, 395
PaJaMo experiments, 647–648
Palindromes, 102
Parallel pleated sheets, 49
Pardee, Arthur, 647–648
p arm, 162
Partial diploids, 200

Passenger strand, 609
Passner, Jonathan, 450
Pauling, Linus, 29, 46–47, 49, 58b, 639
Paulson, James R., 172–173
Pausing, in transcription
  suppression by elongation factors, 487
  transcription pause sites and, 432
Pausing model, primase and, 288b, 289f
PCNA (proliferating cell nuclear antigen), in eukaryotic DNA replication, 294, 294f
PCR (polymerase chain reaction), 137, 138f
Peptide(s), 29
Peptide bonds, 29–31
  amino acid linkage by, 29f, 29–30, 30f
  peptide chain directionality and, 30, 31f
Peptide deformylase, 688
Peptide mapping, 653
Peptidoglycan, 189
Peptidyl transferase activity, peptide bond formation and, 680–683, 682f, 683f
Periplasmic space, 190
Permissive temperature, 196
Peroxisome proliferator-activated receptor (PPAR), 502
Perry, Robert, 554
Perutz, Max, 51
Petri dishes, 192
pH
  electrophoresis and, 35–36, 36f
  isoelectric, 36
Phage. See Bacteriophage(s)
Phage Group, 226
Phage lysate, 228
Phagemids, 249b
Phage T7, DNA size in, 82t
Phenotype, 4
  determination of number of genes responsible for, 200
  mutant, display of, 195–196
  typographic conventions for, 194
Phenylalanine, 27, 28f
  abbreviations for, 29t
Philadelphia chromosome, 164b–165b, 165f
Phosphodiester bonds, 8
Phosphorus, van der Waals radius of, 45t
(6-4) photoproducts, DNA damage due to, 319, 320f, 321
Photoreactivation, 333
(6-4) phytolase, 335, 335f
Phytolase, DNA repair by, 333–335, 334f, 335f
Pickard, Craig S., 469
Piwi interacting RNAs (piRNAs), 617
Plaque(s), formed by bacteriophages, 228–229, 229f
Plaque assay, 229, 229f
Plasmid(s), 92
  basic replicon and, 210
  drug resistance, 93b
  fertility factor. See F plasmid(s)
  recombinant, 128
  relaxed, 210
  stringent, 210
  synthesis in recipient cell, 203–204
  as vectors, insertion of DNA fragments into, 128–129, 129f
  virulence, 93b
Plasmid pBR322, DNA size in, 82t
Plating, to detect auxotrophs, 194
Pleated sheets, parallel and antiparallel, 49
Point mutations, 196, 197f
Pol δ, in eukaryotic DNA replication, 294, 298–299, 299, 299f, 300
Pol ε, in eukaryotic DNA replication, 299, 299f, 300

polA gene, DNA synthesis and, 280
pol gene, LTR retrotransposons and, 397, 398f
pol protein coding region, retroviral, 255–256, 256f
Poly(U), direction of poly(Phe) synthesis by, genetic code and, 650–652
Poly(Phe), synthesis of, direction by poly(U), genetic code and, 650–652
Polycistronic mRNA, 435
Polydeoxyribonucleotides, 8, 9f
Polymerase chain reaction (PCR), 137, 138f
Polymerase I transcription release factor (PTRF), 534, 534f
Polymorphisms, 215
  restriction fragment length, 216–218, 217f
  single nucleotide, 215
  tandem repeat, 215, 216f, 218b
Polyomaviridae, 250, 251f
Polyomaviruses, 250, 251f, 252f, 252–253
Polypeptide(s), 30
  direction of growth of, protein synthesis and, 652–653, 653f
  of DNA polymerase III holoenzyme, 283
  flexibility of, 25
  folding pattern of, 43–46
  interaction of, mutants in study of, 199, 199f
  nascent, processing and folding of, 688f, 688–689
  primary structure of, determination of tertiary structure by, 53–56, 54f
  selenocysteine and pyrrolysine as building blocks for, 645–646, 646f
  sequences of, determination from nucleic acid sequences, 39, 41, 41f
Polypeptide chain(s)
  folding of, 49, 49f
  initiation of, 30S subunit in, 665–667, 667f
Polypeptide chain termination signals, protein synthesis and, 656f, 656–657
Polypeptide folding motifs, 50, 50f
Poly(A) polymerase, 586–587
Polyproteins, retroviral, 255–256
Polypyrimidine tracts, 570
Polyribonucleotides, 8, 10f
Polyribosomes (polysomes), 553
Poly(A) sites
  locating, 585, 585f
  RNA polymerase II and, 587f, 587–588
Poly(A) tails
  pre-mRNA synthesis and, 552–553, 553f, 554f
  synthesis of, coupling of transcription termination of RNA polymerase II and, 584–587, 585f, 586f
Porins, gram staining and, 190
Positive-negative selection technique, 369, 369f, 370f, 371
Positive regulation, 436, 436f
POT1 protein, 178, 178f
Potts, Percival, 325
PPAR (peroxisome proliferator-activated receptor), 502
Precursor m-RNA (pre-mRNA)
  multiple molecule production for single pre-mRNA and, 566–567, 567f
  nascent, cap attachment to, 556
  synthesis of, 552–553, 554f
    in eukaryotes, 552, 552f
    poly(A) tails and, 552–553, 553f, 554f
Preinitiation complexes, 480
  formation of
    general transcription factors and, 482f, 482–483, 483f, 486, 486f
    RNA polymerase I catalyzed eukaryotic transcription and, 533, 533f

"Premature release," 290
Pre-mRNA (precursor m-RNA)
  multiple molecule production for single pre-mRNA and, 566–567, 567f
  nascent, cap attachment to, 556
  synthesis of, 552–553, 554f
    in eukaryotes, 552, 552f
    poly(A) tails and, 552–553, 553f, 554f
Pre-RISC complex, 609
PriA protein, in homologous recombination, 364
Pribnow, David, 417
Pribnow box, 417
Primary structure of proteins, 36–41
  BLAST program and, 41
  DNA sequencing for determination of, 39, 41, 41f
  Edman degradation for determination of, 36–39, 37f, 38f, 40f
  tertiary structure determination by, 53–56, 54f
Primary transcript, 410
Primer, coordination of, DnaB helicase and, 288b–289b, 289f
Primer strand, 133, 133f
Priming loop model, primase and, 289b, 289f
Procapsids, 233
Processed pseudogenes, 401
Processivity
  of DNA polymerase III, 281
  measurement of, 282b
Prokaryotes, DNA of, 81
Proliferating cell nuclear antigen (PCNA), in eukaryotic DNA replication, 294, 294f
Proline, 27–28, 28f
  abbreviations for, 29t
Promoter(s), 322
  in bacterial transcription
    initiation stage of, 413, 416f–418f, 417–418
    RNA polymerase holoenzyme binding to, 418–420, 419f, 420f
  closed, 485b, 485f
  formation of, TFIIB and, 484f, 484b–485b, 485f
  Mediator interaction as, 522, 522f
  open, 485b, 485f
  proximal, 488
  regulatory, 488–490, 489f
    in eukaryotic transcription, 487–490, 488f, 489f
  reporter genes lacking, 391b
  strong, 418
  TATA core, TFIID or TBP binding to, general transcription factors and, 480–482, 480f–482f
  types of, RNA polymerase III catalyzed eukaryotic transcription, 535–537, 536f
Promoter escape, 425
Prophages, 228
Prophase, of mitosis, 156–157
Prophase I, meiotic, 159
Prophase II, meiotic, 162
Protein(s). *See also specific proteins*
  α-amino acids and
    abbreviations representing, 28–29, 29t
    peptide bonds linking, 29f, 29–30, 30f
    structure of, 25–28, 26f–28f
  denatured state of, 53–54, 54f
  fibrous, 25
  folding of, zipping and assembly model of, 55
  globular, 25
  initiator, 275
  intrinsically disordered, 53
  native state of, 53
  peptide bonds in, 29–31, 29f–31f
  purification of. *See* Protein purification
  renaturation of, 54f, 54–56
  in retroviral RNA, 255–256
  structure of. *See* Protein structure
  synthesis of. *See* Protein synthesis
Protein-protein interactions, yeast two-hybrid assay and, 518b–519b, 518f
Protein purification, 31–36
  affinity chromatography for, 35, 35f
  column chromatography for, 32, 32f
  electrophoresis for, 35–36, 36f
  gel filtration for, 33, 34f
  ion-exchange chromatography for, 33, 33f
Protein release factors, protein synthesis and, 685, 687
Protein structure, 36–64, 41f, 42f
  primary, 36–41
    BLAST program and, 41
    DNA sequencing for determination of, 39, 41, 41f
    Edman degradation for determination of, 36–39, 37f, 38f, 40f
    tertiary structure determination by, 53–56, 54f
  quaternary, 43, 56–57, 57f–62f, 58b–61b, 60–63
  secondary, 42–43, 46–50
    α-helix, 46–49, 47f
    β-conformation, 48f, 49
    combinations of, 50, 50f
    loops and turns in, 49, 49f
    prediction of, 50, 50f
  supersecondary, 50, 50f
  tertiary, 43, 50–56
    determination by primary structure, 53–56, 54f
    intrinsically disordered proteins in, 53
    molecular chaperones and, 56
    x-ray crystallography of, 50–53, 51f, 52f
  weak noncovalent bonds and, 43–46
    hydrogen, 43–44, 44f, 49
    hydrophobic interaction and, 44–45, 45f
    ionic bonds and, 44
    van der Waals interaction and, 45t, 45–46, 46f
Protein synthesis, 623–691
  aminoacyl-tRNA and
    binding of, to multiple codons, 659–660, 660f
    coding specificity of, 658–659, 659f
  aminoacyl-tRNA synthetases and
    cognate distinction by, 641, 643–645, 644f–646f
    proofreading functions of, 639–640, 640f
    selenocysteine and pyrrolysine as building blocks for polypeptides and, 645–646, 646f
    valyl-tRNA and valyl-AMP hydrolysis by, 640–641, 641f
  chain termination stage of, 664
  elongation stage of, 664, 677–685
    codon-anticodon helix sensing and, 679, 680f
    EF-Ts and, 679–680, 681f
    EF-Tu•GTP•aminoacyl-tRNA ternary complex and, 677, 679
    elongation factors and, 677, 678f
    hybrid-states translocation model and, 684f, 684–685
    ribosomes and, 680–683, 682f, 683f
  genetic code and, 657t, 657–658, 658f
    synthetic messengers confirming, 654–656, 656f
  initiation stage of, 662, 664–676
    archaeal translation initiation pathway and, 676
    eukaryotic methionine initiator tRNA and, 671, 673, 673f
    initiator and elongator methionine tRNAs and, 664–665, 665f, 666f
    scanning mechanism and, 673–674, 675f
    70s subunit in, 667–668, 669f
    Shine-Dalgarno sequence and anti-Shine-Dalgarno sequence and, 668–671, 670f, 671f
    start codons and, 664
    30s subunit in, 665–668, 667f, 669f
    translation initiation factor phosphorylation in eukaryotes and, 676, 676t
  mRNA and
    codons in, 648f, 648–650
    direction of polypeptide growth and, 652–653, 653f
    direction of reading of, 653–654, 654f
    poly(U) direction of poly(Phe) synthesis and, 650–652
    programming of ribosomes by, 647–648
    trinucleotide promotion of aminoacyl-tRNA molecules to ribosomes and, 654, 655f
  nascent polypeptide processing and folding, 688f, 688–689
  polypeptide chain termination signals and, 656f, 656–657
  recycling stage of, 664, 687–688
  ribosomes and, 625–628, 660–662
    bacterial, structure of, 625, 625f
    eukaryotic, structure of, 625–628, 626f, 627f
    high resolution determination of structure of, 661f, 661–662, 662f
    subunit structure and function and, 660–661, 661f
  signal sequence and, 689–691, 690f
  termination stage of, 685, 687
    mutant tRNA molecules and, 687
    protein release factors and, 685, 687
  tRNA and, 628–639
    amino acid attachment to, 628–629, 630f, 631, 631f, 633f, 633–634, 634f
    CCA$_{OH}$ at ends of, 631, 632f, 633
    cloverleaf secondary structures of, 635–636, 636f–638f, 638
    folding of, 638, 639f
    yeast, sequencing of, 635
Proteome, human, 592
Proto-oncogenes, 318
Prototrophs, 192
Provirus model, 257
Prp8, 579
Pseudogenes, processed, 401
Pseudoknots, in RNA, 105, 106f
Psoralen, interstrand cross-linking and, 330f, 330b–331b
Psoriasis, psoralen treatment for, 331
Ptashne, Mark, 510–511, 514, 515, 516, 517
PTRF (polymerase I transcription release factor), 534, 534f
Punnett, Reginald C., 186
Purine, in DNA, 6, 6f
Puromycin, peptidyl transferase activity and, 681–682, 682f
Pyrimidine, in DNA, 6, 6f
Pyrococcus furiosus, DNA polymerase of, 137
Pyrrolysine, as building block for polypeptides, 645–646, 646f

Q

q arm, 162
Quaternary structure of proteins, 43, 56–57, 57f–62f, 58b–61b, 60–63

R

Rad51, 365–366, 366f
  in homologous recombination, 374

Radiation-induced DNA damage, 319–321
  ionizing radiation and, 321
  ultraviolet light and, 319, 320f, 321
Radioactive tracers, 67b
Radioactivity assays, of enzyme catalyzed reactions, 65–66, 67b
Radman, Miroslav, 348
Ramakrishnan, Venkatraman, 661, 679
RAR (retinoid acid receptor), 502
Rasmussen, Eric B., 556
Rat1, 589–590
Rayment, Ivan, 384
R2Bm element, 398, 399f
RDE-1 gene, 604–605
RDE-4 gene, 604
Reactive oxygen species, DNA damage due to, 324, 324f
Reannealing, of DNA, 88–89, 89f
RecA
  in homologous recombination, 362–363, 363f
  regulation of SOS response by, 348–350, 349f, 350f
RecBCD protein, in homologous recombination, 361–362, 362f
Recessive alleles, 3
Recognition elements, 643
Recognition helix, 446
Recombinant DNA technology, applications of, 20–22
Recombinant plasmids, 128
Recombinases, 291, 291f, 293
  in homologous recombination, 374
Recombination, homologous. *See* Homologous recombination
Recycling stage, of protein synthesis, 664, 687–688
Reece, Richard J., 510–511
Regulated exons, 566–567, 567f
Regulation, positive and negative, 436, 436f
Regulatory enzymes
  control of committed steps in biochemical pathways by, 72f, 72–73, 73f
  sigmoidal kinetics and allosteric effectors and, 73f, 73–74, 74f
Regulatory inactivation of DnaA (RIDA), 279b
Regulatory promoters, 488–490, 489f
  in eukaryotic transcription, 487–490, 488f, 489f
Regulatory subunits, 74
Regulons, 458
Relaxed circle, of circular DNA molecules, 04
Relaxed plasmids, 210
Release factors (RFs), 664
  bacterial, 685, 687
Renaturation
  of DNA, 88–89, 89f
  of proteins, 54f, 54–56
Replicating form I (RFI), 91
Replicating form II (RFII), 91
Replication bubble, 267
Replication factor C, in eukaryotic DNA replication, 294
Replication forks, 269
  collapse of, bacterial, homologous recombination replication restart after, 360f, 360–364, 361f
  enzymes acting at, 280
  meeting of, 291, 291f
Replication protein A (RPA), 90
Replicators, 275
Replicon, bacterial DNA replication and, 275
Replisomes, 287
Reporter genes, 391b, 391f
Repressible enzymes, 438
Repressors, 436

Research Collaboratory for Structural Bioinformatics Protein Data Bank, 52
Res gene, 387–388
Resolution, 364
Resolvases, 364
  in replicative transposition, 388
Restriction endonucleases
  DNA characterization using, 123–125, 124f, 125t
  restriction map construction using, 125–128, 126f, 127f
  types of, 124
Restriction fragment length polymorphisms (RFLPs), 216–218, 217f
Restriction maps, of DNA, constructing, 125–128, 126f, 127f
Retinoid acid receptor (RAR), 502
Retinoid X receptor (RXR), nuclear receptors forming heterodimers with, 502
Retroviruses, 255–257, 256f, 258f
Reverse genetics, 613
Reverse transcriptase
  complementary DNA synthesis and, 259b, 259f
  retroviral replication and, 255–257, 256f, 258f
Reverse transcription, 255
Revertants, 196
Reznikoff, William S., 385
RF1, 685–686
RF2, 685–686
RF3, 685
RF(s). *See* Release factors (RFs)
RFI (replicating form I), 91
RFII (replicating form II), 91
RFLPs (restriction fragment length polymorphisms), 216–218, 217f
RGR gene, 520
Rho utilization (*rut*) site, 433
Ribonucleases (RNAses), 122
Ribonucleic acid. *See* RNA
Ribonucleosides, 6, 6f, 7, 7f
Ribonucleotides, 7
Ribosomal complex, dissembly of, in bacteria, 687–688
Ribosomal recycling factor, 664
Ribosome(s), 19, 19f, 20f
  aminoacyl-tRNA transport to, 677, 679
  archaeal, structure of, 663b
  bacterial, structure of, 625, 625f
  eukaryotic, structure of, 625–628, 626f, 627f, 663b, 663f
  programming of, by mRNA, 647–648
  protein synthesis and, 625–628, 680–683, 682f, 683f
  structure of, 660–662
    in eukaryotes, 625–628, 626f, 627f, 663b, 663f
    high resolution determination of, 661f, 661–662, 662f
    subunit structure and function and, 660–661, 661f
Ribosome-nascent chain complex, 691
Riboswitches, 672b, 672f
Ribozymes, 65
Rich, Alexander, 100, 638
Richmond, Timothy J., 168
RIDA (regulatory inactivation of DnaA), 279b
Rifamycin, 422b, 422f
rIIB gene, 648, 648f
rInr (rRNA initiator), 531
RISC (RNA-induced silencing complex), 606
RISC loading complex, 609, 609f
RNA
  crystal structures for, 105, 106f
  DNA differentiated from, 5, 5f

  exogenous, double-stranded, RNA interference triggered by, 601–610
    Dicer proteins and, 607, 607f
    discovery in *Caenorhabditis elegans*, 602–605, 603f
    in vitro studies of, 605–606, 606f
    RISC loading complex and, 609, 609f
    transposon activation blockage by, 610, 610f
    viral replication blockage by, 609–610, 610f
  functions of, 104
  heterogeneous nuclear, 552
    poly(A) tails in, 553, 553f
  isolation of, 114–115
  messenger. *See* mRNA (messenger RNA)
  microRNAs, blockage of mRNA translation and degradation and, 614f, 614–615, 616f, 617, 617f
  piwi interacting, 617
  as primer for Okazaki fragment synthesis, 273, 273f, 274f
  self-splicing, 580f, 580b–581b, 581f
  single-stranded, in bacteriophages, 239f, 239–240, 240f
  small interfering, 606
  structure of. *See* RNA structure
  transfer. *See* tRNA (transfer RNA)
RNA binding site, 430
RNA-DNA hybrid binding site, of transcription elongation complex, 429–430
RNA editing, 591f, 591–592
RNAi. *See* RNA interference (RNAi)
RNA-induced silencing complex (RISC), 606
RNA interference (RNAi)
  as investigational tool, 612–613
  transitive, 610–612
    negative regulator of, 612
    spread throughout organism, 610–612, 611f
  triggered by exogenous double-stranded RNA, 601–610
    Dicer proteins and, 607, 607f
    discovery in *Caenorhabditis elegans*, 602–605, 603f
    in vitro studies of, 605–606, 606f
    RISC loading complex and, 609, 609f
    transposon activation blockage by, 610, 610f
    viral replication blockage by, 609–610, 610f
RNA interference deficient (*rde*) mutants, 604–605
RNA polymerase(s)
  archaeal, 407, 471, 475f, 475b–476b
  bacterial, 407
    crystal structures of, 420–424, 421f, 423f, 424f
    detection and removal of incorrectly incorporated nucleotides by, 432, 432f
    scrunching of, 425–427, 426f
    size of, 411
    structure of, 410–411, 411t, 412f
  eukaryotic, 407
    crystal structures of, in yeast, 472, 473f, 474
    distinguishing by sensitivities to inhibitors, 470f, 470–471, 471f
    subunits of, 471–472, 472t, 477f
    synthetic capacities of, 474
    types of, 468–469, 469f, 470t
  movement of, 430b, 430f, 431f
  nuclear, use of term, 469, 470t
  phage T7 and, 234, 234f, 237f, 237t, 237–238
  reaction catalyzed BY
    requirements for, 408, 409f, 410, 410f
    RNA polymerase structure and, 410–411, 411t, 412f
  reaction catalyzed by, in bacterial transcription, 408–412
  tryptophan concentration and, 457f, 457–458

INDEX **745**

RNA polymerase holoenzyme, bacterial
  binding to promoter, 418–420, 419f, 420f
  structure of, 410–411, 411t, 412–413, 413f
RNA polymerase I
  eukaryotic, 469
  eukaryotic transcription catalyzed by, 528, 530–534
    pre-initiation complex formation and, 533, 533f
    RNA polymerase I structure and, 532
    rRNA synthesis and, 528, 530f, 530–531, 531f, 532
    rRNA transcription unit promoter and, 531–532, 532f
    termination of, 534, 534f
    transcription elongation complex formation and, 533–534
  structure of, RNA polymerase I catalyzed eukaryotic transcription and, 532
RNA polymerase II, 479–480, 551–592
  cap formation and, 553–558
    cap attachment to nascent pre-mRNA chains and, 556
    CTD phosphorylation and, 558
    at 5'-ends, 553–556, 555f, 556f
    eukaryotic pathway for, 556–557, 557f
  CTD in, 486
  eukaryotic, 469
    crystal structures of, 472, 473f, 474, 477f
    subunits of, 471
  gene concept and, 592
  general transcription factor requirement of, for eukaryotic transcription, 478–480, 479t
  human proteome and, 592
  poly(A) sites and, 587f, 587–588
  poly(A) tail synthesis, coupling of transcription termination with, 584–587, 585f, 586f
  pre-mRNA synthesis and, 552–553, 554f
    in eukaryotes, 552, 552f
    poly(A) tails and, 552–553, 553f, 554f
  reactivation by TFIIS, 487, 487f
  RNA editing and, 591f, 591–592
  spliceosomes and, 575–584
    catabolic site of, 577, 578f, 579
    export and slicing as coupled processes and, 584
    self-splicing RNA and, 580f, 580b–581b, 581f
    snRNPs and, 575–577, 576f
    splice site selection regulation and, 579, 582f, 582–583
    splicing as cotranscriptional, then posttranscriptional process, 583f, 538b–584b
  splicing and, 558–575
    amino acid coding region interruption by noncoding regions and, 559–560, 561f, 562, 563f
    combinations of splicing patterns within individual genes and, 568f, 568–569, 569f
    as cotranscriptional, then posttranscriptional process, 583f, 583b–584b
    exons and introns and, 564t, 564–566, 565f
    multiple molecule production for single pre-mRNA and, 566–567, 567f
    sequences required for, 569–570, 570f, 571f
    splice site selection regulation and, 579, 582f, 582–583
    splicing intermediaries resembling lariats and, 571–572, 572f–574f
    transesterification reactions in, 574f, 574–575
    viral evidence for, 558–559, 559f

TFIIB interaction with, in yeast, 484b, 484f
transcription termination of
  allosteric changes and exonuclease and, 589–591, 590f
  coupling of poly(A) tail synthesis and, 584–587, 585f, 586f
  site of, 588–589, 589f
RNA polymerase III
  eukaryotic, 469
  structure of, 535
  eukaryotic transcription catalyzed by, 535–537
    promoter types and, 535–537, 536f
    RNA polymerase III transcript structure and, 535
RNA polymerase IV, eukaryotic, 469
RNA polymerase V, eukaryotic, 469
RNA-recognition motifs (RRMs), 579, 582, 582f
RNAses (ribonucleases), 122
RNA strands, bacterial, release at intrinsic and Rho-dependent terminators, 432–434, 433f, 434f
RNA structure, 104–105, 106f
  function and, 104
  hairpin (stem-loop), 105
  secondary, 104f, 104–105, 105f
  stick figure representations of, 10f, 11
  tertiary, 105, 106f
RNA Tie Club, 631
"RNA world," 105–107
Roberts, Richard, 558–559
Roeder, Robert G., 468–469, 479
Rolling circle replication, 203, 243, 245
Rosenberg, Barnett, 328b
Rossmann, Michael, 642b
Rossmann-fold, 642f, 642f
Rough endoplasmic reticulum, 627, 627f
Rous, Peyton, 257
Rous sarcoma virus, 257
RPA (replication protein A), 90
RPA genes, 471
Rpb1, 486
RPB genes, 471
RPC genes, 471
Rpd3 deacetylase complex, 525
rpoB gene, 422
RRF-1, 611
RRF-2, 611
RRF-3, 611
RRMs (RNA-recognition motifs), 579, 582, 582f
rRNA initiator (rInr), 531
rRNA synthesis
  auxiliary transcription factors required go, 533, 533f
  in nucleolus, 532
  RNA polymerase I catalyzed eukaryotic transcription and, 528, 530f, 530–531, 531f, 532
  of 5S rRNA, 535
rRNA transcription unit, 531f, 531–532
rRNA transcription unit promoter, RNA polymerase I catalyzed eukaryotic transcription and, 531–532, 532f
RSC complex, 524–525
Rupert, Claude S., 333
rut (rho utilization) site, 433
Rutter, William J., 468–469
RuvAB protein, in homologous recombination, 363–364, 364f
RuvC protein, in homologous recombination, 364
Ruvkun, Gary, 612, 614
RXR (retinoid X receptor), nuclear receptors forming heterodimers with, 502
RXR•RAR response element, 502, 503f

RXR•TR response element, 502, 503f
RXR•VDR response element, 502, 503f

S

S (Svedbergs), 117
Sabitini, David D., 689
Saccharomyces cerevisiae, 211–214, 212f. See also Yeast
  cell division in, 211, 212f
  haploid and diploid stages of, 213–214, 214f
  kinetochore of, 174–175
  notations, conventions, and terminology used with, 212–213, 213t
  telomerase in, 304
S-adenosylmethionine, methionine conversion to, 554–556, 555f
SAGO-1, 611
SAGO-2, 611
Sal box, 534, 534f
Salmonella typhimurium, Ames test and, 331–333
Sanger, Frederick, 8, 139, 141, 664, 665
Saunders, Edith Rebecca, 186
Scaffold model, for chromatin structure, 172–173, 173f
Scanning model, 673
Schimmel, Paul, 640, 641, 643, 644
Schizosaccharomyces pombe, cell division in, 211–212
Schuster, Stephan C., 145b
Scott, Matthew, 497
Scrunching model, 425–427, 426f
SDS-PAGE (sodium dodecyl sulfate-polyacrylamide gel electrophoresis), 121–122
Secondary structure of proteins, 42–43, 46–50
  α-helix, 46–49, 47f
  β-conformation, 48f, 49
  combinations of, 50, 50f
  loops and turns in, 49, 49f
  prediction of, 50, 50f
Segregation, law of, 3
Selective markers, 206
Selectivity factor 1 (SL1), 533, 533f
Selenocysteine, as building block for polypeptides, 645–646, 646f
Semiconservative model of replication, 265–267, 265f–267f
Semidiscontinuous model of replication, 270, 270f
S1 endonuclease, 123
Sense strand, of bacterial DNA, 410, 410f
Sepharose beads, 143b
Sequence hypothesis, 11, 624
Serine, 27, 27f
  abbreviations for, 29t
Setlow, Richard, 340
70S initiation complex, 666, 667–668, 669f
Sex chromosomes, 163
Sex pilus, 202–203, 203f
Shapiro, James A., 381
Sharp, Philip A., 558–559, 571, 605
Shigella, drug resistant, 93b
Shine, John, 669
Shine-Dalgarno sequence, 668–671, 670f, 671f
Short interspersed nucleotide elements (SINEs), 398, 401
Shotgun sequencing, 141f, 141–142
Sickle cell anemia, 58f, 59f, 58b–60b
Side chains, of amino acids, 26–28, 26f–28f
  hydrophilic, 48–49
  hydrophobic, 44–45, 48–49
sid-1 gene, 612
Sigma factor, 413, 414t, 414b–415b, 416

σ⁵⁴, 414, 415
σ⁷⁰ family
  bacterial, conserved domains of, 424–425, 425f
  conserved domains and, 424–425, 425f
σ⁷⁰ polymerase holoenzyme
  bacterial, escape of, initiation stage of, 425
  escape of, 425
σ subunits, of *Bacillus subtilis*, 414b
Sigmoidal kinetic curves, 73f, 73–74, 74f
Signal hypothesis, 689
"Signaling release," 290
Signal recognition particle (SRP), 689–690
Signal sequence, protein synthesis and, 689–691, 690f
SII, 487, 487f
Silencers, in eukaryotic transcription, 491
Silencing, spreading of, 612
Simian virus 40 (SV40)
  DNA of, 250
    size of, 82t
  early transcription unit in, 490, 490f
  genome of, 250, 251f
  as model for eukaryotic DNA replication, 293–295, 294f, 295f
  replication cycle of, 250, 252f, 252–253
SINEs (short interspersed nucleotide elements), 398, 401
Single nucleotide polymorphisms (SNPs), 215
Single-stranded DNA binding proteins (SSBs), 90, 91f
SIN3A, 527
SiRNAs (small interfering RNAs), 606
Sister chromatids, 157, 364
  crossovers between, 367
Site-directed mutagenesis, 138f, 138–139
60S subunit, in protein synthesis, 626–627
Skolnick, Mark, 217–218
SL1 (selectivity factor 1), 533, 533f
Sleeping Beauty transposon, 396–397
Sliding clamp, 283
  tethering of polymerase holoenzyme to DNA by, 285, 285f
Small interfering RNAs (siRNAs), 606
Smith, Hamilton O., 124, 127, 141
Smith, Michael, 138
Smithies, Oliver, 368
Smooth endoplasmic reticulum, 627, 627f
Snf1 kinase, 513
SNPs (single nucleotide polymorphisms), 215
SnRNPs, 575–577, 576f
Sodium dodecyl sulfate-polyacrylamide gel electrophoresis (SDS-PAGE), 121–122
Somatic cells, 155
  genetics of, for genetic mapping, 219
Song, Ok-kyu, 518b
SOS response to DNA damage, 347–353
  DNA polymerase synthesis induced by, 351–352
  human DNA polymerases and, 352t, 352–353
  RecA and LexA regulation of, 348–350, 349f, 350f
  translesion DNA synthesis and, 347–348
Southern blotting (Southern transfer), 130–132, 131f
Spacefill structures, 45
S phase, 156
Spliceosomes, 575–584
  catabolic site of, 577, 578f, 579
  export and slicing as coupled processes and, 584
  self-splicing RNA and, 580f, 580b–581b, 581f
  snRNPs and, 575–577, 576f
  splice site selection regulation and, 579, 582f, 582–583
  splicing as cotranscriptional, then posttranscriptional process, 583f, 583b–584b

3′-Splice sequence, 570, 570f
5′-Splice sequence, 570, 570f
3′-Splice site, 569–570, 570f, 571f
5′-Splice site, 569–570, 570f, 571f
Splicing, 558–575
  alternative, 566–567, 567f
  amino acid coding region interruption by noncoding regions and, 559–560, 561f, 562, 563f
  combinations of splicing patterns within individual genes and, 568f, 568–569, 569f
  exons and introns and, 564t, 564–566, 565f
  multiple molecule production for single pre-mRNA and, 566–567, 567f
  sequences required for, 569–570, 570f, 571f
  splicing intermediaries resembling lariats and, 571–572, 572f–574f
  transesterification reactions in, 574f, 574–575
  viral evidence for, 558–559, 559f
Splicing regulatory proteins (SR proteins), 579, 582, 582f
Split gene(s). *See* Splicing
Split gene hypothesis, 562
SPO11, in homologous recombination, 374, 375f
Spontaneous mutagenesis, 195
Squelching, Mediator and, 517
SR (SRP receptor), 690
SRB1 gene, 520
SRB2 gene, 520
SRP (signal recognition particle), 689–690
SRP receptor (SR), 690
SR proteins (splicing regulatory proteins), 579, 582, 582f
SRP•SR, 691
SSBs (single-stranded DNA binding proteins), 90, 91f
30S initiation complex, 666
30s subunit, in protein synthesis, 625, 626f, 665–668, 667f, 669f
Stabilization helix, 446
Stahl, Franklin, 266–267
Start codons, in protein synthesis, 664
Stationary phase, 192
Steitz, Joan, 575, 577, 670
Steitz, Thomas, 134, 283, 450, 642b, 661
Stem-loop structure, of RNA, 105
Stereoisomers, of amino acids, 26, 26f
Sticky ends, 124, 124f
sti gene, 643b
Stillman, Bruce, 298
Streptococcus pneumoniae, transformation experiments and, 11
Streptomyces, chromatin structure in, 152
Stringent plasmids, 210
Strong promoters, 418
Structural genes, 194
Sturtevant, Alfred H., 187–188
Substrates, of enzymes, 65
  enzyme-substrate complexes and, 68f, 68–72, 70f, 71f
Sucrose gradient centrifugation, 117–118, 118f
Sugars, in nucleic acids, 4f, 5f, 5–6
Suicide enzymes, direct alkyl group removal by, 335–337, 336f
Sulfolobus shibatae, RNA polymerase from, 475b, 475f, 476b
Sulfur, van der Waals radius of, 45t
Sulston, John, 614
Supercoils, 94–96
Superhelix, 94–96
Supersecondary structure of protein, 50, 50f
Suppressor(s), 196

Suppressor-sensitive mutants, 196
Sutherland, Earl W., 449
SV 40. *See* Simian virus 40 (SV 40)
Svedberg, Theodor, 117
Svedbergs (S), 117
SWI/SNF chromatin remodeling complex, 523–525
Sylenocysteyl-tRNA, formation of, bacterial pathway for, 647, 647f
Symplekin, 587
Synapsis, 159
Synthesis-dependent strand annealing, 365–366, 366f
Synthetic elements, 389, 390f
Szostak, Joseph, 300–303

T
TAFs (TBP-associated factors), 481–482, 482f
Tail formation, in bacteriophages, 233
Tandem repeat polymorphisms, 215, 216f, 218b
Target-primed reverse transcription mechanisms, non-LTR retrotransposons and, 398–399, 399f, 400f
Tata, Jamshed R., 468
TATA-binding protein (TBP)
  binding to TATA core promoter, general transcription factors and, 480–482, 480f–482f
  crystal structure of, 480–481, 481f
TATA box, 488
  core promoter for protein-coding genes and, 476, 478, 478f
TATA core promoter, TFIID or TBP binding to, general transcription factors and, 480–482, 480f–482f
Tatum, Edward L., 4, 201
TBP. *See* TATA-binding protein (TBP)
TBP-associated factors (TAFs), 481–482, 482f
TBP•TATA box complexes, 482
T DNA, 93b
Telomerase
  end-replication problem and, 305–307, 306f, 307f
  nucleotide repeat addition to chromosome ends and, 303, 305f
Telomeres, 175–179, 177f, 178f
  addition of nucleotide repeats to chromosome ends and, 303, 305f
  in aging and cancer, 304b
  formation of, in eukaryotes, 300–303, 301f, 302f
Telophase, of mitosis, 157
Telophase I, meiotic, 162
Telophase II, meiotic, 162
Temin, Howard, 255, 257
Temperate phages, 228, 241–248
  λ, 241f–244f, 241–245
  P1, 245f, 245–246, 246b–247b, 248
  plaques produced by, 229
Temperature
  melting, 86
  nonpermissive, 196
  permissive, 196
Temperature-sensitive (Ts) mutants, 196
Templates, for DNA replication, 18, 18f
Template strand, 133, 133f
Terminal inverted repeats, 381
Terminal redundancy, of T-even bacteriophages, 230, 231f
Terminase, 233
Termination, of RNA polymerase I catalyzed eukaryotic transcription, 534, 534f
Termination *(Ter)* sites, 291, 291f

Termination stage
  of bacterial DNA replication, 275
    replication forks and, 291, 291f
    terminus utilization substance and, 291, 291f
  of bacterial transcription
    intrinsic, 433
    Rho-dependent, 433–434, 434f
    RNA strand release at intrinsic and Rho-dependent terminators and, 432–434, 433f, 434f
  of protein synthesis, 685, 687
    mutant tRNA molecules and, 687
    protein release factors and, 685, 687
Terminus utilization substance (Tus), 291, 291f
Ter (termination) sites, 291, 291f
Tertiary structure of proteins, 43, 50–56
  determination by primary structure, 53–56, 54f
  intrinsically disordered proteins in, 53
  molecular chaperones and, 56
  x-ray crystallography of, 50–53, 51f, 52f
Tetrads, 159
Tetrahymena thermophila
  rRNA synthesis in, 580f, 580b–581b, 581f
  telomeres of, 176–177, 300–305, 301f
Tetramerization helix, of Lac repressor subunit, 446, 447f
T-even phages, 226, 230f–232f, 230–234, 233t, 235f–236f
  assembly of, 234b
TFIIA, 479, 479t
  in transcription initiation, 482f, 482–483
TFIIB, 479, 479t
  in transcription initiation, 482f, 483, 483f, 484b–485b
TFIIB recognition elements (BRE), 478
TFIID, 479, 479t, 481f, 481–482, 482f
  binding to TATA core promoter, 480–482, 480f–482f
TFIID·TATA box complexes, 482
TFIIE, 479, 479t
TFIIF, 479, 479t
  in transcription initiation, 482f, 483
TFIIH, 479, 479t
  in transcription initiation, 482f, 483
TFIIIA, promoters and, 536, 537f
TFIIIB, promoters and, 536–537, 537f
TFIIIC, promoters and, 536–537, 537f
TFIIS, 487, 487f
Theoretical maximum rate of reaction (V_{max}), 69, 69f
Thermococcus litoralis, DNA polymerase of, 137
Thermus aquaticus
  core RNA polymerase of, 422b, 422f
  polIIIα subunit from, 283
  RNA polymerase e, 407
Thermus thermophilus
  RNA polymerase e, 407
  σ^{70} RNA polymerase holoenzyme crystal structure in, 423, 423f
  30S subunit of, 661
Θ-replication, 243, 244f, 245
Θ structures, 267, 268f
Θ subunit, of core polymerase, 284–285
Thiomethylgalactoside, 439, 439f
Threonine, 27, 27f
  abbreviations for, 29t
Thymine, 8t
  in DNA, 6, 6f
Thymine glycol, 324, 324f
TIF-1A (transcription initiation factor 1A), 532
TIF-1B (transcription initiation factor IB), 533, 533f
Time-of-entry mapping, 206, 207f

Ti plasmid, 93b
titin gene, 564, 566
Tjian, Robert, 494
t-loops, 177, 177f, 178f
T_m (melting temperature), 86
Tnp gene, 387–388, 389
Tn*5* transposon
  structure of, 383, 383f
  transposition mechanism of, 383–387, 384f–388f
Tn*10* transposon
  structure of, 383, 383f
  transposition mechanism of, 383–387, 384f–388f
Tn*3* transposon, replicative transposition of, 387–388, 388f, 389f
Todo, Takeshi, 335
Tonegawa, Susumu, 394–395
Top agar, 229
Topoisomer(s), 95–96, 96f
Topoisomerase(s), 96f–98f, 97–98
Topoisomerase II, chromatin condensation and, 172, 172f
Topoisomerase IV, separation of sister chromosomes and, 291, 291f, 293
Topoisomer separation, gel electrophoresis and, 121, 121f
Topological isomers, 95–96, 96f
Torpedo model, 589–590
TψC arm, of tRNA, 638
TR, 502
Transcription
  bacterial. *See* Bacterial transcription
  definition of, 407
  eukaryotic. *See* Eukaryotic transcription
  pausing in
    suppression by elongation factors, 487
    transcription pause sites and, 432
Transcription activating proteins
  activation domains of. *See* Activation domains
  DNA-binding domains of. *See* DNA-binding domains
Transcription activator proteins, eukaryotic, 492–517
  activation domains in, 511–517
    association with DNA-binding domains, 514f, 514–517, 516f
    of Gal4, 511–512, 512f
    GAL gene regulation and, 512f, 512–513, 513f
    intrinsically disordered nature of, 511
  AP-1 family of, 507b, 507f, 508f
  DNA-binding domain structures of, 496–511
    basic region leucine zipper proteins, 504–506, 504f–506f
    Cys$_4$ zinc finger, 499–502, 500f–503f
    helix-turn-helix, 497–499, 497f–499f
    Zn$_2$Cys$_6$ binuclear cluster, 506, 509–511, 509f–511f
  gene activity and, 493f, 493–494
  purification by DNA affinity chromatography, 494f, 494–495
  transcription machinery recruitment and, 492f, 492–493
  transfection assay to determine ability to stimulation gene transcription, 495–496, 496f
Transcription arrest, 432
Transcription elongation complex
  duplex-binding site of, 428–429
  formation of, RNA polymerase I catalyzed eukaryotic transcription and, 533–534

nucleoside triphosphate and, 427
nucleotides in, numbering of, 428
RNA-DNA hybrid binding site of, 429–430
Transcription factors, general, eukaryotic, 478–486
  preinitiation complex formation and, 482f, 482–483, 483f, 486, 486f
  RNA polymerase II requirement for, 478–480, 479t
  TFIID or TBP binding to TATA core promoter and, 480–482, 480f–482f
Transcription initiation factor 1A (TIF-1A), 532
Transcription initiation factor IB (TIF-1B), 533, 533f
Transcription pause sites, 432
Transcription termination, 432
  bacterial
    intrinsic, 433
    Rho-dependent, 433–434, 434f
    RNA strand release at intrinsic and Rho-dependent terminators and, 432–434, 433f, 434f
  of RNA polymerase II
    allosteric changes and exonuclease and, 589–591, 590f
    coupling of poly(A) tail synthesis and, 584–587, 585f, 586f
    site of, 588–589, 589f
Transcription termination factor 1 (TTF-1), 534, 534f
Transduction, generalized, in bacteriophage P1, 246b, 247f
Transesterification reactions, in splicing, 574f, 574–575
Transfection assay, to determine transcription activator protein ability to stimulation gene transcription, 495–496, 496f
Transfer origin (*oriT*), in F plasmid, 203
Transfer RNA. *See* tRNA (transfer RNA)
Transforming principle, 11–14, 13f
Transgenic mice, 371
Transient excursion model, 426f, 426–427
Transition mutations, 322, 323f
Transition states, 67
Transitive RNA interference, 610–612
  negative regulator of, 612
  spread throughout organism, 610–612, 611f
Translation, definition of, 407
Translation initiation factor
  eukaryotic, 674
    protein synthesis regulation by, 676, 676t
    phosphorylation of, in eukaryotes, 676, 676t
Translation initiation factors (IFs), 662, 664. *See also* Initiation factor *entries*
Translation initiation pathway, archaeal, 676
Translocation(s), 684
  ionizing radiation inducing, 321
Translocation mechanism, 686b, 686f
Translocons, 690
Transposable elements (transposons), 380–402
  activation blockage of, by RNA interference, 610, 610f
  autonomous, 389
  bacterial, 381–391
    composite, 382f, 382–383, 383f
    in experimental studies, 389–390, 390f
    insertion sequences, 381–382, 382f
    Tn3, 387–388, 388f, 389f
    Tn5 and Tn10, 383–387, 384f–388f
  bearing antibiotic resistance genes, 389–390, 390f, 391b
  composite, 382f, 382–383, 383f

eukaryotic, 391–402
　Alu element, 401
　antibody and T-cell receptor gene formation mechanism and, 392, 394–396, 395f–397f
　dormant, restoration to full activity, 396–397
　*h*AT mobile element transposition and, 392, 392f, 394f
　target-primed reverse transcription mechanism of non-LTR elements and, 398–399, 399f, 400f
　transposition mechanisms of, 391–392
　transposition via an RNA intermediate, 397–398, 398f, 399f
　value of retrotransposons and, 401–402
　nonautonomous, 389
　noncomposite, 388
Transposase, 381
Transposons. *See* Transposable elements (transposons)
Transversion mutations, 322, 323f
TRAP170 Mediator, 520
trp leader, 456, 456f
TRF1 protein, 177–178, 178f
TRF2 protein, 177–178, 178f
Trigger factor, 688
Trinucleotides, promotion of molecules to ribosomes by aminoacyl-tRNA and, 654, 655f
tRNA (transfer RNA), 19, 20f, 629
　aminoacyl
　　binding of, to multiple codons, 659–660, 660f
　　coding specificity of, 658–659, 659f
　　formation of, 633f, 633–634, 634f
　　protein synthesis and, coding specificity of, 658–659, 659f
　　trinucleotide promotion of molecules to ribosomes and, 654, 655f
　　tRNA molecule movement through ribosome and, 685
　CCA$_{OH}$ at ends of, 631, 632f, 633
　genes for, number of, 634
　initiator, in eukaryotes, methionine in, 671, 673, 673f
　methionine, initiator and elongator, protein synthesis and, 664–665, 665f, 666f
　mutant, protein synthesis and, 687
　N-Acetyl-Phe-, 684, 685
　protein synthesis and, 628–639
　　amino acid attachment to, 628–629, 630f, 631, 631f, 633f, 633–634, 634f
　yeast, sequencing of, 635
　sylenocysteyl, formation of, bacterial pathway for, 647, 647f
　valyl, hydrolysis of, by aminoacyl-tRNA synthetases, 640–641, 641f
tRNA^{fMet}, 665, 666f
tRNA ligases, aminoacyl, 629, 630f
tRNA synthetases
　alanyl, editing defect in, neurodegenerative disease caused by, 643b
　aminoacyl, 629, 630f
　　cognate distinction by, 641, 643–645, 644f–646f
　　proofreading functions of, 639–640, 640f
　　selenocysteine and pyrrolysine as building blocks for polypeptides and, 645–646, 646f
　　valyl-tRNA and valyl-AMP hydrolysis by, 640–641, 641f
　Ile
　　posttransfer editing and, 642b, 642f
　　proofreading function of, 639–640, 640f

Trombone model of replication, 287, 288f
trpA gene, 453
trpB gene, 453
trpC gene, 453
trpD gene, 453
trpE gene, 453
trpL gene, 455
trpR gene, 454–455
Trp•tryptophan complex, 454–455
Tryptone broth, 192
Tryptophan, 27, 28f
　abbreviations for, 29t
Tryptophan operon, bacterial, 453–458, 454f–457f
Ts (temperature-sensitive) mutants, 196
TTF-1 (transcription termination factor 1), 534, 534f
Tumor suppressor genes, 318
Turnover number, 69
Turns, in proteins, 49, 49f
Tus (terminus utilization substance), 291, 291f
Two-hybrid assay, yeast, 517, 518f, 518b–519b
Twort, Frederick, 225
Ty1 element, 397
Tyrosine, 27, 27f
　abbreviations for, 29t

U

U2AF, 577, 578f
UAG, 687
UAS. *See* Upstream activating sequence (UAS)
UAS$_{GAL}$, 512, 512f, 514
UBF (upstream binding factor), 533, 533f
UGA, 685
Ultraviolet light, DNA damage caused by, 319, 320f, 321
unc-22 gene, 610
Unidirectional replication, 267, 268f
3′-Untranslated region, 435
5′-Untranslated region (5′-leader; 5′-UTR), 435, 673–674
Up element, 418, 418f
Upstream activating sequence (UAS), 492
　in active yeast genes, Mediator association with activators at, 521–523, 522f
Upstream binding factor (UBF), 533, 533f
Upstream nucleotide sequences, 410
Uracil, 8t
　in DNA, 6, 6f
U-rich elements, 585
U1 snRNA, 575–576, 576f, 577, 578f
U2 snRNA, 575–576, 576f, 577, 578f
U4 snRNA, 575–576, 576f, 577, 578f
U5 snRNA, 575–576, 576f, 577, 578f
U6 snRNA, 575–576, 576f, 577, 578f, 579
5′-UTR (5′-untranslated region), 435, 673–674
uvrA gene, nucleotide excision repair and, 341
UvrA protein, nucleotide excision repair and, 341
uvrB gene, nucleotide excision repair and, 341
UvrB protein, nucleotide excision repair and, 341
uvrC gene, nucleotide excision repair and, 341
UvrC protein, nucleotide excision repair and, 341
UvrD protein, mismatch repair and, 345

V

Vaccinia virus strain WR, DNA size in, 82t
Valine, 27, 28f
　abbreviations for, 29t
Valyl-AMP, hydrolysis of, by aminoacyl-tRNA synthetases, 640–641, 641f
Valyl-tRNA, hydrolysis of, by aminoacyl-tRNA synthetases, 640–641, 641f
Van der Waals interactions, in proteins, 45t, 45–46, 46f

Van der Waals radius, 45t
Variable arm, of tRNA, 638
Variable region domains, of immunoglobulin G, 64
Varmus, Harold E., 257
Vassylyev, Dimitry G., 428
VDR (vitamin D receptor), 502
Vectors, plasmids as, insertion of DNA fragments into, 128–129, 129f
Venter, J. Craig, 141
v-fos gene, 507b
Vibrio cholerae, 225
Vidal, Marc, 613
Virions, 224
　adenoviral, 253, 253f
　assembly of, 234
　release of, chronic phages and, 248f, 248–249
Virulence plasmids, 93b
Virulent phages, S Bacteriophage(s), virulent
Viruses. *See also specific viruses*
　animal, 249–259
　　adenoviruses, 253f, 253–254, 254f
　　as models, 249–250
　　polyomaviruses, 250, 251f, 252f, 252–253
　　retroviruses, 255–257, 256f, 258f
　bacterial. *See* Bacteriophage(s); *specific bacteriophages*
　replication blockage by RNA interference in, 609–610, 610f
　sizes of, 224, 224f
　splicing in, 558–559, 559f
Vitamin D receptor (VDR), 502
v-jun gene, 507b
V$_{max}$ (theoretical maximum rate of reaction), 69, 69f

W

Wang, James, 96
Water
　DNA damage by hydrolytic cleavage reactions and, 321–323, 322f, 323f
　hydrophobicity and, 44–45
Watson, James D., 145b, 265, 631
　DNA structure and, 15–19, 16f–18f
Weaver, Warren, 2
Weinzierl, Robert O. J., 475b
Weiss, Samuel B., 468
Werner, Finn, 475b
Werner, Michael, 520–521
Werner syndrome helicase (WRN), 178
Western blotting, 132
Whole genome shotgun sequencing, 141–142
Wild type genes, 195
Wilkins, Maurice, 15
Wobble hypothesis, 659–660, 660f
Woese, Carl, 152
Woolly mammoth, DNA sequencing of, 145b
Writhing, of circular DNA molecules, 94, 95
WRN (Werner syndrome helicase), 178
WT1 gene, 568, 568f

X

X chromosomes, 163
Xeroderma pigmentosum, 343b–344b
　TFIIH and, 483
X-ray(s), DNA damage due to, 321
X-ray crystallography
　atomic structure of nucleosome core particles and, 168–170, 169f, 179f
　tertiary structure of proteins and, 50–53, 51f, 52f
Xrn2, 589, 590

Y

Yanofsky, Charles, 455, 456
Y chromosomes, 163
Yeast. *See also Saccharomyces cerevisiae*
   gene conversion in, 372–373, 374f
   homologous recombination in, 374, 376
   Mediator of, 520–521, 521f
      association with activators at UAS, 521–523, 522f
   as model system, 211
   notations, conventions, and terminology used with, 212–213, 213t
   origins of replication in, 297f, 297–298, 298f
   RNA polymerase II interaction with TFIIB in, 484b, 484f
   TBP of, 480
   TFIIB interaction with RNA polymerase II in, 484b, 484f
   tRNA of, sequencing of, 635
Yeast artificial chromosome, 302
Yeast chromosome IV, DNA size in, 82t
Yeast spores, 213–214
Yeast two-hybrid assay, 517, 518f, 518b–519b
Yokoyama, Shigeyuki, 423, 642b
Yonath, Ada, 661
Young, Richard, 520

Z

Zamecnik, Paul, 628–629, 631
Z-DNA, 100t, 100–102, 101f
Zinder, Norton D., 239, 246b
Zipping and assembly model, 55
zNA polymerase II, RNA editing and, 591f, 591–592
Zn_2Cys_6 binuclear cluster family, of eukaryotic transcription activator proteins, 506, 509–511, 509f–511f